Biochar for Environmental Management

Science, Technology and Implementation

Third Edition

Edited by
Johannes Lehmann and Stephen Joseph

Routledge
Taylor & Francis Group

LONDON AND NEW YORK

earthscan
from Routledge

Designed cover image: © Johannes Lehmann; Stephen Joseph; Örjan Berglund; Timothy Ryan

Third edition published 2024
by Routledge
4 Park Square, Milton Park, Abingdon, Oxon, OX14 4RN

and by Routledge
605 Third Avenue, New York, NY 10158

Routledge is an imprint of the Taylor & Francis Group, an informa business

First edition published by Routledge 2009
Second edition published by Routledge 2015

British Library Cataloguing-in-Publication Data
A catalogue record for this book is available from the British Library

ISBN: 978-1-032-28615-0 (hbk)
ISBN: 978-1-032-28618-1 (pbk)
ISBN: 978-1-003-29767-3 (ebk)

DOI: 10.4324/9781003297673

Typeset in Plantin
by MPS Limited, Dehradun

Access the Support Material: www.routledge.com/9781032286150

Contents

Contributors

Diego Abalos, Department of Agroecology, Aarhus University, Tjele, Denmark, ETSI Agronomos, Universidad Politécnica de Madrid, Ciudad Universitaria, 28040 Madrid, Spain

Samuel Abiven, Département de Géosciences, École normale supérieure, Paris, France

James E. Amonette, Pacific Northwest National Laboratory, Richland, WA 99352, USA, and Center for Sustaining Agriculture and Natural Resources, Washington State University, Wenatchee, WA 98801, USA

Elias Azzi, KTH Royal Institute of Technology, E-100 44 Stockholm, Sweden

Shahla Hosseini Bai, Centre for Planetary Health and Food Security, School of Environment and Science, Griffith University, Nathan, QLD 4111, Australia

Santanu Bakshi, Bioeconomy Institute, Iowa State University, Ames, IA 50011, USA

Ana Catarina Bastos, Department of Environment and Planning, Centre for Environmental and Marine Studies (CESAM), University of Aveiro, 3810-193 Aveiro, Portugal

Luke Beesley, The James Hutton Institute, Aberdeen, AB15 8QH, United Kingdom

Michael Bird, School of Earth and Environmental Science and Centre for Tropical Environmental and Sustainability Science, James Cook University, Cairns, Queensland, 4870, Australia

Nanthi Bolan, School of Agriculture and Environment, The University of Western Australia, Perth, WA 6001, Australia

Catherine E. Brewer, Department of Chemical and Materials Engineering, New Mexico State University, Las Cruces, NM, 88003, USA

Robert C. Brown, Department of Mechanical Engineering and Bioeconomy Institute, Iowa State University, Ames, IA 50011, USA

Tristan R. Brown, Bioeconomy Development Institute, SUNY College of Environmental Science & Forestry, Syracuse, USA

Thomas D. Bucheli, Agroscope, Environmental Analytics Institute for Sustainability Sciences ISS, CH-8046 Zürich, Switzerland

Wolfram Buss, Research School of Biology, Australian National University, 2601 Canberra, Australia

Beatriz Cerqueira Cancelo, Faculty of Biology, Department of Plant Biology and Soil Science, University of Vigo, As Lagoas-Marcosende, 36310, Vigo, Pontevedra, Spain

Marta Camps-Arbestain, New Zealand Biochar Research Centre, Institute of Agriculture and Environment, Massey University, Palmerston North 4442, New Zealand

Leonardo León Castro, Universidade Escuela Superior Politécnica del Litoral, Facultad de Ciencias de la Vida, Guayaquil, Ecuador

Maria Luz Cayuela, Department of Soil and Water Conservation and Waste Management, CEBAS-CSIC, Campus Universitario de Espinardo, 30100 Murcia, Spain

Hanbo Chen, Institute of Eco-environmental Research, School of Environmental and Natural Resources, Zhejiang University of Science & Technology, Hangzhou, 310023, China

Gerard Cornelissen, Department of Environmental Engineering, Norwegian Geotechnical Institute, Oslo, Norway; Department of Plant and Environmental Sciences, Norwegian University of Life Sciences, Ås, Norway; Department of Applied Environmental Sciences (ITM), Stockholm University, 10691 Stockholm, Sweden

Annette Cowie, NSW Department of Primary Industries / University of New England, Armidale, NSW 2351, Australia

Jane Debode, Institute for Agricultural, Fisheries and Food Research (ILVO) Merelbeke, Belgium

Thomas H. DeLuca, Department of Forest Ecosystems & Society, College of Forestry, Oregon State University, Corvallis, Oregon, USA

Xavier Domene, Consolidated Research Group on Soil Protection, CREAF, 08193 Cerdanyola del Vallès, Spain

Leanne Ejack, Natural Resource Science, McGill University, Montreal, Canada

Yunying Fang, Australian Rivers Institute, School of Environment and Science, Griffith University, Nathan 4111, Australia

Michael B. Farrar, Centre for Planetary Health and Food Security, Griffith University, Nathan 4111, Australia

Jenny R. Frank, Environmental and Renewable Resources Division, SUNY Morrisville, Morrisville, New York, USA

Omer Frenkel, Department of Plant Pathology and Weed Research, The Volcani Center, POB 15159, Derech Hamacabim 68, Rishon Lezyion 7505101, Israel

Marta Gallart, CSIRO Agriculture and Food, Acton 2601, ACT, Australia

Si Gao, Department of Environmental Studies, California State University, Sacramento, California, USA

Xiaodong Gao, Department of Earth, Environmental and Planetary Sciences, Rice University, Houston, TX 77005, USA

Manuel Garcia-Perez, Department of Biological Systems Engineering, Washington State University, Pullman, WA 99164, USA

Bruno Glaser, Martin-Luther-University Halle-Wittenberg, Institute of Agricultural and Nutritional Sciences, Halle, Germany

Luise Giani, Soil Science Division, Carl-von-Ossietzky-Universität Oldenburg, Germany

Ellen R. Graber, Institute of Soil, Water and Environmental Sciences, The Volcani Center, Bet Dagan 50250, Israel

Shamim Gul, Department of Botany, University of Balochistan, Saryab Road, Quetta, Balochistan, Pakistan

Michael J. Gundale, Department of Forest Ecology and Management, Swedish University of Agricultural Sciences, Umeå, Sweden

Roger Hegarty, School of Environmental and Rural Science, University of New England, Armidale NSW 2351, Australia

Michael Hardman, Urban Sustainability, University of Salford, Salford M5 4WT, UK

Nguyen Hien, Soils and Fertilizers Research Institute: Hanoi, Vietnam

William C. Hockaday, Department of Geology, Baylor University, Waco TX 76798, USA

Jeffrey Homburg, University of Arizona, Tucson AR, USA

Amit K. Jaiswal, Department of Horticulture and Landscape Architecture, Purdue University, West Lafayette, USA

Simon Jeffery, Centre for Crop and Environmental Science, Agriculture and Environment Department, Harper Adams University, Newport, Shropshire, TF10 8NB, United Kingdom

Keiji Jindo, Agrosystems Research Group, Wageningen University and Research, 6708 PB Wageningen, The Netherlands

Davey L. Jones, School of Natural Sciences, Bangor University, UK

Stephen Joseph, School of Materials Science and Engineering, University of New South Wales, Sydney, NSW 2251, Australia

Claudia Kammann, Department of Applied Ecology, Geisenheim University, Geisenheim, Germany

Niloofar Karimian, CSIRO, Mineral Resources, Clayton South, Victoria 3169, Australia

Edith Kichamu-Wachira, Centre for Planetary Health and Food Security, Griffith University, Nathan 4111, Australia

Markus Kleber, Department of Crop and Soil Science, Oregon State University, Corvallis, OR 97331, USA

Rai S. Kookana, CSIRO - Environment, University of Adelaide, Glen Osmond 5064, Australia

Harn Wei Kua, Department of the Built Environment, College of Design and Engineering, National University of Singapore, 4 Architecture Drive, Singapore 117566

Johannes Lehmann, Soil and Crop Sciences, School of Integrative Plant Science, Cornell University, Ithaca, NY 14853, USA

Manhattan Lebrun, INRA USC1328, LBLGC EA1207, University of Orléans, Orléans, France

Yu Luo, College of Environmental & Natural Resource Sciences, Zhejiang Provincial Key Laboratory of Agricultural Resources and Environment, Zhejiang University, Hangzhou, 310058, China

M. Derek MacKenzie, Department of Renewable Resources, University of Alberta, Edmonton, AB, T6G 2E3, Canada

Josef Maroušek, Institute of Technology and Business, Okružní 517/10, České Budějovice, Czech Republic

Anna Maroušková Institute of Technology and Business, Okružní 517/10, České Budějovice, Czech Republic

Doyle McKey, Centre d'Ecologie Fonctionnelle et Evolutive, University of Montpellier, Place Eugène Bataillon, 34095 Montpellier, France

Kalidas Mainalis, Department of Biological Systems Engineering, Washington State University, Pullman, WA 99164, USA

Ondřej Mašek, UK Biochar Research Centre, University of Edinburgh, Edinburgh EH9 3JN, United Kingdom, United Kingdom

Caroline A. Masiello, Department of Earth, Environmental and Planetary Sciences, Rice University, Houston, TX 77005, USA

Sarah Meale, School of Agriculture and Food Sciences, University of Queensland, Australia

Leônidas C. A. Melo, Soil Science Department, School of Agricultural Sciences, Federal University of Lavras, Lavras, MG 37203-202, Brazil

Thomas R. Miles, T R Miles Technical Consultants Inc., USA

Babak Minofar, Czech Academy of Sciences, Institute of Microbiology, Zamek 136, Nove Hrady, Czech Republic, 37333

Kerry Mitchell, Department of Public Health & Preventive Medicine, St. George's University, Grenada, West Indies

Sohrab Haghighi Mood, Department of Biological Systems Engineering, Washington State University, Pullman, WA 99164, USA

Peter S. Nico, Earth Sciences Division, Lawrence Berkeley National Laboratory, Berkeley, CA 94720, USA

Yong Sik Ok, Korea Biochar Research Center, APRU Sustainable Waste Management Program & Division of Environmental Science and Ecological Engineering, Korea University; International ESG Association (IESGA), Seoul, 06621, Republic of Korea

Negar Omidvar, Centre for Planetary Health and Food Security, Griffith University, Nathan 4111, Australia

Adam O'Toole, Department of Biogeochemistry and Soil Quality, Norwegian Institute of Bioeoconomy (NIBIO), Norway

Nicholas Paul, School of Science, Technology and Engineering, University of the Sunshine Coast, Queensland Australia

Manuel Raul Pelaez-Samaniego, Department of Biological Systems Engineering, Washington State University, Pullman, WA 99164, USA

Joseph J. Pignatello, The Connecticut Agricultural Experiment Station, New Haven, CT 06504-1106, USA

Melissa Rebbeck, Climate and Agricultural Support Pty Ltd, P.O Box 25, Goolwa, 5214, South Australia

Frédérique Reverchon, Red de Diversidad Biológica del Occidente Mexicano, Centro Regional del Bajío, Instituto de Ecología A.C., Pátzcuaro, Michoacán, México

Xaiofan Rong, College of Resources and Environmental Sciences. Nanjing Agricultural University, Nanjing 210095-China

Ruy Anaya de la Rosa, Corporate Carbon, Sydney NSW 2000, Australia

Cornelia Rumpel, CNRS, Institute for Ecology and Environmental Sciences (IEES), Paris, France

Miguel Ángel Sánchez-Monedero, Department of Soil and Water Conservation and Waste Management, CEBAS-CSIC, Espinardo Murcia, Spain

Jens Schneeweiß, Centre for Baltic and Scandinavian Archaeology University of Kiel, Germany

Simon Shackley, School of GeoSciences, University of Edinburgh, Scotland, EH8 3JX, UK

Jessica G. Shepherd, School of GeoSciences, University of Edinburgh, EH9 3FF Edinburgh, UK

Carlos Alberto Silva, Soil Science Department, School of Agricultural Sciences, Federal University of Lavras, Lavras, MG 37203-202, Brazil

Giuseppe di Rauso Simeone, Department of Agricultural Sciences, University of Naples Federico II, Portici, Italy

Balwant Singh, Sydney Institute of Agriculture, School of Life and Environmental Sciences, The University of Sydney, NSW, Australia

Bhupinder Pal Singh, NSW Department of Primary Industries, Elizabeth MacArthur Agricultural Institute, Menangle NSW 2568, Australia

Saran Sohi, UK Biochar Research Centre, School of GeoSciences, University of Edinburgh, The King's Buildings, EH9 3JN, United Kingdom

Otakar Strunecký, Institute of Technology and Business, Okružní 517/10, České Budějovice, Czech Republic

Einar Stuve, Oplandske Bioenergi AS, Norway

Cecilia Sundberg, Department of Energy and Technology, Swedish University of Agricultural Sciences, 75007 Uppsala, Sweden

Sarasadat Taherymoosavi, School of Materials Science and Engineering, University of NSW, Sydney, NSW 2052, Australia

Ehsan Tavakkoli, NSW Department of Industry Skills and Regional Development: Wagga Wagga, NSW, Australia

Wenceslau Teixeira, Embrapa Solos, Rio de Janeiro, Brazil

Janice E. Thies, Department of Crop and Soil Sciences, Cornell University, Ithaca, NY 14853 USA

Lukas Trakal, Department of Environmental Geosciences, Faculty of Environmental Sciences, Czech University of Life Sciences Prague, Kamýcká 129, Suchdol, 165 00, Prague 6, Czech Republic

Minori Uchimiya, U.S. Department of Agriculture, Agricultural Research Service, New Orleans, USA

Lukas Van Zwieten, NSW Department of Primary Industries, Wollongbar NSW 2477 Australia

Frank G. A. Verheijen, Department of Environment and Planning, Centre for Environmental and Marine Studies (CESAM), University of Aveiro, Portugal

Hailong Wang, School of Environmental and Chemical Engineering, Foshan University, Foshan 528000, China

Shengsen Wang, College of Environmental Science and Engineering, Yangzhou University, Yangzhou 225127, China

Zhe Han Weng, Animal Plant & Soil Sciences, Latrobe University, Australia

Joann K. Whalen, Department of Natural Resource Science, McGill University, Montreal, Canada

Thea Whitman, Department of Soil Science, College of Agriculture and Life Sciences, University of Wisconsin-Madison, Madison, WI 53706, USA

Piumi Amasha Withana, Korea Biochar Research Center, APRU Sustainable Waste Management Program & Division of Environmental Science and Ecological Engineering, Korea University, Seoul, 02841; International ESG Association (IESGA), Seoul, 06621, Republic of Korea

Dominic Woolf, Soil and Crop Sciences, School of Integrative Plant Science, Cornell University, Ithaca, NY 14853 USA

Nicole Wrage-Mönnig, Grassland and Fodder Sciences, Faculty of Agricultural and Environmental Sciences, University of Rostock, Germany

Christian Wurzer, UK Biochar Research Centre, School of GeoSciences, University of Edinburgh, EH9 3FF Edinburgh, UK

Andrew R. Zimmerman, Department of Geological Sciences, University of Florida, Gainesville, FL 32611, USA

Preface

The field of biochar has changed quite a bit since we compiled the first and second editions of this book. Research on biochar has accelerated even further, which by now found broad acceptance within the wider scientific community on soil health, climate change mitigation and adaptation, or circular economy. Beyond a soil conditioner, biochar research includes its use in the fertilizer industry and investigations on soil-less potting media have intensified. In addition to land management, biochar is now also used in diverse materials in the construction industry. This book not only reviews recent advances made in our understanding of biochar properties, behavior, and effects in agriculture, environmental management, and material production, but specifically develops fundamental principles and frameworks of biochar science and application.

The book serves as an introduction to biochar for students, scholars, and lay readers, as well as a comprehensive textbook for anyone who wants to gain a deeper understanding of biochar. At the same time, it highlights new insights at the frontier of biochar science, develops new concepts for its investigation and use, and identifies knowledge gaps and future research and development needs. It is intended to provide essential information available to date to land use planners, farmers, homeowners, teachers, consultants, policymakers, regulatory agencies, and project or business developers.

The interest in biochar is still expanding and reaching new sectors, since despite its ancient roots biochar remains a relatively new industry and topic of science for an increasing and broader group of people. Regional and local groups have been founded in many countries, and the international networks of scientists, industry, project developers, and policymakers interested in biochar have advanced the discourse and sustainable development of biochar. These networks develop frameworks for commercializing biochar such as setting standards of what safe biochar is and how it can be used sustainably to address environmental issues while recognizing social and economic constraints. This book contributes to rigorous scientific inquiry and hopes to motivate the development of responsible and sustainable biochar applications. It attempts to lay out the complexity of biochar systems, covering both the detailed science as well as the broad development and policy picture, to outline concepts for further inquiry as well as realistic and achievable implementation.

The book is divided into five main areas: (1) History and fundamentals of biochar investigation, production and use; (2) Basic physical and chemical properties of biochar and their classification; (3) Persistence, changes and movement of biochar in the environment; (4) Plant productivity and environmental processes that are affected by biochar including soil biota, nutrient and carbon transformations and movement, soil remediation, greenhouse gas emissions, soil water and pollutant dynamics in soil (such as organic pollutants, heavy metals, herbicides); (5) Implementation of biochar that requires assessments of biochar contents, its use in commercial products including biochar-based fertilizers and as part of wider biochar systems, greenhouse gas accounting, certification, economics, commercialization, and policy.

We are extremely grateful to the authors who made this book happen, as well as the numerous referees who spent a significant amount of their time giving expert opinions that ensured the high scientific quality of this publication. In particular, we want to thank Rebecca Abney, Mohammad Al-Wabel, Jim Amonette, Elias Azzi, Chumki Banik, Mattia Bartoli, Julia Berazneva, Luke Beesley, Catherine Brewer, Wolfram Buss, Marta Camps-Arbestain, Marialuz Cayuela, Chih-Hsin Cheng, Gerard Cornelissen, Annette Cowie, Daren Daugaard, Thomas DeLuca, Kathleen Draper, Tomek Falkowksy, Xiaodong Gao, Mauro Giorcelli, Ellen Graber, William Hockaday, Shahla Hosseini-Bai, Paul Imhoff, Jim Ippolito, Simon Jeffery, Davey Jones, Claudia Kammann, Thomas Klasson, Rai Kookana, Isabel Lima, Jingdong Mao, Josef Maroušek, Ondrej Masek, Caroline Masiello, Tim McAllister, Leonidas Melo, Tom Miles, Patryk Oleszczuk, Joseph Pignatello, Ghasideh Pourhashem, Yamina Pressler, Peter Quicker, Cornelia Rumpel, Simon Shackley, Bhupinder Pal Singh, Ron Smernik, Cecilia Sundberg, Sonal K. Thengane, Thomas Trabold, Chi-Hwa Wang, Hailong Wang, Andrea Watson, Han Wei, Joann Whalen, Thea Whitman, Dongke Zhang, Andy Zimmerman, Lukas van Zwieten, and several anonymous referees.

Finally and most importantly, we want to thank our families and friends for all their patience with the frenzy of organizing this volume, all the late-night writing, and their full support, without which we would not have been able to put together this book.

Johannes Lehmann
Ithaca

Stephen Joseph
Saratoga

1

Biochar for environmental management

An introduction

Johannes Lehmann and Stephen Joseph

What is biochar?

Biochar is the product of heating biomass in the absence of or with limited air to above 250°C, a process called charring or pyrolysis also used for making charcoal (Chapter 3). The material distinguishes itself from charcoal or other carbon (C) products in that it is intended for use as a soil application or broader for environmental management including building materials and other bio-products. In some instances, the material properties of biochar may overlap with those of charcoal as an energy carrier, but many types of biochar do not easily burn and charcoals are typically not made to address soil or environmental issues (Nomenclature in Box 1.1). An important defining feature of biochars, similar to charcoal, is a certain level of specific organic C forms, called fused aromatic ring structures (Chapters 6 and 7). These structures are formed during pyrolysis and are key to biochar properties for mineralization (Chapter 11) or adsorption (Chapter 10) among others. Therefore, biochar is typically

enriched in C (Figure 1.1), and even more in phosphorus (P), or other metals such as calcium (Ca) or magnesium (Mg), and sometimes even nitrogen (N). The chemical properties of the organic C structure of biochar are fundamentally different from those of the material that the biochar was produced from and depleted in oxygen (O) and hydrogen (H). In contrast, the macro-morphological characteristics of biochars typically resemble those of the starting material, which means

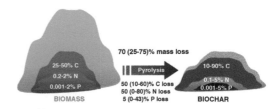

Figure 1.1 *Conversion efficiency of biomass, C, N, and P during pyrolysis (data from Enders et al. (2012); typical losses followed by range in brackets)*

DOI: 10.4324/9781003297673-1

that it typically looks the same, apart from its black color. The intended use as a soil amendment or broader environmental material often requires that biochars do not contain harmful levels of heavy metals (Chapter 21) or organic contaminants (Chapter 22), in keeping with related efforts to make other soil amendments or materials safe for the intended use. Despite these common criteria, it would be wrong to conclude that biochar is a narrowly defined material. In fact, biochars can have very different properties, which have to be recognized, as discussed throughout this book.

Box 1.1 Nomenclature of biochar and related materials in comparison to pyrogenic C structures

Johannes Lehmann, Joseph Pignatello, Michael Bird, Stephen Joseph

The following nomenclature for biochar and related terms has been adopted in this book and may provide guidance for achieving greater clarity more broadly in publications on the subject. In some instances, clarity in conversation may also improve conceptualization and scientific advances, which is intended to promote an understanding of biochar properties and its behavior in the environment.

Biochar: Biochar is the solid product of pyrolysis, designed to be used for environmental management. IBI (2013) defines biochar as: "A solid material obtained from thermochemical conversion of biomass in an oxygen-limited environment. Biochar can be used as a product itself or as an ingredient within a blended product, with a range of applications as an agent for soil improvement, improved resource use efficiency, remediation and/or protection against particular environmental pollution, and as an avenue for greenhouse gas (GHG) mitigation". In addition, to be recognized as biochar according to IBI (2013) or Delinat (2012), the material has to pass several material property definitions that relate both to its value (e.g., H/Corg ratios relate to the degree of charring and therefore mineralization potential in soil) and its safety (e.g., bioavailable heavy metal content). This publication uses the term biochar even when citing publications that use other terms but clearly refer to the use of such materials in the context defined for biochar.

Hydrochar: Hydrochar is the solid product of hydrothermal carbonization (HTC) or liquefaction (sometimes referred to as HTC material) and is distinct from biochar due to its production process and properties (Libra et al., 2011). It typically has higher H/C ratios of above 0.7 mol mol^{-1} (Schimmelpfennig and Glaser, 2012; Kambo and Dutta, 2015) and lower aromaticity than biochar as well as little or no fused aromatic ring structures. Hydrochar is not covered in this publication and is only occasionally discussed in comparison to biochar.

Pyrogenic Carbonaceous Material (PCM): PCM is used here as the umbrella term for all materials that were produced by thermochemical conversion and contain some organic C, such as charcoal, biochar, char, black carbon, soot, and activated carbon. The term refers to the material, does not make a distinction between inorganic or organic components, and does not refer to the C atom.

Pyrogenic Organic Matter (PyOM): PyOM is introduced here as the umbrella term for all materials that were produced by thermochemical conversion by any means that contain no inorganic C or refer to the organic component of PCM. Similar to PCM, the term refers to the material and not the C atom.

Char: Char is defined for the purpose of this publication as the material generated by incomplete combustion processes that occur in natural and man-made fires.

Charcoal: Charcoal is produced by thermochemical conversion from biomass (mainly but not exclusively wood) for energy generation. The term is sometimes used in the context of other uses, e.g., medicine, filtration, separation, etc. If processed further by any form of activation, the use of the term "activated carbon" is proposed.

Activated Carbon: Activated carbon denotes a PCM that has undergone activation, for example, by using steam or the addition of chemicals. It is used in filtration or separation processes, sometimes in restoration, and for specialized experiments in soil (competition, inoculation, etc). "Carbon" in this context should not be abbreviated to "C", since it does not refer to the C atom in activated carbon, but to the material (which also contains atoms other than C). The acronym "AC" for activated carbon will be used in this publication only if needed repeatedly, but otherwise should be spelled out, "activated carbon". Clarification is needed in those instances where biochars were modified after production for which some sources use the term "activation". Such treatment of biochars is typically ill-defined and it should be explained in detail what "activation" of biochars means in a particular study. The use of the term "activated biochar" is discouraged.

Black Carbon: The term black carbon ("carbon" spelled out) is extensively used in the atmospheric, geologic, soil science, and environmental literature to refer to PCM dispersed in the environment from wildfires and fossil fuel combustion. The term should be taken to refer to the entire material, not just the fused ring fraction or the C atom. The use of this term is discouraged (or be used only if absolutely necessary and in the context described here) to avoid confusion with "Black C" which is defined below.

Soot: Soot is a secondary PCM and a condensation product (Chapter 3). Chars, charcoal, biochars, black carbons (and, to a limited extent, also activated carbon), may contain soot, but soot can also be identified as a separate component resulting from gas condensation processes.

Ash: Ash is an operationally defined fraction of biomass or PCM (according to ASTM D1762–84) that typically includes inorganic oxides, chlorides, sulfates, and carbonates (Enders et al., 2012). For the purposes of this publication, the term does not describe the solid residue of combustion which commonly contains some residual organic C.

When referring to the C atoms of the PyOM, the letter C should be used as in "pyrogenic C", abbreviated to PyC. A selection of terms referring to C forms in PyOM relevant to this publication includes:

** "Pyrogenic C" (abbreviated to PyC after first use) refers to the C atom, and not to the material that also contains H, O, N, and ash material (Figure 1.2). PyC includes the (non-inorganic) C atoms that have undergone pyrogenic or thermal transformation, and by this definition only include C present in fused rings, including C on surfaces of fused aromatic C that may also bind to other atoms than C such as C-O/N, non-protonated C, and protonated C. In this publication, the term does not include non-transformed C present in residual carbohydrates or lignin structures, in tars, or in functional groups bound to fused aromatic C such as carboxyl groups. Different methods to quantify PyC typically attempt to capture this C fraction (Chapter 24), but do so with varying success or intentionally capture a portion of it (e.g., only the fused aromatic C without the surface C). When referring to a certain analytically defined fraction, the method should be stated in conjunction with the term PyC (e.g., PyC quantified by CTO-375).

** "Black C" spelled with "C" and not "carbon", is synonymous with PyC. "Black C" should not be abbreviated to BC as this can be confused with biochar (which is in some publications abbreviated to BC; the acronym BC is therefore not used here). PyC should be used preferentially to "black C".

** "Total Organic Carbon" (abbreviated TOC) refers to the entire organic C component of any material, including all thermally altered organic C as well as remaining untransformed organic C. "Total inorganic carbon" (abbreviated TIC) mainly includes carbonate and possibly other compounds such as oxalates (Figure 1.2).

** In some cases, "soot C" is appropriate to indicate the C atom properties of soot (which is a secondary PCM as defined above).

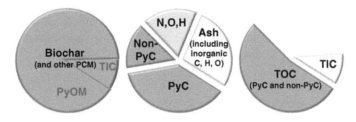

Figure 1.2 *Illustrative sketch of a possible distribution of different C forms and other atoms in biochar (component and acronyms explained in the text)*

A brief history of biochar research and application

Valuing biochar-rich soils and the concept of adding biochar to soil and in potting mixes reaches back several centuries and has found entry into some traditional management concepts in many regions worldwide (Chapter 2). Even though some notable research was done, the historic reports and scientific studies started mostly observational and were initially, in large part, gathered from plant growth responses on former charcoal storage sites (Chapter 2). Biochar application was discussed in major agricultural textbooks (Allen, 1846) and in scientific journals (Anonymous, 1851) and developed into commercial products as a form of "manure" in the mid-1800s, but with varying success as seen for biochar made from peat (Durden, 1849). By the second half of the 19th century, scientific studies on biochar had increased substantially, not the least due to Justus von Liebig's publications (von Liebig, 1847) providing quantitative proof combined with the theoretical underpinning of why biochar may improve nutrient availability. This interest in biochar continued well into the 20th century (e.g., Retan, 1915; Morley, 1927; Verona and Ciriotti, 1935), but most of the research and development ceased by the middle of the 20th century, possibly trailing the development and marketing of inorganic fertilizers. Notable research and development on biochar started again in the 1980s in Japan (Ogawa and Okimori, 2010). The present interest in biochar research and development was mainly motivated by research on Amazonian Dark Earths (also called *Terra Preta de Indio*; Mann, 2002; Marris, 2006). These soils found in the Amazon Basin were created by Amerindian populations several hundred to a few thousand years before the present but maintained their fertility largely due to the high proportion of biochar-type organic matter (Glaser and Birk, 2012). Even though *Terra Preta* soils do not provide a direct analog to biochar management (Lehmann, 2009) and are by far not the only soils containing biochar (Chapter 2), they can be credited for spurring a recent investigation into whether biochar can provide broader soil benefits in its own right (Glaser et al., 2002). In parallel, naturally produced chars from vegetation fires are becoming re-appreciated as the reason for the high fertility attributes of some soils such as those in the U.S. Midwest (Mao et al., 2012).

The term "biochar" was introduced only recently, first as a term to distinguish activated carbon made from fossil fuel and activated carbon made from biomass (Bapat et al., 1999), and shortly thereafter to replace the term "charcoal" as a fuel (Karaosmanoglu et al., 2000) and to distinguish it from coal.

Biochar as the term used in this book and by now more widely accepted globally in the context of a soil amendment, was introduced in 2006 (Lehmann et al., 2006) based on conversations with Peter Read.

Research on *Terra Preta* and naturally occurring chars (often under the term black C or charcoal, Box 1.1) initially dominated the scientific literature on biochar-relevant topics, but in 2008 the number of articles in academic journals on purposeful application of biochar to soil started to increase (Figure 1.3). The term charcoal continues to be used in the context of a soil amendment, but with a decreasing proportion. The publication activity of biochar in the scientific literature now exceeds that in the historically more established subject of compost science (Figure 1.3). Similarly, citations of scientific articles on biochar have risen and are per publication by now about 10 times higher than those on compost, which may be taken as an indicator of scientific interest in biochar research.

In addition to the scientific output, the development of biochar over the past years can also be traced by examining patents, sessions at annual meetings of professional societies, products in the marketplace, or the number of interest groups. Groups of interested stakeholders started to form in 2006 and regional and national groups constituted themselves by 2010. Patents were in appreciable numbers only published after 2010. Biochar has become part of educational curricula and dedicated seminars, and the Intergovernmental Panel on Climate Change (IPCC) has adopted a biochar method in the appendix to the updated guidelines for national greenhouse gas accounting in 2019. The connection with *Terra Preta* soils in the Amazon has provided a narrative that has stimulated a general interest in soils for those who may otherwise have less interest in agriculture.

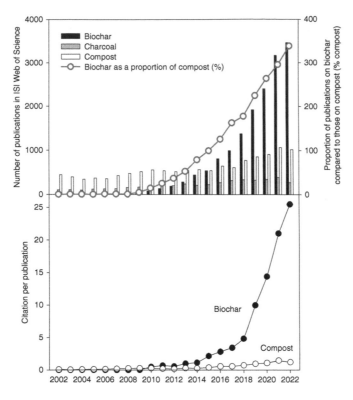

Figure 1.3 *Number of publications and citations per article (in the same year) in scientific journals listed in the ISI Web of Knowledge (http://apps.webofscience.com) with the term biochar in the title of the article in comparison to those with either charcoal or compost or composting in the title (all sources and contexts of the terms were considered, even if charcoal pertains to energy, different from the 2015 edition of this book; not considered were publications that contain relevant research but where the terms were not used in the title, or with the terms black C or pyrogenic C or pyrogenic organic matter even if they are relevant to biochar; the numbers reported here for biochar research are therefore conservative)*

Biochar as a system

In a narrow sense, biochar is the term for a range of materials. But actually, any benefits that the production and use of biochar can generate can often be realized only if biochar is perceived as a systems approach. A wide variety of biomass can be used to produce a wide variety of biochar materials, each with its opportunities and constraints, creating trade-offs and synergies (Jeffery et al., 2015). Some biomass is a valuable commodity for other purposes, such as food and construction wood, or has environmental value for soil protection, shade, or wind breaks. In each specific circumstance, the use or abuse of biomass has to be critically evaluated. When the biomass is heated to a point where pyrolysis occurs, the energy generated by the pyrolysis is sufficient to continue the reaction (Figure 1.4). However, depending on the moisture content of the biomass, the heat

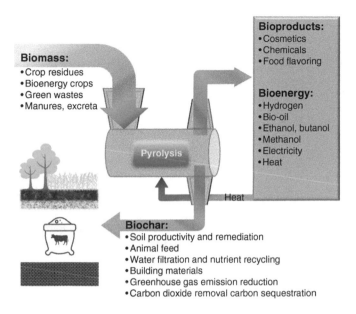

Figure 1.4 *Schematic of a basic biochar system*

can be sufficient to maintain the pyrolysis or require more energy to dry than is contained in the biomass. The rest of the energy that is released can be utilized to produce a wide variety of products, including energy but also other bioproducts such as food flavoring. The energy produced from these volatiles can have various forms owing to the comparatively low temperature used in comparison to combustion or gasification and ranges from heat and electricity to hydrogen gas, converted using microorganisms to ethanol or butanol or using catalysis to methanol or bio-oil. Biochar is a solid product with about a third of the mass yet containing half the C originating from the biomass (Figure 1.1).

For biomass input, bioenergy or bioproduct, and biochar output, many different permutations of the system are possible (Chapters 31–34). Biomass input may not only include plants grown for their sole purpose as feedstock for pyrolysis, but also residues from crop production or food and energy processing as well as manures and sewage. Therefore, the motivation or entry point for a biochar system can be very different. It is useful to distinguish between four broad groups of objectives: improvement of soil or materials, mitigation of soil, water, or air pollution, including climate change, waste management, and energy generation (Figure 1.5).

Biochar use

Biochars may be used as a soil amendment as well as in other ways to address environmental issues, including but not limited to building materials or animal feed. Soil improvement using biochars may target not only (i) crop productivity through improvements of soil nutrient availability (Chapters 8, 16, and 19), soil physical properties and specifically water relations (Chapters 5 and 20), or plant-microbe interactions (Chapters 14, 15, and 23), but also (ii) soil remediation (Chapters 21, 22, and 27). The potential value of biochars in a particular soil is in the

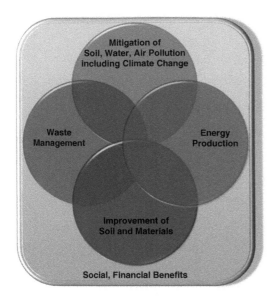

Figure 1.5 *Motivation for applying biochar systems*

first instance related to properties that can also be addressed by additions of other organic matter such as compost or manure, albeit with important nuances. Obviously, not all soil constraints can be addressed with biochar, and if soil properties do not constrain productivity and the soil is very fertile, then biochar additions will likely not improve crop yields (Chapter 13). The fact that biochars can have very different properties depending on the material they were produced from and their production conditions, also changes their utility to address any existing soil constraints. Biochar use in soil-less planting media, as compost additive, in animal feed with subsequent use of the manure in soil (Chapter 29), as admixtures in fertilizers or in green roofs (Chapter 26), as building material (Chapter 28) among others may require very different properties than as a soil conditioner.

The ability to target biochars with very different attributes affords the possibility to design biochars for certain purposes ("fit for purpose", "designer biochar"; Chapter 31) in

a potentially very effective way. For example, biochars made from the same feedstock can have pH values of less than 4 or above 12 (Chapter 8), one being potentially able to address the fertility of soils with pH values that are either too high or too low for optimum productivity. However, the wide variety of biochar materials also hampers effective communication in science, the public, and the marketplace: biochar does not equal biochar to the extent that a common term may even provide a hurdle. Classification is therefore needed to distinguish different biochars (Chapter 9) and it may even prove prudent to develop a nomenclature of different subsets of biochars with different properties and use them to facilitate communication with and between stakeholders.

Benefits derived from biochar use in some way (whether as soil amendment or otherwise) are possibly the defining feature of a biochar system. If the primary objective is to sequester C, a lack of gaining social or financial capital through biochar utilization beyond sequestration may provide a disincentive. Biochar can also be used as a charcoal fuel, and greater energy gains may provide greater reductions in greenhouse gas emissions if no soil benefits can be realized (Gaunt and Lehmann, 2008; Woolf et al., 2010). In fact, reduced greenhouse gas emissions from soils or greater plant growth may need to be achieved for a biochar system to have a preferable emission balance compared to biochar use as a charcoal fuel (Woolf et al., 2010). Revenues from increased crop yields may prove critical for financial viability.

Mitigation of soil, water, and air pollution including climate change

The opportunity to reduce greenhouse gas emissions clearly shows the need to perceive biochar management as a system rather than a material (Chapter 30). The lower mineralization of biochar than the original material that it was produced from (Chapter 11) reduces the CO_2 emissions from the system and is indeed key to climate change mitigation with biochar (Chapter 30). However, CO_2 capture is delivered by plants through photosynthesis, and whether an old-growth forest or a decomposing crop litter is used to produce biochar dramatically changes the C balance (Woolf et al., 2010). Not only the C balance, but the emissions generated or reduced in the entire system determine the life-cycle greenhouse gas budget and include nitrous oxide or methane emissions from soil (Chapter 18) or from the decomposing biomass, as well as from transportation, infrastructure, indirect land use change and others (Chapter 30). A clear distinction between greenhouse gas emissions and carbon dioxide removal is key to understanding tradeoffs and synergies between different practices of biochar systems (Lehmann et al., 2021). Even though the technical or theoretical potential is substantial on a global scale (Woolf et al., 2010) and on par with or exceed many alternatives (Lehmann et al., 2021), the actual mitigation that will be achieved clearly depends on environmental sustainability, social acceptance, technological implementation, and economic competitiveness compared to other mitigation options (Roe et al., 2021). This can not be evaluated without some commercialization at a meaningful scale. Similar to several other agricultural C sequestration strategies (e.g., reducing tillage), certain financial, environmental, and societal benefits can be realized through building soil and animal health or improving properties of, for example, building materials that will create lasting value for crop productivity, clean water or the built environment beyond C sequestration.

Efforts in mitigating pollution of water or air through animal and human excreta may benefit from pyrolysis and retention of nutrients from urine or separated liquids on biochars to generate transportable fertilizers (Krounbi et al., 2021). The ability to densify nutrient-rich manure by drying and pyrolyzing decreases weights by over 90%. Adding biochar to soil to retain nutrients is an additional avenue to reduce water pollution (Chapter 19). It is not clear whether nutrient trading schemes will be able to make use of this mechanism.

Feedstock as waste management

Processing of wastes is a relatively established use of pyrolysis, even if the use of biochar for environmental management is not widespread at present. The generally lower processing temperatures and the higher organic C contents of the solid residue compared to incineration or gasification (Chapter 3) facilitate operations. These have to be weighed against longer processing times in the case of slow pyrolysis and different limitations concerning size and type of installations (Chapter 4). Typically, the range of materials that can be processed by various permutations of pyrolysis technology is wide and includes woody biomass, leaves, grasses, animal and human manures, sludge, or crop residues (nut shells, pits, stones, bagasse, rice hulls, straw, etc.). High moisture contents can make some feedstocks less attractive if there is a need for net energy generation, and beyond a certain threshold of moisture content, more energy is needed for operation than is contained in the feedstock. The choice for a particular type of waste may be more limited by the requirement to produce a biochar that is (i) safe to apply to soil, as well as (ii) appropriate for effectively addressing soil constraints relevant at a project or regional level. Waste management is a common entry point for biochar systems and the economics often dictate the use of materials that need disposal with low or even negative costs sometimes even generated at a single location (Chapters 31–34). Biochar production may be an attractive alternative in those situations where no local disposal is available and the biomass (e.g., yard wastes, animal manures, and human excreta) has to be otherwise transported over long distances. Especially with efforts to close the nutrient and C cycle between urban and agricultural regions, long transportation distances are prohibitive to cost-effective recycling, and it remains to be seen whether biochar technology can provide an alternative. An important opportunity is biochar-based fertilizers made from human excreta and a comprehensive recycling of both urine and feces. In addition, appropriate management of organic wastes can help in (1) mitigating climate change indirectly by decreasing methane and nitrous oxide emissions from landfills; (2) reducing industrial energy use and emissions due to recycling and waste reduction; (3) recovering energy from waste; (4) enhancing C sequestration in forests due to decreased demand for virgin paper; and (5) decreasing energy used in long-distance transport of waste.

Energy production

Pyrolysis is a recognized and long-standing technology to provide energy (Chapters 3 and 4). In addition to heat energy, pyrolysis is also able to generate a variety of high-value liquid and gaseous energy carriers. Furthermore, a range of products can be produced from food flavoring to agrochemicals, fertilizers, cosmetics, medicine, adhesives, and others. In the early 20th century, pyrolysis was the only technology to produce methanol, acetone, or acetic acid, in addition to some liquid fuels (Goldstein, 1981).

Prioritizing energy or other non-solid products will in most cases constitute a tradeoff to biochar production. However, from a life-cycle perspective, maximization of energy generation may be less preferred than weighing soil health and environmental benefits including climate change mitigation against energy generation (Chapters 30 and 31). It is theoretically possible that securing the production base by prioritizing soil fertility in a pyrolysis bioenergy project through biochar additions to the soil will in the long term achieve greater energy gains than maximization of energy generation during pyrolysis.

Bioenergy may on its own not be able to satisfy the growing global energy demand under realistic constraints to biomass production (Kraxner et al., 2013). But it may significantly contribute to a future energy solution (Gelfand et al., 2013), and possibly be competitive for distributed production of liquid or gaseous fuels and bioproducts. Pyrolysis in particular may prove capable of addressing constraints for many bioenergy approaches (Liu et al., 2022) posed by varying availability of feedstock types, either between different locations or different times of the year, because of its versatility in accepting a wide variety of organic materials.

Similar to combustion technology, pyrolysis technology can be operated at different scales, from stoves to cook meals or heat individual homes to large bioenergy installations that generate liquid fuels (Chapter 4). The specific technology solution will need to vary considerably to meet different objectives. The upper limit for the scale of individual pyrolysis reactors will likely remain smaller than that of biomass combustion or fossil fuel-based conversion technologies (Chapter 4). This may mean that pyrolysis may also for this reason be best utilized for distributed energy generation.

Current state of biochar science, development, and deployment

Scientific activity has undoubtedly accelerated significantly over the past five years (Figure 1.3), and the number of research publications can be expected to increase further in the near future. Even though the scientific output is currently high and much information on biochar is now available (Joseph et al., 2021), several critical knowledge gaps are only being filled over time and are identified in each chapter of this book. Of particular note is the lack of a decision tool to identify biochar types suitable to address certain soil, material, and environmental constraints. While a comprehensive tool will only be available with a more mature state of science that considers all or at least most of the permutations of possible biochar properties and application, useful milestones can already be reached at present by identifying those biochars (Chapters 8, 11, 13, 21, and 22) and biochar systems (Chapters 13, 28, 29, and 31) that are sufficiently investigated. The analytical framework is under development for characterizing both the material properties of biochars (Chapter 9) and the systems benefits (Chapters 27–29 and 31). In this context, scientific studies require careful planning to implement valid comparisons of tested biochar applications to either standard biochar available to the global scientific community, a control without adding biochar, or additions of equivalent amounts of crop residues or composts (Jeffery et al., 2015). The alternative to biochar applications typically is not to apply none, but a different type of organic matter, and the alternative to compost applications may not be applications of biochar on its own but together with compost or inorganic fertilizers (Chapter 26). On the one hand, a systems comparison of biochar effects on greenhouse gas emissions from soil (Chapters 18, 30, and 31) may only succeed if it is compared to applications of unpyrolyzed biomass considering the conversion of biomass to biochar (Chapter 3). Mechanistic insights, on the other hand, about how biochar influences greenhouse gas emissions from soil can only be obtained using comparisons on the same mass or C basis. The modification of biochars post-production may deserve particular attention to isolate specific effects by keeping others (e.g., pH) constant (Guerena et al., 2015). Random variation between biochars may often not provide the necessary parameter space to identify the mechanisms by which biochars affect soil processes (Rajkovich et al., 2012). Such necessary refinement of experimental designs and the development of clear and testable hypotheses require prior knowledge. Formulating appropriate expectations can, by now, build on a sufficient body of published research as summarized in this publication.

Despite the impressive increase in the number of scientific studies, relevant knowledge gaps may need to be addressed by experimentation at scale of implementation, notably in the area of life-cycle evaluation of environmental impact and specifically greenhouse gas emissions and carbon dioxide removal (Chapter 31; CDR-fyi, 2022), economic evaluation (Chapters 32 and 33), and production technology (Chapter 4). Commercial-scale production and application will also generate opportunities to address the need for longer-term data sets of biochar effects on crop productivity (Chapter 13) and on off-site impacts through leaching and erosion (Chapters 12 and 19). Fully investigated systems at relevant scales are a prerequisite for regional or global implementation.

The large variety of possible biochar products requires due diligence on the part of research to discover unintended consequences (e.g., Chapters 21–23), and on the part of producers to comply with best management practices and ethical as well as biophysical standards to offer a safe product. The regulatory frameworks must be in place, both to provide incentives and point out limits. Only a rational and considered discussion will ensure that biochar systems develop sustainably.

Important questions arise whether only one motivation or entry point (waste management, mitigation of soil, water, and air pollution including climate change, energy generation, improvement of soil and materials; Figure 1.5) is sufficient to generate the necessary social and financial benefits for a biochar system to operate sustainably; or whether two or even all four value streams are needed. Can, for example, greenhouse gas emission reductions alone be financially viable or socially acceptable, and conversely, is soil improvement socially acceptable even if greenhouse gas emissions are not reduced or even increased? And if several entry points have to generate value, how many opportunities exist that warrant research and development? In addition, tradeoffs may occur between different value streams: a greater biochar production may in the first instance reduce the amount of energy generated (Jeffery et al., 2015) unless the soil is sufficiently improved that biomass and therefore feedstock production increases. Such questions must be answered to develop biochar systems at any particular location. Scaling biochar systems will therefore still need to take local conditions into account, observing sustainability principles (Chapter 31) and working within particular policy (Chapter 33) and social (Chapter 32) frameworks.

References

Allen AB 1846 *The American Agriculturalist; Designed to Improve the Planter, the Farmer, the Stock-Breeder, and the Horticulturalist*, Saxton and Mills, New York, NY.

Anonymous 1851 Charcoal peat. *The Horticultural Review and Botanical Magazine* 1, 422–423.

Bapat H, Manahan SE, and Larsen DW 1999 An activated carbon product prepared from Milo (Sorghum vulgare) grain for use in hazardous waste gasification by ChemChar concurrent flow gasification. *Chemosphere* 39, 23–32.

CDR-fyi 2022 https://medium.com/cdr-fyi/cdr-fyi-2022-year-in-review-d095acd9a1a0

Delinat 2012 *Guidelines for Biochar Production: European Biochar Certificate. Delinat Institute and Biochar Science Network*, Delinat Institut for Ecology and Climate Farming, Arbatz, Switzerland.

Durden EH 1849 On the application of peat and its products, to manufacturing, agricultural, and sanitary purposes. *Proceedings of the Yorkshire Geological Society* 3, 339–366.

Enders A, Hanley K, Whitman T, Joseph S, and Lehmann J 2012 Characterization of biochars to evaluate recalcitrance and agronomic performance. *Bioresource Technology* 114, 644–653.

Gaunt JL, and Lehmann J 2008 Energy balance and emissions associated with biochar sequestration and pyrolysis bioenergy production. *Environmental Science & Technology* 42, 4152–4158.

Gelfand I, et al. 2013 Sustainable bioenergy production from marginal lands in the US Midwest. *Nature* 493, 514–517.

Glaser B, and Birk JJ 2012 State of the scientific knowledge on properties and genesis of

Anthropogenic Dark Earths in Central Amazonia (terra preta de Índio). *Geochimica et Cosmochimica Acta*, 82, 39–51.

Glaser B, Lehmann J, and Zech W 2002 Ameliorating physical and chemical properties of highly weathered soils in the tropics with charcoal – a review. *Biology and Fertility of Soils* 35, 219–230.

Goldstein IS 1981 *Organic Chemicals from Biomass*, CRC Press, Boca Raton, FL.

Guerena D, et al. 2015 Partitioning the contributions of biochar properties to enhanced biological nitrogen fixation in common bean (Phaseolus vulgaris). *Biology and Fertility of Soils* 51, 479–491.

IBI 2013 Standardized product definition and product testing guidelines for biochar that is used in soil. https://biochar-international.org/ibi-biochar-standards/#:~:text=The%20IBI%20Biochar%20Standards%20are,and%20approval%20of%20the%20document, accessed February 20, 2023

Jeffery S, et al. 2015 The way forward in biochar research: targeting trade-offs between the potential wins. *Global Change Biology – Bioenergy* 7, 1–13.

Joseph S, et al. 2021 How biochar works, and when it doesn't: A review of mechanisms controlling soil and plant responses to biochar. *Global Change Biology – Bioenergy* 13, 1731–1764.

Kambo HS, and Dutta A 2015 A comparative review of biochar and hydrochar in terms of production, physico-chemical properties and applications. *Renewable and Sustainable Energy Reviews* 45, 359–378.

Karaosmanoglu F, Isigigur-Ergundenler A, and Sever A 2000 Biochar from the straw-stalk of rapeseed plant. *Energy and Fuels* 14, 336–339.

Kraxner F, et al. 2013 Global bioenergy scenarios – future forest development, and-use implications, and trade-offs. *Biomass and Bioenergy* 57, 86–96.

Krounbi L, Enders A, Gaunt J, Ball M, and Lehmann 2021 Plant uptake of nitrogen adsorbed to biochars made from dairy manure. *Scientific Reports* 11, 15001.

Lehmann J 2009 Terra Preta Nova – Where to from Here? In: Woods WI, et al. (Eds) *Amazonian Dark Earths: Wim Sombroek's Vision*, Springer, Berlin, pp. 473–486.

Lehmann J, Gaunt J, and Rondon M 2006 Biochar sequestration in terrestrial ecosystems – a review. *Mitigation and Adaptation Strategies for Global Change* 11, 403–427.

Lehmann J, et al. 2021 Biochar in climate change mitigation. *Nature Geoscience* 14, 883–892.

Libra JA, et al. 2011 Hydrothermal carbonization of biomass residuals: a comparative review of the chemistry, processes and applications of wet and dry pyrolysis. *Biofuels* 2, 71–106.

Liu T, Miao P, Shi Y, Tang KHD, and Yap PS 2022 Recent advances, current issues and future prospects of bioenergy production: a review. *Science of the Total Environment* 810, 152181.

Mann CC 2002 The real dirt on rainforest fertility. *Science* 297, 920–923.

Marris E 2006 Black is the new green. *Nature* 442, 624–626.

Mao J-D, et al. 2012 Abundant and stable char residues in soils: implications for soil fertility and carbon sequestration. *Environmental Science and Technology* 46, 9571–9576.

Morley J 1927 Why I use charcoal. *The National Greenkeeper* 1, 15.

Ogawa M, and Okimori Y 2010 Pioneering works in biochar research in Japan. *Australian Journal of Soil Research* 48, 489–500.

Rajkovich S, et al. 2012 Corn growth and nitrogen nutrition after additions of biochars with varying properties to a temperate soil. *Biology and Fertility of Soils* 48, 271–284.

Retan GA 1915 Charcoal as a means of solving some nursery problems. *Forestry Quarterly* 13, 25–30.

Roe S, et al. 2021 Land-based measures to mitigate climate change: potential and feasibility by country. *Global Change Biology* 27, 6025–6058.

Schimmelpfennig S, and Glaser B 2012 One step forward toward characterization: Some important material properties to distinguish biochars. *Journal of Environmental Quality* 41, 1001–1013.

Verona O, and Ciriotti P 1935 Azione del carbone sulla vegetazione, Memoria V: il carbone animale. *Bolletino del R. Istituto Superior Agrario di Pisa* 11, 401–420.

von Liebig J 1847 *Chemistry and Physics in Relation to Physiology and Pathology*, Baillière.

Woolf D, Amonette JE, Street-Perrott FA, Lehmann J, and Joseph, S 2010 Sustainable biochar to mitigate global climate change. *Nature Communications* 1, 56.

2

Historical accumulation of biochar as a soil amendment

Bruno Glaser, Doyle McKey, Luise Giani, Wenceslau Teixeira,
Giuseppe di Rauso Simeone, Jens Schneeweiß, Nguyen Hien,
and Jeffrey Homburg

Introduction

Burned clay clasts in small areas of the Australopithecine site of Chesowanja in central Kenya are the earliest evidence of fire associated with humans, about 1.4 million years old (Gowlett, 2016). Fire produces charred residues, which are used by humans in various ways. The first known use of charcoal was for drawing pictures in caves about 38,000 years ago. It began to be used for smelting 4,000 years ago. Its first recorded medicinal use was in 1,500 BCE (i.e., 3500 years ago). However, we argue that the most significant benefits to humans were provided by charcoal that was not produced intentionally, at least not at first, but instead was a felicitous byproduct of people simply living and working in an area over long periods. People use fire for cooking, pottery, and metal production, and the resulting pyrogenic organic matter (PyOM; see Chapter 1 for terminology) accumulates ad persists in the

soil as biochar. Farmers noticed the favorable effects of biochar-enriched soils and began to prefer such soils for agriculture.

This chapter provides a historical overview of how PyOM was produced and accumulated in soils before the relatively recent phase of biochar research and development. We examine whether farmers recognized, and took advantage of, the positive effects of PyOM accumulation on the quality of soils for agriculture. The sections of this chapter are geographically organized, except for a final section that treats in detail a special agricultural practice, soil paring-and-burning. This practice occurs on several continents. Its treatment as a separate geographically cross-cutting section is justified by the features common to all examples, and by the fact that it has escaped the attention of most researchers interested in the human use of fire.

DOI: 10.4324/9781003297673-2

Historical use of biochar in North America and Mesoamerica

Few data exist on the traditional use of biochar for much of North America, although PyOM is a ubiquitous component of soils at many ancient settlements and associated landscapes across the continent. Anthropologists in Mexico have documented charcoal production from pine trees to reduce the weight of fuel transported to villages, where charcoal was used as cooking fuel for large feasts when people gathered. The ash and charcoal remaining from these feasts were commonly spread across agricultural field areas to improve the fertility of their soils (Adams, 1997). However, there are few similar examples where biochar was intentionally produced and used for soil amelioration.

Many open terrestrial ecosystems are fire-maintained, and charcoal and fire-altered organic matter enhance soil functions in these ecosystems (DeLuca and Aplet, 2008; Pingree and DeLuca, 2017). Areas were intentionally burned as a management practice for promoting natural plant productivity for human foraging and for grazing and browsing by wild game such as bison, deer, and elk that were hunted by humans in the past. However, there is little evidence in this region that charcoal was intentionally produced for uses other than as fuel.

Determining if prehistoric fires were the product of human activity or of lightning is difficult, but there are numerous ethnographic and archaeobotanical examples of fires used by humans for landscape management and for increasing the fertility of agricultural soils (Adams, 2004). For example, every 5 to 10 years the Acoma of west-central New Mexico lit fires in four different directions along the way with smoldering juniper bark up to about 20 km from the village. The purpose was not to destroy the land but to renew soil fertility. Other groups,

such as the Zuni and Cochiti of New Mexico and the Sonoran Desert, burned vegetation during fire drives to hunt rabbits, deer, and pronghorn antelope. The fires resulted in large quantities of char being incorporated into the soil (Adams, 2004). By burning, Hohokam farmers in central and southern Arizona cleared vegetation that clogged irrigation canals. Targeted applications of fire would have optimized human access to plant and animal resources, including fuel wood, and burnt chaparral vegetation drew game animals to new shrubby growth (Adams, 2004).

In north-central Montana, prairie fires were set by Native American groups along drive lines in preparation for communal bison hunts (Roos et al, 2018). These fires resulted in char layers in alluvial and colluvial deposits. Radiocarbon dates of these char layers indicate that fire management activities peaked around 1,100–1,650 CE, with the timing of paleofires strongly associated with wetter climate regimes, indicated by the Palmer Drought Severity Index. That is, fires were most common during the wetter periods when the biomass and fuel load of grasses was greatest. The same relationship was reported for a transect study across North America (Glaser and Amelung, 2003). This fire-climate linkage counters arguments often made that human fire management practices were independent of climate regimes in the past.

In North Carolina, natural and human-caused fires could be differentiated by studying the geomorphic settings of fires initiated by modern lightning strikes over the last 3,900 years using char and pollen records preserved in peat deposits in the southern Appalachian Mountains (Delcourt and Delcourt, 1997). Natural fires in forests of oak, pine, chestnut, and hickory caused by lightning strikes were concentrated on ridge summits and upper hill slopes on xeric, west-, southwest-, and south-facing slopes. Lightning strikes on mesic landforms rarely sparked wildfires. Char from weedy plants and cultigens was found in the stratigraphic record of Horse Cove Bog, with plants indicative of human use consisting predominantly of goosefoot, plantain, purslane, sumpweed, and maize (Yarnell, 1976; Delcourt, 1987).

PyOM-rich archeological deposits similar to *Terra Preta* of the Amazon Basin were found in some parts of the southeastern United States, most notably in tree island soils of the Everglades and Upper St. Johns River of Florida and in the Fourche Maline midden mounds of southeastern Oklahoma and southwestern Arkansas. Tree islands in the Everglades and Upper St. Johns River in South Florida were the only landforms suitable for ancient human occupations in these wetlands. Tree islands formed above areas of subsurface limestone bedrock surfaces. Their surfaces are typically elevated about 1 m above the surrounding marshes, and they commonly have a subsurface soil layer cemented with calcium carbonate, that is, calcrete or a petrocalcic horizon (Campbell et al, 1984; Schwadron, 2006; Schwadron et al, 2009; Ardren et al, 2016). These islands are usually vegetated by hammocks of live oaks and cabbage palms, although the tree canopy is rapidly disappearing today due to a variety of causes (e.g., the spread of a deadly fungus, wildfires, and sea level rise). Tree island soils are more than one-meter-thick midden (or trash) deposits composed of black to dark brown, organically enriched soil intermixed with deeply buried and very dense concentrations of vertebrate faunal remains (mainly turtle and snake bones), sherds (fragments of pottery), and other midden debris such as marine shell. Due to the intensity of human occupation over millennia, soil P levels are very high in these midden soils, much higher than in natural soils in the South Florida wetlands. The tree island midden sites suggest that coastal settlements used interior tree islands as special-use sites for hunting, fishing, building fires, and processing food resources as early as 5,000 years ago, and then shifted some settlements into the tree islands permanently about 2,500 years ago (Ardren et al, 2016).

The Bug Hill site is a Fourche Maline midden mound in the Jackfork Valley of the Ouachita Mountains in southeastern Oklahoma. It is an accretional mound composed of organic-rich soil and abundant PyOM, animal bone, lithic tools and debitage, sherds, and other artifacts (Altschul, 1983). The mound summit has a 1.5-m-thick A horizon, about four times the thickness of the natural off-mound area, and the mound soils have basic to slightly alkaline pH values (pH 7–8) that are excellent for bone preservation, unlike the acidic (pH 4.8–5.3) soils off-mound (Figure 2.1). Bug Hill was occupied from the Middle Archaic period through historic times (~5,000 BCE to late 1,800 CE). It was intensively occupied as a base camp for collecting and processing diverse animal and plant food resources over the last three millennia, from the Late Archaic through the early Caddoan period (~1,600 BCE to 1,000 CE). PyOM was concentrated on buried surfaces in the midden mound, on house floors, and in pit features within the mound. Carbonized plant remains included sunflower, sumpweed, pigweed, goosefoot, knotweed, giant ragweed, canary grass, and nutshell species (hickory, walnut, acorns, and pecans). Abundant fuel wood resources were also documented, including red oak, white oak, hickory, ash, black cherry, and elm, and common fuel woods included pine, black walnut, sycamore, dogwood, black gum, maple, hackberry, mulberry, persimmon, plum, honey locust, Osage orange, sweetgum, cottonwood, and willow (Shea and Altschul, 1983).

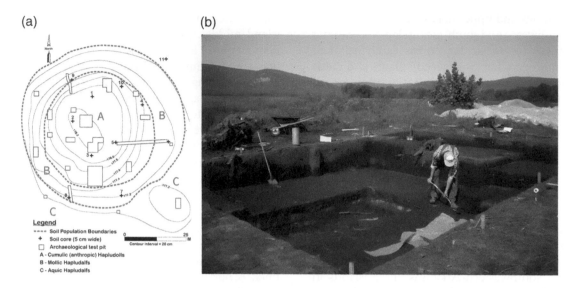

Figure 2.1 *Left: Map of the Bug Hill midden mound site, showing the different soil populations for the mound summit (A soils), mound slopes (B), and off-mound areas (C) (map modified from Johnson (1983)). Right: Charcoal-rich black midden soil in a profile of archeological excavations at Bug Hill in southeastern Oklahoma (with permission by Jeffrey Altschul)*

Historical use of biochar in South America

In Brazil, land covered by old-growth forests and savannas is traditionally cleared for agricultural and habitational purposes using slash-and-burn techniques (Figure 2.1). It has probably been practiced since the beginning of farming practices in Brazil. Some Amazonian indigenous peoples still pile and burn biomass to manage residues and create carbon-rich and fertile soil spots for agricultural purposes. The Kayapo (Hecht, 2009), the Kuikuro (Heckenberger, 2005; Schmidt et al, 2014), and Satere Maues (Steiner et al, 2009) are some of the ethnic groups whose fire management practices have been studied (Figure 2.2).

The arrival of metal axes that replaced stone axes made cutting vegetation more efficient (Carneiro, 1979) and probably facilitated the use of slashing and burning as a

practice for shifting cultivation. Traditional slash-and-burn practices are still employed today. The vegetation is first cut, piled, allowed to dry, and then burned. Burning is usually performed more than once, as incompletely burned charred trunks and branches are piled and reburned. The places where the vegetation is heaped and burned create small spots (*coivaras*) or windrows (*leiras*) rich in ashes and char pieces that become fertile spots. The *coivaras* and *leiras* can be easily identified some years after burning because of the better growth and productivity of crops. The amount of char formed during burning depends mainly on the type and amount of biomass. Secondary forest burns more thoroughly than primary forest. Studies reported that C stock in the char formed during land clearing ranged from 1.6 Mg ha^{-1} to 4.1 Mg ha^{-1}

Figure 2.2 *Areas with old secondary vegetation slashed and burned for agricultural purposes in Upper Rio Negro. Photo: Wenceslau Teixeira*

(Fearnside et al, 1999, 2001, 2007; Graça et al, 1999; Turcios, 2016).

Burning also reduces the weed seed bank and populations of some insects and pathogens (German, 2000). After some years, the effect of increased alkalinity and nutrients from the ashes disappears, and the area is at least temporarily "abandoned". As in other slash-and-burn rotational-agriculture systems, weed invasion is probably at least as important as fertility depletion as a cause for the "abandonment" of fields with PyOM-enriched soils. A new cycle of slash and burn starts in a new area, and the return period for the initial area varies from ten to fifty years.

Return periods are shorter for PyOM-enriched soils (German, 2003). Another way that some small farmers create C-rich and fertile soil is through the sweeping-and-char system (Falcão et al, 2019). Farmers sweep the foliar debris to create heaps, which are burnt periodically, and use the "burned soil", termed *terra queimada*, enriched in ashes and char pieces, to fertilize fruit trees. In Amazonia, char residues are commonly used for vegetable (horticultural and herbs) production in home gardens, and these gardens are in elevated planters to avoid damage caused by domestic animals (Steiner et al, 2003).

For these home gardens, dry woody biomass was collected and a mound was formed, followed by burning. The resulting soils had a strong smell of tars and were left for 3 weeks before planting. Only undemanding crops like manioc, pineapple, and trees are planted directly on these soils. Soils remained fertile for 3 years after this treatment (Steiner et al, 2003). For vegetables and medicinal plants, all available types of organic matter at the settlement were used (bones, wood, leaves, chicken manure), as well as rotten trunks of certain species of trees. This was then reduced in an open fire before it was mixed with the soils that had already been burnt (Steiner et al, 2003).

The best-known of these PyOM-enriched soils are the *Terra Preta de Indio* (usually shortened to *Terra Preta*), soils in the humid tropics of South America that are intensively used by farmers today (Glaser et al, 2001; Glaser and Birk, 2012; Figure 2.3). The recent global interest in biochar research was initiated by the scientific discovery of the genesis of these soils (Glaser et al, 2001, 2002), in which PyOM residues play a major role, accounting for their high and persistent organic matter

Ferralsol yellow very clayey distric – Central Amazon

Anthrosol pretic clayey epieutric– Central Amazon

Anthrosolo pretic epieutric – Central Amazon

Figure 2.3 *The original typical upland yellow dystric Ferralsol in Central Amazonia (left) and the created pretic epieutric Anthrosol (middle and right). Soil type according to the World Reference Base: Pretic Anthrosol (Clayic, Epieutric). Photos: Wenceslau Teixeira*

content and their sustainable fertility. However, until recently there was no strong evidence that *Terra Preta* was in fact intensively used by pre-Columbian cultivators. It is clear by now that both annual and perennial crops were grown by pre-Columbian farmers on these soils (Iriarte et al, 2020).

Hypotheses about the origin(s) of these Anthrosols feature a mix of purposeful and unintentional effects. According to a widely accepted hypothesis (the "midden model"), *Terra Preta* arose as a byproduct of waste accumulation near human settlements. Humans quickly observed the salubrious effects of the soils resulting from habitation sites and used them to grow nutrient-demanding crops in home gardens, or repurposed them as fields (Iriarte et al, 2020). A contemporary analogy illustrates how this process may have played out. Farmers near Iquitos, Peru, began to supplement their income by making charcoal for sale in the city, using some of their land for slash-and-burn farms and some for charcoal production. Increasing pressure on land led them to make farms on old charcoal-production sites. They noticed that crops grew better on these sites, and began to prefer them for opening farms. Such "kiln site agriculture" (Miltner and Coomes, 2015) shows how an initially unintentional byproduct led to an intentional strategy of using biochar for soil amelioration.

Terra Preta is famous for its persistent soil organic matter enrichment and its sustainable fertility (Glaser et al, 2001; Liang et al, 2006). The organic C of Terra Preta is composed of around 35% of PyC, compared to only 14% in the topsoil of surrounding soils (Glaser et al, 2000, 2001). The stock of PyC in *Terra Preta* is approximately 70 times higher than in the surrounding non-anthropic soils (Glaser et al, 2001). The carbonaceous substances in *Terra Preta* show graphitic core structures that act as a sink for cations, such as Ca, and a persistent graphitic structure with primary particles in the 2–8 nm scale that degrade slowly and increase CEC (Jorio et al, 2016). The critical value of about 2–8 nm for graphitic nanocrystallites found for the charred residues in *Terra Preta* implies a structure, different from that of fresh biochar, that reconciles functionality with persistence (Ribeiro-Soares et al, 2013; Archanjo et al, 2014; Jorio et al, 2016).

Historical use of biochar in Europe

Role of biochar in "Urban Dark Earths"

Several published articles on "Urban Dark Earths" in Europe can be found such as the archeological description of the stratigraphy of the Medieval Court of Hoogstraeten in the center of Brussels (Devos et al, 2013). Radiocarbon dating of *Fagus* charcoal revealed an age of around 1000 BP. Although detailed micromorphological investigations were conducted, there is no detailed discussion about the historical use of biochar as a soil amendment in this study.

Cologne constituted one of the political and economic centers of the northernmost frontier (limes) region of the Roman Empire and was at the core of important socio-political upheavals following its collapse. The late Roman and early Medieval periods of the city were examined using an interdisciplinary approach combining zooarchaeology, archaeobotany, and soil micromorphology (Grau-Sologestoa et al, 2023). In Antoniterstrasse and Heumarkt, the first documented urban Dark Earths in Cologne were analyzed, evidencing ongoing settlement activity from late Roman times to the Early Middle Ages. Other macroscopic residues were also registered, such as pottery sherds, charcoal, and bone fragments. The macro-remains spectra from Antoniterstrasse mostly include charred remains, of which 85% were cereal-derived. The most common taxon was emmer (*Triticum dicoccum*), followed by spelt (*Triticum spelta*), oat (*Avena sativa/fatua*), and barley (*Hordeum vulgare*), accompanied by some arable weeds, such as corn-cockle (*Agrostemma githago*), and poppy (*Papaver dubium/rhoeas*). Wild grasses, including brome (*Bromus spec.*), are also frequent, usually occurring as arable weeds. Some ruderal species were present, but no garden plants. This chaff-dominated composition does not reflect the surrounding vegetation, but more likely represents domestic waste. A charred cereal grain from Dark Earth from Antoniterstrasse was radiocarbon dated to 377–535 AD (Grau-Sologestoa et al, 2023). The latter is ubiquitous in all layers containing typical refuse such as bones, eggshells, ashes, charcoal, or other organic archaeobotanical material containing large amounts of charred arable weeds and cereals common for household waste. Fragmented domestic waste, charred chaff, as well as phosphates point to manuring for gardening activity. Moreover, this area, situated close to the Rhine, records minor flooding events which, in addition to the admixture of domestic waste and manure, kept the soil fertile, as is also seen from studies in Brussels (Devos et al, 2013).

In the Wendland region of Northern Germany, a Slavic settlement (10th/11th century AD), revealed a thick black soil (Nordic Dark Earth) with similar properties as *Terra Preta* (Wiedner et al, 2015). The existence of the Nordic Dark Earth in the temperature zone of Europe demonstrates the capability of sandy-textured soils to maintain high soil organic matter contents and nutrient retention over hundreds of years. Knowledge of Nordic Dark Earth was probably an important part of the Viking–Slavic subsistence agriculture system, which could have had a great impact on the development of the Viking age emporia in the 9th/10th century AD.

Role of biochar in Plaggenesch generation

"Esch" means slightly more elevated parts of arable sites and "Plaggen" are bits and pieces of vegetation, its enmeshed roots, and soil material sticking to them shallowly scraped with a hoe or spade (Giani et al, 2014; McKey, 2021). Thus, the name already indicates a man-made influence on the Plaggenesch generation. Plaggenesch cultivation, plaggen management (Blume and Leinweber, 2004), or plaggen agriculture (Giani et al, 2014; Urbanski, 2022) followed hundreds of years of shifting (slash-and-burn) cultivation of forest, which eventually led to its replacement by heathland vegetation. Cultivation of the soils amended by plaggen allowed cropping without fallow (Niemeier, 1939; Behre, 2008). It is assumed that Plaggenesch cultivation started during the Late Neolithic period and the Bronze Age (Simpson et al, 1998) and increased substantially during the early Medieval, approximately 1000 years ago (Blume and Leinweber, 2004).

The introduction of Plaggenesch cultivation was an important agricultural revolution, comparable to the introduction of mineral fertilization (Zoller, 1957). During processing, plaggen were cut in the rural area surrounding the villages, carried to the stables, and enriched with nutrients from cattle excrements. During winter, they were partly decomposed and subsequently spread out onto the fields as an organic-earthy manure in spring.

As Plaggenesch cultivation regularly included activities near the farmhouses, waste (charcoal, remnants of bricks, pottery) was disposed of together with manure and nowadays found as artifacts, confirming the man-made origin of these soils. The main purpose was to gain fertilizer and to enrich soils by using any available type of plaggen material (Eckelmann and Klausing, 1982). Additionally, the lack of bedding material in the stables was overcome and a fundamental soil improvement was achieved (Blume and Leinweber, 2004; Giani et al, 2014). In the field, the plaggen manure decomposed to generate soil organic matter. Over time, the land surface rose by some 10 mm y^{-1} mainly because of the huge amounts of added mineral fraction (Figure 2.4; Driessen and Dudal, 1991). With time, plaggic epipedons grew, regularly reaching thicknesses of 0.6–0.8 m (Eckelmann, 1980), or even up to 1.3 m (Blume and Leinweber, 2004) and 1.5 m (Howard, 2017), resulting in the genesis of a

new human-made soil (Figure 2.4), classified as Plaggic Anthrosols (International Union of Soil Science Working Group World Reference Base, 2015) and Plagganthrepts (Soil Survey Staff, 2014).

Contrary to surrounding soils, Plaggic Anthrosols show improved physical properties (e.g., field capacity, porosity, size, and connection of pores (Niedersächsisches Landesamt für Bodenforschung, 1997; Driessen and Dudal, 2001; Blume and Leinweber, 2004), mainly due to SOM enrichment (Giani et al, 2014). However, the soil organic C (SOC) content of about 25–29 g kg^{-1} (Giani et al, 2014) is similar to that of A horizons of natural sandy soils in NW Germany (Sauer et al, 2007). On the other hand, plowed top horizons of Plaggic Anthrosols have 13–37 g kg^{-1} higher SOC contents than their non-plagged references holding 11–24 g kg^{-1} (Urbanski et al, 2022). Cumulative SOC stocks ranged between 12 and 57 kg m^{-2} for the Plaggic Anthrosol and 7 to 13 kg m^{-2} for the reference soils (Urbanski et al, 2022). Hence, the SOC stocks show an enrichment by a factor of 2–10

Figure 2.4 *Soil profiles from Plagic Anthrosols in NW Germany (Soil of the Year 2013; Giani et al, 2014)) (left), Russia (Hubbe et al, 2007) (middle), and Norway (Schnepel et al, 2014) (right), reproduced with permissions by the publishers*

(Kern et al, 2019; Urbanski et al, 2022), depending on the thicknesses of the plaggic epipedons. In summary, Plaggic Anthrosols are enriched in SOC, but this increase is primarily caused by a large increase in soil volume.

Plaggic Anthrosols can show charcoal fragments of up to 5% of the bulk soil (Hubbe et al, 2007). PyC, identified by molecular markers for the condensed aromatic moieties (Glaser et al, 1998) in Plaggic Anthrosols in northwestern Germany (Kern et al, 2019) and Norway (Acksel et al, 2019) deliver evidence for inputs of combustion residues from ancient fire management or settlements. This part of the SOC content is more persistent (Springob and Kirchmann, 2002; Kuzyakov et al, 2014). It is highly probable that the persistence of this part of the SOC is a result of mineral-organic inter-actions owing to its condensed aromatic structure, which origins either from a pyro-genic (Glaser et al, 1998) or microbial source (Glaser et al, 2008). However, PyC amounts in Plaggic Anthrosols are lower than those in other Anthrosols such as *Sambaquies* and *Terra Preta* and mineralization rates are higher (Kern et al, 2019), pointing to the vulnerability of these soils to land-use change from arable land to forests (Kalinina, 2006).

Historical use of biochar in Eurasia

Most of the area is a boreal forest or forest steppe adjoining it to the south. Natural and human-caused forest fires are an integral part of these biomes. Slash-and-burn, also termed swidden cultivation, entailed the deliberate and controlled burning of forested areas to obtain fertile ash- and PyC-rich soils for grain cultivation (Petrov, 1968). This type of agroforestry was part of a complex eco-nomic system in which hunting also played a major role (Okunev, 1997). In Finland and Russia, it was practiced until the 1930s. Farmers cut down young trees and girdled older trees, then left them to dry and finally burned them. The wood ash served as a fertilizer for the soil, into which the seed was then sown. It is assumed that slash-and-burn cultivation was used early in prehistory, prob-ably since the Bronze Age, and then massively increased with the Iron Age, when better tools became available owing to the availability of iron (Ponomarenko et al, 2019). Biochar plays a key role here, even if it is only a byproduct or co-product of the whole process. Byproduct means an unintentional collateral result of a process performed for some other reason. Co-product leaves open the possibility that there is more than one intentional reason why people cleared forest, e.g., to make open areas for light-demanding crops, to produce ash for crop nutrition, or to produce biochar for soil amelioration. However, it is essential to distinguish between remains of natural forest fires and features of slash-and-burn cultivation, although this is not easy and some ambiguity always remains.

Another tradition that has hardly been studied scientifically so far concerns gardens. In Russia, kitchen or vegetable gardens have always had an important function in the subsistence not only of rural, but also of urban communities, and they were part of the everyday settlement pattern (Schneeweiß, 2021). "Garden archeology" is still a very young field of research and has only recently begun to focus not only on ornamental or flower gardens but also on kitchen gardens. After all, it is still an unquestioned and common practice in the Russian countryside to add to the garden wood ash and the biochar it contains from the banya oven used for sauna and bathing. Furthermore, settlement layers deposited since the Middle Ages have a striking black color, which is likely due to biochar enrichment. According to current archeological and ethnographic knowledge, the traditional direct use of biochar in Russia does not show any peculiarities in comparison to other parts of the world.

Historical use of biochar in Africa

Africa is the cradle of humankind, and there is evidence for the traditional use of biochar in agriculture at least in Benin (Zech et al, 1990), Guinea (Leach and Fairhead, 1995; Fairhead and Leach, 1996; Fraser et al, 2014, 2015, 2016), Liberia (Frausin et al, 2014; Solomon et al, 2016), Sierra Leone (Frausin et al, 2014), and South Africa (Blackmore, 1990).

Anthropologists found a soil cultivation practice in the Kissidougou Prefecture of the Republic of Guinea resembling those that gave rise to *Terra Preta* (Leach and Fairhead, 1995; Fairhead and Leach, 1996). Farmers enriched the soils with excrements of people, domestic animals, and poultry, as well as with residue from crops and fish. Additionally, ashes from wood fires were added over many years (Fairhead and Leach, 2009). In some villages, farmers used soil mounding and raised beds and incorporated unburned and burnt residues into soils over many years, to convert infertile savannah soil into land available for cultivation. This latter technique is related to soil paring-and-burning, discussed in a later section. The villagers confirm that after 3–4 years, the use of these practices produces a mature soil that is easy to cultivate.

Enhanced water infiltration and deeper root growth are observed (Fairhead and Leach, 2009). Pottery findings in "older" soils of former villages and the dark color of the man-made soils are further similarities between these soils and *Terra Preta*. Furthermore, important similarities exist between Amazonia and West Africa, such as the hot and wet climate that produces highly weathered Ferralsols in which biochar enrichment provides the greatest agronomic benefits (Fairhead and Leach, 2009).

Organic matter-rich and fertile soils around former settlements exhibit high similarities to *Terra Preta* and similar soils in Liberia and Benin, especially the composition and persistence of soil organic matter (Zech et al, 1990). One central indigenous perception about the formation of these soils was that "god made the soil, but we made it fertile" (Frausin et al, 2014). In a current indigenous soil management system in West Africa, targeted waste deposition transforms highly weathered, nutrient- and C-poor tropical soils into enduringly fertile, C-rich black African Dark Earths, with similar properties as *Terra Preta* (Solomon et al, 2016).

Historical use of biochar in Asia

Archeological excavations have revealed charred residues in ancient paddy soils. One of these Neolithic agricultural vestiges is the famous Chuodun Site in the lower Yangtze River Delta in eastern China (Hu et al, 2013). Aromatic C is the predominant organic C form in these ancient charred paddy soils, originating mainly from burned rice straw, indicated by solid-state ^{13}C nuclear magnetic resonance spectroscopy (Cao et al, 2006) and PyC analysis as determined by potassium dichromate oxidation (Hu et al, 2013). Radiocarbon dating of carbonized rice grains and other burned plant residues revealed a calibrated age of 5,900 BP (Cao et al, 2006; Hu et al, 2013). Further ancient charred paddy soils were found in the Dinghuishan and Xishanping agricultural sites in Minle County of Gansu in northwestern China (Li et al, 1989, 2007a, b). Due to the absence of such charred soils at a large scale, it can be assumed that these localized deposits were not produced by slash-and-burn agriculture (Hu et al, 2020; Liu et al, 2021). Instead, the production of rice husk biochar directly after harvesting is a common practice in many Asian countries and has been used for soil improvement since the beginning of rice cultivation (Ogawa and Okimori, 2010). Alongside the use of rice husk biochar in paddy soils, Chinese farmers mixed vegetable residues with grass, straw, turf, weeds, and soil and set this mixture on fire (Liebig, 1878). After a few days of burning and smoldering, the black earth produced was used as seedling fertilizer. This may represent an example of soil paring-and-burning techniques as described below.

Kitchen ashes (ash and biochar) are added to toilets after each use to reduce odor and increase the C/N ratio of waste (Polprasert and Koottatep, 2017). Farmers often use the ashes mixed with excreta and powdered lime for composting, which helps to increase pH and facilitate pathogen die-off (Phuc et al, 2006). In Thai Nguyen province (Vietnam), farmers used biochar (charcoal and ash) for vegetables and rice seed germination for at least more than 40 years (Figure 2.5).

In Borneo (East Kalimantan, Indonesia), dark-colored soils have high C contents, high P levels, and improved soil fertility in comparison to adjacent soils, although the origin of these enrichments is still unknown (Sheil et al, 2012; Figure 2.6). Biochar was found in black soils under fruit orchards in Gong Solok and in some soils of the Long Jalan and Lio Mutai study areas. But in many other soils of the region with a deep black color, no biochar was found. At some sites, the C:P ratio of the black soils was around 500–600, but only 150–200 for adjacent soils, indicating inputs into the black soils of C-rich and nutrient-poor material such as wood biochar. However, other black soils showed a C:P ratio of only 100, for instance, the soils in Gong Solok (Figure 2.6), indicating the input

Figure 2.5 *Biochar from open fire cookstoves used in growing vegetables in home gardens in Vietnam (left and bottom right), Rice ash from open fires (cookstoves and on fields) used for rice seed germination in Vietnam (top right). Photos: Hien Nguyen*

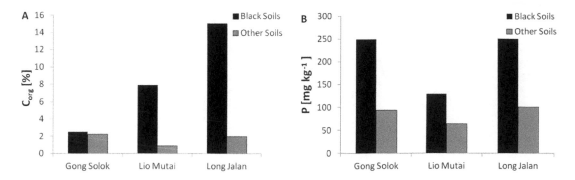

Figure 2.6 *Mean total organic carbon (C_{org}, A) and total P (B) contents of dark-colored and adjacent soils from three sites in Borneo. All pairs of data samples show significant differences ($p < 0.01$) except C_{org} of Gong Solok ($p > 0.5$) (data from Sheil et al, 2012; N = 5)*

of P-rich material such as guano, which is available from nearby caves (Sheil et al, 2012). This source is assumed because high P content in these latter sites is strongly associated with high Ca content, rather than with high contents of Fe and Al (hydr)oxides, as in the other study sides (Sheil et al, 2012).

Dark earths developed on agricultural sites can be distinguished from those developed on settlement sites, owing to the different soil chemical properties and the proximity of some soils to historical settlement sites (Sheil et al, 2012). Ash and biochar were added to the soil, perhaps through partial and repeated burning of secondary forests. This cultivation practice, in combination with swidden agriculture, is also known in areas outside South-East Asia. These black soils found in Borneo are not comparable with *Terra Preta* in

South America (Sheil et al, 2012). This owes mainly to their lower cation exchange capacity, the absence of charcoal or its presence only in low proportions, and the absence of artifacts such as ceramic sherds.

In Japan, charcoal, ashes, and biomass residues have long been used to condition and fertilize soils, absorb odors, and purify water (Nishio, 1998; Yoshizawa et al, 2005). It has also been known for a long time that the input of ash or carbonized materials to a compost pile accelerates decomposition, stimulating bacterial activity and neutralizing acidity (Ogawa and Okimori, 2010). Biochar in composts promotes the production of antibiotic compounds of *Bacillus ssp*, which has a suppressing effect on root diseases and inhibits the growth of some soil-borne pathogens such as *Pythium, Rhizoctonia, Phytophthora* and *Fusarium* (Kobayashi, 2001).

Historical use of biochar in Oceania

In Australia, several man-made soils have been reported with high contents of nutrients and aromatic C, similar to those of *Terra Preta* (Coutts et al, 1976, 1979; Coutts and Witter, 1977). Classified as Cumulic

Anthrosols according to Australian Soil Classification, these soils are developed on sites of Australian Aborigine oven mounds (Downie et al, 2011; Figure 2.7). These mounds are places where Aborigines lived

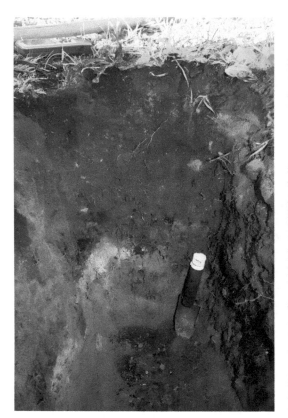

Figure 2.7 *Terra Preta-like soils in Australia (left). In comparison to adjacent sites (right), the dark soils have higher values for pH, cation exchange capacity, and contents of organic and total C, total N, Ca, K, and P (Stephen Joseph, with permission)*

over long periods, usually located above the flood zone of rivers, lakes, or swamps.

The Aborigines used earthen ovens to cook food (Figure 2.8). The debris and refuse were discarded over generations and from these, together with natural sediments, mounds were formed (Coutts, 1976; Coutts et al, 1979). The mounds often contain biochar, burnt clay or stone heat retainers from cooking ovens, animal bones, shells, stone tools, and sometimes Aboriginal burials. There was no agriculture before European colonization. This is a case similar to the one mentioned above, where charcoal was produced as a result of other activities, not for

charcoal production and not intentionally used for agriculture.

Seven Cumulic Anthrosols along the Murray River in SE Australia had radiocarbon ages between 650 ± 30 and $1,609 \pm 34$ BP (Downie et al, 2011). Total C contents were significantly ($p < 0.05$) higher compared to adjacent non-Anthrosols, and solid-state ^{13}C nuclear magnetic resonance spectroscopy showed the predominance of aromatic C. Scanning electron microscopy showed that biochar particles were of plant origin. The surface O/C ratio of a biochar particle was therefore relatively high (0.45 mol mol^{-1}) in contrast to a fresh biochar particle surface

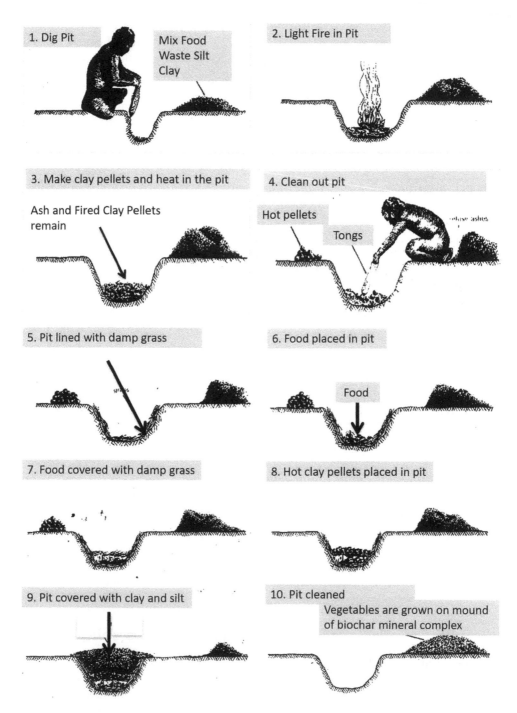

Figure 2.8 *Biochar production of Australian Aborigines (adapted from Coutts et al, 1979)*

($\sim$ 0.15 mol mol^{-1}; Wiedner et al, 2015). Higher values of this ratio are an indicator of biochar aging and are also responsible for the high cation exchange capacity of aged biochar. Total N, P, K, and Ca contents were all significantly higher compared to those in adjacent soils, a result in keeping with the higher cation exchange capacity at a depth of 0.2–0.3 m of 490 mmolc kg^{-1} in contrast to only 180 mmolc kg^{-1} for adjacent soils, and with higher pH values (up to 1.3 units higher) than in adjacent soils. Because the properties of these soils are similar to those of *Terra Preta*, Downie et al. (2011) named them *Terra Preta Australis*. Similar to *Terra Preta*, the development of these soils is a byproduct of subsistence activities and not the result of intentional production for agriculture. Nevertheless, soil properties were improved through the addition of biochar (Downie et al, 2011).

Archeological and radiocarbon evidence indicates that the earliest settlers arrived in the New Zealand archipelago around 1,250–1,300 CE (McFadgen et al, 1994; Higham and Hogg, 1997; Higham et al, 1999). Radiocarbon dating of shells and PyOM particles in Maori gravel soils showed a radiocarbon age of 720 - 470 BP (McFadgen, 1980). In addition to mid-Waikato and Kaiapoi in north Canterbury, one of the best investigated ancient Maori sites is the Waimea Plain near Nelson. Maori gravel soils in the western part of Waimea contain high amounts of charcoal and plant-available nutrients such as P and K (Rigg and Bruce, 1923). The authors suggest that these soils resulted from a long-continued practice of burning wood or scrub (both rich in P and K) taken from the surroundings and incorporated into the soils (Rigg and Bruce, 1923). Besides the fertilization effect of the ash, the blackening effect of biochar may lead to greater absorption of heat, resulting in earlier crop growth (Rigg and Bruce, 1923). Because it has not been shown that Maori added sods, seaweed, or dung into the soils (McFadgen, 1980), it is not accurate to call these soils "Maori Plaggen Soils" as some authors have done. Furthermore, the burning of wood and scrub and incorporation into soil in large amounts (Rigg and Bruce, 1923), seems not to have been a common fertilization practice of prehistoric Maori.

Paring-and-burning systems: A neglected aspect of the interactions of humans and fire

This section treats a set of techniques devised for preparing grassland or heathland soils for cultivation that have the potential to produce biochar. These techniques, grouped as "soil-paring-and-burning", are poorly studied, and the documented occurrences considerably underestimate their present and past geographic distribution. How much biochar they generate, and how much accumulates in soils affected by this practice, is unclear, as is the impact of biochar on the properties of their soils. We discuss factors that could affect variation in the production and accumulation of biochar in these systems and propose research priorities to fill the knowledge gaps.

Introduction to soil-paring systems

Soil-paring is a technique used in many parts of the world to prepare grassland and heathland soils for cultivation. In soil-paring systems, thin sods are cut, the soil left attached to the root system, and the turf and vegetation are piled to form mounds or ridges, which are then covered with soil. In some cases, the mounds or ridges are composting structures: decomposition of the organic matter they contain supplies nutrients for crop growth. Such paring-and-composting systems are described by French authors as "*écobuage compostier*" (Portères, 1972) or "*enfouissage en vert*" (Vennetier, 1966). In other cases, termed paring-and-burning systems, the soil-covered mounds or ridges, once dry, are ignited and then burn slowly over several days. Nutrients for crop growth are supplied by ashes from the partially combusted organic matter. Soil-paring in general, and paring-and-burning techniques in particular, are much more widespread than is generally recognized (Portères, 1972; Sigaut, 1975; McKey, 2021).

For agriculture without inputs of herbicides and fertilizers, soil-paring techniques are well suited to grasslands and heathlands because they solve two big problems of farming in these environments (McKey, 2021). First, they destroy the mat of roots of herbaceous plants that compete with crops. Second, unlike forests, grasslands and heathlands have low-standing biomass that can be mineralized to provide nutrients for crop growth. By soil-paring, farmers can mobilize nutrients for crops by decomposition or combustion not only of the aboveground biomass but also of belowground biomass and soil organic matter. Furthermore, by concentrating soil and biomass from a large area onto a much smaller area of mounds and ridges, farmers can turn very poor soils into soils rich enough to produce crops. Plaggen cultivation, treated in a previous section, is a particular variant of soil-paring in Europe that results in the transfer of fertility at a very large scale (McKey, 2021).

Paring-and-burning techniques: A subset of soil-paring systems

Paring-and-burning in Europe is well described (Portères, 1972; Sigaut, 1975; Olarieta et al, 2011). Descriptions of the practice in other areas are similar, that of the Niari Valley, Congo Republic, being one of the most detailed (Nzila, 1992). In these systems, thin sods bearing herbaceous or heathland vegetation are cut and placed, turf side down and the vegetation facing inward, to form mounds or ridges. Sometimes other vegetation is added, for example, branches of trees and shrubs from surrounding areas (Vennetier, 1966) or even gathered from large areas of forest (Olarieta et al, 2011). The trait common to all systems is that the vegetation is partly buried. Mounds or ridges are covered with soil and then left to dry. The mounds or ridges are designed as veritable ovens, often resembling charcoal kilns, with "chimneys" that can be opened or closed to regulate the intensity of the fire (e.g., Bouteyre, 1958; Vennetier, 1966; Portères, 1972; Sigaut, 1975; Nzila, 1992; Grangeret-Owona, 1997; Olarieta et al, 2011). When mounds or ridges are ignited, they burn slowly over several days. The temperature of the fire varies with depth in the mound or ridge, reaching up to 650°C at the base (Pülschen and Koch, 1990), as does the oxygen supply. The material in the structures after these operations is thus a mix of burnt earth and earth unaffected by heat, ashes, unburnt organic matter, and charred plant material (biochar). Crops are planted on the mound or ridge. Less demanding crops are planted in the areas between mounds or ridges (Nzila, 1992; Michellon et al, 2002), or these spaces are left uncultivated (Vennetier, 1966). In other cases (e.g., in the *guie* system in Ethiopia; Pülschen and Koch, 1990), mounds or ridges are broken up and their material spread over the whole field, which is then planted.

Soil-paring-and-burning was widely practiced in European grasslands and heathlands before commercial fertilizers began to become widely available around the end of the 19th century (Sigaut, 1975). It is also still practiced in numerous locations in Africa and is documented in scattered locations in Asia and South America (Table 2.1, Figure 2.9). Fragmentary old references (Fortune, 1847, re-reported by Liebig, 1878) suggest the practice may have been widespread in northern China. The absence of a widely shared terminology to describe such systems (McKey, 2021) makes searching for reports difficult. While the word *écobuage* is widely used in French to refer to paring-and-burning systems, there is no widely used term in English. Typically, authors give the local name of the practice and describe it as a unique system—e.g., *formiguer* in Spain, *guie* in Ethopia, and *maala* in the Republic of Congo—with only infrequent references to one or two other systems elsewhere. The term "burn-beating" is sometimes used (e.g., Moreau et al, 1998), but should be avoided, as it has also been used to refer to slash-and-burn (without any disturbance of soil). In the absence of easily searchable keywords, it is likely that the recorded occurrences listed in Table 2.1 probably grossly underestimate the distribution of soils affected by this practice.

Whether farmers practice paring-and-composting or paring-and-burning appears to be influenced by climatic factors and soil type (McKey, 2021). In some environments, paring-and-burning techniques confer important advantages relative to paring-and-composting techniques. Under conditions where decomposition rate is constrained by low soil temperature, acidity, biomass rich in tannin and other difficult-to-degrade compounds, waterlogging, or other factors, burning makes the nutrients from organic matter immediately available to crops. Despite the reduction of soil organic matter and decreased cation exchange capacity, paring-and-burning usually results in an increase in available nutrients, particularly P and K. Although some nutrients (particularly N) are lost via volatilization, the covered burns typical of these systems result in lower loss of nutrients to convection than would result from an open running burn. Fire also suppresses weeds, pests, and pathogens. Finally, paring-and-burning is most commonly practiced in heavy, clay-rich, poorly drained soils, and one of its most frequently cited favorable effects is the improvement of soil physical properties, particularly drainage, via the creation of "pseudo sands" formed by the fusion of clay particles into larger aggregates.

Paring-and-burning techniques are labor-intensive, not only for humans but sometimes also for their draft animals. For example, labor needs of *guie* paring-and-burning in the Ethiopian highlands are considerable: "Soil burning is a very tedious practice and includes activities of three to five criss-cross ploughings, breaking clods, soil collection, hollowing and burning and then spreading out the burnt soil. A great amount of human labor and oxen power is expended to accomplish these activities. The total labor requirement on one hectare of barley under GYE [*guie*] is 925 man hours which is more than three times greater than that devoted to a non-GYE plot (280 man hours per hectare). Similarly, oxen power requirement on a GYE plot is nearly twice as much as that on a non-GYE plot: 414 and 220 hours respectively for a pair of oxen for the former and the latter" (Woldekidan, 1985). Paring-and-burning systems are often part of a process of intensification made necessary by demographic pressure or by the necessity to farm marginal land that

Table 2.1 *Known occurrence of soil-paring-and-burning systems, with brief descriptions of their ecological contexts, the practices employed, and the main benefits recognized by authors. This list is not exhaustive and may considerably underestimate actual occurrence. Only fragmentary information is available for some systems. In many cases, paring-and-burning is less practiced today than when the field studies were conducted. Principal references are presented in boldface type. "?" indicates that no information is available*

Country	Site	Ecological context	Description of practices	Main benefits	References
EUROPE					
France, England, Ireland	Large areas of NW Europe	Coastal moors (heathlands), peat bogs, and grasslands of France's Massif Central or the Ardennes of France and Belgium	Archetype of paring-and-burning. Five steps: (i) soil is pared by removing large but thin sods, including root mats, using hoes, shovels, or similar tools; (ii) Sods are dried, turned over grass side down, soil remains attached to roots; (iii) Dried sods arranged into loosely packed mounds, up to 1 m high, to form an oven, with « chimneys » to regulate conditions of the burn; (iv) The ovens are ignited and undergo a slow and incomplete burn over several days; (v) The remains of the collapsed ovens (mounds containing ash, burnt soil, and some charcoal) are scattered over the area to be cultivated.	Increased fertility, weed and pest control, increased pH, improved soil structure (drainage)	Portères (1972); Sigaut (1975); McKey (2021)
Spain	Many regions	Widely used, up until the 1960s, in humid and semi-arid areas, in both acid and calcareous soils.	Termed « formiguer » technique in Catalonia, « hormiguero » or « hormiguero » in other parts of Spain. Similar in most respects to the archetype of paring-and-burning described by Sigaut (1975). Biomass (mostly woody) is piled up and dried, then burned under a soil cover. In semi-arid Mediterranean Catalonia,	Increased fertility (ashes, changes in soil chemical and physical properties). Authors propose that biochar produced in formiguers may affect long-term soil quality. The authors note that	Mestre and Mestres (1949); Olarieta et al (2011)

Table 2.1 *continued*

Country	Site	Ecological context	Description of practices	Main benefits	References
			biomass for formiguers is collected from large areas of forest. In more humid western Spain, soils are richer in organic matter and only vegetation from the immediate surroundings is placed in the mound. The density of formiguers suggested as usually between 260 and 700 ha^{-1}. Volume of each ranged from 0.5 m^3 (when built annually) to 2 m^3 (when built in the first year).	biochar from structures other than formiguers, i.e., charcoal-making kilns, is sold as fertilizer.	
SOUTH AMERICA					
Bolivia	Cochibamba cordillera of the southern Andes, 3500–3900 m	High elevation wet montane grassland (puna grassland).	Conforms quite closely to the archetype of paring-and-burning described by Sigaut (1975); practiced in grasslands that have never been cultivated in memory, or that have lain fallow for at least 8 years. Large sods (0.45–0.5 by 0.25–0.3 by 0.15–0.2 m) dug with shovels or hoes, turned over Sods arranged into combustion ovens 1 m in diameter and 1 m high, spaced about 1.5 m apart. Sods are allowed to dry for several days to several weeks, then ignited. After burning for at least one day and one night, the material is then scattered over the entire plot. Used for growing potatoes.	Elimination of competing plants (and their seed banks), increased soil fertility.	Duval et al (2007)

ASIA

Country	Location	Environment	Practice	Effects/Purpose	References
Bhutan, India	Eastern Himalayas (Bhutan, Meghalaya, Arunachal Pradesh)	Degraded forest (sparse pine woodland); altitude around 1500 m; monsoonal climate with winter frost; sandy, porous podsolic soils, pH 4–6	Variant of *djum* slash-and-burn system. Slash from low pine branches placed in strips along the slope, covered with soil to make ridges and furrows. Crop seeds are planted on the burnt ridges. No mention of charcoal or soil temperatures during burns. Fallows currently 10 years or 5 years (shortened from 20–30 years). At lower elevations, all vegetation is slashed, and seeds are planted directly into the soil-ash complex	Nutrients from the ash; decreased convective loss of nutrients compared to open burns; increased pH; control of weeds and pests	Kerkhoff and Sharma (2006), Mishra and Ramakrishnan (1983a,b)
China	Lowlands of northern provinces	?	"During the summer months, all sorts of vegetable rubbish are collected in heaps by the road-sides, and mixed with straw, grass, parings of turf, &c., which are set on fire and burn slowly for several days ... This manure is not scattered over the land, but reserved for covering the seeds ... This manure is useful mechanically as well as chemically in a stiff soil like that of the lowlands of China, where the seeds are apt to be injured in the process of germination". (Fortune 1847)	Fertilizer; effects on soil structure (noted as being applied to ameliorate "stiff" soils)?	Fortune (1847); re-reported by Liebig (1878)
India	Western coast from the former Bombay Presidency to Kerala, also in Bengal (eastern coast)	Practiced only where annual rainfall > 2500 mm	In the *rab* technique, all kinds of fuel (dried cow manure, grass, leaves, tree branches) are piled and covered with earth, forming mounds 1.5 m high. After burning, the material is spread over the field to form a seedbed for rice or eleusine, both destined for transplantation. In some areas, "clay burning" is conducted, primarily to	Control of weeds and crop pests, improved soil physical properties, increased nutrient availability (ashes; liberation of K from silicate clays).	Mollison (1901); Portères (1972)

Table 2.1 *continued*

Country	Site	Ecological context	Description of practices	Main benefits	References
			improve the physical properties of stiff clay soils, but this requires plant biomass as fuel, so that ashes and charred vegetation are also produced. In other areas, paring and burning are used to render cultivable land with a dense, tough grass cover that hinders plowing. The very top layer (40 mm) of soil is pared and burned to get rid of weeds. After two weeks of dry weather, straw and brushwood were added as fuel. When the fire is started, turf is gradually piled on top of the burning vegetation. Vegetation and turf continue to be added gradually. Earth is scorched rather than burned, and vegetation is charred rather than reduced to ashes (Mollison 1901).		
AFRICA					
Cameroon	Bamenda highlands, various locations	Tropical humid montane climate, around 1000–2300 m; on clayey soils, red, weathered ferralitic soils dominated by kaolinite clays	*Ankara* system: hoes are used to form ridges of plant residue buried under a thin layer of soil; ridges are then burnt when dry. Used primarily by small-scale farmers on localized areas (one or a few ridges). Exists alongside two other methods of residue management: (1) making of compost ridges (vegetation buried in ridges but not burnt); (2) localized surface burning of plant material (predominant where plant residue is plentiful so that burying all of it is time- and labor-	Increased availability of K and P, adequate N; much higher maize yields than with other methods of residue management; intensive heat generation led to reduction of clay fraction, an increase of silt fraction. Farmers see it as a process of concentrating plant nutrients on limited	Osiname and Meppe (1999); Tume et al (2019); Yengoh (2012)

				References
			demanding). Used more frequently on organic matter-rich soils derived from lava basalt than on light-textured soils derived from granite. Cold soil temperatures lead to slow decomposition of organic matter. *Ankara* glows for about 48 hours, generating intense heat (temperatures not measured).	spots to optimize their use in high nutrient-demanding crops. Considered by Osiname & Heppe (1999) as unsustainable: short-term advantages but long-term disadvantages (destruction of organic matter, increased erosion of hill slopes).	
Bamileke country, western province of Cameroon		Practices similar to those of the *Ankara* system (but that term is apparently not used in Bamileke country). Branches poking through the surface of piled vegetation and soil function as "chimneys"; paring-and-burning is always done after a fallow period, never done in the second cycle of cultivation after a fallow (not enough vegetation).	Decreased clay content, lighter, more friable soils; increases in exchangeable bases, P, K; about half of the organic matter burned, the rest remains intact in the burnt ridges. Demanding crops (e.g, cucurbits, taro) grown only on pared-and-burned ridges	Portères (1972); Grangeret-Owona (1997)
Chad	Logone Valley	Seasonal tropical floodplain savanna, flooded 2–3 months yr^{-1}, with grass cover up to 1.8–2 m tall. Young, little-developed soils, mostly coarse silt and fine sand; extensively practiced on silty soils, not on sandy soils	Widespread practice in rice farming. Grass pulled up with roots and soil attached, dried, buried under 0.05–0.1 m soil, formed into ridges spaced every 4–6 m. Fire intensity is regulated with « chimneys », fires burn for several days. The need for large amounts of grass necessitates agreements between villages not to allow wildfire in areas to be cultivated. The Spread of fires set for hunting leads to disputes.	Increased fertility (ashes); "action of heat on the structure of soil" (Bouteyre, 1961)	Bouteyre (1958, 1961); Cabot (1965); Portères (1972)

Table 2.1 *continued*

Country	Site	Ecological context	Description of practices	Main benefits	References
Republic of the Congo	Kukuya Plateau	Grass-dominated savanna, on a plateau around 600–900 m altitude; short dry season, 2 m annual rainfall. Soils iron-rich, nutrient-poor. Soils sandy (fine sand), clay 26–49%	Paring-and-burning is most elaborate on one of the three types of fields, termed *Bvuma*. These are round mounds (1.5 m diameter, 1 m high) or ridges (to 0.8 m high) scattered in fields of the two other types (characterized by less elaborate paring-and-burning). They are reserved for the planting of demanding crops (gourd, potato, tobacco). Farmers bring all kinds of vegetation within several m to the mound before covering it with dirt and igniting it. "Chimneys" are made to regulate the burn.	Increased fertility; ashes from burning confers immediate benefit from the herbaceous cover.	Vennetier (1966); Guillot (1973); Nzila (1992)
	Niari Valley, Mouyondzi Plateau	Grass-dominated savanna, up to 2 m height, with some shrubs. Annual rainfall 1200 mm. Highly weathered ferralitic soil. Practiced on soil with a crumbly structure that facilitates hoeing, and on flat terrain to minimize erosion.	The paring-and-burning technique is termed *maala*. Applied in areas of the highest population density. Grasses are pulled up with root mass and branches of shrubs are cut, then all are piled into mounds or ridges, covered with litter and top layers of soil from areas between mounds or ridges. Half-smothered fire burns for 3–4 days. Temperatures recorded during burns range from 95 to 418°C. Burnt earth (red color) persists for several years. The black color in some parts of the profile indicates incomplete combustion; and carbonized	Increased pH, increased nutrient availability; reduced risk of Mn and Al toxicity; improvement of soil texture (pseudo sands)	Portères (1972); Dzaba (1987); Nzila (1992, 2017); Moreau et al (1998)

Country	Region	Vegetation	Practice	Effects	Reference
Democratic Republic of the Congo	Province de l'Uele (eastern part of the country, south of South Sudan; all following entries are from the western part of the country)	Savanna	vegetation at some levels. Demanding crops are planted on mounds or ridges, with pigeon peas between these structures. *Maala* are cultivated for at least three consecutive years, then left fallow 4–6 years. High grass is pulled up, piled in mounds, dried, and burned. Eleusine is grown, and mixed with sorghum and sesame.		Portères (1972)
	Kwango	?	?	?	Nicolai 1963, as cited in (pp. 411–412)—see Guillot 1973
	Mvuazi; Bakongo country of Bas-Zaïre	?	*Mafuku* technique. Hoed grass and adhering soil are placed in mounds about 1 m high in rectangles of varying size. Soil between mounds hoed and placed over the dry grass. The mound is ignited and burns for 2–7 days. Any unburnt grass is placed on the heap and burned again. Demanding crops are planted in the center of the heap, and less demanding crops at the edges.	Concentration of fertility, improved soil physical properties	Drachoussoff 1965; Ezumah & Okigbo 1980; Edje et al. 1988

Table 2.1 *continued*

Country	Site	Ecological context	Description of practices	Main benefits	References
Rwanda and Burundi	?	?	Sods of grass hoed up, "beaten" with a hoe to remove soil, vegetation piled, dried, and burned. The material is then scattered over the field.	Used to cultivate particular varieties of eleusine.	Portères 1972
Ethiopia	Ethiopian highlands, 2000–3000 m. Annual rainfall 1,100–1,500 mm.	Practiced on restricted areas of hydromorphic heavy clay soils.	Guie system. Fallow lands are plowed (with the ard) two to three times in a criss-cross direction. Sods are collected and piled by hand to form mounds about 1.6 m in diameter and 0.8 m high (800–1600 mounds ha^{-1}). Mounds are ignited once dry and burn for several days to more than 2 weeks. Temperatures reach up to 650° C in the center of the mound. Materials in mounds then spread over the field to form a seedbed. Plots cultivated 2–3 successive years, followed by 7–15 years fallow. Formerly applied to large fields for the cultivation of barley, now used to make small fertile seedbeds for *Eucalyptus* (Mertens et al, 2015).	Increased pH, increased nutrient availability (particularly P), more favorable physical properties (formation of pseudo sands), weed suppression; and carbonized layer are mentioned, but only Amare et al. (2013) mention the potential benefits of PyOM (long-term soil C storage).	Wolde-Yohannes and Wehrmann (1975); Abebe (1981); Woldekidan (1985); Edje et al (1988); Pülschen and Koch (1990); McCann (1995); Amare et al (2013); Mertens et al (2015)
Kenya	Nandi Plateau, western highlands of Kenya	Up to 1500 m altitude.	Applied for the cultivation of Eleusine. The surface of the savanna is pared, and plaques of organic matter and roots are piled, dried, and burned. The residue is then spread over the field and hoed. Although Nandi people are pastoralists, animal manure is never used as fertilizer.		Gourou (1970)

Tanzania, Zambia	Southern Tanzania, northern Zambia	Up to 2000 m altitude. High-altitude grasslands and seasonally flooded marshes	Paring-and-burning is practiced in areas within the miombo woodland zone where cold soil temperatures, poorly drained soil, or both, prevent rapid decomposition of grasses. (In other areas, paring-without-burning [compost mounding] is practiced.) In cold soil areas, the grass is mowed and piled, and thin branches are added to form ridges about 1 m high and 1–2 $tatami$ (2–3 m^2). The grass is covered with soil from the surrounding area, allowed to dry for about one month, then ignited. Burns for several days. In poorly drained soil, the root mat is dug up, overturned, piled into mounds 1–1.5 m in diameter, and left to dry over months. The material is then spread over the whole area before planting.	Increased fertility, increased pH. Although N is lost by ammonia volatilization due to increased alkalinity.	Itani (2002)
"Hautes terres"		About 1500 m altitude; grass-dominated deforested hills. Highly weathered, nutrient-poor ferralitic soils.	Practiced in areas where low soil temperatures slow decomposition. Vegetation is cut, piled into ridges, covered with 0.05–0.1 m soil, allowed to dry, then ignited. Ridges burn for several days. Mounds are sometimes kept intact; in other cases, the material is scattered over the whole field.	Increased soil fertility. Paring-and-burning is seen as a pathway to intensification under high population density. Combined with direct seeding under vegetation cover. Incomplete combustion is mentioned, but no mention of charcoal.	Michellon et al (2001, 2002, 2004)

Table 2.1 *continued*

Country	Site	Ecological context	Description of practices	Main benefits	References
Republic of Guinea	Fouta-Djallon plateau	Practiced on poorly drained soils overlaying lateritic hardpans that are waterlogged in the rainy season, but dry and hard in the dry season. Deforested savannas bearing grass up to 1.3 m tall.	Conforms quite closely to the archetype of paring-and-burning (Sigaut 1975). Locally called *Muki* (also spelled *Mouki*) or *Moki*. Two months before rains, the grass is hoed into thin sods 0.3 m long by 0.2–0.25 m wide. These are left to dry for several weeks. Shortly before rains, sods are grouped into mounds spaced 0.75–2 m apart, depending on vegetation density. The grassy face of sods is turned to the interior of the mound, the root-mat face to the outside. Various fuels are added (dried manure, twigs) and mounds are covered with 50–60 mm of soil. Vents are opened to regulate the conditions of the burn. Mounds are then ignited. After burning, the material is scattered over the field. African rice is grown on the plots for a single year. Plots only have sufficient vegetation for renewed cultivation after 10 to 50 years of fallow.	?	Portères (1972)

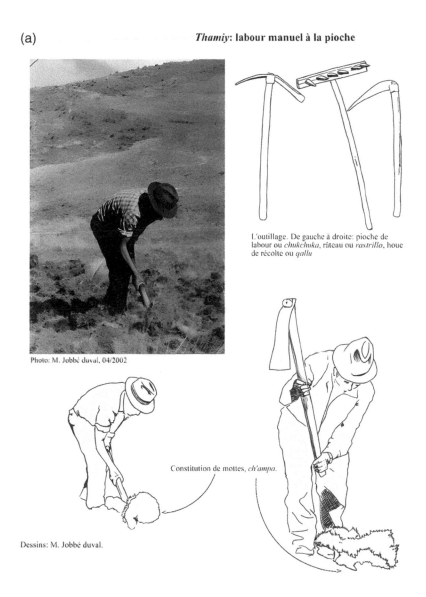

Figure 2.9 *Stages in the chain of operations of soil paring-and-burning as practiced in the Bolivian Andes. (A) Sods are dug up with a mattock or a hoe. (B) Once dried, sods are assembled, grass side down, into "ovens" about 1 m in diameter by 1 m high and ignited. (C) The smoldering sods are surveyed and ovens are modified to regulate combustion. Burning ovens should produce smoke, not flames. (D) After several days of smoldering burn, the burnt clay is reddish in color and the organic matter is incompletely combusted. After cooling, the contents of the ovens are spread over the entire parcel and subjected to paring-and-burning. Photos are taken from Plates 1–4 in Jobbé Duval (2007) and are used by permission. Photo credits: M. Jobbé Duval (A, D), E. Sarrazin (B, C)*

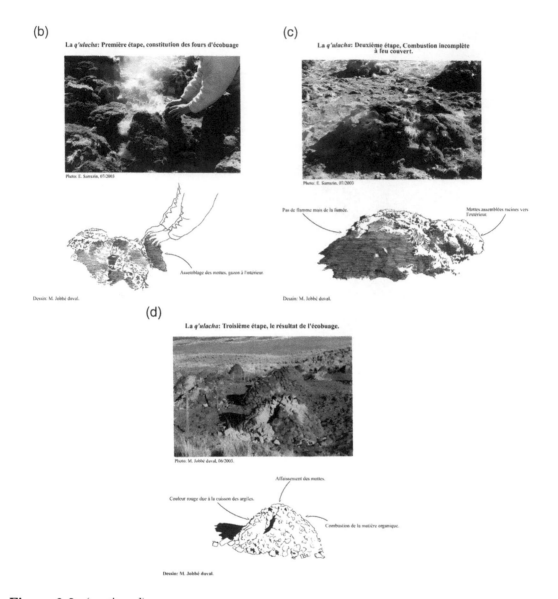

(b) La *q'ulacha*: Première étape, constitution des fours d'écobuage

Photo: E. Sarrazin, 07/2003

Assemblage des mottes, gazon à l'intérieur.

Dessin: M. Jobbé duval.

(c) La *q'ulacha*: Deuxième étape, Combustion incomplète à feu couvert.

Photo: E. Sarrazin, 07/2003

Pas de flamme mais de la fumée.

Mottes assemblées racines vers l'extérieur.

Dessin: M. Jobbé duval.

(d) La *q'ulacha*: Troisième étape, le résultat de l'écobuage.

Photo: M. Jobbé duval, 06/2003.

Affaissement des mottes.

Couleur rouge due à la cuisson des argiles.

Combustion de la matière organique.

Dessin: M. Jobbé duval.

Figure 2.9 *(continued)*

could not be cultivated by other means. In some cases, however, paring-and-burning is applied to make plots of particularly favorable soils on which a few highly demanding crops are grown (e.g., Guillot, 1973; Grangeret-Owona, 1997; Yengoh, 2012).

In many locations, the most recent accounts of paring-and-burning techniques date from work done 25 or more years ago (Table 2.1). The labor-intensive nature of these techniques, along with the increased availability of commercial fertilizers, appear to be leading to their decline (see the section "Filling the knowledge gaps", below).

Paring-and-burning systems and soil biochar

Despite its widespread distribution, soil paring-and-burning is almost completely unknown to most researchers studying the interactions between humans and fire. General treatments of the human dimensions of fire do not even mention the practice (e.g., Bowman et al, 2011; Santín and Doerr, 2016). Two reasons help explain its absence from the literature (McKey, 2021). First, the lack of a shared terminology to characterize systems based on this practice has already been mentioned above. Many cases may yet lie unnoticed, hidden as brief mentions in old documents that are not widely available. Although two French researchers published remarkable syntheses of information available on soil paring-and-burning (*écobuage* in French) in the 1970s (Portères, 1972; Sigaut, 1975), unifying treatment of the subject in French, these papers have gained no traction among Anglophone scholars. Furthermore, even in French, there is confusion around this term. Today, *écobuage* is often used erroneously to mean the setting of running fires to clean pastures. Most French today do not know the word's original meaning (Sigaut and Morlon, 2010). Second, in contrast to the thick organic matter-rich surface horizons of the famed plaggen soils of northern Europe, or the biochar- and potsherd-rich *Terra Preta*, soil paring-and-burning leaves behind no conspicuous vestiges.

Two recent studies of paring-and-burning systems, one experimental (Olarieta et al, 2011), and the other a synthesis paper (Fairhead et al, 2017), independently noted that the documented benefits conferred by paring-and-burning described in the preceding section are all short-term. They postulated that, by generating biochar, paring-and-burning might also confer long-term benefits by causing durable and favorable changes in soil properties such as cation exchange capacity, similar to the benefits conferred by biochar in *Terra Preta*.

However, the amount of biochar actually produced and accumulated in soils subjected to paring-and-burning is unclear. Almost all studies of paring-and-burning soil-preparation techniques were done long before researchers began to recognize the functional importance of PyOM in soils. Although the presence of charred material is often noted in these studies, it is usually given scant attention. Olarieta et al (2011), who devoted some attention to PyOM, found only small quantities of macroscopic biochar, both in the experimental paring-and-burning ovens they incinerated and in soils of the region in which the practice was known to have been frequent about 100 years before. The rarity of macroscopic biochar is not surprising, however, given that, in most cases, herbaceous vegetation dominates the burned biomass.

Resolving the questions about the quantities of biochar produced by paring-and-burning (Olarieta et al, 2011; Fairhead et al, 2017) does not hinge on macroscopic biochar but on microscopic particles that enable biochar to impact soil properties such as cation exchange capacity. However, no studies appear to have quantified the amount of PyC produced by paring-and-burning, or the amount accumulated in soils subjected to this practice over long periods. Besides these two papers (Olarieta et al, 2011; Fairhead et al, 2017), to our knowledge, only one other recent paper (Amare et al, 2013; on the *guie* system of the Ethiopian Highlands), even mentions that paring-and-burning could produce biochar (while noting that they studied only the ash fraction). The study on the *maala* system in the Congo

Republic (Nzila, 1992) is one of the most comprehensive studies of soils under an actual paring-and-burning system. He notes that high values of total exchangeable cations are observed in ash-rich parts of pared-and-burned fields, but that the highest values are observed in parts rich in charred plant debris. However, in soils cultivated 3–4 years after paring-and-burning, values of exchangeable cations are close to those of savanna soils not subjected to paring-and-burning. Whether biochar accumulates and has long-lasting effects is thus unclear.

Filling the knowledge gaps

Determining how much biochar is produced by paring-and-burning, how much has accumulated in soils subjected to paring-and-burning over time, and whether biochar affects important properties of these soils faces several challenges.

The first is to find functioning present-day examples of these systems, which were, with a few exceptions, already abandoned over a century ago in Europe (Sigaut, 1975), and are declining in importance elsewhere. Well-documented studies of ethnographic examples are needed to set benchmarks and to identify geoarchaeological markers for interpreting historical and archeological systems. Guiblais-Starck et al (2020) proposed indicators that could distinguish archeological vestiges of paring-and-burning from other uses of fire in agriculture, but their suggestions should be tested using present-day analogs.

Where they no longer exist, present-day analogs can be recreated experimentally. In Spain, where the practice of making *formiguers* persisted in some places until the 1960s, Olarieta et al (2011) were able to use published descriptions from the 1940s, along with details gathered from interviews of old farmers, to recreate *formiguers*. Their results are highly instructive, but the possibility always remains that experiments do not precisely capture important aspects of the system. Furthermore, experimental studies lack the temporal depth necessary to understand long-term dynamics.

In Africa, it is unclear how many of the examples documented 50 years ago by Portères (1972) still persist today. We have found no reference to paring-and-burning in Chad, Rwanda, Burundi, or the Uele region of the eastern Democratic Republic of the Congo more recent than those cited by that author. For the Republic of Guinea's Fouta-Djallon plateau, the current status of the paring-and-burning system locally called *mouki* is unclear. Boulet and Talineau (1988) wrote that *mouki* was long forbidden (as being destructive) but had been recently re-authorized, whereas Dupuis et al (2007) reported that *mouki* had been forbidden 20 years ago and was at the date of their writing no longer practiced. Studies of paring-and-burning in the Niari Valley of the Congo Republic (Moreau et al, 1998; Nzila et al, 2017) date from the fieldwork for Nzila's PhD thesis, published 30 years ago (Nzila, 1992). Already a few years after Nzila's fieldwork, the *maala* practice had been almost completely abandoned, owing to its labor-intensive nature and the increased availability of fertilizers (B. Barthès, IRD, pers. comm., 2022). In Ethiopia, the replacement of the *guie* system of paring-and-burning by other more efficient systems was already predicted, for similar reasons, by others (Wolde-Yohannes and Wehrmann, 1975; Abebe, 1981; Pülschen and Koch, 1990). Although the practice still persists today, its scale and function have been dramatically altered. Once made for cultivating large fields of barley, it is now used on a smaller scale to make fertile seedbeds for *Eucalyptus* seedlings before their transplantation (Mertens et al, 2015). Paring-and-burning was also reported to be decreasing in frequency relative to open burning and plowing in the Bamileke country of Cameroon (Grangeret-Owona, 1997). Paring-and-burning is now often combined with the application of commercial fertilizer. Opportunities for studying these systems in their traditional forms may thus be rapidly disappearing.

Once suitable systems are found, the second challenge is to document traits that could affect how much biochar they generate and accumulate over time. Other things being equal, the greater the amount of biomass concentrated in small areas for burning, the greater the amount of biochar produced. In most systems, biomass in the mounds and ridges is drawn from vegetation within a radius of a few meters, but sometimes biomass is collected from large areas (up to 25–40 ha) of surrounding forest (Olarieta et al, 2011). Soils subjected to paring-and-burning also vary in their capacity to produce grass biomass. The height of grass in *Hyparrhenia* savannas in the Niari Valley, Republic of Congo, reaches up to 2 m, and this high, dense cover is renewed after 4–6 years of fallow (Nzila, 1992). Where vegetation is less dense, pared-and-burned mounds or ridges are spaced further apart (Drachoussoff, 1965; Portères, 1972; Moreau et al, 1998).

In most cases, grasses account for most of the vegetation burned, but in some cases, branches of trees and shrubs are included (e.g., Mishra and Ramakrishnan, 1983a; Dzaba, 1987; Nzila, 1992; Michellon et al, 2004; Olarieta et al, 2011). The latter certainly produce more biochar. Temperature and oxygen supply during the burn also affect biochar amount and quality. Profiles following burns show great heterogeneity within the burned structures (see for example illustrations in Nzila, 1992, and Olarieta et al, 2011), and observations indicate variation among systems in the relative proportions of ash and PyC produced (Table 2.1).

Finally, while the production of PyOM is initially concentrated in mounds and ridges, these concentrations are then diluted over the landscape. In some systems, this happens immediately, when material in the burned mounds and ridges is scattered over the entire field before planting (Sigaut, 1975, for many European systems; Duval et al, 2007, and references on the *guie* system in Ethiopia in Table 2.1). In other systems, mounds and ridges are not broken up before planting, but the location of mounds and ridges shifts over time: after fallows—a near universal feature in these systems (Table 2.1), mounds or ridges are broken up and new ones constructed in the areas between them (Mollison, 1901; Ezumah and Okigbo, 1980; Grangeret-Owona, 1997; Michellon et al, 2004). Paring-and-burning systems may thus leave a diluted, relatively homogeneous mark on the landscapes affected by them.

Concluding remark

The realization that pre-industrial human activities have already produced biochar-enriched soils in various parts of the world, and those farmers are aware of the value of these soils, has implications for strategies of research on biochar and strategies for its promotion as a component of sustainable development. Where farmers already use biochar-enriched soils, it is important to understand all the facets, ecological, social, and cultural, of the agricultural systems in which these soils are enmeshed (Leach et al, 2012; Bezerra et al, 2019). Such understanding may help us identify ways to improve existing biochar-based systems, rather than "reinventing the wheel", or worse, replacing a functioning, integrative system with a technical "fix" that leads to socially and culturally unacceptable outcomes (Leach et al, 2012).

References

Abebe M 1981 Soil burning in Ethiopia. *Ethiopian Journal of Agricultural Science* 3, 57–74.

Acksel A, et al 2019 Humus-rich topsoils in SW Norway – Molecular and isotopic signatures of soil organic matter as indicators for anthropo-pedogenesis. *Catena* 172, 831–845.

Adams KR 1997 Macrobotanical analyses. In: Homberg JA and Ciolek-Torrello R (Eds) *Agricultural, Subsistence, and Environmental Studies.* Vol. 2. Tucson: SRI Press. pp. 149–177.

Adams KR 2004 Anthropogenic ecology of the North American Southwest. In: Minnis PE (Ed) *People and Plants in Ancient Western North America.* Washington: Smithsonian Books. pp. 167–204.

Altschul JH 1983 Bug Hill: Excavation of a Multicomponent Midden Mound in the Jackfork Valley, Southeast Oklahoma. *New World Research Report of Investigation No. 81-1.* Pollock, Louisiana.

Amare T, Yitaferu B, and Hurni H 2013 Effects of « guie » on soil organic carbon and other soil properties: A traditional soil fertility management practice in the Central Highlands of Ethiopia. *Journal of Agricultural Sciences* 5, 236–244.

Archanjo BS, et al 2014 Chemical analysis and molecular models for calcium–oxygen–carbon interactions in black carbon found in fertile Amazonian Anthrosoils. *Environmental Science and Technology* 48, 7445–7452.

Ardren T, Lowry JP, Memory M, Flanagan K, and Busot, A 2016 Prehistoric human impact on tree island lifecycles in the Florida Everglades. *The Holocene* 26, 772–780.

Behre KE 2008 *Landschaftsgeschichte Nordwestdeutschlands.* Neumünster: Wachholtz.

Bezerra J, et al 2019 The promises of the Amazonian soil: shifts in discourses of Terra Preta and biochar. *Journal of Environmental Policy and Planning* 21, 623–635.

Blackmore AC, Mentis MT, and Scholes RJ 1990 The origin and extent of nutrient-enriched patches within a nutrient-poor savanna in South Africa. *Journal of Biogeography* 17, 463–470.

Blume HP, and Leinweber P 2004 Plaggen Soils: landscape history, properties, and classification. *Journal of Plant Nutrition and Soil Science* 167, 319–327.

Boulet J, and Talineau JC 1988 Eléments de l'occupation du milieu rural et systèmes de production agricole au Fouta Djalon (République de Guinée): tentative de diagnostic d'évolution'. *ORSTOM. Cahiers des Sciences Humaines* 24, 99–117.

Bouteyre G 1958 L'écobuage en culture rizicole dans la région de Kim et Bouko (régions du Mayo-Kebbi et du Logone). Nos Sols. *Haut Commisariat Général de l'Afrique Equatoriale Française Bulletin* 7 et 8, 31–40.

Bouteyre G 1961 Etude pédologique au 1/200.000 de la région de Logone et du Moyen-Chari entre Logone et Bahr-Sara: Moundou, mission 1959, J. Barbery et G. Bouteyre: Koumra, mission 1960, G. Bouteyre.

Bowman DM, et al 2011 The human dimension of fire regimes on Earth. *Journal of Biogeography* 38, 2223–2236.

Cabot J 1965 *Le Bassin du Moyen Logone*. Paris: ORSTOM. pp. 348.

Campbell LJ, Homburg JA, Weed CS, and Thomas PM 1984 Cultural Resources Reconnaissance Survey of Fifty Probability Areas Within the Upper St. Johns River Flood Control Project, Osceola, Brevard and Indian River Counties. *New World Research Report of Investigations No. 84–87*, Pollock, Louisiana.

Cao ZH, et al 2006 Ancient paddy soils from the Neolithic age in China's Yangtze River Delta. *Naturwissenschaften* 93, 232–236.

Carneiro RL 1979 Forest clearance among the Yanomamo, observations and implications. *Antropológica* 52, 39–76.

Coutts PJ, Henderson P, and Fullagar RLK 1979 A preliminary investigation of Aboriginal mounds in north-western Victoria. *Records of the Victorian Archaeological Survey* 9, 1–116.

Coutts PJ, and Witter DC 1977 New radiocarbon dates for Victorian archaeological sites. *Records of the Victorian Archaeological Survey* 4, 59–73.

Coutts PJ, Witter DC, McIlwraith MA, and Frank RK 1976 The mound people of western Victoria: a preliminary statement. *Records of the Victorian Archaeological Survey* 1, 1–54.

Delcourt HR 1987 The impact of prehistoric agriculture and land occupation on natural vegetation. *Trends in Ecology and Evolution* 2, 39–44.

Delcourt HR, and Delcourt PA 1997 Pre-Colombian native American use of fire on Southern Appalachian Landscapes. *Conservative Biology* 11, 1010–1014.

DeLuca TH, and Aplet GH 2008 Charcoal and carbon storage in forest soils of the Rocky Mountain West. *Frontiers in Ecology and the Environment* 6, 18–24.

Devos Y, Nicosia C, Vrydaghs L, and Modrie S 2013 Studying urban stratigraphy: Dark Earth and a microstratified sequence on the site of the Court of Hoogstraeten (Brussels, Belgium). Integrating archaeopedology and phytolith analysis. *Quaternary International* 315, 147e166.

Downie AE, van Zwieten L, Smernik RJ, Morris S, and Munroe PR 2011 Terra Preta Australis: Reassessing the carbon storage capacity of temperate soils. *Agriculture, Ecosystems and Environment* 140, 137–147.

Drachoussoff V 1965 Agricultural change in the Belgian Congo: 1945–1960. Food Research Institute *Studies* 5, 137–201.

Driessen PM, and Dudal R 2001 *The Major Soils of the World*. Zulphen: Koninklijke Wöhrmann BV.

Driessen PM, and Dudal R 1991 *The Major Soils of the World*. Koninklijke Wöhrmann BV.

Dupuis B, Forest F, and Stilmant D 2007 D32. Etude bibliographique: Inventaire des connaissances en termes de variétés cultivées en Guinée, au Mali et au Burkina Faso et inventaire des connaissances en termes de systèmes de culture incluant la production de fonio dans ces trois pays: projet n°. 015403 FONIO. Amélioration de la qualité et de la compétitivité de la filière fonio en Afrique de l'Ouest.' https://agritrop.cirad.fr/562784/

Duval MJ, Cochet H, and Bourliaud JB 2007 L'écobuage andin. Questions sur les origines, l'extension, les modalités et le devenir d'une technique d'ouverture des champs de pomme de terre sur puna humide (Cochabamba, Bolivie). *Techniques and Culture. Revue Semestrielle d'Anthropologie des Techniques* 48-49, 149–188.

Dzaba D 1987 Contribution à l'étude de la dynamique de la fertilité chimique du sol dans un système de culture traditionnel à base d'écobuage' Centre de Recherches Agronomiques de Loudima. In: ORSTOM *Les arbres fixateurs d'azote. L'amélioration biologique de la fertilité du sol.* Paris: ORSTOM. pp 595–627.

Eckelmann W 1980 *Plaggenesche aus Sanden, Schluffen und Lehmen sowie Oberflächenveränderungen als Folge der Plaggenwirtschaft in den Landschaften des Landkreises Osnabrück*. Stuttgart: Schweizerbart'sche Verlagsbuchhandlung.

Eckelmann W, and Klausing C 1982 Plaggenwirtschaft im Landkreis Osnabrück. *Mitteilungen Vereinsgeschichte und Landeskunde Osnabrück* 88, 234–248.

Edje OT, Semoka JMR, and Haule KL 1988 Traditional forms of soil fertility maintenance. In: Wortmann C (Ed) Proceedings of a Workshop on Soil Fertility Research for Bean Cropping Systems in Africa, Addis Ababa, Ethiopia, pp. 7–29.

Ezumah HC, and Okigbo BN 1980 Cassava planting systems in Africa. In: Weber EJ, Toro MJC, and Graham M (Eds) Cassava Cultural Practices: Proceedings of a Workshop. Salvador, Bahia, Brazil, 18–21 March 1980. Ottawa: IDRC. pp. 44–49.

Fairhead J, and Leach M 1996 *Misreading the African Landscape: Society and Ecology in a Forest-Savanna Mosaic.* Cambridge: Cambridge University Press.

Fairhead J, and Leach M 2009 Amazonian dark earths in Africa? In: Woods WI, et al (Eds) *Amazonian Dark Earths: Wim Sombroek's Vision.* Dordrecht: Springer. pp. 265–278.

Fairhead J, et al 2017 Indigenous soil enrichment for food security and climate change in Africa and Asia: A review. In: Sillitoe P (Ed) *Indigenous Knowledge: Enhancing Its Contribution to Natural Resources Management.* Wallingford: CAB International. pp. 99–115.

Falcão NP, et al 2019 Spectroscopic and chemical analysis of burnt earth under Amazonian homegarden systems and anthropic Amazonian dark soils. *Net Journal of Agricultural Science* 7, 1–12.

Fearnside PM, de Alencastro Graça PML, Leal Filho N, Rodrigues FJA, and Robinson JM 1999 Tropical forest burning in Brazilian Amazonia: measurement of biomass loading, burning efficiency and charcoal formation at Altamira, Pará. *Forest Ecology and Management* 123, 65–79.

Fearnside PM, de Alencastro Graça PML, and Rodrigues FJA 2001 Burning of Amazonian rainforests: burning efficiency and charcoal formation in forest cleared for cattle pasture near Manaus, Brazil. *Forest Ecology and Management* 146, 115–128.

Fearnside PM, Barbosa RI, and Graça PML 2007 Burning of secondary forest in Amazonia: biomass, burning efficiency and charcoal formation during land preparation for agriculture in Apiau, Roraima, Brazil. *Forest Ecology and Management* 242, 678–687.

Fortune R 1847 *Three Years Wanderings in the Northern Provinces of China: Including a Visit to the Tea, Silk, and Cotton Countries; with an Account of the Agriculture and Horticulture of the Chinese, New Plants, Etc.* London: J. Murray, No. 34944.

Fraser JA, Leach M, and Fairhead J 2014 Anthropogenic dark earths in the landscapes of Upper Guinea, West Africa: intentional or inevitable? *Annals of the Association of American Geographers* 104, 1222–1238.

Fraser JA, Frausin V, and Jarvis A 2015 An intergenerational transmission of sustainability? Ancestral habitus and food production in a traditional agro-ecosystem of the Upper Guinea Forest, West Africa. *Global Environmental Change* 31, 226–238.

Fraser JA, et al 2016 Cultural valuation and biodiversity conservation in the Upper Guinea forest, West Africa. *Ecology and Society* 21, 36.

Frausin V, et al 2014 God made the soil, but we made it fertile: Gender, knowledge, and practice in the formation and use of African dark earths in Liberia and Sierra Leone. *Human Ecology* 42, 695–710.

German L 2000 Historical contingencies in the coevolution of environment and livelihood: contributions to the debate on Amazonian Black Earth. *Geoderma* 111, 307–331.

German L 2003 Ethnoscientific understandings of Amazonian Dark Earths. In: Lehmann J, Kern DC, Glaser B, and Woods WI (Eds) *Amazonian Dark Earths: Origin, Properties and Management.* Dordrecht: Kluwer. pp. 179–201.

Giani L, Makowsky L, and Müller K 2014 Plaggenesch: Soil of the year 2013 in Germany. A review on its formation, distribution, classification, function and threats. *Journal of Plant Nutrition and Soil Science* 177, 320–329.

Glaser B, and Amelung W 2003 Pyrogenic carbon in native grassland soils along a climosequence in North America. *Global Biogeochemical Cycles* 17, 1064.

Glaser B, and Birk JJ 2012 State of the scientific knowledge on properties and genesis of Anthropogenic Dark Earths in Central Amazonia (terra preta de índio). *Geochimica et Cosmochimica Acta* 82, 39–51.

Glaser B, Haumaier L, Guggenberger G, and Zech W 1998 Black carbon in soils: the use of benzenecarboxylic acids as specific markers. *Organic Geochemistry* 29, 811–819.

Glaser B, and Knorr KH 2008 Isotopic evidence for condensed aromatics from non-pyrogenic sources in soils – implications for current methods for quantifying soil black carbon. *Rapid Communications in Mass Spectrometry* 22, 935–942.

Glaser B, Haumaier L, Guggenberger G, and Zech W 2001 The Terra Preta phenomenon: a model for sustainable agriculture in the humid tropics. *Naturwissenschaften* 88, 37–41.

Glaser B, Lehmann J, and Zech W 2002 Ameliorating physical and chemical properties of highly weathered soils in the tropics with charcoal – a review. *Biology and Fertility of Soils* 35, 219–230.

Glaser B, Balashov E, Haumaier L, Guggenberger G, and Zech W 2000 Black carbon in density fractions of anthropogenic soils of the Brazilian Amazon region. *Organic Geochemistry* 31, 669–678.

Gourou P 1970 *L'Afrique.* Paris: Hachette.

Gowlett JA 2016 The discovery of fire by humans: a long and convoluted process. *Philosophical Transactions of the Royal Society B: Biological Sciences* 371, 20150164.

Graça PM, Fearnside PM, and Cerri CC 1999 Burning of Amazonian forest in Ariquemes, Rondônia, Brazil: biomass, charcoal formation and burning efficiency. *Forest Ecology and Management* 120, 179–191.

Grangeret-Owona I 1997 L'Agriculture Bamiléké vue à travers sa gestion de la fertilité agronomique. *Thèse doctorale, Faculté Universitaire de Sciences Agronomiques de Gembloux*, Belgique. pp. 616.

Grau-Sologestoa I, et al 2023 Animals, crops and dark earth: An interdisciplinary study of urban development from the Late Roman period to the Early Middle Ages in Cologne (Germany), *Environmental Archaeology*, 10.1080/14614103 .2023.2182465.

Guiblais-Starck A, et al 2020 Première identification archéologique d'un écobuage médiéval: le site de Vaudes "Les Trappes" (Aube). *ArcheoSciences* 44, 219–235.

Guillot B 1973 *La Terre Enkou. Recherches sur les Structures Agraires du Plateau Koukouya.* Paris: La Haye, Mouton & Co.

Hecht SB 2009 Kayapó savanna management: fire, soils, and forest islands in a threatened biome. In: Woods WI, Teixeira WG, Lehmann J, Steiner C, and WinklerPrins A (Eds) *Amazonian Dark Earths: Wim Sombroek's Vision.* Dordrecht: Springer. pp. 143–162.

Heckenberger MJ 2005 *The Ecology of Power: Culture, Place, and Personhood in the Southern Amazon, A.D 1000–2000.* New York: Routledge. pp. 404.

Higham TG, and Hogg AG 1997 Evidence for late Polynesian colonization of New Zealand; University of Waikato radiocarbon measurements. *Radiocarbon* 39, 149–192.

Higham T, Anderson A, and Jacomb C 1999 Dating the first New Zealanders: the chronology of Wairau Bar. *Antiquity* 73, 420–427.

Howard J 2017 Anthropogenic soils in agricultural settings. In: Hartemink AE and McBratney AB (Eds) *Progress in Soil Science.* Berlin: Springer. pp. 115–147.

Hubbe A, et al 2007 Evidence of plaggen soils in European North Russia (Arkhangelsk region). *Journal of Plant Nutrition and Soil Science* 170, 329–334.

Hu L, et al 2013 Evidence for a Neolithic Age fire-irrigation paddy cultivation system in the lower Yangtze River Delta, China. *Journal of Archaeological Science* 40, 72–78.

Hu Y, et al 2020 Abundance and morphology of charcoal in sediments provide no evidence of massive slash-and-burn agriculture during the Neolithic Kuahuqiao culture, China. *PLoS One* 15, e0237592.

Iriarte J, et al 2020 The origins of Amazonian landscapes: Plant cultivation, domestication and the spread of food production in tropical

South America. *Quaternary Science Reviews* 248, 106582.

Itani J 2002 Indigenous farming systems in miombo woodlands and surrounding areas in East Africa' (in Japanese). *Asian and African Area Studies* 2, 88–104.

IUSS Working Group WRB 2015 World reference base for soil resources 2014, Update 2015. International soil classification system for naming soils and creating legends for soil maps. World Soil Resources Report No. 106. FAO. https://www.fao.org/3/i3794en/I3794en.pdf

Jobbé Duval M 2007 L'écobuage andin. Questions sur les origines, l'extension, les modalités et le devenir d'une technique d'ouverture des champs de pomme de terre sur puna humide (Cochabamba, Bolivie). *Techniques & Culture* 149–188.

Johnson DL 1983 Geomorphic setting. In: Altschul JA (Ed) *Bug Hill: Excavation of a Multicomponent Midden Mound in the Jackfork Valley, Southeast Oklahoma*. Pollock: New World Research Report of Investigation No. 81-1. pp. 202–219.

Jorio A, et al 2016 Study of carbon nanostructures for soil fertility improvement. In: Jorio A (Ed) *Bioengineering Applications of Carbon Nanostructures*. Berlin: Springer. pp. 85–104.

Kalinina O 2006 Degradation von Plaggeneschen und Konsequenzen für ihre ökologische Bewertung. *Mitteilungen der Deutschen Bodenkundlichen Gesellschaft* 108, 105–106.

Kerkhoff E, and Sharma E 2006 Debating shifting cultivation in the Eastern Himalayas. In: *Farmers' Innovations as Lessons for Policy*. Kathmandu: ICIMOD. p. 92.

Kern J, Giani L, Teixeira W, Lanza G, and Glaser B 2019 What can we learn from ancient fertile anthropic soil (Amazonian Dark Earths, Shell Mounds, Plaggen soil) for soil carbon sequestration? *Catena* 172, 104–112.

Kobayashi N 2001 Charcoal utilization in agriculture (in Japanese). *Nogyo Denka* 54, 16–19.

Kuzyakov Y, Bogomolova I, and Glaser B 2014 Biochar stability in soil: Decomposition during eight years and transformation as assessed by compound-specific ^{14}C analysis. *Soil Biology and Biochemistry* 70, 229–236.

Leach M, and Fairhead J 1995 Ruined settlements and new gardens: Gender and soil-ripening among Kuranko farmers in the forest-savanna transition zone. *IDS Bulletin* 26, 24–32.

Leach M, Fairhead J, and Fraser J 2012 Green grabs and biochar: Revaluing African soils and farming in the new carbon economy. *Journal of Peasant Studies* 39, 285–307.

Li F, Li JY, and Lu Y 1989 New discoveries from Donghuishan Neolithicsite, Minle Gansu (in Chinese). *Agricultural Archaeology* 1, 56–69.

Li JH, Dong YH, and Cao ZH 2007a Distribution and origins of polycyclic aromatic hydrocarbons in a soil profile containing 6000-years old paddy soil (in Chinese). *Acta Pedolog Sinica* 44, 41–46.

Li XQ, Zhou XY, and Zhou J 2007b The earliest archaeobiological evidence of the broadening agriculture In China recorded at Xishanpingsite in Gansu Province. *Science in China Series D-Earth Science* 50, 1707–1714.

Liang B, et al 2006 Black carbon increases cation exchange capacity in soils. *Soil Science Society of America Journal* 70, 1719–1730.

Liebig J 1878 *Chemische Briefe*. Leipzig and Heidelberg: CF Winter`sche Verlagsbuchhandlung.

Liu Y, et al 2021 Rice paddy soils are a quantitatively important carbon store according to a global synthesis. *Communications Earth and Environment* 2, 154.

McCann J 1995 *People of the Plow: An Agricultural History of Ethiopia, 1800–1990*. Madison: University of Wisconsin Press.

McFadgen BG 1980 Maori Plaggen soils in New Zealand, their origin and properties. *Journal of the Royal Society of New Zealand* 10, 3–18.

McFadgen BG, Knox FB, and Cole TL 1994 ^{14}C Radiocarbon calibration curve variations and their implications for the interpretation of New Zealand prehistory. *Radiocarbon* 36, 221–236.

McKey D 2021 Making the most of grasslands and heathlands. Unearthing the links between soil paring-and-burning, plaggen cultivation, and raised-field agriculture. *Revue d'ethnoécologie* 20, 10.4000/ethnoecologie.8120.

Mertens K, et al 2015 Impact of traditional soil burning (guie) on Planosol properties and land-use intensification in southwestern Ethiopia. *Soil Use and Management* 31, 330–336.

Mestre Artigas C, and Mestres Jané A 1949 Aportación al estudio de la fertilización del suelo por medio de hormigueros. *Boletín del Instituto Nacional de Investigaciones Agronómicas* 109, 125–163.

Michellon R, Razakamiaramanana RR, and Séguy L 2001 Developing sustainable cropping systems with minimal inputs in Madagascar: direct seeding on plant cover with « soil smouldering » (écobuage) techniques. *World Congress on Conservation Agriculture Madrid*, 1–5 October, 2001. http://open-library.cirad.fr/files/2/460__1004133629.pdf

Michellon R, et al 2002 Amélioration de la fertilité par écobuage: Influence de la fréquence et de l'intensité de la combustion selon le type de sol de tanety. *Fiche d'essai* 1. http://open-library.cirad.fr/files/2/421__1039388458.pdf

Michellon R, et al 2004 Rapport de Campagne 2002-2003. Région du Vakinankaratra. http://madadoc.irenala.edu.mg/documents/v02846_RAP.pdf

Miltner BC, and Coomes OT 2015 Indigenous innovation incorporates biochar into swidden-fallow agroforestry systems in Amazonian Peru. *Agroforestry Systems* 89, 409–420.

Mishra BK, and Ramakrishnan PS 1983a Slash and burn agriculture at higher elevations in north-eastern India. I. Sediment, water and nutrient losses. *Agriculture, Ecosystems and Environment* 9, 69–82.

Mishra BK, and Ramakrishnan PS 1983b Slash and burn agriculture at higher elevations in north-eastern India. II. Soil fertility changes. *Agriculture, Ecosystems and Environment* 9, 83–96.

Mollison J 1901 *Text Book on Indian Agriculture. Vol. I: Soils, Manures, Implements*. Bombay: Times of India Press.

Moreau R, Nzila JD, and Nyété B 1998 La pratique de l'écobuage maala et ses conséquences sur l'état du sol au Congo',

18e Congrès Mondial des Sols, Symposium no 45. http://natres.psu.ac.th/link/soilcongress/bdd/symp45/6997-t.pdf.

Niedersächsisches Landesamt für Bodenforschung 1997 *Altlastenhandbuch des Landes Niedersachsen: Geologische Oberflächenerkundung*. Berlin: Springer.

Niemeier G 1939 Probleme der bäuerlichen Kulturlandschaft in Nordwestdeutschland. *Deutsche Geographische Blätter* 42, 111–118.

Nishio M 1998 Microbial fertilizers in Japan. National Institute of Agro-Environmental Sciences, Kannodai 3-1-1, Tsukuba.

Nzila JDD 1992 La Pratique de l'Ecobuage dans la Vallée du Niari (Congo). Ses conséquences sur l'évolution d'un sol ferrallitique acide. Document ORSTOM Montpellier, 1992, no 7. Centre ORSTOM de Montpellier. https://horizon.documentation.ird.fr/exl-doc/pleins_textes/divers16-03/36052.pdf

Nzila JDD 2017 Influences de l'écobuage sur la restauration de la productivité des sols argileux acides de la vallée du Niari (Congo RDC). In: Roose E (Ed) *Restauration de la Productivité des Sols Tropicaux et Méditerranéens*. Marseille: IRD Editions. pp. 141–149.

Ogawa M, and Okimori Y 2010 Pioneering works in biochar research, Japan. *Soil Research* 48, 489–500.

Okunev AV 1997 Роль подсечного земледелия в жизни северно-русского общинника во второй половине XIX - начале XX вв. Рябининские чтения 1995 (The role of slash-and-burn agriculture in the life of a northern Russian community member in the second half of the 19th–early 20th centuries. Ryabininsky readings 1995). Петрозаводск: Музей-заповедник «Кижи».

Olarieta JR, Padrò R, Masipa G, Rodriguez-Ochoa R, and Tello E 2011 Formiguers, a historical system of soil fertilization and biochar production? *Agriculture, Ecosystems and Environment* 140, 27–33.

Osiname OA, and Meppe F 1999 Effects of different methods of plant residue management on soil properties and maize yield. *Communications in Soil Science and Plant Analysis* 30, 53–63.

Petrov VP 1968 *Подсечное земледелие (Slash-and-Burn Agriculture)*. Kiev.

Phuc PD, Konradsen F, Phuong PT, Cam PD, and Dalsgaard A 2006 Practice of using human excreta as fertilizer and implications for health in Nghean Province, Vietnam. *Southeast Asian Journal of Tropical Medicine and Public Health* 37, 222.

Pingree MRA, and DeLuca TH 2017 Function of wildfire-deposited pyrogenic carbon in terrestrial ecosystems. *Frontiers in Environmental Science* 30, 1–7.

Polprasert C, and Koottatep T 2017 *Organic Waste Recycling: Technology, Management and Sustainability*. London: IWA Publishing.

Ponomarenko E, Tomson P, Ershova E, and Bakumenko V 2019 A multi-proxy analysis of sandy soils in historical slash-and-burn sites: A case study from southern Estonia. *Quaternary International* 516, 190–206.

Portères R 1972 De l'écobuage comme un système mixte de culture et de production. *JATBA* 19, 151–207.

Pülschen L, and Koch W 1990 The significance of soil burning ("Guie") in Ethiopia with special regard to its effects on the agrestal weed flora. *Journal of Agronomy and Crop Science* 164, 254–261.

Ribeiro-Soares J, Cançado LG, Falcão NPS, Martins Ferreira EH, Achete CA, and Jorio A 2013 The use of Raman spectroscopy to characterize the carbon materials found in Amazonian anthrosoils. *Journal of Raman Spectroscopy* 44, 283–289.

Rigg T, and Bruce JA 1923 The Maori gravel soil of Waimea West, Nelson, New Zealand. *Journal of the Polynesian Society* 32, 85–92.

Roos C, Zedeño MN, Hollenback KL, and Erlick MMH 2018 Indigenous impacts on North American Great Plains fire regimes of the past millennium. *Proceedings of the National Academy of Sciences* 115, 8143–8148.

Santín C, and Doerr SH 2016 Fire effects on soils: the human dimension. *Philosophical Transactions of the Royal Society B: Biological Sciences* 371, 20150171.

Sauer D, et al 2007 Podzol: Soil of the Year 2007. A review on its genesis, occurrence, and functions. *Journal of Plant Nutrition and Soil Science* 170, 581–597.

Schmidt MJ, et al 2014 Dark Earths and the human built landscape in Amazonia: a widespread pattern of anthrosol formation. *Journal of Archaeological Science* 42, 152–165.

Schneeweiß J 2021 Vegetable gardening in the Slavic economy in the early middle ages. Archaeological view and methods of its study. In: Tataurova LV (Ed) *Культура русских в археологических исследованиях: археология Севера России (Culture of Russians in archaeological researches: Archaeology of the North of Russia)*. Omsk: Surgut. pp. 244–248.

Schnepel C, Potthoff K, Eiter S, and Giani L 2014 Evidence of plaggen soils in South-West Norway. *Journal of Plant Nutrition and Soil Science* 177, 638–645.

Schwadron M 2006 Everglades tree islands prehistory: archaeological evidence for regional Holocene variability and early human settlement. *Antiquity* 80, f1–f6.

Schwadron M, Graf MT, Galbraith J, and Chmura G 2009 A Preliminary Study of the Presence of a Carbonate Layer in Tree Island Soil in WCA-3. South Florida Water Management District. Southeast Archaeological Center, National Park Service, Tallahassee, Florida.

Shea A, and Altschul JA 1983 Paleobotanical remains. In: Altschul JA (Ed) Bug Hill: Excavation of a Multicomponent Midden Mound in the Jackfork Valley, Southeast Oklahoma. Pollock, Louisiana; New World Research Report of Investigation No. 81-1. pp 202–219.

Sheil D, et al 2012 Do Anthropogenic Dark Earths occur in the interior of Borneo? Some initial observations from East Kalimantan. *Forests* 3, 207–229.

Sigaut F 1975 *L'agriculture et le feu. Rôle et place du feu dans les techniques de préparation du champ de l'ancienne agriculture européenne*. Paris: Mouton. pp. 320.

Sigaut F, and Morlon P 2010 'Ecobuage', Les Mots de l'Agronomie. https://mots-agronomie.inra.fr/index.php/%C3%89cobuage. accessed 6 October 2021.

Simpson IA, Dockrill SJ, Bull ID, and Evershed RP 1998 Early anthropogenic soil formation at Tofts Ness, Sanday, Orkney. *Journal of Archaeological Science* 25, 729–746.

Soares MB, Cerri CE, Demattê JA, and Alleoni LR 2022 Biochar aging: Impact of pyrolysis temperature on sediment carbon pools and the availability of arsenic and lead. *Science of the Total Environment* 807, 151001.

Soil Survey Staff 2014 Keys to soil taxonomy. 12th Edition, USDA-Natural Resources Conservation Service, Washington DC. accessed at https://www.nrcs.usda.gov/resources/guides-and-instructions/keys-to--soil-taxonomy

Solomon D, et al 2016 Indigenous African soil enrichment as a climate-smart sustainable agriculture alternative. *Frontiers in Ecology and the Environment* 14, 71–76.

Springob G, and Kirchmann H 2002 C-rich sandy Ap horizons of specific historical land-use contain large fractions of refractory organic matter. *Soil Biology and Biochemistry* 34, 1571–1581.

Steiner C, Teixeira WG, Woods WI, and Zech W 2009 Indigenous knowledge about Terra Preta formation. In: Woods WI, Teixeira WG, Lehmann J, Steiner C, and WinklerPrins A (Eds) *Terra Preta Nova: A Tribute to Wim Sombroek*. Berlin: Springer. pp. 193–204.

Steiner C, Teixeira WG, and Zech W 2003 Slash and char: An alternative to slash and burn practiced in the Amazon Basin. In: Glaser B and Woods WI (Eds) *Amazonian Dark Earths Explorations in Space and Time*. Dordrecht: Springer. pp. 183–193.

Tume SJP, Kimengsi JN, and Fogwe ZN 2019 Indigenous knowledge and farmer perceptions of climate and ecological changes in the Bamenda Highlands of Cameroon: Insights from the Bui Plateau. *Climate* 7, 138.

Turcios MM, Jaramillo MM, do Vale Jr JF, Fearnside PM, and Barbosa RI 2016 Soil charcoal as long-term pyrogenic carbon storage in Amazonian seasonal forests. *Global Change Biology* 22, 190–197.

Urbanski L, et al 2022 Legacy of plaggen agriculture: High organic carbon stocks as a result from high carbon input and volume increase. *Geoderma* 406, 115513.

Vennetier P 1966 *Geographie du Congo Brazzaville*. Paris: Gauthier-Villars. p. 125.

Wiedner K, Schneeweiß J, Dippold MA, and Glaser B 2015 Anthropogenic Dark Earth in Northern Germany – The Nordic Analogue to terra preta de Índio in Amazonia. *Catena* 132, 114–125.

Woldekidan B 1985 Resource Allocation in Traditional Agriculture in the Ethiopian Highlands: a Case Study. MSc. Thesis, Australian National University. 97 pp. https://openresearch-repository.anu.edu.au/bitstream/1885/123347/2/b15428941_Woldekidan_Berhanu.pdf

Wolde-Yohannes L, and Wehrmann J 1975 Das Bodenbrennen GUIE in Äthiopien und seine Wirkung auf Boden und Pflanze. *Forstwissenschaftliches Centralblatt* 94, 288–300.

Yarnell RA 1976 Early plant husbandry in easter North America. In: Cleland C (Ed) *Cultural Change and Continuity*. New York: Academic Press. pp. 265–273.

Yengoh GT 2012 Determinants of yield differences in small-scale food crop farming systems in Cameroon. *Agriculture and Food Security* 1, 1–17.

Yoshizawa S, et al 2005 Composting of food garbage and livestock waste containing biomass charcoal. Proceedings of the International Conference and Natural Resources and Environmental Management, Kuching, Sarawak.

Zech W, Haumaier L, and Hempfling R 1990 Ecological aspects of soil organic matter in tropical land use. In: McCarthy P, Clapp CE, Malcolm RL, and Bloom PR (Eds) *Humic Substances in Soil and Crop Sciences. Selected Readings*, Madison: American Society of Agronomy and Soil. pp. 187–202.

Zoller D 1957 Esche und Plaggenböden in Nordwestdeutschland. *Weser-Ems Fachblatt für Land- und Forstwirtschaft* 104, 1614–1629.

3

Fundamentals of biochar production

Robert C. Brown, Santanu Bakshi, and Ondřej Mašek

Introduction

Biochar production cannot be properly discussed without first distinguishing it from char and charcoal. All three forms of carbonaceous material are produced from pyrolysis, the process of heating carbon (C) bearing solid material under oxygen-starved conditions (Chapter 1). Char is defined here as any carbonaceous residue from natural fires. Thus, char is the most general term to employ in scientific descriptions of the products of pyrolysis and fires whether from biomass or other materials. Charcoal is char produced from pyrolysis of animal or vegetable matter in kilns for use in cooking or heating. Biochar is a carbonaceous material produced specifically for application to soil for agronomic or environmental management (Chapter 1). In 2012, the International Biochar Initiative (IBI) released the first Guidelines for "Biochar that is used in Soil" to formally define this carbonaceous product and describe its desired characteristics. Biochar as a soil amendment and C

sequestration agent are relatively new applications requiring significant research and development. However, continuing research is required to understand what constitutes "good" biochar in agronomic and environmental management applications.

Charcoal is readily generated in open fires, whether forest fires or campfires, under oxygen-poor conditions. Thus, charcoal was available to early humankind, who used this byproduct of their campfires to create spectacular cave paintings during the last ice age (Bard, 2001). Charcoal eventually found application in other fields including agronomy, medicine, metallurgy, pyrotechnics, and chemical manufacture. However, its largest application has always been in the preparation of smokeless fuel for cooking, residential heating, smelting, and steel making. The process of charcoal making removes most of the volatile matter responsible for smoke during burning, leaving behind a C-rich fuel with high energy content. Charcoal is a relatively

DOI: 10.4324/9781003297673-3

clean-burning fuel that represented an important innovation in the controlled use of fire. Superficially, charcoal resembles coal, which is also derived from vegetable matter; indeed, the word charcoal may have originally meant "the making of coal" (Encyclopedia Britannica). However, the geological processes from which coal is derived are quite different from charcoal making, resulting in important differences in chemical composition, porosity, and reactivity.

This chapter covers three topics: the history and production of charcoal in traditional kilns; biomass composition and mechanisms of biochar production from plant materials; and chemical insights into factors influencing the yield of biochar. The carbonaceous residue of pyrolysis will be referred to variously as char, charcoal, or biochar depending upon the context of the discussion.

Due to its long history of production, most information on the preparation of carbonaceous material stems from charcoal production. This chapter will therefore review traditional charcoal-making processes as a first step in understanding how to design modern biochar systems. Although C is the major constituent of charcoal, its exact composition and physical properties depend upon the starting material and the conditions under which it is produced. Charcoal contains 65% to 90% solid C, with the balance being volatile matter and mineral matter (ash) (Antal and Grønli, 2003). Nevertheless, other thermochemical pathways are being developed that produce an array of different gaseous, liquid, and solid (char) products while offering improved efficiency, increased quality control, and fewer environmental impacts than traditional charcoal making.

History and production technology of charcoal making

In the early development of pyrolysis, producing charcoal for energy use or metallurgical applications was the sole objective of wood carbonization. Throughout the history of charcoal making, the pyrolysis process has evolved from charcoal pits and mound kilns to modern, fast pyrolysis reactors now employed as potential pyrolysis biorefineries. The first kilns focused on maximizing the production of charcoal at the expense of other pyrolysis products. These early kilns produced copious quantities of acrid smoke, which was highly polluting. By the end of the 18th century, new technologies had been developed to recover and utilize the volatile compounds in the pyrolysis off-gases (Klar, 1925).

The earliest charcoal kilns consisted of temporary pits or mounds, which were simple and inexpensive to construct. Indeed, pit and mound kilns have persisted to the modern era as a result of these virtues. Readily available earth serves as a good insulator and sealant for enclosing the load of wood to be carbonized. Various kinds of brick, metal, and concrete kilns have been introduced to improve the yield of charcoal. All of these operate in batch mode, requiring the periodic charging and discharging of the kiln. A recent innovation in charcoal making is the multiple hearth kiln, which operates continuously, offering energy efficiency and environmental performance advantages compared to batch kilns. Virtually all charcoal kilns employ wood as feedstock although in principle any biomass could be used to produce charcoal.

Pit kilns employ the simplest strategy for controlling access of air and reducing heat loss during carbonization: burying a stack of smoldering wood in the ground (Figure 3.1).

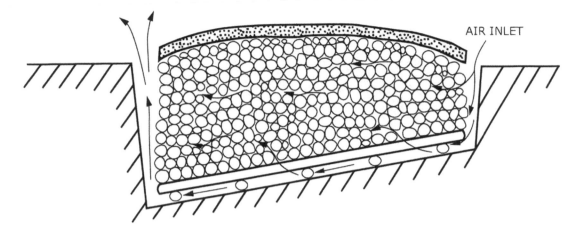

AIR INLET

Figure 3.1 *Large pit kiln (adapted from FAO Forestry Department, 1987)*

Small pit kilns may be only one cubic meter in volume. A small fire is started in the pit and additional wood is added to make a strong fire. At this point, a canopy of branches and leaves is added to support a layer of earth of about 0.2 m. Carbonization may proceed for up to two days before the pit is uncovered and the charcoal is allowed to cool before unloading. Large pit kilns can be 30 m³ or larger and produce 6 tons or more of charcoal per load. Burning in large pit kilns takes place progressively from one end to the other (Figure 3.1). Large pit kilns do not necessarily have higher yields than small pits, but they are more efficient in the use of labor. Pit kilns must be continuously tended, requiring the opening and closing of vent holes in the soil layer, to ensure the right balance between combustion and pyrolysis in the pile. The pit kiln is ideal where the soil is well-drained, deep, and loamy. Charcoal yields are generally very low (theoretically about 30%) (FAO Forestry Department, 1987) and the charcoal is not uniform in quality. The venting of particulate matter and volatile organic compounds into the atmosphere is an obvious disadvantage, as is the potential for soil contamination with

hydrocarbons from condensed pyrolysis liquids.

The mound kiln is essentially an above-ground version of the pit kiln, with the earth mounded up over a stack of wood to control air infiltration and heat loss during carbonization (Figure 3.2). The mound is preferred to a pit when the water table is close to the surface or the soil is hard to work. It is also employed when a permanent site near an agricultural village (which has more scattered wood resources) is preferred to a temporary site located within a timber resource. A typical mound kiln is about 4 m in diameter at the base and 1–1.5 m high in the shape of a flattened hemisphere (see Figure 3.2). Long pieces of fuel wood are stacked vertically against a central post while shorter logs are placed vertically towards the periphery. Gaps between logs are filled with small wood to make a dense pile. It is covered with straw or dry leaves and then a layer of loamy or sandy earth to seal the mound. The center post is removed before lighting, the space serving as both the place to ignite the pile as well as the flue for the smoke to exhaust from the pile. About 6–10 vents at the base of the mount allow control of air infiltration during

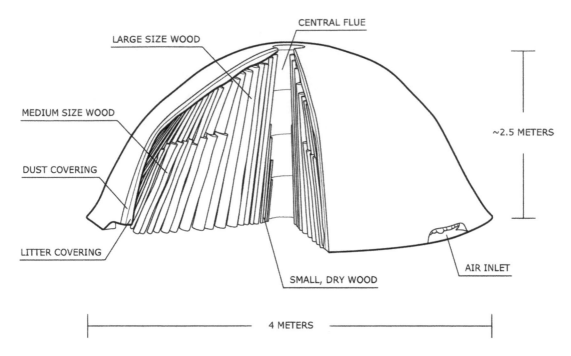

Figure 3.2 *Mound kiln (adapted from FAO Forestry Department, 1987)*

carbonization. Like the pit kiln, the mound kiln has relatively low charcoal yield and, as illustrated in Figure 3.3, it is a source of significant atmospheric pollution.

The brick kiln is an important improvement over traditional pit and mound kilns, producing good quality charcoal at relatively high yields (Figure 3.3) (FAO Forestry Department, 1987). Capital cost is relatively low and labor costs are moderate. The kiln is constructed out of bricks or sometimes other refractory materials such as masonry, cinder blocks, or concrete, which provide good heat insulation and oxygen control, in a hemispherical or beehive shape of 5–7 m diameter set into a foundation of similar refractory material (Figure 3.3).

The kiln typically has two openings diametrically opposite one another and perpendicular to prevailing winds. One opening is used to charge the kiln while the other is used to discharge the charcoal. These openings can either be closed with steel doors or simply bricked over and sealed with mud. Vents are distributed around the base of the kiln to control air infiltration while smoke is exhausted from an "eye" hole at the top of the kiln. Carbonization may occur over 6–7 days, followed by a "purging" stage of 1–2 days during which the perimeter vents are sealed, and finally a cooling stage of 3 days in which the eye hole is also sealed.

Metal kilns made of steel or cast iron originated in Europe in the 1930s and spread to the developing world in the 1960s. Although several variations exist, the transportable metal kiln developed by the Tropical Products Institute (TPI) is illustrative of the main features of this type of kiln (Whitehead, 1980). As shown in Figure 3.4, the TPI kiln consists of two interlocking cylindrical sections and a conical cover with four steam

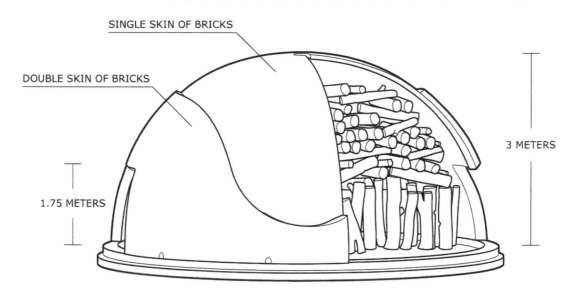

SINGLE SKIN OF BRICKS

DOUBLE SKIN OF BRICKS

3 METERS

1.75 METERS

Figure 3.3 *Brick kiln (adapted from FAO Forestry Department, 1987)*

release ports. The kiln is supported on eight channels projecting radially from the perimeter of the base section. These are designed to serve as air inlets or, when fitted with smoke stacks, to vent smoke out of the kiln. During carbonization, four of the channels are fitted with smoke stacks. The metal kiln has several advantages over traditional or brick kilns. The flow of air into and smoke out of the kiln is readily controlled, which improves charcoal

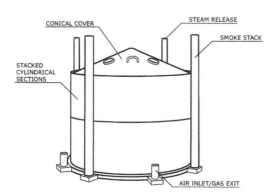

CONICAL COVER
STEAM RELEASE
SMOKE STACK
STACKED CYLINDRICAL SECTIONS
AIR INLET/GAS EXIT

Figure 3.4 *TPI transportable metal kiln (adapted from Whitehead, 1980)*

yield and quality. Unskilled personnel can be quickly trained to operate the kiln and it does not require the constant attention of traditional kilns. Carbonization is complete in three days and all of the charcoal can be recovered from the kiln. The kiln can be operated in areas of high rainfall. The metal kiln does not, however, mitigate the air pollution associated with charcoal making.

The concrete kiln, also known as the Missouri kiln, is a rectangular structure constructed of reinforced concrete or concrete block with steel doors (Figure 3.5). The kiln is designed for mechanized loading and unloading of wood and charcoal. A typical kiln is about 7 m wide and 11 m long with a vault height of 4 m. This gives it a capacity of about 180 m^3 of wood, which is about three times greater than brick kilns. They typically produce 16 tons of charcoal during a 3-week cycle by carrying out carbonization in the temperature range of 260–370°C (Wolk et al, 2007). Yields are higher than for metal kilns because of better thermal insulation and larger volume-to-surface area ratios. Temperature gauges

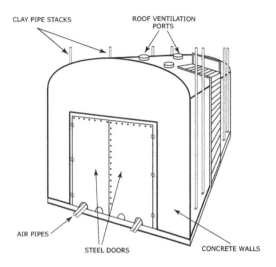

CLAY PIPE STACKS

ROOF VENTILATION
PORTS

AIR PIPES

STEEL DOORS

CONCRETE WALLS

Figure 3.5 *The Missouri-type charcoal kiln (Maxwell, 1976)*

within the kiln also contribute to better yields by allowing hot and cold spots to be identified and corrected by controlling airflow into the kiln. The Missouri kiln is fitted with eight, 0.15 m diameter pipes to serve as chimneys. These can be connected to a central flue and afterburner to mitigate atmospheric emissions of carbon monoxide, volatile organic compounds, and particulate matter (Yronwode, 2000). However, control of emissions from batch-type kilns is difficult because the emissions never reach a steady-state condition. A modernized version of the Missouri kiln, known as Rima Container Kiln (5 m^3 to 40 m^3), commonly used in Brazil, produces a 35% gravimetric yield of charcoal during a 3-h cycle (1 ton of charcoal h^{-1}) by carrying out carbonization in the temperature range of 450–500°C (Vilela et al, 2014).

The main characteristics of some of the most common batch kilns used for commercial charcoal production are summarized in Table 3.1. These can generally be classified according to the mode by which heat is supplied to the reactor. An autothermal kiln admits sufficient air to burn the vapors

released from the pyrolyzing biomass, which directly provides heat to the process. An indirectly heated kiln excludes air from the pyrolyzer, and burns the pyrolysis gases outside the kiln, transferring the heat to the biomass through the metal wall of the kiln.

Continuous reactors were introduced to both eliminate manual loading and unloading of batch kilns, as well as achieve steady operation, which improves charcoal yields and eases the implementation of pollution emission controls. The most prominent of these is the multiple hearth kiln, consisting of a refractory-lined vertical steel shell containing a series of shelves or hearths supported by the walls of the kiln (Figure 3.6). A rotating vertical shaft fitted with rabble arms is located at the center of the shell. As the shaft rotates the rabble arms sweep slowly across the hearths, moving carbonizing wood either radially inward or outward toward penetrations in the hearths where the material drops to the next lower hearth. Air flowing upward through the hollow shaft is mixed with the pyrolysis gases, is re-introduced at the bottom of the reactor, and used as a purge gas. Gases including volatile organic compounds and water vapor released from the carbonizing wood travel counter currently to the flow of biomass in the kiln. Continuous multiple hearth kilns produce an average of 2.75 tons per hour of charcoal by carrying out carbonization over the temperature range of 275–350°C (USEPA, 1995). As a continuous flow reactor, the multiple hearth kiln offers superior control of carbonization time and gas flow, which improves charcoal yields and quality. Continuous processes are also more amenable to pollution control compared to batch processes. Afterburning is estimated to reduce emissions of particulate matter (PM), carbon monoxide (CO), and volatile organic compounds (VOC) by at least 80% (Rolke et al, 1972).

Although charcoal kilns have been continuously improving, they still draw criticism

Table 3.1 *Descriptions of batch kilns for the production of charcoal*

Reactor	Final product	Heating mode	Construction materials/mode of operation	Operation/portability	Raw material used	Loading and discharge methods	Size of the kiln	Process control
Earth Pit/Mound kiln	Biochar	Slow/Autothermal	Earth/Batch	Batch/Built in place	Logs	Manual	Pit: depth: 0.6–1.2 m, Length: 4 m, Capacity: 1–30 m^3 Mound kiln: diameter: 2–15 m, height: 1–5 m	Observing color of produced vapors
Brazilian/Argentine brick kiln	Biochar	Slow/Autothermal	Brick/Batch	Batch/Stationary	Logs	Manual/Mechanical	Diameter: 5–7 m, high: 2–3 m	Observing color of produced vapors
Adam retort	Biochar	Slow/Autothermal	Cinder	Batch/Stationary	Logs	Manual	Capacity: 3 m^3	Observing color of produced vapors
Missouri Kiln	Biochar	Slow/Autothermal	Concrete	Batch/Stationary	Logs	Mechanical	Capacity: 180 m^3, Wide: 7 m, length: 11 m, height: 4 m	Observing color of produced vapors
Black Rock Forest/ Portable metal kiln	Biochar	Slow/Autothermal	Steel	Batch/Stationary	Logs	Manual	Diameter: 2.3 m, Height: 1.7 m	Observing color of produced vapors
Wagon Retort	Biochar/Bio-oil	Slow/Indirect heat	Steel	Batch/Stationary	Logs	Use of wagons	Diameter: 2.5 m, Length: 7.5 m	Direct measurement of temperature

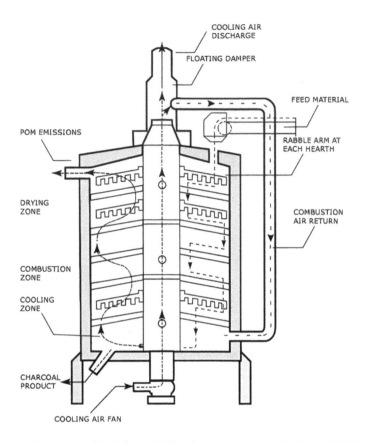

Figure 3.6 *The continuous multiple hearth kiln for charcoal production (USEPA, 1995)*

due to the environmental concerns related to their air emissions (Figure 3.7). Pollutant emissions associated with traditional charcoal making are usually characterized as carbon monoxide (CO), methane (CH_4), non-methane hydrocarbons (NMHC), and total suspended solids (TSPs) (Smith et al, 1999) although "hydrocarbons" is a misnomer because NMHC usually includes methanol, acetic acid, and other oxygenated organic compounds. Carbon dioxide, a potent greenhouse gas, is a major air emission from kilns, it is not an environmental concern as long as old-growth forest is not used as feedstock. Table 3.2 lists emission levels for these pollutants from different kinds of charcoal kilns (Moscowitz, 1978).

Some NMHCs, also referred to as volatile organic compounds (VOCs) are commercially valuable chemical compounds, that can be recovered by distillation. In fact, "wood tar" and "pyroligneous acid" were often the main reasons for operating kilns before the development of petroleum-based chemicals. Destructive distillation of wood produced commercially significant quantities of wood alcohol, mostly acetic acid and methanol (Sjostrom, 1993). Fast pyrolysis, discussed in detail in Chapter 4, was introduced in the 1970s to recover condensable vapors as the primary product with char as a coproduct. The prospect of converting the resulting bio-oil into renewable fuels has increased in recent years.

Figure 3.7 *Operation of a mound kiln showing the heavy smoke emitted during the carbonization process (Source: Weald and Downland Open Air Museum)("Charcoal Making, Weald and Downland Open Air Museum")*

Table 3.2 *Air emissions per kilogram biomass from different kinds of charcoal kilns (Adapted from Moscowitz, 1978)*

	CO (g kg^{-1})	CH$_4$ (g kg^{-1})	NMHC[1] (g kg^{-1})	TSP[2] (g kg^{-1})
Uncontrolled batch	106–336	12.7–57.7	7–95	197–598
Low control batch	24–27	6.6–8.6	1–9	27–89
Controlled continuous	8.0–8.9	2.2–2.9	0.4–3.0	9.1–30

Notes
[1] NMHC – non-methane hydrocarbons (includes recoverable methanol and acetic acid)
[2] TSP – total suspended particulates

Fundamental concepts of thermal decomposition of biomass substrates

The major constituents of fibrous biomass are cellulose, hemicellulose, and lignin with smaller quantities of organic extractives (pectins and proteins) and inorganic minerals (Vassilev et al, 2010). These constituents can vary considerably among different kinds of biomass or even within a species depending upon soil type, climate, and time of harvest.

Consequently, the characteristics of biochar can vary significantly among feedstocks. It is important to understand the constituents of the biomass and the reaction mechanisms to produce biochar. The large variety of potential feedstocks for biochar production makes possible biochars with diverse characteristics (Table 3.3), which determines their suitability

Table 3.3 Organic composition, elemental (ultimate), and proximate analysis of typical examples of biomass (% dry basis) (Cell-cellulose, Hem-hemicellulose, Lig-lignin, VM-volatile matter, FC-fixed carbon)

Selected biomass	Organic composition			Elemental (ultimate) analysis				Proximate analysis			References
Species	Cell	Hem	Lig	C	H	O	N	VM	FC	Ash	
Woody crops											
Poplar	49.0	25.6	23.1	48.5	5.9	43.7	0.5	82.3	16.3	1.3	Toor et al (2011); Jenkins and Ebeling (1985)
Pine	41.7	20.5	25.9	47.3	6.2	42.4	0.3	78.4	18.9	2.7	Toor et al (2011); Wang and Brown (2013)
Red oak	42.2	33.1	20.2	45.2	6.4	47.7	0.1	87.7	11.6	0.7	Pettersen (1984); Rowell et al (2013); Wang and Brown (2013)
Eucalyptus wood	45.0	19.2	31.3	48.4	6.1	45.3	0.1	80.7	18.9	0.5	Yip et al (2010)
Eucalyptus bark	37.4	19.2	28.0	50.4	5.6	43.7	0.3	68.1	27.0	4.0	Yadav et al (2002); Yip et al (2010)
Eucalyptus leaf	37.1	14.3	34.1	59.3	6.8	32.4	1.3	74.6	21.7	3.7	Chilcott and Hume (1984); Yip et al (2010)
Herbaceous crops											
Switchgrass	36.2	21.7	21.2	44.8	6.6	44.4	0.6	76.7	14.4	3.7	Imam and Capareda (2012); Yoshida et al (2008)
Miscanthus	40.2	22.4	24.4	48.1	5.4	42.2	0.5	75.1	17.9	3.1	McKendry (2002)
Agricultural residues											
Corn stover	37.1	24.2	18.2	43.7	5.6	43.3	0.6	75.2	19.3	5.6	Toor et al (2011); Jenkins (1985)
Rice straw	32.0	35.7	22.3	48.2	6.2	44.1	0.8	74.7	15.2	10.1	Wannapeera et al (2008)
Wheat straw	41.3	30.8	7.7	49.7	7.1	39.3	0.8	70.2	25.0	4.8	Tumuluru et al (2011); McKendry (2002)
Coconut shell	14.0	32.0	46.0	71.4	5.4	25.1	5.3	71.8	12.4	15.8	Rout (2013)
Rice husk	30.8	15.0	26.4	31.6	4.3	47.4	0.7	65.3	10.0	24.6	Herrera et al (2022)

Aquatic species

Micro algae	7.1	16.3	42.5	6.8	28.0	6.6	71.0	12.4	16.6	Ververis et al (2007); Wang and Brown (2013)
Macro algae	25.7	13.8	32.9	4.8	35.6	2.5	67.1	9.3	29.1	Trinh et al (2013); Ross et al (2008)
Waste biomass										
Saw dust	33.7	22.6	46.5	5.6	45.7	2.1	77.0	19.9	0.8	Sinağ et al (2011); Park et al (2010)
Manure (swine)	15.1	19.9	47.3	5.9	20.1	4.6	77.0	1.1	22.3	Xiu et al (2010); Cao et al (2011)
Manure (cattle/dairy)	31.4	13.9	37.9	5.5	25.6	3.0	67.0	10.8	22.3	Liao et al (2005); Otero et al (2011)
Municipal solid waste	37.6	8.4	47.6	6.0	32.9	1.2	69.6	9.8	20.6	Lamborn (2009); Kathirvale et al (2004)
Food waste	49.5	7.4	49.6	6.9	32.2	3.4	82.0	13.1	6.3	Lamborn (2009); Caton et al (2010)
Yard trimmings	26.2	11.7	57.7	18.4	22.3	0.1	62.8	30.7	6.5	Lamborn (2009); Miskolczi (2013)
Pure cellulose	100.0	0.0	44.6	6.4	49.0	0.0	92.8	6.1	1.1	Rutkowski and Kubacki (2006); Sanchez-Silva et al (2012)

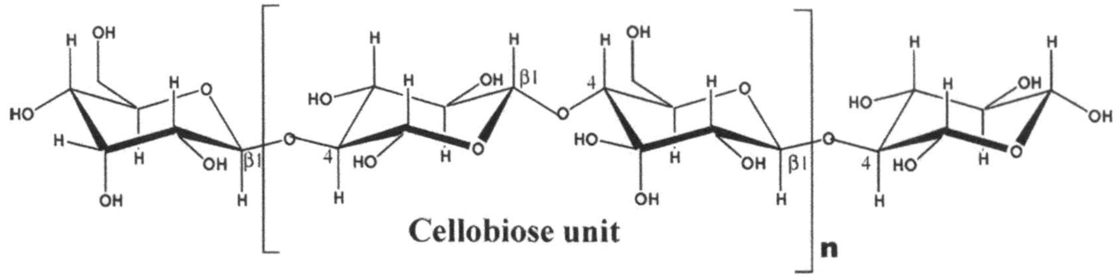

Figure 3.8 *Chemical structure of cellulose (Adopted from Mohan et al, 2006)*

for such applications as soil amendment, carbon sequestration, and/or bioremediation (Spokas et al, 2012).

Cellulose is a linear condensation polymer of β-(1–4)-D-glucopyranose connected by ether linkages known as glycosidic bonds (Figure 3.8) (O'Sullivan, 1997; Mohan et al, 2006). The repeating unit of the cellulose polymer is cellobiosan, which consists of two anhydroglucose units. The number of glucose units in a cellulose chain is known as the degree of polymerization (DP). The average DP for native cellulose is on the order of 10,000. The coupling of adjacent cellulose molecules by hydrogen bonds and van der Waal's forces results in a parallel alignment that gives cellulose a crystalline structure. Cellulose exists as sheets of glucopyranose rings lying in a plane with successive sheets stacked on top of each other to form three-dimensional particles that aggregate into elementary fibrils with crystalline widths of 4–5 nm. This crystalline microfibril arrangement makes cellulose more resistant to thermal decomposition than hemicellulose.

Hemicellulose includes a large number of heteropolysaccharides built from hexoses (D-glucose, D-mannose, and D-galactose), pentoses (D-xylose, L-arabinose, and D-arabinose), and deoxyhexoses (L-rhamnose or 6-deoxy-L-mannose and rare L-fucose or 6-deoxy-L-galactose) linked by glycosidic bonds (Sjostrom, 1993). Small amounts of uronic acids (4-O-methy-D-glucuronic acid, D-galacturonic acid, and D-glucuronic acid) are also present. Hemicelluloses are characterized as a condensation polymer with the removal of water molecules with every linkage both in the D configuration of the six-membered pyranoside forms and in the L configuration of five-membered furanoside forms (Bajpai, 2018). Hardwoods are rich in xylans such as O-acetyl-(4-O-methylglucurono) xylan and contain small amounts of glucomannan. Softwoods are rich in glucomannans such as O-acetyl-galactoglucomannan and smaller amounts of xylans such as arabino-(4-O-glucurono) xylan. Softwood hemicelluloses have more mannose and galactose units and fewer xylose units and acetylated hydroxyl groups than hardwood hemicelluloses.

Short side chains distinguish hemicellulose from cellulose (Figure 3.9). Due to its more open structure compared to cellulose, hemicellulose is more hygroscopic, attracting water molecules (Benaimeche et al, 2020). The chemical and thermal stability of hemicelluloses is lower than that of cellulose due to its lack of crystallinity and lower degree of polymerization, which is only 100–200 (Sjostrom, 1993).

Lignin, a phenylpropane-based polymer, is the largest non-carbohydrate fraction of lignocellulose (Sjostrom, 1993). It is constructed of three monomers: coniferyl alcohol,

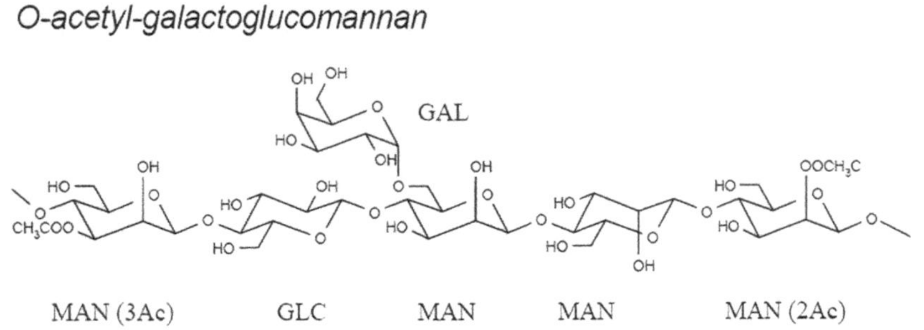

O-acetyl-galactoglucomannan

MAN (3Ac)   GLC   MAN   MAN   MAN (2Ac)

Figure 3.9 *Structural formula for common hemicellulose found in softwoods (GAL = galactose, GLC = glucose, MAN = mannose, Ac = acetyl group, XYL = xylose, GLcA = methylglucuronic acid, ARA = arabinose) (Hartman, 2006)*

sinapyl alcohol, and *p*-coumaryl alcohol, each of which has an aromatic ring with different substituents (Figure 3.10). Softwood lignin contains a higher fraction of coniferyl phenylpropane units (guaiacyl lignin) while hardwood lignin is a co-polymer of both coniferyl and sinapyl phenylpropane units (guaiacylsyringyl lignin). Lignin has an amorphous structure, which leads to a large number of possible interlinkages between individual units. Ether bonds (mainly β-O-4 aryl ether bonds) predominate between lignin units and covalent bonds exist between lignin and polysaccharides (Donaldson et al, 2017). Unlike

cellulose, lignin cannot be depolymerized to its original monomers.

Bundles of elementary cellulose fibrils are embedded in a matrix of hemicellulose with a thickness of 7–30 nm. Lignin is located primarily on the exterior of microfibrils where it covalently bonds to hemicellulose (Klein and Snodgrass, 1993). Lignin impregnates the cell wall, reduces the pore sizes, shields the polysaccharides, and contributes to the recalcitrance of lignocellulose (Saxena and Brown, 2005).

Plant materials also contain other organic compounds collectively known as "extractives" in the engineering literature. These include resins, fats and fatty acids, phenolics, and phytosterols, among other chemical compounds. Extractives are classified as either hydrophilic or lipophilic depending on whether they are soluble in water or organic solvents, respectively. Resin is often used to describe lipophilic extractives, except for phenolic substances. Extractives can influence gaseous emission profiles during pyrolysis but they are not thought to substantially influence charcoal yield because of their low concentrations.

The inorganic content of biomass includes the major elemental nutrients N, P, and K as well as smaller amounts of S, Cl, Si, alkaline earth metals, transition metals, and various

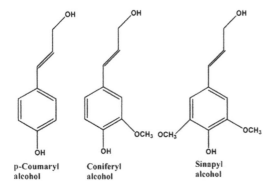

p-Coumaryl alcohol   Coniferyl alcohol   Sinapyl alcohol

Figure 3.10 *Monomers from which lignin is assembled*

Table 3.4 *Influence of heating rate on pyrolysis of cellulose in a thermogravimetric analyzer with nitrogen as sweep gas, flow rate unspecified (Gupta and Lilley, 2003)*

Heating rate ($°C\ min^{-1}$)	Enthalpy of pyrolysis ($J\ kg^{-1}$)	Onset temperature of pyrolysis (°C)	Temperature of maximum decomposition rate (°C)
5	+780	314	345
10	+498	337	360
30	+455	350	383
50	+440	362	396

trace elements. That part of the inorganic content remaining after oxidation of the biomass at high temperature is known as ash.

Cellulose, hemicellulose, and lignin have distinctive thermal decomposition behaviors that depend upon heating rates. Pyrolysis of cellulose is initiated at higher temperatures as the heating rate is increased (Gupta and Lilley, 2003) (Table 3.4). At very low heating rates, typical of muffle furnaces or traditional charcoal kilns, cellulose decomposition begins at temperatures as low as 250°C (Williams and Besler, 1996).

Hemicellulose is the first to decompose, beginning at 220°C and substantially completed by 315°C (Figure 3.11). Cellulose does not start to decompose until about 315°C. If volatiles are quickly removed from the reaction zone, cellulose is mostly converted to condensable organic vapors and aerosols once 400°C is attained. High pressures and the absence of ventilation promote char and gas-forming reactions at the expense of condensable organic vapors. Although lignin begins to decompose at 160°C, it is a slow, steady process extending to 900°C and

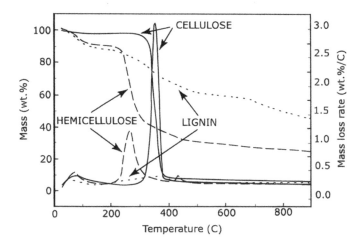

Figure 3.11 *Thermogravimetric analysis of the pyrolysis of cellulose, hemicellulose (xylan), and lignin at a constant heating rate ($10°C\ min^{-1}$) with N (99.9995%) sweep gas at 120 mL min^{-1} using a thermogravimetric analyzer (adapted from Yang et al, 2007)*

yielding a solid residue approaching 40 wt% of the original sample.

Recent research has begun to unravel the mechanisms by which plant polymers thermally decompose. Both cellulose and hemicellulose decompose by the competing reactions of glycosidic bond breaking and pyranose ring scission. In the absence of alkali and alkaline earth cations, glycosidic bond breaking dominates and anhydrosugars are the dominant primary products of cellulose and hemicellulose pyrolysis (Patwardhan et al, 2009, 2011). However, even the relatively small amounts of alkali and alkaline earth metals occurring naturally in most biomass as part of the ash catalyze pyranose ring scission, reducing the yield of anhydrosugars and increasing the yields of light (C2-C3) oxygenated organic compounds, non-condensable gases, and char, offering opportunities for enhanced biochar C sequestration potential (Patwardhan et al, 2010; Mašek et al, 2019).

The depolymerization of polysaccharides forms a melt of anhydrosugars, which has been directly observed during the pyrolysis of cellulose (Dauenhauer et al, 2007). Melted anhydrosugars can either evaporate and escape from the reaction zone or polymerize to non-volatile oligomers. Although the vapor pressures of anhydrosugars are very small compared to many other pyrolysis products, at pyrolysis temperatures they are high enough under well-ventilated conditions for much of the anhydrosugars to escape as vapors, which are relatively stable. However, polymerization is also favored by elevated temperatures. The resulting oligosaccharides are non-volatile and cannot escape the reaction zone. Instead, they dehydrate to biochar, water, and light gases (Bai et al, 2013). Pyrolysis of polysaccharides proceeds as a sequence of two steps (Figure 3.12 for cellulose), each with competing pathways that determine the extent of char formation from carbohydrates. Conditions that promote biochar formation include the presence of ash (Buss et al, 2019) in the biomass, the rate of heating, and the level of ventilation within the reactor.

Among the light oxygenates from polysaccharide decomposition are alcohols, aldehydes, carboxylic acids, and ketones. Decomposition products also include furans including furfural (2-furaldehyde) and 2, 5-dimethylfuran (DMF). Some of these compounds have been commercially recovered from biomass, particularly carboxylic acid, and furfural from the acetyl functionality and pentose sugars, respectively, found in hemicellulose (Rutherford et al, 2004).

Lignin, similar to the polysaccharide in lignocellulosic biomass, also depolymerizes upon rapid heating. Some researchers hypothesize that lignin partially depolymerizes into large fragments that are thermally ejected from the pyrolysis zone, allowing them to be recovered in bio-oil. More consistent with the recent experimental evidence is the depolymerization of lignin to highly substituted phenolic monomers and dimers that are sufficiently volatile to

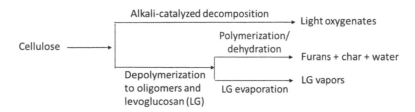

Figure 3.12 *Mechanism of cellulose pyrolysis. LG-levoglucosan*

evaporate (Patwardhan et al, 2011). The side chains of these phenolic compounds include a wide range of functionality, including very reactive methoxy and vinyl groups. In a recent study using an optically accessible reactor, Tiarks et al (2019) concluded that observations of aerosol ejection in laboratory experiments were an artifact of the experimental method rather than a realistic representation of pyrolyzing biomass in larger reactors. These experiments pyrolyzed extracted lignin that was only heated from the undersurface of the sample. The extracted lignin particles were observed to melt and coalesce, a phenomenon not observed for lignocellulosic biomass particles. Because the melt was heated from below, devolatilization occurred within the melt rather than at the surface, resulting in violent gas ejection, which carried droplets of lignin oligomers into the gas stream. By dispersing the extracted lignin particles within an inert matrix, simulating the dispersion of lignin in lignocellulosic biomass, coalescence into a monolithic melt was circumvented and ejection of liquid droplets was no longer observed (Tiarks et al, 2019). Instead, aerosols were observed to form in the vapor phase above the sample, which was attributed to condensation reactions of volatile monomers and dimers in the gas phase to yield non-volatile oligomers that nucleate to aerosols – the yellowish, acrid smoke of pyrolyzing biomass that is recovered as a black, viscous liquid sometimes known as tar but better characterized as phenolic oligomers.

The catalytic activity of naturally occurring alkali and alkaline earth metals (AAEM) present in the biomass can also play an important role in promoting rapid depolymerization of lignin. Biomass pretreatment with ferrous sulfate can promote AAEM passivation by sulfate anions while lignin depolymerization is catalyzed by the ferrous cations (Rollag et al, 2022). This pretreatment has been found to increase pyrolytic sugar production severalfold while preventing agglomeration of biochar during pyrolysis.

Lignin is the source of most of the biochar from biomass pyrolysis. This is because lignin depolymerizes relatively slowly, allowing reactive lignin fragments to repolymerize and subsequently dehydrate to biochar, water, and non-condensable gases. As with cellulose, char formation from lignin can be suppressed, although not eliminated, by rapidly heating the biomass to high temperatures and rapidly transporting the product vapors out of the pyrolysis zone.

Charcoal yields and properties

Traditional charcoal making goes through three successive stages characterized by the color of smoke emitted from the kiln. Biomass is dried in the first step, indicated by the heavy white smoke of condensing water. In the second step, the biomass pyrolyzes, yielding yellow smoke consisting of organic compounds. During the final stage, little smoke is produced (typically transparent or bluish color), indicating that the carbonization process has been completed. At this point, the kiln chimneys are closed to prevent fresh air from entering the kiln, which would encourage combustion (Toole et al, 1961).

In some kilns, water and condensable tars can be recovered as liquid. The water comes from both the moisture in the biomass feedstock as well as "produced water" from chemical reactions accompanying pyrolysis. It is important to distinguish between these kinds of water when performing material balances on pyrolysis processes (Table 3.5).

Table 3.5 *Example of stoichiometry and mass fraction (% w/w.) for wood carbonization at 400°C. Adapted from Antal and Grønli (2003) on information from Klason (1914)*

Reaction	Biomass	→	CO_2 (gas)	+	CO (gas)	+	Water (liquid)	+	Tars (liquid)	+	Biochar (solid)
Molar ratio	$C_{84}H_{120}O_{56}$	→	5 CO_2	+	3 CO	+	28 H_2O	+	$C_{28}H_{34}O_9$	+	$C_{48}H_{30}O_6$
Mass fraction (%)	100	→	11	+	4	+	25	+	25	+	35

Table 3.6 *Biochar yields (dry weight basis for different kinds of batch kilns) (Kammen and Lew, 2005)*

Kiln type	Biochar yield (%)
Pit	13–30
Mound	2–42
Brick	13–33
Portable steel (TPI)	19–31
Concrete (Missouri)	33

The biochar yield $\eta_{biochar}$ from a kiln is given by:

$$\eta_{biochar} = (m_{biochar}/m_{biomass}) \times 100 \quad [3.1]$$

where $m_{biochar}$ is the dry mass of biochar from the kiln and $m_{biomass}$ is the dry mass of biomass loaded into the kiln.

Although reported yields range widely for a given type of kiln (Table 3.6), in general, brick and steel kilns yield more biochar than pit and mound kilns, and concrete kilns are expected to have the highest yields among batch kilns.

Different kinds of biomass have markedly different biochar yields (Li et al, 2017; Phounglamcheik et al, 2020; Štefanko and Leszczynska, 2020) (Table 3.7). The ash content of the biomass is a major determinant of the mass yield of biochar (Li et al,

Table 3.7 *Biochar yields (dry weight basis for different kinds of biomasses) (Li et al, 2017)*

Biomass type	Biochar yield (%)
Sweetgum	9.0
Beech bark	24.6
Acacia	10.7
Acacia bark	21.7
Corn stover	21.7
Red oak	13.5
Switchgrass	9.5
Yellow poplar	7.4
Red maple	7.0
Beech	10.6
Loblolly pine	10.0

2017). Although the biochar yield described by Equation 3.1 is of some practical application, it is not an appropriate measure of the amount of C produced from biomass since it does not account for the ash content of the biomass feedstock and biochar product.

A more meaningful measure of carbonization efficiency is the fixed carbon yield:

$$\eta_{\text{fixed Carbon}} = \frac{m_{\text{biochar}}}{m_{\text{biomass}}} \frac{c_{\text{fixed Carbon}}}{1 - b_{\text{ash}}} \qquad [3.2]$$

where $c_{fixed\ carbon}$ is the fixed carbon content of biochar as measured by ASTM Standard D 1762–84 and b_{ash} is the ash content of the dry biomass. This represents the conversion of ash-free organic mass in the feedstock into ash-free fixed carbon (Antal et al, 2000). An ideal kiln would have a fixed carbon yield equal to the solid elemental C yield as predicted by the thermodynamic equilibrium. For example, the pyrolysis of cellulose at 500°C and 1 MPa should have a fixed carbon

yield of 48%, as illustrated in Figure 3.13 (calculated using the chemical equilibrium software package STANJAN), adapted from Bishnu et al (2001).

In fact, biochar yields from biomass are considerably less than theoretical expectations. Traditional charcoal kilns can have carbonization efficiencies as low as 8% by mass (Straka, 1985). This can arise from the infiltration of O_2 with air into the kiln, which oxidizes biochar to CO and CO_2 and greatly reduces equilibrium yields of C (Figure 3.13).

However, even in the absence of O_2, low biochar yields can result if vapors and gases are removed from the reaction zone before thermodynamic equilibrium can be attained. Although it is often assumed that biochar is the result of solid-phase reactions in which de-volatilized biomass leaves behind a carbonaceous residue (primary pyrogenic carbonaceous matter, PCM), in fact, biochar is also formed by dehydration of the primary products of pyrolysis (secondary PCM), as

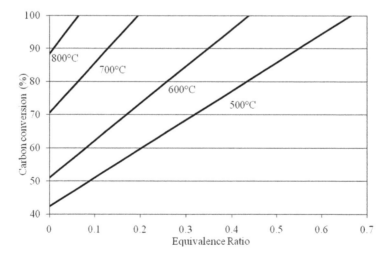

Figure 3.13 *Carbon conversion for gasification of cellulose at 1 MPa as a function of equivalence ratio (fraction of stoichiometric O2 requirement for theoretical complete combustion) calculated with STANJAN chemical equilibrium software. Note that fixed carbon yield (%) equals 100 - carbon conversion (%)*

previously described. This secondary PCM is as chemically reactive as the primary PCM despite differences in its formation (Chen et al, 1997). The escape of pyrolytic vapors prevents the attainment of thermodynamic equilibrium by the original reactants, which favors high biochar yields. Biochar yield of pyrolyzing cellulose can vary from a few percent to almost 20% simply by controlling the venting of vapors during TGA experiments (Varhegyi et al, 1988; Milosavljevic et al, 1996). Klason (1914) recognized the importance of primary and secondary reactions in carbonization almost one hundred

years ago but this fact has yet to be fully exploited in charcoal or biochar manufacture.

The existence of primary and secondary reactions in biochar production helps explain two otherwise difficult-to-understand phenomena. These are the effects of pressure on resulting yields and the report of both endotherms and exotherms during wood pyrolysis.

According to thermodynamic calculations, the pyrolysis of cellulose or wood should not be strongly influenced by pressure (Figure 3.14). In fact, studies dating back as far as the pioneering research by Klason

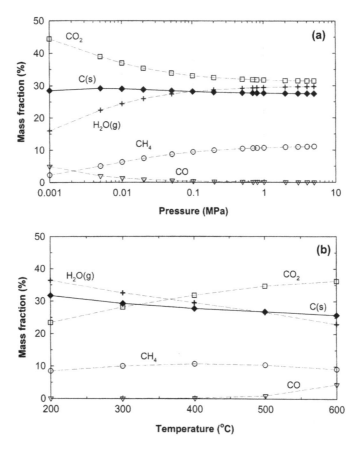

Figure 3.14 *Chemical equilibrium products of cellulose pyrolysis (a) Effect of pressure at 400°C; (b) Effect of temperature at 1 MPa (adapted from Antal and Grønli, 2003)*

(1914) have claimed significant effects of pressure on charcoal yields although others have reported otherwise (Frolich et al, 1928). The question was taken up by Mok and Antal (1983) in tubular flow reactors who demonstrated that charcoal yields increased from around 10 wt% to over 20 wt% as the pressure was increased from 0.1 MPa to 2.5 MPa (Figure 3.15a). They also discovered that the effect was dependent upon the rate at which the reactor was purged with inert gas (Figure 3.15a). This latter observation led them to suggest that pressure is a kinetic rather than a thermodynamic effect: high pressures prolong the intraparticle residence time of pyrolyzing vapors as well as increase the rates of decomposition reactions that allow a closer approach to the expectations of thermodynamic equilibrium. Sweep gas removes vapors before they have a chance to decompose and deposit secondary PCM.

Researchers have variously suggested enthalpies of pyrolysis that have ranged from endothermic (Kung and Kalelkar, 1973) to exothermic (Roberts, 1970). Mok and Antal (1983) used tubular flow reactors embedded in a differential scanning calorimeter to measure the heat of pyrolysis as a function of pressure and purge gas flow (Figure 3.15b). They found the heat of pyrolysis was endothermic at low pressures and exothermic at high pressures. Furthermore, the pressure at which the process transitioned from endotherm to exotherm was dependent on purge gas flow, with low flow rates moving the transition to lower pressures. They attributed the endotherm to the devolatilization of levoglucosan and the exotherm to the *in-situ* carbonization of levoglucosan (Figure 3.12). The ability to control pyrolysis not only improves solid yields but also decreases the energy inputs of biochar reactors (Figure 3.15).

Porosity is an important property of biochar. Although the vascular structure of plant materials contributes to large pores, most of biochar's high surface area derives from nanopores created during the heating process. Porosity is a complex function of

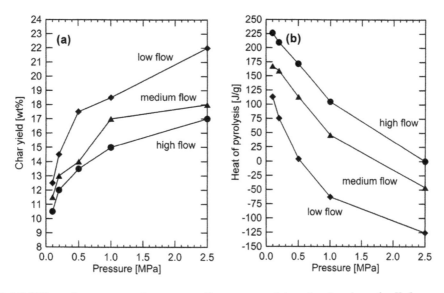

Figure 3.15 *Effect of pressure and purge gas flow rate on (a) carbonization of cellulose and (b) heat of pyrolysis for cellulose (Mok and Antal, 1983)*

heating temperatures, heating rates, and heating times. The origin of the porosity of biochar is due to the quick release of gases during the decomposition of lignin and the reaction of aromatic condensation with increasing temperature (Chen et al, 2012; Zhao et al, 2017; Tomczyk et al, 2020).

Under conditions of slow pyrolysis, aliphatic C in pyrolyzing biomass must first be converted into fused-ring, aromatic C before porosity can develop (Rutherford et al, 2004). For cellulose, this transformation of aliphatic C does not occur below 250°C while for lignin, which already contains significant amounts of aromatic C, temperatures closer to 300°C are required to convert its aliphatic C. At higher temperatures (400–500°C), the porosity of biochar begins to increase drastically as aromatic C is gradually lost, more amorphous C moves into crystalline C, and more volatiles are removed which generates more pore spaces (Leng et al, 2021). Thus, it would appear that the fused-ring structures of aromatic C provide a matrix in which pores can be created. Also, higher temperatures can provide more activation energy used to develop microporosity in biochar (Leng et al, 2021),

It is well known that the pore size distribution is a significant parameter of characterizing the pore structural heterogeneity and internal structure of biochar (Hu et al, 2008; Leng et al, 2021).

Pyrogenic C (PyC) is categorized as either graphitizable or non-graphitizable PyC (Byrne and Marsh, 1995). In both cases, C atoms are arranged in fused hexagonal rings stacked as small crystallites. However, upon heating to high temperatures, the crystallites of graphitizable PyC reorient themselves into parallel sheets of C atoms after passing through the so-called "mesophase" (Marsh and Diez, 1994) known as graphite, which destroys the porosity of the material. In non-graphitizable PyC, the crystallites are randomly oriented and strongly cross-linked to one another (Oberlin, 1984) which resist reorientation and/or graphitization upon heating and preserve porosity. Pyrogenic C derived from pyrolyzing biomass is non-graphitizable (i.e., graphite cannot be made from biomass pyrolysis since the process does not involve the mesophase) and is thought to be associated with the high O content of the starting material (Franklin, 1951; Marsh and Diez, 1994).

Safety measures during biochar production

Biochar production generates gaseous and dust emissions. Carbon monoxide in the gases is toxic and along with H_2 and other flammable gases in the mixture represents an explosion hazard (FAO Forestry Department, 1987). Tars and pyroligneous acids in the exhaust gases are both irritants to the eyes and inhalation hazards to the skin. To prevent exposure to these hazards, safety precautions including personal protective equipment (PPE) are recommended while working in the kiln or pit to avoid

prolonged exposure (Olujimi et al, 2016; Pramchoo et al, 2017).

Biochar making produces fine solid dust particles (<10 μm) that if inhaled can be injurious to health and eye irritant along with allergenic responses and asthma (Manisalidis et al, 2020). The particle size of the biochar product is typically much larger than this, but fines are generated during grinding/crushing for applications such as the production of charcoal briquettes or biochar pellets (e.g., for biochar-based fertilizers, Chapter 26). Appropriate PPE equipment is highly recommended during handling biochar (Chapter 25). If biochar is incorrectly produced or not sufficiently cooled before exposure to air, it may catch fire depending on the function of time after production (Zhao et al, 2014). Appropriate mitigation measures include storing fresh biochar in the absence of air and spraying it with water to reduce its temperature below its autoignition temperature in air (Chapter 25). The addition of water also reduces the dust problems (Prostański, 2013).

Research and development priorities

Biochar is an effective low-cost C sequestration agent produced via the thermochemical conversion of a variety of biomasses. Since the composition of the biomass and production techniques affect the yield and quality of biochar, these factors and their interactions must be carefully considered in designing biochar production facilities to obtain maximum benefits. Despite the long history of intentional charcoal making and growing interest in biochar as a soil amendment and carbon removal agent, commercial production practices are often haphazard or employ rough rules of thumb at best with the result that biochar profitability is often problematic. Commercial prospects for biochar would be improved by research that includes (1) optimizing biochar production techniques to reduce costs and maintain the homogeneity of biochar when produced at a large scale, (2) determining appropriate production parameters and techniques that produce biochar with maximum soil benefits and C sequestration, and (3) developing techno-economic and life-cycle analyses of biochar production and application to evaluate economic and environmental impacts.

References

Antal MJ, et al 2000 Attainment of the theoretical yield of carbon from biomass. *Industrial and Engineering Chemical Research* 39, 4024–4031.

Antal MJ, and Grønli M 2003 The art, science, and technology of charcoal production. *Industrial and Engineering Chemical Research* 42, 1619–1640.

Bai X, Johnston P, and Brown RC 2013 An experimental study of the competing processes of evaporation and polymerization of levoglucosan in cellulose pyrolysis. *Journal of Analytical and Applied Pyrolysis* 99, 130–136.

Bajpai P 2018 Wood and fiber fundamentals. In: Bajpai P (Ed) *Biermann's Handbook of Pulp and Paper (Third Edition)*. Amsterdam: Elsevier. pp19–74.

Bard E 2001 Extending the calibrated radiocarbon record. *Science* 292, 2443–2444.

Benaimeche O, Seghir NT, Sadowski Ł, and Mellas M 2020 The utilization of vegetable fibers in cementitious materials. In: Hashmi S, and Choudhury IA (Eds) *Encyclopedia of Renewable and Sustainable Materials*. Oxford: Elsevier. pp649–662.

Bishnu PS, Hamiroune D, and Metghalchi M 2001 Development of constrained equilibrium codes and their applications in nonequilibrium thermodynamics. *Journal of Energy Resources Technology* 123, 214–220.

Buss W, Jansson S, and Mašek O 2019 Unexplored potential of novel biochar-ash composites for use as organo-mineral fertilizers. *Journal of Cleaner Production* 208, 960–967.

Byrne JF, and Marsh H 1995 Introductory overview. In: Patrick JW (Ed) *Porosity in Carbons – Characterization and Applications.* New York: Halsted Press. pp2–48.

Cao X, Ro KS, Chappell M, Li Y, and Mao J 2011 Chemical structures of swine-manure chars produced under different carbonization conditions investigated by advanced solid-state 13C nuclear magnetic resonance (NMR) spectroscopy. *Energy and Fuels* 25, 388–397.

Caton PA, Carr MA, Kim SS, and Beautyman MJ 2010 Energy recovery from waste food by combustion or gasification with the potential for regenerative dehydration: A case study. *Energy Conversion and Management* 51, 1157–1169.

Charcoal making, Weald and Downland Open Air Museum [WWW Document], n.d. URL http://www.wealddown-schools.org.uk/Furtherinformation/information-charcoal burning.htm (accessed 10.10.13).

Chen G, Yu Q, and Sjöström K 1997 Reactivity of char from pyrolysis of birch wood. *Journal of Analytical and Applied Pyrolysis* 40–41, 491–499.

Chen Y, Yang H, Wang X, Zhang S, and Chen H 2012 Biomass-based pyrolytic polygeneration system on cotton stalk pyrolysis: Influence of temperature. *Bioresource Technology* 107, 411–418.

Chilcott MJ, and Hume ID 1984 Digestion of Eucalyptus andrewsii foliage by the common ringtail possum, Pseudocheirus peregrinus. *Australian Journal of Zoology* 32, 605–613.

Dauenhauer PJ, Dreyer BJ, Degenstein NJ, and Schmidt LD 2007 Millisecond reforming of solid biomass for sustainable fuels. *Angewandte Chemie International Edition* 46, 5864–5867.

Donaldson L, Nanayakkara B, and Harrington J 2017 Wood growth and development. In: Thomas B, Murray BG, and Murphy DJ (Eds) *Encyclopedia of Applied Plant Sciences (Second Edition).* Oxford: Academic Press. pp203–210.

FAO Forestry Department 1987 Simple technologies for charcoal making. FAO Forestry Paper 41. Rome.

Franklin RE 1951 Crystalline growth in graphitizing and non-graphitizing carbons. *Proceedings of the Royal Society of London. Series A. Mathematical and Physical Sciences* 209, 196–218.

Frolich PK, Spalding HB, and Bacon TS 1928 Destructive distillation of wood and cellulose under pressure. *Industrial and Engineering Chemistry* 20, 36–40.

Gupta AK, and Lilley DG 2003 Thermal destruction of wastes and plastics. In: Andrady AL (Ed) *Plastics and the Environment.* Hoboken: Wiley-Interscience. pp629–696.

Hartman J 2006 Hemicellulose as barrier material. *Stockholm: Royal Institute of Technology.* Licentiate thesis.

Herrera K, Morales LF, Tarazona NA, Aguado R, and Saldarriaga JF 2022 Use of biochar from rice husk pyrolysis. *Part A: Recovery as an Adsorbent in the Removal of Emerging Compounds. ACS Omega* 7, 7625–7637.

Hu S, Xiang J, Sun L, Xu M, Qiu J, and Fu P 2008 Characterization of char from rapid pyrolysis of rice husk. *Fuel Processing Technology* 89, 1096–1105.

Imam T, and Capareda S 2012 Characterization of bio-oil, syn-gas and bio-char from switchgrass pyrolysis at various temperatures. *Journal of Analytical and Applied Pyrolysis* 93, 170–177.

Jenkins BM, and Ebeling JM 1985 Thermochemical properties of biomass fuels. *California Agriculture* 39, 14–16.

Kammen DM, and Lew DJ 2005 *Review of technologies for the production and use of charcoal.* Renewable and Appropriate Energy Laboratory Report. Golden, CO.

Kathirvale S, Muhd Yunus MN, Sopian K, and Samsuddin AH 2004 Energy potential from municipal solid waste in Malaysia. *Renewable Energy*, 29, pp559–567.

Klar M 1925 *The Technology of Wood Distillation.* London: Chapman and Hall Ltd.

Klason P 1914 Versuch einer Theorie der Trockendestillation von Holz. I. *Journal für Praktische Chemie* 90, 413–447.

Klein GL, and Snodgrass WR 1993 Cellulose. In: Macrae R, Robinson RK, and Saddler MJ (Eds) *Encyclopedia of Food Science, Food Technology and Nutrition*. London: Academic Press. pp758–767.

Kung HC, and Kalelkar AS 1973 On the heat of reaction in wood pyrolysis. *Combustion and Flame* 20, 91–103.

Lamborn J 2009 Characterisation of municipal solid waste composition into model inputs. In: Proceedings of the Third International Workshop Hydro-Physico-Mechanics of Landfills. Braunschweig, Germany.

Leng L, et al 2021 An overview on engineering the surface area and porosity of biochar. *Science of the Total Environment* 763, 144204.

Li W, Dang Q, Brown RC, Laird D, and Wright MM 2017. The impacts of biomass properties on pyrolysis yields, economic and environmental performance of the pyrolysis-bioenergy-biochar platform to carbon negative energy. *Bioresource Technology* 241, 959–968.

Liao W, et al 2005 Effects of hemicellulose and lignin on enzymatic hydrolysis of cellulose from dairy manure. *Applied Biochemistry and Biotechnology* 124, 1017–1030.

Manisalidis I, Stavropoulou E, Stavropoulos A, and Bezirtzoglou E 2020 Environmental and health impacts of air pollution: A Review. *Frontiers in Public Health* 8, 14.

Marsh H, and Diez MA 1994 Mesophase of graphitizable carbons. In: Shibaev VP, and Lam L (Eds) *Liquid Crystalline and Mesomorphic Polymers*. New York: Springer. pp231–257.

Mašek O, et al 2019 Potassium doping increases biochar carbon sequestration potential by 45%, facilitating decoupling of carbon sequestration from soil improvement. *Scientific Reports* 9, 5514.

Maxwell W 1976 Stationary source testing of a Missouri-type charcoal kiln. USEPA Report Number 907-9-76-001. Kansas City, MO.

McKendry P 2002 Energy production from biomass (part 1): Overview of biomass. *Bioresource Technology* 83, 37–46.

Milosavljevic I, Oja V, and Suuberg EM 1996 Thermal effects in cellulose pyrolysis: Relationship to char formation processes. *Industrial and Engineering Chemistry Research* 35, 653–662.

Miskolczi N 2013 Co-pyrolysis of petroleum based waste HDPE, poly-lactic-acid biopolymer and organic waste. *Journal of Industrial and Engineering Chemistry* 19, 1549–1559.

Mohan D, Pittman CUJ, and Steele PH 2006 Pyrolysis of wood/biomass for bio-oil: A critical review. *Energy and Fuels* 20, 848–889.

Mok WSL, and Antal MJ 1983 Effects of pressure on biomass pyrolysis. II. Heats of reaction of cellulose pyrolysis. *Thermochimica Acta* 68, 165.

Moscowitz C 1978 *Source Assessment: Charcoal Manufacturing – State of the Art*. Cincinnati: USEPA.

O'Sullivan AC 1997 Cellulose: The structure slowly unravels. *Cellulose* 4, 173–207.

Oberlin A 1984 Carbonization and graphitization. *Carbon* 22, 521–541.

Olujimi OO, Ana GREE, Ogunseye OO, and Fabunmi VT 2016 Air quality index from charcoal production sites, carboxyheamoglobin and lung function among occupationally exposed charcoal workers in South Western Nigeria. *SpringerPlus* 5, 1546.

Otero M, Lobato A, Cuetos MJ, Sánchez ME, and Gómez X 2011 Digestion of cattle manure: Thermogravimetric kinetic analysis for the evaluation of organic matter conversion. *Bioresource Technology* 102, 3404–3410.

Park DK, Kim SD, Lee SH, and Lee JG 2010 Co-pyrolysis characteristics of sawdust and coal blend in TGA and a fixed bed reactor. *Bioresource Technology* 101, 6151–6156.

Patwardhan PR, Brown RC, and Shanks BH 2011 Understanding the fast pyrolysis of lignin. *ChemSusChem* 4, 1629–1636.

Patwardhan PR, Satrio JA, Brown RC, and Shanks BH 2009 Product distribution from fast pyrolysis of glucose-based carbohydrates. *Journal of Analytical and Applied Pyrolysis* 86, 323–330.

Patwardhan PR, Brown RC, and Shanks BH 2011 Product distribution from the fast pyrolysis of hemicellulose. *ChemSusChem* 4, 636–643.

Patwardhan PR, Satrio JA, Brown RC, and Shanks BH 2010 Influence of inorganic salts on the primary pyrolysis products of cellulose. *Bioresource Technology* 101, 4646–4655.

Pettersen RC 1984 The chemical composition of wood. In: Rowell R (Ed) *The Chemistry of Solid Wood, Advances in Chemistry*. Washington: American Chemical Society. pp2–57.

Phounglamcheik A, et al 2020 Effects of pyrolysis conditions and feedstocks on the properties and gasification reactivity of charcoal from woodchips. *Energy and Fuels* 34, 8353–8365.

Pramchoo W, Geater AF, Jamulitrat S, Geater SL, and Tangtrakulwanich B 2017 Occupational tasks influencing lung function and respiratory symptoms among charcoal-production workers: A time-series study. *Safety and Health at Work* 8, 250–257.

Prostański D 2013 Use of air-and water spraying systems for improving dust control in mines. *Journal of Sustainable Mining* 12, 29–34.

Roberts AF 1970 A review of kinetics data for the pyrolysis of wood and related substances. *Combustion and Flame* 14, 261–272.

Rolke RW, et al 1972 *Afterburner Systems Study*. Emeryville. Shell Development Company.

Rollag SA, Lindstrom JK, Peterson CA, and Brown RC 2022 The role of catalytic iron in enhancing volumetric sugar productivity during autothermal pyrolysis of woody biomass. *Chemical Engineering Journal* 427, 131882.

Ross AB, Jones JM, Kubacki ML, and Bridgeman T 2008 Classification of macroalgae as fuel and its thermochemical behaviour. *Bioresource Technology* 99, 6494–6504.

Rout TK 2013 *Pyrolysis of Coconut Shell*. Rourkela: National Institute of Technology.

Rowell RM, Pettersen R, and Tshabalala 2013 Chapter 3: Cell Wall Chemistry, In: *Handbook of Wood Chemistry and Composites*. Second Edition, Boca Raton: CRC Press.

Rutherford DW, Wershaw RL, and Cox LG 2004 Changes in composition and porosity occurring during the thermal degradation of wood and wood components. Scientific Investigations Report 5292. United States Geological Survey. Reston, VA.

Rutkowski P, and Kubacki A 2006 Influence of polystyrene addition to cellulose on chemical structure and properties of bio-oil obtained during pyrolysis. *Energy Conversion and Management* 47, 716–731.

Sanchez-Silva L, López-González D, Villaseñor J, Sánchez P, and Valverde JL 2012 Thermogravimetric–mass spectrometric analysis of lignocellulosic and marine biomass pyrolysis. *Bioresource Technology* 109, 163–172.

Saxena IM, and Brown RM 2005 Cellulose biosynthesis: Current views and evolving concepts. *Annals of Botany* 96, 9–21.

Sınağ A, Uskan B, and Gülbay S 2011 Detailed characterization of the pyrolytic liquids obtained by pyrolysis of sawdust. *Journal of Analytical and Applied Pyrolysis* 90, 48–52.

Sjostrom E 1993 *Wood Chemistry: Fundamentals and Applications*, 2nd ed. San Diego: Academic Press.

Smith KR, et al 1999 Greenhouse gases from small-scale combustion devices in developing countries, Charcoal-making kilns in Thailand. United States Environmental Protection Agency/600/R-99/109. Washington, DC.

Spokas KA, et al 2012 Biochar: A synthesis of its agronomic impact beyond carbon sequestration. *Journal of Environmental Quality* 41, 973–989.

Straka TJ 1985 FAO Forestry Department. Industrial Charcoal Making. FAO Forestry Paper 63. Rome.

Tiarks JA, Dedic CE, Meyer TR, Brown RC, and Michael JB 2019 Visualization of physicochemical phenomena during biomass pyrolysis in an optically accessible reactor. *Journal of Analytical and Applied Pyrolysis* 143, 104667.

Tomczyk A, Sokolowska Z, and Boguta P 2020 Biochar physicochemical properties: Pyrolysis temperature and feedstock kind effects. *Reviews in Environmental Science and Bio/Technology* 19, 191–215.

Toole AW, et al 1961 *Biochar Production, Marketing and Use.* Madison: University of Wisconsin Forest Products Laboratory, USDA Forest Service.

Toor SS, Rosendahl L, and Rudolf A 2011 Hydrothermal liquefaction of biomass: A review of subcritical water technologies. *Energy* 36, 2328–2342.

Trinh TN, et al 2013 Comparison of lignin, macroalgae, wood, and straw fast pyrolysis. *Energy and Fuels* 27, 1399–1409.

Tumuluru JS, Sokhansanj S, Hess J, Wright C, and Boardman R 2011 A Review on biomass torrefaction process and product properties for energy applications. *Industrial Biotechnology* 7, 384–401.

Uroić Štefanko A, and Leszczynska D 2020 Impact of biomass source and pyrolysis parameters on physicochemical properties of biochar manufactured for innovative applications. *Frontiers in Energy Research* 8, 138.

USEPA 1995 Charcoal, Section 10.7 in AP-42, Compilation of Air Pollutant Emission Factors, Fifth Edition. Research Triangle Park, NC.

Varhegyi G, Antal MJ, Szekely T, Till F, and Jakab E 1988 Simultaneous thermogravimetric-mass spectrometric studies of the thermal decomposition of biopolymers. 1. Avicel cellulose in the presence and absence of catalysts. *Energy and Fuels* 2, 267–272.

Vassilev SV, Baxter D, Andersen LK, and Vassileva CG 2010 An overview of the chemical composition of biomass. *Fuel* 89, 913–933.

Ververis C, et al 2007 Cellulose, hemicelluloses, lignin and ash content of some organic materials and their suitability for use as paper pulp supplements. *Bioresource Technology* 98, 296–301.

Vilela AO, Lora ES, Quintero QR, Vicintin RA, and Souza TPS 2014 A new technology for the combined production of charcoal and electricity through cogeneration. *Biomass and Bioenergy* 69, 222–240.

Wang K, and Brown RC 2013 Catalytic pyrolysis of microalgae for production of aromatics and ammonia. *Green Chemistry* 15, 675–681.

Wannapeera J, Worasuwannarak N, and Pipatmanoami S 2008 Product yields and characteristics of rice husk, rice straw and corncob during fast pyrolysis in a drop-tube/fixed-bed reactor. *Songklanakarin Journal or Science and Technology* 30, 393–404.

Whitehead WDJ 1980 The construction of a transportable charcoal kiln. *Rural Technology Guide* 13.

Williams PT, and Besler S 1996 The influence of temperature and heating rate on the slow pyrolysis of biomass. *Renewable Energy* 7, 233–250.

Wolk RH, Lux S, Gelber S, and Holcomb FH 2007 Direct carbon fuel cells: Converting waste to electricity. ERDC-CERL Fuel Cell Program Report number ERDC/CERL TR-07-32. Champaign, IL.

Xiu S, Shahbazi A, Shirley V, and Cheng D 2010 Hydrothermal pyrolysis of swine manure to bio-oil: Effects of operating parameters on products yield and characterization of bio-oil. *Journal of Analytical and Applied Pyrolysis* 88, 73–79.

Yadav KR, Sharma R, and Kothari R 2002 Bioconversion of eucalyptus bark waste into soil conditioner. *Bioresource Technology* 81, 163–165.

Yang H, Yan R, Chen H, Lee DH, and Zheng C 2007 Characteristics of hemicellulose, cellulose and lignin pyrolysis. *Fuel* 86, 1781–1788.

Yip K, Tian F, Hayashi J, and Wu H 2010 Effect of alkali and alkaline earth metallic species on biochar reactivity and syngas compositions during steam gasification. *Energy and Fuels* 24, 173–181.

Yoshida M, et al 2008. Effects of cellulose crystallinity, hemicellulose, and lignin on the enzymatic hydrolysis of miscanthus sinensis to monosaccharides. *Bioscience, Biotechnology and Biochemistry* 72, 805–810.

Yronwode P 2000 *From the hills to the grills*. Missouri Resources Magazine Spring Issue.

Zhao MY, Enders A, and Lehmann J 2014 Short- and long-term flammability of biochars. *Biomass and Bioenergy* 69, 183–191.

Zhao SX, Ta N, and Wang XD 2017 Effect of temperature on the structural and physicochemical properties of biochar with apple tree branches as feedstock material. *Energies* 10, 1293.

4

Biochar production technology

Ondřej Mašek, Robert C. Brown, and Santanu Bakshi

Introduction

Biochar is a carbonaceous solid product obtained from a process called pyrolysis, i.e., thermochemical decomposition of organic matter, such as wood, straw, etc., in the absence or presence of restricted amounts of oxygen (O_2). The extent to which pyrolysis products burn depends on the equivalence ratio defined as the ratio between the number of moles of O_2 admitted into the reactor and the stoichiometric ratio of moles of O_2 required for complete combustion of the biomass. When the equivalence ratio becomes equal to zero (complete absence of O_2), the process is referred to as pyrolysis. Pyrolysis can be an endothermic or exothermic reaction depending on the temperature of the reactants, becoming increasingly exothermic as the reaction temperature decreases (Spokas et al, 2012). For equivalence ratios of less than 0.15, the process is known as flaming pyrolysis because some of the volatiles produced during pyrolysis are oxidized in a visible flame while the biochar remains intact. For

equivalence ratios of 0.15–0.3, the process is characterized as gasification, whereby some of the volatile gases and solids are oxidized to carbon monoxide (CO), carbon dioxide (CO_2), and water (H_2O). The exothermic energy released by partial oxidation supports the endothermic pyrolysis reactions at higher temperatures. Pyrolysis is a promising technology for producing persistent carbon (C) for sequestration and the production of an energy carrier known as bio-oil or pyrolysis oil, suitable as feedstock for producing second-generation transportation fuels (Bridgwater and Peacocke, 2000; Huber, 2008; Granatstein et al, 2009; Mason et al, 2009; Woolf et al, 2010). On the other hand, the carbonization process focuses on maximizing the biochar yield. This chapter will discuss the most relevant carbonization technologies.

Wood carbonization processes have been known for millennia and were practiced for charcoal production (to be used in the

DOI: 10.4324/9781003297673-4

smelting of metals) as well as the production of tars and liquid products (wood vinegar). These processes have particular applications, such as, in wood preservation or the production of chemicals (Brown, 1917; Klark, 1925; Emrich, 1985). These applications peaked in the 19th and early 20th centuries with industrial dry distillation plants producing charcoal and chemicals. The technology subsided in the 20th century with the discovery and exploitation of crude oil resources and their refining that could produce the same products at a larger scale and lower cost. At present, charcoal production is mostly concentrated in countries with abundant biomass resources and cheap labor, with only a small number of facilities operating in industrialized countries. In 2018, the world production of charcoal was estimated to be more than 53 Mt (Nabukalu and Gieré, 2020), and most of it was destined for iron smelting and domestic use (cooking, barbecue). Because current production technologies typically yield only 20% w/w of the original biomass as charcoal on average, it can be estimated that at least 300 Mt of dry biomass is processed to produce the world's supply of charcoal annually. This scenario presents an inefficient use of biomass, as most of the energy contained in the biomass is wasted. Therefore, for sustainable development of the biochar industry at scale, enhanced pyrolysis technologies that offer an optimal compromise between high yield and good quality biochar are required. Furthermore, it is critical that gaseous and liquid co-products are valorized, e.g., for producing heat or electricity, or even recovery of chemical products. As most current charcoal production practices do not meet these requirements, their processes simply cannot be adopted for widespread sustainable biochar production.

Presently, biochar production is still on a much smaller scale compared to charcoal fuel production, with a global capacity estimated at over 100,000 t of biochar in 2022 but is growing rapidly. For example, in the EU, biochar production capacity has grown from 35,000 t in 2021 to 65,000 t in 2022, and a similar growth rate is expected in the future (EBI, 2022).

Biomass pyrolysis from antiquity to today

The pyrolysis process for the production of charcoal, tar, and wood vinegar has been in use since prehistory, with the first records of use in smelting dating back to at least five thousand years (James, 1972), but possibly even further (Radivojević et al, 2010). Besides smelting, reports of the use of charcoal and wood vinegar date back at least four thousand years (Day et al, 2021). The production and use of charcoal became widespread in the Bronze and Iron Ages, where in the absence of mineral coal charcoal was the key fuel for metal smelting (Deforce et al, 2021). Throughout medieval times the technologies used for charcoal production gradually transitioned from pit kilns to earth mound kilns and later brick kilns. These were then replaced by retorts in the 19th century when the demand for charcoal and its co-products increased dramatically (Emrich, 1985). The charcoal industry started to decline after the 1st World War as chemical and metallurgical industries moved to new fossil resources. Biomass pyrolysis technology re-emerged only in the second half of the 20th century with the World Oil Crisis sparking interest in alternative fuels, such as bio-oil. Pyrolysis was also developed as a waste processing technology, reducing the volume of waste for disposal to landfill.

Interest in biomass pyrolysis for producing solid products, biochar, came about only at the end of the first decade of the 21st century with increasing climate change concerns and the search for new mitigation solutions. Several new technologies specifically designed for biochar production have been developed, building on the retort and kiln designs used a hundred years earlier. Some of the key processes are discussed in the following section below.

A delicate balance between scientific discoveries, the development of new materials and products, technological improvements, and market forces has shaped biomass pyrolysis' long and convoluted evolution. The following is a list of important milestones in the evolution of pyrolysis technology:

70AD	Gaius Plinius Secundus the Elder (Historia Naturalis) described different uses of liquids and tars in char production process (Emrich, 1985).
1653	Johann Rudolf Glauber confirmed that the acid in pyroligneous water was the same acid in vinegar (Klark, 1925; Emrich, 1985).
1792	England commercialized illuminating gas manufactured from wood (Klark, 1925).
1819	The first pyrolysis oven to transfer heat through its metal walls was designed by Carl Reichenbach (Klark, 1925).
1850	Horizontal retorts (approximately 1 m diameter and 3 m long) were used mainly by Germany, England, and Austria, while the French developed vertical retorts made portable by Robiquete (Klark, 1925).
1870	The rise of the celluloid industry and the manufacture of smokeless powder increased the demand for acetone produced by wood distillation (Klark, 1925).
1850	The wood distillation industry began to expand (Klark, 1925).
1920–1950	The rise of the petroleum industry caused a decline in wood distillation (Klark, 1925).
1970	The World Oil Crisis gave rise to the need for alternative liquid fuels.
1970–90s	Development of new pyrolysis reactors occurred side by side with the understanding of the fundamentals of biomass pyrolysis reactions (Mottocks, 1981; Scott et al, 1984, 1988; Evans et al, 1987a, b; Piskorz et al, 1988; Boroson et al, 1989a; Bridgwater et al, 1999, b, 1994).
1980–90s	Several types of pyrolysis technologies (fast, flash, vacuum, and ablative) reach commercial or near commercial status (Roy et al, 1985; Freel et al, 1990, 1994; Yang et al, 1995; Roy et al, 1997; Bridgwater et al, 2001).
1990s	New crude bio-oil-based products (e.g., bio-lime, slow-release fertilizers, road de-icers, wood preservatives, glues, sealing materials, bio-pitches, hydrogen, browning agents, hydroxyacetaldehyde, phenol-formaldehyde resins) were developed (Underwood, 1990; Underwood and Graham, 1991; Chum and Kreibich, 1993; Oehr, 1993; Radlein, 1999; Roy et al, 2000; Freel and Graham, 2002).
2000s	New bio-oil-based refinery concepts are proposed (Czernik et al, 2002; Bridgwater, 2005; Huber and Dumesic, 2006; Elliott, 2007; Helle et al, 2007; Mahfud et al, 2007; van Rossum et al, 2007; Jones et al, 2009).

| 2005 | The idea of using pyrolysis charcoal for storing C and soil improvement emerges. Intermediate pyrolysis reactors started to be used for the combined production of bio-oil and char (Hornung et al, 2005; Garcia-Perez et al, 2007; Ingram et al, 2008) |
| Since 2010 | A number of new producers offering pyrolysis units designed for biochar production have emerged and grown rapidly, with at least 10 technology suppliers in Europe (EBI, 2022) and similar numbers in North America, Australia, and Asia, especially China and Japan. |

Types of biochar production reactors

Numerous thermochemical pathways are used to produce solid, gaseous, and liquid products, and as a result, thermochemical reactors come in various designs and sizes. The choice of the most suitable process depends on the application of interest. The product distribution of biochar, gas, and liquid depends mainly on the composition of the feedstock and the operating conditions of the process (residence time of solids and gases, temperatures, pressure, the content of oxidizing agent in the reactor) (Shafizadeh, 1982). Table 4.1 summarizes thermochemical conversion technologies and their

Table 4.1 *Biomass thermochemical conversion technologies and product distribution*

Thermochemical	Temperature (°C)	Other defining parameters	Gas	Liquid	Solid	Major intended product
Carbonization	200–400	Air tightness, residence time, materials	60–75	3–5	10–35	Charcoal; solid fuel and industrial input
Pyrolysis (for bio-oil)	400–700	Heating rates, residence time, particle size, gas flow rates	20–40	40–70	10–25	Bio-oil; chemical products and fuels
Pyrolysis (for biochar)	300–700	Residence time, heating rates	40–75	0–15	20–50	Biochar; soil amendment, carbon sequestration, and bioremediation
Gasification	500–1500	Oxidizing media, equivalence ratio	85–95	0–5	5–15	Syngas; gaseous fuel for heat and power, and gas to liquid
Hydrothermal processing	200–700	Elevated pressures, solvent type, and ratio (e.g., water)	0–90	0–80	0–60	Various chemical products
Combustion	1500–2000	Excess air for complete combustion	95	0	5	Energy converted to heat and power

main features (this should be viewed as a qualitative comparison rather than a quantitative reference on product distribution).

A variety of technologies capable of producing biochar have been developed, and several classification schemes can be used. Most often, pyrolysis reactors are classified according to the heating rate achieved in the reactor. Low heating rates in early reactor designs required several hours to produce biochar. These "slow pyrolysis reactors" often have heating rates well below $100°C\ min^{-1}$, and some carbonization systems work at heating rates below $100°C\ hr^{-1}$. Most continuous pyrolysis technologies, e.g., drum, auger, and rotary kilns, operate with heating rates close to or even somewhat above $100°C\ min^{-1}$. Historically, slow pyrolysis reactors (also known as carbonization reactors) were further classified as kilns or retorts (Emrich, 1985), depending on their mode of operation. Kilns were used in traditional charcoal making without particular regard to recovering the accompanying liquid fractions. The term retort was used to describe pyrolizers that recovered gaseous and liquid co-products (Emrich, 1985). However, the distinction is not as clear nowadays, as demonstrated by, for example, rotary kiln units that recover gases and liquids. The currently most commonly used technologies for biochar production are described below.

Batch carbonization units

Due to their simplicity and low cost, batch carbonization (slow pyrolysis) units have been prevalent in the charcoal industry and are also considered for biochar production. Operation of these slow pyrolysis units is typically carried out in individual batches, processing one batch at a time. Sometimes in the case of larger operations, several units can be operated simultaneously in a staggered manner, where each unit is at a different stage of the cycle. This can approximate a continuous production process and result in higher efficiency due to minimal or no downtime. Furthermore, there is the possibility to use gases and vapors generated by one unit for preheating another unit, increasing the overall process efficiency. Due to the intermittent nature of the process, these operations can be labor-intensive and require working in hazardous conditions.

A special type of slow pyrolysis process that requires minimum equipment is the so-called flame curtain pyrolysis. It owes its name to the fact that volatiles released from biomass during the process can react with air and therefore burn and form flames above the solid biomass and biochar. The special shape of the kilns restricts air access to the biochar at the bottom of the kiln while promoting the mixing of air with hot vapors escaping the bed of pyrolyzing biochar. Combusting the vapors in a flame above the biochar bed provides the heat needed for the pyrolysis process. Thanks to the extremely low cost and simplicity of operation, this type of unit is suitable for use in rural areas for small-scale biochar production (Cornelissen et al, 2016). Nevertheless, larger industrial units utilizing the flame curtain pyrolysis principle do exist and can offer the advantage of some level of automation. The disadvantage of this process is the inability to utilize the gaseous and liquid co-products, as these get combusted in the process without the possibility of recovery.

Moving bed reactors

Moving bed reactors rely on gravity to move biomass vertically through a pyrolysis zone, often without any moving mechanical parts in the reactor itself (Zhang et al, 2018). The biomass loaded at the top of the reactor moves downward, with the speed of movement controlled by the discharge rate of the biochar at the bottom of the reactor. While this design offers several advantages, such as relative ease of scale-up and feedstock flexibility, they also present challenges related to the uniform movement of material through the reactor without the formation of hot or cold spots, which affect the process efficiency and biochar quality. Some smaller moving bed reactors, such as the multiple-hearth reactor, include mechanical moving parts in the form of trays that distribute biomass in the reactor over several trays vertically stacked along the reactor axis, with biomass cascading down from the top tray towards the bottom where it is discharged. This design facilitates uniform distribution of the material through the reactor and good interaction with hot gases flowing upwards through the reactor.

Auger reactors

Auger reactors, also called screw pyrolyzers, move biomass through a tubular reactor by the action of a rotating screw (Figure 4.1). The heat needed for the pyrolysis process can be supplied either externally through the reactor wall, using a heat carrier, e.g., sand or steel balls, or in some cases, internally from the heated auger inside the reactor. The screw pyrolyzer is attractive for its potential to operate at small to medium scales. Auger

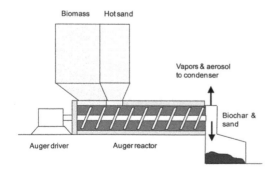

Figure 4.1 *Auger pyrolysis reactor*

pyrolyzers can utilize single, twin, or multiple interlocking screws to achieve desired processing conditions. Individual reactors can also be stacked to achieve a longer overall residence time without the need for an excessively long single reactor or to allow different processing conditions (temperature and residence time) in each stage. One of the first such pyrolyzers was the twin-screw Lurgi-Ruhrgas mixer reactor developed initially for producing town gas or olefins from coal using sand as a heat carrier. The auger pyrolyzer is a popular design for biochar production plants and several suppliers offer units on the market, such as Advanced Biorefinery, Inc., Biomacon, Pyreg GmbH, and Kansai Engineering.

Drum pyrolysis unit

The drum pyrolyzer moves biomass through an externally heated, horizontal cylindrical shell by the action of paddles. No air is intentionally admitted to the drum, although some air enters the voids between feedstock particles. The process takes several minutes for the biomass to transit the drum, which is short compared to traditional batch carbonization processes. The residence time of vapors is long enough for most of them to crack into non-condensable gases, although some tar remains with the gas. Some of the gas is burned in a firebox below the drum to heat the biomass to pyrolysis temperatures. Biomass is dried before entering the drum pyrolyzer to ensure good biochar and gas quality. The drum pyrolyzer of Pacific Pyrolysis is one of the few continuous pyrolyzers that have been employed in the production of biochar.

Rotary kilns

Rotary kilns are another form of continuous pyrolyzer (Arsenault et al, 1980; Bayer and Kutubuddin, 1988). They are similar to drum pyrolyzers in the employment of an externally heated, cylindrical shell except that the shell is oriented at an angle to the horizon and rotated to allow gravity to move the biomass down the length of the kiln. They have similar solids residence times of typically 5–30 min (Boateng, 2008). The advantage over the drum pyrolyzer is the absence of moving parts in the interior. Rotary kilns for biomass pyrolysis have been investigated at low temperatures (350°C) and moderately high temperatures (600–900°C). Klose and Wiest (1999) showed that variations in biomass feed rate and operating temperatures for a rotary kiln pyrolyzer allowed wide control on the relative yields of condensable and non-condensable vapors while biochar yield remained relatively constant in the range of 20–24%. This lack of control over biochar yield suggests that the relatively large volume of a rotary (or drum) kiln does not encourage the interaction of pyrolysis vapors and biochar that produces secondary biochar (Chapter 3). For further reading on the transport phenomena underlying rotary kilns, the reader is referred to Boateng (2008). Examples: Char Technologies, FEECO, Heyl & Patterson, Splainex, VOW/Evensen.

Fast pyrolysis reactors

Fast pyrolysis processes achieve high heating rates (on the order of several hundred °C s^{-1}) through the use of small biomass particles (typically < 2 mm) and reactors are characterized by high mass and heat transfer rates (Mohan et al, 2006). Such reactors are intended to maximize the production of liquids for upgrading to transportation fuels (Bridgwater et al, 1999, 2000, 2001; Czernik and Bridgewater, 2004). Heating of low ash content woody biomass at high heat rates typically yields up to 60–75% w/w bio-oil and approximately 15–25% w/w biochar, depending upon the feedstock and operating conditions. Lower yields are obtained when processing agricultural wastes (wheat straw, corn stover) with higher ash content. Although several kinds of reactors have been designed for fast pyrolysis, the high heat and mass transfer rates obtainable in fluidized beds make them ideal reactors for bio-oil production (Figure 4.2). Changing particle size, reaction temperature, and gas flow rate through the fluidized bed can dramatically alter the distribution of products. Due to relatively high flow rates of gas and low residence time of biochar in the reactor bed

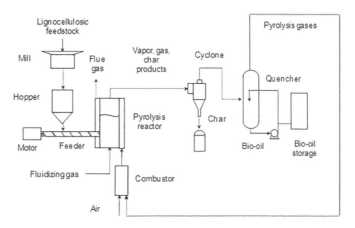

Figure 4.2 *Fluidized bed fast pyrolysis reactor*

compared to the slow pyrolysis process, the produced biochar is distinctively different. However, a systematic comparison is still lacking in the published literature.

Gasification

Aside from pyrolysis, biomass gasification can also produce biochar as a co-product. Unlike in pyrolysis where biomass is heated in the absence of oxygen, in gasification, a small amount of oxygen is injected. As biochars produced by gasifiers are exposed to higher temperatures and an oxidative environment, they are distinct from pyrolysis biochars. Despite this, they can be considered as biochar as they offer C sequestration potential unless combusted. The differences between gasification biochar and pyrolysis biochar are both physical and chemical and stem from the different processing environments compared to pyrolysis, such as a higher processing temperature, and the presence of a gasification agent, such as CO_2 or H_2O (Fryda and Visser, 2015). As a result, only the least reactive carbon is retained in the solid residue together with a large fraction of ash. Consequently, the pH of gasification biochar tends to be higher than that of pyrolysis biochar. Another potential challenge to the safe use of gasification biochar is the contamination with hydrocarbons (tars) that can be higher in gasification biochars, depending on the gasifier type (Rogovska et al, 2012). Despite these challenges, biomass gasifiers can be an important source of biochar due to the relatively widespread use of small-scale biomass gasification units. In certain contexts, gasification may be preferred to pyrolysis due to the higher yield of energy-rich gases that can be used for electricity generation in off-grid locations. Three kinds of gasifiers are suitable for the co-production of producer gas and biochar including updraft, downdraft, and fluidized bed (Brown and Brown, 2014; Figure 4.3).

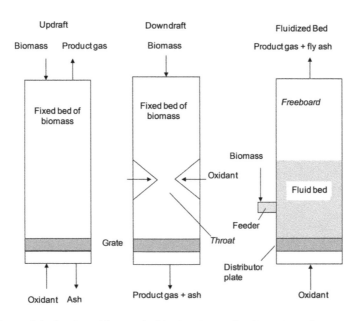

Figure 4.3 *Different kinds of gasifiers suitable for co-production of producer gas and biochar*

Pyrolysis process parameters and modes of operation

Over the past decade, a lot of effort went into establishing relationships between processing conditions, biomass composition, and the yields and properties of biochar relevant to agronomic or environmental management applications. Yields and properties of pyrolysis products are influenced by the composition of the biomass feedstock and several process parameters (Chapter 3). Among these, the highest treatment temperature is the most important one, and newly emerging models based on machine learning have shown promising results in predicting biochar yields and, to some extent, biochar properties. This opens the possibility of controlling operating conditions to improve process efficiency and optimize co-product distribution. Despite the progress made so far, further advances in biochar production are possible. It will require both basic research to understand the mechanisms of biochar formation that can be used to increase yields and biochar stability, and demonstration projects to prove the technical and economic feasibility of large-scale biochar production. This section discusses the key process parameters that affect biochar yields and properties. Classification of thermochemical reactors according to their potential to produce biochar is provided in Table 4.2.

Oxygen content in the thermochemical reactor (pyrolysis, gasification, combustion)

Thermochemical processes can be classified according to the amount of O_2 present: pyrolysis having no O_2; gasification having a sub-stoichiometric amount, and combustion having an excess of O_2. Each can produce biochar, with the amount produced generally decreasing with increasing O_2. If the process is not sufficiently exothermic, an external heat source is required to drive the carbonization process. Alternatively, air can be added to the reactor to release sufficient heat to drive the process (autothermal). Pyrolysis maximizes biochar production if the heat rate is low (slow pyrolysis) and maximizes condensable vapors (bio-oil) production if the heat rate is high (fast pyrolysis). When O_2 is introduced at a low rate or sub-stoichiometric combustion/oxidation rate (typically 0.15–0.3 combustion stoichiometric ratio), the volatile gases and the fixed carbon react with O_2 to form syngas in large quantities. This leaves behind residual biochar (C and ash) that has been exposed to higher temperatures (typically over 800°C) due to the partial oxidation reactions encountered in the gasification process. The resulting biochar product could therefore have a different pore structure and surface functional groups than the biochar resulting from pyrolysis. When the O_2 in the latter process is more than what is required to completely oxidize the volatile matter and fixed carbon, the process is in full combustion mode and the solid residue is mainly ash as most of the carbon would be burnt. Since the kinetics of charcoal combustion is slow, not all the C is completely removed and the solid residue contains some unburned C. The inorganic content and the pH of these materials are typically very high.

Particle size used (logs, large particles, chip, fine particles)

Although pyrolysis is feedstock neutral, feed particle size is an important parameter, as it limits the rate at which heat can be transferred to the material. As mentioned earlier, only small-sized particles (3 mm or less) are

Table 4.2 Key criteria for selecting appropriate thermochemical technology for biochar production

Content of oxygen and temperature	Reactor type	Final products	Heat transfer rate	Particle size (pretreatment)	Mode of operation	Heating method	Construction materials	Portability	Reactor position	Loading mode
Pyrolysis (No oxygen, temperature:350–600°C) Gasification (0.15–0.28 stoichiometric air-biomass ratio, temperature: 700–1200°C) Combustion (stoichiometric air-biomass ratio higher than 1, temperature: over 1500°C)	Fixed bed Fluidized bed Circulating bed Ablative Rotary drums Moving beds Auger	Bio-oil Syngas Biochar Hydrogen Heat Electricity	Slow Fast	Logs; large particles Chips Fine Particles	For Intermittent operation For nearly continuous operation For continuous operation	Heating by direct admission of air to the wood (autothermal) Heating by direct contact of the biomass with furnace gases on the wood Indirect heating Internal radiators Heating through the walls	Earth pits Brickwork Steel	Stationary Semi-portable Portable	Vertical Horizontal	Manual Mechanical loading With rail cars

suitable for fast pyrolysis, while slow pyrolysis reactors can process both small and larger-sized particles. Three main types of processed wood are typically used for biochar production: (1) logs, (2) sawmill chips and pellets, and (3) fine particles.

Logs are an assembly of wood piles that usually consists of material > 1 m in length, but also may range from 0.1 to 0.4 m in length (Toole et al, 1961). Wood that is preferable for a given feed charge should be of the same general size and moisture content. This simplifies the handling of the wood and creates more uniform carbonization. Round wood with a cross-section greater than 0.2 m should be split or cut into smaller pieces (Toole et al, 1961).

Chips and pellets can be produced directly from woody biomass, but transporting them from their source to the pyrolysis plants requires planning, as transporting bulky biomass can incur high transportation costs. The main advantage of processing chips and pellets is their ease of handling.

Fine Particles (< 3 mm diameter) are typically used with fast pyrolysis reactors (fluidized bed reactors, and circulating bed reactors). Such sizes result in increased surface areas required to achieve the necessary high heat transfer rates; they are typically produced by grinding chips. Grinding and pretreatment represent additional energy and thereby cost to the total process cost.

Mode of operation (intermittent operation (batch), nearly continuous, continuous)

Depending upon the mode of operation, pyrolysis reactors can be classified as batch, semi-batch, and continuous.

Batch reactors are typically used for the manufacture of biochar in a process where the recovery of byproducts is often of secondary importance (Klark, 1925). Batch processes normally involve a heating-up period in which the product is produced, followed by a nonproductive cool-down period that prepares the equipment for handling and the next batch operation. In a batch kiln or retort, individual particles remain nearly stationary. These reactors only allow the discharge of biochar after it has been cooled down to the appropriate and safe handling temperatures. Start-up and energy costs to heat and reheat the oven are repetitive and energy-intensive. Also, the utilization of the volatiles formed during the process is difficult for energy recovery, so they are usually released into the atmosphere, causing significant pollution. Batch operations are very common among reactors with small design footprints. Due to the nature of the process, the time required to complete a production cycle can vary from several hours to days, and is labor-intensive, due to the need for loading and offloading of the batch reactors.

Semi-batch operated systems (Carbon Gold, Adams retort/Peter Hirst, Japanese kilns) tend to be portable and make better use of hot ovens. In these systems, heat recovery is accomplished by recycling the hot vapors in between batches. The Carbo Twin Retort, developed by Ekoblok/Carbo Group, is a typical example of a semi-batch operation. The Carbo Twin Retort is a semi-continuous production module developed in The Netherlands in the 1990s consisting of several vessels and an insulated oven. It is operated by placing successive vessels loaded with wood into an oven. As one vessel undergoes pyrolysis in the oven its volatile gases are burnt to heat the freshly loaded incoming vessel in a twin chamber external to the main oven. No external heating source is required, except during start-up. The production capacity is determined by the number of batch runs that can be carried out (Trossero et al, 2008). Some of these systems allow the recovery of liquid products, but most are typically used to produce biochar.

A semi-batch concept similar to the Carbo Twin named POLIKOR and EKOLON has been commercialized by a company called Bioenergy LLC (St Petersburg, Russia). In this technology, the removable retorts are inserted inside a firewood box. The retorts have a special device at the bottom that allows pyrolysis vapors to enter into the combustion chamber for the generation of part of the heat needed to drive the process. The mobile units of this design are marketed as POLYEVKA and KORVET. These semi-portable steel kilns have two advantages: ease of mobility useful for small-scale production and shorter cycles resulting in quicker cooling times (Emrich, 1985).

Continuous operation reactors are designs that ensure a continuous flow of material into a vessel with an axial temperature profile so that the material undergoes drying, preheating, pyrolysis, cooling, and discharge over time and space all in a continuous process. Examples include stationary tunnel kilns with wood-loaded walking boats, directly and/or indirectly heated rotating cylinders, etc. Examples include auger pyrolyzers, such as those commercialized by Pyreg and Biomacon, or rotary kiln pyrolysis units by Mitsubishi Heavy Industries. Continuous pyrolysis plants offer the possibility for a high level of automation, reducing the labor requirement and therefore a popular choice for biochar production in industrialized countries.

Heating method

All pyrolysis reactors require that the biomass particle undergo some form of heating protocol requiring heat exchange between the heat source and the wood particle. Like all heat exchangers, the mode of heat exchange can be important in the economical use of energy as well as product quality. Feed particle dimensions equally influence heat exchange. If the particles are not of the proper dimensions, penetration of heat will be slow and the necessary heating rate will not be reached. Agitation of the feedstock also improves heat exchange significantly. Heat can be transferred from the heat source either by direct admission of air to the biomass (autothermal), heating by direct contact of the biomass with furnace gases, or heating by direct or indirect contact with a solid heat carrier (Figure 4.4).

Partial combustion (autothermal processes)

Partial combustion of pyrolysis vapors and biochar is most common for small-scale operations (Emrich, 1985). Burning part of the raw material with a controlled air inlet provides the energy necessary for the process.

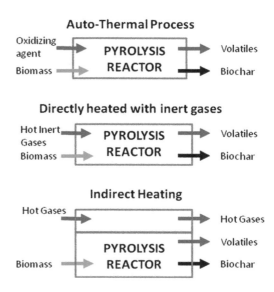

Figure 4.4 *Types of pyrolysis technologies according to the heating method used (Fournier, 2009)*

Carbonization by contact with hot gases

In this arrangement, hot gases from an external source that come into direct contact with the feedstock provide the energy required for carbonization. This method reduces the need for expensive heat transfer surfaces; however, there are costs associated with the heating of the required inert gases to provide the energy needed for the endothermic pyrolysis reactions. Typically, an inferior fuel is used to heat the heat carrier in an external furnace (Klark, 1925). Overall biochar and byproduct yields can be high, making this system suitable for medium to large-scale plant operations (Emrich, 1985).

Indirect heating

With indirect heating, the retort is heated from the outside. Heat is transferred to the biomass by conduction through the reactor walls. The feed charge is contained inside the retort and heated by hot flue gases generated from the combustion of gases released from the retort as the biomass charge to the retort is pyrolyzed. Some systems have used molten salts as an alternative to flue gas for heat carriers (Toole et al, 1961).

Direct contact with solid heat carrier

There are several different ways of using solid heat carriers to supply heat to a pyrolysis process. Most of these rely on the internal circulation of the heat carrier between hot zones (e.g., combustion chamber) and the pyrolysis zone where the heat is required. Sand, metal or ceramic balls, and biochar are among the materials used as solid heat carriers acting like a "heat flywheel". Combustion of pyrolysis co-products, or other fuels typically present, acts as the heat source to heat the solid medium which, in turn, is then allowed to contact the biomass to initiate pyrolysis (Mašek, 2009). The use of solid heat carriers offers several advantages over other methods. The intimate contact between the heat carrier

and the biomass particles intensifies heat transfer, facilitates rapid heat rates, and improves pyrolysis process efficiency, consequently leading to increased throughput. In some designs, the heat carrier media can act as a tar-cracking catalyst, improving the quality of the pyrolysis gas. The concept of conveying solid heat carrier into and out of the pyrolyzer has been demonstrated at Aston University (UK) and Bioliq, a company using technology developed at Karlsruhe Institute of Technology (KIT) in Germany.

Microwave heating

Unlike conventional heating by conduction and convection, microwave (MW) heating is classified as volumetric heating by an electromagnetic wave that is a result of the interaction of dielectric materials with a microwave field. The heating arises from the displacement that charged particles in the material undergo when subjected to electromagnetic radiation, and the extent to which a material heats up when subjected to microwave radiation is mainly determined by its dielectric properties. On one hand, materials that typically exhibit good dielectric properties are organic carbonaceous materials and water (both of which are present in biomass pyrolysis). Dry biomass, however, can be relatively "transparent" to microwaves, and thus not easily heated by microwaves. Industrial microwave heating is performed at frequencies of 915 MHz and 2.45 GHz to avoid interference with communications and other users of the electromagnetic spectrum.

Because microwave heating is volumetric in nature, it is not limited by heat conduction in the reactor or thermal diffusivity of the biomass particles, thereby resulting in improved heat distribution. Microwave pyrolysis has been explored for the production of fuels (Dominguez et al, 2006; Budarin et al, 2010; Zhao et al, 2010) as well as chemicals and biochar (Mašek et al, 2013; Gronnow et al, 2012).

Portability (stationary, semi-portable, portable)

Stationary units are typically large installations requiring the transport of feedstock from its source, which adds considerably to the cost of a project (Dumesny and Noyer, 1908). Due to transportation and building expenses, stationary reactors are justified only when sufficient amounts of biomass are available for the production of biochar and bio-oil. Stationary units offer the opportunity to utilize pyrolysis co-products for heating and electricity generation as the necessary infrastructure can be installed on-site.

Semi-portable systems combine the beneficial features of stationary and portable reactors. In such designs, some components can be stationary but the most expensive components are portable. For example, the furnace, which generates heat for the pyrolysis reactor, can be stationary, while the reactor and condensers are portable. Portable wagon retorts were typically coupled with a stationary brick furnace in the old wood distillation industry.

Portable or mobile units consist of equipment and accessories that can be easily and rapidly assembled and disassembled using simple tools. This system can be moved to a site that may have more available resources like biomass. Portable units typically do not recover pyrolysis co-products, beyond use in the process itself due to a lack of infrastructure for heat recovery or electricity generation.

Loading mode (manual, mechanical, wagons)

Biomass must be loaded for processing in a fashion that allows free circulation of admitted air or product gases. The location of the air entry and smoke outlet openings and the type

of biomass affect how a kiln is loaded and unloaded to attain the desired circulation.

Manual loading of cordwood and slabs is entered by hand through a door (Bates, 1922; Toole et al, 1961). To use kiln capacity effectively, wood must be stacked to allow combustion gases to circulate freely through the pile (Toole et al, 1961). Stacking wood inside kilns is labor-intensive. All of the logs must be packed as close together as possible with thinner pieces against the wall and thicker logs towards the center (Emrich, 1985).

Mechanical loading using mechanized yard handling equipment has several distinct advantages. Conveyer belts, bucket elevators, and tractor scoops can quickly and efficiently move the discharged biochar from the kiln (Toole et al, 1961). Tractor scoops move larger material, while smaller pieces (chips) are fed into the reactor with a conveyor or bucket elevator (Toole et al, 1961).

Using *wagons* to load and unload a pyrolysis reactor can reduce costs considerably. Wagon cars carry feedstock directly into the oven on a track and then back out of the oven with the biochar product (Figure 4.5). However, maintaining thermally fatigued wagons, i.e., those that are frequently subjected to the extreme temperatures of the ovens contributes to operation costs.

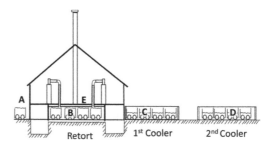

Figure 4.5 *Plan of an American destructive distillation plant in continuous operation. A: car, B: retort, C: first cooler, D: second cooler, E: acetate drying floor (Veitch, 1907)*

Process control

To achieve good yields of high-quality products, reaction conditions must be well controlled. The temperature in the reactor is the most important, although not the only variable, in the control of the pyrolysis process (Zhao et al, 2013). Two ways to control heat conditions of pyrolysis reactors are as follows: (1) measuring and controlling the temperature inside the reactor either manually or with a feedback control system; or (2) observing the color of the vapors produced. Also, suitable (but less commonly used) for units operating in continuous mode with the recovery of heat or bio-oil is the monitoring of the gas composition.

Pressure

Pyrolysis reactors can be operated at atmospheric pressure, in a vacuum, or at high pressure. On one hand, operating a reactor in a vacuum promotes the release of volatiles and thus results in increased production of liquid products enriched with cellulosic sugar fractions like levoglucosan and cellobiosan while decreasing the yield of biochar. On the other hand, high pressure favors secondary, solid-forming, reactions of volatiles, resulting in greater yields of biochar and gases.

Most pyrolysis reactors operate under *atmospheric pressure* conditions for simplicity. Because the creation of a vacuum or high pressure is not necessary, the cost of seals, reactor vessels, and auxiliary equipment is drastically reduced. Reactors that operate in a *vacuum* are more complex due to the necessity of avoiding air leaks.

Operating a reactor under pressure results in limited production of liquids. However, *pressurized* reactors produce much higher yields of biochar and syngas. A high-pressure (flash carbonization) reactor was developed by the University of Hawaii.

Solid and gas contact mode counter-current, co-current, and cross-flow

Heat exchange between the heat source and biomass particles is critical to process efficiency and product quality. Rotary kilns and retorts can either be counter-current or co-current where the combustion gases, i.e., the reaction heat source and the biomass flow, are counter-current or co-current, respectively, within the kiln. In such devices, a reducing atmosphere (a bed of biomass) and an oxidizing atmosphere (the freeboard) co-exist at the same point along the kiln length; the decomposition of the biomass occurs in a reducing atmosphere, releasing volatiles into the freeboard that can be burnt in excess O_2 to provide added heat source (Boateng, 2008). Cross-flow systems can include reactors where there are no discernible regions for the freeboard and bed of biomass, e.g., tunnel kilns, packed beds such as pits, etc. In these systems, the charring occurs in a reactive front.

Pyrolysis atmosphere

Different gaseous atmospheres can be used in the pyrolysis process. A vast majority of biochar research published to date used inert atmosphere, most often nitrogen, due to its good availability, relatively low cost, and good performance. The use of other gaseous atmospheres in the pyrolysis literature is relatively small but on an upward trend. This is motivated by a growing interest in the production of engineered biochar with specific functionalities. Pyrolysis atmosphere plays a significant role in biomass conversion due to its impact on the heat and mass transfer in the reactor, as well as, in many cases, chemically reacting with the products of pyrolysis. Besides inert gases (e.g., N_2 and Ar), reductive gases (mainly H_2 and CH_4)

and oxidizing gases (e.g., steam, O_2 in sub-stoichiometric ratios) have been employed in published studies, to achieve specific effects on the products of pyrolysis. Despite the extensive use of pure inert gases in biochar research, and the opportunities offered by specific reactive gases, most commercial biochar production units use flue gases (from the combustion of pyrolysis co-products) due to their availability on site and minimum cost. The effects of the different flue gas compositions on biochar properties and their comparison to atmospheres used in research studies are not fully understood and present an important area of research and development.

Selected examples of currently used biochar production technologies

A number of technologies have been developed to the point of commercialization and are increasingly being used in the emerging biochar industry. This section provides some examples of the most commonly used processes at different scales.

Continuous processing units

Some of the most commonly used moving bed pyrolysis reactors and their properties that have reached or nearly reached commercialization are listed in Table 4.3. Schemes of the Lurgi and multiple-hearth furnace (Herrenshoff) pyrolysis reactors are shown in Figures 4.6 and 4.7.

Table 4.4 lists some of the most commonly used fast pyrolysis reactors in operation. Schemes for ablative, fluidized bed, circulating fluidized bed, and vacuum pyrolysis reactors are shown in Figure 4.8.

Batch processes – Gasification cookstoves and flame curtain kilns

Cookstoves

Biomass cookstoves are simple units designed for efficient domestic cooking or heating using different biomass feedstock, with low pollutant emissions and primarily intended for small-scale applications in developing countries. There is a wide range of designs available (over 400 types listed in the HEDON database; www.hedon.info/Stoves+Database), depending on the purpose, scale, fuel used, materials of construction, etc. Biochar cookstoves are a special category of cookstoves that, besides cooking, offer the possibility to produce biochar. Most biochar cookstoves operate in batch mode and are based on the so-called Top-Lit Up Draft (TLUD) principle (Figure 4.9). The solid biomass fuel is ignited from the top of the stove; the heat generated establishes a pyrolysis zone or front that moves towards the bottom of the stove. This releases flammable pyrolysis gases that flow upwards and are mixed with secondary air as they reach the top of the stove for sustaining combustion.

The different designs of cookstoves vary in the way air is admitted into the stove, aimed at improving the combustion process to increase fuel use efficiency and minimize emissions. Once the entire batch of fuel is exhausted, the solids accumulated at the bottom of the stove can be quenched (otherwise it continues smoldering) and removed as biochar. The yield of biochar is typically around 20–25%, depending on the feedstock, stove used, and the operator's skills (Table 4.5). To ensure a good yield of

Table 4.3 *Moving bed reactors*

Reactor	Final product	Heat transfer achieved/capacity	Operation/portability	Raw material used	Link of companies commercializing the technology
French/Lambiotte	biochar/bio-oil	Slow/Direct contact with hot gases	Continuous/ Stationary	Cordwood	http://www.lambiotte.com
Herrenshoff Multiple-Hearth Furnace	biochar/heat	Slow/2.5 t/h	Continuous/ Stationary	Chips/fine particles	https://www.pyrocal.com.au
Rotary Drums	biochar/heat	Slow	Continuous/ Stationary	Chips	https://chartechnologies.com/
Auger Reactors	biochar/bio-oil/ heat	Slow/Intermediate	Continuous/ Stationary or mobile	Chips or fine particles	http://www.inbio.net https://www.biogreen-energy.com http://www.egenindustries.com/ https://pyrocore.com/
Pyrovac Design	biochar/bio-oil	Intermediate	Continuous/ Stationary	Chips or fine particles	https://www.corigin.co/ https://pyrovac.com/en/
Paddle Reactor	biochar/bio-oil	Slow/Intermediate	Continuous/ Stationary	Chips	

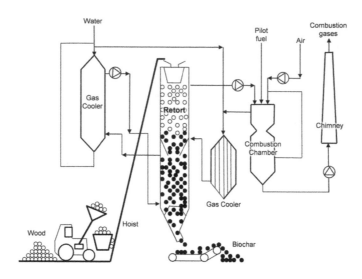

Figure 4.6 *Lurgi process (Garcia-Nunez et al., 2017)*

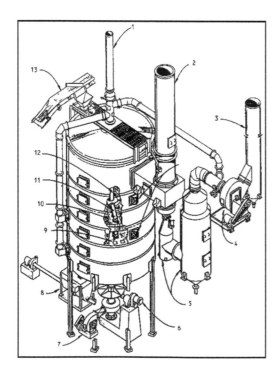

Figure 4.7 *Cross-section of the multiple-hearth furnace (FAO, 1985)*

biochar, identifying the right moment to quench the combustion process to avoid undesirable combustion of residual biochar is important. This can be done by observing the type of flame and its behavior. The pyrolysis gases burn as a yellow-orange flame. The transition to the residual biochar combustion phase is marked by the appearance of a smaller bluish flame. This combustion phase is less intensive but can provide considerable additional heat, at the expense of biochar yield. The type of fuel used in a TLUD stove is to a large extent dictated by its design, as it is uncommon for these stoves to be able to handle a wide range of feedstocks, especially regarding particle size. Small particles can form a bed with high resistance to gas flow and therefore a fan is usually required to boost the supply of primary air. Larger, uniform particles (chips and pellets) are well suited for use in TLUD stoves.

Some cookstoves are specifically designed for biochar production, such as the Anila stove designed by R.V. Ravikumar, of the University of Mysore in India, which has two separate fuel chambers. In the Anila stove, the

Table 4.4 *Fast pyrolysis reactors*

Reactor	Final product	Construction materials/operation mode	Portability	Raw material used	Link of companies commercializing the technology
Fluidized bed	bio-oil and biochar	Steel/continuous	Stationary/ mobile	Fine particles	http://www.dynamotive.com/ technolopgy/ http://www.agri-therm.com/ http://www.avellobioenergy. com/ http://www.bioware.com.br
Circulating bed reactors	bio-oil	Steel/continuous	Stationary	Fine particles	http://www.ensyn.com/tech.htm http://kior.com/
Ablative reactors	bio-oil	Steel/continuous	Stationary	Chips	http://btg-btl.com/

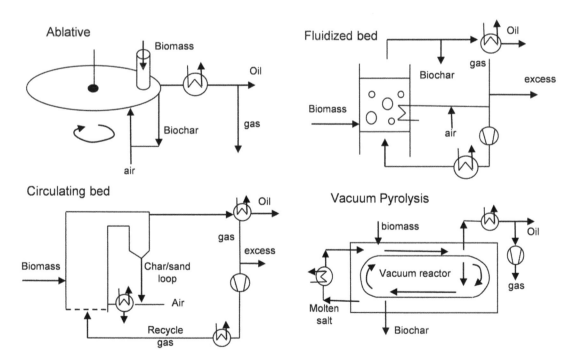

Figure 4.8 *Scheme of fast pyrolysis reactors (ablative, fluidized bed, a circulating bed, and a vacuum pyrolysis reactor)*

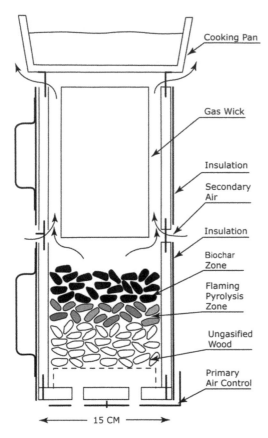

Cooking Pan

Gas Wick

Insulation

Secondary Air

Insulation

Biochar Zone

Flaming Pyrolysis Zone

Ungasified Wood

Primary Air Control

15 CM

Figure 4.9 *Top-lit updraft (autothermal) gasification cookstove*

fuel in the inner combustion chamber undergoes flaming pyrolysis or complete combustion, and the heat generated in this process heats the biomass in the surrounding outer container. This biomass undergoes allothermal pyrolysis without direct contact with a flame. The gases produced in the outer container enter the bottom of the inner combustion chamber where they fuel and further sustain the combustion process. This approach offers several advantages, as it does not require careful monitoring of the burning process to identify when to quench the biochar, thus eliminating the risk of accidental biochar combustion. In addition, as the combustion and pyrolysis processes are physically separated, the stove can use different feedstock in each chamber and thus produce biochar even from materials that would otherwise be difficult to process. One considerable limitation of cookstoves for biochar production is their capacity. Depending on the feedstock used, biochar cookstoves typically produce only between 100–300 g of biochar per batch, i.e., per cooking. The limited quantity of biochar is, therefore, more suitable for spot application by smallholders

Table 4.5 *Selected biochar cookstoves*

Stove type	Country	Fuel	Biohar yield (%)
Champion TLUD	India	chunky biomass (chips, pellets, twigs, briquettes, etc.)	~20
Sampada	India	chunky biomass (chips, pellets, twigs, briquettes, etc.)	20–25
Vesto	Swaziland	chunky biomass (chips, pellets, twigs, briquettes, etc.)	25
MJ Biomass Gas Stove	Indonesia	chunky biomass (chips, pellets, twigs, briquettes, etc.)	30–35
LuciaStoves	Italy	chunky biomass (chips, pellets, twigs, briquettes, etc.)	~30
Anila	India	combination of fuels in two separate compartments	20–30
BMC Rice Husk Gas Stove	Philippines	rice husk	17–35
MJ Rice Husk Gas Stove	Indonesia	rice husk	30

(Torres-Rojas et al, 2011), rather than general spreading on fields.

Flame curtain kilns

Due to its simplicity in terms of operation and equipment (a conical-shaped metal kiln or soil pit), flame curtain pyrolysis has gained a lot of interest in terms of biochar production in the last few years (Schmidt et al, 2014). Unlike the cookstoves discussed above, flame curtain pyrolysis kilns are solely used for biochar production, and such units cannot be used indoors for heating purposes. Flame curtain pyrolyzers, such as the "Kon-Tiki" or "Moki" kilns are appealing for biochar production on a household or community scale in locations without the infrastructure necessary to support more advanced technologies, or where the costs of such technologies are prohibitive.

To ensure the necessary anoxic conditions in the biochar layer, the flame curtain kiln design relies on oxygen from ambient air being consumed in the flame above the biomass bed (Figure 4.10), and therefore cannot reach the biochar at the bottom of the kiln. As a portion of the energy released by the combustion of pyrolysis products in the flame is returned to the surface of the biomass feedstock and promotes further pyrolysis (Torero, 2016; Morrisset et al, 2021), the pyrolyzing biomass is heated from both the hot layer of biochar below and from heat transfer associated with the flame sheet above. The relative contributions of these two heat transfer mechanisms vary with time as the biomass biochars (Cornelissen et al, 2016), until the release rate of pyrolysis gases and volatiles drops to the point where the flame curtain can no longer be sustained across the whole surface (Emberley et al, 2017). As this would allow ingress of air to the biochar layer and therefore its combustion, additional feedstock needs to be added to provide more fuel that generates pyrolysis gases, or the process needs to be quenched, depending on how full the kiln is at this point. In addition to preventing the ingress of O_2, maintaining a constant flame sheet limits the emission of pyrolysis products (e.g., CO and CH_4) to the atmosphere as these are oxidized in the flame to CO_2 and H_2O. The successful optimal operation of the flame curtain pyrolysis reactor, producing quality biochar while minimizing environmental emissions then relies on achieving an appropriate rate of application of the feedstock in the layering process (Jayakumar et al, 2023).

A successful example of the deployment of the flame curtain pyrolysis kilns is the MOKI kiln developed by the late Dr. Makoto Ogawa in Japan that is used in the Cool Vege project (https://coolvege.com/).

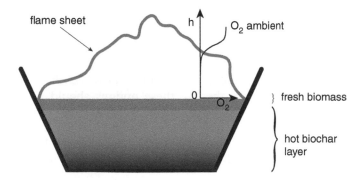

Figure 4.10 *Flame curtain pyrolysis kiln*

Conclusions and recommendations

The history of biomass pyrolysis confirms that a balanced investment in the creation of new knowledge (science), in the design, testing, and scale-up of new technologies, and in the development of new products (market) is needed to build a competitive biochar industry.

The main hurdles facing the deployment of biochar production technologies are the possible negative environmental impacts, lack of heat recovery, and limited control of the quality and yield of the biochar produced. This, together with the lack of well-developed markets and mature technologies for the production, refining, and commercialization of bio-oil-derived pyrolytic products favors the development of slow pyrolysis units that produce biochar and heat. To fully assess the potential of such units for the co-production of quality biochar and heat or power, it is necessary to develop a good understanding of the energy and mass balance of such systems. Although a new body of work has emerged in this area in the past decade, much more is needed to enable efficient deployment of biochar production systems in different geographic and economic contexts and at different scales. Future progress in biochar production technologies will require a balanced investment in the discovery of new knowledge, the development of new technologies, and the commercialization of new products.

Some specific goals for advanced biochar manufacture should include the following:

• Continuous feed pyrolyzers to improve energy efficiency and reduce pollution emissions associated with batch kilns.
• Exothermic operation without air infiltration to lessen energy inputs and increase biochar yields.
• Recovery of co-products to reduce pollution emissions and improve process economics.
• Control of operating conditions to improve biochar properties and allow changes in co-product yields.
• Improve feedstock flexibility making possible the efficient conversion of not only wood but also herbaceous feedstocks, manure, and agro-industrial waste into biochar.

The pyrolysis industry must be well-planned to ensure that long-term sustainability goals are met. Interconnection with other industries in the growing bioenergy and biorefinery sector and energy consumers as well as with a state or national household supply program is critical for success. Programs aimed at substituting fossil fuels in rural communities could utilize the heat generated from the pyrolysis volatiles and render biochar production economically and environmentally friendly. Finally, with the wider deployment of carbon capture and sequestration technologies (CCS) and especially bioenergy with CCS (BECCS), integration of biochar production with CCS facilities and CO_2 transport networks is likely to become viable, and these options should be considered in the planning stages of new, especially larger installations, as such combined facilities would achieve much higher levels of atmospheric CO_2 removal.

References

Arsenault RH, et al 1980 Pyrolysis of agricultural residues in a rotary kiln. In: Jones JL, et al (Eds) *Thermal Conversion of Solid Wastes and Biomass*. Washington: American Chemical Society. pp337–350.

Bates JS 1922 *Distillation of hardwoods in Canada. Forestry Branch-Bulletin #74*. Canada: Department of the Interior.

Bayer E, and Kutubuddin M 1988 Thermochemical conversion of lipid-rich biomass to oleochemical and fuel. In: Bridgwater AV, and Kuester JL (Eds) *Research in Thermochemical Biomass Conversion*. Amsterdam: Elsevier Applied Science. pp518–530.

Boateng AA 2008 *Rotary Kilns: Transport Phenomena & Transport Processes* 1st Edition, Elsevier, Butterworth-Heinemann Publishers, Amsterdam; Oxford; Boston, ISBN: 978-0-7506-7877-3

Boroson ML, Howard JB, Longwell JP, and Peters WA 1989a Product yield and kinetics from the vapor phase cracking of wood pyrolysis tar. *AIChE Journal* 35, 120–128.

Budarin VL, Clark JH, Lanigan BA, Shuttleworth P, and Macquarrie DJ 2010 Microwave assisted decomposition of cellulose: a new thermochemical route for biomass exploitation. *Bioresource Technology* 101, 3776–3779

Bridgwater, AV (2005, December). Fast pyrolysis based biorefineries. In *230th ACS National Meeting*. American Chemical Society.

Bridgwater AV, and Peacocke GVC 1994 Engineering development in fast pyrolysis for bio-oils. Proceedings from the Biomass Pyrolysis Oil Properties and Combustion meeting Sept. 26–28 Estes Park, CO, 110–127.

Bridgwater AV, Meier D, and Radlein D 1999 An overview of fast pyrolysis of biomass. *Organic Geochemistry* 30, 1479–1493.

Bridgwater AV, and Peacocke GVC 2000 Fast pyrolysis processes for biomass. *Renewable and Sustainable Energy Reviews* 4, 1–73.

Bridgwater AV, Czernik S, and Piskorz J 2001 An overview of fast pyrolysis. In: Bridgwater AV (Ed) *Progress in Thermochemical Biomass Conversion*. Hoboken: Blackwell Sciences. pp977–997.

Brown NC 1917 *The Hardwood Distillation Industry in New York*, The New York State College of Forestry Syracuse University, Syracuse.

Brown RC, and Brown TR 2014 *Biorenewable Resources: Engineering New Products from Agriculture*, 2nd Edition. Ames, IA: WIley Blackwell.

Chum HL, and Kreibich RE 1993 Process for preparing phenol formaldehyde resin products derived from fractionated fast-pyrolysis oils. U.S. Patent 5,091,499.

Cornelissen G, et al 2016 Emissions and char quality of flame-curtain "Kon Tiki" kilns for farmer-scale charcoal/biochar production. *PLoS One* 11, e0154617.

Czernik S, and Bridgwater AV 2004 Overview of applications of biomass fast pyrolysis oil. *Energy and Fuel* 18, 977–997.

Czernik S, French R, Feik C, and Chornet E 2002 Hydrogen by catalytic steam reforming of liquid by products from biomass thermochemical processes. *Industrial and Engineering Chemical Research* 41, 4209–4215.

Day GS, Drake HF, Zhou H-C, and Ryder MR 2021 Evolution of porous materials from ancient remedies to modern frameworks. *Communications Chemistry* 4, 114.

Deforce K, Groenewoudt B, and Haneca K 2021 2500 years of charcoal production in the Low Countries: The chronology and typology of charcoal kilns and their relation with early iron production. *Quaternary International* 593–594, 295–305.

Dominguez A, Menendez JA, Inguanzo M, and Pis JJ 2006 Production of bio-fuels by high temperature pyrolysis of sewage sludge using conventional and microwave heating. *Bioresource Technology* 97, 1185–1193.

Dumesny P, and Noyer J 1908 *Wood Products, Distillates and Extracts, Part I; and the Chemical Products of Wood Distillation, Part II, Dyeing and Tanning Extracts from Wood*, Scott, Greenwood & Son. The Oil and Colour Trates Journal Offices & Broadway, Ludgate Hill EC, London.

EBI 2022 European Biochar Market Report 2021/2022.

Elliott DC 2007 Historical developments in hydro-processing bio-oils. *Energy and Fuels* 21, 1792–1815.

Emberley R, Inghelbrecht A, Yu Z, and Torero JL 2017 Self-extinction of timber. *Proceedings of Combustion Institute* 36, 3055–3062.

Emrich W 1985 *Handbook of Charcoal Making*, Springer, Dordrecht.

Evans RJ, and Milne TA 1987a Molecular characterization of the pyrolysis of biomass: I. Fundamentals. *Energy and Fuels* 1, 123–137.

Evans RJ, and Milne TA 1987b Molecular characterization of the pyrolysis of biomass: II, Applications. *Energy and Fuels* 1, 311–319.

FAO 1985 Industrial Charcoal Making, FAO Forestry Paper 63. Rome, Italy: FAO, United Nations.

Fournier J 2009 Low temperature pyrolysis for biochar systems. Presentation made to the Conference Harvesting Clean Energy, January 25, 2009.

Freel BA, and Huffman DR 1994 Applied bio-oil combustion. Proceedings from the Biomass Pyrolysis Oil, Properties and Combustion Meeting, September 26–28, Estes Park, Colorado. pp309–315.

Freel BA, Graham RG, and Huffman DR 1990 *The Scale-Up and Development of Rapid Thermal Processing (RTP) to Produce Liquid Fuels from Wood*, Ontario Ministry of Energy Report, Toronto.

Freel BA, and Graham RG 2002 Bio-oil preservatives. U.S. Patent 6,485,841, filed Oct. 30, 1998 and issued Nov. 26, 2002.

Fryda L, and Visser R 2015 Biochar for soil improvement: evaluation of biochar from gasification and slow pyrolysis. *Agriculture* 5, 1076–1115.

Garcia-Perez M, Adams TT, Goodrum JW, Geller DP, and Das KC 2007 Production and fuel properties of pine chip bio-oil/biodiesel blends. *Energy and Fuels* 21, 2363–2372.

Garcia-Nunez, JA, Pelaez-Samaniego, MR, Garcia-Perez, ME, Fonts, I, Abrego, J, Westerhof, RJM, & Garcia-Perez, M 2017 Historical developments of pyrolysis reactors: A review. *Energy & Fuels* 31, 5751–5775. 10.1021/acs.energyfuels.7b00641.

Granatstein D, Kruger C, Collins H, Garcia-Perez M, and Yoder J 2009 Use of biochar from the pyrolysis of waste organic material as a soil amendment. *Final Project of Interagency Agreement* C0800248.

Gronnow MJ, et al 2012 Torrefaction/biochar production by microwave and conventional slow pyrolysis – comparison of energy properties. *Global Change Biology Bioenergy* 5, 144–152.

Helle S, Bennett NM, Lau K, Matsuio JH, and Duff SJB 2007 A kinetic model for the production of glucose by hydrolysis of levoglucosan and cellobiosan from pyrolysis oils. *Carbohydrate Research* 342, 2365–2370.

Hornung A, Bockhorn H, Appenzeller K, Roggero CM, and Tumiatti W 2005 Plant for the thermal treatment of material and operation process thereof. US Patent Number: 6,901,868.

Huber GW 2008 Breaking the chemical and engineering barriers to lignocellulosic bio-fuels: next generation hydrocarbon biorefineries. *A research roadmap for making lignocellulosic bio-fuels a practical reality*, www.ecs.umass.edu/biofuels/Images/Roadmap2–08.pdf, accessed 2013.

Huber GW, and Dumesic JA 2006 An overview of aqueous phase catalytic process for production of hydrogen and alkanes in a biorefinery. *Catalysis Today* 111, 119–132.

Ingram L, et al 2008 Pyrolysis of wood and bark in an auger reactor: physical properties and chemical analysis of the produced bio-oils. *Energy and Fuels* 22, 614–625.

James TGH 1972 Gold technology in ancient Egypt. *Gold Bulletin* 5, 38–42.

Jayakumar A, et al 2023 Systematic evaluation of pyrolysis processes and biochar quality in the operation of low-cost flame curtain pyrolysis kiln for sustainable biochar production. *Current Research in Environment and Sustainability* 5, 100213.

Jones SB, et al 2009 Production of gasoline and diesel from biomass via fast pyrolysis, hydrotreating and hydro-cracking, a design case. US Department of Energy, prepared under Contract DE-AC05-76RL01830, PNNL-18284 Rev. 1

Klark M, and Rule A 1925 *The Technology of Wood Distillation*, Chapman & Hall Ltd, London.

Klose W, and Weist W 1999 Experiments and mathematical modeling of maize pyrolysis in a rotary kiln. *Fuel* 78, 65–72.

Mahfud FH, Ghijen F, and Heeres HJ 2007 Hydrogenation of fast pyrolysis oil and model compounds in a two-phase aqueous organic system using homogeneous ruthenium catalysts. *Journal of Molecular Catalysis A: Chemical* 264, 227–236.

Mašek O 2009 Allothermal gasification: review of recent developments. In: Badeau J-P, and Levi A (Eds) *Biomass Gasification: Chemistry, Processes and Applications*. Hauppauge: Nova Science Publishers. pp271–287.

Mašek O, et al 2013 Microwave and slow pyrolysis biochar – comparison of physical and functional properties. *Journal of Analytical and Applied Pyrolysis* 100, 41–48.

Mason LC, Gustafson R, Calhoun J, Lippke BR, and Raffaeli N 2009 Wood to energy in Washington. The College of Forest Resources, University of Washington. Report to the Washington State Legislature, www.ruraltech.org/pub/reports/2009/wood_to_energy/index.asp.

Mohan, D, Pittman, CU, and Steele, PH 2006 Pyrolysis of wood/biomass for bio-oil: A critical review. *Energy & Fuels* 20, 848–889. 10.1021/ef0502397.

Morrisset D, Hadden RM, Bartlett AI, Law A, and Emberley R 2021 Time dependent contribution of char oxidation and flame heat feedback on the mass loss rate of timber. *Fire Safety Journal* 120, 103058.

Mottocks TW 1981 Solid and gas phase phenomena in the pyrolytic gasification of wood. M.Sc. Thesis, Department of Mechanical and Aerospace Engineering, Princeton University, Princeton, NJ.

Nabukalu C, and Gieré R 2020 Global charcoal consumption and the question of energy security, Kleinman Center for Energy Policy [WWW Document]. URL https://kleinmanenergy.upenn.edu/news-insights/global-charcoal-consumption-and-the-question-of-energy-security/.

Oehr KH, Scott DS, and Czernik S 1993 Method of producing calcium salts from biomass, U.S. Patent No. 5,264,623, filed January 4, 1993, issued Nov. 23.

Piskorz J, Radlein D, Scott DS, and Czernik S 1988 Liquid products from the fast pyrolysis of wood and cellulose. In: Bridgwater AV, and Kuester JL (Eds) *Research in Thermochemical Biomass Conversion*, Phoenix: Elsevier Applied Science, Ltd. pp557–571.

Radivojević M, et al 2010 On the origins of extractive metallurgy: new evidence from Europe. *Journal of Archaeological Science* 37, 2775–2787.

Radlein D 1999 The production of chemicals from fast pyrolysis bio-oils. In: Bridgwater AV (Ed) *Fast Pyrolysis of Biomass: A Handbook*. Newbury: CPL Press. pp164–188.

Rogovska N, Laird D, Cruse RM, Trabue S, and Heaton E 2012 Germination tests for assessing biochar quality. *Journal of Environmental Quality* 41, 1014–1022.

Roy C, Lemieux S, de Caumia B, and Pakdel H 1985 Vacuum pyrolysis of biomass in a multiple heat furnace. *Biotechnology and Bioenegy Sym.* No 15, 107.

Roy C, Blanchette D, Korving L, Yang J, and de Caumia B 1997 Development of a novel vacuum pyrolysis reactor with improved heat transfer potential. In: Bridgwater AV, and Boocock DGB (Eds) *Developments in Thermochemical Biomass Conversion*. London: Blackie Academic and Professional. pp351–367.

Roy C, Lu X, and Pakdel H 2000 Process for the production of phenolic rich pyrolysis oils for use in making phenol-formaldehyde resol resin. U.S. Patent No. 6,143,856, filed Feb. 5, 1999, issued Nov 7, 2000.

Scott DS, Piskorz J, and Radlein D 1988 The effect of wood species on composition of products obtained by the Waterloo Fast Pyrolysis Process. Proceedings from the Canadian Chemical Engineering Conference, Toronto, 10.

Schmidt H, Taylor P, Eglise A, and Arbaz C 2014 Kon-Tiki flame curtain pyrolysis for the democratization of biochar production. Biochar Journal 1, 14–24.

Shafizadeh F 1982 Introduction to pyrolysis of biomass. Journal of Analytical and Applied Pyrolysis 3, 283–305.

Spokas KA, et al 2012 Biochar: A synthesis of its agronomic impact beyond carbon sequestration. Journal of Environmental Quality 41, 973–989.

Toole AW, et al 1961 Biochar production, marketing and use. Forest Products Laboratory, Madison, WI, USDA-Forest Service, University of Wisconsin, Report July 1961.

Torero J 2016 Flaming ignition of solid fuels. In: Hurley MJ, et al (Eds) SFPE Handbook of Fire Protection Engineering. New York: Springer. pp633–661.

Torres-Rojas D, Lehmann J, Hobbs P, Joseph S, and Neufeldt H 2011 Biomass availability and biochar production in rural households of Western Kenya. Biomass and Bioenergy 35, 3537–3546.

Trossero M, Domac J, and Siemons R 2008 Industrial Biochar Production, TCP/CRO/3101 (A) Development of a Sustainable Biochar Industry, FAO, Zagreb.

Underwood GL 1990 Commercialization of fast pyrolysis products. In: Hogan E, Robert J, Grassi G, and Bridgwater AV (Eds) Biomass Thermal Processing. Newbury: CPL Press. pp226–228.

Underwood GL, and Graham RG 1991 Methods of producing fast pyrolysis liquids for making a high browning liquid smoke composition. U.S. Patent No. 5,039,537, filed June 6, 1991 and issued August 4, 1992.

van Rossum G, Kersten SRA, and Van Swaaij WPM 2007 Catalytic and non-catalytic gasification of pyrolysis oil. Industrial and Engineering Chemistry Research 46, 3959–3967.

Veitch FP 1907 Chemical Methods for Utilizing Wood, Including Destructive Distillation, Recovery of Turpentine, Rosin, and Pulp, and the Preparation of Alcohols and Oxalic Acid. Washington DC: U.S. Department of Agriculture, Bureau of Chemistry Circular No. 30.

Woolf D, Amonette JE, Street-Perrott A, Lehmann J, and Joseph S 2010 Sustainable biochar to mitigate global climate change. Nature Communications 1, 1–56.

Yang J, Tanguy PA, and Roy C 1995 Heat transfer, mass transfer and kinetic study of the vacuum pyrolysis of a large used tire particle. Chemical Engineering Science 50, 1909–1922.

Zhang X, et al 2018 Application of biomass pyrolytic polygeneration by a moving bed: Characteristics of products and energy efficiency analysis. Bioresource Technology 254, 130–138.

Zhao L, Cao X, Mašek O, and Zimmerman A 2013 Heterogeneity of biochar properties as a function of feedstock sources and production temperatures. Journal of Hazardous Materials 256-257, 1–9.

Zhao X, et al 2010 Microwave pyrolysis of corn stalk bale: a promising method for direct utilization of large-sized biomass and syngas production. Journal of Analytical and Applied Pyrolysis 89, 87–94.

5

Characteristics of biochar

Physical and structural properties

Catherine E. Brewer

Introduction

The goal of this chapter is to present an overview of the physical properties of biochars as stand-alone materials, before their application and their interactions with other materials. This is to provide anchoring points between the complexity of feedstock-process combinations and the complexity of soil-water-plant systems. Sections are organized by property with explanations of property definitions, methods to characterize that property, ranges of that property typical for the biochars being considered, and connections between feedstock/process and that property. Throughout the chapter, an attempt is also made to summarize the recent and future research into biochar physical properties.

Particle size

Particle size terminology and measurement

Particle size characterization aims to answer several questions: How large are the biochar particles? What shapes are the biochar particles? How similar are the sizes of biochar particles within a sample? The methods used to answer those questions must balance the need for precision with the resources that can be expended to perform the measurement.

Individual particles can be measured in terms of maximum length in a specific direction, ratios of dimensions such as aspect ratio (length/width), and similarity to specific geometries, such as rectangular cubes, cylinders, or spheres. Such characterization of individual particles is useful in cases where models are being developed, such as modeling of a feedstock particle shrinking during pyrolysis (Chaurasia, 2019) or the penetration of oxygen into a particle during

DOI: 10.4324/9781003297673-5

oxidation treatment (Boigné et al, 2022). For most applications, however, size exclusion methods, such as sieving and settling, provide the information needed for decision-making and material handling. Biochar producers and handlers need to be aware of the hazards and local regulations related to very small particles, specifically the inhalable particulate matter of <10 μm (PM_{10}) and <2.5 μm ($PM_{2.5}$), and the potential for dust explosions. For example, the U.S. Environmental Protection Agency has set limits on the amount of PM_{10} and $PM_{2.5}$ for air quality standards in units of μg m^{-3} averaged over specified periods (U.S. E.P.A., 2021). Dust mitigation measures, such as water sprays, additional ventilation, cyclones, and filtration systems, may be needed, in particular around equipment such as mills and sieves.

Particle size relation to feedstock and reaction conditions

The size distribution of biochar particles is closely related to the size distribution of the feedstock particles. Unless post-treatments such as granulation or pelletization are applied (see below), the sizes of the feedstock particles represent the maximum sizes of the resulting biochar particles as particles shrink during pyrolysis and break apart with handling. The extent of these changes depends on the processing conditions. Downie et al (2009) performed a series of slow pyrolysis experiments with woodchips and sawdust of the same wood at different temperatures to measure the relationship between feedstock particle size distribution and biochar particle size distribution. Biochar particle sizes decrease with the highest treatment temperature (Figure 5.1); some of the changes are attributed to changes in biochar mechanical properties (friability) at higher temperatures (Downie et al, 2009).

Particle size distribution influences, and is influenced by, pyrolysis heating rate and reaction time. Biomass and biochar are, in general, poor thermal conductors. Whereas metals have thermal conductivities in the range of tens to hundreds of W m^{-1} K^{-1}, the thermal conductivities of lignocellulose materials are approximately 0.1–0.2 W m^{-1} K^{-1} (Bergman and Lavine, 2017). This dictates an inverse relationship between feedstock particle size and achievable heating rate. This relationship is complicated by anisotropy: differences in properties as heat moves in different directions through the particles (Larfeldt et al, 2000). The anisotropy of biomass is the result of the differences in plant cell biochemical composition in different kinds of cells and in different parts of the cells. To reach the heating rates characteristic of fast pyrolysis (hundreds of °C s^{-1}), feedstock particles are typically reduced to sizes of a few mm to hundreds of μm. Particles that are too big are "undercooked" at the end of the very short reaction times (seconds) (Brewer et al, 2012). When heating rates are in the °C min^{-1} range and residence times are an hour or two, particles in the cm range for the smallest dimension can be processed "completely". For this reason, biochar particles from fast pyrolysis reactor systems are much smaller than those of biochars from slow pyrolysis and traditional kiln reactor systems: not just because the higher heating rates can result in explosive fracturing of feedstock into smaller particles; smaller particles were needed to make those high heating rates possible.

Attempts to model shrinkage to enable biochar property tailoring are limited by the assumptions of where volume loss occurs (Mettler et al, 2012). In "uniform conversion" and "volumetric decomposition" models, volume is lost uniformly with surface area exposure to the atmosphere. Since most surface area comes from internal pores (see below), the models predict loss of material

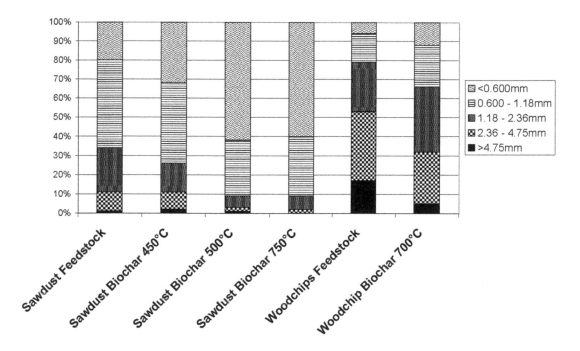

Figure 5.1 *Biochar particle size distribution based on initial wood particle size (woodchips vs. sawdust) and slow pyrolysis temperature (up to 750°C). The heating rate was 5–10°C min⁻¹. Dimensions refer to screen size (Downie et al, 2009)*

internally with little change in the outer dimensions. In "shrinking unreacted particle" and "shrinking core" models, volume loss is external and the bulk particle shrinks. In real pyrolysis, both things happen simultaneously (Gentile et al, 2017). Recently, advances in synchrotron X-ray imaging have enabled in-situ observation of biomass pyrolysis (Barr et al, 2021). For almond and walnut shells, biomass bed shrinkage started between 250–300°C as particles became smaller and packed together. The temperature of the fastest shrinkage rate was affected by the type of nut shells and pre-treatment. Almond shells tended to shrink externally more than walnut shells. Shells pre-treated with a base to reduce lignin content shrank at lower temperatures (~380°C) compared to untreated or water-soaked samples (~410°C) (Barr et al, 2021).

Porosity and density

Porosity and density definitions and measurements

Many of the most interesting impacts of biochar properties in applications are related to the "non-biochar" parts of biochar particles: the pores. Pores determine the bulk density of biochar and the interactions of biochar with other solids, liquids, and gases. To have a meaningful discussion of porosity

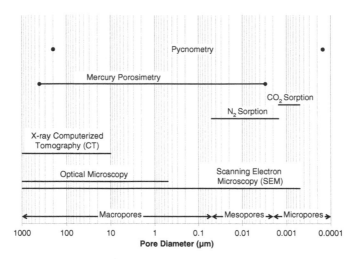

Figure 5.2 *Methods for characterization of material porosity based on pore sizes. Adapted from (Brewer et al, 2014), reprinted with permission of Elsevier*

and density, some clarification of terminology is needed. Material and soil scientists define pore sizes differently based on the pore size distributions most relevant to their areas of investigation. Since this chapter's focus is on the biochar itself as a material, the pore size classifications used here are those from material science: pores smaller than 2 nm (based on the pore's opening diameter), between 2 and 50 nm, and larger than 50 nm are micropores, mesopores, and macropores, respectively. Close attention also needs to be paid to the difference between *intra*particle porosity, the spaces *within* particles, *inter*particle porosity, the spaces *between* particles, and total porosity, the combination of inter- and intra-particle porosity in the bulk material. Different measurements are needed to detect different pore sizes (Figure 5.2; Brewer et al, 2014; Singh et al, 2017).

Density is the relationship between the mass of a material relative to the volume that material occupies. Different types of density are based on how the volume is measured (Figure 5.3): bulk density is a measure of the volume of the whole container holding

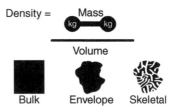

Figure 5.3 *Comparison of volumes in the definition of density for biochar particles, adapted from (Brewer et al, 2014) with permission of Elsevier*

the biochar particles, envelope density of the particle only (as if the particle were not porous), and skeletal or true density of only the solid portion of the particle. Bulk density is the property reported most often because it relates to material handling and is the easiest to measure. Skeletal density is the property most related to biochar chemical composition.

Characteristics of pores in biochars

Biochars contain all three size ranges of pores. Most of the pore volume is found in the macropores, while most of the surface area is found in the micropores. The size and shape of the macropores in biochars come primarily from the size and shape of the biomass cells (~1–100 μm). During pyrolysis, the "innards" of the cells volatilize and escape while the cell wall molecules condense into aromatic structures. Further heating intensity (time and temperature) and activation reactions cause smaller pores, mostly in the micropore size range, to form in the cell wall remnants. At some point, the walls between the pores become so perforated that the smaller pores connect (Antal and Grønli, 2003). This results in the observation of pore size groupings rather than a smooth continuum of pore sizes (Marsh and Rodriguez-Reinoso, 2006). The outlines of cell walls can be seen using scanning electron micrography of pine and poplar wood chars produced at 450 and 550°C (Figure 5.4; Zhang and You, 2013). Pore size distributions of biochars in comparison to the pore size distribution in the original feedstocks can be examined using mercury porosimetry (Figure 5.5; Zhang and You, 2013).

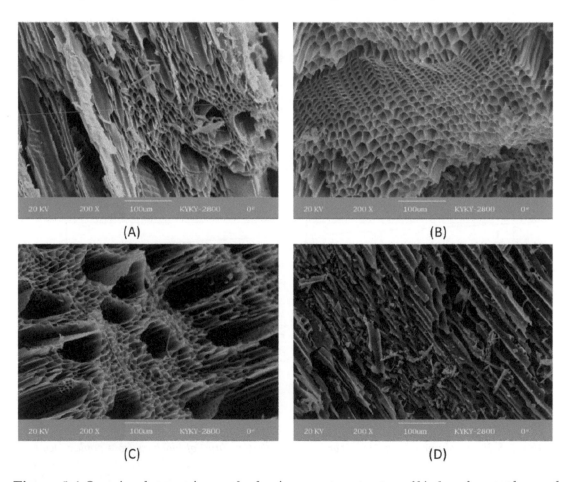

Figure 5.4 *Scanning electron micrographs showing macropore structure of biochars from poplar wood produced at 450°C (A) and 550°C (C) and from pine wood produced at 450°C (B) and 550°C (D), reprinted with permission from (Zhang and You, 2013). Copyright 2023 American Chemical Society*

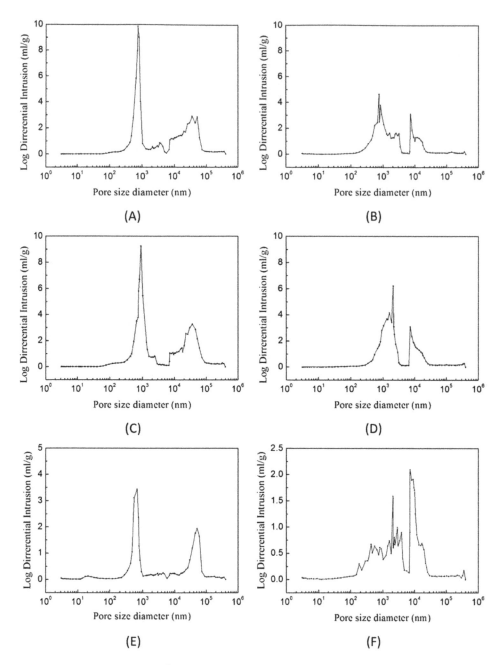

Figure 5.5 *Pore volume (mL g⁻¹) vs. pore size distribution (nm) of biochars from poplar wood produced at 450°C (A) and 550°C (C) and from pine wood produced at 450°C (B) and 550°C (D), and of the poplar (E) and the pine (F) feedstocks, reprinted with permission from (Zhang and You, 2013). Copyright 2013 American Chemical Society*

Assuming that the characterization method is capable of measuring the range of pore sizes present, biochars can have pore volumes up to a few $cm^3 g^{-1}$. Pore volumes reported for gas adsorption techniques, limited to the micro and meso pore range, result in pore volumes of <0.1–0.2 $cm^3 g^{-1}$ (Weber and Quicker, 2018). In general, biochar porosity increases with pyrolysis temperature and time (Fu et al, 2012), mostly within the 300–600°C range (Yang et al, 2021). The heating rate can impact pore formation if the rates of biopolymer melting and decomposition, condensation of C structures, and escape of volatiles have similar orders of magnitude. In this case, "melting" may result in larger pores with smoother walls, while efficient escape of volatiles and formation of stable condensed carbon structures enable more small pores (Angın, 2013; Sun et al, 2012).

Relationships between density and porosity

Intraparticle porosity, as a proportion of the volume, is calculated from knowledge of the skeletal density (ρ_s) and the envelope density (ρ_e) (Equation 5.1). Total porosity is calculated from knowledge of the skeletal density and the bulk density (ρ_b) (Equation 5.2). When all three densities are known, interparticle porosity in a bed of particles can be estimated by the difference between bulk and intraparticle porosity (Brewer et al, 2014) (Equation 5.3).

$$Intraparticle\ porosity\ (\%) = 1 - \frac{\rho_e}{\rho_s} \quad [5.1]$$

$$Total\ porosity\ (\%) = 1 - \frac{\rho_b}{\rho_s} \quad [5.2]$$

$$Interparticle\ porosity\ (\%) = Total\ porosity$$
$$- Intraparticle\ porosity$$
$$[5.3]$$

The skeletal or true density of biochars increases with reaction temperature as the sugar and lignin molecules (approximately 1.2–1.5 $g\ cm^{-3}$) are converted into amorphous aromatic C structures (1.5–2 $g\ cm^{-3}$) towards the skeletal density of graphite (2.25 $g\ cm^{-3}$) (Brewer et al, 2014). In this way, skeletal density can provide information about the temperature reached within the reactor for a given feedstock. Since soil particles and inorganic salts in biochars have higher particle densities than C-based materials, biochars with higher ash contents tend to have higher skeletal densities (Brewer et al, 2009). Skeletal density is also higher when all of the pores are accessible to measurement since closed but hollow pores add apparent volume without adding mass (Brewer et al, 2014).

While skeletal density is consistently related to biomass and reaction temperature, relationships between feedstock, reaction conditions, bulk density, and total porosity are more complicated. Skeletal density and bulk density can be inversely related: as the chemical structure condenses with increasing pyrolysis temperature (resulting in higher skeletal density), more porosity forms with continued heating (resulting in a lower envelope or bulk density) (Weber and Quicker, 2018; Pastor-Villegas et al, 2006). Bulk densities of biochar are often in the range of 0.1–0.5 $g\ cm^{-3}$ depending on how tightly packed the particles are within the container. As a first approximation, the bulk density of wood-derived biochar is 80% that of its dry feedstock (Abdullah and Wu, 2009).

Density and porosity connections to biochar applications

The relatively low bulk density of biochar means that biochar requires substantial volume for transport and additional attention to avoid erosion losses. On the surface of a field, biochar may be fluidized more easily than similarly-sized soil particles and erode with strong winds. In water, biochar tends to float, creating challenges for keeping aqueous suspensions evenly mixed and for keeping biochar in the correct location in a field (Major et al, 2010). As with measurement techniques, the utilization of biochar pores must consider size range. For example, a large micropore volume may be advantageous for gas phase interactions but provide little benefit for microorganism habitat or plant-available water retention. Likewise, biochars with large pores and small surface areas may be better for water management (Ibrahim and Horton, 2021) than for cation exchange capacity and nutrient retention.

Surface area

Surface area terminology and measurements

Surface area is useful for gauging the expected quantity of biochar interactions with the environment. The higher the surface area, the more sites there are for interactions with other molecules. The challenge for interpreting surface area numbers is that those numbers change depending on the molecule used to probe the surface. Some probe molecules, such as nitrogen gas (N_2) or carbon dioxide (CO_2), interact similarly with many types of surface chemistries. This "physical adsorption" interaction means that the "whole" surface area is measured compared to "chemical adsorption" where only certain types of chemical sites result in attachment. Even for "physisorption" methods, mass transfer differences still create differences in the reported surface area number. For example, N_2-probed surface areas tend to be smaller than those probed by CO_2 because CO_2 can diffuse into smaller pores under measurement conditions (Mukherjee et al, 2011). Comparing surface areas between materials, therefore, requires careful attention to the method used.

Surface areas of biochars

Surface area is typically reported in values of $m^2 \, g^{-1}$ and is calculated from models of how molecules cover a surface. For N_2 sorption at 77 K using the Brunauer-Emmet-Teller (BET) model (Brunauer et al, 1938), biomass feedstocks have surface areas around 0.5–1 $m^2 \, g^{-1}$. Surface areas for biochars increase from that of biomass to hundreds of $m^2 \, g^{-1}$ with greater heating intensity (time and temperature) (Weber and Quicker, 2018). The increase in surface area is not linear with temperature but rather corresponds to the decomposition and volatilization of the biopolymers. Carbons made below the decomposition temperature of cellulose (~350°C) usually have surface areas similar to their feedstocks (Brewer et al, 2014). Eventually, the development of pores and surface area reach a maximum, above which no additional pores form or smaller pores collapse into larger pores and surface area decreases. For non-activated biochars, past studies suggest this maximum occurs somewhere around 600–750°C (Weber and Quicker, 2018). Some biochars and activated carbons can have surface areas of a few thousand $m^2 \, g^{-1}$ from the many very small pores between and among aromatic

C sheets (Marsh and Rodriguez-Reinoso, 2006). Higher heating rates, such as those seen with fast pyrolysis, can result in lower surface areas as the volatilization process is more "explosive", creating larger pores and channels. In general, biochars with higher ash contents have lower surface areas as the micropores contributing to the surface area form more within the organic C fraction rather than the mineral fraction (Ronsse et al, 2013).

Biochar activation to increase surface area

For applications where more surface area is desired, biochar producers may add a second processing step: activation. The purpose of activation is the creation of (access to) small pores in the C structure. Activation is accomplished through heating to a relatively high temperature (often >700°C) in combination with or without the addition of a chemical agent that enables controlled oxidation and gasification of the C material. (note that "activation" here refers to changes in physical properties; "activation" in biochar marketing is sometimes used to refer to the incorporation of plant nutrients and/or microorganisms into biochar before application to soils (Chapter 1).)

Higher temperatures decompose the less-stable C structures, including the tars and volatiles that had condensed and blocked pore openings during the initial stages of pyrolysis. Some researchers have used microwave irradiation to enable large (>200°C) increases in temperature for activation in short time durations with less energy than conventional heating (Selvam and Paramasivan, 2022). High surface areas (450–800 $m^2 g^{-1}$) can be generated using microwave-assisted pyrolysis (Li et al, 2016). Heating of the biochar in the presence of steam or CO_2 enhances the effect of temperature through carbon-CO_2 and carbon-water gasification processes (Awasthi, 2022; Kołtowski et al, 2017). Steam tends to create more mesopores while CO_2 tends to create more micropores (He et al, 2021).

Activation with a chemical agent typically involves soaking the biochar in a solution of the desired acid, base, salt, or metal, then drying and heating the biochar. Examples of chemical agents include sulfuric acid, phosphoric acid, potassium hydroxide, potassium carbonate, sodium hydroxide, sodium (bi)carbonate, and zinc chloride (Saletnik et al, 2019). The use of a chemical agent, while effective at creating porosity and surface area, can require rinse steps before the activated biochar is ready for use. Some activation agents are used to impact the specific types of chemical sites on the surface more than to increase the total surface area (Marsh and Rodriguez-Reinoso, 2006). Some examples are hydrochloric acid, nitric acid, hydrogen peroxide, and phosphoric acid (Yang et al, 2021). For biochars with higher ash contents, an increase in measured surface area after an acid rinse "activation" may be an effect of removing mineral content blocking pores.

Mechanical properties

Mechanical strength and durability definitions and tests

Mechanical strength of biochar is important when biochar needs to support a load, such as in an adsorption column, or when the particle size distribution of biochar needs to be maintained (the opposite of "grindability"). Strength is defined based on the type of force applied: "push" (compression), "pull" (tension), "rub" (shear), or "hit" (impact). Commonly reported units are compressive strength (in MPa), crushing strength (in kg m^{-2}), and Hardgrove Grindability Index (HGI), where the particle size reduction of a sample in a mill is compared to the particle size reduction of a reference coal. The higher the HGI, the more grindable the sample is (Weber and Quicker, 2018). When biochar or biochar pellet durability measurements are important, methods are borrowed from the animal feed and solid biofuel sectors (García et al, 2021). The pellet durability index is measured by exposing the pellets to standardized sequences of handling (tumbling, mixing, conveying) and then quantifying the number of broken pellets or the amount of fines (particles smaller than a certain screen size).

Biochar mechanical strength and grindability

Biochar mechanical strength is related to the mechanical strength of the feedstock and the biochar density (inverse of porosity). In general, carbonization makes biomass harder and more brittle as well as less elastic (Zickler et al, 2006). Biochars made from dense biomass with low ash and high lignin contents (nut shells, certain woods) tend to be stronger. As porosity increases with the heating of the biomass, the mechanical strength decreases until the density reaches a point where strength increases with increasing temperature (Weber and Quicker, 2018). Biochar strength decreased until 600°C and was lower for biochars made at faster heating rates (Kumar et al, 1999).

Management of biochar physical properties

Control of biochar physical properties is possible within the constraints of available biomass, equipment, and (as with all engineered processes) time and money. Biochar producers need to balance what they can reasonably procure, measure, and adjust to target certain applications.

Impacts of location on feedstock and system options

The low density, high moisture content, and ash content of biomass mean that transportation is often the limiting factor for feedstock selection. Similarly, the low bulk density of biochar hinders economical transportation. Design of a biochar production system must first identify the biomass available nearby (typically within tens of km), the seasonality of the supply, and the ability to collect feedstocks. Whenever possible, on-site conversion is desirable. Next, producers must identify the target mix of products: biochars for one or more applications and co-products. Heat, power, nutrient delivery, monetized ecosystem services, and avoided disposal costs are often needed to achieve feasible economics (Struhs et al, 2020). Feedstocks with lower ash and moisture contents are easier to handle. Feedstocks with consistent particle sizes and compositions make operation simpler as fewer quality control measurements and adjustments to process parameters are needed. For these reasons, such feedstocks are also less plentiful and more expensive because there is more demand for them. In some cases, feedstock availability and product distribution from the chosen reactor may dictate biochar properties. Up-front efforts to quantify and characterize the feedstocks are critical to project success.

Relationship to target applications

Maximizing the benefits of biochar requires match-making between biochar properties and application needs (Wu et al, 2020). While most recent research has been devoted to the chemical properties of biochars for applications, especially for fertilizer (Chapter 26) and pollution remediation applications (Chapter 27), there have been some examples of cases when the physical properties were specifically important. Very small biochar particles (nanobiochars) have been shown to affect the growth and metabolism of microorganisms (Liu et al, 2019, Guo et al, 2022) and interactions within the rhizosphere (Zhang et al, 2022) differently than the corresponding non-milled biochars. Nanobiochars complexed with metals have different mobility in the soil than "pristine" biochars (Gong et al, 2022). Changing non-magnetic biochar into magnetic biochar through the addition of iron oxides or zero valence iron can make spent biochar easier to recover from water (Hu et al, 2022, Zhou et al, 2017) or from soil (Duan et al, 2022), as well as affect the electrostatic chemistry of the biochar surfaces.

Property modification methods

When biochar physical properties immediately after pyrolysis are not ideal for an application,

biochar producers should consider post-production modifications. Particle size distribution can be modified by crushing, grinding, granulation, and pelletization. Crushing is a function of pressure forces that cause the material to fracture at weaker points. This can be accomplished simply with a shovel or the bucket of a backhoe, or in a more precise manner between large plates or rollers. Grinding or milling uses shear forces (abrasion) to fracture and separate particles; cutting mills, ball mills, hammer mills, and grinders are all unit operations that can be used. Granulation is the process of building up larger particles using binder materials and coating processes to stick smaller particles together. Many activated carbons are formulated from powders using granulation. Recent work has shown that different kinds of biochar can be granulated using common binders in a drum granulator (Briens and Bowden-Green, 2020). Pelletization uses binders and substantial compression forces to press smaller particles together, usually into cylinders or beads (Kim et al, 2019). Porosity and surface area can be modified by physical or chemical activation, removal of minerals through rinsing, treating biochars with minerals (Zhao and Zhou, 2019), and densification. Strength and durability can be modified by either combining stronger feedstocks with weaker feedstocks before conversion or by adding strengthening materials after conversion.

Recommendations for future research

Overcoming challenges for pelletization and granulation

While research on activated carbon granulation, charcoal briquetting, and biomass densification has been ongoing for several decades,

little data and even fewer predictive models are available for the production of biochar pellets or granules. For granulation, more information is needed on the hydrophobic-hydrophilic properties of biochars relative to the ability of binders to agglomerate biochars to the appropriate granule size and durability (Briens and Bowden-Green, 2020). From there, more information is needed on how granulation impacts biochar handling and applications, such as for field application with fertilizers or for green roofs (Liao et al, 2022). For pelletization, some experimental data is available on the relationships between biochar properties (feedstock, particle size distribution, and pyrolysis temperature) and pelletization parameters (compression ratio, moisture content, and binder) for determining the pellet density, durability, and exiting pressure (pressure needed to push a formed pellet out of the die; this increases exponentially as pellets become longer and can lead to mill clogging/failure if the pressure becomes too high) (Mohammadi, 2021). Early experiments indicate that biochar does not behave in the same way as raw biomass or torrified biomass, that there is a minimum moisture content of ~30% for pellets to remain intact, and that higher pyrolysis temperatures may favor easier pelletization (Riva et al, 2021). Granulation and pelletization have the potential to address some of the physical property issues of biochar: low density, low mechanical strength, variability in particle size and shape, and fines or dust (Mohammadi, 2021). Future research should focus on understanding the engineering parameters for these processes so that scale-up can be easier and more successful.

Particle size control before or after feedstock conversion?

Many biomass feedstocks require at least some amount of particle size separation and modification to reach the desired particle size distribution. For biochars, this separation and modification can occur before or after pyrolysis to create the target particle size distribution. Particle size modification before pyrolysis can increase the uniformity of heat

and mass transfer within the reactor, potentially improving biochar quality control and reducing the time and energy for processes like drying. Particle size modification post-pyrolysis, however, may require less energy due to the higher friability of the biochar, especially if very fine particles are needed for the target application. Multiple options need to be compared to identify the ideal product handling sequence relative to time, cost, biochar quality, and safety. Researchers should consider ongoing and expanding research efforts with feedstock handling, including programs such as the Biomass Feedstock National User Facility at Idaho National Laboratory (Idaho National Laboratory, 2022), to transfer best practices for biochar process scale-up and implementation.

Role in biochar classifications and certifications

In the most recent biochar standards published by the International Biochar Initiative, the required properties for reporting are mostly related to chemical composition and potential toxicity (International Biochar Initiative, 2015). Particle size distribution by dry sieving (>50 mm through <0.5 mm) and moisture content are required declarations; the measurements must occur after any activation, weathering, or any other processes that affect particle size. Measurement of surface area by N_2 adsorption is an optional reported property. In the most recent European Biochar Certificate (EBC) guidelines (European Biochar Foundation, 2022), bulk density (based on dry matter <3 mm), mineral additives, water holding capacity, maximum pyrolysis temperature, and moisture content are required to be reported. Measurement of BET surface area and pore size distribution by N_2 adsorption are recommended but are not required due to the analysis cost and the difficulty in relating surface area directly to application performance. Both standards include notes about best management practices for biochars related to physical properties such as flammability, self-heating, and dust. The focus of standards on properties needed for transportation and handling and safety (particle size, density, moisture content) is likely to continue as the "value" of biochar for a particular application based on physical properties is not as well understood. Future research that enables the prediction of performance based on physical properties, such as nanobiochar content or biochar strength, may create a need for other physical property measurements to be added to standards and certifications.

References

Abdullah H, and Wu H 2009 Biochar as a fuel: 1. Properties and grindability of biochars produced from the pyrolysis of mallee wood under slow-heating conditions. *Energy and Fuels* 23, 4174–4181.

Angın D 2013 Effect of pyrolysis temperature and heating rate on biochar obtained from pyrolysis of safflower seed press cake. *Bioresource Technology* 128, 593–597.

Antal MJ, and Grønli M 2003 The art, science, and technology of charcoal production. *Industrial Engineering and Chemical Research* 42, 1619–1640.

Awasthi MK 2022 Engineered biochar: A multifunctional material for energy and environment. *Environmental Pollution* 298, 118831.

Barr MR, et al 2021 Towards a mechanistic understanding of particle shrinkage during biomass pyrolysis via synchrotron X-ray microtomography and in-situ radiography. *Scientific Reports* 11, 2656.

Bergman TL, and Lavine AS 2017 Appendix A. Thermophysical Properties of Matter. In: TL Bergman (Ed) *Fundamentals of Heat and Mass Transfer.* Hoboken, NJ: Wiley. pp899–927.

Boigné E, Bennett NR, Wang A, and Ihme M 2022 Structural analysis of biomass pyrolysis and oxidation using in-situ X-ray computed tomography. *Combustion and Flame* 235, 111737.

Brewer CE, et al 2014 New approaches to measuring biochar density and porosity. *Biomass and Bioenergy* 66, 176–185.

Brewer CE, et al 2012 Extent of pyrolysis impacts on fast pyrolysis biochar properties. *J. Environmental Quality* 41, 1115–1122.

Brewer CE, Schmidt-Rohr K, Satrio JA, and Brown RC 2009 Characterization of biochar from fast pyrolysis and gasification systems. *Environmental Progress & Sustainable Energy* 28, 386–396.

Briens L, and Bowden-Green B 2020 A comparison of liquid binders for drum granulation of biochar powder. *Powder Technology* 367, 487–496.

Brunauer S, Emmett PH, and Teller E 1938 Adsorption of gases in multimolecular layers. *Journal of the American Chemical Society* 60, 309–319.

Chaurasia A 2019 Modeling of downdraft gasification process: Part I - Studies on shrinkage effect on tabular, cylindrical and spherical geometries. *Energy* 169, 130–141.

Downie A, Crosky A, and Munroe P 2009 Physical Properties in Biochar. In: J Lehmann and S Joseph (Eds) *Biochar for Environmental Management: Science and Technology.* Earthscan: London. pp13–32.

Duan L, et al 2022 Zero valent iron or Fe_3O_4-loaded biochar for remediation of Pb contaminated sandy soil: Sequential extraction, magnetic separation, XAFS and ryegrass growth. *Environmental Pollution* 308, 119702.

European Biochar Foundation 2022 European Biochar Certificate – Guidelines for a Sustainable Production of Biochar Version 10.1. Last Modified 10 January 2022.

https://www.european-biochar.org/media/doc/2/version_en_10_1.pdf.

Fu P, et al 2012 Evaluation of the porous structure development of chars from pyrolysis of rice straw: Effects of pyrolysis temperature and heating rate. *Journal of Analytical and Applied Pyrolysis* 98, 177–183.

García R, et al 2021 Residual pyrolysis biochar as additive to enhance wood pellets quality. *Renewable Energy* 180, 850–859.

Gentile G, et al 2017 A computational framework for the pyrolysis of anisotropic biomass particles. *Chemical Engineering Journal* 321, 458–473.

Gong H, Zhao L, Rui X, Hu J, and Zhu N 2022 A review of pristine and modified biochar immobilizing typical heavy metals in soil: Applications and challenges. *Journal of Hazardous Materials* 432, 128668.

Guo S, et al 2022 Ball-milled biochar can act as a preferable biocompatibility material to enhance phenanthrene degradation by stimulating bacterial metabolism. *Bioresource Technology* 350, 126901.

He M, et al 2021 Critical impacts of pyrolysis conditions and activation methods on application-oriented production of wood waste-derived biochar. *Bioresource Technology* 341, 125811.

Hu X, et al 2022 One-pot synthesis of iron oxides decorated bamboo hydrochar for lead and copper flash removal. *Industrial Crops and Products* 187, 115396.

Ibrahim A, and Horton R 2021 Biochar and compost amendment impacts on soil water and pore size distribution of a loamy sand soil. *Soil Science Society of America Journal* 85, 1021–1036.

Idaho National Laboratory 2022 *Biomass Feedstock National User Facility*. Idaho National Laboratory, accessed 22 November 2022. https://bfnuf.inl.gov/.

International Biochar Initiative 2015 Standardized Product Definition and Product Testing Guidelines for Biochar that Is Used in Soil, Version 2.1. Last Modified November 2015, accessed 03 March 2023. https://biochar-international.org/standard-certification-training/biochar-standards/.

Kim J, Song J, Lee S-M, and Jung J 2019 Application of iron-modified biochar for arsenite removal and toxicity reduction. *Journal of Industrial and Engineering Chemistry* 80, 17–22.

Kołtowski M, Charmas B, Skubiszewska-Zięba J, and Oleszczuk P 2017 Effect of biochar activation by different methods on toxicity of soil contaminated by industrial activity. *Ecotoxicology and Environmental Safety* 136, 119–125.

Kumar M, Verma BB, and Gupta RC 1999 Mechanical properties of acacia and eucalyptus wood chars. *Energy Sources* 21, 675–685.

Larfeldt J, Leckner B, and Melaaen MC 2000 Modelling and measurements of heat transfer in charcoal from pyrolysis of large wood particles. *Biomass and Bioenergy* 18, 507–514.

Li J, et al 2016 Biochar from microwave pyrolysis of biomass: A review. *Biomass and Bioenergy* 94, 228–244.

Liao W, Drake J, and Thomas SC 2022 Biochar granulation, particle size, and vegetation effects on leachate water quality from a green roof substrate. *Journal of Environmental Management* 318, 115506.

Liu X, Tang J, Wang L, Liu Q, and Liu R 2019 A comparative analysis of ball-milled biochar, graphene oxide, and multi-walled carbon nanotubes with respect to toxicity induction in Streptomyces. *Journal of Environmental Management* 243, 308–317.

Major J, Lehmann J, Rondon M, and Goodale C 2010 Fate of soil-applied black carbon: downward migration, leaching and soil respiration. *Global Change Biology* 16, 1366–1379.

Marsh H, and Rodriguez-Reinoso F 2006 Characterization of Activated Carbons. In: *Activated Carbon*. Elsevier: Amsterdam. pp143–182.

Mettler MS, Vlachos DG, and Dauenhauer PJ 2012 Top ten fundamental challenges of biomass pyrolysis for biofuels. *Energy and Environmental Science* 5, 7797–7809.

Mohammadi A 2021 Overview of the benefits and challenges associated with pelletizing Biochar. *Processes* 9, 1591.

Mukherjee A, Zimmerman AR, and Harris W 2011 Surface chemistry variations among a series of laboratory-produced biochars. *Geoderma* 163, 247–255.

Pastor-Villegas J, Pastor-Valle JF, Rodríguez JMM, and García MG 2006 Study of commercial wood charcoals for the preparation of carbon adsorbents. *Journal of Analytical and Applied Pyrolysis* 76, 103–108.

Riva L, et al 2021 Considerations on factors affecting biochar densification behavior based on a multiparameter model. *Energy* 221, 119893.

Ronsse F, Van Hecke S, Dickinson D, and Prins W 2013 Production and characterization of slow pyrolysis biochar: influence of feedstock type and pyrolysis conditions. *Global Change Biology Bioenergy* 5, 104–115.

Saletnik B, et al 2019 Biochar as a multifunctional component of the environment—a review. *Applied Sciences* 9, 1139.

Selvam S M, and Paramasivan B 2022 Microwave assisted carbonization and activation of biochar for energy-environment nexus: A review. *Chemosphere* 286, 131631.

Singh B, Camps Arberstain M, and Lehmann J (Eds) 2017. *Biochar: A Guide to Analytical Methods*. Boca Raton: CRC Press.

Struhs E, Mirkouei A, You Y, and Mohajeri A 2020 Techno-economic and environmental assessments for nutrient-rich biochar production from cattle manure: A case study in Idaho, USA. *Applied Energy* 279, 115782.

Sun H, Hockaday WC, Masiello CA, and Zygourakis K 2012 Multiple controls on the chemical and physical structure of biochars. *Industrial Engineering and Chemistry Research* 51, 3587–3597.

U.S. Environmental Protection Agency 2021 National Ambient Air Quality Standards. Last Modified February 2021, accessed 24 February 2022. https://www.epa.gov/criteria-air-pollutants/naaqs-table.

Weber K, and Quicker P 2018 Properties of biochar. *Fuel* 217, 240–261.

Wu P, et al 2020 Visualizing the emerging trends of biochar research and applications in 2019: a scientometric analysis and review. *Biochar* 2, 135–150.

Yang C, Liu J, and Lu S 2021 Pyrolysis temperature affects pore characteristics of rice straw and canola stalk biochars and biochar-amended soils. *Geoderma* 397, 115097.

Zhang J, and You C 2013 Water holding capacity and absorption properties of wood chars. *Energy and Fuels* 27, 2643–2648.

Zhang X, et al 2022 Nanobiochar-rhizosphere interactions: Implications for the remediation of heavy-metal contaminated soils. *Environmental Pollution* 299, 118810.

Zhao Z, and Zhou W 2019 Insight into interaction between biochar and soil minerals in changing biochar properties and adsorption capacities for sulfamethoxazole. *Environmental Pollution* 245, 208–217.

Zhou Z, et al 2017 Sorption performance and mechanisms of arsenic(V) removal by magnetic gelatin-modified biochar. *Chemical Engineering Journal* 314, 223–231.

Zickler GA, Schöberl T, and Paris O 2006 Mechanical properties of pyrolysed wood: a nanoindentation study. *Philosophical Magazine* 86, 1373–1386.

6

Characteristics of biochar

Micro- and nano-chemical properties and interactions

Sohrab Haghighi Mood, Kalidas Mainalis, Manuel Raul Pelaez-Samaniego, and Manuel Garcia-Perez

Introduction

Biochar has received growing attention over the last decade due to its potential use in various applications, including environmental remediation, as raw material for activated carbon, and additive for construction materials. The capacity of biochar for these and other targeted uses relates to its unique micro- and nano-properties (i.e., surface area, porous structure, surface functional groups, size of polyaromatic rings, mineral and ash content, and capacity to transport electrons) (Rajapaksha et al, 2016; Wang et al, 2019). Micro-structures are features at scales between 500 nm and 500 μm (at the cell wall structure level) (McDonald-Wharry et al, 2016). The term nano-structure describes structural details at the scale from a few angstroms to a few nanometers (which is in the range of the size of graphene sheets and pore size) at which biochar particles possess micro- and nano-structures that e.g., play a role in electron transfer between bacteria and iron oxide particles (Figure 6.1). By selecting appropriate feedstock, carbonization (i.e., equipment, processing temperature, heating rate, and residence time), and post-carbonization and activation conditions (e.g., physical activation, chemical activation, and biological modification), it is possible to control biochar's micro- and nano-properties. This chapter aims to describe the connection between feedstock composition, carbonization conditions, and biochar structure at the micro- and nano-level scale. The link between these structures and adsorption, reaction, and electrochemical phenomena relevant to the use of biochar for environmental remediation are discussed.

DOI: 10.4324/9781003297673-6

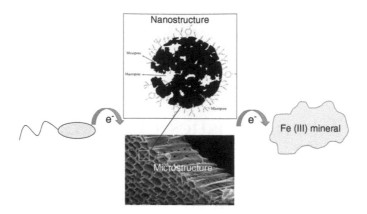

Figure 6.1 *Biochar micro- and nano-structures and their role as electron shuttles in redox reactions (adapted in part from Kappler et al, 2014; Lee et al, 2017)*

Biochar formation and properties

Formation mechanisms

Biochar's most crucial micro- and nano-properties associated with its environmental services are as follows: proximate composition (volatile matter, fixed carbon, and ash), surface area, pore size distribution, surface functional groups, size of polyaromatic ring systems, and z-potential (Leng et al, 2020). By choosing a suitable feedstock, pretreatment, and carbonization or activation conditions, or by formulating products through blending with other materials, it is possible to obtain biochars with desirable properties (Antal et al, 2003; Chen et al, 2011; Kastner et al, 2012; Duan et al, 2019; Leng et al, 2020; Pelaez-Samaniego et al, 2022).

Several feedstocks (e.g., agricultural and forest residues, animal wastes, and sewage sludge) have been used to produce biochar (Khawkomol et al, 2021; Papageorgiou et al, 2021). While agricultural residues (e.g., wheat straw) generally have high cellulose, hemicellulose, and ash content, woody feedstocks are typically rich in lignin

(Volpe et al, 2018). These differences in feedstock composition and morphology influence the hardness and mechanical properties of the resulting biochars.

When lignocellulosic materials are heated to temperatures over 250°C under inert conditions, the weakest ether bonds in cellulose, hemicellulose, and lignin break and depolymerized oligomeric molecules are formed (Paris et al, 2005). These molecules can further crack to form volatile compounds or crosslink to form biochar. At carbonization temperatures of 250–550°C, these oligomers are liquids forming a viscoelastic suspension with the reacting biopolymers and biochar. This suspension is called metaplast (Dufour et al, 2012; Kersten et al, 2013; Montoya et al, 2017; Pecha et al, 2017) (Figure 6.2). This viscoelastic material is responsible for the softening, swelling, shrinking, and resolidification of biomass and biochar particles (Caposciutti et al, 2019). The glass transition of amorphous cellulose is estimated to occur between 243°C and 307°C (Wooten et al, 2004), and the melting point

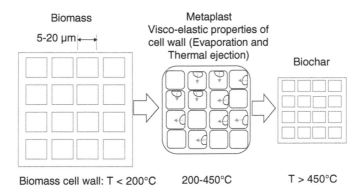

Figure 6.2 *Evolution of biochar micro-structure during pyrolysis. Particle shrinking due to the formation of metaplast with viscoelastic properties at processing conditions*

of levoglucosan and cellobiosan (main monomeric and dimeric products of cellulose pyrolysis) are 180°C and 240°C, respectively (Wooten et al, 2004), which explains the formation of the viscoelastic suspension at these temperatures. The surface area of biochars at temperatures in which the metaplast exists (<500°C) is small (less than 20 m² g⁻¹). The biomass particle therefore shrinks during carbonization with increasing pyrolysis temperature (Figure 6.2).

When reaching the gas-metaplast interface, volatile bubbles formed inside the liquid intermediate generate micro-explosions responsible for aerosol formation (droplet diameter of 1–4 µm) (Teixeira et al, 2011). The formation of these liquid intermediates has been assessed by conducting torrefaction or pyrolysis of softwood and removing these intermediates with dichloromethane (DCM) that disappear with greater pyrolysis temperature (Figure 6.3). Some of the bubbles in the

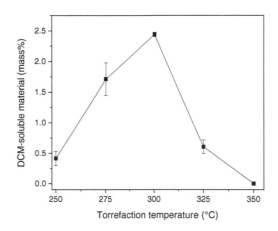

Figure 6.3 *Evolution of the yield of dichloromethane (DCM) soluble material (lignin liquid intermediates) isolated from torrefied wood as a function of processing temperatures (Error bars correspond to STD from 3 replicates) (Pelaez-Samaniego et al, 2014; with permission)*

metaplast harden during solid cooling, forming pores and cavities on biochar wall surfaces (Dufour et al, 2011). Biochars produced below 350°C generally contain leachable organic molecules (mainly oligomers).

Evaporation and thermal ejection are the main mechanisms responsible for the removal of volatile products (Terrell et al, 2020). The boiling point of levoglucosan, cellobiosan, and cellotriosan are 304°C, 581°C, and 792°C, respectively (Wooten et al, 2004). Boiling points of volatile pyrolysis products, sugars, and polycyclic aromatic hydrocarbons have been reported elsewhere (Oja et al, 1998; Oja et al, 1999; Fonts et al, 2021). Figure 6.4 shows the estimated boiling point for lignin oligomers (Terrell et al, 2020). According to Cao et al (2022), the yield of ejected aerosols from cellulose pyrolysis increases from 29% w/w at a heating rate of 1°C s^{-1} to 57% w/w when the heating rate is increased to 1000°C s^{-1}. The yield of evaporated products from cellulose decreased from 43 wt.% at a heating rate of 1°C s^{-1}, to 28% w/w when the heating rate is increased to 1000°C s^{-1}. Montoya et al (2017) developed a mathematical model that considers bubble formation dynamics inside the liquid intermediate. The authors estimated thermal ejection aerosol yield from ligno-cellulosic materials between 10 and 25% w/w, depending on the heating rate.

The solid chemical structure of biochars gradually evolves as a function of pyrolysis temperature (Figure 6.5). At 700°C, the formation of small oxygenated graphene sheets with defects occurs (Smith et al, 2017). Heteroatoms (O and N) play a critical role in crosslinking. As the temperature further increases, biomass-derived oligomers form liquid intermediates (amorphous phase) that further crosslink, form aromatics, and condense into distorted polyaromatic graphene sheets with defects. These polyaromatic rings have been referred to as "graphene sheets" or "aromatic layers", but no consistent definition exists on the size required for polyaromatic rings to be called "graphene". Above 600°C, graphene sheets grow laterally and organize (McDonald-Wharry et al, 2016). Over 1000°C, distorted graphene arranges and forms stacks of parallel graphene units detectable by TEM and XRD. These stacks often compose 2–4 layers of a few nanometers across the

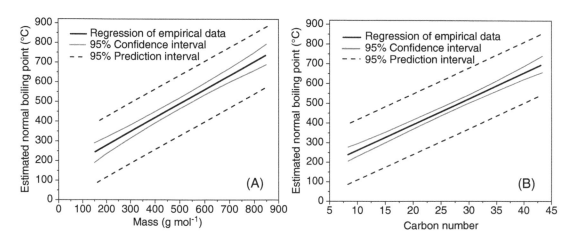

Figure 6.4 *Estimated boiling and condensation point of individual lignin-derived oligomeric molecules as a function of (A) their mass and (B) carbon number (adapted from Terrell et al, 2020; with permission)*

Cellulose ($C_{12}H_{20}O_{10}$)   C300 ($C_{12}H_{18}O_9$)   C400 ($C_{22}H_{10}O_5$)

C500 ($C_{24}H_9O_4$)   C600 ($C_{32}H_{12}O_3$)   C700 ($C_{37}H_{12}O_3$)

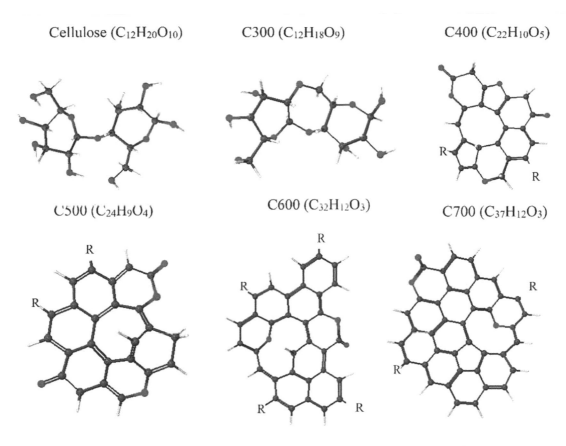

Figure 6.5 *Evolution of the nano-structure of biochar made from cellulose as a function of processing temperature (Smith et al, 2017; with permission)*

interlayer and are known as graphitic crystallites, turbostratic crystallites, parallel layer groups, graphite microcrystals, and nanocrystals (McDonald-Wharry et al, 2016).

The model of Kercher and Nagle (Kercher et al, 2003) can describe the evolution of the graphene packages in materials pyrolyzed up to 1400°C. The model considers that the number of graphene packets and their thickness remains constant while growing laterally at the expense of the amorphous carbons. The graphene domains increase in length from 1–4 nm at pyrolysis around 1000°C to 4–8 nm above 2000°C. The number of graphene layers

per stack also increases from 2–3 to 4–6 (McDonald-Wharry et al, 2016).

A physical model can be drawn that shows the gradual evolution of biochar nano-structure as the carbonization temperature increases (Figure 6.6). Biochars derived from lignocellulosic materials are often called "non-graphitizing" because the randomly oriented stacks held apart by strong crosslinks resist long-range reorganization into true crystalline graphite, even at pyrolysis temperatures as high as 3000°C. The non-graphitability of biomass-derived biochars is due to the formation of non-hexagonal rings, induced curvatures, arbitrary crosslinking, and random

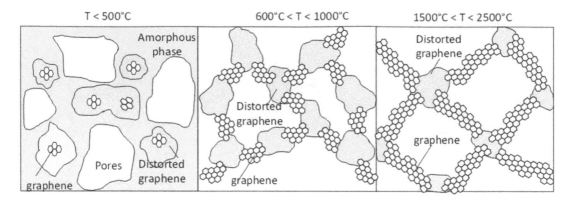

Figure 6.6 *Evolution of biochar nano-structures as a function or pyrolysis temperature evolution. Adapted from (McDonald-Wharry, 2016; with permission)*

graphene stack orientations that occur in the solid phase or in the metaplast early in the carbonization process (McDonald-Wharry et al, 2016).

Immediately after the carbonization process, some biochars have a "wood smoke-like odor", a "wet and sticky" appearance, and a leachable oily phase. The presence of leachable oils on the biochar surface is often due to contact of biochar and vapors during cooling with condensing vapors. Oxidation of biochar surfaces (with O_2, O_3, and H_2O_2) to increase surface oxygenated functional groups could also result in the formation of leachable compounds (Smith et al, 2015). Oxidation breaks some of the weak bridges linking small polyaromatic rings with the biochar matrix and the small clusters produced can be soluble in organic solvents or water (Smith et al, 2015).

Effect of minerals

Agriculture and forest feedstocks typically have clays, organic matter, salts, and chemical fertilizers. For example, K minerals in rice husk are in the form of archerite (KH_2PO_4), chlorocalcite ($KCaCl_3$), kalicinite ($KHCO_3$), pyrocopoite ($K_2MgP_2O_7$),

struvite-K ($KMgPO_4 \cdot 6\ H2O$), and sylvite (KCl) (Xu et al, 2017). Al is present in the form of kaolinite ($Al_2Si_2O_5(OH)_4$) and bentonite ($(Na,Ca)_{0.3}(Al,Mg)_2Si_4O_{10}(OH)_2 \cdot n(H_2O)$) (Rawal et al, 2016). During pyrolysis, the alkali and alkaline earth cations form coordination bonds with oxygenated biomass groups (Kuzhiyil et al, 2012), modifying the thermal degradation mechanism of cellulose, hemicellulose, and lignin. Alkaline and alkaline earth cations thus act as crosslinking agents, increasing biochar yields. Their removal leads to biomass melting and the destruction of biochar micro-structures (Nan et al, 2019). Minerals also influence the unique arrangement of aromatic structures during carbonization (Sun et al, 2013; Xu et al, 2017; Nan et al, 2019). Raman spectra studies on the effect of minerals suggest that inherent minerals tend to drive the formation of disordered structures in biochar (Nan et al, 2019), resulting in biochar prone to oxidation (Chapter 11). Si and P play an essential role in the arrangement of the C structure. Si blends with C and forms Si-encapsulated C with a dense network, positively impacting biochar persistence (Guo et al, 2014). The formation of C-O-P, or C-P could be responsible for increasing C retention when P is present (Li et al, 2014).

The relative content of ash in biochar increases with the carbonization temperature (Suliman et al, 2016). Most of the minerals form crystalline deposits on biochar surfaces. The most common minerals found in biochar are sylvite (KCl), quartz (SiO_2), amorphous silica, calcite ($CaCO_3$), hydroxyapatite ($Ca_5(PO_4)_3OH$), $Ca_2P_2O_7$, Mg_2PO_4OH, magnesia calcite ($MgCO_3$), and dolomite ($CaMg(CO_3)_2$) (Xu et al, 2017). Other salts, in smaller quantities, include phosphates, sulfates, nitrates, and oxides of Ca, Mg, Al, Ti, Mn, Zn, and Fe. When the carbonization temperature increases from 450 to 550°C, iron oxide (Fe_2O_3) changes its structure from bulk micro-sized ferromagnetic crystals to a nanoscale supramagnetic state (Rawal et al, 2016). The pH of most biochars is typically higher than 7 due in part to the influence of alkali- and alkaline earth-bearing minerals.

Effect of acids

Strong acids (H_2SO_4, H_3PO_4) react with alkalines to form stable salts deactivating acids' catalytic properties. Strong acids above the quantity needed to neutralize the salts (Kuzhiyil et al, 2012) contribute to the acidification of the liquid intermediate, catalyzing dehydration reactions and the formation of higher yields of biochar. Also, when acids are added in excess, biomass melts, losing its microscopic structure. Dehydration corresponds to the E-1 elimination (unimolecular elimination reaction rate determined by the dissociation of the leaving group from the carbocation) and proceeds via acid catalysis (Scheirs et al, 2001; Mamleev et al, 2009).

N-containing compounds

N in biochar can be in five forms: aminated functionalities, pyridinic, pyrrolic, graphitic, and oxynitrides (Inagaki et al, 2018; Wan et al, 2020). N-doped biochar can be prepared by pyrolyzing either N-rich biomass (in-situ process) or by supplying N-rich materials as an external source (ex-situ process). In situ, functionalization approaches (self-doping) involve N-doped biochars from N-rich biomass sources (Matsagar et al, 2021). Carbohydrate-containing N functional groups such as glucosamine or chitosan are used as C and N-precursors (Zhao et al, 2010; Titirici et al, 2012). The in-situ approach is a straightforward process and has great potential to get high N content and a stable product. Ex-situ processes involve the carbonization of C and N sources. Different inorganic chemicals (e.g., ammonia gas, ammonium salts, nitric acids, etc.) and organic chemicals (urea, melamine, etc.) have been used (Liu et al, 2011). N-containing gases decompose at high temperatures, generating highly reactive radicals which react with the carbonaceous materials to form N-containing functionalities (Jeon et al, 2020). The introduction of exogenous non-C atoms (e.g., N, S, and B) into ordered sp^2-hybridized graphene changes the electrochemical properties of the π-electron network, creating imbalanced electroactive states (Wan et al, 2020). P-type doping forms when pyridinic N and pyrrolic N donate one and two π-electrons to the adjacent sp^2-hybridized C-conjugated system (Wan et al, 2020). The positive holes on N-doped biochars strongly attract electron-rich rings (Wan et al, 2020). Graphitic N is the most conductive and electronegative N-dopant (Wan et al, 2020). The presence of N could lead to a denser electron cloud, improving charge transfer rates for capacitors (Wan et al, 2020).

Effect of co-composting

Biochar positively changes the dynamics of composting organic wastes and the content of water-soluble nutrients during composting (Chapter 26; Czekała et al, 2016; Awasthi et al, 2017). When biochar is co-composted with organic wastes, minerals and organic agglomerates are formed on the biochar surface, which modifies biochar physico-chemical properties (surface oxidation, fragmentation to smaller particles, redox Fe and Mn nanoparticles on the surface of biochars, adsorption of organic and inorganic compounds). These agglomerates contain N and C functional groups and mixed-valence iron oxide nanoparticles. Pores are formed at the interface between the biochar matrix and the organo-mineral aggregates (Archanjo et al, 2017). These phenomena could explain, at least in part, the positive effects of biochar on biomass co-composting (Chapter 26).

Biochar characterization methods

Characterization of biochar at the micro- and macro-level requires different techniques that intend to visualize the morphology of the material, the bulk elemental composition, the elemental composition at the surface, the presence of functional groups, structure, and mechanical properties, among others (Table 6.1).

Table 6.1 *Methods for biochar characterization*

Characterization method	Property and characteristics analyzed	References
Techniques for morphology analysis		
Scanning electron microscopy (SEM)	Topography and biochar micro-structure.	(Omidi et al, 2017)
Transmission electron microscopy (TEM)	Topography of biochar nano-structure. Biochars produced at less than 1000°C seem to result in similar TEM pictures. Generally interpreted as 1 nm randomly oriented regions of 2–4 layer thick graphene stack along with curved and faceted graphene layers and onion-like multi-layered structure.	(Rhim et al, 2010; Yang et al, 2019)
Atomic force microscopy (AFM)	Study of biochar interaction with polymers in composite materials.	(Heath et al, 2019)
Techniques for analysis of molecular structure and functional groups		
Electron paramagnetic resonance (EPR) or electron spin resonance (ESR)	This technique examines the content of free radicals in biochar. Persistent free radicals (PFRs) typically have 10^{18} unpair spins/grams.	(Fang et al, 2014, Fang et al, 2015)
^{13}C Nuclear magnetic resonance spectroscopy (NMR)	Can be used to measure molecular structure, functional groups, and polyaromatic ring system size. Biochar formed at high temperatures is mainly formed by sp2-bonded	(Wang et al, 2013; McBeath et al, 2014; Smith et al, 2017)

Table 6.1 *continued*

Characterization method	Property and characteristics analyzed	References
	aromatic structures. Chars produced below 350°C contain more than 70% of the C as aromatic O-aryl groups. Chars over 600°C almost all in the form of aromatic C. ^{13}C NMR with long-range dipolar dephasing were used to determine cluster size.	
Raman spectroscopy	Chemical functionalities and graphitization index. Spectra display two broad bands: 1350 cm^{-1} (D band, associated with disorder (defects, distortions, edges of graphite crystals, around sp^2-C) and 1580 cm^{-1} (G band, sp2-bonded C with structures like graphite)	(Smith et al, 2016; Ayiania et al, 2020)
Neutron Scattering	Consistent with the structure of chars composed of 3-fold coordinated sp2-bonded carbon with a small portion of 5–7 rings inducing curvature. In chars prepared at 650, 800, 1000°C major peaks are interpreted as being consistent with ordered graphene sheets with lateral extends of 10–12 Å for biochars produced at 1200°C the lateral extends around 20 Å.	(Acharya et al, 1999, Petkov et al, 1999)
Vibrating sample magnetometer	Magnetic properties at room temperature (magnetic hysteresis loops).	(Yang et al, 2018)
X-ray diffraction (XRD) and scattering	Crystallographic structure of biochar. 400°C<T<700°C broad 002 reflections attributed to the graphene/polyaromatic interlayer spacing (d from 0.39 to 0.37 nm). Biochar produced at 1000°C aside from the 002 reflection (around $2\theta = 25°$ using Cu K α radiation), 100 and 110 broad reflections: appear around $2\theta = 44°$ and 78° related to the two-dimensional in-plane structure of graphene layers. The literature sometimes reports crystallite size (La and Lc), although the stacks lack three-dimensional crystalline order.	(Keiluweit et al, 2010; McDonald-Wharry et al, 2016)
Fourier transform infrared (FTIR) (normally using attenuated total reflection – ATR)	A quick and economical technique used to obtain an infrared spectrum of biochar, based on the transition of its functional groups and aromatization.	(Nair et al, 2022)

Table 6.1 *continued*

Characterization method	Property and characteristics analyzed	References
Gas physisorption analysis	Used to estimate surface area, pore structure, average pore diameter, and pore volume.	(Suliman, 2016)
Interface and electrochemical properties		
Zeta-potential	Indicates the electrical charge of a particle suspended in a liquid.	(Suliman, 2016)
X-Ray photoelectron spectroscopy (XPS)	Chemical and electronic states on a surface-surface composition of biochar, the valence state of chemical elements.	(Smith et al, 2016; Ayiania et al, 2020; Liu et al, 2020)
Solid-state cyclic voltammetry (SSCV)	Cyclic voltammograms are used to measure the redox characteristics of systems (including biochar-based cathods-anods). Voltage sweeps up and down and current is measured.	(Joseph et al, 2015)
Electron energy loss spectrometer (EELS), selective area electron diffraction (SAED)	Material exposed to a bean of electrons of known kinetic energy. Some electrons undergo inelastic scattering (losing energy), which is interpreted based on the cause of this loss. SAED (or SAD) is used with TEM. The technique (using electronic diffraction) is used to measure crystal orientation, lattice constants and examine defects.	(Archanjo et al, 2017; Joseph et al, 2018)
Techniques for elemental composition		
Elemental analysis	Biochr's C, H, N, S, and O content. Rapid decrease in O content of biochar produced between 200 and 500°C. In biochar produced at or above 700°C the C content is higher than 80% w/w. At 1000°C, C content is higher than 90% w/w.	(Ronsse et al, 2013)
Inductively coupled plasma (ICP) spectrometry	Metal content in biochar's ash.	(Onorevoli et al, 2018)
Energy dispersive X-Ray spectroscopy (EDX)	Elemental composition on the surface of biochar (usually coupled with SEM).	(Chia et al, 2012; Archanjo et al, 2017)
Physical and mechanical properties		
Conductivity, dynamic elastic modulus, modulus, mechanical strength, hardness	Massive increase (5 orders of magnitude) in electrical conductivity and a moderate increase in light reflectance for biochars produced between 400°C and 1000°C. The dynamic elastic modulus of cherry wood decreases by an order of magnitude to a	(Antal et al, 2003; Rhim et al, 2010; McDonald-Wharry et al, 2016)

Table 6.1 *continued*

Characterization method	Property and characteristics analyzed	References
	minimum at 330°C, it gains all its lost modulus at around 900°C. Loose in modulus and compressive strength when heated till 400°C, significant increase happening at 800°C. Modulus and hardness decrease at temperatures over 1000°C due to the partial removal of strong crosslinks and the growth of soft graphite-like components.	
Analysis of decomposition products		
Laser desorption mass spectra, temperature-programmed oxidation	Electrons impact biochars produced at 440°C and generate predominantly ions attributed to CO, CO_2, and benzene. Temperature-programmed oxidation: Uses H/C ratios as a measure of the average size of the graphene-like layers. Average length estimates from H/C ratios increase from 0.7 nm to 5.7 nm in biochars produced at 600°C and 1000°C, respectively.	(Braadbaart et al, 2008; Schneider et al, 2010)
Pyrolysis-gas chromatography with mass spectrometry (Py-GC/MS)	Composition of the volatile fraction. Biomass stability scores.	(Ghysels et al, 2022)

Micro- and nano-molecular features of biochar

Biochars can contain many different oxygenated and nitrogenated functional groups on graphene sheets (Figure 6.7). Hydrophobic graphene-rich surfaces are considered suitable sites for the adsorption of non-polar moieties (Lawrinenko et al, 2015). Functional groups with heteroatom are generally polar and play an essential role in controlling the interactions between molecules interactions through hydrogen bonding, ion-dipole, and dipole-dipole (Boehm, 1994; Gai et al, 2014). α or π electrons (OH, NH_2, OR or C(C=O)R) are often classified as electron donors. Groups with empty orbitals ((C=O)OH, (C=O)H, or NO_2) are electron acceptors.

Hydroxyl and carboxyl acidic functional groups are often negatively charged (Lawrinenko et al, 2015) and contribute to biochar cation exchange capacity (CEC) (Cheng et al, 2008; Beesley et al, 2011; Mukherjee, Zimmerman et al, 2011; Guo et al, 2014; Takaya et al, 2016; Banik et al, 2018; Fidel et al, 2018; Xu et al, 2019). Carboxyl groups are strong Bronsted acids and phenols, while carbonyl groups are weak acids. Oxygenated functional groups could come from the lignocellulosic feedstock or can be introduced through post-carbonization biochar oxidation (Cheng et al, 2008; Smith et al, 2015). Together with acidic and base

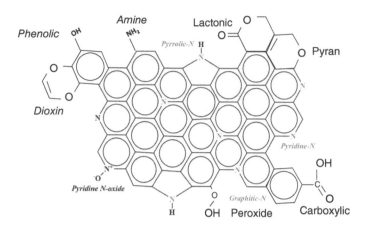

Figure 6.7 *Common N and O functional groups on biochar surfaces*

organic groups, minerals control overall biochar pH. In biochar particles, acid and basic sites are within microns of each other (Amonette and Joseph, 2009).

Basic groups such as amides, aromatic amines, and pyridines are positively charged (Shakoor et al, 2021). At pH below pK$_a$, pyridinic functional groups are positively charged (Lawrinenko et al, 2015). The presence of oxonium heterocycles, N heterocycles (pyridinium groups), and condensed aromatic carbon could increase biochar anion exchange capacity (AEC) (Lawrinenko et al, 2017). Changes in biochar surface properties when added to soils are discussed in Chapter 10.

Biochar surface interactions with other molecules

Pesticides and other persistent organic pollutants such as pharmaceuticals and residues of personal care products can adversely affect human health. Biochar has proven to be an efficient material for removing and stabilizing some of these pollutants (Cui et al, 2011; Bian et al, 2013; Inyang et al, 2015). Biochar performance in specific applications is mediated by its sorption properties, pH, capacity to induce precipitation, catalytic properties, ability to generate free radicals or transport electrons, and interaction with microbes (Worch, 2012).

While physical adsorption occurs through van der Waals interactions (dipole-dipole interactions, dispersion forces, and induction forces), chemical adsorption involves bonds between adsorbate and adsorbent (Inyang et al, 2015). The following phenomena are responsible for biochar interactions with other molecules: (1) diffusion and partitioning, (2) hydrophobic interactions, (3) aromatic π-π interactions, (4) hydrogen bonding, (5) electrostatic interactions, (6) pore filling, (7) cation exchange, (8) induced precipitation, (9) formation of surface complexes, (10) interaction with amine groups, (11) simultaneous adsorption and catalytic degradation, and (12) microorganism mediated (Table 6.2; Inyang et al, 2015; Tan et al, 2016; Krasucka et al, 2021; Haghighi Mood et al, 2022).

Table 6.2 *Mechanisms associated with pollutant removal using biochar*

Mechanism	Schematic	Pollutants removed (references)
Diffusion and partitioning: Contaminant removal is controlled by the solubilization of pollutants into the "viscous liquid intermediates". Linear adsorption isotherms.	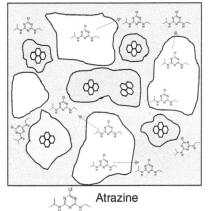 Atrazine	e.g., atrazine (Sun et al, 2011)
Hydrophobic interactions: Entropic effect originated from the disruption of water molecules. Sorbed apolar molecules and water compete for biochar's low polarity surface.	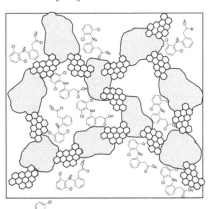 Dichlofenac	e.g., dichlofenac, trichloroethylene (Murphy et al, 1994)
Non-covalent p - electron donor-acceptor (EDA): Aromatic p-systems serve as electron acceptors. Poly-condensed aromatic rings serve as p-donors binding electron-withdrawing molecules. Minerals such as Fe, Mg, Si, K, and Ca may participate in cation p-interaction with biochar PAHs	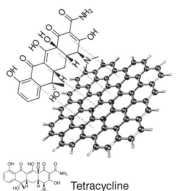 Tetracycline	e.g., tetracycline (Hunter et al, 1990; Hunter et al, 2001; Keiluweit et al, 2010)

Table 6.2 *continued*

Mechanism	Schematic	Pollutants removed (references)
Electrostatic interactions: Organic cations will sorb on negatively charged surfaces. Anionic sorbates will bind to positively charged sites (e.g., mineral-rich biochars or biochars containing N). pH controls biochar surface charge.	 Rhodamine	e.g., rhodamine (Teixidó et al, 2011)
Pore filling: Controlled by physical adsorption. Micropores (<2 nm) and small mesopores (2–20 nm) contribute to most of the surface area. Non-linear sorption kinetic with Langmuir-like sorption isotherm.	 Paracetamol	e.g., CH_4, N_2, paracetamol (Ahmad et al, 2014; Tran et al, 2020)
Ion exchange: Exchange of biochar-free mobile ions by ions in solution with similar charges.	 ● Metal Ions ◐ Exchangable Ions	e.g., heavy metals (Tan et al, 2015)

Table 6.2 *continued*

Mechanism	Schematic	Pollutants removed (references)
Surface complexation: Oxygen-containing functional groups participate in esterification reactions that are solution pH dependent.	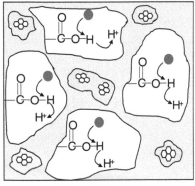 ● Metal Ions	e.g., heavy metals (Wang et al, 2020)
Induced precipitation: Formation of insoluble compounds by reaction with counter ions on the surface. In heavy metals precipitation with ions released from minerals (OH^-, CO_3^{2-}, PO_4^{3-}, SO_4^{2-}, SiO_4^{4-}). Precipitation of heavy metals is controlled by the solubilization of counter ions. Low-temperature biochars release ions more easily than those produced at high temperatures. Ash in biochar produced at low temperatures is more amorphous than the most stable crystals obtained at high temperatures.	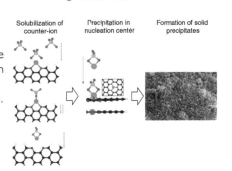	e.g., Cu, Zn, Cd, and PO_4^{3-} can also be removed by induced precipitation. Some of the precipitates formed include: $PbCO_3$, $Pb_3(CO_3)_2(OH)_2$, $Pb_5(PO_4)_3(OH, Cl)$ (Xu et al, 2017; Ayiania et al, 2019; Mood et al, 2020, Haghighi Mood et al, 2021)
Complex mechanisms involving chemical reactions: H_2S is solubilized in a basic water layer followed by oxidation of the HS- ions with O_2 to form elemental S.	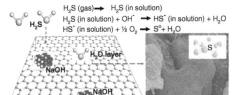	e.g., H_2S (Liu et al, 2013; Pelaez-Samaniego et al, 2020)

The removal of ash in biochar to increase pore transport and accessibility is a common strategy to improve phenanthrene, carbaryl, and atrazine biochar adsorption capacity (Zhang et al, 2013, Tan et al, 2015).

Together with biochar properties, adsorption rate and equilibrium also depend on the physicochemical properties of the adsorbent (e.g., molecular size, solubility, pK_a) and the environment where the adsorption and chemical reaction occur (solution chemistry, pH, concentration of pollutants, temperature) (Inyang et al, 2011). In general, negatively charged pristine biochar surfaces at neutral pH do not attract negative ions (e.g., nitrate and phosphate). The pH affects the protonation and deprotonation of surface functional groups, causing surface charge changes (Liu et al, 2020). The protonation of biochar surface functional groups depends on the dissociation constant of each of the groups, which is affected by the temperature and the nature of the surrounding chemical environment (pH) (Chen et al, 2015). The pH at which the number of positive and negative surface charges equals zero is called zero-point charge (pHzpc) (Li et al, 2013; Li et al, 2016).

Redox reactions

Redox reactions play an important role in element cycles in soil (Joseph et al, 2015). Carbon-based materials act as redox catalysts via three mechanisms: (1) directly interacting with the pollutant via inherent reactive-active moieties (RAMs); (2) producing highly reactive oxygen species (ROS) to degrade the organic contaminant into harmless substances (Wan et al, 2020); and (3) acting as electron shuttles.

Direct interaction with pollutants via inherent reactive-active moieties (RAMs)

RAMs are typically found on two types of molecules: (i) electron-rich functional groups; and (ii) inherent environmentally persistent free radicals (EPFRs). Amino ($-NH_x$) and phenolic ($-OH$) groups with abundant confined electrons can contribute to the degradation of pollutants. Meanwhile, ketonic ($C=O$) groups have high electron density and nucleophilic nature (Wan et al, 2020). Due to the removal of oxygenated functional groups, biochars produced at temperatures above 700°C tend to have low electron paramagnetic resonance (EPR) signals. Oxygen moieties are very important for the free movement of electrons responsible for biochar paramagnetic properties (Wan et al, 2020). RAMs are typically present along structural defects on carbon matrices. The synergism between transition metals such as Ni, Cu, Zn, and Fe (traditional catalysts) and phenols could result in enhancements of EPFRs and, consequently, RAMs on biochar (Fang et al, 2015). The values of EPFR density for common biochars are, for example: (i) pine needles: 13.7×10^{18} spins per g of biochar produced at 550°C; (ii) wheat straw: 28.6×10^{18} spins per g for biochar produced at 500°C; and (iii) maize straw: 30.2×10^{18} spins per g for biochar produced at 500°C, and rice husk: 17.1×10^{18} spins per g for biochar produced at 500°C (Wan et al, 2020).

Production of highly reactive oxygen species (ROS)

Wet oxidation processes happen due to the presence of oxidants such as oxygen, ozone, and hydrogen peroxide, which are converted into ROS (e.g., •OH). Typically, the oxidants are solubilized into an aqueous environment, followed by the formation of ROS, which reacts with dissolved organics. Some electron donor functional groups such as phenolics -OH can facilitate the conversion of H_2O_2 into ROS (Wan et al, 2020). The critical role of quinone/hydroquinone pairs on the redox capacity of biochar electrodes was studied elsewhere (Klüpfel et al, 2014). N-doped biochars (especially those with high content of pyridinic groups) have high oxidative activities (Tews et al, 2021). Pyridinic N possesses unpaired electrons that act as confined radicals to directly capture peroxide molecules. The integrated adsorption-activation process lowers the energy barrier required to dissociate the O-O peroxide bond (Wan et al, 2020).

Electron shuttles

While the electron-donating capacity of biochars has been associated with its content of phenolic groups, electron-accepting capacity is controlled by quinone-hydroquinone functional groups and the presence of π-electron systems (Klüpfel et al, 2014). Biochar could play an important role as an electron shuttle in direct interspecies electron transfer (DIET) abiotic redox processes (Yuan et al, 2018). Oxidized biochars (rich in quinones) often function as electron acceptors, which is important for organic matter oxidative degradation. Reduced biochar (rich in phenols) could serve as electron donors for nitrate and magnetite reduction (Yuan et al, 2018). Biochars produced between 400 and 500°C have the highest content of quinone structures (Klüpfel et al, 2014; Sun et al, 2018).

According to Sun et al (2018), redox cycling of functional groups contributes between 78 and 100% of total electron transfer. Condensed aromatic structures are electroactive, thus allowing electrons to be transferred across the conjugated π-electron systems. In biochars produced at temperatures from 650 to 800°C, rich in nanosized graphene, direct electron transfer of carbon matrices accounts for between 87 and 100% of electron transfer (Sun et al, 2018).

Biochar has the potential to influence methanogenesis. It has been found that, compared with non-biochar control, biochar produced from rice straw and manure accelerated methanogenesis 10.7–12.3 fold (Yuan et al, 2018). Electronic syntrophy exists between methanogens and Geobacteraceae. Biochars favor methanogenesis by facilitating direct interspecies electron transfer between methanogens and Geobacteraceae. Biochar concentrations of 5–10 g L^{-1} stimulated both the rate and the extent of microbial reduction of the Fe (III) oxyhydroxide mineral ferrihydrite (15 mM) by *Shewanella oneidensis* (Kappler et al, 2014). Biochar acted as an electron shuttle. The biochar facilitated the biotic reduction of ferrihydrite from magnetite (Fe_3O_4) to siderite ($FeCO_3$). Electron snorkeling of biochar and pyrogenic carbon and their effect on methanogenesis in artificial microcosm and bioelectrical peat incubations has been tested (Sun et al, 2021). The presence of pyrogenic C produced from forest fires was able to reduce the post-fire methane production from incubated peatland soils by 13–24% (Sun et al, 2021). Biochar acted as an electron snorkel (electron shuttle) facilitating extracellular electron transfer and soil microbial respiration responsible for methane production suppression.

Outlook

The use of biochar for environmental services is a relatively new field of study. Here we suggest a physical model of a biochar-soil aggregate including the phenomena discussed in this chapter that may be relevant for specific cases (Figure 6.8). Not all the phenomena shown in Figure 6.8 are important for a targeted application (e.g., adsorption and oxidation of paracetamol). Some of these phenomena may be present but may not control the process, others will occur and will compete with the phenomena of interest. To develop rational strategies to engineer biochars for targeted services, it is important to identify the controlling mechanisms (phenomena) (e.g., mass transfer rate, surface adsorption rate, chemical reaction rate, transfer of electrons) and how biochar physicochemical properties at the micro- or nano-scales are directly associated with the phenomena of interest. Once this information is collected, researchers will need to work on the design and production of a biochar with the desired micro- and nano-properties and on the identification of the best conditions at which the biochar should be used to achieve desired outcomes.

Although there are many studies describing biochars' removal mechanisms for specific environmental applications (Chapter 27), more quantitative studies contrasting results of mathematical models and experimental studies are expected to be published in

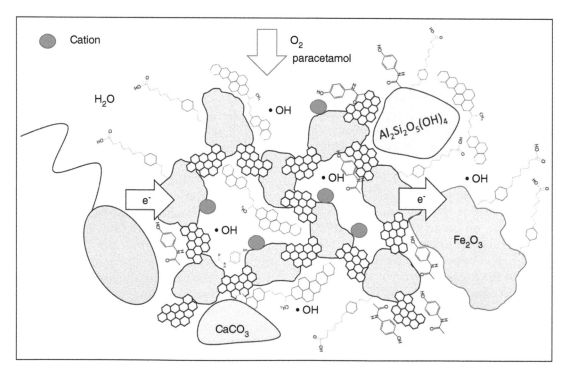

Figure 6.8 *Schematic depicting the complex micro- and nano-structure of biochar-soil aggregates and phenomena associated with biochar environmental services*

the coming years. These new mathematical models need to link adsorption rate and equilibrium equations with biochar nano- and micro-structural features. In the meantime, more studies are needed to describe and model the interactions between biochars and microbes. Chemical and enzymatic reactions mediated by the transfer of electrons, by the formation of reactive oxygen species, or by the interaction of pollutants with inherent reactive-active moieties need to be better studied. A major challenge facing this field is the need to develop realistic models representing the gradual evolution of biochar nano- and micro-structure as a function of feedstock composition and biochar processing conditions. Experimental data to build required biochar physical and mathematical models

are necessary. More studies are also needed to understand the interactions between biomass, clay, and mineral species and their effect on biochar micro- and nano-structure (Rawal et al, 2016). Although advancing our understanding and controlling the phenomena associated with biochar uses in environmental services is important, equally important is to develop products optimized for other targeted markets such as agriculture and construction. Additionally, case-specific biochar market analyses are needed to identify the required properties of engineered biochar products. Finally, more research and development is needed to advance from the formulation of new products at laboratory and bench scale to pilot demonstration, and commercialization projects.

References

Acharya M, et al 1999 Simulation of nanoporous carbons: A chemically constrained structure. *Philosophical Magazine B* 79, 1499–1518.

Ahmad M, et al 2014 Biochar as a sorbent for contaminant management in soil and water: A review. *Chemosphere* 99, 19–33.

Amonette JE, and Joseph S 2009 Characteristics of biochar: microchemical properties. In: Lehmann J, and Joseph S (Eds) *Biochar for Environmental Management: Science and Technology.* London: Earthscan. pp33–52.

Antal MJ, and Grønli M 2003 The art, science, and technology of charcoal production. *Industrial and Engineering Chemistry Research* 42, 1619–1640.

Archanjo BS, et al 2017 Nanoscale analyses of the surface structure and composition of biochars extracted from field trials or after co-composting using advanced analytical electron microscopy. *Geoderma* 294, 70–79.

Awasthi MK, et al 2017 Evaluation of biochar amended biosolids co-composting to improve the nutrient transformation and its correlation as a function for the production of nutrient-

rich compost. *Bioresource Technology* 237, 156–166.

Ayiania M, et al 2019 Production and characterization of H_2S and PO_4^{3-} carbonaceous adsorbents from anaerobic digested fibers. *Biomass and Bioenergy* 120, 339–349.

Ayiania M, et al 2020 Deconvoluting the XPS spectra for nitrogen-doped chars: An analysis from first principles. *Carbon* 162, 528–544.

Banik C, Lawrinenko M, Bakshi S, and Laird DA 2018 Impact of pyrolysis temperature and feedstock on surface charge and functional group chemistry of biochars. *Journal of Environmental Quality* 47, 452–461.

Beesley L, and Marmiroli M 2011 The immobilisation and retention of soluble arsenic, cadmium and zinc by biochar. *Environmental Pollution* 159, 474–480.

Bian R, et al 2013 Biochar soil amendment as a solution to prevent Cd-tainted rice from China: Results from a cross-site field experiment. *Ecological Engineering* 58, 378–383.

Boehm HP 1994 Some aspects of the surface chemistry of carbon blacks and other carbons. *Carbon* 32(5), 759–769.

Braadbaart F, and Poole I 2008 Morphological, chemical and physical changes during charcoalification of wood and its relevance to archaeological contexts. *Journal of Archaeological Science* 35, 2434–2445.

Cao J, Yu Y, and Wu H 2022 Contributions of thermal ejection and evaporation to the formation of condensable volatiles during cellulose pyrolysis. *Energy and Fuels* 36, 1939–1947.

Caposciutti et al 2019 Experimental investigation on biomass shrinking and swelling behaviour: Particles pyrolysis and wood logs combustion. *Biomass and Bioenergy* 123, 1–13.

Chen WH, and Kuo PC 2011 Torrefaction and co-torrefaction characterization of hemicellulose, cellulose and lignin as well as torrefaction of some basic constituents in biomass. *Energy* 36(2), 803–811.

Chen Z, Xiao X, Chen B, and Zhu L 2015 Quantification of chemical states, dissociation constants and contents of oxygen-containing groups on the surface of biochars produced at different temperatures. *Environmental Science and Technology* 49, 309–317.

Cheng C-H, Lehmann J, and Engelhard MH 2008 Natural oxidation of black carbon in soils: Changes in molecular form and surface charge along a climosequence. *Geochimica et Cosmochimica Acta* 72, 1598–1610.

Chia CH, et al 2012 Imaging of mineral-enriched biochar by FTIR, Raman and SEM–EDX. *Vibrational Spectroscopy* 62, 248–257.

Cui L, et al 2011 Biochar amendment greatly reduces size Cd uptake in a contaminated paddy soil: a two-year field experiment. *BioResources* 6, 2605–2618.

Czekała W, et al 2016 Co-composting of poultry manure mixtures amended with biochar – The effect of biochar on temperature and C-CO2 emission. *Bioresource Technology* 200, 921–927.

Duan W, Oleszczuk P, Pan B, and Xing B 2019 Environmental behavior of engineered biochars and their aging processes in soil. *Biochar* 1, 339–351.

Dufour A, et al 2012 In situ analysis of biomass pyrolysis by high temperature rheology in relations with 1H NMR. *Energy and Fuels* 26, 6432–6441.

Dufour A, Ouartassi B, Bounaceur R, and Zoulalian A 2011 Modelling intra-particle phenomena of biomass pyrolysis. *Chemical Engineering Research and Design* 89, 2136–2146.

Fang G, et al 2014 Key role of persistent free radicals in hydrogen peroxide activation by biochar: implications to organic contaminant degradation. *Environmental Science and Technology* 48, 1902–1910.

Fang G, Liu C, Gao J, Dionysiou DD, and Zhou D 2015 Manipulation of persistent free radicals in biochar to activate persulfate for contaminant degradation. *Environmental Science and Technology* 49, 5645–5653.

Fidel RB, Laird DA, and Spokas, KA 2018 Sorption of ammonium and nitrate to biochars is electrostatic and pH-dependent. *Scientific Reports* 8, 17627.

Fonts I, et al 2021 Thermodynamic and physical property estimation of compounds derived from the fast pyrolysis of lignocellulosic materials. *Energy and Fuels* 35, 17114–17137.

Gai X, et al 2014 Effects of feedstock and pyrolysis temperature on biochar adsorption of ammonium and nitrate. *PLoS ONE* 9, e113888.

Ghysels S, et al 2022 Biochar stability scores from analytical pyrolysis (Py-GC-MS). *Journal of Analytical and Applied Pyrolysis* 161, 105412.

Guo J, and Chen B 2014 Insights on the molecular mechanism for the recalcitrance of biochars: interactive effects of carbon and silicon components. *Environmental Science and Technology* 48, 9103–9112.

Haghighi Mood S, Pelaez-Samaniego MR, and Garcia-Perez M 2022 Perspectives of engineered biochar for environmental applications: a review. *Energy and Fuels* 36, 7940–7986.

Haghighi Mood, S, et al 2021 Nitrogen and magnesium Co-doped biochar for phosphate adsorption. *Biomass Conversion and Biorefinery.* 10.1007/s13399-021-01404-1.

Heath GR, and Scheuring S 2019 Advances in high-speed atomic force microscopy (HS-AFM) reveal dynamics of transmembrane channels and transporters. *Current Opinion in Structural Biology* 57, 93–102.

Hunter CA, Lawson KR, Perkins J, and Urch CJ 2001 Aromatic interactions. *Journal of the Chemical Society, Perkin Transactions 2,* 651–669.

Hunter CA, and Sanders JKM 1990 The nature of.pi.-.pi. interactions. *Journal of the American Chemical Society* 112, 5525–5534.

Inagaki M, Toyoda M, Soneda Y, and Morishita T 2018 Nitrogen-doped carbon materials. *Carbon* 132, 104–140.

Inyang M, and Dickenson E 2015 The potential role of biochar in the removal of organic and microbial contaminants from potable and reuse water: a review. *Chemosphere* 134, 232–240.

Inyang M, et al 2011 Enhanced lead sorption by biochar derived from anaerobically digested sugarcane bagasse. *Separation Science and Technology* 46, 1950–1956.

Jeon IY, Noh HJ, and Baek JB 2020 Nitrogen-doped carbon nanomaterials: synthesis, characteristics and applications. *Chemistry – An Asian Journal* 15, 2282–2293.

Joseph S, et al 2015 The electrochemical properties of biochars and how they affect soil redox properties and processes. *Agronomy* 5, 322–340.

Joseph S, et al 2018 Microstructural and associated chemical changes during the composting of a high temperature biochar: Mechanisms for nitrate, phosphate and other nutrient retention and release. *Science of the Total Environment* 618, 1210–1223.

Kappler A, et al 2014 Biochar as an electron shuttle between bacteria and Fe(III) minerals. *Environmental Science and Technology Letters* 1, 339–344.

Kastner A, et al 2012 Catalytic esterification of fatty acids using solid acid catalysts generated from biochar and activated carbon. *Catalysis Today* 190, 122–132.

Keiluweit M, Nico PS, Johnson MG, and Kleber M 2010 Dynamic molecular structure of plant biomass-derived black carbon (biochar). *Environmental Science and Technology* 44, 1247–1253.

Kercher AK, and Nagle DC 2003 Microstructural evolution during charcoal carbonization by X-ray diffraction analysis. *Carbon* 41, 15–27.

Kersten S, and Garcia-Perez M 2013 Recent developments in fast pyrolysis of ligno-cellulosic materials. *Current Opinion in Biotechnology* 24, 414–420.

Khawkomol S, et al 2021 Potential of biochar derived from agricultural residues for sustainable management. *Sustainability* 13, 8147.

Klüpfel L, Keiluweit M, Kleber M, and Sander M 2014 Redox properties of plant biomass-derived black carbon (biochar). *Environmental Science and Technology* 48, 5601–5611.

Krasucka P, et al 2021 Engineered biochar – A sustainable solution for the removal of antibiotics from water. *Chemical Engineering Journal* 405, 126926.

Kuzhiyil N, Dalluge D, Bai X, Kim KH, and Brown RC 2012 Pyrolytic sugars from cellulosic biomass. *ChemSusChem* 5, 2228–2236.

Lawrinenko M, Jing D, Banik C, and Laird DA 2017 Aluminum and iron biomass pretreatment impacts on biochar anion exchange capacity. *Carbon* 118, 422–430.

Lawrinenko M, and Laird DA 2015 Anion exchange capacity of biochar. *Green Chemistry* 17, 4628–4636.

Lee J, Kim KH, and Kwon EE 2017 Biochar as a catalyst. *Renewable and Sustainable Energy Reviews* 77, 70–79.

Leng L, et al 2020 Nitrogen containing functional groups of biochar: An overview. *Bioresource Technology* 298, 122286.

Li F, Cao X, Zhao L, Wang J, and Ding Z 2014 Effects of mineral additives on biochar formation: carbon retention, stability, and properties. *Environmental Science and Technology* 48, 11211–11217.

Li R, et al 2016 Enhancing phosphate adsorption by Mg/Al layered double hydroxide functionalized biochar with different Mg/Al ratios. *Science of the Total Environment* 559, 121–129.

Li W, et al 2013 Molecular level investigations of phosphate sorption on corundum (α-Al2O3) by 31P solid state NMR, ATR-FTIR and quantum chemical calculation. *Geochimica et Cosmochimica Acta* 107, 252–266.

Liu H, Liu Y, and Zhu D 2011 Chemical doping of graphene. *Journal of Materials Chemistry* 21, 3335–3345.

Liu JY, Chang HY, Truong QD, and Ling YC 2013 Synthesis of nitrogen-doped graphene by pyrolysis of ionic-liquid-functionalized graphene. *Journal of Materials Chemistry C* 1, 1713–1716.

Liu J, et al 2020 Preparation, environmental application and prospect of biochar-supported metal nanoparticles: A review. *Journal of Hazardous Materials* 388, 122026.

Mamleev V, Bourbigot S, Le Bras M, and Yvon J 2009 The facts and hypotheses relating to the phenomenological model of cellulose pyrolysis: Interdependence of the steps. *Journal of Analytical and Applied Pyrolysis* 84, 1–17.

Matsagar BM, Yang RX, Dutta S, Ok YS, and Wu KCW 2021 Recent progress in the development of biomass-derived nitrogen-doped porous carbon. *Journal of Materials Chemistry A* 9, 3703–3728.

McBeath AV, Smernik RJ, Krull ES, and Lehmann J 2014 The influence of feedstock and production temperature on biochar carbon chemistry: A solid-state 13C NMR study. *Biomass and Bioenergy* 60, 121–129.

McDonald-Wharry JS, Manley-Harris M, and Pickering KL 2016 Reviewing, combining, and updating the models for the nanostructure of non-graphitizing carbons produced from oxygen-containing precursors. *Energy and Fuels* 30, 7811–7826.

Montoya J, Pecha B, Janna FC, and Garcia-Perez M 2017 Single particle model for biomass pyrolysis with bubble formation dynamics inside the liquid intermediate and its contribution to aerosol formation by thermal ejection. *Journal of Analytical and Applied Pyrolysis* 124, 204–218.

Montoya J, Pecha B, Janna FC, and García-Pérez M 2017 Identification of the fractions responsible for morphology conservation in lignocellulosic pyrolysis: Visualization studies of sugarcane bagasse and its pseudo-components. *Journal of Analytical and Applied Pyrolysis* 123, 307–318.

Mood, SH, Ayiania, M, Jefferson-Milan, Y, and Garcia-Perez, M 2020 Nitrogen doped char from anaerobically digested fiber for phosphate removal in aqueous solutions. *Chemosphere* 240, 124889.

Mukherjee A, Zimmerman AR, and Harris W 2011 Surface chemistry variations among a series of laboratory-produced biochars. *Geoderma* 163, 247–255.

Murphy EM, Zachara JM, Smith SC, Phillips JL, and Wietsma TW 1994 Interaction of hydrophobic organic compounds with mineral-bound humic substances. *Environmental Science and Technology* 28, 1291–1299.

Nair RR, Mondal MM, and Weichgrebe D 2022 Biochar from co-pyrolysis of urban organic wastes—investigation of carbon sink potential using ATR-FTIR and TGA. *Biomass Conversion and Biorefinery* 12, 4729–4743.

Nan H, et al 2019 Interaction of inherent minerals with carbon during biomass pyrolysis weakens biochar carbon sequestration potential. *ACS Sustainable Chemistry and Engineering* 7, 1591–1599.

Oja V, and Suuberg EM 1998 Vapor pressures and enthalpies of sublimation of polycyclic aromatic hydrocarbons and their derivatives. *Journal of Chemical and Engineering Data* 43, 486–492.

Oja V, and Suuberg EM 1999 Vapor pressures and enthalpies of sublimation of d-Glucose, d-Xylose, cellobiose, and levoglucosan. *Journal of Chemical and Engineering Data* 44, 26–29.

Omidi M, et al 2017 Characterization of biomaterials. In: Tayebi L, and Moharamzadeh K (Eds) *Biomaterials for Oral and Dental Tissue Engineering*. Amsterdam: Elsevier. pp97–115.

Onorevoli B, et al 2018 Characterization of feedstock and biochar from energetic tobacco seed waste pyrolysis and potential application of biochar as an adsorbent. *Journal of Environmental Chemical Engineering* 6, 1279–1287.

Papageorgiou A, Azzi ES, Enell A, and Sundberg C 2021 Biochar produced from wood waste for soil remediation in Sweden: Carbon sequestration and other environmental impacts. *Science of the Total Environment* 776, 145953.

Paris O, Zollfrank C, and Zickler GA 2005 Decomposition and c of wood biopolymers—a microstructural study of softwood pyrolysis. *Carbon* 43, 53–66.

Pecha MB, Montoya JI, Ivory C, Chejne F, and Garcia-Perez M 2017 Modified pyroprobe captive sample reactor: characterization of reactor and cellulose pyrolysis at vacuum and atmospheric pressures. *Industrial and Engineering Chemistry Research* 56, 5185–5200.

Pelaez-Samaniego MR, et al 2022 Biomass carbonization technologies. In: Mohan D, Pittman CU, and Mlsna TE (Eds) *Sustainable Biochar for Water and Wastewater Treatment*. Amsterdam: Elsevier. pp39–92.

Pelaez-Samaniego MR, Perez JF, Ayiania M, and Garcia-Perez T 2020 Chars from wood gasification for removing H2S from biogas. *Biomass and Bioenergy* 142, 105754.

Pelaez-Samaniego MR, Yadama V, Garcia-Perez M, Lowell E, and McDonald AG 2014 Effect of temperature during wood torrefaction on the formation of lignin liquid intermediates. *Journal of Analytical and Applied Pyrolysis* 109, 222–233.

Petkov V, et al 1999 High real-space resolution measurement of the local structure of Ga12xInxAs using X-ray diffraction. *Physical Review Letters* 83, 4089–4092.

Rajapaksha AU, et al 2016 Engineered/designer biochar for contaminant removal/immobilization from soil and water: potential and implication of biochar modification. *Chemosphere* 148, 276–291.

Rawal A, et al 2016 Mineral–biochar composites: molecular structure and porosity. *Environmental Science and Technology* 50, 7706–7714.

Rhim YR, et al 2010 Changes in electrical and microstructural properties of microcrystalline cellulose as function of carbonization temperature. *Carbon* 48, 1012–1024.

Ronsse F, van Hecke S, Dickinson D, and Prins W 2013 Production and characterization of slow pyrolysis biochar: influence of feedstock type and pyrolysis conditions. *Global Change Biology Bioenergy* 5, 104–115.

Scheirs J, Camino G, and Tumiatti W 2001 Overview of water evolution during the thermal degradation of cellulose. *European Polymer Journal* 37, 933–942.

Schneider GF, Calado VE, Zandbergen H, Vandersypen LMK, and Dekker C 2010 Wedging transfer of nanostructures. *Nano Letters* 10, 1912–1916.

Shakoor MB, Ye ZL, and Chen S 2021 Engineered biochars for recovering phosphate and ammonium from wastewater: A review. *Science of the Total Environment* 779, 146240.

Smith M, Ha S, Amonette JE, Dallmeyer I, and Garcia-Perez M 2015 Enhancing cation exchange capacity of chars through ozonation. *Biomass and Bioenergy* 81, 304–314.

Smith M, et al 2016 Structural analysis of char by Raman spectroscopy: improving band assignments through computational calculations from first principles. *Carbon* 100, 678–692.

Smith MW, Pecha B, Helms G, Scudiero L, and Garcia-Perez M 2017 Chemical and morphological evaluation of chars produced from primary biomass constituents: cellulose, xylan, and lignin. *Biomass and Bioenergy* 104, 17–35.

Suliman W, et al 2016 Modification of biochar surface by air oxidation: role of pyrolysis temperature. *Biomass and Bioenergy* 85, 1–11.

Sun K, et al 2013 Impact of deashing treatment on biochar structural properties and potential sorption mechanisms of phenanthrene. *Environmental Science and Technology* 47, 11473–11481.

Sun K, Keiluweit M, Kleber M, Pan Z, and Xing B 2011 Sorption of fluorinated herbicides to plant biomass-derived biochars as a function of molecular structure. *Bioresource Technology* 102, 9897–9903.

Sun T, et al 2018 Simultaneous quantification of electron transfer by carbon matrices and functional groups in pyrogenic carbon. *Environmental Science and Technology* 52, 8538–8547.

Sun T, et al 2021 Suppressing peatland methane production by electron snorkeling through pyrogenic carbon in controlled laboratory incubations. *Nature Communications* 12, 4119.

Takaya CA, Fletcher LA, Singh S, Anyikude KU, and Ross AB 2016 Phosphate and ammonium sorption capacity of biochar and hydrochar from different wastes. *Chemosphere* 145, 518–527.

Tan XF, et al 2016 Biochar-based nanocomposites for the decontamination of wastewater: A review. *Bioresource Technology* 212, 318–333.

Tan X, et al 2015 Application of biochar for the removal of pollutants from aqueous solutions. *Chemosphere* 125, 70–85.

Teixeira AR, et al 2011 Aerosol generation by reactive boiling ejection of molten cellulose. *Energy and Environmental Science* 4, 4306–4321.

Teixidó M, Pignatello JJ, Beltrán JL, Granados M, and Peccia J 2011 Speciation of the ionizable antibiotic sulfamethazine on black carbon (biochar). *Environmental Science and Technology* 45, 10020–10027.

Terrell E, Carré V, Dufour A, Aubriet F, Le Brech Y, and Garcia-Pérez M 2020 Contributions to lignomics: Stochastic generation of oligomeric lignin structures for interpretation of MALDI–FT-ICR-MS results. *ChemSusChem* 13(17), 4428–4445.

Tews I, et al 2021 Nitrogen-doped char as a catalyst for wet oxidation of phenol-contaminated water. *Biomass Conversion and Biorefinery*, 10.1007/s13399-020-01184-0.

Titirici MM, White RJ, Falco C, and Sevilla M 2012 Black perspectives for a green future: hydrothermal carbons for environment protection and energy storage. *Energy and Environmental Science* 5, 6796–6822.

Tran HN, et al 2020 Innovative spherical biochar for pharmaceutical removal from water: insight into adsorption mechanism. *Journal of Hazardous Materials* 394, 122255.

Volpe R, Zabaniotou AA, and Skoulou V 2018 Synergistic effects between lignin and cellulose during pyrolysis of agricultural waste. *Energy and Fuels* 32, 8420–8430.

Wan Z, et al 2020 Sustainable remediation with an electroactive biochar system: mechanisms and perspectives. *Green Chemistry* 22, 2688–2711.

Wang L, et al 2019 Mechanisms and reutilization of modified biochar used for removal of heavy metals from wastewater: A review. *Science of the Total Environment* 668, 1298–1309.

Wang S, et al 2020 Biochar surface complexation and Ni(II), Cu(II), and Cd(II) adsorption in aqueous solutions depend on feedstock type. *Science of the Total Environment* 712, 136538.

Wang T, Camps-Arbestain M, and Hedley M 2013 Predicting C aromaticity of biochars based on their elemental composition. *Organic Geochemistry* 62, 1–6.

Wooten JB, Seeman JI, and Hajaligol MR 2004 Observation and characterization of cellulose pyrolysis intermediates by 13C CPMAS NMR. A new mechanistic model. *Energy and Fuels* 18, 1–15.

Worch E, 2012 *Adsorption Technology in Water Treatment: Fundamentals, Processes, and Modeling*. Berlin: De Gruyter.

Xu D, Cao J, Li Y, Howard A, and Yu K 2019 Effect of pyrolysis temperature on characteristics of biochars derived from different feedstocks: A case study on ammonium adsorption capacity. *Waste Management* 87, 652–660.

Xu X, et al 2017 Indispensable role of biochar-inherent mineral constituents in its environmental applications: A review. *Bioresource Technology* 241, 887–899.

Yang F, et al 2019 Porous biochar composite assembled with ternary needle-like iron-manganese-sulphur hybrids for high-efficiency lead removal. *Bioresource Technology* 272, 415–420.

Yang Y, Chen N, Feng C, Li M, and Gao Y 2018 Chromium removal using a magnetic corncob biochar/polypyrrole composite by adsorption combined with reduction: reaction pathway and contribution degree. *Colloids and Surfaces A: Physicochemical and Engineering Aspects* 556, 201–209.

Yao Y, Gao B, Chen J, and Yang L 2013 Engineered biochar reclaiming phosphate from aqueous solutions: mechanisms and potential application as a slow-release fertilizer. *Environmental Science and Technology* 47, 8700–8708.

Yuan HY, et al 2018 Biochar modulates methanogenesis through electron syntrophy of microorganisms with ethanol as a substrate. *Environmental Science and Technology* 52, 12198–12207.

Zhang P, Sun H, Yu L, and Sun T 2013 Adsorption and catalytic hydrolysis of carbaryl and atrazine on pig manure-derived biochars: Impact of structural properties of biochars. *Journal of Hazardous Materials* 244-245, 217–224.

Zhao L, et al 2010 Sustainable nitrogen-doped carbonaceous materials from biomass derivatives. *Carbon* 48, 3778–3787.

7

Characteristics of biochar

Macromolecular properties

William C. Hockaday, Markus Kleber, and Peter S. Nico

Introduction

For more than a decade, researchers on all continents have been engaged in an initiative to simultaneously produce bioenergy, remove excess carbon (C) from the atmosphere, and improve soil and water quality. At the center of this "Charcoal Vision" (Laird, 2008) is biochar technology (Lehmann, 2007a; Lehmann and Joseph, 2009) – the attempt to design integrated agricultural biomass-bioenergy systems that build soil quality and increase productivity, while reducing atmospheric CO_2 (Lehmann, 2007b).

As interest in the matter spread across the research community, the semantic distinction between biochar technology as such and its solid product became blurry, even though a successful biomass-bioenergy system would generate products of equal or even greater significance than biochars: energy and a cleaner atmosphere. A convention evolved of treating the product of pyrolysis as an organic compound with common and generalizable properties. This practice threatens to be detrimental to the success of the "Charcoal Vision" (Laird, 2008), because research efforts tell us that additions of charred organic matter are not automatically a benefit to the receiving agroecosystem (Spokas et al, 2012). Consequently, those interested in moving biochar technology forward are increasingly emphasizing the fact that "there is a fundamental need to recognize that biochar does not refer to one singular product" (Spokas et al, 2012) and that "not all biochars are created equal and that biochars should be designed with special characteristics for use in specific environmental or agronomic settings" (Ippolito et al, 2012).

It has become evident that matching biochar properties with the problem at hand, or "designing" biochar for a specific application, is the road to success for biochar technology. In the paragraphs to follow, we describe the organic macromolecular features that may differ among biochars and identify the major controls acting on the development of these features.

DOI: 10.4324/9781003297673-7

The dynamic molecular model of biochar

Generalizing the effects of thermal treatment on the molecular structure of plant biomass is difficult for several reasons. To begin with, the composition of plant material is not homogeneous, not even within the same plant. Pyrolysis techniques as they are practiced in the field are highly variable and often difficult to control precisely. Heat treatment regimes, reactor technology, and other practical aspects of the pyrolysis process all influence the composition of the final product.

To understand biochar properties, many researchers have adopted a strategy of taking a given organic biomass precursor material through temperature sequences of pyrolytic treatments. By doing so, gradients of change and variations in molecular structure can be observed as a function of heat treatment temperature. Work of this kind has been done for isolated organic compounds such as lignin and cellulose (Bacon and Tang, 1964; Tang and Bacon, 1964) and for chemically more complex plant biomass feedstocks (Baldock and Smernik, 2002; Paris et al, 2005; Chen et al, 2008; Uchimiya et al, 2011; Kloss et al, 2012).

Informed by the growing data pool resulting from such work, the scientific community has converged towards a "dynamic continuum" molecular model for the macromolecular structure of biochar (Paris et al, 2005; Amonette and Joseph, 2009; Keiluweit et al, 2010). The adjective "dynamic" emphasizes the fact that the polymeric components of biomass undergo a range of molecular transformations and rearrangements as the level of applied thermal energy varies. Due to the discrete or quantized energy requirement of reaction activation energies and phase transitions, the transitions in biochar properties are often abrupt (Figure 7.1), with the distinct possibility that a biochar made at a given temperature may have many different properties compared to a biochar from the identical feedstock that was prepared at only slightly lower or higher temperatures.

The dynamic continuum biochar model reflects the sequence in which biochar-forming processes take place as a function of charring temperature. At low temperatures, plant biomass is mainly dehydrated and remains otherwise unaltered. As charring intensity increases, organic macromolecules, e.g., cellulose, lignin, and hemicellulose, are lost and isolated aromatic rings (transition biochar, Figure 7.1) begin to form. Eventually, recognizable biopolymers are lost, and isolated aromatic molecules with 2 and 3 rings are formed (amorphous or disordered biochar). When the temperature is raised further, small and somewhat "defective" sheets of condensed aromatic rings stack up to form so-called turbostratic crystallites. These are small three-dimensional structures that consist of 3 to 5 stacked C sheets with a vertical height (L_c) of 1–2 nm and a lateral extension (L_a) of 2–5 nm (Franklin, 1951; Yen et al, 1961; Heidenreich et al, 1968; Lu et al, 2001; Kercher and Nagle, 2003; Bourke et al, 2007). These dimensions depend on many factors such as feedstock, heat treatment temperature, and mineral content of the feedstock. As long as significant admixtures of amorphous materials are present in addition to turbostratic crystallites, we call such biochars "composite" (Figure 7.1).

When all amorphous organic C has either been volatilized or disordered clusters of aromatic rings are converted into polycyclic aromatic crystallites, the biochar is designated as "turbostratic" or "carbonized" (Bourke et al, 2007). Biochars do not typically fall into this latter region as they either require prolonged residence in the pyrolysis

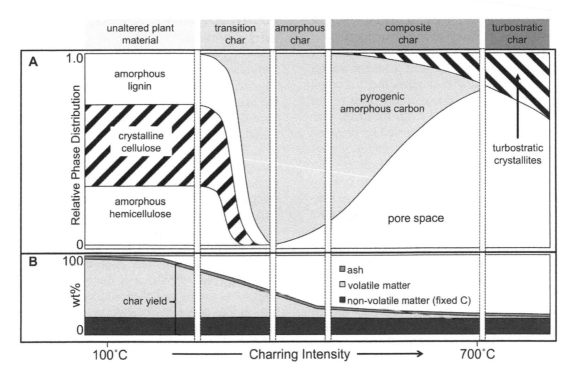

Figure 7.1 *Illustration of the dynamic continuum molecular structure of biochar across a charring gradient and schematic representation of four biochar categories and their individual phases. (A) Physical and chemical characteristics of organic phases. Exact temperature ranges for each category are controlled by both charring conditions (i.e., temperature, duration, and atmosphere) and relative contents of plant biomass components (i.e., hemicellulose, cellulose, and lignin). (B) Biochar composition as inferred from gravimetric analysis. Yields, volatile matter, fixed carbon, and ash contents are averaged across wood and grass chars. Relative contributions above 700°C are estimates (from Keiluweit et al, 2010; with permission from the publisher)*

retort or heating temperatures (HTT) over about 700°C.

While the continuum biochar model and variations in feedstock properties explain why "biochar materials are diverse" (Sohi, 2012), the question remains how knowledge about biochar properties can be used to "match char properties to the right situation" in the soil ecosystem (Sohi et al, 2010). A solution to this challenge requires the identification of ecologically relevant biochar properties as well as the identification of best

practices for measuring them. To identify such properties, we recall from the dynamic biochar model that, depending on heat treatment temperature, biochar may contain three broad categories of organic materials. These are (i) dried or torrefied plant materials; (ii) a "soft", amorphous, flexible organic phase (sometimes operationally designated as "volatile matter"); and (iii) turbostratic crystallites of rigidly stacked, "glassy" C sheets.

The overwhelming majority of biochars made for environmental applications will contain very little torrefied plant material because the purpose of the charring process is to transform such materials into one of the other two phases. We thus concentrate on how to identify and distinguish the functional importance of amorphous compounds versus those of polycondensed C in biochars. Our conceptual approach is informed by previous work which examined the adsorption of non-ionic organic contaminants to biochars (Sun et al, 2012). In this work, variations in polarity and functional group chemistry, relative proportions of aliphatic C, the extent of crystallite growth, and the resulting BET-N$_2$ surface area were successfully used to explain the adsorption of multifunctional phthalate esters. We will adopt this concept to examine how modern analytical techniques can help to both identify the assemblage of functional groups associated with a given species of biochar and determine the relative proportions of amorphous versus crystalline components within biochar.

Aromatic rings, sheets, and crystals

Definition of aromaticity

Thermal treatment of biomass, when carried out in an inert atmosphere (i.e., in the absence of an electron acceptor such as oxygen (O) that would promote combustion and conversion to carbon dioxide (CO$_2$)) induces chemical change. Eventually, new solid phases with increasing proportions of C are formed. The resulting "carbonization" of the precursor materials is a complex process in which many reactions such as dehydrogenation, hydrogen (H) transfer, and isomerization take place concurrently. Of particular importance for the final properties of biochar is the formation of rings of C atoms and their condensation and growth into larger sheets and stacks. When C atoms are assembled into rings with C=C double bonds, overlap of p-orbitals may occur and π-electrons can become delocalized (Figure 7.2) creating "aromatic molecules".

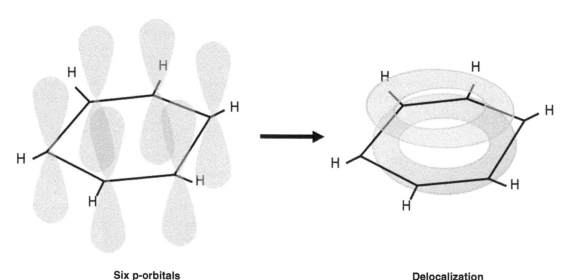

Six p-orbitals   **Delocalization**

Figure 7.2 *Molecular orbitals forming a pi (π)-system containing delocalized electrons. Single bonds made from electrons in the line between the C nuclei are called sigma (σ)-bonds (figure on the left). The two rings above and below the plane of the molecule on the right represent one of three bonding molecular orbitals. Two delocalized electrons can be found anywhere within those rings*

Aromatic molecules were originally called such because many of them have a pronounced odor. The modern definition for aromaticity (IUPAC, 2006) is "the concept of spatial and electronic structure of cyclic molecular systems displaying the effects of cyclic electron delocalization which provide for their enhanced thermodynamic stability (relative to acyclic structural analogs) and tendency to retain the structural type in the course of chemical transformations". Delocalization is a result of orbital mixing across multiple atoms via overlapping atomic orbitals, and p orbitals in the case of C. This creates new molecular orbitals, some of which are at lower energy levels resulting in shorter bonds and greater bond strength compared to arrangements of 6 C atoms without delocalization, such as in the hypothetical cyclic molecule cyclohexatriene. Here it is necessary to emphasize that not all planar, cyclic arrangements of C atoms are automatically aromatic – they must have a closed loop of (4n+2) π-electrons in the cyclic arrangement of p orbitals (Hückel, 1931) for delocalization to occur [n is an integer, 0,1,2,3, ...]. Following this rule, only π-systems containing 2, 6, 10, 14, ... electrons exhibit the enhanced stability that is associated with aromaticity.

Functional consequences of aromaticity and resonance

In addition to their enhanced thermodynamic stability, aromatic rings are remarkable for other reasons. The thermodynamic properties of a six-C molecule vary as a function of molecular architecture (assuming formation from elemental constituents and combustion in O_2) and benzene, in contrast to glucose and hexane, requires an energy input to be formed from its elemental constituents (Table 7.1).

Aromatic rings have a much greater heat of combustion compared to the carbohydrate glucose (Table 7.1), demonstrating that thermodynamically stable aromatic organic compounds can well be significant energy sources for such decomposer organisms that are equipped with the catabolic toolbox to overcome the resonance-induced energy barrier.

Electron delocalization has consequences for the ability of the aromatic ring to engage in adsorption processes within soils and sediments. The fact that the ring is composed of elements with similar electronegativities [2.5 for C and 2.2 for H on the scale of Allred and Rochow (1958)] indicates a relatively even charge distribution across the molecule and a resulting lack of polarity. Consequently, bonding interactions of apolar aromatic ring structures with minerals and organic matter in

Table 7.1 *Properties of compounds with 6 carbons. NOSC = nominal oxidation state of carbon. Data from Dickerson and Geis (1976)*

	Molecular formula	NOSC	Molar mass (g mol^{-1})	Melting point (°C)	Enthalpy of formation (kJ mol^{-1})	Enthalpy of combustion (kJ mol^{-1})	Enthalpy of combustion (kJ g^{-1})
Glucose	$C_6H_{12}O_6$	0	180	146–150	−1271	−2805	−15.6
Hexane	C_6H_{14}	−2.3	86	−94	−198	−4160	−48.3
Benzene	C_6H_6	−1	78	5.5	+83	−3273	−41.9

terrestrial environments have long been considered restricted to weak, nonspecific, hydrophobic interactions (Keiluweit and Kleber, 2009). However, the orientation of the electron-carrying π-orbitals above and below the frame of C atoms generates a structure that is slightly depleted in electron density (δ^+) within the plane of the C atoms, but somewhat enriched in electron density (δ^-) in the π-orbitals below and above the plane (Figure 7.3).

A consequence of the slightly polar charge distribution in the aromatic ring is the resulting ability to interact with other molecules through a variety of specific mechanisms of considerably greater strength than conventional hydrophobic interactions (Keiluweit and Kleber, 2009). In addition, the polarity of aromatic ring systems may be modified by either adding or withdrawing electron density to the system. Generally, substituents with electron-donating properties and additional fused π-systems increase the electron density in π-systems and therefore promote their electron-donor strength. Conversely, electron-withdrawing substituents and heteroatoms render the π-system electron deficient, thus allowing it to function as a π-electron acceptor. As a result, the

ability of polycondensed aromatic structures to engage in mechanisms such as electron-donor acceptor (EDA) interactions may vary greatly. EDA interactions are a mode of bonding that involves the "donation" of a π-electron by an electron-rich donor to an electron-deficient acceptor. The addition of heteroatoms can withdraw electron density from aromatic π-systems with a low degree of condensation and a high proportion of electronegative O atoms (high O/C atomic ratio). Such compounds would be expected to occur in the amorphous biochars that form at relatively low temperatures (Keiluweit et al, 2010). As polycondensed sheets grow larger with increasing charring temperature or duration of exposure to temperature (Rutherford et al, 2012) they also tend to become more continuous (or less "defective"), with the consequence that electronegative heteroatoms and functional groups are mainly found attached to the edges of sheets. It has been suggested that as a result of this geometric arrangement, π-electrons may find themselves pulled towards the edges of graphene sheets to form a region of high electron density next to the edge, while the center of the sheet will become depleted in electron density (McDermott and McCreery, 1994; Zhu and Pignatello, 2005). As a consequence of such electron distribution, aromatic sheets in biochar may have the ability to both accept and donate π-electrons, depending on the extent to which their edges are decorated with O-containing functional groups (Klüpfel et al, 2014).

The previous paragraphs have emphasized that the extent to which the C in biochar has been transformed into energetically stable aromatic ring structures contributes decisively to both the fate and reactivity of biochar in the environment. Consequently, there is great interest in methods to measure or estimate with confidence the:

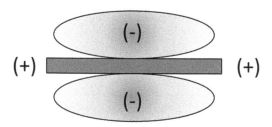

Figure 7.3 *Sideview of an aromatic ring. The dark rectangle represents the benzene ring. The π-system above and below the benzene ring leads to a "quadrupole" charge distribution: negative signs indicate areas of negative polarity, positive signs indicate molecular regions of positive polarity*

1 aromaticity sensu stricto, i.e., the degree to which C rings in biochar exhibit electron delocalization and the associated enhanced thermodynamic persistence relative to the precursor biomass;
2 the extent to which aromatic rings are fused into larger polycondensed units (aromatic condensation); and
3 the proportion of C in condensed ring structures with reference to total C.

The term "aromaticity" has on occasion been used to include all of the above. This requires the reader to use caution when quantitative estimates of "aromaticity" are encountered.

Measuring aromaticity

Aromaticity (rings showing electron delocalization), degree of polycondensation (structures with few rings versus such with many), and the relative proportion of ring C as a fraction of total C (the extent to which the feedstock has been carbonized) can be assessed with the help of several methods. X-ray diffraction techniques and nuclear magnetic resonance (NMR) spectroscopy have seen the most frequent use. One-dimensional NMR spectra provide functional group information which can be used to estimate the stochastic average number of C atoms in an aromatic cluster and indices for either circular or linear catenation. Information from X-ray diffraction provides three-dimensional information about turbostratic crystallites, including sheet diameter, stacking height, and interlayer distance. From these principal limitations, it follows that NMR is more powerful in investigating smaller aromatic structures in low-temperature biochars with low crystallinity, while X-ray diffraction becomes more meaningful with increasing crystallinity of the carbonaceous phase.

X-ray diffraction

The first analytical technique used to investigate the structure of charred organic materials was X-ray diffraction, with significant work done already in the first half of the 20th century (Warren, 1934; Franklin, 1950, 1951). The technique utilizes the ability of crystals to diffract radiation with wavelengths in the same order of magnitude as the spacings between the atoms that make up the diffracting planes in crystal lattices. From the physical principle behind X-ray diffraction follows that the technique is unable to measure the extent of electron delocalization, i.e., aromaticity sensu stricto. It is, however, capable of measuring the vertical (repeat) distance between individual polycondensed sheets [d] and their stacking height (L_c) as well as the lateral extension of polycondensed sheets (L_a) (Figure 7.4).

The quality of X-ray diffraction data improves with the homogeneity of the crystal structure, which tends to be marginal in low-temperature biochars. While traditional approaches of data analysis (Diamond, 1957)

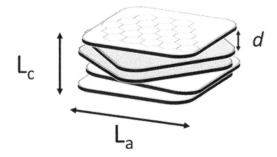

Figure 7.4 *Idealized representation of a turbostratic stack (individual sheets rotated and tilted with reference to each other) of 5 carbon sheets with a diameter of aromatic sheets L_a, stacking height L_c, and interlayer distance d. For a more detailed molecular rendering including alkyl linkages between individual sheets compare Figure 7.9*

are presently being revived with some promising initial results (Sultana et al, 2010, 2011), two ways of interpreting X-ray diffractograms of biochars have attained some popularity and are frequently combined. These are the gamma band (γ-band) method to estimate the relative proportion of aromatic C in a given sample (Yen et al, 1961) and the derivation of the d; L_a and L_c parameters from the intensities of hkl 002 and 100 (10) or 110 (10) reflections in the diffractogram using the Scherrer equation (Lu et al, 2001; Kercher and Nagle, 2003).

The dimensions of individual turbostratic crystallites [d; L_a and L_c] are given by Bragg's law

$$d = \frac{n\lambda}{2\sin\theta} \qquad [7.1]$$

where n = an integer; λ = the wavelength of the radiation used (in angstrom, varies as a function of anode material such as Cu, Co); d = the spacing between the planes in the atomic lattice; θ = the angle between the incident ray and the scattering planes, and by the Scherrer equation (Patterson, 1939):

$$L = \frac{\lambda K}{\beta \cos\theta} \qquad [7.2]$$

where L = the mean crystallite dimension in angstrom units along a line normal to the reflecting plane, λ = the wavelength of the radiation used (varies as a function of anode material such as Cu, Fe); K = a constant near unity; β = the width of a peak at half-height (FWHM) expressed in radians of 2θ (to obtain radians, measure 2θ in degrees and then multiply by $\pi/180$).

For the determination of the height L_a to which individual sheets are stacked in a turbostratic crystal, the 002 reflection is used (Figure 7.5) and K was suggested to be 0.89;

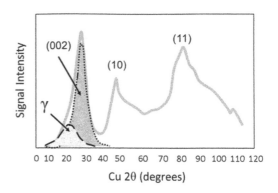

Figure 7.5 *Example for an X-ray diffractogram showing diffraction signals of char prepared from coal at 1200°C. The figure illustrates the close proximity of the γ-band to the hkl 002 band. The two-dimensional (10) and (11) signals superimposed on the three-dimensional (100) and (110) lines. Cu indicates a copper anode was chosen to generate X-rays with a wavelength of 1.54 Ångstrøm (0.154 nm). θ is the angle of incidence between the X-ray source and the sample surface. Figure redrawn after Lu et al (2001)*

for the determination of the lateral C sheet diameter L_c, both 100 (10) and 110 (11) signals can be used, with a recommended K value of 1.84 (Lu et al, 2001; Kercher and Nagle, 2003). However, while the d, L_a, and L_c parameters provide information about the extent to which C crystallites have grown during the pyrolysis process, they cannot be considered true proxies for the whole of aromatic C. Especially in low-temperature biochars, significant proportions of aromatic C may exist that have not been fused into larger sheets and arranged in crystallites capable of diffracting X-rays.

In low-temperature biochars, a second X-ray technique known as electron energy loss spectroscopy (EELS) has been applied to study the growth of aromatic domains. Yoo et al (2018) offered a physical interpretation of the multiple backscattering (bulk plasmon resonance) portion of the EELS spectrum (20–30 eV). They reasoned that multiple resonances originate from the scattering of ejected electrons propagating through multiple concatenated aromatic C comprising the small clusters of aromatic rings. Applying a quantum mechanical (Drude) model, the average plasmon resonance energy can be related to an average C-C distance (nm) as a general measure of order and disorder in low-temperature biochars (Yoo et al, 2018). Multiple phase transitions or structure rearrangements exist between 300°C and 700°C (consistent with Figure 7.1). Through a combination of EELS C-C bond distances and X-ray diffraction (L_a, 100 and d, 002 values) the aspect ratio (length and width) of turbostratic crystallites can therefore be obtained (Yoo et al, 2018). Pinewood biochars formed by fast pyrolysis at temperatures ≥700°C can then be identified as having preferential aromatic crystallite growth along a longitudinal axis, as opposed to a radial growth pattern, thus producing linear catenation rather than annular catenation. The aspect ratio of crystallite growth has direct implications for macromolecular structure, but effects on physical, chemical, and biological properties of biochar remain unclear, and likely only hold relevance for high-temperature biochars (≥700°C).

In the coal industry, the so-called "γ-band" approach to X-ray diffraction analysis is popular to estimate the proportion of aromatic C with reference to total C. The γ-band in the diffractogram (Figure 7.5) is thought to result from non-aromatic C chains arranged in some sort of order, and the rationale of the γ-band approach is that a proxy for aromaticity can be obtained by relating the peak intensities of the "aromatic" 002 C peak to the "aliphatic" γ-band intensity. The technique has seen some use in the characterization of charcoal and biochar, as

well (Mochidzuki et al, 2003; Bourke et al, 2007; Nguyen and Lehmann, 2009). In the γ-band approach, aromaticity is defined as the proportion of C in aromatic rings to total C (Yen et al, 1961) and calculated as:

$$f_a = \frac{C_A}{C} = \frac{A_{002}}{A_{002} + A_\gamma} \qquad [7.3]$$

with f_a = proportion of aromatic C (as % of total C); C = total C; C_A = C in aromatic rings (more precise: C in planar sheets); A_{002} = area of the hkl 002 band in an X-ray diffractogram (the 002 band representing the C organized in vertical stacks of aromatic sheets); A_γ = area of the γ-band in an X-ray-diffractogram representing "saturated structure as in contrast to the hkl 002 band which represents aromatic structure" (Yen et al, 1961).

The y-band method to estimate the proportion of aromatic C can be considered a standard analytical procedure for the characterization of coal and industrial charcoal (Ergun and Tiensuu, 1959; Yen et al, 1961; Schoening, 1982, 1983; Schwager et al, 1983; Lu et al, 2000, 2001; Mochidzuki et al, 2003; Bourke et al, 2007; Saikia, 2010). However, when the technique is applied to biochar (Nguyen and Lehmann, 2009) the user should be mindful of the fact that the method originates from the investigation of asphaltene and coal-type materials. These may contain aliphatic chains with a significant ability to arrange in a somewhat ordered fashion and may thus be better at generating the γ-band reflection in the X-ray diffractogram than the non-aromatic C compounds in biochars. It is doubtful that the γ-band represents the exact same molecular forms of C across samples from different feedstocks. Estimates of total aromaticity in biochars based on the γ-band method should thus be considered as rough approximations.

^{13}C NMR spectroscopy

The ^{13}C NMR spectroscopy of solids became available as an analytical technique for the investigation of organic matter more than 3 decades ago (Barron et al, 1980; Barron and Wilson, 1981; Hatcher et al, 1981). Since then, ^{13}C-NMR has evolved into a standard method for the assessment of changes in the composition of organic compounds, including those changes that occur during thermal treatment. Notable early NMR investigations of the chemical transformations in charred plant biomass include work by Almendros et al (1992, 2003), Knicker et al (1996), and Baldock and Smernik (2002). Recent examples of NMR observations of multistep biochar temperature series can be found in publications by Nguyen et al (2010; oak wood, corn stover); Sun et al (2011; festuca grass and ponderosa pine); McBeath et al, (2011; chestnut wood) and Cao et al (Cao et al, 2012; maple wood).

As an example, cotton seed meal, a lignocellulosic plant biomass, underwent slow pyrolysis at five treatment temperatures, for which ^{13}C NMR spectra of the uncharred and charred materials were obtained (He et al, 2021; Figure 7.6). Each peak represents C atoms in specific chemical environments (i.e., functional groups such as CH_3; OCH; or COO). NMR measurements can be done under a variety of experimental conditions, including multiple ^{1}H-^{13}C cross-polarization (MultiCP) and direct polarization (DP) with magic angle spinning (MAS) (Figure 7.6). Both experiments can be repeated with a spectral editing technique known as dipolar dephasing, which detects only the non-protonated C as well as protonated C that undergo motion, specifically, translation or rotation on the timescale of the NMR experiment (milliseconds). Cross-polarization experiments produce more signal per scan than the DP technique, but it has problems observing all

the (^{13}C-) C atoms in a sample. The MulitCP experiment improves C observability by enhancing the detection of the fastest and slowest relaxing signals. Nevertheless, CP techniques may underrepresent polycondensed aromatic ring systems where aromatic C atoms near the center of a crystallite are several bond lengths removed from ^{1}H atoms. The DP technique is better at observing all C that is present, but it involves much longer experiment times than the CP procedure. The proportion of total C observed by either CP or DP declines with increasing HTT from 87% (CP) and 89% (DP) in a 250°C biochar to 8% (CP) and 57% (DP) in a 700°C biochar (McBeath et al, 2011). Estimates of aromaticity based on CP information are generally lower than those obtained based on DP information (McBeath et al, 2011).

For these reasons, the DP technique is considered the better choice for quantitative examinations of the macromolecular structure of biochars (Cao et al, 2012). DP experiments on biochars produced at temperatures > 500°C detected higher proportions of non-protonated aromatic C than the MultiCP (Figure 7.6; He et al, 2021). Aromatic C appears in the ^{13}C NMR spectral region (Figure 7.6) between 110 and 165 ppm (ppm scale denoting chemical shift compared to a reference compound), with the region between 110 and 145 ppm considered to represent aromatic C (also called "aryl-C") and the region from 145 to 165 ppm ("O-aryl") representing aromatic rings with O or nitrogen (N) containing substituents (Knicker and Ludemann, 1995). By convention, the aromaticity of an organic matter sample is expressed in terms of relative signal intensity, i.e., the signal area between 110 and 165 ppm is integrated and reported as a fraction of the total signal area. When peaks are exceptionally broad and in the presence of spinning side bands,

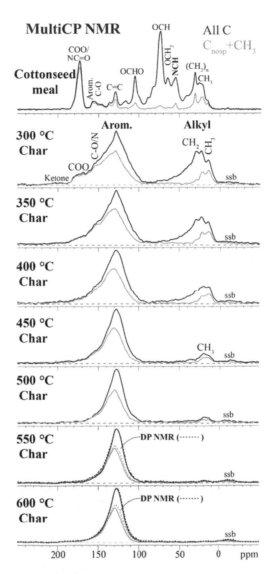

Figure 7.6 *^{13}C MultiCP MAS NMR spectra (black lines) recoupled dipolar dephasing spectra (dashed lines) showing non-protonated C (C_{nonp}) plus mobile groups such as CH_3. Dash-dot lines are quantitative DP MAS NMR spectra after dipolar dephasing. Cotton seed meal is a non-food biomass waste. Biochars were produced by slow pyrolysis at the indicated treatment temperatures. Spectra are scaled to match the magnitude of the highest peak. Spinning side bands (ssb) are signals from the modulation of the magnetic field at the spinning frequency. Figure from He et al (2021)*

mathematical deconvolution and peak fitting procedures are preferred over assigning a fixed chemical shift region as aromatic C (McBeath et al, 2011).

Spectral editing NMR techniques provide greater molecular detail on the dimension of structural domains in biochar. Techniques such as the recoupled long-range C-H dipolar dephasing (Figure 7.6) allow the unraveling of details regarding the evolution of aromatic compounds with increasing HTT (Cao et al, 2012). Therefore, aromatic rings decorated with O-containing functional groups (aromatic C-O) can be distinguished from aromatic rings at the edges of polycondensed domains (aromatic C-H) and aromatic rings within polycondensed aromatic domains (non-protonated aromatic C-C) (Figure 7.7). This information is useful to estimate the potential functionalities of a biochar made under given pyrolysis conditions. For the

pyrolysis conditions and feedstock chosen by Cao et al (2012), we find that in biochars produced at 450°C (Figure 7.7) only about half of the aromatic C is connected to an H or an O, indicating that this aromatic C is either part of rather small clusters of aromatic rings or in polycondensed ring structures with some sort of "defect". At 700°C (Figure 7.7), only a small fraction of the aromatic C is decorated with either H or O, indicating that these C atoms must be located inside larger ring domains. Biochars formed at 700°C had an average number of fused aromatic C per sheet of > 37 C, or > 17 aromatic rings (Brewer et al, 2009). Similarly, an average cluster size of 40 aromatic C was reported for oak biochar treated at 600°C (Nguyen et al, 2010).

Another methodological innovation is based on the principle that delocalized π-electrons in aromatic rings are free to circulate around the aromatic ring. Such movement

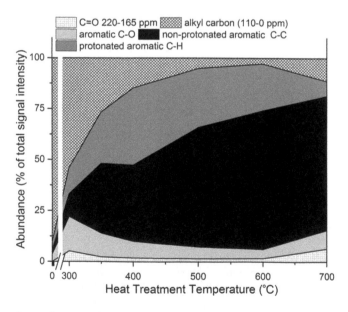

Figure 7.7 *Proportions of aromatic compounds in maple wood as determined by quantitative DP/NMR analysis (protonated aromatic C-H = ring C bonded to an H; non-protonated aromatic C-C = rings surrounded by other aromatic rings; aromatic C-O = ring C bonded to an O-containing functional group) (data taken from Table 2 in Cao et al, 2012)*

can be induced by an external magnetic field. As electrons move, they induce a weak electromagnetic field that is associated with the aromatic ring. More condensed (larger and purer) ring systems produce larger "ring currents" and stronger magnetic fields, which can be assessed through changes in electrical conductivity as well as through variations in the ^{13}C NMR aromatic peak position and line width (Freitas et al, 1999, 2001). Similar trends in line broadening, peak position, and relaxation rates can be observed for ^{13}C-labeled probe molecules adsorbed to organic solids (Smernik, 2005; Smernik et al, 2006). The ^{13}C NMR peak position of the ^{13}C-labeled benzene probe molecule is shifted systematically downfield when adsorbed to biochars of increasing HTT (McBeath and Smernik, 2009; McBeath et al, 2011). This phenomenon, where isotopically-labeled probe molecules "inherit" the NMR behavior (i.e., spin dynamics) of the chemical environment in which they are adsorbed, can be used as a promising method for studying the physical and chemical properties of the interior pores of biochars. The selection of probe molecules of varying physical and chemical properties opens new avenues for studying the various chemical and physical domains and how they might interact with dissolved chemical species. For example, the NMR chemical shift of the ^{13}C-labeled benzene has been used as a proxy for the degree of aromatic condensation based on the principle that ring currents increase with the lateral dimension (L_a) of the C sheet (Smernik et al, 2006; McBeath and Smernik, 2009; McBeath et al, 2011).

Growth of aromatic domains in biochars

Aromaticity is a central determinant of biochar functionality (Chapters 8 and 10) and persistence (Chapter 11) in the environment. Therefore, successful attempts at producing biochars that are tailored for specific functions and applications will depend on the ability to control the pyrolysis process in a way that the desired level of aromaticity can be achieved. Numerical information regarding the dimensions of polycondensed domains in biochars is mainly available from X-ray diffraction techniques and from NMR spectroscopy. This means that, when we develop a tentative process model for the growth of polycondensed aromatic structures in the pyrolysis process, we must attempt to reconcile data from very heterogeneous observational and experimental environments. The picture that emerges (Table 7.2) reflects this diversity of charring conditions and observational techniques. Fortunately, there are robust trends that could inform a concerted research effort to describe the two- and three-dimensional structure of biochars. First, we note that the degree of polycondensation increases with heat treatment temperature (HTT). Second, where HTT and the method of observation were identical (Brewer et al, 2009; Cao et al, 2012) but feedstock was varied we find differences in aromatic domain size that support the role of feedstock in controlling aromaticity. We conclude that the numerical data published to date (Table 7.2) support previous findings that three factors should be particularly relevant for controlling aromaticity. These are a type of feedstock (Antal and Grønli, 2003; Bourke et al, 2007), maximum heat treatment temperature (Paris et al, 2005; Keiluweit et al, 2010), and the duration of the heat treatment (Rutherford et al, 2012). Controlling only one of these factors (for example, HTT) is not likely to yield reproducible product quality when biochars are to be made to end-user specifications.

Non-aromatic biochar carbon

The abundance of non-aromatic biochar C declines sharply with increasing heat

Table 7.2 *Aromatic domain size estimated from ^{13}C NMR spectroscopy and from X-ray diffraction as a function of biomass type, pyrolysis conditions, and heat treatment temperature. Given as (i) sheet diameter L_a in angstrom (10 Å = 1 nm), (ii) number of rings, or (iii) number of C atoms within aromatic clusters*

Biomass type	Pyrolysis conditions	Technique to estimate aromatic domain size	Heat treatment temperature (°C)						Reference
			300	400	500	600	700	greater 700	
Switchgrass	Slow pyrolysis	^{13}C-NMR; ^{1}H-^{13}C Long-Range Recoupled Dipolar Dephasing			> 23 carbons => **7 rings**				Brewer et al. (2009)
Corn residue, Oakwood	Slow pyrolysis	DP ^{13}C NMR	18–19 carbons ≈ **4–5 rings**		37–40 carbons **13 rings**				Nguyen et al (2010)
Chestnut wood	N$_2$ continuous flow, ramp rate 50°C hr^{-1}, held at max T for 5 hr	^{13}C-NMR; ring current method		no larger than coronene = **7 ring**	increasing in size			structures larger than **19 ring** dominating above 700°C	McBeath et al (2011)
Maple wood	N$_2$ continuous flow, ramp rate 25°C min^{-1}, held at max T for 2 hr	^{13}C-NMR; ^{1}H-^{13}C Long-Range Recoupled Dipolar Dephasing	**1–2 rings**	18–20 carbons ≈ **4–5 rings**	40 carbons ≈ **13 rings**	64 carbons ≈ **22–23 rings**	74 carbons ≈ **27 rings**		Cao et al (2012)
Polyvinylidene chloride (C$_2$H$_2$Cl$_2$)$_n$	N-atmosphere; held for 2 hr at max T= 1000°C	XRD in Guinier geometry, L$_a$ determination using two-dimensional {hk} bands and Warrens formula, subsequent fitting of calculated curves						1000°C; mean diameter of 16 Å corresponding to **30–36 rings** in circular catenation	Franklin (1950)
Fiberboard made from southern yellow pine, northern pine, and hardwood	50°C hr^{-1} to 110°C; 3 hr dwell; 15°C hr^{-1} to 210°C; 30°C hr^{-1} to 400°C; 15°C hr^{-1} to 600°C; 50°C hr^{-1} to 1000°C	XRD, L$_a$ determination using {100} data in Scherrer equation, # of rings estimated assuming hexagonal geometry of aromatic cluster			42–43 Å, equivalent to about **225 rings** in circular catenation			greater 60 Å at 800°C growing to over 80°C at 1400°C, corresponding to **> 10^3 rings**	Kercher and Nagle (2003)

Sample	Conditions	Method	Results	Reference
Ponderosa pine wood	Inconel crucible covered with lid, held at max T for 1 hr	X-ray diffraction, L_a determination using {100} data in Scherrer equation, # of rings estimated assuming hexagonal geometry of aromatic cluster	19.3 Å ≈ 49 rings 18.3 Å ≈ 44 rings 20.4 Å ≈ 54 rings	Diffractograms published in Keiluweit et al (2010)
Festuca grass			13.8 Å ≈ 25 rings 19.7 Å ≈ 51 rings 28.5 Å ≈ 106 rings	
Charred vegetation residues from Japanese Andosols	Uncontrolled vegetation fire	X-ray diffraction; theoretical profiles of PAH-mix fitted to measured {11} band	mean L_a between 12.6 and 13.7 Å suggesting the most abundant (34–44%) condensed structures were in the **19 ring** size category. Range observed was 9.2 to 19.2 Å, corresponding to **14–52 rings**	Sultana et al (2010)
Japanese cedar	Porcelain crucible with lid, held at T = 400°C for 1 hr		most abundant (34%) at 7.2 Å ≈ **7 rings** Range observed was 4.5 Å to 14.4 Å ≈ **4 – 25 rings,**	Sultana et al (2011)
Loblolly pine wood	Open crucible for 15 min, N_2 flow 1 L min^{-1} 183°C s^{-1} to 300°C; 216°C s^{-1} to 400°C; 249°C s^{-1} to 500°C; 316°C s^{-1} to 600°C; 383°C s^{-1} to 700°C; 448°C s^{-1} to 800°C; 518°C s^{-1} to 900°C; 649°C s^{-1} to 1000°C;	X-ray diffraction, d-spacing using {002} and L_a using {100} data in Scherrer eqn. for ordered C; for disordered C - electron energy loss spectroscopy (EELS) with °C carbon K-edge resonances in the Drude model for C-C bond length	Layer coherence length, L_a decreased to 400°C then increased 20 Å 21 Å 22 Å 23 Å at 800 °C to 31 Å at 1000 °C	Yoo et al (2018)

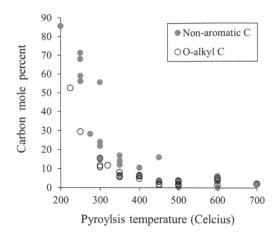

Figure 7.8 *Non-aromatic (solid) and O-alkyl C (open symbols) in biochars decline with pyrolysis treatment temperature. Values were obtained by quantitative direct polarization ^{13}C NMR spectroscopy (data from Brewer et al, 2009; McBeath and Smernik, 2009; Nguyen et al, 2009; McBeath et al, 2011; Wiedemeier et al, 2014)*

treatment temperature (Brewer et al, 2009, 2011; McBeath and Smernik, 2009; Nguyen and Lehmann, 2009; McBeath et al, 2011; Wiedemeier et al, 2014; Figure 7.8). Furthermore, the majority of non-aromatic C consists of O-substituted alkyl C coming from the celluloses and hemicelluloses in the cell walls of plant biomass. According to the general trend in Figure 7.8, less than 10% of biochar C is non-aromatic at temperatures ≥ 400°C. Cellulose pyrolysis reactions begin at 330°C, reach a maximum at 370°C and end around 390°C (Mok and Antal, 1983), which explains the rapid decrease in O-alkyl C abundance (Figure 7.8) for pyrolysis temperatures between 300 and 400°C. Therefore, cellulose reaction kinetics and thermodynamics are likely important controls on the abundance of non-aromatic C in biochars.

The importance of non-aromatic C to biochar properties is underscored by the results of the so-called "proximate analysis". ·roximate analysis is an operational proce-·ure to assay fundamental properties related

to chemical, thermal, or biological persistence. There is growing evidence that volatile matter and persistence assays are measuring the non-aromatic fraction of biochar C. In this section, we provide examples of studies that combined volatile matter analyses and biochar persistence assays with molecular spectroscopy analyses to quantify aromatic and non-aromatic C.

Non-aromatic carbon and chemical stability of biochars

Resistance to acid dichromate oxidation is a common proxy for chemical stability (Kaal and Rumpel, 2009). The degree of de-alkylation of molecules detected by pyrolysis GCMS showed positive linear correlations with the proportion of C that resists acid dichromate oxidation ($r^2 = 0.70$, $p < 0.01$ for benzene/toluene ratio; $r^2 = 0.92$, $p < 0.001$ for benzonitrile/methyl-benzonitrile ratio) (Kaal and Rumpel, 2009). These findings were significant because they demonstrated that variation in the chemical resistance of biochar is related to the relative abundance of non-aromatic C, specifically, the ratio of alkyl C/aromatic C.

Non-aromatic carbon and thermal stability of biochars

The thermal resistance of biochars can be assessed using a modified, low-temperature version (Joseph et al, 2010) of the ASTM D1762–84 standard chemical analysis method (ASTM = American Society for Testing Materials, www.astm.org). Applying this method to biochars of various feedstocks and heat treatment temperatures (Enders et al, 2012), the organic H/C ratio is positively correlated with the thermally labile (i.e., volatile) fraction (H:C_{org} = 0.0149 * volatile % - 0.152; r^2 = 0.658). The H content, and therefore, the H/C ratio is a bulk measure of the non-aromatic C. The thermally resistant (i.e., fixed carbon) C fraction follows a 1:1 linear relationship with aromatic C as measured by ^{13}C NMR spectroscopy (Brewer et al, 2011). ^{13}C NMR spectra of thermally labile fractions of biochar by spectral subtraction of biochar differed in volatile matter abundance (McClellan-Maaz et al, 2021). The thermally labile (volatile) matter was dominated by alkyl C (70 mole %) and phenolic C. Therefore, the abundance of non-aromatic C in biochar is an important control on the thermal resistance of biochar C.

Non-aromatic carbon and biological persistence of biochars

Biological mineralization of biochar (Chapter 11) typically shows a strong statistical correlation with non-aromatic C abundance (Singh et al, 2012), such as those observed with chemical and thermal stability proxies. Calorimetry revealed that the quotient of total energy, ΔE, and activation energy ($\Delta E / E_a$) was a predictor of microbial respiration of biochar-C in soil (Harvey et al, 2016) – a principle termed the energetic return on investment. The consistency of

chemical, thermal, energetic, and biological data in identifying the mineralizability of non-aromatic biochar-C leads to questions about the nature of the non-aromatic C in biochar and its role in the structure of biochar particles. The following passages deal with these topics.

Role of non-aromatic C in the structural dimensions of biochar

A pervading concept in several decades of literature on coal and charcoal pyrolysis posits that, within individual growing C sheets, clusters of fused aromatic rings are connected, or cross-linked, by alkyl (-CH_n-) and carbonyl (-C(O) -) side chains (Fletcher et al, 1992). Figure 7.9 illustrates this view of the biochar chemical structure and illustrates the inferred dimensions of aromatic clusters with non-aromatic sidechains.

In contrast to the fused aromatic ring structures which grow in dimension with increasing pyrolysis temperature (Table 7.2), the abundance and dimension of non-aromatic domains decline with pyrolysis temperature. Pyrolysis GCMS analyses of laboratory-generated peat charcoals suggest that the abundance of alkyl-substituted aromatic clusters decreased as the charring temperatures increased (Almendros et al, 2003; Kaal and Rumpel, 2009). This was ascribed to "side chain burn off" and the growth of aromatic ring clusters via dehydration and condensation of alkyl domains at temperatures above 400°C (Kaal and Rumpel, 2009). At temperatures ≥ 400°C, the biochar structure is dominated by aromatic C (Figure 7.7), while alkyl domains (i.e., side chains) comprise < 10% of biochar-C, declining to <5% of biochar-C at temperatures above 500°C (Figure 7.8). Quantitative ^{13}C NMR with dipolar dephasing has been applied to biochars generated across a range of temperatures (Table 7.2). In addition to estimating the dimensions of the aromatic ring clusters (C atoms per cluster), stoichiometric

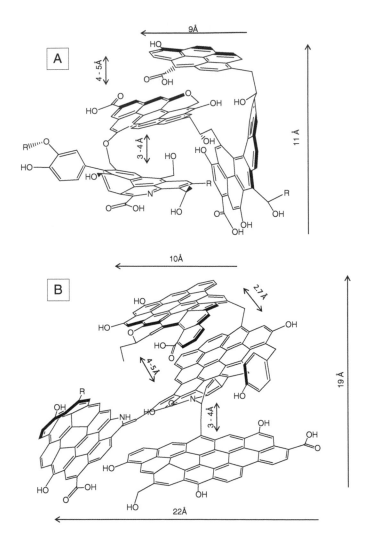

Figure 7.9 *Model chemical structures for composite biochars formed at 400°C (A) and 500°C (B). The proposed structures reconcile the molecular information from* ^{13}C *NMR, ESR, and XRD analyses. The dimensions of aromatic ring clusters are consistent with the literature compiled in Table 7.2, while the single aromatic rings are consistent with relatively slow* ^{1}H-^{13}C *dipolar dephasing (observed by Mao et al, 2012). The molar ratios H/C, O/C, N/C are (0.65,0.20, 0.0091) and (0.47, 0.13, 0.01) for the 400°C and 500°C structures, respectively. Molecular geometries are derived from 3D energy minimization calculations (MM2) performed in ChemBio3D Ultra* ® *software (Perkin Elmer Inc., Waltham, MA, USA)*

considerations make it possible to constrain the average dimensions of non-aromatic side chains (Solum et al, 1989; Mao et al, 2012). Aromatic cluster size, side-chain dimension, and elemental stoichiometry obtained from combustion analysis, XRD, NMR, and electron spin resonance information can be reconciled within a unified concept (Figure 7.9).

The proposed model structures (Figure 7.9) depict an average structural domain comprised of 3 condensed aromatic clusters with non-aromatic side chains for composite biochars formed at heat treatment temperatures of 400°C and 500°C. The XRD-based estimates of aromatic crystallite lateral dimensions, L_a (from Table 7.2), increase from ~7 Å at 400°C HTT to ~16 Å at 500°C HTT. The energy minimization of the 400°C biochar structure converged to a semi-parallel placement of the aromatic crystallites with an interlayer spacing between C atoms ranging from 3.3. to 6 Å, consistent with published d(002) spacings from XRD analysis (Bourke et al 2007; Keiluweit et al, 2010).

Free radicals in biochar

Free radicals are an important characteristic of the biochar structure and need to be considered with respect to its formation, properties, and reactivity. The importance of free radical abundance and the associated property of paramagnetism within biochar are not fully understood. However, free radicals are known to be sites of surface reactivity and thus affect cross-linking and/or polymerization reactions and surface oxidation (Feng et al, 2004), having implications for biological activity (Liao et al, 2014), mechanical strength (Zhou et al, 2013) and redox properties (Klüpfel et al, 2014). An electron spin resonance (ESR) spectroscopy study of biomass and biochars identified three major types of radicals, based on the magnetic moment of the unpaired electrons (g-factors) (Trubetskaya et al, 2016). Uncharred lignocellulosic biomass including grass, wood, and rice husk contained mostly O-centered radicals (g = 2.0026–2.0028), whereas alkyl C-centered radicals (g = 2.0027–2.0028) and aryl C-centered radicals (g = 2.0027–2.0031) were clearly formed during fast pyrolysis at high temperatures (1000°C, 1250°C, and

1500°C). Furthermore, two types of aryl-C centered radicals can be distinguished by having either (i) a narrow ESR signal which implies a narrow grouping of radical energy states consistent with distinct, electronically similar C radicals potentially residing in small sectors of structurally disorganized regions, or (ii) a broad signal which implies a variety of available energies likely due to electronic coupling with delocalized electron systems such as those found in the larger aromatic components of the biochar (Bourke et al, 2007). The narrow type of ESR signals is more common in biochars formed at lower temperatures (400–700°C) while the broader type becomes increasingly important with higher charring temperatures (700–1000°C). However, the two ESR signals representing extensive, organized, and small, disorganized π-systems are not mutually exclusive. Both types were detected in charcoals formed from corn cobs at 750°C, sucrose at 950°C, and wheat straw at 1000°C (Bourke et al, 2007; Trubetskaya et al, 2016). Alternative causes of the broader ESR signal are radicals in non-aromatic regions, including N- and O-centered radicals in nitro, nitroxy, quinone, and semiquinone groups (Trubetskaya et al, 2016 and references therein).

The stable free radicals in biochar originate from both the uncharred biomass and the pyrolysis process. Radicals originating from lignocellulosic biomass are mainly O-centered, whereas pyrolysis reactions generate mainly C-centered radicals. Pyrolysis reactions include free radical formation and recombination. The formation of free radicals is initiated during thermal bond breaking and depolymerization of cellulose, hemicellulose, and lignin. The primary pyrolysis products are generally low in molecular weight (< 200 da) and include free radical and non-radical species (Evans and Milne, 1987). Free radical recombination can occur by reaction of primary radical species with other primary radicals, or with more stable (lignocellulosic) pyrolysis residues through electron abstraction.

The concentration of unpaired electrons varies with temperature and duration of heat treatment. The number of spins reaches a maximum in charcoals generated at HTT between 400°C and 700°C and decreases progressively toward zero at HTT near 1000°C (Lewis and Singer, 1981; Emmerich et al, 1991). Rapid heating (fast pyrolysis) followed by rapid cooling on timescales < 1 second can extinguish radical recombination reactions, resulting in higher concentration of stable free radicals in biochar.

Heteroatom functional groups in biochar

As discussed above, aromatic C is the most obvious and on a mass basis frequently dominant feature of biochar that is formed during the charring process. However, the initial biomass starting materials are dominated by O and N-containing functional groups. These groups include alcohols, phenols, carboxylic acids, amides, amines, carbonyls (aldehyde and ketone), and heterocycles. While many of these are lost during the charring process through dehydration and rearrangements, even materials produced at temperatures as high as 900°C, still contain detectable quantities of O- and N-containing functional groups (Sevilla and Fuertes, 2010).

Once biochar is introduced to the soil environment, one of the first modifications to their organic molecular structure is functionalization of the surface with O-containing functional groups (Chapter 10). For example, biochar retrieved two years after application to soil showed an O/C ratio that had increased from 0.2 in the fresh material to 0.75 using X-ray photon spectroscopy (XPS) analyses (Joseph et al, 2010). This changing O/C ratio was associated with a loss

in aromatic type C and an increase in O-containing functionality from absorption of organic matter and/or oxidation of the biochar surface. The associated surfaces also showed increased N-containing functionality including proteins/peptides, ammonium, and C-N groups. On timescales of hundreds of years, biochar-C becomes extensively oxidized, with carboxyl groups (~17 mole % of PyC-C) being the second most abundant C functional group after aromatic C (~74 mole % of PyC-C) (Mao et al, 2012). Charred residues from Amazonian Terra Preta and Iowa Mollisols showed close spatial proximity of aromatic protons to carboxyl C using two-dimensional NMR heteronuclear correlation spectroscopy (Mao et al, 2012). The spectroscopic data strongly indicate that the abundant carboxyl groups are directly bonded to the aromatic rings of biochar. Based upon their analyses, the Terra Preta chars had a H/C ratio of ~0.5 and an O/C ratio of ~0.4, while the Iowa Mollisol chars had H/C and O/C ratios of ~0.5 and ~0.2, respectively.

Analytical assessments of the molecular properties of biochars must consider the physical regions of the biochars that are probed by a particular technique. The relevant analytical approaches can be roughly broken into those that provide bulk data and surface-sensitive data. Bulk data are those that represent an ensemble average of the properties within the entire sample under investigation, independent of location within the three-dimensional structure of the biochar materials.

Bulk data tend to be statistically more robust than surface data but suffer from the inclusion of potentially "unrelated" material, such as mineral ash. Bulk data are most useful for understanding the mass balance of a charring, decomposition, or soil amendment process. Surface-sensitive data are those that are dominated by some region of the biochar that is near the solid surface. The definition of surface is further complicated by the three-dimensional structure of the biochars, how it is prepared, e.g., ground or not, and differing probe depth of analytical techniques, from a few nanometers for XPS to a few microns for attenuated total reflectance – Fourier transform infrared spectroscopy (ATR-FTIR). Surface data are disproportionally important for understanding the instantaneous reactivity of the biochar material but suffer from issues of statistical representativeness and reproducibility.

By far the two most common bulk analytical approaches used to assess the structure of biochars are elemental ratios (H/C and O/C) and NMR. Information from bulk elemental ratios is frequently understood in the context of "van Krevelen plots" that define a space determined by the O/C mole ratio on the abscissa and the H/C mole ratio on the

ordinate. Because there is some amount of inorganic H, C, and O within produced biochars, when used in this context these ratios should strictly be considered as H_{org}, C_{org}, and O_{org}, although there is a large body of literature that does not explicitly make that distinction. An example van Krevelen plot (Figure 7.10) shows the position of several series of biochars and the direction in which dehydration and decarboxylation alter the position of a material in van Krevelen space.

Compilations of elemental ratios of biochars from multiple starting materials under a variety of conditions have been offered by several authors (Schimmelpfennig and Glaser, 2012), and are summarized in Figure 7.10. In general, organic H/C molar ratios decrease with increasing heat treatment temperature from around ~1.5 to a level significantly below 0.5 for pure compounds such as lignin and cellulose as well as

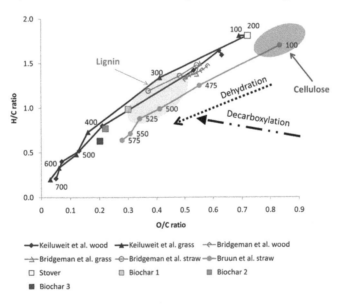

Figure 7.10 *Example van Krevelen Diagram (adapted from Brewer et al, 2011, and Schimmelpfennig and Glaser, 2012) showing H/C vs O/C ratios of several temperature series of biochars along with areas of pure cellulose and lignin. Stover, Biochar 1, 2, and 3 are the corn stover starting material and subsequent biochars produced in Brewer et al (2011). Dehydration and decarboxylation lines indicate the direction those processes move on a van Krevelen diagram*

for more complex plant biomass. Similarly, O/C ratios decline with the duration and intensity of heat treatments. These robust trends render the van Krevelen plot and elemental analysis an excellent tool for obtaining a reliable initial estimate of general biochar characteristics. Data from CHNO elemental analyses can be used to consider the molar elemental ratios as empirical formulae, $C_cH_hN_nO_o$, where the subscripts represent the molar quantities of the element. The molar quantities are then amenable to treatment by traditional organic chemistry concepts. For instance, the molar H/C ratio is a commonly used measure of the degree of unsaturation (the formation of C=C double bonds) in organic polymers.

Since the stoichiometry of a saturated species can be represented as C_cH_{c+2}, the H deficiency (z) relative to a saturated species can be calculated by Equation 7.4. The estimation of H deficiency also facilitates the calculation of double bond equivalents - a quantity that is particularly suited to the classification of aromatic materials. Double bond equivalents (DBE) are defined as the number of double bonds plus the number of rings (e.g., benzene has 4 DBE), and can be calculated from elemental data by Equation 7.5. The DBE value can then be used to calculate the aromatic fraction, f_a, by normalizing DBE to the molar C and N (c + n) content (Equation 7.6). The aromatic fraction calculated from elemental CHN data can be directly compared to the aromatic fraction (sp^2 hybridized C) measured by ^{13}C NMR, according to Equation 7.7. The biochar H/C ratios in the range below 0.5 correspond to f_a values from approximately 0.8 to 1.

$$z = -2c + h \qquad [7.4]$$

$$DBE = \frac{-z}{2} + \frac{n}{2} + 1 \qquad [7.5]$$

$$f_a = \frac{DBE}{(c + n)} \qquad [7.6]$$

$$f_a = \frac{NMR\ peak\ area\ 110 - 220\ ppm}{NMR\ peak\ area\ 0 - 220\ ppm} \qquad [7.7]$$

Fourier transform infrared spectroscopy has been used extensively to understand the development and alterations in functional group chemistry of biochars. Several features appear to be common to the FTIR analyses. These include that even low-temperature pyrolysis processes (~300°C) can induce loss of resolution (i.e., an increase in the overall number of partially overlapping signals) in the fingerprint region, ~1500 to 1100 cm^{-1}, of the FTIR spectra, a result indicative of an increase in the diversity of functional groups. Even at this relatively low temperature, given sufficient time, ~24 hrs, the intensity of peaks representative of aromatic moieties increases (Figure 7.11; Rutherford et al, 2012). Investigations of this kind allow inference of the kind that carbonyl functionality as represented by the signal in the 1700–1750 cm^{-1} range reaches a minimum after ~8 hr of heating at low to moderate temperatures.

Charring for short times at temperatures below ~400–500°C generates an increase in aromatic functionality as determined by aromatic C=C stretching bands at ~1600 cm^{-1} and by aromatic C-H bending at ~885, 815, and 750 cm^{-1} in the FTIR spectra. Also seen are strong signals associated with carbonyl functionality as determined by peaks in the 1700–1750 cm^{-1} region (Keiluweit et al, 2010; Figure 7.12). Biochars generated below ~400–500°C continue to show strong absorbance in the 3500–3100 cm^{-1} range, indicating the continued presence of hydroxyl functionality. Biochars formed above these temperatures show rapid decreases in these two areas, indicative of proportional loss of

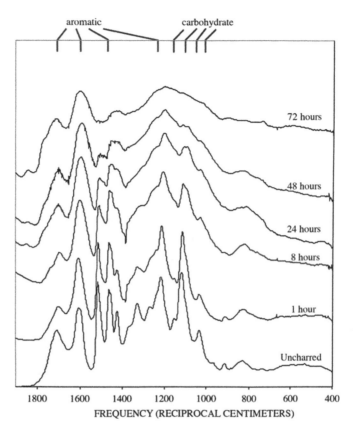

Figure 7.11 *FTIR spectra of pine wood heated at 300°C for up to 168 hr (from Rutherford et al, 2012; with permission from the publisher)*

O-containing functionality. In general, the FTIR spectra become relatively featureless. This, however, does not mean that the materials are not absorbing IR light, but rather that the presence of distinct O-containing functionality is diminished. Raman spectra, based upon the adsorption and scattering of infrared wavelengths, exhibit similar loss of O- and alkyl functionality with increasing heat treatment energy. Additionally, Raman signals of biochar samples yield broad peaks centered at wavelengths $1350\ cm^{-1}$ (D-band) corresponding to aromatic crystallites > 6 fused benzene rings, and $1590\ cm^{-1}$ (G-band) for aromatic ring-breathing vibrations (Li and Li, 2006). The intensity ratio of the D/G signal ratio tends to increase with the heat flux (energy/area) applied to biochar particles (Yu et al, 2018).

FTIR spectra correlate well with surface O-containing functionality, carboxylic acids, anhydrides, and organic carbonates, as determined by titration for biochars formed at temperatures up to ~400°C (Reeves et al, 2008; Reeves, 2012). Biochars produced at 550°C had dramatically lower O-containing functionality than those produced at 400°C as determined by Boehm titration of biochars made from eucalyptus, wood and leaves, paper sludge, poultry litter, and cow manure (Singh et al, 2010). Peaks representing O-

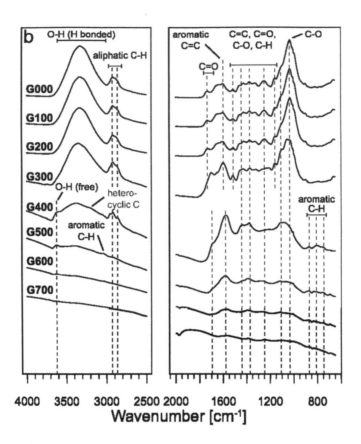

Figure 7.12 *Loss of functionality at higher charring temperatures. Temperature series of ATR-FTIR spectra for biochars from festuca grass (with permission from Keiluweit et al, 2010)*

containing functionality, mostly carboxylic groups, are reduced in the 350–500°C temperature range (Harvey et al, 2012).

Near-edge X-ray absorption fine structure (NEXAFS) spectroscopy both in bulk and in associations with Scanning Transmission X-ray Microscopy (STXM) has been used to compare estimates for biochar aromaticity with respective information from NMR (Heymann et al, 2011; Wiedemeier et al, 2014). The imaging capability of the STXM identified a distinct decrease in aromatic C and a proportional increase in phenolic and carboxylic C as a function of distance from the center of natural biochar particles isolated from Brazilian soil (Lehmann et al, 2005). NEXAFS also showed

a strong increase in aromatic functionality in biochars made from either starting material at 400°C as compared to 300°C (Keiluweit et al, 2010). Concurrent with the increase in aromaticity, a loss in O-containing carboxylate and O-alkyl functional groups was observed. Bulk NEXAFS analyses of a series of reference materials including biochars from wood and grass as well as soot, shale, bituminous, and lignite coals, and different soils showed total aromaticity of ~40% of the biochar samples as well as the two soil samples. However, the NEXAFS spectra did not allow for complete distinction between the wood and grass-derived biochars or between the biochars and fossil fuel-derived soot. Correlations between

NEXAFS spectral deconvolution results and NMR ^{13}C (DPMAS) were also mixed, showing relatively good correlations for aromatic C ($r^2 = 0.63$, $p < 0.05$) and a lower correlation ($r^2 = 0.49$, $p < 0.05$) for aromatic C / O-alkyl C ratio.

Conclusions

Several mechanisms control the evolution of macromolecular features in thermally altered organic materials. An important factor is the highest temperature achieved during the pyrolysis process. Documented effects are also associated with feedstock properties (chemical makeup, biomass density, particle size, particle shape, etc.) and with the heat exposure time or residence time within the pyrolysis retort. The influence of temperature ramp rates is not well understood. Other factors such as mineral admixtures from soil adhering to biomass feedstocks are likely to influence the evolution of organic macromolecular structure during the pyrolysis process as well but have so far not been examined in detail. Our review of published data (Table 7.2) illustrates that it will not be possible to predict the molecular "outcome" of a specific pyrolytic treatment for a given organic feedstock based on heat treatment temperature alone. We conclude that the technological basis of "making (bio-)chars to purpose" has yet to be developed. An alternative route to "match (bio-)char properties to the right situation" (Sohi et al, 2010) may be the establishment of meaningful biochar characterization protocols with a high-throughput capability that would facilitate the selection of biochars for specific applications based on their functionalities.

References

Allred AL, and Rochow E 1958 A scale of electronegativity based on electrostatic force. *Journal of Inorganic and Nuclear Chemistry* 5, 264–268.

Almendros G, González-Vila FJ, Martin F, Fründ R, and Lüdemann HD 1992 Solid-state NMR-studies of fire-induced changes in the structures of humic substances. *Science of the Total Environment* 118, 63–74.

Almendros G, Knicker H, and González-Vila FJ 2003 Rearrangement of carbon and nitrogen forms in peat after progressive thermal oxidation as determined by solid-state ^{13}C- and ^{15}N-NMR spectroscopy. *Organic Geochemistry* 34, 1559–1568.

Amonette JE, and Joseph S 2009 Characteristics of biochar: microchemical properties. In: Lehmann J, and Joseph S (Eds) *Biochar for Environmental Management – Science and Technology*. Sterling: Earthscan. pp 33–52.

Antal MJ, and Grønli M. 2003 The art, science, and technology of charcoal production. *Industrial and Engineering Chemistry Research* 42, 1619–1640.

Bacon R, and Tang MM 1964 Carbonization of cellulose fibers—II. Physical property study. *Carbon* 2, 221–222.

Baldock JA, and Smernik RJ 2002 Chemical composition and bioavailability of thermally altered *Pinus resinosa* (Red pine) wood. *Organic Geochemistry* 33, 1093–1109.

Barron PF, and Wilson MA 1981 Humic soil and coal structure study with magic-angle spinning C-13 CP-NMR. *Nature* 289, 275–276.

Barron PF, Wilson MA, Stephens JF, Cornell BA, and Tate KR 1980 Cross-polarization C-13

NMR-spectroscopy of whole soils. *Nature* 286, 585–587.

Bourke J, et al 2007 Do all carbonized charcoals have the same chemical structure? 2. A model of the chemical structure of carbonized charcoal. *Industrial Engineering Chemistry Research* 46, 5954–5967.

Brewer CE, Schmidt-Rohr K, Satrio JA, and Brown RC 2009 Characterization of biochar from fast pyrolysis and gasification systems. *Environmental Progress and Sustainable Energy* 28, 386–396.

Brewer CE, Unger R, Schmidt-Rohr K, and Brown RC 2011 Criteria to select biochars for field studies based on biochar chemical properties. *Bioenergy Research* 4, 312–323.

Cao XY, et al 2012 Characterization of wood chars produced at different temperatures using advanced solid-state C-13 NMR spectroscopic techniques. *Energy and Fuels* 26, 5983–5991.

Chen BL, Zhou DD, and Zhu LZ 2008 Transitional adsorption and partition of nonpolar and polar aromatic contaminants by biochars of pine needles with different pyrolytic temperatures. *Environmental Science and Technology* 42, 5137–5143.

Diamon R 1957 X-ray diffraction data for large aromatic molecules. *Acta Crystallographica* 10, 359–364.

Dickerson RE, and Geis I 1976. Chemistry, matter and the universe – An integrated approach to general chemistry, WA Benjamin Inc, Menlo Park.

Emmerich FG, Rettori C, and Luengo CA 1991 ESR in heat treated carbons from the endocarp of Babassu coconut. *Carbon* 29, 305–311.

Enders A, Hanley K, Whitman T, Joseph S, and Lehmann J 2012 Characterization of biochars to evaluate recalcitrance and agronomic performance. *Bioresource Technology* 114, 644–653.

Ergun S, and Tiensuu V 1959 Interpretation of the intensities of X-rays scattered by coals. *Fuel* 38, 64–78.

Evans RJ, and Milne TA 1987 Molecular characterization of the pyrolysis of biomass. *Energy and Fuels* 1, 123–137.

Feng JW, Zheng S, and Maciel GE 2004 EPR Investigations of charring and char-air interaction of cellulose, pectin, and tobacco. *Energy and Fuels* 18, 560–568.

Fletcher TH, Solum MS, Grant DM, and Pugmire RJ 1992 Chemical structure of char in the transition from devolatilization to combustion. *Energy and Fuels* 6, 643–650.

Franklin RE 1950 The interpretation of diffuse X-ray diagrams of carbon. *Acta Crystallographica* 3, 107–121.

Franklin RE 1951 Crystallite growth in graphitizing and non-graphitizing carbons. *Proceedings of the Royal Society of London. Series A. Mathematical and Physical Sciences* 209, 196–218.

Freitas JCC, Bonagamba TJ, and Emmerich FG 1999 C-13 High-resolution solid-state NMR study of peat carbonization. *Energy and Fuels* 13, 53–59.

Freitas JCC, Bonagamba TJ, and Emmerich FG 2001 Investigation of biomass- and polymer-based carbon materials using C-13 high-resolution solid-state NMR. *Carbon* 39, 535–545.

Harvey OR, Herbert BE, Kuo LJ, and Louchouarn P 2012 Generalized two-dimensional perturbation correlation infrared spectroscopy reveals mechanisms for the development of surface charge and recalcitrance in plant-derived biochars. *Environmental Science and Technology* 46, 10641–10650.

Harvey OR, et al 2016 Discrimination in degradability of soil pyrogenic organic matter follows a return-on-energy-investment principle. *Environmental Science and Technology* 50, 8578–8585.

Hatcher PG, Schnitzer M, Dennis LW, and Maciel GE 1981 Aromaticity of humic substances in soils. *Soil Science Society of America Journal* 45, 1089–1094.

He Z, Guo M, Fortier M, Cao X, and Schmidt-Rohr K 2021 Fourier transform infrared and solid state ^{13}C nuclear magnetic resonance spectroscopic characterization of defatted cottonseed meal-based biochars. *Modern Applied Science* 15, 108–118.

Heidenreich RD, Hess WM, and Ban LL 1968 A test object and criteria for high resolution electron microscopy. *Journal of Applied Crystallography* 1, 1–19.

Heymann K, Lehmann J, Solomon D, Schmidt MWI, and Regier T 2011 C 1s K-edge near edge X-ray absorption fine structure (NEXAFS) spectroscopy for characterizing functional group chemistry of black carbon. *Organic Geochemistry* 42, 1055–1064.

Hückel E 1931 Quantentheoretische Beiträge zum Benzolproblem. *Zeitschrift für Physik* 70, 204–286.

Ippolito JA, Laird DA, and Busscher WJ 2012 Environmental benefits of biochar. *Journal of Environmental Quality* 41, 967–972.

IUPAC 2006 Compendium of chemical terminology. 2nd edn (the 'Gold Book'). Compiled by McNaught AD and Wilkinson A. Blackwell Scientific Publications: Oxford.

Joseph SD, et al 2010 An investigation into the reactions of biochar in soil. *Australian Journal of Soil Research* 48, 501–515.

Kaal J, and Rumpel C 2009 Can pyrolysis-GC/MS be used to estimate the degree of thermal alteration of black carbon? *Organic Geochemistry* 40, 1179–1187.

Keiluweit M, Nico PS, Johnson, MG, and Kleber M 2010 Dynamic molecular structure of plant biomass-derived black carbon (biochar). *Environmental Science and Technology* 44, 1247–1253.

Keiluweit M, and Kleber, M 2009 Molecular-level interactions in soils and sediments: The role of aromatic π-systems. *Environmental Science & Technology* 43, 3421–3429.

Kercher AK, and Nagle DC 2003 Microstructural eution during charcoal carbonization by X-ray diffraction analysis. *Carbon* 41, 15–27.

Kloss S, et al 2012 Characterization of slow pyrolysis biochars: effects of feedstocks and pyrolysis temperature on biochar properties. *Journal of Environmental Quality* 41, 990–1000.

Klüpfel L, Keiluweit M, Kleber M, and Sander M 2014 Redox properties of plant biomass-derived black carbon (biochar). *Environmental Science and Technology* 48, 5601–5611.

Knicker H, and Lüdemann HD 1995 N-15 and C-13 CPMAS and solution NMR studies of N-15 enriched plant material during 600 days of microbial degradation. *Organic Geochemistry* 23, 329–341.

Knicker H, Almendros G, Gonzalez-Vila, FJ, Martin F, and Lüdemann HD 1996 C-13- and N-15-NMR spectroscopic examination of the transformation of organic nitrogen in plant biomass during thermal treatment. *Soil Biology and Biochemistry* 28, 1053–1060.

Laird DA 2008 The charcoal vision: a win win win scenario for simultaneously producing bioenergy, permanently sequestering carbon, while improving soil and water quality. *Agronomy Journal* 100, 178–181.

Lehmann J 2007a Bio-energy in the black. *Frontiers in Ecology and the Environment* 5, 381–387.

Lehmann J 2007b A handful of carbon. *Nature* 447, 143–144.

Lehmann J, and Joseph S 2009 *Biochar for Environmental Management – Science and Technology*. Sterling: Earthscan.

Lehmann J, et al 2005 Near-edge X-ray absorption fine structure (NEXAFS) spectroscopy for mapping nano-scale distribution of organic carbon forms in soil: application to black carbon particles. *Global Biogeochemical Cycles* 19, GB1013.

Lewis IC, and Singer LS 1981 Electron-spin resonance and the mechanism of carbonization. In: Walker PL, and Thrower PA (Eds) *Chemistry and Physics of Carbon*. New York: Marcel Dekker 17. pp 1–88.

Li X, and Li CZ 2006 Volatilisation and catalytic effects of alkali and alkaline earth metallic species during the pyrolysis and gasification of Victorian brown coal. Part VIII. Catalysis and changes in char structure during gasification in steam. *Fuel* 85, 1518–1525.

Liao S, Pan B, Li H, Zhang D, and Xing B 2014 Detecting free radicals in biochars and determining their ability to inhibit the germination and growth of corn, wheat and rice seedlings. *Environmental Science and Technology* 48, 8581–8587.

Lu L, Sahajwalla V, and Harris D 2000 Characteristics of chars prepared from various pulverized coals at different temperatures using drop-tube furnace. *Energy and Fuels* 14, 869–876.

Lu L, Sahajwalla V, Kong C, and Harris D 2001 Quantitative X-ray diffraction analysis and its application to various coals. *Carbon* 39, 1821–1833.

Mao JD, et al 2012 Abundant and stable char residues in soils: implications for soil fertility and carbon sequestration. *Environmental Science and Technology* 46, 9571–9576.

McBeath AV, and Smernik RJ 2009 Variation in the degree of aromatic condensation of chars. *Organic Geochemistry* 40, 1161–1168.

McBeath AV, Smernik RJ, Schneider MPW, Schmidt MWI, and Plant EL 2011 Determination of the aromaticity and the degree of aromatic condensation of a thermosequence of wood charcoal using NMR. *Organic Geochemistry* 42, 1194–1202.

McClellan-Maaz TM, Hockaday WC, and Deenik JL 2021 Biochar volatile matter and feedstock effects on soil nitrogen mineralization and soil fungal Colonization. *Sustainability* 13, 2018–2033.

McDermott MT, and McCreery, RL 1994 Scanning tunneling microscopy of ordered graphite and glassy carbon surfaces: Electronic control of quinone adsorption. *Langmuir* 10, 4307–4314.

Mochidzuki K, et al 2003 Electrical and physical properties of carbonized charcoals. *Industrial and Engineering Chemistry Research* 42, 5140–5151.

Mok WSL, and Antal MJ 1983 Effects of pressure on biomass pyrolysis. 2. Heats of reaction of cellulose pyrolysis. *Thermochimica Acta* 68, 165–186.

Nguyen BT, and Lehmann J 2009 Black carbon decomposition under varying water regimes. *Organic Geochemistry* 40, 846–853.

Nguyen BT, Lehmann J, Hockaday WC, Joseph S, and Masiello CA 2010 Temperature sensitivity of black carbon decomposition and oxidation. *Environmental Science and Technology* 44, 3324–3331.

Paris O, Zollfrank C, and Zickler GA 2005 Decomposition and carbonisation of wood biopolymers - a microstructural study of softwood pyrolysis. *Carbon* 43, 53–66.

Patterson AL 1939 The Scherrer formula for X-ray particle size determination. *Physical Review* 56, 978–982.

Reeves JB 2012 Mid-infrared spectroscopy of biochars and spectral similarities to coal and kerogens: what are the implications. *Applied Spectroscopy* 66, 689–695.

Reeves JB, McCarty GW, Rutherford DW, and Wershaw RL 2008 Mid-infrared diffuse reflectance spectroscopic examination charred pine wood, bark, cellulose, and lignin: Implications for the quantitative determination of charcoal in soils. *Applied Spectroscopy* 62, 182–189.

Rutherford DW, Wershaw RL, Rostad CE, and Kelly CN 2012 Effect of formation conditions on biochars: Compositional and structural properties of cellulose, lignin, and pine biochars. *Biomass and Bioenergy* 46, 693–701.

Saikia BK 2010 Inference on carbon atom arrangement in the turbostatic graphene layers in Tikak coal (India) by X-ray pair distribution function analysis. *International Journal of Oil Gas and Coal Technology* 3, 362–373.

Schimmelpfennig S, and Glaser B 2012 One step forward toward characterization: some important material properties to distinguish biochars. *Journal of Environmental Quality* 41, 1001–1013.

Schoening FRL 1982 X-ray structural parameter for coal. *Fuel* 61, 695–699.

Schoening FRL 1983 X-ray structure of some South African coals before and after heat-treatment at 500 degrees C and 1000 degrees C. *Fuel* 62, 1315–1320.

Schwager I, Farmanian P, Kwan JT, Weinberg VA, and Yen TF 1983 Characterization of the microstructure and macrostructure of coal-derived asphaltenes by nuclear magnetic resonance spectrometry and X-ray diffraction. *Analytical Chemistry* 55, 42–45.

Sevilla M, and Fuertes AB 2010 Graphitic carbon nanostructures from cellulose. *Chemical Physics Letters* 490, 63–68.

Singh BP, Cowie AL, and Smernik RJ 2012 Biochar carbon stability in a clayey soil as a function of feedstock and pyrolysis temperature. *Environmental Science and Technology* 46, 11770–11778.

Singh BP, Hatton BJ, Singh B, Cowie AL, and Kathuria A 2010 Influence of biochars on nitrous oxide emission and nitrogen leaching from two contrasting soils. *Journal of Environmental Quality* 39, 1224–1235.

Smernik RJ 2005 A new way to use solid-state carbon-13 nuclear magnetic resonance spectroscopy to study the sorption of organic compounds to soil organic matter. *Journal of Environmental Quality* 34, 1194–1204.

Smernik RJ, Kookana, RS, and Skjemstad JO 2006 NMR characterization of C-13-benzene sorbed to natural and prepared charcoals. *Environmental Science and Technology* 40, 1764–1769.

Sohi S, Krull E, Lopez-Capel E, and Bol R 2010 A review of biochar and its use and function in soil. *Advances in Agronomy* 105, 47–82.

Sohi SP 2012 Carbon storage with benefits. *Science* 338, 1034–1035.

Solum MS, Pugmire RJ, and Grant DM 1989 Carbon-13 solid-state NMR of Argonne-premium coals. *Energy & Fuels* 3, 187–193.

Spokas KA, et al 2012 Biochar: a synthesis of its agronomic impact beyond carbon sequestration. *Journal of Environmental Quality* 41, 973–989.

Sultana N, Ikeya K, Shindo H, Nishimura S, and Watanabe A 2010 Structural properties of plant charred materials in Andosols as revealed by X-ray diffraction profile analysis. *Soil Science and Plant Nutrition* 56, 793–799.

Sultana N, Ikeya K, and Watanabe A 2011 Partial oxidation of char to enhance potential interaction with soil. *Soil Science* 176, 495–501.

Sun K, et al 2012 Polar and aliphatic domains regulate sorption of phthalic acid esters (PAEs) to biochars. *Bioresource Technology* 118, 120–127.

Sun K, Keiluweit M, Kleber M, Pan ZZ, and Xing BS 2011 Sorption of fluorinated herbicides to plant biomass-derived biochars as a function of molecular structure. *Bioresource Technology* 102, 9897–9903.

Tang MM, and Bacon R 1964 Carbonization of cellulose fibers—I. Low temperature pyrolysis. *Carbon* 2, 211–214.

Trubetskaya A, et al 2016 Characterization of free radicals by electron spin resonance spectroscopy in biochars from pyrolysis at high heating rates and at high temperatures. *Biomass and Bioenergy* 94, 117–129.

Uchimiya M, Wartelle LH, Klasson KT, Fortier CA, and Lima IM 2011 Influence of pyrolysis temperature on biochar property and function as a heavy metal sorbent in soil. *Journal of Agricultural and Food Chemistry* 59, 2501–2510.

Warren BE 1934 X-ray diffraction study of carbon black. *The Journal of Chemical Physics* 2, 551–555.

Wiedemeier DB, et al 2014 Aromaticity and the degree of aromatic condensation of chars. *Organic Geochemistry* 78, 135–143.

Yen TF, Erdman JG, and Pollack SS 1961 Investigation of the structure of petroleum asphaltenes by X-ray diffraction. *Analytical Chemistry* 33, 1587–1594.

Yoo S, Kelley SS, Tilotta DC, and Park S 2018 Structural characterization of Loblolly pine derived biochar by X-ray diffraction and energy loss spectroscopy. *ACS Sustainable Chemistry and Engineering* 6, 2621–2629.

Yu J, et al 2018 Influence of temperature and particle size on structural characteristics of chars from Beechwood pyrolysis. *Journal of Analytical and Applied Pyrolysis* 130, 127–134.

Zhou XY, et al 2013 Properties of formaldehyde-free environmentally friendly lignocellulosic composites made from poplar fibres and oxygen-plasma-treated enzymatic hydrolysis lignin. *Composites Part B Engineering* 53, 369–375.

Zhu D, and Pignatello JJ 2005 Characterization of aromatic compound sorptive interactions with black carbon (charcoal) assisted by graphite as a model. *Environmental Science & Technology* 39, 2033–2041.

8

Characteristics of biochar

Nutrient properties

Joann K. Whalen, Leanne Ejack, Shamim Gul, and Leonardo León Castro

Introduction

When biochar is used for agricultural soil improvement or land remediation, it contributes to the soil nutrient supply. Biochar is a source of macronutrients and micronutrients that are essential for plant nutrition. Biochars prepared from diverse feedstocks have variable nutrient properties, due to the unique chemistry of each feedstock. Nutrient transformations on the biochar surface and within its matrix, independent of biochar interactions with the environment, are influenced by the temperature and duration of the pyrolysis process, as well as pre- and post-pyrolysis treatments. The same feedstock can be converted into biochar with differing chemical and physical attributes, which makes it difficult to predict the biochar nutrient properties. The goal is to clarify how feedstock and pyrolysis conditions affect the nutrient characteristics of biochar, so that biochar products with specific nutrient profiles and nutrient sorption capacities can be made for agriculture and environmental uses.

This chapter describes the inherent elemental composition and physico-chemical properties that affect the biochar nutrient content (environmental reactions that alter the nutrient supply from biochar, such as leaching and biological reactions in soil-plant systems, are explained in Chapters 12, 13, and 16). Here, the focus is on how feedstocks and pyrolysis conditions alter the total and bioavailable nutrient content of biochar. Also considered is how biochar surfaces and internal matrices contribute to nutrient exchange and nutrient transformations, considering the influence of surface functional groups, ion exchange capacity, precipitation reactions, pore volumes, and pH effects. Physical and chemical treatments can be applied pre- and post-pyrolysis to enhance the biochar nutrient supply. The conclusion proposes future areas of research that will enhance feedstock management and the pyrolysis process to generate biochar with desired nutrient properties.

DOI: 10.4324/9781003297673-8

Total nutrient content

Feedstock

The nutrient content of biochar is not identical to the nutrient concentration of the original feedstock, but the feedstock type is a general indicator of the elemental composition of biochar. Biochar derived from plant-based materials like shells, husks, and wood has a lower total N and total P content than biochar derived from manure, biosolids, and food waste (Hossain et al, 2020; Ippolito et al, 2020). There is 2.5 times more total N and double the total S content in biochar produced from manure and biosolids (2.4 g total N kg^{-1}; 0.9 g total S kg^{-1}) than biochar made with wood residues, with intermediate N and S concentrations in biochar derived from crop and grass residues (Figure 8.1). Biochar made from manure and biosolids has elevated total P, Ca, and Mg concentrations relative to biochar produced from other feedstocks (Figure 8.1). The total K concentration follows a different pattern: pyrolyzed grasses > manure/biosolids > crop residue > wood residue (Figure 8.1). Generally, the biochar nutrient concentration is correlated with the nutrient content of the initial feedstock.

Wood, crop, and grass residues contain modest amounts of N and S, compared to manures, biosolids, food wastes, microalgal biomass, and insect biomass, because these materials tend to have a higher protein content than plant materials. Insect biomass contains chitin, a long-chain polymer of N-acetylglucosamine containing 6–11% N, the second most abundant polysaccharide in nature after cellulose. Biochar from cricket chitin had 0.9–1.1 g total N kg^{-1}, which is nearly the same as the N content in the initial feedstock (Różyło et al, 2022), possibly because chitin is thermally stabilized by amine-N bonding to glucose polymers. The

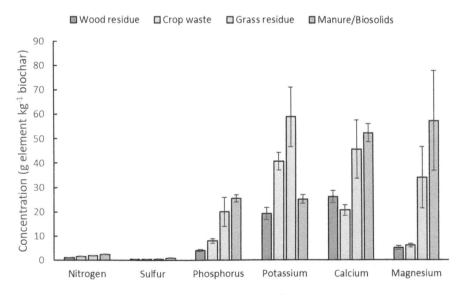

Figure 8.1 *Macronutrient concentrations (g element kg^{-1}) in biochar produced from four types of feedstock (data from Ippolito et al, 2020)*

protein-rich microalgae *Chlorella vulgaris* contains 56% w/w protein (0.79 g total N kg^{-1}) and produces a biochar with 0.42–0.70 g total N kg^{-1}, depending on pyrolysis conditions (Maliutina et al, 2018). Another study found that Chlorella-based algal biomass produced biochar with 1.1 g total N kg^{-1}, similar to the feedstock; the biomass can be applied directly as a soil fertilizer because it contained ≥0.2% w/w of P, K, Ca, Mg, and Fe (Chang et al, 2015).

The elevated P concentration in biochar of grass residues, manures, and biosolids is a result of physiologically driven P accumulation in these materials. Grass residue feedstocks include leguminous plants that require energetic P compounds like adenosine triphosphate (ATP) for growth, seed production, and biological nitrogen fixation. Animal and human diets (and the associated food wastes) contain more P than can be assimilated for metabolic needs, so the excess P is excreted and accumulates in the manures and biosolids. Since biochar may contain up to 20–25 g P kg^{-1}, the P applied in P-rich biochar needs to be considered in soil nutrient management plans for agri-environmental protection on farms.

Grasses like sugarcane, cereals, bamboo, and other natural and cultivated grasses use K to build strong stems that resist diseases, water stress, and lodging from high winds and heavy rainstorms. Non-edible stems and leaves, therefore, contain most of the K remaining in the plant at physiological maturity (i.e., much of the N and P are removed in the grain, seeds, and fruit at harvest). Hence, crop residues and grasses used as feedstocks in thermochemical conversion are naturally rich in K. When animals and humans consume diets with an excess of K, or with excess Ca and Mg, these water-soluble nutrients are excreted into the manure and biosolids, which explains the elevated K, Ca and Mg in manure and biosolids biochar.

Within the biochar matrix, K, Ca, and Mg salts are distributed in pores, associated with organic acids and chelates, or bound to ion exchange sites (Moradi-Choghamarani et al, 2019). When biochar is wetted, these macronutrient cations solubilize and diffuse out of the biochar matrix, or exchange from charged surfaces into aqueous solution.

Micronutrients – e.g., B, Cl, Cu, Fe, Mo, Mn, and Zn – are essential for microbial, plant, and animal growth. As micronutrients are present in feedstocks, biochar contains trace amounts, from a few micrograms to milligrams of each micronutrient kg^{-1} of pyrolyzed material. Biochar produced from manure and biosolids is often enriched with ~ 1 g kg^{-1} of Cu, Fe, and Zn because of the incomplete assimilation of micronutrients from animal diets (Xiao et al, 2018; Hossain et al, 2020). Biochar made from sugarcane filter cake, farmyard manure, and rice husk contained from 15–115 mg Cu kg^{-1}, 0.5–13 g Fe kg^{-1}, and 131–681 mg Mn kg^{-1} with 105–402 mg Zn kg^{-1} and met the micronutrient requirements of a maize test crop (Choudhary et al, 2021). Pyrolysis eliminates antibiotics and organic contaminants from poultry manures and retains most of the essential macronutrients and micronutrients, producing a biochar that could be recycled for agriculture and land reclamation (Dróżdż et al, 2020). However, there is a fine line between micronutrient sufficiency and toxicity in plants. Therefore, land managers must know the biochar micronutrient content and avoid higher application rates that oversupply micronutrients to plants.

Pyrolysis conditions

Pyrolysis conditions, namely the furnace temperature, heating rate, duration, pressure, and reaction atmosphere, influence the total nutrient content of biochar. Of these factors, the pyrolysis temperature is the most important. In general, higher pyrolysis temperatures are associated with the loss of volatile elements (H, O, N, and S) that have high vapor pressure and form gases that evaporate from solid materials.

Biochar from the same feedstock often has a lower N concentration when the pyrolysis temperature increases. Nitrogen concentration declined by about 0.1% for every 100°C increase in pyrolysis temperature (Figure 8.2), independent of the heating rate, duration of pyrolysis, reaction atmosphere, and pressure. Pyrolysis temperatures ≥300°C transform inorganic N salts into N-containing gases (NO, N_2O, NO_2, N_2, NH_3) and increase the reaction of heterocyclic N compounds (pyridine-N and pyrrolic-N) with H radicals to form NH_3 gas (Zheng et al, 2013; Moradi-Choghamarani et al, 2019). Although the N concentration tends to be lower in the biochar than the original feedstock, the remaining N can be conserved by binding to C in aromatic compounds, i.e.,

forming pyrrolic-N or pyridinic-N structures. For example, increasing the pyrolysis temperature from 350 to 700°C produced 7-fold more aromatic C=N in 6-membered oxygenated ring structures and doubled the amount of quaternary aromatic N (Torres-Rojas et al, 2020). High-temperature pyrolysis of feedstock with a higher N concentration was more likely to produce *de novo* heterocyclic aromatic N because pyrolysis at ≥500°C provides enough energy to incorporate N atoms into an aromatic C ring structure.

The S contained in feedstock is susceptible to volatilization at higher pyrolysis temperatures. Wheat straw lost 16% of its S content after 1 h pyrolysis at 200°C and as much as 50% of the initial S content after 4 h pyrolysis at 200°C, but the S content was 45–50% lower than the initial concentration when the pyrolysis temperature was set at 400°C or 600°C, regardless of the duration of pyrolysis (Figure 8.3). Similarly, lignosulfonate that contained 128 g S kg^{-1}

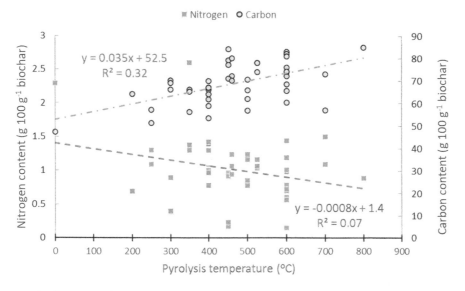

Figure 8.2 *As the pyrolysis temperature increases, there is a decline in the volatile nitrogen concentration (g N 100 g^{-1} biochar) and a slight increase in the non-volatile carbon content (g C 100 g^{-1} biochar) (data from Li et al, 2019; Das et al, 2021)*

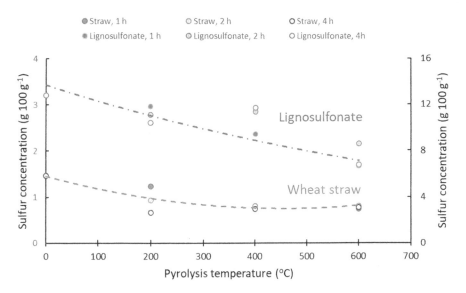

Figure 8.3 *An increase in pyrolysis temperature is associated with lower sulfur concentration (g S 100 g^{-1} biochar) in wheat straw biochar, on the primary y-axis, and lignosulfonate biochar, plotted on the secondary y-axis. There is virtually no difference in the S concentration due to the duration of the pyrolysis reaction (data from Zhang et al, 2015)*

lost more S due to the pyrolysis temperature than the time that the feedstock was retained in the pyrolyzer (Figure 8.3). Pyrolyzed biomass contains inorganic sulfur in reduced forms such as FeS_2, but it also contains organic sulfur compounds that vary in their thermostability. Pyrolysis of bituminous coal produces a char that contains more thiophene > sulfoxide > sulfone > mercaptan or thiophenol (Tian et al, 2018). The many aromatic rings in thiophene-bound sulfur make it relatively non-volatile, compared to other forms of organic sulfur. The chemistry of inorganic and organic sulfur compounds in feedstocks, as well as their bond strength when subjected to pyrolysis, needs more study.

Despite uncertainty about the amount of N and S volatilized during pyrolysis, the loss of these elements is probably unavoidable because some of the N and S in inorganic and organic compounds will transform into gas upon heating. Since there is mass loss

when a feedstock is pyrolyzed, the change in nutrient mass between the feedstock and the biochar (on an equivalent mass basis) will be more accurate than comparing the nutrient concentration of the feedstock and the resulting biochar (Enders et al, 2012).

Non-volatile elements within the feedstock tend to concentrate when the material is pyrolyzed at a higher temperature because more mass is lost from the feedstock. Biochar has a higher P, K, and Ca concentration when it is produced at higher pyrolysis temperatures; Mg dynamics are uncertain due to limited experimental measurements (Figure 8.4). Atmospheric aerosol particles emitted from combustion sources (e.g., coal-burning plants, and automobiles) contain P, K, Ca, Mg, micronutrients, trace metals, and other volatiles. Therefore, a mass balance approach (e.g., Schimmelpfennig and Glaser, 2012) should be used to determine the quantities of macronutrients and

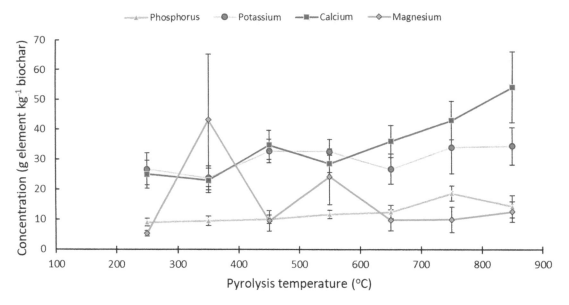

Figure 8.4 *An increase in pyrolysis temperature produces biochar with a greater P, K, and Ca concentration, but variable Mg concentration (data from Ippolito et al, 2020)*

micronutrients from the original feedstock that are retained after pyrolysis.

The heating rate refers to the incremental increase in temperature needed to reach the target pyrolysis temperature. Heating rates <100°C min^{-1} are characteristic of slow pyrolysis, which has greater biochar yield compared to a faster heating rate. Faster heating rates are favored when the goal is to generate more syngas and bio-oil than biochar. Heating rates determine the biochar yield but cause minimal change to the nutritive content. Angin (2013) produced safflower seed cake biochar by heating the feedstock at rates of 10, 30, and 50°C min^{-1}; they had virtually the same nutrient concentrations. Fast pyrolysis with a heating rate >120°C min^{-1} did not affect the C, N, and P concentrations of biochar, but there were generally greater total S, K, Ca, and Mg concentrations in biochar produced by slow pyrolysis (heating rate <120°C min^{-1}; Ippolito et al, 2020). The variation in S, K, Ca, and Mg

concentrations could be affected by the type of feedstock used for fast or slow pyrolysis, rather than the heating rate per se. Additional work is needed to confirm how biochar nutrient properties are affected by the heating rate. Such studies would ideally vary the heating rate of the same feedstock to reach a constant pyrolysis temperature, as such data are scarce.

Besides pyrolysis temperature and heating rate, some data is available on other factors relating to pyrolysis conditions and nutrient content. The duration of pyrolysis, or furnace residence time, tends to co-vary with the pyrolysis temperature because faster reactions occur at higher temperatures. For example, the loss of volatile S from wheat straw was affected more by the pyrolysis temperature than the duration of pyrolysis from 1–4 h (Figure 8.4). When seven types of wood and grass residues were left in the pyrolyzer at 450–700°C for 1–16 h, there were only minor changes (±10%) in the C, N, P, and K concentration of the biochar

(Li et al, 2019). Altering the reaction atmosphere by adding ≤2% O_2 in the inert gas (i.e., N_2, Ar, or CO_2) influenced the reaction of volatile organic compounds and produced oxygenated organic products on biochar surfaces, but it had no detectable effect on the C and N concentration of wood and grass residues (Li et al, 2019). Activating gases contained in steam or from adding CO_2 can remove tar from biochar pores, thus increasing porosity in the nanometer range (Chemerys et al, 2020). However, there are few reported changes in the nutrient concentration of biochar associated with reaction atmosphere and pressure.

Summary of feedstock and pyrolysis impacts on biochar nutrient content

Total nutrients in biochar depend principally upon the feedstock selected and the pyrolysis temperature. Feedstock type accounts for twice as much of the variation in the total N, P, and K concentration in biochar as the pyrolysis temperature, but these factors are of similar importance in determining the total S, Ca, and Mg concentration in biochar (Figure 8.5). In summary, the total nutrient concentration of biochar depends upon the feedstock, which contains inorganic and organic nutrients. During the pyrolysis process, some nutrients will be lost as volatile gases or as aerosol particles. Nutrients remaining in the biochar may be physically transformed with respect to their location, associating with functional groups on reactive surfaces or with pores that form during the thermochemical processing of the feedstock. Another possibility is that nutrients react chemically to form precipitates or become incorporated into aromatic structures within the pyrolyzed organic matter. Pyrolysis leads to a redistribution of nutrients from the original feedstock, which has implications for the biological availability of those nutrients.

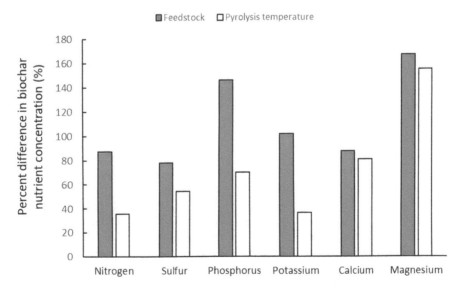

Figure 8.5 *Difference in biochar nutrient concentration as a function of the feedstock type and pyrolysis temperature, reported in a global meta-data analysis. Proportional difference (in %) is plotted as the {(highest nutrient concentration − lowest nutrient concentration) ÷ [(highest nutrient concentration + lowest nutrient concentration)/2]} × 100 (data from Ippolito et al, 2020)*

Available nutrients

The agronomic value of soil amendments like biochar is determined from the proportion of available nutrients that plants absorb through ion transporters. The total nutrient concentration in biochar includes the inorganic ions (often found as salts, precipitates, chelated complexes, and in exchange sites) and the organic compounds that must be biochemically or biologically transformed into cations and anions before they can be assimilated into living biomass. Total nutrient concentration in biochar is assessed by elemental (ultimate) analysis and X-ray fluorescence of the ash. However, available forms are surface-absorbed or dissolvable nutrients, recovered by extracting the biochar with water, a salt solution of low to medium ionic strength, a chelating agent, or a mildly acidic solution that mimics the natural acidity in the plant root zone. The available nutrients in biochar represent the ions that are absorbed by plants and other organisms or that enter rapidly-cycling pools of soil organic matter, gradually releasing nutrients over time to support a yield benefit in the years following the initial biochar application (Major et al, 2010).

Knowing the available nutrient concentration in biochar is important for plant nutrition. Biochar mixed with soil and other growth substrates is a source of macronutrients and micronutrients, increasing the fertility of the growth media (Chapter 13), as shown in Box 8.1 and 8.2. The available nutrients in biochar reduce the need for other plant fertilizers. This has led to calculations of the available N, P, and K supplied by various biochars, including those made from wood and crop residues, and pig and cattle manure (Ippolito et al, 2020). Still, soil test analyses are needed to confirm that biochar application increased the concentration of available N, P, and K under field conditions, ideally with tests done more than once to report dynamic changes in the nutrient supply following biochar application that is observed with time (Major et al, 2010). Another way to validate the available N, P, and K supply from biochar is to compare the plant nutrient removal from biochar-amended and unamended growth media. Such plant tests can also determine if crops absorb any non-essential trace elements or other non-nutritive substances from the biochar.

Environmental managers need to know the available nutrients in biochar to protect water, air, and soil quality. Biochar piles must be protected from rainfall and water infiltration during storage, as available nutrients (notably N and P) within the biochar are susceptible to leaching and runoff. Airborne emission of nutrients occurs during pyrolysis, but also when biochar is spread with farm-scale equipment. Windborne losses of up to 30% of the biochar mass occurred during biochar loading, transportation, and spreading of 5.6 t ha^{-1} of hardwood biochar on an agricultural field with a tractor-drawn lime spreader (Ejack et al, 2021). This represents an unintentional release of biochar nutrients into the environment. Finally, environmental losses of N and P are subject to guidelines and regulations in many jurisdictions, meaning that the available N and P in biochar must be counted in nutrient management schemes. Biochar users should keep a record of the available macronutrient and micronutrient concentrations in biochar products that are stored, transported, and used for agriculture and industrial applications, for environmental protection and compliance.

Box 8.1 Biochar research in Khyber Pakhtunkhwa, Pakistan

The *Acacia nilotica* is a native tree found in the riparian zones of Sindh and Punjab provinces of Pakistan. The wood of this tree is used for large-scale biochar production via slow pyrolysis in underground kilns (known as Bhatti in the local language). The biochar is then sold throughout Pakistan for cooking barbecue (Bar BQ) food. Due to its ready availability, this commercial biochar was tested as an agricultural amendment in field-based experiments that study the impact of biochar on soil quality and crop yield. Improvements in soil fertility and crop growth under field conditions were reported for grain, vegetable, and herb crops (Hameeda et al, 2019; Achakzai et al, 2022; Ullah et al, 2022).

One example comes from a field study on an experimental farm in the Buner District, Khyber Pakhtunkhwa, Pakistan. Slow-pyrolysis biochar produced from *A. nilotica* at 450°C had 568 g kg^{-1} of organic C and 9.5 g kg^{-1} of N, and was applied to field plots at rates of 0, 5, 10, 15, and 20 t ha^{-1}. Applying this biochar as the sole soil amendment or co-applying biochar with synthetic NPK fertilizer gave a significant 4 to 59% increase in maize grain yield. There was a positive linear relationship between the biochar application rate and maize grain yield (Figure 8.6; Ullah et al, 2022).

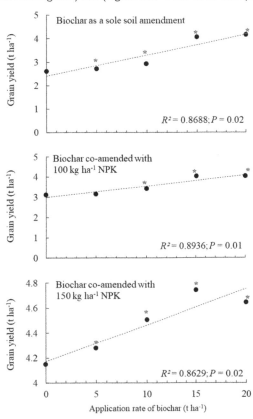

Figure 8.6 *Relationship between the application rate of Acacia nilotica wood-derived slow pyrolysis biochar with maize grain yield. Asterisks on data points indicate significantly greater values than the control without biochar (P ≤ 0.05; data from Ullah et al, 2022)*

In Pakistan, a farmer could purchase a medium-quality biochar that contains broken pieces and fragments for less than \$0.1 kg^{-1}. This is substantially less than the price of synthetic NPK fertilizer, which costs about \$0.7 kg^{-1}. Therefore, biochar may be an affordable option for farmers, who could use biochar as a soil amendment and partial substitute for synthetic NPK fertilizer in their agricultural fields.

Box 8.2 Biochar as an agricultural commodity on family farms in La Sabana, Chone, Ecuador

In Ecuador, biochar was tested as a soil amendment to improve the cultivation of sweet potato (*Ipomoea batatas* L.) with purple pulp. In the La Sabana, Chone area of Manabí province, farmers grow multiple crops on landholdings that are < 2 ha on average. Sweet potato, cassava, peanuts, and rice are the main cash crops. Farmers often fertilize these crops with synthetic NPK fertilizer. The continual increase in fertilizer prices every year, together with a general loss of soil fertility in the area have motivated farmers to consider using other soil amendments (Figure 8.7).

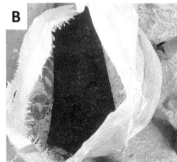

Figure 8.7 *Examples of biochar use. (A) Farmer preparing the soil with biochar for the next growing cycle on a one-hectare family farm located in Manabi province, Ecuador. (B) Biochar used as a soil amendment in Manabi province, Ecuador. (C) Farmer Vicente Quijije showing his sweet potato for sale in Manabi province, Ecuador. The crop was grown in biochar-amended soil (photographs courtesy of Leonardo León Castro, 2022, with permission)*

Currently, the farmers work together with the National Institute for Agricultural Research (INIAP) to study the effect of organic amendments to improve soil fertility and improve production. The experimental farm is located at 1° 09' 52" S, 80° 23' 18" W. In 2022, 12.5 ha of land on the experimental farm was planted with the INIAP-Toquecita variety sweet potato. Biochar produced with rice husk containing 3 g N kg^{-1}, 10 g K kg^{-1}, 230 g Mg kg^{-1}, 250 g Ca kg^{-1}, 960 g Si kg^{-1} was applied to large-scale field plots. To date, the results show an increase in the marketable sweet potato yield of 18.4 t ha^{-1} of tubers, which represents a 15% improvement in yield, compared to the control without biochar (Cobeña, unpublished data).

The production of I kg of sweet potato tubers in the conventional agriculture system is $0.16 and the retail price is $0.20, which is not economically sustainable. Using biochar, the farmers can reduce their production costs by up to 25%. Furthermore, farmers have observed good foliar development and a deeper root system, which helps sweet potatoes obtain energy through photosynthesis and acquire sufficient water and nutrients. This is leading to a system of sweet potato production with a minimum of chemical inputs. Testing with additional organic amendments is underway on farms. The farmers involved in these studies are planning to obtain organic certifications to increase their economic opportunities, particularly in the international export market.

Feedstock

Newly-formed biochar contains surface-adsorbed available nutrients and pore-associated minerals that are soluble in water and other mild extractants. Biochars with higher total nutrient concentrations generally have more available nutrients. Biochar from manure and biosolids feedstock are enriched in NH_4^+, extractable P and Ca relative to plant-based biochars (Figure 8.8). About 20% of the total K in biochar is extractable, and the available K concentration is greater in crop residue biochar (9 g kg^{-1}) and grass residue biochar (16 g kg^{-1}) than manure or biosolids biochar (4 g kg^{-1}) and wood biochar (2 g kg^{-1}; Ippolito et al, 2020). Greater available Mn, Zn, and Cu concentrations are reported in biochars from manures and biosolids than in other biochar types, whereas crop residue biochar had a higher available Fe concentration than biochar

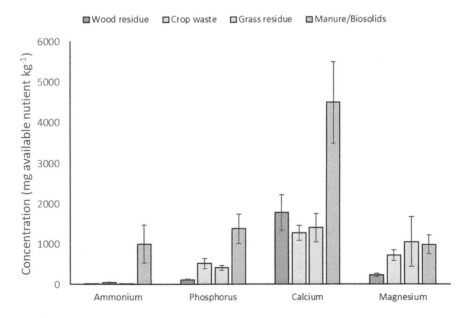

Figure 8.8 *Water-soluble and extractable macronutrient concentrations (mg available nutrient kg^{-1}) in biochar produced from four types of feedstock (data from Ippolito et al, 2020)*

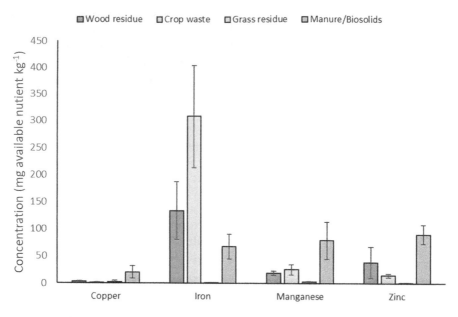

Figure 8.9 *Water-soluble and extractable micronutrient concentrations (mg available nutrient kg⁻¹) in biochar produced from four types of feedstock (data from Ippolito et al, 2020)*

from other feedstock types based on a global meta-data analysis (Figure 8.9). Grass residue biochar has an order of magnitude less available Cu, Fe, Mn, and Zn than other biochar (Figure 8.9).

Biochar is reported to contain available N as nitrate (NO_3^-), but it is unlikely there will be much NO_3^- in newly-formed biochar because pyrolysis is a high-temperature oxygen-free process. However, biochar does contain surface-absorbed NH_4^+ that is available for microbial and plant assimilation (Spokas et al, 2012; Wang et al, 2015), and this NH_4^+ could be biologically transformed into NO_3^- ions. A biochar containing NO_3^- has probably been colonized by ammonia oxidizers (archaea and bacteria) that converted NH_4^+ to NO_3^- in a well-oxygenated environment, i.e., during storage. Plant-based biochars contain ≤ 200 mg available N kg⁻¹,

so an agronomic application of 20 t ha⁻¹ of biochar supplies 4 kg available N ha⁻¹. This is insignificant with respect to the N fertilizer recommendation of 100–200 kg N ha⁻¹ for commodity crops such as maize and wheat. Consequently, the soil fertility improvement following biochar application is generally attributed to changes in soil physico-chemical and biological processes rather than the available nutrients supplied by biochar (Chapters 10, 15, and 16).

Pyrolysis conditions

As the pyrolysis temperature increases, the available NH_4^+ concentration declines (Figure 8.11), likely due to NH_3 volatilization or the integration of NH_3 into aromatic N compounds. The heating of feedstock will release some of the total S in gaseous

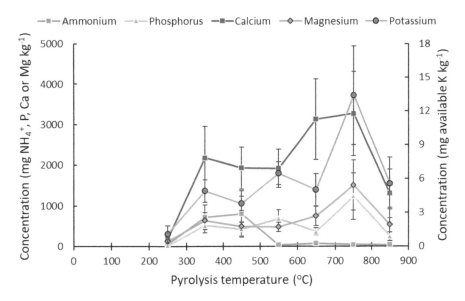

Figure 8.10 *An increase in pyrolysis temperature, up to 700°C produces biochar with more available P, K, Ca, and Mg concentration, but the available NH$_4^+$ concentration declines after 450°C. The global decline in available nutrient concentrations at 850°C may be the result of thermochemical reactions to unavailable forms or aerosol/solid particle emissions. (data from Ippolito et al, 2020)*

products. However, pyrolysis at 200–400°C of powdered seaweed (*Caulerpa peltata*) containing sulfate polysaccharides released sulfate salts like $Mg_2K_2(SO_4)_3$, which dissolve in water and are available S. Increasing the pyrolyzer temperature to 650–700°C will transform the sulfuric acid salt to CaS crystals (Zhang et al, 2020), which might be thermostable or a transient state if CaS was integrated into aromatic compounds that formed around 700–800°C. The concentration of available P, K, Mg, and Ca generally increased when the pyrolysis temperature was 350–800°C (Figure 8.10). A similar pattern was observed for micronutrients, based on available Fe, Mn, Zn, and Cu (Figure 8.11). The sharp decline in available nutrient concentrations when the temperature was >

800°C may indicate that available P and the available cations were emitted as aerosol particles at higher pyrolysis temperatures. This interpretation is consistent with the fact that P compounds volatilize at 760°C, resulting in a decline in total P concentrations at pyrolysis temperatures > 800°C (Knicker, 2007). However, the variability in available nutrient concentrations of biochar may reflect uncertainty in the global meta-data analysis. Further study of available nutrient concentrations in biochar, as a function of pyrolysis temperature, is needed to clearly differentiate between available nutrient fluctuations caused by decomposition of organic forms, thermochemical transformations, and aerosol/solid particle emissions.

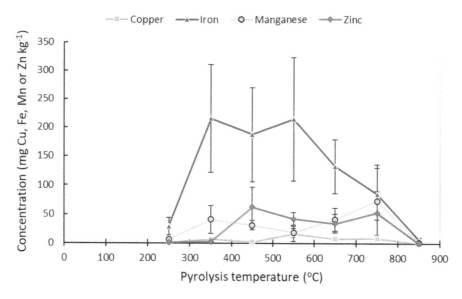

Figure 8.11 *Available micronutrient concentration in biochar is greatest between 350–700°C than at lower and higher pyrolysis temperatures. (data from Ippolito et al, 2020)*

Surface chemistry and nutrient properties

A key component of nutrient availability is the form of the nutrient within and on the matrix, here, the biochar pores and surfaces. Adsorption-desorption kinetics of nutrients from the biochar matrix into the surrounding aqueous solution depends on physical and chemical factors. Biochar surfaces are physically uneven because of the variable particle size geometry, compaction, and porosity of the feedstock. Chemically, biochar surface heterogeneity is the result of hydrophobic, hydrophilic, acidic, and basic functional groups that are created as a function of the chosen feedstock and pyrolysis conditions. Hydrophobic biochar surfaces repel water. This results in less leaching, by at least an order of magnitude, of NH_4^+, dissolved N and dissolved P from hydrophobic pine wood biochar (pyrolyzed at 600 ± 50°C under limited oxygen for 8 hours) with

high hydrophobicity (contact angle value >90°) compared to hydrophilic pine wood biochar (made by heating the hydrophobic biochar at 300°C in ambient air for 8 hours; Hong et al, 2020). Water repellency at the surface can partially explain nutrient retention in hydrophobic biochar, but nutrient adhesion is also affected by biochar mesoporosity and redox-active sites on the surface of a biochar containing an organic coating (Hagemann et al, 2017). More non-exchangeable nutrients sorb to wood-based biochar because water is repulsed by the hydrophobic lignin residues and lipophilic extractive like fats, waxes, terpenoids, and higher aliphatic alcohols that remain after pyrolysis at 450°C for 1–2 h (Usevičiūtė and Baltrėnaitė, 2020). This suggests that hydrophobic biochars have a stable nutrient content with a low proportion of exchangeable

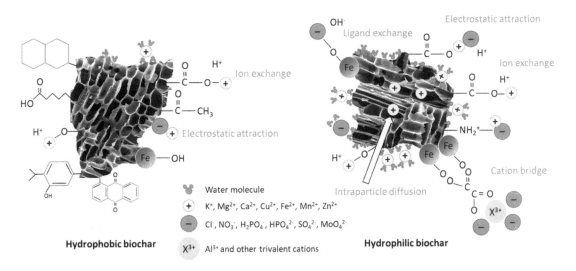

Figure 8.12 *Hydrophobic biochar repels water. Hence, nutrient cations and anions held in exchange sites and through electrostatic attraction are slowly exchanged from the surface. A hydrophilic biochar offers additional sorption capacity through ligand exchange and cation bridges, as well as the possibility of intraparticle diffusion of ions into micro- and nano-sized pores. Hydrophilic biochar has more capacity to retain nutrients that reach the biochar surface through mass transfer and diffusion processes and will exchange cations and anions in an aqueous solution*

nutrient ions, relative to hydrophilic biochars (Figure 8.12). This needs to be tested in equilibrium solutions to determine the nutrient exchange rate of biochars with diverse surface reactivity, at varying reaction times.

The surface charge of a hydrophilic biochar determines its capacity to sorb anions and cations, and this may be deduced from the presence of acidic and basic functional groups. Many, but not all, of the acidic functional groups are oxygenated (e.g., carboxyl, hydroxyl, phenol, phosphate, and phosphonate functional groups). Grass- or manure-based biochars with more oxygenated functional groups are expected to sorb more cations through ion exchange, electrostatic forces, and complexation than wood-based biochar that contains fewer oxygenated functional groups and more aromatic structures (Hassan et al, 2020). Hydrogen ions (H^+) that dissociate from the acidic,

oxygenated functional group exchange with cations from the surrounding environment. Surface acidity of 32–1067 mmol kg^{-1} in wood-based biochar made from pine, poplar, and willow chips at 400°C and 550°C was related to the presence of hydroxyl or phenol groups, which conferred 9–85 $mmol_{c(+)}$ kg^{-1} of exchange capacity (Calvelo Pereira et al, 2015). Cations can also bind to inorganic carbonates in the ash fraction of biochar (Calvelo Pereira et al, 2015). For example, despite its low CEC of 45 $mmol_{c(+)}$ kg^{-1}, kraft lignin biochar adsorbed nearly twice as much Cr (VI) than cow dung biochar with a CEC of 1410 $mmol_{c(+)}$ kg^{-1} prepared under the same condition; this was attributed to cation sorption by minerals in the micro-nano pore structures of kraft lignin biochar (Hu et al, 2022). Further information about pyrolysis temperature and reaction condition effects on the types of oxygenated functional groups, porosity, and specific surface area of biochar is given in Chapters 5, 6, and 7. The nature of biochar CEC being the result of surface binding and other sorption processes within pores makes it challenging to develop predictive models of the cation adsorption-desorption kinetics of biochar.

It is also important to acknowledge that aromatic structures in biochar can attract and bind to cations from aqueous solution through cation-π binding. The electron-donating capacity of biochar is associated with the conjugated π-system created by electroactive quinoid functional groups and polycondensed aromatic sheets that form during pyrolysis (Klüpfel et al, 2014). Grass, wood, and nutshell biochars produced at 400–700°C had a higher capacity to accept and donate electrons, up to 0.2 mmol of electrons per gram of biochar (Klüpfel et al, 2014), but this may underestimate their electron-donating capacity. Electron donation of up to 7 mmol of electrons per gram of biochar was reported for pinewood biochar pyrolyzed at 400°C, and reached ~10–15 mmol of electrons per gram with grass-based biochars (Prévoteau et al, 2016). Electron donation is either associated with more oxygen functional groups (e.g., hydroxyl or ether) linked to the formation of aromatic compounds and a loss of hydroxyl groups through the dehydration of lignin-derived phenols and alcohols at higher pyrolysis temperatures, or with the formation of fused aromatic compounds and shuttling of electrons through the biochar matrix (Sun et al, 2017). The ability of biochar to donate electrons for chemical reactions and to microbial metabolic pathways probably influences the chemical form and amount of available nutrients in the biochar. This topic requires further study.

Biochar may sorb anions to positively charged sites on basic surface oxides and basic functional groups such as amines, hydrazines, and N-containing heterocycles. Oxonium functional groups with +1 charge provide pH-independent anion exchange capacity (AEC); positively-charged pyridinic functional groups and non-specific proton adsorption by condensed aromatic rings are responsible for pH-dependent AEC (Lawrinenko and Laird, 2015). Pyrolyzed albumin, alfalfa meal, maize stover and cellulose (500°C or 700°C for 1 h) had AEC from 6.0–280 $mmol_{c(-)}$ kg^{-1}, whereas wood, crop residue, and nutshell biochars produced at ≥500°C with various pyrolysis conditions had AEC of 3.5–300 $mmol_{c(-)}$ kg^{-1} (Lawrinenko and Laird, 2015; Bakshi et al, 2016). AEC increases when surfaces are protonated. Greater AEC is expected in acidic than alkaline biochar, and AEC should differ between newly produced biochars and biochars that were exposed to soil, water, and other environmental conditions.

Biochar pH and nutrient properties

Biochar tends to be alkaline, although pyrolysis of some wood residue can produce a biochar with pH<6 (Nguyen and Lehmann, 2009; Gezahegn et al, 2019). Like other biochar properties, the degree of alkalinity depends on feedstock and pyrolysis temperatures. In general, wood residue biochar is more acidic than biochar made from crop and grass residues, manure, and biosolids (Ippolito et al, 2020; Tomczyk et al, 2020). The alkalinity in biochar is related to higher ash content, since basic cations like K, Mg, and Ca are associated with carbonate, oxide, chloride, and other anions that consume H^+ (Yuan et al, 2011).

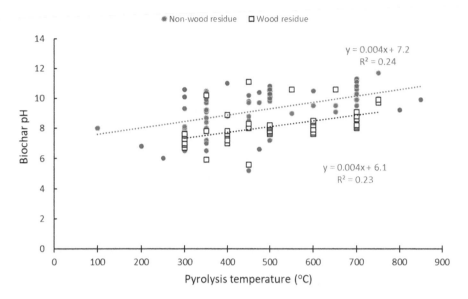

Figure 8.13 *Biochar pH increases at higher pyrolysis temperatures. Biochar made from wood residue tends to be more acidic than biochar from crop and grass residues, manure, and biosolids (data from Gezahegn et al, 2019; Ippolito et al, 2020; Tomczyk et al, 2020)*

Increasing the pyrolysis temperature also raises the biochar pH (Figure 8.13). High-temperature pyrolysis reduces acidity by volatilization and dehydroxylation of inorganic and organic matrices, which eliminates low molecular weight acids and removes H^+ from the biochar surface (Hass et al, 2012; Bardestani and Kaliaguine, 2018). Simultaneously, more Ca-, Mg-, Na-, and K-bearing oxides, hydroxides, and carbonates are formed, along with silicate, iron, and aluminum oxides (Cao and Harris, 2010; Hass et al, 2012). Selecting a non-wood feedstock and higher pyrolysis temperature should produce biochar with more available cations in the ash fraction and on CEC sites and lower AEC and fewer available anions.

The pH of biochar impacts its usefulness for environmental applications such as soil liming and buffering. For example, biochar has a liming effect when used as a soil amendment. Biochars containing carbonates, silicates, oxygen-containing functional groups, and other alkali organic and inorganic compounds that consume H^+ can directly reduce acidity. Biochars made from wood, shells, crop residues, and manures have a liming potential of 2–21% $CaCO_3$-eq (Singh et al, 2017). One advantage of biochar, compared to lime ($CaCO_3$), is that biochar prevents re-acidification of soil by increasing the soil pH buffering capacity (Shi et al, 2019), discussed further in Chapters 10 and 11. Oxygenated functional groups on biochar surfaces and metal oxides in the ash fraction contribute variable charge, which seems to be responsible for the soil pH buffering capacity (Nelson and Su, 2010). The pH-dependent charge of biochar makes it effective in sorbing ions and molecules from the aqueous phase of contaminated sediments and waste streams, although other chemical reactions (e.g., precipitation, complexation) that occur on biochar surfaces and within pores are important for the retention of some pollutants (Chapter 19).

Pre-pyrolysis and post-pyrolysis treatments that change biochar nutrient properties

Treatments that occur before, during, and after pyrolysis alter the nutrient properties of biochar. Physical treatments can change the humidity, size, and oxidation state of the feedstock or biochar. Chemical modification can change the surface reactivity or pH, or add nutritive substances such as micronutrients to create a biochar-based fertilizer. Although one could mix feedstocks with contrasting nutrient contents (e.g., nutrient-rich manure with nutrient-poor wood residue) to create a biochar-based fertilizer with a nutrient profile that matches the crop nutrient requirements (Novak et al, 2014), this option has seldom been reported in the literature. Consequently, most understanding of biochar nutrient properties after physical or chemical alteration comes from studies of a single feedstock or a mixture of similar feedstock types (Bardestani and Kaliaguine, 2018).

Biological treatment of the feedstock is rarely considered a pre-pyrolysis treatment because lignocellulosic biomass resists biological degradation. Pyrolysis has more commonly been used as the first step in processing plant-based residues, followed by recovery of the syngas and pyrolytic sugars for bioethanol fermentation or upgrading (Luque et al, 2016). Still, there is interest in using biochar as a carrier for delivering selected microbiota (inoculants) to growing environments, mixing biochar with biologically active fertilizing materials for agronomic and horticultural use (Chapter 26), and in combining biochar with biological materials for environmental and industrial applications (Chapter 27).

Physical treatments to modify biochar nutrient properties

Finer biochar particles and biochar particles that retain more water are expected to hold more dissolved nutrients in their pores, leading to the eventual sorption of these nutrients through ion exchange and other reactions. For example, sugar maple wood biochar (pyrolyzed at 370°C) that was sieved (0.06–0.5 mm) after pyrolysis had up to 2.5-fold greater water retention capacity than ground biochar (0.06–0.5 mm), likely because sieved particles were more elongated than ground particles and had greater interpore volume (Liao and Thomas, 2019). Smaller particle sizes should expose more biochar surfaces to aqueous solution, allowing for greater exchange of available nutrients, but this needs to be confirmed experimentally.

Pre-pyrolysis steam treatment causes carbonization and physical disruption of the feedstock. For example, pre-pyrolysis steam explosion (2 min at 210°C and 2.5 MPa) of straw residue disrupted bonds between lignin and hemicellulose, and shattered xylem vessels, parenchyma cells, and fibrous structures (Chen et al, 2019). After the straw was pyrolyzed (500°C under N_2 gas for wheat, rice, corn, oilseed rape, and cotton straws), the biochar from steam-exploded residue had less specific surface area and lower pore volume, but 2–3 times greater AEC than biochar from untreated straw residue (Chen et al, 2019). This implies that there were more positively charged functional groups on straw residue biochar after pre-pyrolysis steam treatment, although the mechanism still needs to be explained.

Oxidation is another physical treatment that can be applied during pyrolysis or as a post-pyrolysis treatment. The presence of O_2 in the reaction atmosphere lowers the charcoal yield, possibly because it interferes with repolymerization of volatile matter and generates higher ash content (Li et al, 2019). Thus, oxidation is more common as a post-pyrolysis treatment. However, conifer wood biochar (spruce, fir, and pine pyrolyzed at 475°C and 100 kPa) that was subsequently exposed to mild air oxidation at 200°C had more oxygen surface functional groups (up to 6.8 mmol g^{-1}), especially carboxylic acids, potentially increasing CEC of the biochar (Bardestani and Kaliaguine, 2018). High-temperature steam activation increases biochar pH, as demonstrated for chicken manure biochar (pyrolyzed at 350°C) that was pH 9.9 without treatment and reached pH 11.9 with steam activation (3 mL min^{-1} for 45 min at 800°C; Hass et al, 2012). These post-pyrolysis treatments increase the surface reactivity and make biochar more alkaline, which should boost the CEC. The major drawback of steam treatment is the loss of biochar, based on ~70% mass loss of conifer wood biochar during the post-pyrolysis steam treatment (Bardestani and Kaliaguine, 2018). Techno-economic analysis is needed to document the energy requirement and costs (operating, equipment) of oxygen and steam treatments before implementing these processes at an industrial scale.

Chemical treatments to modify biochar nutrient properties

Biochar nutrient properties are modified by chemically treating the feedstock before drying and heating. Several chemical treatments were developed to conserve N compounds in the feedstock during pyrolysis. Adding CaO reduces total N losses because

solid phase reactions form a thermostable CaC_xN_y product, while treating feedstocks with Na or K acetate increases the soluble Na phosphates and K phosphates, adding to the available P concentration in biochar (Buss et al, 2022). Chicken manure feedstock wetted with $CaCl_2$ and $FeCl_3 \cdot 6H_2O$ solutions produced biochar with slightly higher total N and available N (NH_4^+) concentrations than biochar from untreated feedstock, presumably due to co-precipitation with amine-N compounds (Xiao et al, 2018). However, pre-pyrolysis acidification of poultry manure feedstock with concentrated H_3PO_4 or concentrated HNO_3 reduced the total N content of the resulting biochar (pyrolyzed at 300°C for 2 h; Sahin et al, 2017). This was attributed to an increase in the proportions of NH_3 and amine-N compounds in poultry manure, N forms are vulnerable to volatilization upon heating.

Mixing refined minerals, mineral by-products, and ground rocks with the feedstock changes the total and available nutrient concentration in the biochar. Rice husk was pretreated with solutions of $AlCl_3$, $FeCl_3$, $MnCl_2$, $ZnCl_2$, and $MgCl_2$ (67% w/v) before pyrolysis (600°C for 1 hour in an ambient atmosphere) to increase the content of positively-charged metal oxyhydroxides and oxides to control the P desorption from biochar (Sornhiran et al, 2022). Greater P retention in biochar was associated with Al silicates > Fe silicates > Mn silicates > Zn silicates > Mg silicates formed during pyrolysis. Another option is to treat the biochar with nutrient solutions following pyrolysis. For example, leather tannery waste biochar containing 1 g total N kg^{-1} was soaked in a solution containing 1.1 g Cu (II) kg^{-1}, 1.3 mg Mn (II) kg^{-1}, and 1.2 g Zn (II) kg^{-1}. The micronutrients are released gradually from the impregnated biochar into water and soilless media during germination and vegetative growth tests with cucumber (Skrzypczak et al, 2022). Other methods of combining biochar with nutrient solutions, organic fertilizers, and nutritive materials to create biochar-based nutrient amendments are described in Chapter 26.

Biochar sorptive capacity can be improved by pre-pyrolysis treatments. One approach is to increase the oxygenated functional groups, i.e., hydroxyl and carboxyl groups on biochar, which provides more CEC. Practically, this is done by soaking the feedstock in a strong acid like H_3PO_4 to remove metals and introduce acid functional groups (Rajapaksha et al, 2016). Poultry manure biochar treated with concentrated H_3PO_4 before pyrolysis (300°C) had twice as much water-soluble P as the untreated biochar, but there was no change in water-soluble K, Ca, or Mg concentrations (Sahin et al, 2017). Another option is to treat the feedstock with alkali solutions of KOH or NaOH, but this does not consistently alter the surface area and sorption capacity of the resultant biochars (Wang and Wang, 2019), suggesting a feedstock-specific effect of these pre-pyrolysis treatments.

Metal salts or metal oxides mixed with the feedstock are effective in changing the sorption capacity of biochar, specifically for anions. For example, metal oxyhydroxide surface coatings were developed by wetting crop residues (alfalfa meal, corn stover, cellulose) with 1% (w/w) of solutions containing 1 M $FeCl_3$ or 1 M $AlCl_3$ before slow pyrolysis (500°C and 700°C, N_2 atmosphere). Metal salt treatment increased the AEC in alfalfa biochar to 29–530 $mmol_{c(-)}$ kg^{-1}, which was 2 to 5 times greater than untreated alfalfa biochar (Lawrinenko et al, 2017). The modest change in the AEC of corn stover and cellulose biochars illustrated the dependence of biochar AEC on the distribution of metal oxyhydroxides and their coordination with oxygen functional groups during pyrolysis. Metal-coated biochars also

exhibit desirable catalytic properties and those containing Fe_3O_4 are magnetic, which allows them to be recycled from wastewater treatment effluent (Wang and Wang, 2019).

Modifying the surface charge of biochar in the post-pyrolysis period is another option. This may be done by adding acid and alkali solutions, applying organic solvents or surfactants, or coating biochars with metal salts and metal oxides (Rajapaksha et al, 2016). However, treating biochars with acid and alkali solutions may have the unintended consequence of dissolving minerals in the ash fraction. Poultry manure biochar treated with concentrated H_3PO_4 after pyrolysis had twice as much water-soluble P and Ca as the untreated biochar, but similar water-soluble K and Mg (Sahin et al, 2017). Dissolved minerals that re-precipitate as salts could block access to biochar pores, limiting water infiltration and contact between dissolved nutrients and pore surfaces. Models of metal sorption to biochar reveal the importance of binding to external and internal surfaces (i.e., within pores; Wu et al, 2022). Since nutrient sorption to biochar is a function of surface charge and porosity, it remains important to study how post-pyrolysis treatments affect nutrient diffusion and mass transfer processes into pores.

Many of the post-pyrolysis chemical treatments to modify biochar surface charge (e.g., acid and alkali solutions, solvents, and metals) are expensive and generate undesirable byproducts and wastes. An alternative approach is to use H_2O_2, an inexpensive oxidant that increases oxygenated functional groups before decomposing into water and oxygen. Birch wood biochar (pyrolyzed at 450°C) that was sieved to powder (<1 mm) or granules (1–3 mm), then stirred with 30% H_2O_2 had an increase in CEC to 280–300 $mmol_{c(+)}$ kg^{-1} and more acidic oxygenated functional groups on the surface (Chemerys et al, 2020). The average pore volume of the treated birch wood biochar was 2.3–2.5 nm, which is a favorable size for the sorption of nutrients through ion exchange, complexation, and electrostatic attraction. Post-pyrolysis treatment with H_2O_2 merits further investigation for improving the CEC and nutrient exchange properties of biochar.

Conclusions and future research directions

The elemental composition of biochar is related principally to the nutrient content in the original feedstock and the pyrolysis temperature. Future work should include routine measurements of nutrients relevant to plant growth. Total N concentration is frequently measured, as well as total P, K, S, Ca, and Mg concentrations. Much less information is available about total Cu, Fe, Mn, and Zn concentrations, and virtually no data exists on the total B, Cl, and Mo in biochars. Besides these 13 essential nutrients, some plants also need Co, Ni, Na, Sr, Si, and V, and few reports exist on these elements in biochar.

Increasing the pyrolysis temperature transforms minerals and organic compounds into ions that bind to surface exchange sites, precipitate as salts or complex with organic acids and other chelates. However, nutrient transformations are uncertain at >700°C. High-temperature pyrolysis may cause nutrient volatilization, nutrient binding in de novo aromatic compounds (e.g., heterocyclic aromatic N, thiophene containing S), and the formation of reduced nutrient-rich salts like CaS and FeS. Further study is needed to resolve the effect of higher pyrolysis temperature on biochar nutrient properties.

There is considerable scope for research on the pyrolysis conditions and treatments that control the available nutrient concentration in biochar. Available nutrients are retained on charged surfaces associated with functional groups: CEC is linked to negatively-charged acidic functional groups and π electrons; AEC is related to positively-charged sites on oxonium and basic functional groups. Such charged surfaces exist within the interpore volume at the nanometer scale (0.6–2 nm) and may be a key location for the sorption of nutrient ions, based on experimental evidence and a pore volume and surface diffusion model (Gong et al, 2019; Wu et al, 2022). Pre-pyrolysis and post-pyrolysis treatments should be considered for their capacity to develop surface charge as well as pores in the micro to nano size range that retain nutrients through diffusion and mass transfer mechanisms. Decision-support tools may help determine whether direct pyrolysis of feedstock produced biochar with the desired nutrient concentrations and sorption capacity (Figure 8.14).

Finally, robust techno-economic analysis and circular economy calculation tools must be used to scale the experimental findings to operational levels. Whilst a set of pyrolysis conditions and treatments may be effective in producing biochar with the desired nutrient properties at the bench scale, they may not be viable at an industrial scale due to high energy requirements, lack of suitable equipment, loss of biochar during processing or transfers, use of costly chemicals, generation of hazardous byproducts and so on. The cost of production will likely be the largest constraint to designing biochar with specific nutrient properties, but this remains to be confirmed.

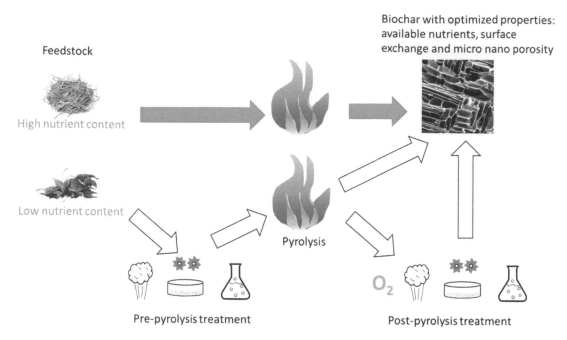

Figure 8.14 *Flow diagram of direct pyrolysis of nutrient-rich feedstock, and possible transformations of a nutrient-poor feedstock through pre-pyrolysis treatment or post-pyrolysis treatment of the biochar to create an optimized product. Pre-pyrolysis treatment could include exposure to steam, grinding, and sieving, or application of chemical additives. Post-pyrolysis treatments could include physical and chemical modifications, as well as oxygen activation. The value of the final biochar product must become higher than the cost of pre- or post-pyrolysis treatments to be viable*

References

Achakzai AG, et al 2022 Evaluation of the effect of biochar-based organic fertilizer on the growth performance of fennel and cumin plants for three years. *Environmental Pollutants and Bioavailability* 34, 374–384.

Angin D 2013 Effect of pyrolysis temperature and heating rate on biochar obtained from pyrolysis of safflower seed press cake. *Bioresource Technology* 128, 593–597.

Bakshi S, Aller DM, Laird DA, and Chintala R 2016 Comparison of the physical and chemical properties of laboratory- and field-aged biochars. *Journal of Environmental Quality* 45, 1627–1634.

Bardestani R, and Kaliaguine S 2018 Steam activation and mild air oxidation of vacuum pyrolysis biochar. *Biomass and Bioenergy* 108, 101–112.

Buss W, et al 2022 Mineral-enriched biochar delivers enhanced nutrient recovery and carbon dioxide removal. *Communications Earth and Environment* 3, 67.

Calvelo Pereira R, Camps-Arbestain M, Vazquez Sueiro M, and Macia-Agullo JA 2015 Assessment of the surface chemistry of wood-derived biochars using wet chemistry, Fourier transform infrared spectroscopy and X-ray photoelectron spectroscopy. *Soil Research* 53, 753–762.

Cao XD, and Harris W 2010 Properties of dairy-manure-derived biochar pertinent to its potential use in remediation. *Bioresource Technology* 101 5222–5228.

Chang Y-M, Tsai W-T, and Li M-H 2015 Chemical characterization of char derived from slow pyrolysis of microalgal residue. *Journal of Analytical and Applied Pyrolysis* 111, 88–93.

Chemerys V, Baltrenaite-Gediene E, Baltrenas P, and Dobele G 2020 Influence of H_2O_2 modification on the adsorptive properties of birch-derived biochar. *Polish Journal of Environmental Studies* 29, 579–588.

Chen X-J, Lin Q-M, Rizwan M, Zhao X-R, and Li G-T 2019 Steam explosion of crop straws improves the characteristics of biochar as a soil amendment. *Journal of Integrative Agriculture* 18, 1486–1495.

Choudhary TK, Khan KS, Hussain Q, and Ashfaq M 2021 Nutrient availability to maize crop (*Zea mays* L.) in biochar amended alkaline subtropical soil. *Journal of Soil Science and Plant Nutrition* 21, 1293–1306.

Das SK, Ghosh GK, Avasthe RK, and Sinha K 2021 Compositional heterogeneity of different biochar: Effect of pyrolysis temperature and feedstocks. *Journal of Environmental Management* 278, 111501.

Dróżdż D, Wystalska K, Malińska K, Grosser A, Grobelak A, and Kacprazak M 2020 Management of poultry manure in Poland - Current state and future perspectives. *Journal of Environmental Management* 264, 110327.

Ejack L, Whalen JK, Major J, and Husk BR 2021 Biochar application on commercial field crops using farm-scale equipment. *Canadian Biosystems Engineering* 63, 6.1–6.8.

Enders A, Hanley K, Whitman T, Joseph S, and Lehmann J 2012 Characterization of biochars to evaluate recalcitrance and agronomic performance. *Bioresource Technology* 114, 644–653.

Gezahegn S, Sain M, Thomas SC 2019 Variation in feedstock wood chemistry strongly influences biochar liming potential. *Soil Systems* 3, 26.

Gong H, Tan ZX, Zhang L, and Huang QY 2019 Preparation of biochar with high absorbability and its nutrient adsorption–desorption behaviour. *Science of the Total Environment* 694, 133728.

Hagemann N, et al 2017 Organic coating on biochar explains its nutrient retention and stimulation of soil fertility. *Nature Communications* 8, 1089.

Hameeda, et al 2019 Biochar and manure influences tomato fruit yield, heavy metal accumulation and concentration of soil nutrients under groundwater and wastewater irrigation in arid climatic conditions. *Cogent Food and Agriculture* 5, 1576406.

Hass A, et al 2012 Chicken manure biochar as liming and nutrient source for acid appalachian soil. *Journal of Environmental Quality* 41, 1096–1106.

Hassan M, et al 2020 Influences of feedstock sources and pyrolysis temperature on the properties of biochar and functionality as adsorbents: A meta-analysis. *Science of the Total Environment* 744, 140714.

Hong N, Cheng Q, Goonetilleke A, Bandala ER, and Liu A 2020 Assessing the effect of surface hydrophobicity/hydrophilicity on pollutant leaching potential of biochar in water treatment. *Journal of Industrial and Engineering Chemistry* 89, 222–232.

Hossain MZ, et al 2020 Biochar and its importance on nutrient dynamics in soil and plant. *Biochar* 2, 379–420.

Hu Y, Li P, Yang Y, Ling M, and Li X 2022 Preparation and characterization of biochar from four types of waste biomass under matched conditions. *BioResources* 17, 6464–6475.

Ippolito JA, et al 2020 Feedstock choice, pyrolysis temperature and type influence biochar characteristics: a comprehensive meta-data analysis review. *Biochar* 2, 421–438.

Klüpfel L, Keiluweit M, Kleber M, and Sander M 2014 Redox properties of plant biomass-derived black carbon (biochar). *Environmental Science and Technology* 48, 5601–5611.

Knicker H 2007 How does fire affect the nature and stability of soil organic nitrogen and carbon? A review. *Biogeochemistry* 85, 91–118.

Lawrinenko M, and Laird DA 2015 Anion exchange capacity of biochar. *Green Chemistry,* 17, 4628–4636.

Lawrinenko M, Jing DP, Banik C, and Laird DA 2017 Aluminum and iron biomass pretreatment impacts on biochar anion exchange capacity. *Carbon* 118, 422–430.

Li WW, et al 2019 Effects of temperature, heating rate, residence time, reaction atmosphere, and pressure on biochar properties. *Journal of Biobased Materials and Bioenergy* 13, 1–10.

Liao WX, and Thomas SC 2019 Biochar particle size and post-pyrolysis mechanical processing affect soil pH, water retention capacity, and plant performance. *Soil Systems,* 3, 14.

Luque L, et al 2016 Comparison of ethanol production from corn cobs and switchgrass following a pyrolysis-based biorefinery approach. *Biotechnology for Biofuels* 9, 242.

Major J, Rondon M, Molina D, Riha SJ, and Lehmann J 2010 Maize yield and nutrition during 4 years after biochar application to a Colombian savanna oxisol. *Plant and Soil* 333, 117–128.

Maliutina K, Tahmasebi A, and Yu JL 2018 Pressurized entrained-flow pyrolysis of microalgae: Enhanced production of hydrogen and nitrogen-containing compounds. *Bioresource Technology* 256, 160–169.

Moradi-Choghamarani F, Moosavi AA, and Baghernejad M 2019 Determining organo-chemical composition of sugarcane bagasse-derived biochar as a function of pyrolysis temperature using proximate and Fourier transform infrared analyses. *Journal of Thermal Analysis and Calorimetry* 138, 331–342.

Nelson PN, and Su N 2010 Soil pH buffering capacity: A descriptive function and its application to some acidic tropical soils. *Australian Journal of Soil Research* 48, 201–207.

Nguyen BT, and Lehmann J 2009 Black carbon decomposition under varying water regimes. *Organic Geochemistry* 40, 846–853.

Novak JM, Cantrell KB, Watts DW, Busscher WJ, and Johnson MG 2014 Designing relevant biochars as soil amendments using lignocellulosic-based and manure-based feedstocks. *Journal of Soils and Sediments* 14, 330–343.

Prévoteau A, Ronsse F, Cid I, Boeckx P, and Rabaey K 2016 The electron donating capacity of biochar is dramatically underestimated. *Scientific Reports* 6, 32870.

Rajapaksha AU, et al 2016 Engineered/ designer biochar for contaminant removal/ immobilization from soil and water: Potential and implication of biochar modification. *Chemosphere* 148, 276–291.

Różyło K, Jędruchniewicz K, Krasucka P, Biszczak W, and Oleszczuk P 2022 Physicochemical characteristics of biochar from waste cricket chitin (*Acheta domesticus*). *Molecules* 27, 8071.

Sahin O, et al 2017 Effect of acid modification of biochar on nutrient availability and maize growth in a calcareous soil. *Soil Use and Management* 33, 447–456.

Schimmelpfennig S, and Glaser B 2012 One step forward toward characterization: Some important material properties to distinguish biochars. *Journal of Environmental Quality* 41, 1001–1013.

Shi R-Y, Li J-Y, Ni N, and Xu R-K 2019 Understanding the biochar's role in ameliorating soil acidity. *Journal of Integrative Agriculture* 18, 1508–1517.

Singh B, Dolk MM, Shen Q, and Camps-Arbestain M 2017 Biochar pH, electrical conductivity and liming potential. In: Singh B, Camps-Arbestain M, and Lehmann J (Eds) *Biochar: A guide to analytical methods.* Clayton, Australia: CSIRO Publishing. pp23–38.

Skrzypczak D, et al 2022 Tannery waste-derived biochar as a carrier of micronutrients essential to plants. *Chemosphere* 294, 133720.

Sornhiran N, Aramrak S, Prakongkep N, and Wisawapipat W 2022 Silicate minerals control the potential uses of phosphorus-laden mineral-engineered biochar as phosphorus fertilizers. *Biochar* 4, 2.

Spokas KA, et al 2012 Biochar: A synthesis of its agronomic impact beyond carbon sequestration. *Journal of Environmental Quality* 41, 973–989.

Sun T, et al 2017 Rapid electron transfer by the carbon matrix in natural pyrogenic carbon. *Nature Communications* 8, 14873.

Tian B, et al 2018 Correlation between bond structures and volatile composition of Jining bituminous coal during fast pyrolysis. *Fuel Processing Technology* 179, 99–107.

Tomczyk A, Sokolowska Z, and Boguta P 2020 Biochar physicochemical properties: Pyrolysis temperature and feedstock kind effects. *Reviews in Environmental Science and Bio-Technology* 19, 191–215.

Torres-Rojas D, et al 2020 Nitrogen speciation and transformations in fire-derived organic matter. *Geochimica et Cosmochimica Acta* 276, 179–185.

Ullah H, et al 2022 Synergistic effects of *Acacia* prunings-derived biochar and nitrogen application on the mineral profile of maize (*Zea mays* L.) grains. *Sustainability* 14, 2995.

Usevičiūtė L, and Baltrėnaitė E 2020 Methods for determining lignocellulosic biochar wettability. *Waste and Biomass Valorization* 11, 4457–4468.

Wang B, Lehmann J, Hanley K, Hestrin R, and Enders A 2015 Adsorption and desorption of ammonium by maple wood biochar as a function of oxidation and pH. *Chemosphere* 138, 120–126.

Wang JL, and Wang SZ 2019 Preparation, modification and environmental application of biochar: A review. *Journal of Cleaner Production* 227, 1002–1022.

Wu JW, Wang T, Shi N, and Pan W-P 2022 Insight into mass transfer mechanism and equilibrium modeling of heavy metals adsorption on hierarchically porous biochar. *Separation and Purification Technology* 287, 120558.

Xiao R, et al 2018 Biochar produced from mineral salt-impregnated chicken manure: Fertility properties and potential for carbon sequestration. *Waste Management*, 78, 802–810.

Yuan J-H, Xu R-K, and Zhang H 2011 The forms of alkalis in the biochar produced from crop residues at different temperatures. *Bioresource Technology* 102, 3488–3497.

Zhang CT, et al 2020 Evolution of the functional groups/structures of biochar and heteroatoms during the pyrolysis of seaweed. *Algal Research – Biomass Biofuels and Bioproducts* 48, 101900.

Zhang J, Liu J, and Liu RJ 2015 Effects of pyrolysis temperature and heating time on biochar obtained from the pyrolysis of straw and lignosulfonate. *Bioresource Technology* 176, 288–291.

Zheng H, et al 2013 Characteristics and nutrient values of biochars produced from giant reed at different temperatures. *Bioresource Technology*, 130, 463–471.

9

A biochar classification system and associated test methods

Balwant Singh, James E. Amonette, Marta Camps-Arbestain, and Rai S. Kookana

Introduction

In this chapter, a biochar classification system is presented as it relates to its use as a soil amendment. It builds on a previous classification system (Camps-Arbestain et al, 2015) and two fundamental biochar standards/ guidelines: "Standardized product definition and product testing guidelines for biochar that is used in soil" (IBI, 2015) (aka IBI Biochar Standards) and "Guidelines for sustainable production of biochar: European Biochar Certificate" (EBC, 2012-2022) (aka EBC Guidelines). The scope of this classification system is constrained to materials with properties that satisfy the criteria for biochar as defined by either the IBI Biochar Standards or the EBC Standards (Annex I). Both the IBI Standards and the EBC Guidelines are product-quality certifications but do not constitute a certificate for the issuance of carbon credits. The classification system will enable stakeholders and commercial entities to (i) assess the climate-impact (CI) value based on biochar properties, soil temperature, and production factors including carbon efficiency, emissions, and bioenergy; (ii) identify the most suitable biochar to fulfill the requirements for a particular soil and/or land-use; and (iii) distinguish the application of biochar for specific niches (e.g., soilless agriculture). This classification system is based on the best current knowledge and will need to be improved as new data and knowledge become available in the scientific literature. The use of biochar in materials (e.g., cement, concrete, asphalt) is excluded from the scope of this chapter.

DOI: 10.4324/9781003297673-9

Biochar classification system

The main thrust of this classification system is the direct or indirect effects that biochar provides from its application to soil or other applications of biochar (e.g., environmental) as depicted in Figure 9.1. The potential effects of biochar from soil and other applications are classified into different categories (Figure 9.1) and provided classes for: (i) climate impact; (ii) fertilizer value; (iii) liming value; (iv) particle size; and (v) binding organic agrochemicals in the following sections.

Climate-impact value and classification

The impact of biochar on climate depends on many factors (Woolf et al, 2010; Cowie et al, 2015; Verheijen et al, 2015; Amonette et al, 2021; Chapter 30). Historically, the primary focus has been on the C-storage properties of biochar with the result that the persistence of biochar C in soil is moderately well understood (Lehmann et al, 2015; Chapter 11) and it has been used as one of the criteria for the classification of biochars

(Camps-Arbestain et al, 2015). A full representation of the climate impact, however, requires a life cycle assessment (LCA) approach that starts with biomass C and compares the release of greenhouse gases and aerosols (GHGAs) from the biomass during biochar production and its subsequent application to soil, with the GHGAs released when the same biomass is used in other pathways. A full LCA of each biochar and biomass combination is impractical for most classification purposes. However, a limited assessment using readily available inputs can capture the essential elements related to the production and soil deployment of biochar and thereby provide a more realistic estimate of its climate impact (CI) than reliance on C-storage properties alone.

Here, an LCA-based biochar classification system (Figure 9.2) is presented that considers C efficiency of biochar production and soil permanence (CESP), biomass harvest, pre-processing and conversion emissions (PEMI), and production of bioenergy to offset fossil energy (ENER) in the biochar

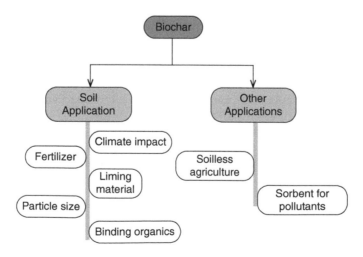

Figure 9.1 *A classification system of biochar based on its effects from soil and other applications*

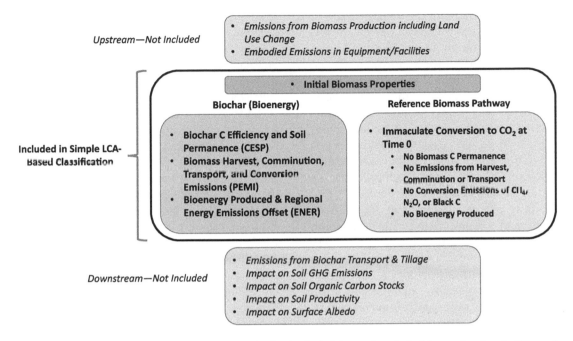

Figure 9.2 *Overview of processes and variables included in and excluded from the simple life cycle assessment (LCA)-based classification of biochar*

pathway. To focus on the relative differences among biochars, all aspects of biochar production are compared against a single, hypothetical, and easily quantifiable biomass reference pathway that is termed here "immaculate conversion to CO_2", for which no biomass C permanence exists, no other GHGA emissions are generated, and no bioenergy is produced. To keep the calculations simple, neither upstream emissions from biomass production (including land-use change) nor embodied emissions associated with the manufacture of equipment and facilities are considered. In addition, any biomass supply chain will have to avoid the following (Woolf et al, 2010): clearance of natural undeveloped lands; conversion of land from production of food crops to biomass; biomass extraction rates that engender loss of soil function; and use of contaminated waste biomass. Downstream

emissions from biochar transport and tillage, and the impact of biochar amendments on soil GHG emissions, organic C stocks, productivity, and surface albedo are excluded. These downstream parameters either contribute little to the overall climate impact (Woolf et al, 2010), are highly location-dependent, or are not sufficiently well understood to include in the limited LCA analysis.

As the primary CI criterion for classification, the average mass of GHGAs released into the atmosphere is calculated over a 100-year period per unit mass of biomass C that was converted at the start of Year 1. Separate 100-year CI values are computed for each portion of the biochar pathway (i.e., $CI_{CESP}100$, $CI_{PEMI}100$, and $CI_{ENER}100$) and then added to obtain the 100-year Biochar Production Climate Impact ($CI_{PROD}100$). All CI values have units of tons of CO_2 equivalents per ton of biomass C (t CO_2e t^{-1}

biomass C). To obtain the average mass of GHGAs over the 100-year period all CIs are calculated on an annual basis (Klasson and Davison, 2002, for a thorough discussion of the dynamic LCA approach) to account for changes in the atmospheric loading of the GHGAs and the annual impacts are then averaged. A major advantage of these $CI_{XXXX}100$ parameters is that, when multiplied by the radiant efficiency of CO_2, they yield radiative forcing per ton of biomass C (Myhre et al, 2013), and thus are a fundamental (if limited) measure of CI associated with conversion of the biomass. The CI calculations are available as a workbook in the Supplementary Material that includes a user portal to allow calculations and classification determinations tailored to individual biochar production situations.

Carbon efficiency and soil persistence (CESP)

The persistence of biochar in soil is predominantly a function of the intrinsic biochar properties and soil temperature (Chapter 11). The biochar property used here for CI classification purposes is the atomic ratio of hydrogen (H) to organic C in the biochar (H/C_{org}), which estimates the degree to which the biochar C approaches a graphite-like structure (Wang et al, 2013). Values of H/C_{org} are best calculated directly from ultimate analysis results, although they can be estimated with lower accuracy from proximate analysis results (Klasson, 2017). Persistence in the soil is given by the permanence factor (F_{PERM}), which is a fractional version of the BC_{+100} parameter (i.e., the proportion of organic carbon (C_{org}) in the biochar that persists in the soil for more than 100 years) used in earlier work. To calculate F_{PERM} values as a function of biochar H/C_{org} and soil temperature, two regression equations are used (Lehmann et al, 2015)

relating BC_{+100} to H/C_{org} and biochar mineralization rate to temperature (Q_{10}) and coupled with temperature adjustments of mineralization rates (Woolf et al, 2021). A contour plot (Figure 9.3) shows the results of this calculation for the full range of H/C_{org} ratios (0.10 to 0.70) considered relevant and useful for C storage purposes and for the mean annual soil temperatures encountered in world soils (0 to 40°C). This plot can be used to obtain a graphical estimate of F_{PERM} in lieu of the workbook calculator in the Supplementary Material. For classification purposes, the F_{PERM} values are separated into five ranges (designated by F_p1 through F_p5, with the F_p5 class having the highest F_{PERM} values). Although the F_{PERM} values are based on correlations of BC_{+100} and H/C_{org} using a two-pool model (Lehmann et al, 2015), the corresponding single-pool rate constant (k = $-\ln(F_{PERM})/100$) is used to calculate the nominal biochar half-lives on which the F_{PERM} classification boundaries are based. These classification half-lives represent a conservative estimate of the persistence of biochar in soil (Lehmann et al, 2015). The workbook calculator (Supplementary Material) allows one to calculate F_{PERM} and related kinetic parameters (single-pool rate constant, half-life, BC_{+100}, and mean residence time) for any combination of H/C_{org} and soil temperature within these limits. In addition to H/C_{org} and temperature, other soil factors, such as moisture, texture, mineralogy, and pH (Lehmann et al, 2015; Chapter 11) can influence the persistence of biochar in soil. These other soil factors are not accounted for in the current estimates of F_{PERM}.

While important, F_{PERM} is agnostic concerning the method by which the biochar is produced. In the first of the three LCA-based CI estimates, $CI_{CESP}100$, the C efficiency of production (i.e., C_{eff}, the fraction of biomass C incorporated into the biochar) is combined with the C efficiency of storage

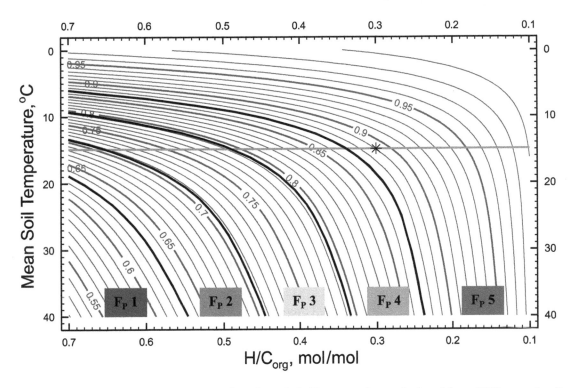

Figure 9.3 *Classification of biochars based on their F_{PERM} values calculated from H/C_{org} and soil temperature for the expected ranges (H/C_{org} from 0.1 to 0.7, soil temperature from 0°C to 40°C) relevant to biochar C storage in soil. Boundaries between the five classes (F_p1 through F_p5) are set by the half-life of biochar C in soil. A horizontal line marks the global mean soil temperature of 14.9°C. An asterisk marks the F_{PERM} value used in biochar production scenarios discussed below*

in soil (i.e., F_{PERM}, the fraction of biochar C remaining after 100 years in soil). As detailed in the Supplementary Material, the calculation considers biogenic CO_2 emissions during the conversion of biomass to biochar and during storage in soil (emissions of fossil-CO_2 before and during conversion and of other biogenic GHGAs during conversion are considered in calculations for $CI_{PEMI}100$). After adjusting both biogenic emission sources on an annual basis for the uptake of atmospheric CO_2 by other Earth processes (such as ocean absorption, photosynthesis, and weathering of silicate rock), a 100-year mean value can be computed for

tons of CO_2 emitted per ton of biomass C. A similar set of calculations for the biomass reference pathway (immaculate combustion to CO_2) yields a reference 100-year mean (2.33 t CO_2 t^{-1} biomass C) which is then subtracted from that calculated for the biochar pathway to yield the value for $CI_{CESP}100$.

A contour plot (Figure 9.4) shows $CI_{CESP}100$ values (obtained using the workbook calculator in the Supplementary Material) for the relevant ranges of C_{eff} and F_{PERM}. Also shown are the five ranges used to classify the biochars in terms of their $CI_{CESP}100$ value, with CI_{CESP} 5 having the greatest climate-mitigation impact and

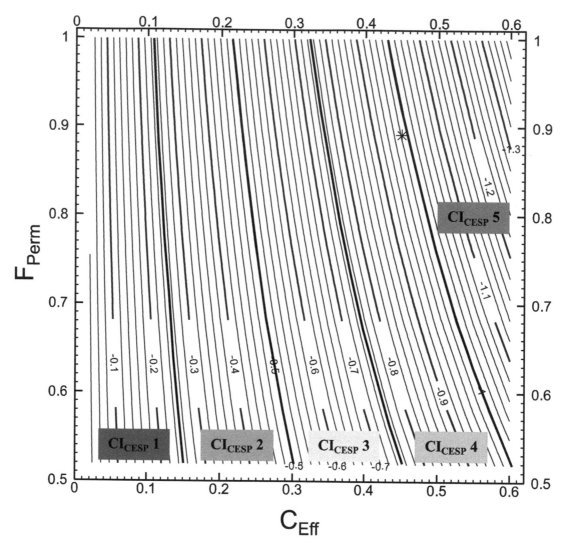

Figure 9.4 *Classification of biochars based on their $CI_{CESP}100$ values calculated from C_{eff} and F_{PERM} for the expected ranges (C_{eff} from 0.02 to 0.60, F_{PERM} from 0.52 to 1.00) relevant to biochar production and C storage in soil. An asterisk marks the $CI_{CESP}100$ value for biochar production scenarios discussed below*

CI_{CESP} 1 the lowest. Classification ranges are listed in Table 9.1.

Production emissions (PEMI)

The second LCA-based CI estimate used for classification, $CI_{PEMI}100$, addresses biogenic non-CO_2 GHGA emissions during biomass conversion to biochar and fossil-CO_2 emissions associated with biomass transport, comminution, and conversion. Because the reference biomass pathway emissions will always be nil (by definition), values for

Table 9.1 Classes for F_{perm} for the CI components ($CI_{CESP}100$, $CI_{PEMI}100$, and $CI_{ENER}100$), and for the Biochar Production CI ($CI_{PROD}100$) assigned using the limited LCA-based classification system

Factor	Special classes			Classification quintiles				
	U (Unacceptable)	? (Not Rated)	0 (No Benefit)	1 (Least Benefit)	2	3	4	5 (Most Benefit)
Biochar soil permanence factor, F_{PERM}	U	--	--	F_p1	F_p2	F_p3	F_p4	F_p5
Range	$F_P < 0.500$	--	--	$0.500 \leq F_P < 0.630$	$0.630 \leq F_P < 0.707$	$0.707 \leq F_P < 0.794$	$0.794 \leq F_P < 0.871$	$F_P \geq 0.871$
Criteria	$C\,t_{1/2} < 100\ y$	--	--	$100\ y \leq C\,t_{1/2} < 150\ y$	$150\ y \leq C\,t_{1/2} < 200\ y$	$200\ y \leq C\,t_{1/2} < 300\ y$	$300\ y \leq C\,t_{1/2} < 500\ y$	$C\,t_{1/2} \geq 500\ y$
Carbon efficiency and soil permanence, $CI_{CESP}100$	--	--	--	$CI_{CESP}\,1$	$CI_{CESP}\,2$	$CI_{CESP}\,3$	$CI_{CESP}\,4$	$CI_{CESP}\,5$
Range (t CO_2e_{100} t^{-1} biomass C)	--	--	--	$-0.25 < CI \leq 0$	$-0.50 < CI \leq -0.25$	$-0.75 < CI \leq -0.50$	$-1.00 < CI \leq -0.75$	$CI \leq -1.00$
Production emissions, $CI_{PEMI}100$	U	?	--	$CI_{PEMI}\,1$	$CI_{PEMI}\,2$	$CI_{PEMI}\,3$	$CI_{PEMI}\,4$	$CI_{PEMI}\,5$
Range (t CO_2e_{100} t^{-1} biomass C)	$CI \geq +1.00$	$CI = 0$	--	$+0.80 \leq CI < +1.00$	$+0.60 \leq CI < +0.80$	$+0.40 \leq CI < +0.60$	$+0.20 \leq CI < +0.40$	$0 \leq CI < +0.20$
Production bioenergy, $CI_{ENER}100$	--	--	0	$CI_{ENER}\,1$	$CI_{ENER}\,2$	$CI_{ENER}\,3$	$CI_{ENER}\,4$	$CI_{ENER}\,5$
Range (t CO_2e_{100} t^{-1} biomass C)	--	--	$CI = 0$	$-0.25 < CI < 0$	$-0.50 < CI \leq -0.25$	$-0.75 < CI \leq -0.50$	$-1.00 < CI \leq -0.75$	$CI \leq -1.00$
Biochar production climate impact, $CI_{PROD}100$	U	?	--	$CI_{PROD}\,1$	$CI_{PROD}\,2$	$CI_{PROD}\,3$	$CI_{PROD}\,4$	$CI_{PROD}\,5$
Range (t CO_2e_{100} t^{-1} biomass C)	$CI > 0$	$CI_{PEMI}100 = 0$	--	$-0.30 < CI \leq 0$	$-0.60 < CI \leq -0.30$	$-0.90 < CI \leq -0.60$	$-1.20 < CI \leq -0.90$	$CI \leq -1.20$

$CI_{PEMI}100$ will always be positive and therefore contribute to climate change. To assess CI of biogenic emissions during conversion, measurements of production-emission factors (i.e., g X emitted kg^{-1} dry biomass) for CH_4, N_2O, and black-C particulate aerosols (BlkC) are required. In contrast, production-emission factors for fossil-CO_2 emissions are usually calculated from the amount of fossil fuel consumed per mile or per hour of operation using standard emission factors (USEPA 2023), normalized per unit of dry biomass transported, comminuted, or converted, and then added to yield a single fossil-CO_2 production-emission factor. As detailed in the Supplementary Material, when the relevant production-emission factors are available, $CI_{PEMI}100$ values are determined by multiplying each production-emission factor by the appropriate emission-impact factor (e.g., 0.00127 for fossil CO_2, 0.125 for CH_4, 0.581 for N_2O, and 5.96 for BlkC all assuming a biomass C content of 0.50) and then summing the four products.

Acceptable values of $CI_{PEMI}100$ range from 0 to 1.00 t CO_2e t^{-1} biomass C (Table 9.1). This range is broken into quintiles, with the highest-ranking quintile (CI_{PEMI} 5) being 0 to 0.20 t CO_2e t^{-1} biomass C (indicating the smallest contribution to climate change) and the lowest-ranking quintile (CI_{PEMI} 1) ranging from 0.80 to 1.00 t CO_2e t^{-1} biomass C. If no emissions data are collected, the value of $CI_{PEMI}100 = 0$, and it is classified as "Not Rated" (shown as "?" in the workbook). Values of $CI_{PEMI}100$ greater than 1.00 t CO_2e t^{-1} biomass C are assigned a classification of "U" indicating a biochar production method that has unacceptable emission levels.

Production bioenergy (ENER)

The third LCA-based CI estimate used to classify biochar, $CI_{ENER}100$, considers the net energy recovered during the biomass conversion process that is used to offset fossil energy. As with $CI_{PEMI}100$, the reference biomass pathway emissions are nil. In contrast to $CI_{PEMI}100$, however, the value of $CI_{ENER}100$ will never be positive. If bioenergy is recovered and used to displace fossil energy, it will enhance the beneficial CI of the biochar. To estimate $CI_{ENER}100$, the maximum energy available for recovery is obtained as the difference between the energy stored in the biomass and that stored in the biochar, both on a dry mass basis. This value is multiplied by the relative energy recovery efficiency (i.e., RER_{eff}, the fraction of the potential energy that is recovered and used to displace fossil energy) and then the energy used to dry the biomass is subtracted (the biochar is bone dry) to obtain the net bioenergy recovered (and used). To complete the estimate of $CI_{ENER}100$, the net bioenergy recovered is expressed per unit of biomass C, multiplied by the C intensity of the fossil energy being displaced, and then multiplied by the mean fraction of CO_2 remaining in the atmosphere over 100 years (i.e., 0.521). The energy stored in the biomass is the lower heating value (LHV), and that in the biochar is the higher heating value (HHV), both of which can either be determined experimentally or estimated using the results of an ultimate analysis (Hosokai et al, 2016; Qian et al, 2020). Similarly, the RER_{eff} may be determined experimentally (Mason et al 2021), or a generic value may be used (e.g., Woolf et al, 2010, set $RER_{eff} = 0.75$). Values for the C intensity of the fossil energy will vary with each situation and region. In the Supplementary Material workbook, where further details of the $CI_{ENER}100$ calculation can be found, some values for C intensity of the primary energy supply globally and in the United States are supplied, as well as for the C intensity of specific fossil fuels.

The classification quintiles for $CI_{ENER}100$ are identical to those for $CI_{CESP}100$, with the highest class (CI_{ENER} 5) being assigned for values less than -1.00 t CO_2e t^{-1} biomass C and the lowest class (CI_{ENER} 1) being assigned to values between 0 and -0.25 t CO_2e t^{-1} biomass C (Table 9.1). There are no unacceptable values for $CI_{ENER}100$ and a value of 0 is given a "No Benefit" classification.

Biochar production climate impact (PROD)

When $CI_{CESP}100$, $CI_{PEMI}100$, and $CI_{ENER}100$ values are summed, they yield the biochar production CI ($CI_{PROD}100$), which estimates the CI of biochar for classification purposes. Classification quintiles (Figure 9.5, Table 9.1) range from 0 to -0.30 t CO_2e t^{-1} biomass C in the lowest class (CI_{PROD} 1) to < -1.20 t CO_2e t^{-1} biomass C in the highest class (CI_{PROD} 5). If no emissions data are reported, the $CI_{PROD}100$ classification is set to "Not Rated" (shown as "?"). Reporting the classification values as CI_{PROD} X (A, B-C-D) is recommended, where X is the CI_{PROD} class and A, B, C, and D, are respectively, the classes

assigned for F_{PERM}, $CI_{CESP}100$, $CI_{PEMI}100$, and $CI_{ENER}100$. For example, CI_{PROD} 4 (Fp5, 2-5-4) would represent a biochar with a $CI_{PROD}100$ value of -1.07 (Class 4), F_{PERM} value of 0.891 (Class Fp5), $CI_{CESP}100$ value of -0.33 (Class 2), $CI_{PEMI}100$ value of $+0.13$ (Class 5) and a $CI_{ENER}100$ value of -0.87 (Class 4). If further detail is needed, one can always report the full suite of $CI_{XXXX}100$ and F_{PERM} values.

Climate impact estimates for three biochar production scenarios

To demonstrate the application of the limited LCA-based classification system and to show the relative impacts of the three CI components, three hypothetical biochar production scenarios are considered. These use the same woody biomass feedstock and have identical C_{eff} (0.45) and F_{PERM} (0.891, corresponding to H/C_{org} of 0.3 and soil temperature of 14.9°C) values, and hence identical $CI_{CESP}100$ values, but differ in their

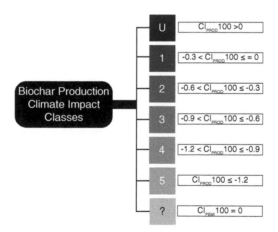

Figure 9.5 *Classification of biochars in terms of their estimated climate impact of production based on the limited LCA-based approach. Ranges are in units of t CO_2e t^{-1} biomass C. A classification of U indicates the CI of the biochar is unacceptable and its production detracts from climate change mitigation. A classification of "?" indicates that no rating can be made because production emissions have not been considered. The estimated benefit to climate change mitigation increases with the class value, with Class 5 providing the highest benefit*

production emissions and production bioenergy parameters. The three production scenarios are low-tech biochar production without emission controls or bioenergy capture (e.g., flame-cap kiln), modern biochar production with bioenergy production but minimal production emission controls (e.g., pyrolytic gasifier), and modern biochar production with bioenergy production and full production emission controls (e.g., a high-temperature gasifier). The production emission and bioenergy parameters for each scenario are based on published (low-tech) or proprietary data shared by manufacturers (modern) (Table 9.2). The full parameter sets for the three scenarios (Table 9.2) show large differences in production emissions with the low-tech scenario having the highest emissions of CH_4 and BlkC and lowest emissions of N_2O and fossil-CO_2.

Using these input parameters, component CI values are assigned biochar production CI classes for each scenario using the workbook calculator (Supplementary Material). The results (Table 9.2) cover the entire range of CI classes with the low-tech scenario being classified as unacceptable (U), the modern with bioenergy as CI_{PROD} 2, and the modern with bioenergy and full emission controls as CI_{PROD} 5. The key differences are in the $CI_{PEMI}100$ values where the unacceptably large value for the low-tech scenario (+2.12) completely overwhelms the $CI_{CESP}100$ value of −0.98, that for the modern with bioenergy scenario (+0.97) cancels the $CI_{CESP}100$ value but the $CI_{ENER}100$ value of −0.51 yields a net benefit, and that for the modern with bioenergy and full production-emissions controls yields a very low $CI_{PEMI}100$ value of +0.13 (16 times smaller than for the low-tech scenario) and a robust $CI_{PROD}100$ value (−1.37). Although CH_4 is also important, the emission factor for BlkC dominates the $CI_{PEMI}100$ calculation underscoring the

need to produce biochar with minimal particulate emissions (for both climate mitigation and public-health reasons).

To show the dynamic nature of the CI calculations and to help visualize the determination of the 100-year mean CI, the annualized CI_{PROD} data for the three scenarios are plotted with a zero-fill representation (Figure 9.6a). Values of CI_{PROD} greater than zero are in the "carbon-positive" region and therefore accelerate climate change relative to the biomass reference scenario, whereas those below zero are "carbon negative" (i.e., they mitigate climate change). For each scenario, the shaded area is summed (carbon-positive areas retaining a positive sign and carbon-negative areas retaining a negative sign) and then divided by 100 to obtain the $CI_{PROD}100$ result. The other three panels (Figure 9.6b, 9.6c, and 9.6d) provide annualized scenario-specific data for the component CIs and CI_{PROD}. As shown in Figures 9.6a and 9.6c for the modern with bioenergy scenario, the cross-over point between carbon positive and carbon negative for the CI_{PROD} curve represents a carbon-payback period that, in this instance, is within the 10 years considered fully sustainable (DeHue et al, 2007; DeHue 2013). In stark contrast, the carbon payback period for the low-tech scenario is a century (Figure 9.6b), which is clearly unsustainable. The modern scenario with bioenergy and full emission controls does not have a carbon payback period (Figure 9.6d) and therefore this scenario is fully sustainable at the time of biochar production.

The main inference to be drawn from the incorporation of a limited LCA-based approach into the classification of the CI of biochars is that the conditions by which the biochar is produced matter greatly. The biochar properties (F_{PERM}) of the three scenarios were identical, yet widely divergent $CI_{PROD}100$ values were obtained depending

Table 9.2 *Parameters used in the limited LCA-based approach to classifying biochars for the three examples of biochar production discussed and the results of applying the approach*

Parameter	Biochar production method		
	Low-tech w/o energy	Modern w/energy	Modern w/energy and full emissions control
	--- Common Parameters ---		
Biomass C content (mass fraction)		0.50	
Biomass H_2O content (mass fraction wet basis)		0.20	
Biomass lower heating value (LHV) (GJ t^{-1} dry biomass)		18.0	
Biochar C content (mass fraction)		0.80	
Biochar H/C_{org} (mol mol^{-1})		0.3	
Biochar H_2O content (mass fraction wet basis)		nil	
Biochar higher heating value, HHV (GJ t^{-1} dry biochar)		31.0	
Soil mean annual temperature (°C)		14.9	
Biochar soil permanence factor, F_{PERM} (fraction)		0.891	
Biochar production carbon efficiency,^{a} C_{eff} (fraction)		0.45	
Carbon intensity of primary energy supply^b (kg CO_2 GJ^{-1})		64.11	
Alternative biomass fate		Immaculate conversion to CO_2	
	--- Production-specific parameters ---		
Relative energy recovery efficiency, RER_{eff} (fraction)	nil	0.75	0.75

Table 9.2 *continued*

Parameter	Biochar production method		
	Low-tech w/o energy	Modern w/energy	Modern w/energy and full emissions control
	------- Production-specific emission factors[c] -------		
Fossil CO_2 emissions for biomass transport, comminution, and conversion (g CO_2 kg^{-1} dry biomass)	25	81	44
Biogenic methane emissions (g CH_4 kg^{-1} dry biomass)	4.107	0.073	0.043
Biogenic nitrous oxide emissions (g N_2O kg^{-1} dry biomass)	0.0088	0.1850	0.1019
Biogenic black carbon emissions (g black carbon kg^{-1} dry biomass)	0.2640	0.1247	0.0015
	------ 100-year climate impact (t CO_2e_{100} t^{-1} biomass C) ------		
C efficiency and soil permanence, $CI_{CESP}100$	−0.98	−0.98	−0.98
Production emissions, $CI_{PEMI}100$	2.12	0.97	0.13
Production bioenergy, $CI_{ENER}100$	0.00	−0.51	−0.51
Biochar production climate impact, $CI_{PROD}100$	1.14	−0.53	−1.37
Biochar production climate impact Class	U	2	5

Notes

[a] Calculated by $(C_{BC}*m_{BC}) / (C_{BM}*m_{BM})$ where C_{XX} and m_{XX} are, respectively, the C content (g C / g XX) and total mass (g XX) of the biomass (XX = BM) converted and biochar produced (XX = BC)

[b] Data for World used by Woolf et al (2010); US mean, minimum, and maximum values for 2020 are 46.87, 29.90, and 70.55 kg CO_2 GJ^{-1}, respectively, derived from https://www.eia.gov/environment/emissions/state/excel/table6.xlsx; Additional representative values for stationary combustion of specific fossil fuels are given in the Supplementary Material

[c] Amonette et al (2023)

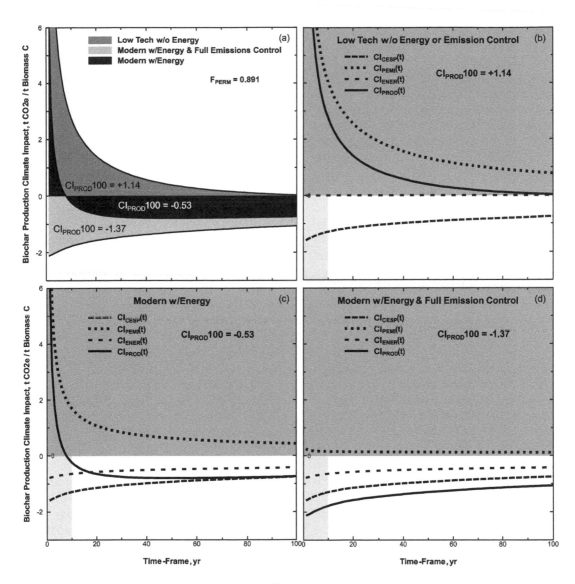

Figure 9.6 *Annualized biochar production CI data for low-tech, modern with bioenergy, and modern with bioenergy and full emission controls scenarios: (a) zero-fill representation of CI_{PROD} data together with mean $CI_{PROD}100$ values; (b), (c), and (d) line plots of annualized CI component and CI_{PROD} data for each scenario. Dark shading above the zero line indicates the C-positive CI region. Light shading below the zero line during first ten years indicates the C-payback period during which the annualized CI_{PROD} value must become (and remain) negative for a scenario to be considered fully sustainable*

on GHGA emissions and production of bioenergy. Although not addressed in the scenarios, $CI_{CESP}100$ values scale directly with C_{eff}, which is far more important to the result than differences in F_{PERM} (Figure 9.7a). In the absence of bioenergy production, a biochar produced at a C_{eff} of 0.45 will have a 3-fold larger climate benefit (i.e., $CI_{CESP}100$) than one produced at a C_{eff} of 0.15. Given the current parameters included in the limited LCA-based approach, differences in C_{eff} are the next most important factor (after GHGA emissions) in determining the $CI_{PROD}100$ and become the most important factor once GHGA emissions are fully controlled (Figure 9.7b).

The ability of the workbook calculator (Supplementary Material) to classify biochars easily for their production-related CI and to discern, in general, how the individual CI components contribute to the whole CI will provide insights that help biochar production methods evolve more quickly towards those that maximize the beneficial CI of biochar production. The workbook calculator itself is expected to evolve to include a broader LCA parameter set, further improving its accuracy and utility. However, while instructive and helpful for classification purposes, the limited LCA-based approach is not a substitute for a full LCA when applied to specific circumstances and considering different biomass feedstocks and reference pathways. Full LCAs are recommended for biochars whose $CI_{PROD}100$ values classify them as U, ?, 1, and even 2, to clarify whether they truly have value for mitigation of climate change.

Fertilizer value and classification

The fertilizer value

The concentration of various nutrients in biochars is largely influenced by the type of feedstock and pyrolysis conditions (Chapter 8; Singh et al, 2010; Ippolito et al, 2020), whereas nutrient availability in biochars is related to the nature of the chemical compounds in which the nutrient occurs (Chapter 16; Wang et al, 2012a,b). Soil properties such as pH may also influence the availability of nutrients from biochar in soils. In the proposed classification system, the concentration of major nutrients present in a biochar determines the potential fertilizer "grade" of the biochar and is represented by an abbreviated code (based on the weight of the element or corresponding oxide, depending on regional regulatory body, and expressed as a percentage of the total dry weight of the compound or mixture). The availability of the nutrient is reported separately. In the classification, six major nutrient elements, i.e., nitrogen (N), phosphorus (P), potassium (K), sulfur (S), calcium (Ca), and magnesium (Mg) have been considered. In the US and parts of Europe, the N, P, K, S, Ca, and Mg values are often expressed as % total N, % P_2O_5, % K_2O, % S, % CaO, and % MgO, respectively, in fertilizers and soil amendments. For biochars, the use of the same type of six-number code for nutrient concentration, as well as a separate value based on the availability of the nutrients is proposed (Table 9.3). As biochar nutrient concentrations are usually smaller but application rates are generally much greater than for fertilizers, the nutrient concentrations should be reported to two decimal points.

The fertilizer classification system

In addition to providing the fertilizer grade for these six nutrients, each biochar can be classified into a fertilizer class (Figure 9.8). For simplification purposes, the present classification does not consider N because of its low availability (see the next section) and Ca that is present in considerable concentrations

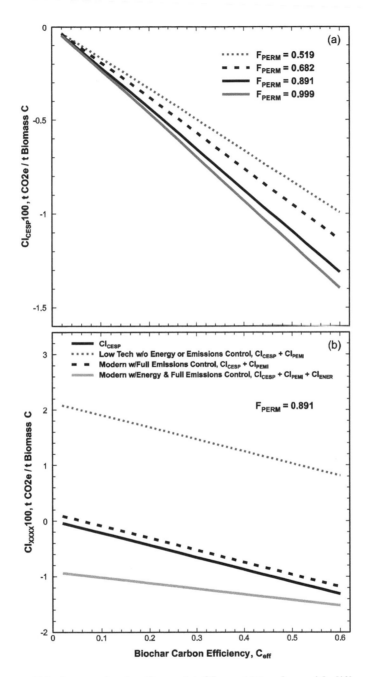

Figure 9.7 *Impact of biochar production C_{eff} on (a) $CI_{CESP}100$ values with different F_{PERM} values, and (b) $CI_{PROD}100$ values calculated when $F_{PERM} = 0.891$ for three biochar production scenarios having high and low production emissions of GHGAs with and without production of bioenergy*

Table 9.3 *Total concentration of N, P (as P_2O_5), K (as K_2O), S, Mg (as MgO), and Ca (as CaO) and their corresponding available fractions in different biochars.* [1] *In the last column, fertilizer class for each of the biochars are given in the last column*

Feedstock[2]	HHT (°C)	H/C_{org}	[3]N_T (%)	[4]N_A/N_T	P_2O_{5-T} (%)	P_A/P_T	K_2O_T (%)	K_A/K_T	S_T (%)	S_A/S_T	MgO_T (%)	Mg_A/Mg_T	CaO_T (%)	Ca_A/Ca_T
Pine wood	450	0.52	0.35	0.09	0.10	0.21	0.34	1.00	<d.l.[5]	n.a.[6]	0.18	0.34	0.74	0.52
Pine wood	550	0.37	0.48	0.08	0.11	0.20	0.35	1.00	<d.l.	n.a.	0.19	0.18	0.85	0.44
Eucalyptus wood	350	0.66	0.40	0.02	0.50	0.30	0.62	1.00	<d.l.	n.a.	0.19	0.40	0.59	0.35
Willow wood	350	0.55	1.36	0.02	0.81	0.26	0.88	1.00	0.41	0.10	0.45	0.70	4.02	0.90
Biosolid + Eucalyptus wood[7]	450	1.00[7]	1.85	0.17	10.95	0.42	0.86	0.52	0.31	0.44	0.53	0.75	2.95	1.00
Biosolid + Eucalyptus wood	550	0.82[8]	1.66	0.05	11.60	0.36	0.84	0.54	0.21	0.35	0.52	0.79	3.22	0.85
Willow wood	550	0.39	1.78	0.01	0.97	0.38	1.29	1.00	0.35	0.75	0.72	0.68	6.00	0.98
Cattle manure	550[9]	0.72	1.08	0.04	3.35	0.66	3.91	0.96	0.40	0.31	2.00	0.85	3.13	0.95
Poultry litter	550[9]	0.55	3.77	0.05	7.60	0.81	4.03	1.00	0.48	0.76	1.50	0.95	8.91	0.99
Eucalyptus wood	450	0.61	0.59	0.05	0.30	0.45	0.43	0.66	0.02	0.88	0.20	0.48	1.04	0.89
Eucalyptus wood	550	0.46	0.60	0.03	0.37	0.39	0.50	0.64	0.02	0.63	0.18	0.38	1.25	0.69
Rice husk	550	0.38	0.71	0.01	0.55	0.24	1.20	0.92	0.01	0.09	0.15	0.52	0.43	0.67
Rice husk	700	0.19	0.62	0.01	0.54	0.30	1.35	0.95	0.003	0.03	0.16	0.56	0.31	0.53
Wheat straw	550	0.36	1.05	0.03	0.49	0.23	1.40	0.82	0.02	0.22	0.04	0.18	0.37	0.40
Wheat straw	700	0.25	1.02	0.01	1.20	0.31	2.02	0.87	0.02	0.48	0.10	0.28	0.60	0.46
Switch grass	400	0.65	0.49	0.01	0.33	0.31	0.11	0.47	0.01	n.a.	0.15	0.48	0.26	0.45
Switch grass	550	0.43	0.54	0.01	0.40	0.30	0.13	0.48	0.003	0.13	0.12	0.35	0.24	0.56

Notes

[1] Total P, K, Mg, and Ca were determined using the modified dry-ash method following a wet digestion (Enders and Lehmann, 2012); total N and S were determined using a total elemental analyzer; available N was measured using 6 M HCl (Wang et al, 2012a); available P using 2% formic acid (Wang et al, 2012b, after Rajan et al, 1992; and AOAC, 2005); available K, S, Mg, and Ca were determined after extracting with 1 M HCl, following the method used for the liming equivalence determination (Rayment and Lyons, 2011); Total concentrations are expressed on oven-dry mass basis of biochars.

[2] Sources: Wang et al (2012a,b), Camps-Arbestain et al (2015), and Singh et al (2017a).

[3] "T" = total;

[4] "A" = available;

[5] "<d.l." = below detection limit;

[6] "n.a." = non-available where data is not available or unreliable;

[7] H_{org}/C_{org} (molar ratio) = 0.67;

[8] H_{org}/C_{org} = 0.55;

[9] "a" = activated.

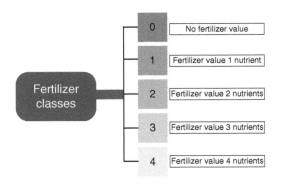

Figure 9.8 *A classification system of biochar based on its fertilizer value in terms of supplying phosphorus (P), potassium (K), sulfur (S), and magnesium (Mg). Minimum levels for available P_2O_5, K_2O, S, and MgO are based on the needs to fulfill the demand of an average corn crop (grain) considering a maximum biochar dose of 10 t ha^{-1}*

in most biochars but to some extent linked to biochars' liming value.

The following steps need to be followed to classify a specific biochar based on its fertilizer value.

Step 1 Available concentrations of P, K, S, and Mg in the biochar should be known. For this, the total concentration of each element in the biochar needs to be multiplied by its corresponding available fraction, as shown in the example shown in Table 9.4.

Step 2 In order to classify a particular biochar based on its fertilizer value, the expected yield and nutrient removal for a specific crop needs to be considered and the fertilizer value of the biochar is expressed in terms of its ability to meet the nutrient removal demand of the crop. Here, corn crop is taken as an example, which is one of the major crops worldwide. The removal of P, K, S, and Mg by corn at the end of the growing season (based on an average yield of 13.5 t ha^{-1}) is estimated to be approximately 103 kg P_2O_5 ha^{-1} (45 kg P ha^{-1}), 54 kg K_2O ha^{-1} (45 kg K ha^{-1}), 17.6 kg S ha^{-1}, and 33.8 kg MgO ha^{-1} (20.4 kg Mg ha^{-1}) (Table 9.5; Havlin et al, 2005). These estimations of nutrient removals do not consider soil properties and soil-plant interactions. More information on nutrient removal by different crops in various countries and regions is

Table 9.4 *Total concentration of P, K, S, and Mg (expressed as a percentage on dry mass basis), the corresponding available fraction, and the corresponding available concentration (expressed as a percentage on dry mass basis) of biochar produced from a mixture of biosolids and eucalyptus wood (50% DW each) at 550°C. The third column is obtained by multiplying column 1 (total concentration of a specific element in oxide form in biochar expressed as a percentage on dry mass basis) × column 2 (fraction of the total content that is available)*

Total P_2O_5 (%)	Available fraction (Avail. P/total P)	Available P_2O_5 (%)
11.60	0.36	4.18
Total K_2O (%)	Available fraction (Avail. K/total K)	Available K_2O (%)
0.84	0.54	0.45
Total S (%)	Available fraction (Avail. S/total S)	Available S (%)
0.21	0.35	0.07
Total MgO (%)	Available fraction (Avail. Mg/total Mg)	Available MgO (%)
0.52	0.79	0.41

Table 9.5 *Nutrient removal[1] or uptake values (kg t^{-1}) for selected agricultural crops (Havlin et al, 2005)*

Crop	N	P_2O_5	K_2O	S	MgO	CaO
Wheat (grain)	19.5	12.7	8.4	1.1	4.6	0.8
Wheat (straw)	8.2	2.1	14.2	2.7	3.6	2.0
Barley (grain)	18.1	8.9	8.0	2.2	2.8	0.8
Barley (straw)	6.8	5.2	21.9	0.9	0.8	2.5
Corn (grain)	12.5	7.6	4.0	1.3	2.5	0.7
Corn (stover)	8.3	2.1	14.6	1.2	4.5	1.7
Alfalfa	26.5	6.9	27.3	3.3	5.0	16.9
Soybean (grain)	39.1	19.5	18.6	4.8	3.4	5.5
Soybean (stover)	10.1	4.2	10.1	1.4	1.7	4.8
Rice (grain)	13.0	6.7	3.6	-	-	-
Rice (straw)	8.3	2.7	21.0	-	-	-
Onion	2.7	2.8	2.9	1.1	0.2	0.9
Potato (sweat)	2.2	2.3	6.4	0.3	0.4	0.3
Potato (white)	2.7	3.1	5.8	0.2	0.4	0.2
Spinach	4.5	3.1	3.3	0.4	0.8	1.5
Tomato	2.7	2.1	4.4	0.3	0.4	0.2
Cotton (seed+lint)	22.0	20.0	13.0	1.7	4.0	2.0
Sugarbeet	4.5	1.0	8.7	0.6	1.9	-
Sugarcane	2.0	1.0	3.4	0.2	0.5	-

[1]   An example of nutrient removal would be a corn (grain) crop with an average yield of 13.5 t ha^{-1} removes 12.5 × 13.5 = 169 kg N, 7.6 × 13.5 = 103 kg P_2O_5, 4.0 × 13.5 = 54 kg K_2O, 1.3 × 13.5 = 17.6 kg S, 2.5 × 13.5 = 34 kg MgO and 0.7 × 13.5 = 9.5 kg CaO

available on the website of the International Plant Nutrition Institute (http://www.ipni.net/article/IPNI-3296). However, end-users are encouraged to use local data when available, because nutrient removal by a particular crop depends on the growing conditions, cultivar, and soil properties.

Step 3 The fertilizer value of the biochar is dose-dependent. Therefore, the present classification system should be adaptable to different dosages. If a biochar application rate of 10 t ha^{-1} is intended to be used on a given soil, then the (rounded) nutrient concentrations in the biochar needed to meet the hypothetical crop (i.e., corn) demand will be: available P_2O_5 = 1.00%, available K_2O = 0.55%, available S = 0.20%, and available MgO = 0.35% (% of the biochar dry weight). If the biochar application rate was 1 t ha^{-1}, then the nutrient concentration in the biochar will obviously need to be 10 times greater to meet that specific nutrient crop demand.

According to the present classification, biochar will not have any fertilizer value if, when at an application rate of 10 t ha^{-1}, it cannot completely fulfill the hypothetical demand of corn for at least one of the four above-mentioned nutrients. In other words, if biochar has values of available P_2O_5 < 1.03%, available K_2O < 0.54%, available S < 0.18%, and available MgO < 0.34% (each given in percentage of total biochar weight) – which at a dose of 10 t ha^{-1} would correspond to approximately 44 kg P ha^{-1}, 42 kg K ha^{-1}, 20 kg S ha^{-1} and 21 kg Mg ha^{-1} –, this biochar is considered to have no fertilizer value for classification purposes. This always assumes that a biochar is applied at a dose ≤ 10 t ha^{-1}; doses above this value have not been considered to classify biochars for their fertility value. Biochars with the fertilizer class 0 can still provide a considerable amount of nutrients, particularly when amended at doses > 10 t ha^{-1}. Biochar users are encouraged to make use of the information provided on the fertilizer grade for N, P, K, S, Ca, and Mg of a specific biochar (and the corresponding availability) (examples in Table 9.3) together with available information on soil fertility, so that the needs for a specific crop are adequately satisfied or balanced with other sources of fertilizer where needed.

Step 4. How are biochars with fertilizer value classified further? An example is provided below in Table 9.6, where the amounts of available P, K, S, and Mg added through a biochar applied at a dosage ranging from 1 to 10 t ha^{-1} are calculated.

Data in Table 9.6 show that when this particular biochar is applied at a rate of 3 t ha^{-1} or above it fulfills the P requirement of a corn crop, and at the application rate of 9 t ha^{-1} it additionally meets the crop requirement for Mg. Therefore, based on the classification system, this biochar has a fertilizer value of 2 P_{3t} Mg_{9t}. The number "2" in the label of this biochar indicates the number of nutrients having substantial fertilizer value that is contained in this biochar, and the indices in subscript indicate the application rate of the biochar at which crop demand is fulfilled for the indicated nutrient. This information should help end-users make an informed decision about the

Table 9.6 *The amount of available P (as P_2O_5), K (as K_2O), S, and Mg (as MgO) added to the soil when the biochar produced from a mixture of biosolids and Eucalyptus wood (50% DW each) produced at 550°C is applied at rates ranging from 1 to 10 t biochar ha^{-1}. The shadowed areas indicate where the nutrient requirement for corn (considered as an example) is fulfilled. Corn grain yield was assumed to be 13.5 t ha^{-1} and nutrients removed by the crop were 103 kg P_2O_5, 54 kg K_2O, 17.6 kg S, and 34 kg MgO (Table 9.5)*

Available nutrient supplied[1]	Biochar application rates (t ha^{-1})									
	1	2	3	4	5	6	7	8	9	10
P_2O_5 (kg ha^{-1})	41.8	83.6	125.4	167.2	209	250.8	292.6	334.4	376.2	418
K_2O (kg ha^{-1})	4.5	9.0	13.5	18.0	22.5	27.0	31.5	36.0	40.5	45.0
S (kg ha^{-1})	0.7	1.4	2.1	2.8	3.5	4.2	4.9	5.6	6.3	7.0
MgO (kg ha^{-1})	4.1	8.2	12.3	16.4	20.5	24.6	28.7	32.8	36.9	41.0

Notes
[1] To convert P to P_2O_5, multiply the P value by 2.291; to convert K to K_2O multiply by 1.205, and to convert Mg to MgO multiply by 1.658

Table 9.7 *Fertilizer classes of biochars (listed in Table 9.3) that meet partial nutrient requirements of corn (13.5 t grain ha^{-1}), with uptake or removal of 45 kg P ha^{-1} (103 kg P$_2$O$_5$ ha^{-1}), 45 kg K ha^{-1} (54 kg K$_2$O ha^{-1}), 18 kg S ha^{-1}, 20 kg Mg ha^{-1} (34 kg MgO ha^{-1}). Subscripts indicate the dose of biochar at which the specific nutrient requirement of corn is met (HHT highest heating temperature)*

Feedstock[1]	HHT (°C)	H/C$_{org}$ (mol mol^{-1})	Fertility class	Nutrient(s)
Pine wood	450	0.52	0	None
Pine wood	550	0.37	0	None
Eucalyptus wood	350	0.66	1	K$_{9t}$
Willow wood	350	0.55	1	K$_{7t}$
Biosolid + Eucalyptus wood	450	1.00	2	P$_{3t}$ Mg$_{9t}$
Biosolid + Eucalyptus wood	550	0.82	2	P$_{3t}$ Mg$_{9t}$
Willow wood	550	0.39	3	K$_{5t}$ S$_{6t}$ Mg$_{7t}$
Cattle manure	550	0.72	3	K$_{2t}$ P$_{5t}$ Mg$_{2t}$
Poultry litter	550	0.55	4	K$_{2t}$ P$_{2t}$ S$_{5t}$ Mg$_{3t}$
Eucalyptus wood	450	0.61	0	None
Eucalyptus wood	550	0.46	0	None
Rice husk	550	0.38	1	K$_{5t}$
Rice husk	700	0.19	1	K$_{4t}$
Wheat straw	550	0.36	1	K$_{5t}$
Wheat straw	700	0.25	1	K$_{3t}$
Switch grass	400	0.65	0	None
Switch grass	550	0.43	0	None

[1] Sources: Wang et al (2012a, b), Camps-Arbestain et al, 2015, and Singh et al (2017a).

application rate of biochar and ensure no excess nutrients are added when the requirements for more than one element are met.

This classification system thus considers the following fertilizer classes: (i) Class 0 (available P$_2$O$_5$ < 1.03%, available K$_2$O < 0.54%, available S < 0.18%; available MgO < 0.34%); (ii) Class 1 (biochar with a fertilizer value for one nutrient out of P, K, S, Mg), (iii) Class 2 (biochar with a fertilizer value for two nutrients), (iv) Class 3 (biochar with a fertilizer value for three nutrients), and (v)

Class 4 (biochar with a fertilizer value for four nutrients). No other nutrients are considered in this classification. Table 9.7 provides the classification of biochars listed in Table 9.3 based on this classification system.

Aromatic and heterocyclic N-ring structures are formed as a direct result of pyrolyzing (Almendros et al, 1990, 2003). These compounds are generally persistent in decomposition and thus conversion of N to forms that are plant-available is often negligible (Chapter 6; Almendros et al, 2003; Yao et al, 2010). The

6 M HCl-hydrolysable N in biochar represented the easily-mineralizable N pool better than either total N or mineral N (2 M KCl extractable) (Wang et al, 2012a). Although the use of a 6 M HCl-extractable N represents an oversimplification of the mechanisms through which N in biochar is released (as it only assumes acid hydrolytic reactions), the method, as further modified by Camps-Arbestain et al (2017), is proposed here until a better methodology is developed. The N fertilizer grade of biochar based on the present classification thus considers the use of this reagent to estimate plant-available N in biochar.

Less than 5% of the organic C in biochars is mineralized in a single year and more typical are rates of less than 1% for biochars with a molar H/Corg ratio of <0.6 (Chapter 11; Woolf et al, 2021 for an online calculator). Assuming a co-mineralization of the N with the C, the amount of N mineralized from a pine wood biochar produced at 450°C (total N values reported in Table 9.3) per year would be 0.2 kg N t^{-1} biochar, whereas that from a poultry litter biochar produced at 550°C (total N values reported in Table 9.3) would be 1.9 kg N t^{-1} biochar, assuming a mineralization rate of 5%. Values of "available N" in fresh biochar as estimated using 6 M HCl were of similar magnitude (0.3 and 1.9 kg N t^{-1} of biochar, respectively; Table 9.3).

Relationship between nutrient availability in biochar and recommended extraction methods

NITROGEN

Considering a minimum requirement of 168 kg available N ha^{-1} to fulfill the annual demand for corn (grain) (Table 9.5), more than 500 and 80 t ha^{-1} of each biochar, respectively, would need to be applied. Practically speaking, however, the maximum amount of each biochar that could be applied to fertilize corn would be 16 t ha^{-1} for the pine-wood biochar and about 1.35 t ha^{-1} for the poultry-litter biochar, based on their K contents. The corresponding amounts of N associated with these application rates would be 2.9 kg ha^{-1} and 2.6 kg ha^{-1}, respectively, which is less than 2% of the annual demand (168 kg ha^{-1}) for corn. Even with potatoes, a crop requiring substantial amounts of K and N (Table 9.5), the levels of P supplied by most biochars would limit application rates to levels with minimal N value (e.g., only 2 t ha^{-1} of the poultry-litter biochar would be enough to supply the P needs of potatoes and this would supply only 4% of the N required). The N fertilization value of biochar is expected to be negligible under realistic application scenarios. Therefore, N is not included in the proposed classification system.

PHOSPHORUS

The availability of nutrients that are present in the ash fraction of biochar (e.g., P and K) is expected to be greater than that of heterocyclic N because it depends primarily on solubility rather than the C mineralization rate. The pH and presence of chelating substances will have a strong influence on ash-nutrient availability (Chapters 8 and 16). Phosphorus in biochar is mainly present as phosphate salts, the nature of these being determined by feedstock composition. Calcium- and Mg-phosphate compounds have been found to dominate in biochars produced from cow manure while Al-phosphate compounds dominate in biochars produced from alum-treated biosolids.

In general, there is a gradual increase in the total P concentration of biochars with increasing pyrolysis temperature (Li et al, 2019). In addition to the enrichment of P, there is a substantial speciation transformation in P forms in biochar during the pyrolysis process. Organic P species, such as phytates, phospholipids, and nucleic acids, are converted to inorganic forms of P with pyrolysis, and this conversion is nearly complete at 450°C (Li et al, 2018). Phytate was converted to inorganic P in manure and plant wastes pyrolyzed at 350°C, whereas inorganic orthophosphate (PO_4^{3-}) was the sole species in manure biochars produced at ≥500°C and pyrophosphate ($P_2O_7^{4-}$) species persisted in plant biochars up to 650°C (Uchimiya and Hirodate, 2014). Water-soluble forms of P are transformed to much less water-soluble P compounds (e.g., hydroxyapatite and oxyapatite) with increasing pyrolysis temperature (Bottezini et al, 2021), but with no decreased but often rather increased P extractability using relevant extraction procedures (Li et al, 2018) and therefore increased availability to plants.

The coordinating cations of phosphate in biochar have strong implications on P availability (Wang et al, 2012b). The IBI Biochar Standards recommend the use of P extractable with 2% formic acid (Wang et al, 2012b; Rajan et al, 1992; AOAC, 2005) to estimate plant-available P. The present classification system also proposes its use to determine the P fertilizer grade of biochars. Available P of different biochars (e.g., P extracted with formic acid) can reach up to 5% (reported as P_2O_5, equivalent to 2.2% P) of the total mass of biochar (Table 8.3). When using a biochar with 4% available P_2O_5, less than 3 t ha^{-1} (Table 8.4) is needed to apply 45 kg available P ha^{-1} (Table 9.6). In fact, biochars from biosolids and cattle manure feedstocks can be as effective as commercial P fertilizers (e.g., calcium dihydrogen phosphate) in increasing shoot yield per unit available P applied (on a formic acid-extractable P basis) (Wang et al, 2012b).

Citric and formic acids extract most amounts of P from the studied biochars, however, the P extracted by these weak acids was not correlated with the P uptake by ryegrass plants (Rose et al, 2019). Much stronger correlations of plant P uptake were observed with the P extracted by water and Bray 2 (0.03 M NH_4F and 0.1 M HCl) reagent. Bray-1 (0.025 M HCl + 0.03 M NH_4F) was found to be the most appropriate extractant for measuring available P for immediate to medium-term, and Mehlich-3 (0.2 M CH_3COOH + 0.015 M NH_4F + 0.001 M EDTA + 0.25 M NH_4NO_3) as the most appropriate for determining long-term available P in raw poultry litter and biochars made at temperatures of 300–600°C (Li et al, 2018). Neutral ammonium citrate with water or Mehlich 1 (0.05 M HCl 0.02 M H_2SO_4) was recommended to extract available P from sewage sludge biochars pyrolyzed at a temperature < 300°C and 2% citric acid for biochars made at higher temperatures (i.e., >300°C) (Figueiredo et al, 2021). Clearly, there are variable and contrasting observations about the availability of P in different biochars, which is probably related to different P compounds formed during the pyrolysis process. It is possible that a single extractant may not be suitable to measure available P in biochars produced from different feedstocks pyrolyzed at different temperatures. While further research is needed in this respect, using a 2% formic acid reagent is recommended to determine available P until a better extractant is found.

Potassium

Potassium in biochar is readily released to solution (Yao et al, 2010) due to the high solubility of K-containing salts, and total K has generally been sufficiently equivalent to plant-available K for characterization (IBI, 2015). This is true as long as the digestion method used does not also solubilize K-bearing aluminosilicate minerals, which may also be present in some biochars. The digestion method proposed by the IBI Biochar

Standards and here – uses a modified dry ashing procedure (Enders and Lehmann, 2012) – will not completely dissolve K-bearing mica or feldspar but may extract a considerable fraction of interlayer K from micaceous minerals (mica, illite) if these materials are abundant; this should be taken into consideration in biochars containing substantial amounts of soil particles. The K extractable with 1 M HCl, following the method used for the determination of the liming equivalence (Rayment and Lyons, 2011), as described in the Test Method's section, is suggested here as an alternative to the modified dry ashing although it has not been evaluated for extraction of K from micaceous or feldspathic minerals. Values of total K and HCl-extractable K are generally equivalent except for the biochars produced from biosolids (Table 8.4).

Sulfur

Sulfur in plant tissue is found in various forms i.e., C-bonded S, ester-S, and sulfate-S (Kok, 1993). C-bonded S thermally decomposes at temperatures below 450°C, while ester-S tends to accumulate and eventually convert to sulfate-S (Blum et al, 2013); sulfate-S is thermally the most stable form of S (Knudsen et al, 2004; Khalil et al, 2008) (Chapter 8). The presence of K, Ca, Cl, and silicate influences the retention or release of S during thermal decomposition; K- and Ca-based additives (e.g., CaO) favor the retention of S through the formation of K_2SO_4 and $CaSO_4$, whereas Cl and silicates promote the release of S by influencing the thermal stability of the formed K_2SO_4 and $CaSO_4$ (Khalil et al, 2008). Sulfate-S in biochar produced from biosolids at 550°C was found to be in a non-crystalline phase (Yao et al, 2010). Such a form of S is readily available to plants as it easily dissolves (Yao et al, 2010; Blum et al, 2013). The method proposed here to determine available S is the

measurement of extractable sulfate-S in 1 M HCl (Table 9.3) following the method used for the determination of the liming equivalence (Rayment and Lyons, 2011).

Calcium and Magnesium

Calcium and Mg are mostly retained in the biochar if pyrolyzed at temperatures < 500°C. During slow pyrolysis, they started to evolve at temperatures above 600°C, whereas during fast pyrolysis of pine sawdust, 10 to 20% of Ca and Mg were evolved at temperatures below 550°C (Okuno et al, 2005). The recovery, however, also depended on the presence of other compounds such as silicates, because the formation of Ca-silicates is more thermo-dynamically favorable than Mg-silicates (Okuno et al, 2005). More than 98% of the total Ca and Mg in cornstalk biochars were extractable with 1 M HCl (Xu and Sheng, 2011). However, this extractant will not recover Ca- and Mg-bearing aluminosilicates if present, and is thus considered adequate for the determination of available Ca and Mg. Therefore, Ca and Mg extractable with 1 M HCl (Table 9.4), following the method used for the determination of the liming equivalence (Rayment and Lyons, 2011), is recommended here.

Liming value and classification

The ash fraction of biochar is generally enriched with inorganic non-crystalline (amorphous) and poorly to well-crystallized (mineral) constituents (Singh et al, 2010; Yuan et al, 2011; Kloss et al, 2012). These originate from the feedstock or the presence of some diluents (e.g., soil, clay minerals) mixed with the feedstock. The inorganic constituents of the ash fraction of biochar are usually metal (mostly alkali and alkaline metals) carbonates, silicates, phosphates, sulfates, chlorides, and oxides (Singh et al,

2010; Vassilev et al, 2013a), some of which have considerable liming value (Vassilev et al, 2013b) (Chapter 8). Biochars with liming properties offer an additional value as a soil conditioner for acidic soils. Liming values of a range of biochars are reported as a proportion of $CaCO_3$ equivalent (% $CaCO_3$-

eq) (Table 9.8). Based on the observed range for $CaCO_3$-eq, biochars have been classified into the following four classes: (i) Class 0 ($CaCO_3$-eq < 1%); (ii) Class 1 (1 ≤ $CaCO_3$-eq < 10%); (iii) Class 2 (10 ≤ $CaCO_3$-eq < 20%); and (iv) Class 3 ($CaCO_3$-eq ≥ 20%) (summary of classes in Figure 9.9).

Table 9.8 *The values of H/C_{org}, pH, ash content, and $CaCO_3$ equivalences ($CaCO_3$-eq) of biochars produced from different feedstocks and different highest heating temperatures (HHT)*

Class	Feedstock	HHT (°C)	H/C_{org} (mol mol^{-1})	pH (H_2O)	Ash content (%)	$CaCO_3$-eq (%)	Ref.
0	Eucalyptus (wood)	400	0.56	6.9	3.5	-0.9	1
0	Eucalyptus (wood)	400a	0.52	7.7	3.7	-0.3	1
0	Sawdust	450	n.d.	5.9	1.2	0.5	2
0	Pine (wood)	400	0.71	6.9	3.7	0.7	3
0	Eucalyptus (wood)	550	0.37	8.8	3.3	0.7	1
1	Pine (wood)	550	0.55	7.9	4.1	1.8	3
1	Cow manure	550a	0.62	8.9	75.7	4.3	1
1	Poplar (wood)	550	0.56	8.8	65.0	6.6	3
1	Poultry litter	550a	0.48	10.3	44.4	8.6	1
1	Willow (wood)	400	0.63	7.5	5.7	9.4	3
2	Corn stover	350	0.68	8.9	9.8	11.0	5
2	Poultry litter	550	n.d.	9.6	41.3	13.0	2
2	Biosolids + wood chips	550	0.41	8.0	51.1	15.1	4
2	Food waste	600	n.d.	11.3	59.8	17.0	2
2	Manure + wood chips	450	0.54	10.0	38.4	17.9	4
3	Paper mill waste	550	0.32	8.2	n.d.	29.0	6
3	Tomato green waste	550	n.d.	12.1	56.2	33.0	7
3	Paper sludge	550a	n.d.	9.2	65.4	40.9	1
3	Paper mill waste	400	n.d.	8.0	51.6	67.0	2
3	Paper mill waste	500	n.d.	9.6	56.3	80.0	2

Sources: [1]Singh et al (2010), [2]Krull et al 2012, [3]Calvelo-Pereira et al (2011), [4]Wang et al (2012a), [5]Herath et al (2013); [6]Van Zwieten et al (2010); [7]Smider and Singh (2014).
"a" stands for activated; "n.d." stands for not determined.

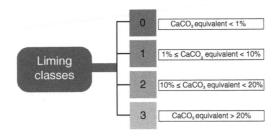

Figure 9.9 *Liming classes for biochars based on CaCO₃ equivalent in biochars*

The test method for the determination of the liming equivalence is described in detail by Singh et al (2017b). It should be noted that the test method provides an approximate measure of the acid-neutralizing capacity of biochars and not the lime content in the biochar; some of the salts of base cations in biochars may neutralize a part of the acidity in the test procedure. Additionally, it has been observed that the rapid titration procedure

---

Box 9.1 Interacting effect of type of soil, type of biochar, and dose on final soil pH 2 weeks after application of biochar to soil

The change in pH of a specific soil with the addition of a specific biochar will depend not only on the liming value of the biochar used, but also on the pH-buffering capacity of the specific soil. Additionally, biochar particle size, soil moisture, and the degree of mixing will affect the change in soil pH from biochar application. A tested Umbrisol is dominated by short-range ordered Al oxyhydroxides, which have an important role in buffering the soil pH around 4.2, whereas the Podzol has a very small pH-buffering capacity and thus responds more readily to the amendment than the Umbrisol, even though both soils have low pH values (Figure 9.10).

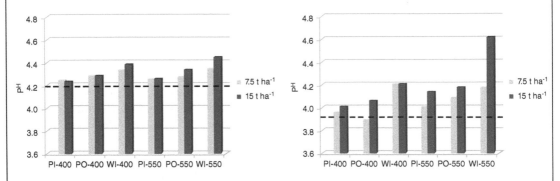

Figure 9.10 *Soil pH values of an Umbrisol (left) and a Podzol (right) amended with biochars from pine (PI), poplar (PO), and willow (WI) produced at 400 and 550°C, added at two doses (7.5 and 15 t ha⁻¹), and incubated for 2 weeks (details of some of these biochars are provided in Table 9.7). Dashed lines represent the pH value of the soil before the amendment (Camps-Arbestain et al, 2015)*

may slightly overestimate the liming potential of biochars compared to slower titration using a dilute acid (Singh et al, 2010). Biochars have the potential to store acidity on surface functional groups and equilibrium might not be achieved in the short time recommended in the rapid titration procedure for determining the liming value.

Particle-size classification

Biochar application in soils generally reduces soil bulk density, increases porosity, decreases hydraulic conductivity, and increases available water (Chapters 5, 8, 20). However, the effects of biochar on soil water retention, drainage, and other physical properties have been reported to be biochar-, dose- and soil-dependent (Herath et al, 2013; Blanco-Canqui, 2017; Chapter 20), which makes it difficult to recommend a specific biochar to ameliorate soil physical properties. For example, a greater decrease in soil bulk density and a greater increase in soil porosity from biochar addition were found in coarse-textured soils than in fine-textured soils (Blanco-Canqui, 2017). In addition to soil texture, biochar particle size distribution, shape, and internal structure may affect soil physical properties, such as hydraulic conductivity and soil water storage.

There are conflicting reports on the effects of biochar particle size on soil water retention and plant available water (Chapter 20), with greater increases in soil water retention observed from coarse biochar than fine biochar in some studies (Liu et al, 2017; de Duarte et al, 2019; Edeh and Mašek, 2021). However, in other studies, either a fine biochar had been found more effective than a coarse biochar (Liao and Thomas, 2019) or the biochar particle size had shown no significant effect on plant available water in biochar-amended soils (Danso et al, 2020; Zhang et al, 2021). Biochar's effects on soil water retention and hydraulic conductivity have been ascribed to the modification in porosity and hydrophobicity caused by different particle sizes of biochars in contrasting textured soils. Clearly, to develop specific recommendations, more research is needed on the effects of biochar particle size on soil physical properties under field conditions.

For textural classes of biochars, a classification system is proposed based on a ternary plot (Figure 9.11). The corresponding test method consists of progressive dry sieving with 50 mm, 25 mm, 16 mm, 2 mm, 1 mm, and 0.5 mm sieves. The ternary plot classifies biochars in different textural classes, i.e., (i) powder (> 50% biochar fraction < 2 mm); (ii) kernel (> 50% biochar fraction between 2 and 16 mm); (iii) lump (> 50% biochar fraction >16 mm); or (iv) blended (other particle sizes) (Figure 9.11). Information on the fractions < 2 mm and > 16 mm is offered in two additional ternary plots. Other measurements provided in the present classification system (e.g., to assess the value of specific biochars as a substrate in soilless agriculture; next section) may also be useful to evaluate the effects of specific biochar on soil physical properties. These measurements are (i) air-filled porosity; (ii) water-holding capacity; and (iii) wettability.

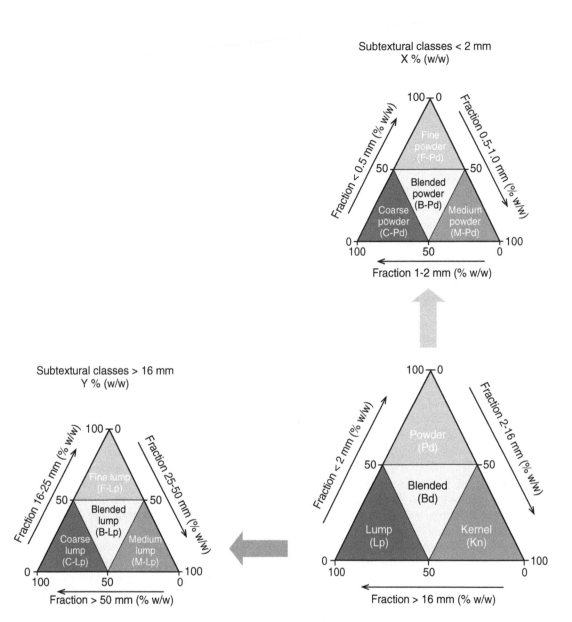

Figure 9.11 *Textural classes for biochars. The large triangle is used to provide the textural class for a specific biochar (e.g., powder, kernel, lump, blended). The small triangles provide additional information on the powders (< 2 mm) and lumps (size > 16 mm). "X" and "Y" stand for the percentages of the powder (< 2 mm) and lump (> 16 mm) fractions, respectively*

Reduced efficacy of agrochemicals in biochar-amended soils

Biochars can act as strong adsorbents for organic chemicals, such as pesticides and industrial organic contaminants (Kookana, 2010; Lian and Xing, 2017; Lima et al, 2022; Chapter 23), sometimes with tens of thousands of times greater sorption capacity than that of soils (Yang and Sheng 2003, Yang et al, 2006, Yu et al, 2009). Weakly sorbed herbicides (e.g. 2,4-D) were sorbed to biochars with high specific surface areas (SSA) (>400 $m^2 g^{-1}$) as much as to activated carbon (AC) (Kearns et al, 2014). Therefore, biochars can potentially influence the bioavailability and, in turn, the efficacy of pesticides used in controlling target weeds and pests (Graber et al, 2011; Kookana et al, 2011). Similarly, biochar can serve as an effective sorbent to minimize the adverse impact of contaminants on the environment (Kearns et al, 2019; Wang et al, 2020).

Application rates of pesticides to soils often need to be adjusted according to the organic matter content of soils, to compensate for the potential deactivation effect of organic C in soils (Chapter 23). Similarly, owing to the strong binding of pesticides to biochars, the addition of biochars to soils also leads to a degree of deactivation of soil-applied pesticides. In fact, numerous studies have now reported that the efficacy of herbicides can be significantly compromised in freshly biochar-amended soils (Yang et al, 2006; Xu et al, 2008, Nag et al, 2011; Graber et al, 2011, 2012; Gámiz et al, 2017). When biochar produced at > 600°C was applied at 1% w/w to soil, it impacted nematode control and nearly twice the recommended rate of 1,3-dichloropropene nematicide application was needed (Graber et al, 2011). In soils amended with the same amount of wheat straw biochar produced at 450°C, the dose required to reduce weed biomass by 50% increased by 1.6 and 3.5

times for trifluralin and atrazine, respectively (Nag et al, 2011). These studies demonstrate that in biochar-amended soils a substantially increased rate of pesticide and herbicide application may be required compared to that in an unamended soil to achieve the same level of pest control. This is likely to have adverse environmental and economic consequences.

The deactivation effect of biochar on herbicides is likely to diminish with time due to the aging of biochar in soils. Three years after biochar application, herbicide sorption in biochar-amended soils was similar to that in the unamended soils (Martin et al, 2012). Besides, the effects of biochar may be different on pre- or post-emergence herbicide applications to soils (Gámiz et al, 2017). This, however, may require a pre-screening of biochars in terms of their potential impact on the deactivation of soil-applied pesticides.

To classify various biochars, in terms of their potential adverse impact on herbicide or pesticide efficacy, one needs to relate their effectiveness as a sorbent to measurable properties of biochars. Akin to ACs, the SSA of biochars is an important property that determines its adsorptive capacity. It is now well established that the nature and properties of biochars are strongly affected by the feedstock (e.g., wood or grass), pyrolytic temperature, and production method (Chapters 5 and 8). Consistent with their SSA, wood feedstock biochars showed a much greater capacity to adsorb pesticides and other organic compounds than biochars produced from grasses (Yu et al, 2006, Bornemann et al, 2007).

The condensed aromatic structure of biochar is an important property from the standpoint of binding of organic chemicals (Chapters 22 and 27). While NMR can be a timeconsuming technique for assessing

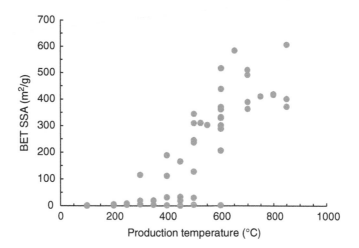

Figure 9.12 *BET specific surface area (SSA) for biochars (n = 50) produced from various feedstocks at different pyrolysis temperatures. Based on data compiled from Chun et al, 2004; Brown et al, 2006; Chen et al, 2008; Bornemann et al, 2007; Zhang et al, 2011; Kearns et al, 2014; Li et al, 2017*

condensed aromatic C contents in biochars, the elemental H/C_{org} ratio can be used as an indicator of their condensed structure (Kookana et al, 2011; Wang et al, 2013; Chapter 11), which depends more on the temperatures and duration of pyrolysis and less so on the biochar feedstock (Kearns et al, 2014). Despite a considerable variation in the SSA of various feedstocks at any given production temperature of biochars (especially in the 300–500°C range), a sigmoidal dependence of SSA on temperature is apparent for a wide variety of biochars (Figure 9.12). The largest increase in SSA is noted in the

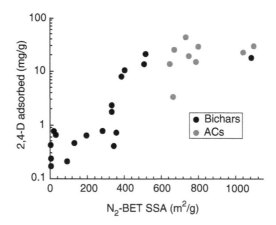

Figure 9.13 *Amount of 2,4-D adsorbed on biochars and activated carbons (ACs) at an equilibrium concentration of 100 μg/L plotted against the BET specific surface area (SSA). Adapted, with permission, from Kearns et al (2014)*

temperature range of 400–600 °C, whereas it tends to plateau below and above this range.

The ability of biochars to remove 2,4-D herbicide from water compares with the activated carbons engineered for water treatment (Figure 9.13). It is worth keeping in mind that 2,4-D is a highly polar herbicide (weak acid) and is typically difficult to remove from water. The amounts of 2,4-D adsorbed onto biochars (q100 µg L^{-1}) and ACs and the BET SSA indicate that some biochars with high SSAs have sorption capacity on par with ACs designed for water treatment. On the other hand, the biochars with low SSAs possess several orders of magnitude lower sorption capacity than ACs. More recently, it has been observed that biochars generated from updraft gasifiers under conditions of simultaneous co-pyrolysis thermal air activation show an even greater ability to adsorb herbicides, such as 2,4-D and simazine, from surface waters containing dissolved organic matter (Kearns et al, 2019). Clearly, biochars have the potential to be as effective sorbents as engineered ACs for water treatment.

Biochar classification for binding pesticides and other organic contaminants from soil and water

Based on the above discussion, the following classification scheme for various biochars is proposed in terms of their effectiveness in binding organic chemicals such as pesticides from soils and waters. In the case of soils, the effectiveness is measured against the inherent sorption capacity of unamended soil, whereas in the case of water treatment, ACs are used as a benchmark to classify their relative efficiency in the removal of contaminants. Based on the data on herbicide sorption (Kearns et al, 2014, 2019), the SSA of biochar measured by the BET-N_2 adsorption technique (which is primarily a function of production temperature

and feedstock, Figure 9.12; Chapter 5), is used as the key property of the biochar in our classification scheme. The activation of biochars such as by thermal air activation is not taken into consideration in this scheme. The cut-off temperature and feedstock types likely to be associated with each category are also indicated in the scheme.

Highly effective sorbent: SSA >400 m^2 g^{-1} or with >10,000 times greater sorption affinity (per unit mass of C) than that of soil organic C or > 100 times greater sorption in a biochar-amended soil (1% w/w) than in an unamended soil. These biochars are likely to be derived from woody feedstock and produced at high pyrolysis temperatures (> 500°C). In the case of water treatment, biochars in this category would be considered nearly similar to ACs in terms of their ability to remove contaminants from water (Figure 9.13).

Moderately effective sorbent: SSA ranging from 10–400 m^2 g^{-1} or with 1,000 times greater sorption affinity than soil organic carbon or at least ten times greater sorption in a biochar-amended soil (1% w/w) than in an unamended soil. These biochars are likely to be derived from woody or grassy feedstocks at pyrolysis temperatures in the range of 300- 500°C. In the case of water treatment, the biochars in this category would be considered 10 times less effective in terms of their ability to remove contaminants from water than ACs (Figure 9.13).

Poorly effective sorbent: SSA <10 m^2 g^{-1} or with 100 times greater sorption affinity than that of soil organic C or less than two times higher sorption in a biochar-amended soil (1% w/w) than in an unamended soil. These biochars are likely to be derived from grasses, crop residues, or animal wastes at low pyrolysis temperatures <300°C. In the case of water treatment, the biochars falling in this category would be considered 100 times less effective than ACs in the removal of contaminants from water (Figure 9.13).

Suitability of biochar as a substrate in potting mixes and soilless agriculture

The use of biochar as a substrate for potting mixes and soilless, hydroponic plant production may provide growers with a cost-effective, renewable, and environmentally responsible recycling method, and supplement substrate and fertilizer requirements. Carbonized materials have been used as soilless substrates in Japan since before the 1970s and in Brazil since the early 1980s (Ikeda, 1985; Kämpf and Jung, 1991; Chapter 26). In recent years, biochar use as a partial or full substitute for peat (a non-renewable resource) in soil-less media has been evaluated in several studies (Banitalebi et al, 2019; Zulfiqar et al, 2019; Jindo et al, 2020). Biochar substrates can potentially increase total porosity (both macro- and micro-pores) and water-holding capacity and promote microbial biomass and activity. However, plant responses to biochar-based media can be highly variable since biochar characteristics are highly variable depending on the feedstock and production conditions. High salt content and alkalinity in biochar are the major concerns for its use in soil-less growing media (Banitalebi et al, 2019; Zulfiqar et al, 2019). Based on the existing research at this stage, biochar could be suggested as an effective additive, partially replacing peat for improving the characteristics of the commercial growing media (Jindo et al, 2020).

The specific mixture of biochar and substrate will likely need to comply with an established standard (e.g., Australian Standards - Potting Mixes; AS 3743–2003, Standards Australia International). The following test methods are proposed to help assess the suitability of a specific biochar as a substrate to be used in either potting mixes or hydroponics: (i) air-filled porosity; (ii) water-holding capacity; (iii) wettability; (iv) pH; (v) electrical conductivity; and (vi) dissolved organic C. In addition to these methods, information on available nutrients, as provided in the fertilizer value of the biochar (including its elemental composition), will help the users adjust their fertigation systems.

Test methods

Test methods as needed to establish the carbon storage value

Organic carbon

The use of C_{org} in place of total C is recommended by the IBI and EBC biochar standards and guidelines for the determination of the amount of C present in biochar, as biochar may also contain inorganic C in the form of carbonate precipitates. Carbonates tend to gradually dissolve in soils under temperate climate conditions or evolve as CO_2 when added to acid soils. Organic C is obtained by determining total C using an elemental analyzer and subtracting inorganic C. Inorganic C can be determined using either (i) a titration method (DIN 51726, ISO 925), in which evolved CO_2 is trapped in an alkaline solution after treating the sample with acid (Bundy and Bremner 1972), or (ii) a manometric method in which the pressure evolved by the CO_2 released after adding a dilute acid to biochar contained in a closed vessel is measured (ASTM D4373-21).

Molar H/C$_{org}$ ratio

Hydrogen is determined using an elemental analyzer. C$_{org}$ is determined as described above. The ratio is expressed on a molar basis.

Molar H$_{org}$/C$_{org}$ ratio

As for C in biochar, H will need to be corrected if there is an important contribution from inorganic H, such as in the case of alum-rich biochar (Wang et al, 2013). A pretreatment of biochar with 10% HF followed by a thorough rinse and oven drying is recommended before determining the H$_{org}$ and C$_{org}$ content of biochar samples.

Test methods as needed to establish the fertilizer value

Total phosphorus, potassium, sulfur, magnesium and calcium

Based on the current knowledge, determining the total content of these elements using the modified dry-ash method following a wet digestion is recommended (Enders and Lehmann, 2012) until better methods are developed. In brief, 0.2–0.5 g of biochar is weighed into a digestion tube and placed into a muffle furnace at 500°C for at least 8 hours. Then 5 mL of HNO_3 is added into the tube and processed at 120°C until dryness is reached. Then the tubes are allowed to cool before adding a mixture of HNO_3-H_2O_2. Samples are then placed back into a pre-heated block, processed at 120°C to dryness, solubilized, and filtered. The digest solution is diluted with deionized water. The elemental composition can be determined using common analytical techniques.

Total nitrogen and sulfur

Determination through dry combustion using a total elemental analyzer.

Available nitrogen

The fraction of N that is hydrolyzable with 6 M HCl is considered here to represent available N (Camps-Arbestain et al, 2017, after Wang et al, 2012a), despite the drawbacks described above. Acid hydrolysis using 6 M HCl involves cleaving ester and amide linkages. It thus preferentially hydrolyses carbohydrates, proteinaceous material, and ester-bound biopolymers (e.g., cutin and suberin), leaving behind a residual OM enriched with alkyl and aromatic structures (Kaal and Rumpel, 2009). Acid hydrolysis is conducted using a modification of the method of Pansu and Gautheyrou (2006). The total N content of the untreated biochar and the non-hydrolyzable residue is determined using an elemental analyzer and hydrolyzable N is determined by difference.

Available phosphorus

The IBI Biochar Standards (IBI, 2015) proposes extraction of P with 2% formic acid (Wang et al, 2012b; after Rajan et al, 1992; and AOAC, 2005) as a method to estimate plant-available P. Formic acid is a monocarboxylated organic acid and thus has a low complexation ability. This reagent solubilizes phosphate by decreasing the solution pH and buffering it at a pH value <3. A sonication step has been added to the existing formic acid extraction method to extract P from biochars to favor the dispersion of the hydrophobic biochar particles (Wang et al, 2012b). The sonication step might only be useful for biochar samples with hydrophobic properties or sparingly soluble P compounds richer in Al and Fe.

Available potassium

Two methods are proposed as valid for this purpose. One is the total content using the modified dry ashing procedure (Enders and Lehmann, 2012), as described above.

Alternatively, the determination of K solubilized in 1 M HCl, as in the liming equivalence determination (Rayment and Lyons, 2011), is considered a valid option. For this, 1 M HCl is added to a known mass of biochar and allowed to stand overnight followed by mechanical shaking for 2 hours (see below). The K in the extract is determined by common analytical techniques.

Available SO_4-S, calcium, and magnesium

The method proposed is the same as described for K, using a 1 M HCl extraction. Sulfate-S, Ca, and Mg will be determined by common analytical techniques.

Test methods as needed to establish the liming value

Liming equivalence

The procedure involves adding 100 ml of 1 M HCl to a known mass of biochar (5.0 g; use 2.5 g for samples with more than 30% carbonate) in a wide-mouth plastic bottle and swirling it occasionally for an hour at room temperature. Allow it to stand overnight, and then shake it on a horizontal shaker for 2 hours. The $CaCO_3$-eq is then determined by titrating the biochar-acid suspension with 0.5 M NaOH to bring the equilibrium solution pH to 7.0 (Singh et al, 2010; Rayment and Lyons, 2011). The reaction of acid is not selective for calcite. Accordingly, the results are reported in terms of calcium carbonate equivalent (g $CaCO_3$-eq kg^{-1}).

Test methods as needed to establish the particle-size classification

The method involves progressive dry sieving with 50 mm, 25 mm, 16 mm, 8 mm, 2 mm, 1 mm, and 0.5 mm standard sieves. Regional authorities however may use different sieve mesh units.

Test required for the classification for binding pesticides and other organic contaminants

Specific surface area

Specific surface area should be measured according to DIN ISO 9277 (BET) or ASTM D6556-21 method. In these methods, N_2 is used as the adsorption gas. Biochar should be dried at 40°C and ground to < 3.15 mm particle size for the analysis. Samples should be degassed under vacuum for about 2 hours keeping the degassing temperature of 150°C. The multipoint (5–7 evenly spaced data points between 0.05 and 0.5 p/p_0) BET method should be used for the adsorption.

Potting mixes and hydroponics class

Air-filled porosity

This can be measured using a modified version of the Australian Standards methods for potting mixes (AS 3743–2003 - Standards Australia International, 2003). The modification consists of soaking the substrate material in water for 24 hours before measuring air-filled porosity to ensure thorough wetting, given the water-repellence of biochar.

Water-holding capacity

This can be measured using a modified version of the German Standard E DIN ISO 14238-2012. The modification consists of soaking the substrate material in water for 24 hours before measuring water-holding capacity to ensure thorough wetting, given the water repellence of biochar.

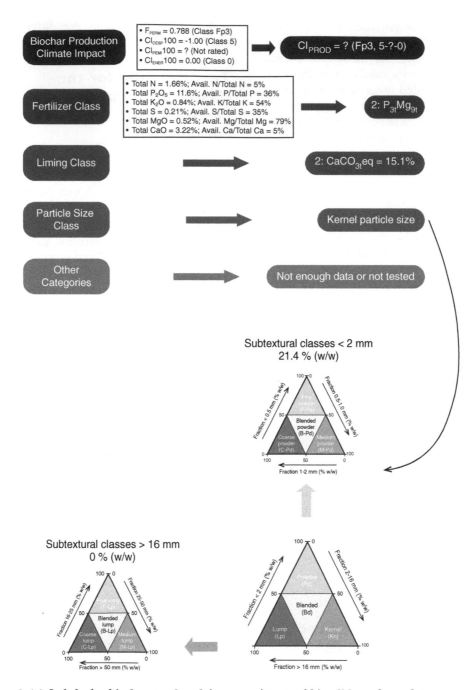

Figure 9.14 *Label of a biochar produced from a mixture of biosolids and eucalyptus wood (50% each; DW basis) at 550°C based on the classification system described in this chapter (Camps-Arbestain et al, 2015)*

Wettability

The method proposed to measure biochar wettability is the molarity of ethanol droplet (MED) (Roy and McGill, 2002). Once the suitable molarity – for which the ethanol drops do not penetrate within 10 seconds – is identified, the contact angle is calculated using the average value of the detected and the immediate high molarities.

pH

Either the method described in the IBI Biochar Standards or that in the EBC Guidelines is suitable as long as the methodology used is reported. The IBI approach follows the pH analysis procedures as outlined in section 04.11 of the US Composting Council and US Department of Agriculture (2001) adapted for the use with biochar (Rajkovich et al, 2012): a 1:20 (w:v) solution of biochar:deionized water equilibrated for 1.5 hours before measurement. The EBC approach follows the DIN ISO 10390 method (ISO 10390) using $CaCl_2$.

Electrical conductivity

Either the method described in the IBI Biochar Standards or that in the EBC Guidelines is suitable as long as the methodology used is reported. The IBI approach follows the procedure as outlined in Section 04.10 of the US Composting Council and US Department of Agriculture (2001) adapted for the use with biochar (Rajkovich et al, 2012) where a 1:20 (w:v) solution of biochar:deionized water is equilibrated for 1.5 hours before measurement. The EBC approach follows the method of the BGK (Federal quality community compost) analog to DIN ISO 11265 where a 1:10 (w:v) solution of biochar:deionized water is equilibrated for 1 hour before measurement (ISO 11265).

Dissolved organic carbon

100 ml of distilled water is added to 10 g of biochar and regularly stirred at 50°C for 24 hours and then centrifuged and filtered (Lin et al, 2012). Dissolved organic C (DOC) is determined by using a DOC analyzer (e.g., Shimatzu TOC).

Product labeling and documentation

Information from this classification should be used for labeling a biochar. An example of biosolids pyrolyzed at 550°C (Wang et al, 2012a; described in Tables 9.3 and 9.7) is shown in Figure 9.14.

Supplementary Material

Supplementary data (Table S9.1: List of properties and associated test methods for biochars) can be found at www.routledge.com/9781032286150.

Supplementary material (calculator) can be found at www.routledge.com/9781032286150.

References

Almendros G, Gonzalez-Vila FJ, and Martin F 1990 Fire-induced transformation of soil organic matter from an oak forest; an experimental approach to the effects of fire on humic substances. *Soil Science* 149, 158–168.

Almendros G, Knicker H, and Gonzalez-Vila FJ 2003 Rearrangement of carbon and nitrogen forms in peat after progressive thermal oxidation as determined by solid-state ^{13}C and ^{15}N-NMR spectroscopy. *Organic Geochemistry* 34, 1559–1568.

Amonette JE, et al 2021 *Biomass to Biochar: Maximizing the Carbon Value. Report from virtual workshop held April-September 2021.* Center for Sustaining Agriculture & Natural Resources, Washington State University,

Pullman, WA. https://csanr.wsu.edu/biomass2biochar/

Amonette JE, Yorgey GG, and Schiller S 2023 *Technical Assessment of Potential Climate Impact and Economic Viability of Biochar Technologies for Small-Scale Agriculture in the Pacific Northwest. PNNL-33894.* Center for Sustaining Agriculture & Natural Resources, Washington State University, Pullman, WA.
75 pp. https://csanr.wsu.edu/publications/technical-assessment-of-potential-climate-impact-and-economic-viability-of-biochar-technologies-for-small-scale-agriculture-in-the-pacific-northwest/ Accessed 4 September 2023.

AOAC (2005) Chapter 2. Fertilizers. In: Horwitz W, and Latimer Jr JW (Eds) *Official Methods of Analysis of AOAC International.* Maryland: AOAC International. pp 1–40.

ASTM D4373-21 2021 Standard Test Method for Rapid Determination of Carbonate Content of Soils. https://www.astm.org/d4373-21.html Accessed 19 August 2023.

ASTM D6556-21 2021 Standard Test Method for Carbon Black—Total and External Surface Area by Nitrogen Adsorption. https://www.astm.org/d6556-21.html Accessed 19 August 2023.

Banitalebi G, Mosaddeghi MR, and Shariatmadari H 2019 Feasibility of agricultural residues and their biochars for plant growing media: Physical and hydraulic properties. *Waste Management* 87, 577–589.

Blanco-Canqui H 2017 Biochar and soil physical properties. *Soil Science Society of America Journal* 81, 687–711.

Blum SC, et al 2013 Sulfur forms in organic substrates affecting S mineralisation in soil. *Geoderma* 200-201, 156–164.

Bornemann LC, Kookana RS, and Welp G 2007 Differential sorption behaviour of aromatic hydrocarbons on charcoals prepared at different temperatures from grass and wood. *Chemosphere* 67, 1033–1042.

Bottezini L, et al 2021 Phosphorus species and chemical composition of water hyacinth biochars produced at different pyrolysis temperature. *Bioresource Technology Reports* 14, 100684.

Brown RA, et al 2006 Production and characterization of synthetic wood chars for use as surrogates for natural sorbents. *Organic Geochemistry* 37, 321–333.

Bundy LG, and Bremner JM 1972 A simple titrimetric method for determination of inorganic carbon in soils. *Soil Science Society of America Journal* 36, 273–275.

Calvelo Pereira R, et al 2011 Contribution to characterisation of biochar to estimate the labile fraction of carbon. *Organic Geochemistry* 2, 1331–1342.

Camps-Arbestain M, et al 2015 A biochar classification system and associated test methods. In: Lehmann J, and Joseph S (Eds) *Biochar for Environmental Management: Science, Technology and Implementation.* 2nd edition. London: Taylor and Francis. pp165–194.

Camps-Arbestain M, et al 2017 Available nutrients in biochar. In: Singh B, Camps-Arbestain M, and Lehmann J (Eds) *Biochar: A Guide to Analytical Methods.* New York: CRC Press. pp109–125.

Chen B, Zhou D, and Zhu L 2008 Transition adsorption and partition of nonpolar and polar aromatic contaminants by biochars of pine needles with different pyrolytic temperatures. *Environmental Science and Technology* 42, 5137–5143.

Chun Y, Sheng G, Chiou CT, and Xing B 2004 Compositions and sorptive properties of crop residue-derived chars. *Environmental Science and Technology* 38, 4649–4655.

Cowie A, et al 2015 Biochar, carbon accounting and climate change. In: Lehmann J, and Joseph S (Eds) *Biochar for Environmental Management: Science, Technology and Implementation.* 2nd edition. London: Taylor and Francis. pp763–794.

Danso EO, et al 2020 Soil structure characteristics, functional properties and consistency limits response to corn cob biochar particle size and application rates in a 36-month pot experiment. *Soil Research* 58, 488–497.

De Duarte SJ, Glaser B, and Cerri CEP 2019 Effect of biochar particle size on physical, hydrological and chemical properties of loamy and sandy tropical soils. *Agronomy* 9, 1–15.

Dehue B, et al 2007 Sustainability reporting within the RTFO: Framework report, Second draft. May 2007, PBIONL062365, Ecofys, Utrecht NL. https://biomass.ucdavis.edu/files/2013/10/10-18-2013-Ecofys-sustainabilityreporting-May07.pdf accessed 22 November 2022.

Dehue B 2013 Implications of a 'carbon debt' on bioenergy's potential to mitigate climate change. *Biofuels Bioproducts Biorefining* 7, 228–234.

DIN 51726, June 2004 – Testing of solid fuels – Determination of the carbonate carbon dioxide content. din-51726-testing-of-solid-fuels-determination-of-the-carbonate-carbon-dioxide-content Accessed 19 August 2023.

EBC 2012-2022 *European Biochar Certificate – Guidelines for a Sustainable Production of Biochar.* Carbon Standards International (CSI), Frick, Switzerland. (http://european-biochar.org). Version 10.2 from 8th Dec 2022. Accessed 3 January 2023.

Edeh IG, and Masek O 2021 The role of biochar particle size and hydrophobicity in improving soil hydraulic properties. *European Journal of Soil Science* 73, e13138.

Enders A, and Lehmann J 2012 Comparison of wet digestion and dry-ashing methods for total elemental analysis of biochar. *Communications in Soil Science and Plant Analysis* 43, 1042–1052.

Figueiredo CC, et al 2021 Assessing the potential of sewage sludge-derived biochar as a novel phosphorus fertilizer: influence of extractant solutions and pyrolysis temperatures. *Waste Management* 124, 144–153.

Gámiz B, et al 2017 Biochar soil additions affect herbicide fate: Importance of application timing and feedstock species. *Journal of Agricultural and Food Chemistry* 65, 3109–3117.

Graber ER, et al 2011 Sorption, volatilization and efficacy of the fumigant 1,3-dichloropropene in a biochar-amended soil. *Soil Science Society America Journal* 75, 1365–1373.

Graber ER, et al 2012 High surface area biochar negatively impacts herbicide efficacy. *Plant and Soil* 353, 95–106.

Havlin JL, et al 2005 Soil Fertility and Fertilizers. *An Introduction to Nutrient Management*, 7th edition. Upper Saddle River, New Jersey: Prentice Hall.

Herath HMSK, Camps-Arbestain M, and Hedley M 2013 Effect of biochar on soil physical properties in two contrasting soils: An Alfisol and an Andosol. *Geoderma* 209-210, 188–197.

Hosokai S, et al 2016 Modification of Dulong's formula to estimate heating value of gas, liquid and solid fuels. *Fuel Processing Technology* 152, 399–405.

IBI 2015 *Standardized Product Definition and Product Testing Guidelines for Biochar That Is Used in Soil-Version 2.1.* International Biochar Initiative. www.biochar-international.org, accessed on 2 January 2023.

Ikeda H 1985 Soilless culture in Japan. *Farming Japan* 19, 36–42.

Ippolito JA, et al 2020 Feedstock choice, pyrolysis temperature and type influence biochar characteristics: A comprehensive meta-data analysis review. *Biochar* 2, 421–438.

ISO 925 2019 Solid mineral fuels — Determination of carbonate carbon content — Gravimetric method. https://www.iso.org/standard/75880.html Accessed 19 August 2023.

ISO 14238 2012 Soil quality — Biological methods — Determination of nitrogen mineralization and nitrification in soils and the influence of chemicals on these processes. https://www.iso.org/standard/56033.html Accessed 19 August 2023.

ISO 10390 2005 Soil quality — Determination of pH. https://www.iso.org/standard/40879.html Accessed 19 August 2023.

ISO 11265 1994 Soil quality — Determination of the specific electrical conductivity. https://www.iso.org/standard/19243.html Accessed 19 August 2023.

ISO 9277 2010 Determination of the specific surface area of solids by gas adsorption — BET method. https://www.iso.org/standard/44941.html Accessed 19 August 2023.

Jindo K, Sánchez-Monedero MA, and Mastrolonardo G 2020 Role of biochar in promoting circular economy in the agriculture sector. Part 2: A review of the biochar roles in growing media, composting and as soil amendment. *Chemical and Biological Technologies in Agriculture* 7, 16.

Kaal J, and Rumpel C 2009 Can pyrolysis-GC/MS be used to estimate the degree of thermal alteration of black C? *Organic Geochemistry* 40, 1179–1197.

Kämpf AN, and Jung M 1991 The use of carbonized rice hulles as an horticultural substrate. *Acta Horticulturae* 294, 271–284.

Kearns JP, et al 2019 High temperature co-pyrolysis thermal air activation enhances biochar adsorption of herbicides from water. *Environmental Engineering Science* 36, 710–723.

Kearns JP, et al 2014 2,4-D adsorption to biochars: Effect of preparation conditions on equilibrium adsorption capacity and comparison with commercial activated carbon literature data. *Water Research* 62, 20–28.

Khalil RA, Seljeskog M, and Hustad JE 2008 Sulfur abatement in pyrolysis of straw pellets. *Energy and Fuels* 22, 1789–1795.

Klasson KT, and Davison BH 2002 A General Methodology for Evaluation of Carbon Sequestration Activities and Carbon Credits. ORNL/TM-2002/235. https://info.ornl.gov/sites/publications/Files/Pub57401.pdf Accessed 08 August 2023.

Klasson KT 2017 Biochar characterization and a method for estimating biochar quality from proximate analysis results. *Biomass and Bioenergy* 96, 50–58.

Kloss S, et al 2012 Characterisation of slow pyrolysis biochars: Effects of feedstocks and pyrolysis temperature on biochar properties. *Journal of Environmental Quality* 41, 990–1000.

Knudsen JN, et al 2004 Sulfur transformations during thermal conversion of herbaceous biomass. *Energy and Fuels* 18, 810–819.

Kok LJ 1993 *Sulfur Nutrition and Assimilation in Higher Plants*. The Hague: SPB Academic Publishing.

Kookana RS 2010 The role of biochar in modifying the environmental fate, bioavailability, and efficacy of pesticides in soils: A review. *Australian Journal of Soil Research* 48, 627–637.

Kookana RS, et al 2011 Biochar application to soil: Agronomic and environmental benefits and unintended consequences. *Advances in Agronomy* 112, 103–143.

Krull ES, et al 2012 *From Source to Sink: A National Initiative For Biochar Research*. Climate Change Research Program, Department of Agriculture, Fisheries and Forestry, Australia.

Lehmann J, et al 2015 Persistence of biochar in soil. In: Lehmann J, and Joseph S (Eds) *Biochar for Environmental Management: Science, Technology and Implementation*. 2nd edition. London: Taylor and Francis. pp235–282.

Li F, et al 2019 Effects of biochar amendments on soil phosphorus transformation in agricultural soils. *Advances in Agronomy* 158, 131–172.

Li S, Lu J, Zhang T, Cao Y, and Li J 2017 Relationship between biochars' porosity and adsorption of three neutral herbicides from water. *Water Science and Technology* 75, 482–489.

Li W, et al 2018 Transformation of phosphorus in speciation and bioavailability during converting poultry litter to biochar. *Frontiers in Sustainable Food Systems* 2, 20.

Lian F, and Xing B 2017 Black carbon (biochar) in water/soil environments: Molecular structure, sorption, stability, and potential risk. *Environmental Science and Technology* 51, 13517–13532.

Liao W, and Thomas S 2019 Biochar particle size and postpyrolysis mechanical processing affect soil pH, water retention capacity, and plant performance. *Soil Systems* 3, 14.

Lima JZ, et al 2022 Biochar-pesticides interactions: An overview and applications of wood feedstock for atrazine contamination. *Journal of Environmental Chemical Engineering* 10, 108192.

Lin Y, et al 2012 Water extractable organic carbon in untreated and chemical treated biochars. *Chemosphere* 87, 151–157.

Liu Z, et al 2017 Biochar particle size, shape, and porosity act together to influence soil water properties. *PloS One* 12, 1–19.

Martin SM, et al 2012 Marked changes in herbicide sorption–desorption upon ageing of biochars in soil. *Journal of Hazardous Materials* 231–232, 70–78.

Mason J, et al 2021 Innovative Microscale Biomass Gasifier Combined Cooling, Heating, and Power System. Final Project Report. Energy Research & Development Division, California Energy Commission. CEC-500-2021-026. https://www.energy.ca.gov/sites/default/files/2021-05/CEC-500-2021-026.pdf accessed 05 August 2023.

Myhre G, et al 2013 Anthropogenic and Natural Radiative Forcing. In: Stocker TF, et al (Eds) *Climate Change 2013: The Physical Science Basis. Contribution of Working Group I to the Fifth Assessment Report of the Intergovernmental Panel on Climate Change.* Cambridge: Cambridge University Press. pp659–740.

Nag SK, et al 2011 Poor efficacy of herbicides in biochar-amended soils as affected by their chemistry and mode of action. *Chemosphere* 84, 1572–1577.

Okuno T, et al 2005 Primary release of alkali and alkaline earth metallic species during the pyrolysis of pulverised biomass. *Energy and Fuels* 19, 2164–2171.

Pansu M, and Gautheyrou J 2006 *Handbook of Soil Analysis – Mineralogical, Organic and Inorganic Methods.* Heidelberg: Springer-Verlag.

Qian C, et al 2020 Prediction of higher heating values of biochar from proximate and ultimate analysis. *Fuel* 265, 116925.

Rajan SSS, et al 1992 Extractable phosphorus to predict agronomic effectiveness of ground and unground phosphate rocks. *Nutrient Cycling in Agroecosystems* 32, 291–302.

Rajkovich S, et al 2012 Corn growth and nitrogen nutrition after additions of biochars with varying properties to a temperate soil. *Biology and Fertility of Soils* 48, 271–284.

Rayment GE, and Lyons DJ 2011 *Soil Chemical Methods – Australasia.* Collingwood: CSIRO Publishing.

Rose TJ, et al 2019 Phosphorus speciation and bioavailability in diverse biochars. *Plant and Soil* 443, 233–244.

Roy JL, and McGill WB 2002 Assessing soil water repellency using the molarity of ethanol droplet (MED) test. *Soil Science* 167, 83–97.

Singh B, Singh BP, and Cowie AL 2010 Characterisation and evaluation of biochars for their application as a soil amendment. *Australian Journal of Soil Research* 48, 516–525.

Singh B, et al 2017a Biochar pH electrical conductivity and liming potential. In: Singh B, Camps-Arbestain M, and Lehmann J (Eds) *Biochar: A Guide to Analytical Methods.* Boca Raton: CRC Press/Taylor Francis Group. pp23–38.

Singh B, Camps-Arbestain M, and Lehmann J (Eds) 2017b *Biochar: A Guide to Analytical Methods.* Boca Raton: CRC Press/Taylor Francis Group.

Smider B, and Singh B 2014 Agronomic performance of a high ash biochar in two contrasting soils. *Agriculture, Ecosystems and Environment* 191, 99–107.

Uchimiya M, and Hiradate S 2014 Pyrolysis temperature-dependent changes in dissolved phosphorus speciation of plant and manure biochars. *Journal of Agricultural and Food Chemistry* 62, 1802–1809.

US Composting Council and US Department of Agriculture 2001 Test Methods for the Examination of Composting and Compost, (TMECC), Thompson WH (Ed) http://compostingcouncil.org/tmecc/, Accessed January 2012

USEPA 2023 Emission Factors for Greenhouse Gas Inventories. https://www.epa.gov/system/files/documents/2023-03/ghg-emission-factors-hub.xlsx, Accessed August 2023.

Van Zwieten L, et al 2010 Effects of biochar from slow pyrolysis of papermill waste on agronomic performance and soil fertility. *Plant and Soil* 327, 235–246.

Vassilev SV, et al 2013a An overview of the composition and application of biomass ash. Part 1. Phase-mineral and chemical composition and classification. *Fuel* 105, 40–76.

Vassilev SV, et al 2013b An overview of the composition and application of biomass ash. *Part 2. Potential utilization, technological and ecological advantages and challenges. Fuel* 105, 19–39.

Verheijen FGA, et al 2015 Biochar sustainability and certification. In: Lehmann J, and Joseph S (Eds) *Biochar for Environmental Management: Science, Technology and Implementation.* 2nd edition. London: Taylor and Francis. pp795–812.

Wang T, Camps-Arbestain M, and Hedley M 2013 Predicting C aromaticity of biochars based on their elemental composition. *Organic Geochemistry* 62, 1–6.

Wang T, et al 2012a Chemical and bioassay characterization of nitrogen availability in biochar produced from dairy manure and biosolids. *Organic Geochemistry* 51, 45–54.

Wang T, et al 2012b Predicting phosphorus bioavailability from high-ash biochars. *Plant and Soil* 357, 173–187.

Wang B, et al 2020 Assessment of bioavailability of biochar-sorbed tetracycline to *Escherichia coli* for activation of antibiotic resistance genes. *Environmental Science and Technology* 54, 12920–12928.

Woolf D, et al 2010 Sustainable biochar to mitigate global climate change. *Nature Communications* 1, 56.

Woolf D, et al 2021 Greenhouse gas inventory model for biochar additions to soil. *Environmental Science and Technology* 55, 14795–14805.

Xu C, Liu W, and Sheng GD 2008 Burned rice straw reduces the availability of clomazone to barnyardgrass. *Science of the Total Environment* 392, 284–289.

Xu M, and Sheng CD 2011 Influences of the heat-treatment temperature and inorganic matter on combustion characteristics of cornstalk biochars. *Energy and Fuels* 26, 209–218.

Yang YN, and Sheng GY 2003 Enhanced pesticide sorption by soils containing particulate matter from crop residue burns. *Environmental Science and Technology* 37, 3635–3639.

Yang YN, Sheng GY, and Huang M 2006 Bioavailability of diuron in soil containing wheat- straw-derived char. *Science of the Total Environment* 354, 170–178.

Yao FX, et al 2010 Simulated geochemical weathering of a mineral ash-rich biochar in a modified Soxhlet reactor. *Chemosphere* 80, 724–732.

Yu XY, Ying GG, and Kookana RS 2006 Sorption and desorption behaviors of diuron in soils amended with charcoal. *Journal of Agricultural and Food Chemistry* 54, 8545–8550.

Yu XY, Ying GG, and Kookana RS 2009 Reduced plant uptake of pesticides with biochar additions to soil. *Chemosphere* 76, 665–671.

Yuan JH, Xu RK, and Zhang H 2011 The form of alkalis in the biochar produced from crop residues at different temperatures. *Bioresource Technology* 102, 3488–3497.

Zhang G, et al 2011 Sorption of simazine to corn straw biochars prepared at different pyrolytic temperatures. *Environmental Pollution* 159, 2594–2601.

Zhang J, Amonette JE, and Flury M 2021 Effect of biochar and biochar particle size on plant-available water of sand, silt loam, and clay soil. *Soil and Tillage Research* 212, 104992.

Zulfiqar F, et al 2019 Challenges in organic component selection and biochar as an opportunity in potting substrates: A review. *Journal of Plant Nutrition* 42, 1386–1401.

10

Aging of biochar in soils and its implications

Joseph J. Pignatello, Minori Uchimiya, and Samuel Abiven

Introduction

This chapter addresses natural or simulated aging processes on biochars and naturally occurring chars and their effects on sorption and reactivity towards organic and inorganic compounds. The subject matter was reviewed in Chapter 10 of the second edition of this book (Pignatello et al, 2015b), and the present chapter will revise and update the findings reported therein and discuss new findings. The term soil will refer to vadose zone soils unless otherwise stated. Some of the discussion focuses on the effects of organic matter (OM) found in soil in either solid or dissolved form on pyrogenic carbonaceous matter (PCM) properties and behavior. Recognizing that the nature of OM continues to be a subject of vigorous debate, for the sake of brevity we merely refer to it here as OM, 'soil organic matter' (SOM), or 'dissolved organic matter' (DOM) and leave it up to the reader to check the cited references in the text, if they wish, as to source and method of isolation and handling.

The biochars referred to were made by slow pyrolysis, with the final heat treatment temperature indicated where necessary. Other chapters in the book will address mineralization (Chapter 11), effects induced by aging on priming of OM decomposition (Chapter 17), microbial communities (Chapter 14), greenhouse gas emissions (Chapter 18), soil-plant interactions (Chapter 15), and C, P, N cycling (Chapter 16). Long-term biochar persistence *per se* is an important issue regarding carbon sequestration potential and greenhouse gas emissions (Leng et al, 2019), but is covered elsewhere in the book (Chapter 30).

Since Chapter 10 of the 2nd edition (Pignatello et al, 2015b), some aspects of the subject matter have been reviewed (Mia et al, 2017; Wang et al, 2020). Chapter 10 of the 2nd edition concluded that an understanding of the chemical alteration of biochar in soil is constrained by the analytical methods in use. Consequently, we lacked a complete understanding of the processes and mechanisms

DOI: 10.4324/9781003297673-10

leading to the chemical alteration of biochar in soil, including oxidation and functionalization; how mineral substances and organic substances interact; and how alterations induced by aging affect sorption and reactions with small molecules. Since then, the literature has expanded and many mechanistic aspects of aging and its consequences have come more clearly into focus.

We begin by dealing with the keystone issue of how best to determine aging effects experimentally.

Experimental approaches for determining aging effects and their merits and shortcomings

There has been a lot of discussion on the proper method to evaluate aging (Wang et al, 2020). The relevant literature is listed in Table 10.1 of the Appendix to this chapter. Methods for biochar aging include artificial laboratory methods that are supposed to reproduce observed aging features, or field/incubation studies, where biochar samples are aged naturally.

Artificial aging methods

Artificial aging methods are based on the characteristics of charcoal pieces found in soil, usually after decades to centuries (Cohen-Ofri et al, 2006). These methods mainly try to reproduce oxidation, physical fragmentation, temperature, or biological degradation.

Oxidation methods try to mimic the observations made on charcoal pieces with oxidized surfaces. Acidic oxidation uses strong oxidizing acids at temperatures above 70°C for several hours to represent the aging of biochar in an acidic soil (Anyanwu et al, 2018) or the effect of acid rain (Chang et al, 2019). A mixture of HNO_3 and H_2SO_4 at 70°C for 6 hours is the most common approach (Liu et al, 2013; Qian and Chen, 2014; Ghaffar et al, 2015; Fan et al, 2018; Chang et al, 2019), but longer exposure also exists (Naisse et al, 2013). Alkaline oxidation was developed to reproduce alkaline environments (Braadbaart et al, 2009). This method uses NaOH at a high concentration above 50°C for several hours, and is usually followed by chemical oxidation with H_2O_2 (Lawrinenko et al, 2016; Fan et al, 2018), without any clear rationale presented. Certain authors propose much longer contact times, such as 147 days (Braadbaart et al, 2009).

Chemical oxidation with H_2O_2 is one of the most commonly used methods. It is among the strongest chemical oxidation treatments with the lowest OM recovery. It involves placing biochar material in contact with H_2O_2 for several days at a constant temperature. The protocol of Cross and Sohi (2013) is the most frequently used, the aging lasting 2 days at 80°C (Cross and Sohi, 2013; Qian and Chen, 2014; Chang et al, 2019; Tan et al, 2020a). This chemical oxidation usually produces the expected effect, an increase in O content at surfaces. It is, however, difficult to evaluate to what age these methods correspond to. Cross and Sohi's method aims to simulate 100 years of aging for biochar under field conditions. Comparison with samples that stayed several decades in the soil is lacking. Strong oxidation with HCl or $K_2Cr_2O_7$ with H_2SO_4 for 12 hours was not enough to reproduce char from the Holocene (Naisse et al, 2013). Some studies have used mild chemical oxidation such as the repeated addition of dilute H_2O_2 solutions over several days (Yang et al, 2021a). Quan et al (2020) proposed photochemical oxidation using UV irradiation of the char for 6 hours. Wang (Wang et al, 2020) used a cyclic regime of UV and simulated solar radiation at 63°C for 400 hours in a soilless system. This method would correspond to the effect of sunlight oxidation when biochar is placed on the soil surface.

Exposure to specific temperatures is often used as a method to age biochar. The approach mostly used in the literature is from Cheng and Lehmann (2009), where biochar was exposed to temperatures of −22°C, 4°C, 30°C and 70°C for 6 to 12 months under saturated and unsaturated conditions. In some cases, the sample is poisoned or irradiated to avoid biological decomposition. Many other combinations of temperature and duration, with or without O_2, exist (Wang et al, 2017; Tang et al, 2021; Yang et al, 2021a). In the most recent studies, the duration is shorter (1–2 months) and the authors focus on the higher temperatures (Tang et al, 2021). Surprisingly, temperature seems to have a less important effect than the duration itself. The rationale behind this temperature treatment is not entirely clear. It may represent decomposition conditions without soil, thus allowing the recovery of all the material. However, constant temperature treatments are not conditions experienced in nature.

Physical cycling treatments such as freeze-thawing and wet-drying cycles are more directly related to field conditions and are the most applied aging methods. In most cases, the freeze-thaw cycle consists of applying −20°C, followed by thawing back to room temperature (Frišták et al, 2014; Naisse et al, 2015; Oleszczuk and Kołtowski, 2018; Xu et al, 2018; Cao et al, 2019; Tan et al, 2020a). The main variations are the duration and number of these cycles. Classically, the cycle lasts about a day and is repeated more than 20 times

(Frišták et al, 2014). To our knowledge, the importance of the cycle occurrence was not specifically studied, so it is difficult to say if the studies using these methods are comparable or not. The wet and drying method is often used in the same studies as freeze-thawing, following the same approach of multiple cycles over several weeks (Wang et al, 2017; Xu et al, 2018; Tan et al, 2020a). Usually, the conditions are set to two water-holding capacity levels, one near the potential and the other near dryness. The duration of the phases usually ranges from several days to several weeks. The outcome of these treatments is the fractionation of the material into smaller particles but also changes in porosity. These methods are clearly reproductions of events in nature, but it is difficult to relate the intensity of these approaches with what would be happening in nature. In general, more than 20 cycles of freeze-thawing would correspond to several winters in a row.

Leaching or, more generally, mixing biochar into water is also proposed as an aging method. The biochar can be mixed for 1 to 24 hours (Spokas et al, 2014; Chen et al, 2021a) at different stirring speeds, with or without increased temperature (Hale et al, 2011). This leaching technique is also often used as a pre-treatment for other aging treatments, mostly to remove soluble ash and volatile compounds produced during pyrolysis (Chen et al, 2021a). The impact of this treatment is usually weak on the chemistry of biochar, but important on the particle size. The heating of the water does not play a very clear role.

Biological aging in the absence of soil has also been proposed (Cheng et al, 2006; Hale et al, 2011; Oleszczuk and Kołtowski, 2018; Quan et al, 2020). It consists of applying an inoculum and nutrition media to the biochar and incubating for several weeks to months. The most commonly used method is the one proposed by Hale et al (2011), where a microbial inoculum extracted from local soils is added and incubated for 18 days at 30°C. Other authors propose longer experiments (for example 1 year in Quan et al, 2020). This method mimics the biological aspect of aging, with the advantage that particles can be easily recovered without soil mineral coating on them. However, the method is difficult to reproduce since it relates to a specific soil inoculum. In addition, the inoculum of a soil represents only a small proportion of the microorganisms living in it.

Most of these laboratory methods have the advantage of being repeatable and relatively easy to set up. Chemical oxidation is the most commonly used and probably the most comparable. They can be compared to archeological samples as well as relatively recent biochar additions in the field. It still remains difficult to evaluate how strong this artificial aging can be. Oxidation techniques seem to lead to relatively large C losses (above 25% (Cross and Sohi, 2013; Naisse et al, 2013)), corresponding to several centuries of degradation. Many regard strong chemical oxidation as "overkill", at least for representing short- or intermediate-term aging (Wang et al, 2020). It is not clear whether laboratory and natural conditions affect the external surfaces and cores of particles in the same way. Physical artificial methods are quite efficient at reproducing fragmentation of the particles, and discriminate well among the different biochars according to their pyrolysis temperature and feedstocks. Many studies propose to compare chemical, physical, and biological conditions, and often find a stronger aging (oxidation mainly) with chemical approaches. This comparison is difficult to evaluate since the strength of the artificial alteration cannot be normalized to the number of years in the field, for example. In addition, all real aging processes take place simultaneously, and, in particular, the increase of fragmentation by physical processes may give access to more surfaces in nature that can then be oxidized further. Several authors propose to combine methods of aging (Meng et al, 2020; Chen et al, 2021a).

Natural aging

Natural aging under controlled or natural conditions combines the advantages of realism and concurrence of different processes. Incubations in soil under constant temperature and moisture conditions have been often used (Cheng et al, 2006; Anyanwu et al, 2018; Kumar et al, 2018; Ren et al, 2018b; He et al, 2019; Deng et al, 2020). In these studies, the authors try to optimize biological decomposition, holding temperature between 20 and 30°C and moisture around 70% of maximum water holding capacity, conditions inspired by other types of fresh OM decomposition where biological processes are the most important. This is questionable for biochar, which for the most part strongly resists microbial oxidation for decades. These constant conditions applied in incubations may result in less fragmentation than in the field. It is not very surprising to note that short incubations lead to very little change.

The most realistic option for achieving all processes of aging taking place concurrently is field aging. Field aging is often carried out over longer times than incubations, usually for several years (Martin et al, 2012; Sorrenti et al, 2016; Dong et al, 2017; Rechberger et al, 2017, 2019; Ren et al, 2018a; Haider et al, 2020) up to 9 years (Yi et al, 2020). The main problem is the recovery of the biochar particles in the soil. One option is to collect the soil, sieve it, and handpick the samples. The biochar in this case is selected by size, and fragments might not be systematically considered. Another option consists of placing the biochar in mesh litter bags, with or without soil

(Ascough et al, 2018; Campos et al, 2021). This method allows easier recovery of the material but may induce local conditions that are slightly different from the field conditions. This bias is probably very small compared to all the methods described above. Field aging leads to a large list of changes, from porosity to surface oxidation to changes in pH (Dong et al, 2017; Ascough et al, 2018).

Few studies have systematically compared field-aged with laboratory-aged results. One such study (Bakshi et al, 2016) compared field-aged (≥3 yr) biochars of different types and sources with laboratory-aged biochars.

The laboratory-aged biochars were subjected to strong chemical oxidation (acidic concentrated H_2O_2 at 40°C for 1 month), followed by exchange with Ca^{2+} and incubation with compost DOM. Field- and lab-aging caused similar increases in volatile matter fraction, CEC, AEC, specific surface area, and similar modifications in O-containing surface functional groups. Because the properties of a given biochar depend on precursor and pyrolysis conditions, and because such properties will affect the outcome of aging, it seems unlikely that a single lab protocol will be found that closely mimics field conditions for all biochars.

Changes in physical-chemical properties of biochars due to aging

Physical-chemical alteration refers to changes in particle size, elemental content, oxidation state, acidity, ion exchange capacities, proportion of aliphatic and aromatic contents, degree of aromatic condensation, changes in porosity and surface area, and preferential loss of C-containing fragments from the biomass body. Physical-chemical alteration can occur by biotic or abiotic processes, or both, and can affect soil water-holding capacity, nutrient retention, and interactions with pollutants and other small molecules and ions.

Chapter 10 of the 2nd edition (Pignatello et al, 2015a) reviewed what was known at the time about feedstock characteristics, pyrolysis conditions, and soil physical and biological processes that favored fragmentation of biochar particles. It pointed out that little was known about solubilization and colloid formation of biochars and their transport in the soil column. The chapter identified the consequences of aging in soil on the surface and the chemical properties of biochar known at the time. Since then, many more papers have appeared that focus on the aging of biochars

and chars under environmental conditions. Such alterations are well described and cited in two recent reviews (Mia et al., 2017; Wang et al., 2020). In general, aging leads to loss of biochar volatile matter, a gain of natural OM through adsorption, changes in mineral content, and increases through biotic and abiotic oxidation in O content, acidity, carboxylic content, the contents of various other O functional groups, hydrophilicity (O/C or (O+N)/C ratios), and CEC. Less clear are the trends in surface and pore dimensions (can increase or decrease), H/C ratio (can increase or decrease), and the effects of feedstock composition and pyrolysis conditions of the original biochar. Here we will focus on highlights and the results of studies published since the reviews.

Transport of particles and colloids in the soil profile

At the landscape level, PCM preferentially accumulates at depth in the soil profile (Soucémarianadin et al, 2019). It is likely that colloidal fractions contribute to this accumulation; however, large particles can still be found in deep horizons (Carcaillet, 2001). Since PCM can only enter the soil from above, this preferential accumulation must be due to both the persistence of the material to microbial degradation and its transport in the profile (Chapter 12). However, this accumulation probably takes place over centuries, which cannot really be captured with current experimental designs. Fewer studies observed lateral movement of biochar. Up to 36% of biochar could be transported by runoff with a single rain event, indicating the potential importance of spatial redistribution after application (Bellè et al, 2020). In addition, researchers found preferential erosion of PCM as compared to rest of the organic C at the level of the meter in a watershed slope (Rumpel et al, 2007, 2009). It indicates that biochar can easily travel and accumulate in the landscape, which is probably beneficial in terms of C storage, but detrimental for agronomical purposes (Chapter 12).

Colloids can be defined as suspended particles between 1 nm and 10 μm (Lead and Wilkinson, 2006). Biochar colloids are formed around biochar particles, but comprise all sorts of materials including clay, metal oxides, and organic molecules. For example, nanoparticles extracted from pecan shell biochar by either alkaline extraction or sonication in water gave both hard and soft (malleable) particles; both C-rich and C, O, and Ca-containing particles (with $CaCO_3$ crystals); both amorphous and "onion-skin" C-rich particles; and traces of carbon nanotubes (Yi et al, 2015). The capacity of biochar to form colloids is mainly related to its surface properties and particle size, and therefore is directly related to its aging. Biochar colloids can be formed during pyrolysis when the feedstock already has this particle size, as well as in the soil during its aging. Mechanical crushing can

take place at any time, for example, with farmers' tools, while breakdown by weathering requires time and constraints; thus, the production of colloids from biochar is likely to become more important over time.

Because of the diversity of materials that can contain biochar colloids, it is not possible to generalize colloidal properties. Usually, colloids tend to be less aromatic and richer in O functional groups as a result of aging (Liu et al, 2018). Consequently, they have a more negative zeta potential (electrokinetic potential) than the parent biochar. The differences tend to grow over time. Oxidation makes colloids more hydrophilic (Ma et al, 2021), and new functional groups may appear. However, as described in the section about aging techniques, it seems especially complicated to collect weathered colloids that went through natural aging. Such experimental designs do not yet exist. Aging also promotes the solubilization of biochar molecular material (Abiven et al, 2011; Jaffé et al, 2013), which, in turn, can form colloids in the soil.

Biochar colloids can travel in the environment more easily than larger particles. Because such transport is difficult to observe in soils, biochar colloid dynamics have been experimentally observed mainly in sand or sandy soil columns under saturated flow conditions (Liu et al, 2016, 2018, 2019; Schiedung et al, 2020). Transport is inhibited when the ionic strength is high or the particles are coated with metal oxides. The transport of particles is promoted when the pH is high and with high sand content (Schiedung et al, 2020).

Effects of short-term aging on physical-chemical properties

Biochars undergo "aging" from the moment they are removed from the reactor and exposed to air. The carbonization process leaves the solid matrix with "dangling bonds" that exist as C- or O-centered unpaired electrons (free radicals, C· or C-O·) or C-centered unshared electron pairs (singlet or triplet carbenes, C:) (Pignatello et al, 2017, and references therein). Many dangling bonds simply self-couple on cooling, while others react with O_2. The rate of chemisorption of oxygen rapidly declines but continues at a significant pace for weeks or more (Antal and Grønli, 2003). Chemisorption of O generates O-centered radicals, such as C-OO· and C-O·, as well as a variety of "stable" O functional groups, such as C=O, C-OH, C-O-C, C-OO-C, and C-OOH. Some of the C-OO-C, C-OOH, and C-OO· groups liberate H_2O_2 when the biochar subsequently comes into contact with water. The liberated H_2O_2 can back-react with the solid, generating reactive O species (ROS), such as hydroperoxyl (HO_2·) and hydroxyl (HO·) radicals (Yang et al, 2017). The H_2O_2 generated on contact with water declines rapidly over the first few hours of storage in air, but quantifiable amounts appear even after 30 days of storage in air. Storage in a vacuum, however, preserves the H_2O_2-generating capacity when subsequently submerged in water.

Some matrix-free radicals can persist for a long time. Persistent free radical (PFR) concentrations in biochars, chars, and related PCM can range from 10^{16} to 10^{19} spins·g^{-1}. PFR in wildfire chars (~10^{19} spins g^{-1}) can persist for years (Sigmund et al, 2021). PFRs owe their persistence to extensive π-delocalization of spin into aromatic rings, or by their fixed location in the matrix that sterically prevents reaction with O_2 and other molecules or their coupling with other radicals. It has been proposed that PFR participate in redox chemistry of biochars with small molecules and microbial cells (Luo et al, 2021). PFR in inhaled soot microparticles have human health implications due to the ROS they may generate in tissues (Balakrishna et al, 2009; Khachatryan et al, 2011).

Experiments with chars (500°C) from pure lignin or cellulose models avoided the participation of minerals and metal ions in subsequent reactivity (Yang et al, 2017). The lignin biochar was generally more reactive than the cellulose biochar with respect to both the generation and decomposition of H_2O_2. Freshly prepared lignin and cellulose biochars contained 6×10^{19} and 8×10^{18} spins g^{-1}, respectively. Within a few days of storage in air, the free radical content declined by 58% and 20%, respectively, and gained more O-centered character. The radical levels and O-character remained stable over the remaining 30-day aging period.

Researchers conducting aging experiments on biochar should be mindful of the above short-term changes occurring after biochar preparation. Most studies do not report the length of time between production and experimentation, nor how the sample was stored. It is possible that storage in a tightly sealed container with little air headspace would consume the O_2 and suspend chemisorption of O_2, later to resume after the sample is withdrawn for experimentation.

Redox behavior

Biochars possess reversible electron exchange capacity, quantifiable as electron-acceptor capacity (EAC) or electron-donor capacity (EDC) (Klüpfel et al, 2014; Chacón et al, 2017; Yuan et al, 2017, 2022; Sun et al, 2017, 2018, 2021). EAC and EDC capacities can be quantified by electrochemical analysis or chemical titration using a pair of electron donor and acceptor (d-a) solute molecules, such as ABTS/ZiV (Klüpfel et al, 2014), ferrocyanide-ferricyanide (Prévoteau et al, 2016), or Ti(III)/ferricyanide (Xin et al, 2019) (Equations 10.1 and 10.2):

$$a + BC \rightarrow (BC)_{ox} + a^- \text{(to determine EDC)}$$
[10.1]

$$d + BC \rightarrow (BC)_{red} + d^+ \text{(to determine EAC)}$$
$$[10.2]$$

where (ox) and (red) represent the resulting oxidized or reduced form of the biochar. The functional groups of biochar involved in electron storage and release are not established for certain, but many think that quinone and hydroquinone, lactone and reduced lactone, aromatic sheets, and PFR may be involved. Values of EDC and EAC up to several milli-equivalents per gram (meq g^{-1}) have been reported.

Little is known about the effects of aging on EDC and EAC of biochars (Sun et al, 2017). Pure lignin biochar (500°C) exposed to air for 0.5 hours had about equal amounts of EAC and EDC (~1.5 meq g^{-1}), whereas the same biochar exposed to air for 180 d gained EAC (to ~3.5 meq g^{-1}) and lost EDC (to ~0.3 meq g^{-1}), presumably by giving up stored electrons to O_2 (Li et al, 2020a).

The EAC and EDC behavior of biochars implies that they can act as sources of oxidizing or reducing equivalents for anthropogenic chemicals. In fact, biomass biochars are known to be intrinsically redox-active toward some organic compounds (see paragraph below). Reactions occur predominantly in the sorbed state (Yang et al, 2017). The mechanism was thought to occur by reduction of O_2 to H_2O_2, followed by decomposition of the H_2O_2 upon back-reaction with the char to give HO· and other ROS. However, recent studies indicate more complex behavior and sensitivity to aging.

Studies report the effects of short-term and longer-term aging on biochar/char reactivity. The reactivity of lignin and cellulose chars (500°C) towards aqueous p-nitrophenol declined gradually with storage in air over 30 days, but by only half the initial rate (Yang et al, 2017). Again, lignin biochar was more reactive than cellulose biochar. Reactions of lignin biochar with a series of para-substituted phenols (-H, -NO$_2$, -CHO, -OCH$_3$) (Li et al, 2020a) gave both oxidized and reduced byproducts whose yields depended on whether the phenol was exposed to the biochar under anoxic or oxic conditions, and how long the biochar had been stored in air (0.5 hours or 180 d). The behavior was found to be consistent with the changes in EDC and EAC mentioned above. Under oxic conditions, linear free energy relationships were observed between the rate constant and either the Hammett electronic substituent constant or the standard electrode potential of the phenol. Oxygen present during aging: (i) incorporates O into surface groups which suppresses sorption of the phenol by making the surface more hydrophilic; (ii) annihilates sites at which phenol reduction occurs; and (iii) reacts with reducing sites to give H_2O_2, the precursor of HO· which can attack the phenol. Thus, depending on reaction conditions and aging, biochars have both oxidizing and reducing capabilities that can be attributed to a combination of direct-reacting sites and HO· originating from H_2O_2 reduction.

It is not clear that direct-reacting sites include PFR sites. The PFR concentrations are too low to react stoichiometrically and can be involved only if they are regenerated. In non-engineered systems, regeneration is plausible if PFR was playing the role of oxidant toward contaminant and transferring its acquired electrons to the bulk oxidant O_2. Nonradical sites are also responsible, in part, for H_2O_2 decomposition; in fact, H_2O_2 pretreatment- depleted sites that were reactive toward p-nitrophenol.

By virtue of their EDC and EAC, biochars have been suggested to act as electron shuttles between chemical species and between chemical species and microbes (Kappler et al, 2014; Saquing et al, 2016; Yu et al, 2015) and evidence exists for *Geobacter species* (Sun et al, 2021). To support this function, the matrix must be electrically conducting over 'long' distances, since cells cannot penetrate pores smaller than ~ 1 μm diameter, where most of the surface area and sorption sites lie. The electrical conductivity of produced biochars is quite low unless the final pyrolysis temperature exceeds about 700°C, and then it increases steeply with temperature (Xu et al, 2013), but also increases to 800°C have been found (Sun et al, 2017). The influence of aging on electron shuttling involving microbes has received little or no attention (Sun et al, 2017).

Effects on biochar physical-chemical properties of long-term aging in soilless systems

Adsorption of natural OM strongly affects biochar physical-chemical properties. Adsorption of a DOM to wood biochars (300, 500, 700°C) increased the H/C and O/C ratios, reduced the N_2 BET specific surface area, shifted the zeta potential to more negative values, and decreased the ash content by assisting the solubilization of metal oxides (Wu and Chen, 2019). DOM adsorption apparently renders the biochar surface similar to the surface of DOM alone.

Artificial chemical aging of a soilless wood biochar (500°C) can be achieved by treatment with mild or strong chemical oxidation (MCO or SCO) (Yang et al, 2021a). MCO entailed exposure to humidified O_2 (60°C for 180 days) or treatment with dilute H_2O_2 daily for 10 days at 20°C. SCO entailed treatment with hot nitric acid or hot ammonium persulfate. Relative to the raw biochar, all treatments increased the O/C ratio, carboxyl acidity, lactonic acidity, and CEC, especially hot nitric acid. The two MCO treatments modestly increased micropore (CO_2) surface area and porosity, especially in the supermicropore region; whereas SCO slightly decreased micropore surface area and

porosity. All treatments shifted the zeta potential-pH curve (pH 3-7) "downward" to more negative values corresponding to more negative surface charge, in the order of dilute H_2O_2 < humidified O_2 < hot HNO_3 ~ hot persulfate.

Soilless aging of willow or sewage sludge biochars (500, 700°C) for up to 18 months in air at various temperatures between −20 and 90°C led to biochar oxidation and a loss of C (Siatecka et al, 2021, 2022). The effects became more pronounced with increasing temperature and were essentially complete by 12 months. Biochars produced at lower temperatures were more prone to abiotic oxidation (Siatecka et al, 2022).

Photochemical transformations in soilless biochar systems have also been observed (Wang et al, 2020). A cyclic regime of UV and simulated solar radiation at 63°C for 400 hours led to small increases in O/C, small decreases in N/C, small changes in the Fourier transform infrared (FTIR) spectrum, modest changes in the O1s x-ray photoelectron spectrum (XPS), and small decreases in N_2 surface area (Li et al, 2020b). Light does not penetrate well below a few millimeters in the soil column and does not reach beyond a few nanometers of the particle 'skin'. Solar radiation of colloidal biochar particles generates ROS, such as HO· and singlet O, that can attack the particle surface (Wang et al, 2020).

Effects on physical-chemical properties of long-term aging in soil media

Usually, the aging of biochar in soil is carried out for a year but periods of up to 7 years have been reported (Tan et al, 2020b; Williams et al, 2019). Typically, aging in soil leads to a decrease in pH (de la Rosa et al, 2018; Mukherjee et al, 2014), an increase in hydrophilicity (as O/C or (N+O)/C ratio) (Ren et al, 2016; Cao et al, 2017; Tan et al, 2020b), and an increase in CEC (Mukherjee et al, 2014). An increase in AEC was observed in one study (Mukherjee et al, 2014). Often there is a decrease in total C (de la Rosa et al, 2018) and a decrease in the ratio of aliphatic to aromatic C (Singh et al, 2016). These and similar observations support the biphasic model of biochar degradation in soil, in which a minor, more aliphatic 'volatile' fraction is mineralized more rapidly than the rest (Mukome et al, 2014). Interestingly, thermal indices measuring biochar recalcitrance and biodegradability of hardwood biochar (450°C) in a sandy clay loam aged for 7 years did not differ significantly between the aged and non-aged samples (Williams et al, 2019).

The ash content is often found to decrease with aging (Ren et al, 2016), possibly due to assisted dissolution of some metal oxides by DOM (Wu and Chen, 2019). However, one study observed an increase in the ash content (Mukherjee et al, 2014). SEM images showed colonization by microbes and extensive coating by OM (Mukherjee et al, 2014).

The effects of natural aging on surface area and small-pore porosity are mixed: some find increases (Cao et al, 2017), while others find decreases (Mukherjee et al, 2014; Ren et al, 2016). Regardless, the changes are not typically great. The increase in N_2 BET SSA and porosity of a rice hull biochar (500 °C) mixed with each of three different soils for 1 and 13 months were attributed to the leaching of minerals, which opens pore entrances (Cao et al, 2017). The aging of biochar in soil quickly leads to a decrease in the N_2 BET surface area due to fouling; experiments proved that the foulant, likely DOM, is transferred through the aqueous phase (Teixido et al, 2013).

Long-term chemical and biological decomposition of biochar in soil may involve early steps of oxidation of aliphatic groups connecting polyaromatic moieties, forming oxidized functional groups at the breaking points (Mia et al, 2017).

Effects of biochar aging on sorption of organic compounds

Sorption is central to the fate and bio-availability of organic compounds in soil. Aging in soil typically leads to a reduction in the soil-water distribution ratio K_d, depending in a complex way on feedstock, pyrolysis conditions, solute structure, and aging conditions (Pignatello et al, 2015b). Aging may also alter the shape of the sorption isotherm—the curve relating the sorbed and dissolved concentrations—typically making it less non-linear. According to evidence available at the time, the main cause of sorption attenuation in soil media was the action of DOM, which plays the dual role of blocking access to internal pores and competing for sorption "sites". Both processes could rationalize the trend toward greater linearity, assuming DOM preferentially affects the smallest pores and the strongest sites. A secondary cause of sorption suppression was suggested to be an increase in biochar O content, which makes the surface more hydrophilic. A third possible cause, though less conclusive, was pore blockage by sorption of metal cations and deposition of soil mineral phases on biochar. Studies published since that review confirm overwhelmingly that fouling by OM is the principal cause of sorption intensity changes, with oxidation of the surface playing a secondary role except for organocation contaminants.

Effects of surface oxidation in soilless systems

Recent studies have confirmed that the chemical oxidation of biochars typically reduces the sorption of neutral compounds by making the surface more hydrophilic. Sorbed water molecules, by aggregating around O functional groups, crowd out the molecules of interest. Storage of lignin biochar (500°C) in ambient air for up to 180 days incorporated O and suppressed sorption and sorption rate of a series of substituted phenols (Li et al, 2020a). Likewise, MCO (daily addition of dilute H_2O_2 for 10 d, or incubation in moist O_2 atmosphere at 60°C for 180 days), as well as SCO (hot nitric acid or ammonium persulfate) of a wood biochar (500°C) increased surface polarity and CEC, but decreased sorption and sorption rate of naphthalene (Yang et al, 2021a).

Organocations respond oppositely than neutral compounds to chemical oxidation. Sorption of the dicationic herbicide, paraquat (N,N'-dimethyl-4,4'-bipyridinium) increased when the biochar was oxidized either under MCO, SCO, or brief post-pyrolysis hot air oxidation (Yang et al, 2021a). The sorption K_d of paraquat for the oxidized aged biochars was also highly pH-dependent, increasing by as much as 1000-fold from pH 4 to pH 10. The Langmuir sorption capacity of paraquat over a wide range of CEC was predicted by a multi-term equation that includes a term for CEC plus a term for the Langmuir sorption capacity of naphthalene as a reference, standing in for nonspecific driving forces that may also be a function of aging. Interestingly, the CEC term of the best-fit correlation was proportional to the *square* of the CEC. The observed adsorption and the adsorption predicted by the multiterm equation are well correlated (Figure 10.1). This square dependence on CEC was attributed to the bidentate interaction of paraquat with pairs of charges on the surface.

Evidently, the bidentate interaction is more energetic than the monodentate interaction and thus dominates sorption until the sorption capacity is exceeded. Paraquat sorption as a function of soilless aging of a softwood biochar (300°C, at different times for up to 360 days in moist warm air) showed a good *linear* correlation between the Langmuir sorption capacity of paraquat and the CEC (Shi et al, 2015). However, their correlation included only four datapoints, and the CEC—normally pH-dependent—was measured at uncontrolled pH.

Effects of organic matter coatings in soilless systems

Soil organic matter, including lipids and polysaccharides, deposit on the surfaces of biochars and natural chars during weathering in soil. Adsorption of added DOM or its surrogates typically increases with decreasing pH, increasing ionic strength, and increasing polyvalent metal ion concentration, as expected considering hydrophobic effects, diffuse double layer theory, and metal ion bridging (Pignatello et al, 2015b). DOM fractions of greater polarity, aromaticity, and molecular weight preferentially sorb to biochars. DOM blocks pore entrances beginning with the micropores (<2 nm effective width) and progressing to small mesopores (2–4 nm). Few if any studies have dealt with the effects of aging on sorption of DOM since that review (Pignatello et al, 2015b).

Several groups have studied the influence of organic substances—either the OM of the soil, or the authigenic DOM of the biochar itself (pyrogenic DOM (PyDOM))—on sorption. Sorption of the insecticide chlorantraniliprole to biochar pre-incubated in different aqueous soil extracts was reduced by

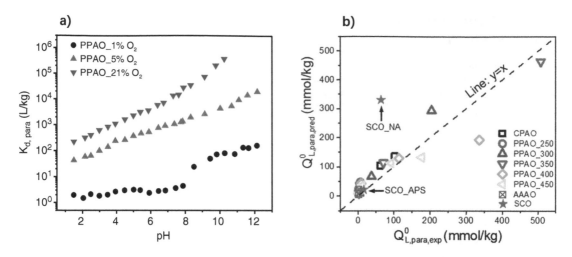

Figure 10.1 *Data for paraquat sorption to a softwood biochar (500°C) oxidized under different conditions showing the profound effect of introducing acidic groups into the biochar through chemical oxidation. (a) Effect of pH on the biochar-water sorption distribution ratio K_d; PPAO: brief post-pyrolysis hot-air oxidation at the indicated headspace O_2 concentration. (b) Predicted vs experimental Langmuir sorption capacity. Capacity was predicted by: $Q^0_{L,para} = (0.12) \cdot Q^0_{L,nap} + (0.0023) \cdot CEC^2$ ($R^2 = 0.91$), where $Q^0_{L,naph}$ is the Langmuir sorption capacity for the reference compound, naphthalene, and CEC is the cation exchange capacity using the ammonium acetate method (Sparks, 1996) at the same pH as sorption (pH 7.4). PPAO: at 21% headspace O_2 concentration and the indicated T, °C; CPAO: pyrolysis in the presence of 1% or 2% air; AAAO: ambient oxidation in warm, moist O_2 for 6 months or addition of dilute hydrogen peroxide daily for 10 days; SCO: strong chemical oxidation using hot nitric acid or peroxymonosulfuric acid (Yang et al, 2021a; with permission)*

up to 85% (Wang et al, 2015). Wheat root exudates promoted the desorption of atrazine from a pig manure biochar (300°C, 700°C) and soils amended with that biochar (Ren et al, 2018b). Prior aging of the biochar-soil mixture for 30 days increased the atrazine desorption rate and the rapidly desorbing fraction by the exudates due to OM coatings gained during aging that reduced sorption capacity.

Sorption of naphthalene, phenanthrene, and 1,3-dinitrobenzene by a maize straw biochar (300 and 500°C) increased after aqueous extraction of the PyDOM, suggesting that PyDOM had blocked the biochar's micropores (Luo et al, 2017). Moreover,

sorption by biochars allowed to take up DOM, as well as biochars aged in soil for 30 days, decreased and their isotherms became more linear relative to the control. Phenanthrene was the most affected. By contrast, aging the biochars in soil pre-treated with H_2O_2, which removed SOM, had no effect on the sorption of 1,3-dinitrobenzene and actually *enhanced* sorption of naphthalene and phenanthrene (Figure 10.2).

Aging of biochars in soil systems

The majority of studies over the past several years (Table 10.2 in the electronic appendix)

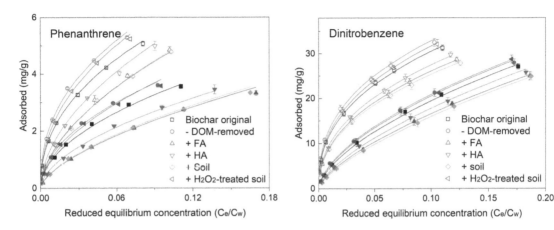

Figure 10.2 *Sorption of phenanthrene and 1,3-dinitrobenzene by maize straw biochar (300 or 500°C) as affected by: removal of the pyrogenic DOM fraction ("PyDOM"); sorption of OM obtained from alkaline extracts; aging in soil for 30 days, or aging in soil pre-treated with hydrogen peroxide (Luo et al, 2017; with permission). This figure shows the strong effects of the addition or removal of OM to the surfaces of biochar on the sorption of chemical compounds*

show that aging of biochar incorporated into soil leads to sorption attenuation: oxyfluorofen after 1–6 months of aging (Wu et al, 2019); atrazine and phenanthrene after up to 2 years (Ren et al, 2018a); fomesafen after 1–6 months (Khoram et al, 2017); diethyl phthalate after 30 days (Zhang et al, 2016); terbuthylazine and 4-chloro-2-methylphenoxyacetic acid after up to 2 years (Trigo et al, 2014); phenanthrene after 7 months (Kumari et al, 2014); and sulfamethazine after mild or severe artificial aging (Teixido et al, 2013). However, one study reports the opposite trend (indaziflam and fluoroethyldiaminotriazine after up to 2 years (Trigo et al, 2014)), and another reports mixed results depending on feedstock and pyrolysis conditions (atrazine and phenanthrene after 90 days (Ren et al, 2018c)). In one case, sorption first increased, then decreased with aging time (Ren et al, 2018a).

A decline in sorption often parallels a decrease in surface area and an increase in polarity index (e.g., Ren et al, 2018a,c). Aging appears to have a greater impact on

biochars made at or below 500°C, which ordinarily have a lower surface area and porosity and higher functional group density, possibly because such biochars are better able to sorb DOM (Teixeido et al, 2013; Kumari et al, 2014; Zhang et al, 2016) or minerals that can block pores (Kim et al, 2021; Kumari et al, 2014). In some cases, the decrease in sorption with aging trends with an increase in leachability (formesafen, Khorram et al, 2017) or bioavailability to plants (acetochlor, Li et al, 2018); fomesafen, Khorram et al, 2017).

Compared to incorporation in the soil, burial of biochar contained in a mesh bag or box seems to give greater variability in the observed effects—some studies show a decline (coumarin, up to 15 mo, Gámiz et al, 2021), others enhancement (mesotrione, 7 mo, Gámiz et al, 2021; picloram, terbuthylazine, and imazamox, 6 mo, Gámiz et al, 2019b), and still others no change (9-anthroic acid, up to 12 mo, Kim et al, 2021). It is not clear whether this variability is a consequence of particle separation methodology or the structure of the test compounds (Trigo et al, 2014), as all compounds showing sorption enhancement or no change possess multiple functional groups, and most compounds were predominantly anionic at the pH of the soil. Increases in sorption of mesotrione, picloram, terbuthylazine, and imazamox in a soil-biochar mixture after aging may be due to metals or other sites provided by the SOM film on the biochar (Gámiz et al, 2019a,b).

The aging effect on sorption can be manifested quite rapidly—even within the timeframe of a typical sorption experiment (Teixido et al, 2013). Sorption to aged hardwood biochars with agricultural soil at 0, 1, or 2% levels under either a mild (2 d, 20°C) or more severe (28 days, 40°C) condition before spiking the antibiotic sulfamethazine did not match predicted curves based on the isotherms of the compound on the individual components. Sorption in the mixtures was dramatically overpredicted (by up to 316 times) by the sum of sorption to the individual components, indicating a strong aging effect even within the sorption experimental time. The biochar-water isotherms, obtained by mathematical subtraction of soil-water isotherms from the total, became more linear, as well. Biochars either mixed with soil or placed in a membrane bag suspended in the soil suspension showed reduced N_2-BET SSA after aging, implying that fouling of the biochar surface by DOM was responsible for the aging effect on sulfamethazine sorption. Likewise, the sorption of pyrene in soil-biochar mixtures after different aging regimes was significantly less than predicted by the sum of sorption to individual components (Zhang et al, 2013). The results should serve as a warning to researchers that sorption to biochar in a soil mixture can be suppressed even for the so-called "aging time zero" sample due to fouling by OM during the sorption equilibration period.

Effects of aging on reactions and sorption of metals

The fundamentals of metal speciation, fate, and aging were covered in depth in Chapter 10 of the Second Edition of this book (Pignatello et al, 2015a). Topics included interfacial sorption mechanisms, thermodynamics, and influence of nanoporous structures. The environmental fate of metals can be predicted from fundamental concepts. A recent review summarizes how such concepts relate to plant and microbial availability of metals with and without soil amendments (Uchimiya et al, 2020). This section will focus on knowledge gaps identified in the Second Edition (Pignatello et al, 2015a), including the lack of: spectroscopic and mechanistic information on hetero-aggregation between biochar and (clay) minerals; evidence confirming the hypothesis that oxidative aging (a popular proxy for biochar aging in soils) occurs from surface to core of the biochar particle; methodology for measuring O/C changes arising from oxidation, as opposed to natural organic matter (NOM) sorption; studies exploring biochar's effects on redox and photo-redox reactivity of metals; and information on how weathering influences the reversibility of metal sorption on biochar.

Effects of biochar aging on redox and photo-redox reactions of metals

Photochemistry
Similar to NOM, biochar and PyDOM can undergo both direct and indirect photolysis.

Direct photolysis chemically transforms the light-absorbing reactants within the biochar matrix, such as the carbonaceous structure itself, PyDOM, or the Fe(III) components. Indirect photolysis produces reactive radicals and other redox-active intermediates (e.g., $^1O_2$, $OH^•$, e_{aq}^-, $O_2^{-•}$, H_2O_2, $RCOO^•$, Fe(II)) that can subsequently undergo reactions or initiate chain reactions. In indirect photolysis, biochar or metals can act as a photosensitizer to convert ground-state molecules to excited-state molecules (e.g., $^3O_2 \rightarrow {}^1O_2$) for secondary photochemistry. Probe compounds (quenchers) are typically used to examine the roles of reactive intermediates, e.g., furfuryl alcohol for $^1O_2^•$, chloroethanol for e_{aq}^-. Heterogeneous photochemistry is particularly relevant to biochar in contact with metal (hydr)oxides. As described previously (Pignatello et al, 2015a), biochar and other particulate matter entering the soil matrix become rapidly coated with metal (hydr)oxides, clays, plant/microbial exudates, and native SOM. Biochar and oxide surfaces can act as semiconductors to form reducing and oxidizing surface sites (conduction-band electrons e_{cb}^-, and holes h^+, respectively). Examples of semiconductor minerals with photocatalytic activity in the visible region include (band gap in eV, equivalent wavelength in nm): PbO (2.76, 449), CdS (2.4, 516), α-Fe_2O_3 (2.34, 530), CdO (2.2, 563), and CuO (1.7, 729) (Stumm et al, 1996). PyDOM and other chromophores sorbed on Fe and Mn oxides can shift the photo-excitation wavelength range and induce ligand-to-metal charge transfer, which potentially can lead to reductive dissolution of the metal and oxidation of PyDOM, although this remains to be be demonstrated. In summary, light-catalyzed reductive dissolution of metal hydroxides (as a coating material on biochar or as a component of aggregates) proceeds within the semiconducting particle and in surface complexes. In addition, PyDOM can complex with Fe(III) and Mn(III/IV); photolysis of such dissolved or particulate metal complexes is likely to play a major role in PyDOM decomposition and influence the overall fate of biochar. In aquatic environments, PyDOM is expected to selectively bind metals. Resulting changes in metal speciation will ultimately impact the fate and transport of metals as well as NOM in a given aquatic environment (Kaal et al, 2022). Those areas are currently underexplored, although many reports have emerged on the design of biochar composites (Mian et al, 2018) for potential industrial applications, including those dealing with photocatalysis. Industrial application of carbonaceous materials (e.g., as a reductant, electrode, catalyst) is by no means new (Antal and Grønli, 2003), and analogous concepts on carbon black and activated carbon are widely available in the legacy literature, and some have met with commercial success.

Information on biochar-mediated photochemistry of metals is sparse, and most of the published studies focused on PyDOM, composites, and materials for industrial applications. Similar to natural soil DOM, PyDOM binds to metals leading to alteration in their photosensitivity (Song et al, 2021). PyDOM (<0.45 μm) induced photoreduction of Hg(II) to Hg^0(aq) via the generation of superoxide ion, $O_2^{-•}$ (Li et al, 2020c). The authors concluded that the composition of PyDOM in biochar filtrate (mainly of low-molecular-weight, carboxylate-rich aromatic clusters, Mao et al, 2012) made it more photoreactive than alkaline extracts (Li et al, 2020c). However, their biochar filtrate was prepared by sonication and likely contained biochar nanoparticles (NPs) (Yi et al, 2015). Superoxide ($O_2^{-•}$) formed by photolysis of PyDOM was proposed to be due to its mineral (silica) content based on comparison with an HCl- and HF-washed control (Fu et al, 2016). Again, their PyDOM sample was prepared by sonication and likely contained biochar NPs; the relative reactivity of PyDOM, as opposed to colloidal particles, is unknown in those experiments. In addition, treatment with concentrated acids causes side reactions such as hydrolysis and coagulation. In summary, a direct comparison of different literature sources on PyDOM reactivity is a challenge due to the lack of a consistent definition of PyDOM and how to isolate it.

Redox reactions

As has been reviewed extensively (Husson, 2013), C is one of the most redox-versatile elements in soil. In oxic soils, E_h varies over a wider range than pH, and changes in E_h signal changes in environmental parameters, including microbial community composition. Much like other C sources such as plant residues and compost, biochar added to soils can alter the transient E_h-pH curve and thereby affect associated environmental processes (Figure 10.3), by shuttling electrons

(Sun et al, 2017; Yu et al, 2022) that can lead to mineral dissolution (Wu et al, 2020), by altering the proportion of Fe oxide phases (Sun et al, 2017; Giannetta et al, 2020), and by serving as the carbon source for the soil microbiome.

The redox reactivity of biochar (Wu et al, 2021b) and pyDOM (Xu et al, 2021) has been widely investigated; see section above on Redox Behavior. However, most redox experiments (e.g., for microbial respiration, Xu et al, 2016) have used biochar as a catalyst in the presence of a bulk reductant (e.g., bacteria, sulfide, dithionite), and not as a direct reductant. Redox mediation (often called electron shuttle or redox catalysis) is a well-known concept in the environmental sciences and biogeochemistry fields; examples of redox catalysts include stable model quinones (anthraquinone disulfonate, AQDS), phenazines, stable Fe complexes, and other reversible electron donor/acceptor couples (Lovley et al, 1996; Newman et al, 2000;

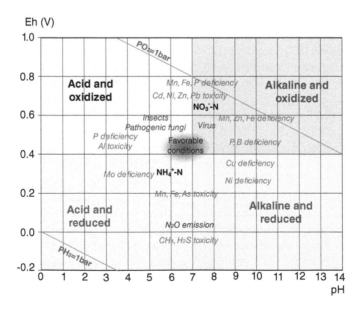

Figure 10.3 *Influence of E_h-pH on soil biogeochemistry (Husson, 2013; with permission). Biochar can alter E_h and pH to shift dominant processes*

Uchimiya et al, 2009). Proposed mechanisms for biochar as a redox catalyst are often based on the AQDS analogy attributing semiquinone radicals detectable by electron spin resonance (ESR) spectroscopy (Xu et al, 2016). Compared to AQDS, juglone, and other dissolved quinones, quinone groups on biochar are expected to be less accessible to microbes; as a result, biochar can only enhance redox reactions by a few to several folds (Xu et al, 2016). Biochar was able to reduce Cr(VI) to Cr(III) if the biochar was modified to contain heterocyclic nitrogen (Zhu et al, 2020) or zero-valent iron (Oh et al, 2021). Unactivated biochar reportedly catalyzes Cr(VI) reduction by ferrihydrite (Xu et al, 2022). Direct reduction of Cr(VI) by biochar was observed only at low pH (2–4) (Xu et al, 2019). These studies indicate that unactivated biochar does not contain sufficient concentration of reducing sites or redox lability to reduce Cr (VI) directly under environmentally relevant conditions. As a result, a large number of studies have emerged on biochar composites with ferric hydroxides, zero-valent iron, and other more redox-active surfaces. In addition to electron shuttling, metal oxide surfaces (including TiO_2, goethite, and kaolinite) are known to catalyze the reduction of Cr(VI) and other contaminants by dissolved reductants (Strathmann et al, 2003). Different environmental surfaces and chelating agents will change the reduction potentials and reactivities of transition metal ions (Orsetti et al, 2013). No information is available to estimate the thermodynamic stability or redox lability of Fe(II) sorbed on biochar, nor other environmentally relevant half-reactions in the presence of biochar, PyDOM, or biochar NPs and colloids. In summary, biochar may act as a redox catalyst by (i) shuttling electrons between bulk reductant and terminal electron acceptor; and (ii) forming reactive Fe(II) and other more redox-labile dissolved/sorbed species via surface binding. Finally, the observable redox reactivity of biochar towards Cr or other metals does not support a claim that biochar is the dominant reductant or oxidant in amended soils. Various soil components can participate in redox reactions (Figure 10.3) either directly (e.g., plant and microbial exudates composed of quinone and other electron-donating or accepting structures (Perlinger et al, 1996; Uchimiya, 2020), Figure 10.3) or indirectly (e.g., by changing the speciation and associated reduction potential of Fe(II) and other electron-donors prevalent in anoxic and sub-oxic soils). A meta-analysis of the redox literature may elucidate the contribution of biochar in the overall soil redox process in particular scenarios.

Effects of aging on metal-biochar interactions

A meta-analysis of the effects of aging on physicochemical properties of biochar (pH, C/H, O) considered 19 experiments of aging by incubation with live soil, 15 by chemical oxidation (multiple different reagents), and 9 by freeze-thaw cycling (Li et al, 2019). No statistically significant effects of freeze-thaw were observed. Incubation had greater impacts compared to oxidation on pH and C/H but the opposite on O content. Despite the limited number of studies considered, this finding suggests the importance of microbially-induced aging of biochar.

Numerous reports investigated how different biochar aging methods—with great emphasis on chemical oxidation due to its convenience—impact the "adsorption" of Pb^{2+} and other metal ions (Tan et al, 2020a; Zhao et al, 2021). However, as described in detail in the Second Edition (Pignatello et al, 2015a) and elsewhere, the fate of metals in soils or biochar-soil mixtures cannot be explained by simple "adsorption" of the nominal metal ion. Upon exposure to environmental media, whether biochar is present or not, metal ions can be hydrolyzed, form organic or inorganic complexes, sorb, precipitate, coprecipitate with other metal oxides, etc. (Stumm et al, 1996; Uchimiya et al, 2020). Changes in pH alone will alter the distribution of different metal species and the total dissolved concentration, which is regarded as a measure of bioavailability. Oxidation of biochar by O_3, photo-Fenton, oxidizing acids, or other traditional laboratory routes is expected to produce dissolved and particulate phases of biochar enriched in O functionality (Sacko et al, 2022). "Endogenous" heavy metals in the ash portion of biochars, such as those prepared from sewage sludge and industrial wastes, will undergo acid, ligand, and reductive dissolution, similar to other minerals in soil. The bioavailability of endogenous heavy metals in biosolids soil amendment has long been a concern regarding root crops such as carrots (especially for domestic gardening), as investigated extensively by Chaney and co-workers (Codling et al, 2015; Holmgren et al, 1993). Pyrolysis is expected to concentrate and stabilize heavy metals that may be present in the biomass feedstock; however, reliable spectroscopic evidence of endogenous heavy metal forms is scarce due to their typically low concentrations (before and after pyrolysis) and poor spectral resolution (Chanaka Udayanga et al, 2019). The following paragraphs will review heavy metal fate in biochar-amended soils and comment on whether their fate can be predicted by first principles.

A meta-analysis of 257 experiments concluded that pH was the primary parameter controlling the "exchangeable" fraction of Cu, Cd, Pb, or Zn of biochar-amended soils, rather than the extent of aging or the biochar pyrolysis temperature (Yuan et al, 2021). The extraction method in data sources primarily relied on 1:10 mixtures of soil and 10 mM $CaCl_2$ solution. Biochar decreased the soluble (presumed bioavailable) fraction of heavy metals in acidic soils but increased it in alkaline soils. Comparing pots with field experiments, biochar addition stabilized heavy metals in pots but mobilized them in the field. Heavy metal bioavailability tended to decrease in short-term (<1 year) but increase in long-term (>1 year) aging experiments. These observations suggest that aging effects may be due to a combination of metal speciation and biochar physicochemical alteration. Most of the aging experiments covered in the meta-analysis study (Yuan et al, 2021) employed spiked heavy metals, which may be the most labile (bioavailable) and experimentally convenient form, but do not represent the heavy metal as it exists in the biochar itself or the environment (Uchimiya et al, 2020).

Fine textured soil increased the available metal fraction in biochar-amended soils, relative to medium and coarse soil fractions (Yuan et al, 2021). Heavy metals are known to have more harmful effects on microorganisms in light-textured soils, especially those with low organic C and clay contents (Giller et al, 1998). In general, biochar and other soil amendments cause maximum effects in sandy soils of low CEC and total organic carbon content (TOC), which lack an intrinsic ability to attenuate perturbations (Uchimiya et al, 2020). Collectively, the literature shows that pH change is typically the primary factor influencing heavy metal bioavailability in biochar-amended soils (Yuan et al, 2021), and microbial aging impacts the long-term bioavailability of metals in biochar-amended soils (Li et al, 2019). The fate of heavy metals in unamended soils is well known to be controlled by the master variables: pH, ionic strength, and TOC. A meta-analysis using machine learning of Cd, Cr, Cu, Pb, Ni, or Zn spiked on 1105 different soil types concluded pH to be the primary descriptor, followed by CEC and initial concentration

(Yang et al, 2021b). The chemical oxidation of biochars is an insufficient indicator to predict the long-term effects of biochars on metals. Biochar colloids prepared by sonication sorbed up to about 4-fold greater amounts of Cr and Cd than bulk biochar under acidic conditions, presumably due to the higher surface area and greater functional group exposure of the colloids (Qian et al, 2016). These findings are also consistent with frequently reported observations that nanoparticulate forms are more reactive than bulk forms of the same material (Nurmi et al, 2005).

Ferrihydrite and other poorly crystalline Fe(III) (hydr)oxides are reactive, metastable phases composed of particles a few nanometers in dimension (Erbs et al, 2008). Over time they convert to stable minerals like goethite and hematite. Metastable ferric (hydr)oxides will form on biochar when Fe(II) species undergo oxidation and the Fe(III) products subsequently hydrolyze. The formation of amorphous phases, followed by the transformation of metastable phases will be part of the aging process of biochar. Metastable phases include various forms of NPs and their aggregates (De Yoreo et al, 2015). Those metastable phases can exist within soil aggregates containing biochar, and their spatiotemporal transition will contribute to the aging of biochar in soil. Organic amendments, regardless of type (biochar, compost, or a combination), increased the fraction of iron complexed to NOM (relative to ferrihydrite) in the fine sand compared to the fine silt and clay fractions of soil in a short-term (<1 year) field experiment (Giannetta et al, 2020). Such size-dependent distribution of metal-DOM complexes could influence the fate of heavy metal contaminants, e.g., by decreasing amorphous Fe (hydr)oxides and other reactive surfaces from finer soil fractions. Based on a few simple dissolution kinetics experiments, it has been claimed that PyDOM is more reactive than OM from alkaline extracts toward the dissolution of iron minerals (Wu et al, 2021a); however, no detailed experiments were conducted to determine the underlying cause, e.g., redox- vs. complexation-dominated mineral dissolution (Rakshit et al, 2009).

Conclusions and outlook

Both field and artificial aging methods have been employed in biochar aging studies. Field aging is realistic, but gives only the net impact of multiple processes, requires long experimental times, has an inherent problem of representative recovery if the biochar is incorporated, and deposition of OM and minerals or metals can mask changes occurring on the biochar body surface itself. Artificial aging saves time and permits better control over variables and conditions. Artificial methods include strong and mild chemical oxidation, incubation with a microbial inoculum, and simulated diagenesis (Chen et al, 2021b). But there is no consensus on what combination mimics the field situation.

A key question, which is not always expressed in papers, is the objective of the aging experiment. When the aging study is done to address agronomic purposes or remediation purposes, an aging period of a couple of equivalent years is adapted and well in line with the needs. However, when addressing the persistence of biochar in soil, these approaches are unadapted, and only decadal to centennial aging is useful. In this case, the aging should be seen as a multi-process event, nonlinear with time, and with tight links to field conditions. This second aspect remains still only partially covered by the current methods.

Both macroscopic and colloidal PCM accumulate preferentially at depth in the soil column, but natural transport takes decades and is not easy to simulate or model. Lateral transport during rainfall events takes place at a quicker pace than vertical transport according to limited studies. Colloidal particles can be formed during pyrolysis, handling, mechanical crushing in the field, or chemical weathering. Their properties may differ from those of the parent material. Little work has been done on aging effects, and naturally-aged particles are difficult to obtain.

Biochars undergo "aging" as soon as they are exposed to ambient air due to the chemisorption of O_2 and water molecules. Subsequent contact with water liberates H_2O_2, PyDOM, and small organic molecules. Persistent free radicals and other sites on the biochar serve to store electrons and initiate chemical reactions with small molecules associated with the particles. More research is needed to fully understand the effects of aging on electron donor-acceptor properties of biochar and biochar reactivities towards small molecules, as well as the proposed ability of biochars to act as electron shuttles among molecules and cells.

Studies published since Chapter 10 in the 2nd Edition of this book confirm that sorption of DOM to biochar increases surface polarity, decreases surface area, and increases surface negative charge. DOM may also assist in the dissolution of minerals. Biotic or abiotic oxidation has positive effects on surface polarity, CEC, acidity, and incorporation of O groups. Due to light shielding, photochemical aging of biochar particles operates from the outside in and is important only for particles at the soil surface. Photochemical transformations are more important for colloidal biochar and PyDOM in natural waters and more work should be done in this area.

Chemical oxidation typically reduces the sorption of neutral compounds due to the increased surface hydrophilicity (i.e., competition by water molecules), but increases organo-cation sorption due to increased CEC. The aging of biochars in soil systems has generally negative impacts on the sorption of neutral molecules due to fouling of the surface by DOM. This conclusion was reached in Chapter 10 of the 2nd Edition and confirmed overwhelmingly in studies published since. Fouling can happen almost immediately.

The literature review above on the effects of aging on metals indicates that the first principles described in the second edition (Pignatello et al, 2015a) sufficiently predicted simple experimental setup, i.e., spiked heavy metals for short-term laboratory experiments using oxidative aging methods widely used within the biochar community. In complex field experiments involving aged metallic phases and various biogeochemical inducers of aging under non-equilibrium, deviations become prominent and systems analysis is necessary. For example, longer-term effects of biochar on heavy metal bioavailability could be explained simply by pH shifts, which are known to control the speciation and solubility of metals. Similar to nutrient availability in soils, microbially-induced dissolution of heavy metals must be taken into account, in addition to purely chemical metal-biochar interactions. Metastable phases will impact the long-term aging of biochar as the coating material or participants in heteroaggregation.

Supplementary Online Appendix Tables

Table 10.1 *Methods of aging.*

Table 10.2 *Effect of biochar aging in soil on sorption of organic compounds.*

References

Abiven S, Hengartner P, Schneider MPW, Singh N, and Schmidt MWI 2011 Pyrogenic carbon soluble fraction is larger and more aromatic in aged charcoal than in fresh charcoal. *Soil Biology and Biochemistry* 43, 1615–1617.

Antal MJ, and Grønli M 2003 The art, science, and technology of charcoal production. *Industrial and Engineering Chemistry Research* 42, 1619–1640.

Anyanwu IN, et al 2018 Influence of biochar aged in acidic soil on ecosystem engineers and two tropical agricultural plants. *Ecotoxicology and Environmental Safety* 153, 116–126.

Ascough PL, et al 2018 Dynamics of charcoal alteration in a tropical biome: A biochar-based study. *Frontiers in Earth Science* 6, 61.

Bakshi S, Aller D, Laird DA, and Chintala R 2016 Comparison of the physical and chemical properties of laboratory and field-aged biochars. *Journal of Environmental Quality* 45, 1627–1634.

Balakrishna S, et al 2009 Environmentally persistent free radicals amplify ultrafine particle mediated cellular oxidative stress and cytotoxicity. *Particle and Fibre Toxicology* 6, 11.

Bellè S-L, et al 2020 Key drivers of pyrogenic carbon redistribution during a simulated rainfall event. *Biogeosciences* 18, 1105–1126.

Braadbaart F, Poole I, and van Brussel AA 2009 Preservation potential of charcoal in alkaline environments: An experimental approach and implications for the archaeological record. *Journal of Archaeological Science* 36, 1672–1679.

Campos P, Knicker H, Miller AZ, Velasco-Molina M, and De la Rosa JM 2021 Biochar ageing in polluted soils and trace elements immobilisation in a 2-year field experiment. *Environmental Pollution* 290, 118025.

Cao T, et al 2017 Surface characterization of aged biochar incubated in different types of soil. *BioResources* 12, 6366-6377.

Cao Y, Jing Y, Hao H, and Wang X 2019 Changes in the physicochemical characteristics of peanut straw biochar after freeze-thaw and dry-wet aging treatments of the biomass. *BioResources* 14, 4329–4343.

Carcaillet C 2001 Soil particles reworking evidences by AMS ^{14}C dating of charcoal. *Comptes Rendus de l'Académie Des Sciences - Series IIA - Earth and Planetary Science* 332, 21–28.

Chacón FJ, Cayuela ML, Roig A, and Sánchez-Monedero MA 2017 Understanding, measuring and tuning the electrochemical properties of biochar for environmental applications. *Reviews in Environmental Science and Bio/Technology* 16, 695–715.

Chanaka Udayanga WD, et al 2019 Insights into the speciation of heavy metals during pyrolysis of industrial sludge. *Science of the Total Environment* 691, 232–242.

Chang R, Sohi SP, Jing F, Liu Y, and Chen J 2019 A comparative study on biochar properties and Cd adsorption behavior under effects of ageing processes of leaching, acidification and oxidation. *Environmental Pollution* 254, 113123.

Chen X, Lewis S, Heal KV, Lin Q, and Sohi SP 2021a Biochar engineering and ageing influence the spatiotemporal dynamics of soil pH in the charosphere. *Geoderma* 386, 114919.

Chen Y, et al 2021b Effects of simulated diagenesis and mineral amendment on the structure, stability and imidacloprid sorption properties of biochars produced at varied temperatures. *Chemosphere* 282, 131003.

Cheng CH, Lehmann J, Thies JE, Burton SD, and Engelhard MH 2006 Oxidation of black carbon by biotic and abiotic processes. *Organic Geochemistry* 37, 1477–1488.

Cheng CH, and Lehmann J 2009 Ageing of black carbon along a temperature gradient. *Chemosphere* 75, 1021–1027.

Codling EE, Chaney RL, and Green CE 2015 Accumulation of lead and arsenic by carrots grown on lead-arsenate contaminated orchard soils. *Journal of Plant Nutrition* 38, 509–525.

Cohen-Ofri I, Weiner L, Boaretto E, Mintz G, and Weiner S 2006 Modern and fossil charcoal: Aspects of structure and diagenesis. *Journal of Archaeological Science* 33, 428–439.

Cross A, and Sohi SP 2013 A method for screening the relative long-term stability of biochar. *Global Change Biology - Bioenergy* 5, 215–220.

de la Rosa JM, Rosado M, Paneque M, Miller AZ, and Knicker H 2018 Effects of aging under field conditions on biochar structure and composition: Implications for biochar stability in soils. *Science of the Total Environment* 613-614, 969–976.

Deng Y, Huang S, Dong C, Meng Z, and Wang X 2020 Competitive adsorption behaviour and mechanisms of cadmium, nickel and ammonium from aqueous solution by fresh and ageing rice straw biochars. *Bioresource Technology* 303, 122853.

De Yoreo JJ, Gilbert PUPA, Sommerdijk NAJM, Penn RL, Whitelam S, Joester D, Zhang H, Rimer JD, Navrotsky A, Banfield JF, Wallace AF, Michel FM, Meldrum FC, Cölfen H, and Dove PM 2015 Crystallization by particle attachment in synthetic, biogenic, and geologic environments. *Science* 349. https://www.science.org/doi/10.1126/science.aaa6760

Dong X, Li G, Lin Q, and Zhao X 2017 Quantity and quality changes of biochar aged for 5 years in soil under field conditions. *Catena* 159, 136–143.

Erbs JJ, Gilbert B, and Penn RL 2008 Influence of size on reductive dissolution of six-line ferrihydrite. *Journal of Physical Chemistry C* 112, 12127–12133.

Fan Q, et al 2018 Effects of chemical oxidation on surface oxygen-containing functional groups and adsorption behavior of biochar. *Chemosphere* 207, 33–40.

Frišták V, Friesl-Hanl W, Pipíška M, Micháleková BR, and Soja G 2014 The Response of artificial aging to sorption properties of biochar for potentially toxic heavy metals. *Nova Biotechnologica et Chimica* 13, 137–147.

Fu H, et al 2016 Photochemistry of dissolved black carbon released from biochar: Reactive oxygen species generation and phototransformation. *Environmental Science and Technology* 50, 1218–1226.

Gámiz B, Velarde P, Spokas KA, and Cox L 2019a Dynamic effect of fresh and aged biochar on the behavior of the herbicide Mesotrione in soils. *Journal of Agricultural and Food Chemistry* 67, 9450–9459.

Gámiz, B, Velarde, P, Spokas, KA, Celis, R, and Cox, L 2019b Changes in sorption and bioavailability of herbicides in soil amended with fresh and aged biochar. *Geoderma* 337, 341–349.

Gámiz B, López-Cabeza R, Velarde P, Spokas KA, and Cox L 2021 Biochar changes the bioavailability and bioefficacy of the allelochemical coumarin in agricultural soils. *Pest Management Science* 77, 834–843.

Ghaffar A, et al 2015 Effect of biochar aging on surface characteristics and adsorption behavior of dialkyl phthalates. *Environmental Pollution* 206, 502–509.

Giannetta B, et al 2020 Iron speciation in organic matter fractions isolated from soils amended with biochar and organic fertilizers. *Environmental Science and Technology* 54, 5093–5101.

Giller KE, Witter E, and McGrath SP 1998 Toxicity of heavy metals to microorganisms and microbial processes in agricultural soils: A review. *Soil Biology and Biochemistry* 30, 1389–1414.

Haider G, et al 2020 Mineral nitrogen captured in field-aged biochar is plant-available. *Scientific Reports* 10, 1–12.

Hale SE, Hanley K, Lehmann J, Zimmerman AR, and Cornelissen G 2011 Effects of chemical, biological, and physical aging as well as soil addition on the sorption of pyrene to activated carbon and biochar. *Environmental Science and Technology* 45, 10445–10453.

He E, et al 2019 Two years of aging influences the distribution and lability of metal(loid)s in a contaminated soil amended with different biochars. *Science of the Total Environment* 673, 245–253.

Holmgren GGS, Meyer MW, Chaney RL, and Daniels RB 1993 Cadmium, lead, zinc, copper, and nickel in agricultural soils of the United States of America. *Journal of Environmental Quality* 22, 335–348.

Husson O 2013 Redox potential (Eh) and pH as drivers of soil/plant/microorganism systems: A transdisciplinary overview pointing to integrative opportunities for agronomy. *Plant and Soil* 362, 389–417.

Jaffé R, Ding Y, Niggemann J, Vähätalo AV, Stubbins A, Spencer RGM, Campbell J, and Dittmar T 2013 Global charcoal mobilization from soils via dissolution and riverine transport to the oceans. *Science* 340, 345–347.

Kaal J, Pérez-Rodríguez M, and Biester H 2022 Molecular probing of DOM indicates a key role of spruce-derived lignin in the DOM and metal cycles of a headwater catchment: Canspruce forest dieback exacerbate future trends in the browning of central European surface waters? *Environmental Science and Technology* 56, 2747–2759.

Kappler A, et al 2014 Biochar as an electron shuttle between bacteria and Fe(III) minerals. *Environmental Science and Technology Letters* 1, 339–344.

Khachatryan L, Vejerano E, Lomnicki S, and Dellinger B 2011 Environmentally Persistent Free Radicals (EPFRs). 1. Generation of reactive oxygen species in aqueous solutions. *Environmental Science and Technology* 45, 8559–8566.

Khorram MS, et al 2017 Effects of aging process on adsorption-desorption and bioavailability of fomesafen in an agricultural soil amended with rice hull biochar. *Journal of Environmental Sciences (China)* 56, 180–191.

Kim J, et al 2021 Sorption of anthracene ($C_{14}H_{10}$) and 9-anthroic acid ($C_{15}H_{10}O_2$) onto biochar-amended soils as affected by field aging treatments. *Chemosphere* 273, 129670.

Klüpfel L, Keiluweit M, Kleber M, and Sander M 2014 Redox properties of plant biomass-derived black carbon (biochar). *Environmental Science and Technology* 48, 5601–5611.

Kumar A, et al 2018 Biochar aging in contaminated soil promotes Zn immobilization due to changes in biochar surface structural and chemical properties. *Science of the Total Environment* 626, 953–961.

Kumari KGID, Moldrup P, Paradelo M, and de Jonge LW 2014 Phenanthrene sorption on biochar-amended soils: Application rate, aging, and physicochemical properties of soil. *Water, Air, and Soil Pollution* 225, 2105.

Lawrinenko M, Laird DA, Johnson RL, and Jing D 2016 Accelerated aging of biochars: Impact on anion exchange capacity. *Carbon* 103, 217–227. 10.1016/j.carbon.2016.02.096.

Lead JR, and Wilkinson KJ 2006 Aquatic colloids and nanoparticles: Current knowledge and future trends. *Environmental Chemistry* 3, 159.

Leng L, Huang H, Li H, Li J, and Zhou W 2019 Biochar stability assessment methods: A review. *Science of the Total Environment* 647, 210–222.

Li Y, et al 2018 Effects of biochars on the fate of acetochlor in soil and on its uptake in maize seedling. *Environmental Pollution* 241, 710–719.

Li H, Lu X, Xu Y, and Liu H 2019 How close is artificial biochar aging to natural biochar aging in fields? A meta-analysis. *Geoderma* 352, 96–103.

Li J, et al 2020a Reaction of substituted phenols with lignin char: Dual oxidative and reductive pathways depending on substituents and conditions. *Environmental Science and Technology* 54, 15811–15820.

Li N, et al 2020b Evaluation of biochar properties exposing to solar radiation: A promotion on surface activities. *Chemical Engineering Journal* 384, 123353.

Li L, et al 2020c Dissolved black carbon facilitates photoreduction of Hg(II) to Hg(0) and reduces mercury uptake by lettuce (*Lactuca sativa* L.). *Environmental Science and Technology* 54, 11137–11145.

Liu Z, Demisie W, and Zhang M 2013 Simulated degradation of biochar and its potential environmental implications. *Environmental Pollution* 179, 146–152.

Liu L, Liu G, Zhou J, Wang J, and Jin R 2019 Cotransport of biochar and *Shewanella oneidensis* MR-1 in saturated porous media: Impacts of electrostatic interaction, extracellular electron transfer and microbial taxis. *Science of the Total Environment* 658, 95–104.

Liu Z, et al 2016 Impacts of biochar concentration and particle size on hydraulic conductivity and DOC leaching of biochar–sand mixtures. *Journal of Hydrology* 533, 461–472.

Liu Z, et al 2018 Effect of freeze-thaw cycling on grain size of biochar. *PLOS ONE* 13, e0191246.

Lovley DR, Coates JD, Blunt-Harris EL, Phillips EJP, and Woodward JC 1996 Humic substances as electron acceptors for microbial respiration. *Nature* 382, 445–448.

Luo K, et al 2021 A critical review on the application of biochar in environmental pollution remediation: Role of persistent free radicals (PFRs). *Journal of Environmental Sciences* 108, 201–216.

Luo L, Lv J, Chen Z, Huang R, and Zhang S 2017 Insights into the attenuated sorption of organic compounds on black carbon aged in soil. *Environmental Pollution* 231, 1469–1476.

Ma P, Yang C, Zhu M, Fan L, and Chen W 2021 Leaching of organic carbon enhances mobility of biochar nanoparticles in saturated porous media. *Environmental Science: Nano* 8, 2584–2594.

Mao JD, et al 2012 Abundant and stable char residues in soils: Implications for soil fertility and carbon sequestration. *Environmental Science and Technology* 46, 9571–9576.

Martin SM, Kookana RS, Van Zwieten L, and Krull E 2012 Marked changes in herbicide sorption–desorption upon ageing of biochars in soil. *Journal of Hazardous Materials* 231–232, 70–78.

Meng Z, et al 2020 Transport and transformation of Cd between biochar and soil under combined dry-wet and freeze-thaw aging. *Environmental Pollution* 263, 114449.

Mia S, Dijkstra FA, and Singh B 2017 Long-term aging of biochar: A molecular understanding with agricultural and environmental implications. *Advances in Agronomy* 141, 1–51.

Mian MM, and Liu G 2018 Recent progress in biochar-supported photocatalysts: Synthesis, role of biochar, and applications. *RSC Advances* 8, 14237–14248.

Mukherjee A, Zimmerman AR, Hamdan R, and Cooper WT 2014 Physicochemical changes in pyrogenic organic matter (biochar) after 15 months of field aging. *Solid Earth* 5, 693–704.

Mukome FND, Kilcoyne ALD, and Parikh SJ 2014 Alteration of biochar carbon chemistry during soil incubations: SR-FTIR and NEXAFS investigation. *Soil Science Society of America Journal* 78, 1632–1640.

Naisse C, et al 2013 Can biochar and hydrochar stability be assessed with chemical methods? *Organic Geochemistry* 60, 40–44.

Naisse C, et al 2015 Effect of physical weathering on the carbon sequestration potential of biochars and hydrochars in soil. *Global Change Biology - Bioenergy* 7, 488–496.

Newman DK, and Kolter R 2000 A role for excreted quinones in extracellular electron transfer. *Nature* 405, 94–97.

Nurmi JT, et al 2005 Characterization and properties of metallic iron nanoparticles: Spectroscopy, electrochemistry, and kinetics. *Environmental Science and Technology* 39, 1221–1230.

Oh SY, Seo YD, Rajagopal R, and Ryu KS 2021 Removal of chromate and selenate in natural water using iron-bearing mineral-biochar composites. *Environmental Earth Sciences* 80, 240.

Oleszczuk P, and Kołtowski M 2018 Changes of total and freely dissolved polycyclic aromatic hydrocarbons and toxicity of biochars treated with various aging processes. *Environmental Pollution* 237, 65–73.

Orsetti S, Laskov C, and Haderlein SB 2013 Electron Transfer between Iron Minerals and Quinones: Estimating the Reduction Potential of the Fe(II)-Goethite Surface from AQDS Speciation. *Environmental Science &*

Technology 47, 14161–14168. https://pubs.acs.org/doi/10.1021/es403658g

Perlinger JA, Angst W, and Schwarzenbach RP 1996 Kinetics of the reduction of hexachloroethane by juglone in solutions containing hydrogen sulfide. *Environmental Science and Technology* 30, 3408–3417.

Pignatello JJ, Mitch WA, and Xu W 2017 Activity and reactivity of pyrogenic carbonaceous matter toward organic compounds. *Environmental Science and Technology* 51, 8893–8908.

Pignatello JJ, Uchimiya M, Abiven S, and Schmidt MWI 2015a Evolution of black carbon properties in soil. In: Lehmann J, and Joseph S (Eds.) *Biochar for Environmental Management: Science, Technology, and Implementation.* London, UK: Taylor & Frances. pp195–234.

Pignatello JJ, Uchimiya M, Abiven S, and Schmidt MWI 2015b Evolution of biochar properties in soil. In: Lehmann J, and Josephs S (Eds.) *Biochar for Environmental Management.* Earthscan. pp195–234.

Prévoteau A, Ronsse F, Cid I, Boeckx P, and Rabaey, K 2016 The electron donating capacity of biochar is dramatically underestimated. *Scientific Reports* 6, 32870.

Qian L, and Chen B 2014 Interactions of aluminum with biochars and oxidized biochars: Implications for the biochar aging process. *Journal of Agricultural and Food Chemistry* 62, 373–380.

Qian L, et al 2016 Effective removal of heavy metal by biochar colloids under different pyrolysis temperatures. *Bioresource Technology* 206, 217–224.

Quan G, et al 2020 Effects of laboratory biotic aging on the characteristics of biochar and its water-soluble organic products. *Journal of Hazardous Materials* 382, 121071.

Rakshit S, Uchimiya M, and Sposito G 2009 Iron(III) bioreduction in soil in the presence of added humic substances. *Soil Science Society of America Journal* 73, 65–71.

Rechberger MV, et al 2017 Changes in biochar physical and chemical properties: Accelerated biochar aging in an acidic soil. *Carbon* 115, 209–219.

Rechberger MV, et al 2019 Enhanced Cu and Cd sorption after soil aging of woodchip-derived biochar: What were the driving factors? *Chemosphere* 216, 463–471.

Ren X, Sun H, Wang F, and Cao F 2016 The changes in biochar properties and sorption capacities after being cultured with wheat for 3 months. *Chemosphere* 144, 2257–2263.

Ren X, Sun H, Wang F, Zhan P, and Zhu H 2018a Effect of aging in field soil on biochar's properties and its sorption capacity. *Environmental Pollution* 242, 1880–1886.

Ren X, Wang F, Cao F, Guo J, and Sun H 2018b Desorption of atrazine in biochar-amended soils: Effects of root exudates and the aging interactions between biochar and soil. *Chemosphere* 212, 687–693.

Ren X, Yuan X, and Sun H 2018c Dynamic changes in atrazine and phenanthrene sorption behaviors during the aging of biochar in soils. *Environmental Science and Pollution Research* 25, 81–90.

Rumpel C, et al 2007 Composition and reactivity of morphologically distinct charred materials left after slash-and-burn practices in agricultural tropical soils. *Organic Geochemistry* 38, 911–920.

Rumpel C, Ba A, Darboux F, Chaplot V, and Planchon O 2009 Erosion budget and process selectivity of black carbon at meter scale. *Geoderma* 154, 131–137.

Sacko O, et al 2022 Sustainable green chemistry: Water-soluble ozonized biochar molecules to unlock phosphorus from insoluble phosphate materials. *ACS Agricultural Science and Technology* 2, 69–78.

Saquing JM, Yu YH, and Chiu PC 2016 Wood-derived black carbon (biochar) as a microbial electron donor and acceptor. *Environmental Science and Technology Letters* 3, 62–66.

Schiedung M, Bellè SL, Sigmund G, Kalbitz K, and Abiven S 2020 Vertical mobility of pyrogenic organic matter in soils: A column experiment. *Biogeosciences* 17, 6457–6474.

Shi K, Xie Y, and Qiu Y 2015 Natural oxidation of a temperature series of biochars: Opposite effect on the sorption of aromatic cationic herbicides. *Ecotoxicology and Environmental Safety* 114, 102–108.

Siatecka A, Różyło K, Ok YS, and Oleszczuk P 2021 Biochars ages differently depending on the feedstock used for their production: Willow- versus sewage sludge-derived biochars. *Science of the Total Environment* 789, 147458.

Siatecka A, and Oleszczuk P 2022 Mechanism of aging of biochars obtained at different temperatures from sewage sludges with different composition and character. *Chemosphere* 287, 132258.

Sigmund G, et al 2021 Environmentally persistent free radicals are ubiquitous in wildfire charcoals and remain stable for years. *Communications Earth and Environment* 2, 68.

Singh B, Fang Y, and Johnston CT 2016 A Fourier-transform infrared study of biochar aging in soils. *Soil Science Society of America Journal* 80, 613–622.

Song F, et al 2021 Novel insights into the molecular-level mechanism linking the chemical diversity and copper binding heterogeneity of biochar-derived dissolved black carbon and dissolved organic matter. *Environmental Science and Technology* 55, 11624–11636.

Sorrenti G, Masiello CA, Dugan B, and Toselli M 2016 Biochar physico-chemical properties as affected by environmental exposure. *Science of the Total Environment* 563, 237–246.

Soucémarianadin L, et al 2019 Pyrogenic carbon content and dynamics in top and subsoil of French forests. *Soil Biology and Biochemistry* 133, 12–15.

Sparks DL 1996 *Methods of soil analysis. Part 3.* Madison: Chemical Methods American Society of Agronomy, Soil Science Society of America.

Spokas KA, et al 2014 Physical disintegration of biochar: An overlooked process. *Environmental Science anf Technology Letters* 1, 326–332.

Strathmann TJ, and Stone AT 2003 Mineral surface catalysis of reactions between FeII and oxime carbamate pesticides. *Geochimica et Cosmochimica Acta* 67, 2775–2791.

Stumm W, and Morgan JJ 1996. *Aquatic chemistry.* New York: Wiley-Interscience.

Sun T, et al 2017 Rapid electron transfer by the carbon matrix in natural pyrogenic carbon. *Nature Communications* 8, 14873.

Sun T, et al 2018 Simultaneous quantification of electron transfer by carbon matrices and functional groups in pyrogenic carbon. *Environmental Science and Technology* 52, 8538–8547.

Sun T, et al 2021 Suppressing peatland methane production by electron snorkeling through pyrogenic carbon in controlled laboratory incubations. *Nature Communications* 12, 4119.

Tan L, et al 2020a Effect of three artificial aging techniques on physicochemical properties and Pb adsorption capacities of different biochars. *Science of the Total Environment* 699, 134223.

Tan L, et al 2020b Changes in biochar properties in typical loess soil under a 5-year field experiment. *Journal of Soils and Sediments* 20, 340–351.

Tang W, Jing F, Laurent ZBLG, Liu Y, and Chen J 2021 High-temperature and freeze-thaw aged biochar impacts on sulfonamide sorption and mobility in soil. *Chemosphere* 276, 130106.

Teixido M, et al 2013 Predicting contaminant adsorption in black carbon (biochar)-amended soil for the veterinary antimicrobial sulfamethazine. *Environmental Science and Technology* 47, 6197–6205.

Trigo C, Spokas KA, Cox L, and Koskinen WC 2014 Influence of soil biochar aging on sorption of the herbicides MCPA, nicosulfuron, terbuthylazine, indaziflam, and fluoroethyldiaminotriazine. *Journal of Agricultural and Food Chemistry* 62, 10855–10860.

Uchimiya M 2020 Proton-coupled electron transfers of defense phytochemicals in sorghum (*Sorghum bicolor* (L.) Moench). *Journal of Agricultural and Food Chemistry* 68, 12978–12983.

Uchimiya M, and Stone AT 2009 Reversible redox chemistry of quinones: Impact on biogeochemical cycles. *Chemosphere* 77, 451–458.

Uchimiya M, et al 2020 Chemical speciation, plant uptake, and toxicity of heavy metals in agricultural soils. *Journal of Agricultural and Food Chemistry* 68, 12856–12869.

Wang H, et al 2017 Effects of atmospheric ageing under different temperatures on surface properties of sludge-derived biochar and metal/metalloid stabilization. *Chemosphere* 184, 176–184.

Wang L, et al 2020 Biochar aging: Mechanisms, physicochemical changes, assessment, and implications for field applications. *Environmental Science and Technology* 54, 14797–14814.

Wang TT, et al 2015 Suppression of chlorantraniliprole sorption on biochar in soil–biochar systems. *Bulletin of Environmental Contamination and Toxicology* 95, 401–406.

Williams EK, Jones DL, Sanders HR, Benitez GV, and Plante AF 2019 Effects of 7 years of field weathering on biochar recalcitrance and solubility. *Biochar* 1, 237–248.

Wu Y, and Chen B 2019 Effect of fulvic acid coating on biochar surface structure and sorption properties towards 4-chlorophenol. *Science of the Total Environment* 691, 595–604.

Wu C, et al 2019 Sorption, degradation and bioavailability of oxyfluorfen in biochar-amended soils. *Science of the Total Environment* 658, 87–94.

Wu P, et al 2020 Contrasting impacts of pH on the abiotic transformation of hydrochar-derived dissolved organic matter mediated by δ-MnO_2. *Geoderma* 378, 114627.

Wu H, et al 2021a Mechanisms for the dissolved biochar promoted iron dissolution and consequential chromium release. *Science of the Total Environment* 796, 148923.

Wu W, et al 2021b Determination of instinct components of biomass on the generation of Persistent Free Radicals (PFRs) as critical redox sites in pyrogenic chars for persulfate activation. *Environmental Science and Technology* 55, 7690–7701.

Xin D, Xian M, and Chiu PC 2019 New methods for assessing electron storage capacity and redox reversibility of biochar. *Chemosphere* 215, 827–834.

Xu WQ, Pignatello JJ, and Mitch,WA 2013 Role of black carbon electrical conductivity in mediating hexahydro-1,3,5-trinitro-1,3,5-triazine (RDX) transformation on carbon surfaces by sulfides. *Environmental Science and Technology* 47, 7129–7136.

Xu S, et al 2016 Biochar-facilitated microbial reduction of hematite. *Environmental Science and Technology* 50, 2389–2395.

Xu Z, Xu X, Tsang DCW, and Cao X 2018 Contrasting impacts of pre- and post-application aging of biochar on the immobilization of Cd in contaminated soils. *Environmental Pollution* 242, 1362–1370.

Xu X, Huang H, Zhang Y, Xu Z, and Cao X 2019 Biochar as both electron donor and electron shuttle for the reduction transformation of Cr(VI) during its sorption. *Environmental Pollution* 244, 423–430.

Xu W, Walpen N, Keiluweit M, Kleber M, and Sander M 2021 Redox properties of pyrogenic dissolved organic matter (pyDOM) from biomass-derived chars. *Environmental Science and Technology* 55, 11434–11444.

Xu Z, et al 2022 Direct and indirect electron transfer routes of chromium(VI) reduction with different crystalline ferric oxyhydroxides in the presence of pyrogenic carbon. *Environmental Science and Technology* 56, 1724–1735.

Yang J, Pignatello JJ, Pan B, and Xing B 2017 Degradation of p-Nitrophenol by lignin and cellulosec: H_2O_2-mediated reaction and direct reaction with the solids. *Environmental Science and Technology* 51 8972–8980.

Yang Y, Duan P, Schmidt-Rohr K, and Pignatello JJ 2021a Physicochemical changes in biomass chars by thermal oxidation or ambient weathering and their impacts on sorption of a hydrophobic and a cationic compound. *Environmental Science and Technology* 55, 13072–13081.

Yang H, et al 2021b Predicting heavy metal adsorption on soil with machine learning and mapping global distribution of soil adsorption capacities. *Environmental Science and Technology* 55, 14316–14328.

Yi P, Pignatello JJ, Uchimiya M, and White JC 2015 Heteroaggregation of cerium oxide nanoparticles and nanoparticles of pyrolyzed biomass. *Environmental Science and Technology* 49, 13294–13303.

Yi Q, et al 2020 Temporal physicochemical changes and transformation of biochar in a rice paddy: insights from a 9-year field experiment. *Science of the Total Environment* 721, 137670.

Yu L, Yuan Y, Tang J, Wang Y, and Zhou S 2015 Biochar as an electron shuttle for reductive dechlorination of pentachlorophenol by *Geobacter sulfurreducens. Scientific Reports* 5, 16221.

Yu W, Chu C, and Chen B 2022 Enhanced microbial ferrihydrite reduction by pyrogenic carbon: Impact of graphitic structures. *Environmental Science and Technology* 56, 239–250.

Yuan C, et al 2021 A meta-analysis of heavy metal bioavailability response to biochar aging: Importance of soil and biochar properties. *Science of the Total Environment* 756, 144058.

Yuan J, Wen Y, Dionysiou DD, Sharma VK, and Ma X 2022 Biochar as a novel carbon-negative electron source and mediator: Electron exchange capacity (EEC) and environmentally persistent free radicals (EPFRs): a review. *Chemical Engineering Journal* 429, 132313.

Yuan Y, et al 2017 Applications of biochar in redox-mediated reactions. *Bioresource Technology* 246, 271–281.

Zhang W, Sun H, and Wang L 2013 Influence of the interactions between black carbon and soil constituents on the sorption of pyrene. *Soil and Sediment Contamination: An International Journal* 22, 469–482.

Zhang X, et al 2016 Effect of aging process on adsorption of diethyl phthalate in soils amended with bamboo biochar. *Chemosphere* 142, 28–34.

Zhao Y, Li Y, Fan D, Song J, and Yang F 2021 Application of kernel extreme learning machine and Kriging model in prediction of heavy metals removal by biochar. *Bioresource Technology* 329, 124876.

Zhu S, et al 2020 Enhanced transformation of Cr(VI) by heterocyclic-N within nitrogen-doped biochar: Impact of surface modulatory Persistent Free Radicals (PFRs). *Environmental Science and Technology* 54, 8123–8132.

11

Persistence of biochar

Mechanisms, measurements, predictions

*Johannes Lehmann, Samuel Abiven, Elias Azzi, Yunying Fang,
Bhupinder Pal Singh, Saran Sohi, Cecilia Sundberg, Dominic Woolf,
and Andrew R. Zimmerman*

Introduction

Persistence is a core property of biochar, with wide-ranging implications for climate change mitigation and adaptation, as well as its broader soil health effects as outlined throughout this book: (i) the net CO_2 emissions from biomass converted to biochar are reduced which offers opportunities for climate change mitigation (Chapter 31); and (ii) any benefits of the presence of biochar in soil continues for a longer period, such as effects on nutrient (Chapters 8, 10, 16, 19) and water availability (Chapter 20) or mitigation of agrochemicals or toxins (Chapters 21–22, 27). The enduring presence of biochars may not always translate into a continuation of any initial positive or negative effects, as their properties can change during exposure to soil (Chapter 10), such as loss of their acid-neutralizing ability (Chapter 8) or ability to adsorb PAH (Chapter 22). However, a desirable property may also emerge as in the case of surface oxidation and the development of cation retention (Chapter 10).

Persistence is defined here as the presence of biochar as opposed to its mineralization to CO_2, by any mechanism (abiotic and biotic) (Box 11.1). Persistence is, therefore, different from aspects of transport (Chapter 12), but it can be affected by transport in the soil profile or landscape (Figure 11.1). Persistence applies to any use of biochar, beyond its application to soil (such as in building materials, Chapter 28). However, this chapter only discusses biochar use in soil, even though some of the basic principles may apply to other uses, as well.

DOI: 10.4324/9781003297673-11

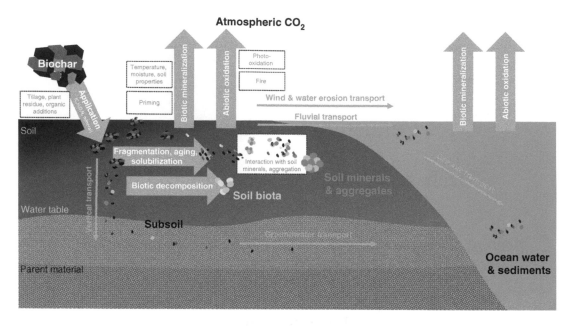

Figure 11.1 *Overview of the mechanisms that influence biochar persistence. Mineralization is a biotic process (by soil microbes and fauna) that together with abiotic oxidation (e.g., by photo-oxidation or fire) converts biochar C to CO_2, compared to decomposition, used here to describe biotic processes to transform biochar into other organic compounds (Box 11.1). This chapter focuses on how the composition of biochar influences mineralization, oxidation, and decomposition (large grey arrows), and how changes in properties (fragmentation, aging, solubilization), management (tillage, plant residue management, organic additions), soil properties, and environmental conditions (temperature, moisture) influence these (dashed boxes). Transport (thin arrows) is a different process from persistence and is only considered in this chapter as it influences mineralization*

Box 11.1 Terminology for quantification of persistence

The term "persistence" is used here deliberately, rather than terms such as stability or recalcitrance, as the term persistence makes no inferences about why biochar remains present (Lehmann et al, 2015). It is typically measured through assays of the CO_2 evolved (for which we use the term "mineralized") or the changes in amounts remaining in soil over time (the latter only if physical movement can be avoided). A numerical value is commonly modeled by assuming an exponential decay, with the resultant dynamic expressed as decay rate, mean residence time (MRT, equivalent to mean lifetime), half-life, or turnover time, which can also be applied to biochar. These are not synonymous but are mathematically related (Six and Jastrow, 2002). An apparent decay rate is the exponent (k, as a function of environmental conditions) in the exponential decay function and has a unit of 1/time:

$$\text{Biochar}_{(at\ time\ t)} = \text{biochar}_{(at\ time\ 0)}e^{-kt} \qquad [11.1]$$

MRT is the inverse of the decay rate ($1/k$) and is the average time that biochar is present. Half-life is the time that elapses before half of the biochar C mineralizes and can be obtained by multiplying the MRT by the natural logarithm of 2. For computing the turnover time where biochar is applied regularly, information about the stock of biochar is required. It is calculated by dividing the stock at equilibrium by the loss per unit time under steady state (applicable for pyrogenic organic matter (PyOM) inputs through regular vegetation fires or for regular biochar additions such as with biochar-based fertilizers, not for one-time biochar additions).

Heterogeneous composite materials such as biochar and natural organic matter are typically composed of a mixture of individual compounds or groups of compounds, here called "fractions", each with different rates of decay. This may necessitate the assignment of multiple exponential functions to describe the overall decay process, using distinct (although usually conceptual) "pools". For biochar, since more persistent fractions seem to predominate than in plant litter, simplification may be possible when considering long time scales. Such equations can be solved mathematically to yield an estimate for the decay rate 'k', provided that the assumption of no interaction between C pools and no transfer of decomposition products to other C pools can be made. Although this assumption cannot hold for all soil organic C forms, it may be used to conservatively estimate the persistence of biochar. The alternate approach is interconnected multi-pool modeling typified by soil organic C models such as Century (Parton and Rasmussen, 1994) and RothC (Coleman et al, 1997), in which material entering a pool as the product of decomposition from one or more other pools is accounted for, and the status of each pool is re-assessed at each successive calculation "time step" (dynamic simulation). This has not been sufficiently attempted for biochar (Foereid et al, 2011; Woolf and Lehmann, 2012; Mondini et al, 2017; Lefebvre et al, 2020; Pulcher et al, 2022).

Mechanisms of mineralization

Biotic processes

Even though biochars have greater persistence than the biomass they are produced from, they can still be mineralized to CO_2 by microorganisms (Potter, 1908; Shneour, 1966), incorporated into their biomass (Kuzyakov et al, 2014), or decomposed to other organic materials (Wengel et al, 2006), as are all organic residues in soil. Pyrogenic C (PyC), which makes up a major proportion of biochars (see Chapter 1 for nomenclature), is chemically different from other natural organic matter and the decomposer community typically lacks or does not produce the full suite of enzymes required to decompose the multitude of thermally altered organic phases produced by pyrolysis to the same extent as other natural organic matter (as discussed above; and Chapter 14).

Possibly, microorganisms preferentially colonize some biochar (Lehmann et al, 2011; Pokharel et al, 2020) and such physical proximity may increase the biotic mineralization of biochar (Farrell et al, 2013; Luo et al, 2017), which may be a function of the amount of easily mineralizable C in biochars (Luo et al, 2017) among other properties such as pH, nutrient and moisture availability, or electrochemical properties (Joseph et al, 2021). Some soil fauna groups, such as earthworms, have been shown to preferentially ingest soil containing fire-derived chars (Topoliantz and Ponge, 2005) or biochar (Van Zwieten et al, 2010) which may increase mineralization directly by soil fauna or indirectly by microorganisms in the faunal gut. Mesofauna may ingest biochar that is associated with a greater population of microorganisms (Domene et al, 2015). Earthworms may also physically disperse biochar in the soil and decrease biochar particle sizes, but whether these processes decrease or enhance its mineralization is not known (Ameloot et al, 2013).

Similarly, some evidence exists for preferential exploration of biochars by roots and root hairs (Vanek and Lehmann, 2015). Roots can exude protons or low-molecular acids, thus changing the chemical environment and the biological activity in the rhizosphere, but also causing fragmentation of biochar particles. Whether these processes will have a net positive or negative effect on biochar mineralization, is not known.

Chemical processes

Even though the activity of soil biota is likely to be a major pathway of mineralization of biochar, abiotic processes can directly lead to CO_2 evolution from biochar or may facilitate subsequent biotic mineralization. However, it should be noted that these processes typically occur simultaneously and may greatly differ between different types of biochars.

Inorganic carbonates contained in certain biochars (Enders et al, 2012) may be dissolved through dissolution reactions (Farrell et al, 2013). If present in significant quantities, the CO_2 released through the dissolution process can influence the estimation of mineralization rates using the natural C-13 abundance and 2-pool mixing model approaches

(Singh et al, 2012a; Bruun et al, 2013). In addition, these inorganic carbonates would not be included in the prediction of persistence (e.g., by using H/C_{org} ratios, outlined below), but are often not analytically separable in mineralization studies. It would be worthwhile to explore whether accounting for inorganic C separately from total C would improve mineralization predictions (Singh et al, 2012a; Farrell et al, 2013). Dissolution of organic C from biochar is also important for the subsequent mineralization of dissolved organic C, especially from low-temperature biochar (Bostick et al, 2018).

Abiotic oxidation of PyC surfaces may initially be important (Cheng et al, 2006; Wang et al, 2017; Chapter 10) and may increase or decrease subsequent biotic metabolization. Short-term abiotic mineralization of organic C to CO_2 has been calculated as a third (Zimmermann et al, 2012) to half (Zimmerman, 2010) or more (Bruun et al, 2013) of total mineralization over the first few months. Incubations of fresh biochar without soil or microbial inoculants also showed initial evolution of CO_2, which were attributed to various processes, including abiotic reactions with water (Spokas and Reicosky, 2009) and desorption of CO_2 (Bruun et al, 2013). Maintaining sterile conditions is challenging even in the laboratory, and abiotic processes may require additional study, since they appear to be significant at least over the short term which may also affect long-term prediction of persistence.

As one type of abiotic oxidation, photo-oxidation may be important for biochars that remain on the soil surface, as was shown for uncharred plant litter (King et al, 2012). Skjemstad et al (1996) analytically defined PyC as the fraction of organic matter that is resistant to photo-oxidation, which suggests that the PyC fraction of biochars may be less prone to mineralization by this process than most uncharred organic matter. Before adding biochars to the soil, photo-oxidation was found to reduce their mineralization (Wang et al, 2017). Isolated dissolved organic C from biochar was found to be highly susceptible to photo-oxidation and associated mineralization (Bostick et al, 2020). Only limited information on this process currently exists for biochar and its leachates in soil (Chapter 10).

Physical processes

Biochars may physically disintegrate in soil over several years to very small sizes (Spokas et al, 2014), thereby increasing the opportunities for mineralization on the one hand and interactions with soil minerals that reduce mineralization on the other. No PyOM particles with a diameter greater than 50 μm could be found in a humid tropical upland soil thirty years after forest fires (Nguyen et al, 2008) pointing at disintegration over decadal time scales. In ecosystems with regular inputs of fire-derived PyOM, particles are typically smaller than 53 μm (88% of PyOM, Skjemstad et al, 1996). Fragmentation may in general be a result of frost, temperature and moisture changes, salt weathering, solubilization, roots, or mechanical stress through soil tillage. Physical disintegration, as mentioned for chemical processes, may be analogous to the weathering processes of minerals, which then allow biotic mineralization to occur, with important differences discussed below.

Biochar properties and persistence

This section discusses the material properties that affect biochar persistence. This discussion should be seen in the context of how the environment (such as soil moisture and temperature among others, as discussed below) affects biochar mineralization (Figure 11.1), as the persistence of any organic material added to soil is a result of soil ecosystem properties, rather than simply a result of the properties of the added organic material itself (Schmidt et al, 2011).

Organic composition of biochar

The principle underpinning biochar persistence (i.e., its slower microbial mineralization than unpyrolyzed biomass) in the environment is that microorganisms more easily utilize organic C forms that require less activation energy for their metabolization. The relevant change in material characteristics is that pyrolysis creates mineral-like properties which can be seen as a progressive transformation towards the highly crystalline and well-ordered C mineral graphite, including (i) the growth of graphene-like sheets that are much larger than those poly-condensed ring systems that have been reported as products of biological metabolism (and microorganisms, therefore, lack adaptation to utilize those as a source of energy); (ii) the assembly of turbostratic stacks with crystalline character; and (iii) the creation of unlimited molecular diversity of poly-condensed structures as a function of varying pyrolysis temperature and feedstock (Chapters 6 and 7).

These structural changes during charring are not homogeneous and vary considerably on a nanometer scale (Chapter 6) and depend on the biomolecules they are produced from (Knicker, 2011). The described ring structures have, on average, relatively small cluster sizes that increase with greater pyrolysis temperatures: 18–40 C atoms were found in fresh biochars made from oak wood and corn residues at 350°C and 600°C using nuclear magnetic resonance (NMR) spectroscopy (Nguyen et al, 2010), 25 to 52 C atoms in biochars made from chestnut wood between 500°C and 700°C determined by adsorption of ^{13}C-labeled benzene (McBeath et al, 2011), and 20 or more C atoms in char residues in a Midwestern Mollisol and an Amazonian Dark Earth (Mao et al, 2012). These clusters are linked in larger PyC assemblages (Mao et al, 2012) which remain in particulate form for long periods (Liang et al, 2008) and can also be seen to gain some degree of spatial ordering at temperatures around 600°C at the nano-scale (Kercher and Nagle, 2003; Nguyen et al, 2010). How important the nano-scale spatial assemblage is for the persistence of biochars remains unclear but could be relevant as different formations of onion-shape or fullerene-type structures are known to exist in PyC and possess different structural stabilities (Hata et al, 2000; Harris, 2005; Paris et al, 2005;

Bourke et al, 2007; Cohen-Ofri et al, 2007). Within a given study, the net effect typically is a greater persistence of biochars that have been pyrolyzed at higher temperatures or for longer durations.

Despite the dominance of aromatic ring structures, biochars also contain varying amounts of other organic compounds (Baldock and Smernik, 2002; Czimczik et al, 2002) that may mineralize more rapidly over relatively short periods (a few weeks to months) such as low molecular weight compounds and those containing aliphatic C (Cheng et al, 2006; Hilscher et al, 2009; Nguyen et al, 2010). The proportion of non-aromatic PyOM-C forms generally decreases with greater pyrolysis temperatures (Keiluweit et al, 2010; Nguyen et al, 2010; McBeath et al, 2014; Wiedemeier et al, 2015). However, considerable chemical variation exists even in biochars produced at the same pyrolysis temperature (as reflected in varying elemental ratios, Figure 11.2), likely due to a combination

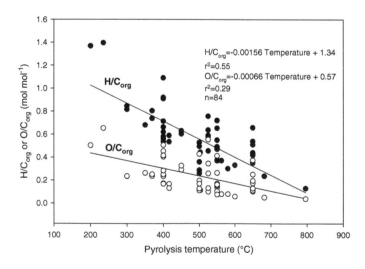

Figure 11.2 *Relationship between the temperature during biochar production and organic composition of biochars using elemental ratios (data for which long-term mineralization experiments are available in Figure 11.5; some variation also caused by H and O that can be part of inorganic structures)*

of other pyrolysis conditions (e.g., duration, air flow, homogeneity in large reactor vessels) and feedstock properties (e.g., ash content), in addition to experimental conditions under which the mineralization was examined (see below). Understanding the molecular composition and indeed general properties of biochar is therefore important for quantifying and predicting the mineralization of different biochar in soil.

Nitrogen composition of biochar

During pyrolysis, nitrogen (N) is volatilized or retained as organic N in proportions approximately equivalent to organic C (Torres-Rojas et al, 2020). During the transformation of biomass to biochar, the retained N is primarily present as 6-ring aromatic and some 5-ring PyC structures and the amount of N in the original biomass does not seem to influence that transformation, while pyrolysis temperature has a large effect. The forms of N are more important for the magnitude of biochar N and C mineralization than the amount of N or C/N ratio (Torres-Rojas et al, 2020). This is different from the mineralization of N and C in an unpyrolyzed litter which typically shows a good correlation with the amount of N or the C/N ratio.

Mineral composition of biochar

The mineral composition of biomass (including but not limited to Ca, Mg, K and their oxides, as well as clay minerals that are intentionally added or occur as part of biomass handling; Xu et al, 2017) and the resulting biochar may also influence biochar mineralization through two different processes: (i) by how these minerals influence the organic composition (degree of aromaticity or formation of fused aromatic ring structures, etc.) during pyrolysis, and

(ii) how these minerals influence microbial mineralization directly. Ca additions to biomass before pyrolysis, for example, decreased the proportion of aromatic C in biochar (Nan et al, 2021). Microbial mineralization studies that unambiguously identify the effects of differences in mineral composition on persistence only exist to a limited extent, with chemical and thermal tests being used more often (Nan et al, 2018; Yang et al, 2018a). Thus, at present, conclusive evidence of the effect of mineral composition on microbial mineralization is limited. Microbial mineralization of biochar made from rice straw decreased by 20–90% when pyrolyzed with kaolin, calcite, or calcium phosphate (Li et al, 2014). This decrease may at least in part be explained by a 33–43% greater formation of aromatic C during pyrolysis in the presence of these minerals (Li et al, 2014). It is not clear, however, how much interactions between soil minerals and biochars on their own would have reduced mineralization (as discussed below).

Particle size of biochar

Biochars consist of particles, and their sizes are a result of the particle size of the biomass, pyrolysis conditions, and any crushing after pyrolysis (Bruun et al, 2012). The particulate nature of biochar also affects its persistence in several ways. Soil biotic and abiotic processes will only have access to the surface of biochar particles and can only oxidize and mineralize its surfaces or leachate from those surfaces. It can therefore be expected that a larger particle size will to some extent increase biochar persistence. Indeed, smaller particles were shown to mineralize more rapidly than larger particles in the absence of clay minerals (Zimmerman, 2010). Many of the currently available mineralization experiments using different particle sizes (Sigua et al, 2014; Chen et al, 2019) do not allow a distinction between an increase in mineralization of soil organic matter or the biochar in the mixture. Regardless of their size, biochar particles also have internal pore surfaces as a result of the cell structure of the biomass they are produced from (Chapter 5). Liquids and gases (Chapter 8), as well as some microorganisms and their exoenzymes (Chapter 14), may or may not have access to these internal surfaces, depending upon their pore diameters and connectivity. In addition, the size of biochar particles will decrease in soil over time through physical disintegration, as discussed above.

Similar to unpyrolyzed biomass (Lehmann and Kleber, 2015), smaller molecules and, therefore, also smaller and more oxidized biochar particles are expected to interact with mineral surfaces or to be incorporated into soil aggregates to a greater extent, which typically increases persistence. In addition, biochar particle size also affects water retention depending on soil texture (Chapter 20) and therefore microbial mineralization. Moreover, biomass particle size affects the aromaticity of biochar (Manyà et al, 2014), and will change C mineralization (Deng et al, 2021). On balance, it is therefore not clear whether a deliberate choice of particle size of a biochar amendment will effectively manage its persistence.

Effects of environment and soil management on persistence

Soil temperature

In general, the activity of all chemical reactions, including organic matter decomposition, increases at higher temperatures. Microbial mineralization of organic matter typically increases with temperature to a greater extent for those organic materials, such as biochar, that mineralize more slowly, expressed as the so-called Q10 temperature coefficient (i.e., the increase in mineralization rate with a temperature increase of 10°C) (Davidson and Jannsens, 2006; with important modifications depending on organic matter interactions with soil and adaptation of the microbial community). Therefore, biochar mineralization may be expected to be more sensitive to temperature changes in soil than unpyrolyzed organic matter if all other factors are kept constant (which is close to impossible) (Fang et al, 2017). Fang et al (2014b) found that, although Q10 values were not influenced by the type of biochar produced from a woody biomass source at 450 or 550°C when incubated in soil, interactions between soil minerals and biochar can reduce temperature sensitivity (Singh et al, 2015) as was pointed out for unpyrolyzed organic matter (Davidson and Jannsens, 2006). In contrast, an increased Q10 of mineralization of corn stover biochar from 1.2 to 1.6 (at 10–20°C) with increasing pyrolysis temperature from 350 to 600°C (Nguyen et al, 2010) may serve as an indication of the magnitude of changes in temperature sensitivity of biochar mineralization without significant interactions with minerals (the incubations were done in sand). Q10 values for the incubation of an aged fire-derived char on glass beads similarly lay at 1.7 (for 20°C (15–25°C); Zimmermann et al, 2012). Calculations based on C stock changes yielded a Q10 of 3.4 at temperatures between 5 and 15°C (Cheng et al, 2008), which is identical to results from oak wood biochar pyrolyzed at 350°C when interpolated to the same temperature (Nguyen et al, 2010). Combining these values, Q10 values decreased non-linearly with the increasing temperature of mineralization (Figure 11.3(a)).

Prediction of mineralization rates at temperatures different from an experimentally

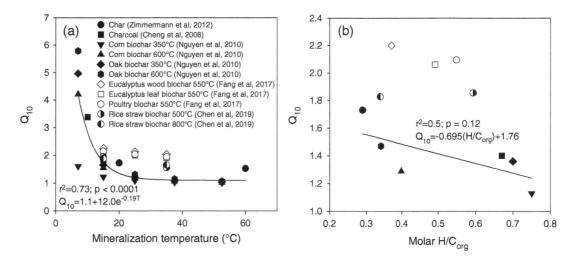

Figure 11.3 *Q10 values of biochar mineralization (a) with increasing temperature, and (b) as a function of molar H/C_{org} ratios of biochars (for Q10 at a mineralization temperature of 20°C; since H/C_{org} ratios for aged biochars refer to their initial property before aging (Fang et al, 2017), these and the incubations in glass beads (Chen et al, 2019) are excluded from the regressions*

determined one, therefore, requires the use of different Q10 values that can be calculated using the experiment temperature (Texp) and the targeted temperature (Ttarget) (Woolf et al 2021), as follows:

$$Q10 = (1.1 * (T_{exp} - T_{target})$$
$$- 63.1579 * e^{-0.19*Texp}$$
$$+ 63.1579 * e^{-0.19*Ttarget})/(T_{exp} - T_{target})$$
[11.2]

This non-linear relationship was found to be related, to a certain extent, to the H/C_{org} ratios of the biochars (Figure 11.3(b)). However, factors beyond material properties are also important. This is illustrated by the different Q10 values found for incubations of identical organic materials (Feng and Simpson, 2008) and biochar (Fang et al, 2014b) in different soils.

The Q10 values experimentally established so far may be biased towards the more

easily mineralizable fraction of biochars, as the incubation results necessarily reflect the properties of the C that evolve during earlier stages of mineralization. Aged biochars obtained from storage sites after 130 years (Cheng et al, 2008), naturally aged chars (Zimmermann et al, 2012), artificially aged biochar (Chen et al, 2021), or aged after multi-year incubation (Fang et al, 2017) indeed show slightly higher temperature sensitivity than fresh biochars with a greater proportion of easily mineralizable C (Figure 11.3). This difference is less pronounced at lower mineralization temperatures.

Moisture

Moisture is a major factor controlling the mineralization of any organic C in soil. It is therefore not surprising that a model of biochar mineralization was very sensitive to variations in moisture and resulted in poor matches between observed and predicted mineralization, especially during dry seasons (Foereid et al, 2011). On the other hand, saturated soil water conditions (and possibly concurrent reduction in O_2 availability) did not significantly decrease the mineralization of biochars that were pyrolyzed at a greater temperature (600°C vs 350°C) over a period of one year compared to unsaturated conditions (Nguyen and Lehmann, 2009). For biochars made from oak or corn stover at 350°C, however, mineralization decreased by half under saturated compared to unsaturated conditions. In addition, interactions between moisture and temperature under field conditions and their effects on biochar mineralization in the short term or adaptations in the long term have not been investigated thus far.

Aggregation

Aggregation and entrapment of organic matter inside aggregates (Tisdall and Oades, 1982) or occlusion in small pores that are inaccessible to microorganisms or their exoenzymes (Kaiser and Guggenberger, 2008) confer greater persistence to otherwise easily decomposable organic C. In addition, organic matter itself promotes aggregation (Six et al, 2004). Therefore, it is not surprising that biochar increases soil aggregation, on average, by 16% for the global dataset examined so far (Islam et al, 2021). Aggregation is increased to a greater extent in neutral or acidic as well as loamy soils than in coarser (Islam et al, 2021) or finer textured soils (Ouyang et al, 2013) by biochar made from wood at higher pyrolysis temperature (> 600°C; Islam et al, 2021). Attesting to a possible protection mechanism of biochar by aggregation, biochar materials have indeed been observed inside aggregates (Brodowski et al, 2005; Lehmann et al, 2005; Liu et al, 2020; Burgeon et al, 2021), but typically to a greater extent in free light fractions outside aggregates (Glaser et al, 2000;

Murage et al, 2007; Dharmakeerthi et al, 2015) and in so-called heavy or mineral-associated fractions (Glaser et al, 2000; Liang et al, 2008). Chars from land clearing using fires were macroscopically visible for about 30 years after deposition with particles (>50 μm) present over centennial (Nguyen et al, 2008; Burgeon et al, 2021) and even millennial time scales (Glaser et al, 2000; Lehmann et al, 2005; Liang et al, 2008).

These particle sizes are too large to fit into the nanometer-size pores of soil minerals that would reduce access by microorganisms. Therefore, occlusion in small pores may not be as important a process conferring persistence to biochar as it is for unpyrolyzed organic matter, as shown for maize straw and its biochar in a Vertisol (Rahman et al, 2018). However, one caveat to our current understanding is that physical fractionation techniques developed for organic matter other than biochar with potentially different specific densities, sizes, and hydrophobicity may not be suitable for biochar.

Another aspect that may need to be considered: the particulate nature of PyOM acts in a way that may be described as "self-aggregation". The physical disconnection of the interior of a biochar particle with virtually unchanged chemical characteristics over millennia from the decomposer community surrounding the particle might reduce mineralization rates (Liang et al, 2008). Therefore, smaller biochar particles mineralize faster than larger ones, as discussed above.

Interaction with soil minerals

Interactions with soil minerals are an important mechanism for the long-term persistence of any organic matter in soil. These processes are, on average, less important for the persistence of biochar than for the corresponding unpyrolyzed organic matter (Woo et al, 2016), albeit likely more important than aggregation (discussed above). Due to the surficial oxidation of PyOM during aging, its high loading with negatively charged functional groups, and possibly due to its radical content and electrochemical properties (Chapter 10), opportunities for interactions with positively charged minerals are, in principle, large. Such interactions could either occur with dissolved mineral elements including Al, Mn, Fe, and Ca, or with soil minerals such as Fe or Al oxides or phyllosilicates. Direct spectroscopic evidence for both processes exists (Nguyen et al, 2008; Joseph et al, 2010, 2013; Chia et al, 2012; Yang et al, 2016; Jing et al, 2022). Finer soil texture typically reduces the mineralization of biochar, by up to half in the presence of kaolinite (Yang et al, 2018b) and greater proportions of clay (Bruun et al, 2013; Yang et al, 2022).

The type of soil minerals is equally important for biochar mineralization that is typically reduced by short-range order soil minerals. Soil PyC contents were positively correlated with short-range order minerals in fire-prone Hawaiian soils (Cusack et al, 2012). The MRT of wood biochar determined in an incubation experiment was 22–35% greater in an Oxisol (with a greater proportion of short-range order soil minerals) than in both a Vertisol and an Inceptisol from Australia (Fang et al, 2014a). Mineralization of biochar made from pine wood was reduced by half over six months when incubated in an andesitic soil with greater amounts of short-range order clay minerals than in a granitic soil; whereas there was no difference in mineralization for uncharred wood (Santos et al, 2012). Mineralization of biochars can therefore be greatly influenced by the type of soil it is added to, which may even exceed the importance of pyrolysis temperature (Herath et al, 2015).

It is probable that biochar-mineral interactions are qualitatively and quantitatively different from those operating on unpyrolyzed organic matter (Lavalle et al, 2019). This requires further study recognizing the time dependence of biochar surface quality and quantity (Chapter 10).

Soil properties

Other soil properties than those related to mineral interactions, aggregation, or priming (as discussed elsewhere in this chapter) also affect the persistence of biochar, including redox conditions, pH, or composition and abundance of biota. Higher soil pH was shown to increase the mineralization of biochars (Cheng et al, 2008; Luo et al, 2011). Both a lower toxicity of free metals (e.g., Al) for microorganisms and a lower stabilization of biochar by short-range order oxides may contribute to this lower persistence, outweighing any stabilizing effects of Ca-bridging that is more likely to occur in soil with alkaline pH values. On the other hand, low soil pH may also increase the abiotic dissolution of inorganic C in biochars in the short term (Sheng et al, 2016).

Tillage

Tillage of soil typically leads to greater mineralization of soil organic C through a variety of processes including aeration, destruction of aggregates, or desorption from mineral surfaces. On the one hand, one may expect a greater acceleration of mineralization by tillage for biochar than other organic matter if a specific mechanism of biochar persistence is its particulate nature (as outlined above), because tillage may break open biochar particles to expose the interior that would otherwise be protected. On the other hand, the opposite may be hypothesized since the mineralization of biochar without mineral interactions or aggregate protection that is typically reduced by tillage is lower than that of unpyrolyzed residues. Consequently, 8–18 years of tilling of several soil types caused virtually no decrease in char from vegetation fires after concurrent fire suppression at two locations in Australia (Skjemstad et al, 2004). Similarly, annual plowing of a bare fallow did not significantly reduce PyC contents, while total organic C decreased by 33% over 55 years (Vasilyeva et al, 2011).

Aggregate stability after biochar additions (Islam et al, 2021) is not more or may even be less vulnerable to tillage than after other organic amendments. Tilling a Luvisol in Nigeria that was amended with hardwood biochar pyrolyzed at 580°C decreased aggregate stability by 18% compared to a reduced tillage regime but decreased aggregate stability by 22% when poultry manure had been added (Agbede, 2021; both amendments had greater aggregate stability compared to an unamended control regardless of tillage). Therefore, until further evidence is available, it may be assumed that proportional increases in mineralization of biochar by tillage are not higher than those documented for other soil organic C. This also means that the estimated mineralization of fire-derived chars in untilled soils must be corrected for the effect of tillage.

Plant residue, organic input, and priming of biochar

The input of organic matter by plants, such as roots, root exudates, or leaf litter, as well as any other organic matter inputs, such as animal manure, green manures, or composts, will change the mineralization of biochar in the soil through a variety of processes (Chapter 17). This effect is often called "priming". In principle, biochar mineralization may be increased (positive priming) or decreased (negative priming) through processes such as co-metabolism or substrate switching, respectively, among others (Hamer et al, 2004; Cui et al, 2017; Zimmerman and

Ouyang, 2019). These effects should be considered when extrapolating biochar mineralization rates from incubation results without plant or biomass inputs to the application of biochar in agricultural soil.

In many instances, biochar mineralization was reduced (negative priming), when plant-derived organic matter was added to biochar (Cui et al, 2017; Zimmerman and Ouyang, 2019). At present, a conservative assumption is that plant residue or biomass input does not diminish the persistence of biochar (Liang et al, 2010; Dharmakeerthi et al, 2015; Jiang et al, 2016).

Reburning of biochar

In-situ burning, such as through vegetation fires or crop residue burning, may lead to CO_2 emissions from biochar. Burning of naturally produced chars in subsequent fires has been proposed as a possible reason for low char accumulation observed in natural ecosystems (Ohlson and Tryterud, 2000; Czimczik et al, 2005). However, an experimental fire was found to only cause a 7% loss of chars buried 20 mm into the organic forest floor (Santin et al, 2013), and less than 8% when attempting to maximize combustion by placing chars on the soil (Saiz et al, 2014), indicating very low rates of reburning. Any incorporation of biochar into mineral soil would probably even reduce such values, as temperatures during experimental fires reached only less than 50°C at depths greater than 20 mm (Bradstock and Auld, 1995). In modern soil management that likely includes the incorporation of biochar and no significant crop residue burning in fields, this process may therefore be less important than in natural ecosystems where most of the char is initially deposited in organic horizons, on the soil surface, or even remain on living or dead aboveground woody vegetation (Santín et al, 2015). Such exposed PyOM may be oxidized to a significant extent by fire soon after deposition (Doerr et al, 2018). Over the long term, however, most PyC is found in the subsoil in natural ecosystems (Koele et al, 2017), which is unaffected by fire.

Erosion and leaching of biochar

Biochar can move in the landscape by erosion (wind and water) or leaching (Chapter 12; Figure 11.1). During transport (in air or water), mineralization of biochar may increase, but likely to a lesser extent than unpyrolyzed biomass, as proportions of PyC in total organic C are typically greater in subsoils or the ocean than in topsoils (Koele et al, 2017). For both unpyrolyzed and pyrolyzed biomass, mineralization typically decreases with soil depth (Rumpel and Kögel-Knabner, 2011; Soucémarianadin et al, 2019) and upon deposition in lake or ocean sediments (Masiello and Druffel,

1998), with erosion typically being seen as increasing organic C sequestration at a regional or global scale (Berhe et al, 2007). Biochar persistence may therefore not decrease after transport (Abney and Berhe, 2018).

The solubility of biochar and its persistence are likely related to some extent. On the one hand, soluble organic carbon forms are mineralized to a larger extent. On the other hand, soluble compounds may leach into the subsoil where they can be adsorbed and are more persistent. Leaching of organic C from oak-wood biochar ranged from 7–0.2% with increasing pyrolysis temperature 250–650°C (Bostick et al, 2018) and 4–11% for ryegrass biochar (Schiedung et al, 2020), being typically higher for grass- than wood-derived biochars (Mukherjee and Zimmerman, 2013). Most of this release occurs during the first rain (85%; Schiedung et al, 2020) but can continue over time (Abiven et al, 2011).

Approaches to quantification of biochar persistence

The persistence of biochar here refers to how much biochar remains in the soil environment after a certain period. The amount lost is a result of various processes discussed above, including both biotic and abiotic processes. It cannot be predicted solely by chemical analyses of the biochar itself. Certain chemical tests can, however, be correlated to experimentally determined persistence in soil. These tests are covered in detail under Prediction of Long-Term Persistence below. In the current section, only the quantification of persistence in the soil itself is covered, using laboratory incubations, field experiments, chronosequences, global and regional PyC budgets, the use of aged biochars, and quantification of biochar C.

Only experiments with biochar additions to soil (in the laboratory or the field), where the production conditions or the properties of the biochar are known, can be used to directly predict biochar persistence.

Identifying biochar-C for assessing persistence

Two principal approaches can be used to directly quantify biochar persistence: (i) measuring how much biochar remains in soil after some time, or (ii) measuring how much CO_2 evolved over a while. Measurements of biochar-derived CO_2 emissions are more sensitive when biochar-C losses are low, while changes of biochar-C in the soil can be challenging to detect against what usually is over 90% biochar-C remaining after only months or years. On the other hand, regular measurements of CO_2 are costly with automatic chamber setups or manual sampling, and irregular field measurements may not capture variations of emissions due to changing moisture and temperature (while continuous measurements are possible in laboratory incubations). In soil or where plants are present, these measurements of CO_2 have to be combined with isotopic techniques to ascertain whether the CO_2 came from the applied biochar, from mineralization of already existing soil organic matter, or from root respiration that both greatly exceed CO_2 derived from biochar C. In incubations without plants or natural organic matter (Zimmerman, 2010; Whitman et al, 2013), no isotopes are needed, but microbial and abiotic processes might be strongly biased. The most common isotopic technique uses ^{13}C, either capitalizing on natural differences in the isotopic signature of biochar and soil organic matter (Major et al, 2010) or on experimental labeling using isotope enrichment (Whitman and Lehmann, 2015) or depletion (Fang et al, 2018). In the laboratory, the radioisotope ^{14}C can be used to more sensitively quantify sources of C (Kuzyakov et al, 2014). For ^{13}C isotope partitioning calculations, absolute enrichments should be used with strongly enriched biochars (Maestrini et al, 2014; Schiedung et al, 2020). In contrast, the so-called delta values can be used with weakly enriched biochars (Chalk and Smith, 2022). The isotope labeling is usually found to be relatively uniform at the bulk soil scale, with a possible exception of inorganic carbonates (Chalk and Smith, 2022), a consideration in evaluating the technique.

In principle, measurements of biochar remaining in soil are easier to perform after long periods, when large amounts of biochar

are mineralized, or in field experiments, where continuous measurements of CO_2 are challenging to perform logistically. Simple quantification of total C without the need for isotopes suffices if incubations are done in a substrate that has no C (Whitman et al, 2013). However, in soil or with plants, either isotopes have to be used, as explained above, or the remaining biochar has to be quantified using other techniques (Chapter 24). These include using biomarkers that are specific to biochar, and thermal and chemical methods to distinguish biochar-derived C from metabolized biochar or ingress of native soil C or roots (Bird et al, 2017).

It should be noted that one-time measurements of biochar remaining in soil pose constraints on calculating MRT of what is a highly non-linear decay, as discussed below. That constraint often makes measurements of biochar or biochar-C remaining less attractive. Taking more than one destructive measurement is often not possible without compromising the integrity of the plot.

A combination of isotope techniques and biomarkers affords additional insights into mechanisms of decay, either by isolating the original biochar or any decomposition products such as microbial biomass (Liang et al, 2010) or their metabolites (Santos et al, 2012; Farrell et al, 2013; Kuzyakov et al, 2014). Specifically, biochar degradation products are not necessarily mineralized (Bird et al, 2017). In addition, such organic C forms that are not in biochar anymore may behave differently than C in biochar. Being able to distinguish these fractions will likely be important to building dynamic models to provide better predictions of biochar turnover and sequestration (Woolf and Lehmann, 2012).

Field experiments

Field studies allow quantification of biochar turnover under the most realistic conditions, including differential climate, soil type, constant organic C input, soil management (e.g., tillage), presence of plants, etc. However, adding biochar in the field poses constraints on distinguishing mineralization from other pathways by which biochar may disappear from the topsoil, where it is typically applied to. Transport of biochar through erosion or leaching can pose larger losses than mineralization (Major et al, 2010) and should not be misinterpreted as mineralization. Ignoring the transport of biochar may therefore underestimate persistence if the remaining biochar or biochar C is quantified. Similarly, litterbags studies of biochar (e.g., Wardle et al, 2008) that are popular in forest ecology and agroforestry research to quantify the decomposition of leaf (and occasionally root) litter, face the issue of not being able to distinguish mineralization of biochar to CO_2 from decomposition to other organic matter or simply transport out of the experimental bags (Lehmann and Sohi, 2008).

Conversely, measurements of biochar-derived CO_2 emissions at the soil surface may overestimate persistence if a portion of the biochar is eroded or leached. More so than laboratory studies, field studies restrict the number of comparisons between biochar types due to cost and space constraints, which is shown by the typically low number of treatments for field studies (Major et al, 2010; Zimmermann et al, 2012; Weng et al, 2017). In addition, using isotopically labeled materials or aged biochar is more challenging in the field than in laboratory studies, as the large amounts of biochar needed restrict plot sizes (Whitman and Lehmann, 2015).

Laboratory incubations

Laboratory studies allow for much greater control over experimental conditions than field studies to investigate the effects of different environments and biochar properties. However, their shortcomings are their typically limited duration (even though, to date, some of the longest published observations with purposeful addition of biochar with known properties stem from laboratory incubation studies) and absence of litter input, macrofauna, plants or soil management, which also affect microorganisms, water and temperature dynamics. Expectedly, variations in calculated MRT, often ranging three orders of magnitude, are large between and within studies that include a wide variety of different biochars and experimental conditions, even when carried out at the same incubation temperature (Table 10.1 in Lehmann et al, 2015). Much of this variation can be explained by different biochar properties and experimental conditions (including soil properties, soil biota, etc.). One important advantage of incubation studies is the opportunity to utilize so-called aged biochars, as outlined below.

Chronosequences

Chronosequences substitute space for time by sampling soils that received biochar-type PyC at different times in the past (Bird et al, 1999; Preston and Schmidt, 2006; Hammes et al, 2008; Nguyen et al, 2008; Vasilyeva et al, 2011; Alexis et al, 2012). They are also called "false time series" as they are not actually derived from sampling the same location repeatedly over time, but from sampling different locations that received PyC at different times in the past; chronosequences have historically been used to study various processes beyond biochar and soil management (Huggett, 1998). One challenge of chronosequences is to identify a sufficiently large number of sites that have received the same amount and type of biochar inputs, under near-identical environmental conditions (vegetation, soil type, climate) and management (tillage, cropping system). The assessment is also restricted to those PyOMs that are typically produced by fires rather than deliberate additions of a range of biochar types. The advantage of this approach is that biochar mineralization can theoretically be examined over longer periods of hundreds (Hammes et al, 2008; Nguyen et al, 2008) or thousands of years (Preston and Schmidt, 2006, calculated from Gavin et al, 2003) than would ordinarily be quantifiable through field observations using researcher-managed biochar additions. Similar to field studies, significant challenges arise to either estimate or exclude the physical transport of biochar. Since erosion can be the major pathway explaining the disappearance of biochar from soil (Chapter 12), neglecting to consider transport will result in erroneous estimates of biochar mineralization (Nguyen et al, 2008). Chronosequences may therefore, at best, provide a lower estimate of mineralization and should not be used unless erosion and leaching rates are known.

Assessment of aged biochars

Assessment of aged biochars is intended to be used to simulate what mineralization rates of biochars would look like in hundreds or thousands of years that cannot be directly quantified.

Aged biochars can either be obtained from naturally aged deposits in the environment or by artificial aging in the laboratory using various approaches. Naturally aged biochars have been obtained from storage sites of historic charcoal production (Cheng et al, 2008; Calvelo Pereira et al, 2014), isolated char fractions of fire-prone soils (Shindo, 1991), or collected on soil surfaces (Zimmermann et al, 2012). Also, incubations of soils with large proportions of PyC have been used to approximate mineralization rates and can be compared to adjacent soils with low or no PyC (Cheng et al, 2008; Liang et al,

2008; Knicker et al, 2013). One challenge with this approach is to experimentally separate unpyrolyzed organic C forms adsorbed to biochar or PyOM particles, as these are not easily distinguishable from the PyOM.

Aging of biochar in the laboratory may deploy chemical oxidants, thermal oxidation, photo-exposure, biological incubation, or physical aging (Cross and Sohi, 2013; Yang et al, 2022; Zeba et al, 2022). Such aging may have three principal effects on biochar persistence: (i) reduction of any easily mineralizable fraction of biochars (mainly biological aging); (ii) increased access to internal surfaces (mainly physical aging); and (iii) weakening of aromatic structures (mainly chemical aging). The first effect would reduce bias through typically large short-term mineralization of a non-PyC fraction of biochar for the extrapolation of MRT. The latter two may resolve whether short-term mineralization experiments overestimate the persistence of biochar over the long term. The unchanged bulk chemical composition of remaining PyOM in soils over decadal (McBeath et al, 2013), centennial (Schneider et al, 2011) and millennial (Liang et al, 2008) time scales does not provide evidence for important biological or chemical aging effects (not to be confused with changes in surface properties, which can be very large, Chapter 10). The decreasing particle sizes of biochar in soil (see discussion above) may indicate that physical aging (e.g., by freeze-and-thaw) may be more important than biological and chemical aging in the long term.

One important issue with natural or laboratory aging of biochars is that it often isolates biochars in advance of mineralization studies or performs aging without the mineral matrix that is known to increase persistence by surface interactions, organic matter adsorption, or aggregation as outlined above. Thus, both approaches lead to underestimation of biochar persistence. Some evidence for the importance of mineral interactions is provided by greater C-14 ages of the most easily chemically oxidizable PyC fraction (Krull et al, 2006).

As a consequence of the highly variable effects of very different aging techniques, some studies find that aging increases persistence, while others report decreases in persistence. For example, chemical, physical, and biological aging decreased the persistence of low-temperature biochars (350°C) but had little influence on high-temperature biochars (550°C) (Zeba et al, 2022). In contrast, biological aging decreased mineralization by more than half over 180 days (Yang et al, 2022). It is therefore doubtful whether any particular aging method provides unequivocal information about long-term persistence in soil (Li et al, 2019).

Global or regional budgets

Budgets of PyC production through fires and PyC stocks remaining in soils and sediments can be used to estimate PyC persistence at a regional or global scale (Masiello, 2004). These calculations assume a long-term steady state in regional or global PyC stocks, i.e., production must be equal to loss. For a range of sites across northern and eastern Australia for which production rates and stocks were measured, MRTs of between 1300 and 2600 years were calculated (Lehmann et al, 2008). On a global scale, the MRT of PyC was estimated at about 1,000 years using a simple budget approach (Bird et al, 2015) and at least 5,387 years using Earth system modeling approaches across all ecosystems (Bowring et al, 2022).

How to deal with inorganic C in biochar

Biochars can contain variable quantities of inorganic C, ranging from no detectable contents to 68% of total C (Enders et al, 2012; Wang et al, 2014; Calvelo Pereira et al, 2017). Most biochars made from wood or crop residues contain small amounts of inorganic C (<0.4% carbonate-C, Wang et al, 2014). In contrast, biochars made from manures or industrial wastes (paper sludge, etc.) can contain large proportions of inorganic C, typically highest at intermediate pyrolysis temperature of about 500–600°C. Inorganic C is likely to evolve over the short term except in alkaline soil (Spokas, 2010). However, additions of biological carbonates to biomass before pyrolysis generated biochar that is more persistent against chemical and thermal oxidation (Zhang et al, 2021). It remains to be seen whether inorganic C in biochar (added before or produced during pyrolysis) can confer greater persistence to biochar in soils. Some evidence also suggests that C sequestration as inorganic C may be increased in the long term by biochar (Dong et al, 2019).

Prediction of long-term persistence from feedstock type, production conditions, and biochar properties

Biochar persistence can be estimated by extrapolating the above-mentioned field and laboratory assessments of mineralization to the long-term and can be related to biochar properties, feedstock type, and production conditions.

Short-term measurements and long-term predictions

Strictly speaking, any experiments that use modern biochar to apply to soil will constitute a "short-term" assessment relative to most expectations of information about "long-term" mineralization rates. This would apply to both laboratory incubation experiments and field experiments. The reason is that the time period that we seek information for (at least 100 years for climate change mitigation needs) exceeds the time period of our experiments (typically <3 years). This requires extrapolation beyond the period of observation, which increases uncertainty and requires multiple approaches to instill confidence in calculated MRTs. In addition to mathematical extrapolations, there are some opportunities to extend the time horizon of observation by: (i) exposing biochar to higher temperatures during what will likely be a laboratory study and therefore accelerate decomposition that may be mathematically corrected by using estimates of temperature sensitivities; or (ii) utilizing aged biochars to estimate long-term mineralization rates, as described above. While field experiments are preferred because of their realistic environmental and management conditions, they do not allow these two strategies to be used effectively, and are typically constrained to testing very few biochar types to keep workloads manageable. In addition, the variability of environmental conditions such as soil moisture and temperature make extrapolations with regular mathematical models challenging, especially in strongly seasonal climates (Maestrini et al, 2014). Laboratory experiments offer more flexibility as to what biochars and experimental conditions are used and allow the application of equations that assume constant conditions in short-term data. It is not fully evident, however,

whether laboratory incubation experiments consistently over- or under-estimate mineralization, as seen from experiments with unpyrolyzed litter (Bonan et al, 2013).

Extrapolation methods

Mineralization of plant litter is commonly modeled using a single exponential function (Equation 11.3; here expressed as proportions of initial biochar-C). However, the large difference in mineralizability of low-molecular acids and N-rich volatile compounds on the one hand, and the fused aromatic C forms in biochars on the other (Chapters 6 and 7), warrants the use of multiple pools, each with their own mineralization rate (Figure 11.4(a,b)). To reduce complexity, typically a 2-pool model (Equation 11.4) is used. The differences between MRT calculated using either single or multiple pools are large (Zimmerman and Gao, 2013; Bird et al, 2015; Equation 11.5), and the number and duration of data available determine the options for computing MRT, as discussed below.

$$(1-\text{pool}): \text{Biochar} - C_{(\text{at time } t)} (\% \text{ of initial biochar} - C)$$
$$= 100\, e^{-k(1/yr)\, t\ (yr)} \qquad [11.3]$$

$$(2-\text{pool}): \text{Biochar} - C_{(\text{at time } t)} (\% \text{ of initial biochar} - C)$$
$$= \text{Pool1}\,(\% \text{ of biochar} - C_{(\text{at time } 0)})\, e^{-k1\ (1/yr)\ t}$$
$$+ \text{Pool2}\,(\% \text{ of biochar} - C_{(\text{at time } 0)})\, e^{-k2\ (1/yr)\ t}$$
$$[11.4]$$

[whereby Pool1+Pool2=100%; note opportunities and limitations of setting this constraint]

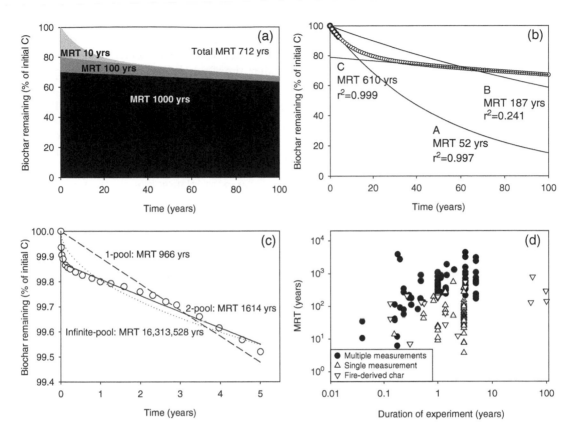

Figure 11.4 *Conceptual and measured data, demonstrating the effects of different methods of calculation of mean residence time (MRT). (a) Conceptual illustration of mineralization of biochar containing different C pools, each with different MRTs (three fractions of 20, 10, and 70% with 10, 100, and 1000 years MRT, respectively, resulting in an average MRT of 712 years); (b) Three different methods of calculating MRT using a 1-pool equation (Biochar-C remaining = $C_0\, e^{-k*yrs}$) and annualized data from panel (a), shown as symbols: (A) C_0 set to 100%, and k determined by fitting over the first 5 years of data, (B) C_0 and k determined by fitting over 100 years of data, (C) C_0 and k determined by fitting over the 50 last years of data. (c) Three different methods of calculating MRT from measured mineralization of wood biochar pyrolyzed at 550°C (Singh et al, 2012a) using either a 1-pool (dashed line; biochar-C = $100e^{-0.00104t}$), a 2-pool (solid line; biochar-C = $0.13e^{-58t} + 99.87e^{-0.00065t}$) or an infinite-pool (Zimmerman, 2010) power model (dotted line; biochar-C = $100-((100(1-e^{0.00081})/(1-0.63))t^{(1-0.627)})$. (d) Temperature-adjusted MRT related to the duration of the experiment using: (i) 2-pool model results of multiple measurements over the duration of experiment; (ii) 1-pool model results of a single measurement after the experiment; or (iii) experiments with fire-derived chars (irrespective of calculation) (Lehmann et al, 2015; n = 156)*

(3 − pool): Biochar

$$- C_{(at\ time\ t)}\ (\%\ of\ initial\ biochar - C)$$

$$= Pool1\ (\%\ of\ biochar - C_{(at\ time\ 0)})\,e^{-k1\ (1/yr)\ t}$$

$$+ Pool2\,(\%\ of\ biochar$$

$$- C_{(at\ time\ 0)})\,e^{-k2\ (1/yr)\ t}$$

$$+ Pool3\,(\%\ of\ biochar$$

$$- C_{(at\ time\ 0)})\,e^{-k3\ (1/yr)\ t}$$

[11.5]

[whereby Pool1+Pool2+Pool3=100%]

Theoretically, it may be desirable to apply a model not just with two but with multiple C pools (Zimmerman, 2010; Bai et al, 2013; Bird et al, 2015; Herath et al, 2015), recognizing that biochar C is composed of a continuum of PyC forms (Preston and Schmidt, 2006) with progressively slower mineralization rates as some of the smaller molecules in biochar are mineralized and a greater proportion of fused aromatic C forms remain. For a mineralization experiment conducted by Singh et al (2012a), the 1-pool equation calculated an MRT of 966 years, the 2-pool equation an MRT of 1614 years, and a model with infinite pools ('power model' introduced by Zimmerman, 2010) an MRT of 16,313,528 years (Figure 11.4(c)). It seems indispensable to utilize at least two pools to adequately describe the mineralization dynamics (Woolf et al, 2021). Alternatively, it may be possible to account for initial rapid mineralization in different ways, such as fitting curves by omitting the initial period of rapid mineralization (Figure 11.4(b)). The inclusion of additional pools into the calculation may provide a better approximation of long-term mineralization but requires longer incubation experiments. Extrapolation beyond measured data may also benefit from using fractions that are actually measured, such as proportions of PyC in biochar or volatile matter.

Even with calculating a single metric (such as an MRT or BC_{100}, as shown in the next section) from multiple pools, individual pool sizes and decomposition rates should be listed to allow verification and calculation of other metrics (such as biochar-C remaining at other times). It is also crucial to report the fitting algorithm used, the initial guess of the parameters, and the boundaries imposed, because fitting 2-pool exponential models is not a linear problem, and algorithms do not necessarily converge to the global optimum (unlike for linear problems).

An extrapolation beyond the period covered by the available data, without scrutinizing the adequacy of the models used for a specific data set, poses undeniable challenges to the assessment of biochar mineralization. This is illustrated by a very large range in MRT values calculated by the three models shown in Figure 11.4(c). In addition, the variability in temperature and moisture conditions in field experiments provides challenges to extrapolation (Maestrini et al, 2014). Dynamic models should be developed that include environmental factors as well as microbial parameters, which are currently only available in incipient form (Foereid et al, 2011; Woolf and Lehmann, 2012; Mondini et al, 2017; Lefebvre et al, 2020; Pulcher et al, 2022).

The duration of an experiment also affects the results. Longer experiments that allowed for the use of a 2-pool model include those calculating the highest estimates for MRT; in comparison, MRT estimates below 200 years were only obtained from those experiments with a duration of less than one year (Figure 11.4(d)). An incubation experiment of 8.5 years (Kuzyakov et al, 2014) calculated an MRT that was twice as long as the one calculated for the first 2.5 years of the same experiment (Kuzyakov et al, 2009). Therefore, shorter periods of observation typically underestimate the persistence of biochar (Zimmerman and Gao, 2013; Fang et al, 2017).

The number of observations poses another constraint to data requirements in addition to the duration of observations: multiple measurements over time are needed to parameterize mineralization models. However, often only one or two measurements are made if the remaining biochar rather than CO_2 is quantified (Nguyen and Lehmann, 2009; Whitman et al, 2013), and using such data results in low estimates of MRT (Singh et al, 2012b).

Biochar properties

The prediction of biochar mineralization will only be fully aligned with prevailing theory if a comprehensive model is available that considers biochar material properties, environmental factors (including soil properties), and microbial dynamics. The use of biochar properties as a practical solution will at best generate an estimate that can and should, over time, be improved through more sophisticated approaches. The requirements for an appropriate proxy for mineralizability to be used by biochar practitioners and as part of more comprehensive modeling include (i) sufficiently low cost to allow routine measurements for research, monitoring and verification; (ii) relatively rapid analyses (ideally within hours); (iii) repeatability; (iv) robustness to different biochar properties and analytical capabilities; (v) strong and preferably linear relationship with mineralization rates; (vi) availability in different analytical laboratories; and (vii) ideally representing a specific chemical property rather than an operational definition.

A range of methods can be used to characterize biochar properties and relate those to biochar persistence as quantified by mineralization experiments (discussed above). These methods include quantification of biochar elemental composition (total C, C_{org}, H, O, N, ash, etc.), their ratios (H/C, O/C, H/C_{org}, O/C_{org}), organic functional group composition (C=C bonds, etc.; using

NMR, Py-GC-MS, NEXAFS, BPCA, and others), thermal stability (ASTM proximate analyses, thermogravimetry, bomb calorimetry, etc.), reduction (hydrogen pyrolysis), or chemical oxidation (dichromate, hydrogen peroxide, etc.) (Singh et al, 2017; Wang et al, 2022). It should be emphasized that these characterization methods cannot be used to quantify persistence directly. These biochar properties can only be used to predict persistence by correlating them to field or laboratory mineralization experiments.

The material properties most likely responsible for the relative persistence of pyrolyzed OM compared to unpyrolyzed OM are the fused aromatic C forms discussed above and in Chapters 6 and 7. Plant residues do not contain fused aromatic C forms, which are created by pyrolysis and increase in proportion with higher pyrolysis temperature and duration (Chapter 6). In addition, feedstock properties influence the C forms in biochars, e.g., due to their mineral contents (Chapter 6). Total aromaticity may be quantified by using spectroscopic techniques such as NMR spectroscopy which has been shown to correlate well with mineralization for individual studies (Singh et al, 2012a). However, correlations including multiple studies may be hampered by the use of different NMR analytical approaches, e.g., cross-polarization (CP) versus direct polarization (DP) techniques. Quantification of aromaticity via ^{13}C NMR spectra may be compromised by overlapping spinning sidebands and low polarization observability (Amin et al, 2016). Further, cluster sizes of aromatic C can vary significantly between biochars independent of aromaticity (McBeath and Smernik, 2009; Nguyen et al, 2010), and may be more important for persistence than aromaticity alone (Mao et al, 2012).

Less expensive and more widely accessible characterization methods are measurements of atomic O/C or H/C ratios (Figure 11.5(a-c)

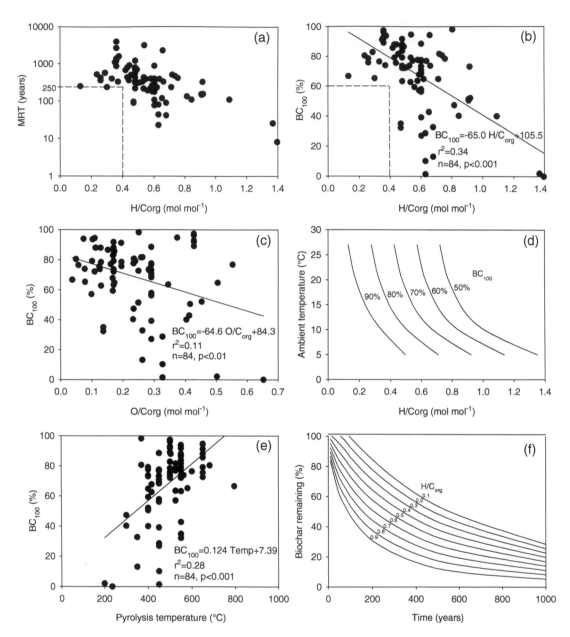

Figure 11.5 *Relationship between biochar elemental ratios or pyrolysis temperature and either MRT or the amount of biochar C remaining after 100 years (BC_{100}). (a,b) relationship between H/C_{org} and MRT or BC_{100}; (c) relationship between O/C_{org} and BC_{100}; (d) effect of H/C_{org} and ambient temperature to which biochar is exposed to on BC_{100}; (e) relationship between pyrolysis temperature and BC_{100}; (f) effect of time of interest and H/C_{org} on biochar C remaining (at a global average temperature of 14.9°C; only data that allowed the use of a 2-pool model of experiments conducted for one or more years using biochars from pyrolysis (not combustion or fire) with available information for biochar properties from Woolf et al, 2021)*

that are valid proxies for aromaticity (Wang et al, 2013) and have been correlated to mineralization rates (Spokas, 2010; Budai et al, 2013; Lehmann et al, 2021; Woolf et al, 2021). For some feedstocks that are rich in alum (aluminum sulfates in sludges) or ash, additional correction for inorganic C, H, and O may be needed (Enders et al, 2012; Wang et al, 2013), and the use of molar H/C_{org} rather than H/C ratios has by now been well established (Woolf et al, 2021) while organic H and O is not regularly assessed. Even more commonly accessible are measurements of volatile and fixed carbon that are often called proximate analyses, a protocol adopted from the coal and charcoal industry (ASTM, 2007). For low-ash biochars, volatile or fixed carbon contents may be sufficiently correlated with O/C or H/C ratios (Enders et al, 2012) and have been useful in predicting biochar mineralization (Zimmerman, 2010), as have other thermal oxidative techniques such as R50 (Harvey et al, 2012). Oxidative techniques often remove the most easily mineralizable organic C and results therefore correlate well with short-term mineralization, but may not be the most robust techniques to predict long-term mineralization, for which information about the non-oxidizable C is needed. Recent method development has made this connection, such as by establishing a relationship between C removable by hydrogen pyrolysis and cluster sizes of fused aromatic ring structures (McBeath et al, 2015).

Measured MRT (converted to an incubation temperature of 14.9°C) consistently exceeded 250 years for biochars with H/C_{org} ratios below 0.4 for the data available to date, which include both field and laboratory experiments (Figure 11.5(a)). This means that more than 60% of the initial C in biochar will remain after 100 years (BC_{100}, Figure 11.5(b)). Most biochars produced by slow pyrolysis above 500°C will have an H/C_{org} ratio below 0.4 (Enders et al, 2012; Schimmelpfennig and

Glaser, 2012). The correlation of BC_{100} with the H/C_{org} ratio is better than with the O/C_{org} ratio (Figure 11.5(b,c)).

Multiple factors beyond biochar properties determine mineralization as discussed above, making the correlations across studies still poor (Figure 11.5). Therefore, data sets from individual studies provide better control for different experimental conditions. Consequently, the correlation between H/C_{org} and BC_{100} was $r^2 = 0.96$ recalculated for ten biochars investigated by Singh et al (2012a), compared to $r^2 = 0.34$ for the global data set in Figure 11.5(b). Much of the variation, in addition to biochar properties, stems from explainable differences in experimental site or soil conditions. A difference in H/C_{org} between 0.24 and 0.5 is equivalent to increasing mineralization temperature from 5 to 20°C and decreasing BC_{100} from 90 to 70% (Figure 11.5(d)). This illustrates that the material properties of biochars alone will not be able to predict all of their mineralization dynamics as is evident from basic theory and known dynamics of soil organic C (Lehmann et al, 2015). Further model development is therefore desirable.

Some basic trends can, however, be discerned. Unpyrolyzed organic matter has H/C_{org} ratios well above 1 (Baldock and Smernik, 2002; Enders et al, 2012: 1.4–1.6), indicating that charring confers, at a minimum, an order of magnitude decrease in mineralization relative to organic residues. Direct experimentation with pyrolyzed (at comparatively low temperatures of 350–450°C) and unpyrolyzed organic matter point to a 1.5 order of magnitude reduction in mineralization (Baldock and Smernik, 2002; Santos et al, 2012). Since different residues (e.g., leaves vs. wood) already possess very different initial mineralization rates without charring (Santos et al, 2012; Whitman et al, 2013), the difference in absolute reduction of mineralization conferred by charring (e.g., BC_{100}) is smaller for those unpyrolyzed organic residues that show a lower mineralization rate. This has implications for life cycle emission balance models (Chapter 30).

Production conditions and feedstock properties

Both production conditions and feedstock properties determine biochar properties, and therefore should, in principle, have a relationship with mineralization data and can be used to predict biochar persistence under certain circumstances. Usually, the prediction of persistence is poorer using production conditions than biochar properties (shown for pyrolysis temperature in Figure 11.5(e)). Feedstock properties alone cannot be used because of the greater importance of pyrolysis temperature in dictating biochar properties. However, production conditions constrained by feedstock properties can generate broad classes of persistence of biochars that are acceptable for use in greenhouse gas emissions reduction accounting (Ogle et al, 2019; Lehmann et al, 2021; Woolf et al, 2021).

Along with pyrolysis temperature (Figure 11.5(e)), the type of pyrolysis technology employed is closely related to biochar properties and may be important for constraining its persistence. Some production processes such as hydrothermal carbonization (producing hydrochar which does not have the same properties as biochar), flash pyrolysis using pressure, fires, gasification, and many emergent thermochemical technologies have to be evaluated for their biochar properties to ascertain persistence that may be longer or shorter than predicted from data in Figure 11.5. Additives and mixtures of feedstocks may create biochars with properties that are not easily predictable from the individual feedstock properties alone (Taherymoosavi et al, 2016). Similarly, post-pyrolysis modifications including, but not limited to, composting, steam activation, and ball-milling, may change the persistence of biochars and have to be evaluated separately.

Persistence over what time?

Predictions of persistence over different time frames will generate very different results. The predicted proportion of biochar-C remaining decreases more with longer experimental duration during earlier than later stages (Figure 11.5(f)). This reduction in predicted BC_{100} is significant compared to differences between biochars of varying properties. Between 100 and 200 years, the C remaining of a biochar with a H/C_{org} ratio of 0.4 decreases from 80% to 64%, which is equivalent to the decrease in persistence due to an increase in the H/C_{org} ratio from 0.4 to 0.6 (Figure 11.5(f)). What time frames are useful to evaluate depends on the intent, which typically depends upon the requirements of carbon markets or national greenhouse gas accounting (discussed in Chapters 30, 31, and 34).

The sensitivity to changes in predicted MRT as a result of differences in biochar properties also differs whether biochar-C remaining is considered after 100 or 500 years. Over 100 years, the difference in biochar-C remaining (BC_{100}) between biochars with varying properties (here shown for H/C_{org}) is lower than over 500 years (for which lines are further apart in Figure 11.5(f)). Experiments with aged biochars should generate more information on how robust these projections are that reach well beyond the periods for which the experiments were conducted.

Conclusions and recommendations

Pyrolyzing organic matter decreases its mineralization by at least one and a half orders of magnitude under otherwise identical environmental conditions (e.g., soil temperature or moisture, soil properties, decomposer community). Although biochar turnover is clearly influenced by many of these environmental conditions, the relative decrease in mineralization is significantly related to biochar's organic composition (such as H/C_{org} or aromaticity). This relationship between biochar properties and persistence depends on how the biomass is pyrolyzed (Chapters 6 and 7), mainly by variations in pyrolysis temperature and duration, the properties of the biomass, including but not limited to its ash contents. This specific variation is now becoming better predictable through assessing relevant material properties of biochars such as aromaticity, aromatic condensation, and atomic ratios of organic C, H, and O.

However, variation between sites of application, experimental conditions (e.g., incubation temperature; incubation media; moisture, field vs. laboratory experiments), and extrapolation approaches to calculate MRT add a layer of complexity to interpretations, and future modeling efforts should incorporate all sources of variation. Recommendations for appropriate scrutiny and interpretation of experimental data and study results include the following:

- Consider biochar variability: biochar products are chemically and physically different. The behavior of one specific biochar product does not allow for predicting the behavior of other biochar products without recognizing their differences.
- Compare apples to apples: for estimating the extent to which pyrolyzing biomass increases persistence in soil, the mineralization of fresh biochar added to soil should only be compared to the mineralization of the corresponding unpyrolyzed biomass under identical experimental conditions. Misconceptions commonly arise by comparing fresh biochar to: (i) mineralization of soil organic C that has already been decomposed from a much larger source of biomass, is already interacting with soil minerals and that typically includes chars from vegetation fires; (ii) mineralization of unpyrolyzed biomass of a different feedstock; and (iii) mineralization of unpyrolyzed biomass under different experimental conditions, different incubation temperatures or soil types (e.g., sand vs iron-rich clays) that may have larger effects than charring (and future research will need to clarify whether all effects are similar irrespective of charring).
- Apply the correct model: calculation of MRT and other measures of mineralization from fresh biochars must use at least a 2-pool model. A 1-pool model is unacceptable and, at best, gives minimum estimates. Further improvements in modeling are a high priority.
- Distinguish between physical movement and mineralization: erosion and leaching (Chapter 12) can be significant pathways that lead to the disappearance of biochar in field experiments but should not be confused with mineralization. In several instances, 'mobility' has been confused with 'lability', and 'stationary' in the landscape with 'stable' against mineralization. In fact, eroded organic matter buried in lake or ocean sediments or organic matter in subsoils may mineralize to a lesser degree than organic matter in topsoils.

Accounting for biochar transport, especially through erosion in field mineralization studies, is difficult; erosion should be prevented, and leaching should be accounted for.

- Attribute mineralization to the correct source: not only biochar contribute to CO_2 evolution from soil, and the difference in CO_2 evolved from plots that received biochar and from those that did not, may not be additive due to priming of both biochar and soil organic C. Therefore, isotopes must be used to distinguish CO_2 or remaining soil C from biochar and other sources. Inorganic C dynamics (release and possible precipitation) further complicate assessments.

Experimental data from field trials are still scarce and a coordinated international effort is needed to stage comparative trials in sufficient locations and over long time scales (Table 11.1). Future experiments on biochar persistence, whether in the field or laboratory, may need to include the following components: (i) use of C-13 or C-14 isotope enrichment to distinguish CO_2 derived from biochar versus from other sources, preferably greater than what is possible by natural abundance; (ii) experimental periods exceeding 1 year; (iii) a sufficient number of measurements over time to allow a 2-pool model to be applied; (iv) testing of multiple types of biochar of known composition; and (v) adequate comparison to uncharred organic matter.

Whether biochar is applied on the soil surface or incorporated, and the types of minerals present in soil, are both known to affect the mineralization of unpyrolyzed

Table 11.1 *Recommended research priorities to predict biochar persistence in soil*

Research priorities	Target knowledge gap	Priority	Challenges
Field studies	Long-term mineralization across different climatic regions; unknown variation from laboratory experiments	High	Quantification of losses other than mineralization is difficult to avoid or quantify; extrapolation is challenged by variations in soil moisture and temperature
Soil properties	Mechanism and magnitude of influence by soil minerals, biota, pH	High	In addition to laboratory incubations, this requires long-term field assessment with accompanying mechanistic studies
Modeling	Simulation of turnover mechanisms, inclusion in standard soil organic carbon models	High	Knowledge gaps in quantitative responses to environmental and soil conditions and decomposer dynamics for parameterization of the model (including interactions among them and with biochar properties)
Soil application	Variation in environmental conditions; photo-oxidation; tillage effects; erosion	Medium	This often requires large amounts of biochar due to the mechanization of application and field trials which are expensive (especially with isotopic labeling)
Feedstock properties	Moisture, mixtures, additions (minerals, etc.)	Medium	Many different options exist, difficult to standardize

organic matter. There are several reasons to expect differential responses with pyrolyzed residues, but the net effect on biochar persistence is not sufficiently quantified. Comprehensive experiments and modeling that account for all processes controlling biochar mineralization, will provide a step forward, not only for predicting the biochar remaining in soil after certain periods, but even more so for understanding biochar mineralization and the interactions between different factors controlling the persistence of biochar. Filling the knowledge gaps that have the greatest sensitivity for understanding biochar mineralization will guide agricultural policy, C trading approaches, and prediction of future C management strategies.

References

Abiven S, Hengartner P, Schneider MP, Singh N, and Schmidt MW 2011 Pyrogenic carbon soluble fraction is larger and more aromatic in aged charcoal than in fresh charcoal. *Soil Biology and Biochemistry* 43, 1615–1617.

Abney RB, and Berhe AA 2018 Pyrogenic carbon erosion: Implications for stock and persistence of pyrogenic carbon in soil. *Frontiers in Earth Science* 6, 26.

Agbede TM 2021 Effect of tillage, biochar, poultry manure and NPK 15-15-15 fertilizer, and their mixture on soil properties, growth and carrot (*Daucus carota* L.) yield under tropical conditions. *Heliyon* 7, e07391.

Alexis MA, Rasse DP, Knicker H, Anquetil C, and Rumpel C 2012 Evolution of soil organic matter after prescribed fire: A 20-year chronosequence. *Geoderma* 189-190, 98–107.

Ameloot N, Graber ER, Verheijen FGA, and De Neve S 2013 Interactions between biochar stability and soil organisms: Review and research needs. *European Journal of Soil Science* 64, 379–390.

Amin FR, et al 2016 Biochar applications and modern techniques for characterization. *Clean Technologies and Environmental Policy* 18, 1457–1473.

ASTM 2007 'ASTM D1762-84 Standard Test Method for Chemical Analysis of Wood Charcoal'. West Conshohocken, PA: ASTM International.

Bai M, et al 2013 Degradation kinetics of biochar from pyrolysis and hydrothermal carbonization in temperate soils. *Plant and Soil* 372, 375–387.

Baldock JA, and Smernik RJ 2002 Chemical composition and bioavailability of thermally altered *Pinus resinosa* (Red pine) wood. *Organic Geochemistry* 33, 1093–1109.

Berhe AA, Harte J, Harden JW, and Torn MS 2007 The significance of the erosion-induced terrestrial carbon sink. *BioScience* 57, 337–346.

Bird MI, Moyo C, Veendaal EM, Lloyd J, and Frost P 1999 Stability of elemental carbon in a savanna soil. *Global Biogeochemical Cycles* 13, 923–932.

Bird MI, et al 2015 The pyrogenic carbon cycle. *Annual Review of Earth and Planetary Science* 43, 273–298.

Bird MI, et al 2017 Loss and gain of carbon during char degradation. *Soil Biology and Biochemistry* 106, 80–89.

Bonan GB, Hartmann MD, Parton WJ, and Wieder WR 2013 Evaluating litter decomposition in earth system models with long-term litterbag experiments: An example using the Community Land Model version 4 (CLM4). *Global Change Biology* 19, 957–974.

Bostick KW, Zimmerman AR, Wozniak AS, Mitra S, and Hatcher PG 2018 Production and composition of pyrogenic dissolved organic matter from a logical series of laboratory-generated chars. *Frontiers in Earth Science* 6, 43.

Bostick KW, et al 2020. Photolability of pyrogenic dissolved organic matter from a thermal series of laboratory-prepared chars. *Science of the Total Environment* 724, 138198.

Bourke, J., et al 2007 Do all carbonized charcoals have the same chemical structure? 2. A model of the chemical structure of carbonized charcoal. *Industrial and Engineering Chemistry Research* 46, 5954–5967.

Bowring SP, Jones MW, Ciais P, Guenet B, and Abiven S 2022 Pyrogenic carbon decomposition critical to resolving fire's role in the Earth system. *Nature Geoscience* 15, 135–142.

Bradstock RA, and Auld TD 1995 Soil temperatures during experimental bushfires in relation to fire intensity: Consequences for legume germination and fire management in South-Eastern Australia. *Journal of Applied Ecology* 32, 76–84.

Brodowski S, Amelung W, Haumeier L, Abetz C, and Zech W 2005 Morphological and chemical properties of black carbon in physical soil fractions as revealed by scanning electron microscopy and energy-dispersive X-ray spectroscopy. *Geoderma* 128, 116–129.

Bruun EW, Ambus P, Egsgaard H, and Hauggaard-Nielsen H 2012 Effects of slow and fast pyrolysis biochar on soil C and N turnover dynamics. *Soil Biology and Biochemistry* 46, 73–79.

Bruun S, Clauson-Kaas S, Bubolska L, and Thomsen IK 2013 Carbon dioxide emissions from biochar in soil: Role of clay, microorganisms and carbonates. *European Journal of Soil Science* 65, 52–59.

Budai A, et al 2013 Biochar carbon stability test method: An assessment of methods to determine biochar carbon stability', IBI Document, Carbon Methodology, accessed at https://biochar-international.org/wp-content/uploads/2018/06/IBI_Report_Biochar_Stability_Test_Method_Final.pdf

Burgeon V, Fouché J, Leifeld J, Chenu C, and Cornelis JT 2021 Organo-mineral associations largely contribute to the stabilization of century-old pyrogenic organic matter in cropland soils. *Geoderma* 388, 114841.

Calvelo Pereira R, et al 2014 Detailed carbon chemistry in charcoals from pre-European Māori gardens of New Zealand as a tool for understanding biochar stability in soils. *European Journal of Soil Science* 65, 83–95.

Calvelo Pereira R, Camps Arbestain M, Wang T, and Enders A 2017 Inorganic carbon. In: Singh B, Camps Arbestain M, and Lehmann J (Eds) *Biochar: A Guide to Analytical Methods*. Boca Raton: CRC Press/Taylor and Francis. pp51–63.

Chalk P, and Smith CJ 2022 ^{13}C methodologies for quantifying biochar stability in soil: A critique. *European Journal of Soil Science* 73, e13245.

Chen G, Wang X, and Zhang R 2019 Decomposition temperature sensitivity of biochars with different stabilities affected by organic carbon fractions and soil microbes. *Soil and Tillage Research* 186, 322–332.

Chen, G., et al 2021 Priming, stabilization and temperature sensitivity of native SOC is controlled by microbial responses and physicochemical properties of biochar. *Soil Biology and Biochemistry* 154, 108139.

Cheng CH, Lehmann J, Thies JE, Burton SD, and Engelhard MH 2006 Oxidation of black carbon by biotic and abiotic processes. *Organic Geochemistry* 37, 1477–1488.

Cheng CH, Lehmann J, Thies JE, and Burton SD 2008 Stability of black carbon in soils across a climatic gradient. *Journal of Geophysical Research* 113, G02027.

Chia CH, et al 2012 Analytical electron microscopy of black carbon and microaggregated mineral matter in Amazonian Dark Earth. *Journal of Microscopy* 245, 129–139.

Cohen-Ofri I, Popovitz-Niro R, and Weiner S 2007 Structural characterization of modern and fossilized charcoal produced in natural fires as determined by using electron energy loss spectroscopy. *Chemistry- A European Journal* 13, 2306–2310.

Coleman K, et al 1997 Simulating trends in soil organic carbon in long-term experiments using RothC-26.3. *Geoderma* 81, 29–44.

Cross A, and Sohi SP 2013 A method for screening the relative long-term stability of biochar. *Global Change Biology Bioenergy* 5, 215–220.

Cui J, et al 2017 Interactions between biochar and litter priming: A three-source ^{14}C and δ^{13}C partitioning study. *Soil Biology and Biochemistry* 104, 49–58.

Cusack DF, Chadwick OA, Hockaday WC, and Vitousek P 2012 Mineralogical controls on soil black carbon preservation. *Global Biogeochemical Cycles* 26, GB2019.

Czimczik CI, Preston CM, Schmidt MWI, Werner RA, and Schulze E-D 2002 Effects of charring on mass, organic carbon, and stable carbon isotope composition of wood. *Organic Geochemistry* 33, 1207–1223.

Czimczik CI, Schmidt MWI, and Schulze ED 2005 Effects of increasing fire frequency on black carbon and organic matter in Podzols of Siberian Scots pine forests, *European Journal of Soil Science* 56, 417–428.

Davidson EA, and Jannsens IA 2006 Temperature sensitivity of soil carbon decomposition and feedbacks to climate change. *Nature* 440, 165–173.

Deng B, et al 2021 Feedstock particle size and pyrolysis temperature regulate effects of biochar on soil nitrous oxide and carbon dioxide emissions. *Waste Management* 120, 33–40.

Dharmakeerthi RS, Hanley K, Whitman T, Woolf D, and Lehmann J 2015 Organic carbon dynamics in soils with pyrogenic organic matter that received plant residue additions over seven years. *Soil Biology and Biochemistry* 88, 268–274.

Doerr SH, Santin C, Merino A, Belcher CM, and Baxter G 2018 Fire as a removal mechanism of pyrogenic carbon from the environment: Effects of fire and pyrogenic carbon characteristics. *Frontiers in Earth Science* 6, 127.

Domene X, Hanley K, Enders A, and Lehmann J 2015 Short-term mesofauna responses to soil additions of corn stover biochar and the role of microbial biomass. *Applied Soil Ecology* 89, 10–17.

Dong X, Singh BP, Li G, Lin Q, and Zhao X 2019 Biochar increased field soil inorganic carbon content five years after application. *Soil and Tillage Research* 186, 36–41.

Enders A, Hanley K, Whitman T, Joseph S, and Lehmann J 2012 Characterization of biochars to evaluate recalcitrance and agronomic performance. *Bioresource Technology* 114, 644–653.

Fang Y, Singh B, Singh BP, and Krull E 2014a Biochar carbon stability in four contrasting soils. *European Journal of Soil Science* 65, 60–71.

Fang Y, Singh BP, and Singh B 2014b Temperature sensitivity of biochar and native carbon mineralisation in biochar-amended soils. *Agriculture, Ecosystems and Environment* 191, 158–167.

Fang Y, Singh BP, Matta P, Cowie AL, and Van Zwieten L 2017 Temperature sensitivity and priming of organic matter with different stabilities in a Vertisol with aged biochar. *Soil Biology and Biochemistry* 115, 346–356.

Fang Y, Singh BP, Luo Y, Boersma M, and Van Zwieten L 2018 Biochar carbon dynamics in physically separated fractions and microbial use efficiency in contrasting soils under temperate pastures. *Soil Biology and Biochemistry* 116, 399–409.

Farrell M, et al 2013 Microbial utilisation of biochar-derived carbon. *Science of the Total Environment* 465, 288–297.

Feng X, and Simpson MJ 2008 Temperature responses of individual soil organic matter components. *Journal of Geophysical Research-Biogeosciences* 113, G03036.

Foereid B, Lehmann J, and Major J 2011 Modeling black carbon degradation and movement in soil. *Plant Soil* 345, 223–236.

Gavin DG, Brubaker LB, and Lertzman KP 2003 An 1800-year record of the spatial and temporal distribution of fire from the west coast of Vancouver Island, Canada. *Canadian Journal of Forest Research* 33(4), 573–586.

Glaser B, Balashov E, Haumaier L, Guggenberger G, and Zech W 2000 Black carbon in density fractions of anthropogenic soils of the Brazilian Amazon region. *Organic Geochemistry* 31, 669–678.

Hamer U, Marschner B, Brodowski S, and Amelung W 2004 Interactive priming of black carbon and glucose mineralisation. *Organic Geochemistry* 35, 823–830.

Hammes K, Torn MS, Lapenas AG, and Schmidt MWI 2008 Centennial black carbon turnover observed in a Russian steppe soil. *Biogeosciences Discussion* 5, 661–683.

Harris PJF 2005 New perspectives on the structure of graphitic carbons. *Critical Reviews in Solid State and Materials Sciences* 30, 235–253.

Harvey OR, et al 2012 An index-based approach to assessing recalcitrance and soil carbonsequestration potential of engineered black carbons (Biochars). *Environmental Science and Technology* 46, 1415–1421.

Hata T, Imamura Y, Kobayashi E, Yamane K, and Kikuchi K 2000 Onion-like graphitic particles observed in wood charcoal. *Journal of Wood Science* 46, 89–92.

Herath HMSK, et al 2015 Experimental evidence for sequestering C with biochar by avoidance of CO_2 emissions from original feedstock and protection of native soil organic matter. *Global Change Biology Bioenergy* 7, 512–526.

Hilscher A, Heister K, Siewert C, and Knicker H 2009 Mineralisation and structural changes during the initial phase of microbial degradation of pyrogenic plant residues in soil. *Organic Geochemistry* 40, 332–342.

Huggett RJ 1998 Soil chronosequences, soil development, and soil evolution: A critical review. *Catena* 32, 155–172.

Islam MU, Jiang F, Guo Z, and Peng X 2021 Does biochar application improve soil aggregation? A meta-analysis. *Soil and Tillage Research* 209, 104926.

Jiang X, Haddix ML, and Cotrufo MF 2016 Interactions between biochar and soil organic carbon decomposition: Effects of nitrogen and low molecular weight carbon compound addition. *Soil Biology and Biochemistry* 100, 92–101.

Jing, F., Sun, Y., Liu, Y., Wan, Z., Chen, J. and Tsang, D.C., 2022. Interactions between biochar and clay minerals in changing biochar carbon stability. *Science of The Total Environment*, 809, 151124.

Joseph SD, et al 2010 An investigation into the reactions of biochar in soil. *Australian Journal of Soil Research* 48, 501–515.

Joseph SD, et al 2013 Shifting paradigms: Development of high-efficiency biochar fertilizers based on nano-structures and soluble components. *Carbon Management* 4, 323–343.

Joseph S, et al 2021 How biochar works, and when it doesn't: A review of mechanisms controlling soil and plant responses to biochar. *Global Change Biology - Bioenergy* 13, 1731–1764.

Kaiser K, and Guggenberger G 2008 Mineral surfaces and organic matter. *European Journal of Soil Science* 54, 219–236.

Keiluweit M, Nico PS, Johnson MG, and Kleber M 2010 Dynamic molecular structure of plant biomass-derived black carbon (biochar). *Environmental Science and Technology* 44, 1247–1253.

Kercher AK, and Nagle DC 2003 Microstructural evolution during charcoal carbonization by X-ray diffraction analysis. *Carbon* 41, 15–27.

King JY, Brandt LA, and Adair EC 2012 Shedding light on plant litter decomposition: Advances, implications and new directions in understanding the role of photodegradation. *Biogeochemistry* 111, 57–81.

Knicker H, González-Vila FJ, and González-Vázquez R 2013 Biodegradability of organic matter in fire-affected mineral soils of Southern Spain. *Soil Biology and Biochemistry* 56, 31–39.

Knicker H 2011 Pyrogenic organic matter in soil: Its origin and occurrence, its chemistry and survival in soil environments. *Quaternary International* 243, 251–263.

Koele N, et al 2017 Amazon Basin forest pyrogenic carbon stocks: First estimate of deep storage. *Geoderma* 306, 237–243.

Krull ES, Swanston CW, Skjemstad JO, and McGowan JA 2006 Importance of charcoal in determining the age and chemistry of organic carbon in surface soils. *Journal of Geophysical Research* 111, G04001.

Kuzyakov Y, Subbotina I, Chen H, Bogomolova I, and Xu X 2009 Black carbon decomposition and incorporation into microbial biomass estimated by 14C labeling. *Soil Biology and Biochemistry* 41, 210–219.

Kuzyakov Y, Bogomolova I, and Glaser B 2014 Biochar stability in soil: Decomposition during eight years and transformation as assessed by compound-specific ^{14}C analysis. *Soil Biology and Biochemistry* 70, 229–236.

Lavallee JM, et al 2019 Selective preservation of pyrogenic carbon across soil organic matter fractions and its influence on calculations of carbon mean residence times. *Geoderma* 354, 113866.

Lefebvre D, et al 2020 Modelling the potential for soil carbon sequestration using biochar from sugarcane residues in Brazil. *Scientific Reports* 10, 19479.

Lehmann J, and Sohi S 2008 Comment on "fire-derived charcoal causes loss of forest humus". *Science* 321, 1295.

Lehmann J, et al 2005 Near-edge X-ray absorption fine structure (NEXAFS) spectroscopy for mapping nano-scale distribution of organic carbon forms in soil: Application to black carbon particles. *Global Biogeochemical Cycles* 19, GB1013.

Lehmann J, et al 2008 Australian climate–carbon cycle feedback reduced by soil black carbon. *Nature Geoscience* 1, 832–835.

Lehmann J, et al 2011 Biochar effects on soil biota – a review. *Soil Biology and Biochemistry* 43, 1812–1836.

Lehmann J, and Kleber M 2015 The contentious nature of soil organic matter. *Nature* 528, 60–68.

Lehmann J, et al 2015 Persistence of biochar in soil. In: Lehmann J, and Joseph S (Eds) *Biochar for Environmental Management: Science, Technology and Implementation*. Oxford: Routledge. pp 235–282.

Lehmann J, et al 2021 Biochar in climate change mitigation. *Nature Geoscience* 14, 883–892.

Li F, Cao X, Zhao L, Wang J, and Ding Z 2014 Effects of mineral additives on biochar formation: carbon retention, stability, and properties. *Environmental Science and Technology* 48, 11211–11217.

Li H, Lu X, Xu Y, and Liu H 2019 How close is artificial biochar aging to natural biochar aging in fields? A meta-analysis. *Geoderma* 352, 96–103.

Liang B, et al 2008 Stability of biomass-derived black carbon in soils. *Geochimica et Cosmochimica Acta* 72, 6069–6078.

Liang B, et al 2010 Black carbon affects the cycling of non-black carbon in soil. *Organic Geochemistry* 41, 206–213.

Liu S, et al 2020 Mineral-ions modified biochars enhance the stability of soil aggregate and soil carbon sequestration in a coastal wetland soil. *Catena* 193, 104618.

Luo Y, Durenkamp M, De Nobili M, Lin Q, and Brookes PC 2011 Short term soil priming effects and the mineralisation of biochar following its incorporation to soils of different pH. *Soil Biology and Biochemistry* 43, 2304–2314.

Luo Y, Lin Q, Durenkamp M, Dungait AJ, and Brookes PC 2017 Soil priming effects following substrates addition to biochar-treated soils after 431 days of pre-incubation. *Biology and Fertility of Soils* 53, 315–326.

Maestrini B, et al 2014 Carbon losses from pyrolysed and original wood in a forest soil under natural and increased N deposition. *Biogeosciences Discussions* 11, 1–31.

Major J, Lehmann J, Rondon M, and Goodale C 2010 Fate of soil-applied black carbon: downward migration, leaching and soil respiration. *Global Change Biology* 16, 1366–1379.

Manyà JJ, Ortigosa MA, Laguarta S, and Manso JA 2014 Experimental study on the effect of pyrolysis pressure, peak temperature, and particle size on the potential stability of vine shoots-derived biochar. *Fuel* 133, 163–172.

Mao J-D, et al 2012 Abundant and stable char residues in soils: Implications for soil fertility and carbon sequestration. *Environmental Science and Technology* 46, 9571–9576.

Masiello CA 2004 New directions in black carbon organic geochemistry. *Marine Chemistry* 92, 201–213.

Masiello CA, and Druffel ERM 1998 Black carbon in deep-sea sediments. *Science* 280, 1911–1913.

McBeath AV, and Smernik RJ 2009 Variations in the degree of aromatic condensation of chars. *Organic Geochemistry* 40, 1161–1168.

McBeath AV, Smernik RJ, Schneider MP, Schmidt MWI, and Plant EL 2011 Determination of the aromaticity and the degree of aromatic condensation of a thermosequence of wood charcoal using NMR. *Organic Geochemistry* 42, 1194–1202.

McBeath AV, Smernik RJ, and Krull ES 2013 A demonstration of the high variability of chars produced from wood in bushfires. *Organic Geochemistry* 55, 38–44.

McBeath AV, Smernik RJ, Krull ES, and Lehmann J 2014 The influence of feedstock and production temperature on biochar carbon chemistry: A solid-state ^{13}C NMR study. *Biomass and Bioenergy* 60, 121–129.

McBeath AV, Wurster CM, and Bird MI 2015 Influence of feedstock properties and pyrolysis conditions on biochar carbon stability as determined by hydrogen pyrolysis. *Biomass and Bioenergy* 73, 155–173.

Mondini C et al 2017 Modification of the RothC model to simulate soil C mineralization of exogenous organic matter., *Biogeosciences* 14, 3253–3274.

Mukherjee A, and Zimmerman AR 2013 Organic carbon and nutrient release from a range of laboratory-produced biochars and biochar–soil mixtures. *Geoderma* 193, 122–130.

Murage EW, Voroney P, and Beyaert RP 2007 Turnover of carbon in the free light fraction with and without charcoal as determined using the ^{13}C natural abundance method. *Geoderma* 138, 133–143.

Nan H, et al 2018 Interaction of inherent minerals with carbon during biomass pyrolysis weakens biochar carbon sequestration potential. *ACS Sustainable Chemistry and Engineering* 7, 1591–1599.

Nan H, et al 2021 Pyrolysis temperature-dependent carbon retention and stability of biochar with participation of calcium: Implications to carbon sequestration. *Environmental Pollution* 287, 117566.

Nguyen B, and Lehmann J 2009 Black carbon decomposition under varying water regimes. *Organic Geochemistry* 40, 846–853.

Nguyen B, Lehmann J, Kinyangi J, Smernik R, and Engelhard MH 2008 Long-term black carbon dynamics in cultivated soil. *Biogeochemistry* 89, 295–308.

Nguyen B, Lehmann J, Hockaday WC, Joseph S, and Masiello C 2010 Temperature sensitivity of black carbon decomposition and oxidation. *Environmental Science and Technology* 44, 3324–3331.

Ogle SM, et al 2019 *Refinement to the 2006 IPCC Guidelines for National Greenhouse Gas Inventories*. In: Calvo Buendia E, et al (Eds) Intergovernmental Panel on Climate Change: Switzerland; Vol. IV.

Ohlson M, and Tryterud E 2000 Interpretation of the charcoal record in forest soils: Forest fires and their production and deposition of macroscopic charcoal. *The Holocene* 10, 519–525.

Ouyang L, Wang F, Tang J, Yu L, and Zhang R 2013 Effects of biochar amendment on soil aggregates and hydraulic properties. *Journal of Soil Science and Plant Nutrition* 13, 991–1002.

Paris O, Zollfrank C, and Zickler GA 2005 Decomposition and carbonisation of wood biopolymers—a microstructural study of softwood pyrolysis. *Carbon* 43, 53–66.

Parton WJ, and Rasmussen PE 1994 Long-term effects of crop management in wheat-fallow: II. CENTURY model simulations. *Soil Science Society of America Journal* 58, 530–536.

Pokharel P, Ma Z, and Chang SX 2020 Biochar increases soil microbial biomass with changes in extra-and intracellular enzyme activities: A global meta-analysis. *Biochar* 2, 65–79.

Potter MC 1908 Bacteria as agents in the oxidation of amorphous carbon. *Proceedings of the Royal Society of London B* 80, 239–250.

Preston CM, and Schmidt MWI 2006 Black (pyrogenic) carbon: A synthesis of current knowledge and uncertainties with special consideration of boreal regions. *Biogeosciences* 3, 397–420.

Pulcher R, Balugani E, Ventura M, Greggio N, and Marazza D 2022 Inclusion of biochar in a C dynamics model based on observations from an 8-year field experiment. *SOIL* 8, 199–211.

Rahman MT, Guo ZC, Zhang ZB, Zhou H, and Peng XH 2018 Wetting and drying cycles improving aggregation and associated C stabilization differently after straw or biochar incorporated into a Vertisol. *Soil and Tillage Research* 175, 28–36.

Rumpel C, and Kögel-Knabner I 2011 Deep soil organic matter—a key but poorly understood component of terrestrial C cycle. *Plant and Soil* 338, 143–158.

Saiz G, et al 2014 Charcoal re-combustion efficiency in tropical savannas. *Geoderma* 219-220, 40–45.

Santin C, Doerr SH, Preston C, and Bryant R 2013 Consumption of residual pyrogenic carbon by wildfire. *International Journal of Wildland Fire* 22, 1072–1077.

Santín C, Doerr SH, Preston CM, and González-Rodríguez G 2015 Pyrogenic organic matter production from wildfires: A missing sink in the global carbon cycle. *Global Change Biology* 21, 1621–1633.

Santos F, Torn MS, and Bird JA 2012 Biological degradation of pyrogenic organic matter in temperate forest soils. *Soil Biology and Biochemistry* 51, 115–124.

Schiedung M, Bellè SL, Sigmund G, Kalbitz K, and Abiven S 2020 Vertical mobility of pyrogenic organic matter in soils: A column experiment. *Biogeosciences* 17, 6457–6474.

Schimmelpfennig S, and Glaser B 2012 One step forward toward characterization: Some important material properties to distinguish biochars. *Journal of Environmental Quality* 41, 1001–1013.

Schmidt MWI, et al 2011 Persistence of soil organic matter as an ecosystem property. *Nature* 478, 49–56.

Schneider M, Lehmann J, and Schmidt MWI 2011 Charcoal quality does not change over a century in a tropical agro-ecosystem. *Soil Biology and Biochemistry* 43, 1992–1994.

Sheng Y, Zhan Y, and Zhu L 2016 Reduced carbon sequestration potential of biochar in acidic soil. *Science of the Total Environment* 572, 129–137.

Shindo H 1991 Elementary composition, humus composition, and decomposition in soil of charred grassland plants. *Soil Science and Plant Nutrition* 37, 651–657.

Shneour EA 1966 Oxidation of graphite carbon in certain soils. *Science* 151, 991–992.

Sigua GC, et al 2014 Carbon mineralization in two ultisols amended with different sources and particle sizes of pyrolyzed biochar. *Chemosphere* 103, 313–321.

Singh BP, Cowie AL, and Smernik RJ 2012a Biochar stability in a clayey soil as a function of feedstock and pyrolysis temperature. *Environmental Science and Technology* 46, 11770–11778.

Singh N, Abiven S, Torn MS, and Schmidt MWI 2012b Fire-derived organic carbon in soil turns over on a centennial scale. *Biogeosciences* 9, 2847–2857.

Singh BP, et al 2015 In situ persistence and migration of biochar carbon and its impact on native carbon emission in contrasting soils under managed temperate pastures. *PLoS One* 10, e0141560.

Singh B, Camps-Arbestain M, and Lehmann J (Eds) 2017 *Biochar: A Guide to Analytical Methods.* Brisbane: CRC Press, CSIRO Publishing.

Six J, and Jastrow J 2002 Organic matter turnover. In: Lal R (Ed) *Encyclopedia of Soil Science.* New York, USA: Marcel Dekker. pp 936–942.

Six J, Bossuyt H, Degryze S, and Denef K 2004 A history of research on the link between (micro) aggregates, soil biota, and soil organic matter dynamics. *Soil and Tillage Research* 79, 7–31.

Skjemstad JO, Clarke P, Taylor JA, Oades JM, and McClure SG 1996 The chemistry and nature of protected carbon in soil. *Soil Research* 34, 251–271.

Skjemstad JO, Spouncer LR, Cowie B, and Swift RS 2004 Calibration of the Rothamsted organic carbon turnover model (RothC ver. 26.3), using measurable soil organic carbon pools. *Australian Journal of Soil Research* 42, 79–88.

Soucémarianadin L, et al 2019 Heterogeneity of the chemical composition and thermal stability of particulate organic matter in French forest soils. *Geoderma* 342, 65–74.

Spokas KA 2010 Review of the stability of biochar in soils: Predictability of O: C molar ratios. *Carbon Management* 1, 289–303.

Spokas KA, and Reicosky DC 2009 Impacts of sixteen different biochars on soil greenhouse gas production. *Annals of Environmental Science* 3, 179–193.

Spokas KΛ, et al 2014 Physical disintegration of biochar: An overlooked process. *Environmental Science and Technology Letters* 1, 326–332.

Taherymoosavi S, Joseph S, and Munroe P 2016 Characterization of organic compounds in a mixed feedstock biochar generated from Australian agricultural residues. *Journal of Analytical and Applied Pyrolysis* 120, 441–449.

Tisdall JM, and Oades JM 1982 Organic matter and water-stable aggregates in soils. *Journal of Soil Science* 33, 141–163.

Topoliantz S and Ponge J F 2005 Charcoal consumption and casting activity by *Pontoscolex corethrurus* (Glossoscolecidae). *Applied Soil Ecology* 28, 217–224.

Torres-Rojas D, et al 2020 Nitrogen speciation and transformations in fire-derived organic matter. *Geochimica et Cosmochimica Acta* 276, 179–185.

Van Zwieten L, et al 2010 Effects of biochar from slow pyrolysis of papermill waste on agronomic performance and soil fertility. *Plant and Soil* 327, 235–246.

Vanek SJ, and Lehmann J 2015 Phosphorus availability to beans via interactions between mycorrhizas and biochar. *Plant and Soil* 395, 105–123.

Vasilyeva NA, et al 2011 Pyrogenic carbon quantity and quality unchanged after 55 years of organic matter depletion in a Chernozem. *Soil Biology and Biochemistry* 43, 1985–1988.

Wang T, Camps-Arbestain M, and Hedley M 2013 Predicting C aromaticity of biochars based on their elemental composition. *Organic Geochemistry* 62, 1–6.

Wang T, et al 2014 Determination of carbonate-C in biochars. *Soil Research* 52, 495–504.

Wang R, et al 2017 Photooxidation of pyrogenic organic matter reduces its reactive, labile C pool and the apparent soil oxidative microbial enzyme response. *Geoderma* 293, 10–18.

Wang H, Nan Q, Waqas M, and Wu W 2022 Stability of biochar in mineral soils: Assessment methods, influencing factors and potential problems. *Science of the Total Environment* 806, 150789.

Wardle DA, Nilsson MC, and Zackrisson O 2008 Fire-derived charcoal causes loss of forest humus. *Science* 320, 629.

Weng H, et al 2017 Biochar built soil carbon over a decade by stabilizing rhizodeposits. *Nature Climate Change* 7, 371–376.

Wengel M, Kothe E, Schmidt CM, Heide K, and Gleixner G 2006 Degradation of organic matter from black shales and charcoal by the wood-rotting fungus Schizophyllum commune and release of DOC and heavy metals in the aqueous phase. *Science of the Total Environment* 367, 383–393.

Whitman T, Hanley K, Enders A, and Lehmann J 2013 Predicting pyrogenic organic matter mineralization from its initial properties and implications for carbon management. *Organic Geochemistry* 64, 76–83.

Whitman T, and Lehmann J 2015 A dual-isotope approach to allow conclusive partitioning between three sources. *Nature Communications* 6, 8708.

Wiedemeier DB, et al 2015 Aromaticity and degree of aromatic condensation of char. *Organic Geochemistry* 78, 135–143.

Woo SH, Enders A, and Lehmann J 2016 Microbial mineralization of pyrogenic organic matter in different mineral systems. *Organic Geochemistry* 98, 18–26.

Woolf D, and Lehmann J 2012 Modelling the long-term response to positive and negative priming of soil organic carbon by black carbon. *Biogeochemistry* 111, 83–95.

Woolf D, et al 2021 A greenhouse gas inventory model for biochar additions to soil. *Environmental Science and Technology* 55, 14795–14805.

Xu X, et al 2017 Indispensable role of biochar-inherent mineral constituents in its environmental applications: A review. *Bioresource Technology* 241, 887–899.

Yang F, Zhao L, Gao B, Xu X, and Cao X 2016 The interfacial behavior between biochar and soil minerals and its effect on biochar stability.

Environmental Science and Technology 50, 2264–2271.

Yang Y, et al 2018a Effect of minerals on the stability of biochar. *Chemosphere* 204, 310–317.

Yang F, et al 2018b Kaolinite enhances the stability of the dissolvable and undissolvable fractions of biochar via different mechanisms. *Environmental Science and Technology* 52, 8321–8329.

Yang Y, et al 2022 Biochar stability and impact on soil organic carbon mineralization depend on biochar processing, aging and soil clay content. *Soil Biology and Biochemistry* 169, 108657.

Zeba N, Berry TD, Panke-Buisse K, and Whitman T 2022 Effects of physical, chemical, and biological ageing on the mineralization of pine wood biochar by a Streptomyces isolate. *PloS one* 17, e0265663.

Zhang J, et al 2021. Biological calcium carbonate with a unique organic–inorganic composite structure to enhance biochar stability. *Environmental Science: Processes and Impacts* 23, 1747–1758.

Zimmerman A 2010 Abiotic and microbial oxidation of laboratory-produced black carbon (biochar). *Environmental Science and Technology* 44, 1295–1301.

Zimmerman AR, and Gao B 2013 The stability of biochar in the environment. In: Ladygina N, and Rineau F (Eds) *Biochar and Soil Biota*. Boca Raton, USA: CRC Press. pp 1–40.

Zimmerman AR, and Ouyang L 2019 Priming of pyrogenic C (biochar) mineralization by dissolved organic matter and vice versa. *Soil Biology and Biochemistry* 130, 105–112.

Zimmermann M, et al 2012 Rapid degradation of pyrogenic carbon. *Global Change Biology* 18, 3306–3316.

12

Biochar transport in terrestrial ecosystems

Fate and impact

Cornelia Rumpel

Introduction

Depending on their formation conditions, most biochar types are characterized by high stability against microbial decomposition, and consequently, high persistence approaching most probably several centuries (Chapter 11). Biochar may be subject to slow mineralization by microorganisms and in addition, it can also be affected by processes leading to its loss from the site of deposition at much shorter timescales. Indeed, biochar and associated nutrients or micropollutants may be transported by water, wind, or animals. Wind and water may be able to induce long-distance transport of biochar and other pyrogenic carbonaceous material (PMC) through landscapes and also may lead to its ultimate storage in ocean sediments (Coppola et al, 2022).

As fresh biochar is characterized by an absence of interaction with soil minerals at the time of application (Czimczik and Masiello, 2007), it is prone to export by leaching or erosion processes, similar to those affecting free particulate organic matter derived from plant material such as crop residues or composts. In the first few years after its application, removal of biochar off-site may be more important in terms of quantitative biochar fluxes than microbial degradation (Major et al, 2010; Foereid, et al, 2011). Transport can occur via vertical mobilization within the soil profile as well as off-site horizontal or lateral export.

Transport processes of biochar have several controls (Figure 12.1) and they are likely altered by biochar weathering after its field exposure. They may be controlled by biochar's inherent physical and chemical properties, such as particle size and density (Chapter 5) and hydrophobicity and sorptive properties (Chapters 6 and 10). In addition, biochar susceptibility to transport is likely affected by pedoclimatic conditions, such as soil type, climate, slope gradient, and biological activity.

The mode of biochar application may play an important role in its removal from the site of deposition. Examples of mode of

DOI: 10.4324/9781003297673-12

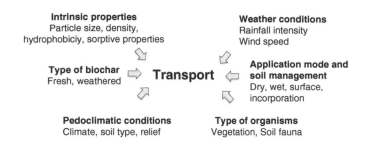

Figure 12.1 *Factors affecting biochar transport*

application and soil conditions favoring biochar transport include the application of biochar as slurry, which may favor vertical transport, particularly in sandy soils, while dry biochar application to clayey soils may favor off-site transport by wind or water erosion. On biologically active sites, biochar may be subject to vertical transport by soil fauna independent of its mode of application. Under humid climate conditions, water erosion may be the main process for off-site biochar export, especially in terrain with steep slopes. In dry environments, wind erosion may be an important transport mechanism. Agricultural management operations may influence biochar transport processes favoring (1) its mobilization (i.e., plowing); and (2) its retention, for example through mixture with other (organic) materials.

Processes leading to material removal from soil may interact with each other, and they are scale-dependent. Therefore, the movement and fate of biochar and other organic matter types, including PMC, in the environment is a complex issue (Lal, 2003; Kuhn et al, 2009). Transport of biochar away from the site of its deposition

does not systematically lead to loss of persistence (i.e., resistance to degradation), as these processes may or may not affect its susceptibility to biological, physical, and/or chemical degradation processes leading to its mineralization. Thus, biochar carbon that is subjected to transport processes may still be withheld from the atmosphere in the medium/long-term, but if lost from the site, it (1) may not anymore be accounted for through carbon credits; and (2) does not lead anymore to soil improvement at the site of its deposition. Depending on the transport processes, biochar will be deposited at various places and contrasting distances from the removal site. Off-site effects of biochar will depend on the deposition site and should be specifically evaluated and taken into account when considering the overall effects of biochar transport.

This chapter summarizes published scientific literature on the physicochemical and biological processes affecting biochar removal from the site of its deposition and its controls. A conceptual overview of the role of biochar transport for its fate in the environment will be presented.

Biochar transport processes, directions, and agents

Physical processes affecting biochar after its deposition are drying and re-wetting or

freezing and thawing processes, which may not only affect biochar friability, surface chemistry, and residence time (Spokas et al, 2014, Naisse et al., 2015) but may also facilitate vertical biochar transport through the formation of pores and/or cracks.

Transport rates at the plot scale

In agricultural experiments, significant downward transport of biochar with transport rates of up to 30 mm a^{-1} and accumulation of biochar in deeper layers was observed under high rainfall in arable soil and grassland soil (Major et al, 2010; Felber et al, 2014). Due to these high transport rates, biochar removal from the upper soil layer can be extremely elevated, accounting for 40–70% of the applied biochar in arable systems under temperate climate conditions (Dong et al, 2017; Nguyen et al, 2009; Haefele et al, 2011).

Transport rates are strongly affected by soil type (Singh et al, 2015) in combination with climate or occurrence of heavy rainfall events (Haefele et al, 2011). During a one-year field experiment where lateral movement was prevented, downward migration of biochar was contrasting for an Arenosol, a Cambisol, and a Ferralsol (Singh et al, 2015). The highest recovery of biochar in the application layer 0–0.08 m was noted for the Cambisol, while migration of 45–50% of the applied biochar was recorded for the two other soil types. Strong differences in migration depth and intensity were observed. In the Ferralsol, all removed material accumulated within 0.5 m depth, while the other two soils showed deposition of small amounts of biochar in the 0.3–0.5 m layer. In the Arenosol, only 80% of the biochar was recovered in the 0–0.5 m profile. After the integration of mineralization losses, these data suggested that 18% of the biochar was lost by vertical migration below 0.5 m depth in this soil type. The losses through vertical migration exceeded mineralization losses for the Arenosol and the Ferralsol but not the Cambisol (Singh et al, 2015). The authors concluded that it is important to understand the interaction mechanisms between biochar and soil to maximize the long-term carbon stabilization potential of biochar.

Transport mechanisms at the plot scale

Several processes can explain biochar transport into deeper soil layers. The simplest explanation is that particles move in the soil column through larger pores or are transported in the water column of these pores. Biochar can have a wide range of diameters (Chapter 5) and accordingly, soil macro- and mesopores in the mm to large μm range may be possible pathways. This mechanism may play a more important role in soils with higher proportions of macropore volume (e.g., sandy mineral soils, organic soils), in preferential flow pathways, and vertic soils with the temporary formation of cracks. In contrast, soils having a high abundance of meso- and micropores and or large amounts of reactive minerals may favor biochar retention (Singh et al, 2015). Biochar particle properties that influence their mobility are size, surface charge (measured as ζ potential) properties, and hydrophilicity. Moreover, pH, ionic strength, and water saturation of the transport media control biochar mobility (Zhang et al, 2010). Biochar retention may be favored in unsaturated media with low pH and high ionic strength, which is attributed to weakened electrostatic repulsion between particles. Larger particles (μm range) may be more affected by mechanical filtration whereas smaller particles may be more affected by solution chemistry. The preferential retention of larger particles could result in a vertical stratification of particle sizes (Zhang et al, 2010).

Vertical biochar transport is also dependent on the form of biochar-C, as, for example, solid particles are transported less effectively than dissolved forms. Indeed, the latter is important for biochar transport especially right after its deposition (Santos et al, 2022). The water extractable organic C (WEOC) originating from 46 biochars with contrasting feedstocks and pyrolysis conditions indicated that WEOC may represent $0.5–40$ mg g^{-1} biochar, the amount being influenced by pyrolysis conditions and feedstock (Liu et al., 2019). Due to the production of condensable vapors (Bridgwater, 2012), fast pyrolysis biochars have higher WEOC concentrations than slow pyrolysis biochars (Bruun et al, 2012; Liu et al., 2017). It has been pointed out that the WEOC originating from biochar contains truly dissolved organic C as well as colloidal material defined as particles with a size range between 0.45 and 1 μm (Wagner et al, 2017). The WEOC compounds composing truly dissolved organic matter consist of small aromatic clusters with a high degree of carboxylic and phenol groups conferring a high polarity and reactivity to this material, which is consequently highly reactive and susceptible to transport in aqueous media (Liu et al., 2019). A recent study indicated that the de-condensation of aromatic compounds could be involved in the WEOC recovered from biochar-amended soils (Braun et al, 2020). WEOC material may be mineralized, accumulate in subsoil, or may be exported to aquatic systems. A recent 2-year laboratory study found that pyrolysis temperature and soil properties may interact to control the vertical transport of dissolved compounds released from biochar (Santos et al, 2022). In a column experiment, Schiedung et al (2020) showed that the vertical mobility of pyrogenic carbon (PyC) after its deposition is limited to a small fraction and was affected by soil type and its degree of oxidation. In particular, the adsorption of PyC occurred in loamy soil, while in sandy soil PyC migration also mobilized native soil organic C (Schieldung et al,

2020). Below 0.3 m depth, biochar accumulation may increase water retention and improve root growth in sandy soils (Bruun et al, 2014). However, PyC accumulations were mainly found in subsoil horizons below the A horizon (0.05–0.1 m depth) in soils rich in short-range order minerals (Diekow et al, 2005; Rumpel et al., 2006), whereas sandy soils did not show any presence of PyC below the A horizon (Alexis et al, 2012). As discussed above, biochar transport and its accumulation in different parts of the soil profile is most probably strongly dependent on soil type and pedoclimatic conditions. These should be determined to evaluate field application rates. Field trials with contrasting agricultural practices are needed to evaluate biochar burial and its importance for long-term carbon sequestration (Button et al, 2022).

In recent years, the behavior of colloidal biochar nanoparticles has received attention, despite their small portion in fresh biochar. Indeed, these small particles are reactive and may be subject to horizontal leaching into surface water similar to PyC in a fire-impacted watershed (Hockaday et al, 2007). In addition, these particles may be released through biochar aging (Sorrenti et al, 2016; de la Rosa, 2018; Chapter 10). Water-dispersible biochar colloids had a high persistence and could form and remain in natural waters and soil solutions, suggesting a risk for long-distance migration (Fang et al, 2020). This poses a possible environmental threat as biochar may be intrinsically enriched in pollutants, such as polycyclic aromatic hydrocarbons, potentially toxic elements, dioxins, volatile organic compounds, and emerging contaminants (e.g., persistent free radicals, metal cyanide) (Han et al, 2022). Biochar may also have a high sorption affinity for environmental contaminants (Haider et al, 2022). Mobile biochar colloids could thus lead potentially to the widespread dissemination of pollutants (Hmeed et al, 2021). The mobility of colloidal biochar nanoparticles in saturated soils, such as those in paddy rice systems, can be significant depending on both their intrinsic properties and the environmental conditions (ionic strength, pH, and natural organic matter) of the site of deposition (Chen et al, 2017).

Bioturbation by organisms has frequently been suggested as a possible mechanism behind the transport of PCM (Skjemstad et al, 1999; Carcaillet 2001; Haefele et al, 2011) and may thus also affect biochar mobility. In recent years some of these hypotheses were tested in laboratory experiments (e.g., Elmer et al, 2015; Ali et al, 2022). The organisms that could influence biochar transport are earthworms and termites, known as bioengineers to be strongly involved in bioturbation. In particular, earthworms were shown to incorporate particulate organic matter deep into mineral soils (Don et al, 2008). It has therefore been suggested that earthworms might act as vehicles for the delivery of biochar into deeper soil (Elmer et al, 2015). A laboratory experiment with *Lumbricus terrestris*, a species belonging to the anecic ecological category that establishes vertical burrows, showed that eight biochar types were removed actively from the soil surface through bioturbation (Elmer et al, 2015). Biochar removal from the site of deposition ranged between 18 and 100% for the eight biochars under study with preferential removal of aged biochar and the least removal of fast-pyrolysis biochar from hardwood sawdust. Although the dependence of these processes on biochar application mode was not studied, these data indicate that earthworms may be important transport vectors in biologically active soils in humid environments. The importance of different earthworm species from other ecological categories for biochar incorporation

into the mineral soil and their role in controlling biochar fate under field conditions has yet to be investigated.

In arid environments, earthworms are absent, and termites may be involved in the transport of biochar. This hypothesis was tested by Ali et al (2022), who showed that depending on the termite species, biochar particles could be transported more than 4 cm in 10 days in a horizontal direction. They showed a preference for biochar enriched with cattle slurry, presumably due to nutritional benefits.

Together, the evidence from field and laboratory work indicates that initial mean PyC and biochar transport rates at the plot scale are in the order of a few mm up to several cm or m per year. This does not mean, however, that all biochar is subject to these transport rates. In fact, the transport of biochar is strongly affected by its intrinsic properties (Table 12.1), and the environmental factors at the site of its deposition. The transport speed and intensity likely change over time as the species of mobilized biochar change (Figure 12.2). Whereas particulate fractions may be transported preferentially directly after the application of biochar, the increased oxidation, solubility, and fragmentation of particles together with preferential retention of larger particles in the upper soil horizons will favor the transport of dissolved biochar, colloidal material, and nanoparticles to greater depths over time. The transport of biochar is strongly affected by its ability to be integrated into soil aggregates or to interact with soil minerals. Therefore, the effect and controls of these associations on biochar mobility in the soil must further be determined and evaluated quantitatively. Also, the role of soil fauna as transport agents for biochar needs to be further explored (Domene, 2016).

Table 12.1 *Processes affecting loss of different biochar fractions through vertical transport at profile scale and off-site transport in horizontal and lateral direction*

Scale	Process	Preferentially affected fraction
Soil profile (vertical movement)	Infiltration	Small particle size/colloidal
	Bioturbation	All fractions
	Colloidal transport	Colloidal
	Leaching	Soluble fraction
Off site (horizontal movement)	Interrill erosion	All fractions
	Rill or gully erosion	All fractions
	Sheet erosion	All fractions
	Tillage erosion	All fractions
	Wind erosion	Light fraction, small particle size
	Rain splash	Light fraction
	Waterflow	Soluble fraction

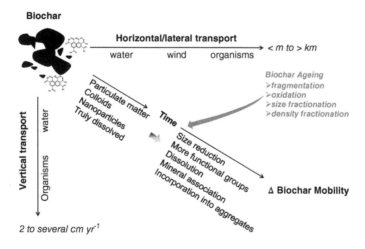

Figure 12.2 *Conceptual sketch of the vertical and horizontal transport of biochar in time and space*

Horizontal and lateral transport of biochar through landscapes

PCM is affected by wind and water erosion leading to its horizontal/lateral redistribution in landscapes and its export to waterbodies and ultimate deposition in freshwater and marine sediments (Coppola et al, 2022). Off-site loss of aromatic material such as PyC and biochar by erosion may be more important than previously thought. Field experiments, as well as modeling studies, have confirmed that erosion may lead to a major export flux of these materials from their site of deposition, affecting their stocks as well as their residence time in the soil system (Major et al, 2010; Foereid et al, 2011; Abney and Berhe, 2018). As erosion mainly concerns the upper soil surface, where organic matter (and biochar) concentrations are highest, eroded sediments are usually enriched in carbon (and biochar) (Rumpel et al, 2006). While wind erosion occurs mainly in arid and semi-arid areas and affects the lightest particles deposited on the soil surface (Webb et al, 2019), water erosion may occur in all climatic zones after strong rainfall events. In semi-arid regions, both processes may interact (Tuo et al, 2016).

Transport rates

Rainfall experiments with PyC under field conditions have shown that PyC is prone to water erosion and extremely mobile immediately after its deposition, with erosion affecting up to 60% of the added PyC (Rumpel et al, 2009; Bellè et al, 2021). Under the controlled conditions of a rainfall experiment, in total, runoff erosion led to the export of up to 19 g PyC m^2 h^{-1} during the first rainfall after its deposition (Rumpel et al, 2009). A field experiment under tropical conditions indicated that up to 45% of the applied biochar could have been subject to lateral transport during one year, corresponding to a vertical transport rate of 0.12 g PCM m^2 h^{-1} (Obia et al, 2017). The authors suggested that the export rates depended on soil type, with greater lateral transport in an Arenosol with a loamy fine sand texture than in an Acrisol with a sandy loam texture (Obia et al, 2017).

Eroded particulate organic matter from agricultural operations is an emerging air quality issue (Pattey and Qiu, 2012), as it was shown that dust originating from intensive agricultural systems is a major contributor to airborne particulate matter <10 μm (PM_{10}) (Madden et al, 2010), which makes up the "thoracic fraction" penetrating the human respiratory system beyond the larynx. Indeed, PM_{10} emissions from agricultural soils amended with biochar may be of great concern, because of the health risk of inhalation of biochar itself and associated organic or inorganic contaminants (Gelardi et al, 2019). Aerosol release of biochar in agricultural settings can occur during its application to soil or as the result of natural and mechanical disturbance, such as wind or tillage events. Biochar-amended soils might have the potential to increase these emissions (Li et al, 2018) because, at sites with bare soil surfaces, biochar could be more affected by wind erosion than mineral-associated or occluded organic matter types due to its low particle densities and the contribution of very fine particles (Ravi et al, 2016). Indeed, wind tunnel experiments showed that biochar application increased the emissions of PM_{10} by up to 400% by direct removal even at wind speeds below threshold values for soil erosion (Ravi et al, 2020). At higher wind speeds, emissions from biochar increased by 300% even when they did not contain an inhalable fraction (Ravi et al, 2020). However, although quantitatively potentially important, organic matter erosion by wind has largely been neglected up to now and no quantitative information on biochar removal is available for sites on which wind and water erosion occur jointly.

Mechanisms of horizontal and lateral biochar transport

Horizontal and lateral biochar transport by wind and water involves several processes (Figure 12.3) operating at several temporal scales and depending on the intensity of the erosion events (i.e., wind speed or rainfall intensity).

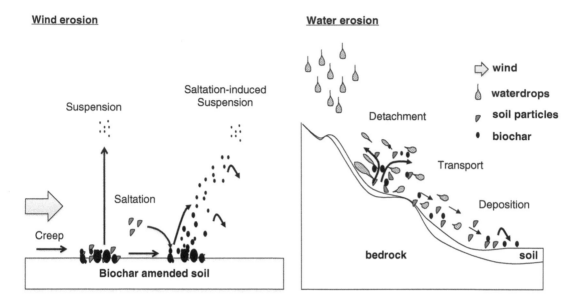

Figure 12.3 *Processes occurring during wind and water erosion of biochar amended soils*

Runoff water transports the finest, lightest particles deposited at the soil surface (Table 12.1). Detachment of particles generally occurs by rainsplash, due to the physical force of raindrops hitting the soil surface (Figure 12.3; Torri and Poesen, 1992). Raindrops impacting the soil surface detach sediments and project them in all directions. A significant part of the detached material may be propelled over a distance of 0.5 m or greater. Thus, locally, rain splash itself can lead to the net transport of soil particles (Kinnell, 1990). For biochar, the kinetic energy of the drops hitting the ground may lead to size fractionation and thus formation of colloidal particles, which are especially susceptible to water transport (Guggenberger et al, 2008). Most mineral particles affected by these processes are redistributed within the plot. However, some soil components, which have a very low settling velocity in water, may be transported over greater distances. This could be the case with biochar. Because of its low density and higher hydrophobicity, it could be easily carried away by runoff water (Wang et al, 2013a; Lee et al, 2018; Peng et al, 2019). Biochar erosion 8 months after its application during heavy rainfall simulation was found to be proportional to its application rate and was dependent on its particle size with coarse particles being less prone to erosion loss (Li et al, 2020). The colloidal nature of the finest biochar particles may also allow them to stay in suspension for longer-range transport (Mulleneers et al, 1999). Indeed, eroded sediments collected at several scales throughout a watershed were found to be more enriched in PCM than other organic matter types (Rumpel et al, 2006; Chaplot et al, 2005). In addition to its particle size (Table 12.1), the horizontal and lateral PCM transport was found to be dependent on its feedstock and also the soil type that received the addition (Bellè et al, 2021). Quantification of material removed by splash erosion showed that this process greatly increased PCM removal from amended plots affected by runoff erosion (Rumpel et al, 2009).

Aeolian transport of soil particles and biochar can be classified into four physical processes (Figure 12.3): creep, during which particles >500 μm are dragged on the soil surface, saltation, which affects particles with a size between 70–500 μm, short-term suspension affecting particles between 20–70 μm size and long-term suspension of particles < 20 μm (Bagnold, 1974; Kok et al, 2012). Saltation-induced particle emission occurs when saltation-sized particles are lifted from the ground and transported at short distances (Bagnold, 1974). When the saltation particle hits the ground, it can collide with other particles and lead to the abrasion of fine material, which may be subject to suspension. Saltation was found to greatly enhance biochar-derived emissions (Ravi et al, 2020), probably because it breaks easily by surface abrasion due to its high friability (Spokas et al, 2014; Le et al, 2020).

Although erosion processes are often thought to lead to biochar loss, this loss may only represent its removal from a given location and its re-deposition elsewhere rather than its mineralization and return of its C to the atmosphere. In fact, water, as well as wind, erosion continuously leads to the removal and deposition of material. In particular, organic matter mobilized by erosion processes, such as creep, saltation, rain splash, and rainwash, travels only short distances and may accumulate in depositional crusts. Export of eroded biochar into the fluvial system may occur through rill or gully erosion, which are non-selective processes leading to soil removal by water flow in rills or gullies. During this transport, eroded organic particles may be subject to degradation processes leading to their mineralization. Quantitative information on the various transport processes is necessary to close the global PCM cycle (Abney et al, 2018; Coppola et al, 2022). This may reduce the uncertainty of global carbon models (Friedlingstein et al, 2014). It must be considered that a high proportion of eroded PCM may not reach the ocean and may instead be mineralized or transported and deposited in terrestrial basins, where its impact and fate need to be assessed (Figures 12.3 and 12.4).

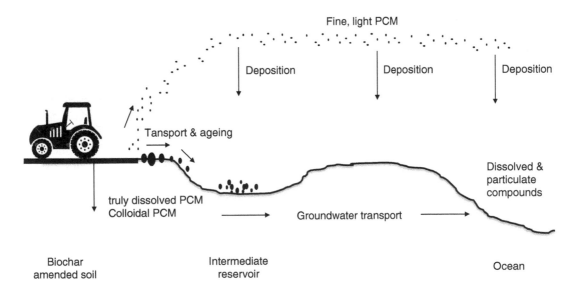

Figure 12.4 *Transport of biochar at the land-water interphase*

Biochar removal, similar to the removal of any organic matter type from its site of formation through erosion, may be affected by many factors such as soil properties, topography, surface cover, and soil wetness. For example, a modeling study with data collected during a rainfall experiment suggested that biochar may be susceptible to export from soils with limited infiltration capacity (Wang et al, 2013b). Moreover, intrinsic biochar properties are important factors influencing its erosion susceptibility. PCM may be subjected to transport selectivity, i.e., leading to its preferential export compared to other organic matter types (Rumpel et al, 2006; Obia et al., 2017; Pyle et al, 2017). However, preferential erosion of PCM with regard to other organic matter types was not reported in all studies (e.g., Güerena et al, 2015). Rather than transport selectivity, rapid PCM removal could be related to the fact that the PCM was deposited on top of the soil surface and therefore is concentrated at the soil surface, where it may be removed before other soil organic matter types.

Wind speed, rainfall intensity, and duration were found to have important effects on organic matter mobilization by erosion processes (Tiefenbacher et al, 2021; Ravi et al, 2014), although these parameters interact strongly with soil moisture and vegetation cover (Meng et al, 2018; Ziadat and Taimeh, 2013). Low-intensity rainfall events preferentially remove particulate organic matter due to low transport capacity (Martinez-Mena et al, 2012) and may lead to biochar loss in sloping terrain. Low wind speeds can remove particles <70 μm which may be present in fresh biochar and/or produced by biochar weathering. High rainfall events generate significant overland runoff that increases sediment and associated organic matter transport, leading to greater net PCM removal. Wind speeds exceeding the minimum threshold velocity leading to saltation (Shao et al, 1993) will enhance biochar-derived emissions through abrasion and particle suspension (Ravi et al, 2016). Overland flow is a key factor controlling the supply and transport of sediment to and by streams (Dietrich and Dunne, 1978). It was also found to control the redistribution of PCM in fire-affected landscapes (Abney and Berhe, 2018; Bellè et al, 2021). These sporadic events may affect both freshly deposited biochar as well as older biochar components already associated with soil particles as a result of the aging process (Czimczik and Masiello, 2007; Liang et al, 2008). In contrast, long-distance aeolian transport affects only the smallest biochar particles < 20 μm having the lowest settling velocities (Gao and Wu, 2014).

Biochar intrinsic properties, in addition to pedoclimatic conditions (see above), determine its susceptibility to wind and water erosion (Table 12.1). Important characteristics are particle size, porosity, and hydrophobicity of biochar, which depend on production conditions and may thus be controllable to some extent. Moreover, the biochar application mode can affect biochar erosion losses (Blanco-Canqui, 2020). Indeed, biochar incorporation into the soil reduces its erosion losses as compared to surface application (Sadeghi et al, 2016). Other strategies to reduce fine particle losses due to (wind) erosion are pelletizing and moistening PCM before application (Maienza et al, 2017; Silva et al, 2015). However, the reduction of particle loss may occur only in the short term, and pelletizing was shown to have negative effects on plant growth and production (Maienza et al, 2017). Co-application of compost or vermicompost and biochar may be another strategy to reduce its losses, possibly through enhancement of aggregate formation (Ngo et al, 2016).

The susceptibility of biochar to transport may change with biochar aging. Indeed, fragmentation and biotic or abiotic oxidation reactions may reduce biochar particle size and generate water-soluble materials from solid biochar after field exposure (Spokas et al, 2014; Liu et al, 2017; Chen et al, 2023, Figure 12.4). The export of small particles and dissolved organic matter from biochar applied to soils might constitute an important loss mechanism, which could be key for linking terrestrial and marine C cycles (Jaffé et al, 2013). The importance of horizontal and lateral versus vertical transport in this process is not completely understood for soils containing PCM. While the common paradigm is that erosion events are probably dominating biochar exports because of the relatively small contribution of biochar-derived WEOC (Braun et al, 2020), vertical transport to the aquatic system was found to dominate losses in the White Nile watershed of Lake Victoria (Güerena et al, 2015). The main export in truly dissolved form may occur for older PCM (Dittmar et al, 2012a). PCM can be mobilized from soils to the water systems in truly dissolved form several decades after its generation (Dittmar et al, 2012b), and export of older PCM may make up much of the PCM transported by rivers as shown by radiocarbon dating (Ziolkowski and Duffel, 2010). Interactions with other soil particles and soil biota in particular gain importance over time and alter biochar transport properties and mechanisms. The major pathway is via lateral and horizontal transport mainly by water whereas vertical transport in the soil profile is likely quantitatively less important and slower, but it may change soil properties and biochar residence times (Figure 12.2). Most transport processes act selectively and lead to the fractionation of biochar moieties along lateral and vertical transport pathways.

Fate of transported biochar

Biochar erosion and deposition could substantially influence its fate and the C budget of both biochar-amended soils and depositional soils or sediments (Figure 12.4). The biochar compounds removed from the site of deposition in particulate or soluble form may be exposed to physical, chemical, and or biological degradation processes. Indeed, 75–90% of condensed aromatic C were lost through photooxidation when biochar leachates were exposed to photoirradiation in a solar simulator (Bostick et al, 2020). The photooxidation was found to change the composition of dissolved particles released from biochar. It could make them more susceptible to biotic and abiotic degradation processes (Fu et al, 2016). Only a (small) proportion of particulate PyC and PCM may reach the ocean sediments, which are considered the ultimate sink (Masiello, 2004; Copolla et al, 2022).

Biochar could be strongly affected by erosion processes and removed from soils, especially in terrain with substantial topography or where exposed to strong winds. This could lead to a net loss of C (and potentially fertility) for the soils amended with biochar, but a net increase of C (and potentially fertility) in soils or sediments where the eroded biochar is re-deposited and accumulates. In some cases, aeolian transport of biochar particles <2.5 μm may compromise the effects that biochar could have on climate change mitigation (Genesio et al, 2016). Indeed, it causes a positive direct radiative forcing (Bond et al, 2013), and also causes indirect radiative forcing through changing the albedo at the site of deposition (Genesio et al, 2016).

Re-deposition of transported biochar may occur close to the site of mobilization or after longer transport to depositional areas within the watershed or into river-, lake- or ocean sediments further afield (Figure 12.3). Eroded particulate PyC was found to be mainly redistributed within a small flat watershed (Pyle et al, 2017; Hanke et al, 2017) and even in sloping land, a small proportion of particulate PyC left the watershed compared to the amount eroded from the soil at the plot scale (Chaplot et al., 2005). The place of re-deposition will depend on the physicochemical properties of the PCM particles, the features of the landscape and watershed in which the biochar is applied, and the magnitude and frequency of the events leading to biochar erosion. Much of the transported PCM may be stored in intermediate reservoirs, such as waterlogged soils, sediments, and groundwaters, although the actual accumulation and turnover rates of PCM are unknown (Coppola et al, 2022). Very few studies have addressed the potential effectiveness of biochar transport processes, pathways, and likely quantities involved, and therefore its fate after removal from the site of deposition is largely unknown. It is important to consider that the burial of eroded biochar at depositional sites is likely to lead to its enhanced preservation due to unfavorable conditions for microbial activity (Lal, 2003; Berhe et al, 2012). Preservation would be particularly enhanced at oxygen-deprived depositional sites such as lake sediments, river and coastal sediments, and ocean sediments (Figure 12.3). Indeed, high amounts of condensed aromatic carbon similar to biochar were found in mollic horizons of European floodplains (Rennert et al, 2021) and in the

surface sediments of Chinese wetlands (Wang et al, 2014). Despite the importance of these reservoirs for long-term storage of PCM eroded from the soil, its depositional and decomposition rates and storage have been poorly quantified (Coppola et al, 2022). Also, the preservation of dissolved aromatic carbon originating from biochar in the deep ocean may be important in the long term (Dittmar and Paeng, 2009; Ziolkowski and Druffel, 2010; Fang et al, 2021). Tidal pumping and groundwater discharge for the continental runoff of dissolved aromatic compounds may be important processes determining its further export to ocean sediments (Dittmar et al, 2012a). Its fate in the ocean may depend on the rate at which condensed aromatic compounds originating from biochar are cycled through the surface ocean's photic zone because these moieties are more photolabile than other dissolved organic matter (Stubbins et al, 2012; Bostick et al, 2020).

Vertically transported soluble material originating from biochar may also reach the river network via groundwater (Figure 12.4), which was shown to be an important pathway of PyC export to the aquatic systems (Dittmar et al, 2012a; Güerena et al, 2015). When biochar is subject to vertical transport down the soil profile, it may also efficiently accumulate at deep soil horizons (Brodowski et al., 2006) depending on soil type and its ability to protect organic matter from decomposition and/or subsequent transport (see above). As the vertical transport rates and river exports of dissolved aromatic moieties originating from biochar may be significant globally (Bowring et al, 2022; Jones et al, 2020), the controls of vertical transport of the various biochar types and forms need to be further studied and included into earth system models (Santos et al, 2022).

Outlook

Despite the importance of transport processes for biochar losses from the site of its application, its movement in the environment, and its potential transformation during transport, few studies have been carried out to quantify these processes and only some have compared biochar losses with losses of uncharred organic matter. It seems clear that biochar is particularly prone to erosion compared to mineral soil particles due to its specific physical and chemical properties. However, while some insights may be inferred from studies on naturally-produced PCM, such as from natural fires, horizontal and vertical transport of biochar in agricultural settings have rarely been assessed quantitatively up to now. Particular emphasis should be given to the different chemical biochar forms, which may be subjected to transport (truly dissolved, particulate, colloidal). Soil fauna, which could be important for biochar movement, has rarely been assessed as a transport vector for biochar under field conditions. Very few data are available concerning the distance traveled as well as the timescale of these processes. Modeling studies are required to integrate the wealth of information generated by process-oriented controlled experiments to hierarchize the various factors that control biochar transport and to connect processes with the evolution of biochar properties over time. Particular focus should be given to biochar fate in the different pedoclimatic environments by integrating their biological, chemical, and physical properties. This could help to give recommendations concerning agricultural practices to minimize biochar loss from the site of its deposition.

References

Abney RB, and Berhe AA 2018 Pyrogenic carbon erosion: Implications for stock and persistence of pyrogenic carbon in soil. *Frontiers in Earth Science* 6, 26.

Alexis MS, Rasse DP, Knicker H, Anquetil C, and Rumpel C 2012 Evolution of soil organic matter after prescribed fire: A 20-year chronosequence. *Geoderma* 189-190, 98–107.

Ali M, Masood N, Muhammad Rashad Javeed HMR, Al-Ashkar I, F. Almutairi K, Liu L, Aqeel Sarwar M, Rajendran K, and EL Sabagh A 2022 Termites Improve the Horizontal Movement of Carbonized Particles: A Step towards Sustainable Utilization of Biochar. *Phyton* 91, 2235–2248.

Bagnold R 1974 *The physics of blown sand and desert dunes* (1st ed). Netherlands: Springer.

Bellè S-L, et al 2021 Key drivers of pyrogenic carbon redistribution during a simulated rainfall event'. *Biogeosciences* 18, 1105–1126.

Berhe AA, et al 2012 Persistence of soil organic matter in eroding versus depositional landform positions. *Journal of Geophysical Research* 117, 10.1029/2011JG001790.

Blanco-Canqui H 2020 Does biochar improve all soil ecosystem services? *Global Change Biology Bioenergy* 13, 291–304.

Bond TC, et al 2013 Bounding the role of black carbon in the climate system: A scientific assessment. *Journal of Geophysical Research: Atmospheres* 118, 5380– 5552.

Bostick KW, et al 2020 Photolability of pyrogenic dissolved organic matter from a thermal series of laboratory-prepared chars. *Science of the Total Environment* 724, 138198.

Bowring SPK, Jones MW, Ciais P, Guenet B, and Abiven S 2022 Pyrogenic carbon decomposition critical to resolving fire's role in the Earth system. *Nature Geoscience* 15(2), 135–142.

Braun M, Kappenberg A, Sandhage-Hofmann A, and Lehndorff E 2020 Leachable soil black carbon after biochar application. *Organic Geochemistry* 143, 103996.

Bridgwater, AV 2012 Review of fast pyrolysis of biomass and product upgrading. *Biomass and Bioenergy* 38, 68–94.

Brodowski S, John B, Flessa H, and Amelung W 2006 Aggregate-occluded black carbon in soil. *European Journal of Soil Science* 57, 539–546.

Bruun EW, Petersen C, Strobel BW, and Hauggaard-Nielsen H 2012 Nitrogen and carbon leaching in repacked sandy soil with added fine particulate biochar. *Soil Science Society of America Journal* 76, 1142–1148.

Bruun EW, Petersen CT, Hansen E, Holm JK, and Hauggaard-Nielsen H. 2014 Biochar amendment to coarse sandy subsoil improves root growth and increases water retention. *Soil Use and Management* 30, 109–118.

Button ES, et al 2022 Deep-C storage: Biological, chemical and physical strategies to enhance carbon stocks in agricultural subsoils. *Soil Biology and Biochemistry* 170, 108697.

Carcaillet C, and Talon B 2001 Soil carbon sequestration by Holocene fires inferred from soil charcoal in the dry French Alps. *Arctic Antarctic and Alpine Research* 33, 282–288.

Chaplot V, Rumpel C, and Valentin C 2005 Water erosion impact on soil and carbon redistributions within uplands of South-East Asia. *Global Biogeochemical Cycles* 19, GB4004.

Chen M, Wang D, Yang F, Xu X, Xu N, and Cao X 2017 Transport and retention of biochar nanoparticles in a paddy soil under environmentally-relevant solution chemistry conditions. *Environmental Pollution* 230, 540–549.

Chen X, et al 2023 Rapid simulation of decade-scale charcoal aging in soil: Changes in physicochemical properties and their environmental implications. *Environmental Science and Technology* 57, 128–138.

Coppola AI, Wagner S, Lennartz ST, Seidel M, Ward ND, Dittmar T, Santín C, and Jones MW 2022 The black carbon cycle and its role in the Earth system. *Nature Reviews Earth & Environment* 3, 516–532.

Czimczik CI, and Masiello CA 2007 Controls on black carbon storage in soils. *Global Biogeochemical Cycles* 21, GB3005.

de La Rosa JM, et al 2011 Contribution of Black Carbon in recent sediments of the Gulf of Cadiz. Applicability of different quantification methodologies. *Quaternary International* 243, 264–272.

de la Rosa JM, Rosado M, Paneque M, Miller AZ, and Knicker H 2018 Effects of aging under field conditions on biochar structure and composition: Implications for biochar stability in soils. *Science of The Total Environment* 613-614, 969–976.

Diekow J, et al 2005 Carbon and nitrogen stocks in physical fractions of a subtropical Acrisol as influenced by long-term no-till cropping systems and N fertilization. *Plant and Soil* 268, 319–328.

Dietrich WE, and Dunne T 1978 Sediment budget for a small catchment in mountainous terrain. *Zeitschrift für Geomorphologie Supplement* 29, 191–206.

Dittmar T, and Paeng J 2009 A heat-induced molecular signature in marine dissolved organic matter. *Nature Geoscience* 2, 175–179.

Dittmar T, Peng J, Gihring TM, Suryaputra IGNA, and Huettel M 2012a Discharge of dissolved black carbon from a fire-affected intertidal system. *Limnology and Oceanography* 57, 1171–1181.

Dittmar T, et al 2012b Continuous flux of dissolved black carbon from a vanished tropical forest biome. *Nature Geoscience* 5, 618–622.

Domene X 2016 A critical analysis of meso-and macrofauna effects following biochar supplementation. In: Ralebitso-Senior T, and Orr C (Eds) *Biochar Application*. Amsterdam: Elsevier. pp 268–292.

Don A, Steinberg B, Schöning I, Pritsch K, Joschko M, Gleixner G, and Schulze E-D 2008 Organic carbon sequestration in earthworm burrows. *Soil Biology and Biochemistry* 40, 1803–1812.

Dong X, Li G, Lin Q, and Zhao X 2017 Quantity and quality changes of biochar aged for 5 years in soil under field conditions. *Catena* 159, 136–143.

Elmer WH, Lattao CV, and Pignatello JJ 2015 Active removal of biochar by earthworms (Lumbricus terrestris). *Pedobiologia-Jounral of Soil Ecology* 58, 1–6.

Fang J, Cheng L, Hameed R, Jin L, Wang D, Owens G, and Lin D 2020 Release and stability of water dispersible biochar colloids in aquatic environments: Effects of pyrolysis temperature, particle size, and solution chemistry. *Environmental Pollution* 260, 114037.

Fang Y, et al 2021 Particulate and dissolved black carbon in coastal China seas: Spatiotemporal variations, dynamics, and potential implications. *Environmental Science and Technology* 55, 788–796.

Felber R, Leifeld J, Horák J, and Neftel A 2014 Nitrous oxide emission reduction with greenwaste biochar: Comparison of laboratory and field experiments. *European Journal of Soil Science* 65, 128–138.

Foereid B, Lehmann J, and Major J 2011 Modeling black carbon degradation and movement in soil. *Plant and Soil* 345, 223–236.

Friedlingstein P, et al 2014 Uncertainties in CMIP5 climate projections due to carbon cycle feedbacks. *Journal of Climate* 27, 511– 526.

Fu H, et al 2016 Photochemistry of dissolved black carbon released from biochar: Reactive oxygen species generation and phototransformation. *Environmental Science and Technology* 50, 1218–1226.

Gao X, and Wu H 2014 Aerodynamic properties of biochar particles: Effect of grinding and implications. *Environmental Science and Technology Letters* 1, 60–64.

Gelardi, DL, Li C, and Parikh SJ 2019 An emerging environmental concern: Biochar-induced dust emissions and their potentially toxic properties. *Science of the Total Environment* 678, 813–820.

Genesio, L, Vaccari, FP, and Miglietta, F 2016 Black carbon aerosol from biochar threats its negative emission potential. *Global Change Biology* 22, 2313–2314.

Güereña DT, et al 2015 Terrestrial pyrogenic carbon export to fluvial ecosystems: Lessons learned from the White Nile watershed of East Africa. *Global Biogeochemical Cycles* 29, 1911–1928.

Guggenberger G, et al 2008 Storage and mobility of black carbon in permafrost soils of the forest tundra ecotone in Northern Siberia. *Global Change Biology* 14, 1367–1381.

Haefele SM, et al 2011 Effects and fate of biochar from rice residues in rice-based systems. *Field Crops Research* 121, 430–440.

Haider FU, et al 2022 Biochar application for remediation of organic toxic pollutants in contaminated soils; An update. *Ecotoxicology and Environmental Safety* 248, 114322.

Hameed R, Lei C, Fang J, and Lin D 2021 Co-transport of biochar colloids with organic contaminants in soil column. *Environmental Science and Pollution Research* 28, 1574–1586.

Han H, et al 2022 Contaminants in biochar and suggested mitigation measures – a review. *Chemical Engineering Journal* 429, 132287.

Hanke, UM, Reddy, CM, Braun, ALL, Coppola, AI, Haghipour, N, McIntyre, CP, Wacker, L, Xu, L, McNichol, AP, Abiven, S, Schmidt, MWI, and Eglinton, TI 2017 What on Earth Have We Been Burning? Deciphering Sedimentary Records of Pyrogenic Carbon. *Environmental Science & Technology* 51, 12972–12980. doi: 10.1021/acs.est.7b03243.

Hockaday WC, Grannas AM, Kim S, and Hatcher PG 2007 The transformation and mobility of charcoal in a fire-impacted watershed. *Geochimica et Cosmochimica Acta* 71, 3432–3445.

Jaffé R, et al 2013 Global charcoal mobilization from soils via dissolution and riverine transport to the oceans. *Science* 340, 345–347.

Jones MW, Coppola AI, Santín C, Dittmar T, Jaffé R, Doerr SH, and Quine TA 2020 Fires prime terrestrial organic carbon for riverine export to the global oceans. *Nature Communications* 11, 2791.

Kinnell PIA 1990 Modelling erosion by rain-impacted flow. *Catena supplement*, 17, 55–66.

Kok JF, Parteli EJR, Michaels TI, and Karam DB 2012 The physics of wind-blown sand and dust. *Reports on Progress in Physics* 75, 106901.

Kuhn NJ, Hoffmann T, Schwanghart W, and Dotterweich M 2009 Agricultural soil erosion and global carbon cycle: controversy over? *Earth Surface Processes and Landforms*, 34, 1033–1038.

Lal R 2003 Soil erosion and the global carbon budget. *Environment International* 29, 437–450.

Le H, Valenca R, Ravi S, Stenstrom MK, and Mohanty SK 2020 Size-dependent biochar breaking under compaction: Implications on clogging and pathogen removal in biofilters. *Environmental Pollution* 266, 115195.

Lee CH, et al 2018 In-situ biochar application conserves nutrients while simultaneously mitigating runoff and erosion of an Fe-oxide-enriched tropical soil. *Science of the Total Environment* 619–620, 665–671.

Li C, Bair DA, and Parikh SJ 2018 Estimating potential dust emissions from biochar amended soils under simulated tillage. *Science of The Total Environment* 625, 1093–1101.

Li, Y, Feng, G, Tewolde, H, Yang, M, and Zhang, F 2020 Soil, biochar, and nitrogen loss to runoff from loess soil amended with biochar under simulated rainfall. *Journal of Hydrology* 591, 125318.

Liang B, et al 2008 Stability of biomass-derived black carbon in soils. *Geochimica et Cosmochimica Acta* 72, 6069–6078.

Liu Z, Dugan B, Masiello CA, and Gonnermann HM 2017 Biochar particle size, shape, and porosity act together to influence soil water properties. *PLoS One* 1–19, 10.1371/journal.pone.0179079.

Liu C-H, Chu W, Li H, Boyd SA, Teppen BJ, Mao J, Lehmann J, and Zhang W 2019 Quantification and characterization of dissolved organic carbon from biochars. *Geoderma*, 335, 161–169.

Madden, NM, Southard, RJ, and Mitchell, JP 2010 Soil water and particle size distribution influence laboratory-generated PM_{10}. *Atmospheric Environment* 44, 745–752.

Maienza A, et al 2017 Impact of biochar formulation on the release of particulate matter and on short-term agronomic performance. *Sustainability* 9, 1131.

Major J, Lehmann J, Rondon M, and Goodale C 2010 Fate of soil-applied black carbon: downward migration, leaching and soil respiration. *Global Change Biology* 16, 1366–1379.

Martínez-Mena M, et al 2012 Organic carbon enrichment in sediments: Effects of rainfall characteristics under different land uses in a Mediterranean area. *Catena* 94, 36–42.

Masiello CA 2004 New directions in black carbon organic geochemistry. *Marine Chemistry* 92, 201–213.

Meng Z, et al 2018 Interactive effects of wind speed, vegetation coverage and soil moisture in controlling wind erosion in a temperate desert steppe, Inner Mongolia of China. *Journal of Arid Land* 10, 534–547.

Mulleneers HAE, Koopal LK, Swinkels GCC, Bruning H, and Rulkens WH 1999 Flotation of soot particles from a sandy soil sludge. *Colloids and Surfaces A: Physicochemical and Engineering Aspects* 151, 293–301.

Naisse C, et al 2015 Effect of physical weathering on the carbon sequestration potential of biochars and hydrochars in soil. *Global Change Biology Bioenergy* 7, 488–496.

Ngo PT, Rumpel C, Janeau J-L, Dang D-K, Doan TT, and Jouquet, P 2016 Mixing of biochar with organic amendments reduces carbon removal after field exposure under tropical conditions. *Ecological Engineering* 91, 378–380.

Nguyen BT, et al 2009 Long-term black carbon dynamics in cultivated soil. *Biogeochemistry* 92, 163–176.

Obia A, Børresen T, Martinsen V, Cornelissen G, and Mulder J 2017 Vertical and lateral transport of biochar in light-textured tropical soils. *Soil and Tillage Research* 165, 34–40.

Pattey E, and Qiu G-W 2012 Trends in primary particulate matter emissions from Canadian agriculture. *Journal of the Air and Waste Management Association* 62, 737–747.

Peng XY, Tong XG, Hao LT, and Wu FQ 2019 Applicability of biochar for limiting interrill erosion and organic carbon export of sloping cropland in a semi-arid area of China. *Agriculture, Ecosystems and Environment* 280, 68–76.

Pyle LA, Magee KL, Gallagher ME, Hockaday WC, and Masiello CA 2017 Short-term changes in physical and chemical properties of soil charcoal support enhanced landscape mobility. *Journal of Geophysical Research: Biogeosciences* 122, 3098–3107.

Ravi S, D'Odorico P, Over TM, and Zobeck TM 2014 On the effect of air humidity on soil susceptibility to wind erosion: The case of air-dry soils. *Geophysical Research Letters* 31, L09501.

Ravi S, Sharratt B, Li J, Olshevski S, Meng Z, and Yang J 2016 Particulate matter emissions from biochar-amended soils as a potential tradeoff to the negative emission potential. *Scientific Reports* 6, 35984.

Ravi S, Li J, Meng Z, Zhang J, and Mohanty S 2020 Generation, resuspension, and transport of particulate matter from biochar-amended soils: A potential health risk. *GeoHealth* 4, e2020GH000311.

Rennert T, et al 2021 Does soil organic matter in mollic horizons of central/east European floodplain soils have common chemical features? *Catena* 200, 105192.

Rumpel C, et al 2006 Preferential erosion of black carbon on steep slopes with slash and burn agriculture. *Catena* 65, 30–40.

Rumpel C, Ba A, Darboux F, Chaplot V, and Planchon O 2009 Erosion budget of pyrogenic carbon at meter scale and process selectivity. *Geoderma* 154, 131–137.

Sadeghi SH, Hazbavi Z, and Harchegani MK 2016 Controllability of runoff and soil loss from small plots treated by vinasse-produced biochar. *Science of the Total Environment* 541, 483–490.

Santos, F, Bird, JA, and Asefaw Berhe, A 2022 Dissolved pyrogenic carbon leaching in soil: Effects of soil depth and pyrolysis temperature. *Geoderma*, 424, 116011.

Schiedung M, Bellè S-L, Sigmund G, Kalbitz K, and Abiven S 2020 Vertical mobility of pyrogenic organic matter in soils: a column experiment. *Biogeosciences* 17, 6457–6474.

Shao Y, Raupach MR, and Findlater PA 1993 Effect of saltation bombardment on the entrainment of dust by wind. *Journal of Geophysical Research: Atmospheres* 98, 12719–12726.

Silva FC, Borrego C, Keizer JJ, Amorim JH, and Verheijen FGA 2015 Effects of moisture content on wind erosion thresholds of biochar. *Atmospheric Environment* 123, 121–128.

Singh BP, et al 2015 In situ persistence and migration of biochar carbon and its impact on native carbon emission in contrasting soils under managed temperate pastures. *PLoS ONE* 10, e0141560.

Skjemstad JO, Taylor JA, Janik LJ, and Marvanek SP 1999 Soil organic carbon dynamics under long-term sugarcane monoculture. *Australian Journal of Soil Research* 37, 151–164.

Sorrenti G, Masiello CA, Dugan B, and Toselli M 2016 Biochar physico-chemical properties as affected by environmental exposure. *Science of the Total Environment* 563–564, 237–246.

Spokas KA, Novak JM, Masiello CA, Johnson MG, Colosky EC, Ippolito JA, et al 2014 Physical disintegration of biochar: An overlooked process. *Environmental Science and Technology Letters* 1, 326–332.

Stubbins A, Niggemann J, and Dittmar T 2012 Photo-lability of deep ocean dissolved black carbon. *Biogeosciences* 9, 1661–1670.

Tiefenbacher A, et al 2021 Antecedent soil moisture and rain intensity control pathways and quality of organic carbon exports from arable land. *Catena* 202, 105297.

Torri D, and Poesen J 1992 The effect of soil surface slope on raindrop detachment. *Catena supplement* 19, 561–577.

Tuo D, Xu M, Gao L, Zhang S, and Liu S 2016 Changed surface roughness by wind erosion accelerates water erosion. *Journal of Soils and Sediments* 16, 105–114.

Wagner S, Ding Y, and Jaffé R 2017 A new perspective on the apparent solubility of dissolved black carbon. *Frontiers in Earth Science* 5, 75.

Wang DJ, Zhang W, Hao XZ, and Zhou DM 2013a Transport of biochar particles in saturated granular media: Effects of pyrolysis temperature and particle size. *Environmental Science and Technology* 47, 821–828.

Wang C, Walter MT, and Parlange J-Y 2013b Modeling simple experiments of biochar erosion from soil. *Journal of Hydrology* 499, 140–145.

Wang Q, Zhang P-J, Liu M, and Deng Z-W 2014 Mineral-associated organic carbon and black carbon in restored wetlands. *Soil Biology and Biochemistry* 75, 300–309.

Webb NP, Chappell A, Edwards BL, McCord SE, Van Zee JW, Cooper BF, Courtright EM, Duniway MC, Sharratt B, Tedela N, and Toledo D 2019 Reducing Sampling Uncertainty in Aeolian Research to Improve Change Detection. *Journal of Geophysical Research: Earth Surface* 124, 1366–1377.

Zhang W, et al 2010 Transport and retention of biochar particles in porous media: Effect of pH, ionic strength, and particle size. *Ecohydrology* 3, 497–508.

Ziadat FM, and Taimeh AY 2013 Effect of rainfall intensity, slop, land use and antecedent soil moisture on soil erosion in an arid environment. *Land Degradation and Development* 24, 582–590.

Ziolkowski LA, and Druffel ERM 2010 Aged black carbon identified in marine dissolved organic carbon. *Geophysical Research Letters* 37, L16601.

13

Plant productivity with biochar applications to soils

Simon Jeffery, Frank G. A. Verheijen, Diego Abalos, and Ana Catarina Bastos

Introduction

Feeding a growing world population without further degrading the Earth's environmental systems is a crucial challenge for humanity (Mueller et al, 2012). The vast majority (99%) of calories for human consumption come from terrestrial sources, with less than 1% provided by marine and freshwater ecosystems (Pimentel et al, 2010). Recent studies suggest that crop production will need to approximately double by 2050 to keep pace with projected demands from population growth, dietary changes, urbanization, and increasing competition for land from bioenergy use (Foley et al, 2011). In this context, how can we increase food production, while also supporting a positive trajectory towards the UN Sustainable Development Goals (Hunter et al, 2017)?

Agriculture is a major contributor to climate change, a problem that is likely to be further exacerbated if it is expanded or intensified to increase food production (Tilman et al, 2011). Several management and policy

options are being investigated and discussed (Chabbi et al, 2017; Lynch et al, 2021), e.g., increased resource use efficiency, promotion of "sensible diets", reduction of food waste, advanced crop varieties including genetically modified crops, and closing "yield gaps" on underperforming lands. Given the magnitude of the challenge, and the difficulties of implementing most of the mentioned options, the application of novel measures by which higher yields can be achieved with little or no damage to the environment is critically needed.

One such proposed measure is the use of biochar as a soil amendment. Biochar application to soils can potentially aid mitigation of climate change by sequestering C (Chapter 30). Concomitant with this posited beneficial effect on the net greenhouse gas balance, the most commonly reported effects of biochar application are increased crop production and yield (Jeffery et al, 2011; Spokas et al, 2012). In some instances, biochar application to soils has been shown to increase yields

DOI: 10.4324/9781003297673-13

by over 400% using biochar applied at 4 t ha^{-1} co-applied with fertilizer (Cornelissen et al, 2013). However, a closer look at the published literature shows that such high yield increases are more of an exception than the rule, and a wide variety of crop yield responses, including no effects or negative responses, are reported (Jeffery et al, 2017a; Ye et al, 2020; Bai et al, 2022). This large variability highlights the need for a mechanistic understanding of the effects of biochar application to soils on crop yields to allow for robust predictions of the likely effects. Furthermore, it is necessary to document current gaps in our understanding of biochar's potential impacts on crop yield (and other ecosystem services), both positive and negative, to support evidence-based policymaking.

Biochar has been compared to other forms of organic matter. However, despite years of research, there is still a poor understanding of the general relationship between soil organic matter and crop yields (e.g., Loveland and Webb, 2003; Oldfield et al, 2019). The same is true regarding the interaction between biochar and crop productivity (Spokas et al, 2012). It is generally accepted that it is the turnover of soil organic matter that is important for soil fertility, and not merely its presence (Janzen, 2006). Organic matter can improve crop yields directly by supplying and exchanging nutrients (by decomposing and by its cation exchange capacity) and water (by retaining water onto the organic matter). Further to this, organic matter can contribute to soil fertility indirectly by providing a soil structure that is conducive to seedling emergence (friability providing a good soil-to-seed contact) and root development (resistance and resilience to compaction), that improves water infiltration and drainage (waterlogging stops microbial decomposition), and that reduces erosion (loss of seed or crop). It can also harbor plant growth-promoting bacteria and fungi. However, much of the organic portion of biochar (as opposed to the inorganic, or ash, portion) has long residence times in soil (Chapter 11), and so it seems likely that biochar will affect soil fertility and productivity differently to soil organic matter, including at different spatial and temporal scales. Initial steps have been made towards achieving a mechanistic understanding of the range of effects of biochar addition to soil on crop yields, but much work remains to be done and many questions remain unresolved.

From an agronomic viewpoint, the potential benefit of biochar application must be considered a consequence of its effects on enhancing soil productivity, which is determined by the entire spectrum of soil properties. Among these properties are: physical attributes, such as the size, continuity, and tortuosity of pores, aggregate stability, and texture, which together determine soil structure; chemical properties, such as organic matter content and composition, nutrient stocks and availability, mineralogy, pH, salinity, and bioavailable contents of potentially toxic compounds (organic and inorganic); biological attributes, such as the abundance, activity, and functional diversity of microbial biomass and soil fauna (Cassman, 1999). Policies permitting biochar application to soils in commercial contexts have been developed in numerous countries. However, due to the complex abiotic and biotic interactions that take place in the soil ecosystem, positive outcomes cannot be assumed. More research is needed to best identify optimum biochar types for different cropping systems and environmental combinations. This would maximize the potential for biochar to help achieve global food security while concurrently aiding with the mitigation of climate change. By enhancing crop yields, biochar use also has the potential to reduce the amount of land required for food production, freeing up potential areas to be more focused on the provision of other key ecosystem services, including C sequestration (Werner et al, 2022).

Introduction to biochar meta-analyses

Biochar affects crop yields, with significant increases, decreases, or no effect, all being repeatedly reported (Lehmann and Joseph, 2015; Gonzaga et al., 2018; Jeffery et al, 2017a; Ye et al, 2020; Wu et al, 2022). Interactions between the soil system, climate and biochar type, application rate, and placement determine the impacts, or lack thereof, when biochar is applied to the soil. As such, variable results are expected across studies, and extrapolation of expected effects from single studies is problematic owing to the wide range of biochar-soil-plant-climate combinations. Furthermore, as different metrics are often used, cross-study interpretation can be difficult. Therefore, experimental data must be combined into a single framework to facilitate interpretation.

Meta-analysis (MA) represents a robust and objective means of combining data from multiple experimental studies into a single analysis. Data from different studies that adhere to pre-defined experimental quality criteria are combined into a single database that can then be subject to statistical interrogation. Meta-analytical techniques have been developed over the past few decades allowing a range of metrics and approaches to be included (Hoffman, 2015). In short, they allow disparate results from contrasting studies, including when quantified using different metrics, to be standardized and combined with a single unified metric - usually the "effect size".

Numerous MAs have been published in recent years, covering several agronomic and environmental impacts consequent to biochar's application to soil (Table 13.1). Here, results from published meta-analyses are combined and analyzed to identify key generalizations and potential mechanisms underlying soil functions and processes that have consequences for plant growth and crop yields when biochar is applied to soils. Most of these studies used untreated biochar, with the majority using a single application of a relatively large amount of biochar (≥ 10 t ha^{-1}).

Plant productivity following biochar application to soil

Positive yield impacts, on average, with biochar application were initially confirmed in the meta-analysis of Jeffery et al (2011), which reported a grand mean result of +11%, based on 177 pairwise comparisons from 16 studies. That analysis also showed variable crop responses to different biochar types, produced from different feedstocks, or under different conditions. It also demonstrated differential responses of different crops to the same biochar, demonstrating the complexity of the issue and the caution with which general findings, such as the grand mean, should be extrapolated.

Since 2011, numerous MAs have been published that have included analysis of yield response (Table 13.1). These have subsequently been reviewed, along with other MAs published on biochar by Jeffery (2018) who identified 18 MAs, nine of which explored yield effects, and Schmidt et al (2021) who reviewed 26 meta-analyses published on biochar in agriculture, three were focused specifically on yield and plant productivity (Table 13.1).

Across all identified MAs, grand mean yield effects are in the region of 10 – 17%. The lowest grand mean result was by Ye et al (2020) at 10%, and the highest was reported by Wu et al (2022) at 16%. However, the range of results of categorical analyses (i.e.,

Table 13.1 *Meta-analyses aimed at investigating biochar interactions with plant production and crop yield*

	Title	Authors and year	Journal	Topic(s) covered	Number of studies/ pairwise comparisons
1	A quantitative review of the effects of biochar application to soils on crop productivity using meta-analysis	Jeffery et al (2011)	Agriculture, Ecosystems and Environment	Biochar impacts on crop yield	16/177
2	Biochar and its effects on plant productivity and nutrient cycling: a meta-analysis	Biederman and Harpole (2013)	Global Change Biology Bioenergy	Biochar impacts on crop yield and biogeochemical cycles (only very poorly covered)	114/371
3	Biochar's effect on crop productivity and the dependence on experimental conditions—a meta-analysis of literature data	Liu et al (2013)	Plant and Soil	Biochar impacts on crop yield	103/880
4	Heterogeneous global crop yield response to biochar: a meta-regression analysis	Crane-Droesch et al (2013)	Environmental Research Letters	Biochar impacts on crop yield	84/365
5	Biochar and forest restoration: a review and meta-analysis of tree growth responses	Thomas and Gale (2015)	New Forests	Biochar impacts on tree growth	17/420
6	Biochar amended soils and crop productivity: a critical and meta-analysis of literature	Baidoo et al (2016)	International Journal of Development and Sustainability	Biochar impacts on crop yield	27/NA
7	Biochar boosts tropical but not temperate crop yields	Jeffery et al (2017a)	Environmental Research Letters	Biochar impacts on crop yield	111/1135
8	A meta-analysis of effects of biochar properties and management practices on crop yield	Jing et al (2017)	Scientia Agricultura Sinica	Biochar impacts on crop yield	97/819

9	Toward a better assessment of biochar–nitrous oxide mitigation potential at the field scale	Verhoeven et al (2017)	Journal of Environmental Quality	Biochar impacts on crop yield and N_2O	40/122
10	Combined effects of biochar properties and soil conditions on plant growth: a meta-analysis	Dai et al (2020)	Science of the Total Environment	Soil conditions and plant growth	153/1254
11	Biochar effects on crop yields with and without fertilizer: a meta-analysis of field studies using separate controls	Ye et al (2020)	Soil Use and Management	Biochar and fertilizer interaction effects on crop yield	56/264
12	Biochar-based fertilizer effects on crop productivity: a meta-analysis	Melo et al (2022)	Plant and Soil	Biochar and fertilizer interaction effects on crop yield	40/148
13	Impact of biochar amendment on soil hydrological properties and crop water use efficiency: a global meta-analysis and structural equation model	Wu et al (2022)	Global Change Biology Bioenergy	Biochar and soil hydrology and crop water use efficiency	81/337
14	Combined effects of biochar and fertilizer applications on yield: A review and meta-analysis	Bai et al (2022)	Science of the Total Environment	Biochar and fertilizer interaction effects on crop yield	57/627

analyzing by soil, crop, biochar type, etc.) is wider, with a 70% average yield increase observed in tropical soils when nutrient biochar was applied compared to an average −17% yield penalty when biochar is applied to soils with pH > 8 (Jeffery et al, 2017a). Furthermore, MAs indicated the expected effect size, along with a measure of confidence, if a random biochar with certain characteristics was applied in another specified but random soil-plant-climate system. They do not provide an indication of the

idealized situation if a specific biochar was chosen to be applied to a soil to ameliorate a known issue – such as high pH biochar being applied to a low pH soil, where it will have a liming effect. In those situations, the effect size could be expected to be higher than those so far discussed here. The current state of the art of matching biochar types with specific soil types is discussed in Chapter 9. Nevertheless, MAs represent a powerful tool to identify possible mechanisms through differential analysis of MA databases.

Possible mechanisms underlying plant productivity effects

Direct effects

Biochars can provide plant macro and micro-nutrients. Different biochars have different plant nutrient profiles (e.g., N, P, K, Ca, Mg, Fe, and Cu) which can be largely predicted based on feedstock choice and pyrolysis conditions (Ippolito et al, 2020). This means that there is potential to design biochars to help ameliorate low fertility in soils through the provision of a variety of plant-available nutrients to support or enhance crop growth.

Nitrogen

The amount of N in biochars is relatively low. The use of stable isotope ^{15}N tracers has shown that at least a portion of N in biochar is plant-available (Taghizadeh-Toosi et al, 2012; Jeffery et al, 2017b; Craswell et al, 2021). This may be in the form of ammonium that is adsorbed to the biochar surface (Taghizadeh-Toosi et al, 2012). Biochar can also function as a nitrate exchanger by capturing and releasing nitrate and some other reactive N forms at exchange sites, while the retention has not been shown to have sufficient longevity to function as slow-release fertilizers (Rasse et al, 2022). Increased Nitrogen Use Efficiency (NUE) has been observed following

biochar application i.e., by reducing losses of fertilizer N through nitrate leaching or gaseous emissions (Borchard et al, 2019). Increases of ~6–7% NUE have been observed for maize following the application of a maize straw biochar (pyrolysis 500°C for 2 hours) when applied at rates equivalent to 12 and 24 t ha^{-1}. NUE can be negatively affected with biochar applied at high rates, e.g., 48 t ha^{-1} of biochar resulted in a ~2% reduction in NUE (Ma et al, 2020).

Potassium

At the global level, K is as limiting as N and P for plant productivity in terrestrial ecosystems (Sardans and Peñuelas, 2015), particularly in natural ecosystems. Most biochars are relatively rich in K, which forms a large part of the ash fraction post pyrolysis and can remain included within the C matrix (Novak et al, 2018). The amount depends to a large extent on the K already present in the feedstock. These are plant-available and impact plant growth (Mia et al, 2014; Wang et al, 2017), including by potentially increasing plant water use efficiency (Grzebisz et al, 2013), and plant community composition (van de Voorde et al, 2014; Oram et al., 2014).

Phosphorous

Several studies have shown that soluble and plant-available P increase in the soil after biochar application (Li et al, 2017, 2019; Yang et al, 2021). The amount of P in biochars is highly variable depending on the feedstock (Li et al, 2017) but generally decreases at high pyrolysis temperatures (Bruun et al, 2017). The availability of P in soil after application of biochar is determined by: (i) the quantity of soluble P in the biochar; (ii) the capacity of biochar to enhance soil P availability by influencing soil pH, complexation, and metabolism; (iii) the P adsorbed by the biochar, thereby improving P retention in soil and affecting P assimilation in plants (Yang et al, 2021); and (iv) potential interactions with K salt or rock powder when added to the feedstock, pre-pyrolysis, which have been shown to greatly increase plant P availability in nutrient-rich biochars made from manure or sewage sludge (Buss et al, 2020).

While the low initial P availability means that biochars are not suitable as P starter fertilizers, they can be a valuable tool to maintain an appropriate level of available P in the soil (Kuligowski et al, 2010), for example when using organic manures (Wang et al, 2015). Through pre-treatments of feedstocks, it may be possible to create biochars with more favorable nutrient stoichiometries, acting as P fertilizers (Buss et al, 2022).

Increased plant-available water (PAW)

Water can often be limiting to crop production. Biochar has been shown to increase the water-holding capacity of soils (Teixeira et al, 2021), especially medium-textured soils with lower proportions of silt and clay (Chapter 20). Not all soil water may be available to plants because it can be bound tightly in micro or nano-pores inside biochar particles. However, a large proportion of pore sizes within biochar fragments exist at sizes relevant for plant available water (PAW; Jeffery et al, 2015; Batista et al, 2018; Werdin et al, 2020) and sufficient pore connectivity is present for biochar particles to interact with soil hydrology after soil application (Lu and Zong, 2018). A meta-analysis of 37 studies found that, on average, biochar increased PAW by 29% on average (mean application rate = 10 t ha^{-1}), but was most effective at improving soil water properties in coarse-textured soils with application rates between 30 and 70 t ha^{-1}, although it varied by particle size, specific surface area, and porosity (Edeh et al, 2020). In addition, soil structure improvements resulting from biochar addition, i.e., lower bulk density and improved aggregation, also improved root growth (Xiang et al, 2017) or resulting build-up of soil organic carbon (Blanco-Canqui et al, 2020), impacting soil physicochemical properties. This may increase infiltration capacity at the soil surface or throughout the biochar-amended part of the soil profile (Abrol et al, 2016; Gholamahmadi et al, 2023). As a consequence, a larger proportion of rain can enter the soil, where it may subsequently be retained more effectively by the biochar. Therefore, the effect of biochar on total soil water storage, i.e., the soil sponge function, needs to consider both infiltration, saturated and unsaturated hydraulic conductivity determining infiltration depth and retention (Verheijen et al, 2019). The extra water that enters the soil equals less water that runs off the soil (if the land is not flat), which may also reduce erosion of soil particles and the export of nutrients. The strongest effects on PAW are generally seen in soils of a medium texture.

Research to date suggests that biochar has the potential to increase crop yields of rain-fed agroecosystems, which constitute 80% of global agriculture by area, and so may play a crucial role in achieving food security. In irrigated systems, the increased water-holding capacity of some biochars may reduce irrigation costs and problems associated with irrigation such as increased salinity and salinization. However, there remains a paucity of evidence in this regard – as with most other effects, the impact of biochar on the water-holding capacity of soil is dependent on the physical and chemical properties of the biochar and the soil, and optimum combinations are not yet well identified. Long-term effects are also less well known. Biochars can weather over time in soils, which may affect their impacts on hydrology as they are broken down into smaller particles and their surface chemistries may change. Therefore, revisiting field studies that were installed, for example, >10 years ago, may provide insights as to whether such changes are starting to occur.

Hydrophilic versus hydrophobic biochar

Porosity alone is insufficient to ensure that biochar interacts with PAW retention, as some biochars can exhibit hydrophobic properties (Jeffery et al, 2015). The degree of biochar hydrophobicity is mostly determined by the pyrolysis temperature, with low temperatures resulting in more hydrophobic biochar, although the precise outcome depends on both feedstock and the pyrolysis process (Mao et al, 2019). Biochar hydrophobicity, quantified as thermally-labile aliphatic compounds, disappeared at a pyrolysis temperature of 500°C or higher (Zornoza et al, 2016). Increasing the residence time of the feedstock in the pyrolysis chamber may have a similar effect. Hydrophobic soils have reduced infiltration rates, which means that rainfall intensity often easily exceeds the soil infiltration rate, leading to runoff and soil erosion, including the export of nutrients. Soil hydrophobicity is a transient property; it disappears after sufficient contact time with water, for example following prolonged rainfall, and it returns when soils dry

out again. This can occur over extended timeframes in soils but the longevity of hydrophobic effects in biochars exhibiting this property remains unknown. Nevertheless, considering that soils are expected to dry out more strongly with climate change until 2050 (IPCC, 2019), it may be wise to avoid adding hydrophobic biochar to such soils. Current biochar certification systems or standards - including as part of C credit systems - may benefit from including quantification of biochar hydrophobicity as a reported characteristic.

Input of phytotoxic compounds into the soil

While biochar can be a valuable source of growth-limiting soil nutrients, particularly in nutrient-poor soils, an increasing number of studies have reported negative effects on agricultural produce, at typical rates for agronomic applications. One of the possible ways through which biochar can negatively impact seed germination, crop growth, and productivity is by supplying potentially bioavailable phytotoxic compounds, ranging from salts to metals or polycyclic aromatic hydrocarbons (PAHs) and volatile matter. Total contents, distribution, and bioavailability of phytotoxic components in biochars are mainly a function of feedstock and pyrolysis conditions, with woody feedstocks generally having lower phytotoxic potential, compared to manure or sludge feedstocks (Hilber et al, 2017a,b; Godlewska et al, 2021). Various studies reported the total and bioavailable concentrations of metals and PAHs in a range of biochars produced within the 200–600°C temperature range, while their bioavailability in biochar was often lower than in the unpyrolyzed feedstocks (Koppolu et al, 2003; Freddo et al, 2012; Hale et al, 2012; Zielińska and Oleszczuk, 2016; Hilber et al, 2017a,b; Wang et al, 2017; Liu et al, 2018). In biochars from "low toxicity provenance" feedstocks (e.g., wood or crop residues), such contents were comparable to soil and ambient background levels within the short-to-medium term (Freddo et al, 2012; Hale et al, 2012; de La Rosa et al, 2016; Hilber et al, 2017a,b). Overall, their application to the soil within typical rates for agronomic use (Jeffery et al, 2017a) is unlikely to induce metal or PAH-induced phytotoxicity in the short-term, although it remains unclear whether it can lead to increased soil total concentrations in the long-term (de La Rosa et al, 2016; Hilber et al, 2017a,b; de Rendese et al, 2018) (Chapters 21 and 22).

Upon application to soil, different chemical groups with phytotoxic properties desorb from the biochar matrix to different extents, with specific desorption rates and bioavailability-related factors that may vary in time, according to local soil environmental conditions and meteorological factors. Crop responses to these chemical fractions depend on a combination of factors, including crop type, specific sensitivity and stage of development, soil type (particularly, textural class), soil organic matter contents, and biochar type and application rates (Godlewska et al, 2021). It highlights the need for complementing analytical with plant effect-based approaches in biochar characterization procedures, for a wide range of crops and environmental combinations. For the majority of woody, crop residue or manure biochars produced between 300–600°C, effects on seed germination, root development, growth, or yield of various staple crops were often stimulatory at applications of up to 100 t ha^{-1}, with increasing phytotoxicity at higher rates, mainly due to over-liming and increased electrical conductivity (Oleszczuk et al, 2013; Stefaniuk et al, 2016; Jeffery et al, 2017; Intani et al, 2019).

Indirect effects

Increased soil pH

On average, soils have a lower pH than biochars. The main natural factors that affect soil pH are the parent material, i.e., limestone and basalt can have a high pH while many sandstones and igneous parent materials have a low pH, as well as the water balance, i.e., with a threshold from more acidic to more alkaline soils where potential evapotranspiration begins to exceed mean annual precipitation (Slessarev et al, 2016). Biochars are generally alkaline (Chapter 8). For example, Jeffery et al (2017a) found a median biochar pH of 9.0 in a global meta-analysis. However, some biochars have even higher pH values, e.g., some manure feedstock biochars, or lower pH values when, for example, using lower pyrolysis temperatures in combination with steam activation (Ippolito et al, 2016). Therefore, biochars generally increase soil pH and thereby the (macro)nutrient availability, by the "liming effect", the level of which differs between different biochar types (Domingues et al, 2017). This is suggested to be one of the main mechanisms of how biochar increases crop yield globally (Jeffery et al, 2017), although in specific circumstances other mechanisms, such as water availability, nutrient inputs, or soil hardness impacts, may be more important.

A sizable proportion of global farmland requires periodic application of agricultural lime to maintain the soil pH in the optimal range; although the pH tolerance of crops varies widely, most crops prefer a soil pH in the range of 5.5–7.0, depending on their (micro)nutrient requirements. As agricultural soils generally acidify over time due to nutrient and proton exchange between the crop and soil, many soils require additives such as lime, to raise the soil pH and keep it in line with crop requirements. Biochar applications may help replace or augment liming applications, thereby providing a cost saving where biochar is cheaper than other liming products, while potentially providing one or more other benefits for plant growth. However, soils that are already at the higher end of the optimal pH range may suffer reduced nutrient availability, and hence yield penalties when biochar is applied, i.e., over-liming, especially manure feedstock biochars at high application rates to already neutral to high pH soils. To ensure the successful use of biochar within a system, current biochar certification systems or standards (Chapter 9) - including as part of C credit systems (Chapter 30) - should provide the consumer with detailed information on how their product will affect the pH of a wide range of soil types under varying land use and soil management, and on how long the effect will last considering the water balance of the field in question.

Stimulation of mycorrhizae and mycorrhizal colonization

Mycorrhizal fungi are important for crop productivity as they form associations with plant roots in the majority of crop species and can increase both nutrient and water uptake. Beneficial effects on mycorrhizal fungi are one of the most often hypothesized mechanisms for observed increases in crop yields (Warnock et al, 2007, 2010; Chapter 14).

Most studies that have investigated the effects of biochar on mycorrhiza have found a strong positive effect of mycorrhizal abundance following biochar application to soil (Ishii and Kadoya, 1994; Vaario et al, 1999). Nishio (1996) stated that "the idea that the application of charcoal stimulates indigenous arbuscular mycorrhizal fungi in soil and thus promotes plant growth is relatively well-known in Japan". Nishio (1996) reported that biochar did not stimulate alfalfa when grown in sterilized soil, but when grown in unsterilized soil containing native mycorrhizal fungi, alfalfa growth was increased by a factor of approximately 1.8. This suggests that it is the interaction between the biochar and the soil mycorrhizal fungi that leads to

positive effects on yield, and not just interactions between biochar and the crop itself.

One of the main reasons for the observed beneficial effects is thought to be the structure of biochar (Chapters 5 and 20), which is highly porous, and so provides protection to extra-radical fungal hyphae which can grow into the pore space of biochar particles and also that such fungal hyphae can sporulate into the micropores of the biochar where there is reduced competition and grazing from other microbes (Saito and Marumoto, 2002). However, Warnock et al (2007) also hypothesized that biochar may have beneficial effects on mycorrhizal fungi due to indirect effects on other soil microbes as well as potentially leading to plant–fungus signaling interference and detoxification of allelochemicals on biochar particle surfaces (Chapter 14). However, while there is a growing body of research on biochar and mycorrhizal interactions, the functionality and dynamics of their interactions, and the associated mechanisms are far from being fully elucidated (Gujre et al, 2021).

Stimulation of biological nitrogen fixation

Symbiotic biological nitrogen fixation (BNF) is a vital source of N for maintaining and improving soil fertility. The global addition of N provided to soils by legumes is estimated to be in the region of 26×10^6 tons N (Herridge et al, 2008). Biochar amendment has been shown to improve BNF in legumes (Nishio, 1996; Rondon et al, 2007) and increase the number of root nodules (Xiu et al, 2021). It has also been shown to increase the proportion of legumes in plant communities in grasslands (van de Voorde et al, 2014; Jeffery et al, 2022).

Several mechanisms through which biochar might affect BNF have been proposed. Biochar application can lead to the immobilization of mineral N (Rondon et al, 2007; Borchard et al, 2019) which may lead to reduced N availability, which is known to stimulate BNF. Furthermore, biochar amendment has been shown to increase P bioavailability (Rondon et al, 2007). Increased P availability after biochar amendment has been correlated with increased BNF in several legumes (Nishio and Okano, 1991; Rondon et al, 2007; Tagoe et al, 2008). In addition, biochar often contains large amounts of nutrients such as K, Ca, and Mg which may be beneficial to legumes and so lead to increased BNF. Increased legume competitivity in grasslands is driven largely by K availability from the ash fraction of the biochar (Oram et al, 2014), which stimulated BNF in red clover (*Trifolium pratense*; Mia et al, 2014). Other potential mechanisms favoring BNF include increased soil pH after biochar application (Ogawa and Okimori, 2010), increased availability of micronutrients (B and Mo) (Rondon et al, 2007), or changes in root growth (Xiu et al, 2021).

Microbial effects

Biochar can provide refugia for mycorrhizal fungi, but also for other microorganisms (Chapter 14). This is hypothesized to be one of the main mechanisms by which biochar can lead to increases in microbial biomass compared to soils without biochar addition (Lehmann et al, 2011). As well as increasing microbial biomass, biochar has been found to increase microbial efficiency (Lui et al, 2020), which is the ratio of respired CO_2 to microbial biomass (Jiang et al, 2016). An increase in CO_2 production from soils is an indicator of increased turnover of soil organic matter (or of increased root-derived inputs). The breakdown of soil organic matter and the associated release of nutrients is a very important factor governing soil fertility, and as such is likely to affect crop productivity. There is some evidence that biochar addition to soil can lead to a positive priming effect (Chapter 17) whereby soil

organic matter is turned over at an accelerated rate (i.e., positive priming) (Wang et al, 2016; Rasul et al, 2022). However, this effect seems to be relatively short-lived (Keith et al, 2011; Wang et al, 2016) and so its contribution to increased nutrient mineralization may be temporary.

Mitigation of bioavailable contaminants in soil

Biochar can be an effective remediation strategy for saline (Ndiate et al, 2022) or acidic soils (van Zwieten et al, 2015), as well as for soils contaminated with a range of organic (Kaur et al, 2020; Haider et al, 2021; Ogura et al, 2021) and inorganic (O'Connor et al, 2018; Peng et al, 2018) compounds, thus indirectly contributing to improving crop yield. Impaired seed germination or root elongation and other negative physiological and metabolic effects can reduce biomass production in contaminated soils. Several mechanisms have been identified that underlie biochar-based contaminant immobilization in soils, often interacting synergistically. These include adsorption, absorption, complexation, cation exchange, and electrostatic interactions (Chapters 21, 22, and 27). Soil-environment combinations, including site-specific spatial and temporal conditions, contaminant type and biochar properties (mainly pore size distribution, specific surface area, and chemistry), application rate, and depth are determining factors for the short-to-medium term effectiveness of biochar in reducing contaminant bioavailability to crops in vulnerable soils (O'Connor et al. 2018; Uchimiya et al, 2011; Nguyen et al, 2023). The potential to produce tailor-made biochars for specific remediation applications, possibly integrating biochar production with waste, soil, and crop management approaches (Krahn et al, 2023), is a promising sustainable option for restoring and/or improving crop productivity and diversity in contaminated

342 SIMON JEFFERY ET AL.

agroecosystems. At present, it remains unclear which contributing factors and associated mechanisms can again lead to the desorption of the contaminant from the biochar matrix in the long term. Contaminant desorption over time may likely be influenced, at least partially, by those same aforementioned factors, combined with physical, chemical, and biological biochar aging processes in soil (Hilber et al, 2017a; Wang et al, 2020; Chapter 10).

Disease resistance

Crop diseases can lead to reduced crop yields, so anything that can increase a crop's resistance to diseases has the potential to increase crop yields. Biochar has been shown to interact with plant immune response systems in a variety of ways (Chapter 23). However, the generality of the effects across crops with the same biochar type remains to be determined for a wide range of biochar. Nevertheless, matching biochar types with particular crops that suffer from particular pests or diseases has the potential to reduce the economic and environmental costs of crop production in terms of reduced pesticide use.

Plant water use efficiency

Biochar generally increases the water use efficiency (WUE) of most crops (Gao et al, 2020; Wu et al, 2022; Chapters 15, 20). However, the current available data show wide variation. For example, Gao et al (2020) found an overall effect of around 20%, while Wu et al (2022) reported an overall effect of around 5%.

Biochar interactions with WUE varied strongly by the scale of measurement - i.e., leaf or whole plant - as well as soil, biochar, and management factors. Generally, whole plant effects were stronger for soils with lower pH and biochar with lower C contents. Importantly, WUE effects appear to be associated with moderate biochar application rates, 10–20 t ha^{-1}, with no significant effects observed at lower/higher application rates. Although more research is needed to identify the various potential mechanisms and optimal biochar-soil-crop combinations, the observed general increase in WUE may be considered a useful tool to help improve drought resistance in agriculture under predicted increases in drought stress (IPCC, 2019). This is particularly true for the mid-latitude region in the northern hemisphere, where WUE has been assessed to have decreased by 48% (Yu et al, 2017), while wheat yields have been predicted to decrease by around 15% due to drought stress in Africa and southern Asia by 2050 (Pequeno et al, 2021). Increased K availability following biochar applications to soils may function to ameliorate some of this water stress due to improved plant WUE (Grzebisz et al, 2013).

Longevity of biochar effects

There is growing evidence that biochar's positive impacts on plant productivity and crop yields can be sustained over significant periods. For example, yield increases were still observed in a maize plantation in Kenya 10 years (20 growing seasons) after biochar was first applied (Kätterer et al, 2019). And grape yields remained elevated over 4 seasons in an experiment in Tuscany, Italy (Genesio et al, 2015). In a meta-analysis of 57 papers, consistent positive yield increases were observed more than three years after initial application (Bai et al, 2022). However, it is unlikely that biochar will always drive yield effects over the long term, dependent on the mechanisms that biochar is interacting with to drive those yield effects. For example, nutrient inputs into the soil such as K in the ash fraction are likely to be relatively short term, but impacts on soil water retention and reduced bulk density are likely to be longer lived.

Matching biochar to field conditions

Besides soil and biochar characteristics, biochar application rate, crop type, and requirements, other site-specific environmental and climatological factors may determine plant productivity responses to biochar application. Considering that soil and biochar characteristics can interact, the same biochar can induce a wide range of effects on plant production – from largely positive to neutral or negative – when applied to different soils, even at comparable application ranges. In the same way, plant production in any specific field can be increased by one biochar applied at a rate of 10 t ha^{-1}, and be reduced by the same biochar at rates of up to 50 t ha^{-1} (e.g., Mia et al, 2014), constituting a "maximum biochar loading capacity" (BLC) of that system (i.e., the maximum amount of biochar that can be added to any specific soil-crop system before negative impacts are observed) (Verheijen et al, 2015). However, whether the BLC of a soil is the same if biochar is applied as a large amount in one application, versus repeated applications at lower rates across multiple years, remains to be determined. The distribution in which biochar is applied across the field can be variable – there is no requirement to apply biochar evenly to drive effects. For example, large yield increases were reported when biochar was applied within the rooting zone of a crop only (Cornelissen et al, 2013). This allows the achieving of higher application rates around the plant than would have been possible if biochar were evenly distributed across a field. Such an approach is likely most useful when the amount of biochar available is restricted, for example, due to limited feedstock availability as is the case for many subsistence farmers.

When biochar is matched to the soil conditions and crop requirements to ensure optimum nutrient provision or PAW, it is more likely for yield benefits to be maximized in that system, with trade-offs kept to a minimum. Furthermore, the poorer the starting conditions and so the larger the yield gap (i.e., the difference between actual and potential yields), the stronger plant production improvements may be achieved if biochar is applied (Jeffery et al, 2017a). Conversely, when the starting conditions are suitable and fields are producing yields close to their potential maximum (i.e., a small yield gap), plant production improvements are less pronounced, or not observed (Jeffery et al, 2017a). Despite this being the case, biochar application may still be a preferred option if assessed on a case-by-case basis given reducing input costs, such as liming, fertilization, crop resilience to abiotic stress (e.g., drought), irrigation and pest control management, thereby providing an economic incentive. If the economic benefits do not outweigh the cost of biochar purchase, production, and application, land users may be incentivized via a grant system (carbon credits) due to potential environmental co-benefits, such as reduced N leaching or N_2O emissions. Furthermore, some of the potential economic incentives for biochar use may accrue over the long term. For example, longer-term soil fertility maintenance or crop yield resilience to extreme weather events, which are predicted to increase due to climate change, may also occur due to biochar reducing soil degradation process that otherwise may have occurred if biochar had not been applied, such as by reducing soil erosion (Gholamahmadi et al, 2023).

The type of biochar is another key factor in the soil-biochar matching process, since crop yields can be improved by one biochar and impaired by another within the same system. According to the type of feedstock, biochars can be classified as "structural" (e.g., wood mill shavings, forest residues, crop residues) or "nutritional" (e.g., manures or sewage residues). Low fertility soils - more commonly found in the tropics than in the temperate zone - were found to have a more than 3 times stronger increase in plant productivity for nutritional than structural biochars (Jeffery et al, 2017a). However, more fertile and well-managed soils did not show any difference in crop yield response for either nutritional or structural biochars.

Considering the aforementioned, it is imperative that current biochar certification systems or standards - including when used as part of C credit systems - provide the consumer with suitable information on the likely impacts of the application of their product to soil and the consequences for plant production for a wide range of soil, crop, land use, and soil management combinations (Verheijen et al, 2012, 2015). Developing the evidence required to allow

optimal matching of biochars to the field-specific conditions by considering, for example, nutrient vs. structure biochars, or biochars with a low H/Corg ratio to minimize N_2O fluxes, etc., will provide a useful first steppingstone in developing a sustainable biochar system. Purpose-made biochars, for instance, by blending nutrient and structure feedstocks, may be the next step in the matching process, with a wider consideration of biochar and soil nutrient dynamics and crop demands (Chapter 26). Further matching steps may include expected future changes in precipitation distribution, plant-available soil water, water-use efficiency, biological N fixation, mitigation of bio-available contaminants in the soil, mycorrhizal colonization, disease pressures, microbial efficiency, and others. In this light, the concept of BLC of any given soil-biochar-crop combination can be applied depending also on the desired goals.

Biochar-based fertilizers and mineral composites

There has been growing interest in recent years in the use of biochars that are produced as composites with fertilizer or minerals (Joseph et al, 2013; Melo et al, 2022; Wang et al, 2022). These represent a very wide range of materials with different characteristics as the highly heterogeneous nature of biochars is increased further by mixing with different materials either pre or post pyrolysis, impregnation, and encapsulation (Wang et al, 2022). More information on the different approaches is included in Chapter 26.

Some mixtures are created through co-composting biochar with compost – a practice that is claimed to "charge" the biochar with nutrients (Chapter 26). Some evidence suggests that plant growth improvements through the addition of such biochars to soil may be

mediated through nitrate available bound to exchange sites within the biochars (Kammann et al, 2015). These biochars have been shown to have stronger yield impacts than constituent parts alone when using co-composted biochar (Agegnehu et al, 2016). Other materials that have been used to enrich biochar with fertilizer include cattle urine, mineral fertilizers, and mineral clay composites. These have also been shown to have beneficial yield impacts greater than the component parts of the composites (Schmidt et al, 2017). However, they have also been shown to have negative effects in some situations (Buss et al, 2016).

In a meta-analysis of biochar-based fertilizers, a grand mean of a 10% increase in crop productivity was found compared to fertilizer alone (Melo et al, 2022). The strongest interactions were when biochar was combined with N fertilizer (~8% increase) or when combined with NPK fertilizer (~15% increase). They concluded that biochar-based fertilizers can increase crop productivity by an average of 15% when added to soils that are not responsive to conventional fertilizers.

Limitations and further considerations

Many pure-biochar yield effects are observed with application rates over 10 t ha^{-1}. At the current costs of biochar, that is unlikely to be a cost-effective approach for many farmers and growers unless farmers are financially incentivized via a grant system (carbon credits; carbon-sink trading).

Many countries have restrictions on the amount and type of biochar that can be applied to soil. In the UK, currently, a maximum of 1 t ha^{-1} year^{-1} of biochar may be applied, and that biochar must be made from virgin woody feedstocks. Therefore, it may take several years to more than a decade for crop yield effects to be realized, and it is unknown whether the slow addition of biochar to the soil over the years will build up to the same effect as when that amount of biochar is applied in one go. In Germany, there is no restriction on the amount of biochar that can be applied, but it must be about 80% C, which means that only biochar from woody feedstocks is likely to be feasible. Neither of these policies seems to be based on a strong evidence base.

Currently, no regulation exists regarding biochar application to soil as a soil and environmental management or climate adaptation and mitigation tool, despite many of the environmental and public health considerations and quality criteria are, or should be, comparable to those involving compost or sewage sludge, for which regulatory and incentive schemes are already in place. This policy is now under construction within the EU at a general and national level, but its final configuration is still "undefined" (Verde and Chiaramonti, 2021) and it is much less developed elsewhere (Chapter 31).

Conclusions and outlook

Biochar has been repeatedly demonstrated to have the potential to increase crop yields and plant productivity. However, the results are variable, and positive effects should not always be assumed. In soils that are already producing near their maximum potential output, biochar application will likely have little to no effect, although input savings might be made (irrigation, fertilizers, pesticides) or a greater resilience to climatic extreme events may be achieved, depending on the conditions. In acidic, nutrient-poor, coarse-textured soils, biochar application can be expected to have a more generally positive effect, with large yield increases likely in some cases. Current indications are that expected increases in crop drought stress by 2050 and 2100 may increase the positive effect of biochar on crop yields through increased plant water supply and water use efficiency, depending on conditions.

In the tropics, biochars that are produced from nutrient-rich feedstocks, such as manures, digestates, and green waste composts, generally cause larger yield increases than biochars produced from less nutrient-rich feedstocks such as wood. However, in the temperate zone, this pattern was not observed. Identification of optimum biochar types for particular soil-plant-climate-environment combinations remains a key goal if the potential benefits of biochar to enhance crop yield and aid global food security are to be realized.

References

Abrol V, et al 2016 Biochar effects on soil water infiltration and erosion under seal formation conditions: Rainfall simulation experiment. *Journal of Soils and Sediments* 16, 2709–2719.

Agegnehu G, Bass AM, Nelson PN, and Bird MI 2016 Benefits of biochar, compost and biochar–compost for soil quality, maize yield and greenhouse gas emissions in a tropical agricultural soil. *Science of the Total Environment* 543, 295–306.

Bai, X, Zhang, S, Shao, J, Chen, A, Jiang, J, Chen, A and Luo, S 2022 Exploring the negative effects of biochars on the germination, growth, and antioxidant system of rice and corn. *Journal of Environmental Chemical Engineering* 10, 107398.

Baidoo I, Sarpong DB, Bolwig S, and Ninson D 2016 Biochar amended soils and crop productivity: A critical and meta-analysis of literature. *International Journal of Development and Sustainability* 5, 414–432.

Batista EM, et al 2018 Effect of surface and porosity of biochar on water holding capacity aiming indirectly at preservation of the Amazon biome. *Scientific Reports* 8, 10677.

Biederman LA, and Harpole WS 2013 Biochar and its effects on plant productivity and nutrient cycling: A meta-analysis. *Global Change Biology Bioenergy* 5, 202–214.

Blanco-Canqui H, et al 2020 Soil carbon increased by twice the amount of biochar carbon applied after 6 years: Field evidence of negative priming. *Global Change Biology Bioenergy* 12, 240–251.

Borchard N, et al 2019 Biochar, soil and land-use interactions that reduce nitrate leaching and N_2O emissions: A meta-analysis. *Science of the Total Environment* 651, 2354–2364.

Bruun S, et al 2017 The effect of different pyrolysis temperatures on the speciation and availability in soil of P in biochar produced from the solid fraction of manure. *Chemosphere* 169, 377–386.

Buss W, Graham MC, Shepherd JG, and Mašek O 2016 Risks and benefits of marginal biomass-derived biochars for plant growth. *Science of the Total Environment* 569, 496–506.

Buss W, Bogush A, Ignatyev K, and Masek O 2020 Unlocking the fertilizer potential of waste-derived biochar. *ACS Sustainable Chemistry and Engineering* 8, 12295–12303.

Buss W, et al 2022 Highly efficient phosphorus recovery from sludgse and manure biochars using potassium acetate pre-treatment. *Journal of Environmental Management* 314, 8.

Cassman, KG 1999 Ecological intensification of cereal production systems: Yield potential, soil quality, and precision agriculture. *Proceedings of the National Academy of Sciences* 96, 5952–5959.

Chabbi A, et al 2017 Aligning agriculture and climate policy. *Nature Climate Change* 7, 307–309.

Cornelissen, G, Martinsen, V, Shitumbanuma, V, Alling, V, Breedveld, G, Rutherford, D, Sparrevik, M, Hale, S, Obia, A and Mulder, J 2013 Biochar Effect on Maize Yield and Soil Characteristics in Five Conservation Farming Sites in Zambia. *Agronomy* 3, 256–274.

Crane-Droesch A, Abiven S, Jeffery S, and Torn MS 2013 Heterogeneous global crop yield response to biochar: A meta-regression analysis. *Environmental Research Letters* 8, 044049.

Craswell ET, Chalk PM, and Kaudal BB 2021 Role of ^{15}N in tracing biologically driven nitrogen dynamics in soils amended with biochar: A review. *Soil Biology and Biochemistry* 162, 108416.

Dai Y, Zheng H, Jiang Z, and Xing B 2020 Combined effects of biochar properties and soil conditions on plant growth: A meta-analysis. *Science of the Total Environment* 713, 136635.

De la Rosa JM, et al 2016 Assessment of polycyclic aromatic hydrocarbons in biochar and biochar-amended agricultural soil from Southern Spain. *Journal of Soils and Sediments* 16, 557–565.

De Resende MF, et al 2018 Polycyclic aromatic hydrocarbons in biochar amended soils: long-term experiments in Brazilian tropical areas. *Chemosphere* 200, 641–648.

Domingues RR, et al 2017 Properties of biochar derived from wood and high-nutrient biomasses with the aim of agronomic and environmental benefits. *PloS One* 12, 0176884.

Edeh IG, Mašek O, and Buss W 2020 A meta-analysis on biochar's effects on soil water properties–New insights and future research challenges. *Science of the Total Environment* 714, 136857.

Foley, JA, Ramankutty, N, Brauman, KA, Cassidy, ES, Gerber, JS, Johnston, M, Mueller, ND, O'Connell, C, Ray, DK, West, PC, Balzer, C, Bennett, EM, Carpenter, SR, Hill, J, Monfreda, C, Polasky, S, Rockström, J, Sheehan, J, Siebert, S, Tilman, D, and Zaks, DPM 2011 Solutions for a cultivated planet. *Nature* 478, 337–342.

Freddo, A, Cai, C and Reid, BJ 2012 Environmental contextualisation of potential toxic elements and polycyclic aromatic hydrocarbons in biochar. *Environmental Pollution* 171, 18–24.

Gao Y, et al 2020 Effects of biochar application on crop water use efficiency depend on experimental conditions: A meta-analysis. *Field Crops Research* 249, 107763.

Genesio, L, Miglietta, F, Baronti, S and Vaccari, FP 2015 Biochar increases vineyard productivity without affecting grape quality: Results from a four years field experiment in Tuscany. *Agriculture, Ecosystems & Environment*, 201, 20–25.

Gholamahmadi B, et al 2023 Biochar impacts on runoff and soil erosion by water: A systematic global scale meta-analysis. *Science of the Total Environment*, 161860.

Godlewska P, Ok YS, and Oleszczuk P 2021. The dark side of black gold: Ecotoxicological aspects of biochar and biochar-amended soils. *Journal of Hazardous Materials* 403, 123833.

Gonzaga MIS, et al 2018 Positive and negative effects of biochar from coconut husks, orange bagasse and pine wood chips on maize (*Zea mays* L.) growth and nutrition. *Catena* 162, 414–420.

Grzebisz W, Gransee A, Szczepaniak W, and Diatta J 2013 The effects of potassium fertilization on water-use efficiency in crop plants. *Journal of Plant Nutrition and Soil Science*, 176, 355–374.

Gujre N, et al 2021 Sustainable improvement of soil health utilizing biochar and arbuscular mycorrhizal fungi: A review. *Environmental Pollution* 268, 115549.

Haider FU, et al 2021 Phytotoxicity of petroleum hydrocarbons: Sources, impacts and remediation strategies. *Environmental Research Letters* 197, 111031.

Hale, SE, Lehmann, J, Rutherford, D, Zimmerman, AR, Bachmann, RT, Shitumbanuma, V, O'Toole, A, Sundqvist, KL, Arp, HPH, and Cornelissen, G 2012 Quantifying the total and bioavailable polycyclic aromatic hydrocarbons and dioxins in biochars. *Environmental Science & Technology* 46, 2830–2838.

Herridge DF, Peoples MB, and Boddey RM 2008 Global inputs of biological nitrogen fixation in agricultural systems. *Plant and Soil* 311, 1–18.

Hilber I, et al 2017a Bioavailability and bioaccessibility of polycyclic aromatic hydrocarbons from (post-pyrolytically treated) biochars. *Chemosphere* 174, 700–707.

Hilber I, et al 2017b The different faces of biochar: Contamination risk versus remediation tool. *Journal of Environmental Engineering and Landscape Management* 25, 86–104.

Hoffman JI 2015 *Biostatistics for medical and biomedical practitioners*. Cambridge: Academic Press.

Hunter MC, et al 2017 Agriculture in 2050: Recalibrating targets for sustainable intensification. *Bioscience* 67, 386–391.

Intani K, Latif S, Islam MS, and Müller J 2019 Phytotoxicity of corncob biochar before and after heat treatment and washing. *Sustainability* 11, 30.

IPCC 2019 Climate Change and Land: an IPCC special report on climate change, desertification, land degradation, sustainable land management, food security, and greenhouse gas fluxes in terrestrial ecosystems: Chapter 1: Framing and Context.

Ippolito JA, et al 2016 Designer, acidic biochar influences calcareous soil characteristics. *Chemosphere* 142, 184–191.

Ippolito, J., Cui, L, Kammann, C, Wrage-Mönnig, Ne, Estavillo, J., Fuertes-Mendizabal, T, Cayuela, ML, Sigua, G, Novak, J, Spokas, Kt, and Borchard, N 2020 Feedstock choice, pyrolysis temperature and type influence biochar characteristics: a comprehensive meta-data analysis review. *Biochar* 2, 421–438.

Ishii, T, & Kadoya, K 1994 Effects of charcoal as a soil conditioner on citrus growth and vesicular-arbuscular mycorrhizal development. *Journal of the Japanese Society for Horticultural Science* 63, 529–535.

Janzen HH 2006 The soil carbon dilemma: Shall we hoard it or use it?. *Soil Biology and Biochemistry* 38, 419–424.

Jiang X, Denef K, Stewart CE, and Cotrufo MF 2016 Controls and dynamics of biochar decomposition and soil microbial abundance, composition, and carbon use efficiency during long-term biochar-amended soil incubations. *Biology and Fertility of Soils* 52, 1–14.

Jeffery S, et al 2017a Biochar boosts tropical but not temperate crop yields. *Environmental Research Letters*, 12, 053001.

Jeffery S, 2018 Biochar application to soil for climate change mitigation and crop production. *Aspects of Applied Biology* 139, 125–132.

Jeffery S, et al 2015 Biochar application does not improve the soil hydrological function of a sandy soil. *Geoderma* 251, 47–54.

Jeffery S, et al 2017b Initial biochar effects on plant productivity derive from N fertilization. *Plant and Soil* 415, 435–448.

Jeffery S, et al 2022 Biochar application differentially affects soil micro-, meso-macro-fauna and plant productivity within a nature restoration grassland. *Soil Biology and Biochemistry* 174, 108789.

Jeffery S, Verheijen, FGA, van der Velde M, and Bastos AC 2011 A quantitative review of the effects of biochar application to soils on crop productivity using meta-analysis. *Agriculture, Ecosystems and Environment* 144, 175–187.

Joseph S, et al 2013 Shifting paradigms: development of high-efficiency biochar fertilizers based on nano-structures and soluble components. *Carbon Management* 4, 323–343.

Kätterer T, et al 2019 Biochar addition persistently increased soil fertility and yields in maize-soybean rotations over 10 years in sub-humid regions of Kenya. *Field Crops Research* 235, 18–26.

Kammann, CI, Schmidt, HP, Messerschmidt, N, Linsel, S, Steffens, D, Müller, C, Koyro, HW, Conte, P, and Joseph, S 2015 Plant growth improvement mediated by nitrate capture in co-composted biochar. *Scientific Reports*, 5.

Kaur V, and Sharma P 2020 Application of biochar as an adsorbent and its significance on berseem (*Trifolium alexandrinum*) growth parameters in farm soil contaminated with PAH. *Journal of Soil Science and Plant Nutrition* 20, 806–819.

Keith A, Singh B, and Singh BP 2011 Interactive priming of biochar and labile organic matter mineralization in a smectite-rich soil. *Environmental Science and Technology* 45, 9611–9618.

Koppolu, L, Agblevor, FA and Clements, LD 2003 Pyrolysis as a technique for separating heavy metals from hyperaccumulators. Part II: Lab-scale pyrolysis of synthetic hyperaccumulator biomass. *Biomass and Bioenergy* 25, 651–663.

Krahn KM, et al 2023 Sewage sludge biochars as effective PFAS-sorbents. *Journal of Hazardous Materials* 445, 130449.

Kuligowski K, Poulsen TG, Rubaek GH, and Sorensen P 2010 Plant-availability to barley of phosphorus in ash from thermally treated animal manure in comparison to other manure based materials and commercial fertilizer. *European Journal of Agronomy* 33, 293–303.

Lehmann, J , and Joseph, S eds., 2015 *Biochar for environmental management: Science, technology and implementation.* Routledge.

Lehmann J, et al 2011 Biochar effects on soil biota–a review. *Soil Biology and Biochemistry* 43, 1812–1836.

Li FY, et al 2019 Effects of biochar amendments on soil phosphorus transformation in agricultural soils. *Advances in Agronomy* 158, 131–172.

Li XX, et al 2017 Plant availability of phosphorus in five gasification biochars. *Frontiers in Sustainable Food Systems* 1, 2.

Liu X, et al 2013 Biochar's effect on crop productivity and the dependence on experimental conditions—a meta-analysis of literature data. *Plant and Soil* 373, 583–594.

Liu Z, et al 2018 Characteristics and applications of biochars derived from wastewater solids. *Renewable Sustainable Energy Reviews* 90, 650–664.

Liu Z, et al 2020 Greater microbial carbon use efficiency and carbon sequestration in soils: amendment of biochar versus crop straws. *Global Change Biology Bioenergy* 12, 1092–1103.

Lu S, and Zong Y 2018 Pore structure and environmental services of biochars derived from different feedstocks and pyrolysis conditions. *Environmental Science and Pollution Research* 25, 30401–30409.

Loveland P, and Webb J 2003 Is there a critical level of organic matter in the agricultural soils of temperate regions: a review. *Soil and Tillage Research* 70, 1–18.

Lynch J, Cain M, Frame D, and Pierrehumbert R 2021 Agriculture's contribution to climate change and role in mitigation is distinct from predominantly fossil CO_2-emitting sectors. *Frontiers in Sustainable Food Systems* 4, 518039.

Ma R, et al 2020 Different rates of biochar application change ^{15}N retention in soil and ^{15}N utilization by maize. *Soil Use and Management* 36, 773–782.

Mao J, Zhang K, and Chen B 2019 Linking hydrophobicity of biochar to the water repellency and water holding capacity of biochar-amended soil. *Environmental Pollution* 253, 779–789.

Melo LCA, Lehmann J, Carneiro JSDS, and Camps-Arbestain M 2022 Biochar-based fertilizer effects on crop productivity: A meta-analysis. *Plant and Soil* 472, 45–58.

Mia S, et al 2014 Biochar application rate affects biological nitrogen fixation in red clover conditional on potassium availability. *Agriculture, Ecosystems and Environment* 191, 83–91.

Mueller, ND, Gerber, JS, Johnston, M, Ray, DK, Ramankutty, N and Foley, JA 2012 Closing yield gaps through nutrient and water management. *Nature* 490, 254–257.

Ndiate NI, et al 2022 Soil amendment with arbuscular mycorrhizal fungi and biochar improves salinity tolerance, growth, and lipid metabolism of common wheat (*Triticum aestivum* L.). *Sustainability* 14, 3210.

Nishio M, and Okano S 1991 Stimulation of the growth of alfalfa and infection of mycorrhizal fungi by the application of charcoal. *Bulletin of National Grassland Research Institute* 45, 61–71.

Nishio M 1996 *Microbial fertilizers in Japan.* Japan: ASPAC Food & Fertilizer Technology Center.

Nguyen, TB, Sherpa, K, Bui, XT, Nguyen, VT, Vo, TDH, Ho, HTT, Chen, CW and Dong, CD 2023 Biochar for soil remediation: A comprehensive review of current research on pollutant removal. *Environmental Pollution* 337, 122571.

Novak JM, Johnson MG, and Spokas KA 2018 Concentration and release of phosphorus and potassium from lignocellulosic-and manure-based biochars for fertilizer reuse. *Frontiers in Sustainable Food Systems* 2, 54.

O'Connor D, et al 2018 Biochar application for the remediation of heavy metal polluted land: A review of in situ field trials. *Science of the Total Environment* 619-620, 815–826.

Ogawa M, and Okimori Y 2010 Pioneering works in biochar research, Japan. *Soil Research* 48, 489–500.

Ogura AP, et al 2021 A review of pesticides sorption in biochar from maize, rice, and wheat residues: Current status and challenges for soil application. *Journal of Environmental Management* 300, 113753.

Oldfield EE, Bradford MA, and Wood SA 2019 Global meta-analysis of the relationship between soil organic matter and crop yields. *Soil* 5, 15–32.

Oleszczuk P, Jośko I, and Kuśmierz M 2013 Biochar properties regarding to contaminants content and ecotoxicological assessment. *Journal of Hazardous Materials* 260, 375–382.

Oram NJ, et al 2014 Soil amendment with biochar increases the competitive ability of legumes via increased potassium availability. *Agriculture, Ecosystems and Environment* 191, 92–98.

Peng X, Deng Y, Peng Y, and Yue K 2018 Effects of biochar addition on toxic element concentrations in plants: A meta-analysis. *Science of the Total Environment* 616-617, 970–977.

Pequeno DN, et al 2021 Climate impact and adaptation to heat and drought stress of regional and global wheat production. *Environmental Research Letters* 16, 054070.

Pimentel D, et al 2010 Will limited land, water, and energy control human population numbers in the future? *Human Ecology* 38, 599–611.

Rasul M, Cho J, Shin HS, and Hur J 2022 Biochar-induced priming effects in soil via modifying the status of soil organic matter and microflora: A review. *Science of the Total Environment* 805, 150304.

Rasse DP, et al 2022 Enhancing plant N uptake with biochar-based fertilizers: Limitation of sorption and prospects. *Plant and Soil* 475, 213–236.

Rondon MA, Lehmann J, Ramírez J, and Hurtado M 2007 Biological nitrogen fixation by common beans (*Phaseolus vulgaris* L.) increases with bio-char additions. *Biology and Fertility of Soils* 43, 699–708.

Saito M, and Marumoto T 2002 Inoculation with arbuscular mycorrhizal fungi: The status quo in Japan and the future prospects. *Diversity and Integration in Mycorrhizas* 273–279.

Sardans J, and Peñuelas J 2015 Potassium: A neglected nutrient in global change. *Global Ecology and Biogeography* 24, 261–275.

Schmidt HP, et al 2021 Biochar in agriculture–A systematic review of 26 global meta-analyses. *Global Change Biology Bioenergy* 13, 1708–1730.

Schmidt HP, Pandit BH, Cornelissen G, and Kammann CI 2017 Biochar-based fertilization with liquid nutrient enrichment: 21 field trials covering 13 crop species in Nepal. *Land Degradation and Development* 28, 2324–2342.

Slessarev EW, et al 2016 Water balance creates a threshold in soil pH at the global scale. *Nature* 540, 567–569.

Spokas, KA, Cantrell, KB, Novak, JM, Archer, DW, Ippolito, JA, Collins, HP, Boateng, AA, Lima, IM, Lamb, MC, McAloon, AJ, Lentz, RD, & Nichols, KA 2012 Biochar: A Synthesis of Its Agronomic Impact beyond Carbon Sequestration. *Journal of Environmental Quality* 41, 973–989.

Stefaniuk, M, Oleszczuk, P and Bartmiński, P 2016 Chemical and ecotoxicological evaluation of biochar produced from residues of biogas production. *Journal of Hazardous Materials* 318, 417–424.

Taghizadeh-Toosi A, Clough TJ, Sherlock RR, and Condron LM 2012 Biochar adsorbed ammonia is bioavailable. *Plant and Soil* 350, 57–69.

Tagoe SO, Horiuchi T, and Matsui T 2008 Effects of carbonized and dried chicken manures on the growth, yield, and N content of soybean. *Plant and Soil* 306, 211–220.

Teixeira WG, Verheijen FGA, and de Oliveira Marques JD 2021 Water holding capacity of biochar and biochar-amended soils. In Manya JJ, and Gasco G (Eds.) *Biochar as a Renewable-based Material: With Applications in Agriculture, the Environment and Energy.* Singapore: World Scientific. pp 61–83.

Tilman, D, Balzer, C, Hill, J and Befort, BL 2011 Global food demand and the sustainable intensification of agriculture. *Proceedings of the National Academy of Sciences* 108, 20260–20264.

Thomas SC, and Gale N 2015 Biochar and forest restoration: A review and meta-analysis of tree growth responses. *New Forests* 46, 931–946.

Uchimiya M, et al 2011 Influence of pyrolysis temperature on biochar property and function as a heavy metal sorbent in soil. *Journal of Agricultural and Food Chemistry* 59, 2501–2510.

Vaario LM, et al 1999 In vitro ectomycorrhiza formation between *Abies firma* and *Pisolithus tinctorius*. *Mycorrhiza* 9, 177–183.

van de Voorde TF, et al 2014 Soil biochar amendment in a nature restoration area: Effects on plant productivity and community composition. *Ecological Applications* 24, 1167–1177.

van Zwieten L, et al 2015 Enhanced biological N_2 fixation and yield of faba bean (*Vicia faba* L.) in an acid soil following biochar addition: Dissection of causal mechanisms. *Plant and Soil* 395, 7–20.

Verde SF, and Chiaramonti D 2021 *The biochar system in the EU: The pieces are falling into place, but key policy questions remain*. Florence: European University Institute.

Verheijen FGA, et al 2019 The influence of biochar particle size and concentration on bulk density and maximum water holding capacity of sandy vs sandy loam soil in a column experiment. *Geoderma* 347, 194–202.

Verheijen FGA, et al 2015 Biochar sustainability and certification. In: Lehmann J, and Joseph S (Eds.) *Biochar for Environmental Management: Science, Technology and Implementation*. London: Routledge. pp 793–810.

Verhoeven E, et al 2017 Toward a better assessment of biochar–nitrous oxide mitigation potential at the field scale. *Journal of Environmental Quality* 46, 237–246.

Wang, Y, Ji, H and Gao, C 2015 Differential responses of soil bacterial taxa to long-term P, N, and organic manure application. *Journal of Soils and Sediments* 16, 1046–1058.

Wang J, Xiong Z, and Kuzyakov Y 2016 Biochar stability in soil: Meta-analysis of decomposition and priming effects. *Global Change Biology Bioenergy* 8, 512–523.

Wang C, Wang Y, and Herath HMSK 2017 Polycyclic aromatic hydrocarbons (PAHs) in biochar–Their formation, occurrence and analysis: A review. *Organic Geochemistry* 114, 1–11.

Wang, L, O'Connor, D, Rinklebe, J, Ok, YS, Tsang, DCW, Shen, Z and Hou, D 2020 Biochar aging: Mechanisms, physicochemical changes, assessment, and implications for field applications. *Environmental Science & Technology* 54, 14797–14814.

Wang, L, Ok, YS, Tsang, DCW, Alessi, DS, Rinklebe, J, Mašek, O, Bolan, NS and Hou, D 2022 Biochar composites: Emerging trends, field successes and sustainability implications. *Soil Use and Management* 38, 14–38.

Warnock DD, Lehmann J, Kuyper TW, and Rillig MC 2007 Mycorrhizal responses to biochar in soil–concepts and mechanisms. *Plant and Soil* 300, 9–20.

Warnock DD, et al 2010 Influences of non-herbaceous biochar on arbuscular mycorrhizal fungal abundances in roots and soils: Results from growth-chamber and field experiments. *Applied Soil Ecology* 46, 450–456.

Werdin J, et al 2020 Biochar made from low density wood has greater plant available water than biochar made from high density wood. *Science of the Total Environment* 705, 135856.

Werner C, Lucht W, Gerten D, and Kammann C 2022 Potential of land-neutral negative emissions through biochar sequestration. *Earth's Future* 10, e2021EF002583.

Wu W, et al 2022 Impact of biochar amendment on soil hydrological properties and crop water use efficiency: A global meta-analysis and structural equation model. *Global Change Biology Bioenergy* 14, 657–668.

Xiang Y, Deng Q, Duan H, and Guo Y 2017 Effects of biochar application on root traits: A meta-analysis. *Global Change Biology Bioenergy* 9, 1563–1572.

Xiu L, et al 2021 Biochar can improve biological nitrogen fixation by altering the root growth strategy of soybean in Albic soil. *Science of the Total Environment* 773, 144564.

Yang L, et al 2021 Effects of biochar addition on the abundance, speciation, availability, and leaching loss of soil phosphorus. *Science of the Total Environment* 758, 143657.

Ye L, et al 2020 Biochar effects on crop yields with and without fertilizer: A meta-analysis of field studies using separate controls. *Soil Use and Management* 36, 2–18.

Yu Z, et al 2017 Global gross primary productivity and water use efficiency changes under drought stress. *Environmental Research Letters* 12, 014016.

Zielinska A, and Oleszczuk P 2016 Effect of pyrolysis temperatures on freely dissolved polycyclic aromatic hydrocarbon (PAH) concentrations in sewage sludge-derived biochars. *Chemosphere* 153, 68–74.

Zornoza R, et al 2016 Stability, nutrient availability and hydrophobicity of biochars derived from manure, crop residues, and municipal solid waste for their use as soil amendments. *Chemosphere* 144, 122–130.

14

Biochar effects on the abundance, activity, and diversity of the soil biota

Janice E. Thies, Xavier Domene, Stephen Joseph, Xaiofan Rong, and Johannes Lehmann

Introduction

Pioneering research in Japan has shown that biochar can strongly affect soil biota behavior and composition (Ogawa et al, 1983), aspects of which have been reviewed recently (Joseph et al, 2021; Schmidt et al, 2021). The literature that examines the effects of biochar on microbial and faunal communities or the effects of adding biochar inoculated with microorganisms into the soil has increased substantially in recent years (Dai et al, 2021). Similarly, basic soil-biochar-plant interactions have been shown to be influenced by biochar amendments and lead to a variety of changes in soil biotic communities, such as increases in microbial biomass and significant changes in microbial community composition (Palansooriya et al, 2019).

Soil microbial communities are highly responsive to many edaphic, climatic, and management factors, including organic matter inputs (Thies and Grossman, 2023), such as biochar, whose properties will also change over time. The biochar aging process depends on the initial soil properties, climatic conditions, and local agronomic practices. Very few long-term studies have been carried out to determine how biochar affects microbial community dynamics over time. However, analogs exist in ancient Anthrosols, such as the Amazonian Dark Earths (O'Neill et al, 2009; Grossman et al, 2010) or African Dark Earths (Camenzind et al, 2018) that have been investigated for their microbial populations hundreds to thousands of years after the biochar was added. Most studies have commonly been conducted under controlled laboratory conditions to identify mechanisms. However, multi-year studies of purposeful biochar additions are now available.

Amending soil with biochar changes the local chemical and physical environment, which has several immediate and downstream influences on microbial communities. Changes in soil pH, Eh, and the release of toxic compounds (e.g., PAHs) are all affected by biochar (McCormick et al, 2013). Biochar

DOI: 10.4324/9781003297673-14

itself may also serve as a habitat for soil microbes (Guerena et al, 2015) and provide cellular resources, such as C and P (Ameloot et al, 2013). Biochar amendment may also enhance the colonization of roots by mycorrhizal fungi (Vanek and Lehmann, 2015), enhance biological N_2 fixation (Guerena et al, 2015), and alter food web interactions (Zhu et al, 2023). Lastly, its ability to participate in electron exchange reactions will enhance some soil processes, such as nitrification or methane production (Sun et al, 2021), but the CEC of biochar may also lead to the binding of signal molecules, thus interfering with some soil processes (Masiello et al, 2013). The effects of biochar on the abundance, activity, and diversity of soil organisms are thus extensive.

Responses of the soil microbial community to biochar addition exhibit high variability with few consistent response patterns between studies (Thies et al, 2015). However, as more

information has come to hand, several data reviews (meta-analyses) have emerged that cover many aspects of the effects of biochar on microbial life and function, including biochar effects on bacterial and fungal community composition, diversity, and richness (Li et al, 2020a; Zhang et al, 2018; Singh et al, 2022; Wang et al, 2023; Xu et al, 2023), microbial biomass (Zhou et al, 2017; Li et al, 2020a), microbial respiratory activity (Liu et al, 2016), enzyme activities (Pokharel et al, 2020; Feng et al, 2023) and N cycling (Liu et al, 2018; Xiao et al, 2019). These have helped to identify the main factors driving changes in microbial characteristics in biochar-amended soils and have led to a better understanding of how biochar might be prepared and used to achieve specific outcomes.

The potential mechanisms shown in Figure 14.1 for driving changes in soil biotic communities are not mutually exclusive and

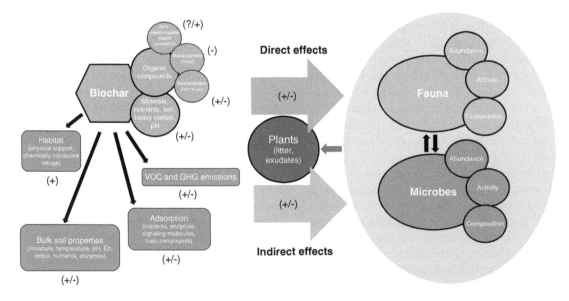

Figure 14.1 *Direct and indirect effects of biochar on soil biota abundance, activity, and composition, including through changes in soil properties and plant activity, as well as interactions between microbes, fauna, and plants (VOC, volatile organic compounds; GHG, greenhouse gasses; effects of biota on biochar also occur (Chapters 10, 11) but are not considered here)*

are likely to act concurrently. Thus, it is a challenge to disentangle causal mechanisms for observations made following the addition of biochar to soil. Caution must also be exercised in data interpretation because the adsorption characteristics of biochar can interfere with many methods used to characterize microbial populations and their activities, especially those that are based on soil extractions.

This chapter synthesizes the most important effects of biochars on soil biota, including the abundance, activity, and community composition of fauna and microorganisms, to identify what direct and indirect mechanisms are playing key roles in affecting soil organisms.

Biochar as a potential habitat and substrate for soil microorganisms

The porous structure of biochar (Chapter 5), its high internal surface area, and its ability to adsorb soluble organic matter (Chapters 17, 19, 23), gases, and inorganic nutrients (Chapters 8, 10, 17) could provide a favorable habitat for microbes to colonize, grow and reproduce, particularly once biochar is aged in soil. In the short term, some studies (Anderson et al, 2011; Chen et al, 2013) suggest that fresh biochar may be a highly selective habitat relative to non-pyrolyzed organic carbon (C) substrates. After three years in field soil, the most readily available C resources will have been mineralized, and microbial colonization of the biochar surface appears to become increasingly sparse (Quilliam et al, 2013). However, high colonization rates have been reported on fire-derived char after several hundred years in soil (Zackrisson et al, 1996) and on biochar particles from Amazonian Dark Earths (Tsai et al, 2009) or in African Dark Earths for fungi (Camenzind et al, 2018). Colonization clearly changes over time as biochar particles age in soil and their surface properties change (Liu et al, 2019).

Some components of biochar, particularly recondensed volatiles, are microbial substrates that may influence many soil processes and characteristics. Soil microbial populations can be affected by both the quality (Chapters 5–8) and quantity of the biochar added to the soil. Biochar qualities strongly depend on the feedstock and pyrolysis conditions used (Chapters 3, 5–8). Flash carbonization (Deenik et al, 2010) and low-temperature pyrolysis leave residual bio-oils and other re-condensed derivatives on the biochar surfaces (Chapters 3, 6, 8). In addition, biochar produced at low temperatures (400–500°C) may contain significant quantities of labile organic matter, which can have a strong influence on colonizing communities of microorganisms and those living in the vicinity of biochar particles (Azeem et al, 2023).

These condensed pyrolysis residuals may serve as substrates for microbial growth and metabolism (Azeem et al, 2023), depending on their composition, but they may also be toxic to plants (McClellan et al, 2007), and possibly toxic to many microbes (Wang et al, 2015; Zhu et al, 2017), at least in the short term. On the other hand, certain microbes may thrive on these residuals and use them as C, nutrient, and energy sources (Maestrini et al, 2014; Azeem et al, 2023). However, these substrates tend to break down relatively quickly and do not determine community composition over the longer term.

Microbial populations that establish themselves on the surfaces of fresh biochar may either be those that can express the enzymes necessary to break down the substrates available or inhabit the space for other reasons (e.g., moisture, nutrients, pH, Eh). Substrates not normally found in soil will take longer to be completely metabolized because microbes either lack the metabolic pathways or the necessary pathways need to be upregulated and thus, there are fewer organisms and associated enzymes able to metabolize them. Organisms colonizing fresh biochar that has residual condensates on its surfaces are likely to differ substantially from those colonizing biochar that has aged in soil. The dominant resources of later colonizing organisms will be the organic C and inorganic nutrients that become adsorbed to the biochar surfaces after any bio-oils, and other volatile and semi-volatile compounds that may have re-condensed on the biochar surface during pyrolysis are gone. Thus, as the surface characteristics change over time, there will be a succession of organisms colonizing these surfaces.

Biochar can alter both the quality and quantity of C substrates available in the soil and, through its adsorptive properties, affect the availability of key nutrient elements. The presence of biochar in soil may enhance the adsorption of DOC (Zhu et al, 2017), dissolved organic N (DON) (Guerena et al, 2013), inorganic nutrients, various gasses, as well as potentially toxic compounds, such as pesticides (Chapter 23), heavy metals (Chapter 21) and polycyclic aromatic hydrocarbons (Chapter 22), all of which can influence the abundance, composition, and activity of soil organisms (Figure 14.1).

Effects of biochar on microbial activity

Soil respiration

The effect of biochar on soil respiration is highly variable, with some studies reporting that biochar amendment reduces respiratory activity of microorganisms (Maestrini et al, 2014). However, this also depends on the length of time since biochar was applied. The reduced respiratory activity could indicate that biochar is: (i) limiting the metabolic activity of biochar-colonizing microorganisms potentially through the lower substrate or nutrient availability, or dilution (Chapter 17); (ii) changing population structure; (iii) increasing C use efficiency (Zhou et al, 2017; Giagnoni and Renella, 2022); (iv) increasing CO_2 fixation by chemolithotrophs associated with biochar particles; (v) decreasing population abundance (Wang et al, 2015); or (vi) some combination of these responses, where many of these potential mechanisms remain unexplored (Chapter 17). One mechanism, the potential toxicity of some biochars, at least in the short-term, has not been explored fully (Wang et al, 2015; Zhu et al, 2017).

Biochar amendment had no significant effect on soil CO_2 fluxes when averaged across 395 paired observations despite increasing soil organic C content by 40% (Liu et al, 2016). Soil CO_2 fluxes vary by land use type, biochar application rate, soil texture, and pH, with the highest CO_2 fluxes from soils receiving biochar made from manure or crop residues. Whereas, biochar amendment led to a significant increase in CO_2 emissions when averaged across 73 studies, with CO_2 emissions greater in finely-textured, acidic soils (Atilano-Camino et al, 2022).

The metabolic quotient (qCO_2) is the microbial respiration rate (CO_2-C) per unit microbial biomass C, with lower values indicating an increase in C use efficiency of the population. A significant reduction in qCO_2 was observed from a range of pot and laboratory studies but was not reduced significantly in the field trials examined (Zhou et al, 2017). Over all experiment types, qCO_2 was reduced by 16% on average, indicating a significant increase in C use efficiency when biochar was applied. Reduced respiratory activity relative to microbial biomass is a common finding that is in part related to improved soil physical and chemical conditions leading to increased C-use efficiency within the microbial community (Giagnoni and Renella, 2022).

Soil enzyme activity

Bacteria and fungi rely on the extracellular enzymes they produce to degrade substrates into smaller molecules that can then be taken up into their cells and used for primary and secondary metabolism (Thies and Grossman, 2023). As such, surfaces become very important regardless of these being those of a soil aggregate, plant root, particle of clay, soil organic matter, or biochar. Biochar will affect activity of extracellular enzymes differently depending on the molecular locations on these proteins that interact with the biochar surface and the nature of the local chemical environment, such as a change in pH. If the enzyme active site is exposed, functional, and free to interact with its milieu, the same or even increased activity could occur. However, if the active site is obscured or the enzyme is denatured, decreased activity would result (Lammirato et al, 2011). Certain classes of enzymes are more active and others less so in biochar-amended soils (Zhang et al, 2019; Pokharel et al, 2020, 2021), likely based on their interactions with soil particles and their molecular composition and folding characteristics in relation to how (or whether) they become adsorbed to biochar surfaces or are denatured by the chemical environment surrounding the biochar particles. Also, various compounds contained in bio-oils and

condensates on the biochar surface may act as enzyme inhibitors (competitive, non-competitive, or uncompetitive) thus altering enzyme function and turnover.

A wide range of effects on enzyme activities has been observed (Figure 14.2). In some studies, biochar application significantly increased the activities of C- and N-acquiring enzymes by 9 and 15%, respectively, but did not significantly affect the activity of P-acquiring enzymes (acid, neutral, or alkaline phosphatase) (Chen et al, 2022a). Further analysis of the ratios of enzyme activities important to the functioning of the C, N, and P cycles, respectively, indicated that soil microbial metabolism was limited by C relative to N and P under biochar amendment and that wood biochar caused the strongest microbial C limitation, followed by crop residue biochar. Microbial C limitations will hinder ecosystem services provided by microbial communities. Thus, adding easily mineralizable C and N sources to support microbial growth and crop production with biochar amendment may be warranted.

In another summary analysis, the activities of N- and P-cycling enzymes increased in biochar-amended soils by 14 and 11%, respectively, across 43 studies, while the activity of C-cycling enzymes decreased by 6% overall (Zhang et al, 2019). Increased activity of N and P cycling enzymes was most pronounced where the biochars were prepared under low temperatures using feedstocks with high nutrient contents and when applied to acidic or neutral soils. C-cycling enzyme activities were suppressed where biochars were prepared under high temperatures, where they were applied at high rates, and where the feedstocks were herbaceous or lignocellulosic materials. Results suggested inhibition of C-cycling enzymes was mainly related to the adsorption and inhibition effects of the biochars added.

The effects of biochar on soil enzyme

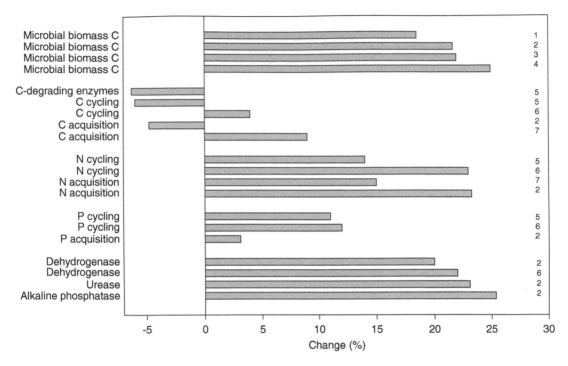

Figure 14.2 *Change in microbial assessment variables with biochar additions. Sources: 1, Xu et al (2023); 2, Pokharel et al (2020); 3, Liu et al (2016); 4, Zhou et al (2017); 5, Zhang et al (2019); 6, Liao et al (2022); 7, Chen et al (2022a)*

activities can also vary widely by soil type, biochar characteristics, and enzyme type (Pokharel et al, 2020, 2021). However, most studies indicate no significant negative effects of biochar on enzyme activities. Rather, biochar application has been found to significantly increase urease (23%), alkaline phosphatase (25%), and dehydrogenase (20%) activities in receiving soils. Increases observed in urease and dehydrogenase activities were more pronounced with biochars prepared at lower temperatures (350–550°C), with a lower C/N ratio and higher pH (>10).

Biochar pyrolysis temperature was also the dominant factor controlling C, N, and P cycling enzyme activities in biochar-amended soils across 129 studies (Liao et al, 2022).

Low-temperature biochars (<500°C) increased the activities of most enzymes involved in microbes acquiring C (4%), N (23%), and P (12%) as well as the activity of dehydrogenase (22%). While C-acquiring enzymes were more sensitive to soil and experimental conditions, N- and P-acquiring enzyme activities increased consistently when low-temperature biochar was used regardless of soil or experimental conditions. In contrast, high-temperature biochars (≥ 500°C) did not have a significant influence on the activities of C, N, or P-acquiring enzymes or dehydrogenase activity. Other factors influencing enzyme activities were biochar application rates, soil C content, soil pH, and texture (Liao et al, 2022).

Nitrogen cycling

Microorganisms play key roles in N cycling in soil where they control mineralization and immobilization reactions, nitrification and denitrification reactions, as well as nitrogen (N) fixation. Adsorption of ammonium (NH_4), nitrate (NO_3), DON, or DOC to biochar may lead to important changes in soil N cycling processes that are discussed in Chapters 15–17. Nitrogen adsorption to biochar may also alter N uptake by plants and soil biota and the oxidation of NH_4 to NO_3 by nitrifying bacteria (Chapter 16) or subsequent NO_3 leaching (Chapter 19) or reduction to N_2 by denitrifying bacteria, with the potential for reducing nitrous oxide (N_2O) release (Chapter 17). Lastly, the fixation of N_2 into NH_3 by N-fixing bacteria is frequently enhanced by biochar amendment (Chapter 16).

Abundance, diversity, and community structure of organisms interacting with biochar

Soil microbial and fauna populations are extremely diverse and composed of assemblages of bacteria, archaea, fungi, algae, protozoa, nematodes, arthropods, and a diversity of invertebrates, spanning a range of sizes, functions, and response times. Interactions among existing biota populations, soil chemical and physical properties, and root exudates (that vary temporally and in response to plant needs in a given environment, as well as among different plant species) will determine the composition of soil biota.

Microbial biomass

Microbial biomass C is most frequently measured by the use of chloroform fumigation and extraction (MBC_{FE}) or by measuring and characterizing phospholipid fatty acids (MBC_{PFLA}). While PLFA analysis can be used to determine MBC, its main purpose is to identify major groups of organisms in a sample. Biochar amendments increased both MBC_{FE} and MBC_{PLFA}, however, the increase in MBC_{FE} was greater than that measured using PLFAs (Li et al, 2020a). This insight suggests matching techniques across studies for better future comparisons.

Increases in MBC_{FE} of between 18–25% across hundreds of observations have been reported as a result of biochar soil amendment (Liu et al, 2016; Zhou et al, 2017; Li et al, 2020a; Pokharel et al, 2020). Increased MBC was significant in field experiments, but not in laboratory incubations. MBC was also higher in N-fertilized and vegetated soils. Alkaline biochars produced at lower temperatures (350–550°C) with lower C/N ratios (from more nutrient-rich feedstocks) resulted in higher microbial biomass, supporting observations made on the effects of biochar on enzyme activities and suggesting that the microbial populations are responding to an increase in resources (both C and other nutrients), as well as increases in soil pH.

Bacteria and archaea

Abundance

Bacterial biomass has been measured with some frequency and is often found to increase in biochar-amended soils, particularly in the short-term (Zhou et al, 2017; Li et al, 2020a; Pokharel et al, 2020). Gram-negative bacteria and actinomycetes appear to respond most quickly and consistently to changes caused by adding biochar to soil. These populations contain many members that are fast-growing opportunists that can respond quickly in terms of growth and metabolism when sources of energy, C, and nutrients become available. Dead microbes would also provide an easily mineralizable source of needed cell constituents and this would be reflected in a short-term increase in surviving populations that can take advantage of the change in availability of organic C and key nutrients. Once this organic matter is consumed, a new equilibrium would be reached and differences between biochar-amended and unamended soils would decrease, as has been observed in longer-term studies (Castaldi et al, 2011; Quilliam et al, 2012; Ventura et al, 2014).

Bacterial diversity

Several recent meta-analyses have examined changes in bacterial diversity in response to biochar soil amendment (Li et al, 2020a; Singh et al, 2022; Wang et al, 2023; Xiang et al, 2023). Common findings among these studies are that biochar soil amendment has a positive effect on overall bacterial diversity as measured by the Shannon or Chao1 indices, with some variations across the different meta-analyses. General findings suggest:

- Bacterial diversity increases when biochar is applied.
- The soil organic C/N ratio was the leading factor influencing the bacterial Shannon and Chao1 indices following biochar addition (Xiang et al, 2023).
- Biochar made from herbaceous feedstocks, but not from wood, manure, or ligno-cellulosic waste, significantly increased bacterial diversity.
- Biochar application significantly increased bacterial diversity in coarse and medium-textured soils to a similar extent, but not in fine-textured soils.
- Increases in total microbial diversity were greater in acidic and sandy soils with low soil organic carbon content (Li et al, 2020a).
- Lower-temperature biochar (<500°C) had a larger effect in increasing diversity compared with higher-temperature biochar, although biochar produced at all temperatures significantly increased diversity.
- Bacterial diversity was significantly increased at medium and low application rates but not at high (>80 t ha^{-1}) rates with no difference between the medium and low rates (<39 t ha^{-1}).
- Impacts of biochar on functional diversity (i.e., C substrate richness consumed by soil microbes) increased with higher biochar application rates, which might relate to increased genetic richness (Xu et al, 2023).
- Biochar applications increased the ratios of soil fungi to bacteria (F/B) and the ratios of Gram-positive bacteria to Gram-negative bacteria (G$^+$/G$^-$) (Zhang et al, 2018).
- Biochars produced at low temperatures added to soils with lower pH and nutrients lead to a greater increase in F/B ratios. Residue biochars applied in dryland soils increased the ratios of G$^+$ bacteria to G$^-$ bacteria the most (Zhang et al, 2018).

Changes in biochar-associated microbial community composition are not consistent across studies (Li et al, 2020a) and are in general more challenging to capture and interpret than abundance. Higher diversity as measured by the Shannon (Li et al, 2020a, Singh et al, 2022, Xiang et al, 2023) and Chao1 (Chen et al, 2013; Wang et al, 2023) indices are often found. At the highest application rate of 40 t biochar ha^{-1}, abundance of Burkholderiaceae, Acidobacteria (Gp 3), and Anaerolineaceae (Chloroflexi) and members of the Gammaproteobacteria, Acidobacteria, and Chloroflexi generally increased; whereas, significant decreases in members of the families Hydrogenophilaceae, Methylophilaceae, Chlorobi, Nitrospira, Planctomycetes, Bacteroidetes, and a general decrease in Betaproteobacteria were observed. With biochar additions to pepper plants (Kolton et al, 2011), 92–95% of the bacterial population was affiliated with the phyla Bacteroidetes (increased from 12 to 30%), Proteobacteria (decreased from 71 to 47%), Actinobacteria, and Firmicutes. Many species among these phyla have proven to possess plant growth-promoting properties. Among the genera enriched in biochar-amended soil were *Chitinophaga* and *Cellvibrio* which degrade chitin and cellulose, and *Hydrogenophaga* and *Dechloromonas* shown to degrade a range of aromatic compounds. Biochar-amended potting mixes showed a preponderance of bacterial genera with plant growth-promoting (PGP) abilities, among them were *Pseudomonas*, *Brevibacillus*, *Streptomyces*, *Mesorhizobium*, and *Bacillus* species (Graber et al, 2010). Enrichment of PGP species may be responsible for the increase in growth observed in greenhouse-grown tomato and pepper in these experiments. Low doses of some compounds associated with biochar that would be toxic at higher concentrations, may have stimulated plant growth at low doses (hormesis) (Chapter 23).

Fungal abundance and diversity

Fungal community responses to biochar soil amendment are more variable than those of bacteria, with some summaries even indicating no effect (Wang et al, 2023) or a negative effect (Chen et al, 2013) on fungal

diversity, while others suggest a modest increase in fungal diversity (Singh et al, 2022). Some general findings include:

*   Biochar application significantly increased fungal diversity compared to the control in both greenhouse and field studies but not in the laboratory.
*   Biochar from lignocellulosic waste and herbaceous feedstocks significantly increased fungal diversity compared to the control, while there was no effect from wood biochar. For biochar from animal waste, there was insufficient information to assess fungal biodiversity.
*   Fungal diversity was significantly increased in fine-textured soils, but not in medium- or coarse-textured soils.
*   Biochar prepared under both high and low pyrolytic temperatures (>500°C or <500°C) significantly increased fungal diversity compared to the control, with effects being more pronounced with biochar prepared at lower temperatures.
*   Fungal diversity was significantly increased at any application rate, but more so at lower application rates (<39 t ha^{-1}) where it increased by 43%.

Saprophytic fungi

Saprophytic fungi, as decomposers, are particularly important, as they may influence the persistence and modification of biochar materials in soil. In contrast to bacteria, fungi have a hyphal, invasive growth habit (aptly likened to tunneling machines; Wessels, 1999), which gives them access to the interior of materials. Fungi also have a wide range of enzymatic capabilities, which further highlights the need to study fungi as decomposers of biochar. For example, fungi (*Trichoderma* and *Penicillium* spp) can contribute to the depolymerization of coal (hard coal, sub-bituminous coal, and lignite) via the production of enzymes such as Mn-peroxidase and phenoloxidase (Laborda et al, 1999). Fungal laccase was also shown to degrade biochar (Hockaday, 2006).

Biochar particles may be argued to serve as a habitat for soil fungi, more so than for bacteria, since hyphae can efficiently explore their interiors. Hyphae are specifically adapted for growing on solid surfaces and invading substrates and can exert large penetrative mechanical forces. Fungi have evolved a high degree of metabolic versatility that also enables them to thrive in a variety of environments that are too harsh for other organisms. As biochar ages in soil, pores may become occluded with organic matter, which may block colonization by bacteria. Fungal hyphae, however, could explore occluded pores and benefit from metabolizing their contents. Fungal colonization of biochar will depend highly on the nature of the biochar itself as well as on the amount and nature of organic molecules that sorb to it in soil. Sorbed organic compounds may serve as significant sources of C and energy for soil microbes. In an approximately 100-year-old biochar, filamentous growth of unidentified microbes was visualized inside the aged biochar particles (Hockaday et al, 2007). These filaments were about 4 μm in diameter, and therefore likely to be fungi rather than actinomycetes. Given that fungi can inhabit both the exterior and the interior of biochar particles, questions regarding the community composition and identity of these (presumably) saprophytic fungi arise. On the other hand, a high pH of biochar may favor bacteria over fungi (McCormack et al, 2013), and the net effects may not always be clear.

Mycorrhizal fungi

Interest in the interactions between mycorrhizae and biochar is likely due to three reasons: the ubiquity and ecological importance of mycorrhizae, their sensitivity to management, and their responsiveness to biochar amendment. First, mycorrhizal fungi are ubiquitous and are key components in virtually all biomes (Treseder and Cross, 2006). Therefore, it is important to understand how any soil additive, including biochar, may affect their performance. Arbuscular mycorrhizal fungi (AMF) colonize most of the important crop species (e.g., corn, rice, wheat), so they are also important with regard to agroecosystem productivity and sustainability. Second, mycorrhizae are sensitive to management interventions (Schwartz et al, 2006), such as adding biochar, and it is tempting to speculate on the possible synergistic effects of mycorrhizal inoculation and biochar application in enhancing soil quality and plant growth. Applying biochar to soil stimulated the colonization of crops by AMF. For example, the mycorrhizal inoculation potential of wheat in Australia increased even two years after biochar application (Solaiman et al, 2010). Likewise, root colonization of sorghum increased when biochar from apple wood sawdust was added in a short-term (4 weeks) greenhouse experiment (LeCroy et al, 2013). On the other hand, inoculation with AM fungi provided no benefit for tomato growth and yield with or without the addition of biochar (Nzanza et al, 2012). Lastly, the majority of the studies reported in the literature show a strongly positive effect of biochar on mycorrhiza abundance (Warnock et al, 2007; Solaiman et al, 2010; Le Croy et al, 2013; Guerena et al, 2015; Vanek and Lehmann, 2015), which prompts questions about underlying mechanisms.

Several potential mechanisms may be responsible for biochar effects on mycorrhizae (Warnock et al, 2007; LeCroy et al, 2013; Vanek and Lehmann, 2015). Suggested mechanisms include physical, chemical, and biological interactions;

these are, of course, strongly interrelated and would be acting concurrently (Figure 14.1).

Microorganisms including hyphae and spores of AMF can be visible on extracted biochar particles, which suggests that AMF colonize biochar particles (Saito, 1989). It was suggested that the porous nature of the biochar particles or that decreased competition from saprophytes (for which this habitat was presumed to be less suitable) could have contributed to this (Saito, 1989; Saito and Marumoto, 2002). AMF root colonization increased in the presence of ground biochar as opposed to non-ground material, attributed to the porous nature of biochar (Ezawa et al, 2002); however, high application rates of 30% (v/v) were used. AMF distinctly preferred biochar pores without an apparent nutrient or moisture benefit (Vanek and Lehmann, 2015). Therefore, surface phenomena and micropore habitats play an important role in affecting microbial abundance and composition, and thereby likely interactions with plant roots.

Mycorrhizal fungi and the C supply from their hosts can react sensitively to these variables. Adsorption of inhibitory compounds or their sequestration and the slow release of positive signaling molecules are examples of how biochar could interfere with root-microbe or other signal exchanges (Masiello et al, 2013). AM fungi can respond to several chemical signals (such as flavonoids, sesquiterpenes, and strigolactones), which alter root growth or branching (Akiyama et al, 2005). While a likely mechanism, little direct evidence has been provided for biochar signal interference among AM fungi in the soil. However, biochar can adsorb compounds used in intercellular signaling and regulation of gene expression among bacteria (Masiello et al, 2013), and reduce DOC in soil that then changes signaling between roots and bacteria (Del Valle et al, 2020).

Soil fauna

Compared to other aspects of biochar research, relatively little direct data are available yet on the effects on soil faunal communities. Biochar application can have a dual impact on faunal communities, influencing them through both direct and indirect mechanisms (Figure 14.1). The first category of impacts resulting from applying biochar includes the release of usable organic matter fractions, salts, pollutants, or volatile organic compounds (VOCs). These substances can directly influence soil conditions and have implications for the ecological functions performed by soil fauna, since biochar can be ingested by earthworms (Topoliantz and Ponge, 2003, 2005; Elmer et al, 2015; Ferreira et al, 2021), collembolans, enchytraeids (Domene et al, 2015a,b), and nematodes (Li et al, 2020b). The second category involves changes in soil habitat conditions such as moisture, pH, and aeration, nutrients, or sorption of toxins or signaling molecules, as well as shifts in biological groups interacting with soil fauna. These changes in habitat conditions and biological interactions can ultimately affect the ecological functions carried out by soil fauna (Domene, 2016).

Soil fauna populations are bottom-up regulated by microorganisms and plants, as the energy flow within food webs occurs through either detritus-based or plant-based energy channels (Bardgett, 2005). The detritus channel encompasses microbial feeder fauna, including fungivores and bacterivores, as well as detritivores (consuming plant or animal decaying materials). In contrast, the plant-based channel involves herbivore animals that feed on living plant roots. In addition to any direct effects, any effects by biochar on these two channels would indirectly influence soil fauna (Figure 14.1). Conversely, any impact of biochar on soil fauna would impact microorganisms and plant productivity through a top-down regulation (Chapter 23).

While biochar unquestionably impacts soil fauna communities, certain soil fauna also influence biochar (Chapters 10, 11, 16). First, detritivores can alter the fate of biochar in soil by fragmenting it, inoculating it with microorganisms, charging it with soil enzymes, enclosing it in fecal pellets or casts, or redistributing it in the soil profile (Domene et al, 2016). This is evident in the altered shape of biochar particles found in earthworm casts due to abrasion in their digestive tract (Ferreira et al, 2021). Secondly, fauna can also influence beneficial biochar properties, potentially altering their capacity to impact soil functioning. For example, earthworms can enhance the capacity of biochar to stimulate plant productivity as well as the abundance of soil collembola and fungal biomass (Garbuz et al, 2020). Fauna can also charge biochar with enzymes after passing it through their digestive tract (Sanchez-Hernandez et al, 2018). Conversely, earthworms reduce the capacity of biochar to mitigate greenhouse gas emissions (Augustenborg et al, 2012; Wang et al, 2021; Wu et al, 2021; Gong et al, 2022) and bioavailability of heavy metals (Huang et al, 2020, Noronha et al, 2022).

Earthworms

By far the most information about biochar effects on soil fauna is available for earthworms. Interestingly, both attraction and avoidance can be observed. Avoidance is caused by the release of potentially harmful substances such as NH_3 (Liesch et al, 2010),

PAHs (Prodana et al, 2019; Shi et al, 2021), and heavy metals (Amaro, 2013; Shi et al, 2023) from certain biochars. A high mortality was observed after introducing poultry litter biochar at 2.4% (equivalent to 22 t ha^{-1}) (Liesch et al, 2010). Additionally, the coarse texture of some biochars has been also claimed to cause avoidance in both anecic and endogeic earthworms (Sanchez-Hernandez, 2019).

Preference by earthworms for biochar has been linked to factors such as increased soil moisture (Tammeorg et al, 2014) and pH (Topoliantz and Ponge, 2005; Van Zwieten et al, 2010; Busch et al, 2011; McCormack et al, 2013), which seem to make these mixtures more inviting. Other authors have suggested a potential trophic relationship between earthworms and biochar. Given earthworms' habit of ingesting soil particles, it is not surprising that they might also ingest biochar (Lehmann et al, 2011; Weyers and Spokas, 2011; Ameloot et al, 2013). Yet, the mechanisms for such preference remain unclear and could be driven by the nutritive value of microbial communities on biochar surfaces, its liming properties, or perhaps a more passive mechanism (Domene et al, 2016). Studies have offered conflicting insights, and while some authors have observed mere particle manipulation, with earthworms pushing aside char particles rather than ingesting them (Topoliantz and Ponge, 2003), others have demonstrated an active consumption even when alternative food sources were available (Elmer et al, 2015). The latter has been attributed to the acquisition of calcium that could serve in the production of the mucilaginous gel used to coat earthworm castings, as this response was unrelated to microbial abundance. In addition, a capacity for the selection of biochar particles with lower pollutant burden by earthworms has also been demonstrated (Ferreira et al, 2021).

The earthworm's trophic interest in biochar, if true, should be associated with improved performance (growth, reproduction, or survival), but this is not shown in the available literature. Growth promotion seems to be a delicate balance, with concentrations below 10% failing to yield positive effects (Topoliantz and Ponge, 2005; Liesch et al, 2010; Prodana et al, 2019; Ferreira et al, 2021; Shi et al, 2023) and concentrations above this threshold often resulting in negative impacts (Li et al, 2011; Gomez-Eyles et al, 2011; Elliston and Oliver, 2020). Some studies even point to growth inhibition at concentrations below 10%, coupled with decreased activity of antioxidant enzymes and genotoxicity (Sanchez-Hernandez et al, 2019; Anyanwu et al, 2020; Huang et al, 2020; Shi et al, 2021; Jia et al 2023). The discourse is similar regarding the effects of biochar on earthworm survival. Concentrations above 10% tend to reduce their abundance (Anyanwu et al, 2018), while concentrations below this threshold typically do not negatively impact earthworm survival (Prodana et al, 2019; Sanchez-Hernandez et al, 2019; Elliston and Oliver, 2020, Huang et al, 2020; Ferreira et al, 2021), with few exceptions (Liesch et al, 2010; Jia et al, 2023; Shi et al, 2023). However, feedstock and pyrolysis temperature influence the extent of the effects. For example, earthworm mortality increased above 10% for corncob biochar, below 8% with sewage sludge biochar, and below 5% for cow manure biochar, for the latter increasing with pyrolysis temperature (350, 550, and 750°C) (Shi et al, 2021). Additionally, no effects on reproduction had been observed at 10% and greater rates of apple wood biochar (Li et al, 2011). All these studies give rise to the question of what the underlying causes are of the negative effects of some biochars. While the mechanisms remain elusive, a spectrum of possibilities has been proposed: bioavailable and toxic compounds generated during pyrolysis (Gale et al, 2016; Jia et al, 2023), the role of particle size that increases the bioavailability

of compounds such as PAHs (Prodana et al, 2019), decomposition end products such as NH_3 released from easily mineralizable organic N (Domene et al, 2008; Liesch et al, 2010; Edwards and Arancon, 2022), the reduction of soil water availability that affects the aqueous layer on earthworms and therefore impairing their O_2 exchange (Sun et al, 2022), or excessive pH increases (Xu et al, 2022).

The abundance of earthworms ranges from the preference for biochars in the field (Van Zwieten et al, 2010; Kamau et al, 2019), no effects (Tammeorg et al, 2014) to negative effects (Dermiyati et al, 2018; Briones et al, 2020; Xu et al, 2022; Shi et al, 2023). Intriguingly, contradicting results are often found between controlled laboratory settings and field studies. Such inconsistences might be at least partly shaped by factors such as field aging of biochar which may mitigate negative effects seen in laboratory experiments as seen for *Aporrectodea caliginosa* exposed to 30 t ha^{-1} of spruce chip biochar (Tammeorg et al, 2014). This could be attributed to the toxicity reduction experienced by some fresh biochar after aging in soil (Weyers and Spokas et al, 2011; Anyanwu et al, 2018; Whalen et al, 2021; Jeffery et al, 2022; Honvault et al, 2023). Over a significantly extended temporal scale, Brazilian Anthropogenic Dark Earths situated within mature forests exhibited higher earthworm taxa richness than the nearby reference soils as well as a distinct community (Conrado et al, 2023).

Enchytraeids

Enchytraeids, also known as potworms, are the taxa with the highest proportion of edaphic species (Anthony et al, 2023). Often underestimated, their role is pivotal in decomposition and mineralization within diverse soil environments (Cole et al, 2000; Coleman et al, 2004). Due to their translucent bodies, we know of their ability to ingest biochar particles (Domene et al, 2016). As detritivores, they ingest both mineral and organic particles (Coleman and Wall, 2014) and can either be affected by biochar or affect biochar properties.

Studies in soils after wildfires have reported adverse impacts on this group linked to pH increases (Liiri et al, 2007; Lundkvist, 1998; Nieminen, 2008; Nieminen and Haimi, 2010), as enchytraeids are expected to thrive better in acidic soils (McCormack et al, 2013). In laboratory environments, the existing body of work largely fails to confirm detrimental effects on survival or reproduction at rates commonly employed for biochar application (Domene et al, 2015b). Instances of negative outcomes have been attributed to drastic pH increases or the presence of PAHs (Marks et al, 2014). These findings mirror earlier observations, revealing robust enchytraeid activity in soil enriched with char following wildfires, as evident from the abundance of fecal pellets (Topoliantz et al, 2006).

Field experiments show inconsistent results regarding the impact of biochar on enchytraeid populations. The introduction of *Miscanthus* biochar at rates of 25–50 t ha^{-1} led to increasing enchytraeid populations over two years, accompanied by an upsurge in microarthropods, particularly collembola (Briones et al, 2020). Curiously, earthworm populations experienced a decline. The reasons behind this positive response remain unclear, with isotope measurements discarding changes in soil detritus- or plant-based energy channels and pointing towards indirect influences, such as soil structure or physicochemical conditions alterations. Conversely, hay-derived biochar added at a rate of 10 t ha^{-1} to a forage crop over three years did not affect enchytraeid populations or other fauna (Jeffery et al, 2022).

Microarthropods

Early studies indicated that soil enriched with char after wildfires or with charcoal from charcoal production exhibits a high abundance of microarthropod fecal pellets (Bunting and Lundberg, 1987) and higher collembola abundance and richness (Uvarov et al, 2000). Collembolans are a diverse group that are also found to ingest biochar (Domene et al, 2016).

Laboratory tests using collembolans found no evidence of biochar avoidance (Amaro et al, 2013), and avoidance was only noted under laboratory conditions at high application rates of above 25 or even 225 t ha^{-1} (Conti et al, 2018; Gruss et al, 2019) or only for short periods of time of several weeks (Domene et al, 2015b). Biochar also showed no significant impact on survival when applied in field experiments even at higher rates of above 225 t ha^{-1} (Domene et al, 2015a; Conti et al, 2018). Similarly, reproduction was generally unaffected at low application rates (Hale et al, 2013; Gruss et al, 2019) or only moderately at high application rates (Domene et al, 2015a). Negative effects at low application rates have been reported for specific biochars such as those made by gasification from grape marc (Conti et al, 2018). These divergent results are not easily linked to specific parameters even though they are often suggested to be a result of increases in pH (Marks et al, 2014; Gruss et al, 2019), PAHs, and heavy metals (Conti et al, 2018). On the other hand, the existence of a trophic mechanism driving collembola behavior and performance is also plausible (Lehmann et al, 2011), and evidence includes observations of reduced avoidance of high concentrations of corn stover biochar when microbial biomass was high (Domene et al, 2015b) and instances of wood and sewage sludge biochars enhancing collembola reproduction (Marks et al, 2014). Yet, biochar is also predicted to decrease collembola populations in acidic soils due to anticipated reductions in fungal biomass stemming from increased soil pH (McCormack et al, 2013). Collembolans consume fungi and alterations in fungal communities due to pH changes may therefore directly impact their abundance. The avoidance of collembolans in soil-biochar mixtures of already alkaline soils may partially be explained by this mechanism (Marks et al, 2014). However, this might mostly be true for strictly fungivorous taxa, as collembolans are a diverse group with varied food preferences, nowadays considered to be more omnivorous than previously thought (Feng et al, 2019).

Field studies show mixed results, mostly reporting no effects at agronomic application rates in temperate soils (Domene et al, 2014; Pressler et al, 2017; Andrés et al, 2019; McCormack et al, 2019) and even at higher rates in subtropical and neutral tropical soils (Prober et al, 2014; Dermiyati et al, 2018). While collembola community indicators such as density, richness, and diversity remained unaffected in soils at former pre-industrial charcoal kiln sites, more epigeic species were found in croplands and grasslands and more euedaphic species in forests (Pollet et al, 2022) likely related to soil moisture (Ferrín et al, 2023).

Clear positive impacts on microarthropod abundance and diversity have been observed in field studies (Eo et al, 2018; Briones et al, 2020; Garbuz et al, 2021). However, these effects are not consistently attributed to changes in soil pH (McCormack et al, 2013). It is possible that relevant pH shifts have simply not been detected in bulk soil but do operate at the spatial scale of the organism (Lehmann et al, 2015).

Field-related adverse effects of biochar on soil microarthropods (Conti et al, 2018; Marks et al, 2016) frequently reflect the toxicity witnessed in laboratory experiments (Marks et al, 2014). However, the underlying mechanisms are not well established. Over time, populations usually recover from these effects, but specific indicators may still change, especially at high application rates (Llovet et al, 2021; Xu et al, 2022). Six years after applying biochar to cereal crops, faunal feeding rates recovered, yet collembola and nematode diversity stayed lower, especially with the highest rate of 50 t ha^{-1} (Llovet et al, 2021). These effects were not linked to pH or trophic shifts but were related to higher PAH contents in the biochar. Conversely, in a poplar plantation, increased pH coincided with reduced microarthropod abundance and diversity, affecting mites and collembolans, and decreasing fine root biomass, while microbial biomass was unchanged (Xu et al, 2022).

While laboratory experiments suggest that biochar might affect soil microarthropod populations by changing microbial levels, this connection lacks definitive confirmation in most of the field studies currently published (McCormack et al, 2019; Llovet et al, 2021; Xu et al, 2022), with specific cases being an exception (Garbuz et al, 2021).

Nematodes

Nematodes, commonly referred to as roundworms, encompass diverse trophic levels, ranging from bacterivores and fungivores to plant feeders, predators, and omnivores, therefore playing diverse roles in soil, functioning as decomposition regulators and even acting as plant pathogens (Coleman and Wall, 2014). In soils enriched with char due to fires or charcoal production, an increase in nematode abundance has often been observed (Uvarov et al, 2000; Matlack et al, 2001) but not always (Soong et al, 2017). Some of these studies also demonstrated nematode community shifts favoring bacterivores but the connection with biochar is not straightforward and more related to disturbances caused by burning (Matlack et al, 2011; Soong et al, 2017). This contrasts with some demonstrated effects after biochar addition in laboratory studies and to arable land outlined below.

Under highly controlled laboratory conditions, adverse effects on the bacterivore *Caenorhabditis elegans* have been linked to the suppression of an anticancer gene connected to resistance against carcinogenic PAHs when exposed to a hydrochar and ADE (Chakrabarti et al, 2011), as well as oxidative stress stemming from free radicals within the biochar at very high application rates (Lieke et al, 2018). In contrast, a preference for peanut shell biochar over other substrates has been observed in this species and attributed to the emission of unidentified volatile compounds (Li et al, 2020b).

Greenhouse mesocosm studies including plants indicate that biochar can either increase or decrease nematode abundance (Fox et al, 2014; Al-Fraih, 2015; Liu et al, 2020a), associated with trophic group shifts, though mechanisms have not been demonstrated and trends are contradictory. For instance, after introducing *Miscanthus* biochar at 1–2% (equivalent to 26–52 t ha^{-1}), bacterivorous nematodes increased while omnivorous nematodes decreased, alongside elevated pH and increased abundance of growth hormone-producing bacteria, possibly linked to enhanced plant growth (Al-Fraih, 2015). However, in another study, two years after applying 60 t ha^{-1} biochar to tropical acidic soil, bacterivores, fungivores, and herbivores declined, with increased pH, PAHs, crop yield, and microbial abundance suggesting intricate interactions (Liu et al, 2020a). Similarly, field studies display mixed outcomes, with no change being observed from months to several years after single biochar additions to neutral and alkaline soils (Biederman and Harpole, 2013; Zhang et al, 2013; Pressler et al, 2017; McCormack et al,

2019; Domene et al, 2021) or even after repeated biochar additions (Jeffery et al, 2022). Positive impacts on nematode density have been also noted in both alkaline and acidic soils, even years later and at lower rates (Eo et al, 2018; Liu et al, 2020b; Llovet et al, 2021). Exceptions include a study with *Prosopis juliflora* biochar added to a tropical maize crop, causing negative effects at rates of 5–10 t ha^{-1} (Kamau et al, 2019).

While in most cases, total nematode abundance is not affected by biochar, community shifts are common. However, they do not show a consistent pattern, similar to the results from laboratory studies. The suppression of herbivore nematodes found in some studies with increases of fungivores in others, holds agricultural significance due to their roles as pests and vectors of plant diseases (Fry, 1982). However, the mechanisms driving the varying impact of biochar remain unexplained.

Increases in fungivore nematodes can coincide with the suppression of pathogenic fungi, as demonstrated in a tropical acidic soil with rice husk addition (Eo et al, 2018). Such effects could align with the proposed pH-mediated enhancement of bacteria over fungi with biochar additions (McCormack et al, 2013). An alternative explanation is that biochar may make bacteria less accessible to microbivore nematodes (Mulder and Vonk, 2011; Gul et al, 2015; Liu et al, 2020a), while fungivore nematodes can readily prey on unprotected fungi. This could result in a higher ratio of fungivore to bacterivorous nematodes in biochar-treated soils (Liu et al, 2020a). Finally, either increased microbial biomass or increased plant growth (Chapter 13) in the presence of biochar could amplify the energy flux to soil ecosystems through microbivores and herbivores (Zhu et al, 2023), and elevate fungivore/bacterivore ratios (Oladele et al, 2021).

A variety of mechanisms have been proposed for how biochar may affect herbivore nematodes. Greater abundance may result from the adsorption and reduction of decomposition products (Domene et al, 2021), such as NH_3 thought to be toxic to herbivore nematodes (Nahar et al, 2006; Oka, 2010). Conversely, the suppression of herbivorous nematodes might occur if certain types of biochars amplify microbial abundance, therefore fostering antagonistic taxa (Jatala, 1986; Siddiqui and Mahmood, 1999). The frequently observed inverse relationship between herbivore and microbivore nematodes (Rahman et al, 2014; Cole et al, 2021; Oladele et al, 2021; Liu et al, 2020b, Domene et al, 2021) is consistent with this explanation. Other reasons for decreases in herbivore nematodes include pH increases (Matlack, 2001; Cole et al, 2021) and the release of toxic compounds (Huang et al, 2015). An interference of biochar with the detection of plant exudates by herbivore nematodes, curtailing their ability to find plant roots, may play a role (Huang et al, 2015). Biochar-induced systemic resistance, potentially caused by ethylene release from specific biochars, has been also offered as a mechanism of herbivore suppression (George et al, 2016). Finally, an enhanced contribution of fungi to C transfer after biochar addition to tropical acidic soil favoring fungivore over herbivore nematodes has been shown (Zhu et al, 2023), aligning with the aforementioned observations.

A clear effect of the application rate also affects the variation in abundance or trophic group composition of nematodes. Both herbivores and predators thrived at low rates (5 t ha^{-1}) while herbivores were suppressed and microbivores gained prominence at relatively high rates (30 t ha^{-1}) (Domene et al, 2021). The same biochar, in a different study and soil, increased both herbivores and fungivores at a high rate (50 t ha^{-1}) compared to a low rate (12 t ha^{-1}) (Llovet et al, 2021). Similarly, at 10% addition (~260 t ha^{-1}) in a tropical tomato crop, biochar suppressed herbivores and promoted fungivores, a pattern not observed at 5% or 15% rates (Ningsi et al, 2022). The abundance of omnivorous and predator taxa increased at the lowest rate and increased in certain bacterivorous taxa at higher rates as shown for a rice husk biochar in a tropical acidic soil tested at 5, 20, and 40 t ha^{-1} (Van Sinh et al, 2022).

Macroarthropods

Macroarthropods are a diverse group that includes larger insects, spiders, ants, and termites (Coleman and Wall, 2014). They play a pivotal role in the establishment of soil structure, bioturbation, and overall functioning, particularly through the activity of ants and termites (alongside earthworms). Around charcoal kilns, lower arthropod diversity has been observed, potentially tied to factors such as altered bulk density, soil organic matter levels, and pH (Fontodji et al, 2009). On the other hand, ant fecal pellets increased in char-enriched soils after wildfires (Topoliantz et al, 2006).

In the scarce laboratory experiments available, intriguing findings emerge. Isopod preference for soil-biochar mixtures has been shown for *Porcellio scaber* and *Porcelionides pruinosus* (Amaro et al, 2013; Madžarić et al, 2018), although in the first study it was restricted to the initial months. Such preference coincided with reduced feeding and growth (Madžarić et al, 2018). In contrast, termites such as *Coptotermes formosanus* actively avoided soils amended with rice straw biochar due to high pH (Chen et al, 2022b), suggesting potential biochar applications in termite control.

Under field conditions and over very long periods, the richness and abundance of termites decreased, while the richness of millipedes and snails increased in ADE, likely linked to their higher Ca levels (Demetrio et al, 2020).

Purposeful biochar additions yielded mixed results, with no clear patterns. Short-term effects on macroarthropod communities were minimal in temperate crop soil, except for the ant species *Tetramorium caespitum* (Castracani et al, 2015), a species associated with disturbed habitats, attracted by the more elevated temperatures in the biochar-treated plots. Over longer periods, macroarthropod populations were not affected by char, contrary to the positive effects observed on earthworms (Kamau et al, 2019). Macroarthropods may possess greater mobility than earthworms and therefore less directly rely on biochar as a food source. In contrast, decreased macroarthropod abundance can be found, as shown for Cuculionidae (Coleoptera) and Chironomidae larvae (Diptera) in a poplar plantation after the introduction of wood biochar, accompanied by a decline in microarthropod abundance (Xu et al, 2022).

The intricate nature of interactions between biochar and this fauna group is evident. The diverse effects seen by different macroarthropod groups can be attributed to their extensive taxonomic diversity, relatively soil-independent lifestyle, and elevated ecological roles. These factors collectively complicate understanding of the underlying mechanisms.

Outlook and future research

Amending soil with biochar should aim to enable the soil biota to carry out their key ecosystem functions to assure long-term soil health and sustained crop production. This involves a careful selection of the feedstock and pyrolysis conditions to avoid detrimental effects on soil biota. Among the many other aims of adding biochar to soil, it is important to ensure sustained functioning of the soil biota so that critical ecosystem functions are maintained, and to avoid inadvertently favoring plant-antagonistic organisms, such as pathogens.

It remains difficult to predict how biota in a given soil system will respond to biochar amendments. To move our understanding of underlying mechanisms forward, we must expand work with well-characterized biochars, in defined soil-plant systems. A systematic approach should enable us to identify dominant factors controlling the diversity of observations currently being reported across different systems and help us to use biochar in these systems to its best advantage.

In the future, the effects of biochar on various soil biota groups, their diversity, and functioning need to be carefully considered. While there is a solid database emerging for microbial communities, the study of soil fauna significantly lags – this should be a focus of future biochar research on soil biota communities. We highlight some important avenues for future studies:

1   Studies need to address specific mechanistic hypotheses, thus moving beyond mere observational assessment of effects. Only in this way will the underlying causes of biochar effects be understood, and thus enable clear management recommendations to be made.
2   The full parameter space (e.g., feedstock, production temperature, application rate, time since application, soil nutrient status, and ecosystem type) still needs to be explored to understand biochar effects on soil biota more fully. At this point, generalizations remain difficult.
3   Reporting of negative or neutral effects is equally important; perhaps there has been biased reporting towards positive effects.

4 In order to understand what effects biochar has on soil biota, response variables need to be examined in a more differentiated way and using realistic application rates. This entails measuring different phases of the life cycle of biota, e.g., for AM fungi the extraradical and the intraradical phases, sporulation, and the fungal community composition in the root and in the soil.

5 The majority of studies to date have focused on the effects of biochar on soil biota, chiefly in terms of abundance measures and correlations. However, there is a shortage of studies that examine how the soil biota community, as affected by biochar, affects plant growth. This is difficult to address experimentally since the tools for examining this are limited.

6 Finally, the above discussion has centered mostly on the effects of biochar on individual soil biota, but soil biota are part of a food web with interactions in the soil microbiome. At the soil ecosystem level, effects on several levels of the trophic system may be enlightening.

We are now at a stage in biochar research where purely phenomenological studies are succeeded by studies with a more mechanistic focus; at the same time, studies examining a range of conditions (biochar, environmental, biotic communities) are still needed, since they are the raw material for emerging syntheses and meta-analyses.

References

Akiyama K, Matsuzaki K-I, and Hayashi H 2005 Plant sesquiterpenes induce hyphal branching in arbuscular mycorrhizal fungi. *Nature* 435, 824–827.

Al-Fraih AM 2015 Effect of soil amendments and root containment on nematode populations in organic greenhouse tomatoes in the Netherlands. Master Dissertation. Wageningen: Wageningen University, The Netherlands.

Amaro A 2013 Optimised tools for toxicity assessment of biochar-amended soils. Master Dissertation. Aveiro: Departamento de Biologia, Universidade de Aveiro, Portugal.

Ameloot N, Graber ER, Verheijen FGA, and De Neve S 2013 Interactions between biochar stability and soil organisms: Review and research needs. *European Journal of Soil Science* 64, 379–390.

Anderson CR, et al 2011 Biochar induced soil microbial community change: Implications for biogeochemical cycling of carbon, nitrogen and phosphorus. *Pedobiologia* 54, 309–320.

Andrés P, et al 2019 Belowground biota responses to maize biochar addition to the soil of a Mediterranean vineyard. *Science of the Total Environment* 660, 1522–1532.

Anthony MA, Bender SF, and van der Heijden MGA 2023 Enumerating soil biodiversity. *Proceedings of the National Academy of Sciences* 120, 33, e2304663120.

Anyanwu IN, et al 2018 Influence of biochar aged in acidic soil on ecosystem engineers and two tropical agricultural plants. *Ecotoxicology and Environmental Safety* 153, 116–126.

Anyanwu IN, Onwukwe DJ, and Anorue CO 2020 In vivo genotoxicity of rice husk biochar on *Eudrilus eugeniae* in soil. *Bulletin of Environmental Contamination and Toxicology* 105, 650–655.

Atilano-Camino MM, Canizales Laborin AP, Ortega Juarez AM, Valenzuela Cantú AK, and Pat-Espadas AM 2022 Impact of soil amendment with biochar on greenhouse gases emissions, metals availability and microbial activity: A meta-analysis. *Sustainability* 14, 15648.

Augustenborg CA, et al 2012 Biochar and earthworm effects on soil nitrous oxide and carbon dioxide emissions. *Journal of Environmental Quality* 41, 1203–1209.

Azeem M, et al 2023 Biochar-derived dissolved organic matter (BDOM) and its influence on soil microbial community composition, function, and activity: A review. *Critical Reviews in Environmental Science and Technology* 53, 1912–1934.

Bardgett E 2005 *The Biology of Soil: A Community and Ecosystem Approach*. Oxford: Oxford University Press.

Biederman LA, and Harpole WS 2013 Biochar and its effects on plant productivity and nutrient cycling: A meta-analysis. *Global Change Biology Bioenergy* 5, 202–214.

Briones MJI, Panzacchi P, Davies CA, and Ineson P 2020 Contrasting responses of macro-and meso-fauna to biochar additions in a bioenergy cropping system. *Soil Biology and Biochemistry* 145, 107803.

Bunting BT, and Lundberg J 1987 The humus profile – concept, class and reality. *Geoderma* 40, 17–36.

Busch D, Kammann C, Grünhage L, and Müller C 2011 Simple biotoxicity tests for evaluation of carbonaceous soil additives: Establishment and reproducibility of four test procedures. *Journal of Environmental Quality* 40, 1–10.

Camenzind T, et al 2018 Arbuscular mycorrhizal fungal and soil microbial communities in African Dark Earths. *FEMS Microbial Ecology* 94, fiy033.

Castaldi S, et al 2011 Impact of biochar application to a Mediterranean wheat crop on soil microbial activity and greenhouse gas fluxes. *Chemosphere* 85, 1464–1471.

Castracani C, et al 2015 Biochar–macrofauna interplay: Searching for new bioindicators. *Science of the Total Environment* 536, 449–456.

Chakrabarti S, Kern J, Menzel R, and Steinberg CEW 2011 Selected natural humic materials induce and char substrates repress a gene in *Caenorhabditis elegans* homolog to human anticancer P53. *Annals of Environmental Science* 5, 1–6.

Chen J, et al 2013 Biochar soil amendment increased bacterial but decreased fungal gene abundance with shifts in community structure in a slightly acid rice paddy from Southwest China. *Applied Soil Ecology* 71, 33–44.

Chen Y, et al 2022b The effect of amending soils with biochar on the microhabitat preferences of *Coptotermes formosanus* (Blattodea: Rhinotermitidae). *Ecotoxicology and Environmental Safety* 232, 113240.

Chen Z, et al 2022a Ecoenzymatic stoichiometry reveals stronger microbial carbon and nitrogen limitation in biochar amendment soils: A meta-analysis. *Science of the Total Environment* 838, 156532.

Cole EJ, et al 2021 Soil nutrient and nematode community changes in response to hardwood charcoal application. *Communications in Soil Science and Plant Analysis* 52, 917–925.

Cole L, Bardgett RD, and Ineson P 2000 Enchytraeid worms (Oligochaeta) enhance mineralization of carbon in organic upland soils. *European Journal of Soil Science* 51, 185–192.

Coleman DC, and Wall DH 2014 Soil fauna: Occurrence, biodiversity, and roles in ecosystem function. In: Paul EA (Ed) *Soil Microbiology, Ecology, and Biochemistry*, 4th Edition. Cambridge (USA): Elsevier Academic Press. pp 111–149.

Coleman DC, Crossley DA Jr, and Hendrix PF 2004 *Fundamentals of Soil Ecology*. Cambridge (USA): Elsevier Academic Press.

Conrado AC, et al 2023 Amazonian earthworm biodiversity is heavily impacted by ancient and recent human disturbance. *Science of the Total Environment* 895, 165087.

Conti FD, Visioli G, Malcevschi A, and Menta C 2018 Safety assessment of gasification biochars using *Folsomia candida* (Collembola) ecotoxicological bioassays. *Environmental Science and Pollution Research* 25, 6668–6679.

Dai Z, et al 2021 Association of biochar properties with changes in soil bacterial, fungal and fauna communities and nutrient cycling processes. *Biochar* 3, 239–254.

Deenik JL, McClellan T, Uehara G, Antal MJ, and Campbell S 2010 Charcoal volatile matter content influences plant growth and soil nitrogen transformations. *Soil Science Society of America Journal* 74, 1259–1269.

Del Valle I, et al 2020 Soil organic matter attenuates the efficacy of flavonoid-based plant-microbe communication. *Science Advances* 6, eaax8254.

Demetrio WC, et al 2020 A "Dirty" footprint: Macroinvertebrate diversity in Amazonian Anthropic Soils. *Global Change Biology* 27, 4575–4591.

Dermiyati E, et al 2018 Soil fauna population during the maize (*Zea mays* l.) growth with the addition of organonitrophos, inorganic fertilizer and biochar. *IOP Conference Series Earth and Environmental Science* 215, 012003.

Domene X, Ramírez W, Mattana S, Alcaniz JM, and Andrés P 2008 Ecological risk assessment of organic waste amendments using the species sensitivity distribution from a soil organisms test battery. *Environmental Pollution* 155, 227–236.

Domene X, Mattana S, Hanley K, Enders A, and Lehmann J 2014 Medium-term effects of corn biochar addition on soil biota activities and functions in a temperate soil cropped to corn. *Soil Biology and Biochemistry* 72, 152–162.

Domene X, Enders A, Hanley K, and Lehmann J 2015a Ecotoxicological characterization of biochars: Role of feedstock and pyrolysis temperature. *Science of the Total Environment* 512-513, 552–561.

Domene X, Hanley K, Enders A, and Lehmann J 2015b Short-term mesofauna responses to soil additions of corn stover biochar and the role of microbial biomass. *Applied Soil Ecology* 89, 10–17.

Domene X 2016 A critical analysis of meso-and macrofauna effects following biochar supplementation. In: Ralebitso-Senior K, and Orr CH (Eds) *Biochar Application: Essential Soil Microbial Ecology*. Amsterdam: Elsevier Science. pp 268–292.

Domene X, Mattana S, and Sánchez-Moreno S 2021 Biochar addition rate determines contrasting shifts in soil nematode trophic groups in outdoor mesocosms: An appraisal of underlying mechanisms. *Applied Soil Ecology* 158, 103788.

Edwards CA, and Arancon NQ 2022 Effects of agricultural practices and chemicals on earthworms. In: Edwards CA, and Arancon NQ (Eds) *Biology and Ecology of Earthworms*. New York: Springer. pp 413–465.

Elmer WH, Lattao CV, and Pignatello JJ 2015 Active removal of biochar by earthworms (*Lumbricus terrestris*). *Pedobiologia* 58, 1–6.

Elliston T, and Oliver IW 2020 Ecotoxicological assessments of biochar additions to soil employing earthworm species *Eisenia fetida* and *Lumbricus terrestris*. *Environmental Science and Pollution Research* 27, 33410–33418.

Eo J, Park KC, Kim MH, Kwon SI, and Song YJ 2018 Effects of rice husk and rice husk biochar on root rot disease of ginseng (*Panax ginseng*) and on soil organisms. *Biological Agriculture and Horticulture* 34, 27–39.

Ezawa T, Yamamoto K, and Yoshida S 2002 Enhancement of the effectiveness of indigenous arbuscular mycorrhizal fungi by inorganic soil amendments. *Soil Science and Plant Nutrition* 48, 897–900.

Feng L, et al 2019 What is the carcass-usage mode of the Collembola? A case study of *Entomobrya proxima* in the laboratory. *Insects* 10, 67.

Feng J, et al 2023 Trade-offs in carbon-degrading enzyme activities limit long-term soil carbon sequestration with biochar addition. *Biological Reviews* 98, 1184–1199.

Ferreira T, et al 2021 Earthworm-biochar interactions: A laboratory trial using *Pontoscolex corethrurus*. *Science of the Total Environment* 777, 146147.

Ferrín M, et al 2023 Trait-mediated responses to aridity and experimental drought by springtail communities across Europe. *Functional Ecology* 37, 44–56.

Fontodji JK, Mawussi G, Nuto Y, and Kokou K 2009 Effects of charcoal production on soil biodiversity and soil physical and chemical properties in Togo, West Africa. *International Journal of Biological and Chemical Sciences* 3, 870–879.

Fox A, Kwapinski W, Griffiths BS, and Schmalenberger A 2014 The role of sulfur-and phosphorus-mobilizing bacteria in biochar-induced growth promotion of *Lolium perenne*. *FEMS Microbiology Ecology* 90, 78–91.

Fry WE 1982 *Principles of Plant Disease Management*. London: Academic Press.

Gale NV, Sackett TE, and Thomas SC 2016 Thermal treatment and leaching of biochar alleviates plant growth inhibition from mobile organic compounds. *PeerJ* 4, e2385.

Garbuz S, Camps-Arbestain M, Mackay A, DeVantier B, and Minor M 2020 The interactions between biochar and earthworms, and their influence on soil properties and clover growth: A 6-month mesocosm experiment. *Applied Soil Ecology* 147, 103402.

Garbuz S, Mackay A, Camps-Arbestain M, DeVantier B, and Minor M 2021 Biochar amendment improves soil physico-chemical properties and alters root biomass and the soil food web in grazed pastures. *Agriculture, Ecosystems and Environment* 319, 107517.

George C, Kohler J, and Rillig MC 2016 Biochars reduce infection rates of the root-lesion nematode *Pratylenchus penetrans* and associated biomass loss in carrot. *Soil Biology and Biochemistry* 95, 11–18.

Giagnoni L, and Renella G 2022 Effects of biochar on the C use efficiency of soil microbial communities: Components and mechanisms. *Environments* 9, 138.

Gomez-Eyles JL, Sizmur T, Collins CD, and Hodson ME 2011 Effects of biochar and the earthworm *Eisenia fetida* on the bioavailability of polycyclic aromatic hydrocarbons and potentially toxic elements. *Environmental Pollution* 159, 616–622.

Gong X, et al 2022 Cattle manure biochar and earthworm interactively affected CO_2 and N2O emissions in agricultural and forest soils: Observation of a distinct difference. *Frontiers of Environmental Science and Engineering* 16, 1–13.

Graber ER 2010 Induction of systemic resistance in plants by biochar, a soil-applied carbon sequestering agent. *Phytopathology* 100, 913–921.

Grossman J, et al 2010 Amazonian anthrosols support similar microbial communities that differ distinctly from those extant in adjacent,
unmodified soils of the same mineralogy. *Microbial Ecology* 60, 192–205.

Gruss I, Twardowski JP, Latawiec A, Medyńska-Juraszek A, and Królczyk J 2019 Risk assessment of low-temperature biochar used as soil amendment on soil mesofauna. *Environmental Science and Pollution Research* 26, 18230–18239.

Guerena D, et al 2013 Nitrogen dynamics following field application of biochar in a temperate North American maize-based production system. *Plant and Soil* 365, 239–254.

Guerena D, et al 2015 Partitioning the contributions of biochar properties to enhanced biological nitrogen fixation in common bean (*Phaseolus vulgaris*). *Biology and Fertility of Soils* 51, 479–491.

Gul S, Whalen JK, Thomas BW, Sachdeva V, and Deng H 2015 Physicochemical properties and microbial responses in biochar-amended soils: Mechanisms and future directions. *Agriculture, Ecosystems and Environment* 206, 46–59.

Hale SE, et al 2013 Short-term effect of the soil amendments activated carbon, biochar, and ferric oxyhydroxide on bacteria and invertebrates. *Environmental Science and Technology* 47, 8674–8683.

Hockaday WC 2006 The organic geochemistry of charcoal black carbon in the soils of the University of Michigan Biological Station, PhD Dissertation. Columbus: Ohio State University, USA.

Hockaday WC, Grannas AM, Kim S, and Hatcher PG 2007 The transformation and mobility of charcoal in a fire-impacted watershed. *Geochimica et Cosmochimica Acta* 71, 3432–3445.

Honvault N, et al 2023 Positive or neutral effects of biochar-compost mixtures on earthworm communities in a temperate cropping system. *Applied Soil Ecology* 182, 104684.

Huang W, Ji H, Gheysen G, Debode J, and Kyndt T 2015 Biochar-amended potting medium reduces the susceptibility of rice to root-knot nematode infections. *BMC Plant Biology* 15, 267.

Huang C, Wang W, Yue S, Adeel M, and Qiao Y 2020 Role of biochar and *Eisenia fetida* on metal bioavailability and biochar effects on earthworm fitness. *Environmental Pollution* 263, 114586.

Jatala P 1986 Biological control of plant-parasitic nematodes. *Annual Review of Phytopathology* 24, 453–489.

Jeffery S, et al 2022 Biochar application differentially affects soil micro-, meso-macro-fauna and plant productivity within a nature restoration grassland. *Soil Biology and Biochemistry* 174, 108789.

Jia H, et al 2023 Significant contributions of biochar-derived dissolved matters to ecotoxicity to earthworms (*Eisenia fetida*) in soil with biochar amendment. *Environmental Technology and Innovation* 29, 102988.

Joseph S, et al 2021 How biochar works, and when it doesn't: A review of mechanisms controlling soil and plant responses to biochar. *Global Change Biology Bioenergy* 13, 1731–2021.

Kamau S, Karanja NK, Ayuke FO, and Lehmann J 2019 Short-term influence of biochar and fertilizer-biochar blends on soil nutrients, fauna and maize growth. *Biology and Fertility of Soils* 55, 661–673.

Kolton M, et al 2011 Impact of biochar application to soil on the root-associated bacterial community structure of fully developed greenhouse pepper plants. *Applied and Environmental Microbiology* 77, 4924–4930.

Laborda F, Monistrol IF, Luna N, and Fernandez M 1999 Processes of liquefaction/solubilization of Spanish coals by microorganisms. *Applied Microbiology and Biotechnology* 52, 49–56.

Lammirato C, Miltner A, and Kaestner M 2011 Effects of wood char and activated carbon on the hydrolysis of cellobiose by β-glucosidase from *Aspergillus niger*. *Soil Biology and Biochemistry* 43, 1936–1942.

LeCroy C, Masiello CA, Rudgers JA, Hockaday WC, and Silberg JJ 2013 Nitrogen, biochar, and mycorrhizae: Alteration of the symbiosis and oxidation of the char surface. *Soil Biology and Biochemistry* 58, 248–254.

Lehmann J, et al 2011 Biochar effects on soil biota – a review. *Soil Biology and Biochemistry* 43, 1812–1836.

Lehmann J, Kuzyakov Y, Pan G, and Ok YS 2015 Biochars and the plant soil interface. *Plant and Soil* 395, 1–5.

Li D, Hockaday WC, Masiello CA, and Alvarez PJJ 2011 Earthworm avoidance of biochar can be mitigated by wetting. *Soil Biology and Biochemistry* 43, 1732–1737.

Li X, Wang T, Chang SX, Jiang X, and Song Y 2020a. Biochar increases soil microbial biomass but has variable effects on microbial diversity: A meta-analysis. *Science of the Total Environment* 749, 141593.

Li J, et al 2020b Integration of behavioural tests and transcriptome sequencing of *C. elegans* reveals how the nematode responds to peanut shell biochar amendment. *Science of the Total Environment* 707, 136024.

Liao X, Kang H, Haidar G, Wang W, and Malghani S 2022 The impact of biochar on the activities of soil nutrients acquisition enzymes is potentially controlled by the pyrolysis temperature: A meta-analysis. *Geoderma* 411, 115692.

Liesch AM, Weyers SL, Gaskin JW, and Das KC 2010 Impact of two different biochars on earthworm growth and survival. *Annals of Environmental Science* 4, 1–9.

Lieke T, Zhang X, Steinberg CE, and Pan B 2018 Overlooked risks of biochars: Persistent free radicals trigger neurotoxicity in *Caenorhabditis elegans*. *Environmental Science and Technology* 52, 7981–7987.

Liiri M, Ilmarinen K, and Setala H 2007 Variable impacts of enchytraeid worms and ectomycorrhizal fungi on plant growth in raw humus soil treated with wood ash. *Applied Soil Ecology* 35, 174–183.

Liu S, et al 2016 Response of soil carbon dioxide fluxes, soil organic carbon and microbial biomass carbon to biochar amendment: A meta-analysis. *Global Change Biology Bioenergy* 8, 392–406.

Liu Q, et al 2018 How does biochar influence soil N cycle? A meta-analysis. *Plant and Soil* 426, 211–225.

Liu Z, et al 2019 The responses of soil organic carbon mineralization and microbial communities to fresh and aged biochar soil amendments. *Global Change Biology Bioenergy* 11, 1408–1420.

Liu T, et al 2020a Biochar exerts negative effects on soil fauna across multiple trophic levels in a cultivated acidic soil. *Biology and Fertility of Soils* 56, 597–606.

Liu X, et al 2020b Soil nematode community and crop productivity in response to 5-year biochar and manure addition to yellow cinnamon soil. *BMC Ecology* 20, 1–13.

Llovet A, et al 2021 Long-term effects of gasification biochar application on soil functions in a Mediterranean agroecosystem: Higher addition rates sequester more carbon but pose a risk to soil faunal communities. *Science of the Total Environment* 801, 149580.

Lundkvist H 1998 Wood ash effects on enchytraeid and earthworm abundance and enchytraeid cadmium content. *Scandinavian Journal of Forest Research* 13, 86–95.

Madžarić S, Kos M, Drobne D, Hočevar M, and Kokalj AJ 2018 Integration of behavioral tests and biochemical biomarkers of terrestrial isopod *Porcellio scaber* (Isopoda, Crustacea) is a promising methodology for testing environmental safety of chars. *Environmental Pollution* 234, 804–811.

Maestrini B, Herrmann AM, Nannipieri P, Schmidt MWI, and Abiven S 2014 Ryegrass-derived pyrogenic organic matter changes organic carbon and nitrogen mineralization in a temperate forest soil. *Soil Biology and Biochemistry* 9, 291–301.

Marks EA, Mattana S, Alcañiz JM, and Domene X 2014 Biochars provoke diverse soil mesofauna reproductive responses in laboratory bioassays. *European Journal of Soil Biology* 60, 104–111.

Marks EAN, Mattana S, Alcañiz JM, Pérez-Herrero E, and Domene X 2016. Gasifier biochar effects on nutrient availability, organic matter mineralization, and soil fauna activity in a multi-year Mediterranean trial. *Agriculture, Ecosystems and Environment* 215, 30–39.

Masiello CA, et al 2013. Biochar and microbial signaling: Production conditions determine effects on microbial communication.

Environmental Science and Technology 47, 11496–11503.

Matlack GR 2001 Factors determining the distribution of soil nematodes in a commercial forest landscape. *Forest Ecology and Management* 146, 129–143.

McClellan AT, Deenik J, Uehara G, and Antal M 2007 Effects of flash carbonized macadamia nutshell charcoal on plant growth and soil chemical properties. *American Society of Agronomy Abstracts*, 3–7 Nov., New Orleans, LA.

McCormack S, Ostle N, Bardgett RD, Hopkin DW, and VanBergen AJ 2013 Biochar in bioenergy cropping systems: Impacts on soil faunal communities and linked ecosystem processes. *Global Change Biology Bioenergy* 5, 81–95.

McCormack SA, et al 2019 Soil biota, carbon cycling and crop plant biomass responses to biochar in a temperate mesocosm experiment. *Plant and Soil* 440, 341–356.

Mulder C, and Vonk JA 2011 Nematode traits and environmental constraints in 200 soil systems: Scaling within the 60–6000 μm body size range. *Ecology* 92, 2004–2004.

Nahar MS, et al 2006 Differential effects of raw and composted manure on nematode community, and its indicative value for soil microbial, physical and chemical properties. *Applied Soil Ecology* 34, 140–151.

Nieminen J, and Haimi J 2010 Body size and population dynamics of enchytraeids with different disturbance histories and nutrient dynamics. *Basic Applied Ecology* 11, 638–644.

Nieminen JK 2008 Labile carbon alleviates wood ash effects on soil fauna. *Soil Biology and Biochemistry* 40, 2908–2910.

Ningsi F, et al 2022 Use of biochar to control root-feeding soil nematodes on Muna local tomatoes variety. *Journal of Tropical Soils* 27, 37–48.

Noronha FR, Manikandan SK, and Nair V 2022 Role of coconut shell biochar and earthworm (*Eudrilus euginea*) in bioremediation and palak spinach (*Spinacia oleracea* L.) growth in cadmium-contaminated soil. *Journal of Environmental Management* 302, 114057.

Nzanza B, Marais D, and Soundy P 2012 Effect of arbuscular mycorrhizal fungal inoculation and biochar amendment on growth and yield of tomato. *International Journal of Agriculture and Biology* 6, 965–969.

Ogawa M, Yambe Y, and Sugiura G 1983 Effects of charcoal on the root nodule formation and VA mycorrhiza formation of soybean. *The Third International Mycological Congress (IMC3) Abstracts*. Tokyo. p.578.

Oka Y 2010 Mechanisms of nematode suppression by organic soil amendments—a review. *Applied Soil Ecology* 44, 101–115.

O'Neill B, et al 2009 Bacterial community composition in Brazilian anthrosols and adjacent soils characterized using culturing and molecular identification. *Microbial Ecology* 58, 23–35.

Oladele SO, Adeyemo A, Awodun M, Adegaye A, and Ingold M 2021 Impact of biochar amendment on soil nematode communities in a West African rain-fed rice cropland. *Nematology* 24, 159–170.

Palansooriya KN, et al 2019 Response of microbial communities to biochar-amended soils: A critical review. *Biochar* 1, 3–22.

Pokharel P, Ma Z, and Chang SX 2020 Biochar increases soil microbial biomass with changes in extra-and intracellular enzyme activities: A global meta-analysis. *Biochar* 2, 65–79.

Pokharel P, Ma Z, and Chang SX 2021 Correction to: Biochar increases soil microbial biomass with changes in extra-and intracellular enzyme activities: A global meta-analysis. *Biochar* 3, 715.

Pollet S, et al 2022 Limited effects of century-old biochar on taxonomic and functional diversities of collembolan communities across land-uses. *Soil Biology and Biochemistry* 164, 108484.

Pressler Y, Foster EJ, Moore JC, and Cotrufo MF 2017 Coupled biochar amendment and limited irrigation strategies do not affect a degraded soil food web in a maize agroecosystem, compared to the native grassland. *Global Change Biology Bioenergy* 9, 1344–1355.

Prober SM, Stol J, Piper M, Gupta VVSR, and Cunningham SA 2014 Enhancing soil biophysical condition for climate-resilient restoration in mesic woodlands. *Ecological Engineering* 71, 246–255.

Prodana M, et al 2019 Influence of biochar particle size on biota responses. *Ecotoxicology and Environmental Safety* 174, 120–128.

Quilliam RS, et al 2012 Nutrient dynamics, microbial growth and weed emergence in biochar amended soil are influenced by time since application and reapplication rate. *Agriculture Ecosystems and Environment* 158, 192–199.

Quilliam RS, Glanville HC, Wade SC, and Jones DL 2013 Life in the 'charosphere' - does biochar in agricultural soil provide a significant habitat for microorganisms? *Soil Biology and Biochemistry* 65, 287–293.

Rahman L, Whitelaw-Weckert MA, and Orchard B 2014 Impact of organic soil amendments, including poultry-litter biochar, on nematodes in a Riverina, New South Wales, vineyard. *Soil Research* 52, 604–619.

Saito M 1989 Charcoal as a micro-habitat for VA mycorrhizal fungi and its practical implication. *Agriculture, Ecosystems and Environment* 29, 341–344.

Saito M, and Marumoto T 2002 Inoculation with arbuscular mycorrhizal fungi: The status quo in Japan and the future prospects. *Plant and Soil* 244, 273–279.

Sanchez-Hernandez JC 2018 Biochar activation with exoenzymes induced by earthworms: A novel functional strategy for soil quality promotion. *Journal of Hazardous Materials* 350, 136–143.

Sanchez-Hernandez JC, Ríos JM, Attademo AM, Malcevschi A, and Cares XA 2019 Assessing biochar impact on earthworms: Implications for soil quality promotion. *Journal of Hazardous Materials* 366, 582–591.

Schmidt H-P, et al 2021 Biochar in agriculture –A systematic review of 26 global meta-analyses. *Global Change Biology Bioenergy* 13, 1708–1730.

Schwartz MW, et al 2006 The promise and the potential consequences of the global transport of mycorrhizal fungal inoculum. *Ecology Letters* 9, 501–515.

Solaiman ZM, Blackwell P, Abbott LK, and Storer P 2010 Direct and residual effect of biochar application on mycorrhizal root colonisation, growth and nutrition of wheat. *Australian Journal of Soil Research* 48, 546–554.

Shi Z, et al 2021 Effects of biochar and thermally treated biochar on *Eisenia fetida* survival, growth, lysosomal membrane stability and oxidative stress. *Science of the Total Environment* 770, 144778.

Shi Z, Wen M, Zhao Y, and Wang C 2023 Vermitoxicity of aged biochar and exploring potential damage factors. *Environment International*, 172, 107787.

Siddiqui ZA, and Mahmood I 1999 Role of bacteria in the management of plant parasitic nematodes: A review. *Bioresource Technology* 69, 167–179.

Singh H, Northup BK, Rice CW, and Vara Prasad PV 2022 Biochar applications influence soil physical and chemical properties, microbial diversity, and crop productivity: A meta-analysis. *Biochar* 4, 8.

Soong JL, Dam M, Wall DH, and Cotrufo MF 2017 Below-ground biological responses to pyrogenic organic matter and litter inputs in grasslands. *Functional Ecology* 31, 260–269.

Sun T, et al 2021 Suppressing peatland methane production by electron snorkeling through pyrogenic carbon in controlled laboratory incubations. *Nature Communications* 12, 4119.

Sun D, et al 2022 Does micro-sized pyrogenic carbon made in lab affect earthworm mortality in restrained water content? *Applied Soil Ecology* 177, 104540.

Tammeorg P, et al 2014 Effects of biochar on earthworms in arable soil: Avoidance test and field trial in boreal loamy sand. *Agriculture, Ecosystems and Environment* 191, 150–157.

Thies JE, Rillig MC, and Graber ER 2015 Biochar effects on the abundance, activity and diversity of the soil biota. In: Lehmann J, and Joseph S (Eds) *Biochar for Environmental Management*. London: Routledge. pp327–389.

Thies JE, and Grossman J 2023 The soil habitat and soil ecology. In: Uphoff N, and Thies JE (Eds) *Biological Approaches to Regenerative Soil Systems*. Boca Raton: CRC Press. pp69–83.

Topoliantz S, and Ponge JF 2003 Burrowing activity of the geophagous earthworm *Pontoscolex corethurus* (Oligochaeta: Glossoscolecidae) in the presence of charcoal. *Applied Soil Ecology* 23, 267–271.

Topoliantz S, and Ponge JF 2005 Charcoal consumption and casting activity by *Pontoscolex corethurus* (Glossoscolecidae). *Applied Soil Ecology* 28, 217–224.

Topoliantz S, Ponge JF, and Lavelle P 2006. Humus components and biogenic structures under tropical slash-and-burn agriculture. *European Journal of Soil Science* 57, 269–278.

Treseder KK, and Cross A 2006 Global distributions of arbuscular mycorrhizal fungi. *Ecosystems* 9, 305–316.

Tsai SM, et al 2009 The microbial world of terra preta. In: Woods WI, Teixeira W, Lehmann J, Steiner C, and WinklerPrins A (Eds) *Amazonian Dark Earths: Wim Sombroek's Vision*. Berlin: Springer. pp 299–308.

Uvarov AV 2000 Effects of smoke emissions from a charcoal kiln on the functioning of forest soil systems: A microcosm study. *Environmental Monitoring and Assessment* 60, 337–357.

Van Sinh N, Kato R, Linh DTT, Phuong NTK, and Toyota K 2022 Influence of rice husk biochar on soil nematode community under upland and flooded conditions: A microcosm experiment. *Agronomy* 12, 378.

Van Zwieten L, et al 2010 Effects of biochar from slow pyrolysis of papermill waste on agronomic performance and soil fertility. *Plant and Soil* 327, 235–246.

Vanek S, and Lehmann J 2015 Phosphorus availability to beans via interactions between mycorrhizas and biochar. *Plant and Soil* 395, 105–123.

Ventura M, et al 2014 Effect of biochar addition on soil respiration partitioning and root dynamics in an apple orchard. *European Journal of Soil Science* 65, 186–195.

Wang Z, et al 2015 Reduced nitrification and abundance of ammonia-oxidizing bacteria in acidic soil amended with biochar. *Chemosphere* 138, 576–583.

Wang W, et al 2021 How do earthworms affect the soil organic carbon fractions and CO_2

emissions after incorporation of different maize straw-derived materials. *Journal of Soils and Sediments* 21, 3632–3644.

Wang M, et al 2023 Meta-analysis of the effects of biochar application on the diversity of soil bacteria and fungi. *Microorganisms* 11, 641.

Warnock DD, Lehmann J, Kuyper TW, and Rillig MC 2007 Mycorrhizal responses to biochar in soil - concepts and mechanisms. *Plant and Soil* 300, 9–20.

Wessels JGH 1999 Fungi in their own right. *Fungal Genetics and Biology* 27, 134–145.

Weyers Sl, and Spokas KA 2011 Impact of biochar on earthworm populations: A review. *Applied and Environmental Soil Science* 2011, 541592.

Whalen JK, and Benslim H 2021 Earthworm populations are stable in temperate agricultural soils receiving wood-based biochar. *Pedosphere* 31, 398–404.

Wu Y, Liu J, Shaaban M, and Hu R 2021 Dynamics of soil N_2O emissions and functional gene abundance in response to biochar application in the presence of earthworms. *Environmental Pollution* 268, 115670.

Xiang Y, et al 2023 Biochar addition increased soil bacterial diversity and richness: Large-scale evidence of field experiments. *Science of the Total Environment* 893, 164961.

Xiao Z, et al 2019 The effect of biochar amendment on N-cycling genes in soils: A meta-analysis. *Science of the Total Environment* 696, 133984.

Xu H, et al 2022 Differential responses of soil arthropods to the application of biogas slurry and biochar in a coastal poplar plantation. *European Journal of Soil Biology* 113, 103447.

Xu W, et al 2023 Global meta-analysis reveals positive effects of biochar on soil microbial diversity. *Geoderma* 436, 116528.

Zackrisson O, Nilsson M-C, and Wardle DA 1996 Key ecological function of charcoal from wildfire in the Boreal forest. *Oikos* 77, 10–19.

Zhang L, Jing Y, Xiang Y, Zhang R, and Lu H 2018 Responses of soil microbial community structure changes and activities to biochar addition: A meta-analysis. *Science of the Total Environment* 643, 926–935.

Zhang L, Xiang Y, Jing Y, and Zhang R 2019 Biochar amendment effects on the activities of soil carbon, nitrogen, and phosphorus hydrolytic enzymes: A meta-analysis. *Environmental Science and Pollution Research* 26, 22990–23001.

Zhang X-K, et al 2013 Soil nematode response to biochar addition in a Chinese wheat field. *Pedosphere* 23, 98–103.

Zhou H, et al 2017 Changes in microbial biomass and the metabolic quotient with biochar addition to agricultural soils: A meta-analysis. *Agriculture, Ecosystems and Environment* 239, 80–89.

Zhu X, Chen B, Zhu L, and Xing B 2017 Effects and mechanisms of biochar-microbe interactions in soil improvement and pollution remediation: A review. *Environmental Pollution* 227, 98–115.

Zhu B, et al 2023 Biochar enhances multifunctionality by increasing the uniformity of energy flow through a soil nematode food web. *Soil Biology and Biochemistry* 183, 109056.

15

Biochar effects on plant ecophysiology

Claudia Kammann and Ellen R. Graber

Introduction

The central aim of this chapter is to evaluate biochar effects from a plant-centered perspective (for plant nutrition see Chapter 15). Overall, there is a shortage of detailed fundamental studies on the impact of biochar on plant ecophysiology with abundant studies tending to focus on agronomic yield. Plants are often considered passive-reacting and enduring, due to their "rooted to the spot" nature, while they can actively modify unfavorable conditions, e.g., by root exudation, root foraging, rhizobia interactions, or N_2-fixing symbioses; they can sense and prepare for current and coming stresses, both abiotic and biotic (Larcher, 2003; Lambers et al, 2008). A well-known example is autumnal leaf senescence, triggered by temperature and photo-periodicity, where plants re-allocate valuable nutrients to perennial tissue before they shed their leaves. Plants can outcompete other plants, e.g., by outgrowing other individuals to gain better access to light, but they may also aid each other, e.g., by emitting volatile chemical signals that warn about stresses they experience. The genetic heritage of wild plants is to master stresses, compete for resources, resist attacks, and produce offspring. Agronomic species may differ since humans select certain traits, promote their distribution, and protect their growth. Understanding the ecophysiology of crop and wild plants concerning biochar effects at the single-organism level (autecology) and the canopy or community level (synecology) can elucidate plant reactions to biochar; synecology and biochar is still a severely neglected topic in biochar research.

DOI: 10.4324/9781003297673-15

Biochar and plant-soil water relations

Biochar has been shown to alleviate environmental stress connected to the plant water household, namely drought and salinity stress. Since the last edition of this book, the number of studies investigating the effects of biochar on plant physiology, biochemistry, growth, and yield parameters has increased considerably (e.g., Haider et al, 2015; El Nahhas et al, 2021; Mahmoud et al, 2022; Sofy et al, 2022). To extract water from the soil, a plant needs to exert a greater force (suction) than the soil retaining the water. This suction force is provided by the water vapor pressure deficit of the atmosphere (Figures 15.1A and 1452). Plants can regulate their water loss by (i) modifying their water uptake via changes in root architecture, depth of rooting and root exudates or symbiotic partner acquisition; (ii) by reducing transpiring surfaces by altering shoot and leaf area development and form; and (iii) by modifying biochemical, morphological, osmotic and stomatal controls over transpiration (Figures 15.1A and 15.3). Thus, biochar amendment-mediated improvements in the soil water potential, plant-available water or root morphology, root tips, and surface size will increase or prolong the water delivery to plants.

Belowground: Biochar effects on plant-available water and water-conserving mechanisms

A coarse-textured soil with a low SOM concentration has a much lower ability to hold water against gravity than a fine-textured or SOM-rich soil (Figure 15.2). The highest increases in the amount of plant-available water after biochar addition are usually reported for coarse-textured soils (Omondi et al, 2016). Generally, water percolates rapidly through pores >50 μm, moves slowly in pores between 10 and 50 μm in diameter, and is held against gravity by capillary forces in pores smaller than 10 μm. Plant-available water is largely defined as the amount of water between −0.033 MPa (field capacity) and −1.5 MPa (permanent wilting point), where the pores are usually too small to be drained by plant-atmosphere suction forces. Meta-analysis confirms that the amount of plant-available water increases on average by 15% with biochar use, most prominently in coarse-textured, sandy soils (by 24.3%) or with low SOM contents (Omondi et al, 2016); increases were favored by residue- and wood- over sludge biochars and by production temperatures >500°C. In sandy soils with added biochar, the amount of water held at −1.5 MPa also increased (Omondi et al, 2016), but the water amount at field capacity (also called maximum water holding capacity (WHC)) increased to a larger degree, depending on biochar properties. However, in most laboratory and controlled settings, high biochar amendment rates (4–8% w-w; 40–80 Mg ha^{-1}, or more) were needed to achieve significant increases in plant-available water (Omondi et al, 2016). In rain-fed agriculture in dry climates, however, even a modest gain in plant available water may make the difference. Conservation farming approaches using root zone application of biochar with fertilizers (Schmidt et al, 2017; Grafmüller et al, 2022) enables the use of lower amounts of biochar (0.7 to 4 t ha^{-1}) and still reap benefits as described by Cornelissen et al (2013). Increases in plant available water were observed in vineyard soils in Tuscany, Italy, with a wood-biochar application at 22 and 44 Mg ha^{-1} (Genesio et al, 2015). Interestingly, grape yield increases with biochar were higher when the growing

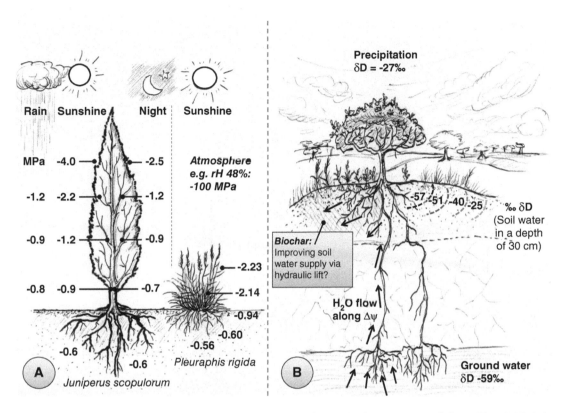

Figure 15.1 *The soil-plant-atmosphere continuum. A: vertical water potential profiles (MPa) in plants with different growth forms. The tree species Juniperus scopulorum (arid regions in the western part of the US) was measured on a rainy and sunny day, and the night following a sunny day; the C4 tussock grass Pleuraphis rigida ('Big Galleta') was measured at the time of maximum transpiration (re-drawn and modified after Larcher, 2003). B: schematic sketch of 'Hydraulic lift': Roots of trees or shrubs that reach (ground) water take up, lift, and release the water to the soil if the surrounding soil water potential (Ψ) is more negative than that of the shoots, e.g., when transpiration is low (at night). In the temperate tree species Acer saccharum, the isotopic water (deuterium) signature (‰ δD) indicated that most of the xylem water was derived from the groundwater; in the forbs in the tree shade ('water parasitism'), the proportion of lifted groundwater was the larger the closer the vegetation was to the tree (re-drawn and modified after Schulze et al, 2005). Biochar effects in this context are within the 'unknown' category of under-researched topics*

season was drier, without reducing grape must quality (Genesio et al, 2015).

The amount of plant-available soil water is also determined by the amount of water that infiltrates into soils upon precipitation. Particularly in sloped landscapes, considerable amounts of water can be lost as run-off water otherwise. With biochar application, runoff, and soil erosion were on average reduced by 8% and 30% in tropical soils, and by 28% and 9% in temperate soils, with subtropical regions figuring in between both (Gholamahmadi et al, 2023).

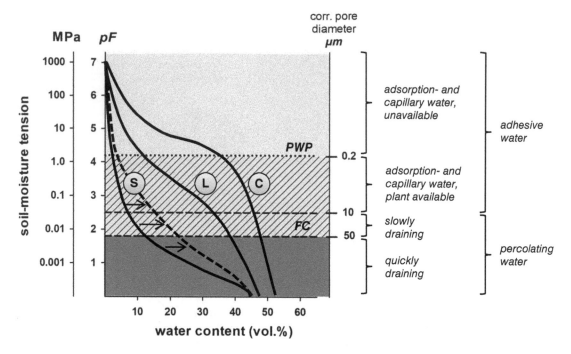

Figure 15.2 *Schematic illustration of the relationship between soil moisture tension, pore diameter, and volumetric water content. The volumetric soil water content (x-axis) in a sandy (S), loamy (L) or clay (C) soil is plotted against the soil suction (matric potential) expressed in MPa (0.1 MPa = 1 bar) or as the logarithmic value of cm water head or pF value (left y-axes); the corresponding mean pore diameter of the capillaries (e.g., of soil-biochar mixtures) and the resulting water availability to the 'average' plant is indicated on the right y-axis. The dashed curve exemplifies the shift that can occur in sandy soil mixed with biochar schematically. PWP = permanent wilting point of most plants; FC = field capacity (roughly equal to WHC = water holding capacity of a disturbed soil sample). Biochar has been shown to improve plant-available water in sandy soils (meta-analysis: Omondi et al, 2016)*

Erosion reduction was double in vegetated compared to non-vegetated soils with biochar application, as the rhizosphere generally promotes soil aggregate formation and stability, one of the main mechanisms for reduced erosion and improved water infiltration. Physically, the pore size distribution and pore connectivity of biochar particles (and their change over time) will partly determine its water retention and delivery characteristics, but overall, soil aggregation, reduced erosion and run-off, and retained soil moisture are more important for soil biota

and soil structure related to plant water supply (Genesio et al, 2015). Moreover, many pores in biochar are smaller than 0.2 μm or are dead ends, unable to fill with water. Hence, the observed increases in plant available water can therefore partly be attributed to the indirect positive effects of biochar on fine root turnover and root exudation, soil biota and thus soil aggregation, water infiltration, reduced soil erosion or water run-off (Omondi et al, 2016; Xiang et al, 2017; Gholamahmadi et al, 2023). Biochar amendments (10 Mg ha^{-1}) significantly increased SOC build-up over 6 years beyond biochar-C addition itself with effects on cumulative water infiltration (Blanco-Canqui et al, 2020).

Green roofs with their artificial "soil" substrates provide extreme growth conditions for most plants, where water retention and supply are crucial characteristics. Here, biochar amendments to green roof substrates (5–30% v/v) have been shown to improve plant-available water and water-plus-nutrient retention (Cao et al, 2014; Chen et al, 2018; Gan et al, 2021; Tan et al, 2023). Reducing rainfall runoff with green roofs aids urban stormwater management (Gan et al, 2021; Liao et al, 2022a, b), while improved plant growth on green roofs helps to reduce the surface temperatures by increasing evapotranspiration (Tan et al, 2023) to combat the urban heat island effect. The topic of designing and optimizing biochar use in green roof substrates for increasing plant water supply deserves more research attention (Liao et al, 2022a, b) in our rapidly heating climate.

Belowground: Biochar effects on root growth and root water acquisition

The root system of a plant develops according to its species-specific morphology and local soil conditions with considerable plasticity. After germination and before senescence, roots grow continuously, foraging for water, nutrients, and eventually symbionts, as only the non-suberized younger root tips and elongation zones including root hairs can usually take up water (Larcher, 2003; Lambers et al, 2008). Temperature impacts root development. A darkening by biochar added was shown to increase soil temperature by (max.) 0.7°C and CO_2 efflux by (max.) 14.3% in the first vegetation period (June – Sept.) after biochar

application at rates of 5 and 10 Mg ha^{-1} to the surface of a southern Finland boreal forest soil (Zhu et al, 2020), but not in the second year after mosses and lichens had overgrown the surface-added biochar (Palviainen et al, 2018). In general, nutrients are better retained through additions of biochar addition (Chapters 8, 10, 12, 16). However, to our knowledge, root morphology changes in response to biochar addition have not yet been investigated as a function of soil temperature or nutrient retention in boreal soils, where the reported temperature increase may have impacted both root growth and soil nutrient transformations (Chapter 16).

Urban trees, with their restricted rooting space and compacted, water- and salt-stressed soil environment often suffer and show restricted growth, reduced stress resilience, and premature death (Schuett et al, 2022). Experiments regarding urban tree substrates (not soils) amended with biochar demonstrated large increases in fine root development and proliferation (Embrén 2016; Heinrich and Saluz, 2017; Stockholm Stad Handbook, 2017). An experiment was carried out by adding biochar only to certain sectors of the large planting pots. Tree-root proliferation was observed especially in sectors of pots that had been amended with biochar in root-steering tests, but only if the tree roots had been soil- and clay-free at the start (Heinrich and Saluz, 2017). The use of biochar in macadam substrates, which are structured gravel-substrate 'soils' (Embrén, 2016; Saluz et al, 2022) definitely deserves much more research attention. Urban green maintenance problems are expected to worsen exponentially with accelerating global warming, and biochar use in green roofs (Chen et al. 2018) and urban tree substrates also provides an urban C-sink potential (Azzi et al, 2022) with a yet unknown magnitude. Since such urban biochar uses lack critical scientific evidence, they are not reflected in the scientific literature for CDR (carbon dioxide removal) assessments (Smith et al, 2023).

Roots that sense dry soil conditions enhance their abscisic acid (ABA) production which induces stomatal closure and reduces cell wall extensibility (Lambers et al, 2008; Marschner, 2012). With assimilates preferentially exported to the roots, root growth is promoted compared to the shoot and root architecture changes. ABA suppresses ethylene biosynthesis, thus suppressing side-root branching and root hair formation. Roots elongate until they reach moister patches, at which time ABA production declines, ethylene synthesis recovers, and roots branch and form more root hairs to exploit the moisture (Figure 15.3). Severe drought, salinity, or herbivory stress, however, can also trigger surplus ethylene

synthesis in two peaks, where the first peak induces gene expressions involved in coping with the stress, but the second, larger one often is detrimental, damaging the plant that produces the ethylene (Glick et al, 2007). Other physiological, phytohormonal, and enzymatic responses under osmotic stress concern reactive oxygen species (ROS) scavenging, osmolyte regulation, reduced photosynthetic C inflow, and water use efficiency.

Biochar may impact this hormonal regulation cycle in several ways (Figure 15.3). It can (I) retain more plant-available water, as discussed above; (II) reduce soil tensile strength and hence allow greater penetration depths and denser top- or subsoil root branching (Bruun et al, 2014; Genesio et al, 2015; Xiang et al, 2017); (III) directly improve K nutrition via its minerals (Ippolito et al, 2020), which is physiologically relevant for abating both drought and salt stress; (IV) theoretically

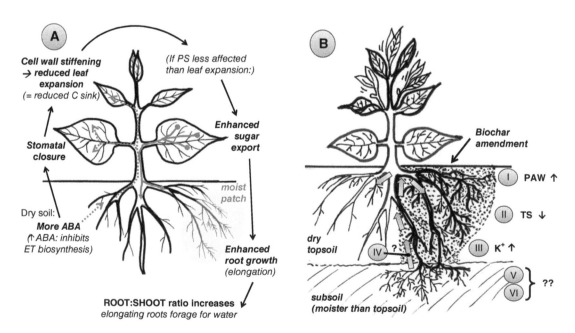

Figure 15.3 *Conceptual scheme of (A) source-sink C partitioning and phytohormone signaling under drought stress and (B) biochar impact pathways I – VI (explanation see text) (modified after Lambers et al, 2008)*

reinforce nocturnal hydraulic lift, making use of an existing deep root system (see text below; Figure 15.1B); (V) support beneficial soil rhizobia, either by favoring their growth directly or via modified biochar as a carrier for microbe inoculation (Sofy et al, 2022); or (VI) it can have phytohormonal effects, thanks to the multitude of chemical substances it may carry (Graber et al, 2014).

Root penetration depths, root morphology including branching patterns, and root hair development can increase with biochar amendments. Root biomass, root length, and the number of root nodules in symbiotic plants increased on average by 32%, 52%, and 25% with biochar amendment, respectively (Xiang et al, 2017). Arbuscular mycorrhizal root colonization per unit length of root did not increase, but since the root length itself increased, the overall root system colonization was higher (Xiang et al, 2017). The overall change in morphological root parameters that the authors reported suggested a better root-length development and soil exploration, indicating either reduced drought stress, better root growth with reduced tensile strength, or both. Root biomass increased on average by 32%, most pronounced with gramineous feedstock biochars applied at a high rate (>40 t ha^{-1}) (Zou et al, 2021). While Xiang et al (2017) reported alternatively, the lowest (Xiang et al, 2017) or highest (Zou et al, 2021) root biomass increase was found with biochar addition to perennials and trees; however, the number of studies on perennials in meta-analyses are currently still small. Overall, all biochar-induced increases in root system parameters (including N_2 fixation, Chapter 14) support the build-up of soil organic C and N beyond "just" the biochar-C addition (Chapter 17).

The use of biochar in subsoils or furrows connecting sub- and topsoils may enable plant root system access to nutrients and water from the subsoil. Wheat roots penetrated deeper, and showed more branching and soil-volume exploitation when biochar was added to the subsoil, resulting in significantly improved wheat straw and grain yields (Bruun et al, 2014). In on-farm trials in Germany, a modified deep-plow tractor device enabled vertical biochar trickle-down injection when the subsoil compacted plow sole at a depth of about 0.30 m was broken up (Figure 15.4). Digging up those profiles revealed preferential root growth of *Vicia faba* green-manure plants down biochar-amended furrows (Figure 15.4), improved biomass growth and nodulation, and reduced penetration resistance (L. Kohl, personal communication, March 5, 2022). The concept of using biochar to break up plow pans to better include agricultural subsoils warrants future research attention.

Belowground: Biochar 'unknowns' for future research

Another unexplored area in which biochar may ameliorate drought stress is via the phenomenon of hydraulic redistribution (HR). HR is the transport of water from wet to dry soil layers, upward or downward, through plant roots (Figure 15.1B). In temperate and, in particular, in arid environments, where perennial plants can have roots that reach tens of meters deep (Schulze et al, 2005), water can move downwards through roots when available in excess at the surface, but also upwards from deep, moister soil horizons, driven by the suction force differential of the respective soil water potentials. The mechanism can help neighboring shallower-rooting species to grow. Tree-root-facilitated HR supplied 49% and 14% of the tree and grass transpiration in a semiarid riparian woodland in the US, respectively (Lee et al, 2021). In a savanna ecosystem, however, trees

Figure 15.4 *Deep plow injection of biochar by L. Kohl (Germany) for better root subsoil exploration through a hardened plow pan (from left to right): (A,B) self-made tractor-pulled plow with biochar injector vessels for trickling down biochar into the deep-plow furrows (injection in spring 2021); (C,D) two dug-up biochar furrows showed improved faba bean root growth into the subsoil along the biochar furrows in July 2021 (Photos: Lucas Kohl, with permission)*

largely use hydraulically lifted water for themselves (Barron-Gafford et al, 2017). A second mechanism in dry farming systems is the movement of water vapor in soils driven by water potential differences (Ao et al, 2016). During nocturnal cooling, water diffuses upwards from deeper, warmer, wetter layers and re-condenses in the topsoil (thermo-condensation). Together with mist condensation and dewfall, this is sometimes sufficient to sustain plants for dryland farming (e.g., in the Namib Desert). In theory, biochar particles could provide an extra trap for water vapor during thermo-condensation. Water vapor sorption by two different biochars

(Cornelissen et al, 2013) supports this idea (Figure 15.3). However, to our knowledge, no mechanistic study exists on biochar and hydraulic redistribution or thermo-condensation. The topics warrant further attention, given that (i) biochar has been shown to alleviate osmotic stresses physiologically (next section); (ii) agroforestry systems are discussed as a novel C-sink and CDR farming method for increasing agricultural system resilience, where deeper-rooting trees lift water and nutrients and reduce wind speed for annual crops; and (iii) semi-arid farmland areas expand with accelerating global heating.

Above- and belowground: Effects of biochar application on plant physiology under osmotic stress

Potassium is essential for the regulation of stomatal movements and the plants' osmotic potential; K deficiency reduces stomatal aperture, photosynthesis, and phloem-assimilate loading, while it increases dark respiration and thus C loss and declines water use efficiency (WUE) (Marschner, 2012). Also, sufficient Mg nutrition (chlorophyll; ribosomal structures) is mandatory for alleviating osmotic stresses. Biochar provides K, Ca, and Mg as part of its mineral ash fraction (Ippolito et al, 2020) and hence it can alleviate osmotic stress. Plant WUE increased by 19% and leaf WUE by 20% with biochar addition (Gao et al, 2020). Positive effects were most prominent in alkaline soils and with herbal biochars (that have a higher ash content), supporting the idea of a plant nutrition-based mechanism of biochar amendments. Besides drought, osmotic stress can also be caused by salts which adversely affect the growth of most crop plants (Hafez et al, 2020; Mehmood et al, 2020). Globally, 424 Mha of topsoil

(0–0.3 m) and 833 Mha of subsoil (0.3–1.0 m) are salt-affected, which represents 4% and 9% of the total land area, respectively (FAO, 2021). Of global salt-affected topsoils, 85% are saline soils with high amounts of soluble salts (with electrical conductivity in the saturated soil extract of >2 dS m^{-1}), 10% are sodic soils with exchangeable Na contents above 15%, and 5% both, i.e. saline-sodic soils (subsoils: 62% saline, 24% sodic and 14% saline-sodic). More than two-thirds of salt-affected soils are found in arid and semi-arid climatic zones (FAO, 2021). These stresses may be alleviated by biochar application. The mechanisms include enhanced salt tolerance through a decreased Na in the soil solution, adsorbed to exchange sites, and increasing the soils' cation exchange capacity, depending on the soil and biochar properties. All of the above can reduce the Na uptake by plants and thus subsequently lower the burden of Na in plant root and shoot tissue (Hafez et al, 2019, 2020; Ibrahim et al, 2020

Mehmood et al, 2020; Farhangi-Abriz and Ghassemi-Golezani, 2021). Physiologically, biochar soil amendments therefore frequently improved parameters related to (stress-impacted) plant growth and yield-forming physiology, such as the activity or production of ROS (oxidative stress) scavenging enzymes, or the upregulation of genes that are related to stress defense mechanisms (Mehmood et al, 2020). Most studies, however, report biochar amendment effects on plant physiology and biochemistry without deeper cause-and-effect investigations.

While drought or salt stress negatively impacts the following biochemical or physiological parameters, biochar use in soils usually improves them; in most studies, biochars made from crop residues at low pyrolysis temperatures were used to good effect, with amendment rates between 5–30 Mg ha^{-1} (Chapter 13). Biochar amendment significantly lowered the ROS scavenging enzymatic activities, e.g., of superoxide dismutase, catalase, ascorbate peroxidase, and peroxidase. However, in certain experiments or without soil, the activity of these stress-related enzymes was upregulated (together with the gene expression producing these enzymes) under moderate and strong salinity stress (Mehmood et al, 2020). In both instances, the biochar effects benefited plant growth under stress. Biochar use usually lowered the levels of ABA while it often increased the levels of plant phytohormones such as indole-3-acetic acid and gibberellic acid that are responsible for coordinated plant-organ growth (only the former), germination (only the latter), cell elongation and rapid shoot and root growth. Subsequently, under salt stress, biochar uses increased chlorophyll a and b concentrations in leaves and enhanced net photosynthetic assimilation rates, improved the WUE (see above), plant nutrient uptake (except Na), and ultimately, via the above-mentioned physiological effects, crop biomass

and yields (Hafez et al, 2020; Ibrahim et al, 2020; Mehmood et al, 2020; Ekinci et al, 2022). Regarding the biochar dosage, there seems to be a U-shaped or a levelling-off dose-response curve where "more biochar" is not always the better option for stress alleviation and optimum biomass yield, despite improved plant nutrition. Biomass was largest with 2.5% (w/w) rather than 5% biochar addition, with shoot and root dry weight biomass increases of 373% and 200% compared to the control, respectively, after the addition of an urban waste biochar (550°C, N content of 14 mg g^{-1}) with and without salt stress (Ekinci et al, 2022). Biochar treatment under saline conditions significantly lowered sucrose, proline, H$_2$O$_2$, and malondialdehyde by 33%, 14%, 49%, and 18% in cabbage seedlings compared to the control. Plant-nutritional imbalances were considerably reduced and nutrient concentrations other than those of Na and Cl increased due to biochar addition. Interestingly, for nutrition, the 5% biochar addition was more effective than 2.5%, while the latter was the most effective dose for most of the enzymatic activities, the leaf relative water content, and ABA concentrations (Ekinci et al, 2022). The results indicate that the effects of biochar addition on salt stress in plants were not purely nutritional in nature.

Within the published literature that investigates biochar for reducing drought or salt-related stress responses in plants, a multitude of studies tested additional alleviating additives or modifications of biochar. Interestingly, the combined use of one (or several) additional alleviating measures on top of biochar frequently produced better results than using either the biochar or the other alleviating measure alone. This included the following successful combinations: (i) foliar spraying of salicylic acid; (ii) glycine betaine of the crop plants; (iii) K and silicone nanoparticle application with drip irrigation, each treatment respectively combined with the soil application of a low-temperature (350°C) rice husk or corn stalk biochar (Hafez et al, 2020, 2021; Mahmoud et al, 2022); (iv) chitosan coating of a 450°C wheat straw biochar (Mehmood et al, 2020); (v) biochar-based nutritional nanocomposites with FeO and ZnO, respectively MgO and MnO, precipitated onto the biochar surfaces (Farhangi-Abriz and Ghassemi-Golezani, 2021; Rahimzadeh and Ghassemi-Golezani, 2022); or (vi) the combination of biochar use with inoculation or concomitant application of beneficial plant-growth promoting rhizobia (PGPR; microbes: Nehela et al, 2021; gibberellic acid producing fungi: Adhikari et al, 2022). However, reviews, meta-analyses, or even systematic investigations into such combinations, including systematic variations of biochar feedstocks and properties are lacking to date. More systematic knowledge about food security improvements of such combinations may improve the assessment of the global potential for land- or rather calorie-neutral biochar carbon removal (Werner et al, 2022).

Biochar and the important role of plant-microbe interactions

A brief introduction to the rhizosphere microbiome

Biochar may act as a carrier or as a synergistic component when added together with PGPR (Chapter 14) as a stress-alleviating or remediation technology for plants, as outlined above. However, the positive effects of biochar-carrier PGPR combinations are not guaranteed. Among the positive examples, biochar used as a carrier for *Enterobacter cloacae* (an opportunistic bacterial pathogen able to fix N_2 that is also used as PGPR) increased soil total N content and plant N uptake by 34% and 15%, and soil nitrogenase activity by 253% compared to the zero control while the relative abundance of other N_2-fixing PGPR like *Burkholderia* and *Bradyrhizobium* significantly increased (Gou et al, 2023). A biochar-based inoculant by combining birch wood biochar with an effective, pre-selected strain of *Bacillus altitudinis* successfully enhanced plant performance under heavy metal (e.g., Cu) stress (Kumar et al, 2022). In a pot study with tomato and cadmium spiking, biochar amendments reduced the Cd uptake by stimulating Cd-resistant PGPR strains in the tomato rhizosphere (Zhou et al, 2022).

Most, if not all, rhizobacteria and rhizosphere fungi are prolific producers of metabolites that inhibit the growth or activity of competing microorganisms (Kong and Liu, 2022). They also produce a variety of volatile organic compounds (VOCs) that participate in long-distance communication in the rhizosphere (Junker and Tholl, 2013) and can induce systemic resistance in plants while simultaneously promoting plant growth. Microbial VOCs have been found to serve as signaling molecules that mediate short- and long-distance interactions between microbes and plant roots, thanks, in part, to their mobility in both the soil's gaseous and aqueous phases. Since biochars added to soil can cause significant changes in the structure of the rhizosphere microbial community (Kolton et al, 2011), there is likely a concomitant change in the profile of bacterial VOC metabolites in the rhizosphere. Moreover, many biochars themselves release numerous small organic molecules and VOCs (Spokas et al, 2010; Graber et al, 2014; Buss and Mašek, 2016), some of which are the same or similar in structure to microbial VOCs and may have similar direct impacts on plant growth and health. Biochars can also adsorb VOCs (Graber et al, 2011; Buss and Mašek, 2016) and may cause substantial changes to the profile of VOCs in the subsurface. Direct investigations of changes in microbial metabolites as a result of biochar additions that impact plant physiology, as well as studies of the impact of biochar-borne chemicals on plant signaling, are still needed; our knowledge has not advanced much since the 2nd edition of this book.

Biochar effects on phytohormonal signaling – the example of ethylene

Ethylene (C_2H_4) is a simple chemical carbohydrate often perceived as the "bud break", "senescence" or "fruit ripening" phytohormone. While the first, smaller synthesis peak helps plants to cope by upregulating gene expression related to the stressor, the second, larger synthesis peak is mostly harmful to the stressed plants (Glick et al, 2007). To counteract the harmful effects of the second peak, plants can have bacterial partners (PGPR) that contain the enzyme 1-aminocyclopropane-1-carboxylic acid (ACC) deaminase that grows in the rhizosphere, but also on or within plant tissue (Orozco-Mosqueda et al, 2020). The molecules' cleavage prevents the formation of ethylene, which reduces the senescence-like negative effects of drought, salinity, or heavy metal stress in plants and improves parameters/indicators (Orozco-Mosqueda et al, 2020; Glick and Nascimento, 2021). Ethylene can also be produced and consumed in soils by microbiota (Zechmeister-Boltenstern and Smith, 1998). Biochar has been shown to emit ethylene, absorb ethylene, or have no activity with respect to ethylene (Spokas et al, 2010; Fulton et al, 2013). Of 12 tested

biochars and one activated carbon for their ethylene emission potentials under various conditions, it became clear that (i) some dry biochars can emit C_2H_4; (ii) wetted biochars tend to emit more C_2H_4; and (iii) biochar mixed with soil at field capacity emitted, on average, higher amounts than biochars (mechanisms responsible for ethylene emission from biochar were not evaluated) (Spokas et al, 2010). Biochar-derived ethylene emissions probably decline within days to weeks after biochar production or soil incorporation (Fulton et al, 2013). On the other hand, biochar particles may serve as a growth matrix or stimulant for certain microbiota. Microbes associated with biochar matrices may produce or consume ethylene (Zechmeister-Boltenstern and Smith, 1998); however, we are still not aware of any research associated with this topic. However, it has been shown that biochar can enhance, promote or otherwise aid in the growth of plant growth promoting rhizobia that produces ACC deaminase, downregulating the plants' excess (second-peak) ethylene synthesis, thereby reducing the negative impact of drought, salinity, or pathogen attacks to plants. Various experiments in the greenhouse and field showed that biochars applied together with PGPRs resulted in greater crop yield (Danish and Zafar-ul-Hye, 2019; Zafar-ul-Hye et al, 2019; Danish et al, 2020).

A few studies have evaluated the use of biochar with in-vitro approaches to adsorb ethylene and prevent growth damage of the seedlings due to high ethylene concentrations. Root development and proliferation were considerably improved with biochar or activated carbon amendment (Di Lonardo et al, 2013). Biochar in in-vitro growth media for strawberry and pitaya improved seedling and root growth with biochar, attributed to ethylene adsorption by the biochar (Santos et al, 2023). The major points are that (i) in-vitro production of propagated plant material can be improved with biochar use and this use pathway deserves more research attention; and (ii) biochar removed rather than emitted ethylene, which may also occur in soils after initial ethylene emissions of production-fresh biochar (Spokas et al, 2010).

In summary, the topic of biochar and the phytohormone ethylene, including indirect effects through the microbiota growing at the root-biochar interface, is still under-explored. The combined use of biochar soil

amendment and PGPR with ACC deaminase offers interesting prospects for combatting plant stresses by mitigating the effects of the second ethylene peak. Methodically, the use of transgenic plants or microbes that are either insensitive or unable to produce a certain phytohormone or enzyme, or plants that use a certain pathogen defense mechanism, may be helpful to gain further insights into the complex topic of biochar effects with regard to phytohormonal regulation of plant growth.

Karrikins: A role in biochar mediated effects on plant responses?

Karrikire a family of plant growth regulators discovered in smoke from burning plant material (Nelson et al, 2012; Banerjee et al, 2019) that are potent in breaking the dormancy of seeds of many species adapted to environments that regularly experience fire and smoke. It has recently been shown that biochars can contain such karrikins or karrikin-like compounds (Kochanek et al, 2016; Yao et al, 2017; Lekhak et al, 2023). In a systematic test of a range of biochars, produced at around 550°C to 650°C from different feedstocks via three distinct commercial-scale production technologies, karrikinolide was detected in all six biochars but the substance was abundant only in two (Kochanek et al, 2016). The liquid byproduct "smoke water" of one technology was also high in karrikinolide. Dose-response assays of seed germination were conducted for two model species that require karrikinolide to

break their dormancy. The high-karrikin biochars were effective, as was the smoke water, even when highly diluted (Kochanek et al, 2016). Root- and shoot-growth-promoting effects of wheat-straw biochar (1–3 mg ml^{-1}) were observed on wheat germination and seedling development and several enzyme activities during seed germination; their wheat straw biochar (unfortunately, the production temperature and procedure were not reported) carried karrikin-like compounds and beta-sitosterol (Lekhak et al, 2023). The majority of germination studies that do not explicitly explore karrikins and their mechanisms mostly reported more or less neutral effects of biochar on seed germination, sometimes slightly positive, sometimes slightly negative (Busch et al, 2012, 2013; Rogovska et al, 2012). However, to date, most studies have focused on crops rather than plant species adapted to periodic fires.

The effects of biochar use at the plant community level

There is a lack of research on the effects of biochar application on the plant community composition, plant species diversity, or more

general synecology effects within multispecies ecosystems. Biochar use can deliver benefits such as the absorption and retention of agrochemicals, pesticides, and P- or nitrate pollution in contaminated or over-fertilized soils, thereby also preventing contaminants' drainage into waterways (Chapters 19, 21, 22). Contaminated areas are sites where species-rich mixtures are often sown or vegetation succession can be monitored since crop plants cannot be grown. Thus, the scarcity of studies where biochar effects on plant community composition, diversity indices, or succession were investigated is somewhat surprising.

However, some results exist. Typically, legumes are promoted by the addition of biochar (Major et al, 2005; Van de Voorde et al, 2014; Schimmelpfennig et al, 2015). In the first vegetation period after the addition of herbaceous biochar produced at 400°C and 600°C at a rate of 10 t ha^{-1} to an abandoned grassland, legumes were significantly promoted after two and 18 different grassland species were sown (7 grasses, 7 forbs, and 4 legumes) (Van de Voorde et al, 2014). However, this may be a transient phenomenon, lasting only two years (Jeffery et al, 2022). In an extensively N-fertilized, species-rich grassland ecosystem in Mid-Germany, grass biomass was lowest in the biochar plots (67 vs. 50% of the total yield) while the forbs increased (31 vs. 47% of the total yield) (Schimmelpfennig et al, 2015). Here, again, the effect was transient (S. Schimmelpfennig, unpublished data).

In some situations, no changes in community composition are observed (Bonin et al, 2018). In others, plant productivity, species richness, and diversity were enhanced 9 months after biochar additions of up to 10% by volume to a subtropical landfill cover soil (Chen et al, 2018). Ecological studies involving vegetation surveys and plant community and functional group composition in response to biochar use are severely lacking although they may offer additional insights into the mechanisms of biochar effects and benefits for biodiversity restoration measures.

Biochar from a plant perspective: An ongoing quest

The sections above illustrate that addressing biochar effects from a plant ecophysiological perspective may shed additional light on the mechanisms responsible for the broad variety of reported plant responses to biochar use in soils. It seems promising to combine biochar, or modified biochar, with other abiotic or biotic stress-alleviating measures that are already known (or suspected) to have beneficial effects. In particular concerning drought-, soil contaminants- or salt stress, biochar may offer considerable potential for land-neutral carbon dioxide removal strategies with yield benefits in a world where food security becomes more and more affected by heat waves, drought, aridity, and spreading of saline-sodic soils with accelerating global heating. Applying the strengths of combined approaches involving biochar soil application that reduce soil erosion, strengthen plant water supply, alleviate ROS, improve photosynthetic C uptake, and root growth, or enhance plant community diversity (in natural or semi-natural ecosystems) may help to increase SOC accumulation in addition to the biochar-C addition. When the plant (most importantly: root) biomass input increases over time or is better retained against decomposition, biochar additions may over time provide 'secondary' stress alleviation via SOC benefits. An improved mechanistic and plant ecophysiological understanding may help shape advanced biochar use with economic and CDR paybacks. However, a large number of under-researched aspects of biochar effect on plant ecophysiology still exist, such as biochar use in hydraulic lift research, for allowing plant roots to reach and exploit the subsoil, for extreme urban environments in tree and roof substrates that would generally profit from an improved systematic understanding of the effects of well-defined biochars on plant physiology, biochemistry, gene expression and regulation.

References

Adhikari A, Khan MA, Imran M, Lee K-E, Kang S-M, Shin JY, Joo G-J, Khan M, Yun B-W, and Lee I-J 2022 The combined inoculation of curvularia lunata AR11 and biochar stimulates synthetic silicon and potassium phosphate use efficiency, and mitigates salt and drought stresses in rice. *Frontiers in Plant Science* 13, 816958. 10.3389/fpls.2022.816858.

Ao Y, Han B, Lu S, and Li Z 2016 Internal evaporation and condensation characteristics in the shallow soil layer of an oasis. *Theoretical and Applied Climatology* 125, 281–293.

Azzi ES, Karltun E, and Sundberg C 2022 Life cycle assessment of urban uses of biochar and case study in Uppsala, Sweden. *Biochar* 4, 18.

Banerjee A, Tripathi DK, and Roychoudhury A 2019 The karrikin 'calisthenics': Can compounds derived from smoke help in stress tolerance? *Physiologia Plantarum* 165, 290–302. 10.1111/ppl.12836.

Barron-Gafford GA, et al 2017 Impacts of hydraulic redistribution on grass-tree competition vs facilitation in a semi-arid savanna. *New Phytologist* 215, 1451–1461.

Blanco-Canqui H, Laird DA, Heaton EA, Rathke S, and Acharya BS 2020 Soil carbon increased by twice the amount of biochar carbon applied after 6 years: field evidence of negative priming. *Global Change Biology Bioenergy* 12, 240–251.

Bonin CL, et al 2018 Perennial biomass crop establishment, community characteristics, and productivity in the upper US Midwest: effects of cropping systems seed mixtures and biochar applications. *European Journal of Agronomy* 101, 121–128.

Britt-Marie Alvem, Traffic Office, Rebecka Grönjord, Sweco Architects 2017 Planting beds in Stockholm city – a handbook, City of Stockholm. https://www.biochar.info/docs/urban/Planting_beds_in_Stockholm_2017.pdf

Bruun EW, Petersen CT, Hansen E, Holm JK, and Hauggaard-Nielsen H 2014 Biochar amendment to coarse sandy subsoil improves root growth and increases water retention. *Soil Use and Management* 30, 109–118.

Busch D, Kammann C, Grünhage L, and Müller C 2012 Simple biotoxicity tests for evaluation of carbonaceous soil additives: Establishment and reproducibility of four test procedures. *Journal of Environmental Quality* 41, 1023–1032.

Busch D, Stark A, Kammann CI, and Glaser B 2013 Genotoxic and phytotoxic risk assessment of fresh and treated hydrochar from hydrothermal carbonization compared to biochar from pyrolysis. *Ecotoxicology and Environmental Safety* 97, 59–66.

Buss W, and Mašek O 2016 High-VOC biochar—effectiveness of post-treatment measures and potential health risks related to handling and storage. *Environmental Science and Pollution Research* 23, 19580–19589. 10.1007/s11356-016-7112-4.

Cao CTN, Farrell C, Kristiansen PE, and Rayner JP 2014 Biochar makes green roof substrates lighter and improves water supply to plants. *Ecological Engineering* 71, 368–374.

Chen HM, et al 2018 Biochar increases plant growth and alters microbial communities via regulating the moisture and temperature of green roof substrates. *Science of the Total Environment* 635, 333–342.

Chen XW, et al 2018 Effects of biochar on the ecological performance of a subtropical landfill. *Science of the Total Environment* 644, 963–975.

Cornelissen G, et al 2013 Biochar effect on maize yield and soil characteristics in five conservation farming sites in Zambia. *Agronomy* 3, 256–274.

Danish S, et al 2020 Drought stress alleviation by ACC deaminase producing *Achromobacter xylosoxidans* and *Enterobacter cloacae*, with and without timber waste biochar in maize. *Sustainability* 12, 6286.

Danish S, and Zafar-ul-Hye M 2019 Co-application of ACC-deaminase producing PGPR and timber-waste biochar improves pigments formation, growth and yield of wheat under drought stress. *Scientific Reports* 9, 5999.

Di Lonardo S, et al 2013 Biochar successfully replaces activated charcoal for in vitro culture of two white poplar clones reducing ethylene concentration. *Plant Growth Regulation* 69, 43–50.

Ekinci M, Turan M, and Yildirim E 2021 Biochar mitigates salt stress by regulating nutrient uptake and antioxidant activity, alleviating the oxidative stress and abscisic acid content in cabbage seedlings. *Turkish Journal of Agriculture and Forestry* 46 (1). 10.3906/tar-2104-81.

El Nahhas N, et al 2021 Biochar and jasmonic acid application attenuates antioxidative systems and improves growth, physiology, nutrient uptake and productivity of faba bean (*Vicia faba* L.) irrigated with saline water. *Plant Physiology and Biochemistry* 166, 807–817.

Embrén B 2016 Planting urban trees with biochar, *the Biochar Journal (tBJ)* Arbaz, Switzerland. ISSN 2297-1114, 44–77. www.biochar-journal.org/en/ct/77.

FAO (Food and Agriculture Organization of the United Nations) 2021 *Global Map of Salt-Affected Soils*. GSAS Map V1.0. Eds: Italy: FAO, Rome.

Farhangi-Abriz S, and Ghassemi-Golezani K 2021 Changes in soil properties and salt tolerance of safflower in response to biochar-based metal oxide nanocomposites of magnesium and manganese. *Ecotoxicology and Environmental Safety* 211, 111904. 10.1016/j.ecoenv.2021.111904.

Fulton W, Gray M, Prahl F, and Kleber M 2013 A simple technique to eliminate ethylene emissions from biochar amendment in agriculture. *Agronomy for Sustainable Development* 33, 469–474.

Gan L, Garg A, Wang H, Mei GX, and Liu JQ 2021 Influence of biochar amendment on stormwater management in green roofs: experiment with numerical investigation. *Acta Geophysica* 69, 2417–2426.

Gao Y, et al 2020 Effects of biochar application on crop water use efficiency depend on experimental conditions: a meta-analysis. *Field Crops Research* 249, 107763.

Genesio L, Miglietta F, Baronti S, and Vaccari FP 2015 Biochar increases vineyard productivity without affecting grape quality: results from a four years field experiment in Tuscany. *Agriculture, Ecosystems and Environment* 201, 20–25.

Gholamahmadi B, et al 2023 Biochar impacts on runoff and soil erosion by water: A systematic global scale meta-analysis. *Science of the Total Environment* 871, 161860.

Glick BR, et al 2007 Promotion of plant growth by bacterial ACC deaminase. *Critical Reviews in Plant Sciences* 26, 227–242.

Glick BR, and Nascimento FX 2021 Pseudomonas 1-aminocyclopropane-1-carboxylate (ACC) deaminase and its role in beneficial plant-microbe interactions. *Microorganisms* 9, 2467.

Gou ZC, Zheng HY, He ZQ, Su YJ, Chen SJ, Chen H, Chen G, Ma NL, and Sun Y 2023 The combined action of biochar and nitrogen-fixing bacteria on microbial and enzymatic activities of soil N cycling. *Environmental Pollution* 317, 120790. 10.1016/j.envpol.2022.120790.

Graber ER, Tsechansky L, Khanukov J, and Oka Y 2011 Sorption, volatilization and efficacy of the fumigant 1,3-dichloropropene in a biochar-amended soil. *Soil Science Society of America Journal* 75, 1365–1373.

Graber ER, Frenkel O, Jaiswal AK, and Elad Y 2014 How may biochar influence severity of diseases caused by soilborne pathogens? *Carbon Management* 5, 169–183.

Grafmüller J, Schmidt H-P, Kray D, and Hagemann N 2022 Root-Zone amendments of biochar-based fertilizers: yield increases of white cabbage in temperate climate. *Horticulturae* 8, 307.

Hafez EM, Alsohim AS, Farig M, Omara AE-D, Rashwan E, and Kamara MM 2019 Synergistic effect of biochar and plant growth promoting rhizobacteria on alleviation of water deficit in rice plants under salt-affected soil. *Agronomy* 9, 847. 10.3390/agronomy9120847.

Hafez EM, et al 2020 Differences in physiological and biochemical attributes of wheat in response to single and combined salicylic acid and biochar subjected to limited water irrigation in saline sodic soil. *Plants-Basel* 9, 21.

Haider G, et al 2015 Biochar but not humic acid product amendment affected maize yields via improving plant-soil moisture relations. *Plant and Soil* 395, 141–157.

Haider G, Steffens D, Müller C, and Kammann CI 2016 Standard extraction methods may underestimate nitrate stocks captured by field-aged biochar. *Journal of Environmental Quality* 45, 1196–1204.

Heinrich A, und Saluz AG 2017 Versuche im Garten- und Landschaftsbau 2017: Lenkung der Baumwurzeln in Stadtbaumsubstraten. Hrsg. Forschungsgesellschaft Landschaftsentwicklung Landschaftsbau e.V., Versuche in der Landespflege. Gemeinsame Veröffentlichung der Forschungsinstitute des Deutschen Gartenbaues, Ausgabe 2017.

Ibrahim MEH, Ali AYA, Zhou G, Elsiddig AMI, Zhu G, Nimir NEA, and Ahmad I 2020 Biochar application affects forage sorghum under salinity stress. *Chilean Journal of Agricultural Research* 80, 317–325. 10.4067/s0718-58392020000300317.

Ippolito JA, et al 2020 Feedstock choice, pyrolysis temperature and type influence biochar characteristics: a comprehensive meta-data analysis review. *Biochar* 2, 421–438.

Jeffery S, et al 2022 Biochar application differentially affects soil micro-, meso-macro-fauna and plant productivity within a nature restoration grassland. *Soil Biology and Biochemistry* 174, 108789.

Junker RR, and Tholl D 2013 Volatile organic compound mediated interactions at the plant-microbe interface. *Journal of Chemical Ecology* 39, 810–825.

Kochanek J, Long RL, Lisle AT, and Flematti GR 2016 Karrikins identified in biochars indicate post-fire chemical cues can influence community diversity and plant development. *PLoS ONE* 11 (8), e0161234. 10.1371/journal.pone.0161234.

Kolton M, et al 2011 Impact of biochar application to soil on the root-associated bacterial community structure of fully developed greenhouse pepper plants. *Applied and Environmental Microbiology* 77, 4924–4930.

Kong ZY, and Liu HG 2022 Modification of rhizosphere microbial communities: a possible mechanism of plant growth promoting rhizobacteria enhancing plant growth and fitness. *Frontiers in Plant Science* 13, 920813.

Kumar A, et al 2022 Biofertilizer based on biochar and metal-tolerant plant growth promoting rhizobacteria alleviates copper impact on morphophysiological traits in *Brassica napus* L. *Microorganisms* 10, 2164.

Lambers H, Chapin IIIFS, and Pons TL 2008 *Plant Physiological Ecology*. Berlin, Heidelberg, New York: Springer.

Larcher W 2003 *Physiological Plant Ecology: Ecophysiology and Stress Physiology of Functional Groups*. Berlin Heidelberg, Germany: Springer.

Lee E, et al 2021 Convergent hydraulic redistribution and groundwater access supported facilitative dependency between trees and grasses in a semi-arid environment. *Water Resources Research* 57, e2020WR028103.

Lekhak B, Dubey A, and Verma AK 2023 Molecular docking of compounds present in pyrolyzed biomass products with the karrikin receptor and its impact on seed germination in *Triticum aestivum*. *Journal of Plant Growth*

Regulation 42, 465–480. 10.1007/s00344-021-10567-0.

Liao WX, Drake J, and Thomas SC 2022a Biochar granulation enhances plant performance on a green roof substrate. *Science of the Total Environment* 813, 152638.

Liao WX, Drake J, and Thomas SC 2022b Biochar granulation, particle size, and vegetation effects on leachate water quality from a green roof substrate. *Journal of Environmental Management* 318, 115506.

Mahmoud AWM, et al 2022 Nanopotassium, nanosilicon, and biochar applications improve potato salt tolerance by modulating photosynthesis, water status, and biochemical constituents. *Sustainability* 14, 723.

Major J, Steiner C, DiTommaso A, Falcão NPS, and Lehmann J 2005 Weed composition and cover after three years of soil fertility management in the central Brazilian Amazon: compost, fertilizer, manure and charcoal applications. *Weed Biology and Management* 5, 69–76.

Marschner P 2012 *Marschners Mineral Nutrition of Higher Plants*. Amsterdam, The Netherlands: Elsevier Academic Press.

Mehmood S, et al 2020 Chitosan modified biochar increases soybean (*Glycine max* L.) resistance to salt-stress by augmenting root morphology, antioxidant defense mechanisms and the expression of stress-responsive genes. *Plants-Basel* 9, 25.

Nehela Y, Mazrou YSA, Alshaal T, Rady AMS, El-Sherif AMA, Omara AE-D, Abd El-Monem AM, and Hafez EM 2021 The integrated amendment of sodic-saline soils using biochar and plant growth-promoting rhizobacteria enhances maize (Zea mays L.) resilience to water salinity. *Plants* 10, 1960. 10.3390/plants10091960.

Nelson DC, Flematti GR, Ghisalberti EL, Dixon K, and Smith SM 2012 Regulation of seed germination seedling growth by chemical signals from burning vegetation. *Annual Review of Plant Biology* 63, 107–130.

Omondi MO, et al 2016 Quantification of biochar effects on soil hydrological properties using meta-analysis of literature data. *Geoderma* 274, 28–34.

Orozco-Mosqueda MdC, Glick BR, and Santoyo G 2020 ACC deaminase in plant growth-promoting bacteria (PGPB): an efficient mechanism to counter salt stress in crops. *Microbiological Research* 235, 126439.

Palviainen M, et al 2018 Effects of biochar on carbon and nitrogen fluxes in boreal forest soil. *Plant and Soil* 425, 71–85.

Rahimzadeh S, and Ghassemi-Golezani K 2022 Biochar-based nutritional nanocomposites altered nutrient uptake and vacuolar H +-Pump activities of dill under salinity. *Journal of Soil Science and Plant Nutrition* 22, 3568–3581. 10.1007/s42729-022-00910-z.

Rogovska N, Laird D, Cruse RM, Trabue S, and Heaton E 2012 Germination tests for assessing biochar quality. *Journal of Environmental Quality* 41, 1014–1022.

Saluz AG, Bleuler M, Krähenbühl N, and Schönborn A 2022 Quality and suitability of fecal biochar in structurally stable urban tree substrates. *Science of the Total Environment* 838, 156236.

Santos LS, et al 2023 Biochar as an alternative to improve the in vitro environment for Pitaya (*Hylocereus undatus* Haw) and strawberry (*Fragaria x ananassa* Duch) growing. *African Journal of Agricultural Research* 19, 226–234.

Schimmelpfennig S, Kammann C, Moser G, Grünhage L, and Müller C 2015 Changes in macro- and micronutrient contents of grasses and forbs following Miscanthus x giganteus feedstock, hydrochar and biochar application to temperate grassland. *Grass and Forage Science* 70, 582–599.

Schmidt HP, Pandit BH, Cornelissen G, and Kammann CI 2017 Biochar-based fertilization with liquid nutrient enrichment: 21 field trials covering 13 crop species in Nepal. *Land Degradation and Development* 28, 2324–2342.

Schuett A, Becker JN, Groengroeft A, Schaaf-Titel S, and Eschenbach A 2022 Soil water stress at young urban street-tree sites in response to meteorology and site parameters. *Urban Forestry and Urban Greening* 75, 13.

Schulze E-D, Beck E, and Müller-Hohenstein K 2005 *Plant Ecology*. Berlin, Heidelberg, Germany: Springer.

Smith SM, Geden O, Nemet G, Gidden M, Lamb WF, Powis C, Bellamy R, Callaghan M, Cowie A, Cox E, Fuss S, Gasser T, Grassi G, Greene J, Lück S, Mohan A, Müller-Hansen F, Peters G, Pratama Y, Repke T, Riahi K, Schenuit F, Steinhauser J, Strefler J, Valenzuela JM, and Minx JC 2023 The state of carbon dioxide removal - 1st Edition. Available at: https://www.stateofcdr.org

Sofy M, Mohamed H, Dawood M, Abu-Elsaoud A, and Soliman M 2022 Integrated usage of arbuscular mycorrhizal and biochar to ameliorate salt stress on spinach plants. *Archives of Agronomy and Soil Science* 68, 2005–2026.

Spokas KA, Baker JM, and Reicosky DC 2010 Ethylene: potential key for biochar amendment impacts. *Plant and Soil* 333, 443–452.

Tan KH, and Wang JS 2023 Substrate modified with biochar improves the hydrothermal properties of green roofs. *Environmental Research* 216, 114405.

Van de Voorde TFJ, Bezemer TM, Van Groenigen JW, Jeffery S, and Mommer L 2014 Soil biochar amendment in a nature restoration area: effects on plant productivity and community composition. *Ecological Applications* 24, 1167–1177.

Werner C, Lucht W, Gerten D, and Kammann C 2022 Potential of land-neutral negative emissions through biochar sequestration. *Earths Future* 10, e2021EF002583.

Xiang Y, Deng Q, Duan H, and Guo Y 2017 Effects of biochar application on root traits: a meta-analysis. *Global Change Biology - Bioenergy* 9, 1563–1572. 10.1111/gcbb.12449.

Yao L, Naeth MA, and Mollard FPO 2017 Ecological role of pyrolysis by-products in seed germination of grass species. *Ecological Engineering* 108, 78–82. 10.1016/j.ecoleng.2017.08.018.

Zafar-ul-Hye M, Danish S, Abbas M, Ahmad M, and Munir TM 2019 ACC deaminase producing PGPR *Bacillus amyloliquefaciens* and *Agrobacterium fabrum* along with biochar improve wheat productivity under drought stress. *Agronomy-Basel* 9, 343.

Zechmeister-Boltenstern S, and Smith KA 1998 Ethylene production and decomposition in soils. *Biology and Fertility of Soils* 26, 354–361.

Zhou XG, et al 2022 Biochar amendment reduces cadmium uptake by stimulating cadmium-resistant PGPR in tomato rhizosphere. *Chemosphere* 307, 136138.

Zhu XD, et al 2020 Short-term effects of biochar on soil CO_2 efflux in boreal Scots pine forests. *Annals of Forest Science* 77, 59.

Zou ZH, et al 2021 Response of plant root growth to biochar amendment: a meta-analysis. *Agronomy-Basel* 11, 2442.

16

Biochar effects on soil nutrient transformations

Thomas H. DeLuca, Michael J. Gundale, M. Derek MacKenzie, Si Gao, and Davey L. Jones

Introduction

Biochar application to agricultural and forest soils is known to influence soil fertility and plant production (Chapter 13). Plant productivity and soil fertility are directly influenced by nutrient availability, which is a product of nutrient transformations in the soil environment. Biochar is also known to represent a persistent form of ecosystem C that remains in the soil for long periods compared to other amendments (Chapter 11). For these reasons, there is a great deal of interest in how biochar applications to soil can influence nutrient transformations and plant availability while increasing net C storage in the soil ecosystem. Although increasing evidence suggests that biochar addition to soil may enhance plant production in a variety of natural and agricultural environments (Lehmann and Rondon, 2006; Atkinson et al, 2010; Jeffery et al, 2011; Gao et al, 2019; Hossain et al, 2020), the direct influence of biochar on soil nutrient cycling is inconsistent and remains somewhat of an enigma. This is partially due to the variation in the soils, crops, and biochar amendments that are used in the experiments, and the vast predominance of short-term, pot-based studies in the literature (Chapter 13).

The purpose of this chapter is to summarize several general mechanisms through which biochar affects nutrient availability to plants, and to specifically evaluate the effect biochar has on nutrient cycling and specific transformations for several key nutrients. We explore some of the knowns and unknowns regarding how biochar influences soil nutrient transformations, which are likely to have both short- and long-term impacts on plant productivity in forest and agricultural landscapes. We specifically focus on the influence of biochar additions to soil on transformations of N, P, S, and micronutrients, Cu, Fe, Mn, and Zn, and explore the implications for modification of

DOI: 10.4324/9781003297673-16

these cycles in terms of plant availability of nutrients and their long-term budgets across a range of ecosystems. Throughout our review, we attempt to differentiate between the short-term and long-term effects of biochar on ecosystem processes.

Some general mechanisms by which biochar influences nutrient turnover and transformations

The application of biochar to agricultural and forest soils has been found to increase the bioavailability and uptake of many nutrients in plants (Glaser et al, 2002; Lehmann et al, 2003; Steiner et al, 2007; Nelson et al, 2011; Jeffery et al, 2011; Gao et al, 2017; Gao et al, 2019; Gao and DeLuca 2020). While some mechanisms causing increased nutrient availability have been extensively described and summarized (Atkinson et al, 2010; Joseph et al, 2021), less research has been conducted on the influence of biochar on specific nutrient cycling mechanisms (Gorovtsov et al, 2020). For instance, numerous studies have described high concentrations of available nutrients on the surface of newly created biochar made over a wide range of temperatures and oxidation conditions, and from a range of feedstocks, suggesting that biochars themselves can have fertilization effects over short time scales (Jeffery et al, 2011). As an example, the direct contribution of NH_4^+ salts from newly formed biochar has been described in numerous studies (see Chapter 8; Gundale and DeLuca 2006a, Spokas et al, 2012). Considerably less attention has been given to the effect of biochar on specific transformations, i.e. indirect alterations of the N cycle via the addition of polyaromatic hydrocarbons (PAHs), alteration of soil pH, microbial colonization of biochar, and alteration of soil moisture conditions (Dutta et al, 2017; Xiao et al, 2019; Razzaghi et al, 2020). These indirect alterations can influence nitrification, biological N-fixation, N-mineralization, nitrification, and gaseous N losses (Clough and Condron 2010; Karim et al, 2019; Gorovtsov et al, 2020).

The influence of biochar on nutrient transformations has consequences for the long-term effect of biochar on plant productivity and nutrient stocks (Figure 16.1), and therefore has important implications for the viability and sustainability of biochar as a climate change mitigation strategy (Lehmann 2007; Roberts et al, 2010). In the following section we identify three general mechanisms through which biochar may influence nutrient cycles: (1) Increase in the nutrient pool and the turnover of available organic nutrients, (2) Alteration of soil physical and chemical properties, and (3) Modification of the soil microbial community and its function.

Increase in the nutrient pool and the turnover of available organic nutrients

A primary mechanism by which biochar may accelerate nutrient cycling over long time scales is by serving as a short-term source of highly available nutrients (Figure 16.1), which become incorporated into living biomass and rapidly mineralizing soil organic nutrient pools (Jeffery et al, 2011). As described above and in detail in Chapter 8, new, unweathered biochar, especially that generated from nutrient-rich material can be a source of highly available nutrient salts that provide a direct short-term source of nutrition to plants (Atkinson et al, 2010; Piash et al, 2022). During the pyrolysis

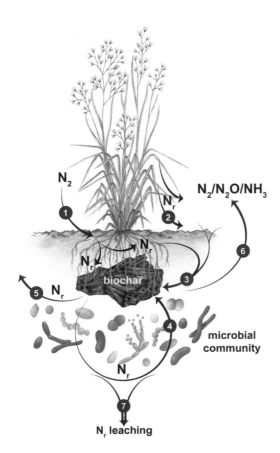

Figure 16.1 *A conceptual model for the influence of biochar on nutrient (Nr) turnover in the soil environment. Biochar can influence many nitrogen transformation processes, including (1) biological N_2 fixation, (2) plant N inputs, (3) direct adsorption of reactive N (N_r), (4) mineralization and immobilization via the soil microbial community, (5) N_r availability for plant uptake, (6) gaseous N losses, and (7) N_R leaching*

process, heating causes some nutrients to volatilize (e.g., N as NO_x, S as SO_2), especially at the surface of the material, while other nutrients become concentrated in the remaining biochar (Gundale and DeLuca 2006a; Nelson et al, 2011; Guo et al, 2021). Feedstock, pyrolysis temperature,

the time a material is held at a given temperature, oxygen availability, and the heating rate directly influence the surface chemistry of biochar (Gundale and DeLuca 2006b; Atkinson et al, 2010; Ippolito et al, 2020). Some specific elements are disproportionately lost to the atmosphere, retained in persistent organic forms, or liberated as soluble oxides during the heating process, affecting the chemical composition of ash residues on the biochar surface (Chan and Xu 2009). For wood-derived biochars, C begins to volatilize around 100°C, N above 200°C, S above 375°C, and potassium (K) and P between 700°C and 800°C (Neary et al, 2005), whereas the volatilization of magnesium (Mg), calcium (Ca), and manganese (Mn) only occurs at temperatures above 1000°C (Neary et al, 1999; Knoepp et al, 2005). These differences in volatilization temperatures among elements cause shifts in the stoichiometry of biochar elemental concentrations, with total S and N concentrations often decreasing relative to other elements due to their lower volatilization temperatures (Knudsen et al, 2004; Trompowsky et al, 2005). Correspondingly, several nutrient salts accumulate on biochar surfaces, with NH_4^+ and SO_4^{2-} concentrations increasing in low-temperature biochars (< 500 °C) (Knudsen et al, 2004; Gundale and DeLuca 2006b), and NO_3^-, PO_4^{3-}, Ca^{2+}, Mg^{2+}, and trace metals increasing, especially in biochars formed at high temperatures (Gundale and DeLuca 2006b; Chan and Xu 2009; Atkinson et al, 2010; Nelson et al, 2011). Accordingly, higher-temperature biochar also has greater alkalinity compared to low-temperature biochar created from the same feedstock (Ippolito et al, 2020; Guo et al, 2021).

Because soils generally contain a relatively large total pool of most nutrients, biochar additions to soil (especially those from low-nutrient feedstocks) usually provide only a modest contribution to the total soil nutrient capital (Chan et al, 2007). Only a small fraction of the total soil nutrient capital is usually bio-available, meaning that the addition of nutrient salts in biochar surface residues can constitute a significant increase in the bio-available pool of some nutrients (Gundale and DeLuca 2006b; Yamato et al, 2006; Chan et al, 2007, also see Chapter 8). This short-term input of bio-available nutrients can enhance plant productivity (i.e. total biomass) and improve tissue quality, and therefore influence both the quantity and quality of nutrient-containing plant residues returned to the soil (Major et al, 2010). Plant C inputs to the soil occur through root exudation and turnover, and through senescence and death of aboveground tissues. It is also well known that the nutrient concentration of plant litter has a strong control on nutrient mineralization rates (Paul, 2015). Therefore, larger inputs of higher quality plant organic matter to the soil in response to biochar-derived nutrients, likely result in an increase in the available nutrient pool, thereby in theory increasing the total quantity of readily available organic nutrients returned to the soil and available for mineralization (Gul and Whalen, 2016; El-Naggar et al, 2019). This feedback involving higher plant nutrient uptake, a higher return of available organic nutrients to the soil, and higher nutrient mineralization rates could enhance nutrient availability to plants over longer time scales as implied in Figure 16.1. The persistence of accelerated nutrient turnover between plants and soil is likely dependent on the size of the nutrient pool added from biochar, the frequency of its addition (e.g. single dose, multiple doses, or annual), the degree to which nutrient capital is removed from a system during harvesting activities, the degree to which nutrients are fixed into sparingly available organic or mineral pools, the long term losses in nutrient capital through leaching or volatilization that occur at a given site, and the long-term build-up (or decline) of stable, recalcitrant organo-mineral complexes beyond the pure biochar (Borchard et al, 2019; Gao et al, 2019; El-Naggar et al, 2019; Joseph et al, 2021; Zhang et al, 2021).

Alteration of soil physical and chemical properties

In addition to its direct contribution of available nutrients to the soil, biochar has a variety of physical and chemical properties that influence soil nutrient transformations. For a more detailed review of biochar's physical and chemical properties, see Chapters 5 and 6, Atkinson et al (2010), and Ippolito et al (2020). Biochar has a high surface area (Beesley et al, 2011), is highly porous (Keech et al, 2005) (see Figure 16.2), has a variable surface charge and often has a surface residue enriched in alkaline metals (Atkinson et al, 2010). When added to soil, biochar has the potential to alter the physical and chemical properties of soil, which in turn can influence nutrient transformation rates. Due to the porous and alkaline nature of most forms of biochar, applications to soil have often been shown to increase soil water holding capacity, alter gas exchange, increase cation exchange capacity (CEC), increase surface sorption capacity, increase base saturation of acidic mineral soils, and alter soil pH (Glaser et al, 2002; Keech et al, 2005; Ding et al, 2016; Karim et al, 2019). These biochar properties are highly dependent on the temperature (see Figure 16.3 and Chapter 8) and duration of pyrolysis (Glaser et al, 2002; Gundale and DeLuca 2006a; Bornermann et al, 2007; Ippolito et al, 2020), and the feedstock from which biochar is made (Gundale and DeLuca 2006b; Streubel et al, 2011; Ippolito et al, 2020).

Soil micro-organisms require environments with appropriate water potential and redox conditions to carry out their metabolic

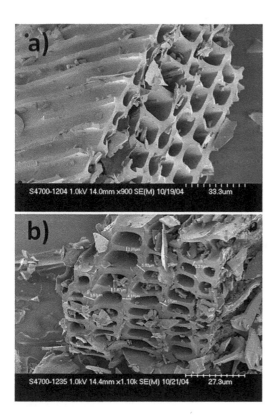

Figure 16.2 *Electron micrographs of a high sorption (a) and low sorption (b) char collected from forest soils in northern Idaho, USA (Brimmer, 2006). The high sorption char (immature char formed in a recent fire) has open pores that follow tracheids whereas the low sorption char (mature char) has many of the pores occluded with organics*

activities (Alexander 1991; Briones 2012). The physical structure of biochar contains a range of larger pore sizes which are influenced by feedstock characteristics (Keech et al, 2005) and pyrolysis conditions (Braghiroli et al, 2020) which can directly influence the water potential and redox environment of soil micro-organisms (Joseph et al, 2010). Micropores, defined by soil scientists as pores with < 30 μm diameter, serve as capillary spaces with high surface area to volume ratios, and can retain water even when soil moisture is strongly depleted (Kammann et al, 2011; Braghiroli et al, 2020), thereby creating moist microsites (Lehmann and Rondon 2006). Biochar also often contains macropores (>75 μm diameter) which can serve as gas exchange channels, thereby influencing the redox environment for soil biota (Joseph et al, 2010; Lehmann et al, 2011). Organic residues decompose much more rapidly under aerobic conditions, and therefore biochar may enhance nutrient mineralization in soils with inherently poor gas exchange properties by increasing soil aeration (Gundale and DeLuca 2006b; Asai et al, 2009). Likewise, several specific nutrient transformations generally require oxygen as an electron acceptor, such as nitrification and sulfur oxidation, which suggests that the physical structure of biochar may increase oxidative transformations in soils with inherently poor gas exchange environments (DeLuca et al, 2006; Asai et al, 2009; Joseph et al, 2010). The highly variable pore size distribution of biochar thus assures the presence of a wide variety of soil microsites with contrasting moisture and redox conditions under variable environmental conditions (Joseph et al, 2010). The addition of biochar to soil may thus intensify microbial or root-associated gross nutrient cycling processes by creating more "microsite opportunities" with *steeper* redox, pH, or nutrient-concentration gradients around or across biochar particles (Briones 2012; Joseph et al, 2013). If these micro-site opportunities are in the presence of great organic (e.g., crop or root residues) inputs, a positive feedback cycle occurs with intensified gross nutrient cycling and improved soil fertility in the long run (Figure 16.1).

Additional mechanisms through which biochar amendments can alter nutrient

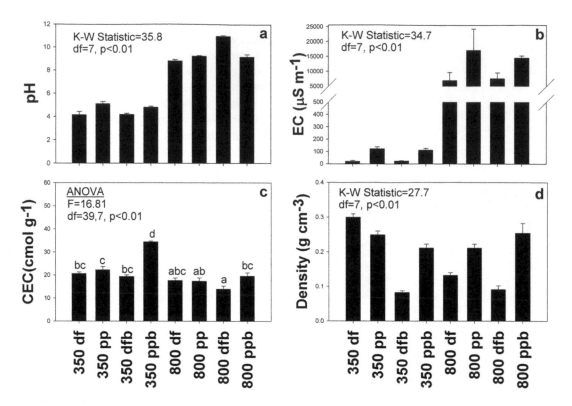

Figure 16.3 *The pH, electrical conductivity (EC), cation exchange capacity (CEC), and density of biochar produced from Douglas-fir or ponderosa pine wood or bark at 350°C or 800°C (redrawn from Gundale and DeLuca, 2006a). Data meeting the assumptions of normality were compared with one-way ANOVA followed by the Student-Neuman-Kuels post hoc procedure where letters indicate pairwise differences. Non-normal data were compared using the Kruskal-Wallis (K-W) statistic*

transformations include: (1) Adsorbing nutrients thereby reducing nutrient loss from the soil (Crutchfield et al, 2010; Ding et al, 2010; Prendergast-Miller et al, 2011; Ventura et al, 2013); (2) Increasing or decreasing fixation of nutrients into insoluble mineral or persistent organic pools (Cui et al, 2011; Nelson et al, 2011); (3) Reducing losses of nutrients (N) via volatilization of NH_3 or transformation to N_2 or N_2O (Prendergast-Miller et al, 2011; Spokas et al, 2012; Arezoo et al, 2012; Borchard et al, 2019; Sha et al, 2019; Liu et al, 2019a); (4) By ameliorating other constraints of

nutrient cycling e.g. in contaminated soils by its adsorptive properties (Figure 16.1). Biochar has been shown to have a transient anion exchange capacity, and moderately high cation exchange capacity that also changes with time in the soil (Brewer et al, 2011). Biochar also can ameliorate soil pH due to the alkaline ash residue commonly associated with biochar as mentioned above. A variety of studies suggests that biochar can simultaneously reduce nutrient leaching and volatilization losses through its influence on soil pH and CEC (Karimi et al, 2020); however, the alkaline nature of some biochars may actually increase NH_3 volatilization in surface soils amended with biochar (Sha et al, 2019). Biochar can harbor a relatively high exchange capacity per unit mass (Atkinson et al, 2010), therefore its addition to some soils can increase surface soil exchange capacity. This exchange capacity can act to reduce leaching and volatilization losses (Prendergast-Miller et al, 2011; Spokas et al, 2012; Arezoo et al, 2012; Ventura et al, 2013).

An additional characteristic of biochar that can influence nutrient cycling is its effect on soil solution C chemistry (Figure 16.1) and turnover (see Chapter 17). While wood and crop residue biochar have been shown to contain only a minor fraction of bio-available C (Major et al, 2010; Jones et al, 2011); low-temperature biochar generated from feedstocks with high concentrations of soluble C can yield high rates of dissolved organic matter (Sun et al, 2021) which can lead to nutrient immobilization or enhance N loss through denitrification. Several studies suggest that biochar can function as a strong adsorptive surface for the adsorption of a wide range of C compounds. The high surface area, porous (Figures 16.2 and 16.3), and often hydrophobic nature of biochar direct after production makes it an ideal surface for the sorption of hydrophobic and volatile organic compounds (Cornelissen et al, 2004; Keech et al, 2005; Bornermann et al, 2007; Gundale and DeLuca 2007; Kumar et al, 2020). Numerous studies have shown a reduction in soluble or free phenolic compounds when activated C is added to soils (DeLuca et al, 2002; Wallstedt et al, 2002; Berglund et al, 2004; Keech et al, 2005; MacKenzie and DeLuca 2006; Gundale and DeLuca 2006a; Kumar et al, 2020) or when pyrogenic C is formed during wildfires or prescribed fire and introduced into the soil (DeLuca et al, 2006; Gundale and DeLuca 2006a; MacKenzie and DeLuca 2006;

Brimmer 2006; Bornermann et al, 2007). These sorption reactions may: (1) Reduce the activity of compounds that may be either inhibitory to nutrient transformation specialists, such as nitrifying bacteria (White 1991; Ward et al, 1997; Paavolainen et al, 1998; Kuppusamy et al, 2016); (2) Reduce complexation of nutrient-rich molecules such as proteins into tannin-complexes (Kraus et al, 2003; Gundale et al, 2010); (3) Reduce the concentration of bio-available C in the soil solution that would otherwise enhance the immobilization of inorganic N, P or S (Paul 2015) (Figure 16.1). The interaction of soluble soil C with biochar surfaces is a key mechanism that may influence nutrient availability and transformations (MacKenzie and DeLuca 2006; Nelissen et al, 2012) or may induce the priming of resident soil organic matter (Fiorentino et al, 2019).

Alteration of microbial communities

Biochar additions to soil have the potential to alter soil microbial biomass, the microbial community composition (Gorovtsov et al, 2020; Zhang et al, 2021), and the activity of soil microbes (Gorovtsov et al, 2020), all of which can influence nutrient mineralization from decomposing plant residues, as well as several specific nutrient transformations. For a complete review of biochar's effects on soil microbial communities, see Chapter 14. There are several mechanisms proposed by which biochar can influence soil microbes, including: (1) The porous structure of biochar which may provide a habitat for microbes (Pietikainen and Fritze 1993; Quilliam et al, 2013b; Gorovtsov et al, 2020); (2) Biochar effects on plant growth and associated plant C inputs (Major et al, 2010); (3) Biochar can function as a source of mineral nutrients for microbial use (Rondon et al, 2007); (4) The sorption of microbial signaling compounds or inhibitory plant phenolic compounds by biochar (DeLuca et al, 2006; Ni et al, 2010; Yu et al, 2018); (5) The effect of biochar on soil's physical and chemical properties (Gorovtsov et al, 2020). Although an increasing number of studies have attempted to characterize the relative importance of these factors in determining microbial response to biochar applications, substantial uncertainty remains regarding the mechanisms through which biochar influences soil microbial community properties (Whitman et al, 2019; Wang et al, 2020b; Zhang et al, 2021).

Despite mechanistic uncertainty, several studies have shown that increases in microbial biomass appear to occur in response to soil biochar amendments. Numerous studies have demonstrated an increase in microbial biomass and activity with biochar additions to soil (Woolet and Whitman 2020; Pokharel et al, 2020). Mechanisms for these increases vary, but most are related to the alteration of soil pH, nutrient availability (Wang et al, 2020b), or physical properties. Other studies have shown no significant shift in microbial activity with biochar amendments to soils (Palansooriya et al, 2019). Although soil microbes are the primary driver of organic nutrient mineralization and oxidative or reductive nutrient transformations, these studies suggest that biochar-induced changes in microbial communities likely have consequences for nutrient turnover rates between plants and soil.

In addition to observed shifts in microbial biomass in response to biochar, a variety of studies have shown that microbial community composition can be altered by biochar (Whitman et al, 2019; Zhang et al, 2021), sometimes resulting in an increased abundance of functional groups that have key roles in nutrient cycling and plant nutrient acquisition (Lehmann et al,

2011). Mycorrhizal fungi, which play a key role in extracting nutrients from persistent organic or insoluble mineral pools have been observed to generally increase with biochar additions to soil (Saito 1990; Makoto et al, 2010; Solaiman et al, 2010, (Zhang et al, 2018). The specific relationship between biochar and mycorrhiza is dependent on the nature of the biochar and the chemistry of the soil to which the biochar has been added (Gujre et al, 2021; Xu et al, 2021). Given the specific functional role of mycorrhizas in nutrient acquisition, changes in mycorrhizal biomass and colonization likely influence the flux of nutrients from un-available nutrient pools (i.e., persistent organic matter and insoluble minerals, in particular, P) into biomass and therefore labile organic pools that actively turnover between plants and soil. In addition to mycorrhizas, several specific nutrient transformations have been shown to either increase or decrease in response to soil biochar amendments, and, in some cases, altered transformation rates have been linked to changes in the abundance of specific soil biota (Zhang et al, 2021). An example of this is the observed increase in nitrification rates in biochar-amended forest soils that otherwise demonstrate little or no net nitrification (DeLuca et al, 2006; Gundale and DeLuca, 2007) which has been linked to increased populations of nitrifying bacteria within biochar pore spaces (Ball et al, 2010; described in further detail below). An increasing number of studies have described direct links between biochar amendment and shifts in the microbial community composition and resultant shifts in nutrient transformation rates (Lehmann et al, 2011; Bello et al, 2020; Xu et al, 2021). An overview of some of these direct links is described with regard to individual nutrient transformations below.

Influences of biochar on specific nutrient transformations

As described above, there are a range of mechanisms through which biochar can influence the loss of nutrients from forest or agricultural ecosystems, as well as the gross annual turnover between soils and plants. In the following sections we review specific mechanisms by which biochar influences N, P, S, and some alkaline and trace metal cycles. Biochar always contains some quantity of soluble inorganic nutrients (see Chapter 8) which it readily or slowly delivered to soil; however, in this section we will focus on the influence of biochar on nutrient transformations as opposed to nutrient delivery.

Nitrogen

Nitrogen is the single most limiting plant nutrient in most cold or temperate terrestrial ecosystems (Vitousek and Howarth, 1991), and also frequently limits agricultural productivity. In soils, the majority of N exists in complex organic forms that must be mineralized (converted from organic N to NH_4^+ or NO_3^-) prior to uptake by most agricultural plants, although most plants also have the capacity to take up organic N with or without mycorrhizal symbionts (Paul 2015). Recent studies have demonstrated that the addition of biochar to surface mineral soils may directly or indirectly influence soil N transformations (Nguyen et al, 2017; Liu et al, 2018; Gao et al, 2019). Here we review the evidence for the direct and indirect influences of biochar on ammonification, nitrification, NH_3 volatilization, denitrification, nitrous oxide emission (see also Chapter 18), and N_2-fixation, while providing potential mechanisms that may be driving these transformations.

Ammonification and nitrification

Nitrogen mineralization is the process by which organic N is converted to inorganic forms (primarily NH_4^+ and NO_3^-). The conversion of organic-N to NH_4^+ is generically termed ammonification. This process is driven by a broad consortium of organisms capable of enzymatic denaturation of proteins and the removal of amide groups from organic compounds (e.g., amino acids and amino sugars). Nitrification represents the oxidation of organic N (via heterotrophic organisms) or NH_4^+-N to NO_3^- by autotrophic bacteria and archaea as well as certain fungi (Stevenson and Cole 1999; Leininger et al, 2006). Biochar addition to temperate and boreal forest soils has been found to increase net nitrification rates in soils that otherwise demonstrate little or no net nitrification (Berglund et al, 2004; DeLuca et al, 2006); whereas, there has been little evidence for such an effect in grassland (DeLuca et al, 2006) or agricultural soils (Lehmann et al, 2003; Rondon et al, 2007; Craswell et al, 2021), which may already accommodate an active nitrifying community. Results from the literature have been summarized in Figure 16.4 which is adapted from a meta-analysis specifically focusing on N transformations and active pools resulting from biochar amendment (Liu et al, 2018).

Several studies in forest ecosystems have aimed to understand the mechanisms underlying increased nitrification following biochar addition. Using forest soils with very low inorganic N concentrations, DeLuca et al (2002) showed that the injection of heat-activated biochar into the organic horizon induced a slight stimulation of nitrification, but the injection of glycine with activated C

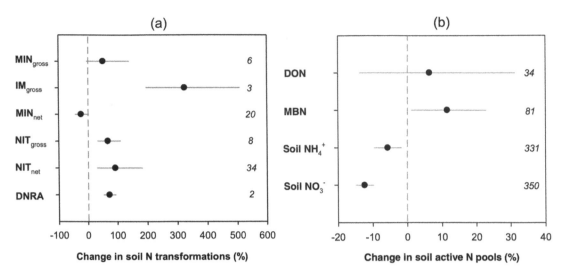

Figure 16.4 *Meta-analysis of the relative changes in soil N transformations (a) and soil active N pools (b) in biochar-amended soils compared to unamended soils. Bars represent 95% confidence intervals. Soil N transformations include gross mineralization (MINgross), gross immobilization of $NH_4^+ - N$ to organic N (IMgross), net mineralization (MINnet), gross nitrification (NITgross), net nitrification (NITnet), and dissimilatory nitrate reduction to ammonium (DNRA). Soil active N pools include dissolved organic nitrogen (DON), microbial biomass nitrogen (MBN), ammonium (NH_4^+), and nitrate (NO_3^-). Adapted from a meta-analysis performed by Liu et al (2018)*

consistently stimulated high rates of nitrification, demonstrating that biochar alleviated the factor limiting nitrification (DeLuca et al, 2002; Berglund et al, 2004). Biochar collected from recently burned forests (MacKenzie and DeLuca 2006; DeLuca et al, 2006) or generated in laboratories under controlled conditions (Gundale and DeLuca 2006a) were found to stimulate net nitrification in laboratory incubations and in short-term (24 hr) nitrifier activity assays. One possible mechanism is that activated carbon-adsorbed organic compounds (and specifically terpenes) either inhibited net nitrification or caused immobilization of NH_4^+ (Sujeeun and Thomas 2017; Bieser et al, 2022). The rapid response of the nitrifier community to biochar additions in soils with low nitrification activity and the lack of a stimulatory effect on

actively nitrifying communities suggest that biochar may be adsorbing inhibitory compounds in the soil environment (Zackrisson et al, 1996) that then allows nitrification to proceed. Similarly, fire induces a short-term influence on N availability, but biochar may act to maintain that effect for years to decades after a fire. It is also possible that the presence of biochar in these forest soils enhances the numbers of ammonia-oxidizing bacteria by creating conditions conducive to their growth, including: increased pH, reduced inhibitory compounds, microsites, redox potential, and external electron transfer (Ball et al, 2010).

In another study seeking to explain char-induced increased nitrification rates in nutrient-poor conifer forests, DeLuca et al (2006) evaluated gross nitrification rates in char-treated and untreated forest soils. Gross nitrification rates in the char-amended forest soils were nearly four times that in the untreated soil, demonstrating the stimulatory effect of char on the nitrifying community rather than reduced immobilization. Wood ash commonly contains high concentrations of metal oxides including CaO, MgO, Fe_2O_3, TiO_2, and CrO (Koukouzas et al, 2007). Exposure of biochar to solubilized ash may result in the retention of these potentially catalytic oxides on active surfaces of the biochar (Le Leuch and Bandosz, 2007). These oxide surfaces may in turn effectively adsorb NH_4^+ or NH_3 and potentially catalyze the photo-oxidation of NH_4^+ (Lee et al, 2005).

In contrast to forested ecosystems, biochar additions in agricultural systems have yielded mixed results, partially based on the variety of feedstocks tested in agricultural trials (see Gao et al, 2019). Biochar additions to agricultural soils have been found to reduce, have no effect, or in some cases increase net N mineralization (Yoo and Kang 2010; Streubel 2011; Güereña et al, 2013; Gao et al, 2017; Gao et al, 2019). However, more consistently, studies have demonstrated an increase in gross N mineralization rates (see Figure 16.4) in agricultural soils with the addition of biochar co-composted with organic residues (Mia et al, 2017; Pokharel et al, 2021; Bieser et al, 2022). Recently, it has also been suggested that co-composting biochar with organic residues produces an organic coating on the outer and inner pore spaces of the biochar, which may explain why biochar retains nutrients and water, as well as stimulating N turnover (Hagemann et al, 2017). Using molecular analyses (TRFLP and 454 pyrosequencing) microbial response to biochar additions was studied in agricultural soils; the presence of *Nitrosovibro* ($NH_4^+ \rightarrow NO_2^-$) was found to decrease in the presence of biochar while

Nitrobacter ($NO_2^- \rightarrow NO_3^-$) was observed to increase in the presence of biochar (Anderson et al, 2012). However, these shifts could have little consequence for nitrification rates as molecular analyses have also demonstrated little or no relationship between ammonia-oxidizing bacteria gene abundance and rates of NO_3^- accumulation (Ducey et al, 2013). Such results emphasize the contrast between the strong positive effects biochar amendment has on forest soils, where little or no net nitrification occurs, compared to a much smaller effect in agricultural soils, that already exhibit inherently high rates of net nitrification and NO_3^- accumulation (e.g., over 113 mg NO_3^--N kg^{-1} in the control; Ducey et al, 2013) before biochar additions. Interestingly, Nelissen et al (2012) reported a significant increase in gross ammonification and nitrification rates in sandy soils amended with maize biochar with the increase in nitrification being attributed to greater substrate availability for autotrophic nitrifying bacteria.

The length of time that biochar resides in the soil environment has also been shown to affect N mineralization potential which may be related to its occlusion with organic matter over time as reported by a couple of studies (Zackrisson et al, 1996; Hagemann et al, 2017). Dempster et al, (2012) found that 1-year-old soils amended with biochar resulted in greater inorganic N accumulation than soils recently amended with biochar in different agronomic soils from both Australia and the UK. This might have significant implications for management practices that are using biochar to retain inorganic N fertilizer on-site. Regular additions of 'fresh' biochar to agricultural systems might be needed to help retain inorganic N fertilizers and this practice may also sequester large amounts of C. In contrast, Novak et al (2010) reported a modest increase in net N mineralization when fresh wood biochar was added to acidic agricultural soils. Alternatively, co-composting biochar with organic residues may solve this problem, by increasing hydrophilicity and nutrient availability (Hagemann et al, 2017), while solving the land application problem as well, given that compost should be easier to spread than dry biochar.

Immobilization

Several studies have shown that respiration rates can increase following biochar additions to soil, suggesting that biochar may either be a direct C source to microbes or have a priming effect on already existing soil organic matter (Wardle et al, 2008; Spokas et al, 2009; Novak et al, 2010). These changes in C availability to microbes, therefore, have the potential to influence nutrient immobilization. The degree to which biochar supplies bio-available C to soil microbes appears to vary substantially among biochars, depending on a variety of factors such as the feedstock biochar is made from (Maaz et al, 2021), the period after which biochar has been added (Nelissen et al, 2015), and potentially also the temperature at which biochar is made (Craswell et al, 2021). Regarding feedstock properties, it has generally been found that some decomposition occurs when fresh biochar is added to the soil (Schneour 1966; Liang et al, 2006; Spokas et al, 2009; Jones et al, 2011), although wood biochar is relatively more persistent (DeLuca and Aplet, 2008). Biochar made from wood or other woody feedstocks are typically N-depleted materials that have the potential to immobilize N; whereas, biochars generated from N-rich feedstocks, such as manures or sewage sludges, may serve as net N mineralization sources (Lehmann et al, 2006; Maaz et al, 2021). However, the degree to which net immobilization or mineralization occurs is strongly dependent on the C chemistry of biochar, which is influenced by the temperature of formation (Gundale et al, 2006a; Nelissen et al, 2015; Maaz et al, 2021). Low-temperature biochars are known to have higher concentrations of residual bio-oils (Steiner et al, 2007; Nelissen et al, 2012; Clough et al, 2013) or surface functional groups (Liang et al, 2006) that can serve as microbial substrates, and hence promote immobilization. Higher temperature biochars, in contrast, contain much higher concentrations of graphene structures, which are much more resistant to microbial metabolism, and hence do not promote immobilization. When biochars do provide a significant concentration of bioavailable C (i.e. < 500°C), immobilization appears to be stimulated (see Chapter 17). Using ^{15}N labeling approaches, Nelissen et al (2015) showed that the addition of a low-temperature biochar immediately stimulated soil ammonium and nitrate immobilization by +4500% and +511%, respectively; however, one year later they found that biochar had a neutral effect on immobilization/mineralization in their soil. This suggests that easily mineralizable C on the surfaces of low-temperature biochar is quickly consumed, and immobilization is short-lived. Once bio-available C fractions are consumed, the remaining more persistent biochar fractions are left behind, which has very little impact on N immobilization (Steiner et al, 2007; Nelissen et al, 2015; Maaz et al, 2021). In summary, while reported effects of biochar on N immobilization have been highly variable, it appears that pyrolysis temperature is of primary importance (<500°C) in controlling C bioavailability that stimulates microbial growth and activity (Fiorentino et al, 2019; Craswell et al, 2021; Xu et al, 2021); whereas, feedstock stoichiometry (i.e., C:N ratio) and time since addition help explain additional variation in nitrogen immobilization rates in response to biochar.

Gaseous nitrogen emissions

Over the past several years, there has been an increasing interest in understanding how biochar influences the gaseous soil N transformations to understand ecosystem N budgets and the effects of biochar management on greenhouse gas (GHG) emissions. Much interest has focused on the influence of biochar on N_2O flux (i.e. it has a global warming effect per molecule that is 298 times greater than CO_2) (Yanai et al, 2007; Spokas et al, 2009; Clough et al, 2010; Cornelissen et al, 2012; Borchard et al, 2019), because of its importance as a greenhouse gas (Hansen et al, 2005) and ozone-depleting substance (Ravishankara et al, 2009). Several studies have also addressed the influence of biochar applications on denitrification and NH_3 volatilization potential to evaluate the influence of biochar on N conservation in agricultural soils (Jones et al, 2012; Taghizadeh-Toosi et al, 2012b; Sha et al, 2019). Nitrous oxide emissions from soil are associated with the processes of nitrification and denitrification, this topic is covered in detail in Chapter 18.

Ammonia volatilization represents a significant pathway for N loss from agroecosystems. For this reason, there has been increasing interest in understanding the role of biochar in soil NH_3 volatilization rates (Steiner et al, 2010, Doydora et al, 2011, Jones et al, 2012, Taghizadeh-Toosi et al, 2012a, 2012b, Chen et al, 2013, Mandal et al, 2018, Dong et al, 2019). A recent meta-analysis emphasized that there is no single unifying pattern for how biochar affects NH_3 volatilization (Sha et al, 2019); however, there are a few noted trends. Ammonia volatilization in agricultural soils is favored at alkaline pH and when high concentrations of NH_4^+ are present, and is reduced in soils with high CEC values (Paul, 2015). Biochar and biochar mixed with ash are known to

temporarily increase soil pH (Glaser et al, 2002; Jones et al, 2012), but usually not to a high enough level to increase NH_3 volatilization. Taghizadeh-Toosi et al (2012a, b) have shown instead that NH_3 is effectively sorbed to the surface of wood biochar, but also demonstrate that it can be desorbed into solution as NH_4^+ thereby reducing N losses to the atmosphere.

Biochar additions to agricultural soils as well as acid forest soils have been found to reduce NH_4^+ concentrations (Le Leuch and Bandosz, 2007; Taghizadeh-Toosi et al, 2012a) which reduces the potential for NH_3 volatilization. Steiner et al (2010) found a clear reduction in NH_3 evolution during poultry litter composting when biochar amendment rates were 20% (w/w). Doydora et al (2011) found 50 – 60% reductions in NH_4^+ available for volatilization when composted poultry litter was cut 1:1 with biochar before incorporation into the soil. This finding is supported to some degree by Jones et al (2012) who found a clear capacity of biochar to adsorb NH_4^+. Furthermore, in field trials, the researchers showed a reduction in NH_3 volatilization at rates of 50 Mg char ha^{-1}, but not at 25 Mg char ha^{-1} (Jones et al, 2012). In agricultural soils, it appears that biochar generally results in a reduced presence of extractable NH_4^+, likely as a result of sorption of soluble NH_4^+ to biochar surfaces (Nguyen et al, 2017). Other analyses have suggested that increasing rates of biochar application result in an increasing rate of NH_3 volatilization (Feng et al, 2022); however, this general observation does not address differences in feedstock or temperature. Wood-based biochar (such as that used in the Jones et al (2012) study described above, may be more likely to decrease NH_3 volatilization compared to N-rich and low-temperature biochars (Sha et al, 2019).

Nitrogen fixation

Biological N_2 fixation historically provided the vast majority of N inflow into agroecosystems (Galloway et al, 2008). Today it is mandatory in low-input agroecosystems where external N inputs are minimal. Although there have been reports of the influence of char on N_2 fixation in leguminous plants for over seventy years (Tyron 1948), results have generally been found to be inconsistent. More broadly, a recent meta-analysis identified 25 studies that had evaluated the influence of biochar on N_2 fixation and reported a 50% increase in total N_2 fixation across the range of studies (Liu et al, 2018). Table 16.1 provides a summary of the results of a collection of studies on the influence of biochar applications on nodulation and N fixation in leguminous plants. Some examples of these findings are described below.

Table 16.1 *Summary of research findings on the influence of biochar on growth, nodulation, and N_2 fixation in leguminous crops. For each study, proportional changes in individual variables were calculated relative to an experimental control. All studies are pot trials with the exception of Quilliam et al. (2013b) which combines the field application of biochar with a growth chamber pot trial*

Biochar type and rate	Response plant	Growth response	Nodulation	Nitrogenase activity or N_2 fixed	Source
Wood biochar 2%	*Pisum sativum*	+37%	+25%	NA	Vantis and Bond, 1950
Wood biochar 4%	*Pisum sativum*	+45%	−11%	NA	Vantis and Bond, 1950
Wood biochar 8%	*Pisum sativum*	+8%	−31%	NA	Vantis and Bond, 1950
Animal biochar 2%	*Pisum sativum*	NS or neutral	−%	NA	Vantis and Bond, 1950
Wood biochar 1% –2%	*Trifolium pretense*	NA	+97%	NA	Turner, 1955
Wood biochar powder 1:1	*P. sativum*	−24%	−39%	NA	Devonald, 1982
Wood bark biochar ~1%	*Medicago sativum*	+70%	NA	+517%	Nishio and Okano, 1991
Wood biochar 3%	*Phaseolus vulgaris*	+25%	NA	+42%	Rondon et al, 2007
Wood biochar 6%	*Phaseolus vulgaris*	+39%	NA	+64%	Rondon et al, 2007
Wood biochar 9%	*Phaseolus vulgaris*	NS	NA	NS	Rondon et al, 2007
Chicken manure biochar ~0.4%	*Glycine max*	+5%	+100%	NA	Tagoe et al, 2008

Table 16.1 *continued*

Biochar type and rate	Response plant	Growth response	Nodulati-on	Nitrogenase activity or N_2 fixed	Source
Chicken manure biochar ~0.8%	*Glycine max*	+41%	+190%	NA	Tagoe et al, 2008
Wood biochar ~2.5%	*Trifolium repens*	Neutral	NA	+250%	Quilliam et al., 2013b
Wood biochar ~5%	*Trifolium repens*	Neutral	−70%	+350%	Quilliam et al., 2013b
Grass biochar (600°C) 10 Mg ha^{-1}	*Trifolium pretense*	+400%	NA	+300%	Van de voorde et al, 2014
Maize stover biochar 15 Mg ha^{-1}	*Phaseolus vulgaris*	+133	+2825%	+1491%	Güereña et al, 2015
Rice straw biochar 15 Mg ha^{-1}	*Phaseolus vulgaris*	+190	+3825%	+2620%	Güereña et al, 2015
sWood biochar 1.1%	*Trifolium repens*	Neutral	NA	−5%	Mia et al, 2018
Wood biochar 10 – 20 Mg ha^{-1}	*Mixed legumes*	−46%	NA	−45%	Mia et al, 2018
Wood biochar 10 Mg ha^{-1}	*Glycine max*	Neutral	+41	NA	Ma et al, 2019
Wood biochar 1.5%	*Glycine max*	+56%	+152%	NA	Yin et al, 2021
Wood biochar 5%	*Glycine max*	+48%	+42%	NA	Yin et al, 2021

NA: Not available; NS: Not significant at $P < 0.05$

In older studies, Vantis and Bond (1950) found that the addition of wood biochar to soils at a rate of 1% (v/v) resulted in a reduction in the number of nodules on clover, but increased the total nodule mass and total N fixed in *Pisum sativum* (L.). However, at higher rates of biochar (greater than 2%), there was no effect or a negative effect of biochar on nodulation (Vantis and Bond, 1950). Turner (1955) found a significant increase in the number of root nodules in clover (*Trifolium pretense* L.) and that 'boiled biochar' further increased nodulation, perhaps due to the removal of inhibitory compounds by pretreatment (Turner, 1955) (this treatment may have influenced phytohormone-like chemicals, see Chapter 15). Investigation of composts with or without biochar added (5% w/w) as a growth medium suggested that the biochar additions resulted in a significant decrease in nodule number and size (Devonald, 1982), however, there is no discussion on pretreatment of the biochar or its polyaromatic hydrocarbon (PAH) content. Studies involving the application of activated carbon to soils have demonstrated a significant inhibitory effect of the amendment on nodulation in *Lotus corniculatus* (L.) (Wurst and van Beersum, 2008). On the other hand, the application of a nutrient-rich biochar (carbonized chicken manure) to silt loam soils in a greenhouse experiment was found to increase nodule number and mass in soybeans (*Glycine max* L.) and increase total N yield (Tagoe et al, 2008). Quilliam et al (2013a) reported that high rates of wood biochar applied to temperate agricultural soils (total applications of 50 and 100 Mg biochar ha^{-1}) significantly reduced total nodulation in clover (*T. repens*), but increased the mass of individual nodules and increased total nitrogenase activity (Quilliam et al, 2013a).

Rondon et al (2007) tested the effect of adding different amounts of wood (eucalyptus) biochar to nodulating and non-nodulating varieties of the common bean (*Phaseolus vulgaris*) and found that biochar significantly increased N$_2$ fixation and bean productivity at application rates of 30 or 60 g biochar kg^{-1} compared to a control, but the highest application rate, 90 g biochar kg^{-1} soil reduced bean productivity (Rondon et al, 2007). Studies suggest that biochar may stimulate N$_2$ fixation as the result of increased availability of alkaline (K, Mg) (Ma et al, 2019) and trace metals (e.g. nickel (Ni), iron (Fe), boron (B), titanium (Ti), and molybdenum (Mo)) (Rondon et al, 2007). Similar findings were reported in a more recent study, where wood waste biochar was found to increase N$_2$ fixation in wild soybeans at application rates of 1.5% w/w, but had no effect at 5% w/w in a pot study involving sandy coastal soils in China (Yin et al, 2021a). In contrast, another recent study reported a negative response of legumes to wood biochar applications in both field and pot trials, an effect found to be exacerbated by field aging of the biochar (Mia et al, 2018).

It is possible that the lack of consistent effects of biochar on legume performance and nodulation (see Table 16.1) is due to differences in nutrient contents of the various types of biochar and their respective potential to adsorb signaling compounds. Nodule formation in leguminous plants is initiated by the release of signaling compounds, often flavonoids (Jain and Nainawatee, 2002). Such polyphenolic compounds are readily sorbed by biochar (Gundale and DeLuca, 2006a; Kumar et al, 2020). This might explain why some studies have shown that activated C reduces nodulation, while low sorption P-rich biochars increase nodulation, which is presumably the result of alleviating the P-limitation of nodulating bacteria with high P demands, such as *Rhizobium* spp. (Rondon et al, 2007). Alternatively, biochar may reduce the presence of environmental stressors (such as salt stress), thereby indirectly increasing the nodulation and performance of legumes (Farhangi-Abriz and Torabian, 2018). These stressors may or may not have been measured in the experiment and inadvertently overlooked as a causal factor.

Numerous studies have been conducted to evaluate the potential for increasing the activity of free-living N$_2$-fixing bacteria in agroecosystems, however, the effect of biochar on free-living N$_2$ fixation has only been directly evaluated in a limited number of papers (Ducey et al, 2013; Liu et al, 2019b; Zhao et al, 2021). Biochar additions to soil likely increase background ethylene production (Spokas et al, 2010; see also chapter 15), which can interfere with the outputs from the acetylene reduction assay to estimate nitrogenase activity if not properly controlled. It is not clear whether background levels of ethylene production were accounted for during some of the incubations (e.g., Liu et al, 2019b). Regardless, several studies have shown that soils amended with biochar show an increase in the abundance of *nif*H, a gene encoding for nitrogenase enzymes in diazotrophic bacteria (Ducey et al, 2013; Liu et al, 2019b; Zhao et al, 2021). Further, field-oriented biochar studies have demonstrated large increases in N$_2$ fixation as measured using isotopic methods (Güereña et al, 2015). It is well understood that excess soluble N in the soil solution reduces N_2 fixation rates in free-living N_2-fixing bacteria (Kitoh and Shiomi, 1991; DeLuca et al, 1996) and available soil P or micronutrients can stimulate N_2 fixation (Chapin et al, 1991). Therefore, it is possible that the activity of free-living N_2-fixing bacteria could be increased by biochar-induced increases in P or trace metal solubility (Lehmann et al, 2003; Steiner et al, 2007) and reduced soluble soil N concentrations (due to immobilization or surface adsorption of NH_4^+). Biochar therefore potentially represents a good carrier or medium for the growth and proliferation of free-living N_2-fixing bacteria. Wood- and cellulose-based biochars are low-N media, yet serve to adsorb soil P (see Chapter 9) and potentially enhance the environment for free living diazotrophs.

Phosphorus

Following N, P tends to be the next major nutrient limiting primary production in most ecosystems. Unlike N, there is little evidence for the direct uptake of organic P by plants, and therefore soil organic matter containing organic P polymers must be enzymatically broken down outside the cell before the uptake of inorganic P ($_{Pi}$). Inorganic P is most commonly taken up by plants in the HPO_4^{2-} or $H_2PO_4^-$ form. Some low molecular weight organic P can be directly taken up by microbial cells (e.g., adenosine phosphates), however, this pathway is probably small in comparison to the uptake of $_{Pi}$. In contrast to N, however, the solubility and rate of diffusion of $_{Pi}$ in soils is typically extremely low due to strong sorption to the mineral phase (e.g., on Fe and Al oxyhydroxide surfaces) and its potential to form mineral precipitates (e.g., Ca-P). In the past decade, biochar additions to soil have been found to have various effects on soil P

availability (Gao and DeLuca, 2016; Gul and Whalen, 2016; Hossain et al, 2020; Ghodszad et al, 2021), and the soil P responses are often found to be a function of biochar characteristics, soil background conditions, the amount and residence time of biochar in soils, and the plant and ecosystem type (Glaser and Lehr, 2019; Gao et al, 2019; Tesfaye et al, 2021). Biochar itself can provide a source of readily available P (see Chapter 8) and can also directly and indirectly influence P behavior in soil by a range of other major mechanisms including: (i) impact on P leaching via influence on soil physical processes and biochar-soil interactions, (ii) alteration in biotic P processes such as enzyme activities and P-solubilizing bacteria, and (iii) formation of organo-mineral complexes that influence soil P solubility.

Release of P from biochar and impacts on P leaching

Biochars have different properties depending on the feedstock they are produced from and depending on the pyrolysis conditions (Chapter 8). Once biochar is applied to soils, many environmental factors can further influence the release of available P from biochar. For example, biochar P release was found to decrease with an increase in soil solution pH (Wang et al, 2015); increase with the existence of certain anions (e.g., Cl^-, SO_4^{2-}); and increase with the residence time of biochar in soils (Pogorzelski et al, 2020).

Although biochar releases some amount of P to soils upon application, there has been little evidence suggesting enhanced soil P leaching. Biochar may reduce soil P leaching loss by directly adsorbing ortho-P in soil solutions via electrostatic attraction, surface anion-exchange capacity, or other mechanisms (Schneider and Haderlein 2016; Dari et al, 2016). Biochar may also indirectly reduce P leaching loss by altering soil hydraulic properties and/or plant P uptake or use efficiency (Zhang et al, 2020a; Razzaghi et al, 2020). In a field-based study, Gao et al (2016) described a significant increase in available soil P under the application of wood biochar at 20 Mg ha^{-1} with or without an organic fertilizer in temperate sandy agricultural soils originated from glacial till parent material. This noted increase in surface soil P availability was closely associated with a reduction of cumulative ortho-P leaching over the growing season, an increase in soil water holding capacity, and a significant increase in crop P concentration and productivity following biochar application (Gao et al, 2016, 2017). However, it is important to note that the capacity of biochar to influence soil P leaching is also dependent on biochar and soil properties (e.g. biochar specific surface area, soil texture) (Bornø et al, 2018), plant and system type (e.g. rooting depth, mycorrhizal associations, P-poor or P-rich ecosystem) (Gao and DeLuca 2020, 2021), and other environmental conditions or management practices (also see Chapter 19). To date, few studies have used field experiments to elucidate the influence of biochar on soil P leaching over multiple growing seasons in agricultural ecosystems (Xie et al, 2021). The fate and behavior of P in subsurface soils in response to biochar addition requires further exploration.

Effect of biochar on phosphatase enzymes and P solubilizing bacteria

Despite the significant amount of functional redundancy in the microbial population, the influence of biochar on the shifts in soil microbial community structure, size, and activity (see Chapter 14) may cause changes in rates of soil biotic P cycling. In the past few years, research on biochar and soil biological P cycling has progressed significantly with the help of a diverse range of molecular tools. Biochar has been observed to influence the mycorrhizal colonization of plant roots which in turn may biologically alter soil P availability and plant P uptake (Chapter 15). Below we provide a few highlights on how biochar may influence the activity of soil enzymes associated with biological P mineralization and P-solubilizing microorganisms associated with P solubilization.

Extracellular phosphatase enzymes produced by soil microorganisms are responsible for soil organic P hydrolysis or the cleavage of P-containing organic compounds releasing inorganic P readily available for microbial and plant use. Numerous lab and field studies have been conducted to examine biochar and soil phosphatase activity in the past few years. Across various soil and biochar types, biochar application to soils, on average, results in an approximately 11% increase in soil phosphatase activity (Zhang et al, 2019). Several synthesis studies suggest that biochar produced at 350 – 600°C can have the most significant positive effect on soil phosphatase activity (Gul and Whalen, 2016; Pokharel et al, 2020). Biochar produced at low to mid pyrolysis temperature generally contains more easily-mineralizable organic P compounds that can serve as substrates for phosphatase (Xu et al, 2016). Soil pH is also a factor highly correlated with shifts in phosphatase activity following biochar additions. For example, alkaline phosphatase activity is often found to be more sensitive to biochar in acidic-to-neutral soils (pH less than 7.5) than alkaline soils, possibly because the short-term liming effect of biochar modified soil pH to an optimal condition favoring enzyme activity (Jin et al, 2016). In a field experiment, Cao et al (2021) found a 24 – 33% increase in alkaline phosphatase activity along with a 0.32–0.50 unit increase in soil pH following straw biochar application to a neutral pH Luvisol under maize monocropping.

Shifts in phosphatase enzyme activity with biochar additions are often found with changes in microbial biomass and activity, but may or may not be closely associated with responses in soil P bioavailability or the relative abundance of P-cycling functional genes (Khadem and Raiesi, 2019; Lu et al, 2020; Yang and Lu, 2022). Evidence suggests that phosphatase activity cannot be used alone to explain soil biological P mineralization patterns in response to biochar because of possible methodological interference due to the direct sorption of phosphatase enzyme or the hydrolysis reaction product onto biochar surface (Swaine et al, 2013; Jindo et al, 2014; Foster et al, 2018). Gao and DeLuca (2018) investigated the relationship between biochar-induced changes in soil biological P cycling dynamics and the abundance of P-cycling functional genes, but found no significant difference in the relative abundance of *phoC* or *phoD* gene (encodes acid and alkaline phosphatase production) between biochar and control despite a significant increase in soil P availability with wood biochar additions. Complementary studies with microbial DNA sequencing further suggested that changes in phosphatase activity were most likely associated with shifts in the community composition, and not the abundance, of *phoC* or *phoD*-harboring microbial community in soils; where soil biological P mineralization can be driven by rare taxa (Wei et al, 2019). For example, biochar was found to result in a specific enrichment in the abundance of *Micromonosporaceae* which possibly played a critical role in facilitating P mineralization in a C-rich, P-poor soil with a high microbial P demand (Tian et al, 2021).

A substantial number of microbial species in soils can also excrete organic acids to dissolve or convert insoluble P into soluble forms. P-solubilizing bacteria (PSB) is the most ubiquitous group responsible for P solubilization which constitutes 1 – 50% of the total microbial population (Sharma et al, 2013). Biochar is generally found to increase PSB abundance in P-poor soils (Deb et al,

2016; Xu et al, 2019). Pyrosequencing evidence in a recent study suggests that rice husk biochar (400°C pyrolysis temperature) applied at 20 Mg ha^{-1} to an acidic soil significantly increased the diversity of PSB and the relative abundance of *Thiobacillus*, *Pseudomonas*, and *Flavobacerium* (all three were genera of PSB), which was predominantly explained by shifts in soil pH and water holding capacity and together contributed to an increase in soil P availability, microbial biomass C and P over two months (Liu et al, 2017). Similarly, in a pot incubation study, biochar produced from forest harvest residues (600°C pyrolysis temperature) applied to soils at 3% (w/w) significantly influenced the soil PSB abundance, for example, the relative abundance of genera *Burkholderia-Paraburkholderia* and *Planctomyces* were increased by 123% and 436% compared to the control (Zhou et al, 2020).

Precipitation, sorption, complexation

A significant component of the P cycle consists of a series of precipitation reactions that influence the solubility of P, ultimately influencing the quantity of P that is available for uptake and actively recycled between plants and microbes. The degree to which these precipitation reactions occur is strongly influenced by soil pH, due to the pH-dependent activities of the ions responsible for precipitation (e.g. Al^{3+}, Fe^{3+}, and Ca^{2+}) (Stevenson and Cole, 1999). In alkaline soils, P solubility is primarily regulated by its interaction with Ca^{2+}, where a cascading apatite mineral pathway develops. In acid soils, P availability is primarily regulated by its interaction with Al^{3+} and Fe^{3+} ions, where highly insoluble Al- and Fe- phosphates form. Biochar may influence the precipitation of P into these insoluble pools by altering the pH, and thus the strength of ionic P interactions with Al^{3+}, $Fe^{3+}$and Ca^{2+}

(Lehmann et al, 2003; Topoliantz et al, 2005) or by sorbing organic molecules that act as chelates of metal ions that otherwise precipitate P (Ghodszad et al, 2021).

Numerous studies have demonstrated that biochar can modify soil pH, normally by increasing pH in acidic soils (Gao et al, 2019). An increase in pH associated with adding biochar to acid soils is due to an increased concentration of alkaline metal (Ca^{2+}, Mg^{2+}, and K^+) oxides in the biochar and a reduced concentration of soluble soil Al^{3+} (Steiner et al, 2007). Adding these alkaline metals, both as soluble salts and associated with biochar exchange sites, is likely the single most significant effect of biochar on P solubility in the short-term, particularly in acidic soils where subtle changes in pH can result in substantially reduced P precipitation with Al^{3+} and Fe^{3+}. In contrast, adding biochar (and associated ash residue) to neutral or alkaline soils may have a limited effect on P availability because adding alkaline metals would only exacerbate Ca-driven P limitations (Gao et al, 2019).

In addition to its effect on soil pH, biochar may also influence the bioavailability of P through several other mechanisms associated with P precipitation, such as biochar-induced surface sorption of chelating organic molecules. Biochar is an exceptionally good surface for sorbing polar or non-polar organic molecules across a wide range of molecular mass (Schmidt and Noack, 2000; Preston and Schmidt, 2006; Bornermann et al, 2007). Organic molecules involved in the chelation of Al^{3+}, Fe^{3+}, and Ca^{2+} ions can potentially be sorbed to hydrophobic or charged biochar surfaces so that, in the long run, organo-biochar or organo-mineral-biochar complexes begin to form over time that may aid in the retention and exchange of soluble P around aged biochar particles (Briones, 2012; Joseph et al, 2013; Gao and DeLuca 2018; Wang et al, 2020a). Examples of chelating compounds include simple organic acids, phenolic acids, amino acids, and complex proteins or carbohydrates (Stevenson and Cole, 1999). The sorption of complexing agents may have a positive or negative influence on P solubility. A clear example of this type of interaction is provided in Figure 16.5. Here, two compounds that have been reported as possible allelopathic compounds released as root exudates from *Centaurea* species: catechin and 8-hydroxy-quinoline (Vivanco et al, 2004; Callaway and Vivanco, 2007) have also been reported to function as potent metal chelates (Stevenson and Cole, 1999; Shen et al, 2001) that may indirectly increase P solubility. Catechin effectively increased P solubility in an alkaline (pH 8.0) calcareous soil and the 8-hydroxy-quinoline increased P solubility when added to an acidic (pH 5.0) Al-rich soil (Figure 16.5). The addition of biochar to these soils eliminated the presence of soluble chelate in the soil system and in turn eliminated the effect of the chelate on P solubility. This interaction may explain the observed reduction in P sorption by ionic resins with increasing biochar application rates in the presence of actively growing *Koleria macrantha* (Gundale and DeLuca, 2007). Such indirect effects of biochar on P solubility would vary with soil type and vegetative cover and underscores the complexity of plant-soil interactions (Makoto and Koike, 2021).

Potassium

After N and P, potassium (K) often represents the next biggest constraint to plant production (Zorb et al, 2014; He et al, 2015). It is well established that a large amount of K contained in biochar (1 - 60 g kg^{-1}) is bioavailable and can provide a useful source of fertilizer K to plants (Limwikran et al., 2018; Liu et al, 2019c; Poormansour et al, 2019; Beusch et al, 2022). In addition, through cation

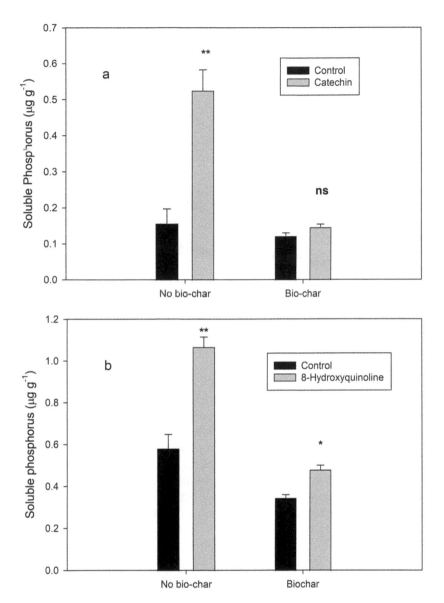

Figure 16.5 *Soluble P leached from columns filled with (a) calcareous soil (pH = 8) amended with catechin alone or with biochar or (b) acid Al rich soil (pH = 6) amended with 8-hydroxy quinoline alone or with biochar (DeLuca, unpublished data). Studies were conducted by placing 30 g of soil amended with 50 mg P kg^{-1} soil as rock phosphate into replicated 50 mL leaching tubes (n = 3). Soils were then treated with nothing (control), chelate, or chelate plus biochar (1% w/w), allowed to incubate for 16 h moist and then leached with 3 successive volumes of 0.01 M CaCl$_2$. Leachates were then analyzed for orthophosphate on a segmented flow Auto Analyzer III. Data were subject to ANOVA by using SPSS*

exchange processes, biochar can aid in the retention of K in coarse-textured soils, reducing leaching (Kuo et al, 2020; Li et al, 2019; Beusch et al, 2022). In most soils, however, biochar often promotes K leaching due to the high amount of K and other salts present in the biochar which can reduce K sorption to the soil (Rens et al, 2018; Palanivell et al, 2019; Krishnan et al, 2021). Biochar addition also typically leads to more efficient use of K fertilizers and increased tissue K concentrations (Biederman et al, 2013). What is less well understood, however, is how biochar directly and indirectly influences K transformations in soil. It is well known that biochar can change the activity and composition of the soil microbial community; however, several studies have now indicated that this may also be associated with an increase in bacteria capable of solubilizing K minerals (Wang et al, 2018; Zhang et al, 2020b). This is most likely mediated by changes in soil pH and the promotion of plant growth-promoting bacteria in the rhizosphere (Zhang et al, 2021). To date, there have been no studies that specifically evaluate the interactions (e.g., weathering rates) between biochar and natural K minerals found in soil (e.g., K-feldspars and feldspathoids, micas, etc.) or the release of K held in clay minerals. In addition, despite the thousands of studies on K-biochar interactions, none have yet to capitalize on the use of K isotopes (^{39}K, ^{40}K, ^{41}K) to discriminate between the uptake of soil- and biochar-derived sources of K by plants. This would be particularly useful in longer-term agronomic field trials to permit the calculation of fertilizer K use efficiency and legacy effects.

Sulfur

There remains a limited number of studies that have focused on the influence of biochar soil amendments on soil S transformations. Sulfur plays an extremely important role in the biochemistry of soils and the physiology of plants (Paul, 2015). Sulfur as a component of two amino acids (cysteine and methionine), is required in protein synthesis and is a fundamental component of energy transformations in all living organisms. Sulfur also represents a source of energy for autotrophic organisms and an alternative electron acceptor for oxidative decomposition under anaerobic conditions (Paul, 2015). It is clear that biochar produced from high-S feedstocks has the potential to release S into the soil solution (Uchimiya et al, 2010, Hu et al, 2021); however, there have been few studies that provide direct evidence for enhanced S oxidation or reduction with biochar applications (Xu et al, 2020, Wang et al, 2021a). Even though the majority of soil S originates from the geologic parent material, most soil S actually exists in an organic state and must be mineralized (converted from organic S to SO_4^{-2}) before plant uptake. Organic S exists as either ester sulfate or as carbon-bonded S, the latter having to be oxidized to SO_4^{-2} before plant uptake (Paul, 2015). With the interest in biochar as an agricultural soil amendment or as an environmental remediation agent, there has been an increasing number of studies conducted that either directly or indirectly address the influence of biochar on S transformations in mineral soils (Marks et al, 2016, Chao et al, 2018, Zhao and Zhang 2021).

One of the earlier studies conducted to directly investigate the influence of biochar on S transformations was conducted with two soil types and four crop residue amendments and performed in PVC columns in the laboratory (Churka Blum et al, 2013). In this study, S, C, and N mineralization were observed following the addition of corn husk biochar to soil compared with fresh residues of corn husks, pea, and rape residues. Although C mineralization and N mineralization were notably low with the biochar amendment, the highest rate of S mineralization for all amendments was observed with the corn husk biochar. The authors conclude that the release of S from the residues is likely a function of the S compounds within the residues and suggests that soluble SO_3^{-2} and SO_4^{-2} are readily liberated from the ester S, allowing for rapid accumulation of inorganic S in soils treated with biochar (Churka Blum et al, 2013). Studies involving pine chip biochar generated in a gasifier also exhibited a net release of organic S and subsequent oxidation of S resulting in a temporary increase in soil SO_4^{-2} in mineral soil mesocosms treated with 50 t biochar ha^{-1} (Marks et al, 2016).

Sulfur mineralization is favored at slightly acid to neutral pH soils, and biochar tends to increase the pH of acidic soils and this effect may indirectly enhance S mineralization (Tabatabai and Al-Khafaji, 1980). Biochar generated from S-rich feedstocks has the potential to release significant amounts of organic and inorganic S into mineral soil (Chao et al, 2018; Hu et al, 2021; Zhao and Zhang, 2021) which may be taken up, oxidized, or reduced depending on the oxidation state of the soils. Organic S tends to absorb to the surface of biochar which may enhance the net mineralization of organic S to SO_4^{-2} or it may remain temporarily adsorbed to the biochar.

Oxidation of reduced mineral forms of S is carried out by both autotrophic (e.g., *Thiobacillus* spp.) and some heterotrophic organisms (e.g., *Pseudomonas* spp.). However, autotrophs obtain their energy from the oxidation of S and therefore tend to be the dominant S-oxidizing organisms in soil (Wainwright, 1984). Sulfur oxidation by acidophilic *Thiobacillus* spp. is not favored by pH increases induced by the addition of biochar. However, different species of Thiobacillus (e.g., *T. thioparus*) can tolerate mildly alkaline conditions and can 'seed' the oxidation process allowing for acid-loving *T. thiooxidans* to transform the

remaining S once the pH drops below 4. Further, these autotrophic organisms have uniquely high requirements for certain trace elements that are in relatively high concentrations in biochar (Chapter 14) and are increased in soil when biochar is added (Rondon et al, 2007). Incubation studies comparing S oxidation rates in slurries containing S-coated bamboo biochar or elemental S with a wetting agent demonstrated more rapid S oxidation in the presence of the biochar, with the pH declining from 6.5 to less than 2.0 (Wu et al, 2020). The authors do not provide a likely mechanism for the observed increase in S oxidation in the presence of bamboo biochar. However, another slurry incubation study involving chalcopyrite (an iron sulfide mineral) demonstrated that biochar addition to the slurry results in slower dissolution of chalcopyrite as a result of surface adsorption of elemental and reduced sulfur (Yang et al, 2020). This effect was more pronounced in low-temperature biochar, suggesting that the surface functional groups are likely involved in S retention and associated reductions in acid production.

Biochar additions to mineral soils may also directly or indirectly affect S sorption reactions and S reduction. As with NO_3^-, non-aged, production-fresh biochar may lack any significant capacity to adsorb SO_4^{2-} (Borchard et al, 2012). Once in a reduced or elemental form, S is more likely to adsorb to the biochar surface (Yang et al, 2020). Accordingly, biochar has been found to be an effective sorbent of H_2S gas associated with landfill extraction wells (Zhang et al, 2017) which may have implications for biochar retention of S in wet mineral soils. The S adsorbed onto the corn stover biochar was found to be readily oxidized to SO_4^{-2} and taken up by crop plants (Cheah et al, 2014). Sulfur is also readily adsorbed to mineral surfaces in the soil environment and particularly to exposed Fe and Al oxides.

Once Fe and Al have been sorbed to biochar surfaces, SO_4^{2-} may interact with the exposed metal oxides. Conversely, organic matter additions to soil have been shown to reduce the extent of SO_4^{2-} sorption in acid forest soils (Johnson 1984), therefore biochar amendments could increase concentrations of S in acid, iron-rich soils. The lack of studies devoted to the evaluation of S transformations following biochar addition to soils calls for additional studies in this area.

Micronutrients

Copper

Most work on the interactions between biochar and copper (Cu) have focused on the ability of biochar to remediate contaminated land (Inyang et al, 2016). Indeed, numerous studies have demonstrated that biochar effectively lowers the bioavailability of Cu, reducing phytotoxicity and metal leaching (Quartacci et al, 2015; Tomczyk et al, 2019). In comparison, much less work has focused on how biochar affects the fate and bioavailability of Cu in non-contaminated soils. Biochar can readily bind Cu^{2+} from soil solution, however, evidence suggests that this process is reversible and does induce plant micronutrient deficiency. Cu release from the biochar surface may also be promoted by the release of complexing agents in root exudates (e.g., citrate). The Cu sorption process is also pH dependent with greater Cu-biochar binding as the soil solution pH increases (Guo et al, 2014). Where excessive amounts of biochar are added, however, and the soil pH rises above 7, this may promote Cu precipitation [$Cu(OH)_{2(s)}$] and reduce plant availability (Gonzaga et al, 2020; Yang et al, 2019). Further, as the biochar ages, the amount of Cu retained on the biochar surface can be expected to fall due to a reduction in CEC and specific surface area (Guo et al,

2014; Hao et al, 2017; Wang et al, 2021b). In terms of Cu cycling in agricultural systems, biochar can promote micronutrient retention in soil and reduce leaching losses, particularly in sandy textured soils (Riedel et al, 2015; Wang et al, 2020c). Biochar may also indirectly support enhanced Cu uptake through the promotion of mycorrhizal and root growth (Gujre et al, 2021). The amount of Cu added to soil in biochar is highly dependent on the biomass feedstock (e.g., wood vs. sewage sludge), its moisture content, pyrolysis temperature, and pyrolysis time (Song et al, 2017). This can result in a wide range of intrinsic biochar Cu contents (1 to 5000 mg kg^{-1}; Hossain et al, 2011; Zielinska et al, 2015; Domingues et al, 2017). Care must be taken, however, to avoid biochar derived from waste streams where high levels of Cu may be present (e.g., wood treated with Cu preservative) as this may lead to Cu toxicity (Lucchini et al, 2014). For a typical forestry or crop-residue-based biochar, an application rate of at 10 t ha^{-1} would equate to a fertilizer dose of 0.5 kg Cu ha^{-1}, which is probably insufficient to rectify any Cu deficiencies or boost crop production. Generally, however, most experiments have reported a positive influence of biochar on Cu crop offtake under non-contaminated conditions (Hunt et al, 2013; Jatav et al, 2018; Chrysargyris et al, 2019; Nzanza et al, 2012), although few differences in foliage Cu content have been reported. It is likely therefore that the increased Cu offtake reflects greater biomass production caused by the removal of other soil constraints by biochar (e.g., low pH, macronutrient deficiency) rather than a direct effect on Cu cycling *per se*. The direct and indirect effects of biochar on plant Cu uptake and microbial Cu cycling, however, remain to be fully elucidated, particularly under field conditions using non-contaminated land and realistic biochar loading rates.

Iron

Iron is required in moderate quantities by plants, however, due to its relative insolubility in most soils, plants have evolved a range of strategies to enhance its solubility and root uptake from soil (Ancuceanu et al, 2015; Tripathi et al, 2018). While much attention has been paid to the chemical modification of biochar using Fe (Wu et al, 2019; Wan et al, 2020), or the co-addition of biochar and Fe-nanoparticles (Su et al, 2016), much less attention has focused on how biochar affects intrinsic microbial Fe cycling and root Fe acquisition. Most of the current evidence surrounds the mechanisms by which biochar directly promotes Fe cycling in paddy soils (Jia et al, 2016). Firstly, biochar can act as an electron shuttle between bacteria and Fe-minerals stimulating the microbial reduction of insoluble Fe-oxyhydroxides under anaerobic conditions leading to increased availability of Fe^{2+} (Kappler et al, 2014; Wang et al, 2017). This reduction in Fe^{3+} may also induce the solubilization and bioavailability of P previously held in Fe-P minerals (Cui et al, 2011). Further, in some soils, biochar has been shown to stimulate the abundance of Fe-reducing bacteria whilst suppressing other microorganisms associated with Fe oxidation (Kappler et al, 2014; Jia et al, 2018). The addition of biochar that is produced under low pyrolysis temperatures can also lead to an increased concentration of DOC leading to the complexation of Fe^{3+} making it more bioavailable to both plants and microorganisms (Wang et al, 2017). While the discussion above mainly relates to waterlogged soils, there is less information available on well-drained aerobic soils. Although the Fe content of biochar can be appreciable in some products ($10-2000$ mg kg^{-1}), it is generally present in an oxidized insoluble form and this has low bioavailability and little fertilizer value. However, studies have indicated that the co-addition of biochar and Fe-fertilizers may be beneficial in alleviating Fe deficiency in some crops (Alburquerque et al, 2015; Ramzani et al, 2016). If excess biochar is added to the soil and the pH becomes too alkaline it may induce deficiency. Similarly, Fe may become bound to the surface of biochar, making it less available to plants (Sorrenti et al, 2016).

Manganese

Manganese is an essential nutrient, required in trace quantities by plants. Manganese deficiency is a common problem for plants, especially in sandy soils, heavily weathered tropical soils, and alkaline soils (Schmidt et al, 2016; Leeper, 1934). In contrast, toxic concentrations of Mn can occur naturally in serpentine soils, or soils impacted by industrial mining activities. Biochar has the potential to both increase Mn availability in Mn-poor soils and reduce toxicity in soils where Mn concentrations exceed plant tolerance. The first mechanism by which biochar can influence Mn availability is by serving as a direct source of associated ash residues (Muhammad et al, 2017). As Mn has an extremely high volatilization temperature (ca. 2000°C), it can be found in high concentrations in biochar ash residues (Smider and Singh, 2014; Bodi et al, 2014), which can help alleviate limitations in Mn-poor soils. In soils where Mn reaches toxic levels, biochar can reduce toxicity through a variety of mechanisms. Firstly, when Mn is a divalent cation, it can interact with biochar surfaces, which typically have an abundance of negative exchange sites, through electrostatic adsorption and ion exchange (Zhong et al, 2020). Additionally, Mn can undergo electron donor-receptor complexation reactions with functional groups on biochar surfaces, such as –OH, –COOH, and C=N, which can reduce solubility and bioavailability (Zhong et al, 2020). Manganese can also undergo precipitation reactions with certain anions, including manganese hydroxide, manganese sulfate, and manganese chloride. These precipitates have the potential to form in the ash residue associated with biochar, potentially reducing their toxicity.

Zinc

Biochar can influence soil Zn dynamics and availability predominantly through cation exchange, sorption, and precipitation. For instance, the presence of negatively charged surface functional groups on biochar can directly contribute to cation exchange capacity and increase the retention of positively charged nutrient ions such as Zn^{2+} in soils. Changes in soil Zn^{2+} dynamics can also be driven by chelation with biochar organic groups (R-COOH, R-OH$^-$) or precipitation onto inorganic groups (CO_3^{2-}, PO_4^{3-}). Biochar can change soil solution pH that will indirectly influence the behavior of Zn^{2+} electrostatic sorption on biochar surface, subsequently influencing Zn immobilization in soils (Houben et al, 2013). In addition, biochar aging in soils can affect soil Zn dynamics. For instance, the oxygen-containing functional groups (e.g., O-H, C=O, and C-O) on biochar surfaces have been found to increase during biochar aging, providing more sorption sites for Zn^{2+} and further influencing Zn mobility in soils over time (Nie et al, 2021).

Future research directions

Biochar has a potentially important role to play in enhancing the biochemical and physical condition of agricultural and forest soils or in remediating lands degraded by extractive practices including mining. In this chapter, we reviewed biochar as a modifier of soil nutrient transformations and discussed the known and potential mechanisms that drive these modifications. Biochar additions to soils may directly or indirectly alter nutrient transformations and, depending on the specific objectives of biochar applications, biochars with different properties might be chosen or even modified to meet those objectives. Biochar applications to agricultural soils along with a nutrient source generally increase NH_4^+ concentration and retention in soil and have often been observed to increase N uptake by crop plants; however, the NH_4^+ availability is not consistently increased by biochar applied without a nutrient source, such as manure. Biochar may also increase gross nitrification across a range of ecosystems and net nitrification in forest soils with otherwise little or no nitrification.

The observed increases in net nitrification in forest soils occur at a level that would have minimal influence on net N leaching and N_2O emissions. Although biochar additions have been observed to increase net ammonification, observations have not been consistent. While P solubility appears to generally increase with biochar additions, this may be primarily a result of direct P addition with the applied biochar or a function of the often observed increase in soil pH with biochar additions to soil. There is a distinct need for studies directed at explaining mechanisms for increased P uptake with biochar additions to agricultural soils. It is possible that biochar additions to soils stimulate mycorrhizal colonization, which may increase P uptake, but when applied with P-rich materials, this effect may be lost. There is a great need for additional studies that mechanistically describe the effect of biochar on soil nutrient transformations, both immediately following application as well as over multiple years or decades. Some key areas that require attention include: (1) Under what conditions does biochar stimulate or reduce N mineralization, nitrification, and immobilization in different ecosystems? (2) Does NH_4^+ adsorption by biochar greatly reduce N availability or does it concentrate N for plant and microbial use? (3) Do all enzymes that adsorb to biochar retain their activity? (4) By what mechanisms does biochar alter S mineralization, oxidation, and reduction? (5) How does biochar influence the dissolution and transformation of trace elements including Cu, Fe, Mn, and Zn? The answers to these questions can only be obtained through rigorous investigation of biochar as a soil conditioner and agricultural amendment. To date, the vast majority of biochar studies have been conducted in soil incubations and greenhouse pot studies. These studies are efficient and highly informative; however, there is an increasing need to emphasize field-based research that incorporates the whole system when evaluating the effect of biochar on nutrient transformations. We also look forward to more studies that integrate microbial measurements and isotopic methods into biochar research, which will facilitate a more mechanistic understanding of how biochar influences nutrient cycling in forest, agricultural, and disturbed soil ecosystems.

References

Alburquerque J, Cabello M, Avelino R, Barron V, del Campillo MC, and Torrent, J 2015 Plant growth responses to biochar amendment of Mediterranean soils deficient in iron and phosphorus. *Journal of Plant Nutrition and Soil Science* 178, 567–575.

Alexander M 1991 *Introduction to Soil Microbiology*. Malabar: Krieger Publishing Company.

Ancuceanu R, Dinu M, Hovane M, Anghel A, Popescu C, and Negre S 2015 A survey of plant iron content- A semi-systematic review. *Nutrients* 7, 10320–10351.

Anderson CR et al 2012 Biochar induced soil microbial community change: Implications for biogeochemical cycling of carbon, nitrogen and phosphorus. *Pedobiologia* 54, 309–320.

Arezoo TT, Clough TJ, Sherlock RR, and Condron LM 2012 Biochar adsorbed ammonia is bioavailable. *Plant and Soil* 350, 57–69.

Asai H et al 2009 Biochar amendment techniques for upland rice production in Northern Laos 1. Soil physical properties, leaf SPAD and grain yield. *Field Crop Research* 111 (1–2), 81–84.

Atkinson CJ, Fitzgerald JD, and Hipps NA 2010 Potential mechanisms for achieving agricultural benefits from biochar application to temperate soils: a review. *Plant and Soil* 337, 1–18.

Ball P, MacKenzie MD, DeLuca TH, and Holben WB 2010 Wildfire and charcoal enhance AOB in forests of the Inland Northwest. *Journal of Environmental Quality* 39, 1243–1253.

Beesley L, Moreno-Jimenez E, Gomez-Eyles JL, Harris E, Robinson B, and Sizmur, T 2011 A review of biochar's potential role in the remediation, revegetation and restoration of contaminated soils. *Environmental Pollution* 159, 3269–3282.

Bello A et al 2020 Microbial community composition, co-occurrence network pattern and nitrogen transformation genera response to biochar addition in cattle manure-maize straw composting. *Science of the Total Environment* 721, 137759.

Berglund LM, DeLuca TH, and Zackrisson, O 2004 Activated carbon amendments of soil alters nitrification rates in Scots pine forests. *Soil Biology and Biochemistry* 36, 2067–2073.

Beusch C, Melzer D, Cierjacks A, and Kaupenjohann M 2022 Amending a tropical Arenosol: increasing shares of biochar and clay improve the nutrient sorption capacity. *Biochar* 4, 16.

Biederman LA, and Harpole WS 2013 Biochar and its effects on plant productivity and nutrient cycling: a meta-analysis. *Global Change Biology Bioenergy* 5, 202–214.

Bieser JMH, Al-Zayat M, Murtada J, and Thomas SC 2022 Biochar mitigation of allelopathic effects in three invasive plants: evidence from seed germination trials. *Canadian Journal of Soil Science* 102 (1), 213–224.

Bodi MB et al 2014 Wild land fire ash: Production, composition and eco-hydro-geomorphic effects. *Earth-Science Reviews* 130, 103–127.

Borchard N, Prost K, Kautz T, Moeller A, and Siemens J 2012 Sorption of copper (II) and sulphate to different biochars before and after composting with farmyard manure *European Journal of Soil Science* 63 (3), 399–409.

Borchard N et al 2019 Biochar, soil and land-use interactions that reduce nitrate leaching and N_2O emissions: A meta-analysis. *Science of the Total Environment* 651, 2354–2364.

Bornemann L, Kookana RS, and Welp G 2007 Differential sorption behavior of aromatic hydrocarbons on charcoals prepared at different temperatures from grass and wood. *Chemosphere* 67, 1033–1042

Bornø ML, Müller-Stöver DS, and Liu F 2018 Contrasting effects of biochar on phosphorus dynamics and bioavailability in different soil types. *Science of the Total Environment* 627, 963–974.

Braghiroli FL, Bouafif H, Neculita CM, and Koubaa A 2020 Influence of pyro-gasification and activation conditions on the porosity of activated biochars: A literature review. *Waste and Biomass Valorization* 11 (9), 5079–5098.

Brewer CE, Unger R, Schmidt-Rohr K, and Brown RC 2011 Criteria to select biochars for field studies based on biochar chemical properties. *BioEnergy Research* 4 (4), 312–323.

Brimmer RJ 2006 *Sorption Potential of Naturally Occurring Charcoal in Ponderosa Pine Forests of Western Montana.* Missoula: MS, University of Montana

Briones AM 2012 The secrets of El Dorado viewed through a microbial perspective. *Frontiers in Microbiology* 3, 10.3389.2012.00239.

Callaway RM, and Vivanco JM 2007 Invasion of plants into native communities using the underground information superhighway. *Allelopathy Journal* 19, 143–151.

Cao D et al 2021 Phosphorus fractions in biochar-amended soil – chemical sequential fractionation, ^{31}P NMR, and phosphatase activity. *Archives of Agronomy and Soil Science* 69, 1–13. 10.1080/03650340.2021.1967327.

Chao X et al 2018 Effect of biochar from peanut shell on speciation and availability of lead and zinc in an acidic paddy soil. *Ecotoxicology and Environmental Safety* 164, 554–561.

Chan KY, Van Zwieten L, Meszaros I, Downie A, and Joseph S 2007 Agronomic values of green waste biochar as a soil amendment. *Australian Journal of Soil Research* 45, 629–634.

Chan KY, and Xu Z 2009 Biochar: nutrient properties and their enhancement. In: Lehmann J, and Joseph S (Eds) *Biochar for Environmental Management.* London: Routledge. pp67–84.

Chapin DM, Bliss LC, and Bledsoe LJ 1991 Environmental regulation of nitrogen fixation in a high arctic lowland ecosystem. *Canadian Journal of Botany* 69, 2744–2755.

Cheah S, Malone SC, and Feik CJ 2014 Speciation of sulfur in biochar produced from pyrolysis and gasification of oak and corn stover. *Environmental Science and Technology* 48 (15), 8474–8480.

Chen CR, Phillips IR, Condron LM, Goloran J, Xu ZH, and Chan KY 2013 Impacts of greenwaste biochar on ammonia volatilisation from bauxite processing residue sand. *Plant and Soil* 367, 301–312.

Churka Blum S, Lehmann J, Solomon D, Caires EF, and Alleoni LRF 2013 Sulfur forms in organic substrates affecting S mineralization in soil. *Geoderma* 200–201, 156–164.

Chrysargyris A, Prasad M, Kavanagh A, and Tzortzakis N 2019 Biochar type and ratio as a peat additive/partial peat replacement in growing media for cabbage seedling production. *Agronomy* 9, 693.

Clough TJ et al 2010 Unweathered wood biochar impact on nitrous oxide emissions from a bovine-urine-amended pasture soil. *Soil Science Society America Journal* 74, 852–860.

Clough TJ, and Condron LM 2010 Biochar and the nitrogen cycle: Introduction. *Journal of Environmental Quality* 39, 1218–1223.

Clough TJ, Condron LM, Kammann C, and Müller C 2013 A review of biochar and soil nitrogen dynamics. *Agronomy* 3, 275–293.

Cornelissen G, Elmquist M, Groth I, and Gustafsson O 2004 Effect of sorbate planarity on environmental black carbon sorption. *Environmental Science and Technology* 38, 3574–3580.

Cornelissen G, Rutherford DW, Arp HPH, Dorsch P, Kelly CN, and Rostad CE 2012 Sorption of pure N_2O to biochars and other organic and inorganic materials under anhydrous conditions. *Environmental Science and Technology* 47, 7704–7712.

Craswell ET, Chalk PM, and Kaudal BB 2021 Role of ^{15}N in tracing biologically driven nitrogen dynamics in soils amended with biochar: A review. *Soil Biology and Biochemistry*, 162, 108416.

Crutchfield EF, Merhaut DJ, McGiffen ME, and Allen EB 2010 Effects of biochar on nutrient leaching and plant growth. *Hortscience* 45 (8), S163–S163.

Cui HJ, Wang MK, Fu ML, and Ci E 2011 Enhancing phosphorus availability in phosphorus-fertilized zones by reducing phosphate adsorbed on ferrihydrite using rice straw-derived biochar. *Journal of Soils and Sediments* 11 (7), 1135–1141.

Dari B et al 2016 Relative influence of soil- vs. biochar properties on soil phosphorus retention. *Geoderma* 280, 82–87.

Deb D, Kloft M, Lessig J, and Walsh S 2016 Variable effects of biochar and P solubilizing microbes on crop productivity in different soil conditions. *Agroecology and Sustainable Food Systems* 40, 145–168.

Dutta T et al 2017 Polycyclic aromatic hydrocarbons and volatile organic compounds in biochar and biochar-amended soil, a review. *GCB Bioenergy* 9, 990–1004.

DeLuca TH, Drinkwater LE, Wiefling BA, and DeNicola DM 1996 Free-living nitrogen-fixing bacteria in temperate cropping systems: influence of nitrogen source. *Biology and Fertility of Soils* 23, 140–144.

DeLuca TH, Nilsson M-C, and Zackrisson O 2002 Nitrogen mineralization and phenol accumulation along a fire chronosequence in northern Sweden. *Oecologia* 133, 206–214.

DeLuca TH, MacKenzie MD, Gundale MJ, and Holben WE 2006 Wildfire-produced charcoal directly influences nitrogen cycling in forest ecosystems. *Soil Science Society America Journal* 70, 448–453.

DeLuca TH, and Aplet GH 2008 Charcoal and carbon storage in forest soils of the Rocky Mountain West. *Frontiers in Ecology and the Environment* 6, 1–7.

Dempster DN, Jones DL, and Murphy DV 2012 Organic nitrogen mineralisation in two contrasting agro-ecosystems is unchanged by biochar addition. *Soil Biology and Biochemistry* 48, 47–50.

Devonald VG 1982 The effect of wood charcoal on the growth and nodulation of peas in pot culture. *Plant and Soil* 66, 125–127.

Ding Y et al 2016 Biochar to improve soil fertility. A review. *Agronomy for Sustainable Development* 36 (2), 36.

Ding Y, Liu YX, Wu WX, Shi DZ, Yang M, and Zhong ZK 2010 Evaluation of biochar effects on nitrogen retention and leaching in multi-layered soil columns. *Water, Air and Soil Pollution* 213 (1–4), 47–55.

Domingues RR et al 2017 Properties of biochar derived from wood and high-nutrient biomasses with the aim of agronomic and environmental benefits. *PLoS One* 12, e0176884.

Dong Y, Wu Z, Zhang X, Feng L, and Xiong Z 2019 Dynamic responses of ammonia volatilization to different rates of fresh and field-aged biochar in a rice-wheat rotation system. *Field Crops Research* 241, 107568.

Doydora SA, Cabrera ML, Das KC, Gaskin JW, Sonon LS, and Miller WP 2011 Release of nitrogen and phosphorus from poultry litter amended with acidified biochar. *International Journal of Environmental Research and Public Health* 8, 1491–1502.

Ducey TF, Ippolito JA, Cantrell KB, Novak JM, and Lentz RD 2013 Addition of activated switchgrass biochar to an aridic subsoil increases microbial nitrogen cycling gene abundances. *Applied Soil Ecology* 65, 65–72.

El-Naggar A et al 2019 Biochar application to low fertility soils: A review of current status, and future prospects. *Geoderma* 337, 536–554.

Farhangi-Abriz S, and Torabian S 2018 Biochar improved nodulation and nitrogen metabolism of soybean under salt stress. *Symbiosis* 74 (3), 215–223.

Feng Y et al 2022 How does biochar aging affect NH_3 volatilization and GHGs emissions from agricultural soils? *Environmental Pollution* 294, 118598.

Fiorentino N et al 2019 Interactive priming of soil N transformations from combining biochar and urea inputs: A 15N isotope tracer study. *Soil Biology and Biochemistry* 131, 166–175.

Foster EJ, Fogle EJ, and Cotrufo MF 2018 Sorption to biochar impacts?-Glucosidase and phosphatase enzyme activities. *Agriculture* 8, 158.

Galloway JN et al 2008 Transformation of the nitrogen cycle: recent trends, questions, and potential solutions. *Science* 320 (5878), 889–892.

Gao S, and DeLuca TH 2016 Influence of biochar on soil nutrient transformations, nutrient leaching, and crop yield. *Advances in Plants and Agriculture Research* 4, 00150.

Gao S, and DeLuca TH 2018 Wood biochar impacts soil phosphorus dynamics and microbial communities in organically-managed croplands. *Soil Biology and Biochemistry* 126, 144–150.

Gao S, and DeLuca TH 2020 Biochar alters nitrogen and phosphorus dynamics in a western rangeland ecosystem. *Soil Biology and Biochemistry* 148, 107868.

Gao S, and DeLuca TH 2021 Influence of fire retardant and pyrogenic carbon on microscale changes in soil nitrogen and phosphorus. *Biogeochemistry* 152, 117–126.

Gao S, Hoffman-Krull K, Bidwell AL, and DeLuca TH 2016 Locally produced wood biochar increases nutrient retention and availability in agricultural soils of the San Juan Islands, USA. *Agriculture, Ecosystems and Environment* 233, 43–54.

Gao S, Hoffman-Krull K, and DeLuca TH 2017 Soil biochemical properties and crop productivity following application of locally produced biochar at organic farms on Waldron Island, WA. *Biogeochemistry* 136, 31–46.

Gao S, DeLuca TH, and Cleveland CC 2019 Biochar additions alter phosphorus and nitrogen availability in agricultural ecosystems: A meta-analysis. *Science of the Total Environment* 654, 463–472.

Ghodszad L et al 2021 Biochar affects the fate of phosphorus in soil and water: A critical review. *Chemosphere* 283, 131176.

Glaser B, Lehmann J, and Zech W 2002 Ameliorating physical and chemical properties of highly weathered soils in the tropics with charcoal – a review. *Biology and Fertility of Soils* 35, 219–230.

Glaser B, and Lehr VI 2019 Biochar effects on phosphorus availability in agricultural soils: A meta-analysis. *Scientific Reports* 9, 1–9.

Gonzaga MIS et al 2020 Aged biochar changed copper availability and distribution among soil fractions and influenced corn seed germination in a copper-contaminated soil. *Chemosphere* 240, 124828.

Gorovtsov AV et al 2020 The mechanisms of biochar interactions with microorganisms in soil. *Environmental Geochemistry and Health* 42 (8), 2495–2518.

Gujre N, Soni A, Rangan L, Tsang DC, and Mitra S 2021 Sustainable improvement of soil health utilizing biochar and arbuscular mycorrhizal fungi: A review. *Environmental Pollution* 268, 115549.

Gul S, and Whalen JK 2016 Biochemical cycling of nitrogen and phosphorus in biochar-amended soils. *Soil Biology and Biochemistry* 103, 1–15

Güereña D, Lehmann J, Hanley K, Enders A, Hyland C, and Riha S 2013 Nitrogen dynamics following field application of biochar in a temperate North American maize-based production system *Plant and Soil* 365, 239–254.

Güereña D et al 2015 Partitioning the contributions of biochar properties to enhanced biological nitrogen fixation in common bean (Phaseolus vulgaris). *Biology and Fertility of Soils* 51 (4), 479–491.

Gundale MJ, and DeLuca TH 2006a Temperature and substrate influence the chemical properties of charcoal in the ponderosa pine/Douglas-fir ecosystem. *Forest Ecology and Management* 231, 86–93.

Gundale MJ, and DeLuca TH 2006b Temperature and source material influence ecological attributes of ponderosa pine and Douglas-fir charcoal. *Forest Ecology and Management* 231 (1–3), 86–93.

Gundale MJ, and DeLuca TH 2007 Charcoal effects on soil solution chemistry and growth of Koeleria macrantha in the ponderosa pine/Douglas-fir ecosystem. *Biol and Fertility of Soils* 43 (3), 303–311.

Gundale MJ, Sverker J, Albrectsen BR, Nilsson MC, and Wardle DA 2010 Variation in protein complexation capacity among and within six plant species across a boreal forest chronosequence. *Plant Ecology* 211 (2), 253–266.

Guo Y, Tang W, Wu JG, Huang ZQ, and Dai JY 2014 Mechanism of Cu(II) adsorption inhibition on biochar by its aging process. *Journal of Environmental Science* 26, 2123–2130.

Guo J et al 2021 Effects of various pyrolysis conditions and feedstock compositions on the physicochemical characteristics of cow manure-derived biochar. *Journal of Cleaner Production* 311, 127458.

Hagemann N et al 2017 Organic coating on biochar explains its nutrient retention and stimulation of soil fertility. *Nature Communications* 8, 1089.

Hansen J et al 2005 Efficacy of climate forcings. *Journal of Geophysical Research* 110, D18104.

Hao H, Jing YD, Ju WL, Shen L, and Cao YQ 2017 Different types of biochar: effect of aging on the Cu(II) adsorption behavior. *Desalin Water Treat* 95, 227–233.

He P et al 2015 Temporal and spatial variation of soil available potassium in China (1990–2012). *Field Crops Research* 173, 49–56.

Hossain MZ et al 2020 Biochar and its importance on nutrient dynamics in soil and plant. *Biochar* 2 (4), 379–420.

Hossain MK, Strezov V, Chan KY, Ziolkowski A, and Nelson PF 2011 Influence of pyrolysis temperature on production and nutrient properties of wastewater sludge biochar. *Journal of Environmental Management* 92, 223–228.

Houben D, Evrard L, and Sonnet P 2013 Mobility, bioavailability and pH-dependent leaching of cadmium, zinc and lead in a contaminated soil amended with biochar. *Chemosphere* 92, 1450–1457.

Hu H, Xi B, and Tan W 2021 Effects of sulfur-rich biochar amendment on microbial methylation of mercury in rhizosphere paddy soil and methylmercury accumulation in rice. *Environmental Pollution* 286, 117290.

Hunt PG, Cantrell KB, Bauer PJ, and Miller JO 2013 Phosphorus fertilization of ryegrass with ten precisely prepared manure biochars. *Transactions ASABE* 56 (6), 1317–1324.

Ippolito JA et al 2020 Feedstock choice, pyrolysis temperature and type influence biochar characteristics: a comprehensive meta-data analysis review. *Biochar* 2 (4), 421–438.

Inyang MI et al 2016 A review of biochar as a low-cost adsorbent for aqueous heavy metal removal. *Critical Reviews in Environ Science and Technology* 46, 406–433.

Jain V, and Nainawatee HS 2002 Plant flavonoids: Signals to legume nodulation and soil microorganisms. *Journal of Plant Biochemistry and Biotechnology* 11, 1–10.

Jatav HS, Singh SK, Singh Y, and Kumar O 2018 Biochar and sewage sludge application increases yield and micronutrient uptake in rice (*Oryza sativa* L.). *Communication in Soil Science and Plant Analysis* 49, 1617–1628.

Jeffery S, Verheijen FGA, van der Velde M, and Bastos AC 2011 A quantitative review of the effects of biochar application to soils on crop productivity using meta-analysis. *Agriculture Ecosystems and Environment* 144 (1), 175–187.

Jia R, Li LN, Qu D, and Mi NN 2016 Enhanced iron(III) reduction following amendment of paddy soils with biochar and glucose modified biochar. *Environmental Science and Pollution Research* 25, 91–103.

Jia R, Qu Z, You P, and Qu D 2018 Effect of biochar on photosynthetic microorganism growth and iron cycling in paddy soil under different phosphate levels. *Science of the Total Environment* 612, 223–230.

Jin Y et al 2016 Manure biochar influence upon soil properties, phosphorus distribution and phosphatase activities: A microcosm incubation study. *Chemosphere* 142, 128–135.

Jindo K et al 2014 Methodological interference of biochar in the determination of extracellular enzyme activities in composting samples. *Solid Earth* 5, 713–719.

Johnson DW 1984 Sulfur cycling in forests. *Biogeochemistry* 1, 29–43.

Jones DL, Murphy DV, Khalid M, Ahmad W, Edwards-Jones G, and DeLuca TH 2011 Short-term biochar-induced increase in soil respiration is both biotically and abiotically mediated. *Soil Biology and Biochemistry* 43, 1723–1731.

Jones DL, Rousk J, Edwards-Jones G, DeLuca TH, and Murphy DV 2012 Biochar-mediated changes in soil quality and plant growth in a three year field trial. *Soil Biology and Biochemistry* 45, 113–124.

Joseph SD et al 2010 An investigation into the reactions of biochar in soil. *Australian Journal of Soil Research* 48 (6–7), 501–515.

Joseph S et al 2013 Shifting paradigms: development of high-efficiency biochar fertilizers based on nano-structures and soluble components. *Carbon Management* 4, 323–343.

Joseph SD et al 2021 How biochar works, and when it doesn't: A review of mechanisms controlling soil and plant responses to biochar. *GCB Bioenergy*, 13 (11), 1731–1764.

Kammann CI, Linsel S, Gossling JW, and Koyro HW 2011 Influence of biochar on drought tolerance of Chenopodium quinoa Willd and

on soil-plant relations. *Plant and Soil* 345 (1–2), 195–210.

Kappler A, Wuestner ML, Ruecker A, Harter J, Halama M, and Behrens S 2014 Biochar as an electron shuttle between bacteria and Fe(III) minerals. *Environ Sci Technol Lett* 1, 339–344.

Karim AA, Kumar M, Mohapatra S, and Singh SK 2019 Nutrient rich biomass and effluent sludge wastes co-utilization for production of biochar fertilizer through different thermal treatments. *Journal of Cleaner Production* 228, 570–579.

Karimi A, Moezzi A, Chorom M, and Enayatizamir N 2020 Application of biochar changed the status of nutrients and biological activity in a calcareous soil. *Journal of Soil Science and Plant Nutrition* 20, 450–459.

Keech O, Carcaillet C, and Nilsson M-C 2005 Adsorption of allelopathic compounds by wood-derived charcoal: the role of wood porosity. *Plant and Soil* 272, 291–300.

Kitoh S and Shiomi N 1991 Effect of mineral nutrients and combined nitrogen sources in the medium on growth and nitrogen fixation of the Azolla-Anabaena association. *Journal of Soil Science and Plant Nutrition* 37, 419–426.

Khadem A, and Raiesi F 2019 Response of soil alkaline phosphatase to biochar amendments: Changes in kinetic and thermodynamic characteristics. *Geoderma* 337, 44–54.

Knoepp JD, DeBano LF, and Neary DG 2005 *Soil Chemistry. Wildland Fire in Ecosystems: Effects of Fire on Soil and Water.* Ogden, UT: U.S. Department of Agriculture, Forest Service, Rocky Mountain Research Station.

Knudsen JN, Jensen PA, Lin WG, Frandsen FJ, and Dam-Johansen K 2004 Sulphur transformations during thermal conversion of herbaceious biomass. *Energy and Fuels* 18, 810–819.

Koukouzas N, Hämäläinen J, Papanikolaou D, Tourunen A, and Jäntti T 2007 Mineralogical and elemental composition of fly ash from pilot scale fluidised bed combustion of lignite, bituminous coal, wood chips and their blends. *Fuel* 86, 2186–2193.

Kraus TEC, Dahlgren RA, and Zasoski RJ 2003 Tannins in nutrient dynamics of forest ecosystems – a review. *Plant and Soil* 256, 41–66.

Krishnan K et al 2021 Mitigating potassium leaching from muriate of potash in a tropical peat soil using clinoptilolite zeolite forest litter compost and chicken litter biochar. *Agronomy* 11, 1900.

Kumar A, Singh E, Khapre A, Bordoloi N, and Kumar S 2020 Sorption of volatile organic compounds on non-activated biochar. *Bioresource Technology* 297, 122469.

Kuppusamy S, Thavamani P, Megharaj M, Venkateswarlu K, and Naidu R 2016 Agronomic and remedial benefits and risks of applying biochar to soil: Current knowledge and future research directions. *Environment International* 87, 1–12.

Kuo YL, Lee CH, and Jien SH 2020 Reduction of nutrient leaching potential in coarse-textured soil by using biochar. *Water* 12, 2012.

Le Leuch LM, and Bandosz TJ 2007 The role of water and surface acidity on the reactive adsorption of ammonia on modified activated carbons. *Carbon* 45, 568–578.

Lee DK, Cho JS, and Yoon WL 2005 Catalytic wet oxidation of ammonia: Why is N_2 formed preferentially against NO_3-? *Chemosphere* 61, 573–578.

Leeper GW 1934 Relationship of soils to manganese deficiency of plants. *Nature*, 134, 972–973.

Lehmann J 2007 Bio-energy in the black. *Frontiers in Ecology and the Environment* 5 (7), 381–387.

Lehmann J et al 2003 Nutrient availability and leaching in an archaeological Anthrosol and a Ferrasol of the Central Amazon basin: Fertilizer, manure, and charcoal amendments. *Plant and Soil* 249, 343–357.

Lehmann J, Gaunt J, and Rondon M 2006 Bio-char sequestration in terrestrial ecosystems – a review. *Mitigation and Adaptation Strategies for Global Change* 11, 403–427.

Lehmann J, Rillig MC, Thies J, Masiello CA, Hockaday WC, and Crowley D 2011 Biochar effects of soil biota – A review. *Soil Biology and Biochemistry* 43, 1812–1836.

Lehmann J, and Rondon M 2006 Bio-Char soil management on highly weathered soils in the humid tropics. In: Uphoff NT, Ball AS, Fernandes E (Eds) *Biological Approaches to Sustainable Soil Systems*. Boca Raton: Taylor and Francis. pp517–530.

Leininger S et al 2006 Archaea predominate among ammonia-oxidizing prokaryotes in soils. *Nature* 442, 806–809.

Liang B et al 2006 Black carbon increases cation exchange capacity in soils. *Soil Science Society America Journal* 70, 1719–1730.

Li Y et al 2019 Effects of biochar-based fertilizers on nutrient leaching in a tobacco-planting soil. *Acta Geochimica* 38, 1–7.

Limwikran T, Kheoruenromne I, Suddhiprakarn A, Prakongkep N, and Gilkes RJ 2018 Dissolution of K Ca and P from biochar grains in tropical soils. *Geoderma* 312, 139–150.

Liu S et al 2017 Rice husk biochar impacts soil phosphorous availability, phosphatase activities and bacterial community characteristics in three different soil types. *Applied Soil Ecology* 116, 12–22.

Liu Q et al 2018 How does biochar influence soil N cycle? A meta-analysis. *Plant and Soil* 426 (1), 211–225.

Liu Q et al 2019a Biochar application as a tool to decrease soil nitrogen losses (NH_3 volatilization, N_2O emissions, and N leaching) from croplands: Options and mitigation strength in a global perspective. *Global Change Biology*, 25(6), 2077–2093.

Liu X et al 2019b Impact of biochar amendment on the abundance and structure of diazotrophic community in an alkaline soil. *Science of The Total Environment* 688, 944–951.

Liu L, Tan Z, Gong H, and Huang Q 2019c Migration and transformation mechanisms of nutrient elements (n p k) within biochar in straw-biochar-soil-plant systems: A review. *ACS Sustainable Chemical Engineering* 7, 22–32.

Lu H et al 2020 Effects of biochar on soil microbial community and functional genes of a landfill cover three years after ecological restoration. *Science of The Total Environment* 717, 137133.

Lucchini P, Quilliam RS, DeLuca TH, Vamerali T, and Jones DL 2014 Increased bioavailability of metals in two contrasting agricultural soils treated with waste wood-derived biochar and ash. *Environmental Science and Pollution Research* 21, 3230–3240.

Ma H et al 2019 Effect of biochar and irrigation on the interrelationships among soybean growth, root nodulation, plant P uptake, and soil nutrients in a sandy field. *Sustainability* 11 (23), 6542.

Maaz TM, Hockaday WC, and Deenik JL 2021 Biochar volatile matter and feedstock effects on soil nitrogen mineralization and soil fungal colonization. *Sustainability* 13, 2018.

MacKenzie MD, and DeLuca TH 2006 Charcoal and shrubs modify soil processes in ponderosa pine forests of western Montana. *Plant and Soil* 287, 257–267.

Major J, Lehmann J, Rondon M, and Goodale C 2010 Fate of soil-applied black carbon: downward migration, leaching and soil respiration. *Global Change Biology* 16 (4), 1366–1379.

Makoto K, and Koike T 2021 Charcoal ecology: Its function as a hub for plant succession and soil nutrient cycling in boreal forests. *Ecological Research*, 36 (1), 4–12.

Makoto K, Tamai Y, Kim YS, and Koike T 2010 Buried charcoal layer and ectomycorrhizae cooperatively promote the growth of *Larix gmelinii* seedlings. *Plant Soil* 327, 143–152.

Mandal S et al 2018 The effect of biochar feedstock, pyrolysis temperature, and application rate on the reduction of ammonia volatilisation from biochar-amended soil. *Science of the Total Environment* 627, 942–950.

Marks EAN et al 2016 Gasifier biochar effects on nutrient availability, organic matter mineralization, and soil fauna activity in a multi-year Mediterranean trial. *Agriculture, Ecosystems and Environment* 215, 30–39.

Mia S, Dijkstra FA, and Singh B 2018 Enhanced biological nitrogen fixation and competitive advantage of legumes in mixed pastures diminish with biochar aging. *Plant and Soil* 424 (1), 639–651.

Mia S, Singh B, and Dijkstra FA 2017 Aged biochar affects gross nitrogen mineralization and recovery: a15N study in two contrasting soils. *Global Change Biology Bioenergy* 9 (7), 1196–1206.

Muhammad, I, Rafiullah, Kaleri, FN, Muhammad, R, and Imran, M 2017 Potential value of biochar as a soil amendment: a review. *Pure and Applied Biology*, 6 (4), 1494–1502.

Neary DG, Klopatek CC, DeBano LF, and Ffolliott PF 1999 Fire effects on belowground sustainability: a review and synthesis. *Forest Ecology and Management* 122, 51–71.

Neary DG, Ryan KC, and DeBano LF 2005 Wildland fire in ecosystems: effects of fire on soil and water. vol Gen. Tech. Rep. RMRS-GTR-42-vol.4. Department of Agriculture, forest Service, Rocky Mountain Research Station.

Nelissen V, Rütting T, Huygens D, Staelens J, Ruysschaert G, and Boeckx P 2012 Maize biochars accelerate short-term soil nitrogen dynamics in a loamy sand soil. *Soil Biology and Biochemistry* 55, 20–27.

Nelissen V, Rutting T, Huygens D, Ruysschaert G, and Boeckx P 2015 Temporal evolution of biochar's impact on soil nitrogen processes – a ^{15}N tracing study. *Global Change Biology Bioenergy* 7 (4), 635–645.

Nelson NO, Agudelo SC, Yuan W, and Gan J 2011 Nitrogen and phosphorus availability in biochar-amended soils. *Soil Science* 176 (5), 218–226.

Nguyen TTN et al 2017 Effects of biochar on soil available inorganic nitrogen: A review and meta-analysis. *Geoderma* 288, 79–96.

Ni JZ, Pignatello JJ, and Xing BS 2010 Adsorption of aromatic carboxylate ions to black carbon (biochar) is accompanied by proton exchange with water. *Environ Science and Technology* 45 (21), 9240–9248.

Nie T et al 2021 Effect of biochar aging and co-existence of diethyl phthalate on the mono-sorption of cadmium and zinc to biochar-treated soils. *Journal of Hazardous Materials* 408, 124850.

Nishio M, and Okano S 1991 Stimulation of the growth of alfalfa and infection of mycorrhizal fungi by the application of charcoal. *National Grassland Research Institute* 45, 61–71.

Novak JM et al 2010 Short-term CO_2 mineralization after additions of biochar and switchgrass to a Typic Kandiudult. *Geoderma* 154, 281–288.

Nzanza B, Marais D, and Soundy P 2012 Effect of arbuscular mycorrhizal fungal inoculation and biochar amendment on growth and yield of tomato. *International Journal of Agricultural Biology* 14, 965–969.

Paavolainen L, Kitunen V, and Smolander A 1998 Inhibition of nitrification in forest soil by monoterpenes. *Plant and Soil* 205, 147–154.

Palanivell P, Ahmed OH, Latifah O, and Abdul Majid NM 2019 Adsorption and desorption of nitrogen phosphorus potassium and soil buffering capacity following application of chicken litter biochar to an acid soil. *Applied Science* 10, 295.

Palansooriya KN et al 2019 Response of microbial communities to biochar-amended soils: a critical review. *Biochar* 1, 3–22.

Paul EA 2015 *Soil Microbiology, Ecology, and Biochemistry*. Boston: Academic Press.

Piash MI, Iwabuchi K, and Itoh T 2022 Synthesizing biochar-based fertilizer with sustained phosphorus and potassium release: Co-pyrolysis of nutrient-rich chicken manure and Ca-bentonite. *Science of the Total Environment* 822, 153509.

Pietikainen J, and Fritze H 1993 Microbial biomass and activity in the humus layer following burning: short-term effects of two different fires. *Canadian Journal of Forest Research* 23, 1275–1285.

Pogorzelski D et al 2020 Biochar as composite of phosphate fertilizer: Characterization and agronomic effectiveness. *Science of the Total Environment* 743, 140604.

Pokharel P, Ma Z, and Chang SX 2020 Biochar increases soil microbial biomass with changes in extra- and intracellular enzyme activities: a global meta-analysis. *Biochar* 2, 65–79.

Pokharel P, Qi L, and Chang SX 2021 Manure-based biochar decreases heterotrophic respiration and increases gross nitrification rates in rhizosphere soil. *Soil Biology and Biochemistry* 154, 108147.

Poormansour S, Razzaghi F, and Sepaskhah AR 2019 Wheat straw biochar increases potassium concentration root density and yield of faba bean in a sandy loam soil. *Communications in Soil Science and Plant Analysis* 50, 1799–1810.

Prendergast-Miller MT, Duvall M, and Sohi SP 2011 Localisation of nitrate in the rhizosphere of biochar-amended soils. *Soil Biology and Biochemistry* 43 (11), 2243–2246.

Preston CM, and Schmidt MWI 2006 Black (pyrogenic) carbon: a synthesis of current knowledge and uncertainties with special consideration of boreal regions. *Biogeosciences* 3, 397–420.

Quartacci MF, Sgherri C, Cardelli R, and Fantozzi A 2015 Biochar amendment reduces oxidative stress in lettuce grown under copper excess. *Agrochemica* 59, 188–202.

Quilliam RS, Jones DL, and DeLuca TH 2013a Biochar application reduces nodulation but increases nitrogenase activity in clover. *Plant and Soil* 366, 83–92.

Quilliam RS, Glanville HC, Wade SC, and Jones DL 2013b Life in the 'charosphere' – Does biochar in agricultural soil provide a significant habitat for microorganisms? *Soil Biology and Biochemistry* 65, 287–293.

Ramzani PMA, Khalid M, Naveed M, Ahmad R, and Shahid M 2016 Iron biofortification of wheat grains through integrated use of organic and chemical fertilizers in pH affected calcareous soil. *Plant Physiology and Biochemistry* 104, 284–293.

Ravishankara AR, Daniel JS, and Portmann, RW 2009. Nitrous oxide (N_2O): the dominant ozone-depleting substance emitted in the 21st century. *Science*, 326 (5949), 123–125.

Razzaghi F, Obour PB, and Arthur E 2020 Does biochar improve soil water retention? A systematic review and meta-analysis. *Geoderma* 361, 114055.

Rens H, Bera T, and Alva AK 2018 Effects of biochar and biosolid on adsorption of nitrogen phosphorus and potassium in two soils. *Water, Air and Soil Pollution* 229, 281.

Riedel T, Hennessy P, Iden SC, and Koschinsky A 2015 Leaching of soil-derived major and trace elements in an arable topsoil after the addition of biochar. *European Journal of Soil Science* 66, 823–834.

Roberts KG, Gloy BA, Joseph S, Scott NR, and Lehmann J 2010 Life cycle assessment of biochar systems: estimating the energetic, economic, and climate change potential. *Environmental Science Technology* 44, 827–833.

Rondon M, Lehmann J, Ramirez J, and Hurtado M 2007 Biological nitrogen fixation by common beans (Phaseolus vulgaris L.) increases with bio-char additions. *Biology and Fertility of Soils* 43, 699–708.

Saito M 1990 Charcoal as a micro habitat for VA mycorrhizal fungi, and its practical application. *Agriculture, Ecosystems and Environment* 29, 341–344.

Schmidt, SB, Jensen PE, and Husted, S 2016 Manganese Deficiency in Plants: The Impact on Photosystem II. *Trends in Plant Science* 21 (7), 622–632.

Schmidt MWI, and Noack AG 2000 Black carbon in soils and sediments: Analysis, distribution, implications, and current challenges. *Global Biogeochemical Cycles* 14, 777–793.

Schneider F, and Haderlein SB 2016 Potential effects of biochar on the availability of phosphorus – mechanistic insights. *Geoderma* 277, 83–90.

Schneour EA 1966 Oxidation of graphite carbon in certain soils. *Science* 151, 991–992

Sha Z et al 2019 Response of ammonia volatilization to biochar addition: A meta-analysis. *Science of the Total Environment* 655, 1387–1396.

Sharma SB, Sayyed RZ, Trivedi MH, and Gobi TA 2013 Phosphate solubilizing microbes: Sustainable approach for managing phosphorus deficiency in agricultural soils. *SpringerPlus* 2, 1–14.

Shen C, Kahn A, and Schwartz J 2001 Chemical and electrical properties of interfaces between magnesium and aluminum and tris-(8-hydroxy quinoline) aluminum. *Journal of Applied Physics* 89, 449–459.

Smider B and Singh B 2014 Agronomic performance of a high ash biochar in two contrasting soils. *Agriculture Ecosystems and Environment,* 191, 99–107.

Solaiman ZM, Blackwell P, Abbott LK, and Storer P 2010 Direct and residual effect of biochar application on mycorrhizal root colonisation, growth and nutrition of wheat. *Australian Journal of Soil Research* 48 (6–7), 546–554.

Song B et al 2017. Evaluation methods for assessing effectiveness of in situ remediation of soil and sediment contaminated with organic pollutants and heavy metals. *Environment International,* 105, 43–55.

Sorrenti G, Masiello CA, and Toselli M 2016 Biochar interferes with kiwifruit Fe-nutrition in calcareous soil. *Geoderma* 272, 10–19.

Spokas KA, Koskinen WC, Baker JM, and Reicosky DC 2009 Impacts of woodchip biochar additions on greenhouse gas production and sorption/degradation of two herbicides in a Minnesota soil. *Chemosphere* 77, 574–581

Spokas KA, Baker JM, and Reicosky DC 2010 Ethylene: potential key for biochar amendment impacts. *Plant and Soil* 333, 443–452.

Spokas KA, Novak JM, and Venterea RT 2012 Biochar's role as an alternative N-fertilizer: ammonia capture. *Plant Soil* 350, 35–42.

Steiner C, Das KC, Melear N, and Lakly D 2010 Reducing nitrogen loss during poultry litter composting using biochar. *Journal of Environmental Quality* 39, 1236–1242.

Steiner C et al 2007 Long term effects of manure, charcoal, and mineral fertilization on crop production and fertility on a highly weathered Central Amazonian upland soil. *Plant and Soil* 291, 275–290.

Stevenson FJ, and Cole MA 1999 *Cycles of the Soil.* 2nd ed. New York: John Wiley and Sons, Inc.

Streubel JD, Collins HP, Garcia-Perez M, Tarara J, Granatstein D, and Kruger CE 2011 Influence of contrasting biochar types on five soils at increasing rates of application *Soil Science Society of America Journal* 75, 1402–1413.

Su HJ, Fang ZQ, Tsang PE, Fang JZ, and Zhao DY 2016 Stabilisation of nanoscale zero-valent iron with biochar for enhanced transport and in-situ remediation of hexavalent chromium in soil. *Environmental Pollution* 214, 94–100.

Sujeeun L and Thomas SC 2017 Potential of biochar to mitigate allelopathic effects in tropical island invasive plants: evidence from seed germination trials. *Tropical Conservation Science* 2017, 10.

Sun Y et al 2021 Roles of biochar-derived dissolved organic matter in soil amendment and environmental remediation: A critical review. *Chemical Engineering Journal* 424, 130387.

Swaine M, Obrike R, Clark JM and Shaw LJ 2013 Biochar alteration of the sorption of substrates and products in soil enzyme assays. *Applied and Environmental Soil Science* 2013, 1–5.

Tabatabai MA, and Al-Khafaji AA 1980. Comparison of nitrogen and sulfur mineralization in soils. *Soil Science Society of America Journal*, 44 (5), 1000–1006.

Taghizadeh-Toosi A, Clough TJ, Sherlock RR, and Condron LM 2012b Biochar adsorbed ammonia is bioavailable. *Plant and Soil* 350, 57–69.

Taghizadeh-Toosi A, Clough TJ, Sherlock RR, and Condron LM 2012a A wood based low-temperature biochar captures NH_3-N generated from ruminant urine-N, retaining its bioavailability. *Plant and Soil* 353, 73–84.

Tagoe SO, Horiuchi T, and Matsui T 2008 Effects of carbonized and dried chicken manures on the growth, yield and N content of soybean. *Plant Soil* 306, 211–220.

Tesfaye F et al 2021 Could biochar amendment be a tool to improve soil availability and plant uptake of phosphorus? A meta-analysis of published experiments. *Environmental Science and Pollution Research* 28, 34108–34120.

Tian J et al 2021 Biochar application under low phosphorus input promotes soil organic phosphorus mineralization by shifting bacterial phoD gene community composition. *Science of The Total Environment* 779, 146556.

Tomczyk A, Boguta P, and Sokolowska Z 2019 Biochar efficiency in copper removal from Haplic soils. *International Journal of Environmental Science and Technology* 16, 4899–4912.

Topoliantz S, Pong J-F, and Ballof S 2005 Manioc peel and charcoal: a potential organic amendment for sustainable soil fertility in the tropics. *Biology and Fertility of Soils* 41, 15–21.

Tripathi DK et al 2018 Acquisition and homeostasis of iron in higher plants and their probable role in abiotic stress tolerance. *Frontiers in Environmental Science* 5, 86.

Trompowsky PM, Benites VDM, Madari BE, Pimenta AS, Hockaday WC, and Hatcher PG 2005 Characterization of humic like substances obtained by chemical oxidation of eucalyptus charcoal. *Organic Geochemistry* 36, 1480–1489.

Turner ER 1955 The effect of certain adsorbents on the nodulation of clover plants. *Annals of Botany* 19, 149–160.

Tyron EH 1948 Effect of charcoal on certain physical, chemical, and biological properties of forest soils. *Ecological Monographs* 18, 82–115.

Uchimiya M, Lima I, Klasson K, and Wartelle L 2010 Contaminant immobilization and nutrient release by biochar soil amendment: Roles of natural organic matter. *Chemosphere* 80 (8), 935–940.

van de Voorde TF, Bezemer TM, Van Groenigen JW, Jeffery S, and Mommer, L. 2014. Soil biochar amendment in a nature restoration area: effects on plant productivity and community composition. *Ecological Applications* 24 (5), 1167–1177.

Vantis JT and Bond G 1950 The effect of charcoal on the growth of leguminous plants in sand culture. *Annals of Applied Biology* 37, 159–168.

Ventura M, Sorrenti B, Panzacchib EG, and Tonon G 2013 Biochar reduces short-term nitrate leaching from A horizon in an apple orchard. *Journal of Environmental Quality* 42, 76–82.

Vitousek PM and Howarth RW 1991 Nitrogen limitation on land and in the sea: How can it occur? *Biogeochemistry* 13, 87–115.

Vivanco JM, Bais HP, Stermitz FR, Thelen GC, and Callaway RM 2004 Biogeographical variation in community response to root allelochemistry: novel weapons and exotic invasion. *Ecology Letters* 7, 285–292.

Wainwright, M. 1984. Sulfur oxidation in soils. *Advances in Agronomy* 37, 349–396.

Wallstedt A, Coughlan A, Munson AD, Nilsson M-C, and Margolis HA 2002 Mechanisms of interaction between *Kalmia angustifulia* cover and *Picea mariana* seedlings. *Canadian Journal of Forest Research* 32, 2022–2031.

Wan XM, Li CY, and Parikh SJ 2020 Simultaneous removal of arsenic, cadmium, and lead from soil by iron-modified magnetic biochar. *Environmental Pollution* 261, 114157.

Wang L, Xue C, Nie XX, Liu Y, and Chen F 2018 Effects of biochar application on soil potassium dynamics and crop uptake. *Plant Nutrition and Soil Science* 181, 635–643.

Wang N, Xue XM, Juhasz AL, Chang ZZ, and Li HB 2017 Biochar increases arsenic release from an anaerobic paddy soil due to enhanced microbial reduction of iron and arsenic. *Environmental Pollution* 220, 514–522.

Wang Y et al 2015 Phosphorus release behaviors of poultry litter biochar as a soil amendment. *Science of The Total Environment* 512–513, 454–463.

Wang, Y et al 2021a Biochar-impacted sulfur cycling affects methylmercury phytoavailability in soils under different redox conditions. *Journal of Hazardous Materials* 407, 124397.

Wang Z et al 2021b Research on the adsorption mechanism of Cu and Zn by biochar under freeze-thaw conditions. *Science of The Total Environment* 774, 145194.

Wang L et al 2020a Biochar aging: mechanisms, physicochemical changes, assessment, and implications for field applications. *Environmental Science and Technology* 54, 14797–14814.

Wang Y, Zheng J, Liu X, Yan Q, and Hu Y 2020b Short-term impact of fire-deposited charcoal on soil microbial community abundance and composition in a subtropical plantation in China. *Geoderma* 359, 113992.

Wang Z et al 2020c Regulation of Cu and Zn migration in soil by biochar during snowmelt. *Environmental Research* 186, 109566.

Ward BB, Courtney KJ, and Langenheim, JH 1997 Inhibition of *Nitrosmonas europea* by monoterpenes from coastal redwood (*Sequoia sempervirens*) in whole-cell studies. *Journal of Chemical Ecology* 23, 2583–2599

Wardle D A, Nilsson MC, and Zackrisson O 2008. Fire-derived charcoal causes loss of forest humus. *Science* 320, 629.

Wei X et al 2019 Rare taxa of alkaline phosphomonoesterase-harboring microorganisms mediate soil phosphorus mineralization. *Soil Biology and Biochemistry* 131, 62–70.

Whitman T et al 2019 Soil bacterial and fungal response to wildfires in the Canadian boreal forest across a burn severity gradient. *Soil Biology and Biochemistry* 138, 107571.

White C 1991 The role of monoterpenes in soil nitrogen cycling processes in ponderosa pine. *Biogeochemistry* 12, 43–68.

Woolet J, and Whitman T 2020 Pyrogenic organic matter effects on soil bacterial community composition. *Soil Biology and Biochemistry* 141, 107678.

Wu C, Shi LZ, Xue SG, Li WC, Jiang XX, Rajendran M and Qian ZY 2019 Effect of sulfur-iron modified biochar on the available cadmium and bacterial community structure in contaminated soils. *Science of The Total Environment* 647, 1158–1168.

Wu C et al 2020. Exploring the recycling of bioleaching functional bacteria and sulfur substrate using the sulfur-covered biochar particles. *Environmental Sciences Europe* 32, 70.

Wurst S, van Beersum S 2008 The impact of soil organism composition and activated carbon on grass-legume competition. *Plant and Soil* 314, 1–9.

Xiao Z et al 2019 The effect of biochar amendment on N-cycling genes in soils: A meta-analysis. *Science of The Total Environment* 696, 133984.

Xie Z et al 2021 Effects of biochar application and irrigation rate on the soil phosphorus leaching risk of fluvisol profiles in open vegetable fields. *Science of The Total Environment* 789, 147973.

Xu G, Zhang Y, Shao H, and Sun J 2016 Pyrolysis temperature affects phosphorus transformation in biochar: Chemical fractionation and ^{31}P NMR analysis. *Science of The Total Environment* 569–570, 65–72.

Xu M et al 2019 Biochar impacts on phosphorus cycling in rice ecosystem. *Chemosphere* 225, 311–319.

Xu X, Sivey JD and Xu W 2020. Black carbon-enhanced transformation of dichloroacetamide safeners: Role of reduced sulfur species. *Science of the Total Environment* 738, 139908.

Xu WH, Whitman WB, Gundale MJ, Chien CC. and Chiu CY. 2021 Functional response of the soil microbial community to biochar applications. *Global Change Biology Bioenergy* 13 (1), 269–281.

Yamato M, Okimori Y, Wibowo IF, Anshiori S, and Ogawa M 2006 Effects of the application of charred bark of Acacia mangium on the yeild of maize, cowpea and peanut, and soil chemical properties in South Sumatra, Indonesia. *Soil Science and Plant Nutrition* 52, 489–495.

Yanai Y, Toyota K, and Okazaki M 2007 Effects of charcoal addition on N20 emissions from soil resulting from rewetting air-dried soil in short-term laboratory experiments. *Soil Science and Plant Nutrition* 53, 181–188.

Yang B, Luo W, Wang X, Yu S, Gan M, Wang J, Liu X, and Qiu G 2020. The use of biochar for controlling acid mine drainage through the inhibition of chalcopyrite biodissolution. *Science of the Total Environment* 737, 139485.

Yang C and Lu S 2022 Straw and straw biochar differently affect phosphorus availability, enzyme activity and microbial functional genes in an Ultisol. *Science of The Total Environment* 805, 150325.

Yang ZY, Xing R, and Zhou WJ 2019 Adsorption of ciprofloxacin and Cu^{2+} onto biochars in the presence of dissolved organic matter derived from animal manure. *Environmental Science and Pollution Research* 26, 14382–14392.

Yin S et al 2021a Biochar Enhanced Growth and Biological Nitrogen Fixation of Wild Soybean (*Glycine max* subsp. soja Siebold andamp; Zucc.) in a Coastal Soil of China. *Agriculture* 11 (12), 1246.

Yin XL et al 2021 Effects of nitrogen-enriched biochar on rice growth and yield, iron dynamics, and soil carbon storage and emissions: A tool to improve sustainable rice cultivation. *Environmental Pollution* 287, 117565.

Yoo G and Kang H 2010 Eff ects of biochar addition on greenhouse gas emissions and microbial responses in a short-term laboratory experiment. *Journal of Environmental Quality* 41, 1193–1202.

Yu L, Duan L, Naidu R, and Semple KT 2018 Abiotic factors controlling bioavailability and bioaccessibility of polycyclic aromatic hydrocarbons in soil: Putting together a bigger picture. *Science of the Total Environment* 613–614, 1140–1153.

Zackrisson O, Nilsson M-C, and Wardle DA 1996 Key ecological function of charcoal from wildfire in the Boreal forest. *Oikos* 77, 10–19

Zielinska A, Oleszczuk P, Charmas B, Skubiszewska-Zieba J, and Pasieczna-Patkowska S 2015 Effect of sewage sludge properties on the biochar characteristic. *Journal of Analytical and Applied Pyrolysis* 112, 201–213.

Zhang H, Voroney RP, Price GW, and White AJ 2017. Sulfur-enriched biochar as a potential soil amendment and fertiliser. *Soil Research* 55, 93–99.

Zhang L, Jing Y, Xiang Y, Zhang R, and Lu H 2018. Responses of soil microbial community structure changes and activities to biochar addition: A meta-analysis. *Science of the Total Environment*, 643, 926–935.

Zhang L, Xiang Y, Jing Y, and Zhang R 2019 Biochar amendment effects on the activities of soil carbon, nitrogen, and phosphorus hydrolytic enzymes: a meta-analysis. *Environmental Science and Pollution Research* 26, 22990–23001.

Zhang L et al 2021 Habitat heterogeneity induced by pyrogenic organic matter in wildfire-perturbed soils mediates bacterial community assembly processes. *The ISME Journal* 15 (7), 1943–1955.

Zhang Q et al 2020a Effects of six-year biochar amendment on soil aggregation, crop growth, and nitrogen and phosphorus use efficiencies in a rice-wheat rotation. *Journal of Cleaner Production* 242, 118435.

Zhang MY, Riaz M, Liu B, Xia H, El-desouki Z, and Jiang CC 2020b Two-year study of biochar: Achieving excellent capability of potassium supply via alter clay mineral composition and potassium-dissolving bacteria activity. *Science of the Total Environment* 717, 137286.

Zhao J et al 2021 Low-pyrolysis-temperature biochar promoted free-living N2-fixation in calcareous purple soil by affecting diazotrophic composition. *Geoderma* 388, 114969.

Zhao B and Zhang T 2021 Effects of biochar and sulfate amendment on plant physiological characteristics, soil properties and sulfur phytoavailability of corn in Calcids soil. *Polish Journal of Environmental Studies* 30, 2917–2925.

Zhong, YC et al. 2020 Effects of aging and weathering on immobilization of trace metals/metalloids in soils amended with biochar. *Environmental Science-Processes and Impacts* 22 (9), 1790–1808.

Zhou C et al 2020 Biochar addition to forest plantation soil enhances phosphorus availability and soil bacterial community diversity. *Forest Ecology and Management* 455, 117635.

Zorb C, Senbayram M, and Peiter E 2014 Potassium in agriculture – status and perspectives. *Journal of Plant Physiology* 171, 656–669.

17

Biochar effects on soil carbon turnover

Thea Whitman, Yunying Fang, and Yu Luo

Introduction

Soil organic matter (SOM) plays critical roles in nutrient cycling and availability, soil structure, soil water dynamics, and, of course, carbon (C) storage (Janzen, 2006). Depending on the specific soil, environmental conditions, properties and application rates of biochar, and timescale, adding biochar to soil can result in net increases or net decreases in SOM stocks. A small change in SOM mineralization could significantly affect SOM stocks and, hence, atmospheric CO_2. Thus, understanding the effects of biochar additions on SOM is critical.

For C management purposes, there is a pressing need to understand and quantify the effects of biochar on non-biochar soil organic C (SOC) (and inorganic soil C) stocks. This is particularly salient when integrating biochar into carbon credit or offsetting schemes, to ensure that the C benefits of biochar are not under- or overestimated. At the same time, while from a climate change mitigation perspective, reducing losses of SOM-C is a key goal, it is also essential to remember that SOM has value not only as a static stock of C, but also as a dynamically cycling pool of nutrients (Janzen, 2006). Any reduction in SOM mineralization would thus be expected to be accompanied by a concomitant reduction in the release of mineral nutrients.

Due to the wide range of biochar properties, the innate diversity of soils and management systems, and the myriad effects of biochar on soil properties (described throughout this book), there are many different mechanisms through which biochar may affect soil C cycling. Thus, the research has moved beyond a purely observational approach – simply adding biochar to soils and quantifying its effects on soil C stocks – to a systematic approach to understanding these effects and the underlying mechanisms. Such an approach results in a deeper understanding of the system, affords greater

DOI: 10.4324/9781003297673-17

predictability, and also has implications for how these effects and their associated biotic and edaphic factors may be incorporated into SOC models.

In this chapter, we offer a brief introduction to biochar interactions with SOM, a description of methods and approaches used to understand biochar–SOM interactions, an overview of the current literature studying these interactions, a consideration of the mechanisms that may drive these interactions, and future research directions.

Biochar interactions with SOM

The observation that the addition of one substrate could alter the decomposition of others traces back at least to Löhnis (1926), and the term "priming" (as well as "negative priming") was employed as early as Bingeman et al (1953). There, it was defined as "a greater loss of soil organic matter, in a soil receiving an organic amendment, than the loss of organic matter in an untreated soil". Since then, the phenomenon has been investigated in many contexts, including priming of SOM by root exudates (Broadbent and Norman, 1947), priming of deep SOM by fresh SOM (Fontaine et al, 2007), and, of course, priming of SOM by charcoal or biochar (Wardle et al, 2008).

The definition of priming has also varied over time, including definitions as broad as "strong short-term changes in the turnover of soil organic matter caused by comparatively moderate treatments of the soil", which includes drying and rewetting effects, fertilizer inputs, and mechanical treatment of soil. For this chapter, we aim to consider all phenomena through which biochar may affect non-biochar C cycling (Figure 17.1). Throughout, we will use the terms "increased or decreased SOC mineralization" rather than "positive or negative priming", primarily because the former are mechanistically-agnostic terms, without the varying and perhaps contrasting preconceptions readers may associate with the term "priming", due to its rich history and wide-ranging applications. Second, while biochar itself is, once added to soil, also "soil organic matter or carbon", for this chapter, we will use the terms SOM or SOC to refer to all soil organic matter or carbon except for the biochar additions themselves, often also referred to as "native SOM or SOC (nSOM or nSOC)" or "non-pyrogenic SOC (npSOC)". Finally, when we use the term "mechanisms", unless otherwise indicated, we are referring specifically to the mechanisms through which biochar is affecting SOM or SOC mineralization.

The litterbag study by Wardle et al (2008), which indicated increased losses of SOM after the addition of charcoal to a boreal forest soil, ignited a flurry of research designed to further investigate this phenomenon, particularly in the context of potential unintended effects of biochar additions on net C storage. Today, the effects of biochar addition on SOC mineralization have been investigated in a wide range of systems, and our mechanistic understanding of the phenomenon is much improved. Still, important unknowns and challenges remain, as discussed in the final section of this chapter.

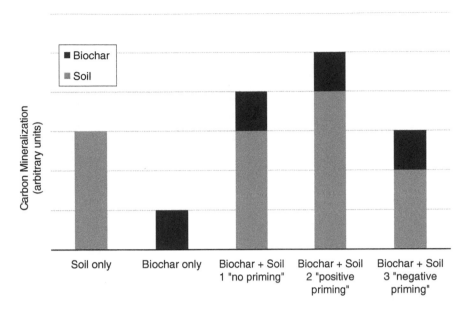

Figure 17.1 *Conceptual illustration of "priming", or increased or decreased mineralization of soil organic C (SOC) with biochar additions. Colored bars represent C sources from soil (brown) or biochar (black) (e.g., contribution to CO_2 flux). The leftmost two bars represent the C mineralization from each when incubated independently. The rightmost three bars represent three scenarios where biochar and soil are incubated together, where Scenario 1 indicates no effects on SOC ("no priming"), Scenario 2 indicates an increase in SOC mineralization ("positive priming"), and Scenario 3 indicates a decrease in SOC mineralization ("negative priming"). We do not illustrate the effects of SOC on biochar mineralization here, because this is not the focus of the chapter. However, interactive priming effects are certainly expected, and are discussed to some extent in Chapter 11. Note that the size of the bars is primarily illustrative and not necessarily reflective of the expected degree of increased or decreased mineralization for each component*

Methods for studying biochar-SOM interactions

Additive methods

The first scientific observations of increased SOC losses after adding biochar generally relied on an additive approach. That is, soil was incubated alone (in the laboratory or the field), biochar was incubated alone, and a mixture of the two was incubated (Figure 17.2). By adding up the C losses in the two one-component treatments and comparing it to the C losses in the two-component treatment, one could infer whether an interactive effect had occurred, where the presence of one C source affected the mineralization of the other. For example, Blanco-Canqui et al. (2020) found that total soil C increased by more than twice the amount of biochar-C added, thus inferring decreased SOC losses due to the biochar additions. However, with such an approach, it is impossible to determine how each source's mineralization has changed. That is, if a net increase in CO_2 emissions is observed, it may

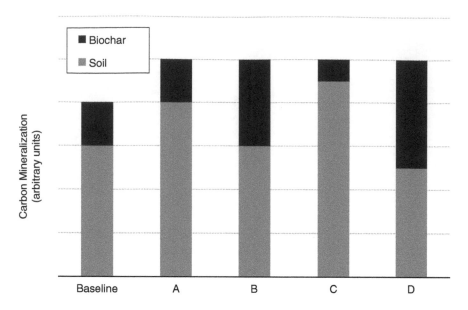

Figure 17.2 *Conceptual illustration of problems with additive methods. Colored bars represent C sources from soil (brown) or biochar (black) (e.g., contribution to CO₂ flux). "Baseline" represents a scenario without an interaction between SOC mineralization and biochar C mineralization. A-D represent four scenarios with the same increase in total C mineralization, but different reasons for this increase. In A, soil C mineralization increases. In B, biochar C mineralization increases. In C, soil C mineralization increases while biochar C mineralization decreases. In D, soil C mineralization decreases while biochar C increases. Despite important differences in the underlying mechanism, these four scenarios would be indistinguishable without using conclusive partitioning methods, such as those that rely on isotopic C source partitioning. Note that the size of the bars is primarily illustrative and not necessarily reflective of the expected degree of increased or decreased mineralization for each component*

be due to increases in biochar mineralization alone, SOC mineralization alone, increases in both, or perhaps even decreases in one while the other increases even more. While creative and thoughtful reasoning could begin to constrain the possibilities (e.g., Cardelli et al, 2016), approaches that conclusively separate the two sources, such as isotopic labeling, have become the gold standard.

Isotopic methods

Isotopically-based differentiation of two (or more) components of a mixed system

remains a workhorse of biogeochemistry (Balesdent et al, 1996). Briefly, C has two stable isotopes (^{13}C and ^{12}C) and one radioisotope commonly found on Earth (^{14}C). The ratios of C isotopes in different materials can be established using isotope ratio mass spectrometry for ^{13}C and ^{12}C or accelerator mass spectrometry or liquid scintillation counting for ^{14}C. The classic mixing model can determine the isotopic composition of a mixture and subsequent partitioning between the two constituent C sources (Equations 17.1–17.3; Figure 17.3A).

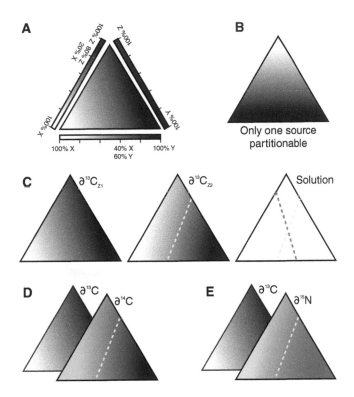

Figure 17.3 *Approaches to partitioning three different C sources. (A) Shading in triangles represents different isotopic ratios for two isotopes (e.g., ^{13}C and ^{12}C, where more ^{13}C is shaded white and more ^{12}C is shaded black). Each of the three points of the triangle represents the isotopic signature of one of three C sources (X, Y, and Z) for a given system. A given shade within the triangle space represents the combined isotopic signature of the three sources. The dashed lines represent the continuum of solutions possible given a combined 3-component isotopic signature. (B) If two C sources have the same isotopic signature (represented by the two black corners), the system can be solved for the fraction the remaining portion represents, but the two identical C sources cannot be separated. (C) If one C source is present in a paired treatment with two different isotopic signatures for that C source (here, $\partial^{13}C_{Z1}$ and $\partial^{13}C_{Z2}$ in the two different shaded triangles), a conclusive solution is possible. (D) A similar approach to that illustrated in (C) is possible if three different isotopes are analyzed – here, ^{13}C, ^{12}C, and ^{14}C (shaded in white, black, and blue). (E) A similar approach to that illustrated in (C) is possible if two different elements with different isotopic signatures are analyzed – here $^{13}C/^{12}C$ (shaded in white/black) and $^{15}N/^{14}N$ (shaded in yellow/green)*

$$f_{BC} + f_{SOC} = 1 \qquad [17.1]$$

$$\delta^{13}C_{BC} \cdot f_{BC} + \delta^{13}C_{SOC} \cdot f_{SOC} = \delta^{13}C_{Total} \quad [17.2]$$

$$f_{BC} = (\delta^{13}C_{Total} - \delta^{13}C_{SOC})/(\delta^{13}C_{BC} - \delta^{13}C_{SOC})$$
$$[17.3]$$

where f_{BC} and f_{SOC} represent the fraction of C contributed by biochar and SOC, respectively, and $\delta^{13}C_{BC}$, $\delta^{13}C_{SOC}$, and $\delta^{13}C_{Total}$ represent the $\delta^{13}C$ signatures for biochar, SOC, and the total mixture, respectively.

A few different approaches have been used to obtain biochar and SOC with distinct isotope ratios. For stable isotopes, there are natural differences in the $^{13}C/^{12}C$ ratio content of C_3 vs. C_4 plants, and the soils that are derived from primarily C_3 or C_4 plant inputs. For example, biochar effects on SOC mineralization were calculated by applying biochar derived from a C_4 plant, *Miscanthus giganteus*, to C_3 soil, as the natural $\delta^{13}C$ signature of SOC in this C_3 soil was 10‰ depleted compared to C_4 *Miscanthus* (Luo et al, 2011). However, while differences of this magnitude have been applied for isotopic partitioning in biogeochemical studies, the relatively small difference between the ^{13}C content of C_3 vs. C_4 plants means that uncertainty can easily eclipse the signal in systems where one source contributes relatively little to the total C in the mixture. Because biochar generally is not very readily mineralized (Chapter 11), and if keeping biochar inputs to ecologically realistic levels, this can be a constraining factor in biochar-SOC interaction studies. Additionally, it is only applicable to systems that can be designed to contain C from different C_3 and C_4 sources. A solution to these challenges is to use biochar that is produced from biomass grown in a ^{13}C-enriched or depleted atmosphere. This allows one to detect the relatively small biochar-derived C fluxes or stocks. Note that it would be extremely difficult to uniformly label the SOC with isotopes since it requires continuous labeling of C inputs to the soil over the entire duration of its development. Thus, the biochar is generally the ^{13}C-labeled source, and the SOC is used at natural abundance levels.

For radioisotope studies, plants grown in a ^{14}C-enriched atmosphere (e.g., Kuzyakov et al, 2009), or plants that have otherwise been enriched (Ji et al, 2019) or depleted (e.g., Tilston et al, 2016; Bird et al, 2017) in ^{14}C can be used to produce biochar, which is distinguishable from SOC with a different ^{14}C signature. Drawbacks of this technique can include stricter regulations for working with radiolabelled materials and appropriate safety precautions, compared to stable isotopes of C and, depending on measurement techniques and levels of enrichment, analyses that can be comparatively expensive.

Three-pool isotopic methods

The approaches described above, using Equations 17.1–17.3, allow one to conclusively separate two different sources of C – for our purposes, biochar-C and SOC. However, many natural and managed ecosystems include three or more C sources of interest – for example, biochar, soils, and plants – that all may interact with each other. Thus, the ability to conclusively separate three different sources is often desirable (Equations 17.4–17.9). For a three-part system, with biochar C, SOC, and plant-derived C:

$$f_{BC} + f_{SOC} + f_{Plant} = 1 \qquad [17.4]$$

$$\delta^{13}C_{BC-1} \cdot f_{BC} + \delta^{13}C_{SOC} \cdot f_{SOC} + \delta^{13}C_{Plant} \cdot f_{Plant}$$
$$= \delta^{13}C_{Total-1} \qquad [17.5]$$

$$\delta^{13}C_{BC-2} \cdot f_{BC} + \delta^{13}C_{SOC} \cdot f_{SOC} + \delta^{13}C_{Plant} \cdot f_{Plant}$$
$$= \delta^{13}C_{Total-2} \qquad [17.6]$$

$$f_{BC} = (\delta^{13}C_{Total-1} - \delta^{13}C_{Total-2})$$
$$/(\delta^{13}C_{BC-1} - \delta^{13}C_{BC-2}) \qquad [17.7]$$

$$f_{SOC} = ((\delta^{13}C_{Total-1} - \delta^{13}C_{Plant})$$
$$-f_{BC} \cdot (\delta^{13}C_{BC-1} - \delta^{13}C_{Plant}))$$
$$/(\delta^{13}C_{SOC} - \delta^{13}C_{Plant}) \qquad [17.8.1]$$

$$f_{SOC} = ((\delta^{13}C_{Total-2} - \delta^{13}C_{Plant})$$
$$-f_{BC} \cdot (\delta^{13}C_{BC-2} - \delta^{13}C_{Plant}))$$
$$/(\delta^{13}C_{SOC} - \delta^{13}C_{Plant}) \qquad [17.8.2]$$

$$f_{Plant} = 1 - f_{BC} - f_{SOC} \qquad [17.9]$$

where f_{BC}, f_{SOC}, and f_{Plant} represent the fraction of C contributed by biochar, SOC, and plant (or other C source) respectively; $\delta^{13}C_{Biochar}$, $\delta^{13}C_{SOC}$, and $\delta^{13}C_{Plant}$ represent the $\delta^{13}C$ signatures for biochar, SOC, and plant (or other C source), respectively; $\delta^{13}C_{BC-1}$ and $\delta^{13}C_{BC-2}$ represent two biochars that are identical except for having different $\delta^{13}C$ signatures, and $\delta^{13}C_{Total-1}$ and $\delta^{13}C_{Total-2}$ represent the $\delta^{13}C$ signature for the combined biochar, SOC, and plant for each of the two ^{13}C-labeled biochars.

Four isotopically-based approaches have been used to separate three (or more) sources (Figure 17.3). First, all three C isotopes can be used – e.g., biochar is enriched in ^{14}C, while the plants are labeled in a ^{13}C-enriched atmosphere, in soil with natural abundances of all isotopes (Cui et al, 2017; Luo et al, 2017) (Figure 17.3D). Second, a dual-label replicated treatment can be used in which one source (e.g., biochar) is produced with two distinct levels of isotopic enrichment, and the experimental treatment with that source is replicated for the two levels of enrichment (Whitman et al, 2015) (Figure 17.3C). Here, again, the expected relative contribution from all three sources needs to be considered, and sufficiently high enrichment is used to allow for the three to be differentiated. Third, in some cases (e.g., if partitioning total mass, rather than CO_2 fluxes), multiple elements could potentially be used, where each source has a different combination of – e.g., $^{12}C/^{13}C$ and $^{14}N/^{15}N$, or $^{12}C/^{13}C$ and $^{16}O/^{18}O$ (Weng et al, 2020) (Figure 17.3E). However, this approach requires the assumption that the stoichiometry of each source does not change throughout the experiment, which in many cases is incorrect. Finally, one ^{13}C-labeled source can be separated from the other two C sources, if the remaining two have the same ^{13}C signature (Figure 17.3B). Using this approach, fresh OM additions to biochar-rich Anthrosols ("*Terra preta* soils") in Brazil were found to be less readily mineralized than in adjacent low-biochar soils with identical mineralogy (Liang et al, 2010). Weng et al (2017) found that ^{13}C-labeled rhizodeposits accumulated more readily in biochar-amended soil than in unamended soil, whereas Fu et al (2022) found that ^{13}C-labeled rhizodeposits accumulated less readily in a biochar-amended soil. However, as illustrated by these studies, this approach is best suited to following fresh organic additions (e.g., litter or root-derived C), not separating biochar-C and SOC mineralization. Importantly, all of these three-component partitioning approaches will still be limited if the contribution of one source is very small.

Methods for isolating biochar-SOC interaction mechanisms

Conclusive partitioning of CO_2 fluxes or C stocks essentially requires isotopic labeling methods. However, moving beyond quantification to understanding biochar-SOC interactions requires the investigation of the mechanisms driving these effects. Because of the multiple effects of biochar on soil properties, it is likely that multiple mechanisms drive the effects of biochar on SOC mineralization in a given system. Thus, two approaches to investigating these mechanisms are to quantify the effect in isolation, or to attempt to control or eliminate the effect of interest. For example, many biochars may change soil pH

(Chapter 8), which, in turn, is known to affect soil microbial community composition (Delgado-Baquerizo et al, 2018) and activity (Malik et al, 2018). To quantify the pH-related effect in isolation, a complementary treatment could be added that produces a pH shift of the same magnitude without the addition (and other effects) of biochar (e.g., using HCl or NaOH). Additional mechanisms that have been simulated without biochar include the effects of dilution and increased surface area (DeCiucies et al, 2018). Alternately, to mask the effect of pH, researchers could adjust the pH of biochar to match that of the soil before initiating the experiment (e.g., Whitman et al, 2014). Similarly, nutrient solutions could be added to all treatments to balance any effects of biochar on specific nutrients, and moisture contents can be normalized across treatments (e.g., DeCiucies et al, 2018). Such approaches allow the researcher to isolate and focus on one specific mechanism at a time. By running sequential experiments with different mechanisms isolated or controlled for, the relative contribution of various mechanisms to biochar-SOC interactions may be estimated. However, some effects can be hard to replicate accurately without biochar additions. Additionally, mechanisms are likely interactive, so, for example, eliminating the effect of biochar on pH may change the response of the soil microbial community to biochar.

Modeling approaches for predicting long-term effects

Complemented by field observations and experimental approaches, models also play a key role in informing our understanding of the likely magnitude of the effects of biochar on SOC mineralization over time and across systems. They may also offer support for or against the likely importance of specific mechanisms. Such models highlight the importance

of understanding which mechanisms are relevant. For example, Woolf et al (2012) modified a version of the Roth C model (Coleman and Jenkinson, 1996) to include biochar additions and the effects of biochar on SOC cycling, based on an understanding of the relevant underlying mechanisms. Thus, increases in SOC mineralization rates were represented by increasing decomposition rate coefficients, whereas decreases in SOC mineralization rates were represented by changing partitioning coefficients (transferring a greater proportion of SOC to the more stable pools). In contrast, Pulcher et al (2022) simulated biochar-induced decreases in SOC mineralization by modifying the decomposition rate coefficient. These decisions affect the fundamental underlying structure of each model, and, hence, define the range of potential outcomes. At the same time, results of models such as these can help identify which poorly constrained parameters or model structures are most influential in determining future C stocks (and, hence, are most important to target for future research).

Mechanistic indicators in microbial communities

Biochar can have marked effects on microbial communities (Chapter 14). The microbial response to biochar additions can further inform our understanding of the likely mechanisms controlling the effects of biochar on SOC mineralization.

Changes in microbial biomass can provide clues as to which mechanisms may be active in a given system. For example, toxicity-related decreases in CO_2 mineralization might be accompanied by decreases in microbial biomass. "Apparent priming" – a term used to refer to increases in CO_2 emissions that are due to increased microbial biomass turnover, rather than to increased non-microbial SOC mineralization

– might be suspected if total increases in mineralized C remain smaller than the total microbial biomass C (Luo et al, 2011). The dilution effect – when microbial mineralization of SOC briefly decreases simply because there is an increase in total available C due to biochar additions – is expected to occur only over short durations before the microbial community has had sufficient time to increase in biomass to take advantage of these increases in available C. Changes in microbial biomass over time may indicate whether this mechanism is occurring. While the observation of changes in microbial biomass alone would not necessarily confirm the occurrence or absence of a specific mechanism, taken together with additional evidence, they may provide support for certain mechanisms over others.

Changes in microbial community composition or other microbial indicators, such as enzyme activities or protein abundances, can also offer information on which mechanisms are relevant in a given system. Changes in SOC mineralization due to biochar-induced changes in pH, toxicity, nutrient status, or moisture in soils could all potentially be reflected in increases or decreases in the abundance of taxa that are characteristic of those specific conditions or effects. For example, in an incubation study, increased SOC mineralization with biochar additions was accompanied by shifts in the abundance of a few specific taxa (Whitman et al, 2021). This suggested that general biological stimulation or "soil conditioning" was not likely the primary active mechanism. At broad phylogenetic scales, shifts in the ratio of bacteria to fungi with biochar additions can also reflect the direction and magnitude of biochar-induced effects on SOC mineralization. For instance, if biochar additions result in fungi being outcompeted by bacteria, this could decrease soil aggregation by fungi, while increasing bacterial SOC mineralization (Chen et al, 2021a).

Trends and mechanisms for increases in SOC mineralization

General trends

While the meta-analysis of 21 studies by Wang et al (2016) finds an overall trend toward net decreases in SOC mineralization, individual studies have observed increases in SOC mineralization of a range of magnitudes. Recent reviews and studies have helped clarify the general trends or conditions under which increased SOC mineralization with the addition of biochar is likely (Maestrini et al, 2014; Wang et al, 2016; Ding et al, 2018; DeCiucies et al, 2018; Han et al, 2020; Rasul et al, 2022). Specifically, increased SOC mineralization is more likely to occur under conditions where: (1) the relative availability of C and nutrient content of SOM is lower (or the relative availability of C and nutrient content of biochar is higher); or (2) shorter timescales are considered. The specific mechanisms by which biochar additions are likely to increase SOC mineralization are explored in detail below.

Specific mechanisms

General biological stimulation or soil conditioning

Because SOC mineralization is driven by microbial activity, any changes to the soil environment as experienced by microbes that support increased activity could also lead to increased SOC mineralization, sometimes termed "soil conditioning" (Zimmerman and Ouyang, 2019). For example, this could include changes to the physical environment that increase oxygen levels in the soil, or either increase or decrease moisture (Chapter 20) to levels optimal for microbial activity (Figure 17.4C). Because biochar can, in some cases, decrease biological exposure to toxic substances such as heavy metals (Chapter 21) or other environmental contaminants (Chapter 27), biochar applications in these circumstances could also increase microbial mineralization of SOC (Figure 17.4A). Alleviation of pH-related constraints offers another important mechanism for increasing microbial mineralization of SOC with biochar additions (Figure 17.4B). Biochar additions can shift soil pH in many systems (Chapters 8 and 13), and pH is a fundamental control of microbial community composition and activity (Rousk et al, 2009; Delgado-Baquerizo et al, 2018). For example, a much stronger relationship between total soil C and C use efficiency (C mineralized per unit C incorporated into microbial biomass) was observed in soils with pH greater than 6.2 (Malik et al, 2018), suggesting that microbial communities in soils with pH below 6.2 were likely growth-limited by acidity. These observations suggest that the amendment of acidic soils with basic biochar could readily be expected to increase microbial growth and activity. Consistent with these observations, biochar additions increased soil pH, resulting in shifts in microbial community composition, and decreases in C use efficiency (Chen et al, 2019). Similarly, biochar additions to acidic and basic soils increased and decreased SOC mineralization, respectively (Sheng and Zhu, 2018). Depending on the exact mechanisms through which soil pH controls microbial activity, these effects might be expected to persist over the same timescales of any pH shifts.

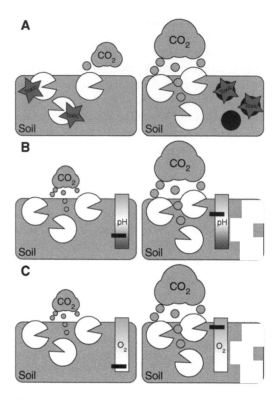

Figure 17.4 *Illustration of three different mechanisms by which biochar may increase SOC mineralization due to changes in the chemical or physical environment. Circles with open mouths represent soil microbes. Black circles represent biochar, with pre-addition on the left, and post-addition on the right. (A) Biochar sorbs or otherwise decreases the bioavailability of toxic substances (represented by stars) such as heavy metals. (B) Biochar shifts pH to a level more favorable for microbial activity. (C) An example of soil conditioning, where biochar increases oxygen availability in soil, increasing microbial activity. Other forms of soil conditioning could include effects on soil moisture or structure*

Effects from nutrients and C in biochar

Biochar can contain meaningful amounts of nutrients (Chapter 8). In cases where microbial activity is limited by nutrients present in biochar, adding biochar could potentially alleviate nutrient limitation and stimulate microbial activity, thus increasing SOC mineralization (Figure 17.5A). However, as discussed later, the converse may be more common, whereby alleviation of nutrient limitation reduces SOC mineralization for nutrient mining. Testing for this mechanism is relatively straightforward, and could be at least approximated by adding nutrients similar in composition to the nutritional content of biochar on their own, and determining if an equivalent effect size is observed. Somewhat conversely, if added biochar has relatively high levels of available C, but low contents of other essential nutrients, microbial "mining" of SOM for these missing nutrients could occur, increasing SOC mineralization at the same time (Figure 17.5B). The degree to which this effect is relevant could be assessed by again adding available nutrients, this time along with the biochar, and determining whether this prevents the increase in SOC mineralization.

Perhaps most importantly, despite large portions of total biochar C being relatively unusable by or unavailable to microbes, there are portions of biochar that are readily mineralizable (Chapter 11) – much of which is present in the water-soluble fraction (Whitman et al, 2014). In a system where C is limiting for microbes, the addition of easily mineralizable biochar C can result in generally increased microbial activity (Figure 17.6A). In this case, microbial biomass and production of extracellular enzymes increase in response to the readily available biochar C fraction, thus increasing the mineralization of easily available SOC. Co-metabolism (Figure 17.6B) may also occur, which is distinguished from general increases in microbial activity in that biochar serves as the primary energy source during the mineralization of certain SOM compounds, although these two effects may be difficult to distinguish experimentally. These mechanisms mean that soils with relatively low SOC and low microbial activities initially, such as those at deeper soil depths (Santos et al, 2021) are more likely to experience boosted microbial activities, increasing SOC losses with biochar additions (Whitman et al, 2021). Increases in SOC mineralization due to easily mineralizable biochar C usually diminish over a timescale of days to weeks, as most of the readily mineralizable C is exhausted. Thus,

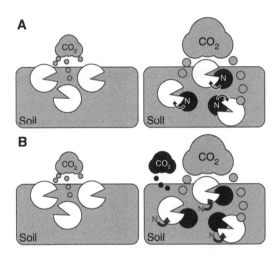

Figure 17.5 *Illustration of two different mechanisms by which the addition of biochar may increase SOC mineralization, due to alleviation of nutrient limitation and microbial mining of SOM. Circles with open mouths represent soil microbes. Black circles represent biochar, with pre-addition on the left, and post-addition on the right. (A) Biochar alleviates a nutrient limitation (here, represented by N, but could be any limiting nutrients), increasing microbial SOC mineralization. (B) Biochar increases nutrient demand (here, represented by N, but could be other limiting nutrients) by microbes, resulting in the mining of soil organic matter for the nutrient and increasing SOC mineralization in the process*

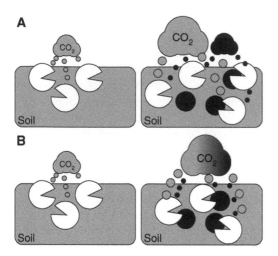

Figure 17.6 *Illustration of two different mechanisms by which adding biochar may increase SOC mineralization. Circles with open mouths represent soil microbes. Black circles represent biochar, with pre-addition on the left, and post-addition on the right. (A) A portion of biochar C is easily mineralized, generally increasing microbial activity thus increasing SOC mineralization. (B) Specific molecules in biochar C allow for co-metabolism, where biochar compounds are mineralized along with SOC and increase microbial SOC mineralization*

understanding what proportion of biochar-C and SOC are readily mineralizable by microbes is critical to understanding the effect of biochar on SOC mineralization. Techniques such as Fourier-transformed ion cyclotron resonance mass spectrometry (FT-ICR MS), have allowed researchers to trace the shift in dissolved organic matter (DOM) chemistry with much higher resolution. Using FT-ICR MS, increased SOC losses with biochar additions could be associated with more persistent DOM components (such as condensed aromatics and tannins) rather than the less persistent components (such as carbohydrates and amino sugars) (Ling et al, 2021). These observations would be consistent with co-metabolism or increased microbial activity mechanisms – if soil already has abundant easily mineralizable C, additions of biochar C are expected to have less of an effect. There may also be counteracting or additive effects of biochar C on SOC mineralization – to the extent that SOM helps hold aggregates in place, thereby protecting SOM from microbial activity, increased degradation of SOM might be predicted to destabilize aggregates, resulting in further losses of SOC. For example, biochar additions decreased the stabilization of root-derived C due to diminished fungal-promoted aggregation, leading to increased SOC mineralization (Chen et al, 2021a).

Trends and mechanisms for decreases in SOC mineralization

General trends

Recent reviews and studies have improved the understanding of the conditions under which decreased SOC mineralization may occur with the addition of biochar (Maestrini et al, 2014; Wang et al, 2016; Ding et al, 2018; DeCiucies et al, 2018; Han et al, 2020; Rasul et al, 2022). Specifically, these effects are more likely to be observed under conditions where: (1) the relative availability of C and nutrient content of SOM is high (or the relative availability of C and nutrient content of biochar is low); (2) biochar has high sorption potential, often related to high surface area; or (3) long timescales (on the order of months to years) are considered. In a meta-analysis of 21 studies, Wang et al. (2016) found biochar decreased net SOC mineralization by a mean of 3.8%, but this mean result across studies encompassed myriad conditions and mechanisms. The specific mechanisms by which biochar may decrease SOC mineralization are explored in detail below.

Specific mechanisms

General biological inhibition

Because organic matter mineralization is driven by microbial and faunal activity, any soil physicochemical changes that impair these organisms, such as shifts in pH or toxicity due to specific elements or compounds, will also be expected to decrease SOC mineralization. Soil pH (Figure 17.7A) is a dominant factor that structures microbial community composition (Lauber et al, 2009; Delgado-Baquerizo et al, 2018; Braus and Whitman, 2021), and it can have important effects on microbial growth and activity (Rousk et al, 2009). If applications of biochar shift pH away from optimal levels for microbial activity in a given soil, this could also decrease SOC mineralization (Fernández-Calviño et al, 2011).

Direct toxicity (Figure 17.7B) could be a second potential source for generally decreased biological activity in soils. Potential properties of biochar with negative effects on soil biota (detailed in Chapter 14) include high salinity, polyaromatic hydrocarbons and free radicals (Liao et al, 2014; Odinga et al, 2020) (Chapter 22), or heavy metals (Chapter 21). Directly assessing microbial biomass and soil fauna populations would help establish whether this is a likely mechanism for decreased SOC mineralization in a given system.

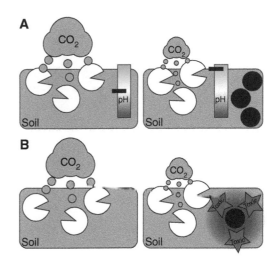

Figure 17.7 *Illustration of two different mechanisms by which the addition of biochar may decrease SOC mineralization, due to changes in the physicochemical environment. Circles with open mouths represent soil microbes. Black circles represent biochar, with pre-addition on the left, and post-addition on the right. (A) Addition of biochar shifts pH to a level less favorable for microbial activity. (B) Addition of biochar is accompanied by toxic substances such as heavy metals*

Effects from nutrients and C in biochar

Changes to the relative availability of nutrients or C after biochar additions to soil can decrease SOC mineralization. For C availability, there are two relatively straightforward mechanisms – substrate switching and the dilution effect – which have a subtle distinction. Briefly, in both of these cases, biochar C replaces SOC for microbial use. The key difference is that microbes preferentially use easily available biochar C over SOC under substrate switching (Figure 17.8A), whereas neither biochar C nor SOC is preferred under the dilution effect (Figure 17.8B). During the dilution effect, the influx of additional C in biochar results – usually only over short

A

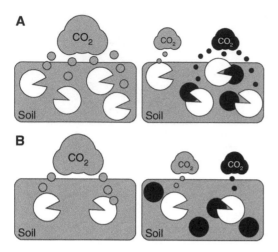

B

Figure 17.8 *Illustration of two different mechanisms by which the addition of biochar may decrease SOC mineralization, due to substrate switching and the dilution effect. Circles with open mouths represent soil microbes. Black circles represent biochar, with pre-addition on the left, and post-addition on the right. (A) Substrate switching: Addition of easily mineralizable biochar C results in decreased SOC mineralization as microbes prefer biochar. (B) Dilution effect: Shortly after additions of easily mineralizable biochar, microbial biomass has not yet sufficiently increased to respond to biochar, and results in a usually temporary reduction in SOC mineralization*

periods before the microbial community has grown to take advantage of the additional C – in microbes and their extracellular enzymes acting on an increased total amount of C without a concomitant increase in microbial biomass or activity. Effectively, the SOC is "diluted" with biochar C. Both of these mechanisms are expected to be relatively short-lived (~days), as the easily mineralizable portion of biochar C that is preferred or equivalent to SOC is rapidly consumed by microbes. Thus, these mechanisms may be inferred in cases where reduced SOC mineralization occurs early but briefly after the

application of biochar, and would be more likely in cases where there are large amounts of readily-mineralizable biochar C, due to either high application rates or the properties of the specific biochar (see Chapter 11 for a discussion of the properties that affect biochar mineralizability). For example, applications of the water-extractable portion of biochar to soil resulted in decreased SOC mineralization which could be due to substrate switching (Chen et al, 2021b).

If added biochar contains limiting nutrients that were previously being "mined" from SOM by microbes, this could decrease SOC mineralization over the short term (days to weeks). This mechanism is invoked in the body of literature that has reported decreased SOC mineralization with mineral N additions (Ramirez et al, 2012). However, although theoretically possible, this mechanism is unlikely to be a dominant one for biochar-altering SOC mineralization. This is because large amounts of readily available mineral N are not commonly found in biochar (Chapter 8). Although there can be a higher relative amount of N than C remaining in biochar as production temperatures increase (Torres-Rojas et al, 2020), large portions of N in organic matter volatilize at temperatures above 200°C (Gray and Dighton, 2006), and N remaining in biochar tends to be found in poorly-mineralizable heterocyclic structures (Torres-Rojas et al, 2020). That said, as outlined in Chapter 16, biochar can affect nutrient availability in many different ways, highlighting the complexities of biochar-soil systems and the need for comprehensive and mechanistic approaches to explain elemental cycling in these systems.

Effects from physiochemical properties of biochar

Biochar is noted for its ability to sorb diverse compounds, including organic and inorganic molecules (Chapters 16, 19, and 21). To the extent that biochar sorption decreases the microbial availability or accessibility of a given molecule, it may also decrease its mineralization. The primary mechanisms can be divided into the sorption of SOC itself, making it less available, and sorption of other molecules that microbes require for SOC mineralization, such as the sorption of microbial products such as enzymes or signaling molecules, and the sorption of microbially-required nutrients.

The sorption of SOC by biochar has been reported to decrease SOC mineralization (Figure 17.9A). These studies include what are effectively anecdotal observations of SOC sorbed on biochar surfaces using nano-scale secondary ion mass spectrometry (nanoSIMS) (Whitman et al, 2014; DeCiucies et al, 2018), as well as more quantitative measurements. For example, quantified DOC sorption potential by biochar also induced reductions in SOC mineralization, although the degree of reduction in SOC mineralization was not directly proportional to the sorptive potential (DeCiucies et al, 2018), indicating multiple mechanisms may be acting simultaneously. The sorptive protection effect of biochar is presumably because sorbed substrates are less accessible to microbially produced exoenzymes than those in their free state (Zimmerman et al, 2004).

Decreased SOC mineralization with biochar additions to soil could also result from increased SOC protection and stabilization mechanisms. For example, if biochar increases stable aggregate formation (Zheng et al, 2018), this could lead to the increased physical protection of SOC from microbial degradation. This mechanism could serve as a potential explanation for decreased mineralization of SOC and added sugarcane residue with the addition of biochar in a Vertisol (Fang et al, 2019), and why biochar affected SOC mineralization differently in different

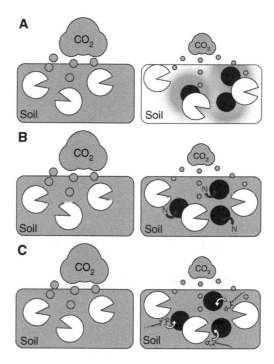

Figure 17.9 *Illustration of three different mechanisms by which the addition of biochar may decrease SOC mineralization. Circles with open mouths represent soil microbes. Black circles represent biochar, with pre-addition on the left, and post-addition on the right. (A) Biochar sorbs SOC, decreasing its microbial availability and resulting in decreased SOC mineralization. (B) Biochar sorbs nutrients, decreasing microbial activity and SOC mineralization. (C) Biochar sorbs microbial molecules, such as acyl-homoserine lactones, impairing microbial communication, activity, and SOC mineralization*

aggregate-size fractions (Zheng et al, 2021). Similarly, if biochar additions increase interactions between SOC and minerals (Chapter 16), this could also lead to increased persistence of SOC. Standard fractionation approaches for quantifying the relative amount of SOC in different (generally methodologically defined) soil fractions may be able to offer evidence for the relevance of these mechanisms.

Biochar can affect nutrient availability – most often assessed in the context of availability to plants, but likely with implications for availability to microbes as well (Chapters 16 and 19). Thus, it stands to reason that the sorption of nutrients by biochar could potentially reduce their availability and, hence, limit microbial activity (Figure 17.9B). That said, interactions between nutrient availability and growth, which are already complex in agricultural systems with a single plant species, are even more difficult to parse when considering entire microbial communities that span domains of life, since different organisms have wide-ranging optima and tolerance levels for diverse environmental conditions. Conditions that might be limiting for one species may be ideal for another, and since there is high functional redundancy within soil microbial communities for the fundamental function of C mineralization, this mechanism may be both unlikely and difficult to conclusively demonstrate. Biochar has also been demonstrated to sorb other microbially-relevant molecules, such as extracellular enzymes (Lammirato et al, 2011) or signaling molecules such as acyl-homoserine lactones (Masiello et al, 2013; Gao et al, 2016) (Figure 17.9C). However, consistently predicting the effect of biochar sorption of microbial molecules on SOC mineralization would likely be difficult, again due to the complexity of microbial communities and their wide-ranging potential activities.

Multiple mechanisms acting simultaneously

Although we considered each mechanism separately in the two sections above, it is essential to remember that multiple mechanisms are likely acting simultaneously to produce observed net effects on SOC mineralization. For example, the common observation of increased SOC mineralization switching to net decreases in mineralization over time is often interpreted as a shift from microbial stimulation or co-metabolism increasing SOC mineralization rates, to sorption of SOC on biochar surfaces decreasing SOC mineralization rates (Zimmerman et al, 2011; DeCiucies et al, 2018). However, there is no reason to think that sorption does not also occur initially – rather, the net effect of all mechanisms is just initially positive. Using approaches such as those outlined in the Methods and Approaches section to distinguish between and quantify the relative importance of each mechanism will help parse and, ultimately, explain the temporal and spatial variation in the net effects of biochar on SOC mineralization.

As we expand system boundaries to include additional C sources, such as plant roots or inorganic soil C, it is essential to recognize that potential interactions between all these components also become increasingly complex. To predict how different plant-biochar-soil systems might be expected to translate to changes in net SOC mineralization (Whitman et al, 2014), the effects of biochar on plants (Chapter 13) and the substantial body of non-biochar priming research must be considered. To predict the effects of biochar on inorganic C, a strong understanding of soil chemistry and the effects of biochar on soil properties (as detailed throughout this book) will be required. Critically, approaches that allow the researcher to conclusively distinguish between the different sources of C in the system are essential to developing a clear understanding of these increasingly complex multi-component systems.

Future directions

Notable progress has been made over the last fifteen years or so in understanding how biochar can affect SOC mineralization. An improved understanding of the mechanisms that are most important has emerged and general trends have been established, such as the typical timescales over which increased vs. decreased SOC mineralization effects are likely to occur. Going forward, we offer several important future research directions:

1 **Improving predictive understanding:** Predictive abilities will be improved as we further deepen our mechanistic understanding of biochar effects on SOC mineralization through targeted experiments and modeling.

2 **Assessing long-term effects:** As is the case for many phenomena, the long-term (decadal to centennial) effects of biochar additions to soil remain less well-characterized, including the effects of repeated amendments over time (e.g., Luo et al, 2018) and the effects of aging (Chapter 10) on SOC-biochar interactions (e.g., Yang et al, 2022) as well as how interactions may be affected by changing climate (e.g., CO_2 levels (Pei et al, 2020) or temperatures (Chen et al, 2021b; Fang et al, 2017)).

3 **Addressing laboratory constraints:** As laboratory incubation conditions (particularly duration) and biochar and soil properties explain a large portion (72%) of variation in SOC response to biochar additions (Ding et al, 2018), an iterative experimental–modeling approach may help make progress in this area.

4 **Improving biological understanding:** The specific organisms involved with biochar-SOC interactions – including microbes, soil fauna, and plant roots – and how they determine or are themselves affected by these interactions remains an area ripe for exploration (e.g., Yu et al, 2020).

5 **Evaluating biochar effects on inorganic C:** Soils contain large stocks of soil inorganic C (SIC), but the cycling of this important stock is often overlooked (Monger, 2014). Despite its perception as a less readily manipulated or affected stock of C (i.e., part of the "slow C cycle"), inorganic C stocks can be altered by soil management decisions. For example, N additions can have important effects on inorganic C cycling (Zamanian et al, 2018). Wang et al (2023) report increased SIC in a calcareous subsoil after biochar application to the topsoil. The effects of biochar on SIC, and the related biological, physical, and chemical mechanisms, should be further studied.

6 **Increasing complexity:** In general, increasing the complexity of systems under investigation to approach realistic scenarios will require thoughtful design, but should yield important advances in our understanding.

7 **Customizing biochar design:** Given the interest in using biochar for C management, leveraging our emerging understanding of which biochar properties and which biochar-soil combinations are most likely to result in increased vs. decreased SOC losses, or formulating

biochar and biochar management systems for specific purposes including SOC management, may become increasingly achievable in the future.

The complexity of soils, ecosystems, and biochar makes the prediction of biochar effects on soil C dynamics challenging. However, through the systematic application of experimentation, observation, and modeling, we are optimistic that the field will continue to progress toward an increasingly predictive understanding of these critical interactions.

References

Balesdent J, and Mariotti A 1996 Measurement of soil organic matter turnover using ^{13}C natural abundance. In: Boutton TW, and Yamasaki SI (Eds) *Mass Spectrometry of Soils*. New York: Marcel Dekker. pp83–111.

Bingeman CW, Varner JE, and Martin WP 1953 The effect of the addition of organic materials on the decomposition of an organic soil. *Soil Science Society of America Journal* 17, 34–38.

Bird MI, et al 2017 Loss and gain of carbon during char degradation. *Soil Biology and Biochemistry* 106, 80–89.

Blanco-Canqui H, Laird DA, Heaton EA, Rathke S, and Acharya BS 2020 Soil carbon increased by twice the amount of biochar carbon applied after 6 years: Field evidence of negative priming. *Global Change Biology Bioenergy* 12, 240–251.

Braus MJ, and Whitman TL 2021 Standard and non-standard measurements of acidity and the bacterial ecology of northern temperate mineral soils. *Soil Biology and Biochemistry* 160, 108323.

Broadbent FE, and Norman AG 1947 Some factors affecting the availability of the organic nitrogen in soil—A preliminary report. *Soil Science Society of America Journal* 11, 264–267.

Cardelli R, Becagli M, Marchini F, and Saviozzi A 2016 Short-term releases of CO_2 from newly mixed biochar and calcareous soil. *Soil Use and Management* 32, 543–545.

Chen L, et al 2019 Competitive interaction with keystone taxa induced negative priming under biochar amendments. *Microbiome* 7, 77.

Chen Z, et al 2021a Biochar decreased rhizodeposits stabilization via opposite effects on bacteria and fungi: diminished fungi-promoted aggregation and enhanced bacterial mineralization. *Biology and Fertility of Soils* 57, 533–546

Chen G., Fang Y, Van Zwieten L, Xuan Y, Tavakkoli E, Wang X, and Zhang R 2021b Priming, stabilization and temperature sensitivity of native SOC is controlled by microbial responses and physicochemical properties of biochar. *Soil Biology and Biochemistry* 154, 108139.

Coleman K, and Jenkinson DS 1996 RothC-26.3 – a model for the turnover of carbon in soil. In: Powlson DS, Smith P, and Smith JU (Eds) *Evaluation of Soil Organic Matter Models. NATO ASI Series*, vol 38. Berlin, Heidelberg: Springer. pp237–246.

Cui J, et al 2017 Interactions between biochar and litter priming: A three-source ^{14}C and δ^{13}C partitioning study. *Soil Biology and Biochemistry* 104, 49–58.

DeCiucies S, Whitman T, Woolf D, Enders A, and Lehmann J 2018 Priming mechanisms with additions of pyrogenic organic matter to soil. *Geochimica et Cosmochimica Acta* 238, 329–342.

Delgado-Baquerizo M, et al 2018 A global atlas of the dominant bacteria found in soil. *Science* 359, 320–325.

Ding F, et al 2018 A meta-analysis and critical evaluation of influencing factors on soil carbon priming following biochar amendment. *Journal of Soils and Sediments* 18, 1507–1517.

Fang Y, Singh BP, Matta P, Cowie AL, and Zwieten LV 2017 Temperature sensitivity and priming of organic matter with different stabilities in a Vertisol with aged biochar. *Soil Biology and Biochemistry* 115, 346–356.

Fang Y, et al 2019 Interactive carbon priming, microbial response and biochar persistence in a Vertisol with varied inputs of biochar and labile organic matter *European Journal of Soil Science* 70, 960–974.

Fernández-Calviño D, Rousk J, Brookes PC, and Bååth E 2011 Bacterial pH-optima for growth track soil pH, but are higher than expected at low pH. *Soil Biology and Biochemistry* 43, 1569–1575.

Fontaine S, et al 2007 Stability of organic carbon in deep soil layers controlled by fresh carbon supply. *Nature* 450, 277–280.

Fu Y, et al 2022 Biochar accelerates soil organic carbon mineralization via rhizodeposit-activated *Actinobacteria*. *Biology and Fertility of Soils* 58, 565–577.

Gao X, et al 2016 Charcoal disrupts soil microbial communication through a combination of signal sorption and hydrolysis. *ACS Omega* 1, 226–233.

Gray DM, and Dighton J 2006 Mineralization of forest litter nutrients by heat and combustion. *Soil Biology and Biochemistry* 38, 1469–1477.

Han L, Sun K, Yang Y, Xia X, Li F, Yang Z, and Xing B 2020 Biochar's stability and effect on the content, composition and turnover of soil organic carbon. *Geoderma* 364, 114184. 10.1016/j.geoderma.2020.114184.

Janzen HH 2006 The soil carbon dilemma: Shall we hoard it or use it? *Soil Biology and Biochemistry* 38, 419–424.

Ji X, et al 2019 Preferential alternatives to returning all crop residues as biochar to the crop field? A three-sources ^{13}C and ^{14}C partitioning study. *Journal of Agricultural and Food Chemistry* 67, 11322–11330.

Kuzyakov Y, Subbotina I, Chen H, Bogomolova I, and Xu X 2009 Black carbon decomposition and incorporation into soil microbial biomass estimated by ^{14}C labeling. *Soil Biology and Biochemistry* 41, 210–219.

Lammirato C, Miltner A, and Kaestner M 2011 Effects of wood char and activated carbon on the hydrolysis of cellobiose by β-glucosidase from Aspergillus niger. *Soil Biology and Biochemistry* 43, 1936–1942. 10.1016/j.soilbio.2011.05.021.

Lauber CL, Hamady M, Knight R, and Fierer N 2009 Pyrosequencing-based assessment of soil pH as a predictor of soil bacterial community structure at the continental scale. *Applied and Environmental Microbiology* 75, 5111–5120.

Liang B, et al 2010 Black carbon affects the cycling of non-black carbon in soil. *Organic Geochemistry* 41, 206–213.

Liao S, Pan B, Li H, Zhang D, and Xing B 2014 Detecting free radicals in biochars and determining their ability to inhibit the germination and growth of corn, wheat and rice seedlings. *Environmental Science and Technology* 48, 8581–8587.

Ling L, et al 2021 Organic matter chemistry and bacterial community structure regulate decomposition processes in post-fire forest soils. *Soil Biology and Biochemistry* 166, 108311.

Löhnis F 1926 Nitrogen availability of green manures. *Soil Science* 22, 253–290.

Luo Y, Durenkamp M, De Nobili M, Lin Q, and Brookes PC 2011 Short term soil priming effects and the mineralisation of biochar following its incorporation to soils of different pH. *Soil Biology and Biochemistry* 43, 2304–2314.

Luo Y, et al 2017 Priming effects in biochar enriched soils using a three-source-partitioning approach: [14]C labelling and ^{13}C natural abundance. *Soil Biology and Biochemistry* 106, 28–35.

Luo Y, Lin Q, Durenkamp M, and Kuzyakov Y 2018 Does repeated biochar incorporation induce further soil priming effect? *Journal of Soils and Sediments* 18, 128–135.

Maestrini B, Nannipieri P, and Abiven S 2014 A meta-analysis on pyrogenic organic matter induced priming effect. *Global Change Biology Bioenergy* 7, 577–590.

Malik AA, et al 2018 Land use driven change in soil pH affects microbial carbon cycling processes. *Nature Communications* 9, 3591.

Masiello CA, et al 2013 Biochar and microbial signaling: production conditions determine effects on microbial communication.

Environmental Science and Technology 47, 11496–11503.

Monger HC 2014 Chapter 3: Soils as generators and sinks of inorganic carbon in geologic time. In: Hartemink AE, and McSweeney K (Eds) *Soil Carbon. Progress in Soil Science.* Switzerland: Springer International Publishing. pp27–36.

Odinga ES, et al 2020 Occurrence, formation, environmental fate and risks of environmentally persistent free radicals in biochars. *Environment International* 134, 105172.

Pei J, et al 2020 Biochar-induced reductions in the rhizosphere priming effect are weaker under elevated CO_2. *Soil Biology and Biochemistry* 142, 107700.

Pulcher R, Balugani E, Ventura M, Greggio N, and Marazza D 2022 Inclusion of biochar in a C dynamics model based on observations from an 8-year field experiment. *Soil* 8, 199–211.

Ramirez KS, Craine JM, and Fierer N 2012 Consistent effects of nitrogen amendments on soil microbial communities and processes across biomes. *Global Change Biology* 18, 1918–1927.

Rasul M, Cho J, Shin H-S, and Hur J 2022 Biochar-induced priming effects in soil via modifying the status of soil organic matter and microflora: A review. *Science of the Total Environment* 805, 150304.

Rousk J, Brookes PC, and Bååth E 2009 Contrasting soil pH effects on fungal and bacterial growth suggest functional redundancy in carbon mineralization. *Applied and Environmental Microbiology* 75, 1589–1596.

Santos F, Rice DM, Bird JA, and Berhe AA 2021 Pyrolysis temperature and soil depth interactions determine PyC turnover and induced soil organic carbon priming. *Biogeochemistry* 153, 47–65. 10.1007/s10533-021-00767-x.

Sheng Y, and Zhu L 2018 Biochar alters microbial community and carbon sequestration potential across different soil pH. *Science of the Total Environment* 622, 1391–1399.

Tilston EL, Ascough PL, Garnett MH, and Bird MI 2016 Quantifying charcoal degradation and negative priming of soil organic matter with a ^{14}C-dead tracer. *Radiocarbon* 58, 905–919.

Torres-Rojas D, et al 2020 Nitrogen speciation and transformations in fire-derived organic matter. *Geochimica et Cosmochimica Acta* 276, 170–185.

Wang J, Xiong Z, and Kuzyakov Y 2016 Biochar stability in soil: meta-analysis of decomposition and priming effects. *Global Change Biology Bioenergy* 8, 512–523.

Wang Y, et al 2023 Inducing inorganic carbon accrual in subsoil through biochar application on calcareous topsoil. *Environmental Science and Technology* 57, 1837–1847.

Wardle DA, Nilsson MC, and Zackrisson O 2008 Fire-derived charcoal causes loss of forest humus. *Science* 320, 629–629.

Weng ZH, et al 2017 Biochar built soil carbon over a decade by stabilizing rhizodeposits. *Nature Climate Change* 7, 371–376.

Weng ZH, et al 2020 Priming of soil organic carbon induced by sugarcane residues and its biochar control the source of nitrogen for plant uptake: A dual ^{13}C and ^{15}N isotope three-source-partitioning study. *Soil Biology and Biochemistry* 146, 107792.

Whitman T, Enders A, and Lehmann J 2014 Pyrogenic carbon additions to soil counteract positive priming of soil carbon mineralization by plants. *Soil Biology and Biochemistry* 73, 33–41.

Whitman T, Zhu Z, and Lehmann J 2014 Carbon mineralizability determines interactive effects on mineralization of pyrogenic organic matter and soil organic carbon. *Environmental Science and Technology* 48, 13727–13734.

Whitman T, and Lehmann J 2015 A dual-isotope approach to allow conclusive partitioning between three sources. *Nature Communications* 6, 8708.

Whitman T, et al 2021 Microbial community shifts reflect losses of native soil carbon with pyrogenic and fresh organic matter additions and are greatest in low-carbon soils. *Applied and Environmental Microbiology* 87, 1–38.

Woolf D, and Lehmann J 2012 Modelling the long-term response to positive and negative priming of soil organic carbon by black carbon. *Biogeochemistry* 111, 83–95.

Yang Y, et al 2022 Biochar stability and impact on soil organic carbon mineralization depend on biochar processing, aging and soil clay content. *Soil Biology and Biochemistry* 169, 108657.

Yu Z, Ling L, Singh BP, Luo Y, and Xu J 2020 Gain in carbon: deciphering the abiotic and biotic mechanisms of biochar-induced negative priming effects in contrasting soils. *Science of the Total Environment* 746, 141057.

Zamanian K, Zarebanadkouki M, and Kuzyakov Y 2018 Nitrogen fertilization raises CO_2 efflux from inorganic carbon: A global assessment. *Global Change Biology* 24, 2819-2817.

Zheng H, Wang X, Luo X, Wang Z, and Xing B 2018 Biochar-induced negative carbon mineralization priming effects in a coastal wetland soil: roles of soil aggregation and microbial modulation. *Science of the Total Environment* 610–611, 951–960.

Zheng T, Zhang J, Tang C, Liao K, and Guo L 2021 Positive and negative priming effects in an Ultisol in relation to aggregate size class and biochar level. *Soil and Tillage Research* 208, 104874.

Zimmerman AR, Chorover J, Goyne KW, and Brantley SL 2004 Protection of mesopore-adsorbed organic matter from enzymatic degradation. *Environmental Science and Technology* 38, 4542–4548.

Zimmerman AR, Gao B, and Ahn M-Y 2011 Positive and negative carbon mineralization priming effects among a variety of biochar-amended soils. *Soil Biology and Biochemistry* 43, 1169–1179.

Zimmerman AR, and Ouyang L 2019 Priming of pyrogenic C (biochar) mineralization by dissolved organic matter and vice versa. *Soil Biology and Biochemistry* 130, 105–112.

18

Biochar influences methane and nitrous oxide emissions from soil

Lukas Van Zwieten, Maria Luz Cayuela, Claudia Kammann,
Stephen Joseph, Nicole Wrage-Mönnig, Annette Cowie,
Niloofar Karimian, and Ehsan Tavakkoli

Introduction: Non-CO$_2$ greenhouse gas emissions and the role of biochar

Soil is a source and sink for greenhouse gases (GHGs). Besides carbon dioxide (CO$_2$), there are direct GHGs: nitrous oxide (N$_2$O) and methane (CH$_4$); and indirect GHGs such as NH$_3$ (Schlesinger et al 1992) and nitric oxide (NO) (Davidson and Kingerlee, 1997). Nitrous oxide is a potent greenhouse gas with a global warming potential (GWP), according to the IPCC Sixth Assessment Report, of 273 times greater than CO$_2$. It is also an important ozone-depleting compound. Its atmospheric concentration has increased from 270 parts per billion by volume (ppbv) in the pre-industrial era to over 330 ppbv in 2021 (Nisbet et al, 2021). Human-induced emissions (both direct and indirect) are dominated by the use of N fertilizers in agriculture (reactive N-N$_r$). This has resulted in a 30% increase in emissions of N$_2$O over the past four decades, to 7.3 Tg annually (Tian et al, 2020). Agriculture is also responsible for around 90% of the global NH$_3$ emissions

(Galloway et al, 2008) with sources being animal manures and NH$_3$ fertilizers (Zhu et al, 2015). The volatilization of NH$_3$ can transport N$_r$ over large distances where its deposition can result in non-point source emissions of N$_2$O via microbial N cycling. Nitrogen oxides (NOx) play a role in the photochemical synthesis of tropospheric ozone (Atkinson, 2000), contributing to radiative forcing and the greenhouse effect. The redeposition of NOx from the atmosphere to soil is also a non-point source for the microbial production of N$_2$O. The flux (production and consumption) of N$_2$O from soil is controlled by biological and abiotic processes including: (i) chemical decomposition of hydroxylamine during autotrophic and heterotrophic nitrification; (ii) chemodenitrification of soil NO$_3^-$ and abiotic decomposition of ammonium nitrate (NH$_4$NO$_3$) in the presence of light, humidity and reactive surfaces; (iii) nitrifier denitrification by the same

DOI: 10.4324/9781003297673-18

nitrifiers; (iv) denitrification by organisms capable of using nitrogen oxides as alternative electron acceptors under oxygen (O_2)-limiting environmental conditions; (v) co-denitrification of organic N compounds with NO; and (vi) NO_3^- ammonification or dissimilatory nitrate reduction to NH_4^+ (Butterbach-Bahl et al, 2013). The role of the widely distributed complete ammonia oxidizer bacteria (comammox), able to individually oxidize ammonia to nitrate via nitrite (Han et al, 2021), and ammonia-oxidizing archaea (AOA) and their role in N_2O production (Wu et al, 2020) have also been studied.

Methane has a GWP of 28 times greater than CO_2 (IPCC Fifth Assessment Report). Methane was found at a concentration of 1893 ppbv in the atmosphere in 2021 (Nisbet et al, 2021), more than double its concentration from the pre-industrial era. Agriculture is the largest anthropogenic source of CH_4 with 145 Tg released into the atmosphere in 2017 (Smith et al, 2021) from sources including enteric fermentation, manure management, rice cultivation, and the in-field combustion of residues such as crop straw. Net CH_4 exchange between soils and the atmosphere is controlled by the ubiquitous methanogenic (methane-producing) *Archaea* and methanotrophic (methane-consuming) bacteria. Methane production takes place in anoxic environments where organic C is microbially

degraded (Kammann et al, 2009). Net CH_4 consumption results from the activity of methanotrophic α- and γ-proteobacteria that use CH_4 as the sole C source and require O_2 to function (Conrad, 2007) and is largely determined by the soil gas diffusivity and by methanotrophic activity. Methane consumption and production processes are antagonistic with the net flux determining whether the ecosystem is a source or sink for CH_4 (Kammann et al, 2009).

A broad base of literature exists on the role of biochar in lowering direct GHG emissions, while information is emerging on the effects on indirect GHG emissions. Biochar can influence GHG emissions from the soil through direct interactions with these gases, including sorption and reaction with biochar structure, as well as modifying soil properties that control many of these processes. These soil properties include soil water content, pH, redox status, availability of substrates needed for GHG production such as readily decomposable C and N_r, soil pH, CEC and the soil composition, and activity of the microbial community, which can influence both direct and indirect GHG emissions. This chapter critically assesses biochar for its ability to modify non-CO_2 soil GHG emissions, elucidates key mechanisms under different agricultural systems, and makes recommendations on key areas for future research and policy development.

Evidence for decreased GHG emissions

A wide range of effects of biochar on N_2O and CH_4 emissions have been found (Table 18.1), ranging from small or non-significant changes to both N_2O and CH_4, to a 49% reduction in N_2O emissions (i.e., -49%). The meta-analyses included data from both dryland and paddy studies (usually

reporting these separately) (incorporating laboratory, glasshouse, and field studies) across a wide range of soil types and climates. Incubation or laboratory studies, which tend to be of shorter-term duration, and often use large application rates of biochar, found greater mitigation of N_2O

Table 18.1 *Summary of meta-analyses of effect sizes of biochar on nitrous oxide and methane emissions. Meta-analyses were selected considering quality criteria according to Grados et al (2022). Only meta-analyses using lnRR as effect size were included*

	System	Effect size N_2O (% change)	*No. N_2O	Effect size CH_4 (% change)	*No. CH_4	Notes/summary of key findings
Ji et al, 2018	Paddy soils glasshouse and field	na	na	−12	77	The results of this synthesis suggest that the role of biochar in soil CH_4 mitigation might be overestimated, particularly in fields where biochar is applied in combination with N fertilizer.
	Paddy soils field only	na	na	−4	35	
	Dryland	na	na	−72	83	
Lee et al, 2021	mixed crops East Asian studies only	−21.1	196			This review analyzed data on the biochar-induced N_2O mitigation affected by experimental conditions, including experimental types, biochar types and application rates, soil properties, and chemical forms and application rates of N fertilizer for East Asian countries.
Liao et al, 2021	Paddy soils studies under 2 years duration	ns	125	−13%	150	The results showed that the application of nitrification inhibitors (NI) and biochar significantly increased rice yield by an average of 9.5% and 9.1%, and simultaneously reduced GWP by 24% and 14%, respectively, mainly through lower CH_4 emissions. Biochar application increased rice yield by 10% while reducing CH_4 emissions, N_2O emissions, and GWP by 18%, 22%, and 24%, respectively, at an experimental duration ≥ two years.
	Paddy soils studies over 2 years duration	−22	63	−18%	70	

Table 18.1 *continued*

System	Effect size N_2O (% change)	*No. N_2O	Effect size CH_4 (% change)	*No. CH_4	Notes/summary of key findings
Verhoeven et al, 2017					
Dryland - mixed crops-field sites- variance weighted	−11.5	70	na	na	Significant emission reductions were observed when weighting by the inverse of the pooled variance (−18.1 to −7.1%) but not when weighting by the number of observations per site (−17.1 to +0.8%), thus revealing a bias in the existing data by sites with more observations. Mean yield increased by 1.7 to 13.8%.
Dryland - mixed crops-field sites- site weighted	−10.8	70	na	na	
paddy -variance weighted	−14	52	na	na	
Paddy- site weighted	−6.1 (ns)	52	na	na	
He et al, 2017	Mix of field and laboratory data	−30.9	371	ns	121	No significant effect in the field when unfertilized but a significant reduction in field N_2O under fertilized systems. The increase in GWP (44%) was driven by increases in CO_2 emissions.
Wu et al, 2019	Overall-field studies	−18.7	182	−9	155	Field studies were continued for at least a full crop season. Greater mitigation of CH_4 was observed in paddy soils vs dryland soils, while dryland cropping had greater mitigation of N_2O emissions. Mitigation of CH_4 emissions was not evidenced in field studies under 0.5 y, with maximum mitigation in studies between 1–2 years duration.
Borchard et al, 2019	Overall, field and laboratory studies	−38	608	na	na	N_2O emission reductions tended to be negligible after one year. Overall, soil NO_3^- concentrations remained unaffected while NO_3^- leaching was

Study	System					Notes
				na	na	reduced by 13% with biochar; greater leaching reductions (−26%) occurred over longer experimental times (30 days). Biochar had the strongest N_2O emission reduction in paddy soils and sandy soils. The greatest effect on N_2O was observed with mineral N fertilizers, followed by urea and organic fertilizers. The lowest impact (−27%) was seen in systems without fertilizer addition.
Zhang et al, 2021a	field only, crop, forest, and pasture	−14.9	195	na	na	No significant change in indirect emissions of NO using 12 paired comparisons. While biochar increased the soil content of NH_4^+ and NO_3^- and increased N cycling (mineralization and nitrification), N leaching was lowered by 11%. Biochar increased the abundance of soil denitrifying/ nitrifying genes (amoA, narG, nirS/nirK+S, and nosZ).
Liu et al, 2019	field, incubation, pot	−32	468	na	na	Biochar made from manure or pyrolyzed at temperatures lower than 350°C shows a weak or insignificant impact on lowering soil N_2O emissions
Feng et al, 2022	field data comparing fresh biochar effects	−18	29	134	15	A focus on the indirect GHG NH_3, with a mitigation of −7.3% NH_3 emission with aged biochar. No effect was observed with fresh biochar. This was suggested due to the lower alkalinity of biochar with aging, and the increase in oxygen-containing functional groups on the biochar
	field data comparing aged biochar effects	ns	29	ns	15	

Table 18.1 *continued*

System	Effect size N₂O (% change)	*No. N₂O	Effect size CH₄ (% change)	*No. CH₄	Notes/summary of key findings
					surface. Note the relatively small number of paired comparisons showing increased CH₄ emissions.
Song et al, 2016					
Field data	−19	55	19	24	Strong trend for greater reduction in N₂O emissions with increasing biochar dose from 10 to above 40 t ha⁻¹. No effect on N₂O after 2–3 years duration.
Laboratory data	−15	122	−18	7	
He et al, 2021					
mix of field and laboratory data- overall	−14.7	143	ns	29	N addition increased N₂O emissions by 288%, but with biochar, this increase was significantly less at 148% compared to the non -fertilized control. N addition did not have a significant effect on CH₄ emissions or CH₄ uptake, noting however a small sample size.
Cayuela et al, 2015					
overall	−49	1375	na	na	A key finding was the importance of the molar H:C$_{org}$ ratio of biochar in determining mitigation of N₂O with a ratio <0.3, indicative of a high degree of aromatic condensation, having the greatest effect (−73%).
field	−28	204	na	na	
incubation	−54	1092	na	na	
glasshouse	ns	70	na	na	

Notes
* No. refers to the number of paired comparisons evaluated in the study
ns = not significant

emissions (−54%) compared to field-based studies, which found emissions reductions of generally less than 30%. Differences in N_2O emissions were also apparent between dryland cropping systems compared to flooded paddy soils. Greater effects have been reported in dryland cropping (−41%) than in paddy fields (−17%) (Liu et al, 2019) confirmed by follow-up analyses (Lyu et al, 2022), with observed mitigation (overall change −18.7%) driven mainly by changes from dryland cropping studies (Wu et al, 2019). Field studies across 6 years in paddy rice showed a negligible change in N_2O fluxes (Wu et al, 2019). Because the effect of biochar on soil N_2O emissions varied and was site- or system-specific, Verhoeven et al (2017) employed both variance-weighted (which is the standard approach) and site-weighted (weighting by the inverse of the number of observations per site) analysis. The second approach allows for control for a potential bias from non-independence in studies with many observations. While significant mitigation of N_2O emissions was observed in dryland cropping systems using both statistical approaches, the site-weighted analysis of N_2O mitigation from paddy fields showed that biochar was not different from the control. Differences were also observed with the presence or absence of added fertilizer (Verhoeven et al, 2017; Borchard et al, 2019; He et al, 2021), with the type of N fertilizer also influencing the mitigation potential of biochar. Greater mitigation potential (−46%) was observed when NH_4NO_3, $(NH_4)_2SO_4$, or KNO_3 were used compared to urea (−34%) or N fertilization where organic amendments were co-applied (−32%) (Borchard et al, 2019). The duration of the experiment, related to the aging of biochar in soil, is an important consideration, with the mitigation effect tending to be somewhat transient (less than 12 months) (Borchard et al, 2019; Feng et al, 2022).

However, when focusing on data from paddy rice production, biochar had little effect on emissions of N_2O for the first 2 years, but significant (−22%) reductions were observed when experiments exceeded 2 years in duration (Liao et al, 2021). Other factors that were commonly assessed in the meta-analyses included the feedstock used to make the biochar, the temperature at which it was pyrolyzed, and the application dose. Biochar made from manure or pyrolyzed at temperatures lower than 350°C had non-significant effects on soil N_2O emissions (Liu et al, 2019), with the molar $H:C_{org}$ ratio, related to feedstock and production temperature, also influencing mitigation potential (Cayuela et al, 2015). Higher temperature biochars, those that are more aromatically condensed (i.e., lower molar $H:C_{org}$ ratio), and made from woody or crop residues were more effective at reducing soil N_2O emissions. Biochar dose showed a strong relationship with mitigation potential (Borchard et al, 2019). Application rates below 10 t ha^{-1} generally showed little potential to mitigate soil N_2O emissions, but application rates exceeding 10 t ha^{-1} had an increasing mitigation effect. Indeed, effects exceeding − 60% were observed when the biochar dose was 40 t ha^{-1} or greater. This leads to the question of whether fertilizer products amended with biochar (representing a lower per-hectare dosage but with a high biochar concentration in close contact with the applied N fertilizer) could be utilized to enhance fertilizer N use efficiency and lower N_2O emissions per quantity of N applied.

Meta-analyses vary in the response of biochar amendment to N_2O emissions when comparing soil properties. For example, amending soils with a pH range of 5.5–7 showed the greatest mitigation in N_2O emissions (Borchard et al, 2019), while the lowest effects were observed in strongly acidic or alkaline soils. While soil properties have been shown to control the efficacy of mitigation (Borchard et al, 2019) soil classification could not determine whether biochar would be effective or not. In analyzing the effects of soil organic C concentration (Borchard et al, 2019), few differences were seen, other than soils with a soil organic C content above 2.4% w/w having a lower effect with biochar application. A few meta-analyses have investigated the role of biochar on both N_2O and NO_x emissions. For example, biochar lowered N_2O emissions (−14%), simultaneously decreasing NO emissions by − 8.3% (Yangjin et al, 2021). However, considering field studies only (Zhang et al, 2021a), no impact of biochar on NO emissions was observed. Similarly, no impact of biochar on NH_3 emissions has been observed (Zhang et al, 2021b; Feng et al, 2022).

Meta-analyses of changes to CH_4 fluxes following biochar application to soil have tended to show greater variability in responses than those for N_2O emissions. These include changes ranging from decreases by 72% (Ji et al, 2018) for incubation studies in aerobic soils with predominant CH_4 consumption to increases of 134% (Feng et al, 2022), but noting this increase was observed from only a limited number of comparisons (15) in a paddy system, with the effect becoming non-significant upon biochar aging. This was assumed to be due to the removal of easily mineralizable C fractions in biochar over time. In a study using incubation, glasshouse, and field studies, for both dryland and paddy soils, no significant effect of biochar on CH_4 flux in the absence of fertilizer was detected, but biochar increased CH_4 emissions by 12% when N-fertilizer was co-applied (He et al, 2017, 2021). Evidence, however, suggests that biochar is likely to present opportunities for lowering CH_4 emissions in rice paddies (Wu et al, 2019) where emissions tend to be much greater than in dryland cropping systems. While an earlier meta-analysis (Ji et al, 2018) reported no effect of biochar in paddy soils when field trial data were used, significant effects (−12%) from incubation and glasshouse studies were detected. In a later analysis that incorporated a greater number of field studies from paddy soils, it was shown that biochar lowered CH_4 emissions in both short (below 2 years) (−13%) and longer-term studies (−18%) (Liao et al, 2021) (Figure 18.1).

Mechanisms for altered emissions of nitrogenous GHGs

While there is general agreement in the literature about the main factors that influence N_2O emissions from soil, the biotic and abiotic mechanism(s) and the degree and duration of response of soil N_2O emissions following biochar application to soil remain elusive (Borchard et al, 2019). Further, changes to, and mechanisms of indirect GHG emissions following biochar amendment, including NO_x and NH_3, remain understudied. Importantly, biochar properties, soil properties, particularly the nature of the farming system (rice paddy vs dryland), and the co-application of chemical fertilizers are key factors that impact soil N_2O emissions following biochar addition. Biochar application to soil may affect N_2O emissions by changing (i) soil physical properties (e.g., gas diffusivity, aggregation, water retention) (Quin et al, 2014); (ii) soil chemical properties such as pH (Obia et al, 2015; Bolan et al, 2022), Eh (Joseph et al, 2015), availability of organic and mineral N and dissolved organic C (Haider et al, 2016; Rasse et al, 2022); and

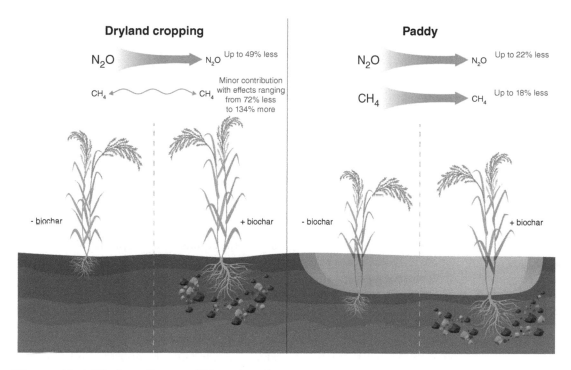

Figure 18.1 *Biochar effects on CH_4 and N_2O emissions from dryland and paddy cropping systems*

(iii) soil biological properties (e.g., microbial community structure, N cycling enzymes) (Cayuela et al, 2013). Such changes in soil properties are associated with the physical characteristics of biochar as well as the functional groups on biochar surfaces, which have been shown to decrease the N_2O/N_2 ratio under denitrifying conditions (Harter et al, 2014). These changes to soil properties and the role of biochar properties in regulating direct and indirect nitrogenous GHG emissions are discussed in detail below.

Biochar limits N availability in soil

Biochar can influence N_2O production by limiting N availability and N-cycling processes. While multiple studies have reported a range of effects of biochar on N_r content in

soil, a meta-analysis (Nguyen et al, 2017) concluded an overall reduction in soil NH_4^+ and NO_3^- content of 10%. Lowering the N_r content in soils can occur via a range of mechanisms including physisorption of N_r onto biochar, and via microbially induced N immobilization processes. Biochars tend to have a higher C:N ratio than their parent materials and contain some low molecular weight organic compounds that can be easily consumed by microorganisms in soil, thus meeting the microbial demand for N which is met by consumption of N_r. Lower temperature biochars, and those made from feedstocks containing a higher content of carbohydrates are more effective at lowering N_r due to their higher contents of easily-mineralizable C (Nguyen et al, 2017). Biochar may also retain soil N through physisorption of NO_3^- and NH_4^+ on biochar surfaces and within pores (Kammann et al, 2015; Haider et al, 2016). A significant correlation between both the CEC and surface area of biochar and NH_4^+ adsorption has been observed (Nguyen et al, 2017). Sorption of NH_4^+ on biochar may increase as negatively-charged functional groups develop on biochar surfaces during oxidation in soil, while surface oxygen groups control the reactivity of biochar (Spokas et al, 2012).

In an experiment with ^{15}N-labeled NH_3, NH_3 adsorbed to biochar was stable in ambient air, but became plant-available when added to soil (Taghizadeh-Toosi et al, 2012). The recovery by plants was larger with acidic biochar. Biochar has also been proposed as an adsorbent for removing NH_3 produced in livestock production systems or during fertilizer production (Ro et al, 2015). It has been suggested that NO_3^- can move into the pore structure of biochar via movement with water (Sohi et al, 2010). Sorption capacities of biochar for NO_3^- have been shown to range between 0.03 to 1.4 mg NO_3^--N g biochar^{-1} (Haider et al, 2016; Yang et al, 2017) with the higher surface area from higher temperature biochars suggested as having greater sorption potential. The chemical oxidation of biochar has been shown to increase the acid group content and, concomitantly, the sorption of NO_3^- (Sanford et al, 2019). Also, aging in agricultural soil or co-composting increased the NO_3^- loading up to 5 mg N g^{-1} of biochar (Kammann et al, 2015; Haider et al, 2016, 2020). The NO_3^- was only released and measurable when repeated 2 M KCl extractions (at room temperatures and 80°C) were performed on the same biochar particles, hence the NO_3^--N loading may be missed in studies when standard methods are used.

High-temperature biochars and protonation by acidification increased NO_3^- sorption by biochars. The sorption of NO_3^- in aqueous solution can be maximized (1.4–1.5 mg N g^{-1}) with biochar derived from high pyrolysis temperatures (600 °C) and low pH (3.5–4) (Fidel et al, 2018). In an experiment evaluating the use of low-molecular-weight organic acids with $NaNO_3$ and either rice husk or miscanthus biochar (700°C), significant sorption of NO_3^- has been shown (Heaney et al, 2020). The study showed that the NO_3^- concentration in the solution went down to near-zero when malic, citric, or oxalic acids were used to acidify the biochar-NO_3^- solution, equaling a sorption of 1.4 mg N g^{-1} of biochar. In all studies mentioned, the (pH-maximized) NH_4^+ sorption was lower than that of NO_3^-. While low-temperature biochars (400°C) showed no significant NO_3^- sorption, aging, co-composting, or chemical oxidation of biochars produced at 600°C and 700°C achieved sorption capacities ranging from 0.50 to 3.97 mg NO_3^--N g biochar^{-1}. It was suggested that the oxygenated functional groups and the increase in positively charged minerals on biochar surfaces contribute to cation bridging, resulting in NO_3^- sorption (Sanford et al, 2019). Acidic root exudates may also facilitate greater sorption of NO_3^- by biochar. Thus it is feasible to consider that upon aging in soil, the sorption capacity of biochars for NO_3^- may increase, contributing to lower N_r content of soil. A lower availability of adsorbed N for microbes is plausible since biochar particles in the nanometer range are not accessible for soil microbes with a diameter in the 0.1 - 1 µm range. Such an effect may also contribute to the observed increase in copy numbers of the denitrifier *nosZ* gene (Harter et al, 2014) promoting complete denitrification to N_2 in response to a shortage of NO_3^- availability to microbes.

Biochar changes bioavailable C supply in soil, affecting biological processes

While the N_r content of soil is a key driver for the production of N_2O, readily decomposable C is also an important controlling factor for microbially mediated N_2O production (Weier et al, 1993). The reduction of NO_3^- by denitrifiers requires a readily available organic C supply with a theoretical 30 moles of C required for the complete denitrification of 24 moles of NO_3^- (Saggar et al, 2013). While it was previously proposed that biochar may limit the availability of organic C in soil (Van Zwieten et al, 2015), hence lower substrate requirement for microbial N cycling processes, a recent meta-analysis (Chagas et al, 2022) using 169 studies concluded that biochar results in an increase in microbial biomass C by 20% and soil mineralizable C by 23% compared to unamended controls. An increase in mineralizable organic C can result in increased heterotrophic respiration, although a wide range of effects (increase and decrease) have been reported in the literature (Chapter 17). An increased respiration may induce localized sites of anaerobiosis to favor processes such as denitrification (Harter et al, 2014), which would under aerobic conditions reduce the activity of O_2-sensitive N_2O reductases. An increase in gene and transcript copy number of the $nosZ$-encoded bacterial N_2O reductase has been observed with biochar addition to compost (Wang et al, 2013) and soil (Harter et al, 2014). It has been suggested that by changing the geochemical properties of soil, in particular the availability of the key electron acceptor (NO_3^-) and donors (NH_4^+, dissolved organic C), biochar causes a shift in the denitrifier community composition by promoting the abundance and activity of N_2O-reducing bacteria (containing a $nosZ$ gene) relative to $nirS$- and $nirK$-containing denitrifiers (Harter et al, 2014).

Impacts of biochar pH on GHG flux

Biochars are generally alkaline in nature and have been shown to increase the pH of acidic soils. The resulting liming value varies depending on feedstock type and pyrolysis temperature (Bolan et al, 2022). The alkalinity of biochar is mainly due to the content of $CaCO_3$, $MgCO_3$, $Mg(OH)_2$ and MgO. Oxygen-containing organic functional groups and fused-ring aromatic structures also contribute to the alkalinity of biochar (Yuan et al, 2011; Li et al, 2013). The N_2O production rate via denitrification and heterotrophic nitrification has been shown to increase at lower soil pH, while the N_2O production rate via autotrophic nitrification decreases at lower soil pH (Zhang et al, 2021b). A meta-analysis investigating the effects of liming acidic soil has shown an increase in the bacterial abundance of N_2O reductase ($NosZ$) and a decrease in the fungi:bacteria ratio, both contributing to a lower $N_2O:N_2$ product ratio of denitrification (Zhang et al, 2022) (i.e., more complete denitrification). The same meta-study also showed that liming may have lowered soil CH_4 emissions and the abundance of methanogens, but it had no effect on CH_4 uptake and abundance of methanotrophs. An observed decrease in soil N_2O emissions was shown to be facilitated by increased $nosZ$ and $nirK$ gene abundance at higher soil pH resulting from biochar amendment (Aamer et al, 2020), thus facilitating the completion of denitrification (i.e., N_2O to N_2). In a field study, fertilized plots with biochar amendment resulted in a significant increase of $nosZ$ gene copy numbers (hence the increased potential for more complete denitrification) compared to the control and the lime treatment (Krause et al, 2018), suggesting that a shift in community composition of *nosZ-II* bearing bacteria was due to more than biochar's liming effect alone, and possibly involved an improvement in the availability of substrates, in particular readily degradable C. To understand the effects of pH in comparison to other changes to soil physicochemical properties, Obia et al (2015) separated the readily degradable C from biochar and showed that it stimulated denitrification, while increased pH promoted complete denitrification in two acidic soils. The same study showed that the increase in soil pH using NaOH also stimulated complete denitrification, suggesting the effect is most likely to be solely due to the pH increase following biochar application to these acidic soils. Contrary to this, however, mimicking the increase in soil pH using $CaCO_3$ did not have any significant impact on N_2O production (Cayuela et al, 2013), suggesting other mechanisms may be at play.

Biochar modulates the soil microbial community

While it has been frequently observed that biochar addition to soil can lower N_2O emissions, this observation has not been consistent across soils. This has been attributed to the microbial process responsible for N_2O production (Ji et al, 2020): While biochar can promote N_2O production by nitrifiers, it usually decreases N_2O production by denitrifiers by facilitating more complete denitrification. For example, N_2O emissions increased by 54% after the addition of biochar in a soil where N conversions were dominated by nitrification and nitrifier denitrification, using stable isotope techniques with the use of the nitrification inhibitor dicyandiamide in mesocosms (Sánchez-García et al, 2014). The biochar was produced by continuous

slow pyrolysis of green waste at 550°C and decreased N_2O emissions by 76% in another soil incubated under the same conditions but dominated by denitrification. A 27% increase in N_2O emissions following the application of woody biochar to soil has also been reported, with the increase attributed to the increased N_2O production by nitrifiers, despite enhanced complete denitrification to N_2 (Wells and Baggs, 2014).

Often, the increase of nitrification following biochar application to soil was in line with an increase in the abundance of nitrifiers in the soil. In a 42-day incubation of soil retrieved from a 3-year field experiment on an acidic oxisol, with rice straw biochar added at 0, 2.25, and 22.5 Mg ha^{-1}, nitrification was increased by the highest application rate of biochar (He et al, 2016). This was attributed to an increase in the gene copy numbers of ammonia-oxidizing archaea (AOA) and ammonia-oxidizing bacteria (AOB) of 10 and 22 times, respectively, compared to the control. A shift towards *Nitrosospira* cluster 3 and related groups was observed. These are associated with more pH-neutral rather than acidic conditions (Stephen et al, 1998). Interestingly, the kinetics changed from zero to first order, probably because there was no longer a limitation to enzyme activity, but rather a limitation in substrate availability (He et al, 2016). The specific surface area of biochar also has an important effect on the microbial community with a higher abundance of AOA and AOB in soils where biochar had a smaller surface area, whereas higher surface areas of biochar promoted complete denitrification to N_2 (Liao et al, 2021). The degradation of DOC in subsoil has been shown to release electrons and thus facilitate complete denitrification (Qin et al, 2017). Biochar can shuttle these electrons, thus promoting complete microbial denitrification to N_2 (Cayuela et al, 2013; Yuan et al, 2019).

On a mechanistic level, there are several suggestions on how biochars may alter conditions that change the microbial community and their function (Table 18.2). The role of biochar on soil aeration is a likely candidate: biochar can increase the water-holding capacity of soils (Chapter 20) and large additions (i.e., 10%) decrease bulk density (Case et al, 2012), making the soils apparently physiologically drier for microbes, which could stimulate nitrification and decrease denitrification. However, when a sandy loam was incubated at a water content corrected for the biochar moisture effect, still a reduction of N_2O emissions of up to 98% was observed with 10% biochar added (Case et al, 2012). On the other hand, some biochars are known to increase the water retention capacity and available water content, particularly in coarse-textured soil (Razzaghi et al, 2020). This could lead to more anaerobic microsites with increased N_2O production from denitrification processes (including nitrifier denitrification), but might still increase nitrifier abundance, as water from the soil may be preferentially retained in biochar pores (Sohi et al, 2010). Furthermore, as discussed above, temporary NO_3^- sorption from soil solution by biochar particles may contribute to reduced N_2O emissions from denitrification or shifts in the $N_2O:N_2$ ratio towards more complete denitrification. The effects may also be more indirect: It has been suggested that biochars may adsorb soluble phenols or terpenes that can inhibit nitrification, or that biochars may introduce microbially toxic compounds into soil (Clough and Condron, 2010).

Biochar affects gas diffusivity, GHG retention, and reactivity in soil

Biochar can change a range of soil physical properties including bulk density (Rassagji et al, 2020), pore connectivity (Quin et al,

2014), and soil hydraulic characteristics (Xiaoqin et al, 2021). While studies linking soil aggregation, soil moisture variation, and their role in N_2O emissions are lacking (Jayarathne et al, 2021), the maximum N_2O emissions tended to occur when lower diffusivity facilitated both anaerobic conditions while still allowing the diffusion of N_2O through the soil. Conversely, increased diffusivity in soil initially promoted then reduced N_2O fluxes. Biochar can also lead to the direct entrapment of N_2O (Cornelissen et al, 2013; Quin et al, 2015; Harter et al, 2016; Yang et al, 2022). Changes in surface functional groups of biochar upon exposure to N_2O, including -O-C = N, pyridine pyrrole, or NH_3 also occur (Quin et al, 2015), indicating a reaction between biochar and N_2O. Nanometer-scale microscopic and spectroscopic analysis revealed that biochars had reacted with soil organic and mineral matter to form redox active organo-mineral clusters. This suggested that two possible mechanisms were occurring. The first involved the oxidation of aromatic and aliphatic compounds by N_2O (which is a strong oxidizing agent) to produce N_2 and ketones (Avdeev et al, 2005). The second mechanism involved the oxidation of Fe_2^+ with the subsequent reduction of N_2O on the redox-active surfaces of the aged biochar. These reactions took place in water-filled pores that had a significant concentration of N_r. The authors noted that the posited reactions are consistent with the research related to the role of Fe_2^+/Fe_3^+ cycling, Eh-pH with N release dynamics, and formation and reduction of N_2O to N_2. The mitigation of N_2O emissions from soil has also been shown to occur via entrapment of N_2O in water-saturated pores of the soil-biochar matrix, with the concurrent stimulation of microbial N_2O reduction to N_2, resulting in an overall decrease of the $N_2O/(N_2O+N_2)$ ratio (Harter et al, 2016).

Table 18.2 *Potential influences of biochar on soil conditions and assumed effects on microbial sources of N_2O*

Biochar impact on soil conditions	Biochar characteristics favoring this	Effect regarding N_2O emissions
Addition of (traces of) N_r	Biochar from manure or co-composted	Increases N_2O production by processes using this as substrate; potential priming effects on mineralization
Increase of dissolved organic carbon	Lower temperature biochars	Increases heterotrophic processes; potentially leads to more anaerobic conditions due to increased respiration, which may favor denitrification; may provide electrons for increased total denitrification to N_2, thus decreasing N_2O emission
Increase of pH	High alkalinity of biochar	Favors bacterial processes if pH is acidic; favors nitrification more than denitrification; increases N_2O reduction to N_2
Improved aeration	High water-holding capacity, large pore volume	Favors aerobic processes, especially nitrification; decreases total denitrification rates, but also increases N_2O to N_2 ratio
Improved electron shuttle capacity	(Limited information available)	Increases complete denitrification to N_2
Less available NH_4^+ in soil due to Adsorption Increased plant uptake Immobilization Increased NH_3 volatilization	Large pore volume, many surface oxygen groups, high CEC	Decreases nitrification-related N_2O production (by nitrification, nitrifier denitrification, nitrification-coupled denitrification)
Less available NO_3^- in soil due to Adsorption Increased plant uptake	High water-holding capacity, large pore volume, many surface oxygen groups, higher production temperatures, field aging of biochar	Decreases N_2O production from denitrification
Increase of soil Fe	Fe-rich feedstocks used	Increases chemodenitrification to N_2
Organic compounds that lower N-cycling (e.g., nitrification inhibitors)	Biochar can be a platform material for natural or synthetic nitrification inhibitors	Decreases N_2O production
Adsorption of inhibitory or toxic compounds to biochar from soil (e.g., terpenes or phenols)	Large pore volume	Increases N_2O production
Adsorption of compounds used for cell-cell signaling	Large surface area	Less well-regulated gene expression among microbes with unclear effect on N_2O emission

Mechanisms for the mitigation of methane production

While meta-analyses (Table 18.1) show a wide range of responses to biochar additions for CH_4 emissions (production or consumption) following biochar application, evidence suggests that the greatest benefit of biochar will be in systems where methanogenesis is dominant- i.e., paddy soils. These responses are likely due to a range of mechanisms associated with both biotic and abiotic processes. Very little work has been carried out on potential abiotic mechanisms for lowering CH_4 production by methanogenic archaea, while more is known on biotic mechanisms, such as the interplay between methanogenic archaea and methanotrophic bacteria (Gao et al, 2022); the latter require O_2 and provide a 'biofilter function' for anaerobically produced CH_4 before it escapes to the atmosphere.

One of the main biotic mechanisms for lowering CH_4 emissions from flooded or waterlogged soils is the stimulation of O_2-requiring methanotrophic bacteria, due to biochar's introduction of oxic or anoxic interfaces into waterlogged soils (Feng et al, 2012; Han et al, 2016; Huang et al, 2019a; Rubin et al, 2020). For example, biochar-amended soils stimulated CH_4 monooxygenase (pmoA) gene copy numbers of methanotrophic bacteria in rice mesocosms, while gene copies of methanogenic archaea were unchanged (Feng et al, 2012). Similarly, increased methanotroph pmoA gene copy numbers, as well as methanotrophic activity, have been shown to occur in rice systems after biochar amendment (Han et al, 2016; Huang et al, 2019a), resulting in lower CH_4 emissions. This mechanism has also been shown to lower cumulative CH_4 emissions by up to 92% from restored wetland soils (Rubin et al, 2020). In all of these studies, however, effective biochar doses were high (i.e., 10% w/w or up to 80 t ha^{-1}). Although the pH increase induced by biochar application may play a role in promoting methanotrophic activity, a reason that was not discussed in these studies (but would be supported by the data of Han et al (2016)) is that biochar increased rice-plant and root system growth, which results in more oxygenated rhizoplane area. At the rhizoplane of aerenchymatous plants, O_2 diffuses from the root system into the surrounding (biochar-amended) soil (Huang et al, 2019b), providing a living habitat for methanotrophic bacteria. Thus, a larger methanotroph-enriched rhizoplane forms a better biofilter barrier against the CH_4 produced in the surrounding anoxic environment, limiting its escape through the rice plants' aerenchyma (a pathway which comprises up to 90% of the CH_4 escape from planted rice paddies).

Municipal solid waste landfills are major sources of CH_4 due to anaerobic breakdown of organic compounds. Landfill cover soils (LCS) are usually designed to act as methanotrophic bacteria-harboring biofilters to reduce or prevent CH_4 emissions from anaerobic landfill (Huang et al, 2021; Chetri et al, 2022a, b; Qin et al, 2022). Biochar amendments considerably improved LCS's biofilter function, even under low O_2 conditions (Reddy et al, 2014). In later studies, preconditioning the added biochar with methanotrophic bacteria (Chetri et al, 2022a, b) increased CH_4 oxidation rates, and reduced the initial lag phase, which usually exists when unmodified biochar is added to landfill cover soils. The pore volume, cation exchange capacity, and surface area of biochar collectively accounted for 79% of the variances in the microbial community structures in LCS (Huang et al, 2021), with pore volume being the most important factor; they also reported that Fe-modified biochars encouraged a bacterial consortium (methanotroph, methylotrophs, and N_2-fixing bacteria) in LCS that significantly improved CH_4 oxidation capacities by up to 26–74%, compared to that of unmodified biochar. In a subsequent study, it was also shown that biochar captured toxic H_2S which in turn improved the performance of methanotrophic bacteria in LCS (Huang et al, 2022). Using hydrophobic biochar in LCS lowered water infiltration and pore clogging, which also assisted the overall methanotrophic biofilter function of the LCS (Qin et al, 2022).

The application of a modified (acidic) biochar (HNO_3, H_2SO_4) at 22.4 t ha^{-1} significantly increased N concentrations in soil and concomitantly lowered CH_4 emissions (Nan et al, 2021). This was attributed to the increase in -OOH groups on the biochar that could oxidize the CH_4 while the decrease in soil pH also inhibited methanogens. The addition of biochar with a high content of O-functional groups can act as the sole electron acceptor in an anaerobic environment (Zhang et al, 2019) which may result

in the biological oxidation of CH_4. They proposed a similar chemical reaction to that of Avdeev et al (2005):

$$CH_4 + 8C - O = 2\ H_2O + 8\ C - OH + CO_2$$

The authors also proposed a biotic mechanism whereby anaerobic methanotrophic archaea (ANME) utilize an extracellular electron transfer mechanism to oxidize CH_4 with direct interspecies electron transfer between ANME-2d and Geobacteraceae; or indirect extracellular electron transfer from ANME-2d to biochar based on interspecies electron transfer between Geobacteraceae and ANME-2d via metabolism of the intermediate acetate (direct electron transfer from ANME-2d to biochar is the thermodynamic premise for biosynthesis of acetate from CH_4 oxidation in this case). The physicochemical properties of biochar have been directly related to their efficacy

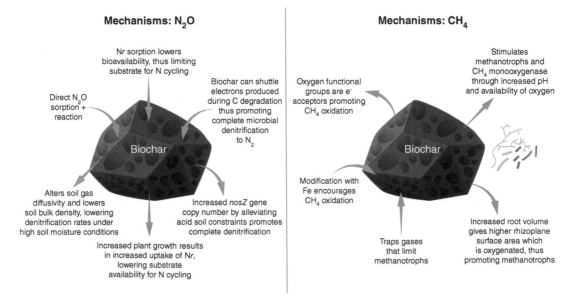

Figure 18.2 *A summary of key mechanisms for lowering emissions of N_2O and CH_4 from soil using biochar. Note, arrows to biochar indicate the direct effect of biochar properties, while arrows away indicate the effect of biochar on the surrounding soil environment*

in inhibiting or improving the rate and extent of the oxidation of CH_4 (Blanca Pascual et al, 2020). Biochars made from woody feedstocks and pyrolyzed at 600°C resulted in the highest CH_4 oxidation rates; however, biochars with high ash concentrations (and higher pH) and electrical conductivity significantly lowered CH_4 oxidation rates. A summary of mechanisms is shown in Figure 18.2.

Engineered biochar products for mitigation of nitrous oxide and methane emissions

Biochars, as proposed in multiple studies (Mandal et al, 2016; Rawat et al, 2022), serve as versatile materials that can be engineered to optimize the mitigation of N_2O and CH_4 emissions from soil. While most studies apply biochar to soils separately to fertilizer N_r, there has been a recent surge in interest in the utilization of biochar as a fertilizer carrier to increase N fertilizer use efficiency and concomitantly lower soil GHG emissions (Rombel et al, 2022). The notion of aligning crop N demand with supply is widely accepted as an effective strategy for both improving fertilizer use efficiency and reducing environmental N losses, including N_2O emissions (Xia et al, 2017). Enhanced efficiency N fertilizers often employ chemical inhibitors, such as nitrification or urease inhibitors, to slow down microbial processing of N. Other approaches include polymer coating technologies for controlled N release. The porous structure and reactive nature of biochar make it an ideal candidate for developing controlled-release N fertilizer products, influencing the soil environment to improve fertilizer use efficiency.

A recent review on N-absorbing and N-enrichment mechanisms in the design of biochar compound fertilizers (BCFs) (Rasse et al, 2022) has demonstrated their potential to boost N-use efficiency, thereby enabling farmers to decrease N application. A particular BCF was found to enhance rice biomass by 67% and fertilizer use efficiency (Chew et al, 2020). This BCF was associated with an 85 mV increase in soil oxidation-reduction potential and a 65 mV increase in the potential difference between rhizosphere soil and the root membrane, thus reducing the free energy needed for root nutrient accumulation. A year-long field study revealed that a BCF made from ground wheat straw biochar mixed with fertilizer and then granulated significantly reduced soil N_2O emissions (Zhou et al, 2021). This N_2O mitigation was linked to a decrease in water-soluble organic N in the soil and the inhibition of soil urease and protease activities. Biochar can react with urea to form cyclic anhydrides, thereby decelerating the release of Nr into the soil, thus reducing the substrate for N_2O production (Dong et al, 2021). During granulation, urea can bind strongly to clay and biochar surfaces, decreasing N leaching which was shown to enhance N-use efficiency, thus lowering fertilizer requirements (Wei et al, 2020).

Biochar has electron-mediating properties in biological redox reactions that can be optimized for mitigating soil GHG emissions. Biochar produced at lower pyrolysis temperatures contains a large concentration of redox-active functional groups and exhibits lower conductivity. However, at higher pyrolysis temperatures, redox-active functional groups gradually decrease, and the C matrices become more electron-conducting, which may facilitate electron transfer. Both the redox potential and electron conductivity of pyrolyzed C are shown to be involved in the CH_4 production process. The C matrices and redox activity could facilitate electron transfer (Sun et al, 2017), while the functional groups might act as electron acceptors or donors, allowing electron exchange with methanogens (Yu et al, 2016). For instance, biochar with the highest electron-donating capacity (0.85 mmol g^{-1}) was found to enhance CH_4 production by 33.3%, while biochar with the electron-accepting ability and a lower electron-donating capacity was shown to inhibit CH_4 generation (Zhang et al, 2021c).

The engineering of biochar to contain other soil-modifying reagents, such as metallic monomers and their respective oxides, has emerged as a promising area of research for its potential to lower soil GHG emissions. Biochar, when modified with Fe, can maintain a reduced soil Eh, fostering the formation of high-concentration root-surface Fe films, and enhancing the persistence of the biochar (Liu et al, 2021). The presence of Fe on the biochar surface acts as a redox mediator in soil, stimulating redox reactions, which leads to the maintenance of a low soil Eh — a condition promoting beneficial processes such as nutrient availability, microbial activity, and organic matter decomposition. In particular, Fe oxides are commonly employed as corrective agents for paddy soils to mitigate greenhouse gas emissions (Li et al, 2022). The extent of CH_4 emissions from paddy soils is intimately linked with the soil Eh, with marked CH_4 generation observed within an Eh range of -150 to -100 mV.

Sodium ferrate-modified biochars have been shown to lower CH_4 and N_2O emissions from soil (Zhou et al, 2022), using a short-term field study in rice. The global warming potential was also lower from this system which increased rice yield. In that study, CH_4 emissions were significantly negatively correlated with yield and the soil labile C pool. N_2O production was significantly negatively correlated with urease activity and significantly positively correlated with the soil C pool management index. The authors concluded that ferrate-modified biochar provides a tool to enhance both production and soil C content. The mechanisms associated with Fe-modified biochar for increasing CH_4 oxidation include increased soil pore volume, cation exchange capacity (CEC), and surface area of the engineered biochar (Huang et al, 2022). Fe-modified biochar encouraged methanotroph, methylotrophs, and N_2-fixing bacteria in the soil with significantly enhanced CH_4 oxidation (by between 26–74%) compared to that of non-modified biochar. The surface oxidation of algal biochars with HNO_3 resulted in the complete inhibition of CH_4 production during batch anaerobic digestion of glucose (Jiang et al, 2022). However, oxidation with H_2O_2 resulted in greater production of CH_4 compared with raw algal biochar. It was suggested that the decrease of the pH to less than the optimum pH for methanogenesis (6.8–7.2) and the generation or dissolution of inhibitory compounds by nitrate and nitro groups after HNO_3 oxidization were the main mechanisms controlling CH_4 production.

Avoided emissions

The use of biomass for biochar diverts that biomass from current uses or disposal options. For some forms of biomass, the current use is associated with significant emissions of greenhouse gases, so the use of biochar can reduce or avoid those emissions. For example, manure-based feedstocks (feedlot manure, poultry litter, and biosolids) are often stockpiled and then spread in fields as a nutrient source. Stockpiled manure releases large quantities of CH_4 and N_2O during storage and handling, as the substrate is moist and rich in N_r and easily metabolizable organic C. Furthermore, once spread onto the soil, substantial quantities of N_r are emitted as N_2O and also as NH_3, an indirect source of N_2O, particularly in wet climates (Van der Weerden et al, 2021).

The direct N_2O emissions from manure vary widely and can be lower than from synthetic fertilizer per unit N applied (Grados et al, 2022). However, due to the relatively higher rates of NH_3 volatilization from manures (Bouwman et al, 2002), the application of unprocessed manures leads to high total emissions of N_2O. Therefore, when manure-based feedstocks are pyrolyzed before application to land, N_2O emissions per unit N applied can be substantially reduced (Ginebra et al, 2022). A field study has shown that raw poultry litter resulted in emissions of 4.9 kg N_2O–N ha^{-1} while at a matching N rate, the same manure, but pyrolyzed, released 1.5 kg N_2O–N ha^{-1} (Van Zwieten et al, 2013).

Current methods for handling manure-based feedstocks often include composting to manage pathogens, reduce volume, and aid handling and product consistency. Composting these materials leads to high emissions of NH_3, and can also release CH_4 and N_2O, especially in warm moist climates (IPCC, 2019). The addition of biochar to compost has been shown to substantially reduce these emissions. For example, biochar amended at 10% (by dry weight) into a poultry litter compost lowered emissions of N_2O by 2–3 fold (Agyarko-Mintah et al, 2017a), with spectral analysis revealing sorption of N on biochar surfaces, likely being a key mechanism for the lower N_2O emissions. In another study, significantly lower emissions of the indirect GHG NH_3 coincided with increases in N retention in the poultry litter compost (Agyarko-Mintah et al, 2017b). The effectiveness of biochar in decreasing N_2O and CH_4 emissions from compost is associated with its absorptive and adsorptive properties, surface N reactions, enhancement of microbial activity, increased aeration, and effects on pH (Yin et al, 2021). A full accounting of GHG emissions has to use a life cycle assessment of biochar systems that are discussed in detail in Chapters 30 and 31.

Conclusion and recommendations

There is now a significant weight of evidence supporting the role of biochar in lowering soil emissions of non-CO_2 GHGs, in particular for N_2O emissions from cropping systems. While increases in CH_4 emissions following biochar amendment have been observed in dryland systems, these tend to be a minor contribution compared to the decrease in CH_4 from flooded and paddy systems. In these paddy soils, evidence from relatively short-term field trials is mounting

that biochar can have benefits in lowering CH_4 emissions by 10–20%. Several research opportunities exist that may enhance technology adoption and climate change mitigation opportunities. These are identified below:

- Meta-analyses (Table 18.1) have shown the mitigation of N_2O emissions from soil occurs generally within the first 2 years following the application of biochar. For this to be recognized in emissions trading, we need reliable estimates of efficacy for specific biochar x soil type x climate, including longevity of the effect.

- Over the past few years, there has been mounting interest in using biochar as a platform material for the development of enhanced-efficiency fertilizers. Through the sorption, porosity, and surface functional groups on biochar, it has been possible to engineer products that increase plant productivity, increase N-use efficiency, and lower soil N_2O emissions. This field of material sciences is likely to result in significant opportunities for commercial investment.

- Many potential mechanisms for lowering emissions of N_2O from soil have been discussed here. Mechanisms associated with modified soil and rhizosphere redox properties that enhance the uptake of N across the root membrane remain an exciting area of investigation, that could provide many opportunities to modify and optimize biochar properties.

- While there are now several studies showing that pyrolysis of manures and the addition of biochar to manure handling and composting systems can lower N_2O and CH_4 emissions, comprehensive studies quantifying the changes to GHG emissions in different contexts are still lacking. These need to be supported by life cycle assessment to determine the system-level effects on net GHG emissions and to inform climate and waste management policy.

- While many studies have measured changes to soil CH_4 emissions, variable responses - from significant decreases to increases in CH_4 emissions from soil - have been reported. Methane emissions from aerobic soils tend to be negligible, with methanotrophy often being dominant. However, opportunities exist in paddy soils where methanogenesis is dominant. Model systems have suggested that biochars with an electron-accepting capacity and lower electron-donating capacity, such as those produced at higher temperatures (ca. 600°C) can lower methanogenesis, but these need to be tested in soil systems and in the field. Nevertheless, this poses exciting opportunities for lowering emissions from paddy soils.

- To understand if stimulation of root growth by biochar can increase methanotrophic activity (and hence reduce net CH_4 emissions from paddy soil), we suggest a further assessment of correlations between root biomass or root length increases and increases of methanotrophic abundance or activity with biochar.

- The physical modification of soil by biochar has remained relatively unstudied, in particular how improvements to soil pore connectivity and bulk density can modify gas diffusion, soil redox conditions, and the generation and consumption of both CH_4 and N_2O. Modification of these soil physical aspects can also affect water flow, potentially leading to greater movement of N_r as well as dissolved N_2O, resulting in non-point source emissions of N_2O.

- While many reports now focus on changes to soil biological properties following biochar amendment, particularly those associated with C and N cycling (and emissions of CH_4 and N_2O), being able to manipulate these changes for the desired outcome remains elusive. This is likely to remain an important topic of research in the coming years.

References

Aamer M, et al 2020 Biochar mitigates the N_2O emissions from acidic soil by increasing the *nosZ* and *nirK* gene abundance and soil pH. *Journal of Environmental Management* 255, 109891.

Agyarko-Mintah E, et al 2017a Biochar increases nitrogen retention and lowers greenhouse gas emissions when added to composting poultry litter. *Waste Management* 61, 138–149.

Agyarko-Mintah E, et al 2017b Biochar lowers ammonia emission and improves nitrogen retention in poultry litter composting. *Waste Management* 61, 129–137.

Atkinson, R 2000 Atmospheric chemistry of VOCs and NOx. *Atmospheric Environment* 34, 2063–2101.

Avdeev V, Ruzankin S, and Zhidomirov G 2005 Molecular mechanism of direct alkene oxidation with nitrous oxide: DFT analysis. *Kinetics and Catalysis* 46, 177–188.

Blanca Pascual M, Sánchez-Monedero MA, Chacón FJ, Sánchez-García M, and Cayuela ML 2020 Linking biochars properties to their capacity to modify aerobic CH_4 oxidation in an upland agricultural soil. *Geoderma* 363, 114179.

Bolan N, et al 2022 Multifunctional applications of biochar beyond carbon storage. *International Materials Reviews* 67, 150–200.

Bouwman AF, Boumans LJM, and Batjes NH 2002 Estimation of global NH_3 volatilization loss from synthetic fertilizers and animal manure applied to arable lands and grasslands. *Global Biogeochemical Cycles* 16, 1024.

Borchard N, et al 2019 Biochar, soil and land-use interactions that reduce nitrate leaching and N_2O emissions: a meta-analysis. *Science of the Total Environment* 651, 2354–2364.

Butterbach-Bahl K, Baggs EM, Dannenmann M, Kiese R, and Zechmeister-Boltenstern S 2013 Nitrous oxide emissions from soils: how well do we understand the processes and their controls? *Philosophical Transactions of the Royal Society B: Biological Sciences* 368, 20130122.

Case SD, McNamara NP, Reay DS, and Whitaker J 2012 The effect of biochar addition on N_2O and CO_2 emissions from a sandy loam soil–the role of soil aeration. *Soil Biology and Biochemistry* 51, 125–134.

Cayuela ML, et al 2013 Biochar and denitrification in soils: when, how much and why does biochar reduce N_2O emissions? *Scientific Reports* 3, 1–7.

Cayuela M, Jeffery S, and Van Zwieten L 2015 The molar H: Corg ratio of biochar is a key factor in mitigating N_2O emissions from soil. *Agriculture, Ecosystems and Environment* 202, 135–138.

Chagas JKM, Figueiredo CCD, and Ramos MLG 2022 Biochar increases soil carbon pools: Evidence from a global meta-analysis. *Journal of Environmental Management* 305, 114403.

Chetri JK, Reddy KR, and Green SJ 2022a Methane oxidation and microbial community dynamics in activated biochar-amended landfill soil cover. *Journal of Environmental Engineering* 148, 04022009.

Chetri JK, Reddy KR, and Green SJ 2022b Use of methanotrophically activated biochar in novel biogeochemical cover system for carbon sequestration: Microbial characterization. *Science of the Total Environment* 821, 153429.

Chew J, et al 2020 Biochar-based fertilizer: Supercharging root membrane potential and biomass yield of rice. *Science of the Total Environment* 713, 136431.

Clough TJ, and Condron LM 2010 Biochar and the nitrogen cycle: introduction. *Journal of Environmental Quality* 39, 1218–1223.

Conrad R 2007 Microbial ecology of methanogens and methanotrophs. *Advances in Agronomy* 96, 1–63.

Cornelissen G, et al 2013 Sorption of pure N_2O to biochars and other organic and inorganic materials under anhydrous conditions. *Environmental Science and Technology* 47, 7704–7712.

Davidson EA, and Kingerlee W 1997 A global inventory of nitric oxide emissions from soils. *Nutrient Cycling in Agroecosystems* 48, 37–50.

Dong D, et al 2021 Mitigation of methane emission in a rice paddy field amended with biochar-based slow-release fertilizer. *Science of the Total Environment* 792, 148460.

Feng Y, Xu Y, Yu Y, Xie Z, and Lin X 2012 Mechanisms of biochar decreasing methane emission from Chinese paddy soils. *Soil Biology and Biochemistry* 46, 80–88.

Feng Y, et al 2022 How does biochar aging affect NH_3 volatilization and GHGs emissions from agricultural soils? *Environmental Pollution* 294, 118598.

Fidel RB, Laird DA, and Spokas KA 2018 Sorption of ammonium and nitrate to biochars is electrostatic and pH-dependent. *Scientific Reports* 8, 17627.

Galloway JN, et al 2008 Transformation of the nitrogen cycle: recent trends, questions, and potential solutions. *Science* 320, 889–892.

Gao J, Liu L, Shi Z, and Lv J 2022 Biochar amendments facilitate methane production by regulating the abundances of methanogens and methanotrophs in flooded paddy soil. *Frontiers in Soil Science* 2, 801227.

Ginebra M, Muñoz C, and Zagal E 2022 Carbon stability and soil N_2O emissions. Pyrolyzed or unpyrolyzed manure? *Journal of Environmental Management* 322, 116095.

Grados D, et al 2022 Synthesizing the evidence of nitrous oxide mitigation practices in agroecosystems. *Environmental Research Letters* 17, 114024.

Haider G, Steffens D, Müller C, and Kammann CI 2016 Standard extraction methods may underestimate nitrate stocks captured by field-aged biochar. *Journal of Environmental Quality* 45, 1196–1204.

Haider G, et al 2020 Mineral nitrogen captured in field-aged biochar is plant-available. *Scientific Reports* 10, 1-12.

Harter J, et al 2014 Linking N_2O emissions from biochar-amended soil to the structure and function of the N-cycling microbial community. *The ISME Journal* 8, 660–674.

Harter J, et al 2016 Soil biochar amendment shapes the composition of N_2O-reducing microbial communities. *Science of the Total Environment* 562, 379–390.

Han X, et al 2016 Mitigating methane emission from paddy soil with rice-straw biochar

amendment under projected climate change. *Scientific Reports* 6, 24731.

Han P, et al 2021 N_2O and NO_y production by the comammox bacterium *Nitrospira inopinata* in comparison with canonical ammonia oxidizers. *Water Research* 190, 116728.

He L, et al 2016 Comparison of straw-biochar-mediated changes in nitrification and ammonia oxidizers in agricultural oxisols and cambosols. *Biology and Fertility of Soils* 52, 137–149.

He Y, et al 2017 Effects of biochar application on soil greenhouse gas fluxes: A meta-analysis. *Global Change Biology Bioenergy* 9, 743–755.

He Y, et al 2021 Antagonistic interaction between biochar and nitrogen addition on soil greenhouse gas fluxes: A global synthesis. *Global Change Biology Bioenergy* 13, 1636–1648.

Heaney N, Ukpong E, and Lin C 2020 Low-molecular-weight organic acids enable biochar to immobilize nitrate. *Chemosphere* 240, 124872.

Huang Y, et al 2019a Methane and nitrous oxide flux after biochar application in subtropical acidic paddy soils under tobacco-rice rotation. *Scientific Reports* 9, 17277.

Huang L, Liang YK, Liang Y, Luo X, and Chen YC 2019b Influences of biochar application on root aerenchyma and radial oxygen loss of *Acorus calamus* in relation to subsurface flow in a constructed wetland. *Huan Jing Ke Xue* 8, 1280–1286.

Huang D, Bai X, Wang Q, and Xu Q 2021 Validation and optimization of key biochar properties through iron modification for improving the methane oxidation capacity of landfill cover soil. *Science of the Total Environment* 793, 148551.

Huang D, Xu W, Wang Q, and Xu Q 2022 Impact of hydrogen sulfide on biochar in stimulating the methane oxidation capacity and microbial communities of landfill cover soil. *Chemosphere* 286, 131650.

IPCC 2019 *Refinement to the 2006 IPCC Guidelines for National Greenhouse Gas Inventories.* Gavrilova et al Emissions from livestock and manure management Volume 4, Chapter 10. In: Buendia C et al (Eds) IPCC, Switzerland.

Jayarathne J, et al 2021 Effect of aggregate size distribution on soil moisture, soil-gas diffusivity, and N_2O emissions from a pasture soil. *Geoderma* 383, 114737.

Ji C, et al 2018 Variation in soil methane release or uptake responses to biochar amendment: a separate meta-analysis. *Ecosystems* 21, 1692–1705.

Ji C, et al 2020 Differential responses of soil N_2O to biochar depend on the predominant microbial pathway. *Applied Soil Ecology* 145, 103348.

Jiang Q, et al 2022 Deciphering the effects of engineered biochar on methane production and the mechanisms during anaerobic digestion: Surface functional groups and electron exchange capacity. *Energy Conversion and Management* 258, 115417.

Joseph S, et al 2015 The electrochemical properties of biochars and how they affect soil redox properties and processes. *Agronomy* 5, 322–340.

Kammann C, Hepp S, Lenhart K, and Müller C 2009 Stimulation of methane consumption by endogenous CH_4 production in aerobic grassland soil. *Soil Biology and Biochemistry* 41, 622–629.

Kammann C, et al 2015 Plant growth improvement mediated by nitrate capture in co-composted biochar. *Scientific Reports* 5, 1–13.

Krause HM, et al 2018 Biochar affects community composition of nitrous oxide reducers in a field experiment. *Soil Biology and Biochemistry* 119, 143–151.

Lee SI, et al 2021 Biochar-induced reduction of N_2O emission from East Asian soils under aerobic conditions: Review and data analysis. *Environmental Pollution* 291, 118–154.

Li X, et al 2013 Functional groups determine biochar properties (pH and EC) as studied by two-dimensional ^{13}C NMR correlation spectroscopy. *PLoS One* 8, e65949.

Li Y, et al 2022 Sources and intensity of CH_4 production in paddy soils depend on iron

oxides and microbial biomass. *Biology and Fertility of Soils* 58, 181–191.

Liao P, et al 2021 Identifying agronomic practices with higher yield and lower global warming potential in rice paddies: a global meta-analysis. *Agriculture, Ecosystems and Environment* 322, 107663.

Liu X, Mao P, Li L, and Ma J 2019 Impact of biochar application on yield-scaled greenhouse gas intensity: a meta-analysis. *Science of the Total Environment* 656, 969–976.

Liu Y, et al 2021 The long-term effectiveness of ferromanganese biochar in soil Cd stabilization and reduction of Cd bioaccumulation in rice. *Biochar* 3, 499–509.

Lyu H, et al 2022 Biochar affects greenhouse gas emissions in various environments: A critical review. *Land Degradation and Development* 33, 3327–3342.

Mandal S, et al 2016 Designing advanced biochar products for maximizing greenhouse gas mitigation potential. *Critical Reviews in Environmental Science and Technology* 46, 1367–1401.

Nan Q, Xin L, Qin Y, Waqas M, and Wu W 2021 Exploring long-term effects of biochar on mitigating methane emissions from paddy soil: a review. *Biochar* 3, 125–134.

Nisbet RA, et al 2021 Atmospheric methane and nitrous oxide: challenges along the path to net zero. *Philosophical Transcripts of the Royal Society A.* 379, 20200457.

Nguyen TTN, et al 2017 Effects of biochar on soil available inorganic nitrogen: a review and meta-analysis. *Geoderma* 288, 79–96.

Obia A, Cornelissen G, Mulder J, and Dörsch P 2015 Effect of soil pH increase by biochar on NO, N_2O and N_2 production during denitrification in acid soils. *PLoS One* 10, e0138781.

Qin S, et al 2017 Irrigation of DOC-rich liquid promotes potential denitrification rate and decreases $N_2O/(N_2O+ N_2)$ product ratio in a 0–2 m soil profile. *Soil Biology and Biochemistry* 106, 1–8.

Qin Y, et al 2022 Methane emission reduction and biological characteristics of landfill cover soil amended with hydrophobic biochar.

Frontiers in Bioengineering and Biotechnology 10, 905466

Quin PR, et al 2014 Oil mallee biochar improves soil structural properties—A study with x-ray micro-CT. *Agriculture, Ecosystems and Environment* 191, 142–149.

Quin P, et al 2015 Lowering N_2O emissions from soils using eucalypt biochar: the importance of redox reactions. *Scientific Reports* 5, 16773.

Rasse DP, et al 2022 Enhancing plant N uptake with biochar-based fertilizers: limitation of sorption and prospects. *Plant and Soil* 475, 213–236.

Rawat N, Nautiyal P, Kumar M, Vimal V, and Karim AA 2022 Engineered biochar: sink and sequestration of carbon. In: Ramola S, Mohan D, Masek O, Méndez A, and Tsubota T (Eds) *Engineered Biochar: Fundamentals, Preparation, Characterization and Applications.* Singapore: Springer Nature Singapore. pp223–235.

Razzaghi F, Obour PB, and Arthur E 2020 Does biochar improve soil water retention? A systematic review and meta-analysis. *Geoderma* 361, 114055.

Reddy KR, Yargicoglu EN, Yue D, and Yaghoubi P 2014 Enhanced microbial methane oxidation in landfill cover soil amended with biochar. *Journal of Geotechnical and Geoenvironmental Engineering* 140, 04014047.

Ro KS, Lima IM, Reddy GB, Jackson MA, and Gao B 2015 Removing gaseous NH_3 using biochar as an adsorbent. *Agriculture* 5, 991–1002.

Rombel A, Krasucka P, and Oleszczuk P 2022 Sustainable biochar-based soil fertilizers and amendments as a new trend in biochar research. *Science of the Total Environment* 816, 151588.

Rubin RL, Anderson TR, and Ballantine KA 2020 Biochar simultaneously reduces nutrient leaching and greenhouse gas emissions in restored wetland soils. *Wetlands* 40, 1981–1991.

Saggar S, et al 2013 Denitrification and $N_2O{:}N_2$ production in temperate grasslands: Processes, measurements, modelling and mitigating negative impacts. *Science of the Total Environment* 465, 173–195.

Sánchez-García M, Roig A, Sánchez-Monedero MA, and Cayuela ML 2014 Biochar increases soil N_2O emissions produced by nitrification-mediated pathways. *Frontiers in Environmental Science* 2, 25.

Sanford J, Larson R, and Runge T 2019 Nitrate sorption to biochar following chemical oxidation. *Science of the Total Environment* 669, 938–947.

Schlesinger WH, and Hartley AE 1992 A global budget for atmospheric NH_3. *Biogeochemistry* 15, 191–211.

Smith P, Reay D, and Smith J 2021 Agricultural methane emissions and the potential for mitigation. *Philosophical Transactions of the Royal Society A* 379, 20200451.

Sohi SP, Krull E, Lopez-Capel E, and Bol R 2010 A review of biochar and its use and function in soil. *Advances in Agronomy* 105, 47–82.

Song X, Pan G, Zhang C, Zhang L, and Wang H 2016 Effects of biochar application on fluxes of three biogenic greenhouse gases: a meta-analysis. *Ecosystem Health and Sustainability* 2, e01202.

Spokas KA, Novak JM, and Venterea RT 2012 Biochar's role as an alternative N-fertilizer: ammonia capture. *Plant and Soil* 350, 35–42.

Stephen J, et al 1998 Analysis of β-subgroup ammonia oxidiser populations in soil by DGGE analysis and hierarchical phylogenetic probing. *Applied Environmental Microbiology* 64, 2958–2965.

Sun T, et al 2017 Rapid electron transfer by the carbon matrix in natural pyrogenic carbon. *Nature Communications* 8, 14873.

Taghizadeh-Toosi A, Clough TJ, Sherlock RR, and Condron LM 2012 Biochar adsorbed ammonia is bioavailable. *Plant and Soil* 350, 57–69.

Tian H, et al 2020 A comprehensive quantification of global nitrous oxide sources and sinks. *Nature* 586, 248–256.

Van der Weerden TJ, et al 2021 Ammonia and nitrous oxide emission factors for excreta deposited by livestock and land-applied manure. *Journal of Environmental Quality* 50, 1005–1023.

Van Zwieten L, et al 2013 Pyrolysing poultry litter reduces N_2O and CO_2 fluxes. *Science of the Total Environment* 465, 279–287.

Van Zwieten L, et al 2015 Enhanced biological N_2 fixation and yield of faba bean (*Vicia faba* L.) in an acid soil following biochar addition: dissection of causal mechanisms. *Plant and Soil* 395, 7–20.

Verhoeven E, et al 2017 Toward a better assessment of biochar–nitrous oxide mitigation potential at the field scale. *Journal of Environmental Quality* 46, 237–246.

Wang C, et al 2013 Insight into the effects of biochar on manure composting: evidence supporting the relationship between N_2O emission and denitrifying community. *Environmental Science and Technology* 47, 7341–7349.

Wei W, et al 2020 Biochar effects on crop yields and nitrogen loss depending on fertilization. *Science of the Total Environment* 702, 134423.

Weier K, MacRae I, and Myers R 1993 Denitrification in a clay soil under pasture and annual crop: losses from ^{15}N-labelled nitrate in the subsoil in the field using C_2H_2 inhibition. *Soil Biology and Biochemistry* 25, 999–1004.

Wells NS, and Baggs EM 2014 Char amendments impact soil nitrous oxide production during ammonia oxidation. *Soil Science Society of America Journal* 78, 1656–1660.

Wu Z, Zhang X, Dong Y, Li B, and Xiong Z 2019 Biochar amendment reduced greenhouse gas intensities in the rice-wheat rotation system: six-year field observation and meta-analysis. *Agricultural and Forest Meteorology* 278, 107625.

Wu L, et al 2020 A critical review on nitrous oxide production by ammonia-oxidizing Archaea. *Environmental Science and Technology* 54, 9175–9190.

Xia L, et al 2017 Can knowledge-based N management produce more staple grain with lower greenhouse gas emission and reactive nitrogen pollution? A meta-analysis. *Global Change Biology* 23, 1917–1925.

Xiaoqin S, Dongli S, Yuanhang F, Hongde W, and Lei G 2021 Three-dimensional fractal characteristics of soil pore structure and their relationships with hydraulic parameters in biochar-amended saline soil. *Soil and Tillage Research* 205, 104809.

Yang J, Li H, Zhang D, Wu M, and Pan B 2017 Limited role of biochars in nitrogen fixation through nitrate adsorption. *Science of the Total Environment* 592, 758–765.

Yang X, et al 2022 Adsorption properties of seaweed-based biochar with the greenhouse gases (CO_2, CH_4, N_2O) through density functional theory (DFT). *Biomass and Bioenergy* 163, 106519.

Yangjin D, Wu X, Bai H, and Gu J 2021 A meta-analysis of management practices for simultaneously mitigating N_2O and NO emissions from agricultural soils. *Soil and Tillage Research* 213, 105142.

Yin Y, et al 2021 Research progress and prospects for using biochar to mitigate greenhouse gas emissions during composting: A review. *Science of the Total Environment* 798, 149294.

Yu L, Wang Y, Yuan Y, Tang J, and Zhou S 2016 Biochar as electron acceptor for microbial extracellular respiration. *Geomicrobiology Journal* 33, 530–536.

Yuan JH, Xu RK, and Zhang H 2011 The forms of alkalis in the biochar produced from crop residues at different temperatures. *Bioresource Technology* 102, 3488–3497.

Yuan H, et al 2019 Biochar's role as an electron shuttle for mediating soil N_2O emissions. *Soil Biology and Biochemistry* 133, 94–96.

Zhang X, et al 2019 Biochar-mediated anaerobic oxidation of methane. *Environmental Science and Technology* 53, 6660–6668.

Zhang L, et al 2021a Effects of biochar application on soil nitrogen transformation, microbial functional genes, enzyme activity, and plant nitrogen uptake: A meta-analysis of field studies. *Global Change Biology Bioenergy* 13, 1859–1873.

Zhang Y, et al 2021b Microbial pathways account for the pH effect on soil N_2O production. *European Journal of Soil Biology* 106, 103337.

Zhang P, et al 2021c The conductivity and redox properties of pyrolyzed carbon mediate methanogenesis in paddy soils with ethanol as substrate. *Science of the Total Environment* 795, 148906.

Zhang HM, et al 2022 Liming modifies greenhouse gas fluxes from soils: A meta-analysis of biological drivers. *Agriculture, Ecosystems and Environment* 340, 108182.

Zhou J, et al 2021 Biochar-based fertilizer decreased while chemical fertilizer increased soil N2O emissions in a subtropical Moso bamboo plantation. *Catena* 202, 105257.

Zhou W, et al 2022 Ferrate-modified biochar for greenhouse gas mitigation: first-principles calculation and paddy field trails. *Agronomy* 12, 2661.

Zhu L, et al 2015 Sources and impacts of atmospheric NH_3: current understanding and frontiers for modeling, measurements, and remote sensing in North America. *Current Pollution Reports* 1, 95–116.

19

Biochar effects on nutrient leaching

Shahla Hosseini Bai, Michael B. Farrar, Marta Gallart,
Frédérique Reverchon, Sarasadat Taherymoosavi, Negar Omidvar,
Edith Kichamu-Wachira, and Stephen Joseph

Introduction

Over the last two decades, biochar has re-emerged as a promising soil ameliorant that can support crop productivity and improve soil quality. Biochar can offer a multitude of benefits to agricultural systems including carbon (C) sequestration, reduction of greenhouse gas emissions, improved soil nutrient retention, increased soil fertility, and reduction of nutrient loss (Hagemann et al, 2017; Shi et al, 2020). In recent years, there has been an increased interest in the capacity for biochar to retain water and nutrients, and the synergistic effects of combining biochar with inorganic or organic fertilizers to increase crop yield and reduce nutrient leaching in agroecosystems (Konczak and Oleszczuk, 2018; Gao et al, 2019a; Hannet et al, 2021; Bai et al, 2022). However, biochar itself can act as a source of leachable nutrients depending on feedstock composition, production process,

and the plant-soil matrix where it is applied (Mukherjee et al, 2011). Understanding the underlying mechanisms by which biochar can act as a source of leachable nutrients or, on the contrary, reduce soil water and nutrient losses and make them plant available will assist the development of slow-release fertilizers suitable for different crops and farming systems. This chapter addresses the following aspects while providing a comprehensive review of the current literature on: (1) biochar as a source of nutrient leaching; (2) biochar capacity to retain nutrients in soil; (3) the potential mechanisms involved in the biochar capacity to retain and control water and nutrient availability; (4) leaching of inorganic and organic nutrients from biochar and; (5) tailoring of biochar to decrease released soluble nutrients, enhance nutrient retention, and minimize leaching.

DOI: 10.4324/9781003297673-19

Biochar as a source of leachable nutrients

Biochars inherently contain some amount of nutrients including NPK, calcium, or magnesium (Mg) in mineral or organic forms, as well as organic C, all of which are potentially leachable (Chapter 6) (Gao and DeLuca, 2016; Alkharabsheh et al, 2021). The extent to which a specific biochar can act as a source of available/leachable nutrients depends largely on the feedstock type and pyrolysis process during manufacturing (El-Naggar et al, 2019). For example, biochar produced from animal manure or biosolids contained more plant-available P than wood-derived biochar (Wang et al, 2012). In particular, biochars that are richer in Ca or Mg phosphate complexes also usually present higher P availability than biochars richer in Al or Fe phosphate complexes because the latter are generally less soluble (Lindsay et al, 1989). The same pattern has been observed for N. Biochar N contents tend to be greater in biochar derived from nutrient-rich organic wastes rather than from wood or grass feedstocks (Figueredo et al, 2017; El-Naggar et al, 2019). Moreover, nutrients present in the feedstock material can be lost via volatilization during pyrolysis, especially when temperatures are high. As an example, decreasing the pyrolysis temperature from 700°C to 350°C increased available P, K, and sulfur (S) concentrations in chicken manure biochar, although the authors of this study recommended low biochar application rates to minimize the risk of excessive dissolved P in leachates (Hass et al, 2012). Phosphate, ammonium, and nitrate released from biochar were detected by Luo et al (2019) when using swine manure biochar for water nutrient removal. Larger concentrations of total and available P and total K in leachates have also been reported with cow dung biochar applications (Guo et al, 2014), confirming the potential of biochar to act as a source of nutrients, especially when applied at higher rates. Therefore, a better understanding of biochar physicochemical properties and their relation with the sorption and desorption mechanisms of mineral or organic compounds is required to optimize the counteracting effects of biochar on leaching.

Biochar capacity to retain nutrients in soil

The ability of biochar to: (1) retain water, mineral nutrients, and inorganic and organic matter; (2) control nutrient dissolution rate; and (3) release them gradually during crop growth, depends on the characteristics of the biochar (e.g., feedstock composition, production conditions and pre-treatment of biomass or post-treatment of biochar). A given set of production characteristics will strongly influence other biochar properties including specific surface area, porosity, abundance of surface functional groups, and aromaticity, all of which are important determinants of nutrient retention capacity. The methods and strategies to improve biochar storage capacity through feedstock selection, modification of pyrolysis conditions (most importantly, pyrolysis temperature), pre-treatment of the biomass (e.g., loading nutrients into biochar pores and onto surfaces), and post-treatment (e.g., acid washing), have all been discussed by several studies (Hagemann et al, 2017; Zheng et al, 2017; Shi et al, 2020; Leng et al, 2021; Schmidt et al, 2021). Here, we briefly report on the key physicochemical properties that determine the capacity of biochar to

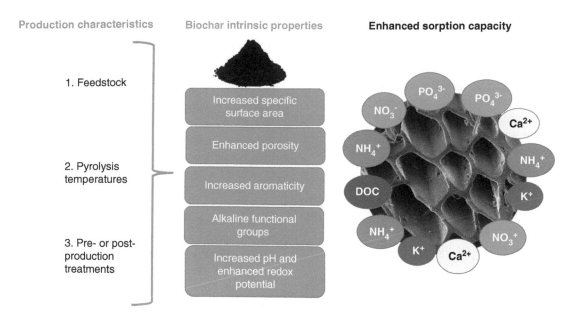

Figure 19.1 *Biochar sorption capacity depends on its physicochemical properties. Adjusting the feedstock material (1), production parameters such as pyrolysis temperature and time (2), and pre- or post-production treatments such as the addition of nanoparticles (3) can help maximize biochar nutrient adsorption and retention capacity, in turn favoring slow release of nutrients and decreasing associated losses*

absorb water and nutrients as summarized in Figure 19.1.

Specific surface area and porosity

Multiple chemical and physical reactions occur on the surface of biochar. Biochar surface area is one of the most important physical properties that can influence biochar nutrient bonding and water-holding capacity. Specific surface area can range from 2 up to 800 m^2 g^{-1} depending on the feedstock biomass, presence of added minerals, processing conditions, and activating agents such as steam, acid bases, and salts (Leng et al, 2021). Biochar with a higher surface area is more effective in adsorbing water, organic and mineral nutrients, and toxic contaminants

(Lehmann, 2007; Ahmad et al, 2014). Biochar produced with lignocellulosic biomass and pyrolysis temperature between 600°C and 700°C display the highest surface areas (Chapter 5; Leng et al, 2021). Biochar surface area increases with increasing temperature until the minerals in the biochar start to melt and block pores (Leng et al, 2021).

The porous nature of biochar can increase net water holding capacity and improve water and nutrient retention when applied to soils (Chapters 8, 10, 20). Biochar pores may become filled or surrounded by mineral nutrients because of chemical reactions and physical incorporation that occurs during pyrolysis (300–500°C) and/or when adding nutrients to biochar as a pre-treatment. High porosity in biochar is generally associated with a high surface area (Leng et al, 2021), and thus its sorption capacity. In particular, micropores are attributed to be the biochar "trapping space" (Chen et al, 2017) and microporosity has been strongly related to the capacity of biochar to act as an efficient sorbent (Ahmad et al, 2014). However, pore structure should also be considered integrally when looking at a biochar sorption capability, since distinct nutrients or compounds would sorb to different pore sizes (Leng et al, 2021). Similar to biochar surface area, porosity is largely influenced by feedstock and pyrolysis temperature (Chapter 5). For example, feedstocks with high lignin content have been shown to produce biochar with high porosity (Sun et al, 2020; Tomczyk et al, 2020), whilst high pyrolysis temperatures have been associated with enhanced microporosity (Chapter 5).

Aromaticity

Biochar possesses an aromatic structure that has been documented to act as π electron donors/acceptors and to form π bonds with mineral nutrients and organic matter (Chapters 7, 8) (Islam et al, 2021). Biochar aromaticity is also highly influenced by pyrolysis temperature and is determined by the ratios of H/C and O/C (Luo et al, 2016). Increases in pyrolysis temperature have been reported to break down the alkyl and ester groups, reduce volatile components and the concentration of both O and N, and hence, change the ratios of H/C and O/C. A highly aromatic structure of biochar is shown by reduced H/C and O/C ratios, suggesting biochar has been highly carbonized (Luo et al, 2016). Biochar aromaticity has been related to its agrochemical sorption capacity (Li et al, 2013) and to Dissolved organic C (DOC) retention (Yang et al, 2019). Moreover, biochars with higher aromaticity have displayed stronger π-π bonds with organic aromatic molecules confirming the potential of biochar for use as a sorbent of organic compounds (Yang et al, 2018).

Surface functional groups

Another key factor moderating the sorption capacity of biochar is biochar surface functionalities (Chapter 10) that can interact with both organic and inorganic compounds and mineral nutrients to retain or release them. Functional groups present on the surface of biochar can directly determine electrostatic interactions that occur on the surface (and within the pores) of the biochar. The electrostatic adsorption, known as anion exchange capacity (AEC) and cation exchange capacity (CEC), is determined as the capacity of biochar to adsorb ionic nutrients and metals (contaminants). Biochar possesses various functional groups on its surface, such as hydroxyls, carbonyls, and carboxyls (Tan et al, 2015). Oxygen-rich carbonaceous functional groups are associated with high CEC of biochar and their concentration defines the affinity of biochar towards metal ions (Bardestani et al, 2018). Depending on feedstock composition, biochar may also contain non-carbonized functional groups, such as iron oxides, that can further enhance the ion exchange capacity of biochar. However, the type and concentration of surface functionalities can change through modification of feedstock composition and production temperature, which affect the absorption capacity of biochar. For example, under specific production temperatures, the presence of more aromatic structures in biochar can increase the negative charge in the orbital (Li et al, 2017). This subsequently enhances the adsorption of ionic species (through ion exchange).

pH

The efficacy of biochar to absorb, stabilize, and enhance the bioavailability of nutrients (ions) is substantially determined by pH. Both biochar pH and pH of the substrate onto which biochar is applied and where the interactions occur (soil and water), have a significant effect on the affinity of organic and inorganic compounds for the biochar surface (Chapter 10). The changes in functional groups through pyrolysis alter the biochar pH. As the pyrolysis temperature increases, O-containing functional groups (i.e., the acidic functional groups such as -OH, C=O, and C-O-C bonds) are destroyed and more alkaline functional groups are formed instead. Moreover, increased pyrolysis temperature induces higher ash content in the produced biochar (Tomczyk et al, 2020). Therefore, biochars produced at relatively high temperatures are often alkaline. Alkaline functional groups have been reported to make biochar into a good electron acceptor or donor and a good sorbent of nutrients and heavy metals (Yang et al, 2019; Islam et al, 2021). Biochars can be designed using specific feedstock and production conditions to control biochar pH by tailoring its redox properties (Chacón et al, 2020). For example, biochars with low ash content and a high number of acidic functional groups typically exhibit low pH (Jia et al, 2020). The charge of the biochar surface functional groups may also be altered with respect to the pH of the solution (and soil), consequently affecting biochar sorption behavior.

Mechanisms involved in the retention of nutrients and water within the biochar

Water, organic and inorganic compounds, and mineral nutrients are adsorbed onto and into biochar through a range of chemisorption and physisorption mechanisms. Sorption mechanisms may include electrostatic attraction, pore filling, partitioning, hydrophobic interaction, H-bonding, ion exchange, complexation, covalent bonding, precipitation, and van der Waals adsorption (Guo et al, 2020; Islam et al, 2021). Bonds can be considered as strong or weak and therefore considered as primary and secondary bonds. Primary bonds include covalent, ionic, and metallic bonds, while dipole–dipole interactions and H-bonding are classified as weaker secondary bonds. To understand how biochar bonds with water and nutrients we first provide an overview of the important sorption mechanisms.

Electrostatic attraction

Electrostatic attraction occurs when a negatively charged atom is attracted by a positively charged atom and vice versa. Adsorption of ionizable organic species (and organic pollutants) on the surface of the biochar occurs through electrostatic attraction, where ionic strength and pH have the greatest effect on interactions between sorbate and sorbent (Guo et al, 2020). For instance, biochar with a negatively charged surface area can sorb nutrients with a positive charge (e.g., NH_4^+) (Nguyen et al, 2017). However, lowering pH values increases the positive charge on the biochar surface leading to improved adsorption of negatively charged organic species (Islam et al, 2021).

Covalent bonding

Covalent bonding (inner-sphere adsorption) is a type in which two or more atoms share electrons equally. The most common type of covalent bond is a single bond in which two atoms share two electrons. Other types of covalent bonds include double and triple bonds. Covalent bonds are often formed between N (in the form of ammonia gas) and organic C in biochar. Covalent bonds are stronger than electrostatic attractions to retain nutrients (Tan et al, 2015). The cations bonded via electrostatic attraction often remain on the outer-sphere layer of biochar particles and can be subsequently lost via leaching (Tan et al, 2015). In computational simulations, covalent bonding is involved in phosphate adsorption onto biochar (Yin et al, 2021). Phosphate interactions through covalent bonding onto Mg-modified biochar were confirmed experimentally by Shin et al (2020) who observed the formation of magnesium phosphate crystals. These findings suggest that covalent bonding is an important mechanism for biochar retention capacity.

Complexation

Complexation is a type of adsorption where metal ions accept electrons from a range of electron-donating functional groups present on the surface of biochar. Therefore, complexation is an important property when trace metals are needed to be retained. Biochars have been suggested to be good sorbents of heavy metals (Chapter 21) because their large specific surface area and abundance of O-containing functional groups favor metal surface complexation (Wang et al, 2020). As for other biochar properties, feedstock and pyrolysis temperature largely influence the capacity of biochar to adsorb toxic elements by complexation. Feedstock, as previously mentioned, will partly determine the abundance of biochar functional groups that can form complexes with metals. On the other hand, biochar produced at lower pyrolysis temperatures often binds metal ions through complexation since low-temperature biochar contains more carboxyl (-COOH), hydroxyl (-OH), and amino (-NH$_2$) groups than those produced under high temperature. The presence of more O-containing functional groups during slow pyrolysis increases the adsorption of metal ions through complexation due to the greater extent of surface oxidation (Li et al, 2017).

Partitioning

Partitioning is a process where organic sorbate molecules diffuse into the organic matter within the noncarbonized portion of biochar, then later in partitioning, these organic compounds solubilize within the matrix of organic matter of biochar to enhance their sorption (Abbas et al, 2018). In biochar specifically, the partitioning of organic compounds occurs on the C amorphous phase containing aliphatic and poly-aromatic compounds such as ketones, sugars, and phenols (Abbas et al, 2018). It has been reported that biochars with higher contents of volatile matter allow greater partitioning of organic molecules and thus, display higher sorption capacities for certain compounds (Liu et al, 2018a; Islam et al, 2021). For example, the application of chicken manure biochar reduced extractable Cd, Pb, and Cu compared with green-waste biochar from artificially contaminated soil (Park et al, 2011). A subsequent experiment involving *Brassica juncea* showed a reduced accumulation of heavy metals in foliar tissue when biochar was applied, most likely due to changes in the partitioning of Cd, Cu, and Pb. The authors suggested a shift of heavy metals from the exchangeable phase to the less bio-available organic bound phase led to decreased plant uptake (Park et al, 2011).

Aging

Biochar undergoes natural aging over time and the aging process modifies biochar physicochemical properties and electrostatic charge characteristics (Chapter 10; Liu and Chen, 2022). Several studies have reported the development of acid groups through oxidation processes during biochar aging under laboratory and field conditions (LeCroy et al, 2013; Mukherjee et al, 2014). The long-term persistence of biochar in soils modifies the dominant processes of how nutrients interact with biochar such as hydrophobic partitioning with surface sorption mechanisms (Jing et al, 2021). For example, an increase in dissolved organic matter (DOM) sorption onto the surface of an aging biochar was described to correspond with a subsequent reduction in the biochar surface area (Heitkötter and Marschner, 2015). Furthermore, once applied to soil, the sorption of soil organic matter (SOM) onto biochar induces an increase in negative surface charges in aging biochar (Mukherjee et al, 2014). Studies have also shown that the application of biochar can be employed for long-term environmental remediation given its capacity to control the bioavailability and mobility of solutes (Yang ct al, 2016; Mingxin et al, 2020). However, once again, the biomass feedstock type and pyrolysis temperature affect biochar properties and are major determinants of the extent of oxidation during the aging process (Heitkötter and Marschner, 2015). Biomass and production characteristics should be carefully evaluated to ensure a lasting beneficial effect before biochar application.

Hydrophobicity

Hydrophobic interaction is important for the adsorption of both water and hydrophobic organic compounds onto biochar (Kinney et al, 2012). Hydrophobic effects are defined as the tendency for nonpolar substances to aggregate in an aqueous solution and eliminate water molecules. Hydrophobic interaction can assist nutrient translocation to plants where biochar has been applied to soil. Hydrophobicity is another biochar characteristic that is temperature-dependent because it is correlated with the presence of aliphatic domains on the surface of biochar and biochar porosity.

Low-temperature biochar can be more hydrophobic due to an accumulation of aliphatic compounds in pores and on biochar surfaces (Joseph et al., 2021) However, hydrophobicity has been demonstrated to disappear when biochar is subjected to water (Das and Sarma, 2015). In another study, aliphatic compounds linked with hydrophobicity were lost as pyrolysis temperatures increased (from 370°C to 620°C) and pores developed in biochar produced from hazelnut shells and Douglas fir chips, leading to more hydrophilic biochar (Gray et al, 2014). Additionally, water molecules are more constrained in small pores, whereas water molecules are more mobile as the pores expand during the pyrolysis process (Conte et al, 2021).

Feedstock composition is also important in moderating biochar hydrophobicity. For example, biochar with high ash content has been associated with an abundance of negative charges on the biochar surface (Tan et al, 2020). Additionally, high-ash biochar is less hydrophobic due to the existence of hydrophilic mineral components in the ash and filling of pores (Hartmann et al, 2010). Aging biochar also influences hydrophobicity: biochar can become more hydrophilic over time, which positively affects ion exchange with the soil solution (Knicker, 2011; Rechberger et al, 2017). Incorporation of hydrophobic biochar can prevent water infiltration and decrease nutrient retention by changing soil sorption properties and potentially increasing leaching (Knicker, 2011).

Hydrophobicity is directly related to the biochar surface area, pore volume, and functional groups such as carboxylic acids on the biochar surface (Mao et al, 2019). Surface area is important in determining biochar hydrophobicity since absorption of water and organic compounds mostly occur through the surface of biochar. Biochar hydrophobicity can also be a function of biochar particle size, where reducing particle size can result in increases in hydrophobicity and further improvement in soil water movement and retention (Gray et al, 2014; Edeh and Masek, 2022). Once water penetrates biochar pore spaces, H_2O molecules interact with pore boundaries according to the inner-sphere mechanism allowing H-bonds (e.g., OH–π) to form on the biochar surface (Jain et al, 2009; Kalra et al, 2020). Water-filled pores within biochar are not only a source of water for plants but can facilitate nutrient retention in the soil profile (Kammann et al, 2015; Hagemann et al, 2017). Finally, biochar hydrophobicity can be also controlled by other parameters, such as soil composition (e.g., the concentration of SOC) where biochar is incorporated into the soil (Mao et al, 2019).

Ligand exchange

Ligand is a type of molecular binding where an ion or functional group binds to a central metal atom which forms a coordination complex. For example, organic matter released from biochar can be used as an organic ligand to bind with metals (e.g., iron or aluminum), thus reducing the amount of iron and aluminum oxides in soil (Ghodszad et al, 2021; Yang et al, 2021). Ammonium may have the potential to undergo ligand exchange with negatively charged functional groups existing on biochar surface (Zhang et al, 2020). Biochar may also adsorb P via ligand exchange, in the form of phosphate ions adsorbed onto the biochar surface with hydroxyl and carboxyl functional groups (Ghodszad et al, 2021). Metal-modified biochar such as Fe(II)-biochar has been proposed

as a means to increase ammonium and phosphate retention while contributing to enhancing P availability (Dong et al, 2020; Wu et al, 2020; Jiao et al, 2021; Xu et al, 2022).

Microbial transformation and immobilization

Soil microorganisms play a major role in transforming nutrients from one form to another (Chapter 16), affecting soil nutrient availability and retention (Liu et al, 2018b). Biochar application can lead to both N microbial immobilization and mineralization. Studies suggest that one-off biochar applications combined with N fertilizer can enhance topsoil N retention by an increase in microbial biomass and incorporation of mineral N into the organic N pool through microbial cycling (Guerena et al, 2013). However, the effects of biochar on microbial biomass are strongly determined by biochar properties (Chapter 14).

Biochar has been shown to alter soil enzymatic activity affecting P availability (Gao et al, 2019b). Biochar can stimulate soil microbial biomass P leading to increased P availability (Gao et al, 2019b). However, increased P availability in response to biochar application may not be always observed (Gul et al, 2015). For example, in acidic soils, originally less P is available for soil microbes to be made available for plant assimilation regardless of biochar presence (Li et al, 2019). Therefore, soil-biochar-microbial interactions have diverse implications on nutrient retention and leaching.

Mechanisms for biochar to bond with N, P, K, Na, and dissolved organic C

Nutrient leaching has become a major global concern and is correlated with the continued

intensification of agricultural production systems. Nutrient leaching is especially concerning in regions where precipitation substantially exceeds evaporation and where irrigation is used to ensure crop yield. Current methods to reduce leaching from agricultural systems include ensuring continuous vegetation cover or use of slow-release or low-solubility fertilizers (Raave et al, 2014). More recently, biochar has been identified as another option to reduce leaching by applying it to soils. Following

biochar application to soil, the bioavailability of macro and micronutrients is affected by several concurrent processes that depend on the combination of soil type, climate, and biochar properties (Bornø et al, 2019). This chapter will explore how and where biochar application to soils in agricultural production systems can moderate leaching pathways for macronutrients, Na, and organic C.

Nitrogen

Several factors significantly influence N loss through the soil profile. The chemical N form, timing, and rate of N-fertilization strongly influence N solubility and mobility in soil ecosystems (Snyder, 1996; Borchard et al, 2019). Nitrate (NO_3^-) is the inorganic N form most prone to leaching as it moves by rapid mass flow, whereas cationic ammonium (NH_4^+) and dissolved organic N (DON) have greater soil retention and are generally transported through diffusion (Giehl et al, 2014).

N cycling after biochar application: Mechanisms and examples

Using biochar as a soil amendment can induce a reduction of N loss through several mechanisms. Biochar interactions with NO_3^- and NH_4^+ control inorganic N leaching (Gao and DeLuca, 2016; Nguyen et al, 2017; Huang et al, 2020). Biochar applications have been shown to decrease soil N inorganic concentrations by up to 10% within the first year of the amendment (Nguyen et al, 2017). The reduction of inorganic N concentrations during the first year of biochar application is mainly driven by the adsorption of NH_4^+ to biochar

thereby reducing availability for transformation in soils expressing net nitrification conditions, and therefore, reducing N available for leaching. The NH_4^+ is eventually released and becomes available for plant uptake. Biochar application also increases the residence time of soil NO_3^- (Bai et al, 2015) due to the NO_3^- retention in biochar pores (Borchard et al, 2019) leading to decreased leaching. However, plant water uptake also affects N retention. For example, plant water transpiration reduces soil water contents and leads to decreased N leaching (Major et al, 2012). NO_3^- leaching has been significantly mitigated when fertilizers are co-applied with biochar, however, the leaching response is still fertilizer rate-dependent (Borchard et al, 2019). A recent meta-analysis has shown that NO_3^- leaching decreases when biochar is combined with fertilizer rates of <150 kg N ha^{-1}, fertilizer rates of > 150 kg N ha^{-1} do not reduce leaching, and fertilizer rates of over 300 kg N ha^{-1} have shown NO_3^- leaching from the system due to the limited biochar capacity to retain NO_3^- on its surface (Borchard et al, 2019).

Biochar can also alter soil N cycling by changing the environmental conditions where soil microbial N transformations occur.

Biochar interferes with the dynamics of soil nitrification-denitrification favoring the N conversion into chemical forms that are less prone to environmental loss (Clough et al, 2013). The addition of biochar drives an increased abundance of microbial N functional genes associated with N conversions, including nitrification and denitrification when soil pH shifts from very acidic to acidic (pH from <5 to 5.5–6.5) (Xiao et al, 2019). For example, where biochar increases pH, microbial conversion of NO_3^- to atmospheric N_2 is stimulated (Clough et al, 2013; Xu et al, 2014). An increased conversion of inorganic N forms into N_2 would result in decreased N_2O emission as well as reduced N leaching (Clough et al, 2013; Xu et al, 2014). Nevertheless, the magnitude of biochar effects on soil microbial N transformations is difficult to predict as it is influenced by multiple factors including soil characteristics, cropping systems, fertilization types, and biochar properties (Xiao et al, 2019).

Despite the benefits of biochar reducing N leaching loss, the rate of application of biochar can strongly influence the portion of N prone to loss. A recent study of increasing biochar rates in a commercial macadamia farm has reported a 3-fold increase in NO_3^- excess when biochar was applied at 30 t ha^{-1} compared with 10 t ha^{-1} rates after 20 months (Asadyar et al, 2021). Generally, decreases in NO_3^- and NH_4^+ leaching after biochar additions to agricultural soils vary from no effect to over 80% N leaching reduction, depending on the biochar type and soil characteristics (Gao and DeLuca, 2016; Borchard et al, 2019; Huang et al, 2020). For example, biochar produced at >500°C decreased NO_3^- leaching due to high NO_3^- retention (Borchard et al, 2019). This is also because biochar produced at high temperatures contains smaller amounts of easily mineralizable C for microbial N immobilization. In addition, the cation exchange capacity (CEC) of biochar determines the NH_4^+ retention, thus higher CEC induces higher N retention (Clough et al, 2013). The biochar feedstock also determines the biochar capacity of nutrient retention. For instance, biochar obtained from lignocellulosic biomass is more effective in reducing NO_3^- leaching than biochar obtained from manure, likely due to the high N content of manures (Borchard et al, 2019).

Soil properties can indirectly affect the level of NO_3^- leaching after biochar application. For example, highly permeable soils with coarse particle sizes and low CEC, such as sandy soils, are more prone to NO_3^- leaching, and biochar applications to these soils can play an important role in decreasing NO_3^- leaching due to increases in soil water-holding capacity (Borchard et al, 2019; Joseph et al, 2021). Additionally, soil pH and organic matter content are other factors driving NO_3^- formation and retention. For example, adding biochar to acidic soils or high organic matter content soils can result in increased N mineralization and nitrification leading to increased NO_3^- concentrations in soil which in turn may affect NO_3^- leaching (Borchard et al, 2019). Biochar can induce net nitrification in natural ecosystems with little or no measurable net nitrification (Clough et al, 2013). This is likely due to decreased consumption of inorganic N and some increases in the actual gross rate of nitrification. Heavy biochar applications can also increase soil permeability, increase soil pH, and aeration, which in turn can increase net nitrification. Usually, biochar does not induce net nitrification in agricultural lands where high net nitrification is experienced naturally (Clough et al, 2013). Hence, both biochar characteristics and soil properties need to be considered when biochar is applied to maximize NO_3^- retention.

Phosphorus

Leached P in groundwater causes eutrophication in aquatic ecosystems where P enrichment causes algae blooms, suffocating aquatic life, decreasing biodiversity, and increasing CH_4 emissions (Bennett and Carpenter, 2002). Soil P can exist in organic and inorganic forms and organic P is generally located close to the soil surface where organic matter is highest (Weihrauch and Opp, 2018). Orthophosphate ions of $H_2PO_4^-$, HPO_4^{2-}, and PO_4^{3-} are bioavailable and can be adsorbed by plant roots (Blum et al, 2018; Weihrauch and Opp, 2018). However, orthophosphates readily precipitate into inorganic forms in both alkaline (Ca-P) and acidic (Fe-P, Al-P) soils and become highly insoluble and immobile (Chapter 16). Additionally, P leaching is more problematic in coarse-textured soils or where excessive P has been applied. This text will examine how biochar application affects: (1) mineralization of organic P; and (2) sorption and precipitation of inorganic P from the soil solution which in turn will moderate P leaching from soils.

Change in the availability of P in soils following biochar application is associated with a variety of physicochemical mechanisms such as altering soil pH, forming alkali-metal phosphate interactions, and providing a net source of P rather than forming what are often strong interactions with soil Al and Fe-oxides. Biochar has been integrated with synthetic fertilizers, used as a bulking agent to improve composts, and, most recently, incorporated into biochar-based fertilizers (BBF) by combination with clays, oxides, minerals, and nutrients (Joseph et al, 2013; Farrar et al, 2019). BBF made at low temperatures increased soil P, K, and Ca availability and total plant P uptake compared with organic fertilizer alone (Farrar et al, 2019). Biochar P adsorption capacity decreases with pyrolysis temperatures above 400°C (Chapter 16). Additionally, BBF can reduce the rate of P diffusion leading to a reduction in P leaching from coarse soils (Melo et al, 2022). Therefore, using biochar and BBFs made from manure feedstocks rather than applying raw manures to agricultural systems can reduce leaching following excessive application.

Potassium

Potassium leaching occurs for a wide range of soil texture groups and is more prevalent in soils with intensive agricultural activity (Jalali and Jalali, 2022). The pathway for K leaching is predominantly to groundwater and the risk of leaching is especially high where soils require frequent irrigation or in tropical regions where high rainfall events are becoming more common (Liu et al, 2021; Jalali and Jalali, 2022). K leaching is highest from permeable sandy soils with low CEC because K cations are attracted to colloidal surfaces with a net negative charge. Leaching of K from soils to groundwater can deplete the soil reserve and lead to plant deficiency, additionally, groundwater high in K should be treated before public supply (World Health Organization, 2009). Leaching of K can be greatly increased by the introduction of H^+ ions to the soil solution following fertilization, decomposition of organic matter, oxidation by weathering, input by precipitation or release from plant roots cause H^+ and Al^{3+} that then adsorb to exchange sites, and the concomitant release of base cations (Ca^{2+}, Mg^{2+}, K^+, and Na^+) (Blum et al, 2018). Alkaline metals (including K) can then be free to leach from soils by percolating water and together with corresponding anions such as NO_3^- and SO_4^{2-} (Blum et al, 2018). Therefore, biochar properties including CEC, pH, and water-holding capacity can play a major role in K leaching.

The concept of "ideal" pyrolysis and feedstock conditions to produce biochar that can simultaneously address leaching and nutrient delivery has not been identified. Additionally, increased K availability can lead to increased leaching and biochar application does not always reduce K leaching (Widowati et al, 2014; Pituello et al, 2015). For example, mixing low and high temperature (3:1) bamboo-biochar has increased soil K concentration (+ 89%) and subsequently led to increased foliar K concentration (+ 25%) compared with organic fertilizer amendments alone (Farrar et al, 2021a),

however on the contrary, applying rice husk, wood or coconut shell biochar reduced K leaching (Widowati et al, 2014). Most recently, biochar compound fertilizers (BCFs) have been found to increase plant K uptake two years after application and without additional organic fertilizer application (Farrar et al, 2021b). BCFs (similar to BBFs) combine biochar with clays, minerals, and nutrients to form novel slow-release products that can be applied at lower rates than pure biochar or fertilizer alone (Joseph et al, 2013; Farrar et al, 2021b). Therefore, mixing biochars with differing physical properties or creating bespoke BCFs can allow farmers to leverage two approaches to reduce K leaching: 1) reducing leaching from the soil by base cation management and 2) reducing the net amount and frequency of fertilizer use. The ability of biochar to moderate cation uptake positively has prodigious significance for the sustainability of intensive agriculture.

Sodium

The leaching of Na^+ is an important process for both the formation and the reclamation of saline-sodic and sodic soils (Blum et al, 2018). Clay soils that contain >15% Na bound to adsorption sites become dispersed and when dry settle into dense layers impeding plant growth (Seelig, 2000). Salinization and sodification are a major threat to arable cropping soils and allowing Na accumulation can adversely affect crop growth and sustainability (Dahlawi et al, 2018). Removal of Na from cation exchange sites is possible using other cations such as calcium (Ca) and subsequent leaching by application of high-quality water or rain. Gypsum is globally accepted

as a useful source of Ca^{2+} to replace exchangeable Na^+, however, gypsum application also increases soil S and Ca which may not always be desirable (Zoca and Penn, 2017; Dahlawi et al, 2018). Biochar application can also decrease Na retention in soils and subsequent adsorption into plants, in turn increasing K, Ca, or Mg through antagonistic effects (Bornø et al, 2019; Farrar et al, 2021). Most recently, an investigation into the potential for rapid assessment of soil Na using hyperspectral methods identified that biochar and BBF application does not impede Na prediction, indicating that biochar and imaging technologies may be used in conjunction to manage salinity issues (Farrar et al, 2023). Careful consideration is required to maximize the benefit of biochar and BBF application in systems with sub-optimal soil base

cation ratios or salinity issues and avoid unwanted deleterious effects such as suppressing cations via antagonistic processes (Dahlawi et al, 2018).

Dissolved organic carbon

DOC is typically low in biochar and can be easily leached (Cheng et al, 2018; Liu et al, 2018c). Biochar pyrolysis temperature is one of the factors affecting biochar DOC content and biochar characteristics to retain DOC. Biochar produced at a low temperature usually contains higher DOC than those produced at higher temperatures (Liu et al, 2018c), additionally, specific surface area and microporosity increase when pyrolysis temperature increases leading to greater DOC adsorption (Cheng et al, 2018). Increased biochar application rate may increase DOC leaching in the short term (Bu et al, 2017). Usually, DOC contents of biochar are not high (Major et al, 2010) and biochar can increase C-C/C-H and COOH groups on biochar surface over time which in turn increases DOC adsorption on biochar surfaces (Darby et al, 2017). DOC on biochar surfaces can weaken biochar-driven changes in soil pH and CEC (Hale et al, 2011; Weng et al, 2017) with implications for nutrient leaching.

Biochar application methods and their implications for nutrient leaching

Biochar can be applied to soils using a variety of approaches (Chapter 25). The method used to apply biochar can dictate the likelihood of nutrient leaching. For example, nitrate leaching magnitude is lower when biochar is mixed evenly at a depth of

0–0.2 m compared with those mixed at depths of 0–0.1 m or 0.1–0.2 m (Li et al, 2018). Regardless of how biochar is applied, biochar is usually co-applied with fertilizers in agroecosystems to supply sufficient nutrient levels for crop yield. Biochar co-application with fertilizers has generally been shown to increase soil nutrient retention even where biochar had only been applied on the soil surface (Bai et al, 2015; Asadyar et al, 2021). Surface application of biochar can reduce nitrification compared with biochar incorporated into the soil profile which can slow soil N losses when biochar has been co-applied with fertilizer (Li et al, 2020). Whilst the effects of biochar characteristics, soil properties, and climatic conditions on nutrient leaching and soil nutrient holding capacity have been widely explored, it is still uncertain how biochar application methods affect soil nutrient holding capacity or nutrient leaching.

Little information is available regarding the effect of biochar application frequency on nutrient leaching. One-off large application of biochar reduced NO_3^- leaching in temperate soils of North-East England but available supply rates need to be considered when recommending the incorporation of large quantities of biochar into the soil (Bell and Worrall, 2011). Contrastingly, more frequent biochar applications were also suggested to minimize NO_3^- leaching (Verheijen et al, 2010). Future research should integrate biochar application frequency into field assessments of nutrient leaching.

Capacity of biochar to retain nutrients

Biochar has been shown to increase plant-available nutrient concentrations in soils when co-applied with fertilizers, hydrophobic organic matter, and inorganic compounds (Bai et al, 2015; Hagemann et al, 2017; Chen et al, 2018; Asadyar et al, 2021; Joseph et al, 2021; Schmidt et al, 2021). By improving soil physicochemical properties such as cation exchange capacity (CEC), biochars made from different feedstocks can enhance plant nutrient availability and soil nutrient retention (Alkharabsheh et al, 2021). The application of biochar at 30 t ha^{-1} in a macadamia orchard increased soil NO_3^- concentration twice above recommended macadamia N requirements after 20 months compared with fertilizer-only application (Asadyar et al, 2021), which highlights

biochar potential to reduce fertilizer inputs and minimize the associated nutrient losses.

Biochar compound fertilizers have been recently designed and investigated across different agricultural systems (Joseph et al, 2013) to further reduce nutrient leaching (Figure 19.2). BCFs usually comprise 20–80% biochar and 5–8% mineral or organic compounds containing N, P, and K (Chew et al, 2020), thus being richer in nutrients than biochar alone whilst showing potential benefits for soil fertility, nutrient loss control, and plant nutrient uptake (Chen et al, 2018; Farrar et al, 2019; Chew et al, 2020; Shi et al, 2020).

The most efficient production method of BCF consists of a granulation process to produce granules of biochar with minerals,

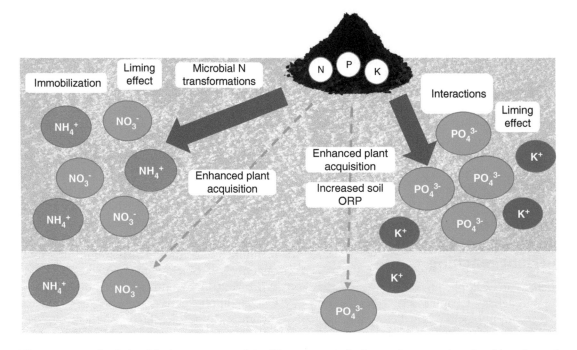

Figure 19.2 *Applying biochar compound fertilizer as a method to reduce nutrient leaching through reduced fertilizer application, slow release of nutrients, and increased plant nutrient uptake*

inorganic NPK, and organic binders (termed here as BNPK) through mechanical extrusion or disc granulation. The increased specific surface area and nutrient adsorption capacity of engineered biochar, including biochar compound fertilizers, enhance nutrient loading and increase nutrient availability for crop growth (Melo et al, 2022; Rasse et al, 2022; Wang et al, 2022) thereby decreasing nutrient leaching. Several studies have reported the positive effects of BCF granules on the availability of soil water, mineral nutrients, and organic matter (Zheng et al, 2017; Chen et al, 2018; Shi et al, 2020). A recent study using soil columns provided evidence of nutrient leaching reduction after the application of biochar-mixed urea granules compared with urea fertilizer (Shi et al, 2020). A significant reduction in N and DOC release rates was reported after 30 days of treatment with biochar-mixed urea granules compared with urea. In a similar study, the application of urea-coated granules with biochar reduced up to 65% nutrient leaching compared with urea alone, acting as a slow-release fertilizer (Chen et al, 2018).

Engineered biochar composites to reduce nutrient leaching

In addition to engineering biochar into BCFs to maximize nutrient content and efficacy, biochar may also be engineered into biochar composites for other specific purposes. Broadly, a BCF is a type of biochar composite with the desired purpose of maximizing plant nutrition. Here, we also briefly discuss biochar composites more broadly where modifications are made to change biochar physicochemical properties and adsorbent behavior.

Biochar composites are made in several ways: (1) pre-treatment or "loading" of feedstock biomass with desired nutrients, metal oxides, hydroxides and/or nano agents before pyrolysis; (2) manipulation of feedstock types and pyrolysis parameters including the highest temperature of treatment (HTT); (3) post-treatment of biochar with aforementioned substances and treated with or without secondary pyrolysis at low temperature; and (4) a combination or permutation of these techniques. This section will focus on exploring some examples of the various methods and the resultant effects on sorption and leaching.

Pre-treatment of feedstock prior to pyrolysis

Pre-treatment substances may be doped into biomass feedstocks via bioaccumulation or blended with feedstocks before pyrolysis. Plants can be grown in high concentrations of specific nutrients and, in turn, bioaccumulate those nutrients (doping), then, when used as a feedstock, the resultant biochar can also display higher specific nutrients. For example, biochar made from Mg-enriched tomato tissue showed better P sorption than the control or Ca-enriched tissues (Yao et al, 2013). Pre-treatment of feedstock by blending metals into biomass before pyrolysis can affect biochar surface area and redox potential (Dieguez-Alonso et al, 2019). Various pyrolysis temperatures and substances have been investigated including blending either aluminum, copper, iron, metal salts including potassium and magnesium chloride and magnesium hydroxide with varying results (Dieguez-Alonso et al, 2019). Biomass pre-treated with magnesium hydroxide and pyrolyzed at 400°C showed the highest phosphate adsorption, while blending with aluminum chloride demonstrated the highest arsenate adsorption (Dieguez-Alonso et al, 2019). The same study showed that K blended biochar produced at 400°C demonstrated no to low release of NO_3^- during initial extraction, indicating the potential of this biochar to reduce NO_3^- leaching (Dieguez-Alonso et al, 2019). Additionally, Fe-blended biochars produced at both low (400°C) and high (700°C) temperatures and K- or Mg-blended biochars (700°C) showed low NO_3^- leaching (Dieguez-Alonso et al, 2019). Importantly, NO_3^- is highly soluble and does not readily adsorb to biochar surfaces without the presence of positively charged sites under acidic conditions, therefore methods such as metal blending with Fe is an important strategy that can increase N retention and reduce leaching (Joseph et al., 2018). Metal blending has been shown to increase microporosity and mesoporosity which then increases biochar adsorption capacity (Dieguez-Alonso et al, 2019). It should also be noted that increasing the content of a specific nutrient via pretreatment may also lead to greater availability following application to soils and therefore greater potential for subsequent leaching.

Pyrolysis parameter manipulation and post-treatment of biochar

The type of biomass feedstock and the pyrolysis temperature are important parameters that influence biochar properties and surface concentration of functional groups (Hassan et al, 2020). Generally, increasing pyrolysis temperature increases biochar surface area, pore size (Chapter 5), ash content, pH, hydrophobicity, and total P and K (Chapter 8) (Pituello et al, 2015; Hassan

et al, 2020). Other factors such as redox are not directly moderated by pyrolysis temperature. For example, Chacon et al (2020) identified that the most important approach to influence the redox properties of biochar is to increase the number of C–OH and C=O groups. Through manipulation of production and modification methods, biochar may be engineered to target specific applications such as electron shuttling or reduction of heavy metals to influence nutrient leaching. It should also be noted that biochar and biochar composites can reduce the leaching of organic matter (Fang et al, 2014).

Freshly produced biochar may be enriched by post-treatment using metal oxides or nanoparticles, clays, and additional nutrient sources and can provide different physico-chemical properties than biochar alone. Biochar surface area and reactivity can be enhanced by metal oxide nanoparticles on pristine biochar and these particles can act as agents to bind with nutrients and pollutants in both soils and water (Tan et al., 2015; Farrar et al, 2019; Mandal et al, 2020). The use of Al-enriched biochars from different feedstocks decreased the leaching of P and N forms (Novais et al, 2018; Yin et al, 2018). Potassium-iron-rice-straw biochar composite promoted PO_4^{3-}, NO_3^-, and NH_4^+ sorption, reducing their leaching when applied to nutrient-poor substrates (Chandra et al, 2020). Adding biochars enriched in nutrients does therefore not automatically translate into increased leaching. Most importantly, the performance and properties of biochar are different if the impregnation occurs before or after pyrolysis (Mandal et al, 2020). There is still a lack of definitive research on biochar composites concerning the effects on soil nutrient leaching/retention in specific agricultural (site-soil-crop) settings.

Conclusions, implications, and further development

Applying biochar to soils affects leaching via different mechanisms and is moderated by biochar properties and soil parameters. Physical and chemical characteristics (including porosity, surface functional groups, and pH) of particular biochar types are important factors affecting nutrient retention and leaching. In recent years, application practices are changing and, more commonly, biochar is being co-applied in combination with fertilizers (Bai et al, 2022). There are opportunities to develop biochar and fertilizer applications by developing BCFs that support sustainable practices aiming to decrease nutrient inputs and nutrient leaching.

Further research needs

Understanding how biochar application methods, application frequencies, and incorporation depth in soil would affect soil nutrient leaching and retention remains a major knowledge gap. Moreover, experiments aimed at determining biochar effects on nutrient leaching and retention need to move from soil-only assessment and incorporate plants in the models, as plant nutrient acquisition is a critical parameter in the assessment of nutrient balances. Ongoing research is still required to engineer specific BCFs to enhance nutrient retention and develop approaches that can be applied in a wide range of soils across cropping systems. Therefore, using biochar and BCFs in high-value, intensive, and irrigated cropping systems will help to minimize leaching. Further research to investigate the incorporation of biochar into potting media blends to reduce nutrient leaching is an important and immediate priority.

References

Abbas Z, et al 2018 A critical review of mechanisms involved in the adsorption of organic and inorganic contaminants through biochar. *Arabian Journal of Geosciences* 11(16), 1–23.

Ahmad M, et al 2014 Biochar as a sorbent for contaminant management in soil and water: a review. *Chemosphere* 99, 19–33.

Alkharabsheh HM, et al 2021 Biochar and its broad impacts in soil quality and fertility, nutrient leaching and crop productivity: A review. *Agronomy* 11(5), 993.

Asadyar L, et al 2021 Soil-plant nitrogen isotope composition and nitrogen cycling after biochar applications. *Environmental Science and Pollution Research* 28, 6684–6690.

Bai SH, et al 2022 Combined effects of biochar and fertilizer applications on yield: A review and meta-analysis. *Science of the Total Environment* 808, 152073.

Bai SH, et al 2015 Wood biochar increases nitrogen retention in field settings mainly through abiotic processes. *Soil Biology and Biochemistry* 90, 232–240.

Bardestani R, and Kaliaguine S 2018 Steam activation and mild air oxidation of vacuum pyrolysis biochar. *Biomass Bioenergy* 108, 101–112.

Bell MJ, and Worrall F 2011 Charcoal addition to soils in NE England: a carbon sink with environmental co-benefits?. *Science of the Total Environment* 409, 1704–1714.

Bennett E, and Carpenter SR 2002 P soup. *World Watch* 15, 24–32.

Blum WEH, Schad P, and Nortcliff S 2018 *Essentials of Soil Science. Soil Formation,*

Functions, Use and Classification (World Reference Base). Clayton South, Australia: CSIRO Publishing.

Borchard N, et al 2019. Biochar, soil and land-use interactions that reduce nitrate leaching and N₂O emissions: a meta-analysis. *Science of the Total Environment* 651, 2354–2364.

Bornø, ML, Müller-Stöver, DS and Liu, F 2019 Biochar properties and soil type drive the uptake of macro- and micronutrients in maize (*Zea mays L.*). *Journal of Plant Nutrition and Soil Science*, 182, 149–158.

Bu X, Xue J, Zhao C, Wu Y, and Han F 2017 Nutrient leaching and retention in riparian soils as influenced by rice husk biochar addition. *Soil Science* 182, 241–247.

Chacón FJ, Sánchez-Monedero MA, Lezama L, and Cayuela ML 2020 Enhancing biochar redox properties through feedstock selection, metal preloading and post-pyrolysis treatments. *Chemical Engineering Journal*, 395, 125100.

Chandra S, Medha I, and Bhattacharya J 2020 Potassium-iron rice straw biochar composite for sorption of nitrate, phosphate, and ammonium ions in soil for timely and controlled release. *Science of the Total Environment* 712, 136337.

Chen Y, Zhang X, Chen W, Yang H, and Chen H 2017 The structure evolution of biochar from biomass pyrolysis and its correlation with gas pollutant adsorption performance. *Bioresource Technology* 246, 101–109.

Chen S, et al 2018 Preparation and characterization of slow-release fertilizer encapsulated by biochar-based waterborne copolymers. *Science of the Total Environment* 615, 431–437.

Cheng H, Jones DL, Hill P, Bastami MS, and Tu CL 2018 Influence of biochar produced from different pyrolysis temperature on nutrient retention and leaching. *Archives of Agronomy and Soil Science* 64, 850–859.

Chew J, et al 2020. Biochar-based fertilizer: supercharging root membrane potential and biomass yield of rice. *Science of the Total Environment* 713, 136431.

Clough TJ, Condron LM, Kammann C, and Müller C 2013 A review of biochar and soil nitrogen dynamics. *Agronomy* 3, 275–293.

Conte P, et al 2021 Recent developments in understanding biochar's physical–chemistry. *Agronomy* 11, 615.

Dahlawi S, Naeem A, Rengel Z, and Naidu R 2018 Biochar application for the remediation of salt-affected soils: Challenges and opportunities. *Science of the Total Environment* 625, 320–335.

Darby I, et al 2017 Short-term dynamics of carbon and nitrogen using compost, compost-biochar mixture and organo-mineral biochar. *Environmental Science and Pollution Research* 23, 11267–11278.

Das O, and Sarmah AK 2015 The love–hate relationship of pyrolysis biochar and water: a perspective. *Science of the Total Environment* 512, 682–685.

Dieguez-Alonso A, et al 2019 Designing biochar properties through the blending of biomass feedstock with metals: Impact on oxyanions adsorption behavior. *Chemosphere* 214, 743–753

Dong D, et al 2020 An effective biochar-based slow-release fertilizer for reducing nitrogen loss in paddy fields. *Journal of Soils and Sediments* 20, 3027–3040.

Edeh IG, and Mašek O 2022 The role of biochar particle size and hydrophobicity in improving soil hydraulic properties. *European Journal of Soil Science* 73, e13138.

El-Naggar A, et al 2019 Biochar composition-dependent impacts on soil nutrient release, carbon mineralization, and potential environmental risk: A review. *Journal of Environmental Management* 241, 458–467.

Fang B, Li X-q, Zhao B, and Zhong L 2014 Influence of biochar on soil physical and chemical properties and crop yields in rainfed field. *Ecology and Environmental Sciences* 23, 1292–1297

Farrar MB, et al 2023 Rapid assessment of soil carbon and nutrients following application of organic amendments. *CATENA* 223, 106928.

Farrar MB, et al 2019 Short-term effects of organo-mineral enriched biochar fertiliser on ginger yield and nutrient cycling. *Journal of Soils and Sediments* 19, 668–682.

Farrar MB, et al 2021a Biochar co-applied with organic amendments increased soil-plant potassium and root biomass but not crop yield. *Journal of Soils and Sediments* 21, 784–798.

Farrar MB, et al 2021b Biochar compound fertilisers increase plant potassium uptake two years after application without additional organic fertiliser. *Environmental Science and Pollution Research International,* 29, 1–15.

Figueredo NAD, Costa LMD, Melo LCA, Siebeneichlerd EA, and Tronto J 2017 Characterization of biochars from different sources and evaluation of release of nutrients and contaminants. *Revista Ciência Agronômica* 48, 395–403.

Gao S, and DeLuca T 2016. Influence of biochar on soil nutrient transformations, nutrient leaching, and crop yield. *Advances in Plants & Agriculture Research* 4, 1–16.

Gao J, et al 2019a Biochar prepared at different pyrolysis temperatures affects urea-nitrogen immobilization and N_2O emissions in paddy fields. *PeerJ* 7, e7027.

Gao S, DeLuca TH, and Cleveland CC 2019b Biochar additions alter phosphorus and nitrogen availability in agricultural ecosystems: a meta-analysis. *Science of the Total Environment* 654, 463–472.

Ghodszad L, Reyhanitabar A, Maghsoodi MR, Lajayer BA, and Chang SX 2021 Biochar affects the fate of phosphorus in soil and water: a critical review. *Chemosphere* 283, 131176.

Giehl RF, and von Wirén N 2014 Root nutrient foraging. *Plant Physiology* 166, 509–517.

Gray M, Johnson MG, Dragila MI, and Kleber M 2014 Water uptake in biochars: the roles of porosity and hydrophobicity. *Biomass Bioenergy* 61, 196e205.

Guerena D, Lehmann J, Hanley K, Enders A, Hyland C, et al 2013 Nitrogen dynamics following field application of biochar in a temperate North American maize-based production system. *Plant and Soil* 365, 239–254.

Gul S, Whalen JK, Thomas BW, Sachdeva V, and Deng H 2015 Physico-chemical properties and microbial responses in biochar-amended soils: Mechanisms and future directions. *Agriculture Ecosystems and Environment* 206, 46–59

Guo M, Song W, and Tian J 2020 Biochar-facilitated soil remediation: Mechanisms and efficacy variations. *Frontiers in Environmental Science* 8, 521512.

Guo Y, Tang H, Li G, and Xie D 2014 Effects of cow dung biochar amendment on adsorption and leaching of nutrient from an acid yellow soil irrigated with biogas slurry. *Water, Air, & Soil Pollution* 225, 1–13.

Hagemann N, et al 2017 Organic coating on biochar explains its nutrient retention and stimulation of soil fertility. *Nature Communications* 8, 1089.

Hale S, Hanley K, Lehmann J, Zimmerman A, and Cornelissen G 2011 Effects of chemical, biological, and physical aging as well as soil addition on the sorption of pyrene to activated carbon and biochar. *Environmental Science and Technology* 45, 10445–10453.

Hannet G, et al 2021 Effects of biochar, compost, and biochar-compost on soil total nitrogen and available phosphorus concentrations in a corn field in Papua New Guinea. *Environmental Science and Pollution Research* 28, 27411–27419.

Hartmann P, Fleige H, and Horn R 2010 Changes in soil physical properties of forest floor horizons due to long-term deposition of lignite fly ash. *Journal of Soils and Sediments* 10, 231e239.

Hass A, et al 2012. Chicken manure biochar as liming and nutrient source for acid Appalachian soil. *Journal of Environmental Quality* 41, 1096–1106.

Heitkötter J, and Marschner B 2015 Interactive effects of biochar ageing in soils related to feedstock, pyrolysis temperature, and historic charcoal production. *Geoderma* 245, 56–64.

Huang LQ, et al 2020 Advances in research on effects of biochar on soil nitrogen and phosphorus. *In IOP Conference Series: Earth and Environmental Science* 424, 012015.

Hassan M, et al 2020 Influences of feedstock sources and pyrolysis temperature on the properties of biochar and functionality as adsorbents: A meta-analysis. *Science of the Total Environment* 744, 140714.

Islam T, Li Y, and Cheng H 2021 Biochars and engineered biochars for water and soil remediation: A review. *Sustainability* 13, 9932.

Jain A, Ramanathan V, and Sankararamakrishnan R 2009 Lone pair ... π interactions between water oxygens and aromatic residues: Quantum chemical studies based on high-resolution protein structures and model compounds. *Protein Science* 18, 481–675.

Jalali M, and Jalali M, 2022 Investigation of potassium leaching risk with relation to different extractants in calcareous soils. *Journal of Soil Science and Plant Nutrition* 22, 1290–1304.

Jia, Y, Hu, Z, Mu, J, Zhang, W, Xie, Z, and Wang, G 2020 Preparation of biochar as a coating material for biochar-coated urea. *Science of The Total Environment*, 731, 139063.

Jiao GJ, et al 2021 Nitrogen-doped lignin-derived biochar with enriched loading of CeO_2 nanoparticles for highly efficient and rapid phosphate capture. *International Journal of Biological Macromolecules* 182, 1484–1494.

Jing F, Liu Y, and Chen J 2021 Insights into effects of ageing processes on Cd-adsorbed biochar stability and subsequent sorption performance. *Environmental Pollution* 291, 118243.

Joseph S, et al 2013 Shifting paradigms: development of high-efficiency biochar fertilizers based on nano-structures and soluble components. *Carbon Management* 4, 323–343.

Joseph S, et al 2018 Microstructural and associated chemical changes during the composting of a high temperature biochar: mechanisms for nitrate, phosphate and other nutrient retention and release. *Science of the Total Environment* 618, 1210–1223.

Joseph S, et al 2021 How biochar works, and when it doesn't: A review of mechanisms controlling soil and plant responses to biochar. *Global Change Biology - Bioenergy* 13, 1731–1764.

Kalra K, Gorle S, Cavallo L, Oliva R, and Chawla M 2020 Occurrence and stability of lone pair-π and OH–π interactions between water and nucleobases in functional RNAs. *Nucleic Acids Research* 48, 5825–5838.

Kammann CI, et al 2015. Plant growth improvement mediated by nitrate capture in co-composted biochar. *Scientific Report* 5, 11080.

Kinney TJ, et al 2012 Hydrologic properties of biochars produced at different temperatures. *Biomass Bioenergy* 41, 34–43.

Knicker H 2011 Pyrogenic organic matter in soil: Its origin and occurrence, its chemistry and survival in soil environments. *Quaternary International* 243, 251–263.

Konczak M, and Oleszczuk P 2018 Application of biochar to sewage sludge reduces toxicity and improve organisms growth in sewage sludge-amended soil in long term field experiment. *Science of the Total Environment* 625, 8–15.

LeCroy C, Masiello CA, Rudgers JA, Hockaday WC, and Silberg JJ 2013 Nitrogen, biochar, and mycorrhizae: Alteration of the symbiosis and oxidation of the char surface. *Soil Biology and Biochemistry* 58, 248–254.

Lehmann J 2007 Bio-energy in the black. *Frontiers in Ecology and the Environment* 5, 381–387.

Leng L, et al 2021 An overview on engineering the surface area and porosity of biochar. *Science of the Total Environment* 763, 144204.

Li F, et al 2019 Effects of biochar amendments on soil phosphorus transformation in agricultural soils. *Advances in Agronomy* 158, 131–172.

Li H, et al 2017 Mechanisms of metal sorption by biochars: biochar characteristics and modifications. *Chemosphere* 178, 466–478.

Li S, Zhang Y, Yan W, and Shangguan Z 2018 Effect of biochar application method on nitrogen leaching and hydraulic conductivity in a silty clay soil. *Soil and Tillage Research* 183, 100–108.

Li, X et al. 2020 Application methods influence biochar–fertilizer interactive effects on soil nitrogen dynamics. *Soil Science Society of America Journal*, 84, 1871–1884.

Li J, Li Y, Wu M, Zhang Z, and Lu J 2013 Effectiveness of low-temperature biochar in controlling the release and leaching of herbicides in soil. *Plant Soil* 370, 333–344.

Li, H, Dong, X, da Silva, EB, de Oliveira, LM, Chen, Y, and Ma, LQ 2017 Mechanisms of metal sorption by biochars: Biochar characteristics and modifications. *Chemosphere*, 178, 466–478.

Lindsay WL, Vlek PL, and Chien SH 1989 Phosphate minerals. *Minerals in soil environments, 2nd edition. SSSA Book Series* 1, 1089–1130.

Liu Y, and Chen J 2022 Effect of ageing on biochar properties and pollutant management. *Chemosphere* 292, 133427.

Liu DL, et al 2021 Characterizing spatiotemporal rainfall changes in 1960–2019 for continental Australia. *International Journal of Climatology* 41, E2420–E2444.

Liu G, et al 2018a Formation and physicochemical characteristics of nano biochar: Insight into chemical and colloidal stability. *Environmental Science & Technology* 52, 10369–10379.

Liu Q, et al 2018b How does biochar influence soil N cycle? A meta-analysis. *Plant and Soil* 426, 211–225.

Liu CH, et al W 2018c Quantification and characterization of dissolved organic carbon from biochars. *Geoderma* 335, 161–169

Luo J, et al 2016 Cassava waste derived biochar as soil amendments: Effects on Kinetics, equilibrium and thermodynamic of atrazine adsorption. *Fresenius Environmental Bulletin* 25, 4607–4617.

Luo L, et al 2019 The characterization of biochars derived from rice straw and swine manure, and their potential and risk in N and P removal from water. *Journal of Environmental Management* 245, 1–7.

Major J, Rondon M, Molina D, Riha SJ, and Lehmann J 2012 Nutrient leaching in a Colombian savanna Oxisol amended with biochar. *Journal of Environmental Quality* 41, 1076–1086.

Major J, Lehmann J, Rondon M, and Goodale C 2010 Fate of soil-applied black carbon: downward migration, leaching and soil respiration. *Global Change Biology* 16, 1366–1379.

Mandal S, et al 2020 Progress and future prospects in biochar composites: Application and reflection in the soil environment. *Critical Review Environmental Science and Technology* 51, 219–271.

Mao J, Zhang K, and B 2019 Linking hydrophobicity of biochar to the water repellency and water holding capacity of biochar-amended soil. *Environmental Pollution* 253, 779–789.

Melo LCA, Lehmann J, Carneiro J, and Camps-Arbestain M 2022 Biochar-based fertilizer effects on crop productivity: a meta-analysis. *Plant and Soil* 472, 45–58.

Mingxin G, Weiping S, and Jing T 2020 Biochar-Facilitated Soil Remediation: Mechanisms and Efficacy Variations. *Frontiers in Environmental Science* 8, 521512.

Mukherjee A, Zimmerman AR, Hamdan R, and Cooper WT 2014 Physicochemical changes in pyrogenic organic matter (biochar) after 15 months of field aging. *Solid Earth* 5, 693–704.

Mukherjee A, Zimmerman AR, and Harris W 2011 Surface chemistry variations among a series of laboratory-produced biochars. *Geoderma* 163, 247–255.

Nguyen TTN, et al 2017 Short-term effects of organo-mineral biochar and organic fertilisers on nitrogen cycling, plant photosynthesis, and nitrogen use efficiency. *Journal of Soils and Sediments* 17, 2763–2774.

Novais SV, Zenero MDO, Barreto MSC, Montes CR, and Cerri CEP 2018 Phosphorus removal from eutrophic water using modified biochar. *Science of the Total Environment* 633, 825–835.

Park JH, Choppala GK, Bolan NS, Chung JW, and Chuasavathi T 2011 Biochar reduces the bioavailability and phytotoxicity of heavy metals. *Plant and Soil* 348, 439–451.

Pituello C, et al 2015 Characterization of chemical–physical, structural and morphological properties of biochars from biowastes produced at different temperatures. *Journal of Soils and Sediments* 15, 792–804.

Raave H, et al 2014 The impact of activated carbon on NO_3^--N, NH_4^+-N, P and K leaching in relation to fertilizer use. *European Journal of Soil Science* 65, 120–127.

Rasse DP, et al 2022 Enhancing plant N uptake with biochar-based fertilizers: limitation of sorption and prospects. *Plant and Soil* 475, 213–236.

Rechberger MV, et al 2017 Changes in biochar physical and chemical properties: accelerated biochar aging in an acidic soil. *Carbon* 115, 209–219.

Schmidt HP, Kammann C, Hagemann N, Leifeld J, Bucheli TD, et al 2021 Biochar in agriculture–A systematic review of 26 global meta-analyses. *Global Change Biology Bioenergy* 13, 1708–1730.

Seelig B 2000 Salinity and sodicity in North Dakota soils. NDSU Extension Service, North Dakota State University of Agriculture and Applied Science, North Dakota EB 57.

Shi W, et al 2020 Biochar bound urea boosts plant growth and reduces nitrogen leaching. *Science of the Total Environment* 701, 1344243.

Shin H, Tiwari D, and Kim DJ 2020 Phosphate adsorption/desorption kinetics and P bioavailability of Mg-biochar from ground coffee waste. *Journal of Water Process Engineering* 37, 101484.

Sun Y, et al 2020 Tailored design of graphitic biochar for high-efficiency and chemical-free microwave-assisted removal of refractory organic contaminants. *Chemical Engineering Journal* 398, 125505.

Tan X, et al 2015 Application of biochar for the removal of pollutants from aqueous solutions. *Chemosphere* 125, 70–85.

Tan Z, Yuan S, Hong M, Zhang L, and Huang Q 2020 Mechanism of negative surface charge formation on biochar and its effect on the fixation of soil Cd. *Journal of Hazardous Materials* 384, 121370.

Tomczyk A, Sokołowska Z, and Boguta P 2020 Biochar physicochemical properties: pyrolysis temperature and feedstock kind effects. *Reviews in Environmental Science and Bio/Technology* 19, 191–215.

Verheijen F, Jeffery S, Bastos AC, Van der Velde M, and Diafas I 2010 Biochar application to soils. A critical scientific review of effects on soil properties, processes, and functions. *EUR*, 24099, 162.

Wang T, Camps-Arbestain M, Hedley M, and Bishop P 2012 Predicting phosphorus bioavailability from high-ash biochars. *Plant and Soil* 357, 173–187.

Wang S, et al 2020 Biochar surface complexation and Ni (II), Cu (II), and Cd (II) adsorption in aqueous solutions depend on feedstock type. *Science of the Total Environment* 712, 136538.

Wang C, et al 2022 Biochar-based slow-release of fertilizers for sustainable agriculture: A mini review. *Environmental Science and Ecotechnology*, 10, 100167.

Weihrauch C, and Opp C 2018 Ecologically relevant phosphorus pools in soils and their dynamics: the story so far. *Geoderma* 325, 183–194.

Weng ZH, et al 2017. Biochar built soil carbon over a decade by stabilizing rhizodeposits. *Nature Climate Change* 7, 371–376.

Widowati W, Asnah A, and Utomo W 2014 The use of biochar to reduce nitrogen and potassium leaching from soil cultivated with maize. *Journal of Degraded and Mining Lands Management* 2, 211.

World Health Organization 2009 *Potassium in Drinking-Water: Background Document for Development of WHO Guidelines for Drinking-Water Quality*. World Health Organization, Geneva.

Wu L, Zhang S, Wang J, and Ding X 2020 Phosphorus retention using iron (II/III) modified biochar in saline-alkaline soils: Adsorption, column and field tests. *Environmental Pollution* 261, 114223.

Xiao Z, et al 2019. The effect of biochar amendment on N-cycling genes in soils: a meta-analysis. *Science of the Total Environment* 696, 133984.

Xu HJ, et al 2014 Biochar impacts soil microbial community composition and nitrogen cycling in an acidic soil planted with rape. *Environmental Science and Technology* 48, 9391–9399.

Xu D, et al 2022 Removal of nitrogen and phosphorus from water by sludge-based biochar modified by montmorillonite coupled with nano zero-valent iron. *Water Science and Technology* 85, 2114–2128.

Yang K, Jiang Y, Yang J, and Lin D 2018 Correlations and adsorption mechanisms of aromatic compounds on biochars produced from various biomass at 700 °C. *Environmental Pollution* 233, 64–70.

Yang F, et al 2021 Stabilization of dissolvable biochar by soil minerals: Release reduction and organo-mineral complexes formation. *Journal of Hazardous Materials* 412, 125213.

Yang X, Zhang S, Ju M, and Liu L 2019 Preparation and modification of biochar materials and their application in soil remediation. *Applied Sciences* 9, 1365.

Yang X, et al 2016 Effect of biochar on the extractability of heavy metals (Cd, Cu, Pb, and Zn) and enzyme activity in soil. *Environmental Science and Pollution Research* 23, 974–984.

Yao Y, et al 2013 Engineered carbon (biochar) prepared by direct pyrolysis of Mg-accumulated tomato tissues: characterization and phosphate removal potential. *Bioresource Technology* 138, 8–13.

Yin Q, Liu M, Li Y, Li H, and Wen Z 2021 Computational study of phosphate adsorption on Mg/Ca modified biochar structure in aqueous solution. *Chemosphere* 269, 129374.

Yin Q, Wang R, and Zhao Z 2018 Application of Mg–Al-modified biochar for simultaneous removal of ammonium, nitrate, and phosphate from eutrophic water. *Journal of Cleaner Production* 176, 230–240.

Zhang D, et al 2020 Utilization of Jujube biomass to prepare biochar by pyrolysis and activation: Characterization, adsorption characteristics, and mechanisms for nitrogen. *Materials* 13, 5594.

Zheng J, et al 2017 Biochar compound fertilizer increases nitrogen productivity and economic benefits but decreases carbon emission of maize production. *Agriculture, Ecosystems and Environment* 241, 70–78.

Zoca SM, and Penn C 2017 An important tool with no instruction manual: a review of gypsum use in agriculture. *Advances in Agronomy* 144, 1–44.

20

Biochar effects on water availability

Xiaodong Gao and Caroline A. Masiello

Introduction

Soil water improvement is one of the first changes observed after soil is amended with biochar. When biochar is made from plant materials it has very high porosity, and its soil application has the potential to improve a variety of soil physical properties related to water, including bulk density, total porosity, pore size distribution, and other pore-related hydraulic properties. In particular, biochar can significantly increase soil water retention and availability, playing an important role in increasing the resilience of agricultural production to severe droughts due to global climate change (Jeffery et al, 2011; Ali et al, 2017).

Understanding the biochar-soil interactions that result in soil water improvements and optimizing them will help the biochar community deliver reliable benefits to farmers, ranchers, and urban land managers. Here we discuss biochar's effects on fundamental soil water properties and provide a framework for selecting and applying biochar to increase soil water availability. For this chapter, the term "biochar" specifically refers to biochar made from plant remains, which is highly porous. When other feedstocks are used, the resulting biochar may have a much lower porosity and may not yield the same hydrologic benefits.

Soil water is critical for soil ecological and microbial processes

Soil water is essential for the survival and growth of plants and soil microbes, and is critical for all soil ecological processes, including organic matter production, carbon and nutrient cycling, as well as soil formation. Soil water is conceptually divided into gravitational water retained in macropores, capillary water existing in meso- and micro-pores, and adsorbed water that adheres to soil organic matter (SOM) and clay particles (Figure 20.1). The size of each soil water pool is controlled by interacting properties.

DOI: 10.4324/9781003297673-20

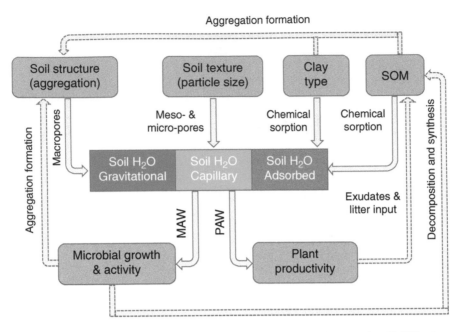

Figure 20.1 *Schematic of soil water pools and their controlling soil properties. PAW: plant available water, MAW: microbially available water (adapted from Moyano et al, 2013; Jabro et al, 2020)*

Plants and microorganisms generally can only extract and use water from the capillary water pool. Furthermore, soil biota can modify the soil water through their contribution to the SOM pool and soil aggerate formation.

The most important characteristics of soil water are the amount of water in the soil (soil water content) and the force holding the water in the porous soil matrix, known as the soil water potential (Jury et al, 1991). Soil water content influences soil properties, such as O_2 and nutrient concentrations, soil temperature, and number of water-filled pores. Soil water potential controls water movement, including the efficiency of plant water extraction, the amount and speed of drainage occurring under gravity, and the movement of solutes after irrigation and rainfall events (Jury et al, 1991).

Water movement and retention processes in the soil such as infiltration, drainage, and evaporation are very complex, often described by mathematic models derived from soil water

theory and experimental data (Simunek et al, 2003; Beven and Germann, 2013). Ongoing climate change significantly increases the frequency of extreme hydrological events, such as extreme precipitation events, flooding, and long-term severe droughts, posing an imminent threat to global food production (Trenberth, 2011; Reichstein et al, 2014). Consequently, the ability of soil to retain enough water to sustain biological activities during droughts, while also providing adequate drainage of water after floods is critical for sustainable agricultural production.

Definitions of important soil water properties

Water is held within the soil matrix by adsorption at the surface of particles and by capillary forces in the pores, influenced by soil texture, structure, SOM contents, and the type of clays present (Figure 20.1). The total amount of water held in soils is sometimes less important than the energy with which soil holds onto that water, because some clay soils can hold water so tightly that it is not available to plants or microbes. Soil water potential is a measure of the energetic state of soil water, indicating how tightly H_2O molecules are bound to the soil matrix, either structurally or chemically, which provides information on the availability of soil water to plants and microbes. Water potential is the sum of several forces, including matric potential (Ψ_m), osmotic potential (Ψ_o), gravitational potential (Ψ_g), and pressure potential (Ψ_p). Among these, the matric potential and osmotic potential are the most important for plants and microbes. Ψ_m depends on soil matric properties, including soil texture, structure, type of particle surfaces, and SOM content. Ψ_o depends on the concentration of soil solutes (both ionic and non-ionic)

that create diffusion gradients for water movement (Marshall et al, 1996).

The soil water characteristic curve (SWCC) describes the relationship between soil water potential and soil water content. The conceptual model of McQueen and Miller (1974) suggests that SWCC can be approximately divided into three regions (Figure 20.2), corresponding to (i) the tightly adsorbed region dominated by molecular bonding mechanisms, e.g., hydrogen-bonding and van der Waals attraction ($\sim -10^6$ to -10^4 kPa), as a thin water film adhering to soil particle surfaces; (ii) the adsorbed film region ($\sim -10^4$ to -100 kPa) dominated by short-range solid-liquid interactions mechanisms, and (iii) the capillary region where water retention mechanisms are controlled by capillary forces (-100 kPa to 0).

Three key measures of soil water were proposed to describe the amount of soil water that is accessible to plants: (i) field

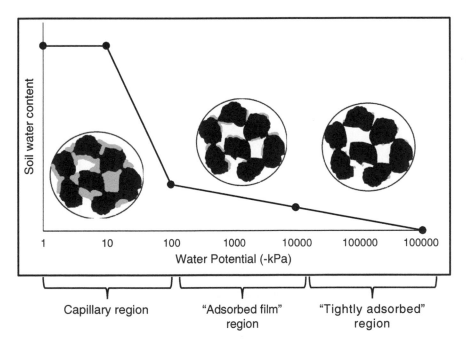

Figure 20.2 *Conceptual model of the soil water characteristic curve (adapted from McQueen and Miller, 1974; Huang et al, 2021)*

capacity (FC), the soil water retained in soil after free drainage, generally defined at −33 kPa suction or −10 kPa for sandy soils (Israelsen and West, 1922; Veihmeyer and Hendrickson, 1931, 1949), (ii) permanent wilting point (PWP), the water content of a soil when plants wilt and cannot recover turgor after watering, generally occurring at −1500 kPa suction (Veihmeyer and Hendrickson, 1928), and (iii) plant available water (PAW), the water content of a soil between FC and PWP (Veihmeyer and Hendrickson, 1927) (Equation 20.1):

$$PAW = FC - PWP \qquad [20.1]$$

Therefore, any practice to increase the FC or decrease the PWP will lead to an increase in available water for plant uptake and agricultural productivity. The PAW is related to many soil properties. Among them, soil texture or particle size distribution is the major determinant of the PAW of a soil. Coarse-textured (sandy) soils generally have both small FC and PWP values, whereas fine-textured (clay) soils have large FC and large PWP values. In both cases, the PAW is relatively low (Figure 20.3). Soils with intermediate texture such as loam soil, have greater differences between the FC and PWP, resulting in the largest PAW for plant growth (Figure 20.3). The PAW described in Figure 20.3 is an estimate that can be improved with species-specific information. This is because plant species have variable water demands depending on their phenology and environmental conditions (Curtis and Claassen, 2008).

Soil water is also critical for microbial activities and growth. Soil microorganisms inhabit the pore network and rely on free water in the pores and water films on surfaces for metabolic and biochemical processes that are essential to their survival (Vos et al, 2013). Low water potential causes osmotic stress that results in a drastic decrease in microbial activities once the soil water potential drops below a threshold point (Manzoni et al, 2012). For instance, the CH_4 oxidation activity of methanotrophs in

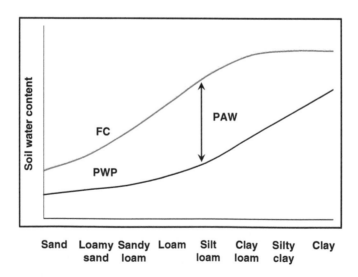

Figure 20.3 *Soil water curves, illustrating the water content at field capacity (FC) and the permanent wilting point (PWP) as a function of the soil texture. Plant available water (PAW) is estimated from the difference between FC and PWP*

landfill soil ceased when the water potential reached the threshold at −1.5 MPa (Spokas and Bogner, 2011). Here, we propose a new concept: microbially available water (MAW), equivalent to PAW for plants. We define MAW as the water content between soil saturation and the threshold that causes microbes to cease their growth. The water-stress threshold for microbial activities varies significantly among the microbial species. Bacteria exhibited water-stress thresholds of −0.6 MPa for gram-negative (Stark and Firestone, 1995) and −2 MPa for gram-positive bacteria (Fierer et al, 2003), whereas fungi generally remain active in much dryer conditions (as dry as −60 MPa), likely due to their filamentous structure (Williams and Hallsworth, 2009).

Soil porosity and pore size distribution are key to retaining soil water and supplying plant-available water

Pore size is critical for soil water retention in the capillary pool, which is available for plants and microorganisms. Soil is a highly complex and heterogeneous matrix of aggregated solid particles with large pore spaces filled with air and water (Babin et al, 2013; Vos et al, 2013). Soil pores are traditionally classified into five different sizes: macropores (> 80 μm) that allow rapid flow of water, mesopores (30–80 μm) where water moves in response to matric potential differences, micropores (5–30 μm) that retain water by capillary forces, ultra-micropores (0.1–5 μm), and nanopores (< 0.1

μm) (Soil Science Society of America, 1996). Water moves quickly through large macropores and is drained by gravity, whereas ultra-micropores and nanopores in the sub-micron range retain water so strongly that it is inaccessible to plants or microbes. Only micropores that contain capillary water are relevant as a source of plant-available water (Kameyama et al, 2019).

Soil aggregation also contributes to soil water retention. Larger pores formed when soil particles aggregate (i.e., inter-aggregate pores between soil agglomerates) will hold water at lower energies than the water present in smaller pores that exist between individual soil grains, also called intra-aggregate pores. The process of aggregation allows clay soils to hold more plant-available water. Aggregation also creates new flow paths for water in the soil, fundamentally altering the soil hydrological system. The bimodal pore size distribution caused by aggregation means that soil with a well-defined structure (i.e., with many aggregates > 250 μm) often exhibits SWCC with a discernible double-hump feature, resulting from the two-pore systems: one for the inter-aggregate pores and other for the intra-aggregate pores (Ghezzehei, 2011). Water-filled pore space and water-film thickness decrease with decreasing soil water potential, with large pores draining out first, followed by meso- and micro-pores (Vos et al, 2013). Precise knowledge of the soil pore size distribution and its ability to control water retention is the key to understanding the soil-water dynamics that develop through biochar-water interactions in soils.

Biochar properties that modulate soil water retention

Biochar is a porous substance that contains air and water in the spaces between solid carbonaceous materials. However, biochar pores are often classified using a different system at much smaller scales because the pore sizes are smaller in biochar than in soil.

Biochar porosity includes macropores (> 50 nm), mesopores (2–50 nm), and micropores (< 2 nm) following IUPAC conventions (Gray et al, 2014). This classification for biochar pores was adopted from gas adsorption equations to assess the porosity of activated carbon, and does not adequately describe large micron-sized pores that are also present in biochar. These micron-sized pores may be the dominant reservoir for water retention and PAW (Gray et al, 2014). Therefore, we will use the Soil Science Society of America (1996) classification system for soil pores to explain the porosity of biochar in this chapter.

The impacts of biochar on soil water retention and other hydraulic properties have been thoroughly studied since the early 2000s. The first report came from Amazonia *Terra preta* soils that were enriched in charcoal and had greater water retention than adjacent soils without charcoal (Glaser et al, 2002). Even though the effect of biochar on soil water properties strongly depends on soil type and biochar properties, the positive effect of biochar on soil water dynamics has been confirmed and corroborated by many subsequent studies during the following decades (e.g., Omondi et al, 2016; Razzaghi et al, 2019; Edeh et al, 2020). Biochar application generally decreases soil bulk density, increases soil porosity and wet aggregate stability, and results in greater soil water retention (Omondi et al, 2016; Blanco-Canqui, 2017; Edeh et al, 2020).

The effect of biochar on soil water properties is a function of both the properties of the soil being amended and of the biochar itself. For example, when biochar particles are larger than soil grains, the addition of biochar creates larger spaces between soil pores. Larger pores allow faster flow of water, increasing the rate of soil drainage and reducing water lost through surface runoff. In addition, larger pores hold water less tightly, allowing more plant access. The amendment scenario where biochar particles are larger than soil particles could reasonably be expected to improve performance in clay soils but may not be beneficial in sandy soils that have naturally high macroporosity. Conversely, a situation where biochar particles are smaller than soil particles could be expected to improve water retention in sandy soil.

Particle size is not the only biochar property that impacts soil water properties: biochar's capacity to improve soil water retention depends on the combination of biochar properties (e.g., porosity and pore size distribution, hydrophobicity, particle size; Chapter 5) and soil properties (e.g., soil texture and aggregation, SOM content) (Figure 20.4), and because of this, it is not possible to characterize the water-improving properties of any particular biochar product on its own.

Increased soil water retention after biochar application can be a direct effect from biochar intrapores (the pores inside biochar particles) or an indirect effect from interpores (the pores between soil-biochar particles). Biochar is a highly porous material that has a direct capacity to adsorb water onto its surface and physically hold water in its pores. Indirectly, biochar particles interact with soil particles to create new agglomerates, creating interpore space at the biochar-soil interface. The macro- and mesopores within the biochar structure and the interparticle pores created by biochar-soil interactions may vary in size and have fundamentally different roles in water retention (Yang and Lu, 2021). In a meta-analysis of soil water properties, biochar application increased FC (+20.4%), PWP (+16.7%), and PAW (+28.5%), but the effects varied significantly among soil types (Edeh et al, 2020). The key biochar factors determining biochar

Figure 20.4 *Parameters including biochar properties (e.g., biochar intrapore volume, pore size distribution, hydrophobicity, particle size), management practices (e.g., biochar application rate), soil properties (e.g., soil texture, aggregation, and SOM content), and climate conditions impact biochar soil application on soil water retention, plant available water (PAW) and microbially available water (MAW)*

performance were biochar particle size, specific surface area, and porosity, indicating that both biochar intrapores and biochar-soil interpores contributed to the improved water properties (Edeh et al, 2020). Therefore, it is not surprising that inconsistent findings were documented in previous research. The heterogeneity between experiments in terms of biochar characteristics and application rates, soil properties, and experimental design all could lead to variable results and limit direct comparisons among different studies (Razzaghi et al, 2019). This points to the need to carefully report several biochar and soil characteristics (Figure 20.4) for more robust meta-analyses and to reduce the uncertainty in predictive models.

Biochar contribution to soil porosity

Soil amended with biochar gains porosity from biochar intrapores and soil-biochar interpores. Interpores are typically larger, and their size is controlled by soil texture and its degree of aggregation. Biochar affects interpore development because there is a difference in the shape and size of biochar particles and soil aggregates. The effects of biochar on soil interpores typically drive the most significant and first-order influence on soil water. Intrapores, which occur inside biochar particles, are typically smaller than interpores and usually exert a second-order influence on soil water retention (Liu et al, 2017).

Biochar's intrapore contribution to soil water retention

Biochar intraporosity and pore size distribution, determined by production conditions and feedstock type, affect its water retention and PAW (Kinney et al, 2012; Brewer et al, 2014, Gao et al, 2017). Greater biochar intraporosity is expected to increase water retention. However, the size of the pores determines whether the water will be available for plant uptake. Biochar contains two types of pores that vary over several orders of magnitude (Chapter 5): pyrogenic nanopores in the range of nanometers (< 30 nm) generated during pyrolysis, and residual macropores in the size range of 1–100 μm that are part of the plant cellular structure in the biomass feedstock (Gray et al, 2014; Gao and Masiello, 2017). Pyrogenic nanopores generated at high temperatures only increase the water storage at water potential < −10 MPa, which is far below the water stress thresholds of most plants (Verheijen et al, 2010). Plant roots can access water from pores > 0.2 μm at the PWP of −1500 kPa and obtain water from pores < 30 μm at the FC of −33 kPa, based on the calculation of capillary forces using Equation 20.2 (Verheijen et al, 2010; Kameyame et al, 2019). Thus, the large micron-sized pores inherited from the feedstock plants often account for about 95% of total intraporosity in biochar, and play a dominant role in retaining PAW (Gao and Masiello, 2017). In other words, feedstock type is more important than pyrolysis temperature for providing effective porosity for PAW. Indeed, this is confirmed by Kameyama et al (2019), who reported greater PAW with wood biochar that had intact cell wall structures and large pores in the order of 10 μm than biochar made from bamboo, rice husk, and poultry manure, regardless of the pyrolysis temperature.

The application of biochar with a large volume of intrapores is expected to increase soil water retention and PAW. Despite the apparent link between biochar total porosity and pore size distribution with soil water retention, there are few reports of the pore size distribution in biochar applied to soil. Biochar application can induce an increase in PAW in clay soil mainly in the matric potential region from −0 to −316 kPa, corresponding to pores > 1 μm (Rasa et al, 2018). The authors attributed the increase in soil porosity and PAW to the additional porosity from biochar intrapores with sizes at 50 and 10 μm diameter. Similarly, the pore size distribution of clay soil before and after biochar application showed the formation of a large proportion of pores with sizes of 2–3 μm and 10–100 μm after biochar application (Sun and Lu, 2014). This coincides with the biochar intrapore size distribution and suggests a direct contribution of biochar intrapores to the increased soil porosity and PAW. However, both studies were conducted in clay soils that cannot rule out the contribution of interpores formed between clay particles and biochar particles with pore sizes similar to the micro-sized biochar intrapores and contribute more PAW (Liu et al, 2017). It would be helpful to repeat such studies with sandy soils, to confirm the PAW contained in biochar intrapores under realistic conditions.

Biochar's particle size determines its contribution to soil interpores

Soil amended with biochar forms interpores between biochar particles and soil particles, which also contribute to soil water retention. In addition to biochar intrapores discussed above, two additional mechanisms by which biochar application might increase soil porosity and thus soil water retention were proposed: (i) modification of the pore system by creating packing or accommodation pores between biochar particles and surrounding soil particles, and (ii) improved persistence of soil pores due to increased aggregate stability (Hardie et al, 2014). Particle size is the single most important biochar parameter that can affect these mechanisms.

Despite the large number of research studies that have examined the impact of biochar on soil water properties, the effect of biochar particle size has received much less research attention. The shape and size of biochar particles can significantly differ from soil particles (Liu et al, 2017). When incorporated into soils, biochar does not only alter the total porosity but also alters the pore size distribution of the soils (Figure 20.5).

The ability of biochar to enhance the soil water properties, particularly in coarse-textured soil, is highly dependent on the biochar particle size. When fine biochar particles are mixed into soils, they can fill the larger pore space between soil particles and aggregates, creating small pores for effective water retention (Figure 20.5). In addition, fine biochar particles have a larger active surface area and combine easily with soil particles to promote aggregation, resulting in higher water retention compared to larger biochar particles (Blanco-Canqui, 2017). However, it is also possible that small biochar particles may clog the openings of existing soil pores, causing a reduction of accessible porosity (Esmaeelnejad et al, 2016).

Fine biochar particles may fill the large pore spaces in coarse-textured soil, shifting the soil pore size distribution to the meso- and micro-pore range, leading to an increase in PAW, as shown in a greenhouse incubation study examining the effect of biochar particle size (< 0.1, 0.1–0.5, 0.5–1, and 1–2 mm) on the physical and hydraulic properties of a sandy loam soil (Alghamdi et al, 2020). The smallest biochar particle size can therefore be responsible for the largest increase in both soil water content at FC and PAW due to an increase in the number of soil mesopores in coarse loamy and sandy soils (Villagra-Mendoza and Horn, 2018).

In addition, biochar application to soil may create new soil aggregates and enhance aggregate stability, resulting in an increase in soil porosity (Verheijen et al, 2010; Obia

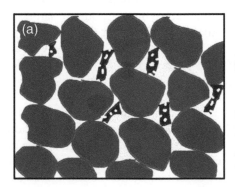 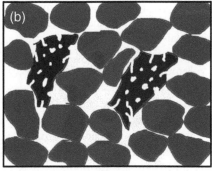

Figure 20.5 *In addition to intrapores (pores inside biochar particles), biochar application creates interpores (pores between biochar particles and soil particles). Biochar particle size determines its impact on soil interpore volumes and pore size distribution. The shape and size of biochar particles differ from soil particles. When biochar particles are smaller than soil particles (a), it creates small interpores for effective water retention; whereas when biochar particles are larger than soil particles (b), adding biochar increases soil total porosity and creates larger interpores, resulting in an increase in soil hydraulic conductivity*

et al, 2016). The binding between soil minerals and biochar particles may increase the inter-particular cohesion, allowing the newly-aggregated soil-biochar agglomerate to resist physical forces. Yang and Lu (2021) reported that biochar addition to a paddy soil (silt loam) significantly promoted soil aggregation and aggregate stability by cementing soil particles together, and thereby created pores ranging from 1 to 10 s of microns, leading to increased soil water retention and PAW. However, some studies have reported a negligible or even negative influence of biochar on aggregate stability (Abel et al, 2013; Hardie et al, 2014; Jeffery et al, 2015). These contradicting results suggest that biochar characteristics and soil properties must be considered as determinants of the soil-biochar interactions, underlining the importance of characterizing both particle types when studying soil water dynamics.

Water repellency of biochar and soil water retention

Soil water repellency not only impacts water entry and distribution in soils but also impacts other biophysical processes including aggregation and microbial activities (Blanco-Canqui, 2017). Greater water repellency generally leads to a larger volume of entrapped air in soils, which may reduce soil water retention and hydraulic conductivity (Glab et al, 2016). Biochar's water repellency or hydrophobicity depends on its surface functional groups, whereas biochar water retention also depends on its porosity and pore size distribution. It is important to characterize biochars for their water repellency because when biochar is applied to soils, there is a risk that the biochar may increase soil water repellency, leading to unexpected hydrologic outcomes. However, biochar will undergo physical, chemical, and biological weathering processes in the environment, and is likely to form a hydrophilic patina layer. Thus, the hydrophobic effect is expected to be transitory and short-term in soils.

Two factors contribute to the water repellency of biochar: feedstock and pyrolysis temperature. Woody feedstocks that contain waxes, lipids, and other water-repellent substances are more likely to be water repellent than non-woody feedstocks because these chemical substances may create a hydrophobic coating on biochar surfaces. Biochar hydrophobicity generally declines at higher pyrolysis temperatures. Biochar produced at 300–400°C has more water repellency than biochar produced at higher temperatures (Kinney et al, 2012; Gray et al, 2014; Hallin et al, 2015; Jeffery et al, 2015; Kameyama et al, 2019; Mao et al, 2019). Hydrophobic functional groups from the feedstock were volatilized upon pyrolysis at > 500°C, making the high-temperature biochar more hydrophilic. Ethanol uptake is often used as an indicator of the relative importance of hydrophobicity vs. porosity in controlling biochar hydrologic behaviors. Low-temperature biochar typically takes up less H_2O than high-temperature biochar, but the same amount of ethanol, as shown for three biochar produced at 370, 500, and 620°C (Gray et al, 2014). This suggests that differences in water uptake are driven by hydrophobicity, not the porosity of the biochar.

We can estimate how biochar hydrophobicity will affect soil water retention with a mathematical model. Water uptake in porous media depends on capillary forces that can be described by the Young-Laplace equation (Gray et al, 2014):

$$P_c = \frac{2\gamma \cos \theta}{r} \qquad [20.2]$$

where P_c is the capillary forces across the liquid-gas interface, γ is the surface tension of water, θ is the contact angle of water which is a function of surface hydrophobicity, and r is the pore radius. When the contact angle is <90°, the surface is defined as hydrophilic, which generates positive capillary forces, driving water into pores; whereas when the contact angle is >90°, the surface is defined as hydrophobic, which generates a negative capillary force and excludes water from entering pores (Gray et al, 2014). In addition, the magnitude of capillary forces shows an inverse linear relationship with the pore radius (Equation 20.2), indicating large pores are less affected by capillary forces than small pores. We speculate that the hydrophobic nature of biochar will affect the water movement into, and retention within biochar intrapores, but has a negligible effect on water dynamics in the larger interpores.

Increased soil water retention in biochar-amended soil does not necessarily translate into increased plant- and microbially-available water

Soil amended with biochar generally has more porosity and greater soil water content, but simply storing more water will not improve the suitability of biochar-amended soil for agriculture. More water retained at FC in the soil does not necessarily equal to more PAW and MAW. It is crucial to determine the impact of biochar on PAW when evaluating the potential benefits of biochar use in agriculture. Many of the early studies that reported greater water retention at FC with biochar application neglected to include the PWP data. It is possible that biochar has increased the FC as well as the PWP, meaning no net change or even a loss

of PAW (Herath et al, 2013). Without these critical details, it is impossible to quantify the increase in PAW resulting from biochar use. One way to overcome this problem is to estimate the SWCC over a range of soil water potentials, as a proxy for PAW in realistic field conditions.

Prior research indicates that the impact of biochar on soil PAW strongly depends on soil texture. The impact of biochar on PAW was first characterized by Tyron (1948), when he reported that biochar addition to soil significantly increased PAW in sandy soil, had no impact in loamy soil, and decreased PAW in clay soil due to an increase at PWP. More recent studies have reported similar results. In a greenhouse study (Aller et al, 2017), biochar had no impact on FC and PWP for silt soils; for clay loam soils, biochar application reduced the water content at FC and had no impact on the water content at PWP, resulting in a decrease in PAW; on the other hand, biochar application increased the water content at FC and had no impact on the water content at PWP for the sandy loam, causing an increase in PAW. Similarly, biochar application significantly increased PAW for a sandy loam soil, while had negligible effects for a silt loam soil and a clay loam soil (Burrell et al, 2016). Although biochar increased the water content at FC, the water held at PWP was also increased by biochar application, resulting in a net zero change in PAW for the silt loam and clay loam (Equation 20.1).

The effect of biochar on soil water retention at high matric potential (at the wet end of the water potential curve, near FC) is generally greater than the lower potential (dry end near the PWP). The increased storage of PAW in biochar-amended soil could be attributed to the direct contribution of biochar's internal meso- and macro-pores and the reorganization of pore systems via interactions between biochar and soil particles; whereas biochar's ability to hold more water at low matric potentials (PWP) is attributed to its pyrogenic nanopores.

Measuring SWCC over a wide range with biochar application is labor-intensive and time-consuming, which is affected by biochar properties, biochar application rate, and soil properties (Figure 20.4). A predictive model was developed to estimate the effect of biochar application on soil porosity, soil water retention, and PAW (Yi et al, 2020). The model used independent measurements of biochar and soil properties to predict water retention in biochar-amened soils. Even though the model was only tested with laboratory data, it will improve our understanding of biochar's impact on water retention and guide us to engineer biochars with desired properties and select biochar application rates to increase soil water retention.

Increasing the biochar application rate for greater soil water retention

Increasing the biochar application rate should increase soil water retention. This is based on the physical nature of biochar as a source of intrapore volume that holds water,

as well as newly-formed soil-biochar agglomerates with greater water retention capacity in the interpore space (e.g., Omondi et al, 2016; Ajayi and Horn, 2016; 2017; Yang and Lu, 2021). However, the marginal benefits are diminished when the biochar application rate is > 60 Mg ha^{-1}, and higher biochar application rates may cause undesirable effects such as a decrease in soil aggregate stability (Ajayi and Horn, 2016; Fu et al, 2019). If the application rate is too low, it may be insufficient to alter the SWCC, whereas if the application rate is too high, it may become too costly for large-scale field applications in commercial fields. For instance, most laboratory studies have used about 1–2% of biochar by weight, which is equivalent to 25–50 Mg ha^{-1} biochar, while some studies applied biochar at rates as high as 100–200 Mg ha^{-1} (Ajayi and Horn 2016; Wong et al, 2017). Since field-based application rates are in the range of 5–20 Mg ha^{-1}, it is important to determine the optimum application rate that would be effective at altering the soil water retention, PAW, and MAW supply. The economically optimal rate for biochar use will vary with soil type and biochar properties, making it crucial to consider site-specific properties and biochar-soil interactions before proceeding with large-scale field deployment.

Soil texture determines biochar's effect on soil water properties

The effect of biochar application on soil water properties is significantly influenced by soil texture. While biochar application increasing water retention in coarse-textured sandy soils has been consistently reported by many studies, there is limited

evidence that biochar improves water retention in fine-textured clay soils (Abel et al, 2013; Ibrahim et al, 2013; Bruun et al, 2014; Laghari et al, 2015; Blanco-Cauqui, 2017; Atkinson, 2018). A thorough understanding of all soil physical and hydraulic properties is needed to evaluate the potentials and limitations for biochar use and management.

Application of biochar to coarse-texture soils (e.g., sand, loamy sand, sandy loam) significantly improved soil physical and hydraulic properties, e.g., reducing soil bulk density, increasing porosity and the stability of soil aggregates (Glab et al, 2016; Blanco-Canqui, 2017; Carvalho et al, 2020). Biochar application generally decreased saturated hydraulic conductivity in coarse-textured sandy soils, increased it in fine-textured soil, and had limited or no effects on medium-textured soils (Bruun et al, 2014; Lim and Spokas, 2016; Blanco-Canqui, 2017; Aller et al, 2017; Edeh et al, 2020). The decrease in saturated hydraulic conductivity in sandy soils was attributed to increased tortuosity due to the filling or clogging of soil macropores with fine biochar particles (Lim and Spokas, 2016). For fine-textured clay soils, the addition of biochar with particle sizes larger than clay particles can create macroporosity and improve aggregation, increasing saturated hydraulic conductivity (Blanco-Canqui, 2017).

In a meta-analysis including 274 studies, biochar on average increased porosity by 8% and PAW by 15% (Omondi et al, 2016).

However, a large portion of the increase was driven from laboratory studies where coarse-textured sandy soils were used. In another meta-analysis study, biochar application had the greatest effect in coarse-textured soils with PAW, FC, and PWP increasing by 33%, 24%, and 22% respectively, whereas the effect of biochar on fine-textured soils was much lower, but still showed an increase of PAW and FC by 9% and 4%, and a decrease of PWP by 0.4% (Edeh et al, 2020). Similarly, biochar applications had a much larger impact on coarse-textured soils with PAW increases by 45% compared to medium-textured and fine-textured soils with PAW increases by 21% and 14%, respectively (Razzaghi et al, 2019). In another study, biochar amended to clay soil had no significant effect on soil water retention characteristics including FC and PAW (Hardie et al, 2014).

In addition, SOM also impacts biochar's ability to improve soil water retention. Biochar increased PAW in a coarse-grained sandy soil with low SOM content (Abel et al, 2013). However, for the soil with high SOM content, which initially possesses a high PAW, no substantial increase in PAW was detected (Abel et al, 2013).

Extrapolating laboratory results for field deployment needs further investigation

There is little evidence from field studies demonstrating that biochar application significantly improves soil water retention in agricultural fields on commercial farms. Contradictory results are often reported and the underlying mechanisms for biochar-mediated alteration of soil moisture remain unclear in field environments. For example, biochar application to two sandy loam and loamy sand agricultural soils in Zambia significantly

increased the total porosity and PAW of the soils after one growing season (Obia et al, 2016). However, another field study showed that biochar application (47 Mg ha^{-1}) had no significant effect on any soil water properties, including FC, PWP, and PAW, thirty months after initial application (Hardie et al, 2014). Similarly, no significant changes in any soil water-related properties (e.g., water retention, saturated hydraulic conductivity, and aggregates stability) were observed after applying biochar to a sandy soil (50 Mg ha^{-1}) for three years (Jeffery et al (2015) and to a silty clay loam soil (10 Mg ha^{-1}) (Wang et al, 2019). These inconsistent results in the literature underline the importance of maintaining the connection between lab-scale experiments, which can provide mechanistic information, and field trials.

The majority of the laboratory and greenhouse studies used soils that were sieved and repacked before the experiments, which will significantly alter the soil structure, pore architecture, and pore size distribution, thus changing the water properties from the field conditions (Hardie et al, 2014). Indeed, sieved, repacked soil bears little resemblance to field soil with regard to its humidity, porosity, and other water properties, which explains why biochar effects on soil water properties are more pronounced in laboratory and greenhouse studies than field studies (Abel et al, 2013; Edeh et al, 2020). Making robust predictions requires well-connected lab, greenhouse, and field trials, where lab trials generate hypothesized mechanisms for further testing in greenhouse and field environments.

Long-term impacts of biochar on soil water properties need further investigation

Soil water properties are generally measured immediately after biochar application or following a short-term incubation. It is now well known that biochar aging impacts its physical and chemical properties and has consequences for its ecosystem service (Gao et al, 2017). However, our knowledge of the long-term impact of biochar on soil water properties in the field remains limited. The physical breakdown of biochar particles, the gradual filling of biochar pores with mineral and biological materials (microbes and plant roots), as well as loss of biochar-soil assemblages through preferential horizontal movement (runoff), will impact the porosity-related properties and long-term hydrological services in biochar-amended soils (Gao et al, 2017).

Biochar aging in the laboratory is a way to mimic biochar natural weathering and assess the long-term impact of biochar on soil water properties. Aller et al (2017) reported that laboratory chemical aging (i.e., exposure to 30% H_2O_2) significantly reduced biochar's capacity to increase FC and PAW in both clay and sandy soils, compared to the fresh biochar. However, these results are not consistent with field observations. In a field trial, Fu et al (2019) reported that freeze-thaw cycles significantly increased the total porosity of the soil with biochar application and thereby improved soil water retention during the melting period. Similarly, field-aged biochar had greater impacts on water retention and PAW of a loamy soil than fresh biochar after 3 years (Paetsch et al, 2018). The authors speculated that mechanical stresses of freeze-thaw cycles during field exposure created new cracks and fractures in the biochar matrix, thereby increasing the pore connectivity of large biochar particles (~10 mm). More field studies are needed to understand the long-term change of biochar in the field and the consequences for soil water properties. It is crucial that these field studies are linked to lab studies that provide a mechanistic understanding of results.

Conclusions and recommendations

Long-term studies are needed to determine how long soil water modifications can persist following biochar application. Most of the published data is from short-term studies (< 2 years). Biochar's physical, chemical, and biological properties are altered with environmental exposure (Chapter 10), likely causing long-term shifts in its hydrologic and ecosystem services on the timescale of years to decades. Determining how to conduct such long-term studies will require creativity on the part of researchers, because the relevant timescale for an aging experiment is from 5–100 years. While it may be possible for researchers to undertake a 5+ year aging experiment, a creative approach is needed for longer experiments. One model for such aging experiments is the chronosequence of 19th-century charcoal kiln studies described by Cheng et al (2008).

More field studies are needed, and future field studies must be well-connected to manipulative lab experiments that reveal mechanisms. To make predictions about key biochar properties and their dynamics in relation to soil water requires the essential mechanistic understanding of what factors are responsible for the field observations. There are large discrepancies between the results from the field studies and those from laboratory or greenhouse studies, which constrains our ability to scale up the laboratory results to realistic environments. More research is needed to connect laboratory, greenhouse, and field trials.

We need a model or decision-support system that can make robust predictions of the probable biochar effects on soil water properties. This will allow us to make better recommendations for biochar use, considering the biochar porosity and suitable application rates, for different climates, soil conditions, as well as crop types.

References

Abel S, et al 2013 Impact of biochar and hydrochar addition on water retention and water repellency of sandy soil. *Geoderma* 202–203, 183–191.

Ajayi AE, and Horn R 2016 Modification of chemical and hydrophysical properties of two texturally differentiated soils due to varying magnitudes of added biochar. *Soil and Tillage Research* 164, 34–44.

Ajayi AE, and Horn R 2017 Biochar-induced changes in soil resilience: effects of soil texture and biochar dosage. *Pedosphere* 27, 236–247.

Alghamdi AG, Alkhasha A, and Ibrahin HM 2020 Effect of biochar particle size on water retention and availability in a sandy loam soil. *Journal of Saudi Chemical Society* 24, 1042–1050.

Ali S, et al 2017 Biochar soil amendment on alleviation of drought and salt stress in plants: a critical review. *Environmental Science and Pollution Research* 24, 12700–12712.

Aller D, Rathke S, Laird D, Cruse R, and Hatfield J 2017 Impacts of fresh and aged biochars on plant available water and water use efficiency. *Geoderma* 307, 114–121.

Atkinson CJ 2018 How good is the evidence that soil-applied biochar improves water-holding capacity? *Soil Use and Management* 34, 177–186.

Babin D, et al 2013 Metal oxides, clay minerals and charcoal determine the composition of microbial communities in matured artificial soils and their response to phenanthrene. *FEMS Microbiology Ecology* 86, 3–14.

Beven K, and Germann P 2013 Macropores and water flow in soils revisited. *Water Resources Research* 49, 3071–3092.

Blanco-Canqui, H 2017 Biochar and soil physical properties. *Soil Science Society of America Journal* 81, 687–711.

Brewer CE, et al 2014 New approaches to measuring biochar density and porosity. *Biomass and Bioenergy* 66, 176–185.

Bruun EW, Petersen CT, Hansen E, Holm JK, and Hauggaard-Nielsen H 2014 Biochar amendment to coarse sandy subsoil improves root growth and increases water retention. *Soil Use and Management* 30, 109–118.

Burrell LD, Zehetner F, Rampazzo N, Wimmer B, and Soja G 2016 Long-term effects of biochar on soil physical properties. *Geoderma* 282, 96–102.

Carvalho ML, De Moraes MT, Cerri CEP, and Cherubin MR 2020 Biochar amendment enhances water retention in a tropical sandy soil. *Agriculture* 10, 62.

Cheng CH, Lehmann J, and Engelhard M 2008 Natural oxidation of black carbon in soils: changes in molecular form and surface charge along a climosequence. *Geochimica et Cosmochimica Acta* 72, 1598–1610.

Curtis MJ, and Claassen VP 2008 An alternative method for measuring plant available water in inorganic amendments. *Crop Science* 8, 2447–2452.

Edeh IG, Mašek O, and Buss W 2020 A meta-analysis on biochar's effects on soil water properties–New insights and future research challenges. *Science of the Total Environment* 714, 136857.

Esmaeelnejad L, Shorafa M, Gorji M, and Hosseini SM 2016 Enhancement of physical and hydrological properties of a sandy loam soil via application of different biochar particle sizes during incubation period. *Spanish Journal Agricultural Research* 14, e1103.

Fierer N, Schimel JP, and Holden PA 2003 Influence of drying-rewetting frequency on soil bacterial community structure. *Microbial Ecology* 45, 63–71.

Fu Q, et al 2019 Effects of biochar addition on soil hydraulic properties before and after freezing-thawing. *Catena* 176, 112–124.

Gao X, et al 2017 Effect of environmental exposure on charcoal density and porosity in a boreal forest. *Science of the Total Environment* 592, 316–325.

Gao X, and Masiello CA 2017 Analysis of biochar porosity by pycnometry. In: Singh B, Camps-Arbestain M, and Lehmann J (Eds) *Biochar: A Guide to Analytical Methods*. Boca Raton: CRC Press. pp132–140.

Ghezzehei TA 2011 Soil Structure. In: Huang PM, Li Y, and Sumner M (Eds.) *Handbook of Soil Sciences: Properties and Processes*. Boca Raton: CRC Press. pp1–18.

Glaser B, Lehmann J, and Zech W 2002 Ameliorating physical and chemical properties of highly weathered soils in the tropics with charcoal – a review. *Biology and Fertility of Soils* 35, 219–230.

Głąb T, Palmowska J, Zaleski T, and Gondek K 2016 Effect of biochar application on soil hydrological properties and physical quality of sandy soil. *Geoderma* 281, 11–20.

Gray M, Johnson MG, Dragila MI, and Kleber M 2014 Water uptake in biochars: the roles of porosity and hydrophobicity. *Biomass and Bioenergy* 61, 196–205.

Hallin IL, Douglas P, Doerr SH, and Bryant R 2015 The effect of addition of a wettable biochar on soil water repellency. *European Journal of Soil Science* 66, 1063–1073.

Hardie M, Clothier B, Bound S, Oliver G, and Close D 2014 Does biochar influence soil physical properties and soil water availability? *Plant and Soil* 376, 347–361.

Herath HMSK, Camps-Arbestain M, and Hedley M 2013 Effect of biochar on soil physical properties in two contrasting soils: an Alfisol and an Andisol. *Geoderma* 209-210, 188–197.

Huang H, et al 2021 Effects of pyrolysis temperature, feedstock type and compaction on water retention of biochar amended soil. *Scientific Reports* 11, 7419

Ibrahim HM, Al-Wabel MI, Usman ARA, and Al-Omran, A 2013 Effect of *conocarpus* biochar application on the hydraulic properties of a sandy loam soil. *Soil Science* 178, 165–173.

Israelsen OW, and West FL 1922 Water holding capacity of irrigated soils. *Utah State Agricultural Experiment Station Bull* 182, 1–24.

Jabro JD, Stevens WB, Iversen WM, Allen BL, and Sainju UM 2020 Irrigation scheduling based on wireless sensors output and soil-water characteristic curve in two soils. *Sensors* 20, 1336.

Jeffery S, et al 2015 Biochar application does not improve the soil hydrological function of a sandy soil. *Geoderma* 251, 47–54.

Jeffery S, Verheijen FGA, van der Velde M, and Bastos AC 2011 A quantitative review of the effects of biochar application to soils on crop productivity using meta-analysis. *Agriculture, Ecosystem and Environment* 144, 175–187.

Jury WA, Gardner WR, and Gardner WH 1991. *Soil Physics*. John Wiley and Sons: New York, USA.

Kameyama K, Miyamoto T, and Iwata Y 2019 The preliminary study of water-retention related properties of biochar produced from various feedstock at different pyrolysis temperatures. *Materials* 12, 1732.

Kinney T, et al 2012 Hydrologic properties of biochars produced at different temperatures. *Biomass and Bioenergy* 41, 34–43.

Laghari M, et al 2015 Effects of biochar application rate on sandy desert soil properties and sorghum growth. *Catena* 135, 313–320.

Lim TJ, Spokas KA, Feyereisen G, and Novak JM 2016 Predicting the impact of biochar additions on soil hydraulic properties. *Chemosphere* 142, 136–144.

Liu Z, Dugan B, Masiello CA, and Gonnermann HM 2017 Biochar particle size, shape, and porosity act together to influence soil water properties. *PLoS One* 12, e0179079.

Mao J, Zhang K, and Chen B 2019 Linking hydrophobicity of biochar to the water repellency and water holding capacity of biochar-amended soil. *Environmental Pollution* 253, 779–789.

McQueen IS, and Miller RF 1974 Approximating soil moisture characteristics from limited data:

Empirical evidence and tentative model. *Water Resources Research* 10, 521–527.

Manzoni S, Schimel JP, and Porporato A 2012 Responses of soil microbial communities to water-stress: results from a meta-analysis. *Ecology* 93, 930938.

Marshall TJ, Holmes JW, and Rose CW 1996 *Soil Physics*. Cambridge: Cambridge University Press.

Moyano FE, Manzoni S, and Chenu C 2013 Responses of soil heterotrophic respiration to moisture availability: An exploration of processes and models. *Soil Biology and Biochemistry* 59, 72–85.

Obia A, Mulder J, Martinsen V, Cornelissen G, and Børresen T 2016 In situ effects of biochar on aggregation, water retention and 'porosity in light-textured tropical soils. *Soil and Tillage Research* 155, 35–44.

Omondi MO, Xia X, Nahayo A, Liu X, Korai PK, and Pan G 2016 Quantification of biochar effects on soil hydrological properties using meta-analysis of literature data. *Geoderma* 274, 28–34.

Paetsch L, et al 2018 Effect of in-situ aged and fresh biochar on soil hydraulic conditions and microbial C use under drought conditions. *Scientific Reports* 8, 6852.

Rasa K, et al 2018 How and why does willow biochar increase a clay soil water retention capacity? *Biomass and Bioenergy* 119, 346–353.

Razzaghi F, Obour PB, and Arthur E 2019 Does biochar improve soil water retention? A systematic review and meta-analysis. *Geoderma* 361, 114055.

Reichstein M, et al 2014 Climate extremes and the carbon cycle. *Nature* 500, 287–295.

Simunek J, Jarvis NJ, van Genuchten MT, and Gardenas A 2003 Review and comparison of models describing non-equilibrium and preferential flow and transport in the vadose zone. *Journal of Hydrology* 272, 14–35.

Soil Science Society of America (SSSA). Glossary of Soil Science Terms. 1996. Soil Science Society of America Inc., Madison WI.

Spokas K, and Bogner J 2011 Limits and dynamics of methane oxidation in landfill cover soils. *Waste Management* 31, 823–832.

Stark JM, and Firestone MK 1995 Mechanisms for soil moisture effects on activity of nitrifying bacteria. *Applied and Environmental Microbiology* 61, 218–221.

Sun F, and Lu S 2014 Biochars improve aggregate stability, water retention, and pore-space properties of clayey soil. *Journal of Plant Nutrition and Soil Science* 177, 26–33.

Trenberth, KE 2011 Changes in precipitation with climate change. *Climate Research* 47, 123–138.

Tryon EH 1948 Effect of charcoal on certain physical, chemical, and biological properties of forest soils. *Ecological Monographs* 18, 81–115.

Veihmeyer FJ, and Hendrickson AH 1927 The relation of soil moisture to cultivation and plant growth. *Proceedings of the First International Congress of Soil Science*. 3, 498–513.

Veihmeyer FJ, and Hendrickson AH 1928 Soil moisture at permanent wilting of plants. *Plant Physiology* 3, 355–357.

Veihmeyer FJ, and Hendrickson AH 1931 The moisture equivalent as a measure of the field capacity of soils. *Soil Science* 32, 181–193.

Veihmeyer FJ, and Hendrickson AH 1949 The application of some basic concepts of soil moisture to orchard irrigation. *Proceedings of Washington State Horticultural Association* 45, 25–41.

Verheijen F, Jeffery S, Bastos AC, van der Velde M, and Diafas I 2010. *Biochar Application to Soils. A Critical Scientific Review of Effects on Soil Properties, Processes and Functions*. Italy: European Commission.

Villagra-Mendoza K, and Horn R 2018 Effect of biochar addition on hydraulic functions of two textural soils. *Geoderma* 326, 88–95.

Vos M, Wolf AB, Jennings SJ, and Kowalchuk GA 2013 Micro-scale determinants of bacterial diversity in soil. *FEMS Microbiology Reviews* 37, 936–954.

Wang D, Li C, Parikh SJ, and Scow KM 2019 Impact of biochar on water retention of two agricultural soils - A multi-scale analysis. *Geoderma* 340, 185–191.

Williams JP, and Hallsworthm JE 2009 Limits of life in hostile environments: no barriers to biosphere function? *Environmental Microbiology* 11, 3292–3308.

Wong JTF, Chen Z, Chen X, Ng CWW, and Wong MH 2017 Soil-water retention behavior of compacted biochar-amended clay: a novel landfill final cover material. *Journal of Soils and Sediments* 17, 590–598.

Yang CD, and Lu SG 2021 Effects of five different biochars on aggregation, water retention and mechanical properties of paddy soil: A field experiment of three-season crops. *Soil and Tillage Research* 205, 104798.

Yi S, Chang NY, and Imhoff PT 2020 Predicting water retention of biochar-amended soil from independent measurements of biochar and soil properties. *Advances in Water Resources* 142, 103638.

21

Biochar and heavy metals

Luke Beesley, Beatriz Cerqueira Cancelo, Michael Hardman, Manhattan Lebrun, Kerry Mitchell, and Lukas Trakal

Introduction: Heavy metals in the environment

Definitions

Heavy metals in soils are derived from both geogenic and anthropogenic sources. In the case of the latter, this may be due to point or diffuse sources as diverse as mining, smelting, industrial processing, waste disposal, fertilizer, and herbicide and pesticide usage (Ross, 1994). In excessive concentration, those heavy metals regarded as the most toxic and environmentally deleterious are cadmium (Cd), chromium (Cr), copper (Cu), mercury (Hg), nickel (Ni), lead (Pb) and zinc (Zn), although several of these, especially those that are transition metals, are nonetheless essential for plant metabolism (e.g., Cu, Ni, Zn). By definition, heavy metals are a group of elements with specific gravities of >5 g cm^{-3} (Ross, 1994) which are both industrially and biologically relevant (Alloway, 1995). Although not heavy metals by chemical definition, the metalloids arsenic (As) and antimony (Sb) are given the status of "risk elements" or "potentially toxic elements" due to their impacts on humans and toxicity to plants after excessive exposure (Moreno-Jiménez et al, 2012). In that case, those heavy metal(loid)s that cause a toxic response to biota or humans resulting in an unacceptable level of environmental risk (Adriano, 2001; Abrahams, 2002; Vangronsveld et al, 2009) may be classed as pollutants. At the ecosystem level, the chemical behavior of heavy metal(loid)s in soils resulting in their mobility and toxicity are complex and, since this book is concerned with "environmental management" we will focus on interactions between biochar and heavy metal(loid)s through an environment lens. This chapter covers the main or "master" mechanisms by which biochars impact metal(loid)s in soils and discusses these via applied examples. Biochar's impacts on metal(loid)s in water is covered elsewhere in this book (Chapter 27).

Exposure and risk

Heavy metal(loid)s in soils and sediments are partitioned into a number of binding phases either (i) incorporated in the solid phase;

DOI: 10.4324/9781003297673-21

(ii) bound to the surface of the solid phase; (iii) bound to ligands in solution; or (iv) as free ions in solution. Only the free ions in solution (i.e., phase (iv)) can be taken up by organisms and, therefore, only the free ions are bioavailable (Di Toro et al, 2001; Thakali et al, 2006). In soils, there is often dis-equilibrium between these four phases due to contrasting geochemical conditions, though the tendency is strongly to equilibrate in the most rapid time. If the concentration of metal ions dissolved in solution decreases (for example, due to plant uptake), then equilibration will again occur by which more metals desorb to increase the amount of metal ions in solution.

To cause a toxic effect, heavy metals must dissolve into a solution, be taken up by an organism, and be transported to cells where a toxic effect can occur. This complex interaction between organisms and contaminants can be described by a simple model known as the source-pathway-receptor model (Hodson, 2010). The source of the pollution is a heavy metal (e.g., Pb), the receptor is a biological organism (e.g., an earthworm), and the pathway is the process that leads to the contaminant being taken up by the organism (e.g., desorption of Pb from the soil surface into the soil solution and diffusion across the gut wall of the earthworm).

Therefore, remediation of heavy metal con-taminated sites can be performed by (i) removing all or part of the source; (ii) eliminating the pathway; or (iii) modifying the exposure of the receptor (Nathanail and Bardos, 2004). Thus, remediation is achieved in heavy metal-polluted environ-ments by reducing the bioavailability of the metals to the receptor organisms. Since heavy metals cannot be degraded or broken down (i.e., the source cannot be depleted in situ), and receptors often cannot be fully or even partially isolated in complex media, such as soils; the only viable option to break the source-pathway-receptor linkage is to disrupt the pathway between the contami-nant and the receptor. Manipulation of bioavailability increasingly forms the basis of risk assessment and classification of polluted areas, rather than absolute concen-trations in soils (Bolan et al, 2008; Durães et al, 2018). Importantly, in the legislative context of most nations, it is this potential to cause harm to humans or ecosystems (the effect) that defines polluted sites and not the presence (concentration) of the contam-inant per se. Therefore, if biochars are to be deployed to heavy metal(loid) contaminated systems, then it is their ability to break the pathway from source to receptor that becomes a focal point of their deployment (Park et al, 2011) (Figure 21.1).

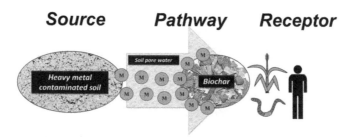

Figure 21.1 *Schematic representation of biochar disrupting the pathway of heavy metal(loid)s (M) from their source to receptor organisms*

Biochar as a remedial amendment

Biochars, in common with other organic amendments, can reduce heavy metal(loid) mobility and bioavailability by various physicochemical means (Bolan and Duraisamy, 2003). The application of organic amendments to soils, from a remedial point of view, has typically been justified by their relatively low cost, compared to "hard" engineering solutions (such as removal and disposal of soils elsewhere) as well as their prevalence as a waste, ordinarily requiring other forms of disposal (burial in landfill, incineration, etc.). The pyrolysis of organic materials to produce biochar increases the surface area and effective cation exchange capacity (CEC) compared to the unpyrolyzed source, and has a lower mineralization rate than unpyrolyzed materials (Chapter 11), theoretically requiring less frequent additions to maintain efficacy than other, more easily mineralizable organic materials, such as composts, manures, etc.

Therefore, the justification for the addition of biochar to metal(loid) contaminated soil is that it can work as a sorbent for metals in solution by establishing a new equilibrium between the concentrations of metals sorbed to surfaces and that in solution. Before this chapter embarks on the detail of the mechanistic, advantageous, and disadvantageous functions of biochar, an important premise should be noted; the same features of biochar that render it suitable for remediation of heavy metal contaminated substrates may at once deem it unsuitable for application, specifically where the desired effect is to increase the bioavailability of metals. The obvious example is Zn, an essential plant nutrient and important element to fortify food and feed but, in excess, a toxin. Rather than considering absolute increases or decreases in heavy metal concentrations in substrates receiving biochars, the emphasis should be placed on bioavailability, mobility, and specific requirements related to land use.

Heavy metal-biochar interactions at the soil-water interface

Direct mechanisms

Direct mechanisms of heavy metal immobilization by biochar include, but are not limited to, fundamental chemical "at-soil surface" processes, such as adsorption and complexation. It is widely acknowledged and discussed that biochars may both mobilize and immobilize heavy metals and As by direct means such as ion exchange, chemical and physical adsorption, precipitation, etc. (Abdelhadi et al, 2017; Soria et al, 2020). These mechanisms are discussed in the following section.

Chemical sorption

During exposure to the atmosphere, such as environmental weathering of freshly produced biochars applied to soils, the oxygenation of biochars' myriad surfaces occurs (Cheng et al, 2006) forming oxygen (O) containing functional groups (e.g., carboxyl, hydroxyl, phenol, and carbonyl groups; Uchimiya et al, 2010, 2011; He et al, 2019). These functional groups induce a negative charge and a high CEC (Sorrenti et al, 2016). CEC first increases, and then decreases, with increasing pyrolysis temperatures (Harvey et al, 2011; Mukherjee et al, 2011;

Wu et al, 2016; Tomczyk et al, 2020); a peak CEC of up to 450 $mmol_c kg^{-1}$ has been shown to occur between 250 and 350°C, depending on the source material. The lower oxygen to carbon (O/C) ratio and reduced abundance of oxygenated (acid) functional groups lowers CEC after higher temperature pyrolysis (Uchimiya et al, 2011; Harvey et al, 2011; Shen et al, 2012). The capacity for metal immobilization demonstrated by lower temperature (<500°C), and faster pyrolysis biochars (Beesley et al, 2010; Beesley and Marmiroli 2011) is therefore, in part, a result of the high CEC of these biochars; biochars with a similar CEC to the soil they are applied to will not immobilize heavy metals as effectively as biochars with greater CEC than the receiving soil (Gomez-Eyles et al, 2011, 2013). Highly weathered acidic tropical soils, low in organic C, whose mineralogy is dominated by kaolinite and iron (Fe)- or aluminum (Al)-oxyhydroxides, yield a low CEC (Schaefer et al, 2008). These soils are more readily phytotoxic than soils from temperate regions due to their inherent inability to retain heavy metals (Melo et al, 2011). In such soils, it is more likely that adding biochars will increase CEC and be more effective in immobilizing heavy metals.

Surface sorption of metals corresponds directly with the release of H^+ ions from biochars, but also of the release of Na, Ca, S, K, and Mg into the solution, which indicates retention of metals on protonated (acidic) functional groups but also metal exchange with other cations (Uchimiya et al, 2010, 2011; Park et al, 2019). Phosphorus- and sulfur- (S)-containing ligands influence the sorption of metal ions such as Pb that have a stronger affinity for phosphates and sulfates (Cao et al, 2009). Biochar surface oxygenated functional groups may impact the oxidation of redox-sensitive metals whilst biochar application to soils also changes soil porosity and modifies soil physical structure, which may

influence microscale redox condition. In these cases, redox-sensitive elements will change their speciation and geochemistry; for instance, As(III) is found in anoxic environments (<100 mV) and is more mobile in soils and toxic than As(V); Cr can be oxidized in aerobic environments (>300–400 mV) and Cr(VI) is more toxic than Cr(III), whilst Cu (I) can also be found under anoxic conditions.

Physical sorption

Aside from a pure ion exchange between biochar surfaces and metals, a non-stoichiometric release of protons and other cations from the surface of biochars can occur; more metals are adsorbed than protons or cations are released and sorption can occur at pH below the point of zero net charge (Sánchez-Polo and Rivera-Utrilla, 2002). The immobilization of metals by biochar cannot, in these instances, be purely attributed to ion exchange alone. Metal sorption to biochars is an endothermic physical process (Liu and Zhang, 2009; Harvey et al, 2011) and an electrostatic interaction between the positively charged metal cations and π-electrons associated with either C=O ligands or C=C of a shared electron 'cloud' on aromatic structures of biochars occurs (Cao et al, 2009; Uchimiya et al, 2010; Harvey et al, 2011).

An increase in the pyrolysis temperature of biochars increases their aromaticity whilst the abundance of oxygenated functional groups decreases (Jung et al, 2016). Thus, increasing pyrolysis temperature increases the proportion of cations sorbed due to 'weak' electrostatic bonding (i.e., cation-π interactions) and decreases the proportion due to stronger chemisorption (i.e., by cation exchange). Therefore, lower temperature pyrolysis should result in effective short-term metal immobilization due to the formation of inner and outer sphere complexes with oxygenated (acid) functional groups, but with time these may diminish in the soil environment.

Thereafter there may be a release of metals back into solution. Higher pyrolysis temperatures result in a negative surface charge that should remain stable for longer, but metals will be weakly (physically) adsorbed to biochar surfaces and immobilization may be more easily reversed (summary in Table 21.1).

Precipitation

Biochar source materials are unlikely to be 100% organic in nature and contain minerals that remain entrained in the biochar matrix after pyrolysis, resulting in a non-organic (or ash) fraction in biochar. Source material mineral contents can range from <1% for woody biomass, up to ~25% for manure or crop residues. Following high-temperature pyrolysis, the ash content of biochars can be up to 50% for manure-derived or 85% for bonemeal-derived biochars. Thus, mineral salts of Na, K, Ca, Mg, P, S, Si, and C are found in abundance in the ash fraction, usually in an oxidized form, their concentrations of which increase with pyrolysis temperature. Uchimiya et al (2010) found Pb phosphate precipitates effective in immobilizing Pb in a broiler litter-derived biochar whilst precipitation of Pb with phosphates contributed to as much as 87% of total Pb sorption to a dairy manure-derived biochar (Cao et al, 2009). Lead-phosphate minerals contributing to sorption in biochars include hydrocerussite and hydroxylpyromorphite (Cao et al, 2011), Pb phosphate, and Pb hydroxyapatite (Chen et al, 2006). Lead-phosphate minerals have a very low solubility so their formation could result in the increased capability of biochars to adsorb higher concentrations of Pb, compared to other divalent cations (Uchimiya et al, 2010; Namgay et al, 2010; Trakal et al, 2011). Precipitation may also occur with other metals such as Cu, Cd, or Zn which precipitate as insoluble phosphate and carbonate salts, mainly at high pH (Lindsay, 2001).

Indirect mechanisms

Indirect mechanisms can also be defined as the effects that biochars have on soil characteristics (physical, biological, and chemical) that then impact heavy metal retention or release. The addition of biochar to soils can, amongst myriad other effects, increase soil pH, microbial biomass, organic C, and water holding capacity which may in turn impact heavy metal retention and release. This section is mainly concerned with, and summarizes the results of, studies examining pH changes induced by biochar additions to soils. Other impacts of biochar addition are briefly covered.

pH changes

It is widely reported that the addition of alkaline biochars to soils resulted in pH increases (Ippolito et al, 2017; Lebrun et al, 2020); metal solubility changes according to pH, generally being lower at higher pH. For metalloids, the geochemistry is somewhat opposing, with higher pH conditions reducing retention (Adriano, 2001). Arsenic solubility and availability increase when pH in soils rises, in most cases, since As binds to positively charged surfaces such as Fe and Mn oxides in soils and anion exchange capacity (AEC) is inversely related to pH (Moreno-Jiménez et al, 2012). Cationic metals (e.g., Cu, Zn, Pb), which are bound to the negatively charged surfaces of soils such as clay minerals and organic matter, increase in solubility as pH decreases because CEC is positively related to pH. When the soil pH increases, metals are increasingly bound to negatively charged surfaces. Contrary to cationic metals, As is released from positively charged soil surfaces when the soil pH is increased; an increase of soil pH has thus been reported to increase As mobility and uptake by organisms (Fitz and Wenzel, 2002).

Studies have reported that soil pore water pH increases after biochar application

Table 21.1 *Selected examples of the influence of pyrolysis temperature on heavy metal sorption capacity, assessed by batch sorption experiments*

Experiment	Biochar preparation	Findings	Reference
Batch aqueous sorption of Pb, solution added at 0.02 g biochar for 20 mL solution	Biochars produced by pyrolysis of sawdust of white spruce, canola straw, wheat straw, and manure pellet at 300°C, 500°C and 700°C	Maximum Pb(II) adsorption increased with increasing pyrolysis temperature. Sorption occurred through multilayer adsorption on the heterogeneous surface, with a finite number of sites.	(Kwak et al, 2019)
Batch aqueous sorption of Cu, solution (20 mL) added to 10 mg biochar	Biochar produced from the pyrolysis of Jerusalem artichoke stalks at 300°C, 500°C and 700°C	Highest maximum adsorption potential (Qm) measured for the low-temperature biochar (300°C, 17 mg g^{-1}). For low-temperature biochar (300°C), sorption took place on a homogeneous surface, while for high-temperature biochar (500°C, 700°C), sorption occurred on a heterogeneous surface. Sorption mainly occurred through the reduction of Cu(II) to Cu (I) (for biochars produced at 300°C and 700°C), surface complexation (importance declines with the increase in pyrolysis temperature), Cu cation-π bonding and precipitation with PO_4^{3-} (for high-temperature biochars).	(Wei et al, 2019)
Batch aqueous sorption of Cd, solution (25 mL) added to 0.05 g biochar	Biochars produced from the pyrolysis of pine cones, pine needles, and pine bark at 300°C to 600°C	Cd sorption increased with the increase in pyrolysis temperature. Sorption mechanisms were identified to be the release of phosphate, the removal of cations, and interaction with functional groups (C=C and COO).	(Park et al, 2019)
Batch sorption of Pb, solution (60 mL) added to 0.03 g biochar	Biochars from cotton straw pyrolyzed at five different temperatures, from 300°C to 700°C	Highest removal efficiency with the biochar produced at 600°C (51 mg g^{-1}). Adsorption mechanism was identified to be ion exchange with Na^+, K^+, Ca^{2+}, and Mg^{2+}, cation-π interactions, and precipitation on the biochar surface, as hydroxypyromorphite and hydrocerussite. Difference of mechanism dominance with pyrolysis temperature: complexation at low temperature, precipitation at high temperature.	(Wang et al, 2021)

Table 21.1 *continued*

Experiment	Biochar preparation	Findings	Reference
Batch sorption of Fe, Ni, Cu, Cr and Pb	Biochars produced from the pyrolysis of bamboo, sugarcane, or neem at 450°C and 550°C	Higher removal efficiency with the high-temperature biochars. Mechanism of sorption was identified to be chemisorption, with exchange of valence electron.	(Singh et al, 2021)
Batch sorption of Pb, solution (20 mL) added to 40 mg biochar	Biochars produced from cotton stalks pyrolyzed between 250°C and 650°C	Pb removal efficiency increased with increasing pyrolysis temperature. Chemisorption identified as the dominant adsorption mechanism: formation of lead precipitates on the biochar surface (principally cerussite), cation exchange with K and Ca, complexation with oxygen functional groups, and cation π interactions with other functional groups (C=C, -CH), and electrostatic interaction.	(Gao et al, 2021)
Batch sorption of Cd, Pb, and Zn, solution (100 mL) added to 10 g biochar	Biochars produced from the pyrolysis of poultry manure at 425°C, 575°C and 725°C	Lowest sorption capacity of the biochar produced at 425°C, highest Zn and Pb sorption capacity for the biochar produced at 575°C, and highest Cd sorption capacity for the biochar produced at 727°C. Sorption mechanisms are identified as cation exchange, surface complexing, precipitation, and electrostatic interaction.	(Sobik-Szołtysek et al, 2021)
Batch sorption of Cu, solution (20 mL) added to 25 mg biochar	Biochars produced from cow manure pyrolyzed at temperatures from 399°C to 700°C	Highest sorption capacity with the biochar produced at 700°C. Primary mechanisms of Cu sorption identified to be co-precipitation (with formation of copper phosphate and copper carbonate) and cation exchange (with K^+, Ca^{2+} and Mg^{2+}), accounting for 93 to 97% of the adsorption. Secondary mechanisms identified as complexation with oxygen functional groups and cation π interactions.	(Zhang et al, 2021)
Batch sorption of Cr, solution (20 mL)	Biochars prepared from pineapple peels, pyrolyzed at	Removal efficiency of Cr decreased with increasing pyrolyzing temperature. Mechanisms if sorption identified as	(Shakya and Agarwal, 2019)

Table 21.1 *continued*

Experiment	Biochar preparation	Findings	Reference
added to 100 mg biochar	temperatures between 350°C and 650°C	precipitation, complexation with oxygen containing functional groups, electrostatic attraction	
Batch sorption of Pb and Cd, solution (30 mL) added to 0.05 g biochar	Biochars made from tobacco stems pyrolyzed at 400°C, 500°C, 600°C and 700°C	Adsorption of Pb and Cd increased with increasing pyrolysis temperature; highest sorption with biochar at 700°C (Pb: 22 mg g^{-1}, Cd: 19 mg g^{-1}). Mechanisms of sorption hypothesized to be precipitation with inorganic salts.	(Wang et al, 2020)
Batch sorption of Fe, Ni, Cu and Zn; solution (50 mL) added to 0.1 g biochar	Biochars made from date palm wastes (fronds and leaves) pyrolyzed 400°C to 600°C	No influence of pyrolysis on the sorption of Cu, Fe, Ni and Zn.	(Sizirici et al, 2021)

to circumneutral and acidic contaminated substrates, explaining changes in metal and As mobility in pore water (Table 21.2). Various other studies report a soil liming effect of biochars, often resulting from alkaline biochars added to very acidic mine soils. Sizmur et al (2011), for example, highlight an extreme example of an increase in soil pH of more than 4 units when nettle-derived biochar was added to a mine soil (pH 2.7). Other authors note more moderate pH increases of 1–2 units (Jones et al, 2012; Li et al, 2018), whilst column leaching studies have shown the pH effect to be reduced over time as much as ~3 units (Beesley et al, 2022).

Dissolved organic carbon (DOC) and trace elements in biochars

Fluctuations in dissolved organic carbon (DOC) concentrations have been measured as a consequence of biochar application to soils, and several studies have previously measured the potential co-mobilization of metals in complexes with leached organic materials. In a study utilizing a recirculating column approach, any co-mobilization of Zn induced by DOC from biochar was mitigated by the strong surface binding of this metal with biochar (Beesley et al, 2022). This is likely to be the case for the majority of soil biochar applications, the exception being where some amorphous organic matter is occluded in biochar pores after pyrolysis. The mechanisms for the co-mobilization of As and soluble organic matter are less clear than for metals in the context of biochars. Ternary complex formation between arsenate and ferric iron complexes of 'humic substance' extracts could be responsible for the increasing As mobility with increasing DOC (Mikutta and Kretzschmar, 2011). Alternatively, DOC may compete with As directly for retention sites on soil surfaces (Fitz and Wenzel, 2002), resulting in an increase in soluble As with increasing concentrations of DOC (Hartley et al, 2009).

Biochars can be sources of (Chapter 8), or enhance the bioavailability of P (Chapter 16). As phosphate is chemically analogous to As (V), increases in P availability result in the release of As from soil surfaces, into solution and uptake into plants via phosphate ion channels (Meharg and Macnair, 1992).

Table 21.2 *Selected examples detailing pH effects of biochars on heavy metal solubility in pore water, assessed by pot tests*

Experiment	Soils and biochars	Extraction procedure	Findings	Reported in reference
Pot trial to determine if biochar is efficient in stabilizing As and Pb	Former mine technosol, very acidic and polluted with As and Pb Hardwood biochar (500°C) applied at 5%	Soil pore water extraction CaCl₂ and NH₄NO₃ extraction	Soil pore water pH increased by 3 units CaCl₂ extractable As and Pb reduced, by 40% and 42%, respectively NH₄NO₃ extractable Pb reduced by 76% Soil pore water Pb concentration decreased by 96% Biochar effective to stabilize Pb.	(Lebrun et al, 2019)
Pot trial to determine the capacity of biochar to stabilize As and Pb	Former mine technosol, very acidic and polluted with As and Pb Hardwood biochar (500°C) applied at 5%	Soil pore water extraction	Soil pore water pH increased by 2.4 units. Soil pore water Pb concentration decreased by 86% Biochar effective in stabilizing Pb	(Lebrun et al, 2021a)
Pot culture experiment to evaluate the potential impacts of biochar on the bioavailability of multiple metals (Cd, Pb, Cu, and Zn)	Farmland soil (fluvo-aquic soil), with basic pH and contaminated by Pb and Zn Rice huck biochar applied at 0% (control), 1%, 2%, and 3%	DTPA-extractable metals	Biochar application decreased the DTPA-extractable Cd, Pb, Cu, and Zn contents. Biochar effective in stabilizing Pb and Zn	(Wang et al, 2021)

Although arsenate is desorbed from soil surfaces by phosphate (Cao et al, 2003), it is not always available for plant uptake since P and As will compete again for the same root transporter. Therefore, As (V) uptake into plants can be avoided by high concentrations of soluble P (Moreno-Jiménez et al, 2012), but if the soluble fraction of As is not taken up by plants, there is a risk it may leach to surface and groundwaters (Fitz and Wenzel, 2002). On the contrary, phosphate is known to precipitate and sorb Pb (Zeng et al, 2017). Phosphate-rich compounds applied to Pb-contaminated soils have also been found to reduce Pb bioavailability (Miretzky and Fernandez-Cirelli, 2008).

Biochars have heavy metals inherent within their structure, derived from their source material, which may be accumulated and concentrated in ash fractions during pyrolysis. Some biochars exceed European topsoil concentrations, suggesting that they may contribute heavy metal loadings if applied to soils (Table 21.3).

Table 21.3 *Summary of range of selected heavy metal(loid) concentrations of biochars extracted by acid (aqua-regia)*

Heavy metal (loid)	Background European topsoil concentrations (mg kg^{-1})[a]	Range of concentrations measured in biochars (mg kg^{-1})	Source publications
As	6	0.01–9	(Hossain et al, 2010; Bird et al, 2012; Freddo et al, 2012)
Cd	0.2	<0.01–8	(Hossain et al, 2010; He et al, 2010; Knowles et al, 2011; Bird et al, 2012; Gascó et al, 2012; Freddo et al, 2012; Van Poucke et al, 2018, 2020)
Cr	22	0.02–230	(Hossain et al, 2010; Bird et al, 2012; Freddo et al, 2012; Van Poucke et al, 2018; Van Poucke et al, 2020; Lebrun et al, 2021a)
Cu	14	<0.01–2100	(Graber et al, 2010; He et al, 2010; Mankasingh et al, 2011; Knowles et al, 2011; Bird et al, 2012; Gascó et al, 2012; Méndez et al, 2012; Freddo et al, 2012; Van Poucke et al, 2018; Van Poucke et al, 2020; Lebrun et al, 2021a)
Pb	16	0.1–196	(Hossain et al, 2010; He et al, 2010; Knowles et al, 2011; Bird et al, 2012; Gascó et al, 2012; Méndez et al, 2012; Freddo et al, 2012; Van Poucke et al, 2018; Van Poucke et al, 2020)
Zn	52	0.7–3300	(Graber et al, 2010; He et al, 2010; Mankasingh et al, 2011; Knowles et al, 2011; Bird et al, 2012; Gascó et al, 2012; Méndez et al, 2012; Freddo et al, 2012; Van Poucke et al, 2018; Van Poucke et al, 2020; Lebrun et al, 2021a)

Notes
[a] Source: Lado et al (2008) based on 1588 samples across 26 EU member states; data reported are median values.

Toxicity to plants (phytotoxicity)

The impacts of biochars on soil properties and resultant phytotoxicity are subject to many factors (Chapter 13), such as soil type and local climate conditions (Gao et al, 2020). In a wide-ranging meta-analysis, soil organic matter, pH, and texture were found to be among the key factors to consider when evaluating the effect of biochars on soil parameters that influence plant growth (Arabi et al, 2021). For example, the same biochar was very efficient in immobilizing Pb, promoting vegetation growth, and reducing plant metal uptake in an acidic former mine technosol (Lebrun et al, 2017), while it had no effect at an alkaline former industrial site (Lebrun et al, 2018c). Similarly, biochar application reduced the mobility of Cd, Pb, and Zn when applied on acidic soil but not to basic soil (Álvarez-Rogel et al, 2018). The content of clay and the presence of co-existing ions in the soil were among the factors influencing the effect of biochar on metals and plants (Guo et al, 2020; summarized in Table 21.4).

Table 21.4 *Selected examples of pot trials where biomass and heavy metal uptake has been measured after soil amendment with biochars alone, or in combination with other organic amendments*

Experiment	Soil	Biochar	Findings	Reported in reference
To determine whether biochar can reduce As and Pb toxicity to allow plant establishment	Former mine soil, acidic and highly contaminated with As and Pb	Pinewood biochar added at 2% and 5% (w/w)	Increased *Salix viminalis*, *Salix alba,* and *Salix purpurea* dry weight with 2% and 5% biochar. Decreased As stem concentration (88%) in *Salix alba* and increased in leaf (41-fold); increased As concentration in the stem of *Salix viminalis*. Increase in leaf Pb concentration of Salix alba and decreased in roots; decreased in Pb leaf and root concentrations of *Salix viminalis*; increased leaf Pb concentration of *Salix purpurea*. Biochar effective at improving soil fertility and plant growth, while stabilizing As and Pb in the root zone	(Lebrun et al, 2017)
To determine the effect of biochar feedstock for the stabilization of As and Pb	Former mine soil, acidic and highly contaminated with As and Pb	Softwood and Pinewood biochars added at 2% and 5% (w/w)	Increase in soil organic matter content and pH. Immobilization of Pb. Increase in *Salix viminalis* and *Populus euramericana* dry weight. Increase in Pb aerial concentrations with biochar. Decrease in As concentration in *Salix viminalis* with pinewood biochar and increase with	(Lebrun et al, 2018b)

Table 21.4 *continued*

Experiment	Soil	Biochar	Findings	Reported in reference
			lightwood biochar. Biochar efficient for the assisted phytostabilization of As and Pb in association with poplar or willow. Biochar feedstock important parameter for As and Pb stabilization	
To determine the effect of biochar particle size on the stabilization of As and Pb	Former mine soil, acidic and highly contaminated with As and Pb	Hardwood biochars, with four particle sizes, added at 2% and 5% (w/w)	Increase in soil pH and immobilization of Pb. Increase in *Salix viminalis* dry weight. Fine biochars decreased As stem concentrations, while coarse biochars decreased As leaf concentration. All biochars increased Pb leaf and stem concentration. Biochar improved soil growing conditions and is efficient for As and Pb stabilization when associated with *Salix viminalis*. Fine biochars induce effects more rapidly than coarse biochars	(Lebrun et al, 2018a)
To determine whether biochar, associated with compost, can stabilize As and Pb	Former mine soil, acidic and highly contaminated with As and Pb	Hardwood biochar added at 5% Compost added at 5%	Increased soil organic matter (biochar + compost > biochar > compost). Decreased As availability (biochar = biochar +compost) and Pb availability (biochar > compost = biochar +compost). Increased soil pH and immobilization of Pb. Increased *Salix viminalis* dry weight (biochar = compost = biochar +compost). Decreased As and Pb root concentrations. Biochar,	(Lebrun et al, 2019)

Table 21.4 *continued*

Experiment	Soil	Biochar	Findings	Reported in reference
			with or without compost, improved soil conditions and plant growth, and is efficient for As and Pb stabilization	
To determine if the association of biochar with red mud (bauxite mining residue) is efficient for the remediation of As and Pb	Former mine soil, acidic and highly contaminated with As and Pb	Hardwood biochar added at 2% Red mud added at 1%	Increased soil pH, Pb immobilization, and increased *Salix dasyclados* growth. Decreased root Pb concentration and increased leaf and stem Pb concentrations. The association of biochar with red mud is effective for the assisted phytostabilization of As and Pb	(Lebrun et al, 2021d)
To determine if biochar, compost and iron sulfate can stabilize As and Pb	Former mine soil, acidic and highly contaminated with As and Pb	Hardwood biochar added at 5% Compost added at 5% Iron sulfate added at 0.15%	Increased soil C, pH, P and K. Immobilization of Pb. Improved *Agrostis capillaris* growth. The association of the three amendments, with Agrostis, is a good option for the remediation of As and Pb	(Nandillon et al, 2021)
To test whether biochar assist the revegetation of mine spoil	Exposed mine spoil materials from five disused/abandoned mine	Rice husk biochar or wheat straw biochar, added at 5% and 10% (w/w)	More than 99% Zn and 97% Pb adsorbed by the biochars. Increased ryegrass mass and reduced Cd assimilation (2–4 fold), As and Sb concentrations. Decreased Al, Cd, Pb, and Zn mobility. Biochar addition can successfully establish a vegetation cover on mine soil and stabilize metals	(Alhar et al, 2021)
To determine if biochar can reduce pollutant uptake to tomato	Metal contaminated site	Maize stalks biochar added at 5 and 10 t ha^{-1}	Increased N, P, and K availability and decreased pollutant availability. Increased concentration of photosynthetic pigments,	(Almaroai and Eissa, 2020)

Table 21.4 *continued*

Experiment	Soil	Biochar	Findings	Reported in reference
			tomato yield, and fruit quality. Decreased Zn, Cu, and Pb concentration in shoots and roots of tomatoes. High doses of biochar decreased Zn, Pb, Cd, and Ni and increased N and K contents in fruits. Biochar application recommended to a metal contaminated soils to improve tomato quality and productivity	
To determine if biochar can restore contaminated soil	Mine waste	Biochar from the distilled waste of lemongrass, added at 1%, 2%, and 4%	Increased soil pH, water holding capacity, organic carbon, microbial biomass C, and nutrient availability. Enhanced palmarosa biomass yield and reduced production of oxidative enzyme (in a dose-dependent manner). The plantation of palmarosa with biochar can restore the soil	(Jain et al, 2020)
To determine the biochar dose allowing soil remediation	Paddy field polluted by wastewater discharged from chemical plant of $ZnSO_4$	Biochar from bamboo powder, added at 1%, 3%, 5%, 7%	Decreased hydrolyzable N and available P. increased soil organic matter, available K, and the activity of urease and polyphenol oxidase. Decreased soil metal concentration. Little effect on *Salix psammophila* growth, except 7% (reduction of growth). The dose of 3% biochar is the best option	(Li et al, 2021)
To evaluate the effect of biochar particle size and application dose	Metal polluted soil	Biochars from oil palm empty fruit bunch, of different	Decreased Pb and Cd solubility. Increased *Brassica juncea* height. Decreased root and shoot	(Samsuri et al, 2020)

Table 21.4 *continued*

Experiment	Soil	Biochar	Findings	Reported in reference
		particle sizes, added at 0.5% and 1%	Cd and Pb concentrations. The application of a fine particle size biochar at 1% is the best option to reduce metal transfer in plants	
To determine the influence of different biochars on plant growth and metal uptake	Basic metal contaminated soil. Acidic metal contaminated soil	Biochars from sewage sludge and pruning trees, added at 6%	No influence of biochar on soil pore water pH No durable and significant effect of biochars on metal mobility. No effectiveness of biochar when applied to basic soil Biochar increased soil pore water pH and decreased metal concentration in soil pore water Biochar is a useful option to reduce metal mobility in acidic soil	(Álvarez-Rogel et al, 2018)

Remediation of contaminated sites: lessons from case examples

Principles of soil remediation

We have discussed the main mechanisms by which biochars interact with metal(loid)s in soils, and how these may be directly and indirectly related to biochars' addition to soils. Now we will discuss these factors regarding principles of contaminated land remediation. Contaminated, industrially impacted, mining and urban lands are often characterized by young, poorly developed soils and a scarcity or absence of vegetation cover (Mench et al, 2003). Re-vegetation of contaminated soils is key to their stabilization and remediation (Arienzo et al, 2004; Ruttens et al, 2006), as the presence of a vegetative cover over bare soil reduces the potential for migration of contaminants to proximal waters or inhalation by receptor organisms (Tordoff et al, 2000), as well as the restoration of the natural cycling of organic matter and nutrients. Barriers to re-vegetation include phytotoxic concentrations of heavy metals. In this context, those which plants may not be able to immobilize at the root level (Pulford and Watson, 2003), and poor functionality (organic matter [cycling], nutrient status, structure of soils, water-holding capacity). To overcome these limitations is the general aim of soil remediation using organic amendments (Fangueiro et al, 2018).

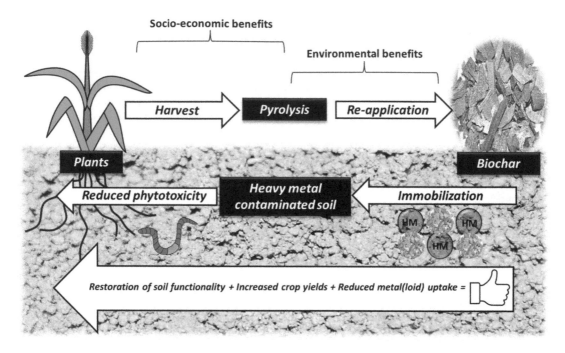

Figure 21.2 *Schematic summary of a remediation system where biochar is deployed to immobilize heavy metals, reduce phytotoxicity, and improve biomass yield. Biomass may be pyrolyzed and re-applied to soils, maintaining a closed system*

Biochars have several well-documented effects on soil health, which should promote functionality and the recovery of degraded land either directly or by indirect mechanisms; liming effects, increased water holding capacity, and improved soil structure, for example (Chapters 13, 16, 20). Many of the documented benefits of biochar addition in this context are only seen when organic or inorganic fertilizers are added together with the biochar amendment, suggesting that biochar alone is often unsuitable as a soil amendment to stimulate re-vegetation (Ye et al, 2020). In fact, some studies report a decrease in plant growth after the amendment of soils with biochar (Beesley et al, 2011, 2013; Li et al, 2021), although others report agronomic benefits when biochar is exclusively added to soils (Shahbaz et al, 2018; Rehman et al, 2019; Almaroai and Eissa, 2020).

Trade-offs between (i) biochar efficiency for adjusting the equilibrium between mobile/bioavailable and stable/complexed heavy metals (toxicity); and (ii) soil functionality is the final aspect to consider in biochar application to contaminated sites. This allows an idealized system to be proposed, summarized below (Figure 21.2), where biochar addition to contaminated soils gives multiple benefits to soil quality which, in turn, are beneficial socially and economically.

Case examples from acid mines and smelter-impacted soils

Skeletal or weakly structured soils supplemented by waste tailings are common at former mine sites (Wong, 2003) where the original soil horizon sequences may be buried deeply below waste. Point and diffuse pollution of soils are also features of areas previously dominated by industrial processes such as smelting and tanning. As such, limitations exist to revegetation of such soils due to unusual stoichiometry of nutrients to pollutants. In a phytotoxicity test using *Phaseolus vulgaris*, Lomaglio et al (2017) demonstrated the ability of wood biochar to increase soil pH and reduce Cd, Pb, and Zn mobility when applied at 2 and 5% to soil in the vicinity of a former smelter site. In addition, they observed an increase in leaf dry weight at the same application rate, while leaf Cd, stem Pb, and leaf Zn concentrations decreased at 2% and 5% application rate. This demonstrated the potential of such biochar to reduce the toxicity of contaminated soil and thus allow revegetation. Similarly increasing doses of wood biochars were applied to soil impacted by industrial waste disposal. In general, only the higher application rates used here (1% and 2%) had beneficial effects on soil pH, pollutant bioavailability, and wheat growth (Ali et al, 2019). Root and shoot metal uptake decreased following biochar amendment, except for the increase in shoot Cu concentration at 0.5% and 1% biochar doses.

Often, to create functional soils at mine sites, some sort of technical soil (technosol) is created by blending mine spoil with other materials. This was the case in several studies on an As and Pb-rich mine spoil in France. Hardwood biochars, added at 2% and 5% to a mine-impacted soil increased *Populus euramericana* dry weight, raised Pb accumulation in aerial tissues, and lowered leaf As concentration (Lebrun et al, 2021a). In a phytotoxicity experiment, biochar (hardwood, 2% w/w) was mixed with iron sulfate, which increased bean dry weight, without affecting metal(loid) plant accumulation, while biochar alone had no effect (Lebrun et al, 2021b); this example illustrates the intelligent application of iron-based materials with biochars for As

contaminated soils. Red mud was also combined with biochar (bamboo, or bark sap oak wood), which increased *Salix triandra* dry weight by 3-fold (leaf), 7-fold (stem), and 15-fold (roots). Lastly, in a study applying biochar, manure, and ochre, in single and combined applications, only the treatments containing manure improved *Agrostis* growth and reduced As and Pb aerial concentrations (Lebrun et al, 2021c). These studies demonstrate the added value of combining biochar (s) with other amendments that can (i) provide specific metal(loid) sorption capacity (e.g., Fe-rich materials for As); and (ii) supply plant-available nutrients to assist revegetation (e.g., manures).

A similar strategy was employed to revegetate Cu-rich mine soil in Spain (Forján et al, 2018a). Mesocosms containing biochar/compost and biochar/technosol ratios applied to soil was 4:11 in both cases, to reflect as closely as possible its application in the field. Furthermore, the treatments were applied on the first 0.15–0.2 m of soil as this would be the way they would be applied. Additionally, seeds of *Brassica juncea* L. (phytoremediator species) were planted and pore water samples were collected at three different depths of the cylinders. The Cu concentrations decreased in pore water samples collected for 11 months, demonstrating Cu stabilization in the settling pond soil with the addition of biochar compost. The main factors favoring the decrease of available Cu concentrations were the increase in pH, the increase of organic carbon content, and the correction of cation exchange capacity (Forján et al, 2018a). In addition, specific surface area and functional groups of the biochar could have a positive effect in reducing the available Cu concentrations (Forján et al, 2018a). On the other hand, both treatments besides increasing carbon content also contributed to a rise in total soil nitrogen, improved nutrient retention, and contributed to soil restructuring. Thus, the treatments improved the physicochemical properties of the soil and favored the establishment of vegetation (Forján et al, 2018b). The biochar-compost treatment therefore improved both immobilization of Cu and plant growth (Forján et al, 2018c). *Brassica juncea* L. had a clear phytostabilizing activity over the Cu, despite not having phytoextraction capacity according to the obtained translocation factor (TF) and transfer coefficient (TC). At low TF values, Cu was not translocated from the root to the shoot, while it was fixed in the roots. This phytostabilization technique proved most appropriate to immobilize metals in mining sites, resulting in less exposure to this toxic element by livestock, wildlife, and human health (Forján et al, 2018b).

The particular issue of calcareous soils

Studies detailing the deployment of biochars into calcareous soils are scarcer than into acidic soils. It is important to remember that association with the exchangeable or carbonate fraction may reduce mobility under specific conditions of alkalinity. However, a reduction in pH, though slight, may cause the release of metals, and as such, metals in these associations are considered potentially mobile.

In a study carried out in a mining region in central Mexico, sediments presented geochemical characteristics generally conducive with limiting metal mobility, including neutral pH, high organic matter content, and very calcareous conditions (Mitchell et al, 2016). While the impact that each of these characteristics has on metal mobility depends principally on pH, a general prevalence of metals associated with the carbonate fraction can be expected in soils with neutral to alkaline pH, which was confirmed through sequential extraction. These associations

with the carbonate fractions are considered temporary as changes in physicochemical characteristics, especially pH, can easily remobilize associated metals.

To evaluate the effect of the amendment, biochar produced from sewage sludge and the non-pyrolyzed source material were applied to the affected sediments. Each week, for 28 days, metals were leached with $CaCl_2$ 0.01 M at pH 5.5, simulating typical conditions, and pH 3.5, simulating more extreme conditions (Houben et al, 2013). Notably, non-pyrolyzed sewage sludge was more effective at reducing metal(loid) mobility at pH 5.5, while biochar was more effective at pH 3.5. Sequential extraction confirmed an increase in the association of Pb and Cd to stable Fe, Mn, and organic matter fractions and a decrease in association with the mobile carbonate fraction (Mitchell et al, 2020). Zinc mobility demonstrates different sorption characteristics when found in complex metal matrices and, in the highlighted study, there were no significant differences in Zn mobility when leached at pH 3.5 or pH 5.5. It is reasonable to assume that the extraction at a more acidic pH, results in a predominance of metal-organic matter associations instead of precipitation, which may occur at higher pH. Sequential extraction showed a reduction of Zn associated with the potentially mobile exchangeable and carbonate-bound fractions. There was an increase in Zn association with more stable fractions in both cases, though association with the organic matter fraction was predominant in sediments amended with unpyrolyzed sludge, likely due to the high affinity of Zn for the major complexing agents found in unmodified organic matter including oxygenated carboxylic and amino groups. On the other hand, associations with the Fe and Mn oxide increased when amended with biochar, likely due to the decreased presence of these functional groups but a proportionate increase of stable Mn oxide sites. Thus, even in calcareous soils and sediments with an inherently elevated buffer capacity, amendment with biochar could reduce metal leachability in the long term.

Specific enhancements of biochars

In the context of the remediation of soils (Chapter 27), improvements or amendments to biochar formulations, such as by co-composting with other organic materials (Chapter 26) have been shown to favorably adjust soil physical-chemical characteristics in metal-contaminated soils (Teodoro et al, 2020). Iron oxides (FeOx) and other metal oxides (Al, Mn, etc.) are effective binding surfaces for metals and metalloids such as As, Hg, Se, Cr, Pb, etc., and are used in the remediation of heavy metal contaminated substrates (Warren et al, 2003; Waychunas et al, 2005). Arsenic is widely known to be immobilized by Fe-rich materials as they provide anion exchange sites (Masscheleyn et al, 1991), so optimizing biochar for metal and As retention may be possible by modifying its characteristics during production or pre-application (Dixit and Hering, 2003) and Fe oxide impregnated sorbents have been widely applied to contaminated waters (Reed et al, 2000; Vaughan and Reed, 2005). Soaking the source material with Fe chloride solution before pyrolysis entrains the Fe oxide into the biochar structure (Chen et al, 2011). Alternatively, the biochar may be soaked in a Fe solution after pyrolysis (Muñiz et al, 2009; Lebrun et al, 2018a). The cost of producing these biochars will be greater than unmodified biochar, so they may only be suitable for specific small-scale applications.

Biochar applications to urban agriculture sites

The application of biochars on an experimental basis to urban agriculture sites has been limited, with previous studies highlighting mixed results (Barrow, 2012; Werner et al, 2019). The potential role of biochar within urban farming practices highlights a smaller scale, and very targeted application of biochars (Song et al, 2020); many urban agriculture sites consist of small pockets of land, and diverse history and pollution legacy (Dennis et al, 2020). Here, biochar presents an opportunity to produce a soil amendment in situ and at the point of usage. With land often at a premium in cities, approaches such as retrofitting urban rooftop farming have gained traction in recent years (Hardman and Larkham, 2014). An obvious and prohibitive barrier to this has been the mass-to-volume ratio of traditional (soil-based) substrates (Orsini et al, 2017), especially when saturated or semi-saturated. The use of biochar on rooftops may be driven by a need for lighter substrates and improved plant-available water (Cao et al, 2014).

Conclusions and recommendations for biochar deployment

Biochars can offer several advantages alone or in combination with other amendments during heavy metal remediation of soils because:

• Biochars have a greater surface area than soils to which they are added and some have higher CEC meaning they are capable of sorbing high concentrations of heavy metals, such as Cd and Zn from the soils they amend. This is especially the case in soils low in organic matter with intrinsically low CEC.

- Biochars are more persistent in soils for longer periods than other commonly applied soil amendments, such as composts and sludges, and may not induce as great deal of co-mobilization effect between easily mineralizable organic fractions and heavy metals. Biochars can also raise the pH of soils, making some nutrients more available to plants, immobilizing some heavy metals, and liming acid soils.
- Biochars can assist in the re-vegetation of some contaminated soils by reducing phytotoxicity and improving the germination of seeds.
- Biochars may be effective at immobilizing metal(loid)s in acidic and calcareous soils, though studies on the latter are scarce.

Remediation strategies for industrial, mine, or urban soils could include biochars for reducing the leaching of heavy metals, decreasing the phytotoxicity of substrates, and assisting revegetation. In the case of As-contaminated sites, and especially where there is a potential that food crops may be cultivated (for example urban allotment sites), a greater degree of caution should be exercised not only

in whether or not to apply biochar, but also in what method of application and how much should be applied. Some biochars also contain elevated concentrations of heavy metals due to their source material. It must also be remembered that most biochars appear sub-optimal as fertilizers if applied alone, so there may be a need to combine them with materials containing available nutrients; co-composting with other organic materials is one such option. At very heavily contaminated and denuded sites, such as former mine areas, particularly where there are surface leachates of heavy metals and unconsolidated soils and wastes, biochars may be useful to restrict the wider impact of contamination beyond site boundaries. By the same principle, after a surface pollution incident, biochars may also be used to reduce the spread of contaminants to groundwater. Combination with other organic materials is likely to be required for effective phytostabilization and remediation. At old industrial sites, there may be sufficient native soil remaining, and remaining nutrients so that biochars can be applied alone to contamination hotspots to restart natural plant successions. In all cases, an intelligent approach should be taken to biochar application to land after some data has been gathered about the specific soil characteristics, and heavy metals present in elevated concentrations, preferably their bioavailability and their wider dispersal.

References

Abdelhadi SO, Dosoretz CG, Rytwo G, Gerchman Y, and Azaizeh H 2017 Production of biochar from olive mill solid waste for heavy metal removal. *Bioresource Technology* 244, 759–767.

Abrahams PW 2002 Soils: their implications to human health. *Science of The Total Environment* 291, 1–32.

Adriano DC 2001 *Trace Elements in Terrestrial Environments*. New York, NY: Springer.

Alhar MAM, Thompson DF, and Oliver IW 2021 Mine spoil remediation via biochar addition to immobilise potentially toxic elements and promote plant growth for phytostabilisation. *Journal of Environmental Management* 277, 111500.

Ali A, et al 2019 Application of wood biochar in polluted soils stabilized the toxic metals and enhanced wheat (*Triticum aestivum*) growth and soil enzymatic activity. *Ecotoxicology and Environmental Safety* 184, 109635.

Alloway BJ 1995 *Heavy Metals in Soils*. Springer Dordrecht.

Almaroai YA, and Eissa MA 2020 Effect of biochar on yield and quality of tomato grown on a metal-contaminated soil. *Scientia Horticulturae* 265, 109210.

Álvarez-Rogel J, Tercero Gómez MDC, Conesa HM, Párraga-Aguado I, and González-Alcaraz MN 2018 Biochar from sewage sludge and pruning trees reduced porewater Cd, Pb and Zn concentrations in acidic, but not basic, mine soils under hydric conditions. *Journal of Environmental Management* 223, 554–565.

Arabi Z, et al E 2021 (Im)mobilization of arsenic, chromium, and nickel in soils via biochar: A meta-analysis. *Environmental Pollution* 286, 117199.

Arienzo M, Adamo P, and Cozzolino V 2004 The potential of *Lolium perenne* for revegetation of contaminated soil from a metallurgical site. *Science of the Total Environment* 319(1), 13–25.

Barrow CJ 2012 Biochar: Potential for countering land degradation and for improving agriculture. *Applied Geography* 34, 21–28.

Beesley L, and Marmiroli M 2011 The immobilisation and retention of soluble arsenic, cadmium and zinc by biochar. *Environmental Pollution* 159, 474–480. 10.1016/j.envpol.2010.10.016

Beesley L, et al 2013 Biochar addition to an arsenic contaminated soil increases arsenic concentrations in the pore water but reduces uptake to tomato plants (*Solanum lycopersicum* L.). *Science of the Total Environment* 454–455, 598–603.

Beesley L, Moreno-Jiménez E, and Gomez-Eyles JL 2010 Effects of biochar and greenwaste compost amendments on mobility, bioavailability and toxicity of inorganic and organic contaminants in a multi-element polluted soil. *Environmental Pollution* 158, 2282–2287.

Beesley L, Trakal L, Hough R, and Mitchell K 2022 Mobility and crop uptake of Zn in a legacy sludge-enriched agricultural soil amended with biochar or compost: insights from a pot and recirculating column leaching test. *Environmental Science and Pollution Research* 29, 83545–83553.

Bird MI, et al 2012 Algal biochar: effects and applications. *Global Change Biology Bioenergy* 4, 61–69.

Bolan NS, and Duraisamy VP 2003 Role of inorganic and organic soil amendments on immobilisation and phytoavailability of heavy metals: A review involving specific case studies. *Australian Journal of Soil Research* 41, 533–555.

Bolan NS, et al 2008 Manipulating bioavailability to manage remediation of metal-contaminated soils. In: Hartemink AE, McBratney AB, and Naidu R (Eds) *Developments in Soil Science* Vol. 32. Amsterdam: Elsevier. pp657–678.

Cao, CTN, et al 2014 Biochar makes green roof substrates lighter and improves water supply to plants. *Ecological Engineering*, 71, 368–374. 10.1016/j.ecoleng.2014.06.017.

Cao X, Ma L, Gao B, and Harris W 2009 Dairy-manure derived biochar effectively sorbs lead and atrazine. *Environmental Science and Technology* 43, 3285–3291.

Cao X, Ma L, Liang Y, Gao B, and Harris W 2011 Simultaneous immobilization of lead and atrazine in contaminated soils using dairy-manure biochar. *Environmental Science and Technology* 45, 4884–4889.

Cao X, Ma LQ, and Shiralipour A 2003 Effects of compost and phosphate amendments on arsenic mobility in soils and arsenic uptake by the hyperaccumulator, *Pteris vittata* L. *Environmental Pollution* 126, 157–167.

Chen S-B, Zhu Y-G, Ma Y-B, and McKay G 2006 Effect of bone char application on Pb bioavailability in a Pb-contaminated soil. *Environmental Pollution* 139, 433–439.

Chen X, et al 2011 Adsorption of copper and zinc by biochars produced from pyrolysis of hardwood and corn straw in aqueous solution. *Bioresource Technology*. 102, 8877–8884.

Cheng C-H, Lehmann J, Thies JE, Burton SD, and Engelhard MH 2006 Oxidation of black carbon by biotic and abiotic processes. *Organic Geochemistry* 37, 1477–1488.

Dennis M, Beesley L, Hardman M, and James P 2020 Ecosystem (dis)benefits arising from formal and informal land-use in Manchester (UK); a case study of urban soil characteristics associated with local green space management. *Agronomy* 10, 552.

Di Toro DM, et al 2001 Biotic ligand model of the acute toxicity of metals. 1. Technical Basis. *Environmental Toxicology and Chemistry* 20, 2383–2396.

Dixit S, and Hering JG 2003 Comparison of arsenic(V) and arsenic(III) sorption onto iron oxide minerals: implications for arsenic mobility. *Environmental Science and Technology* 37, 4182–4189.

Durães N, Novo LAB, Candeias C, and da Silva EF 2018 Distribution, transport and fate of pollutants. In: Duarte AC, Cachada A, and Rocha-Santos T (Eds) *Soil Pollution.* Cambridge: Academic Press. pp29–57.

Fangueiro D, Kidd P, Alvarenga P, Beesley L, and Varennes A 2018 Strategies for soil protection and remediation. In: Duarte AC, Cachada A, and Rocha-Santos T (Eds) *Soil Pollution.* Amsterdam: Elsevier. pp251–281.

Fitz WJ, and Wenzel WW 2002 Arsenic transformations in the soil–rhizosphere–plant system: fundamentals and potential application to phytoremediation. *Journal of Biotechnology* 99, 259–278.

Forján R., Rodríguez-Vila A, Cerqueira B, and Covelo EF 2018a Comparison of compost with biochar versus technosol with biochar in the reduction of metal pore water concentrations in a mine soil. *Journal of Geochemical Exploration* 192, 103–111.

Forján R, et al 2018b Comparative effect of compost and technosol enhanced with biochar on the fertility of a degraded soil. *Environmental Monitoring and Assessment* 190, 610.

Forján R, Rodríguez-Vila A, Pedrol N, and Covelo EF 2018c Application of compost and biochar with *Brassica juncea* L. to reduce phytoavailable concentrations in a settling pond mine soil. *Waste Biomass Valorization* 9, 821–834.

Freddo A, Cai C, and Reid BJ 2012 Environmental contextualisation of potential toxic elements and polycyclic aromatic hydrocarbons in biochar. *Environmental Pollution* 171, 18–24.

Gao L, et al 2021 Impacts of pyrolysis temperature on lead adsorption by cotton stalk-derived biochar and related mechanisms. *Journal of Environmental Chemical Engineering* 9, 105602.

Gao R, et al 2020 High-efficiency removal capacities and quantitative sorption mechanisms of Pb by oxidized rape straw biochars. *Science of the Total Environment* 699, 134262.

Gascó G, Paz-Ferreiro J, and Méndez A 2012 Thermal analysis of soil amended with sewage sludge and biochar from sewage sludge pyrolysis. *Journal of Thermal Analytical Calorimetry* 108, 769–775.

Gomez-Eyles JL, Sizmur T, Collins CD, and Hodson ME 2011 Effects of biochar and the earthworm Eisenia fetida on the bioavailability of polycyclic aromatic hydrocarbons and potentially toxic elements. *Environmental Pollution* 159, 616–622.

Gomez-Eyles JL, et al 2013 Evaluation of biochars and activated carbons for in situ remediation of sediments impacted with organics, mercury, and methylmercury. *Environmental Science and Technology* 47, 13721–13729.

Graber ER, et al 2010 Biochar impact on development and productivity of pepper and tomato grown in fertigated soilless media. *Plant and Soil* 337, 481–496.

Guo X, Liu H, & Zhang J 2020 The role of biochar in organic waste composting and soil improvement: A review. *Waste Management* 102, 884–899. 10.1016/j.wasman.2019.12.003.

Hardman M, and Larkham P 2014 *Informal Urban Agriculture: The Secret Lives of Guerrilla Gardeners.* New York: Springer.

Hartley W, Dickinson NM, Riby P, and Lepp NW 2009 Arsenic mobility in brownfield soils amended with green waste compost or biochar and planted with Miscanthus. *Environmental Pollution* 157, 2654–2662.

Harvey OR, Herbert BE, Rhue RD, and Kuo L-J 2011 Metal interactions at the biochar-water interface: energetics and structure-sorption relationships elucidated by flow adsorption microcalorimetry. *Environmental Science and Technology* 45, 5550–5556.

He E, et al 2019 Two years of aging influences the distribution and lability of metal(loid)s in a contaminated soil amended with different biochars. *Science of the Total Environment* 673, 245–253.

He YD, et al 2010 The fate of Cu, Zn, Pb and Cd during the pyrolysis of sewage sludge at different temperatures. *Environmental Technology* 31, 567–574.

Hodson ME 2010 The need for sustainable soil remediation. *Elements* 6, 363–368.

Hossain MK, Strezov V, Yin Chan K, and Nelson PF 2010 Agronomic properties of wastewater sludge biochar and bioavailability of metals in production of cherry tomato (*Lycopersicon esculentum*). *Chemosphere* 78, 1167–1171.

Houben D, Evrard L, & Sonnet P 2013 Mobility, bioavailability and pH-dependent leaching of cadmium, zinc and lead in a contaminated soil amended with biochar. *Chemosphere* 92, 1450–1457. 10.1016/j.chemosphere.2013.03.055.

Ippolito JA, et al 2017 Biochars reduce mine soil bioavailable metals. *Journal of Environmental Quality* 46, 411–419.

Jain S, et al 2020 Biochar aided aromatic grass [*Cymbopogon martini* (Roxb.) Wats.] vegetation: A sustainable method for stabilization of highly acidic mine waste. *Journal of Hazardous Materials* 390, 121799.

Jones DL, Rousk J, Edwards-Jones G, DeLuca TH, and Murphy DV 2012 Biochar-mediated changes in soil quality and plant growth in a three year field trial. *Soil Biology and Biochemistry* 45, 113–124.

Jung K-W, Kim K, Jeong T-U, and Ahn K-H 2016 Influence of pyrolysis temperature on characteristics and phosphate adsorption capability of biochar derived from waste-marine macroalgae (*Undaria pinnatifida* roots). *Bioresource Technology* 200, 1024–1028.

Knowles OA, Robinson BH, Contangelo A, and Clucas L 2011 Biochar for the mitigation of nitrate leaching from soil amended with biosolids. *Science of the Total Environment* 409, 3206–3210.

Kwak J-H, et al 2019 Biochar properties and lead (II) adsorption capacity depend on feedstock type, pyrolysis temperature, and steam activation. *Chemosphere* 231, 393–404.

Lado, LR, Hengl, T, and Reuter, HI 2008 Heavy metals in European soils: A geostatistical analysis of the FOREGS Geochemical database. *Geoderma* 148, 189–199. 10.1016/j.geoderma.2008.09.020.

Lebrun M, et al 2017 Effect of biochar amendments on As and Pb mobility and phytoavailability in contaminated mine technosols phytoremediated by Salix. *Journal of Geochemical Exploration* 182, 149–156.

Lebrun M, et al 2018a Eco-restoration of a mine technosol according to biochar particle size and dose application: study of soil physico-chemical properties and phytostabilization capacities of *Salix viminalis*. *Journal of Soils and Sediments* 18, 2188–2202.

Lebrun M, et al 2018b Assisted phytostabilization of a multicontaminated mine technosol using biochar amendment: Early stage evaluation of biochar feedstock and particle size effects on As and Pb accumulation of two Salicaceae species (*Salix viminalis* and *Populus euramericana*). *Chemosphere* 194, 316–326.

Lebrun M, et al 2018c Effect of Fe-functionalized biochar on toxicity of a technosol contaminated by Pb and As: sorption and phytotoxicity tests. *Environmental Science Pollution Research* 25, 33678–33690.

Lebrun M, et al 2019 Biochar effect associated with compost and iron to promote Pb and As soil stabilization and *Salix viminalis* L. growth. *Chemosphere* 222, 810–822.

Lebrun M, et al 2020 Effect of different tissue biochar amendments on As and Pb stabilization and phytoavailability in a contaminated mine technosol. *Science of the Total Environment* 707, 135657.

Lebrun M, Miard F, Nandillon R, Morabito D, and Bourgerie S 2021a Effect of biochar, iron sulfate and poultry manure application on the phytotoxicity of a former tin mine. *International Journal of Phytoremediation* 23, 1222–1230.

Lebrun M, et al 2021b Physiological and molecular responses of flax (*Linum usitatissimum* L.) cultivars under a multicontaminated technosol amended with biochar. *Environmental Science Pollution Research* 28, 53728–53745.

Lebrun M, et al 2021c Effects of biochar, ochre and manure amendments associated with a metallicolous ecotype of *Agrostis capillaris* on As and Pb stabilization of a former mine technosol. *Environmental Geochemistry and Health* 43, 1491–1505.

Lebrun M, et al 2021d Effects of carbon-based materials and redmuds on metal(loid) immobilization and growth of *Salix dasyclados* Wimm. on a former mine Technosol contaminated by arsenic and lead. *Land Degradation and Development* 32, 467–481.

Li H, et al 2018 Distribution and transformation of lead in rice plants grown in contaminated soil amended with biochar and lime. *Ecotoxicology and Environmental Safety* 165, 589–596.

Li X, Xiao J, Salam MMA, Ma C, and Chen G 2021 Impacts of bamboo biochar on the phytoremediation potential of *Salix psammophila* grown in multi-metals contaminated soil. *International Journal of Phytoremediation* 23, 387–399.

Lindsay WL 2001 *Chemical Equilibria in Soils*. Caldwell: The Blackburn Press.

Liu Z, and Zhang F-S 2009 Removal of lead from water using biochars prepared from hydrothermal liquefaction of biomass. *Journal of Hazardous Materials* 167, 933–939.

Lomaglio T, et al 2017 Effect of biochar amendments on the mobility and (bio) availability of As, Sb and Pb in a contaminated mine technosol. *Journal of Geochemical Exploration* 182, 138–148.

Mankasingh U, Choi P-C, Ragnarsdottir V 2011 Biochar application in a tropical, agricultural region: A plot scale study in Tamil Nadu, India. *Applied Geochemistry* 26, S218–S221.

Masscheleyn PH, Delaune RD, Patrick WH 1991 Effect of redox potential and pH on arsenic speciation and solubility in a contaminated soil. *Environmental Science and Technology* 25, 1414–1419.

Meharg AA, and Macnair MR 1992 Suppression of the high affinity phosphate uptake system: a mechanism of arsenate tolerance in *Holcus lanatus* L. *Journal of Experimental Botany* 43, 519–524.

Melo LCA, Alleoni LRF, and Swartjes FA 2011 Derivation of critical soil cadmium concentrations for the state of São Paulo, Brazil, based on human health risks. *Human and Ecological Risk Assessment: An International Journal* 17, 1124–1141.

Mench M, et al 2003 Progress in remediation and revegetation of the barren Jales gold mine spoil after in situ treatments. *Plant and Soil* 249, 187–202.

Méndez A, Gómez A, Paz-Ferreiro J, and Gascó G 2012 Effects of sewage sludge biochar on plant metal availability after application to a Mediterranean soil. *Chemosphere* 89, 1354–1359.

Mikutta C, and Kretzschmar R 2011 Spectroscopic evidence for ternary complex formation between arsenate and ferric iron complexes of humic substances. *Environmental Science and Technology* 45(22), 9550–9557.

Miretzky P, and Fernandez-Cirelli A 2008 Phosphates for Pb immobilization in soils: a review. *Environmental Chemistry Letters* 6, 121–133.

Mitchell K et al 2016 Evaluation of environmental risk of metal contaminated soils and sediments near mining sites in Aguascalientes, Mexico. *Bulletin of Environmental Contamination and Toxicology* 97, 216–224. 10.1007/s00128-016-1820-9.

Mitchell K et al 2020 The effect of low-temperature biochar and its non-pyrolyzed composted biosolids source on the geochemical fractionation of Pb and Cd in calcareous river sediments. *Environmental Earth Sciences*, 79. 10.1007/s12665-020-08908-5.

Moreno-Jiménez E, Esteban E, and Peñalosa JM 2012 The fate of arsenic in soil-plant systems. In: Whitacre DM (Eds) *Reviews of Environmental Contamination and Toxicology New York*, New York: Springer. pp1–37.

Mukherjee A, Zimmerman AR, and Harris W 2011 Surface chemistry variations among a series of laboratory-produced biochars. *Geoderma* 163, 247–255.

Muñiz G, et al 2009 Synthesis, characterization and performance in arsenic removal of iron-doped activated carbons prepared by impregnation with Fe(III) and Fe(II). *Journal of Hazardous Materials* 165, 893–902

Namgay T, et al 2010 Influence of biochar application to soil on the availability of As, Cd, Cu, Pb, and Zn to maize (*Zea mays* L.). *Soil Research* 48, 638–647.

Nandillon R, et al 2021 Contrasted tolerance of *Agrostis capillaris* metallicolous and non-metallicolous ecotypes in the context of a mining technosol amended by biochar, compost and iron sulfate. *Environmental Geochemistry and Health* 43,1457–1475.

Nathanail CP, Bardos RP 2004 Remediation Application. In: *Reclamation of Contaminated Land*. Hoboken: John Wiley and Sons, Ltd. pp207–225.

Orsini F, Dubbeling M, de Zeeuw H, and Gianquinto G 2017 *Rooftop Urban Agriculture*. Cham: Springer International Publishing.

Park JH, Choppala GK, Bolan NS, Chung JW, and Chuasavathi T 2011 Biochar reduces the bioavailability and phytotoxicity of heavy metals. *Plant and Soil* 348, 439–451.

Park J-H, et al 2019 Cadmium adsorption characteristics of biochars derived using various pine tree residues and pyrolysis temperatures. *Journal of Colloid and Interface Science* 553, 298–307.

Pulford ID, and Watson C 2003 Phytoremediation of heavy metal-contaminated land by trees—a review. *Environment International* 29, 529–540

Reed BE, Vaughan R, and Jiang L 2000 As(III), As(V), Hg, and Pb removal by Fe-oxide impregnated activated carbon. *Journal of Environmental Engineering* 126, 869–873.

Rehman M, et al 2019 Influence of rice straw biochar on growth, antioxidant capacity and copper uptake in ramie (*Boehmeria nivea* L.) grown as forage in aged copper-contaminated soil. *Plant Physiology and Biochemistry* 138, 121–129.

Ross SM 1994 *Toxic Metals in Soil-Plant Systems*. London: Wiley.

Ruttens A, et al 2006 Phytostabilization of a metal contaminated sandy soil. I: Influence of compost and/or inorganic metal immobilizing soil amendments on phytotoxicity and plant availability of metals. *Environmental Pollution* 144, 524–532.

Samsuri AW, Fahmi AH, Jol H, and Daljit S 2020 Particle size and rate of biochar affected the phytoavailability of Cd and Pb by mustard plants grown in contaminated soils. *International Journal of Phytoremediation* 22, 567–577.

Sánchez-Polo M, and Rivera-Utrilla J 2002 Adsorbent–adsorbate interactions in the adsorption of Cd(II) and Hg(II) on ozonized activated carbons. *Environmental Science and Technology* 36(17), 3850–3854.

Schaefer CEGR, Fabris JD, Ker JC 2008 Minerals in the clay fraction of Brazilian Latosols (Oxisols): A review. *Clay Minerals* 43(1), 137–154.

Shahbaz AK, Iqbal M, Jabbar A, Hussain S, and Ibrahim M 2018 Assessment of nickel bioavailability through chemical extractants and red clover (*Trifolium pratense* L.) in an amended soil: Related changes in various parameters of red clover. *Ecotoxicology and Environmental Safety* 149, 116–127.

Shakya A, and Agarwal T 2019 Removal of Cr (VI) from water using pineapple peel derived biochars: Adsorption potential and re-usability assessment. *Journal of Molecular Liquids*. 293, 111497.

Shen Y-S, Wang S-L, Tzou Y-M, Yan Y-Y, and Kuan W-H 2012 Removal of hexavalent Cr by coconut coir and derived chars – The effect of surface functionality. *Bioresource Technology* 104, 165–172.

Singh E, et al 2021 Pyrolysis of waste biomass and plastics for production of biochar and its use for removal of heavy metals from aqueous solution. *Bioresource Technology* 320, 124278.

Sizirici B, Fseha YH, Yildiz I, Delclos T, and Khaleel A 2021 The effect of pyrolysis temperature and feedstock on date palm waste derived biochar to remove single and multi-metals in aqueous solutions. *Sustainable Environment Research* 31, 9.

Sizmur T, Wingate J, Hutchings T, and Hodson, ME 2011 Lumbricus terrestris L. does not impact on the remediation efficiency of compost and biochar amendments. *Pedobiologia* 54, S211–S216. 10.1016/j.pedobi.2011.08.008.

Sobik-Szołtysek J, Wystalska K, Malińska K, and Meers E 2021 Influence of pyrolysis temperature on the heavy metal sorption capacity of biochar from poultry manure. *Materials* 14, 6566.

Song SJ, et al 2020 New application for assessment of dry eye syndrome induced by particulate matter exposure. *Ecotoxicology and Environmental Safety* 205, 111125.

Soria RI, Rolfe SA, Betancourth MP, and Thornton SF 2020 The relationship between properties of plant-based biochars and sorption of Cd(II), Pb(II) and Zn(II) in soil model systems. *Heliyon* 6, e05388.

Sorrenti G, Masiello CA, Dugan B, and Toselli M 2016 Biochar physico-chemical properties as affected by environmental exposure. *Science of the Total Environment* 563–564, 237–246.

Teodoro M, et al 2020 Application of co-composted biochar significantly improved plant-growth relevant physical/chemical properties of a metal contaminated soil. *Chemosphere* 242, 125255.

Thakali S, et al 2006 Terrestrial Biotic Ligand Model. 2. Application to Ni and Cu toxicities to plants, invertebrates, and microbes in soil.

Environmental Science and Technology 40, 7094–7100.

Tomczyk A, Sokołowska Z, and Boguta P 2020 Biochar physicochemical properties: pyrolysis temperature and feedstock kind effects. *Reviews in Environmental Science and Biotechnology* 19, 191–215.

Tordoff GM, Baker AJM, and Willis AJ 2000 Current approaches to the revegetation and reclamation of metalliferous mine wastes. *Chemosphere* 41, 219–228.

Trakal L, Komárek M, Száková J, Zemanová V, and Tlustoš P 2011 Biochar application to metal-contaminated soil: Evaluating of Cd, Cu, Pb and Zn sorption behavior using single- and multi-element sorption experiment. *Plant Soil and Environment* 57, 372–380.

Uchimiya M, et al 2010 Immobilization of heavy metal ions (CuII, CdII, NiII, and PbII) by broiler litter-derived biochars in water and soil. *Journal of Agricultural and Food Chemistry* 58, 5538–5544.

Uchimiya M, Klasson KT, Wartelle LH, and Lima IM 2011 Influence of soil properties on heavy metal sequestration by biochar amendment: 2. *Copper desorption isotherms*. *Chemosphere* 82, 1438–1447.

Van Poucke R, et al 2018 Chemical stabilization of Cd-contaminated soil using biochar. *Applied Geochemistry* 88, 122–130.

Van Poucke R, et al 2020 Application of biochars and solid fraction of digestate to decrease soil solution Cd, Pb and Zn concentrations in contaminated sandy soils. *Environmental Geochemistry and Health* 42, 1589–1600.

Vangronsveld J, et al 2009 Phytoremediation of contaminated soils and groundwater: Lessons from the field. *Environmental Science and Pollution Research* 16, 765–794.

Vaughan RL, and Reed BE 2005 Modeling As (V) removal by a iron oxide impregnated activated carbon using the surface complexation approach. *Water Research* 39, 1005–1014.

Wang X, et al 2020 Influence of pyrolysis conditions on the properties and Pb^{2+} and Cd^{2+} adsorption potential of tobacco stem biochar. *BioResources* 15(2), 4026–4051.

Wang Z, Xu J, Yellezuome D, and Liu R 2021 Effects of cotton straw-derived biochar under different pyrolysis conditions on Pb (II) adsorption properties in aqueous solutions. *Journal of Analytical and Applied Pyrolysis* 157, 105214.

Warren GP, et al 2003 Field trials to assess the uptake of arsenic by vegetables from contaminated soils and soil remediation with iron oxides. *Science of the Total Environment* 311(1), 19–33.

Waychunas GA, Kim CS, and Banfield JF 2005 Nanoparticulate iron oxide minerals in soils and sediments: unique properties and contaminant scavenging mechanisms. *Journal of Nanoparticle Research* 7, 409–433.

Wei J, et al 2019 Assessing the effect of pyrolysis temperature on the molecular properties and copper sorption capacity of a halophyte biochar. *Environmental Pollution* 251, 56–65.

Werner S, et al 2019 Nutrient balances with wastewater irrigation and biochar application in urban agriculture of Northern Ghana. *Nutrient Cycling in Agroecosystems* 115, 249–262.

Wong MH 2003 Ecological restoration of mine degraded soils, with emphasis on metal contaminated soils. *Chemosphere* 50, 775–780.

Wu W, et al 2016 Influence of pyrolysis temperature on lead immobilization by chemically modified coconut fiber-derived biochars in aqueous environments. *Environmental Science and Pollution Research* 23, 22890–22896.

Ye L, et al 2020 Biochar effects on crop yields with and without fertilizer: A meta-analysis of field studies using separate controls. *Soil Use and Management* 36, 2–18.

Zeng G, et al 2017 Precipitation, adsorption and rhizosphere effect: The mechanisms for Phosphate-induced Pb immobilization in soils—A review. *Journal of Hazardous Materials* 339, 354–367.

Zhang P, Zhang X, Yuan X, Xie R, and Han L 2021 Characteristics, adsorption behaviors, Cu (II) adsorption mechanisms by cow manure biochar derived at various pyrolysis temperatures. *Bioresource Technology* 331, 125013.

22

Organic contaminants in biochar

Wolfram Buss, Christian Wurzer, Jessica G. Shepherd, and
Thomas D. Bucheli

Introduction

For biochar to be permitted for use in agricultural and environmental applications it must meet regulated contaminant threshold values to demonstrate to regulators that it does not pose a risk to human, animal, and environmental health. There are two main types of contaminants in biochar: potentially toxic inorganic elements (Chapter 21) and organic contaminants (discussed below). Organic contaminants can either originate from the pyrolysis process itself or from feedstock materials.

In this chapter, we discuss the formation of organic contaminants during pyrolysis and the removal and destruction of organic contaminants present in feedstock materials. We investigate the link between pyrolysis conditions and contaminant content in biochar and perform a meta-analysis summarizing data on PAH content in biochar. We further provide recommendations on how to minimize the quantity of organic contaminants in biochar by considering feedstock selection, pyrolysis parameters, and post-treatments.

Organic contaminants formed during pyrolysis

Polycyclic aromatic hydrocarbons

Polycyclic aromatic hydrocarbons (PAHs) are organic compounds composed of carbon (C) and hydrogen (H) with at least two condensed aromatic rings (Baek et al, 1991), which can be formed during incomplete combustion processes, including pyrolysis. The PAH ring structure can also be heterocyclic, or substituted, in the case of oxygenated PAHs, nitrogen-containing PAHs, or methylated PAHs (US Department of Health and Human Services, 1995; Weidemann et al, 2017;

DOI: 10.4324/9781003297673-22

Lam et al, 2018). Although there are many different PAHs, typically only 16 unsubstituted homocyclic PAHs are analyzed in environmental samples. These 16 PAHs were identified by the United States Environmental Protection Agency (US EPA) as priority pollutants in the late 1970s (Keith, 1979). PAHs exhibit chronic carcinogenic, mutagenic, and/or teratogenic effects but can also cause acute toxicity to human health and the environment (The Environmental Applications Group LTD, 1990; US Department of Health and Human Services, 1995).

PAHs are formed through two distinct processes during biomass pyrolysis and combustion: low-temperature PAH formation, and pyrosynthesis at higher temperatures (McGrath et al, 2003; Keiluweit et al, 2012). At temperatures below ~600°C, hydrocarbons originating in the biomass are first aromatized (dehydrogenated) and then polymerized to form PAHs (see Figure 22.1 in Bucheli et al, 2015). During pyrosynthesis,

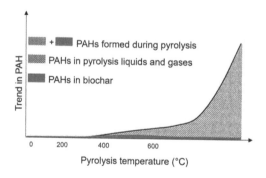

Figure 22.1 *Qualitative relationship of pyrolysis temperature with PAHs formed during pyrolysis (PAH formation minus decomposition), distinguished into PAHs in pyrolysis liquids and gases (vaporized from pyrolysis solids) and PAH content in biochar. PAH content in biochar and link with pyrolysis temperature are qualitative results from a meta-analysis conducted for this Chapter (quantitative data in Figures 22.2 and 22.3)*

single-aromatic radicals or other single aromatics, such as phenol, are formed from lignin, cellulose, and hemicellulose, and subsequently, PAHs are formed through gas-phase reactions of these small aromatic compounds that condense to form larger poly-aromatic compounds (McGrath et al, 2003; Sharma and Hajaligol, 2003).

Such PAHs formed during the production process can subsequently contaminate the biochar. However, quantification of PAHs in biochar can be challenging since it is influenced and possibly biased by the biochar sampling technique and method used for PAH analysis (Hilber et al, 2012; Fabbri et al, 2013; Bucheli et al, 2014, 2015). For example, increasing the biochar extraction time from 6 to 36 hours can increase PAH recovery 20-fold (Buss et al, 2015), and biochar samples of smaller particle size have been shown to have higher PAH content than the same biochar with larger particle size (Hilber et al, 2017b). Therefore, comparing reported biochar PAH concentrations across different studies can be challenging or even misleading where they were quantified using different analytical methodologies. The pyrolysis unit used for biochar production introduces another factor of uncertainty. Small variations in the unit design or inconsistencies during biochar production can result in large variations in PAH content, as these factors lead to temperature fluctuations that cause condensation of pyrolysis vapors, which contain organic contaminants (Buss et al, 2022b). Such effects are not typically obvious and are thus not reported in the scientific literature. This increases the difficulties of comparing data from different studies.

Effect of highest treatment temperature on PAH content in biochar

Investigating the effect of the highest treatment temperature (HTT) on PAH content in biochar has proven very challenging. The thermochemical reactions involving PAHs during pyrolysis are dynamic; PAHs are formed, decomposed, and fused into larger compounds, and can also be incorporated into the biochar carbon structure (Buss et al, 2016). During pyrolysis, the biomass feedstock and emerging gases undergo different reactions depending on the temperature in the pyrolysis unit as it increases towards the HTT. Therefore, the total PAH formation is the sum of net PAH formation integrated over the full temperature profile until the HTT is reached (Buss et al, 2016).

The total amount of PAHs formed typically increases with pyrolysis temperature up to 750°C (Figure 22.1) and probably even beyond (up to ~950°C) (Dai et al, 2014a; Zhou et al, 2014). Total PAH yield during pyrolysis (in absolute amounts, e.g., mg) is not equivalent to the PAH content in residual pyrolysis solids (biochar; relative amounts, e.g., mg kg^{-1}), as most PAHs (typically >99%) are vaporized and are therefore in the gas and liquid products of pyrolysis (Figure 22.1) (Dai et al, 2014b; Fagernäs et al, 2012). There is thus no straightforward link between accumulated PAH yield and PAH content in biochar.

The effect of HTT on PAH content in biochar has been studied extensively, yet no consistent relationship has been reported (Freddo et al, 2012; Hale et al, 2012; Keiluweit et al, 2012; Rogovska et al, 2012; Fabbri et al, 2013; Dai et al, 2014b; Devi and Saroha, 2015; Buss et al, 2016; Wang et al, 2017). To gain a greater understanding of the relevant factors, it is necessary to conduct meta-analyses including multiple studies. However, due to limited comparability of total PAH content in biochar for reasons given above, we standardized PAH content within the same study (mean of PAH content in each study set to 1). These biochars are produced with the same production unit and

PAH extraction techniques. This standardization allows for investigating the relative effect of HTT on PAH content in biochar without the bias introduced by different pyrolysis units and extraction techniques, which can lead to inconsistencies and extreme outliers in the analysis.

The 16 US EPA PAHs were examined in biochars produced from slow pyrolysis from 9 studies containing 23 temperature series (Figures 22.2: Hale et al, 2012; Zheng et al, 2013; Lyu et al, 2016; Madej et al, 2016; Kończak and Oleszczuk, 2018; Wang et al, 2018; Chen et al, 2019; de la Rosa et al, 2019; Adánez-Rubio et al, 2021). Only temperature series with ≥ 3 data points that covered at least a temperature range of 200°C within 350–750°C were used. The mean 16 US EPA PAH content in these 74 biochars was 6.2 ± 24.0 mg kg^{-1} with a large range and a median of 0.8 mg kg^{-1}, which means that threshold values set by the International Biochar Initiative (IBI) (6 mg kg^{-1}) and the European Biochar Certificate (EBC) (4 mg kg^{-1}) can be met by most biochars in these studies. Other studies confirm that biochars can routinely meet these limits (Rombolà et al, 2015; Sigmund et al, 2017; Hilber et al, 2019).

PAH contents at 450, 600, 650, and 700°C are significantly lower than those in biochars produced at 350°C (Figure 22.2). This can be explained by PAH formation and PAH condensation and adsorption that take place simultaneously and together determine the PAH content in biochar. The total PAH content in all pyrolysis products increases with HTT (at least until 750°C), while the thermodynamic conditions at higher HTT favor the association of PAHs with the gas phase (Bucheli et al, 2015; Figure 22.1). Elevated temperatures minimize condensation and adsorption of PAHs to the solid phase (Figure 22.2).

The HTT significantly influenced the content of 2-, 3-, 4-, 5- and 6-ring PAHs in

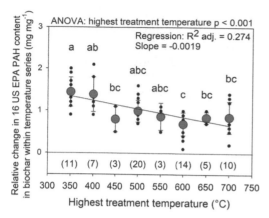

Figure 22.2 *Relationship of total 16 US EPA PAH content in biochar with highest treatment temperature (HTT). Relative change of PAH content within individual temperature series shown where biochars were produced in the same production unit, from the feedstock, and under the same production conditions and extracted using the same extraction technique. The PAH content in each temperature series was standardized setting the mean to 1 and hence, the data illustrate the relative change in PAH concentrations only based on the HTT. Individual data points are shown (black dots) and mean with standard deviation (red dots). Results of one-way ANOVA (p-value) and Tukey's post-hoc test (letters) are displayed above the figure panel and above data points of each temperature, respectively. The numbers in brackets signify the number of samples per temperature. Linear regression shown (adjusted $R^2 = 0.2738$, p-value temperature effect < 0.001). Data based on literature (Hale et al, 2012; Zheng et al, 2013; Lyu et al, 2016; Madej et al, 2016; Kończak and Oleszczuk, 2018; Wang et al, 2018; Chen et al, 2019; de la Rosa et al, 2019; Adánez-Rubio et al, 2021)*

biochars (ANOVA p-values < 0.05, above each panel in Figure 22.3). The regression lines clearly demonstrate that the amount of HMW PAHs in biochar decreased to a larger

extent with HTT than the one of lower molecular weight (LMW) PAHs (more negative slopes of HMW PAHs in Figure 22.3).

Naphthalene has a boiling point of 218°C, therefore, it is predominantly distributed in the gas phase, rather than in the solid phase already at a pyrolysis temperature of 350°C, and its presence in biochar only marginally changes at higher HTTs (Figure 22.3A); the content of naphthalene in biochar is mostly independent of HTT as already hypothesized in the second edition of this textbook (Bucheli et al, 2015). PAHs with 4 rings, for example, have much higher boiling points than naphthalene (384, 404, 438, and 448°C for fluoranthene, pyrene, benzo[a]anthracene and chrysene, respectively (Adánez-Rubio et al, 2021)) and HTT, therefore, reduces their presence in biochar (Figure 22.3C). This effect is confirmed for 5- and 6-ring PAHs (negative slope of regression curves in Figure 22.3D, E). The result clearly demonstrates that HMW PAHs can be efficiently prevented from condensation on biochar via pyrolysis at elevated HTTs (Figure 22.3C-E).

Naphthalene clearly dominates the composition of PAHs in biochar across different pyrolysis units, typically making up 30–80% of total PAHs identified in biochar (Freddo et al, 2012; Kloss et al, 2012; Fabbri et al, 2013; Buss et al, 2022b). In the 74 biochars reported in Figures 22.2 and 22.3, the average naphthalene content was ~40% of total PAHs, in line with the findings of the second edition of this chapter (Bucheli et al, 2015). It is still unclear why the naphthalene content in biochar is so high despite its high boiling point and hence its tendency to be present in the gas rather than solid phase (Buss et al, 2022b). Phenanthrene is often the second most prevalent PAH in biochar, which is in line with the findings of our analyses in this chapter (23.5% of total PAHs in the 74 biochars) (Hilber et al, 2012; Fabbri et al, 2013; Quilliam et al, 2013; Buss et al, 2022b).

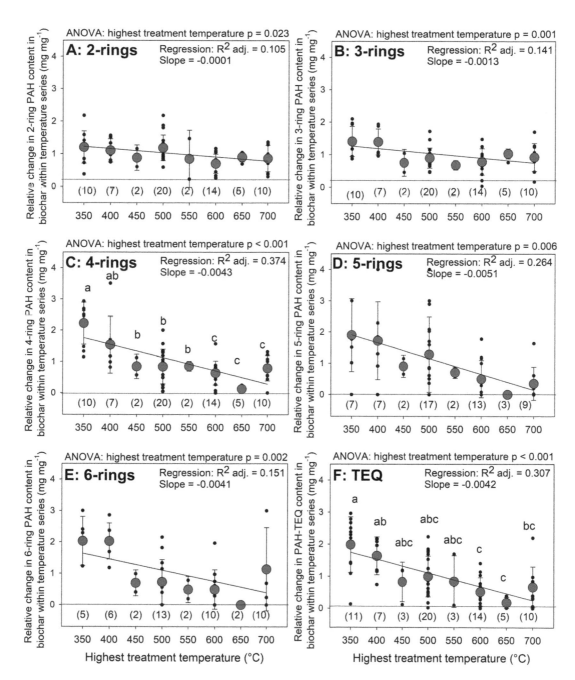

Figure 22.3 *Meta-analysis results showing the relationship of highest treatment temperature and PAH(s) with (A) 2-rings (naphthalene), (B) 3-rings, (C) 4-rings, (D) 5-rings, (E) 6-rings, and (F) toxicity equivalent quantity (TEQ). Relative change of PAH content within individual temperature series shown where biochars were produced in the same production unit, from the same feedstock,*

PAH-related toxicity can be calculated using toxicity equivalent quantity (TEQ) by relating the toxicity of the 16 US EPA PAHs to the toxicity of a reference compound (benzo[a]pyrene (B[a]P) – TEQ of 1). The TEQ of PAHs identified in biochar fluctuates greatly, with values from 0 to 25 mg TEQ kg^{-1} (relative to B[a]P) in a set of 73 biochar samples from different production conditions, with mean and median of 1.4 and 0.05 mg kg^{-1}, respectively (Buss et al, 2022b). The average TEQ of the 16 US EPA PAHs in the 74 biochars described in Figures 22.2 and 22.3 was 0.24 ± 0.82 mg kg^{-1} with a median of 0.006 mg kg^{-1}. The IBI threshold value, in comparison, is 3 mg kg^{-1} (Table 22.3) demonstrating that most biochars easily meet PAH-TEQ threshold values (IBI, 2011).

In contrast to a previous study that investigated the relationship between biochar PAH-TEQ and HTT, our meta-analysis comparing relative changes in each temperature series showed a decrease in TEQ with HTT (Wang et al, 2019; Figure 22.3F). HMW PAHs are less likely to recondense onto biochar at elevated HTTs, resulting in a reduction of PAH-TEQ with HTT. Overall, at an HTT of 350°C, the total PAH content in all pyrolysis products is low, however, the PAHs that are formed (excluding naphthalene) condense more easily since their boiling points are higher than the HTT, resulting in higher total PAH content and higher TEQs (Figures 22.1, 22.2 and 22.3F).

Effect of the pyrolysis unit on PAH content in biochar

The design of the area where pyrolysis solids and vapors are separated has the greatest effect on biochar PAH content (Buss et al, 2022b). Typically, more than 99% of PAHs formed during pyrolysis are produced in the gas phase and, subsequently, can be found in pyrolysis liquids or gases (Figure 22.1). Therefore, condensation of pyrolysis vapors in cold spots of the pyrolysis unit (i.e., the area where pyrolysis solids and vapors are separated) or vapors trapped within the biochar structure can result in highly elevated biochar PAH content (Buss et al, 2022b). This effect is more important for more toxic, higher-molecular weight PAHs because of their higher boiling point compared to naphthalene (boiling point 218°C), for example.

Within the zone of the pyrolysis unit that is at HTT, PAHs that are formed are typically either removed from the pyrolysis solids or incorporated into the biochar-carbon framework. However, this is not the case at lower temperature areas in the pyrolysis unit outside the pyrolysis zone or when the temperature distribution within the pyrolysis zone is uneven. Contact of biochar with pyrolysis vapors produced at >650°C can cause significant contamination with toxic PAHs, as PAH formation increases exponentially at temperatures >650°C (Figure 22.1) (Dai et al, 2014a; Zhou et al, 2014). Such contamination of biochar can result in dramatic outliers in studies e.g., as

under the same production conditions, and extracted using the same extraction technique. The PAH content in each temperature series was standardized setting the mean to 1 and hence the data illustrate the relative change in PAH content only based on the HTT. Individual data points are shown (black dots) and mean with standard deviation (red dots). Results of one-way ANOVA (p-value) and Tukey's post-hoc test (letters) are displayed above the figure panel and above data points of each temperature, respectively. The numbers in brackets signify the number of samples per temperature. Linear regression curve shown. Data based on literature (same as Figure 22.2).

described in Buss et al (2022b). Including these in the calculation of average biochar PAH content to analyze trends in biochar PAH content with HTT can lead to misleading results (Buss et al, 2016).

Pyrolysis under uncontrolled conditions, such as in traditional kilns, typically resulted in the highest PAH content in biochar (Hale et al, 2012; de la Rosa et al, 2014; Wang et al, 2017). Yet simple, "Kon-Tiki" flame curtain pyrolysis produced biochar with low PAH content (Cornelissen et al, 2016). For the production of low-PAH biochar, trapping of pyrolysis vapors and hence PAHs in biochar pores has to be avoided.

Effect of other pyrolysis parameters and feedstock on total PAH content in biochar

Amongst biochar production conditions, the flow of gases within the pyrolysis unit has a clear effect on biochar PAH content – the higher the flow, the lower the PAH content in the biochar. This is the result of the removal (or increased volatilization) of PAHs from biochar during pyrolysis and has been shown in multiple studies (Buss et al, 2016; Madej et al, 2016; Greco et al, 2021). Residence time at the HTT, in contrast, typically has a limited effect on biochar properties, including PAH content, as already reported in the second edition of this chapter (Bucheli et al, 2015; Crombie and Mašek, 2015; Buss et al, 2016).

When comparing the major feedstock classes, grass (e.g., wheat and rice straw) biochar typically has higher PAH content than wood biochar (Hale et al, 2012; Keiluweit et al, 2012; Kloss et al, 2012; Fabbri et al, 2013; Buss et al, 2016; de la Rosa et al, 2019). Comparing pairs of grass and wood biochars produced at the same residence time and HTT and in the same production units across studies (21 pairs that span across the HTT range of 350–700°C; Table 22.1), on average, the total 16 US EPA PAH content in grass-derived biochars

was two times higher than in wood- (soft and hardwood) derived biochar (geometric mean; geometric standard deviation 3.3) (Freddo et al, 2012; Hale et al, 2012; Keiluweit et al, 2012; Kloss et al, 2012; Buss et al, 2016; Madej et al, 2016). Biochars from some feedstocks, such as sewage sludges or manures, can contain PAHs that originate from the feedstock material, however at HTT $\geq$ 500°C, organic contaminants are efficiently removed and instead the PAH content in biochar is dominated by PAHs formed during the production process (Buss, 2021) (discussed above).

Biochars with total PAH contents below EBC and IBI threshold values (~4–10 mg kg^{-1}, Table 3) can routinely and reproducibly be produced under controlled industrial conditions and be considered safe (Rombolà et al, 2015; Sigmund et al, 2017; Hilber et al, 2019).

PAH availability, persistence, and risk to the environment

Most legally binding threshold values are still based on the total content of PAHs in environmental samples. However, to assess their risk in the environment, total PAH content only has limited relevance. It is the bioavailable fraction of PAHs that interacts with the environment and target organisms, causing ecotoxicological effects (Bogolte et al, 2007). One method that assesses this fraction of PAHs in biochar is based on passive sampling using materials such as polyoxymethylene (POM) (Jonker and Koelmans, 2001), polyethylene (PE) (Adams et al, 2007), or silicon rubbers (Estoppey et al, 2016), which mimic the environmental exposure of organisms to the pollutant.

The bioavailable PAH fraction in biochar assessed by POM passive samplers was below the detection limit or below soil or sediment background content (Hale et al, 2012; Mayer et al, 2016; Hilber et al, 2017a). Concentrations of 0.17 – 10 ng L^{-1}, for

Table 22.1 *Total 16 US EPA PAHs content in grass and wood biochar comparing biochars produced under the same production conditions and in the same pyrolysis unit. HTT, highest treatment temperature; RT, residence time*

Feedstocks	HTT (°C)	RT (min)	Ratio PAH content grass: wood	Cited reference
Rice, bamboo and maize vs. redwood	600	150	15	(Freddo et al, 2012)
Grass vs. oak and pine	400	180	0.59	(Hale et al, 2012)
Grass vs. oak and pine	650	180	1.8	(Hale et al, 2012)
Switch grass vs. pine wood	350	480	3.3	(Hale et al, 2012)
Switch grass vs. pine wood	500	480	2.1	(Hale et al, 2012)
Switch grass vs. pine wood	600	480	3.5	(Hale et al, 2012)
Switch grass vs. pine wood	700	480	1.5	(Hale et al, 2012)
Tall fescue straw vs. Ponderosa Pine wood	400	60	1.6	(Keiluweit et al, 2012)
Tall fescue straw vs. Ponderosa Pine wood	500	60	3.6	(Hale et al, 2012)
Tall fescue straw vs. Ponderosa Pine wood	600	60	2.7	(Hale et al, 2012)
Tall fescue straw vs. Ponderosa Pine wood	700	60	0.47	(Hale et al, 2012)
Wheat straw vs. poplar and spruce wood	400	300	0.30	(Kloss et al, 2012)
Wheat straw vs. poplar and spruce wood	460	600	0.90	(Hale et al, 2012)
Wheat straw vs. poplar and spruce wood	525	600	17	(Hale et al, 2012)
Wheat straw vs. softwood pellets	350	10	18	(Buss et al, 2016)
Wheat straw vs. softwood pellets	350	40	11	(Buss et al, 2016)
Wheat straw vs. softwood pellets	650	10	0.73	(Buss et al, 2016)
Wheat straw vs. softwood pellets	650	40	2.4	(Buss et al, 2016)
Miscanthus and wheat straw vs. willow	500	240	0.62	(Madej et al, 2016)
Miscanthus and wheat straw vs. willow	600	240	1.2	(Madej et al, 2016)
Miscanthus and wheat straw vs. willow	700	240	1.5	(Madej et al, 2016)
Geometric mean			2.1	

Table 22.1 *continued*

Feedstocks	HTT (°C)	RT (min)	Ratio PAH content grass: wood	Cited reference
Geometric standard deviation			3.3	
Median			1.8	
Q1			0.9	
Q3			3.5	
Interquartile range			2.6	

example, were reported in a wide range of biochars produced from slow pyrolysis, which is far lower than contents recorded for urban sediments (80–342,000 ng L^{-1}) (Hale et al, 2012).

Generally, LMW PAHs desorbed easier in passive sampler experiments and water extractions than HMW PAHs given their lower biochar-water partitioning coefficients (Hilber et al, 2017a; Chen et al, 2019). There was no consistent effect of biochar HTT on bioavailable biochar PAH content across various studies (Hale et al, 2012; Chen et al, 2019; Krzyszczak et al, 2021). Compared to total PAHs, far fewer studies exist on bioavailable PAHs in biochar and the effect of HTT on bioavailable PAH levels. Therefore, it is not possible to conduct a meaningful meta-analysis at this stage.

Few biochars have been found to have elevated bioavailable ('available') PAH content and in such cases, it is possible that these PAHs are taken up by plants when used in plant experiments (Buss et al, 2015; Hilber et al, 2017a; Wang et al, 2018). Most of the biochars with elevated available PAH content also typically had high total PAH content. Yet a clear correlation could not be found between PAH content in plant tissue and biochar (Wang et al, 2018) and total and available PAHs in biochar, except for LMW

PAHs such as naphthalene and acenaphthylene and when the total PAH content was high (Hilber et al, 2017a). Biochars contaminated by re-condensing vapors showed both elevated contents of total and water-extractable PAHs (water-extractable PAH content in two contaminated biochars: 1.62 and 2.04 µg kg^{-1} PAHs vs. <0.001 µg kg^{-1} PAHs in clean biochar from same feedstock and production conditions) (Buss and Mašek, 2014; Buss et al, 2015). PAHs from these oily condensates were much more accessible to plants and microorganisms (Buss et al, 2015). Condensation and trapping of pyrolysis vapors during pyrolysis therefore needs to be avoided. This highlights the link between total and bioavailable PAHs in biochar, at least in biochars with elevated total PAH content.

Despite the generally high PAH sorption strength of biochar and, hence low bioavailability, dissipation of biochar-based PAHs in soil has generally been found to occur within the first few years after biochar application (Rombolà et al, 2015, 2019; Kuśmierz et al, 2016; de Resende et al, 2018). When bioavailable, PAHs can be degraded in days by the action of various soil organisms, including algae, fungi, and bacteria (Ghosal et al, 2016; Premnath et al, 2021). These results demonstrate that the highest potential risk of PAHs in biochar is within the initial phase after soil application when a small proportion of the PAHs in biochar are bioavailable, and that accumulation of PAHs in soil and hence risk after repeated biochar application is unlikely.

In soil, root exudates or microbial activity can increase the release of PAHs from biochar, which could lead to increased toxicity effects (Oleszczuk and Kołtowski, 2018; Wang et al, 2018). Yet, microbial treatment or co-composting did not have a consistent effect on the bioavailable fraction of biochar PAHs (Hilber et al, 2017a; Oleszczuk and Kołtowski, 2018). Furthermore, the effect of aging on the release of different PAHs in biochar-amended soil was inconsistent in field experiments (Rombolà et al, 2019; Sigmund et al, 2017). Importantly, across various studies, total and available biochar PAH content could not be correlated with acute toxicity in plants, soil, and water organisms, suggesting that compounds other than PAHs were responsible for the toxic effects, if any, observed (Buss and Mašek, 2014; Oleszczuk et al, 2014; Buss et al, 2015; Kołtowski and Oleszczuk, 2015; Oleszczuk and Kołtowski, 2018).

Passive samplers and artificial aging methods do not necessarily reflect biochar field aging in soil and associated PAH changes in biochar (Sigmund et al, 2017), highlighting that PAH-based risk assessments are ideally conducted in the biological system of interest. In a cow's rumen experiment, for example, PAH desorption from biochar was higher than expected, given data from passive samplers; up to 50% of total biochar PAHs were released after 48 hours of exposure (Hilber et al, 2019). Furthermore, the desorption of HMW PAHs was higher than LMW PAHs (Hilber et al, 2019). The cow rumen represents a near "infinite PAH sink" that continuously transports desorbed biochar PAHs away from the biochar, allowing PAH desorption to continue. In

soil, an equilibrium between PAHs in soil solution (desorbed) and PAHs attached to biochar will be reached, limiting further desorption. In simulated lung fluids, PAH release from biochar was very low (0.5–0.75% of totals) resulting in low carcinogenic risk (Liu et al, 2019).

While most biochars are a sink rather than a source of PAHs and are therefore of little concern to human health and the environment (Mayer et al, 2016; Hilber et al, 2017a), biochars with elevated total PAH content (Hilber et al, 2019) should be handled with care and not used in soils or other biological systems.

Chlorinated aromatic compounds

Three groups of chlorinated hydrocarbons are most commonly analyzed in biochars, all of which are considered persistent organic pollutants: PCDDs (known simply as dioxins), PCDFs (known simply as furans) (collectively known as PCDD/F), and PCBs. There are 210 structurally related compounds of dioxins and furans and 209 PCBs, which often co-exist in the environment as the result of incomplete combustion of chlorine-rich materials (Conesa et al, 2009; Environment Agency, 2009). The World Health Organization (WHO) has introduced toxicity equivalent factors (TEF) for 7 dioxins, 10 furans, and 12 PCBs that compare their toxicity to 2,3,7,8-Tetrachlorodibenzodioxin (Environment Agency, 2009).

Dioxins and furans contain oxygen and thus higher quantities are formed during high-temperature processes in the presence of oxygen (combustion) than in its absence (pyrolysis) (Conesa et al, 2009). The highest levels of formation of dioxins and furans were reported during combustion at ~300–400°C (Xhrout et al, 2001; McKay, 2002). However, in the same temperature range (300–500°C) during the pyrolysis of switchgrass, no formation of chlorinated hydrocarbons was detected, confirming that oxygen is required for their formation from this feedstock (Björkman and Strömberg, 1997).

Levels of PCDD/F in biochar are generally below threshold values, in the range of background levels in soil or even below detection limits (Granatstein et al, 2009; Downie et al, 2012; Hale et al, 2012; Wiedner et al, 2013; Lyu et al, 2016; Weidemann et al, 2017). The biochar with the highest dioxin levels (sum of 130 toxic and nontoxic congeners) in five biochars from different feedstocks, for example, was produced from food waste pyrolyzed at 400°C (92 pg g^{-1}), which contained elevated levels of chlorine (2.9%) (Hale et al, 2012). It cannot be excluded that dioxins can be present in biochar because of contamination of the feedstock material, e.g., sewage sludge (Wiedner et al, 2013). However, pyrolysis efficiently removes most organic compounds, including chlorinated hydrocarbons, from biochar (see below).

Biochar PCDD/F content could not be quantified in biochar produced from anaerobically digested sewage sludge pyrolyzed at 550°C, and from softwood pellets and wheat straw produced at both 550 and 700°C (LOQ-PCDF: 0.2 pg g^{-1}, LOQ-PCDD: 0.3 pg g^{-1}) (Weidemann et al, 2017). Only monochlorinated dibenzofuran (MoCDF) was detectable in some of the biochars and its presence was associated with lower HTTs (Weidemann et al, 2017). Total and available PCDD/F content in biochars produced from sawdust decreased with pyrolysis temperature from 360 pg g^{-1} (400°C) to 50 pg g^{-1} (700°C) and 6.8 pg g^{-1} (400°C) to 1.5 pg g^{-1} (700°C), respectively (Lyu et al, 2016). Low carcinogenic risk was associated with

the biochar PCDD/F content (Lyu et al, 2016).

Most importantly, in the few cases where bioavailable dioxin contents were analyzed, it was below the detection limit (Hale et al, 2012). Pyrogenic carbonaceous material is known for its ability to strongly sorb PCDD/F and PCBs, which renders them unavailable and hence they pose a limited risk for the environment (Koelmans et al, 2006) as confirmed in various studies (Downie et al, 2012; Hale et al, 2012; Weidemann et al, 2017; Wiedner et al, 2013; Wilson and Reed, 2012). In summary, the risk of PCDD/Fs and PCBs found in biochar causing adverse effects on human health and the environment is very low (Downie et al, 2012; Hale et al, 2012; Wilson and Reed, 2012; Wiedner et al, 2013; Weidemann et al, 2017).

Volatile organic compounds

Volatile organic compounds (VOCs) are organic compounds with a boiling point below 250°C (which by definition also include naphthalene) (Directive 2004/42/CE of the European Parliament and of the council, 2004). VOCs are highly mobile in the environment as they are typically water soluble and can vaporize at room temperature. They can also be toxic and therefore are potentially high-risk compounds in biochar (Cordella et al, 2012; Buss and Mašek, 2014; Buss et al, 2015).

During pyrolysis, biomass breaks down into a myriad of organic compounds that are impossible to fully assess quantitatively or even qualitatively (Spokas et al, 2011; Cordella et al, 2012; Buss et al, 2020). Such compounds include, for example, organic acids, phenols, short-chain alcohols, and ketones, but also naphthalene. Due to their low boiling point, VOCs are typically vaporized during biochar production (≥350°C), and form part of the pyrolysis liquid fraction (Spokas et al, 2011; Buss et al, 2015). At pyrolysis temperatures of 450 and 600°C, for example, only 0.03–0.08% and 0.01% (respectively) of tars (the heavy portion of pyrolysis liquids) produced during pyrolysis were sorbed to the biochar, demonstrating the near complete separation of pyrolysis vapors and biochar during pyrolysis (Yang et al, 2013).

The quantity of VOCs in biochar and the ease of VOC release decreases with biochar production temperature, as expected (Gundale and DeLuca, 2006; Ghidotti et al, 2017b). In a large set of biochars (n=70), no clear relationship between biochar composition and content of VOCs with feedstock type could be found, yet biochars produced at HTTs <350°C contained more short-chain aldehydes, furans and ketones, and biochars produced at higher pyrolysis temperatures contained more aromatic compounds and longer-chain hydrocarbons (Spokas et al, 2011). However, in a study analyzing water-soluble organic compounds sorbed to biochar, the molecular complexity and degree of aromaticity of VOCs identified decreased with pyrolysis temperature (Ghidotti et al, 2017a). Generally, water-extractable VOC contents in biochar are typically very low (Buss et al, 2015). Similar to other organic contaminants, VOCs are strongly sorbed to the biochar matrix and hence mostly not bioavailable (Dutta et al, 2017). However, few studies exist that explicitly investigate VOC availability in biochar.

As for PAHS, VOCs can be trapped in biochar pores or biochar can be contaminated with pyrolysis liquids because of condensation of liquids in the post-pyrolysis zone of a pyrolysis unit, resulting in larger amounts of VOCs in the resulting biochar (Spokas et al, 2011; Buss et al, 2015). The most abundant VOCs identified in biochar that was contaminated by condensing pyrolysis liquids were: acetic,

formic, butyric, and propionic acids; methanol; phenol; o-, m- and p-cresol and 2,4-dimethylphenol, all with levels >100 μg g^{-1} (Buss et al, 2015).

Organic compounds attached to biochar can be both phytotoxic and plant stimulating, depending on the type and concentration of the compounds present (Bargmann et al, 2013; Ghidotti et al, 2017a; Backer et al, 2018). Typically, VOCs are not present in biochars in quantities that result in toxic effects (Buss and Mašek, 2014; Buss et al, 2015; Ghidotti et al, 2017b). However, in biochars affected by condensation of pyrolysis liquids, the concentrations were high enough to cause phytotoxic effects (Buss et al, 2015) and high potential adverse effects on human health because of biochar handling and storage (Buss and Mašek, 2016). However, due to the corresponding co-contamination of the biochar with PAHs, the toxic effects observed cannot be unambiguously associated with VOCs. Post-pyrolysis condensation can be avoided by active heating of the discharge chamber of the pyrolysis unit (separation of pyrolysis vapors and solids) or good insulation to keep the temperature in the discharge chamber at a similar level to the temperature in the pyrolysis zone.

Post-treatment techniques are very effective in removing VOCs from biochar, such as heating in air at 200°C (Buss and Mašek, 2016). Importantly, VOCs such as LMW organic acids and phenols easily volatilize or leach from soil and are readily degraded by soil microorganisms (Wilson and Jones, 1996; van Schie and Young, 2000). Therefore, even if larger amounts of VOCs are present in biochar, no long-term negative effects on plant growth caused by VOCs are expected and any phytotoxic effects would be mitigated by application of the biochar to soil several weeks before planting. This early application would ensure the desorption of VOC from biochar and subsequent biodegradation (under oxic conditions).

While VOCs in biochar have the potential to cause toxic effects, biochar produced in most pyrolysis units does not contain levels of VOCs significant enough to pose a risk to human health and the environment.

Elevated levels of VOCs are typically also accompanied by elevated PAH levels in biochar, therefore, recommended biochar production practices described in the PAH section of this chapter will also ensure low levels of biochar VOCs.

Organic contaminants originating from feedstocks

While the formation of organic contaminants, such as PAHs, during pyrolysis has received much attention in the literature, the fate of organic contaminants that are already present in feedstock materials has been far less thoroughly investigated. This section summarizes the current knowledge on the removal and decomposition of contaminants originating from biomass feedstock during pyrolysis.

PAHs, VOCs, and chlorinated aromatic compounds

In addition to their formation during pyrolysis, PAHs, VOCs, and chlorinated aromatic compounds can also originate from biochar feedstock material. Various studies have investigated the presence of contaminants before and after pyrolysis to assess contaminant removal efficiency.

In studies investigating sewage sludge biochar, PAHs were efficiently removed (>95%) at an HTT of 500°C (Zielińska and Oleszczuk, 2015; Kong et al, 2019; Moško et al, 2021). This supports the results from our meta-analysis where PAH levels decreased with increasing pyrolysis temperature due to PAH vaporization from pyrolysis solids (Figure 22.2). Similar removal efficiencies were reported for PCBs and dioxins at HTTs of 400–500°C: >99.9 PCB and 90.3% dioxin removal, respectively (Moško et al, 2021), >96% PCB removal (Bridle et

al, 1990) and 97% dioxin removal (Dai et al, 2018). Volatile organic compounds have, by definition, a boiling point below 250°C, so it follows that VOCs in feedstock materials should be removed from biochar to an even higher degree than PAHs and chlorinated aromatic compounds. However, at the publication of this chapter, we are not aware of any studies specifically investigating the removal of VOCs from feedstock materials via pyrolysis.

Contaminants of emerging concern

Contaminants of emerging concern are anthropogenic substances, such as pharmaceuticals, pesticides, or microplastics, which are increasingly identified in the environment at low concentrations (mg to ng per kg). Their presence in the environment is a result of diffuse source pollution from domestic and commercial use. While these contaminants might present a low individual or acute risk, their cumulative effects are still unknown, and repeated or prolonged exposure to emerging contaminants poses concern due to the risk of potentially detrimental effects on ecosystems and human health (Tiedeken et al, 2017). They are ubiquitously present in the environment, and in many cases are persistent, bioaccumulative, or toxic (PBT), and these persistent, mobile, and toxic (PMT) characteristics make them contaminants of emerging concern for scientists and legislators.

Although emerging contaminants are now almost omnipresent in the environment, biochar feedstocks are not regularly analyzed for emerging contaminants and their typical levels are therefore often unknown. Elucidating the presence of individual emerging contaminants in biochars is often not a priority for researchers due to the lower risks they are assumed to pose compared to other contaminants such as heavy metals (Huygens et al, 2019). However, the accumulated risk of hundreds of contaminants and their accompanying degradation products might indicate an additional, currently unknown risk.

Biochar feedstock contamination with emerging contaminants

The pervasive presence of emerging contaminants and the high mobility for many of them, e.g., per- and poly-fluoroalkyl substances (PFAS), implies that a certain level of contamination can be found in most, if not all, biomass used for biochar production, but significant contamination is typically limited to specific feedstocks. Hotspots for emerging contaminants include animal husbandry and hospital wastes due to the spatially concentrated use of pharmaceuticals, and agricultural waste through the application of pesticides and herbicides (Aktar et al, 2009). Pre-treated feedstocks, such as digestate or compost can also contain emerging contaminants; anaerobic digestion and composting are typically inefficient in degrading emerging contaminants (Congilosi and Aga, 2021; Keerthanan et al, 2021).

Wastewater treatment plants are generally considered a main hotspot for emerging contaminants and the resulting sewage sludge is the primary pathway for emerging contaminant introduction into the environment (Keerthanan et al, 2021). Compared to most other feedstocks used in biochar production, sewage sludge can contain very high emerging contaminant concentrations (Moško et al, 2021). Pyrolysis is also studied as one of the most promising sludge recycling technologies regarding phosphorus recovery and receives increasing attention in the search for novel sludge utilization strategies (Buss et al, 2022a). Therefore, contamination issues of sewage sludge biochar represent a current spotlight of legislative interest highlighted by the exclusion of sewage sludge as a permissible biochar feedstock in the fertilizer regulation of the European Union, which was justified by gaps in the scientific knowledge about the removal of organic contaminants during pyrolysis (Huygens et al, 2019). Consequently, and in an attempt to bridge these gaps to the best of our capabilities, in the following paragraphs we focus on sewage sludge as the feedstock source.

Methods for assessing the fate of emerging contaminants during pyrolysis, and their challenges

While the presence of emerging contaminants in some sewage sludge is certain, their fate during pyrolysis is less well-known (Winchell et al, 2022). Available literature is fragmented, with studies either focusing on degradation mechanisms of specific contaminants during pyrolysis or comparison of the initial contaminant level in the feedstock and the final level in the resulting biochar, i.e., the "removal efficiency". Although much is still unknown about the fate of organic contaminants during pyrolysis, HTT is generally considered to be the main determining factor in their removal and degradation (Buss, 2021; Mercl et al, 2021; Moško et al, 2021; Winchell et al, 2022). The residence time is generally considered to be of less importance; a short residence time at HTT (e.g., 5 min) can be sufficient for contaminant removal. The residence time is therefore contingent upon the HTT. The pyrolysis unit design likely also plays a role, homogenous heating of the feedstock is required and hence units with controlled pyrolysis conditions will be favorable. However, to our knowledge, no data exist on the effect of pyrolysis unit design on the efficacy of feedstock-inherent contaminant removal.

Studies that analyze specific degradation pathways of pure compounds are often based on thermogravimetric analysis coupled with analysis of pyrolysis gases (Tian et al, 2017; Madadian and Simakov, 2022). This bottom-up approach allows the identification of temperature-dependent decomposition phases until complete degradation is achieved (e.g., conversion into non-condensable gases). Such studies are limited by their reliance on single or few contaminants and difficulties in linking sequential decomposition phases of contaminants to environmental risks, i.e., the determination of a satisfactory level of contaminant degradation if transformation products are present (Tian et al, 2017).

Studies focusing on removal efficiency are more comprehensive, covering larger numbers of targeted contaminants and all interactions between feedstock and multiple contaminants. However, the results rely on efficient extraction methods to correctly identify the presence or absence of contaminants in the resulting biochar (Ross et al, 2016). A further risk associated with this approach is the potential presence of unknown degradation and transformation products in biochar that are not analyzed as part of the assessment (Neuwald et al, 2021; Hajeb et al, 2022). These constraints might mask the continued presence of contaminants that are only partly degraded or where the extraction efficiency for the parent compounds becomes low, potentially understating the environmental risk of the biochar.

The heterogeneous composition of biochar feedstock materials, such as sewage sludge, and the evolving nature of analytical capabilities result in inconsistent data quality, making comparisons between studies problematic (Mejías et al, 2021). As discussed for PAHs, extraction methods developed for other materials might not be appropriate for use on biochar as they can result in low recoveries (Ross et al, 2016; Hilber et al, 2022). Identifying emerging contaminants in biochar is complicated by differing biochar-sorption strengths, analytical interferences by co-extracted compounds, and therefore varying extraction efficiencies (Oesterle et al, 2020). Recovery rates for emerging contaminants extracted from biochar can range from 2–253% when multiple contaminants are analyzed simultaneously (Vom Eyser et al, 2015). The extraction of emerging contaminants, as well as necessary clean-up and pre-concentration steps for these extracts (Hajeb et al, 2022), require tedious method development for accurate quantification of emerging contaminants on biochar (Oesterle et al, 2020). While analytical methods for emerging contaminant identification and quantification are rapidly evolving (Hajeb et al, 2022), standardized extraction and analysis protocols specifically designed for biochar (and its feedstock material) are urgently needed.

Pharmaceuticals and personal care products

The contaminant class of pharmaceuticals and personal care products (PPCPs) is extremely heterogeneous and includes anti-inflammatories, painkillers, antidepressants, antibiotics, anti-fungal medication, soaps, detergents, and cosmetics (Patel et al, 2019). Of 72 PPCPs that were analyzed in sludge samples, 26 were identified in all 101 samples, with a mean total concentration of 74.4 mg kg^{-1} (d.w.) (McClellan and Halden, 2010).

Degradation studies on individual, pure PPCPs show degradation onset temperatures between 250 and 430°C (Lizarraga et al, 2007; Tian et al, 2017), with degradation often being incomplete before reaching higher temperatures of more than 600°C. The review of thermal decomposition temperatures of 249 pharmaceuticals showed that complete degradation of sulfamethoxazole (SMX) and chloramphenicol (CHL) only occurred at temperatures of 600 and 704°C, respectively (Bean et al, 2016).

Testing degradation of individual PPCPs within sewage sludge, three out of seven antibiotics required an HTT of 600°C to be fully degraded (Tian et al, 2019), comparable to results for phenoxymethylpenicillin (Wang et al, 2021a). Thermal degradation studies of 15 pure PPCPs and co-pyrolysis of all individual 15 PPCPs with synthetic sewage sludge revealed crystallized PPCP remained in biochar produced at an HTT of 600°C (Madadian and Simakov, 2022), demonstrating that although decomposition is typically complete for most compounds, some can breakdown at higher temperatures.

As mentioned before, aside from contaminant degradation, many studies determine contaminant removal based on the presence of the target contaminant in the biochar. The removal of 42 pharmaceuticals in sewage sludge at HTTs between 220 and 620°C demonstrated that an HTT of 420°C already reduced all contaminants to below the detection limit (Mercl et al, 2021), similar to another study where 27 pharmaceuticals were tested (Moško et al, 2021). However, neither study evaluated the extraction efficiency for the target contaminants from the resulting biochars. Therefore, non-detection of the contaminants must be treated with caution.

Differences between removal (i.e., non-detection within biochar) and degradation temperatures can be attributed to the volatilization of certain contaminants before degradation occurs (Suzuki et al, 1978), similar to the way PAHs are vaporized during pyrolysis. These volatilized compounds are subsequently degraded in the secondary combustion stage in commercial pyrolysis units. The boiling point of emerging contaminants is sometimes used as a metric to predict potential volatilization (Moško et al, 2021). Using such metrics, minimum treatment temperatures of around 400–500°C were suggested for the removal of PPCPs (Mercl et al, 2021; Moško et al, 2021). However, it is crucial to understand that these minimum temperatures are specific to the contaminants that were analyzed and do not account for non-targeted contaminants that might still be present. The boiling points of organic contaminants, including PPCPs, can vary greatly and reach up to 821°C, as is the case for the antibiotic chlortetracycline (Puicharla et al, 2014) (Table 22.2). Similarly, degradation temperatures for certain PPCPs can vary widely and may require temperatures above 600°C (Tian et al, 2019). As a result, certain PPCPs may still be present in biochar produced at typical pyrolysis temperatures (500–700°C).

While some studies point to a minimum treatment temperature of around 400–500°C for the removal of PPCPs (Mercl et al, 2021; Moško et al, 2021), degradation temperatures for some PPCPs indicate temperatures of >600°C (Tian et al, 2019). As existing literature covers only a small range of contaminants and due to the varying compound stabilities, more systematic studies on pharmaceuticals with a focus on their thermal stability and any catalytic effects caused by pyrolysis products are required to allow generalized conclusions on PPCP removal during pyrolysis.

Antibiotic-resistance genes

Based on the ubiquitous abundance of antibiotics in wastewater, sewage sludge is a

Table 22.2 *Boiling points of selected pharmaceuticals and personal care products, and the minimum temperature for complete degradation during pyrolysis treatment stated in respective literature*

Analyte	Drug class	Boiling point (°C)*	Reference	Minimum HTT (°C)	Cited reference
5-Fluorouracil	Anticancer	190	ChemSpider	282	(Bean et al, 2016)
Allopurinol	Antigout	423	ChemSpider	379	
Aspirin	Analgesics	140	ChemSpider	370	
Atenolol	Beta-blocker	508	ChemSpider	335	
Chloramphenicol	Antibiotics	645	ChemSpider	704	
Diclofenac	Nonsteroidal anti-inflammatory	412	ChemSpider	260	
Estradiol	Hormone	446	ChemSpider	317	
Fluoxetine	Antidepressant	395	ChemSpider	300	
Gliclazide	Diabetes	489	ChemSpider	429	
Verapamil	Calcium channel blocker	586	ChemSpider	320	
Tylosin	Antibiotics	980	ChemSpider	300	(Tian et al, 2019)
Tetracycline	Antibiotics	738	ChemSpider	300	
Chlortetracycline	Antibiotics	821	ChemSpider	300	
Doxycycline	Antibiotics	762	ChemSpider	600	
Sulfamethazine	Antibiotics	526	ChemSpider	600	
Sulfadiazine	Antibiotics	512	ChemSpider	600	
Phenoxymethylpenicillin	Antibiotics	681	ChemSpider	600	(Wang et al, 2021a)
P-nitrophenol	Fungicide	279	PubChem	600	
Amitriptyline	Antidepressants	398	ChemSpider	420	(Mercl et al, 2021)
Amantadine	Antivirotics	225	ChemSpider	420	
Carbamazepine	Antiepileptic	411	ChemSpider	420	
Ketoprofen	NSAIDs	431	PubChem	420	
Lidocaine	Anesthetics	350	PubChem	420	
Flurazoldine	Veterinary	531	ChemSpider	420	
Tramadol	Analgesics	406	PubChem	420	
Sulfamethoxazole	Antibiotics	482	ChemSpider	420	
Indomethacin	NSAIDs	499	PubChem	420	

Table 22.2 *continued*

Analyte	Drug class	Boiling point (°C)*	Reference	Minimum HTT (°C)	Cited reference
Caffeine	Central nervous system stimulant	178	PubChem	400	(Moško et al, 2021)
Diclofenac	Nonsteroidal anti-inflammatory	412	ChemSpider	400	
Ibuprofen	Nonsteroidal anti-inflammatory	157	PubChem	400	
Triclosan	Antibacterial	120	PubChem	400	
Trimethoprim	Antibiotics	405	ChemSpider	400	
Tramadol	Pain medication	406	PubChem	400	

Notes
* Boiling points may be based on thermodynamic models rather than derived from experimental data

repository for antibiotic-resistance genes (ARGs), threatening the effectiveness of the limited number of known antibiotics (Cui et al, 2022). DNA degrades completely at around 190°C, therefore ARGs should be decomposed at typical pyrolysis temperatures (Karni et al, 2013) and indeed complete gene destruction was demonstrated at HTTs above 300°C (Kimbell et al, 2018). Beta-lactam resistance genes and other ARGs in fermentation sludge and swine manure, for example, were completely degraded during pyrolysis (Zhou et al, 2019; Wang et al, 2021b). Pyrolysis of swine manure and sewage sludge at 650 and 850°C, respectively, left some ARG residues, however, in very low quantities (1,000–100,000-fold reduction in ARG contents from feedstock to biochar) (Zhi et al, 2020). As no study found relevant ARG concentrations in biochar, pyrolysis is a suitable ARG destruction process even at low pyrolysis temperatures of 300°C.

Endocrine-disrupting chemicals
Due to the chemical similarities with pharmaceuticals in the PPCP contaminant group,

endocrine-disrupting chemicals (EDCs) show comparable degradation during pyrolysis. More than 99% of EDCs were removed at HTTs of 400 and 500°C, respectively (Moško et al, 2021). Similar results were observed in degradation studies of nonylphenol in sewage sludge, with significant degradation at 300°C (>90%), and complete degradation at an HTT of 500°C (Ross et al, 2016).

Herbicides and pesticides
In studies that investigated the thermal decomposition of chlorpyrifos, a widely used pesticide, two decomposition stages at 550–650°C and above 650°C were identified (Kennedy and Mackie, 2018; Weber et al 2020), while glyphosate degradation started between 550 and 600°C (Mackie and Kennedy, 2019), demonstrating incomplete degradation at lower pyrolysis temperatures. Analysis of char residues from the combustion of contaminated wood pellets showed that various pesticides can be resistant to thermal degradation even at high temperatures of 750°C in an oxidizing atmosphere (Růžičková et al, 2021).

Per- and poly-fluoroalkyl substances

Per- and poly-fluoroalkyl substances (PFAS) commonly volatilize at temperatures above 450°C and should therefore be removed to the gas phase during pyrolysis at typical biochar HTTs (Longendyke et al, 2022). Indeed, between 500 and 600°C high removal rates are reported for most studied PFAS, except for PFBA (pentafluorobenzoic acid), which has a low boiling point of only 220°C, and therefore should have been removed efficiently (Kundu et al, 2021; Thoma et al, 2022). Complete mineralization of PFAS to fluoride ions has been reported during pyrolysis at 700–800°C (Xiao et al, 2020). Similar to other contaminant classes, uncertainty in evaluating the risk of PFAS contamination in biochar persists, at least at treatment temperatures below 650°C (Winchell et al, 2021).

Given the variety of PFAS species, current literature only covers the fate of key compounds and some of their degradation products (McNamara et al, 2022; Thoma et al, 2022). Additionally, analytical studies are mostly confined to laboratory settings and may not fully account for matrix effects present in commercial pyrolysis units (Yao et al, 2022). Similar to the ongoing analytical challenges faced in other contaminant classes, degradation products of PFAS compounds (or their precursors) are sometimes unknown and the accompanying extraction methods are not optimized for biochar (McNamara et al, 2022). To gain a comprehensive understanding of the fate of PFAS during pyrolysis and avoid potential false negatives, analytical methods for the extraction and analysis must be further developed and applied at a commercial scale, taking into account the strong binding capacity of biochar for many PFAS compounds.

Microplastics

Microplastics in sewage sludge and pine wood were completely degraded during pyrolysis at an HTT of 500°C (Ni et al, 2020; Wang et al, 2021b). All major types of plastic completely degrade at temperatures below 550°C (Sharuddin et al, 2016), which suggests this could be a sensible minimum treatment temperature for plastic contamination of feedstocks. However, so far, no proper method exists for the quantification of microplastics in biochar.

Pyrolysis can be a valuable option for the treatment of microplastic-contaminated waste biomass, as it offers the distinct advantage of eliminating microplastics compared to alternative treatment methods such as composting or anaerobic digestion (Cydzik-Kwiatkowska et al, 2022; Hoang et al, 2022). However, microplastics are known to have a high adsorption capacity for contaminants such as PPCPs, pesticides, herbicides, or heavy metals (Zhao et al, 2022). Therefore, the presence of microplastics in waste biomass may signal the presence of other contaminants too. As a result, the minimum treatment temperature for waste biomass contaminated with microplastics might have to comply with treatment levels for these other contaminant classes as well.

Guidelines and legal aspects of contaminants in biochar

Current guidelines and regulations

Current regulations of contaminants in biochar are restricted to PAHs, PCBs, and PCDD/Fs and do not yet exist for any of the emerging contaminants covered above. Table 22.3 gives an overview of current voluntary biochar quality standards and (inter-)national laws.

Table 22.3 *Relevant guidelines and legislation threshold values for organic contaminants in biochar. The table is based on Meyer et al (2017) with updates from the most recent developments in voluntary biochar quality standards and (inter-)national laws. All values are based on dry matter content. Four and two sub-categories of the EBC and UK BQM are described, respectively. IBI-BS, International Biochar Initiative Biochar Standard; EBC, European Biochar Certificate; BQM, UK Biochar Quality Mandate; EFSA, European Food Safety Authorization; TEQ, toxicity equivalent quantity*

	Unit	IBI-BS	EBC				UK BQM		Germany	Austria	Switzerland	Italy	EU product fertilizing regulation
			Feed	Agro organic	Agro	Basic materials	Standard Gr.	High Gr.	Fertilizer ordinance	Fertilizer ordinance	Fertilizer ordinance	Fertilizer decree #75	
16 US EPA PAHs[1]	mg kg^{-1}	≤300	declaration	<4	<6	–	<20	<20	–	<6	≤4	<6	<6
8 EFSA PAHs[2]	mg kg^{-1}	–	–	<1	–	<4	–	–	–	–	–	–	–
Benzo[e]pyrene	mg kg^{-1}	–	–	<1	–	–	–	–	–	–	–	–	–
Benzo[j]fluoranthene	mg kg^{-1}	–	–	<1	–	–	–	–	–	–	–	–	–
Benzo[a]pyrene TEQ	mg kg^{-1}	≤3	–	–	–	–	–	–	–	–	–	–	–
7/6 US EPA PCBs[3]	mg kg^{-1}	≤1	–	<0.2	<0.2	–	<0.5	<0.5	–	<0.2	–	<0.5	<0.8
PCDDs/Fs TEQ[4]	ng kg^{-1}	≤17	–	<20	<20	–	<20	<20	≤30	≤20	≤20	<9	<20

Notes

[1] PAHs: naphthalene, acenaphthylene, acenaphthene, fluorene, phenanthrene, anthracene, fluoranthene, pyrene, benzo[a]anthracene, chrysene, benzo[b]fluoranthene, benzo[k]fluoranthene, benzo[a]pyrene, indenol[1,2,3-cd]pyrene, dibenzo[a,h]anthracene, benzo[ghi]perylene

[2] EFSA PAHs: benzo[a]pyrene, benzo[a]anthracene, chrysene, benzo[b]fluoranthene, benzo[k]fluoranthene, dibenzo[a,h]anthracene, indeno[1,2,3-cd]pyrene, benzo[ghi]perylene

[3] PCBs: Aroclor 1016, Aroclor 1221, Aroclor 1232, Aroclor 1242, Aroclor 1248, Aroclor 1254, Aroclor 1260

[4] PCDDs/Fs: 2,3,7,8-Tetrachlorodibenzo-p-dioxin, 1,2,3,7,8-Pentachlorodibenzo-p-dioxin, 1,2,3,4,7,8-Hexachlorodibenzo-p-dioxin, 1,2,3,6,7,8-Hexachlorodibenzo-p-dioxin, 1,2,3,7,8,9-Hexachlorodibenzo-p-dioxin, 1,2,3,4,6,7,8-Heptachlorodibenzo-p-dioxin, 1,2,3,4,6,7,8-Octachlorodibenzo-p-dioxin, 2,3,7,8-Tetrachlorodibenzofuran, 1,2,3,7,8-Pentachlorodibenzofuran, 2,3,4,7,8-Pentachlorodibenzofuran, 1,2,3,4,7,8-Hexachlorodibenzofuran, 1,2,3,6,7,8-Hexachlorodibenzofuran, 1,2,3,7,8,9-Hexachlorodibenzofuran, 2,3,4,6,7,8-Hexachlorodibenzofuran, 1,2,3,4,6,7,8-Heptachlorodibenzofuran, 1,2,3,4,5,6,7,8-Octachlorodibenzofuran

Guidelines and legislation threshold values are primarily based on total levels of contaminants rather than available levels and hence their environmental risk. In the case of biochar, total contaminant content does not reflect risk well because the availability of organic contaminants in biochar is very low. In various countries soil and sediment guidelines now open up for bioavailability assessments in second-line risk assessments, and such risk-based guidelines could also be developed for biochars.

A large focus has been placed on setting threshold values for PAHs in biochar. Given the high proportion of naphthalene in the sum of 16 US EPA PAHs in biochar (mean 40%), biochar naphthalene content often determines whether a particular biochar complies with PAH threshold values. Meeting PAH threshold values that include naphthalene can be challenging since the proportion of naphthalene fluctuates highly from 30–80% (Bucheli et al, 2015) or even 11–100% (Buss et al, 2022b) and so far, no particular pyrolysis process parameter or feedstock type has been found to correlate with biochar naphthalene content (as discussed above). Considering the low carcinogenicity of naphthalene (toxicity equivalent factor (TEF) relative to B[a]P of NAP 0.001 vs. TEF dibenzo[a,h]anthracene 5), and low acute toxicity (lethal dose where 50% of individuals are killed (LD 50) for mice and rats 350–9,500 mg kg^{-1} (EU Scientific commission on food)), the use of the sum of 16 US EPA PAHs in legislation can result in biochar being arbitrarily excluded from use despite posing little risk to the environment. Yet, threshold values for the sum of 16 US EPA PAHs are consistently used in regulations and legislation (Table 22.3).

Suggestions for regulation of emerging contaminants

Regulatory limits for contaminants such as PAH, PCBs, dioxins, or PCDD, are well established in legislation and voluntary certification schemes (e.g., EBC, IBI) (Table 22.3). However, most emerging contaminants are still unregulated and miss analytical standard methods, not only in biochar but also in water or soil matrices, and feedstocks such as sewage sludge (Paz-Ferreiro et al, 2018). Identification of relevant organic compounds and their safe levels is difficult at present due to largely unknown individual and cumulative effects in the environment (Huygens et al, 2019). Therefore, maximum limits might not provide a practical assessment of emerging contaminant levels in biochar.

Some authors reported proportional reductions in contaminant levels to reach "satisfactory" levels of removal (Moško et al, 2021). However, due to largely varying base levels of contaminants in different feedstocks and different contaminant-specific risks, this approach can only be temporary. Unfortunately, bioassays also provide little additional information on the cumulative risks of emerging contaminants with repeated exposure (Huygens et al, 2019).

Based on the precautionary principle, a consensus on acceptable contaminant levels of emerging contaminants might therefore only be found in setting a minimum treatment temperature and time for each major reactor type to reach a standardized risk reduction. This approach will require the systematic assessment of specifically selected, thermally stable, and representative contaminants to evaluate their degradation efficiency in various environmental matrices. However, the safety of biochars from specific feedstocks must also be assessed against alternative treatment options. In the case of sewage sludge, pyrolysis can significantly reduce environmental risks compared to the current practice of direct land application, composting, or anaerobic digestion. It might therefore be pragmatic and meaningful to assess the safety of biochar not only against the environmental risks of its application, but also against the environmental risks of alternative treatments.

Comparative Life Cycle Assessment (LCA) can serve as a useful multifactorial decision-making tool to evaluate the risks associated with pyrolysis treatment scenarios in comparison to alternative methods and the current status quo (Huang et al, 2022). In the context of sewage sludge treatment, due to the specific risks of organic contaminants, impact factor categories such as ecotoxity might be the primary focus in assessing different treatment options, rather than solely focusing on global warming potential (GWP) (Tarpani et al, 2020). Despite the current limitations in understanding the fate of emerging contaminants during pyrolysis, LCAs can still provide valuable insights into general trends and high-risk areas that can inform the work of researchers and policymakers alike.

References

Adams RG, et al 2007 Polyethylene devices: Passive samplers for measuring dissolved hydrophobic organic compounds in aquatic environments. *Environmental Science and Technology* 41, 1317–1323.

Adánez-Rubio I, et al 2021 Exploratory study of polycyclic aromatic hydrocarbons occurrence and distribution in manure pyrolysis products. *Journal of Analytical and Applied Pyrolysis* 155, 105078.

Aktar W, Sengupta D, and Chowdhury A 2009 Impact of pesticides use in agriculture: Their benefits and hazards. Interdisciplinary *Toxicology* 2, 1–12.

Anuar Sharuddin SD, Abnisa F, Wan Daud WMA, and Aroua MK 2016 A review on pyrolysis of plastic wastes. *Energy Conversion and Management* 115, 308–326.

Backer R, et al 2018 Getting to the root of the matter: Water-soluble and volatile components in thermally-treated biosolids and biochar differentially regulate maize (Zea mays) seedling growth. *PLoS One* 13, e0206924.

Baek SO, et al 1991 A review of atmospheric polycyclic aromatic hydrocarbons: Sources, fate and behavior. *Water, Air, and Soil Pollution* 60, 279–300.

Bargmann I, et al 2013 Hydrochar and biochar effects on germination of spring barley. *Journal of Agronomy and Crop Science* 199, 360–373.

Bean TG, et al 2016 Evaluation of a novel approach for reducing emissions of pharmaceuticals to the environment. *Environmental Management* 58, 707–720.

Björkman E and Strömberg B 1997 Release of chlorine from biomass at pyrolysis and gasification conditions. *Energy and Fuels* 11, 1026–1032.

Bogolte BT, Ehlers GAC, Braun R, and Loibner AP 2007 Estimation of PAH bioavailability to Lepidium sativum using sequential supercritical fluid extraction – A case study with industrial contaminated soils. *European Journal of Soil Biology* 43, 242–250.

Bridle TR, Hammerton I, and Hertle CK 1990 Control of heavy metals and organochlorines using the oil from sludge process. *Water Science and Technology* 22, 249–258.

Bucheli TD, et al 2014 On the heterogeneity of biochar and consequences for its representative sampling. *Journal of Analytical and Applied Pyrolysis* 107, 25–30.

Bucheli TD, Hilber I, and Schmidt HP 2015 Polycyclic aromatic hydrocarbons and polychlorinated aromatic compounds in biochar. In: Lehmann J, and Joseph S (Eds) *Biochar for Environmental Management: Science and Technology and Implementation*, Second Edition. London: Earthscan Ltd. pp. 595–624.

Buss W, et al 2022a Highly efficient phosphorus recovery from sludge and manure biochars using potassium acetate pre-treatment. *Journal of Environmental Management* 314, 115035.

Buss W 2021 Pyrolysis solves the issue of organic contaminants in sewage sludge while retaining carbon – Making the case for sewage sludge treatment via pyrolysis. *ACS Sustainable Chemistry and Engineering* 9, 1048–1053.

Buss W, et al 2020 Comparison of pyrolysis liquids from continuous and batch biochar production—influence of feedstock evidenced by FTICR MS. *Energies* 14, 9.

Buss W, Graham MC, MacKinnon G, and Mašek O 2016 Strategies for producing biochars with minimum PAH contamination. *Journal of Analytical and Applied Pyrolysis* 119, 24–30.

Buss W, Hilber I, Graham MC, and Mašek O 2022b Composition of PAHs in biochar and implications for biochar production. *ACS Sustainable Chemistry and Engineering* 10, 6755–6765.

Buss W and Mašek O 2016 High-VOC biochar – Effectiveness of post-treatment measures and potential health risks related to handling and storage. *Environmental Science and Pollution Research* 23, 19580–19589.

Buss W and Mašek O 2014 Mobile organic compounds in biochar – A potential source of contamination – phytotoxic effects on cress seed (*Lepidium sativum*) germination. *Journal of Environmental Management* 137, 111–119.

Buss W, Mašek O, Graham M, and Wüst D 2015 Inherent organic compounds in biochar–their content, composition and potential toxic effects. *Journal of Environmental Management* 156, 150–157.

Chen X, Yang L, Myneni SCB, and Deng Y 2019 Leaching of polycyclic aromatic hydrocarbons (PAHs) from sewage sludge-derived biochar. *Chemical Engineering Journal* 373, 840–845.

Conesa JA, et al 2009 Comparison between emissions from the pyrolysis and combustion of different wastes. *Journal of Analytical and Applied Pyrolysis* 84, 95–102.

Congilosi JL and Aga DS 2021 Review on the fate of antimicrobials, antimicrobial resistance genes, and other micropollutants in manure during enhanced anaerobic digestion and composting. *Journal of Hazardous Materials* 405, 123634.

Cordella M, et al 2012 Bio-oils from biomass slow pyrolysis: A chemical and toxicological screening. *Journal of Hazardous Materials* 231–232, 26–35.

Cornelissen G, et al 2016 Emissions and char quality of flame-curtain "Kon Tiki" kilns for farmer-scale charcoal/biochar production. *PLoS ONE* 11, 1–16.

Crombie K and Mašek O 2015 Pyrolysis biochar systems, balance between bioenergy and carbon sequestration. *Global Change Biology Bioenergy* 7, 349–361.

Cui T, et al 2022 Distribution, dissemination and fate of antibiotic resistance genes during sewage sludge processing—A review. *Water, Air, and Soil Pollution* 233, 138.

Cydzik-Kwiatkowska A, Milojevic N, and Jachimowicz P 2022 The fate of microplastic in sludge management systems. *Science of the Total Environment* 848, 157466.

Dai Q, et al 2018 Distribution of PCDD/Fs over the three product phases in wet sewage sludge pyrolysis. *Journal of Analytical and Applied Pyrolysis* 133, 169–175.

Dai Q, et al 2014a Formation of PAHs during the pyrolysis of dry sewage sludge. *Fuel* 130, 92–99.

Dai Q, et al 2014b Temperature influence and distribution in three phases of PAHs in wet sewage sludge pyrolysis using conventional and microwave heating. *Energy and Fuels* 28, 3317–3325.

de la Rosa JM, Paneque M, Miller AZ, and Knicker H 2014 Relating physical and chemical properties of four different biochars and their application rate to biomass production of *Lolium perenne* on a Calcic Cambisol during a pot experiment of 79 days. *Science of the Total Environment* 499, 175–184.

de la Rosa JM, Sánchez-Martín ÁM, Campos P, and Miller AZ 2019 Effect of pyrolysis conditions on the total contents of polycyclic aromatic hydrocarbons in biochars produced from organic residues: Assessment of their hazard potential. *Science of the Total Environment* 667, 578–585.

de Resende MF, et al 2018 Polycyclic aromatic hydrocarbons in biochar amended soils: Long-term experiments in Brazilian tropical areas. *Chemosphere* 200, 641–648.

Devi P, and Saroha AK 2015 Effect of pyrolysis temperature on polycyclic aromatic hydrocarbons toxicity and sorption behaviour of biochars prepared by pyrolysis of paper mill effluent treatment plant sludge. *Bioresource Technology* 192, 312–320.

Directive 2004/42/CE of the European parliament and of the council 2004 on the limitation of emissions of volatile organic compounds due to the use of organic solvents in certain paints and varnishes and vehicle refinishing products and amending Directive 1999/13/EC. Official Journal of the European Union L143, 87–96.

Downie A, et al 2012 Biochar as a geoengineering climate solution: Hazard identification and risk management. *Critical Reviews in Environmental Science and Technology* 42, 225–250.

Dutta T, et al 2017 Polycyclic aromatic hydrocarbons and volatile organic compounds in biochar and biochar-amended soil: A review. *GCB Bioenergy* 9(6), 990–1004.

Environment Agency 2009 Science Report SC050021/Dioxins SGV: Soil guideline values for dioxins, furans and dioxin-like PCBs in soil.

Estoppey N, et al 2016 An in-situ assessment of low-density polyethylene and silicone rubber passive samplers using methods with and without performance reference compounds in the context of investigation of polychlorinated biphenyl sources in rivers. *Science of the Total Environment* 572, 794–803.

Fabbri D, Rombolà AG, Torri C, and Spokas KA 2013 Determination of polycyclic aromatic hydrocarbons in biochar and biochar amended soil. *Journal of Analytical and Applied Pyrolysis* 103, 60–67.

Fagernäs L, Kuoppala E, and Simell P 2012 Polycyclic aromatic hydrocarbons in Birch Wood slow pyrolysis products. *Energy & Fuels* 26, 6960–6970. 10.1021/ef3010515.

Freddo A, Cai C, and Reid BJ 2012 Environmental contextualisation of potential toxic elements and polycyclic aromatic hydrocarbons in biochar. *Environmental Pollution* 171, 18–24.

Ghidotti M, et al 2017a Source and biological response of biochar organic compounds released into water; relationships with bio-oil composition and carbonization degree. *Environmental Science and Technology* 51, 6580–6589.

Ghidotti M, Fabbri D, and Hornung A 2017b Profiles of volatile organic compounds in biochar: insights into process conditions and quality assessment. *ACS Sustainable Chemistry and Engineering* 5, 510–517.

Ghosal D, Ghosh S, Dutta TK, and Ahn Y 2016 Current state of knowledge in microbial degradation of polycyclic aromatic hydrocarbons (PAHs): A review. *Frontiers in Microbiology* 7, 1369.

Granatstein D, Collins H, Garcia-Perez M, and Yoder J 2009 *Use of biochar from the pyrolysis of waste organic material as a soil amendment, Final project report.* Center for Sustaining Agriculture and Natural Resources, Washington State University, Wenatchee, WA.

Greco G, et al 2021 Importance of pyrolysis temperature and pressure in the concentration of polycyclic aromatic hydrocarbons in wood waste-derived biochars. *Journal of Analytical and Applied Pyrolysis* 159, 105337.

Gundale MJ, and DeLuca TH 2006 Charcoal effects on soil solution chemistry and growth of Koeleria macrantha in the ponderosa pine/ Douglas-fir ecosystem. *Biology and Fertility of Soils* 43, 303–311.

Hajeb P, Zhu L, Bossi R, and Vorkamp K 2022 Sample preparation techniques for suspect and non-target screening of emerging contaminants. *Chemosphere* 287, 132306.

Hale SE, et al 2012 Quantifying the total and bioavailable polycyclic aromatic hydrocarbons and dioxins in biochars. *Environmental Science and Technology* 46, 2830–2838.

Hilber I, et al 2017a Bioavailability and bioaccessibility of polycyclic aromatic hydrocarbons from (post-pyrolytically treated) biochars. *Chemosphere* 174, 700–707.

Hilber I, et al 2012 Quantitative determination of PAHs in biochar: A prerequisite to ensure its quality and safe application. *Journal of Agricultural and Food Chemistry* 60, 3042–3050.

Hilber I, Arrigo Y, Zuber M, and Bucheli TD 2019 Desorption resistance of polycyclic aromatic hydrocarbons in biochars incubated in cow ruminal liquid in vitro and in vivo. *Environmental Science and Technology* 53, 13695–13703.

Hilber I, Blum F, Schmidt H-P, and Bucheli TD 2022 Current analytical methods to quantify PAHs in activated carbon and vegetable carbon (E153) are not fit for purpose. *Environmental Pollution* 309, 119599.

Hilber I, Schmidt H-P, and Bucheli TD 2017b Chapter 1: Sampling, storage and preparation of biochar for laboratory analysis. In: Singh B, Arbestain MC, and Lehmann J (Eds) *Biochar: A Guide for Analytical Methods*. Boca Raton: CRC Press. pp 1–8.

Hoang SA, et al 2022 Treatment processes to eliminate potential environmental hazards and restore agronomic value of sewage sludge: A review. *Environmental Pollution* 293, 118564.

Huang C, Mohamed BA, and Li LY 2022 Comparative life-cycle assessment of pyrolysis processes for producing bio-oil, biochar, and activated carbon from sewage sludge. *Resources, Conservation and Recycling* 181, 106273.

Huygens D, et al 2019 Technical proposals for selected new fertilising materials under the Fertilising Products Regulation (Regulation (EU) 2019/1009) – Process and quality criteria, and assessment of environmental and market impacts for precipitated phosphate salts and derivate, Publications Office of the European Union.

International Biochar Initiative (IBI) 2011 Standardized product definition and product testing guidelines for biochar that is used in soil, v. 2.1. November 23, 2015.

Jonker MTO and Koelmans AA 2001 Polyoxymethylene solid phase extraction as a partitioning method for hydrophobic organic chemicals in sediment and soot. *Environmental Science and Technology* 35, 3742–3748.

Karni M, et al 2013 Thermal degradation of DNA. *DNA and Cell Biology* 32, 298–301.

Keerthanan S, Jayasinghe C, Biswas JK,,Vithanage M 2021 Pharmaceutical and personal care products (PPCPs) in the environment: Plant uptake, translocation, bioaccumulation, and human health risks. *Critical Reviews in Environmental Science and Technology* 51, 1221–1258.

Keiluweit M, et al 2012 Solvent-extractable polycyclic aromatic hydrocarbons in biochar: Influence of pyrolysis temperature and feedstock. *Environmental Science and Technology* 46, 9333–9341.

Keith LH 1979 Priority pollutants: I – a perspective view. *Environmental Science and Technology* 13, 416–423.

Kennedy EM, and Mackie JC 2018 Mechanism of the thermal decomposition of chlorpyrifos and formation of the dioxin analog, 2,3,7,8-tetrachloro-1,4-dioxino-dipyridine (TCDDpy). *Environmental Science & Technology* 52, 7327–7333. 10.1021/acs.est.8b01626.

Kimbell LK, Kappell AD, and McNamara PJ 2018 Effect of pyrolysis on the removal of antibiotic resistance genes and class I integrons from municipal wastewater biosolids. *Environmental Science: Water Research and Technology* 4, 1807–1818.

Kloss S, et al 2012 Characterization of slow pyrolysis biochars: Effects of feedstocks and pyrolysis temperature on biochar properties. *Journal of Environmental Quality* 41, 990–1000.

Koelmans AA, et al 2006 Black carbon: The reverse of its dark side. *Chemosphere* 63, 365–377.

Kołtowski M and Oleszczuk P 2015 Toxicity of biochars after polycyclic aromatic hydrocarbons removal by thermal treatment. *Ecological Engineering* 75, 79–85.

Kończak M and Oleszczuk P 2018 Application of biochar to sewage sludge reduces toxicity and improve organisms growth in sewage sludge-amended soil in long term field experiment. *Science of the Total Environment* 625, 8–15.

Kong L, et al 2019 Integrating metabolomics and physiological analysis to investigate the toxicological mechanisms of sewage sludge-derived biochars to wheat. *Ecotoxicology and Environmental Safety* 185, 109664.

Krzyszczak A, Dybowski MP, and Czech B 2021 Formation of polycyclic aromatic hydrocarbons and their derivatives in biochars: The effect of feedstock and pyrolysis conditions. *Journal of Analytical and Applied Pyrolysis* 160, 105339.

Kundu S, et al 2021 Removal of PFASs from biosolids using a semi-pilot scale pyrolysis reactor and the application of biosolids derived biochar for the removal of PFASs from contaminated water. *Environmental Science: Water Research and Technology* 7, 638–649.

Kuśmierz M, et al 2016 Persistence of polycyclic aromatic hydrocarbons (PAHs) in biochar-amended soil. *Chemosphere* 146, 272–279.

Lam MM, Engwall M, Denison MS, and Larsson M 2018 Methylated polycyclic aromatic hydrocarbons and/or their metabolites are important contributors to the overall estrogenic activity of polycyclic aromatic hydrocarbon–contaminated soils. *Environmental Toxicology and Chemistry* 37, 385–397.

Liu X, et al 2019 Release of polycyclic aromatic hydrocarbons from biochar fine particles in simulated lung fluids: Implications for bioavailability and risks of airborne aromatics. *Science of the Total Environment* 655, 1159–1168.

Lizarraga E, Zabaleta C, and Palop JA 2007 Thermal stability and decomposition of pharmaceutical compounds. *Journal of Thermal Analysis and Calorimetry* 89, 783–792.

Longendyke GK, Katel S, and Wang Y 2022 PFAS fate and destruction mechanisms during thermal treatment: A comprehensive review. *Environmental Science: Processes and Impacts* 24, 196–208.

Lyu H, et al 2016 Effect of pyrolysis temperature on potential toxicity of biochar if applied to the environment. *Environmental Pollution* 218, 1–7.

Mackie JC and Kennedy EM 2019 Pyrolysis of glyphosate and its toxic products. *Environmental Science and Technology* 53, 13742–13747.

Madadian E and Simakov DSA 2022 Thermal degradation of emerging contaminants in municipal biosolids: The case of pharmaceuticals and personal care products. *Chemosphere* 303, 135008.

Madej J, Hilber I, Bucheli TD, and Oleszczuk P 2016 Biochars with low polycyclic aromatic hydrocarbon concentrations achievable by pyrolysis under high carrier gas flows irrespective of oxygen content or feedstock. *Journal of Analytical and Applied Pyrolysis* 122, 365–369.

Mayer P, et al 2016 How to determine the environmental exposure of PAHs originating from biochar. *Environmental Science and Technology* 50, 1941–1948.

McClellan K and Halden RU 2010 Pharmaceuticals and personal care products in archived U.S. biosolids from the 2001 EPA national sewage sludge survey. *Water Research* 44, 658–668.

McGrath TE, Chan WG, and Hajaligol MR 2003 Low temperature mechanism for the formation of polycyclic aromatic hydrocarbons from the pyrolysis of cellulose. *Journal of Analytical and Applied Pyrolysis* 66, 51–70.

McKay G 2002 Dioxin characterisation, formation and minimisation during municipal solid waste (MSW) incineration: Review. *Chemical Engineering Journal* 86, 343–368.

McNamara P, et al 2022 Pyrolysis transports, and transforms, PFAS from biosolids to py-liquid. *Environmental Science: Water Research and Technology* 9, 386–395.

Mejías C, et al 2021 Occurrence of pharmaceuticals and their metabolites in sewage sludge and soil: A review on their distribution and environmental risk assessment. *Trends in Environmental Analytical Chemistry* 30, e00125.

Mercl F, et al 2021 Pyrolysis of biosolids as an effective tool to reduce the uptake of pharmaceuticals by plants. *Journal of Hazardous Materials* 405, 124278.

Meyer S, et al 2017 Biochar standardization and legislation harmonization. *Journal of Environmental Engineering and Landscape Management* 25, 175–191.

Moško J, et al 2021 Effect of pyrolysis temperature on removal of organic pollutants present in anaerobically stabilized sewage sludge. *Chemosphere* 265, 129082.

Neuwald IJ, et al 2021 Ultra-short-chain PFASs in the sources of German drinking water: prevalent, overlooked, difficult to remove, and unregulated. *Environmental Science and Technology* 56, 6380–6390.

Ni BJ, et al 2020 Microplastics mitigation in sewage sludge through pyrolysis: The role of pyrolysis temperature. *Environmental Science and Technology Letters* 7, 961–967.

Oesterle P, Lindberg RH, Fick J, and Jansson S 2020 Extraction of active pharmaceutical ingredients from simulated spent activated carbonaceous adsorbents. *Environmental Science and Pollution Research* 27, 25572–25581.

Oleszczuk P, et al 2014 Microbiological, biochemical and ecotoxicological evaluation of soils in the area of biochar production in relation to polycyclic aromatic hydrocarbon content. *Geoderma* 213, 502–511.

Oleszczuk P and Kołtowski M 2018 Changes of total and freely dissolved polycyclic aromatic hydrocarbons and toxicity of biochars treated with various aging processes. *Environmental Pollution* 237, 65–73.

Patel M, et al 2019 Pharmaceuticals of emerging concern in aquatic systems: Chemistry, occurrence, effects, and removal methods. *Chemical Reviews* 119, 3510–3673.

Paz-Ferreiro J, et al 2018 Biochar from biosolids pyrolysis: A review. *International Journal of Environmental Research and Public Health* 15, 956.

Premnath N, et al 2021 A crucial review on polycyclic aromatic hydrocarbons – Environmental occurrence and strategies for microbial degradation. *Chemosphere* 280, 130608.

Puicharla R, et al 2014 A persistent antibiotic partitioning and co-relation with metals in wastewater treatment plant – Chlortetracycline. *Journal of Environmental Chemical Engineering* 2, 1596–1603.

Quilliam RS, et al 2013 Is biochar a source or sink for polycyclic aromatic hydrocarbon (PAH) compounds in agricultural soils? *Global Change Biology Bioenergy* 5, 96–103.

Rogovska N, et al 2012 Germination tests for assessing biochar quality. *Journal of Environment Quality* 41, 1014–1022.

Rombolà AG, et al 2019 Changes in the pattern of polycyclic aromatic hydrocarbons in soil treated with biochar from a multiyear field experiment. *Chemosphere* 219, 662–670.

Rombolà AG, et al 2015 Relationships between chemical characteristics and phytotoxicity of biochar from poultry litter pyrolysis. *Journal of Agricultural and Food Chemistry* 63, 6660–6667.

Ross JJ, et al 2016 Emerging investigators series: Pyrolysis removes common microconstituents triclocarban, triclosan, and nonylphenol from biosolids. *Environmental Science: Water Research and Technology* 2, 282–289.

Růžičková J, et al 2021 The occurrence of pesticides and their residues in char produced by the combustion of wood pellets in domestic boilers. *Fuel* 293, 120452.

Sharma RK, and Hajaligol MR 2003 Effect of pyrolysis conditions on the formation of polycyclic aromatic hydrocarbons (PAHs) from polyphenolic compounds. *Journal of Analytical and Applied Pyrolysis* 66, 123–144.

Sigmund G, et al 2017 Effect of ageing on the properties and polycyclic aromatic hydrocarbon composition of biochar. *Environmental Science: Processes and Impacts* 19, 768–774.

Spokas KA, et al 2011 Qualitative analysis of volatile organic compounds on biochar. *Chemosphere* 85, 869–882.

Suzuki M, Misic DM, Koyama O, and Kawazoe K 1978 Study of thermal regeneration of spent activated carbons: Thermogravimetric measurement of various single components organics loaded on activated carbons. *Chemical Engineering Science* 33, 271–279.

Tarpani RRZ, Alfonsín C, Hospido A, and Azapagic A 2020 Life cycle environmental impacts of sewage sludge treatment methods for resource recovery considering ecotoxicity of heavy metals and pharmaceutical and personal care products. *Journal of Environmental Management* 260, 109643.

The Environmental Applications Group LTD 1990 *The Environmental toxicology of polycyclic aromatic hydrocarbons. Ontario Ministry of the Environment, Minister of Supply and Services Canada*. Ottawa: National Printers.

Thoma ED, et al 2022 Pyrolysis processing of PFAS-impacted biosolids: A pilot study. *Journal of the Air and Waste Management Association* 72, 309–318.

Tian L, Bayen S, and Yaylayan V 2017 Thermal degradation of five veterinary and human pharmaceuticals using pyrolysis-GC/MS. *Journal of Analytical and Applied Pyrolysis* 127, 120–125.

Tian R, et al 2019 Preparation of biochar via pyrolysis at laboratory and pilot scales to remove antibiotics and immobilize heavy metals in livestock feces. *Journal of Soils and Sediments* 19, 2891–2902.

Tiedeken EJ, Tahar A, McHugh B, and Rowan NJ 2017 Monitoring, sources, receptors, and control measures for three European Union watch list substances of emerging concern in receiving waters – A 20 year systematic review. *Science of the Total Environment* 574, 1140–1163.

US Department of Health and Human Services 1995 *Toxicological profile for polycyclic aromatic hydrocarbons*. Atlanta, Georgia.

van Schie PM and Young LY 2000 Biodegradation of phenol: mechanisms and applications. *Bioremediation Journal* 4, 1–18.

Vom Eyser C, et al 2015 Determination of pharmaceuticals in sewage sludge and biochar from hydrothermal carbonization using different quantification approaches and matrix effect studies. *Analytical and Bioanalytical Chemistry* 407, 821–830.

Wang C, Wang Y, and Herath HMSK 2017 Polycyclic aromatic hydrocarbons (PAHs) in biochar – Their formation, occurrence and analysis: A review. *Organic Geochemistry* 114, 1–11.

Wang J, et al 2018 Application of biochar to soils may result in plant contamination and human cancer risk due to exposure of polycyclic aromatic hydrocarbons. *Environment International* 121, 169–177.

Wang J, et al 2019 Polyaromatic hydrocarbons in biochars and human health risks of food crops grown in biochar-amended soils: A synthesis study. *Environment International* 130, 104899.

Wang Q, Zhang Z, Xu G, and Li G 2021a Pyrolysis of penicillin fermentation residue and sludge to produce biochar: Antibiotic resistance genes destruction and biochar application in the adsorption of penicillin in water. *Journal of Hazardous Materials* 413, 125385.

Wang J, et al 2021b Adsorption and thermal degradation of microplastics from aqueous solutions by Mg/Zn modified magnetic biochars. *Journal of Hazardous Materials* 419, 126486.

Weber NH, et al 2020 Products and mechanism of thermal decomposition of chlorpyrifos under inert and oxidative conditions. *Environmental Science: Processes and Impacts* 22, 2084–2094.

Weidemann E, et al 2017 Influence of pyrolysis temperature and production unit on formation of selected PAHs, oxy-PAHs, N-PACs, PCDDs, and PCDFs in biochar – A screening study. *Environmental Science and Pollution Research* 25, 3933–3940.

Wiedner K, et al 2013 Chemical evaluation of chars produced by thermochemical conversion (gasification, pyrolysis and hydrothermal carbonization) of agro-industrial biomass on a commercial scale. *Biomass and Bioenergy* 59, 264–278.

Wilson BK and Reed D 2012 IBI white paper implications and risks of potential dioxin presence in biochar. *International Biochar Initiative*, www.biochar-international.org.

Wilson SC and Jones KC 1996 The fate and behavior of volatile aromatic hydrocarbons in sewage sludge-amended soil. In: *Volatile Organic Compounds in the Environment*. ASTM Special Technical Publication 1261. pp. 119–123.

Winchell LJ, et al 2021 Per- and polyfluoroalkyl substances thermal destruction at water resource recovery facilities: A state of the science review. *Water Environment Research* 93, 826–843.

Winchell LJ, et al 2022 Pyrolysis and gasification at water resource recovery facilities: Status of the industry. *Water Environment Research* 94, 1–20.

Xhrout C, Pirard C, and de Pauw E 2001 De novo synthesis of polychlorinated dibenzo-p-dioxins and dibenzofurans on fly ash from a sintering process. *Environmental Science and Technology* 35, 1616–1623.

Xiao F, et al 2020 Thermal stability and decomposition of perfluoroalkyl substances on spent granular activated carbon. *Environmental Science and Technology Letters* 7, 343–350.

Yang H, et al 2013 Detailed analysis of residual volatiles in chars from the pyrolysis of biomass and lignite. *Energy and Fuels* 27, 3209–3223.

Yao B, et al 2022 The first quantitative investigation of compounds generated from PFAS, PFAS-containing aqueous film-forming foams and commercial fluorosurfactants in pyrolytic processes. *Journal of Hazardous Materials* 436, 129313.

Zhao M, et al 2022 Adsorption of different pollutants by using microplastic with different influencing factors and mechanisms in wastewater: A review. *Nanomaterials* 12, 2256.

Zheng H, et al 2013 Characteristics and nutrient values of biochars produced from giant reed at different temperatures. *Bioresource Technology* 130, 463–471.

Zhi L, et al 2020 Pyrolyzed biowastes deactivated potentially toxic metals and eliminated antibiotic resistant genes for healthy vegetable production. *Journal of Cleaner Production* 276, 124208.

Zhou H, et al 2014 Polycyclic aromatic hydrocarbon formation from the pyrolysis/gasification of lignin at different reaction conditions. *Energy and Fuels* 28, 6371–6379.

Zhou X, et al 2019 Turning pig manure into biochar can effectively mitigate antibiotic resistance genes as organic fertilizer. *Science of the Total Environment* 649, 902–908.

Zielińska A and Oleszczuk P 2015 The conversion of sewage sludge into biochar reduces polycyclic aromatic hydrocarbon content and ecotoxicity but increases trace metal content. *Biomass and Bioenergy* 75, 235–244.

23

How does biochar influence plant biotic stress?

Amit K. Jaiswal, Omer Frenkel, Jane Debode, and Ellen R. Graber

Introduction

General

Adding biochar to soil can affect the response of plants to biotic stresses because the biochar addition changes the plants' chemical, microbial, and physical environments. These changes elicit cascading feedback responses in the plant that can activate or alter plants' innate defenses.

The aim of this chapter is to: (i) briefly review plant pathogens and pests, and the defenses plants have against them; (ii) summarize the impact of biochar on plants' ability to cope with pests and diseases; (iii) critically synthesize current understanding of biochar modes of action; (iv) discuss techniques that may lead to improving biochar ability to help plants contend better with biotic stresses; and (v) identify future research needs.

Phytopathologists conventionally refer to viruses and single-celled organisms such as fungi, oomycetes, and bacteria as 'pathogens'. Multicellular eukaryotic pests include Arthropods (insects, arachnids, myriapods), Mollusks (snails and slugs), and Annelids (nematodes, worms), and are herein referred to as AMA pests. Besides having the potential to cause dramatic damage themselves, AMA pests can additionally serve as vectors of plant pathogens. Higher parasitic plants, invasive climbing plants, and parasitic green algae are also pests that can cause devastating plant disease and destruction, for example, broomrape in tomatoes and mistletoe in a range of host trees.

Plant pests are one of the major challenges for agricultural and horticultural crops. They reduce yield and quality, cause economic losses, and affect food security worldwide (Savary and Willocquet, 2020). Moreover, some plant pathogens produce secondary metabolites that can be harmful to humans and animals (Evidente et al, 2019). Plant pests hamper international trade of plants due to fears of introducing new pests to previously unaffected areas. Climate change is also resulting in an

DOI: 10.4324/9781003297673-23

epidemic of emerging and invasive pests in new areas (Sikes et al, 2018). For this chapter, we broadly subdivide plant pathogens and AMA pests into the environmental compartment in which they live: soilborne (soil-dwelling) versus foliar (above ground). Biochar is employed in the soil compartment where it can come into contact with soil-dwelling organisms and have both direct and indirect impacts on them. This is not the case for organisms that reside above ground, where the effects of biochar can only be indirect. This difference can give us important insights into biochar's mode of action, as well as help suggest ways of promoting biochar efficacy against plant pathogens and AMA pests. The most common use of biochar is in mineral soil, but biochar is also used in soilless cultivation, mainly in peat-based growing media. Even though mineral soils and growing media have very distinct properties, and the modes of action of biochar may be dissimilar in the different systems, the distinction between mineral soil and growing media is not always clearly made in literature, particularly when biochar is tested for its impacts on plant disease resistance. Therefore, when we use the term 'soil' in this chapter, it refers to both mineral soil and growing media.

What are plant defenses against pests and diseases?

Terrestrial plants exhibit a wide array of resistance traits designed for defense against herbivores and pathogens, including structural and chemical constitutive defenses and indirect defenses (Hanley et al, 2007).

Constitutive defense mechanisms are structural and chemical traits that are always expressed within a plant. Structural traits are designed to discourage herbivores from feeding on the plant, either by harming them or retarding their progress (Mortensen, 2013).

Structural defenses include spines, trichomes (plant hairs), thickened tough leaves and stems, and microscopic sand- or needle-like particles inside plant tissues (Mortensen, 2013).

Constitutive chemical traits are metabolites that exist in healthy plants in their biologically active forms or that are quickly formed from existing inactive precursors in response to tissue damage or pathogen attack (Osbourn, 1996). These plant molecules are likely to represent one of the first chemical barriers to pathogens. Pre-formed chemical inhibitors within plants tend to be tissue-specific, with some concentrated in outer cell layers of plant organs and others, like many antifungal compounds, being sequestered inside vacuoles or organelles in healthy plants (Osbourn, 1996). Many constitutive plant metabolites have antifungal activity (Osbourn, 1996).

Plant nutrient status (including macro- and micronutrients) is also a part of constitutive defenses, and can impact both disease development (Soulie et al, 2020) and insect feeding habits (e.g., Lu et al, 2005; Nam et al, 2006; Yermiyahu et al, 2006; Lecompte et al, 2010). Both under- and overly-fertilized plants can be more susceptible or more resistant to pathogen and pest attacks, depending on the circumstances (Baker and Martinson, 1970; Wall et al, 1994; Nam et al, 2006; Xu et al, 2013; Lacomino et al, 2022).

When pests attack plants, indirect defenses can also come into play. Necrosis of cells around the incursion site can occur due to reactive oxygen species, called the hypersensitive response. Plant hypersensitive response curbs the advance of the attacker at the site of the attack and can also cause signal transduction that induces resistance in other unaffected parts of the plant. At the hypersensitive response site, local acquired resistance is developed by reinforcing the cell wall, accumulation of antimicrobial compounds, and expression of genes encoding pathogenesis-related proteins (Vallad and Goodman, 2004). At non-affected parts, induced resistance can occur. Induced resistance is broadly subdivided into two main types: systemic acquired resistance and induced systemic resistance. Systemic acquired resistance involves the synthesis of pathogen-related genes mediated by the accumulation of the phytohormone salicylic acid both locally and systemically (Sticher et al, 1997; Durrant and Dong, 2004; Vlot et al, 2009; Fu and Dong, 2013), and can be triggered by chemical elicitors or environmental triggers (Ayres, 1984; Wiese et al, 2004). Induced systemic resistance is commonly triggered by plant growth-promoting rhizobacteria and fungi, and it relies on the phytohormones ethylene, jasmonic acid, and methyl jasmonate (Van Loon et al, 1998; Vallad and Goodman, 2004; Shoresh et al, 2010; Pieterse et al, 2014). No single hormone controls plant immunity. Instead, plant hormones tend to act interdependently through complex antagonistic or synergistic interactions, referred to as hormonal crosstalk (Robert-Seilaniantz et al, 2011; Shigenaga and Argueso, 2016).

Does biochar application in soil help protect plants from attack by pathogens or AMA pests?

Adding biochar to soil to improve its fertility and help protect against pests is an ancient practice (Ogawa and Okimori, 2010). In some of the earlier studies, *Fusarium* root rot in asparagus seedlings was shown to be suppressed after the amendment of the soil with biochar (Matsubara et al, 2002), and bacterial wilt (*Ralstonia solanacearum*) in tomatoes was shown to be reduced when grown in soil amended with biochar (Nerome et al, 2005). Biochar in the soil helped protect tomato and pepper plants from two foliar fungal pathogens and a foliar mite (Elad et al, 2010). The fact that the biochar location during all stages of plant development was spatially separate from the site of infection pointed to its ability to elicit induced systemic plant responses. The ability of biochar to elicit induced resistance was confirmed by the up-regulation of defense-related genes in the canopy of strawberry plants growing in biochar-amended soils after attacks by various foliar fungal pathogens (Meller Harel et al, 2012).

Since then, there has been a significant increase in studies exploring the effect of biochar on plant disease (Supplementary Table 23.1; www.routledge.com/978103 2286150), and the subject has been reviewed several times (e.g., Bonanomi et al, 2015; Frenkel et al, 2017; Iacomino et al, 2022; Yang et al, 2022). We identified ~120 studies that targeted biochar's influence on plant disease through the closing days of 2022. To date, 71 different pathosystems (pathosystem refers to a given plant pathogen-pest pair) were studied and analyzed (Supplementary Table 23.1). Most of the studies were conducted in greenhouse or growth chamber trials (86%), with only 14% conducted in field conditions. Most studies were conducted in soil (72%) compared to growing media (28%). The vast majority of experiments lasted less than 16 weeks, and even the longest field experiments lasted at most 36–48 months. About 50% of the total studies employed *Solanaceae* crops as their model, with tomato (*Solanum Lycopersici*) making up the vast majority (41% of the total), followed by *S. tuberosum* (4% of total), *Capsicum annum* (4% of total), and

Nicotiana tabacum (2% of total). Other studies involved *Jatropha curas* (6%), *Panax notojinseng* (6%), *Lactuca sativa, Cucumis sativus* (4%), *Morus alba* (3%), *Asparagus officinalis* (3%), and *Glycine max* (3%). An additional 26 species made up the remaining 24% of the studies. Concerning the biochar pyrolysis process, the most common pyrolysis temperature range was 500°C–600°C, followed by 300°C–400°C and 600°C–700°C (Figure 23.1A), and the most common feedstock types were lignin-rich biochars such as wood residues, followed by cellulose-rich biochar such as straw and stalk and crop residues (Figure 23.1B). When summarizing the pathogen types, the most commonly studied organisms were pathogenic fungi (48%), followed by insects (15%), nematodes (13.5%), oomycetes (9%), bacteria (8.5%), and others (6%). Among the fungi, the most studied species were *Fusarium oxysporum* f. sp. *radicis lycopersici,* followed by *Rhizoctonia solani, Botrytis cinerea, Fusarium oxysporum, Alternaria solani, Fusarium solani, Sclerotinia sclerotiorum,* and *Fusarium* spp. Among bacterial pathogens, *Ralstonia solanacearum* was the most investigated species; for nematodes, insects, and oomycetes, the most studied species were *Meloidogyne incognita* and *Epitrix fuscula.* Finally, for parasitic plants, both *Phelipanche aegyptiaca* and *Orobanche crenata* were investigated.

In the global dataset using meta-analysis, disease severity decreased by 47% upon biochar amendment, accompanied by an increase in plant biomass by 44% (Yang et al., 2022. All three biochar feedstock types (food waste, greenhouse waste, straw, and wood) were efficient at reducing disease, with straw-derived biochar demonstrating the highest disease severity reduction. Yet, only greenhouse waste biochars concurrently produced a significant increase in plant growth. Biochars produced at temperatures between 350 to 600°C were effective at both disease control and growth promotion, while biochars produced at both lower and higher temperatures had variable impacts on disease control and plant growth. There was a distinct inverted U-shape in disease response versus biochar application rate curves, with the application of biochars at a rate of 3–5% w/w in the soil resulting in the greatest improvement in both disease suppression and plant growth. The meta-analysis confirmed the "Shifted Rmax" effect (Jaiswal et al, 2015), whereby a given biochar application rate may differentially affect plant growth and susceptibility to disease. In their meta-analysis (Yang et al, 2022), rates of biochar application from <1% up to 5% significantly decreased disease severity, but only the application range between 3 and 5% was significant for improving plant growth. Biochars were found to suppress disease severity in cash crops (vegetables, berries, and tobacco) by 52% and enhance plant growth by 53%, but they had limited influence on plant disease and growth in cereal grains and perennial trees. The effectiveness of biochar in plant disease suppression was higher for foliar pathogens than for soilborne pathogens.

We compartmentalized the dataset in Supplementary Table 23.1 according to soilborne and foliar pathogens. Then we compared the effect of the highest tested biochar concentration on disease severity to the effect of the unamended control and also to the biochar concentration which had the most positive effect on disease suppression (Figure 23.1C; Frenkel et al, 2017). This approach gave insights into the biochar modes of action and also into approaches for commercializing biochar (Figure 23.1D, E).

Biochar effects on 22 foliar disease pathosystems were reported in the literature (Supplementary Table 23.1). In addition, there were 13 insect and mite pathosystems. In most cases, no net negative effect of the

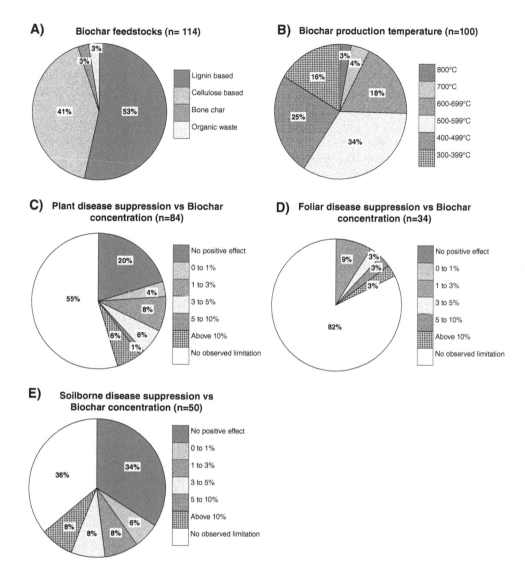

Figure 23.1 *Pie charts presenting the proportion of pathosystem case studies as a function of: (A) biochar feedstock; (B) biochar preparation temperature; (C) the maximal biochar concentration that reduces disease severity in the pathosystem based on all 84 case studies; (D) the maximal biochar concentration that reduces disease severity in 34 foliar pathosystems; (E) the maximal biochar concentration that reduces disease severity in the 50 soilborne pathogens pathosystems. n= number of case studies analyzed. In panes C, D, and E, "No positive effect" means that biochar did not improve disease suppression at any tested concentration, and may have had a negative effect at some or all tested concentrations. The meaning of a defined concentration range is that biochar added at a maximum concentration within this range was observed to improve disease suppression, but above this, it may have had no positive effect or a negative effect. "No observed limitation" means that biochar had a positive effect at all tested concentrations, and no negative effect was observed*

biochar was detected even when the highest applied concentration exceeded 10% biochar (individual studies often without specifying whether values are by weight or by volume). The lowest concentration tested usually had limited positive suppressive activity, and the suppressiveness increased at intermediate concentrations (usually 1–3%). In 82% of the studies, the highest concentration did not have a negative effect on foliar disease suppression (Figure 23.1D). However, in three of the 22 pathosystems, no additional improved suppressiveness was detected compared to the intermediate concentration (usually ≤3%). For foliar AMA pests, control efficacy continued to increase even at concentrations above 10%.

Soilborne pests with biochar additions were studied in 52 pathosystems and included 30 fungal, 3 fungal complexes, 6 oomycetes, 11 nematodes, and two bacterial pathosystems (Supplementary Table 23.1; www.routledge.com/9781032286150). No negative effect of biochar, i.e., disease promotion, was found for nematodes and bacterial pathogens, even at the highest applied biochar concentrations. However, biochar sometimes had a noticeable negative influence on plant resistance to fungal soilborne pathogens. This negative effect was much more serious than for foliar pathogens

(Figure 23.1E), and was clearly pathogen-species dependent. For example, there were few indications of increased disease syndrome at any tested biochar concentrations for *Fusarium* spp., *Sclerotinia sclerotiorum*, and *Cylindrocarpon*. In contrast, *R. solani* pathosystems usually exhibited relative promotion of the plant disease at the highest tested biochar concentrations and at all application rates of 3% w/w and beyond. For the oomycetes *Pythium aphanidermatum* and *P. ultimum*, pathogens that result in fast-acting damping off disease, particularly in young plants, biochar addition was either neutral or disease-promoting at all tested concentrations, even at 1% w/w.

It can thus be concluded that biochar application can help plants protect themselves from attack by pathogens and AMA pests, but it is not a 'one-size-fits-all'. For the most part, the risk of increased disease severity due to biochar addition is greater for diseases caused by soilborne pathogens than by foliar pathogens. A better understanding of the mechanisms by which biochar can improve plant resistance will allow us to predict the potential of biochar to help fight against pathogens and pests (next section). Moreover, additional means of improving biochar safety and efficacy are needed, as elaborated later.

Mechanisms by which biochar can improve plant resistance to disease

Changes in nutritious elements or constitutive defenses

Biochar can supply nutrients (originating from the feedstock) and/or make nutrients more or less available to the plant, usually through changing the pH of the soil

environment. Soil nutrients affected by biochar include nitrogen (N), phosphorus (P), potassium (K), magnesium (Mg), manganese (Mn), iron (Fe), and silicon (Si) (Chapters 8, 16). Considering that plant nutritional status can impact its susceptibility to pests and pathogens (e.g., Baker and Martinson, 1970; Wall et al, 1994; Lu et al,

2005; Nam et al, 2006; Yermiyahu et al, 2006; Lecompte et al, 2010; Xu et al, 2013), the nutrient effect of biochar has been proposed as a possible means by which biochar elicits plant systemic disease resistance responses. Few studies have specifically tested this mechanism, and in only a few pathosystems. In those works, no consistent effect of biochar soil amendment on plant N levels and plant resistance to disease was revealed (Hou et al, 2015; Amery et al, 2021; De Tender et al, 2021). Moreover, no relationships were found between nematode infection rates and any measured shoot nutrient concentrations in carrots growing in soil amended with biochar, except for iron (George et al, 2016). However, there is evidence that biochar soil additions can increase plant tissue Si concentration, which in turn impairs feeding by insect pests (Bakhat et al, 2021; Chen et al., 2019b; Hou et al, 2015). Likewise, alkali-enriched biochars resulted in increased Si concentration in plant tissues and were reported to protect perennial ryegrass against a fungal pathogen that causes gray leaf spots (Wang et al, 2019).

No studies were identified that specifically addressed the impact of biochar on other constitutive plant defenses of plants (e.g., structural such as spines, trichomes, thickened leaves and stem, or chemicals such as phenols, glycosides, lactones, and others). Thus, the impact of biochar on plant constitutive defenses in general, and their role in plant disease development in particular, are woefully under-addressed and would be well served by additional research.

Changes in soil physiochemical properties

Biochar is commonly alkaline and has been frequently observed to increase soil pH (Chapter 8). In addition, both solid and water-soluble parts of biochar are redox active (Chapter 10) and can alter the soil redox potential (Eh). Since many soil pathogens thrive under narrow Eh–pH ranges (Husson, 2013), biochar-induced changes in pH or coupled changes in Eh–pH (Yuan and Xu, 2012; Husson, 2013) in the rhizosphere could strongly alter pathogen viability and virulence. In general, pH and Eh are important drivers of microbial community development, diversity, structure, and activity (Husson, 2013). The high pH and buffer capacity of many biochars could potentially reduce the influence of toxic acids near plant roots and alter the activities of enzymes produced by soilborne pathogens (Bateman, 1970). Changes in water holding capacity of the soil (Zhang and You, 2013), electrical conductivity (Chintala et al, 2014), and temperature (Verheijen et al, 2013; Zhang et al, 2013) due to biochar additions can all alter environmental conditions in the rhizosphere, thus altering niches favored by pathogens and root growth. A greater water-holding capacity may decrease nematode infection rates (George et al, 2016).

Biochar-mediated increases in water-holding capacity could potentially favor pathogens that produce zoospores, such as *Pythium* and *Phytophthora* spp., or bacterial species that can move through water (Fry and Grünwald, 2010). Increasing electrical conductivity and reducing the albedo of soils by biochar (Verheijen et al, 2013; Chintala et al, 2014), can affect root growth and development. Several fungal and oomycetes pathogens perform better in specific conditions of pH, Eh, electrical conductivity, water, and temperature, such that biochar-induced changes in environmental conditions can affect their establishment and survival.

Clearly, such biochar-related changes in soil physical and chemical characteristics and the domino effect they have on soil microbes make it challenging, if not impossible, to tease out their specific contribution to the overall effect of the added biochar. As a result, these remain mainly speculations at this time. We address soil-biochar complex systems feedback in a later section.

Biochar-borne toxic compounds

Several compounds that are known to adversely affect microbial growth and survival have been identified in biochars, including ethylene glycol and propylene glycol, hydroxyl-propionic and butyric acids, benzoic acid and o-cresol, quinones (resorcinol and hydroquinone), and 2-phenoxyethanol (Graber et al, 2010). Methoxyphenols and phenols are formed during the pyrolysis of hemicelluloses and lignin (Faix et al, 1990; McDonald et al, 2000; Lingens et al, 2005). These compounds, along with carboxylic acids, furans, and ketones, are known to inhibit microbial activity (Klinke et al, 2004; Mu et al, 2006; Mun and Ku, 2010) and can theoretically affect soilborne pathogens. *In vitro* experiments have shown variable results (positive, negative, none) for biochar impact on fungal mycelium radial growth and biomass (Graber et al, 2014; Jaiswal et al, 2015; Copley et al, 2015; Akhter et al, 2016; Jaiswal et al, 2017). Likewise, in vitro results for root rot nematodes were variable (Huang et al, 2015; Rahayu and Sari, 2017; Ikwunagu et al, 2019; Arshad et al, 2020; Ebrahimi et al, 2021; Mondal et al, 2021). When direct effects of biochar against nematodes and fungi were tested in soil, rarely were any impacts detected (Rahman et al, 2014; Jaiswal et al, 2015; George et al, 2016; Cao et al, 2018). Therefore, it appears that direct toxicity in the soil is unlikely to be a main or straightforward disease suppression mechanism.

Adsorption of pathogenic enzymes, toxic metabolites, and signaling molecules

Biochars are well-known adsorbents of large and small organic compounds, with adsorption affinities and capacities that can exceed those of typical soil components by several orders of magnitude (Chapters 6, 8) (Graber et al, 2011; Graber and Kookana, 2015). Adsorbed molecules can be extracellular enzymes (Rani et al, 2000; Daoud et al, 2010; Lammirato et al, 2011), microbial signaling molecules (Masiello et al, 2013), plant-plant signaling molecules (Eizenberg et al, 2017), plant-microbe signaling molecules (Gu et al, 2017), and many organic molecules found in the soil subsurface together with added biochar. Considering the adsorbing power of biochar, it was hypothesized that sorption immobilization and deactivation of cellulolytic, pectinolytic, and other cell wall degrading enzymes and toxic metabolites by biochars could reduce their contact with root cell walls, thus helping to protect plants from their ravages (Graber et al, 2014).

In vitro, biochar does adsorb plant pathogenic enzymes, immobilizing and substantially deactivating them (Lammirato et al, 2011; Jaiswal et al, 2018b). This effect was confirmed in plant-soil bioassays both for enzymes and other pathogen toxins (Jaiswal et al, 2018b; Li et al, 2022). In a plant-soil bioassay, adsorption of root exudates by biochar was also confirmed to be instrumental in preventing an outbreak of broomrape, a parasitic plant whose seeds are stimulated to germinate by strigolactones exuded by the host plant (Eizenberg et al, 2017).

These findings are consistent with many others which demonstrate that adsorption of soil compounds on biochar can impact soil processes. However, there are many unknowns regarding the principles that control enzyme or signaling molecule adsorption, immobilization, and deactivation on biochar, and no clear relationship was found between deactivation extent and any biochar physical and chemical characteristic or substrate-product adsorption characteristics (Daoud et al, 2010; Lammirato et al, 2011; Jaiswal et al, 2018b).

We anticipate that adsorption and consequent deactivation of virulence factors by biochar is part of the puzzle responsible for biochar impacts on the severity of soilborne diseases, but not the only part. This is because most biochar dose-disease resistance curves are non-monotonic and reveal the best results at some interim biochar doses. This would not be the case if *only* adsorption and deactivation were important since in that event, increasing biochar dose would increasingly remove virulence factors from near the plant roots.

Soil microbial community structure, diversity, and functioning

Adding biochar to the soil introduces elements and compounds that can influence the surrounding microbial community in the bulk soil and rhizosphere through a variety of mechanisms (Chapter 14). Specifically regarding plant disease, biochar can stimulate chitinolytic bacteria and cellulolytic bacteria that can digest fungal and oomycete cell walls, thus releasing fragments that are known to be active elicitors of plant defense responses (Klarzynski et al, 2000; Wan et al, 2008). Often, biochar amendments substantially increase bacterial taxonomic and functional diversity, microbial activity, and an overall shift in carbon-source utilization (Jaiswal et al, 2017). There is a strong link between the diversity and richness of the soil microbiome and enhanced plant productivity

(Van der Heijden et al, 2008; Wagg et al, 2014), suppression of soilborne pathogens (Garbeva et al, 2004; Mendes et al, 2011; Mendes et al, 2013; Raaijmakers and Mazzola, 2016) and other ecological services (Bell et al, 2005; Wagg et al, 2014; Delgado-Baquerizo et al, 2016).

Biochar was revealed to facilitate a stable, more favorable microbial community and to moderate shifts in the soil and rhizosphere microbial community in the presence of *F. oxysporum* f. sp. *radicis-cucumerinum* (Dror et al, 2022). These effects could be associated with the capacity of biochar to chemically stabilize the root environment (i.e., through maintaining pH) or to provide micro-niches within its unique carbon skeleton that certain bacterial groups can colonize (Abujabhah et al, 2016; Wang et al, 2019).

In bacterial disease, studies also found a link between disease suppression, microbial diversity, and shifts in microbial community structure (Gao et al, 2019; Chen et al, 2022). For the most part, only a few studies revealed a relationship between nematode suppression and soil life (including non-parasitic nematodes) (Rahman et al, 2014; Cao et al, 2018; Ebrahimi et al, 2021).

Many studies suggest that specific groups of beneficial bacteria or fungi are associated with disease suppression (De Tender et al, 2016, 2021; Jaiswal et al, 2018a, 2019; Gao et al, 2019), which potentially opens new opportunities to harness them to reduce diseases. However, the fact that different types of biochars can reduce disease severity suggests that microbial populations are influencing plant disease through more general mechanisms such as increased microbial diversity. It is still very unclear what it is about biochar that results in increased microbial diversity in the rhizosphere.

In-planta molecular mechanisms underlying biochar-mediated induced systemic resistance

Induced systemic resistance related to biochar has been demonstrated in several pathosystems (Supplementary Table 23.1), including evidence of upregulation of defense-related genes governing salicylic acid and jasmonic acid-ethylene pathways and priming (Meller Harel et al, 2012). In the *B. cinerea* –tomato pathosystem, biochar-mediated induced resistance and priming of defense-related gene expression was jasmonic acid-dependent, and correlated with whole plant systemic priming of the early oxidative burst response in a jasmonic acid-dependent manner (Mehari et al, 2015). These characteristics are specific to the systemic acquired resistance pathway, usually involving beneficial rhizosphere microorganisms. In contrast, genes of soybean associated with primary metabolism and salicylic acid and jasmonic acid pathways were downregulated after exposure of soybean to a high concentration of biochar (Copley et al, 2017). This downregulation was implicated in the enhanced susceptibility of soybean to Rhizoctonia foliar blight under high concentrations of biochar (Copley et al, 2017). Global gene expression (microarrays) data for *Arabidopsis thaliana* grown in soil amended with biochar revealed down-regulation of defense-related genes and most categories of secondary metabolites (Viger et al, 2014), but since no pathogen challenge was made, it cannot be known if the general downregulation of defense-involved genes would have resulted in subsequent susceptibility to pathogen attack. Transcriptomic analysis (RNA-seq) of tomato demonstrated that biochar had a priming effect on gene expression and upregulated pathways and genes associated with plant defense and growth, such as

jasmonic acid, brassinosteroids, cytokinins, auxin, and synthesis of flavonoid, phenylpropanoids, and cell wall, while biosynthesis and signaling of the salicylic acid pathway were downregulated (Jaiswal et al, 2020). Biochar was also found to act on gene pathways involved in plant defenses against various AMA pests such as the root-knot nematode (Huang et al, 2015), the white-backed rice hopper (*Sogatella furcifera* Horváth) (Waqas et al, 2018), and the English grain aphid *Sitobion avenae* (Homoptera: *Aphididae*) (Chen et al, 2019a). Finally, biochar had a different impact on the *Botrytis cinerea* resistance of strawberry fruits as compared to the leaves of strawberries, which may indicate a trade-off between plant parts (De Tender et al, 2021). From the evidence, therefore, it is clear that biochar does induce plant systemic defenses, but there is still much to be learned and clarified, including the complex interaction between various plant defense mechanisms and plant growth regulation through hormonal pathways and cross-talk.

Direct versus indirect impacts of biochar on foliar versus soilborne pest and disease suppression

Some of the mechanisms involved in biochar impact on pest-pathogen-plant can operate directly on pests and pathogens that have at least one part of their life cycle below ground (soilborne) (Figure 23.2). In contrast, for foliar pests and pathogens, all relevant mechanisms are expected to have indirect impacts only, because of the spatial separation between biochar and the organisms. This difference can help enhance our understanding of the dominant factors playing a role in biochar's impact on pests and disease (Table 23.1).

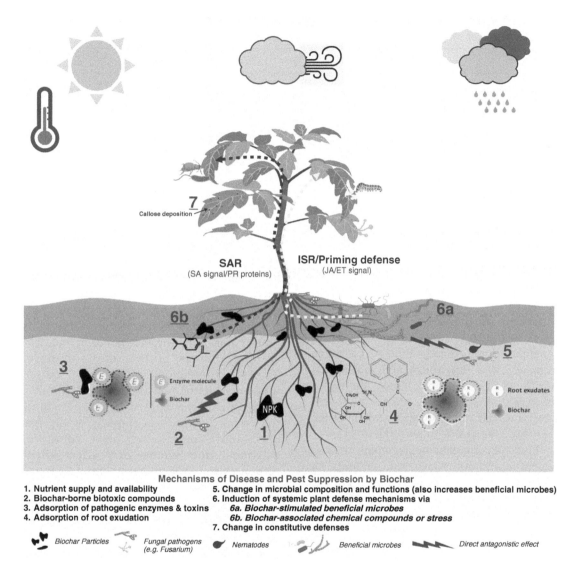

Figure 23.2 *Schematic representation of various mechanisms that may be responsible for biochar suppressiveness of plant diseases and pests. Environmental weather, light, and humidity conditions can also affect the interactions of pests with plants and hence, alter the biochar-plant effects. JA/ET refers to the jasmonic acid/ethylene defense pathway and SA/PR refers to the salicylic acid/pathogenesis-related proteins pathway. This figure was created with BioRender.com (license number GS25GGF9LX)*

Table 23.1 *Type of influence (direct and indirect) each control mechanism may have on soilborne vs. foliar pathogens or AMA pests*

Mechanism	Soilborne	Foliar
Changes in constitutive defenses or nutritious elements	Indirect	Indirect
Soil physiochemical properties	Direct and indirect	Indirect
Biochar-borne toxic compounds	Direct and indirect	Indirect
Adsorption of pathogenic enzymes, toxic metabolites, and signaling molecules	Direct and indirect	Indirect
Soil microbial community structure, diversity, and functioning	Direct and indirect	Indirect

Biochar and system complexity

The severity of diseases caused by plant pathogens traditionally has been considered a function of the interplay between the factors at the three vertices of the 'disease triangle': host susceptibility, pathogen virulence, and environmental conditions (Agrios, 2005). To this triangle, the rhizosphere microbiome can be added as a fourth factor, one that can impinge on and alter the other three factors (Graber et al, 2014).

The soil compartment, in which biochar is most commonly deployed, is itself an archetype of a complex system (Turner, 2021). Soil is comprised of tightly coupled physical, chemical, and biological processes that have all the characteristics of a complex system: (1) constantly changing, (2) tightly coupled, (3) governed by feedback, (4) nonlinear, (5) history-dependent, (6) self-organizing, (7) adaptive, (8) exhibit trade-offs, (9) counterintuitive, and (10) policy resistant (Turner, 2021).

Biochar, while arguably less complex than soil before it is deployed, soon takes on all the complex system characteristics of the host soil compartment. Being foreign to the soil, it may then profoundly influence the complex plant-pathogen-soil environment into which it is placed. By virtue of numerous physical and chemical properties such as nutrient content, water holding capacity, redox activity, adsorption, pH, toxic and hormone-like compound contents, color, and others, it can directly affect all three factors of the disease triangle, as well as indirectly affect them via its influence on the soil microbiome (Graber et al, 2014; Jaiswal et al, 2015). In turn, the direct and indirect impacts of biochar on the environment, host plant, pathogen, and soil microbiome can affect plant responses to biotic stresses caused by pests and pathogens.

In other words, biochar, particularly when used in soil systems, becomes part of an already complex and multi-faceted network of linked micro and macro compartments, where processes in one compartment feedback into others, most commonly in obscure and unknown ways. As a result, often a biochar concentration or a biochar type that benefits one process (e.g., plant growth) may be detrimental to another (e.g., disease resistance), and, therefore, it is quite challenging to elucidate the reason why, and to predict outcomes for any given new system.

Current and future directions: Improving biochar application with respect to plant disease

General considerations

Studies published during the last decade help shape our understanding of biochar impacts on plant disease, along with its potential limitations. Since biochar persists for a long time in soil, we should err on the side of caution when deploying it. We concentrate on two major elements of importance concerning plant disease:

Shifted dose-response curves between plant growth and disease resistance (shifted Rmax)

In specific pathosystems, it has been seen that the range of biochar doses having a positive influence on plant growth is wider relative to the positive effect dose range for disease suppression (Jaiswal et al, 2015). As a result, using high doses of biochar that promote plant growth in the absence of biotic stress pressures may result in increased susceptibility to pathogens or AMA pests when they appear (Figure 23.3). This can be very detrimental to plant performance and result in elevated susceptibility to pathogens (Jaiswal et al, 2014, 2015; Akhter et al, 2015; Frenkel et al, 2017; Waqas et al, 2018), referred to as a shifted Rmax (Jaiswal et al, 2015). From our analysis of the literature here, it appears that the shifted Rmax dilemma is greater when soilborne pathogens are concerned and less problematic for foliar pests.

Variability in biochar characteristics and performance

A major element of concern is the great variability in biochar characteristics and qualities, which leads to yet extra complex feedback in the already highly complicated soil/plant/pathogen system, and to variable performances with respect to plant disease (Jaiswal et al, 2014; Jaiswal et al, 2015; Qiang et al, 2018). Improving biochar consistency will enable safer and more efficient use on a large scale.

Improving biochar safety and utility

Several different approaches for addressing some of these concerns and extending the safe use potential of biochar are reviewed here.

Combining biochar with compost

The beneficial impact of biochar on soil physiochemical and microbiological properties can be combined with additional beneficial organic amendments such as compost to potentially achieve synergistic impacts on plant health and soil properties (Chapter 26).

Composts are known to sometimes exhibit suppressive potential towards a wide variety of diseases caused by soilborne pathogens (Bonanomi et al, 2007; Mehta et al, 2014). Due to biochar's relatively high surface area and porous structure, as well as its enhanced ability to retain water and nutrients, biochar combined with compost may result in synergies not just in terms of improved plant growth but also in terms of controlling plant disease. This suggestion has not yet borne fruit (Schulz and Glaser, 2012). Currently, the number of successful combinations of biochar and compost is not high, and compost alone is still more beneficial to disease suppression than its combination with biochar (Ebrahimi et al, 2016, 2021; Debode et al, 2020; Safaei Asadabadi

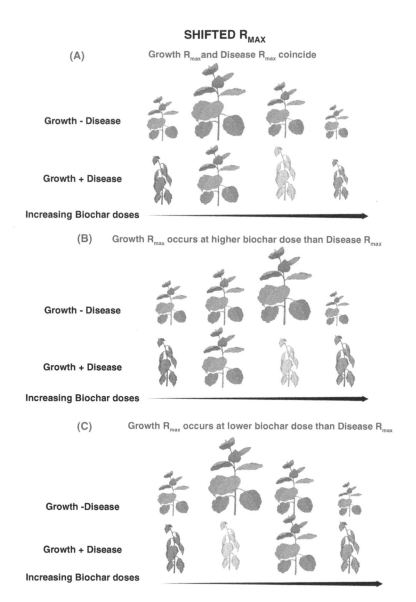

Figure 23.3 *A conceptual graph of the "Shifted Rmax-Effect", where Rmax refers to maximum growth response (G-Rmax) or disease reduction (D-Rmax). The figure shows the three main possibilities: (A) G-Rmax and D-Rmax coincide; (B) G-Rmax occurs at a higher biochar dose than D-Rmax, and (C) G-Rmax occurs at a lower biochar dose than D-Rmax. This figure was created with BioRender.com (license number UQ25GGCBQ8)*

et al, 2021). However, there are many knowledge gaps in the process, and sometimes the results depend on the biochar feedstock type (Akhter et al, 2015).

Inoculating biochar with beneficial organisms

Another approach is inoculating biochar or biochar-compost mixtures with plant growth-promoting rhizobacteria (Rasool et al, 2021), mycorrhizal fungi (Akhter et al, 2015; Arshad et al, 2021), and specific biological control agents (Ma et al, 2023; Postma and Nijhuis, 2019). No clear outcome to the efficacy of such inoculation exists to date (Postma and Nijhuis, 2019; Arshad et al, 2021; Jia et al, 2022; Doherty and Roberts, 2023). We suspect that adding biological control agents to biochar does not favor their adaptation to the soil-rhizosphere environment and that they are easily overwhelmed by the native, well-adapted populations.

One strategy to harness biological control agents is by manipulating the local population of microorganisms (Jin et al, 2022). This could have several advantages over the introduction of non-indigenous biocontrol agents, which are subject to well-known difficulties in the establishment, survival, reproduction, and colonization in the inoculation sites. Indigenous microorganisms are already present in the soil, adapted to the environment, and have a better chance of maintaining stable complex communities. Agricultural practices that support the recruitment and maintenance of beneficial microbial communities in the soil and rhizosphere may thus be more successful than inoculation of non-indigenous biocontrol agents for enhancing the efficiency and sustainability of intensive crop production, as illustrated in the next section.

Preconditioning of biochar amended growing media

Rhizosphere microbial enhancement could be promoted by biochar that is 'activated' by pre-wetting in the soil before planting (Jaiswal et al, 2018a, 2019), much in the same way that manure or compost is often used. This simple approach may result in synergies between biochar and soil moisture that lead to increased microbial abundance, diversity, and activity via pre-planting stimulation of microbial growth and activation of dormant populations of soil microorganisms. These changes may collectively enhance the efficacy of biochar for plant growth promotion and disease suppression, as well as reduce phytotoxicity.

Mineral-biochar complexes and biochar-based fertilizers

Another means of modifying biochar to improve its ease of use and increase its homogeneity involves granulizing it into complexes or fertilizers that have variably been termed "biochar-nanoparticle composites", "biochar compound fertilizer", "biochar-mineral complex (BMC)", or biochar-based fertilizers (Chapter 26) (Joseph et al, 2013; Qian et al, 2014; Ye et al, 2016). While specific methods and protocols to prepare such composite granules vary, the general idea is to combine biochar with a mixture of ground rocks, clays, other minerals, mineral acids, and even organic materials such as molasses or sludge, usually through heating at torrefaction temperatures (Chia et al, 2010, 2014). The idea is to mimic the biochar-mineral complexes found in Amazonian Dark Earths. Promising effects of biochar-mineral complexes on plant growth and microbial changes in the rhizosphere (Wang et al, 2019) should encourage exploration of their utility for suppressing plant disease, which so far has not been reported.

Uses in nurseries and soilless culture

There are still many knowledge gaps about the biochar effect concerning plant diseases. A convenient "pre-soil" stage, such as a nursery system, will encourage more extensive use of biochar in the future, and its benefits will be more easily quantified after gaining valuable experience in the nursery setting. The pyrolysis-biochar platform may also be an attractive and green solution for local nursery wastes. Nurseries demand very strict plant protection protocols since they need to avoid infected seedlings that will spread the pathogens into greenhouses and fields. As a result, adding a "green" defense layer such as biochar in the root plug could be an attractive option. A concentration of biochar at 0.5–3% w/w can induce resistance in the seedlings and encourage microbial diversity in the root plug, thus avoiding disease development and limiting an epidemic outbreak in the greenhouse (Jaiswal et al, 2019). When transferred into the soil, the seedling still carries a small amount of biochar which may provide the growing plants with the beneficial microbial community and other advantages of the biochar for plant growth. The length and intensity of this effect in the field against early-stage plant diseases is a fertile and interesting new area to investigate.

Consistency of biochar as a product for plant protection

One of the barriers to the widespread use of biochar in agriculture for plant protection is its heterogeneous nature: wide varieties of feedstocks are used, feedstock quality can be variable even for a given feedstock type, and different pyrolysis production methods are employed. Altogether, "biochar" can be an inconsistent product.

Such inconsistency is a problem for pest control. In several pathosystems, the disease control efficacy of different biochar types applied at equal concentrations was not the same (Jaiswal et al, 2014, 2015, 2016; Rogovska et al, 2017). Agricultural systems need products that have maximum consistency and homogeneity. In the challenging and complex system in which a farmer operates, yet more unknowns coming from the products in use are the last thing the farmer needs. At a bare minimum, a farmer should be confident that the products used in the soil are homogeneous and reproducible over time.

These considerations require the biochar industry to strive towards: (i) consistent feedstock type and quality characteristics; (ii) stable and reproducible pyrolysis methodology and parameterization; and (iii) consistent and controlled product storage until sale. In other words, biochar needs to be treated as an industrial product, with all the consequent ramifications: (i) a preference for large-scale production systems rather than small-scale production systems; and (ii) product development accompanied by pathosystem–specific research, to enable the industry to give product-specific recommendations for crop and soil types. This would work in a fashion similar to that of fertilizer and pest control companies.

Closing remarks and suggestions for future research

Studies clearly demonstrate that applying biochar to soil can be effective in decreasing plant disease severity and simultaneously increasing plant biomass. Yet, the use of biochar in soil can also have negative impacts on plant resistance to pests and diseases.

There is still much to learn and do in this field. Work on standardizing biochar with respect to feedstock, pyrolysis temperature, and application rate for improving its function in disease control is needed. The influence of biochar on biotic plant stress is a relatively understudied research topic, and thus far, fully 50% of all the studies have been done on plant pathogenic fungi in solanaceous crops. An indirect effect of biochar on plant disease via the soil microbiome is strongly supported by the literature, but there is little attention to understanding the governing characteristics of biochar that account for the changes in microbial communities. It should be noted that:

i   The dominant mode of action may significantly depend on the biochar feedstock, concentration, and pathosystem. For example, biochar produced from a nutrient- or ash-rich (e.g., N or Si) feedstock may work differently compared to one produced from a nutrient- or ash-poor feedstock.

ii The shifted dose-response curves between plant growth and disease resistance (shifted Rmax) appear to be sharper when soilborne pathogens are concerned and less problematic for foliar pests.

iii Several additional modes of action have not yet been studied. For example, no studies were identified that specifically addressed the impact of biochar on constitutive plant defenses of plants other than the plant nutrient content. Few studies have been published concerning the importance of the soil microbiome on biochar-elicited changes in plant responses to AMA pests, whereas this mechanism has been more widely studied for plant pathogens.

iv It is likely that an interplay between several mechanisms is involved in the biochar's effect on plant disease. Biochar and its systems' complexity need to be recognized.

v   There is a major knowledge gap regarding redox and electrical changes in the root zone due to biochar addition and its potential influence on plant root physiology and susceptibility to attack by disease-causing pathogens and pests.

Most of the studies use plant trials to elucidate the mode of action by measuring the correlation between biotic stress resistance and specific plant parameters such as nutrient content, hormones, gene expression, and the rhizosphere microbiome. However, these plant trials do not allow for easy differentiation between the different modes of action involved. Moreover, due to the time and labor intensiveness of these trials, only a limited number of biochars having a large set of differences can be compared with each other. We propose two strategies to address this knowledge gap in the future. First, fast screening tests need to be developed to enable the screening of a high number of biochars as a disease or pest-suppressing agent. Such screening tests have been developed for other applications such as N adsorption (Viaene et al, 2023) or germination (Rogovska et al, 2012), but to the best of our knowledge, no such tests exist for resistance against biotic stress. Second, research is needed to identify which biochar characteristics (e.g., chemical and structural) and soil characteristics (e.g., chemical, structural; mineral soil versus growing media) may predict the disease-suppressive potential of the biochar. Again, such studies have been published for other applications, such as the potential to improve soil physical and chemical properties (Brewer et al, 2011) but not for disease or pest suppression. Once the mechanisms governing biochar as a disease and pest suppressive agent are understood with greater precision, the work on improving biochar by preconditioning, combining with compost, microbial inoculants, minerals, or other technologies will progress more rapidly.

Supplementary Table 23.1 at www.routledge.com/9781032286150.

References

Abujabhah IS, Bound SA, Doyle R, and Bowman JP 2016 Effects of biochar and compost amendments on soil physico-chemical properties and the total community within a temperate agricultural soil. *Applied Soil Ecology* 98, 243–253.

Agrios GN 2005 *Plant Pathology*. Amsterdam: Elsevier.

Akhter A, Hage-Ahmed K, Soja G, Steinkellner S 2015 Compost and biochar alter mycorrhization, tomato root exudation, and development of *Fusarium oxysporum* f. sp *lycopersici. Frontiers in Plant Science* 6, 529.

Akhter A, Hage-Ahmed K, Soja G, and Steinkellner S 2016 Potential of Fusarium wilt-inducing chlamydospores, in vitro behaviour in root exudates and physiology of tomato in biochar and compost amended soil. *Plant and Soil* 406, 425–440.

Amery F, et al 2021 Biochar for circular horticulture: Feedstock related effects in soilless cultivation. *Agronomy* 11, 629.

Arshad U, et al 2021 Combined application of biochar and biocontrol agents enhances plant growth and activates resistance against Meloidogyne incognita in tomato. *Gesunde Pflanzen* 73, 591–601.

Arshad U, Naveed M, Javed N, Gogi MD, and Ali MA 2020 Biochar application from different feedstocks enhances plant growth and resistance against *Meloidogyne incognita* in tomato. *International Journal of Agriculture and Biology* 24, 961–968.

Ayres PG 1984 The interaction between environmental stress injury and biotic disease physiology. *Annual Review of Phytopathology* 22, 53–75.

Baker R and Martinson C 1970 Epidemiology of diseases caused by *Rhizoctonia solani*. In: Parmeter JR (ed) *Rhizoctonia solani, Biology and Pathology*. Berkeley: University of California Press. pp 172–188.

Bakhat HF, et al 2021 Rice husk bio-char improves brinjal growth, decreases insect infestation by enhancing silicon uptake. *Silicon* 13, 3351–3360.

Bateman D 1970 Pathogenesis and disease. In: Parmeter JR (ed) *Rhizoctonia solani, Biology and Pathology*. Berkeley: University of California Press. pp 161–171.

Bell T, Newman JA, Silverman BW, Turner SL, and Lilley AK 2005 The contribution of species richness and composition to bacterial services. *Nature* 436, 1157–1160.

Bonanomi G, Antignani V, Pane C, and Scala E 2007 Suppression of soilborne fungal diseases with organic amendments. *Journal of Plant Pathology* 89, 311–324.

Bonanomi G, Ippolito F, and Scala F 2015 A "black" future for plant pathology? Biochar as a new soil amendment for controlling plant diseases. *Journal of Plant Pathology* 97, 223–234.

Brewer CE, Unger R, Schmidt-Rohr K, and Brown RC 2011 Criteria to select biochars for field studies based on biochar chemical properties. *Bioenergy Research* 4, 312–323.

Cao Y, Gao Y, Qi Y, and Li J 2018 Biochar-enhanced composts reduce the potential leaching of nutrients and heavy metals and suppress plant-parasitic nematodes in excessively fertilized cucumber soils. *Environmental Science and Pollution Research* 25, 7589–7599.

Chen C, Su J, Ali A, and Zhai Z 2022 Cornstalk biochar promoted the denitrification performance and cellulose degradation rate of *Burkholderia* sp. CF6. *Journal of Environmental Chemical Engineering* 10, 106998.

Chen Y, Li R, Li B, and Meng L 2019a Biochar applications decrease reproductive potential of the English grain aphid Sitobion avenae and upregulate defense-related gene expression. *Pest Management Science* 75, 1310–1316.

Chen Y, Shen Y, Meng L, and Li B 2019b The effect of biochar with different feedstock materials on the English grain aphid *Sitobion avenae* Fab.(Hemiptera: Aphididae). *Crop Protection* 124, 104859.

Chia CH, Munroe P, Joseph S, and Lin Y 2010 Microscopic characterisation of synthetic Terra Preta. *Soil Research* 48, 593–605.

Chia CH, Singh BP, Joseph S, Graber ER, and Munroe P 2014 Characterization of an enriched biochar. *Journal of Analytical and Applied Pyrolysis* 108, 26–34.

Chintala R, Mollinedo J, Schumacher TE, Malo DD, and Julson JL 2014 Effect of biochar on chemical properties of acidic soil. *Archives of Agronomy and Soil Science* 60, 393–404.

Copley T, Bayen S, and Jabaji S 2017 Biochar amendment modifies expression of soybean and Rhizoctonia solani genes leading to increased severity of *Rhizoctonia* foliar blight. *Frontiers in Plant Science* 8, 221.

Copley TR, Aliferis KA, and Jabaji S 2015 Maple bark biochar affects *Rhizoctonia solani* metabolism and increases damping-off severity. *Phytopathology* 105, 1334–1346.

Daoud FB-O, Kaddour S, and Sadoun T 2010 Adsorption of cellulase *Aspergillus niger* on a commercial activated carbon: kinetics and equilibrium studies. *Colloids and Surfaces B: Biointerfaces* 75, 93–99.

De Tender C, et al 2021 Biochar-enhanced resistance to *Botrytis cinerea* in strawberry fruits (but not leaves) is associated with changes in the rhizosphere microbiome. *Frontiers in Plant Science* 12, 700479.

De Tender CA, et al 2016 Biological, physicochemical and plant health responses in lettuce and strawberry in soil or peat amended with biochar. *Applied Soil Ecology* 107, 1–12.

Debode J, et al 2020 Has compost with biochar added during the process added value over biochar or compost to increase disease suppression? *Applied Soil Ecology* 153, 103571.

Delgado-Baquerizo M, et al 2016 Microbial diversity drives multifunctionality in terrestrial ecosystems. *Nature Communications* 7, 10541.

Doherty J and Roberts J 2023 Topdressing biochar compost mixtures and biological control organism applications suppress foliar pathogens in creeping bentgrass fairway turf. *Plant Disease*, published online.

Dror B, Amutuhaire H, Frenkel O, Jurkevitch E, and Cytryn E 2022 Identification of bacterial populations and functional mechanisms potentially involved in biochar-facilitated antagonism of the soilborne pathogen Fusarium oxysporum. *Phytobiomes Journal* 6, 139–150.

Durrant WE and Dong X 2004 Systemic acquired resistance. *Annual Review of Phytopathology* 42, 185–209.

Ebrahimi M, Mousavi A, Souri MK, and Sahebani N 2021 Can vermicompost and biochar control Meloidogyne javanica on eggplant? *Nematology* 23, 1053–1064.

Ebrahimi N, et al 2016 Traditional and new soil amendments reduce survival and reproduction of potato cyst nematodes, except for biochar. *Applied Soil Ecology* 107, 191–204.

Eizenberg H, Plakhine D, Ziadne H, Tsechansky L, and Graber ER 2017 Non-chemical control of root parasitic weeds with biochar. *Frontiers in Plant Science* 8, 939.

Elad Y, et al 2010 Induction of systemic resistance in plants by biochar, a soil-applied carbon sequestering agent. *Phytopathology* 100, 913–921.

Evidente A, Cimmino A, and Masi M 2019 Phytotoxins produced by pathogenic fungi of agrarian plants. *Phytochemistry Reviews* 18, 843–870.

Faix O, Meier D, and Fortmann I 1990 Thermal degradation products of wood. *Holz als Roh- und Werkstoff* 48, 281–285.

Frenkel O, et al 2017 The effect of biochar on plant diseases: What should we learn while designing biochar substrates? *Journal of*

Environmental Engineering and Landscape Management 25, 105–113.

Fry WE and Grünwald NJ 2010 Introduction to oomycetes. *The Plant Health Instructor*, 10. 1094/PHI-I-2010-1207-01.

Fu ZQ, and Dong X 2013 Systemic acquired resistance: Turning local infection into global defense. *Annual Review of Plant Biology* 64, 839–863.

Gao Y, Lu Y, Lin W, Tian J, and Cai K 2019 Biochar suppresses bacterial wilt of tomato by improving soil chemical properties and shifting soil microbial community. *Microorganisms* 7, 676.

Garbeva P, van Veen JA, and van Elsas JD 2004 Microbial diversity in soil: Selection of microbial populations by plant and soil type and implications for disease suppressiveness. *Annual Review of Phytopathology* 42, 243–270.

George C, Kohler J, and Rillig MC 2016 Biochars reduce infection rates of the root-lesion nematode Pratylenchus penetrans and associated biomass loss in carrot. *Soil Biology and Biochemistry* 95, 11–18.

Graber E, Frenkel O, Jaiswal AK, and Elad Y 2014 How may biochar influence severity of diseases caused by soilborne pathogens? *Carbon Management* 5, 169–183.

Graber E, Tsechansky L, Khanukov J, and Oka Y 2011 Sorption, volatilization, and efficacy of the fumigant 1, 3-dichloropropene in a biochar-amended soil. *Soil Science Society of America Journal* 75, 1365–1373.

Graber ER and Kookana R 2015 Biochar and retention/efficacy of pesticides. In: Lehmann J, and Joseph S (Eds) *Biochar for Environmental Management: Science and Technology*. London: Earthscan Books. pp 655–678.

Graber ER, et al 2010 Biochar impact on development and productivity of pepper and tomato grown in fertigated soilless media. *Plant and Soil* 337, 481–496.

Gu Y, et al 2017 Application of biochar reduces Ralstonia solanacearum infection via effects on pathogen chemotaxis, swarming motility, and root exudate adsorption. *Plant and Soil* 415, 269–281.

Hanley ME, Lamont BB, Fairbanks MM, and Rafferty CM 2007 Plant structural traits and their role in anti-herbivore defence. *Perspectives in Plant Ecology, Evolution and Systematics* 8, 157–178.

Hou X, Meng L, Li L, Pan G, and Li B 2015 Biochar amendment to soils impairs developmental and reproductive performances of a major rice pest *Nilaparvata lugens* (Homopera: Delphacidae). *Journal of Applied Entomology* 139, 727–733.

Huang W-K, Ji H-L, Gheysen G, Debode J, and Kyndt T 2015 Biochar-amended potting medium reduces the susceptibility of rice to root-knot nematode infections. *BMC Plant Biology* 15, 1–15.

Husson O 2013 Redox potential (Eh) and pH as drivers of soil/plant/microorganism systems: a transdisciplinary overview pointing to integrative opportunities for agronomy. *Plant and Soil* 362, 389–417.

Iacomino G, Idbella M, Laudonia S, Vinale F, and Bonanomi G 2022 The suppressive effects of biochar on above-and belowground plant pathogens and pests: A review. *Plants* 11, 3144.

Ikwunagu E, Ononuju C, and Orikara C 2019 Nematicidal effects of different biochar sources on root-knot nematode (*Meloidogyne* spp.) egg hatchability and control on mungbean (*Vigna radiata* (L.) Wilczek). *International Journal of Entomology and Nematology Research* 4, 1–14.

Jaiswal A, Elad Y, Cytryn E, Graber ER, and Frenkel O 2018a Activating biochar by manipulating the bacterial and fungal microbiome through pre-conditioning. *New Phytologist* 219, 363–377.

Jaiswal AK, et al 2020 Molecular insights into biochar-mediated plant growth promotion and systemic resistance in tomato against Fusarium crown and root rot disease. *Scientific Reports* 10, 1–15.

Jaiswal AK, Elad Y, Graber ER, Cytryn E, and Frenkel O 2016 Soilborne disease suppression and plant growth promotion by biochar soil amendments and possible mode of action. *Acta Horticulturae* 1207, 69–76.

Jaiswal AK, Elad Y, Graber ER, and Frenkel O 2014 *Rhizoctonia solani* suppression and plant growth promotion in cucumber as affected by biochar pyrolysis temperature, feedstock and concentration. *Soil Biology and Biochemistry* 69, 110–118.

Jaiswal AK, et al 2017 Linking the belowground microbial composition, diversity and activity to soilborne disease suppression and growth promotion of tomato amended with biochar. *Scientific Reports* 7, 44382.

Jaiswal AK, Frenkel O, Elad Y, Lew B, and Graber ER 2015 Non-monotonic influence of biochar dose on bean seedling growth and susceptibility to *Rhizoctonia solani*: The "Shifted Rmax-Effect". *Plant and Soil* 395, 125–140.

Jaiswal AK, Frenkel O, Tsechansky L, Elad Y, and Graber ER 2018b Immobilization and deactivation of pathogenic enzymes and toxic metabolites by biochar: A possible mechanism involved in soilborne disease suppression. *Soil Biology and Biochemistry* 121, 59–66.

Jaiswal AK, Graber ER, Elad Y, and Frenkel O 2019 Biochar as a management tool for soilborne diseases affecting early stage nursery seedling production. *Crop Protection* 120, 34–42.

Jia H, et al 2022 Control efficiency of biochar loaded with *Bacillus subtilis* Tpb55 against tobacco black shank. *Processes* 10, 2663.

Jin X, et al 2022 Biochar stimulates tomato roots to recruit a bacterial assemblage contributing to disease resistance against Fusarium wilt. *iMeta* 1, e37.

Joseph S, et al 2013 Shifting paradigms: Development of high-efficiency biochar fertilizers based on nano-structures and soluble components. *Carbon Management* 4, 323–343.

Klarzynski O, et al 2000 Linear beta-1,3 glucans are elicitors of defense responses in tobacco. *Plant Physiology* 124, 1027–1037.

Klinke HB, Thomsen A, and Ahring BK 2004 Inhibition of ethanol-producing yeast and bacteria by degradation products produced during pre-treatment of biomass. *Applied Microbiology and Biotechnology* 66, 10–26.

Lacomino G, Idbella M, Laudonia S, Vinale F, and Bonanomi G 2022 The suppressive effects of biochar on above-and belowground plant pathogens and pests: A review. *Plants* 11, 3144.

Lammirato C, Miltner A, and Kaestner M 2011 Effects of wood char and activated carbon on the hydrolysis of cellobiose by β-glucosidase from *Aspergillus niger*. *Soil Biology and Biochemistry* 43, 1936–1942.

Lecompte F, Abro MA, and Nicot PC 2010 Contrasted responses of *Botrytis cinerea* isolates developing on tomato plants grown under different nitrogen nutrition regimes. *Plant Pathology* 59, 891–899.

Li T, et al 2022 Biochar inhibits ginseng root rot pathogens and increases soil microbiome diversity. *Applied Soil Ecology* 169, 104229.

Lingens A, Windeisen E, and Wegener G 2005 Investigating the combustion behaviour of various wood species via their fire gases. *Wood Science and Technology* 39, 49–60.

Lu Z, Yu X, Heong K, Hu C 2005 Effects of nitrogen content in rice plants and *Nilaparvata lugens* Stal. *Acta Ecologica Sinica* 8, 1838–1843.

Ma M, Taylor PW, Chen D, Vaghefi N, and He J-Z 2023 Major soilborne pathogens of field processing tomatoes and management strategies. *Microorganisms* 11, 263.

Masiello CA, et al 2013 Biochar and microbial signaling: Production conditions determine effects on microbial communication. *Environmental Science and Technology* 47, 11496–11503.

Matsubara Y, Hasegawa N, and Fukui H 2002 Incidence of Fusarium root rot in asparagus seedlings infected with arbuscular mycorrhizal fungus as affected by several soil amendments. *Journal of the Japanese Society for Horticultural Science* 71, 370–374.

McDonald JD, et al 2000 Fine particle and gaseous emission rates from residential wood combustion. *Environmental Science and Technology* 34, 2080–2091.

Mehari ZH, Elad Y, Rav-David D, Graber ER and Harel YM 2015 Induced systemic resistance in tomato (*Solanum lycopersicum*) against *Botrytis cinerea* by biochar amendment involves jasmonic acid signaling. *Plant and Soil* 395, 31–44.

Mehta CM, Palni U, Franke-Whittle I, and Sharma A 2014 Compost: Its role, mechanism and impact on reducing soil-borne plant diseases. *Waste Management* 34, 607–622.

Meller Harel Y, et al 2012 Biochar mediates systemic response of strawberry to foliar fungal pathogens. *Plant and Soil* 357, 245–257.

Mendes R, Garbeva P, and Raaijmakers JM 2013 The rhizosphere microbiome: Significance of plant beneficial, plant pathogenic, and human pathogenic microorganisms. *FEMS Microbiology Reviews* 37, 634–663.

Mendes R, et al 2011 Deciphering the rhizosphere microbiome for disease-suppressive bacteria. *Science* 332, 1097–1100.

Mondal S, Ghosh S, and Mukherjee A 2021 Application of biochar and vermicompost against the rice root-knot nematode (*Meloidogyne graminicola*): An eco-friendly approach in nematode management. *Journal of Plant Diseases and Protection* 128, 819–829.

Mortensen B 2013 Plant resistance against herbivory. *Nature Education Knowledge* 4, 5.

Mu J, Yu ZM, Wu WQ, and Wu QL 2006 Preliminary study of application effect of bamboo vinegar on vegetable growth. *Forestry Studies in China* 8, 43–47.

Mun SP, and Ku CS 2010 Pyrolysis GC-MS analysis of tars formed during the aging of wood and bamboo crude vinegars. *Journal of Wood Science* 56, 47–52.

Nam M, Jeong S, Lee Y, Choi J, and Kim H 2006 Effects of nitrogen, phosphorus, potassium and calcium nutrition on strawberry anthracnose. *Plant Pathology* 55, 246–249.

Nerome M, et al 2005 Suppression of bacterial wilt of tomato by incorporation of municipal biowaste charcoal into soil. *Soil Microorganisms* 59, 9–14.

Ogawa M and Okimori Y 2010 Pioneering works in biochar research, Japan. *Soil Research* 48, 489–500.

Osbourn AE 1996 Preformed antimicrobial compounds and plant defense against fungal attack. *The Plant Cell* 8, 1821.

Pieterse CMJ, et al 2014 Induced systemic resistance by beneficial microbes. *Annual Review of Phytopathology* 52, 347–375.

Postma J and Nijhuis EH 2019 Pseudomonas chlororaphis and organic amendments controlling Pythium infection in tomato. *European Journal of Plant Pathology* 154, 91–107.

Qian L, et al 2014 Biochar compound fertilizer as an option to reach high productivity but low carbon intensity in rice agriculture of China. *Carbon Management* 5, 145–154.

Qiang F, Baoping L, and Ling M 2018 Effects of biochar amendment to soil on life historytraits of Laodelphax striatellus (Hemiptera: Delphacidae) on rice plants. *Chinese Journal of Rice Science* 1, 200.

Raaijmakers JM and Mazzola M 2016 Soil immune responses soil microbiomes may be harnessed for plant health. *Science* 352, 1392–1393.

Rahayu DS and Sari NP 2017 Development of Pratylenchus coffeae in biochar applied soil, coffee roots and its effect on plant growth. *Pelita Perkebunan* 33, 24–32.

Rahman L, Whitelaw-Weckert M, and Orchard B 2014 Impact of organic soil amendments, including poultry-litter biochar, on nematodes in a Riverina, New South Wales, vineyard. *Soil Research* 52, 604–619.

Rani A, Das M, and Satyanarayana S 2000 Preparation and characterization of amyloglucosidase adsorbed on activated charcoal. *Journal of Molecular Catalysis B: Enzymatic* 10, 471–476.

Rasool M, Akhter A, Soja G, and Haider MS 2021 Role of biochar, compost and plant growth promoting rhizobacteria in the management of tomato early blight disease. *Scientific Reports* 11, 6092.

Robert-Seilaniantz A, Grant M, and Jones JD 2011 Hormone crosstalk in plant disease and defense: More than just jasmonate-salicylate antagonism. *Annual Review of Phytopathology* 49, 317–343.

Rogovska N, Laird D, Cruse R, Trabue S, and Heaton E 2012 Germination tests for assessing biochar quality. *Journal of Environmental Quality* 41, 1014–1022.

Rogovska N, Laird D, Leandro L, and Aller D 2017 Biochar effect on severity of soybean root disease caused by Fusarium virguliforme. *Plant and Soil* 413, 111–126.

Safaei Asadabadi R, Hage-Ahmed K, and Steinkellner S 2021 Biochar, compost and arbuscular mycorrhizal fungi: A tripartite approach to combat *Sclerotinia sclerotiorum* in soybean. *Journal of Plant Diseases and Protection* 128, 1433–1445.

Savary S and Willocquet L 2020 Modeling the impact of crop diseases on global food security. *Annual Review of Phytopathology* 58, 313–341.

Schulz H and Glaser B 2012 Effects of biochar compared to organic and inorganic fertilizers on soil quality and plant growth in a greenhouse experiment. *Journal of Plant Nutrition and Soil Science* 175, 410–422.

Shigenaga AM and Argueso CT 2016 No hormone to rule them all: Interactions of plant hormones during the responses of plants to pathogens. *Seminars in Cell and Developmental Biology* 56, 174–189.

Shoresh M, Harman GE, and Mastouri F 2010 Induced systemic resistance and plant responses to fungal biocontrol agents. *Annual Review of Phytopathology* 48, 21–43.

Sikes BA, et al 2018 Import volumes and biosecurity interventions shape the arrival rate of fungal pathogens. *PLoS Biology* 16, e2006025.

Soulie MC, et al 2020 Plant nitrogen supply affects the *Botrytis cinerea* infection process and modulates known and novel virulence factors. *Molecular Plant Pathology* 21, 1436–1450.

Sticher L, Mauch-Mani B, and Métraux, JP 1997 Systemic acquired resistance. *Annual Review of Phytopathology* 35, 235–270.

Turner BL 2021 Soil as an archetype of complexity: A systems approach to improve insights, learning, and management of coupled biogeochemical processes and environmental externalities. *Soil Systems* 5, 39.

Vallad GE and Goodman RM 2004 Systemic acquired resistance and induced systemic resistance in conventional agriculture. *Crop Science* 44, 1920–1934.

Van der Heijden MGA, Bardgett RD, and van Straalen NM 2008 The unseen majority: Soil microbes as drivers of plant diversity and productivity in terrestrial ecosystems. *Ecology Letters* 11, 296–310.

Van Loon L, Bakker P, and Pieterse C 1998 Systemic resistance induced by rhizosphere bacteria. *Annual Review of Phytopathology* 36, 453–483.

Verheijen FG, et al 2013 Reductions in soil surface albedo as a function of biochar application rate: implications for global radiative forcing. *Environmental Research Letters* 8, 044008.

Viaene J, et al 2023 Screening tests for N sorption allow to select and engineer biochars for N mitigation during biomass processing. *Waste Management* 155, 230–239.

Viger M, Hancock RD, Miglietta F, and Taylor G 2014 More plant growth but less plant defence? First global gene expression data for plants grown in soil amended with biochar. *Global Change Biology Bioenergy* 7, 658–672.

Vlot AC, Dempsey DMA, and Klessig DF 2009 Salicylic acid, a multifaceted hormone to combat disease. *Annual Review of Phytopathology* 47, 177–206.

Wagg C, Bender SF, Widmer F, and van der Heijden MGA 2014 Soil biodiversity and soil community composition determine ecosystem multifunctionality. *Proceedings of the National Academy of Sciences* 111, 5266–5270.

Wall P, Neate S, Graham R, Reuter D, and Rovira A 1994 The effect of rhizoctonia root disease and applied nitrogen on growth, nitrogen uptake and nutrient concentrations in spring wheat. *Plant and Soil* 163, 111–120.

Wan J, Zhang X-C, and Stacey G 2008 Chitin signaling and plant disease resistance. *Plant Signaling and Behavior* 3, 831–833.

Wang M, et al 2019 Effect of alkali-enhanced biochar on silicon uptake and suppression of gray leaf spot development in perennial ryegrass. *Crop Protection* 119, 9–16.

Waqas M, et al 2018 Biochar amendment changes jasmonic acid levels in two rice varieties and alters their resistance to herbivory. *PLoS One* 13, e0191296.

Wiese J, Kranz T, and Schubert S 2004 Induction of pathogen resistance in barley by abiotic stress. *Plant Biology* 6, 529–536.

Xu X, Robinson J, and Else MA 2013 Effects of nitrogen input and deficit irrigation within the commercial acceptable range on susceptibility of strawberry leaves to powdery mildew. *European Journal of Plant Pathology* 135, 695–701.

Yang Y, Chen T, Xiao R, Chen X, and Zhang T 2022 A quantitative evaluation of the biochar's influence on plant disease suppress: a global meta-analysis. *Biochar* 4, 43.

Ye J, et al 2016 A combination of biochar–mineral complexes and compost improves soil bacterial processes, soil quality, and plant properties. *Frontiers in Microbiology* 7, 372.

Yermiyahu U, Shamai I, Peleg R, Dudai N, and Shtienberg D 2006 Reduction of *Botrytis cinerea* sporulation in sweet basil by altering the concentrations of nitrogen and calcium in the irrigation solution. *Plant Pathology* 55, 544–552.

Yuan J-H and Xu R-K 2012 Effects of biochars generated from crop residues on chemical properties of acid soils from tropical and subtropical China. *Soil Research* 50, 570–578.

Zhang J and You C 2013 Water holding capacity and absorption properties of wood chars. *Energy and Fuels* 27, 2643–2648.

Zhang Q, et al 2013 Effects of biochar amendment on soil thermal conductivity, reflectance, and temperature. *Soil Science Society of America Journal* 77, 1478–1487.

24

Test procedures for biochar analysis in soils

Michael Bird

Introduction

Biochar is a product of the heating of organic matter under conditions of restricted oxygen. Biochar is a pyrogenic carbonaceous material (PCM) that can exhibit a wide range of pyrogenic carbon (PyC) contents (~20–95%; McBeath et al, 2015), as well as inorganic ash constituents and, in some cases, partially or completely unpyrolysed organic matter. As a result, biochars can vary widely in their chemical composition and stability under natural environmental conditions (Chapters 5–11). The chemical characteristics of biochars therefore overlap with all other materials across the PCM continuum that are variously known as pyrogenic organic matter, char, charcoal, black carbon, soot, microcrystalline graphite, and elemental carbon (Schmidt and Noack, 2000; Zimmerman and Mitra, 2017; see Chapters 1 and 7).

PCM occurs in soils where biomass burning is a natural feature of the ecosystem (savannas, grasslands, and some forests), and indeed can make up more than 50% of total soil organic carbon (SOC) in some soils (e.g., Lehmann et al, 2008; Figure 24.1). PCM can also be abundant in soils where there have been significant historic (Schmidt et al, 2000), or prehistoric (Glaser and Birk, 2012), anthropogenic inputs of PCM to the soil. PCM in a modern soil may therefore have been present in a soil for centuries to millennia. Over time this PCM becomes comminuted into the finer particle sizes (Skjemstad et al, 1999; Nocentini et al, 2010; De la Rosa et al, 2018) and can undergo significant chemical alteration, particularly on particle surfaces (Kaal et al, 2008a; Ascough et al, 2011; De la Rosa et al, 2018). Therefore, the prior existence of PCM in many soils complicates the measurement and isolation of biochar added to soil, because the PCM may or may not be similar, both physically and chemically, to biochar purposefully added to the soil.

Reasons to measure biochar in soil include the verification of biochar carbon

DOI: 10.4324/9781003297673-24

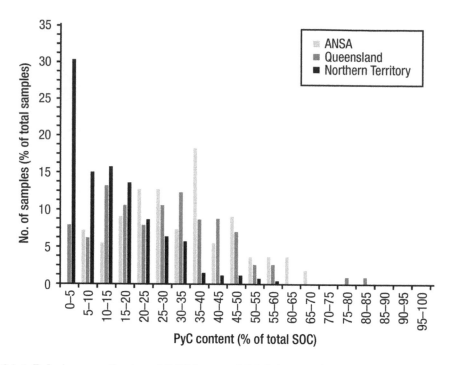

Figure 24.1 *Relative contribution of PCM to total SOC in 452 Australian soils from Queensland (Qld, the Northern Territory (NT), and the Australian National Soil Archive (ANSA), modified from Lehmann et al (2008)*

(C) stocks for C sequestration and C trading as well as to study the dynamics of biochar stocks and fluxes. Routine verification of biochar C stocks is best achieved through analysis of the biochar before addition to the soil and knowledge of application rate (Chapters 11 and 30). Part of the process of verification may require an assessment of the proportion of C that will persist in the soil on at least centennial timescales (Budai et al, 2013), depending on the C trading scheme. This in turn requires field, as well as laboratory, assessment of biochar persistence to predict, rather than monitor, biochar leaching and/or remineralization over time (e.g., Bird et al, 2000; Nguyen et al, 2008;

Vasilyeva et al, 2011; De la Rosa et al, 2018). In addition, understanding the dynamics of, and interactions within, the biochar-plant-soil system under a range of soil, climate, and land management practices commonly requires the quantification of biochar stocks over time and the isolation of biochar for further analysis and characterization (e.g., Major et al, 2010a, 2010b; Liu et al, 2019; Zheng et al, 2022). The major reason for measuring PCM, including biochar PCM in soil is therefore to provide the fundamental research base required to underpin improvements in the application of biochar systems to climate change mitigation and the enhancement of soil fertility and resilience.

Quantification, isolation, or characterization?

Many techniques are now available to quantify, isolate, and characterize PCM, including biochar PCM in soil, usually by measuring the PyC component of PCM. Most techniques were developed outside soil science – in chemistry, materials science, geosciences, and atmospheric sciences. These have been adapted for soil PCM analysis in response to the need to better understand the occurrence and dynamics of PCM in the environment and, more recently, the impact, behavior, and interactions of biochar PCM in soils.

The techniques can be grouped into five major classes. *Physical techniques* are largely non-destructive and rely on a simple difference in density or size as the basis for separating PCM from other soil components. *Chemical oxidation techniques* are destructive and rely on the greater resistance of some components of the PCM continuum to an oxidant than other components of SOC in a sample. *Thermal techniques* are also destructive and rely on the greater resistance of some components of the PCM continuum to decomposition at elevated temperatures relative to other SOC components in a sample.

Spectroscopic techniques are non-destructive. They rely on stimulating a sample with a magnetic field, infrared, or x-ray radiation and measuring a magnetic or photon response to the stimulation in the sample. This can be used to infer the nature and abundance of chemical bonds in a sample, including those that are characteristic of PCM. *Molecular marker techniques* are destructive, decomposing a sample chemically and/or thermally to then measure the abundance of the multiple compounds liberated by decomposition that are known to have derived from PCM.

The techniques range widely in terms of the components of the PCM continuum that are measured, the cost and accessibility of instrumentation, and the level of chemical detail on the nature of the PCM component that is quantified (Table 24.1). No method is 'ideal', and while most methods will produce a generally precise and reproducible analytical result, it is expected that the concentration of PCM reported for the same sample will vary widely between different techniques, and, in some cases, unfortunately between laboratories using the same technique (e.g.,

Table 24.1 *Comparison of PCM techniques. Note: I = Isolation, Q = quantification, C = chemical composition, H = high, M = medium, L = low*

Technique	Advantages	Limitations	IQC	Access	Cost
Physical	Background				
Flotation	• easy and cheap to implement • minimal equipment required	• cannot easily quantify fine material • can be time-consuming	I(Q)	H	L
Density separation	• easy to implement • minimal equipment required	• cannot easily quantify fine material • can introduce contamination • usually requires additional physical or chemical separation steps	I(Q)	H	L

Table 24.1 *continued*

Technique	Advantages	Limitations	IQC	Access	Cost
Chemical					
Dichromate oxidation	• widely used and tested on standards • isolates and quantifies	• requires demineralization • multiple handling steps • no agreed protocol • does not eliminate hydrophobic compounds	IQ	M	M
UV oxidation	• widely tested on standards • has been the basis for calibration of other techniques (e.g., MIR)	• requires specialist equipment • requires demineralization • multiple handling steps • quantification relies on NMR	(I)Q	L	M
Peroxide oxidation	• can provide robust results but only demonstrated on other sample types	• no agreed protocol • untested on soils	IQ	M	M
Chlorite oxidation	• tested on standards • easy to implement • isolates and quantifies	• no agreed protocol • multiple handling steps • does not eliminate hydrophobic compounds?	IQ	M	M
RTO oxidation	• oxidation occurs at room temperature	• untested on soils	I?Q?	M	M
Nitric acid oxidation	• simple and rapid • equipment widely available	• poor isolation of PCM from OC • untested on soils	IQ	M	M
Thermal					
Loss on ignition	• easy and cheap to implement • minimal equipment required	• may require site-specific calibration • assumes no change in SOC	Q	H	L
Chemo-thermal oxidation	• thoroughly tested on standards • simple to implement • tight control over process conditions • quantifies a well-defined component of highly condensed PCM • isolates and quantifies	• issues relating to standardization of oxygen flow/furnace conditions • can involve multiple handling steps • under/over-estimation possible due to matrix effects/charring • not sensitive to less condensed PCM	(I)Q	M	M
TGA-DSC	• straightforward rapid analysis	• interpretation requires specialist knowledge	Q	M	M

Table 24.1 *continued*

Technique	Advantages	Limitations	IQC	Access	Cost
Thermal (continued)					
	• instrumentation widely available • minimal handling required	• incomplete separation of OC and PCM for some samples			
TOT/R	• thoroughly tested on aerosol samples • isolates and quantifies	• requires demineralization • requires specialist equipment and knowledge of incomplete separation of OC and PCM for some samples	Q	L	M
Hydrogen pyrolysis	• quantifies a well-defined component of PCM • tested against standards • minimal handling required • isolates and quantifies	• possibility of trace charring • instrumentation rare but commercially available • not sensitive to less condensed PCM	IQC	L	M
Rock-Eval	• widely used technique • minimal handling required	• does not always result in complete separation of PCM and OC	Q	H	L
MESTA	• also provides information on N and H in PCM • minimal handling required	• instrumentation rare but commercially available • requires specialist knowledge • does not always provide separation between PCM and OC	QC	L	H
ThG	• minimal handling required • instrumentation widely available	• performs poorly on soil	Q	L	M
Spectroscopic					
NMR	• extensive testing on standards • detailed information on organic components including PCM	• requires demineralization • requires specialist knowledge and equipment	QC	L	H
Infrared	• minimal handling required • instrumentation widely available • non-destructive	• relies on calibration using other techniques • site-specific calibration? • interpretation requires specialist knowledge	QC	M	L

Table 24.1 *continued*

Technique	Advantages	Limitations	IQC	Access	Cost
Spectroscopic					*(continued)*
PFL	• non-destructive	• performs poorly on soil	(Q)	M	L
NEXAFS	• detailed chemical finger-printing possible • absorption spectra dependent only on the local bonding environment • non-destructive	• limited research base to date • interpretation requires specialist knowledge	(Q)C	L	H
Molecular marker					
BPCA	• widely tested on standards • agreed protocol • isolates and quantifies • provides a detailed chemical characterization of PCM	• multiple handling steps • requires specialist knowledge and equipment • does not measure most condensed PCM component • oxidation may form non-PCM BPCAs	IQC	L	H
Py GC/MS	• provides detailed information on components of PCM	• requires specialist knowledge and equipment • possibility of charring during analysis • does not measure most condensed PCM component	C	L	H
Levoglucosan	• specific marker for cellulose combustion	• variable rate of production • poor preservation in soil	C	M	H

Schmidt et al, 2001; Hammes et al, 2007; Roth et al, 2012; Zimmerman and Mitra, 2017). This inherent range is demonstrated clearly for standard soil samples analyzed by multiple techniques and laboratories in Figure 24.2. It is also worth noting that except for some physical techniques, most techniques measure only the PyC component of PCM and do not measure components such as H, O, N, S, or ash, hence the term PyC is used to refer to analytical results in this chapter.

Some chemical and thermal techniques can isolate PCM from other organic matter, or from all soil components, and this can be particularly useful as the PCM is thereby available for further characterization, such as partitioning sources using stable isotopes, estimating age with radiocarbon dating, or determining chemical composition with spectroscopy. However, it is rare for all components of the PCM continuum in a sample to be uniquely distinguishable from other SOC using these techniques and hence most methodologies seek to maximize the retention of PCM while minimizing interference from other SOC, or to quantify only a specific component of the PCM continuum. If

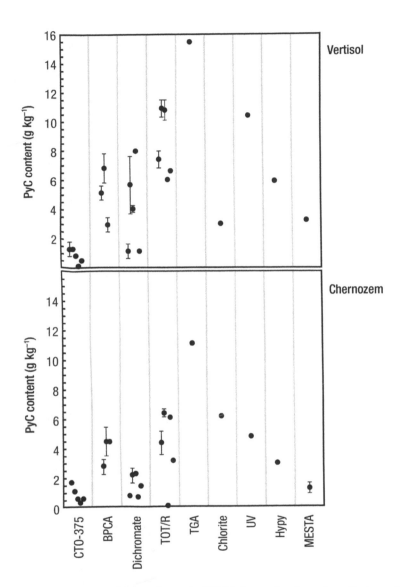

Figure 24.2 *Comparison of PyC abundance in PCM determined on the soil ring trial standards of Hammes et al (2007) including additional results generated using hypy (Meredith et al, 2012) and MESTA (Hsieh, 2007)*

quantification of the broadest range of PCM components is the main aim, then this may come at the cost of some inclusion of other SOCs in the analytical window. If isolation of PCM for further characterization (of isotope composition, surface characteristics, chemical characteristics) is the main aim, then removal of all other SOC may come at the cost of narrowing the range of PCM components that are retained for analysis, generally to the more recalcitrant end of the PCM continuum. Non-destructive methods can potentially enable quantification of the entire continuum of PCM, or a specific sub-set of PCM, with the disadvantage that the PCM is either not separated from the soil for further analysis or cannot be completely separated from the soil. Some non-destructive techniques also rely on calibration using one or more of the destructive techniques or are not calibrated.

Field sampling and laboratory preparation

Sampling and preparation of soil samples are vital steps in any soil-related biochar studies. In this regard, it is worth noting: (i) PCM occurs naturally in many soils, often comminuted to a size and form that is not visibly identifiable; and (ii) biochar added to soil often is, at least initially, 'in the soil' but not 'of the soil', as the soil is usually defined as material <2 mm in size and biochar for agricultural use may be of a larger size class. This has the advantage that in those cases a significant fraction of biochar can be recovered from soil by simple sieving, but also means that care should be taken in sample preparation not to inadvertently break up large fragments, unless all material is to be ground to pass a 2 mm sieve.

Soil is a highly heterogeneous material with characteristics that vary both laterally and vertically, with some soil characteristics (e.g., nutrients) also varying temporally. This variability extends to PCM in soil and must be taken into account in designing a sampling strategy to ensure robust, representative analyses of soil parameters are obtained. This usually involves taking multiple samples from each sampling unit of interest, though these can be amalgamated into fewer, or a single, sample representative of each unit for analysis. Replication of sampling is important as it provides a measure of variability across the sampled area. The number of samples as well as the amount and spatial scale of replication of analyses will depend on the specific aims of each project.

It is often necessary to report soil parameters, including biochar PCM or PyC content, in terms of soil volume rather than weight. While soil samples can be collected using any digging implement, it is preferable to use a solid-walled tube to take a known volume of soil to a known depth so that soil bulk density (g cm^{-3} or kg m^{-3}) can be calculated using the weight of dry soil in a known volume. Soil bulk density can then be used to calculate the abundance, or 'stock', of any soil constituent per unit area. Where sites have been subject to a change in soil structure as a result of a change in soil management, it may be necessary to report concentrations relative to an equivalent soil mass (Ellert and Bettany, 1995).

The usual practice is to decide (i) the number of sampling units that are required to represent the range of variability expected and address the question being asked; (ii) how many samples to take from how many depth interval(s) to adequately represent each sampling unit; and (iii) the degree to which individual samples can be combined into a few number of samples for analysis. The collection of samples should be undertaken using a soil corer or pipe of known volume, generally a minimum 25-mm diameter and preferably larger. The sampler should be driven into the ground to the desired depth, extracted without loss of soil from the bottom, emptied into a labeled bag, and weighed. If samples are combined, they should be well homogenized before sub-sampling. Normally, samples are allowed to air-dry before further preparation. If bulk density is to be determined, a portion of each sample must be dried at 105°C to remove all free water and reweighed. Further preparation may include detailed size fractionation and/or crushing depending on the needs of the study.

Further guidance on soil sampling and preparation can be found in, for example, Major (2009) and Boone (1999). The choice of a particular method, therefore, necessarily depends on the question to be addressed, the components of the PCM continuum that are of interest, the degree and type of chemical characterization required, as well as the number of samples that need to be analyzed, access to the facilities required to undertake the measurements, and the cost of each analysis.

Analytical techniques

Physical separation techniques

These techniques exploit the fact that char (and biochar) PCM usually has an effective density (including unwetted pore space) of <2 g cm^{-3}, thus being less dense than the mineral matrix in which it is found (generally >2 g cm^{-3}), and a proportion total PCM is larger than the mineral matrix. Thus, both flotation and density separation can be very effective at isolating biochar from soil. Schreiner and Brown (1912) used wet sieving followed by density separation using a solution of C tetrachloride and bromoform with a specific gravity of 1.8 g cm^{-3} and hand-picking under a microscope to isolate char PCM from other components in 34 soils with a wide range of characteristics from across the United States.

One of the first PCM samples to be radiocarbon dated was a bread roll which had been turned to char PCM by the eruption that destroyed Pompeii in AD79 (Arnold and Libby, 1949). The recognition that macroscopic PCM could be accurately dated rapidly led to the development of techniques for its isolation from sediments of archeological interest. Flotation techniques, developed for archeological use by Matson (1955), involve agitating a slurry of sediment in water, such that the low density, relatively coarse PCM fraction (along with other plant debris) remains in suspension longer than the mineral fraction. The mineral fraction is briefly allowed to settle, then the water is poured through a sieve that retains the PCM but allows fine mineral material also in suspension to pass through. The procedure is repeated until no more material is retained on the sieve. The material retained on the sieve can be washed, dried and PCM manually handpicked to separate it from any plant remains. The mesh size of the sieve can be adjusted to suit the nature of the project. If only a small amount of material is sufficient, then handpicking a few tens of milligrams from a concentrate under a binocular microscope is possible for particles down to ~100 µm in size (Bird et al, 2002). A range of flotation techniques was reviewed by Kenward et al (1980) and useful refinement was also introduced by Gumerman and Umemoto (1987) who used a fish-tank cleaning siphon to improve the recovery of organic remains from archeological sediments, particularly the recovery of comparatively dense charcoal PCM. A further refinement has been suggested by Bartůněk et al (2017) involving the dissolution of any remaining mineral impurities using 1-butyl-3-methylimidazolium hexafluorophosphate for 1 hour at 200°C to provide a pure isolate of PCM for further analysis.

Possibly the most basic technique is that of Kloss et al (2012) who simply sieved soil samples from a chronosequence of burnt sites in Austria and then handpicked PCM under a binocular microscope, demonstrating that PCM concentration decreased with increasing age since the location was burnt and the maximum concentration also shifted further down the soil profiles. Zackrisson et al (1996) also used hand-picking and sieving to quantify PCM in boreal forest soils and investigate soil-microbe-char interactions, while Nguyen et al (2008) used hand-picked PCM fragments to investigate changes in surface chemistry over a 100-year chronosequence of soils in Kenya.

Further control on the efficiency of isolation, particularly at smaller particle sizes can be achieved by using a range of liquids of different specific gravities. Struever (1968) used zinc chloride ($ZnCl_2$; 1.6 g cm^{-3}) while Bodner and Rowlett (1980) used ferric sulfate ($Fe_2[SO_4]_3.xH_2O$; 1.6 g cm^{-3}) to improve the recovery of PCM from sediments. More recently, sodium polytungstate ($Na_6[H_2W_{12}O_{40}]$; 1.8 g cm^{-3}) has successfully been used to isolate small PCM fragments (Bird et al, 2002) and to separate soot PCM from lithogenic graphite (Veilleux et al, 2009). All of these compounds are provided in solid form and mixed with water, so fine control over the final specific gravity of the solution is possible. Other liquids such as acetone (0.79 g cm^{-3}) and ethyl ether (0.72 g cm^{-3}) can potentially also be used to separate 'fresh' biochar from biochar associated with mineral phases, or biochar from organic detritus, depending on the densities of the materials to be separated.

Density separation using sodium iodide (NaI) or sodium polytungstate with specific gravities ranging from 1.6 to 2.0 g cm^{-3}, commonly in association with additional chemical separations is now employed in a range of fractionation schemes. These approaches are designed to isolate soil C pools of different turnover times that can serve as input to soil C models (e.g., Trumbore and Zheng, 1996; Sohi et al, 2001; Zimmermann et al, 2007). PCM is considered to be the dominant component of the fraction that is low density and resistant to whatever chemical treatment is employed, categorized as the resistant SOC pool.

The advantage of the physical techniques discussed above is that they are low cost, easily implemented, and water-only flotation allows sample processing in the field. In the specific context of biochar added to agricultural soil, which tends to be of comparatively large particle size, it is likely that they can efficiently be used to separate biochar from soil after a period of incorporation and provide a rough measure of the amount of biochar in the soil, or to provide samples for detailed characterization. The drawbacks of the techniques are that the time involved in handpicking biochar fragments increases dramatically as particle size decreases and therefore biochar stocks in soil are likely to be underestimated if the estimate is based solely on the amount of large fragments that are easily separated. This may, however, provide a useful measure of the rate at which biochar fragments are broken up in the soil and transmuted into finer particle size fractions. Care should also be taken to ensure that the choice of chemicals for density separation does not compromise the analyses to be undertaken on the isolated material. Organic solvents can potentially dissolve organic compounds from the biochar and inorganic compounds can be difficult to completely remove from a sample.

Chemical techniques

Dichromate oxidation

Quantification of SOC through oxidation by an acid dichromate reagent was first proposed by Schollenberger (1927) and, with modification, became the basis of the widely used Walkley-Black method (Walkley and Black, 1934). The Walkley-Black method does not fully oxidize all organic C and oxidizes PCM at a slower rate than other forms of organic matter (Piper, 1942). Wolbach and Anders (1989) first exploited this difference in oxidation times to isolate an 'elemental C' PCM fraction from sediments that could then be combusted for quantification and determination of stable C isotope composition (Wolbach et al, 1988). The dichromate oxidation step was performed after decarbonation of the sample with HCl and the removal of silicates using HCl/HF, to provide a concentrate of pure organic matter for oxidation. Lim and Cachier (1996) found that while reaction times required to remove kerogen from lacustrine and marine sediments ranged up to 28 hours, 60 hours at 55°C in a 0.1 M $K_2Cr_2O_7$/2 M sulfuric acid solution provided optimal isolation of 'elemental C' PCM from the widest range of materials, and that this length of oxidation did not greatly reduce the abundance of the elemental C PCM component. Bird and Gröcke (1997) first applied dichromate oxidation to the analysis of laboratory-produced PCM and soil samples. The study demonstrated that char PCM exhibited a wide range of susceptibilities to oxidation and hence dichromate oxidation only quantifies one region of the PCM continuum. The same conclusion was reached by Skjemstad et al (1999) who reported that the oxidation rate was also partly a function of particle size.

Bird and Gröcke (1997) showed that >95% of total organic C was removed from the soil samples after 72 hours of oxidation in 0.1 M $K_2Cr_2O_7$/2 M sulfuric acid solution at 60°C. The residue after this treatment was termed 'oxidation resistant elemental carbon' (OREC, a component of PCM that can be reliably quantified, but potentially contains non-pyrogenic material) and for which the

C isotope composition can be determined to obtain information on the source of the PCM (Figure 24.3). The invariant C isotope composition of the material remaining after periods of oxidation longer than 72 hours confirmed that the removal of organic C was complete after this time. Bird et al (2000) used the technique on soil samples from the Matopos Fire Trials in Zimbabwe and demonstrated that a significant component of PCM had been lost from the 0–50 mm interval of the soils over a 50-year period of fire protection.

Masiello et al (2002) compared the use of 0.1 M $K_2Cr_2O_7$ at 60°C with the use of 0.25 M $K_2Cr_2O_7$ at 23°C on the quantification of PCM from ocean sediments and the determination of the stable C isotope composition and radiocarbon activity of the isolated PCM. The study concluded that reaction rates were faster at higher temperatures and calculated PCM concentrations were similar at both temperatures, but the precision of the radiocarbon measurements was better at 23°C. Song et al (2002) added a solvent extraction and base extraction to the procedure of Lim and Cachier (1996) to also isolate lipids and a so-called humic acid fraction from soils, sediment, and PCM for separate characterization.

Dichromate oxidation under a range of experimental conditions has since been widely implemented for PCM isolation and PyC quantification from soil (e.g., Rumpel et al, 2006; Knicker et al, 2007; Hammes et al, 2007; Meredith et al, 2012; Murano et al, 2021) and for quantification of the resistant component in biochar (Hammes et al, 2007; Calvelo Pereira et al, 2011; Ascough et al, 2011; Meredith et al, 2012; Naisse et al, 2013). It is clear that the technique can provide an internally reproducible and demonstrably useful measure of an operationally defined region of the PCM continuum. It is also a significant advantage of

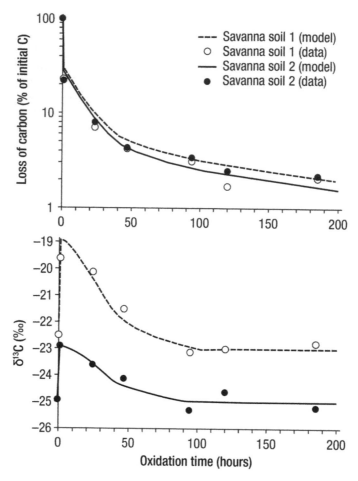

Figure 24.3 *Carbon loss and C isotope composition (δ¹³C value) of two savanna soils subjected to dichromate oxidation (data from Bird and Gröcke, 1997). The data have been modeled as three components. Easily-mineralizable SOC (70%, $T_{1/2}$ = 0.1hrs; soil 1 = −23.5‰, soil 2 = −26‰, slow-mineralizable SOC and/or easily-mineralizable PCM (26%, $T_{1/2}$ = 10 hrs; soil 1 = −18.4‰, soil 2 = −22.5‰) and slow-mineralizable PCM (4%, $T_{1/2}$ = 150hrs; soil 1 = −23‰, soil 2 = −25‰)*

the technique that the material that is isolated is amenable to further analysis by other techniques, including stable isotopes (for source attribution), nuclear magnetic resonance (NMR) spectroscopy (to determine molecular composition), and radiocarbon (to estimate age or residence time).

However, there are several issues relating to the use of the technique that have yet to be

resolved. The first issue is that dichromate oxidation appears not to result in pure PCM. Bird and Gröcke (1997) noted that a small component of organic C survived dichromate oxidation in a sample of Antarctic marine sediment unlikely to contain significant PCM, but this was susceptible to oxidation in a basic peroxide solution. Knicker et al (2007) found that hydrophobic compounds (paraffins and long-chain hydrocarbons) from both plant and soil samples could survive dichromate oxidation, albeit after a much reduced 6-hr oxidation time. Meredith et al (2013) examined the dichromate oxidation residue of a soil sample of Song et al (2002) using hydrogen pyrolysis (Hypy) (see below). The results indicate that 12.2% of the C remaining after oxidation was susceptible to Hypy. GC/MS analysis of the hydropyrolysate showed that of this, 70% were polycyclic aromatic hydrocarbons, potentially of pyrogenic origin, but 31%, by weight of the non-BC fraction identified by Hypy, or 4% of the C identified as PCM by dichromate oxidation, following both solvent and alkali extraction, was composed of hydrophobic long-chain hydrocarbons that are most probably components of non-pyrogenic (microbial) biolipids, and epicuticular waxes. The results indicate the potential for dichromate oxidation analyses to significantly overestimate PCM in a sample.

The second issue, highlighted by the data in Figure 24.2 and discussed by Hammes et al (2007) is that a range of different protocols is used, and this results in a range of values reported for the same soil standards by different laboratories that varies by a factor of six. This is not surprising as comparing the range of oxidation conditions used to examine the same standards (Hammes et al, 2007; Meredith et al, 2012) shows that the concentration of the oxidant ranged from 0.1 to 0.5 M, oxidation times ranged from 10 min to 400 hours and the oxidation temperature ranged from 22.5 to 80°C.

Evident from the discussion above, standard conditions are suggested below, based mainly on those studies that have specifically examined the oxidation kinetics of organic C and PCM in soils and a range of PCM materials (Bird and Gröcke, 1997; Song et al, 2002) with details provided in those papers. These conditions provide optimal removal of non-PCM with minimal removal of PCM, as demonstrated by stable isotope measurements (Bird and Gröcke, 1997).

Step 1   Initial sample preparation may involve preliminary size fractionation or crushing along with drying at 105°C, or freeze-drying to remove water.

Step 2   Removal of carbonate using 6 M HCl at room temperature overnight.

Step 3   Demineralization by digestion in a 2:1 mixture of 22 M HF:6 M HCl (in Teflon), at least overnight at 60°C followed by treatment with 6 M HCl overnight at 60°C to remove fluorides.

Step 4   Removal of hydrophobic organic compounds. Song et al (2002) used soxhlet extraction with 2:3:5 methanol:acetone:benzene, however, it is likely that other solvents in combination with sonication or accelerated solvent extraction may also achieve an effective removal of lipids.

Step 5   If separate quantification of an alkaline ("humic acid" in older literature) extract is required, an overnight room temperature extraction with 0.1 M NaOH can be used, but this is not required for PCM isolation.

Step 6   Removal of non-PCM organic C by digestion in 0.1 M $K_2Cr_2O_7$ / 2 M H_2SO_4 at 55–60°C for 60–72 hours. The exact choice of conditions within these ranges is unlikely to significantly affect the analysis. However, 72 hours represents a more convenient period for routine analysis. The C remaining after this step can be quantified by elemental analysis as PyC.

The dichromate oxidation procedure described above can provide robust operationally defined estimates of PyC abundance in soils and biochars and can be readily implemented in most wet chemistry laboratories. In the specific context of biochar in soils, the residual hydrophobic compounds that may survive the protocol described above are likely to only be a very minor component of the isolated PCM sample. Murano et al (2021) have presented evidence that robust analyses, without the need for an organic extraction step, can be achieved if the $K_2Cr_2O_7$ concentration (Step 6) is increased to 0.3 M, and this approach, avoiding the need for step 4 above, may be appropriate depending on the needs of an individual study.

The technique has the advantages of a considerable and growing literature and the isolated PCM is available for further analyses of chemical structure and isotope composition. A major disadvantage of dichromate oxidation is the multi-step procedure using hazardous chemicals such as HF acid for which substantial health and safety measures are necessary. It should also be noted that poorly condensed PCM (and potentially also PCM subject to environmental degradation) will be oxidized by this technique, and hence the analytical window of technique is biased to measurement of more highly condensed PCM only.

UV photo-oxidation

Skjemstad et al (1993) originally developed ultra-violet (UV) photo-oxidation as a technique to examine the fraction of SOC that is protected by occlusion within mineral aggregates. Skjemstad et al (1996) used a more powerful 2.5 kW photo-oxidizer in conjunction with NMR (see below) and scanning electron microscopy to examine the residues after photo-oxidation. They concluded that up to 30% of C in the residue derived from char pyrogenic C (PyC,) with a significant proportion (up to 88%) of this in the fine (<53 μm) fractions.

Skjemstad et al (1999) used optimized UV photo-oxidation to directly estimate the char

PyC content of soil. The technique involves demineralization of a size-fractionated soil sample with HF to concentrate the organic component, then photo-oxidized in oxygen-saturated water for 2 hours. The CP-MAS ^{13}C NMR signal for the treated samples was calibrated against the Bloch Decay ^{13}C NMR signal from the same sample to correct the aryl-C peak indicative of char PyC for interference from an O-aryl peak derived from residual lignin in the sample. In turn, this enabled the more rapid CP-MAS ^{13}C NMR technique to be used to estimate char PyC in the sample.

As noted by Skjemstad et al (1999) the technique "makes a number of assumptions about the distribution and chemistry of char and lignin, and so cannot be regarded as quantitative". The assumptions are, however, conservative such that the estimate can be considered a minimum for the 'true' char PyC. In the methodological inter-comparison for PyC determination in the soil of Schmidt et al (2001), UV photo-oxidation measured 2–3 times more PyC than other chemical or thermal techniques for most samples. In the Hammes et al (2007) inter-comparison, the technique consistently quantified more PyC for PCM samples than other techniques and produced results at the upper end of the range for soil samples. This is consistent with the comparatively gentle nature of the UV oxidation, and with the ability of ^{13}C NMR to quantify all, not just pyrogenic, aromatic C (Gerke, 2019). The technique was unable to discriminate between PyC in PCM and some interfering materials such as coal.

The main advantage of the technique is the ability to measure aromatic C from a wide range of the PCM continuum. The disadvantages of the technique are that it has only been used in one laboratory, relies on access to specialized equipment, and also relies on calibrations that are likely to vary between samples. UV photo-oxidation has been employed to calibrate other more rapid techniques such as mid-infrared spectroscopy (Janik et al, 2007; see below, and such calibrations have in turn been used to demonstrate the significant contribution (commonly 10–30% and up to 80%) of PyC to the Australian SOC pool (Lehmann et al, 2008).

Peroxide oxidation

Smith et al (1973, 1975) first used an alkaline hydrogen peroxide solution to oxidize organic matter from marine sediments to quantify 'elemental carbon' PyC by infrared spectroscopy. Using this as a basis, several researchers developed basic (NaOH or KOH) peroxide treatments to isolate PCM from a range of sediments in order to quantify changes in PyC abundance in sediments over time, generally combining one or more of an acid treatment (demineralization with hydrofluoric acid and density separation, followed with quantification either by combustion to CO_2 or particle counting (e.g., Rose, 1990; Emiliani et al, 1991). Wolbach and Anders (1989) reported that the basic peroxide reagent was comparatively unstable and abandoned initial experiments in favor of an acid dichromate treatment. However, while this approach has yet to be tested on standard soil materials, a study by Wu et al (1999) found that it did have potential as a technique for PyC quantification in soils. In contrast, Raya-Moreno et al (2017) found that hydrogen peroxide oxidation (33% H_2O_2 replenished several times over a week at room temperature) was not able to adequately separate PyC from SOC. Cross and Sohi (2013) have also employed hydrogen peroxide as a reagent with which to accelerate the aging of biochar samples while Orr et al (2021) found a peroxide oxidation step assisted in the decontamination of PCM prior to radiocarbon dating.

Sodium chlorite oxidation

Simpson and Hatcher (2004) tested a technique where sodium chlorite ($NaClO_2$) in an acetic acid solution is used to oxidize lignin and other polyphenolic compounds that are not PCM, and quantified the PyC component remaining in the sample by ^{13}C NMR. The technique was found to be able to discriminate between PyC and non-PyC, including PyC in soils. Only one laboratory reported results in the ring trial study of Hammes et al (2007), with the results suggesting that char PyC was aggressively attacked by the reagent, but the values reported for the soil standards were comparable to those determined using other oxidation techniques. Hammes et al (2007) concluded the technique shows promise but needs further development. Quantitative ^{13}C NMR is also relatively expensive. De la Rosa et al (2008a) found that the technique performed well on coastal and estuarine sediments, and the technique has also been used to oxidize biochar samples to increase their ability to adsorb nitrate (Sanford et al, 2019).

Ruthenium tetroxide oxidation

Ruthenium tetroxide (RuO_4) (RTO) is a strong oxidant that has long been used to oxidize organic matter, generally at room temperature, for a variety of purposes and particularly for the analysis of coals and hydrocarbons (Berkowitz and Rylander, 1958). The technique has yet to be applied to PyC quantification but several studies have now demonstrated that RTO liberates condensed aromatic rings (3–37 rings in size), some of which may be of PCM origin (e.g., Ikeya et al, 2011), and PyC has been identified in the residual component left after oxidation (Quénéa et al, 2005). The technique has unexplored potential for PyC quantification in soils, with further work required to benchmark the technique against other techniques.

Nitric acid oxidation

Verardo (1997) proposed a simple and rapid nitric acid oxidation technique for quantifying PyC in marine sediments. The technique involves dropwise addition of concentrated nitric acid at 50°C to samples in weighed silver capsules on a hotplate also at 50°C, with evaporation to dryness following each addition. Once the pretreatment is complete, the C content of the sample is determined by elemental analysis. The technique is attractive because of its simplicity and rapidity. However, Bird and Cali (1998) found that a sample of Antarctic marine sediment previously shown to contain <1% of organic C as PyC by dichromate oxidation, returned 50% PyC using the in situ nitric acid digestion technique. Middelberg et al (1999) also concluded that simple nitric acid oxidation significantly overestimated PyC abundance in marine sediments. The technique would require further validation using soil materials before being routinely used for biochar PyC quantification in soil.

Thermal techniques

Loss on ignition

Loss on ignition (LOI) involves heating a sample of known weight in a furnace for a specified length of time and attributing the weight loss observed to the combustion of organic matter in the sample. The LOI has long been used as an approximate measure of SOC content (Ball, 1964). The conditions used in the determination of LOI vary widely from 375–850°C and times ranging from tens of minutes to several hours, depending on the purpose of the analysis. Determination of SOC by LOI is complicated by the fact that clay minerals in particular lose structural water over the same temperature range, and this can be problematic, particularly in clay-rich soils.

The LOI technique has been adapted specifically for use in biochar studies by Koide et al (2011). The methodology employs LOI determination on soil with and without biochar at 550°C for 4 hours in a muffle furnace. The mass of biochar in a soil sample can be calculated from the LOI of the soil sample containing biochar, the LOI of the 'pure' soil (not including the biochar), and the LOI of the pure biochar. This approach requires that a representative sample of soil from before amendment with biochar is available, along with a sample of the biochar added to the soil. It also depends on the LOI of both the soil and the biochar being stable over time. Koide et al (2011) demonstrated that this was the case over the 15-month duration of their study but acknowledged that these assumptions may not hold for all soils or biochars, in the latter case because biochar chemistry is known to change over time, which is an issue for all methods. Raya-Moreno et al (2017) used a combination of LOI measurements on soil and biochar at three temperatures (375°C /18 hours; 550°C/5 hours; 950°C 9 hours) to successfully quantify biochar, soot, and other mineral components respectively in biochar-amended agricultural soils. Nakhli et al (2019) have also demonstrated that paired lower and higher temperature LOI measurements on biochar, soil, and biochar-amended soil can quantify PyC in soils, with the temperatures chosen likely dependent to some degree on the specific soil and biochar used in a study.

LOI is attractive as an inexpensive and easily implemented method of demonstrating carbon additivity in studies where 'before' and 'after' samples of soil and biochar are available, a situation that is usually the case where biochar has been added as a soil amendment.

Chemo-thermal oxidation

Chemo-thermal oxidation (CTO) techniques were originally developed for quantifying char and soot PCM in sediments and aerosols. Winkler (1985) proposed a simple nitric acid digestion followed by the determination of PCM content by LOI of the residue at 450–500°C for three hours, later modified by Laird and Campbell (2000) who lengthened the nitric acid oxidation step, and analyzed PyC in the residue directly using a total organic C analyzer. Cachier et al (1989) developed a simple thermal technique for quantifying soot PyC in aerosol samples that involves decarbonation by exposure to HCl fumes, followed by removal of organic C in a stream of pure oxygen at 340°C for two hours. The PyC remaining is then determined by combustion to CO_2. Kuhlbusch (1995) extended the thermal approach to the quantification of PCM from biomass-burning residues. This technique added an initial acid (HCl and HNO_3) and base (NaOH) extractions, and isolated PyC from the remaining organic C by exposure to a flow of pure oxygen at 340°C for two hours, with final PyC quantification by elemental analyzer.

The technique that has come to be known as CTO-375 was first proposed by Gustaffson et al (1996) and was developed to specifically isolate soot PyC from sediments. In the original technique, organic C was removed by exposure to air in a muffle furnace at 375°C for 24 hours, then decarbonated, and the soot PyC component was determined by an elemental analyzer. Gustafsson et al (2001) validated the technique on a range of sample types that both contained, or were free, of PCM.

Gelinas et al (2001) noted that any thermal technique runs the risk of producing PyC from organic C during the thermal treatment, leading to a positive bias in the amount of PyC reported. In order to minimize the risk of neoformation of PyC, they added additional chemical steps before the thermal treatment. First, the samples are demineralized using HF/HCl, and then organic matter is hydrolyzed using successively stronger and longer treatments with trifluoroacetic acid and HCl. Following these pretreatments, the samples are thermally treated at 375°C in air for 24 hours, with PyC remaining after treatment quantified by an elemental analyzer. While this approach does reduce the risk of charring during the thermal step, it also risks partial loss of the hydrophobic soot component during washing (Elmquist et al, 2004).

There has been considerable testing and validation of the CTO-375 technique since its initial development. Elmquist et al (2004) used a standards addition approach to infer that soot PyC in natural samples may be under-estimated due to enhanced oxidation caused by catalytic contact with chloride and metal cations in the sample. Elmquist et al (2006) detected no PyC in natural char, also due to catalytic reactions with minerals and the greater internal surface area of char PCM compared to soot PCM, which enhances the contact between oxygen and particle surfaces. Agarwal and Bücheli (2011) slightly modified the method by adjusting the temperature ramp rates to 375°C and thoroughly tested the technique for use on organic-rich soil samples.

Comparative studies of CTO-375 and other methods using the same samples (Schmidt et al, 2001; Hammes et al, 2007) have confirmed that the technique reliably quantifies the most condensed component of PCM and can discriminate between PCM and interfering materials, but cannot reliably quantify less condensed PCM, such as found in most biochars and by implication in the soil (Nguyen et al, 2004; Elmquist et al, 2006). Another inter-comparison study that included CTO-375 reported good quantification for pure soot PyC, but a wide range (15-270%) of recoveries for soot spiked into natural soils (Roth et al, 2012). The study also reported systematic differences for the same samples related to different oxygen penetration efficiencies associated with the use of a tube furnace versus a muffle furnace. Murano et al (2021) also reported that while the CTO-375 method was simpler than dichromate oxidation, it overestimated PyC abundances due to the in-situ production of PyC during the thermal treatment of high organic matter soils.

In summary, the CTO-375 technique has the advantages that there is a reasonably clearly defined protocol, minimal handling is involved, the equipment is widely available and process conditions are amenable to precise control. The disadvantages are that the analytical window of the technique is limited to the highly condensed component of PCM and both underestimation due to a variable degree of catalytic oxidation associated with matrix effects and overestimation due to charring are possible. The technique has now been used to quantify PCM in globally distributed sets of soil samples (Bücheli et al, 2004; Nam et al, 2009; Agarwal and Bücheli, 2011; Hamilton and Hartnett, 2013; Qi et al, 2017). The technique has the additional advantage that PyC can be isolated for further analysis of morphology or isotope composition using the technique (e.g., Zencak et al, 2007; Song et al, 2012).

Thermogravimetry – differential scanning calorimetry

Thermogravimetry measures the loss of mass as a sample is heated under controlled conditions, while differential scanning calorimetry measures the energy flux into or out of a sample as its constituent compounds decompose exothermically or endothermically. The techniques have been widely used to characterize minerals and organic compounds both in pure form and in complex mixtures such as soils. The combination of the two techniques enables the relative contributions of mineral and organic components to

be identified and organic components can be quantified based on the temperatures over which they decompose. Lopez-Capel et al (2005) used Thermogravimetry – differential scanning calorimetry (TG-DSC) to identify multiple organic components in SOC fractions and found that several components could be identified corresponding to SOC fractions ranging from labile to refractory.

Leifeld (2007) specifically examined the potential of the technique for quantifying and characterizing PyC in soils as well as pure PCM materials. Samples were progressively heated to 600°C under a constant flow of synthetic air, with peak temperature and 50% burn-off temperatures, peak height, and total heat of reaction were used to estimate the stability of PCM in the samples (Figure 24.4). Samples were also assessed using a standard addition approach. The study found that exotherms at or above 520°C were solely derived from PCM and that PCM in soil could be reliably quantified provided PCM was >3% of SOC. De la Rosa et al (2008b) found good agreement between 'refractory organic matter' including PCM quantified by TG-DSC peaks integrated between 475 and 650°C for burnt soils, and the aromatic component of SOC quantified by NMR techniques, although clear distinction of PCM was not possible. The same approach has successfully been applied to measuring PCM abundance in the soils underlying pre-industrial charcoal kiln sites (Hardy et al, 2017). Hardy et al (2022) extended this approach to determining PyC abundance by comparing the thermograms of the kiln soils to those of the adjacent soils, exploiting the distinct thermal profiles of PCM and non-PCM organic matter to then provide an assessment of PyC distribution across the entire PCM continuum from minimally to highly condensed PyC in the kiln soils.

Only one laboratory participated in the ring trial study of Hammes et al (2007) so it

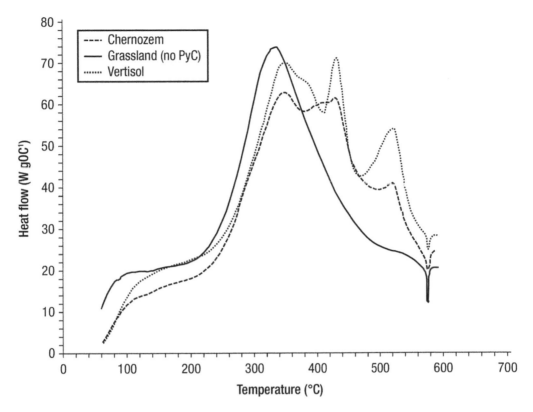

Figure 24.4 *TG-DSC thermograms of Chernozem soil (long-dashed line), Vertisol (short-dashed line), and grassland soil without PyC (solid line) from Leifeld (2007). Lower temperature peaks (<400°C) derive from SOC, and peaks above 400°C, where present, derive from PCM of varying thermal stability*

is not possible to assess the inter-laboratory reproducibility of the technique. Hammes et al (2007) concluded that TG-DSC has potential as it is operationally simple and potentially detects all PyC, however, it may overestimate PyC in low SOC soils and also returned 'false positives' for some non-PyC-containing materials. While the technique does not isolate PyC, Manning et al (2008) have demonstrated that it is possible to determine the C isotope composition CO_2 in-line as it is evolved, and this technique has shown considerable potential for source identification for SOC in soil samples (e.g., De la Rosa et al, 2008b).

Thermal-optical transmittance/reflectance

Thermal-optical transmittance/reflectance (TOT/R) was originally developed for the separate quantification of organic and elemental carbon PyC from atmospheric aerosol samples. The technique heats a sample on a filter paper at several temperatures up to 800°C in gas mixes of varying proportions of He and O_2. CO_2 produced during the analysis is converted to methane and quantified using a flame ionization detector. As charring of the sample is inevitable, the reflectance (or transmission) from a laser is continuously monitored throughout the analysis and is used to correct for the in-situ charring of organic matter, enabling quantification of PyC in the sample (Chow et al, 1993). More recently, the technique has been extended to measuring stable C isotope ratios of both organic C and PyC in aerosol samples (Huang et al, 2006) and to discrimination between char PyC and soot PyC (Han et al, 2007a).

Han et al (2007b) extended the method to soils and sediments by including a demineralization step using HCl and HF, depositing the residue on a filter disc for TOT/R analysis. The study found that the reproducibility of the technique was generally ~±10% and the concentrations for soils and sediments were loosely positively correlated with the CTO-375 results. Zhan et al (2013) used a standards addition approach to confirm that TOT/R could be used to measure PCM in loess samples and also found the results were positively correlated with CTO-375 results on the same samples. Zhan et al (2012) used TOT/R to demonstrate a wide range of PCM abundances (0.02–5.5 g kg^{-1}; 2–37% of SOC), decreasing with depth, in soils of the Loess Plateau in China, and that soot PCM abundance was positively correlated with char PCM in the sample set.

Four laboratories used the TOT/R methodology in the ring trial study of Hammes et al (2007) and reproducibility between laboratories was reasonable, but difficulties were reported in achieving homogeneous distribution of the sample on the filters. The technique also reported substantial 'false positives' for several of the standards known not to contain PyC including shale, melanoidin, and coal. The results on the soil standards were broadly comparable with those generated by other techniques. Further method development to reduce 'false positives' is warranted before the technique can be routinely applied in the context of soil-related biochar studies.

Hydrogen pyrolysis

Hypy is a comparatively recent innovation, originally developed to extract the labile component of sedimentary rocks including coals and oil shales for detailed characterization (e.g., Love et al, 1995). The technique was first tested as a potential tool for PyC quantification and isolation by Ascough et al (2009). Hypy uses pyrolysis at a comparatively slow heating rate (8°C per minute to 550°C) assisted by high hydrogen pressures (>150 bar) with a dispersed sulfide molybdenum catalyst to separate hypy-labile C from hypy-resistant PyC. It has been shown that conversions of ~100% are achieved for thermally labile materials, with the principal product being a dichloromethane-soluble oil. PyC, if present in a sample, is not affected by hypy and can be retrieved from the reactor for characterization. It has also been shown that the hydrocarbon products of hypy are released in high yields (Love et al, 1997, with the advantage that it is possible to separately identify and characterize both the labile and PyC constituents of a sample at a molecular level (Ascough et al, 2010).

Meredith et al (2012) reported the performance of hypy for PyC measurement using all of the ring trial samples examined by Hammes et al (2007). The study found that (i) all interfering materials were successfully removed by hypy except for high-rank coal, which is similar in composition to PyC; (ii) that for all sample matrices, including soils, the hypy results were well within the range of results reported by other techniques; and (iii) the analyses are highly reproducible at ±0.5% or better. The study also concluded that the PyC component that is isolated by hypy can be defined as polyaromatic compounds with a ring size greater than ~7 (coronene, consistent with the boiling point of coronene being 525°C), just below the final hold temperature of 550°C used in the analysis.

In the context of the C sequestration potential of biochar, the PyC component isolated by hypy (SPAC; Stable Polycyclic Aromatic Carbon) is the PyC component likely to persist in the soil on centennial to millennial timescales, as it is known that smaller PAH's of pyrogenic origin are readily microbially degradable (e.g., Juhasz et al, 2000a, b; Seo et al, 2009). Figure 24.9 shows the component of wood-derived biochars that are identified as PyC by hypy in comparison to ^{13}C NMR and BPCA (see below) results for the same materials (Wurster et al, 2013). The aromatic C measured by ^{13}C NMR and BPCA included the polyphenolic compounds, and hence a higher apparent concentration of PyC than hypy, particularly at lower temperatures. The trend to higher measured concentration of PyC with increasing temperature of formation is similar for all techniques. Above ~600°C >90% of C in the biochar is stable under the hypy treatment.

Wurster et al (2012) used mixtures of biochar and a range of organic matrices of differing C isotope composition to test the ability of hypy to both remove all easily mineralizable C and to determine the stable isotope composition of the component of PCM isolated by hypy. The study found that trace contamination amounting to a median value of 0.5% of the original total organic C did survive the hypy treatment. Correction of PyC abundance for this amount of carry-over contamination was straightforward and did not affect the calculation of C isotope composition, provided PyC constituted a minimum of 4% of the total organic carbon in a sample. These conditions are very likely to be met in biochar-related studies.

The hypy technique requires little wet chemistry, is rapid, and provides possibly the most precise definition amongst all techniques of the component of the PyC continuum that is measured. Haig et al (2020) have demonstrated that up to seven samples can be added into a single reactor run, greatly increasing the efficiency of sample preparation. If full quantification of all components of the PyC continuum is required, then the hydropyrolysate is available for further detailed characterization (e.g., Meredith et al, 2013), although single samples must be run if this is the aim. In the context of biochar C sequestration in particular, it is convenient that the definition of hypy resistant PyC (C in polycyclic aromatic compounds with a ring size >~7) is likely to closely equate to the component of biochar that will persist on centennial to millennial timescales. The main limitation of hypy at present is that there are less than a dozen units in operation globally, with only four engaged in research on biochar.

Rock-Eval pyrolysis

'Rock-Eval' pyrolysis was designed for petroleum exploration to automatically screen large sets of rock and sediment samples for hydrocarbon potential without any preliminary treatment. Samples are initially heated under an inert atmosphere to 650°C, followed by heating in an oxidizing atmosphere to 850°C, with evolved hydrocarbons, CO, and CO_2 monitored continuously through the heating process.

Oen et al (2006) first reported the correlation between PyC determined by CTO-375 and Rock-Eval pyrolysis results. Poot et al (2009) assessed the potential of the technique, specifically for quantification using a subset of the ring trial standards of Hammes et al (2007). Two possible measures of PyC were developed (i) the percent of C liberated in the oxidation step (%RC; refractory carbon); and (ii) the temperature at which 50% of the C was oxidized using an oxidation-only protocol ($T_{50\%}$). The study found that both measures correlated well with results obtained using CTO-375 for the char PyC, soot, and aerosol standards of Hammes et al (2007), suggesting the methods quantify a similar thermally-stable fraction of PCM. However, Rock-Eval also measured a significant component of mela-noidin as PyC, suggesting that charring during the heating of the sample is an issue. Rock-Eval was not used on the soil standards from the ring trial, so its applicability to the quantification of PyC in soil remains to be assessed.

Multi-element scanning thermal analysis

Multi-element scanning thermal analysis (MESTA) progressively heats a sample in an enclosed quartz tube from ambient temperature to 800°C at a constant heating rate and under a user-controllable atmosphere. The volatile C, N, and S components in the sample are carried through a high-temperature combustion tube, followed by continuous quantification of their respective oxides as they evolve through a heating cycle (Hsieh, 2007)

Hsieh and Bugna (2008) optimized the technique for PyC using several materials, including some of the ring trial standards of Hammes et al (2007), concluding tentatively that C liberated above 550°C could be attributed to PyC. Similar to several other techniques, MESTA robustly identified PyC in highly condensed materials such as soot PyC but found little PyC in the wood and grass char standards, hence is more suited to analysis of 'resistant' PyC, a component likely to remain in the soil on centennial timescales and hence of particular interest to biochar C sequestration studies. Interestingly, MESTA yielded PyC abundance in the mid-range of reported results for the two-ring trial soil standards, further suggesting potential in soil-related biochar studies.

Thermal gradient method

In the thermal gradient method (ThG) method, a sample is placed in a tube furnace that is heated progressively to 1000°C under a constant flow of oxygen. The progressive loss of C (as CO_2) and water is continuously monitored by infrared detectors (Schwartz, 1995). Roth et al (2012) assessed the potential of the method for PyC quantification and found that the technique produced precise results but underestimated soot-PyC on the one hand and severely overestimated PyC in soil mixtures. While the simplicity of ThG is attractive, further development would be required before routine application of the technique is a possibility.

Spectroscopic techniques

Nuclear magnetic resonance spectroscopy

Nuclear magnetic resonance spectroscopy (NMR) exploits the magnetic properties of specific atomic nuclei to obtain information on the chemical structure and environment of organic molecules. Barron et al (1980) first used ^{13}C cross-polarization NMR with magic angle spinning to non-destructively examine the chemical structure of soil organic matter, demonstrating that alkyl, O-alkyl, and aromatic groups could be resolved based on characteristic chemical shifts. There has since been considerable refinement in the applications of both cross-polarization (CP) and direct polarization (DP) ^{13}C NMR techniques to the study of natural organic matter in soil. Molecular Mixing Models (MMM) simultaneously solve a series of equations to estimate the content of biomolecular components representing the organic materials present in greatest abundance in soil - carbohydrate, protein, lignin, aliphatic material, and PCM (Nelson and Baldock, 2005). In combination with elemental analysis, MMMs can provide semi-quantitative determination of PCM in soils.

The fact that PyC is largely composed of polyaromatic (aryl) C means that NMR can be used to quantify PyC in soils after correction for interferences from lignin O-aryl groups, if present (Skjemstad et al, 1996). Quantification of PyC is generally not achievable on untreated soil due to dilution of the signal and also due to interferences from paramagnetic mineral phases (Smernik et al, 2000). NMR for PyC quantification has generally been achieved on materials initially prepared using other techniques. Skjemstad et al (1999) and Smernik et al (2000) used demineralization and UV oxidation followed by NMR quantification, while Knicker et al (2008) used a combination of demineralization, dichromate oxidation, and NMR to successfully determine PyC in grassland soils from Brazil.

McBeath et al (2011) extended the NMR technique to the determination of the size of aromatic clusters in PyC using measurement of the chemical shift of ^{13}C-labeled benzene adsorbed to char PyC. This technique has yet to be applied to PCM chemically or thermally isolated from soil. NMR on a subset of samples is increasingly used to calibrate more cost-effective techniques such as mid-infrared spectroscopy (MIR; see next section) with MIR then used to determine PyC abundance on much larger soil sample sets (Rossell et al, 2019; De La Rosa et al, 2019; Jiménez-González et al, 2021).

The major advantage of ^{13}C NMR is that it quantifies a range of aromatic cluster sizes, thus including a large portion of the PCM continuum. Additionally, ^{13}C NMR provides detailed information on the nature of any other organic components in a sample. The disadvantages of NMR are that it is comparatively time-consuming and expensive, can potentially quantify non-PyC polycyclic aromatics as PyC (Gerke, 2019), and is generally used in conjunction with one or more other techniques, each introducing its own additional potential biases into the analysis.

Mid-infrared spectroscopy

Mid-infrared spectroscopy (MIR) measures the transmittance or absorbance of infrared radiation from a sample. Individual chemical bonds respond to stimulation at characteristic wavenumbers, enabling quantification by reference to calibration curves generated by comparison with other methods such as NMR. The technique was first exploited in PCM studies of marine sediment by Griffin and Goldberg (1975), using a technique involving basic peroxide oxidation followed by extended grinding to enhance the absorbance of 'elemental carbon' PCM at characteristic wave numbers (Smith et al, 1975). MIR has also long been used to characterize SOC (Skjemstad et al, 1993)

The partial least squares regression technique of Haaland and Thomas (1988) has been developed over two decades for use in characterizing SOC (Janik and Skjemstad, 1995). Janik et al (2007) used MIR to determine particulate organic C and char PyC in soil samples from Austria and Kenya, with quantification enabled by calibration (from 0-11 g char PyC per kilogram of soil) to values derived from UV-oxidation (Figure 24.5). Zimmermann et al (2007) quantified the 'resistant' SOC fraction, using MIR on 111 topsoil samples, after concentration of the resistant SOC fraction by chlorite oxidation of size-fractionated material. Bornemann et al (2008) predicted PyC in 300 soil samples using MIR calibrated using the BPCA technique, and also suggested that MIR prediction of the BPCA marker, mellitic acid, could be used as a measure of PyC condensation. Roth et al (2012) also concluded that MIR could be used to predict char and soot PyC. Michel et al (2009) and Hobley et al (2016) found that MIR could precisely determine char PyC in artificial soil–char mixtures. Various calibrations against other PCM analytical

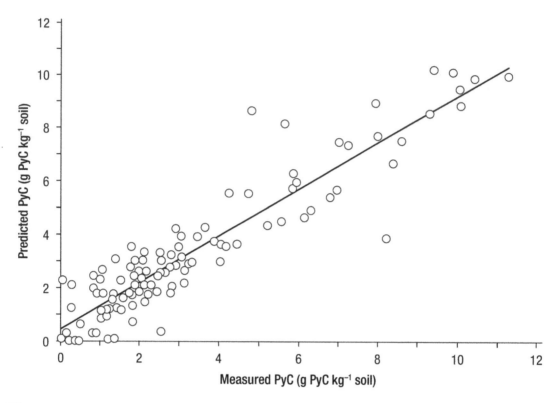

Figure 24.5 *Relationship between MIR predicted PyC and PyC measured by the UV oxidation technique (data from Janik et al, 2007)*

techniques including NMR, hypy, BPCA, and a range of oxidation techniques have been used to measure PyC abundances in soils and sediments locally (Cotrufo et al, 2016) and develop regional to continental maps of PyC abundance in natural soils (Sanderman et al, 2021; Rossell et al, 2019; Jiménez-González et al, 2021).

MIR has the distinct advantages that it is rapid, cheap, and can be used to predict the abundance of multiple soil components, including PyC. The disadvantages of MIR are that there are similarities between the functional groups present in PyC and SOC (Bellon-Maurel and McBratney, 2011) and the technique inherently requires calibration against results generated by an external PyC

quantification technique. There is also no global calibration for PyC in soil (Roth et al, 2012), and for optimal performance a calibration dataset should likely be created that is specific to the soils in a given study area.

Pyrene fluorescence loss

Flores-Cervantes et al (2009) tested the possibility of using pyrene Pyrene fluorescence loss (PFL) to quantify PyC in a range of sample types. This technique capitalizes on the ability of some forms of PyC to sorb polycyclic aromatic compounds and measures the loss of a fluorescence signal from pyrene spiked into a suspension of a solid sample as the pyrene is adsorbed to PyC in the sample. The standard chars and soil samples used in the Hammes et al (2007) ring trial returned significantly higher estimates of PyC by PFL than by the CTO-375 method, but the authors contend that the CTO-375 method only measures the most condensed PyC in a sample, and hence only a fraction of the total PyC in a sample. However, the abundances measured by PFL for soil and char standards were in several cases at least an order of magnitude higher than determined by any techniques used in the Hammes et al (2007) study, and further verification of the technique would be required before routine application to soil biochar studies.

Near edge X-ray absorption fine edge structure spectroscopy

Near edge X-ray absorption fine edge structure (NEXAFS) spectroscopy uses intense, synchrotron-generated, polarized X-rays to probe the surface (10–100 nm) of a sample to obtain absorption spectra that are characteristic of the local bonding environment of atoms. Keiluweit et al (2010) used

NEXAFS and several other techniques to quantify the molecular changes in biochar chemical structure as a function of the temperature of formation, from dominantly amorphous polyaromatic forms at lower temperatures to disordered graphitic crystallites at higher temperatures. Heymann et al (2011) used NEXAFS to analyze the samples used in the PyC ring trial of Hammes et al (2007) and found that NEXAFS could be used to identify characteristic aromatic C in the PyC reference samples. NEXAFS was also successfully used to distinguish PyC from potential interfering organic materials such as kerogen and melanoidin, but was unable to distinguish PyC from C in bituminous coal or lignite due to similarities in the absorbance spectra. Reasonable correlations were obtained with aromatic C abundances as determined by NMR but the authors considered that further research was required before NEXAFS could be considered a quantitative analytical tool and access to instrumentation is currently limited.

Molecular marker techniques

Benzene polycarboxylic acid

Benzene polycarboxylic acids (BPCAs) are a class of compounds composed of a benzene ring with one to six carboxyl groups. Glaser et al (1998) first suggested that BPCAs could be used to quantify PyC, as BPCAs are produced upon oxidation of PyC (Shafizadeh and Sekiguch, 1983) and are a significant component of alkaline ("humic") extracts from SOC, thought to be derived from PyC (Haumaier and Zech, 1995; Figure 24.6). Schneider et al (2010) measured BPCAs in a thermosequence of biochars and demonstrated that the number of carboxyl substitutions per benzene ring increases as the temperature of

formation of the char increases, due to the increasing size of polyaromatic clusters.

Brodowski et al (2005) modified the original method of Glaser et al (1998) to eliminate known methodological artefacts. The current method involves the removal of Fe and Al by 4 M trifluoro acetic acid (TFA) digestion at 105°C followed by conversion of PyC to its constituent BPCAs by nitric acid oxidation at 170°C for 8 hours. An ion exchange resin is used to remove cations, and the extract is then freeze-dried. The BPCAs in the purified extract are measured by gas chromatography of trimethylsilyl derivatives of the BPCAs with flame ionization

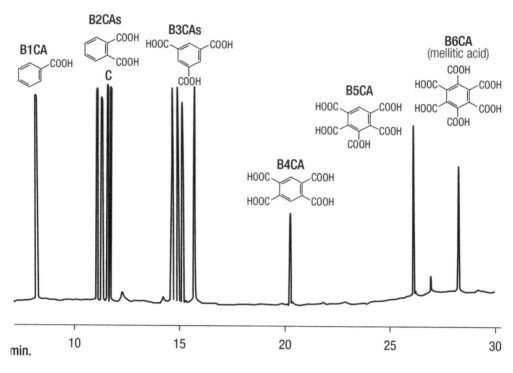

Figure 24.6 *Gas chromatogram of BPCAs with example compounds from B1CA to B6CA, with citric acid standard marked with 'C' (modified from Glaser et al, 1998)*

detection. Both Glaser et al (1998) and Brodowski et al (2005) used citric acid as an internal standard, but Schneider et al (2010) recommend phthalic acid, which is more stable in acid solutions and more similar in structure to BPCAs. It is also possible to quantify BPCAs using high-performance liquid chromatography, with the advantage that no derivatization is required and yield is improved (Dittmar, 2008; Schneider et al, 2011; Wiedemeier et al, 2013). To reduce the dependence of the technique on a lengthy (7-hour) nitric acid digestion in a high-pressure digestion apparatus, Glaser et al (2021) substituted a one-hour microwave-assisted digestion at 190°C. The study indicated that the results from microwave digestion were comparable

to the high-pressure digestion, but larger sample sizes were required.

The BPCA technique has the advantage of providing a measure of the degree of polycondensation (as mellitic acid abundance) of PyC while also capturing a relatively large region of the PyC continuum. The isolated BPCA compounds can also be further analyzed for stable and radiocarbon isotope composition, yielding valuable information on source and age (Glaser and Knorr, 2008; Rodionov et al, 2010; Ziolkowski and Druffel, 2010). The technique has been used, for example, to examine the dynamics of natural biochar aging (Abiven et al, 2011) and has also been found to provide reasonable correlations with MIR spectral data (Bournemann et al, 2008)

The disadvantages of the technique are that Hammes et al (2007) found that three laboratories varied by a factor of two in their analysis of PyC-containing ring trial standards, and also measured significant apparent PyC in samples known not to contain PyC. In the context of biochar C sequestration, the technique has the additional disadvantage that a component of the most condensed aromatic C (likely to persist on centennial timescales) is not affected by the nitric acid oxidation and hence the technique underestimates total PyC abundance, and non-polyaromatic PCM is also not identified. Glaser et al (1998) used a single correction factor of 2.27 to account for this, but Schneider et al (2010) have pointed out that this factor is very likely to vary widely depending on feedstock type and temperature of pyrolysis, and Kaal et al (2008b) also considered this factor too low for aged and partly depolymerized soil PCM.

It also seems to be the case that at least some BPCAs can be formed by mechanisms other than pyrolysis, particularly B3CA and B4CA. Glaser and Knorr (2008) found that up to 25% of BPCAs isolated from soils incubated with an isotope-labeled C source derived from the labeled C source and hence were not of pyrogenic origin but were fungal-derived PCM-like compounds. Chang et al (2018) have likewise identified non-pyrogenic BPCAs in soil fractions. Kappenberg et al (2016) found that multiple forms of non-pyrogenic organic matter yielded BPCAs following exposure to the oxidation treatment to liberate pyrogenic BPCAs. They concluded that only samples containing <5 mg organic carbon should be analyzed and that the abundances of only BPCAs with five and six carboxyl groups should be reported as pyrogenic. This restricts the utility of the technique to samples containing lower organic carbon and/or relatively high PCM abundances.

Pyrolysis gas chromatography – mass spectrometry (Py-GC-MS)

Py-GC-MS thermally decomposes large molecules, which cleave at their weakest points, to form more volatile smaller molecules that can retain a fingerprint of their origin. These molecules are swept into a gas chromatograph where they are separated from each other for quantification with subsequent identification by mass spectrometry. González-Vila et al (2001) demonstrated that useful structural and genetic information could be obtained by Py-GC-MS of char. More recent studies have demonstrated that Py-GC-MS is a powerful tool for detailed examination of the composition of PCM from a range of sources (Song and Peng, 2010; Nocentini et al, 2010; Kaal et al, 2012), the degree of thermal alteration of PCM (Kaal and Rumpel, 2009; De la Rosa et al, 2008; Fabbri et al, 2012), the natural aging of PyC in soil (Kaal et al, 2008b; Calvelo Pereira et al, 2013; Chen et al, 2020), the composition of PyC derived from slash and burn agriculture (Rumpel et al, 2007) and the contribution of compounds derived from PyC to total SOC and the base soluble component of SOC (Kaal et al, 2008a).

Py-GC-MS has the significant advantage that it can provide detailed structural information on complex mixtures of compounds (Figure 24.7). It is most suited to the characterization of PCM itself, or PCM in soil that was formed at temperatures below ~500°C (Kaal et al, 2012) as a significant proportion of PyC is stable under the pyrolysis conditions used and so escapes the analytical window. The disadvantage of Py-GC-MS for studies of biochar in soil is that, at best, the technique provides only a semi-quantitative of PyC abundance that likely misses a substantial fraction of the more condensed PyC of interest for its stability on longer timescales (Rombolá et al, 2016). An additional complication is that small

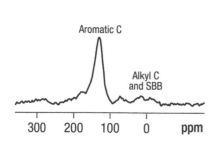
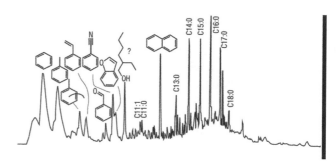

Figure 24.7 *^{13}C CP-MAS-NMR spectrum (left) and total ion pyrogram (right) for fine PCM from an agricultural soil from Laos (FF) after demineralization and dichromate oxidation, showing relative abundance and range of chemical compositions for PyC and aliphatic compounds and aliphatic in the sample (modified from Kaal and Rumpel, 2009)*

amounts of PyC molecules may be produced during the pyrolysis step, before analysis (Sáiz-Jiménez, 1994; Kaal and Rumpel, 2009), which adds ambiguity to detecting PCM by Py-GC-MS, especially in soil with low PyC content.

Levoglucosan

Simoneit et al (1999) demonstrated that levoglucosan – a monosaccharide formed exclusively from the thermal breakdown of cellulose – was present in aerosols and could be used as a specific marker of biomass burning that involved cellulose combustion.

Kuo et al (2008) examined the possibility that the utility of levoglucosan could be extended to quantify PyC in environmental media. They found that the compound was only present in char formed at <350°C and that its abundance in samples formed at identical temperatures varied widely between species. This means the technique cannot be used for quantification, but does provide a useful marker of low-temperature, cellulose-derived PCM in a sample. The results of Knicker et al (2013) suggest that the compound is efficiently microbially degraded and unlikely to persist in the soil for any length of time.

Conclusions

The potential significance of biochar as a tool for climate change mitigation, as well as the enhancement of soil resilience and crop productivity, has focused significant effort over the last decade on the development of a growing number of analytical tools for the quantification and isolation of biochar PCM in soil. Many of the techniques reviewed above were originally developed outside soil science and have been adapted *post-hoc* for

use in biochar-related studies and it is clear that no one technique represents a 'magic bullet'. Each technique targets a different component of the PyC continuum from the relatively labile to the most condensed components of PyC. The selection of a technique should therefore depend on project-specific requirements that also lie along a continuum that must balance the precision, cost, complexity, and availability of one technique

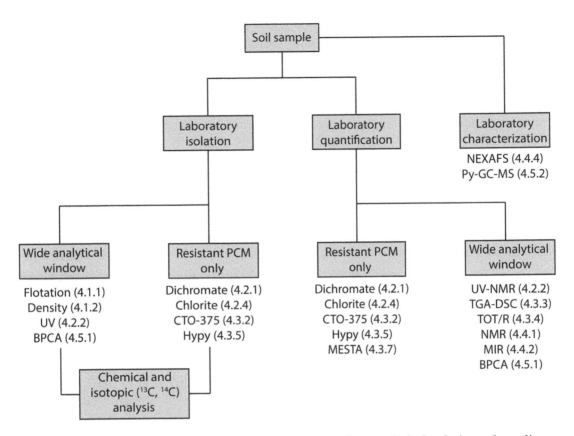

Figure 24.8 *Decision tree to enable selection of the appropriate analytical techniques depending on project requirements. Only those techniques in relatively common usage and that have been validated for soil PCM have been included. Note that many techniques that are primarily used for quantification can also provide information useful for characterization. For further details see the relevant section (given in brackets after the technique)*

against another. The characteristics of each technique are summarized in Table 24.1 and a decision tree is designed to guide the selection of the most appropriate technique for a particular application (Figure 24.8).

It is now abundantly clear from several inter-comparison studies (Schmidt et al, 2001; Hammes et al, 2007; Roth et al, 2012; Meredith et al, 2012) that the available techniques yield widely divergent estimates of PyC on the same standard materials. Comparison of different techniques on project-specific sample sets also

shows that while results are generally correlated, the absolute abundances measured by different techniques for the same samples, are different (Cotrufo et al, 2016; Raya-Moreno et al, 2017; Murano et al, 2021; Hardy et al, 2022). While this means that the results from individual studies may be internally consistent, it becomes difficult to compare results across studies or to extrapolate results to broader geographic regions (Reisser et al, 2016).

Most of the variability between techniques is because different techniques preferentially measure C from different regions of the PyC continuum, but it is also the case that some techniques vary in their ability to uniquely quantify PyC in complex matrices. Zimmerman and Mitra (2017) noted that for many techniques, there is a significant correlation between total organic carbon content and PCM content, which suggests that for some techniques at least, a component of non-PCM carbon is analyzed as PCM carbon, although because PCM contributes C to TOC and the same processes that stabilize TOC also stabilize PCM (Cusack et al, 2012), it may be that some correlation should be expected. The correlations have been clearly demonstrated through experiments for at least dichromate oxidation, chemothermal oxidation, NMR, Hypy, and BPCA analysis as discussed above. Although corrections can and have been developed and applied for some techniques, these uncertainties flow through into measurements by, for example, MIR that rely on calibration data from another technique. It is also worth noting that the 'aging' of biochar and ongoing environmental degradation over time in the field may progressively change the proportion of biochar that is quantified as PyC by most of the techniques discussed here (Kaal et al, 2008b; Calvelo Pereira et al, 2013).

Inter-comparison studies have also demonstrated that there can be a significant range in results reported by different labs for the same standards, using the same technique (Figure 24.2). This inter-laboratory variability has two likely sources. First, several techniques involve multiple pre-treatment steps, with each handling step potentially introducing a bias into the results that is lab-specific. Second, for most techniques, there remains a lack of defined and agreed protocols such that, while the reagents used may be broadly similar, the concentrations, temperatures, and times of contact between

sample and reagent are not. These issues can be addressed to some extent by implementing the routine analysis of standards such as those developed by the black carbon ring trial (Hammes et al, 2007) in each laboratory so that the results of that laboratory can be placed in the broader context. Ideally, protocols would be standardized across all laboratories, but this is unlikely as each technique may have to be varied to optimize performance on specific sample types and depending, for example, on whether isolation, quantification, or characterization is the primary goal. Zimmerman and Mitra (2017) advocate for further standardization of protocols used in different labs and a further round of inter-laboratory comparisons to address some of the issues identified above.

In the specific context of soil-related biochar studies, the fact that biochar added to soil is generally of comparatively large particle size means that low-technology options such as sieving, flotation, and density separation may be sufficient for isolation of biochar from the soil matrix for quantification, particularly in field areas where analytical infrastructure is lacking. For many requirements, these simple techniques may be entirely adequate.

The most promising option for a 'universal', rapid, low-cost technique for quantifying biochar in soil is MIR with the caveat that potentially site-specific calibration is required, which involves the application of another technique with greater selectivity for biochar C to a large subset of samples to develop a calibration. While MIR has found wide application in chemometrics approaches to infer the spatial distribution of natural PyC abundances in soil (Rossell et al, 2019; Jiménez-González et al, 2021; Sanderman et al, 2021) there has yet to be a significant effort to apply the technique to biochar-related soil investigations. At this point, BPCA (all aromatic PyC, except the

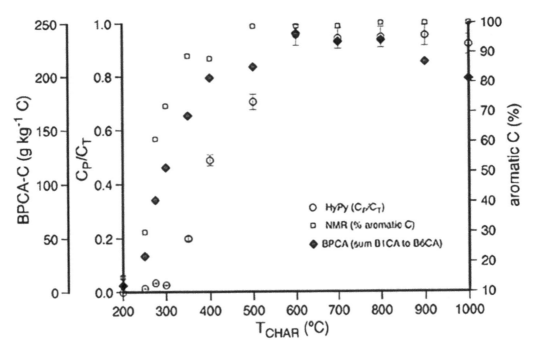

Figure 24.9 *Proportions of biochar determined as PyC by the definitions used in BPCA (section 4.5.1), Hypy (section 4.3.5), and NMR (section 4.4.1) analysis using a thermosequence produced from the same chestnut wood feedstock (redrawn from Wuster et al, 2013)*

most condensed polyaromatic PyC component), hypy (all polyaromatic PyC with ring size greater than ~7), and NMR (all aromatic PyC) hold the most promise as techniques that can provide clearly defined, reproducible measures of the broadest part of the PyC continuum with which to calibrate the MIR technique. These techniques, of course, do not yield the same abundance of PyC in the same samples, particularly in the lower temperature range characteristic of slow pyrolysis biochars, as shown in Figure 24.9. However, these techniques are all amenable to further cross-calibration such that a harmonized and standardized set of techniques to enable the robust determination of PyC abundance in soil-related biochar studies can be achieved.

References

Abiven S, Hengartner P, Schneider MP, Singh N, and Schmidt MW 2011 Pyrogenic carbon soluble fraction is larger and more aromatic in aged charcoal than in fresh charcoal. *Soil Biology and Biochemistry* 43, 1615–1617.

Agarwal T and Bücheli TD 2011 Adaptation, validation and application of the chemothermal oxidation method to quantify black carbon in soils. *Environmental Pollution* 159, 532–538.

Arnold JR and Libby WF 1949 Age determinations by radiocarbon content: Checks with samples of known age. *Science* 110, 678–680.

Ascough PL, et al 2009 Hydropyrolysis as a new tool for radiocarbon pre-treatment and the quantification of black carbon. *Quaternary Geochronology* 4, 140–147.

Ascough PL, et al 2010 Hydropyrolysis: Implications for radiocarbon pre-treatment and characterization of Black carbon. *Radiocarbon* 52, 1336–1350.

Ascough PL, et al 2011 Variability in oxidative degradation of charcoal: Influence of production conditions and environmental exposure. *Geochimica et Cosmochimica Acta* 75, 2361–2378.

Ball DF 1964 Loss-on-ignition as an estimate of organic matter and organic carbon in non-calcareous soils. *Journal of Soil Science* 15, 84–92.

Barron PF, Wilson MA, Stephens JF, Cornell BA, and Tate KR 1980 Cross-polarization ^{13}C NMR spectroscopy of whole soils. *Nature* 286, 585–587.

Bartůněk V, et al 2017 Obtaining black carbon—A simple method for the safe removal of mineral components from soils and archaeological layers. *Archaeometry* 592, 346–355.

Bellon-Maurel V and McBratney A 2011 Near-infrared NIR and mid-infrared MIR spectroscopic techniques for assessing the amount of carbon stock in soils – Critical review and research perspectives. *Soil Biology and Biochemistry* 43, 1398–1410

Berkowitz LM and Rylander PN 1958 Use of ruthenium tetroxide as a multi-purpose oxidant. *Journal of the American Chemical Society* 80, 6682–6684.

Bird MI and Cali JA 1998 A million-year record of fire in sub-Saharan Africa. *Nature* 394, 767–769.

Bird MI and Gröcke DR 1997 Determination of the abundance and carbon isotope composition of elemental carbon in sediments. *Geochimica et Cosmochimica Acta* 61, 3413–3423.

Bird MI, Veenendaal EM, Moyo C, Lloyd J, and Frost P 2000 Effect of fire and soil texture on soil carbon in a sub-humid savanna Matopos, Zimbabwe. *Geoderma* 94, 71–90.

Bird MI, et al 2002 Radiocarbon analysis of the early archaeological site of Nauwalabila I, Arnhem Land, Australia: Implications for sample suitability and stratigraphic integrity. *Quaternary Science Reviews* 21, 1061–1075.

Bodner CC and Rowlett RM 1980 Separation of bone, charcoal, and seeds by chemical flotation. *American Antiquity* 45, 110–116.

Boone RD, Grigal DF, Sollins P, Ahrens RJ, and Armstrong DE 1999 Soil sampling, preparation, archiving, and quality control. In: Robertson GP, Coleman DC, Bledsoe, CS and Sollins P (Eds) *Standard Soil Methods for Long-term Ecological Research*. New York: Oxford University Press. pp 3–28.

Bornemann L, Welp G, Brodowski S, Rodionov A, and Amelung W 2008 Rapid assessment of black carbon in soil organic matter using mid-infrared spectroscopy. *Organic Geochemistry* 39, 1537–1544.

Brodowski S, Rodionov A, Haumaier L, Glaser B, and Amelung W 2005 Revised black carbon assessment using benzene polycarboxylic acids. *Organic Geochemistry* 36, 1299–1310.

Bucheli TD, Blum F, Desaules A, and Gustafsson Ö 2004 Polycyclic aromatic hydrocarbons, black carbon, and molecular markers in soils of Switzerland. *Chemosphere* 56, 1061–1076.

Budai A, et al 2013 Justification for the "Standard test method for estimating biochar stability BC+100". In: Koper T, et al (Eds) *Methodology for Biochar Projects, Version 1.0*, American Carbon Registry, pp 104–129, accessed 8 November 2013.

Cachier H, Bremond MP, and Buat-Ménard P 1989 Determination of atmospheric soot carbon with a simple thermal method. *Tellus B* 41, 379–390.

Calvelo Pereira R, et al 2011 Contribution to characterisation of biochar to estimate the labile fraction of carbon. *Organic Geochemistry* 42, 1331–1342.

Calvelo Pereira R, et al 2013 Detailed carbon chemistry in charcoals from pre-European Māori gardens of New Zealand as a tool for understanding biochar stability in soils. *European Journal of Soil Science* 65, 83–95.

Chang Z, et al 2018 Benzene polycarboxylic acid—A useful marker for condensed organic matter, but not for only pyrogenic black carbon. *Science of the Total Environment* 626, 660–667.

Chen H, Rhoades CC, and Chow AT 2020 Characteristics of soil organic matter 14 years after a wildfire: A pyrolysis-gas-chromatography mass spectrometry Py-GC-MS study. *Journal of Analytical and Applied Pyrolysis* 152, 104922.

Chow JC, et al 1993 The DRI thermal/optical reflectance carbon analysis system: Description, evaluation and applications in US air quality studies. *Atmospheric Environment Part A General Topics* 27, 1185–1201.

Cotrufo MF, et al 2016 Quantification of pyrogenic carbon in the environment: An integration of analytical approaches. *Organic Geochemistry* 100, 42–50.

Cross A and Sohi SP 2013 A method for screening the relative long-term stability of biochar. *Global Change Biology – Bioenergy* 5, 215–220.

Cusack DF, Chadwick OA, Hockaday WC and Vitousek, PM 2012 Mineralogical controls on soil black carbon preservation. *Global Biogeochemical Cycles* 26, GB2019.

De la Rosa JM, et al 2008a Determination of refractory organic matter in marine sediments by chemical oxidation, analytical pyrolysis and solid-state 13C nuclear magnetic resonance spectroscopy. *European Journal of Soil Science* 59, 430–438.

De la Rosa JM, et al 2008b Direct detection of black carbon in soils by Py-GC/MS, carbon-13 NMR spectroscopy and thermogravimetric techniques. *Soil Science Society of America Journal* 72, 258–267.

De la Rosa JM, Rosado M, Paneque M, Miller AZ, and Knicker H 2018 Effects of aging under field conditions on biochar structure and composition: Implications for biochar stability in soils. *Science of the Total Environment* 613, 969–976.

De la Rosa JM, Jiménez-González MA, Jiménez-Morillo NT, Knicker H, and Almendros G 2019 Quantitative forecasting black pyrogenic carbon in soils by chemometric analysis of infrared spectra. *Journal of Environmental Management* 251, 109567.

Dittmar T 2008 The molecular level determination of black carbon in marine dissolved organic matter. *Organic Geochemistry* 39, 396–407.

Ellert BH and Bettany JR 1995 Calculation of organic matter and nutrients stored in soils under contrasting management regimes. *Canadian Journal of Soil Science* 75, 529–538.

Elmquist M, Gustafsson O, and Andersson P 2004 Quantification of sedimentary black carbon using the chemothermal oxidation method: An evaluation of ex situ pre-treatments and standard additions approaches. *Limnology and Oceanography: Methods* 2, 417–427.

Elmquist M, Cornelissen G, Kukulska Z, and Gustafsson Ö 2006 Distinct oxidative stabilities of char versus soot black carbon: Implications for quantification and environmental recalcitrance. *Global Biogeochemical Cycles* 20, GB2009.

Emiliani C, Price DA, and Seipp J 1991 Is the post-glacial artificial? In: Taylor HP O'Neil JR and Kaplan IR (Eds) *Stable-isotope Geochemistry: A tribute to Samuel Epstein.* Geochemical Society Special Publications 3, pp 229–231.

Fabbri D, Torri C, and Spokas KA 2012 Analytical pyrolysis of synthetic chars derived from biomass with potential agronomic application biochar relationships with impacts on microbial carbon dioxide production. *Journal of Analytical and Applied Pyrolysis* 93, 77–84.

Flores-Cervantes DX, Reddy CM, and Gschwend PM 2009 Inferring black carbon concentrations in particulate organic matter by observing pyrene fluorescence losses. *Environmental Science and Technology* 43, 4864–4870.

Gélinas Y, Prentice KM, Baldock JA, and Hedges JI 2001 An improved thermal oxidation method for the quantification of soot/graphitic black carbon in sediments and soils. *Environmental Science and Technology* 35, 3519–3525.

Gerke J 2019 Black pyrogenic carbon in soils and waters: A fragile data basis extensively interpreted. *Chemical and Biological Technologies in Agriculture* 61, 1–8.

Glaser B, Haumaier L, Guggenberger G, and Zech W 1998 Black carbon in soils: The use of benzenecarboxylic acids as specific markers. *Organic Geochemistry* 29, 811–819

Glaser B and Knorr KH 2008 Isotopic evidence for condensed aromatics from non-pyrogenic sources in soils-implications for current methods for quantifying soil black carbon. *Rapid Communications in Mass Spectrometry* 22, 935–942.

Glaser B and Birk JJ 2012 State of the scientific knowledge on properties and genesis of Anthropogenic Dark Earths in Central Amazonia terra preta de Índio. *Geochimica et Cosmochimica Acta* 82, 39–51.

Glaser B, Guenther M, Maennicke H, and Bromm T 2021 Microwave-assisted combustion to produce benzene polycarboxylic acids as molecular markers for biochar identification and quantification. *Biochar* 34, 407–418.

González-Vila FJ, Tinoco P, Almendros G, and Martin F 2001 Pyrolysis-GC-MS analysis of the formation and degradation stages of charred residues from lignocellulosic biomass. *Journal of Agricultural and Food Chemistry* 49, 1128–1131.

Griffin JJ and Goldberg ED 1975 The fluxes of elemental carbon in coastal marine sediments. *Limnology and Oceanography* 20, 456–463. 10.4319/lo.1975.20.3.0456.

Gumerman IV G and Umemoto BS 1987 The siphon technique: An addition to the flotation process. *American Antiquity* 52, 330–336.

Gustafsson Ö, Haghseta F, Chan C, MacFarlane J, and Gschwend PM 1996 Quantification of the dilute sedimentary soot phase: Implications for PAH speciation and bioavailability. *Environmental Science and Technology* 31, 203–209.

Gustafsson Ö, et al 2001 Evaluation of a protocol for the quantification of black carbon in sediments. *Global Biogeochemical Cycles* 15, 881–890.

Haaland DM and Thomas EV 1988 Partial least-squares methods for spectral analyses 1 Relation to other quantitative calibration methods and the extraction of qualitative information. *Analytical Chemistry* 60, 1193–1202.

Haig J, Ascough PL, Wurster CM, and Bird MI 2020 A rapid throughput technique to isolate pyrogenic carbon by hydrogen pyrolysis for stable isotope and radiocarbon analysis. *Rapid Communications in Mass Spectrometry* 3410, e8737.

Hamilton GA, and Hartnett HE 2013 Soot black carbon concentration and isotopic composition in soils from an arid urban ecosystem. *Organic Geochemistry* 59, 87–94.

Hammes K, et al 2007 Comparison of quantification methods to measure fire-derived black/elemental carbon in soils and sediments using reference materials from soil, water, sediment and the atmosphere. *Global Biogeochemical Cycles* 21, GB3016

Han Y, et al 2007a Evaluation of the thermal/optical reflectance method for discrimination between char-and soot-EC. *Chemosphere* 69, 569–574.

Han Y, et al 2007b Evaluation of the thermal/optical reflectance method for quantification of elemental carbon in sediments. *Chemosphere* 69, 526–533.

Hardy B, et al 2017 Evaluation of the long-term effect of biochar on properties of temperate agricultural soil at pre-industrial charcoal kiln sites in Wallonia, Belgium. *European Journal of Soil Science* 681, 80–89.

Hardy B, Borchard N, and Leifeld J 2022 Identification of thermal signature and quantification of charcoal in soil using differential scanning calorimetry and benzene polycarboxylic acid (BPCA) markers. *Soil* 8(2), 1–27.

Haumaier L and Zech W 1995 Black carbon—possible source of highly aromatic components of soil humic acids. *Organic Geochemistry* 23, 191–196.

Heymann K, Lehmann J, Solomon D, Schmidt MW, and Regier T 2011 C 1s K-edge near edge X-ray absorption fine structure NEXAFS spectroscopy for characterizing functional group chemistry of black carbon. *Organic Geochemistry* 42, 1055–1064.

Hobley EU, Brereton AG, and Wilson B 2016 Soil charcoal prediction using attenuated total reflectance mid-infrared spectroscopy. *Soil Research* 551, 86–92.

Hsieh YP 2007 A novel multielemental scanning thermal analysis MESTA method for the identification and characterization of solid substances. *Journal of AOAC International* 90, 54–59.

Hsieh, YP and Bugna GC 2008 Analysis of black carbon in sediments and soils using multi-element scanning thermal analysis MESTA. *Organic Geochemistry* 39, 1562–1571.

Huang L, et al 2006 Stable isotope measurements of carbon fractions OC/EC in airborne particulate: A new dimension for source characterization and apportionment. *Atmospheric Environment* 40, 2690–2705.

Ikeya K, Hikage T, Arai S, and Watanabe A 2011 Size distribution of condensed aromatic rings in various soil humic acids. *Organic Geochemistry* 42, 55–61.

Janik LJ, Skjemstad JO, Shepherd KD, and Spouncer LR 2007 The prediction of soil carbon fractions using mid-infrared-partial least square analysis. *Australian Journal of Soil Research* 45, 73–81.

Janik LJ and Skjemstad JO 1995 Characterization and analysis of soils using mid-infrared partial least-squares 2 Correlations with some laboratory data. *Australian Journal of Soil Research* 33, 637–650.

Jiménez-González MA, et al 2021 Spatial distribution of pyrogenic carbon in Iberian topsoils estimated by chemometric analysis of infrared spectra. *Science of The Total Environment* 790, 148170.

Juhasz AL, Stanley GA, and Britz ML 2000a Microbial degradation and detoxification of high molecular weight polycyclic aromatic hydrocarbons by Stenotrophomonas maltophilia strain VUN 10,003. *Letters in Applied Microbiology* 30, 396–401.

Juhasz AL and Naidu R 2000b Bioremediation of high molecular weight polycyclic aromatic hydrocarbons: A review of the microbial degradation of benzo a pyrene. *International Journal of Biodeterioration and Biodegradation* 45, 57–88.

Kaal J, Martínez-Cortizas A, Niero KG, and Buurman P 2008a A detailed pyrolysis-GC/MS analysis of a black carbon-rich acidic colluvial soil Atlantic ranker from NW Spain. *Applied Geochemistry* 23, 2395–2405.

Kaal J, Brodowski S, Baldock JA, Nierop KG, and Cortizas AM 2008b Characterisation of aged black carbon using pyrolysis-GC/MS, thermally assisted hydrolysis and methylation THM, direct and cross-polarisation ^{13}C nuclear magnetic resonance DP/CP NMR and the benzenepolycarboxylic acid BPCA method. *Organic Geochemistry* 39, 1415–1426.

Kaal J and Rumpel C 2009 Can pyrolysis-GC/MS be used to estimate the degree of thermal alteration of black carbon? *Organic Geochemistry* 40, 1179–1187.

Kaal J, Schneider MP, and Schmidt MW 2012 Rapid molecular screening of black carbon biochar thermosequences obtained from chestnut wood and rice straw: A pyrolysis-GC/MS study. *Biomass and Bioenergy* 45, 115–129.

Kappenberg A, Bläsing M, Lehndorff E, and Amelung W 2016 Black carbon assessment using benzene polycarboxylic acids: limitations for organic-rich matrices. *Organic Geochemistry* 94, 47–51.

Keiluweit M, Nico PS, Johnson MG, and Kleber M 2010 Dynamic molecular structure of plant biomass-derived black carbon biochar. *Environmental Science and Technology* 44, 1247–1253.

Kenward HK, Hall AR, and Jones AKG 1980 A tested set of techniques for the extraction of plant and animal macrofossils from waterlogged archaeological deposits. *Science and Archaeology* 22, 3–15.

Kloss S, Sass O, Geitner C, and Prietzel J 2012 Soil properties and charcoal dynamics of burnt soils in the Tyrolean Limestone Alps. *Catena* 99, 75–82.

Knicker H, Müller P, and Hilscher A 2007 How useful is chemical oxidation with dichromate for the determination of "Black Carbon" in fire-affected soils? *Geoderma* 142, 178–196.

Knicker H, Wiesmeier M, and Dick DP 2008 A simplified method for the quantification of pyrogenic organic matter in grassland soils via chemical oxidation. *Geoderma* 147, 69–74.

Knicker H, Hilscher A, de la Rosa JM, González-Pérez JA, and González-Vila FJ 2013 Modification of biomarkers in pyrogenic organic matter during the initial phase of charcoal biodegradation in soils. *Geoderma* 197, 43–50.

Koide RT, Petprakob K, and Peoples M 2011 Quantitative analysis of biochar in field soil. *Soil Biology and Biochemistry* 43, 1563–1568.

Kuhlbusch TAJ 1995 Method for determining black carbon in residues of vegetation fires. *Environmental Science and Technology* 29, 2695–2702.

Kuo LJ, Herbert BE, and Louchouarn P 2008 Can levoglucosan be used to characterize and quantify char/charcoal black carbon in environmental media? *Organic Geochemistry* 39, 1466–1478.

Laird LD and Campbell ID 2000 High resolution palaeofire signals from Christina Lake, Alberta: A comparison of the charcoal signals extracted by two different methods. *Palaeogeography, Palaeoclimatology, Palaeoecology* 164, 111–123.

Lehmann J, et al 2008 Australian climate-carbon cycle feedback reduced by soil black carbon. *Nature Geoscience* 1, 832–835.

Leifeld J 2007 Thermal stability of black carbon characterised by oxidative differential scanning calorimetry. *Organic Geochemistry* 38, 112–127.

Lim B and Cachier H 1996 Determination of black carbon by chemical oxidation and thermal treatment in recent marine and lake sediments and Cretaceous-Tertiary clays. *Chemical Geology* 131, 143–154.

Liu Z, et al 2019 The responses of soil organic carbon mineralization and microbial communities to fresh and aged biochar soil amendments. *Global Change Biology Bioenergy* 1112, 1408–1420.

Lopez-Capel E, Sohi SP, Gaunt JL, and Manning DA 2005 Use of thermogravimetry-differential scanning calorimetry to characterize modelable soil organic matter fractions. *Soil Science Society of America Journal* 69, 136–140.

Love GD, Snape CE, Carr AD, and Houghton RC 1995 Release of covalently-bound alkane biomarkers in high yields from kerogen via catalytic hydropyrolysis. *Organic Geochemistry* 23, 981–986.

Love GD, McAulay A, Snape CE, and Bishop AN 1997 Effect of process variables in catalytic hydropyrolysis on the release of covalently bound aliphatic hydrocarbons from sedimentary organic matter. *Energy and Fuels* 11, 522–531.

Major J 2009 A guide to conducting biochar trials, International Biochar Initiative, http://www.carbon-negative.us/docs/IBI%20Biochar%20Trial%20Guide%20v1-1.pdf, accessed 8 November 2013

Major J, Rondon M, Molina D, Riha S, and Lehmann J 2010a Maize yield and nutrition during 4 years after biochar application to a Colombian savanna oxisol. *Plant and Soil* 333, 117–128.

Major J, Lehmann J, Rondon M and Goodale C 2010b Fate of soil-applied black carbon: Downward migration, leaching and soil respiration. *Global Change Biology* 16, 1366–1379.

Manning DA, Lopez-Capel E, White ML, and Barker S 2008 Carbon isotope determination for separate components of heterogeneous materials using coupled thermogravimetric analysis/isotope ratio mass spectrometry. *Rapid Communications in Mass Spectrometry* 22, 1187–1195.

Matson FR 1955 Charcoal concentration from early sites for radiocarbon dating. *American Antiquity* 21, 162–169.

Masiello CA, Druffel ERM, and Currie LA 2002 Radiocarbon measurements of black carbon in aerosols and ocean sediments. *Geochimica et Cosmochimica Acta* 66, 1025–1036.

McBeath AV, Smernik RJ, Schneider MP, Schmidt MW, and Plant EL 2011 Determination of the aromaticity and the degree of aromatic condensation of a thermosequence of wood charcoal using NMR. *Organic Geochemistry* 42, 1194–1202.

McBeath AV, Wurster CM, and Bird MI 2015 Influence of feedstock properties and pyrolysis conditions on biochar carbon stability as determined by hydrogen pyrolysis. *Biomass and Bioenergy* 73, 155–173.

Meredith W, et al 2012 Assessment of hydropyrolysis as a method for the quantification of black carbon using standard reference materials. *Geochimica et Cosmochimica Acta* 97, 131–147.

Meredith W, et al 2013 Direct evidence from hydropyrolysis for the retention of long alkyl moieties in black carbon fractions isolated by acidified dichromate oxidation. *Journal of Analytical and Applied Pyrolysis* 103, 232–239.

Michel K, Terhoeven-Urselmans T, Nitschke R, Steffa P, and Ludwig B 2009 Use of near-and mid-infrared spectroscopy to distinguish carbon and nitrogen originating from char and forest-floor material in soils. *Journal of Plant Nutrition and Soil Science* 172, 63–70.

Middelburg JJ, Nieuwenhuize J, and van Breugel P 1999 Black carbon in marine sediments. *Marine Chemistry* 65, 245–252.

Murano H, et al 2021 Quantification methods of pyrogenic carbon in soil with soil as a complex matrix: Comparing the CTO-375 and Cr_2O_7 methods. *Soil Science and Plant Nutrition* 674, 380–388.

Naisse C, et al 2013 Can biochar and hydrochar stability be assessed with chemical methods? *Organic Geochemistry* 60, 40–44.

Nakhli SAA, Panta S, Brown JD, Tian J, and Imhoff PT 2019 Quantifying biochar content in a field soil with varying organic matter content using a two-temperature loss on ignition method. *Science of The Total Environment* 658, 1106–1116.

Nam JJ, Sweetman AJ, and Jones KC 2009 Polynuclear aromatic hydrocarbons PAHs in global background soils. *Journal of Environmental Monitoring* 11, 45–48.

Nelson PN and Baldock JA 2005 Estimating the molecular composition of a diverse range of natural organic materials from solid-state 13C NMR and elemental analyses. *Biogeochemistry* 72, 1–34.

Nguyen TH, Brown RA, and Ball WP 2004 An evaluation of thermal resistance as a measure of black carbon content in diesel soot, wood char, and sediment. *Organic Geochemistry* 35, 217–234.

Nguyen BT, et al 2008 Long-term black carbon dynamics in cultivated soil. *Biogeochemistry* 89, 295–308.

Nocentini C, et al 2010 Charcoal mineralisation potential of microbial inocula from burned and unburned forest soil with and without substrate addition. *Soil Biology and Biochemistry* 42, 1472–1478.

Oen AM, Breedveld GD, Kalaitzidis S, Christanis K, and Cornelissen G 2006 How quality and quantity of organic matter affect polycyclic aromatic hydrocarbon desorption from Norwegian harbor sediments. *Environmental Toxicology and Chemistry* 25, 1258–1267.

Orr TJ, Wurster CM, Levchenko V, Ascough PL, and Bird MI 2021 Improved pretreatment method for the isolation and decontamination of pyrogenic carbon for radiocarbon dating using hydrogen pyrolysis. *Quaternary Geochronology* 61, 101124.

Piper CS 1942 Organic matter. In: Piper CS (ed) *Soil and Plant Analysis: A Laboratory Manual of Methods for the Examination of Soils and the Determination of the Inorganic Constituents of Plants*, The University of Adelaide, Adelaide, Australia, pp 213–229.

Poot A, Quik JT, Veld H, and Koelmans AA 2009 Quantification methods of black carbon: comparison of rock-eval analysis with traditional methods. *Journal of Chromatography A* 1216, 613–622.

Qi F, et al 2017 Pyrogenic carbon in Australian soils. *Science of the Total Environment* 586, 849–857.

Quénéa K, et al 2005 Study of the composition of the macromolecular refractory fraction from an acidic sandy forest soil Landes de Gascogne, France using chemical degradation and electron microscopy. *Organic Geochemistry* 36, 1151–1162.

Raya-Moreno I, Cañizares R, Domene X, Carabassa V, and Alcañiz JM 2017 Comparing current chemical methods to assess biochar organic carbon in a Mediterranean agricultural soil amended with two different biochars. *Science of the Total Environment* 598, 604–618.

Reisser M, Purves RS, Schmidt MWI, and Abiven S 2016 Pyrogenic carbon in soils: A literature-based inventory and a global estimation of its content in soil organic carbon and stocks. *Frontiers in Earth Science* 4, 80.

Rodionov A, et al 2010 Black carbon in grassland ecosystems of the world. *Global Biogeochemical Cycles* 24, GB3013.

Rombola AG, Fabbri D, Meredith W, Snape CE, and Dieguez-Alonso A 2016 Molecular characterization of the thermally labile fraction of biochar by hydropyrolysis and pyrolysis-GC/MS. *Journal of Analytical and Applied Pyrolysis* 121, 230–239.

Rose NL 1990 A method for the extraction of carbonaceous particles from lake sediment. *Journal of Paleolimnology* 3, 45–53.

Roth PJ, et al 2012 Differentiation of charcoal, soot and diagenetic carbon in soil: Method comparison and perspectives. *Organic Geochemistry* 46, 66–75.

Rumpel C, et al 2006 Black carbon contribution to soil organic matter composition in tropical sloping land under slash and burn agriculture. *Geoderma* 130, 35–46.

Rumpel C, et al 2007 Composition and reactivity of morphologically distinct charred materials left after slash-and-burn practices in agricultural tropical soils. *Organic Geochemistry* 38, 911–920.

Sanderman J, et al 2021 Soil organic carbon fractions in the Great Plains of the United States: An application of mid-infrared spectroscopy. *Biogeochemistry* 1561, 97–114.

Sanford JR, Larson RA, and Runge T 2019 Nitrate sorption to biochar following chemical oxidation. *Science of the Total Environment* 669, 938–947.

Sáiz-Jiménez C 1994 Production of alkylbenzenes and alkylnaphthalenes upon pyrolysis of unsaturated fatty acids. *Naturwissenschaften* 81, 451–453.

Schmidt MW, Knicker H, Hatcher PG, and Kögel-Knabner I 2000 Airborne contamination of forest soils by carbonaceous particles from industrial coal processing. *Journal of Environmental Quality* 29, 768–777.

Schmidt MWI and Noack AG 2000 Black carbon in soils and sediments: Analysis, distribution, implications, and current challenges. *Global Biogeochemical Cycles* 14, 777–793.

Schmidt MWI, et al 2001 Comparative analysis of black carbon in soils. *Global Biogeochemical Cycles* 15, 163–167.

Schneider MP, Hilf M, Vogt UF, and Schmidt MWI 2010 The benzene polycarboxylic acid BPCA pattern of wood pyrolyzed between 200°C and 1000°C. *Organic Geochemistry* 41, 1082–1088.

Schneider MP, Smittenberg RH, Dittmar T, and Schmidt MWI 2011 Comparison of gas with liquid chromatography for the determination of benzenepolycarboxylic acids as molecular tracers of black carbon. *Organic Geochemistry* 42, 275–282.

Schollenberger CJ 1927 A rapid approximate method for determining soil organic matter. *Soil Science* 24, 65–68.

Schreiner O and Brown MP 1912 Occurrence and nature of carbonized material in soil. *US Bureau of Soils Bulletin* 90, 5–28.

Schwartz V 1995 Fractionated combustion analysis of carbon in forest soils—new possibilities for the analysis and characterization of different soils. *Fresenius' Journal of Analytical Chemistry* 351, 629–631.

Seo JS, Keum YS, and Li QX 2009 Bacterial degradation of aromatic compounds. *International Journal of Environmental Research and Public Health* 6, 278–309.

Shafizadeh F and Sekiguchi Y 1983 Development of aromaticity in cellulosic chars. *Carbon* 21, 511–516.

Simoneit BR, et al 1999 Levoglucosan, a tracer for cellulose in biomass burning and atmospheric particles. *Atmospheric Environment* 33, 173–182.

Simpson MJ and Hatcher PG 2004 Determination of black carbon in natural organic matter by chemical oxidation and solid-state ^{13}C nuclear magnetic resonance spectroscopy. *Organic Geochemistry* 35, 923–935.

Skjemstad JO, Janik LJ, Head MJ, and McClure SG 1993 High energy ultraviolet photo-oxidation: A novel technique for studying physically protected organic matter in clay-and silt-sized aggregates. *Journal of Soil Science* 44, 485–499.

Skjemstad JO, Clarke P, Taylor JA, Oades JM, and McClure SG 1996 The chemistry and nature of protected carbon in soil. *Australian Journal of Soil Research* 34, 251–271.

Skjemstad JO, Taylor JA, and Smernik RJ 1999 Estimation of charcoal char in soils. *Communications in Soil Science and Plant Analysis* 30, 2283–2298.

Smernik RJ, Skjemstad JO, and Oades JM 2000 Virtual fractionation of charcoal from soil organic matter using solid state ^{13}C NMR spectral editing. *Australian Journal of Soil Research* 38, 665–683.

Smith DM, Griffin JJ, and Goldberg ED 1975 Spectrophotometric method for the quantitative determination of elemental carbon. *Analytical Chemistry* 47, 233–238.

Smith DM, Griffin, JJ, and Goldberg ED 1973 Elemental carbon in marine sediments: A baseline for burning. *Nature* 241, 268–270.

Sohi SP, et al 2001 A procedure for isolating soil organic matter fractions suitable for modelling. *Soil Science Society of America Journal* 65, 1121–1128.

Song J, Peng PA, and Huang W 2002 Black carbon and kerogen in soils and sediments 1 quantification and characterization.

Environmental Science and Technology 36, 3960–3967.

Song J and Peng PA 2010 Characterisation of black carbon materials by pyrolysis-gas chromatography-mass spectrometry. *Journal of Analytical and Applied Pyrolysis* 87, 129–137.

Song J, Huang W, and Peng PA 2012 Stability and carbon isotope changes of soot and char materials during thermal oxidation: Implication for quantification and source appointment. *Chemical Geology* 330-331, 159–164.

Struever S 1968 Flotation techniques for the recovery of small-scale archaeological remains. *American Antiquity* 33, 353–362.

Trumbore SE and Zheng S 1996 Comparison of fractionation methods for soil organic matter 14C analysis: 14C and soil dynamics. Special section. *Radiocarbon* 38, 219–229.

Vasilyeva NA, et al 2011 Pyrogenic carbon quantity and quality unchanged after 55 years of organic matter depletion in a Chernozem. *Soil Biology and Biochemistry* 43, 1985–1988.

Veilleux MH, Dickens AF, Brandes J, and Gélinas Y 2009 Density separation of combustion-derived soot and petrogenic graphitic black carbon: Quantification and isotopic characterization *In IOP Conference Series: Earth and Environmental Science IOP Publishing*, 5, 012010.

Verardo DJ 1997 Charcoal analysis in marine sediments. *Limnology and Oceanography* 42, 192–197.

Walkley A and Black IA 1934 An examination of the Degtjareff method for determining soil organic matter, and a proposed modification of the chromic acid titration method. *Soil Science* 37, 29–38.

Wiedemeier DB, Hilf MD, Smittenberg RH, Haberle SG, and Schmidt MWI 2013 Improved assessment of pyrogenic carbon quantity and quality in environmental samples by high-performance liquid chromatography. *Journal of Chromatography A* 1304, 246–250.

Winkler MG 1985 Charcoal analysis for paleoenvironmental interpretation: A chemical assay. *Quaternary Research* 23, 313–326.

Wolbach WS, Gilmour I, Anders E, Ort, CJ, and Brooks RR 1988 Global fire at the Cretaceous-Tertiary boundary. *Nature* 334, 665–669.

Wolbach WS and Anders E 1989 Elemental carbon in sediments: determination and isotopic analysis in the presence of kerogen. *Geochimica et Cosmochimica Acta* 53, 1637–1647.

Wu Q, Blume HP, Beyer L, and Schleuß U 1999 Method for characterization of inert organic carbon in Urbic Anthrosols. *Communications in Soil Science and Plant Analysis* 30, 1497–1506.

Wurster CM, Lloyd J, Goodrick I, Saiz G, and Bird MI 2012 Quantifying the abundance and stable isotope composition of pyrogenic carbon using hydrogen pyrolysis. *Rapid Communications in Mass Spectrometry* 26, 2690–2696.

Wurster CM, Saiz G, Schneider M, Schmidt MWI and Bird MI 2013 Quantifying pyrogenic carbon from thermosequences of wood and grass using hydrogen pyrolysis. *Organic Geochemistry* 62, 28–32.

Zackrisson O, Nilsson MC, and Wardle DA 1996 Key ecological function of charcoal from wildfire in the Boreal forest. *Oikos* 77, 10–19.

Zencak Z, Elmquist M, and Gustafsson Ö 2007 Quantification and radiocarbon source apportionment of black carbon in atmospheric aerosols using the CTO-375 method. *Atmospheric Environment* 41, 7895–7906.

Zhan C, et al 2012 Spatial distributions and sequestrations of organic carbon and black carbon in soils from the Chinese loess plateau. *Science of the Total Environment* 465, 255–266.

Zhan C, et al 2013 Validation and application of a thermal-optical reflectance TOR method for measuring black carbon in loess sediments. *Chemosphere* 91, 1462–1470.

Zheng N, et al 2022 Can aged biochar offset soil greenhouse gas emissions from crop residue amendments in saline and non-saline soils under laboratory conditions? *Science of The Total Environment* 806, 151256.

Zimmermann M, Leifeld J, and Fuhrer J 2007 Quantifying soil organic carbon fractions by infrared-spectroscopy. *Soil Biology and Biochemistry* 39, 224–231.

Zimmerman AR and Mitra S 2017 Trial by fire: On the terminology and methods used in pyrogenic organic carbon research. *Frontiers in Earth Science* 5, 95.

Ziolkowski LA and Druffel ERM 2010 Aged black carbon identified in marine dissolved organic carbon. *Geophysical Research Letters* 37, L16601.

25

Biochar handling, storage, and transportation

Thomas R. Miles

Introduction

Biochar production has grown significantly in the last several years. Production systems have increased in capacity from a few cubic meters of biochar per day to plants processing more than 1 t hr^{-1} resulting in the movement of many truckloads per week (You et al, 2022; Xia et al, 2023). The qualities of biochars including size, and composition, affect their handling, storage, transportation, and use. Storage is essential for adjusting for differences between production and demand. Production for most thermal processes is continuous while demand may be seasonal for agriculture and landscaping. Transport and handling account for major production costs and must comply with safety standards. The techniques described in this chapter are informed by industry experience and scientific studies where available. Most commercial producers in the US or Europe currently utilize biochars made from woody materials, whereas the largest production of biochar occurs in China using crop residues such as rice husk (Chapter 33). This chapter describes current practices for biochar handling, storage, and transportation. Products and their characteristics are included as they affect the processing, storage, and transportation of biochar. Biochars may be shipped or stored in bulk and processed by densification, granulation, milling, activation, inoculation, suspension in liquids, and blending or processing with other components.

Biochar production

Biochar is produced at small, medium, and industrial scales (Chapter 4). Place-based production is typically small-scale in portable or mobile kilns such as flame-cap (-curtain)

DOI: 10.4324/9781003297673-25

kilns in pits or metal kilns with capacities of 1–3 m³. Feedstocks are usually round wood or branches. Kilns of this kind in North America tend to be used for hand piles and have traditionally been in operation to produce charcoal from forest residues in less accessible areas (Pecha et al, 2021) and still do so in many parts of the world (Cornelissen et al, 2016). These traditional designs are increasingly used in vineyards and orchards to produce biochar from agricultural prunings. The kilns are quenched with water, manure slurries, and soil. The resulting biochar products are chunky and friable. Although several kilns can be operated at the same time, the quantities of biochar produced tend to be small. An individual kiln might produce 0.5–1 t day^{-1} or 3–5 m³ day^{-1} of biochar. One person can tend to more than one kiln, but teams are often present for safety and auxiliary operations. Biochars are used on-site or locally in crops or reforestation. The biochar is handled in sacks, bags, or with small loaders. Producers around the world have become creative in the use of these small kilns and the application of the products.

Larger mobile kilns have been developed for portable use that are based on Air Curtain Incinerators (ACI; Air Burners Inc., airburners.com, CharBoss developed with the US Forest Service) (Schapiro, 2002; Lee 2017). The primary purpose of the ACI is a reduction of large volumes of dry urban, agricultural, or forest residues which is important for the emission balance (Puettmann et al, 2020). Small kilns like the "CharBoss" operate best with fuels up to 100 mm in diameter. In these kilns, biochar forms as a blanket of pyrolysis gas which separates the solid fuel from a curtain of air which burns to pyrolysis gases. Unburned biochar is withdrawn from the fire before it can ignite. It is cooled, quenched, and discharged. Ash is separated or rinsed to ensure

that wood biochars have high C contents. The small versions of these kilns convert about 10 t day^{-1} (0.5–1 t hr^{-1}) to produce 5 m³ biochar day^{-1} (10-hr day). Biochars are used locally and transported in bags or bins. The larger mobile carbonizers can process 8–10 t biomass hr^{-1} yielding 0.5–5 t biochar day^{-1} or 20–25 m³ biochar day^{-1}. Mobile systems usually require a loader and operator. They are permitted by air quality authorities to operate only during daylight hours. A major issue during summer with open kilns (especially air-curtain kilns) is the production of embers which could ignite major fires, which require permission from local fire authorities, as common in Australia. The biochar is wet and chunky in form. It is either bagged or loaded into bulk trailers.

A few pyrolyzers convert chunky wood fuel, husks, or straw in batches at rates of 2 t per batch producing about 500 kg biochar kiln^{-1} or (2 m³ biochar kiln^{-1}) (www.biocharnow.com). Small systems tend to be labor intensive. Batch systems in Europe which are fully automated operate at much larger capacities (www.polytechnik.com). These kilns produce a wide range of sizes which are usually processed for specific markets.

Modular kilns have been developed that can process wood chips, poultry litter, pelletized agricultural residues, or biosolids at approximately 500 kg hr^{-1} (12 t day^{-1}) with yields of 20%–30% w/w, generating 2.5–3.5 t biochar day^{-1} (10–14 m³ day^{-1}). These are stationary units that are sometimes prefabricated in containerized modules with capacities up to 40 t per day^{-1} which are capable of producing 8–12 t biochar day^{-1} (40–60 m³ day^{-1}) or up to one truckload per day. The maximum size of the biochar is usually about 12 mm. The quantity of fines (<2 mm) depends on the feedstock, the processing system, and the way it is operated. These systems are usually delivered with systems to fill bags with a volume of 1.5 m^3. They often include systems to recover heat from the off-gas. Depending on the degree of automation, they can be operated from four man-hours per day to up to five operators in 24 hours. In a typical year, they can produce from 2500 to 4000 t biochar.

Biochars in North America and New Zealand are also recovered from biomass boilers (Spokas et al, 2011). Similarly, biochar is produced and sold from sugar and rice mills, brickworks, and lime kilns in Asia and from bioenergy plants using crop residues in China (Kisor et al, 2010; Danish et al, 2015). Biochars with high C contents are separated from flue gas after they pass through the boiler, and are often reinjected and reburned in the furnace. This biochar can have a higher value when used as biochar than when converted to heat or electricity. It is separated, quenched, and handled in bulk. It is often screened to remove fines that contain mineral ash. The quality and quantity of biochar depend on the feedstock, the design of the boiler, and the operation of the boiler. A 20 MWe biomass plant can produce about 200 kg hr^{-1} (1.3 m^3 hr^{-1}), equivalent to about

4.8 t day^{-1} or 1500–2000 t yr^{-1}. This biochar in developed countries is delivered in bulk in 60–80 m^3 loads directly to farms, compost yards, or soil blenders. Labor is required for post-processing and packaging. Wood processing facilities also recover biochar from boilers and gasifiers which are used to convert wet sawdust to heat to dry lumber. A single sawmill with a 40 GJ burner might produce 200 kg hr^{-1} (1 m^3 hr^{-1}, 24 m^3 day^{-1}). The biochar is similar to the biochar from biomass boilers. It is handled and packaged using similar methods. In Asia, a large portion of the biochar from the energy plants and food and mineral processing are in many instances collected by farmers or a third party from the site in trucks without bagging.

The largest plants in North America and China produce up to 15,000 t biochar yr^{-1} as a byproduct of energy production. Facilities are in design and development to produce 40,000–100,000 t yr^{-1}. These plants will be capable of producing a range of biochar qualities which will likely be processed into a variety of products. Large-scale production >50 t day^{-1} (229 m^3 day^{-1}) is loaded out by truck to offsite storage or use.

Biochar products and quality for handling

Biochars are produced in different ways which affect storage, handling, and transportation. Feedstocks and processing technology determine the ease of product handling and use. The biochar industry is broad and the decisions by many people can affect the quality and suitability of biochar for different purposes, which all affect handling. Feedstock producers can include resource managers, foresters, contractors, woodland owners, farmers, wastewater recycling plants, and the organics recycling

industry, which includes construction and demolition, recycling, composting, forestry, and landscaping. Processors include biochar, biocarbon, and charcoal producers engaged in thermal processing of mainly wood residues, less often crop residues, with other waste processing still being developed (e.g., manures). Biomass energy facilities recover residues with high C contents from biomass boilers or biofuel production. Biomass gasifiers that wood processing industries use to supply heat to dry lumber produce biochar as a by-product. Biochar brokers and re-processors recover the biochar and assure quality and delivery to appropriate markets. Important biochar qualities include volatility, size, density, moisture content, and packaging.

Volatility

Low processing temperatures and short residence times result in biochars with higher contents of volatile matter (Enders et al, 2012). A typical target is to reduce volatile matter to less than 20%. Producers report instability in handling and storage for biochars high in volatile matter, especially if the moisture content is less than 15% (wet base). These biochars tend to be shipped wet and blended with organics for co-composting. Volatile matter can also cause problems in storage and transportation when they come in contact with air. Producers often use steam quenching before cooling and blankets of inert gas such as N_2 to reduce the potential for autoignition (see the section on combustion below).

Size

The particle size of biochars in general resembles the particle size of the feedstock material, with important exceptions. For example, friable feedstock may disintegrate and generate very fine biochar particle sizes. All production methods result in some proportion of fine biochar particle sizes (called "fines" from here on). Some processors screen fines to separate ash and increase the C contents of the biochar product or grind biochar to a desired particle size. Fines with a higher ash content are often diverted to compost and other uses. Fines may have to be combined with some form of dust control to reduce hazards for health, risk of fire, or be pelletized.

Bulk density

Biochar bulk density matters for logistics, application, and transportation costs and methods. Bulk density is the density of the bulk material (e.g., of an entire bag), in contrast to particle density (including the pores in a particle) or true density (i.e., density only of the solid material) (Chapter 5). Bulk density of biochars can vary considerably (Weber and Quicker, 2018) from dry biochars of woody material (e.g., from power production) at 100–200 kg m^{-3}, manure-based biochars at 200–700 kg m^{-3}, to some agricultural residues at 100–150 kg m^{-3} for cereal stover and 350–450 kg m^{-3} for nut shells (Rajkovich et al, 2012). The bulk density of biochars typically decreases with increasing pyrolysis temperature for low-ash feedstocks (Abdullah and Wu, 2009), but can also increase for ash-rich manure feedstocks (Rajkovich et al, 2012). In general, bulk biochars are adjusted with moisture to optimize the weight to volume in transport to about 240 kg m^{-3}. A standard US full truck freight weight limit is 36,287 kg (80,000 lbs) which includes the truck; with a truck weight of 16 t, approximately 20 t are available for a legal commercial load.

The bulk density increases with ash content (Chapter 5) (Table 25.1). Removal of fine ash reduces bulk density which can increase transport costs per unit weight. Biochar recovered from the flue gas of biomass power plants is often light, between 80–160 kg m^{-3} on a dry basis. Producers sometimes add moisture to arrive at an optimum shipping weight. Variations of bulk density between commercial biochars have been reported between 144–192 kg m^{-3} with an average of 160 kg m^{-3} on a dry basis (K Strahl, personal communication, 2022-12-13).

Biochar bulk densities will also be driven by particle size, moisture (Table 25.1), or materials such as nails in recycled wood products.

Table 25.1 *Particle size, moisture, and ash effects on bulk density of industrial biochars (Biochar Solutions). Particle size by sieving; total C content by dry combustion; ash by ASTM D-1762-84; moisture by ASTM D1762-84 (105c)*

Biochar type	Feedstock	Particle size (mm)		Moisture (%)	Ash (%)	Carbon (%)	Bulk density (kg m^{-3})
Biochar particular	Pine wood, western NA[a]	0.5	4.0	6	8	82	125
Biochar particular	Pine wood, western NA	0.5	4.0	35	8	82	169
Biochar mixed granular	Pine wood, eastern NA	0.1	5.0	67	14	78	275
Biochar powder	Pine wood, NA	0.005	0.03	3	5	94	402
Biochar power plant	Pine wood, south-eastern NA	0.1	4.0	18	78	14	467
Biochar power plant	Pine wood, Rocky Mountains, NA	0.1	8.0	4	25	50	140

Notes
[a] North America

Biochar density should be determined for a specific biochar material. A density outside of the expected range is a signal for the producer, user, or other stakeholder to test the material and identify what is causing a deviation from the expected density. Biochar may also be blended with other products which will change characteristics including density.

Mechanical stability

Mechanical stability is important for transport and handling (Weber and Quicker, 2018), as well as any post-pyrolysis processing. Key thereby is compressive strength that typically is lower than in the feedstock that it was produced from (Kumar et al, 1999). However, compressive strength normally does not decrease uniformly with increasing pyrolysis temperature, but may rather increase above a temperature of 600°C (Kumar et al, 1999). Biochars with high compressive strength are usually those made from feedstock that have high lignin contents (Emmerich and Luengo, 1994).

Moisture content

Moisture is the most common method to control dust and the potential for auto-ignition or ignition from other sources. The moisture content of processed biochars can range from 20% to 300% depending on the end use. The EBC (2023) requires moisture contents of 30% which generally prevents dust formation or autoignition (see section below). Biochars are often shipped at 30–50% moisture content, which is still dry to the touch.

Products

Various biochar-based products have evolved for soil applications, including agriculture and horticulture, retail gardens, landscapes, turf, trees, orchards, and vineyards (Chapter 26). These all require different handling, storage, and application methods. Products include biochar on its own, biochar combined with compost, co-composted biochar which contains 5–20% biochar, biochar amended to animal bedding and litter, biochar-based fertilizers which contain 15–25% biochar, and biotic soil amendments which combine biochar with minerals and effective microorganisms (Chapter 26). Biochars are delivered in granulated and liquid forms for seeding. Biochar extracts are used as foliar sprays. Biochar-based nano-fertilizers are in development. Each of these uses has different requirements for biochar particle sizes.

Particle sizes for different products

Processors screen raw biochar into fractions such as "chips" (3–25 mm) for direct application, composting, or constructed soils; "medium" (0.7–3 mm) for top dressing, biofiltration, bio-filler, microbial carriers; "small" (0.3–0.7 mm) for air seeders and irrigation systems; and powder (<0.3 mm) for suspension in water for irrigation and sprays (Table 25.2). Some suppliers use 0.21 mm (70 mesh) as a reference size for seed coating (Zhang et al, 2022), granulation (Bowden-Green and Briens, 2016), or use in "liquid" biochars which are popular with arborists. Producers and consumers have different preferences depending on the feedstock and the application. Some soil blenders prefer biochars sized from 2–6 mm. Others will use everything below 2 mm. Biochars of 3 mm and greater are preferred for direct

applications in soils where fines tend to move by leaching or erosion (Major et al, 2010).

Each size can have different flow characteristics. Some suppliers screen fines to remove larger particles and improve flow characteristics for air seeders and applicators. Nursery workers prefer the less dusty biochars when they are used as substitutes for vermiculite or perlite in growing media.

Biochar-blended products for agriculture and horticulture

Biochar-blended products are beneficial for several reasons. The biochar in such blends contribute to increased moisture and nutrient holding capacity for the long term (Chapter 11), which can be blended or injected into the soil. In comparison, a compost or nutrient product (blood meal, bone meal, fish meal, etc.) will mostly provide nutrients, with compost contributing to organic matter for the short to medium term. Nutrient admixtures will adsorb or precipitate onto the surfaces or in the pores of the biochar particles (Chapter 26), with the intent to be released more gradually in synchrony with plant needs. Biochar as a component of a blend will allow that blend to be more drought tolerant (Chapters 15 and 20), release nutrients more slowly (Chapters 19 and 26), and be darker in color. The factors that will contribute to the commercial success of such blends will not only be the mentioned attributes, but will also include the logistical and economic conditions surrounding all aspects of handling, including packaging, storage, and transportation. While concerns over combustion or flammability may be eliminated, some blends may be biologically active and therefore have a reduced shelf life, or require non-decomposable containers for transport.

Table 25.2 *Biochar size selection (modified after https://biocharnow.com/products/; Adetayo et al, 1995; He et al, 2017, 2018; Gupta and Kua, 2019; Willaredt et al, 2022)*

Applications	Particle sizes (mm)			
	Chip 25–3	Medium 3–0.7	Small 0.7–0.3	Powder <0.3
Constructed soils for tree planting	×	×		
Composting with biochar	×	×		
Construction material	×	×	×	×
Direct placement in soil	×	×	×	×
Biofilter for air or water	×	×	×	×
Carrier for microbial inoculants	×	×	×	×
Top-dressed on soil or grass		×	×	
Biofiller for plastic, concrete, or asphalt		×	×	×
Insertion into soil using air seeder			×	
Irrigation system			×	×
Biochar-based fertilizers			×	×
Seed coating				×
Suspended in water for soil injection				×
Suspended water spray				×

In addition, labeling on the package must conform to local regulations that at present still remain transient. For example, most US states and Canadian provinces require that products labeled "biochar" must specify the feedstock and contain at least 60% C in the biochar (Alexander, 2019). In the EBC (2023) document, no minimum C content is required and the label "biochar" applies to materials with any C content.

Packaging

Packaging varies with transportation and use. Biochars are typically shipped in bulk or bags. A bulk truckload will contain 89–110 m³ or 10–13 t dry weight. A truckload can contain 48–52 bags each with a volume of 1.5–2 m³ bags per load for a total dry weight of 8–9 t. Some producers will not accept orders of less than a truckload. Mixing of an order is not uncommon, for example, shipping 60% in bulk truckloads, 30% in bagged truckloads, and 10% in 1.5 m³ bags (J Strahl, personal communication, December 31, 2022).

Bulk bags are common, also called super sacks or Flexible Intermediate Bulk Containers

Figure 25.1 *Packaging options for biochar products in bulk and retail. (a) bulk bag; (b) bags loaded on pallets; (c) biochar blend for retail market; (a by T Miles; b by K Strahl; c by J Levine; reproduced with permission)*

(FIBCs) that range in size from 1 to 2 m^3. Bulk bags holding 1.5 m^3 are common and can stack easily with 2 bulk bags to a pallet for a pallet volume of 3 m^3.

Biochar products for retail consumers are often packaged in 20 to 28-L bags or 20-L buckets. Bags are shipped on pallets. Granular biochar from 0.5–24 mm can be packaged as pure biochar or as a blended product. Package types include bags holding 14–57 L. As many as 80 bags with 28 L volume each can stack on a pallet. Such bags will be stacked 5 per layer where 16 layers would be a maximum height and 60 units or 12 layers may be more common.

Biochar powder is sometimes defined as biochar that is less than a max size of 0.3 mm (Table 25.1). Typically, biochar powder needs to be dry. Dry powders can flow like liquids. If they are moist, they will clump and not flow. Dry powders present several challenges including airborne dust, thermal combustion, and explosion risk. If producing or using dry powders, the style of packaging can increase the safety and useability of these dry powders. A dry powder can be packaged in a valve-style bag or in a poly-lined bulk bag (Figure 25.1). These styles of packaging will reduce the risk of airborne powders. Producers and users of dry powders should be equipped to manage the risks that come from these products. In addition to the distinct advantage that powders can flow like a liquid or be incorporated into liquids, they allow for a much smaller application rate with higher effectivity in particular applications where fine distribution is needed. In general, however, particle size was found to have minor effects on soil improvement between 2 and 20 mm (Lehmann et al, 2003). By effectively packaging biochar powders, that powder can be applied with less mess and greater safety.

Health and safety

Health and safety concerns have expanded with biochar applications. Biochars and biochar-enhanced products have increased in complexity, in the number of products offered and uses. While initially intended for agriculture, uses now extend to non-soil applications such as additives for building products, concrete, and asphalt (Chapter 28).

The electrochemical properties of biochars have led to application in advanced products for medical uses which will require special handling (Chapter 32). Product distribution has expanded from local use, and local direct sales, to distribution networks for retail garden centers, contractor supply for landscape, turf, and trees, and environmental uses such as structured soils, and stormwater management. Biochars are post-processed to densified, granulated, and bagged biochar-enhanced fertilizers (Chapter 26). Packaging has therefore expanded from bucket size to bulk bags and bulk truckloads.

Health and safety concerns are similar for applications including agriculture, forestry such as wildfire fuel reduction, reforestation, range improvement, growing media for nursery and out planting, revegetation, reclamation of mines and degraded land environment, remediation, erosion control, revegetation remediation compost to reduce persistent herbicides, water quality, stormwater filtration, and water treatment.

Hazardous classifications

Biochars can fall into several hazardous categories. First, and most widely known are auto-ignition characteristics (Zhao et al, 2014; Babu et al, 2020). Secondly, biochars can contain toxic levels of both organic and inorganic compounds (Chapters 21 and 22). In most cases, beneficial use cases may be obtained for applications that immobilize these compounds or mitigate the risk of ignition.

Biochar producers must provide Material Safety Data Sheets (MSDS) for their products for shipping purposes. The SDS or MSDS is often required by customers, and is mostly mandatory for courier or postal shipments. Biochars of all forms must comply with requirements for similar products for occupational health and safety and environmental concerns. United States processors of biosolids

to biochar must comply with the same standards as biosolids application to soil. Workers are typically equipped with Personal Protection Equipment (PPE) for dust as determined by plant site officials.

Contaminants and toxins are typically related to either feedstocks or the use of co-products. Each of the regional and international standards contains limits for toxic components from most biomass feedstocks (International Biochar Initiative; European Biochar Certificate; Australia New Zealand Biochar Industry Group). International Standards and specifications for biocarbon and biochar are under consideration by the International Standards Organization with the collaboration of the regional biochar associations (ISO TC-238 Solid Biofuels, TG1 Biocarbon Working Group, https://www.iso.org/committee/554401.html).

Biochar safety for dust and combustion

A dust-tight conveyance and bag filling station allow producers to produce and bag dry biochar. Most customers do require moisture content to mitigate dust when the biochar is being used. Some biochars require as little as 15% moisture. To be truly air-borne dust-free generally requires 35% moisture, with 30% often recommended as a minimum (EBC, 2023).

The following information is mostly intended for dry and powder biochar products, although anyone working with biomass and biochar materials should be familiar with safe handling practices. In general, dry and dusty products require additional handling practices for safety. Moist and large particles are in general easier to handle than dry and small particles.

When producing or purchasing biochar, one should be familiar with any safety needs for dust, explosion, and combustion (Weber and Quicker, 2018). A basic step is having a Safety Data Sheet (SDS) formerly referred to as a Material Safety Data Sheet (MSDS) available. To ensure that biochar is not auto-thermal, meaning that it will not self-ignite once it is cooled and packaged for sale, or is flammable (Zhao et al, 2014; UN, 2019). In the US, materials are regulated by CFR (2023), which largely takes its test requirements and methods from the UN (2019), whereby biochars fall under Hazard Class 4 (§ 173.124 Class 4, Divisions 4.1, 4.2 and 4.3):

Division 4.1 (Flammable Solid).
Division 4.2 (Spontaneously Combustible Material)
Division 4.3 (Dangerous when wet material)

Flammability is examined with a 250-mm long mold, if the burning reaches a certain speed across the mold or the external flame is applied longer than a certain time (UN, 2019). Biochars are typically not flammable according to such a test. The degree of flammability is promoted by high heating rates, short dwell times, and feedstock impurities that favor the formation of carbon-free radicals during pyrolysis, but dissipate within minutes (Zhao et al, 2014). However, spontaneous combustion is a concern that has to be tested (Figure 25.2). Biochars made at a pyrolysis temperature of 450°C are the most prone to self-heating,

regardless of feedstock (Restuccia et al, 2019). Biochars made at temperatures above 550–600°C are less reactive than the feedstock, with biochars made from rice being more likely to spontaneously combust than those from softwood (Restuccia et al, 2019).

Risks related to airborne dust can be tested using ASTM E1226 (2000). This is an economical and practical way to determine if the dust in the sample has the potential to be explosive. Testing consists of exposing the fine dust in the sample to low-energy igniters inside a 20-liter Siwek explosion chamber to determine the explosion under pressure. In addition, an explosion severity test indicates the severity of the dust explosion by determining the deflagration parameters. For this test, the dust is suspended and ignited in the Siwek chamber and the maximum pressure and the rate of pressure rise are measured.

Volatility for safety concerns

As discussed above, biochars contain volatile matter defined as organic compounds that have boiling points of ≤250°C and due to their volatility are often considered contaminants that can threaten air quality and cause human health concerns. Even low heat treatment can mitigate these concerns also for biochars containing high amounts of volatile matter (Buss and Masek, 2016).

Storage

Storage needs vary among producers. Small-scale producers generally store biochar in covered piles or bags. Larger producers have storage for intermittent or seasonal demand. As an example, one producer stores over 1000 m³ in inventory at any given time between bulk and bagged biochar. Only dry

granular or powdered biochar is stored under a roof. No more than one semi-truck load of dry product is stored at a time to prevent embers from external sources, such as embers from a nearby wildfire from igniting it. Spontaneous combustion due to self-heating is a concern, and has to be assessed for

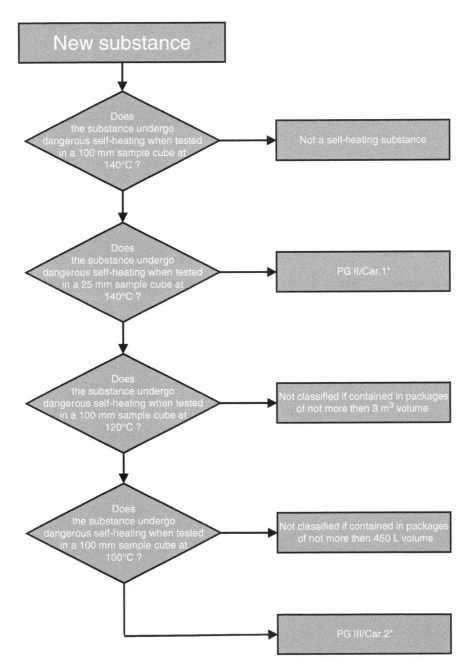

Figure 25.2 *Testing scheme for spontaneous combustion according to UN (2019) (redrawn);* **substances with a temperature for spontaneous combustion higher than 50°C for 27 m³ should not be classified as self-heating substances*

different biochars (see above), and managed, typically by storing biochars moist.

Dry biochars that are burning, are notoriously difficult to extinguish due to their very high combustion temperature. The best way to extinguish them is to drown them out with water. Sprinklers and fire hoses should always be available. One producer uses N_2 to blanket biochar in conveying and loading facilities.

Biochars can be stored outside at up to 50% moisture content (wet basis). One outdoor covered bunker typically stores 1500 m^3. Bulk commercial users (blending companies, greenhouse operators, etc.) usually store biochar outside (70%), in bulk bags (25%) and in bins (5%).

Transport and loading

The easiest way to ship biochar regionally is to utilize local bulk transport. All that is required is a means of loading the biochar into the truck. A bucket loader such as a skid steer or front-end wheel loader can be used as they are readily available and generally utilized in other areas of a biochar plant (Figure 25.3). However, long term it is better to use a hopper, silo, or chip bin. There are two major reasons for this. First, it cuts down on labor and fuel costs. Second, it limits the amount of material lost due to crushing down and spillage.

For longer-distance shipping, bagging is generally required as bulk carriers are usually limited to regional shipping. This is in part due to the lack of consistent backhauls available in bulk commodities. Therefore, shipping on flatbeds or box vans has been identified as a preferred method for shipping biochar over 1000 km in North America (J Levine, personal communication, December 31, 2022).

Most biochars are lightweight and somewhat flowable, causing stability challenges that must be addressed. Biochar is generally shipped in a single layer of 4 bags of 1 m^3 each, or double stacked in super-sacks (Figure 25.1). Both of these can be unstable if they are not correctly prepared. Designing the bags to meet the requirements of the lighter-weight biochars being shipped is more difficult than might be expected. Most bulk

(a)

(b)

Figure 25.3 *Bulk shipping options. (a) Bulk loading 60–75 m^3 trailer; (b) filling a 1.5 m^3 bag (images by K Strahl; reproduced with permission)*

bags deform in some way during handling and transportation, leading to issues either during loading or during unloading. Some expand sideways, resulting in a shipment that may have had 50 mm of space between pallets during loading, being wedged tightly inside the trailer at their destination. Other bag types will be even more unstable. Some bags also have a very short shelf life, starting to deteriorate quickly, compounding these issues. Warpage is another issue that quickly becomes apparent when filling bags. A bag that is not filled or stored correctly may lean to one side or another or develop a rounded layer of compacted biochar on the bottom leading to instability during transportation.

In summary, careful consideration and planning are needed when ordering the bulk bags being used, as well as, in the way that these bags will be filled, stored, and transported.

Quantities and capacities of truckloads

Modes of transport that work for shipping biochar may vary widely. The most basic is bulk transport in a dump truck, end dump, belt trailer, or walking floor. Dump trucks can generally hold between 7.6 and 11.5 m^3 of biochar, while end dumps have a capacity between 19 and 30 m^3. Belt trailers are generally in the range of 57 to 65 m^3, with walking floor trailers being the preferred mode as they can generally hold between 84 and 111 m^3. These trucks or trailers generally have some cracks or gaps in them so wetting down the biochar before transport is advisable to minimize inadvertent losses.

Box trailers are another means of transporting biochar and are usually the most economical method of cross-country transport available. The major courier services will haul biochar in tote bags so long as it is hydrated. These carriers will also offer lift gate delivery, which is necessary for some customers. These carriers will generally utilize smaller trailers with a length of 9–10 m that are pulled behind a truck (sometimes called pup trailers), which limits them to 14 and 16 total pallets, respectively (Sharpe and Rodriguez, 2018). However, for customers with greater needs, they do offer double pup trailer loads, with 2 trailers in a "b-train" configuration, which in total can carry 28 to 32 pallets. Despite their higher cost per distance, this can sometimes make them competitive with 16-m dry van trailers as the total cost per volume transported is lower.

Dry vans with 16 m length are the go-to for road transportation in the US, of the 4.5 million semi-trucks on the road today, over 1.7 million are hauling dry vans. Assuming that each pallet slot has a pallet mass of 23 kg of biochar with 26 slots available, the weight of all 26 pallets would amount to approximately 600 kg.

Early in the development of biochar markets, in about 2009, limiting moisture contents was discussed as a way to maximize the shippable value of biochar. Due to the general light weight of biochar along with the potential for dust and thermal challenges, the addition of moisture can add to the ease of use and safety of biochar without, in most cases, limiting the shippable values. Biochars at or less than 250 kg m^{-3} should not be limited by weight but by volume in commercial freight trucks. Many biochars will have lower bulk densities even with water (Table 25.1). Another way to think about this is if 700 and 900 kg can be packaged per pallet space, a truckload is in a good economic point of efficiency. This general set of assumptions is also true for less than truckload pallet freight as well as bulk freight. While these are generally correct each biochar load should be characterized and managed individually.

Flatbed trucks are highly suitable for shipping biochar, as each row of bags is strapped down individually, mitigating some of the loading concerns of box trucks. They generally fall into three categories:

Category 1 small trucks with flatbeds: hotshots are generally heavy-duty pickup trucks with a trailer that has 5–12 m of usable deck, and can carry up to 12 t of cargo; they usually haul small, time-sensitive less than truckloads to a single customer or location.

Category 2 large trucks flatbeds or step-decks: these trucks have deck space of 12–16 m and can transport up to 22 t of cargo.

Category 3 double trailer setups, such as b-trains: the advantage of this method of transport is that they usually have enough deck space for 30 to 32 pallets, which allows with 3 m^3 of each pallet a total transportation of 98 m^3. In comparison, the most that a 16-m flatbed can accommodate is 80 m^3.

In China and Australia, processing of those biochars to make specific granular or pelleted products using mineral binders and nutrients is considered safe to store, does not cause dust or self-combustion concerns, and is being implemented by several manufacturers (Chapter 27). Similarly, biochar in concentrated liquid products such as foliar sprays or liquid fertilizers to not possess the same constraints.

Conclusions and outlook

Handling, storage, and transportation of biochar can fall back on established infrastructure, equipment, guidelines, regulations, and recommendations from other more established industries such as the charcoal industry among others. However, the wide variety of biochar materials including products containing biochar requires important adaptations of known handling strategies. Particle size is not only important for product quality geared towards specific uses, but also determines handling. In addition, moisture emerges as a critically important aspect of handling to mitigate combustion and dust formation.

Knowledge gaps that need to be addressed include the effects of size distributions and mixtures of sizes on handling, choices of particle sizes for improved flow, and optimization of moisture to facilitate handling without increasing the cost of transport.

References

Abdullah H, and Wu H 2009 Biochar as a fuel: 1. Properties and grindability of biochars produced from the pyrolysis of mallee wood under slow-heating conditions. *Energy and Fuels* 23, 4174–4181.

Adetayo AA, Litster JD, Pratsinis SE, and Ennis BJ 1995 Population balance modelling of drum granulation of materials with wide size distribution. *Powder Technology* 82, 37–49.

Alexander R 2019 Biochar labeling guidance document. Report, accessed at https://biochar-us.org/labeling-guidelines-biochar-products on 2023-8-2.

ASTM E1226 2000 Standard test method for explosibility of dust clouds.

Babu K, et al 2020 A review on the flammability properties of carbon-based polymeric composites: State-of-the-art and future trends. *Polymers* 12, 1518.

Bowden-Green B and Briens L 2016 An investigation of drum granulation of biochar powder. *Powder Technology* 288, 249–254.

Buss W and Mašek O 2016 High-VOC biochar—Effectiveness of post-treatment measures and potential health risks related to handling and storage. *Environmental Science and Pollution Research* 23, 19580–19589.

CFR 2023 Code of federal regulations – Title 49 Subtitle B Chapter I Subchapter C Part 173 Subpart D § 173.124. Accessed at https://www.ecfr.gov/current/title-49/subtitle-B/chapter-I/subchapter-C/part-173/subpart-D/section-173.124 on August 3 2023.

Cornelissen G, et al 2016 Emissions and char quality of flame-curtain "Kon Tiki" kilns for farmer-scale charcoal/biochar production. *PLoS ONE* 11, e0154617.

Danish M, Naqvi M, Farooq U, and Naqvi S 2015 Characterization of South Asian agricultural residues for potential utilization in future 'energy mix'. *Energy Procedia* 75, 2974–2980.

EBC 2023 European biochar certificate. Ithaka Institute, accessed at https://www.european-biochar.org/media/doc/2/version_en_10_3.pdf on 2023-8-1.

Emmerich FG and Luengo CA 1994 Reduction of emissions from blast furnaces by using blends of coke and babassu charcoal. *Fuel* 73, 1235–1236.

Enders A, Hanley K, Whitman T, Joseph S, and Lehmann J 2012 Characterization of biochars to evaluate recalcitrance and agronomic performance. *Bioresource Technology* 114, 644–653.

Gupta S, and Kua HW 2019 Carbonaceous micro-filler for cement: Effect of particle size and dosage of biochar on fresh and hardened properties of cement mortar. *Science of the Total Environment* 662, 952–962.

He X, et al 2017 Evaluation of biochar powder on oxygen supply efficiency and global warming potential during mainstream large-scale aerobic composting in China. *Bioresource Technology* 245, 309–317.

He X, Yin H, Sun X, Han L, and Huang G 2018 Effect of different particle-size biochar on methane emissions during pig manure/wheat straw aerobic composting: Insights into pore characterization and microbial mechanisms. *Bioresource Technology* 268, 633–637.

Kishor P, Ghosh AK, and Kumar D 2010 Use of fly ash in agriculture: A way to improve soil fertility and its productivity. *Asian Journal of Agricultural Research* 4, 1–14.

Kumar M, Verma BB, and Gupta RC 1999 Mechanical properties of acacia and eucalyptus wood chars. *Energy Sources* 21, 675–685.

Lee E, and Han H-S 2017 Air curtain burners: A tool for disposal of forest residues. *Forests* 8, 296.

Lehmann J, et al 2003 Nutrient availability and leaching in an archaeological Anthrosol and a Ferralsol of the Central Amazon basin: Fertilizer, manure and charcoal amendments. *Plant and Soil* 249, 343–357.

Major J, Lehmann J, Rondon M, and Goodale C 2010 Fate of soil-applied black carbon: Downward migration, leaching and soil respiration. *Global Change Biology* 16, 1366–1379.

Pecha B, et al 2021 Biochar production. In: Amonette J, et al (Eds) *Biomass to Biochar – Maximizing the Carbon Value.* Pullman: Washington State University, Center for Sustaining Agriculture and Natural Resources. pp 149–155.

Puettmann M, Sahoo K, Wilson K, and Oneil E 2020 Life cycle assessment of biochar produced from forest residues using portable systems. *Journal of Cleaner Production* 250, 119564.

Rajkovich S, et al 2012 Corn growth and nitrogen nutrition after additions of biochars with varying properties to a temperate soil. *Biology and Fertility of Soils* 48, 271–284.

Restuccia, F, Masek, O, Hadden, R, and Rein G 2019 Quantifying self-heating ignition of biochar as a function of feedstock and the pyrolysis reactor temperatures. *Fuel* 236, 201–213.

Schapiro AR 2002 Use of Air Curtain Destructors for Fuel Reduction. US Department of Agriculture, Forest Service, San Dimas Technology and Development Center, 5100, 0251 1317—SDTDC, accessed at https://airburners.net/tech_docs/usda_fs_techtip0251-1317.pdf on 2023.8.1.

Sharpe B and Rodriguez F 2018 Market analysis of heavy duty commercial trailers in Europe. International Council on Clean Transportation, accessed at https://theicct.org/sites/default/files/publications/EU_Trailer_Market_20180921.pdf on 2023-8-6.

Spokas KA, et al 2011 Qualitative analysis of volatile organic compounds on biochar. *Chemosphere* 85, 869–882.

UN 2019 Recommendations on the transport of dangerous goods – manual of tests and criteria. 7th revision, United Nations, No. E.20.VIII.1, accessed at https://unece.org/fileadmin/DAM/trans/danger/publi/manual/Rev7/Manual_Rev7_E.pdf on 2023-8-3.

Weber K and Quicker P 2018 Properties of biochar. *Fuel* 217, 240–261.

Willaredt M, Peters A, and Nehls T 2022 Predicting water retention curves for binary mixtures–concept and application for constructed technosols. *Hydrology and Earth System Sciences Discussions*, hess-2022-265.

Xia L, et al 2023 Climate mitigation potential of sustainable biochar production in China. *Renewable and Sustainable Energy Reviews* 175, 113145.

You S, et al 2022 Energy, economic, and environmental impacts of sustainable biochar systems in rural China. *Critical Reviews in Environmental Science and Technology* 52, 1063–1091.

Zhang K, et al 2022 Biochar coating is a sustainable and economical approach to promote seed coating technology, seed germination, plant performance, and soil health. *Plants* 11, 2864.

Zhao MY, Enders A, and Lehmann J 2014 Short- and long-term flammability of biochars. *Biomass and Bioenergy* 69, 183–191.

26

Biochar-based fertilizers, co-composting, and growing media

Leônidas C. A. Melo, Carlos Alberto Silva, Miguel Ángel Sánchez-Monedero, Keiji Jindo, Sarasadat Taherymoosavi, and Stephen Joseph

Introduction

Conventional fertilizers were part of the green revolution and have increased food production and quality since the 1960s (Roberts, 2009). Overall, these fertilizers have low efficiency and are usually over-applied due to susceptibility to nutrient losses through leaching, volatilization, denitrification, fixation, erosion, and runoff, generating environmental pollution (Tilman et al, 2002; Rashid et al, 2021). Thus, the development of technologies to produce smart and cost-effective slow-release fertilizers is urgently needed (Calabi-Floody et al, 2018; Rashid et al, 2021), as well as the adoption of agricultural best management practices (Roberts and Johnston, 2015) to reduce nutrient losses, while improving fertilizer agronomic efficiency. Additionally, the application of chemical fertilizers alone does not contribute to soil carbon (C) accumulation in comparison to organic fertilizers (Yang et al, 2016).

Combining biochar with mineral fertilizers has recently been proposed as a new route to produce biochar-based fertilizers (BBF) and enhance nutrient use efficiency from conventional fertilizer sources (Ndoung et al, 2021; Rombel et al, 2022). Biochar seems to have a synergistic effect with fertilizers (Faloye et al, 2017; Ye et al, 2020; Bai et al, 2022), which causes an increase in crop productivity (Chapter 13). This synergism suggests that biochar plays a role in improving nutrient use efficiency from conventional fertilizers that could be commercially explored as BBF due to the low application rates needed (Melo et al, 2022). Moreover, it opens opportunities for recycling organic wastes and producing alternative fertilizers for smallholders in developing countries where the cost of conventional fertilizers is sometimes prohibitive.

The economic feasibility of high application rates of biochar is uncertain and

DOI: 10.4324/9781003297673-26

might be economically viable in high-value crops. For instance, biochar from forest residues applied at 10 t C ha^{-1} covers the average total costs for beets but not for potatoes in Canada (Keske et al, 2020). Biochar was also shown not to be feasible for cereal farmers due to its relatively high cost and limited agronomic benefits for grain production (Dickinson et al, 2015). Adding straw biochar to field crops in China was not economically viable; however, the addition of small amounts together with inorganic nitrogen (N), phosphorus (P), and potassium (K) had a positive return on investment (Clare et al, 2015). Thus, crop yield increases do not offset the costs of biochar application in most scenarios, and C credit prices should be much higher to compensate for biochar costs (Bach et al, 2016). The financial feasibility of 33 relevant publications has been reviewed, and it was concluded that biochar might be financially feasible only in small-scale farming on tropical soils (low fertility) in lower-income countries, but it was suggested that BBF may prove a profitable pathway for the application of biochar in large-scale agriculture, even in higher-income countries, based on yield gains obtained at low application rates of nutrient-enriched biochar (Robb et al, 2020).

Some nutrient-rich feedstocks that are currently misused or challenging to dispose of may be managed using pyrolysis systems by converting them into biochar that can be used as alternative fertilizer sources. This is especially the case for poultry litter-derived biochar, which has been shown to have long-term benefits over and above the provision of available P (Bai et al, 2015). Coffee husk, which has a low bulk density that makes it difficult to return to the field in raw form, when transformed into biochar, can reach up to 10% K (w/w) and may have a high cation

exchange capacity (CEC) depending on the pyrolysis temperature (Domingues et al, 2017). Sewage sludge is also difficult to handle, and its conversion into biochar stabilizes C and nutrients (slow-release) and might provide high levels of P to replace mineral fertilizers (Patel et al, 2020). Moreover, the addition of KCl or K_2SO_4 (Fachini et al, 2021) or even mixing with K-rich feedstocks can enrich sewage sludge-derived biochar to formulate organomineral fertilizers. Slaughterhouse waste (animal bones) in co-pyrolysis with lignocellulose agricultural waste and bioaugmentation with phosphate solubilizing microorganisms can also generate an amendment that can fully replace chemically soluble P fertilizer (Ahmed et al, 2021). Thus, converting waste materials into valuable products through pyrolysis to produce high-value BBFs can create a strong circular economy (see Chapter 31).

The challenge in this area is to conceive new biochar synthesis routes to produce BBFs capable of meeting plant nutritional requirements at low biochar application rates (Melo et al, 2022; Rasse et al, 2022). In this chapter, we aimed to explore (i) the best synthesis routes for producing BBF, including the influence of biochar on BBF properties and the mechanisms of nutrient release from BBF; (ii) the effectiveness of BBF on plant productivity; (iii) co-composting and BBF; (iv) BBF as a constituent of growing media; and (v) BBF potential as a foliar spray and liquid fertilizer. Despite recent advances in BBF synthesis and use, its development is still in its infancy, and many questions remain unanswered regarding the interaction of biochar with mineral fertilizers, the properties of BBFs governing the kinetics of nutrient release, and the effects of BBF as a plant growth promoter (Ndoung et al, 2021). Thus, in this chapter, a ranking of research priorities is also suggested.

Biochar-based fertilizer: Synthesis routes

Common strategies for the preparation of BBF among different studies include direct thermal treatment through pyrolysis of nutrient-rich feedstocks and pre-treatment of feedstock before pyrolysis or post-treatment after pyrolysis (Ndoung et al, 2021; Sim et al, 2021; Marcińczyk and Oleszczuk, 2022; Rombel et al, 2022; Rasse et al, 2022). A step further includes the encapsulation technique aiming to reduce nutrient losses and improve nutrient release mechanisms as well as increase nutrient use efficiency (Sim et al, 2021). Figure 26.1 summarizes the main methods and steps proposed to produce BBFs.

Each BBF production method might be suitable in different situations and generates

BBF with different characteristics regarding the kinetics of nutrient release, nutrient concentration, and role played as a single or multiple nutrient carrier. In addition, each BBF production method might affect the fertilizer production cost. For instance, for direct preparation, coffee husk biochar prepared up to 450°C generates a material rich in available K (Domingues et al, 2017) and of high CEC that is effective in increasing soil CEC (Domingues et al, 2020). Additionally, co-pyrolysis of nutrient-rich waste materials (e.g., sewage sludge) with K minerals can unlock P and improve the fertilizer potential of otherwise low-efficiency fertilizer material (Buss et al, 2020).

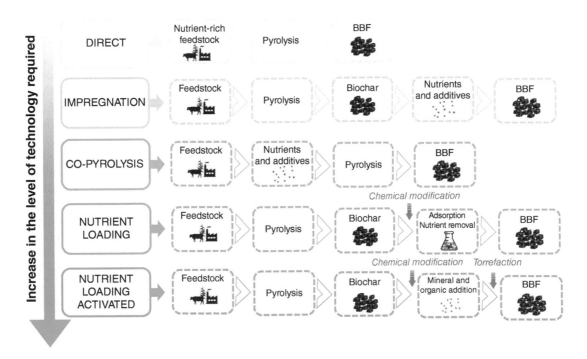

Figure 26.1 *Flow diagram for biochar-based fertilizer production routes*

Usually, the impregnation or blending process does not cause a slow release of nutrients, as has been verified for triple superphosphate (soluble P) incorporation into biochar after pyrolysis, and this BBF type did not show an increase in crop productivity in P-fixing soils in short-term experiments under controlled conditions over soluble P sources (Santos et al, 2019; Pogorzelski et al, 2020). The recovery of P from aqueous solution using biochar has also been intensively studied in the last decade, and most studies recommend the reuse of P-loaded biochar as a fertilizer (Marcińczyk et al, 2022). To efficiently adsorb and remove large amounts of P from aqueous media, biochar must usually be modified by loading cations into feedstock before pyrolysis (Almanassra et al, 2021; Nardis et al, 2022), which should be further tested as fertilizers. P recovered from aqueous solution was used in Mg-impregnated biochar from poultry litter, pig manure, and sewage sludge and evaluated as a fertilizer in a short-term greenhouse experiment using a tropical soil and showed similar or greater agronomic performance than triple superphosphate (Nardis et al, 2020). A biochar prepared from nutritionally Mg-enriched tomato tissues to recover and reuse P from an aqueous solution showed potential as a slow-release P fertilizer (Yao et al, 2013). These findings highlight the need to evaluate the kinetics release and agronomic value of P-loaded BBF under field conditions to show its agronomic efficiency as a sustainable P source.

Other approaches using mixtures of soluble P sources with feedstock before pyrolysis have been shown to increase C conversion efficiency to biochar during pyrolysis, assuring greater C persistence (Zhao et al, 2016; Carneiro et al, 2018). However, depending on the pyrolysis temperature, this co-pyrolysis process transforms P into insoluble P forms, which might reduce the fertilizer efficacy in the short-term due to the restriction of plant P supply (Lustosa Filho et al, 2017), which is reverted in the medium- to long-term, showing similar plant growth as soluble P sources (Carneiro et al, 2021). Bioaugmentation using phosphate-solubilizing microorganisms has also shown promising results using either fungi (Mendes et al, 2014; Ahmed et al, 2021) or bacteria (Leite et al, 2020) to increase the availability of sparingly soluble P compounds in soil. These products, however, still lack consistent results, especially under field conditions, to prove the effectiveness of this approach to supply enough P for plant nutrition, mainly for high P-demanding crops.

Research work has also shown that biochars can be enhanced by coating them with a mixture of manure, minerals, and wood vinegar and then torrefying at low temperatures (approximately 220°C). These organomineral-coated biochars, termed biochar-mineral complexes (BMCs), can be applied at low rates and can significantly change the abundance of growth-promoting microorganisms, especially AM fungi (Blackwell et al, 2015). BMC applied in the first crop cycle at 1.0 and 5.0 ton ha^{-1} and then in the second crop cycle at 0.1 and 0.4 ton ha^{-1} in a barley-sweet corn cropping cycle resulted in a greater yield response than the application of chemical fertilizers for the same input of N and P (Blackwell et al, 2015). This was driven by a difference in the relative abundance of a small number of microbial community members. It was concluded that "overall, a large number of microorganisms appear to be influenced by EB amendment compared with fertilizer use leading to a complex rewiring of community composition and associations".

For N, it is difficult to control the release rate using a BBF approach. It may not be possible to produce concentrated N slow-release BBF based on biochar adsorption properties only (Rasse et al, 2022). Thus, some technologies might emerge aiming to overcome such limitations, such as (i) sorption of NH_3 on functionalized biochar, (ii) biochar-clay composites to improve sorption and release properties for N, and (iii) physical barriers for conventional fertilizers (e.g., coating/encapsulation) (Rasse et al, 2022). Biochars activated either by ultrasonication, or oxidation with hydrogen peroxide or nitric acid were shown to increase the N retention, which was distributed in water-soluble (<15%), hydrolyzable, and non-hydrolyzable N that might progressively become available for plants, acting as slow-release fertilizers (Castejón-Del Pino et al, 2023). To produce an N-based BBF, liquid urea was impregnated onto biochar and applied as a source of N (applied at once) in powder form for the cultivation of common beans and maize in a greenhouse experiment (Barbosa et al, 2022). A fast release of N was observed that caused an increase in soil electrical conductivity (EC) at higher N doses and caused phytotoxicity in common bean, reducing its productivity, but this BBF showed a greater N residual effect (N legacy) and supplied N to maize in second cultivation as opposed to conventional urea (Barbosa et al, 2022). In another study, 17 N-based BBFs were tested for N availability and ammonia loss with N contents ranging from 3% to 38% (Puga et al, 2020b). BBFs with biochar/N ratios (w/w) between 2 and 10 showed a slower N release rate than conventional fertilizers (ammonium sulfate and urea). The most promising mixtures from the former study with slow-release characteristics for N were chosen and assessed for their performance under tropical field conditions (Puga et al, 2020a). Side-dressing application resulted in up to a 12% increase in N use efficiency and up to a 21% increase in maize grain productivity over conventional N fertilizers.

Influence of the physical forms of BBF on the nutrient release rate

Producing N-based BBF with slow-release characteristics is challenging and requires advanced technologies, including granulation, pelletization, or coating. Normally, granulation/pelletization including additives control the nutrient release rate better than in powder form. However, coating/encapsulation can provide a more sustained and controlled release rate of nutrients aiming to produce enhanced efficiency N fertilizers (Sim et al, 2021). Thus, it might be the way forward for producing concentrated N fertilizers. For instance, biochar-coated urea increased N use efficiency (NUE) by 20% compared to conventional urea, mainly due to a reduction in nitrate leaching, while additional biochar application increased the fertilization cost and slightly reduced NUE (Jia et al, 2021). Thus, the coating was ranked as a promising technology to reduce costs and facilitate application while increasing NUE.

An organomineral fertilizer was produced from sewage sludge biochar enriched with KCl or K_2SO_4 using three post-production methods (granules, pellets, and powders) (Fachini et al, 2021). The morphology and physical characteristics of these BBFs were more influenced by the form of the fertilizer (powder, pellet, or granule) than by the source of K (KCl or K_2SO_4). The diffusion and solubility of P in a BBF enriched with triple superphosphate (TSP) or phosphoric acid mixed with MgO was studied as a P fertilizer (Lustosa Filho et al, 2019). The granular form of BBF produced with phosphoric acid and MgO was equivalent to TSP to provide P for maize while preserving higher amounts of P in the granule after cultivation when compared with TSP. However, for BBF produced using TSP and MgO, the granulation caused a slow release that limited P acquisition by roots and reduced maize productivity. Usually, granulation, pelletization, or coating facilitates handling and field application (Chapter 25). Studies in this direction with BBF are still scarce, and there is a range of possibilities; thus, they should definitely advance under field conditions, including economic evaluations aiming to increase farmers' acceptance of this novel technology.

Influence of biochar on BBF properties

Biochar properties are mainly influenced by the feedstock type and pyrolysis temperature (more appropriately, the highest heating temperature – HHT), and these properties might greatly influence the nutrient release behavior from BBF as well as the interaction with soil microorganisms or plant root systems. In a recent meta-analysis of BBF effect on crop productivity, a significant and higher effect size on crop productivity was observed for BBFs designed with more than 30% C in their final composition than for those with less than 30% C when compared with conventional fertilizers (Melo et al, 2022). A significant effect was also observed for BBF produced with biochar with a high HHT (> 400°C). High pyrolysis (> 400°C) temperature biochar is more effective at promoting root length (Xiang et al, 2017), which might help to explain the better effect on crop productivity. The higher C content in BBF might increase porosity, specific surface area, and surface charges owing to a variety of surface functional groups in biochar, but one has to be careful with biochar pH and volatilization of N. Carbon-related properties of biochar are key to designing enhanced efficiency BBFs as well as minerals that are lacking in soil. For instance, the addition of clay and $FeSO_4$ to the surface of bamboo and pyrolysis at different temperatures increased the abundance of microorganisms that made Fe and S more available to plants (Ye et al, 2017). The charge on the surface of a micron or submicron mineral particle will be different from that of the C matrix (Ye et al, 2017). Thus, there is a potential difference between these two areas, which in turn may influence the local biotic and abiotic redox reactions that take place in the soil.

Biochar properties such as C content and charge distribution on biochar surface might increase nutrient use efficiency from BBF compared with conventional fertilizer, especially for N fertilizers due to increased sorption capacity, nitrification, denitrification, and change in the microbial community (Xu et al, 2016). A recent study also demonstrated that electrochemical reactions in the rhizosphere induced by BBF application might reduce the membrane potential of root cells and facilitate nutrient uptake (Chew et al, 2020). The lower membrane potential was further elucidated in another study using rice seedlings fertilized with NPK microsized BBF versus conventional NPK fertilizer. This study showed an increase in nutrient uptake and plant growth and an increase in the potential difference between the soil and epidermal layer, which induced root nutrient accumulation resulting in rice seedling biomass accumulation (Chew et al, 2022). Thus, it is possible that mixtures that increase electron exchange capacity, such as wood ash (Grafmüller et al, 2022), might change the potential at the rhizosphere level and influence nutrient uptake as well as influence changes in the microbial community due to electron transfer kinetics influenced by pyrogenic C (Sun et al, 2017). This is an area that needs further research.

Mechanisms of nutrient release from BBF

Biochar has a porous structure and abundant functional groups along its surface area that interact with nutrients (e.g., NO_3^-, NH_4^+, PO_4^{3-}, and K^+) through different reactions, including physisorption and chemisorption (Sim et al, 2021). Biochar can provide surface area and porosity, as well as a higher capacity to store and donate electrons than regular NPK, and this causes a slower dissolution rate for P and K by forming organo-mineral clusters of these nutrients with C, Si, Al, and Fe (Tahery et al, 2022). However, without the presence of roots, the studied BBF showed no ability to reduce the N dissolution rate, which suggests that the interaction of BBF and the rhizosphere might play a role (Tahery et al, 2022). In fact, only a few studies have evaluated NPK together, as most studies have focused on the individual behavior of these nutrients (Table 26.1).

For phosphate (PO_4^{3-}), most studies indicate its binding with cations (e.g., Ca^{2+}, Mg^{2+}, Fe^{3+}, and Al^{3+}) forming bridges on the biochar surface, and despite this reducing P solubility in water, it still maintains plant availability (Nardis et al, 2020). P fertilizers coated with biochar have shown that P water solubility is regulated by diffusion and swelling or relaxation mechanisms, which can control better P release than simply blending biochar with P fertilizer (Pogorzelski et al, 2020). However, when PO_4^{3-} is co-pyrolyzed with biomass, there are indications of P transformation into insoluble forms, greatly reducing the P release rate (Zhao et al, 2016; Lustosa Filho et al, 2017). Although P transformations due to pyrolysis can limit plant access to P in the short term, there are indications that in the medium- to long-term, plants fertilized with P-based BBFs show similar performance to soluble P fertilizers (Carneiro et al, 2021).

For K, a kinetic study showed that the release of K from BBF occurs in two phases, and rapid dissolution of readily available K occurs on the BBF surface, creating channels that allow water to diffuse into the fertilizer but that also prevent water from entering small pores, greatly reducing the K release rate (Fachini et al, 2022). Then, a second slow-release phase begins where K is physically protected inside BBF pores and slowly diffuses to the surface where it can also interact with negative charges on the biochar surface and limit its fast release to the solution.

Table 26.1 *Main mechanisms of nutrient release from biochar-based fertilizers depending on the production method*

Mechanism of nutrient release	BBF production method				
	Direct	Impregnation	Co-pyrolysis	Nutrient loading	Nutrient loading activated
Physisorption, chemisorption		✓		✓	
Electrostatic attraction				✓	
Diffusion	✓	✓	✓		✓
Dissolution	✓	✓	✓		✓
Precipitation				✓	
Ion exchange				✓	✓
Pore filling		✓			✓

For N, limitations exist in producing highly concentrated N-based BBF only focused on adsorption properties, i.e., surface area, porosity, and biochar functional groups (Rasse et al, 2022). Nitrogen is retained by biochar mainly through ion exchange and electrostatic adsorption physical adsorption (Dai et al, 2020). Thus, the limited adsorption capacity of N by biochar suggests the need for more advanced and integrated techniques to retain and control the release rate of N to develop BBFs of enhanced efficiency. In this regard, a slow release of N can be achieved by integrating the coating of conventional N sources (e.g., urea or ammonium nitrate) using biochar and binders such as stalk, kaolinite, bentonite or sepiolite (Shi et al, 2020; Wang et al, 2022). Encapsulation is usually effective in increasing the control over the release rate of N from urea and other concentrated N sources, as it creates a diffusion barrier that controls nutrient release, but it is a complex and expensive process that limits its large-scale utilization for BBF production (Wang et al, 2022). Encapsulated BBF has mechanisms of nutrient release based on a multistage diffusion model where the binder (e.g., bentonite, zeolite, ligning, and starch) absorbs water and swells at the pores of biochar. Then, dissolved nutrients inside the granule are released slowly through diffusion under a concentration gradient, and when water shortage occurs in the soil, the dehydration of the binder reduces diffusion. Another approach is to impregnate a small amount of molybdenum (Mo) (Mo: biochar ratio of 1:200,000 w/w) into biochar by adsorption and apply it to the soil (Huang et al, 2022). BBF supplied Mo slowly and steadily to Chinese flowering cabbage and increased nitrate assimilation by the plant, which resulted in better vegetable growth and quality.

BBF effectiveness on plant productivity

Although BBF has been claimed to have positive effects on the environment by improving nutrient use efficiency, its effects on crop productivity remain site- and crop-specific when compared with a conventionally fully fertilized control. In a meta-analysis, a 10% increase (grand mean effect) in crop productivity of BBF was observed compared with fertilized control from 148 pairwise comparisons representing 40 independent peer-reviewed articles (Melo et al, 2022). Crop responses to BBF compared with conventional fertilizers range from negative (−18%) to positive (up to +36%) and vary depending on the soil, crop, climate, and BBF preparation method (Table 26.2). Higher crop productivity with higher C contents (> 30%) in the BBF and in soils of higher fertility that are less responsive to chemical fertilizers indicates that BBF could solve specific problems that conventional fertilizers could not. The positive crop productivity response observed with a low BBF application rate (average of 0.9 ton ha^{-1} considering field studies) encourages further studies, especially under field conditions, to evaluate the economic feasibility of using BBF as an enhanced efficiency fertilizer with a multitude of possibilities.

N-based BBF caused a crop yield increase only when clay was included in the mixture for BBF preparation, which likely contributed to improving the N release rate and consequently the N use efficiency (Rasse et al, 2022). Most studies on BBF are still recent and performed under laboratory or greenhouse conditions. Therefore, validation of this technology is needed under field conditions with different crops and soils and using medium- and long-term experiments.

Mechanistic understanding of the effect of BBF on enhancing crop productivity

Table 26.2 *Biochar-based fertilizers by different methods and effects on crop productivity*

BBF	Study	Crop type	BBF preparation method	Crop productivity response (%) *	References
NPK	Field	Sorghum	Mixed thermal treatment	−4	Blackwell et al, 2015
PK	Pot	Sugarcane	Mixed granulation	+18	Borges et al, 2020
P	Pot	Grass; maize; common bean	Co-pyrolysis	−1	Carneiro et al, 2021
NP	Field	Maize	Co-pyrolysis, impregnation and baking	+10	Chen et al, 2021
N	Pot	Maize	Solid mixture impregnation	+4	Dil et al, 2014
NPK	Field	Rice	Coating	+8	Dong et al, 2020
N	Pot	Rapeseed	Coating	+25	Jia et al, 2021
NPK; N	Field	Maize	Solid physical mixture	+1	Kamau et al, 2019
N	Pot	Maize	Impregnation and encapsulating/coating	+17	Khajavi-shojaei and Moezzi, 2020
N	Pot	Oilseed rape	Impregnation	+9	Liao et al, 2020
P	Pot	Maize	Co-pyrolysis	−18	Lustosa Filho et al, 2019
P	Pot	Grass	Co-pyrolysis	+10	Lustosa Filho et al, 2020
P	Pot	Maize	Nutrient-loaded adsorbent	+36	Mosa et al, 2018
P	Pot	Maize	Nutrient-loaded adsorbent	+19	Nardis et al, 2020
N	Field	Maize	Solid physical mixture	+4	Peng et al, 2021
P	Pot	Millet	Mixed granulation/coating	+3	Pogorzelski et al, 2020
N	Field	Maize	Mixed granulation	+9	Puga et al, 2020a
NPK	Field	Rice	Mixed granulation	+21	Qian et al, 2014
P	Pot	Maize	Mixed granulation	+2	Santos et al, 2019
N	Pot	Maize	Mixed granulation	+14	Shi et al, 2020
NPK	Field	Tea	Mixing, extruding; Granulating	+20	Yang et al, 2021
NPK	Field	Green pepper	Impregnation	+16	Yao et al, 2015
S	Pot	Maize	Impregnation	+24	Zhang et al, 2017
N	Field	Tobacco	Impregnation and granulating	+1	Zhang et al, 2021a
N	Field	Maize	Mixed granulation	+5	Zheng et al, 2017

Note
* Calculated as the average of all the treatments (when more than 1 has been tested) compared with the fertilized control (conventional fertilizer)

through plant stimulation compared with conventional fertilizers is still under investigation. In general, it is attributed to the slow-release mechanism for different nutrients as well as an increase in nutrient use efficiency through soil pH that changes the abundance of growth-promoting microorganisms, increased root growth and N uptake, increased mycorrhizal root colonization, increased physical retention of dissolved nutrients and reduced leaching, increased nitrification, improved redox conditions, and changes in abundance of growth-promoting microorganisms, increased P and K availability, and slower diffusion of reactive N to soil solution (Rasse et al, 2022).

Composting and BBF

Composted biochar has attracted attention as a BBF for its potential benefits for soil fertility and plant growth derived from the combined effects and synergies between biochar and compost (Hagemann et al, 2017). The combination of biochar with organic materials, such as composts or manures, is a traditional strategy to benefit from the complementary properties of both substrates (Fischer and Glaser, 2012; Kammann et al, 2016). Biochar provides favorable properties to interact with the nutrient, organic C, and microbial pools supplied by composts, enabling controlled organic matter mineralization and nutrient release. Mixing biochar and compost allows an intimate mixture to optimize their interaction and ensure the loading of biochar with the nutrients supplied by the compost. Mixing was therefore found to be a more effective strategy than applying biochar and compost separately (Rombel et al, 2022). However, the addition of biochar at the early stages of the composting process not only allows for optimizing their interaction but also enhances the composting process and the quality of the end product (Wang et al, 2019; Antonangelo et al, 2021).

Impact of biochar on the composting process

The use of biochar as an additive in organic waste composting has been widely studied in recent years because of its potential role in enhancing the process (Steiner et al, 2015). Both wood and straw biochars have been traditionally used as compost additives due to their suitable physical and physicochemical properties. Application rates in the range of 5 - 20% v/v have been successfully used for a wide range of composting operations, biochar feedstock, and composting materials, even though an application rate of 10% is the most widely used (Godlewska et al, 2017; Sanchez-Monedero et al, 2018). The persistence of wood biochar, especially those prepared at high pyrolysis temperatures, allows a long-lasting effect of biochar during the composting process and after soil application. However, a recent meta-analysis based on 876 observations from 84 studies recommended using straw biochar at an application rate between 10% and 15% to improve the quality of the final composts (Zhou et al, 2022). Most of the straw biochars studied in the literature are based on rice straw, which is a widely available resource in Asian countries with a long-established tradition of agricultural use of biochar. These biochars have a very high content of amorphous microns and nanosized SiO_2, and there is some evidence that these are sites that catalyze the decomposition of organic matter and immobilize heavy metals and toxic organics (Sui et al, 2021). Additionally, it has been shown that biochar can accelerate the removal of polyaromatic hydrocarbons (PAH) in the thermophilic phase when the application rate is higher than 2% v/v, and the resulting biochar-compost has a negligible cancer risk and is safe for soil application (Chen et al, 2022).

The physical and physicochemical properties of biochar, such as its high porosity and large functionalized surface area, make it a suitable habitat for microorganisms (Lehmann et al, 2011). Additionally, biochar can also promote microbial growth in the pile by enhancing its environmental conditions, such as (i) greater aeration and gas exchange diffusion, (ii) improved water retention capacity that can prevent desiccation, (iii) larger sorption capacity for nutrients and easily mineralizable organic fractions that can serve as a source of nutrients and energy to support microbial growth, and (iv) protective role as pH buffer capacity, toxin inhibition and protection from predators. Consequently, biochar has been identified as an efficient composting additive that can enhance bacterial abundance and diversity during composting (Duan et al, 2019).

There is abundant information in the literature on the impact of biochar on compost microbiology (Antonangelo et al, 2021). Biochar is known to enhance the composting temperature and CO_2 evolution during the thermophilic stage (Steiner et al, 2011; Czekała et al, 2016), which is reflected in intense microbial activity due to the increased decomposition of easily mineralizable organic compounds (Khan et al, 2014). Not only the enhanced temperature but also the changes in the physicochemical parameters (C, N, C/N, pH), as a result of biochar in the composting pile, are important drivers for the change in the compost microbial community (Malinowski et al, 2019). Available N and C resources by biochar addition are one of the underlying factors of the abundances of some microbial taxonomies (e.g., Comamonas and Leucobacter,

Qiu et al, 2019). This is especially important for straw biochar, which supplies easily mineralizable C that stimulates microbial biomass (Zhou et al, 2022). As a consequence, composting piles enriched with biochar are characterized by an enhanced abundance of bacteria, fungi, and actinomycetes (Du et al, 2019). The changes in the microbial community during composting occur remarkably during the thermal phase but also in the cooling and maturity stages (Jiang et al, 2019), such as changes in Actinobacteria during maturation in composting piles enriched with biochar (Qiu et al, 2019).

Changes in compost microbiology can affect key microbial processes involved in C and N cycles (Zhang et al, 2021b). Biochar is known to enhance the abundance of several bacteria during the thermophilic phase, such as Solibaccilus, Actinibacteria, and Aspergills, which have high degradation capacity for lignocellulose and lignin (Jiang et al, 2019; Qiu et al, 2019). Biochar also strengthens most enzymatic activities, including extracellular hydrolytic enzymes (b-glucosidase and aryl-sulfatase) as well as lignocellulolytic enzymes (cellulase) (Du et al, 2019). All these enzymes are involved in the degradation of lignocellulose compounds. Biochar is thought to contribute to organic matter decomposition by other complementary mechanisms (Jindo et al, 2016), such as the release of aromatic moieties from the biochar itself (Heymann et al, 2014; Wang et al, 2014) and the adsorption of organic matter onto biochar causes a reduction in mineralization in the composting matrix (Ni et al, 2011). This appears to follow similar dynamics as for priming in soil (Chapter 17).

The changes in microbial dynamics also had an impact on reducing the emissions of volatile organic carbon (Sánchez-Monedero et al, 2019) and methane (Sonoki et al, 2013). The enhanced porosity in the composting matrix by the presence of biochar affects methane emissions (Sonoki et al, 2013). Biochar particle size is a key factor regulating porosity and, consequently, methane emissions. Granular biochar (4-10 mm) reduced the mcrA/pmoA ratio and increased the diversity of bacteria and archaea during composting, causing a 22% reduction in methane emissions, whereas powdered biochar (<1 mm) increased methane emissions by 57% (He et al, 2018), partly caused by differences in bulk density. However, the use of powdered biochar may have negative effects on other GHG emissions during large-scale composting due to the reduced porosity caused by the high proportion of small particles (He et al, 2017). BMC that is added to compost can have a greater effect on microbial communities, compost quality, and plant response than biochar alone (Ye et al, 2016). BMC is produced by treating wood biochar with HHT at 600°C with phosphoric acid and then mixing with a mixture of chicken manure, rock phosphate clay, ilmenite, dolomite, and basalt dust, and when added to compost, BMC interacts synergistically to further increase C, N, total soluble N, nitrate and available K concentrations (Nguyen et al, 2017).

One of the best-known impacts of biochar on the composting process is enhanced N retention in a pile, which increases the agronomical value of biochar-enhanced composts and decreases the environmental impact of the process (reduced gaseous emissions). Biochar can reduce total N losses by approximately 30-36%, as reported by several recent meta-analyses (Zhao et al, 2020; Zhang et al, 2021b). Most of the reduction in the N loss effect is due to reduced NH_3 volatilization and, to a minor extent, reduced N_2O emissions. The mechanisms behind this effect are derived from the physical adsorption of NH_3 in the inner pore space of biochar (Agyarko-Mintah et al,

2017), the retention of NH_4^+ on the large and functionalized surface of biochar (Agyarko-Mintah et al, 2017), the microbial assimilation of NH_4^+ to organic N (Zhou et al, 2022), or nitrification (Sánchez-García et al, 2015; Ye et al, 2016). Straw biochars show a higher N retention capacity than wood biochar as a consequence of the more acidic and functionalized surface (chemical adsorption) in straw biochars compared to the porous and large surface area (physical adsorption) of wood biochars (Zhou et al, 2022), which is also enhanced by oxidizing biochar (Hestrin et al, 2020). The impact of biochar on the microbial processes involved in the N cycle is more complex than that in the C cycle. Consequently, future research aimed at evaluating the interaction of biochar with key microbial communities related to nitrogen transformation should be based on the use of new technologies for high-throughput sequencing of DNA (Sun et al, 2019).

Apart from the impact on N retention, wood biochar does not have a significant impact on the nutritional value of the enriched composts since this type of biochar does not represent a source of nutrients (Casini et al, 2021). However, straw biochars and ash-rich biochars can represent a source of nutrients for composts (Domingues et al, 2017; Zhou et al, 2022). An enhanced K concentration in biochar-amended composts has been observed as a consequence of the direct supply from biochar and from increased K^+ retention facilitated by the functionalized surface of straw biochars (Zhao et al, 2020). At the same time, biochar reduced the inorganic P availability during composting (Vandecasteele et al, 2016), implying that there is an increased microbial transformation of organic P fractions to regulate P availability and prevent the leaching of inorganic P (Wei et al, 2021).

Application of composted BBF for improvement of plant growth: Soil amendment

Apart from the impact of biochar on the composting process, biochar also undergoes intense oxidation and weathering during composting, leading to enhanced carboxyl groups and CEC, which, alongside the sorption of dissolved organic matter on biochar, can enhance the interaction and retention of nutrients (Prost et al, 2013; Wiedner et al, 2015). Co-composting biochar led to the formation of a hydrophilic organic coating on biochar particles, which was identified as a composite of the original compost and biochar particles (Hagemann et al, 2017). This coating was characterized by the presence of nonbiochar organic matter and enrichment with N (mostly NO_3^-) and other nutrients that could explain the enhanced capacity of composted biochar to retain nutrients and stimulate soil fertility.

Several recent reviews and meta-analyses evaluating the use of co-composted biochar as a soil amendment and its impact on soil fertility and plant growth (Agegnehu et al, 2016; Wang et al, 2019; Antonangelo et al, 2021; Rombel et al, 2022) identify multiple benefits, such as improvement in crop yield and soil fertility, reduction in N_2O emissions, and enhancement of C sequestration in soil due to the biochar contained in the compost. The benefits of this type of material were associated with the direct supply of nutrients and the enhanced soil physicochemical properties to retain nutrients (improved soil pH, cation exchange, and water holding capacity).

The addition of 10-15% w/v straw biochar to a composting pile was found to enhance the nutritional value of compost and reduce its ecological risk associated with metal bioavailability (Zhou et al, 2022). A significant increase of 40% in grain yields of cereal grasses in acidic soils treated with co-composted biochar was observed (Wang et al, 2019), but these effects were not significant when the studies were conducted in temperate regions (Agegnehu et al, 2016). In general, the best agronomical performance was observed for rice husk biochar compared to woody biochar, but the impacts of co-composted biochar were mostly related to the properties of the composts, which is in line with previous findings reporting that the impacts of co-composted biochars are mostly driven by the properties of the starting materials of the composts (Kammann et al, 2016; Schmidt et al, 2021).

The interaction of biochar and biochar-blended compost with soil N dynamics and the N use efficiency of native or applied nutrients is one of the key aspects behind the effect of these materials (Ahmad et al, 2021; Rasse et al, 2022). However, other mechanisms may be involved, such as the stimulation of soil microbial processes (Nielsen et al, 2014; Ye et al, 2016), which is affected by the application rate and edaphic conditions (Nielsen et al, 2014; Yan et al, 2021); lower abundance and activity of soilborne pathogens, such as the genera *Mortierella* and *Fusarium*, in soil treated with composted biochar (Bello et al, 2021); salinity reduction by biochar-compost mixtures, facilitating nutrient uptake, such as K^+ (Lashari et al, 2013; Mithu et al, 2022); and the presence of low doses of biostimulants present in biochar (Rasse et al, 2022). It is worthwhile to mention, however, that soil types and biochar characteristics must be considered. While the bacterial soil community is likely more sensitive to the addition of pyrogenic C to acidic paddy soil than the fungal community (Dai et al, 2016), the fungal community can be altered over the long term (3 years) after amendment in alkaline soil (Gao et al, 2021).

Biochar in growing media

Growing media is a solid material, other than soil, that consists of a single matrix or a mixture of two or more matrices aiming at the full and healthy growth of young plants. It is possible, on a smaller scale and in a cost-effective way, to optimize plant growth conditions so that seedlings transplanted to the field are more successful. In the formulation of growing media, materials must provide suitable physiochemical and biological characteristics for plant growth, such as nutrient storage capacity, aeration, friability, CEC, water retention, nutrient adsorption, and nutrient supply in a balanced way; in addition, they must be practical from a commercial point of view (Zulfiqar et al, 2019). Furthermore, it is necessary to eliminate the pathogens and create conditions in the growing media to suppress diseases that can affect the seedlings. There can be no toxic substances (e.g., heavy metals) or excessive amounts of salts in the substrate, high EC, nutritional imbalances, and pH outside the range considered optimal for plant growth must be avoided. Thus, in the formulation of growing media, components must be prioritized that assure optimal conditions and nutrient supply for maximum root growth. Therefore, at the industrial scale, it is possible to create growing media that surpass soils in terms of agronomic efficiency and ability to nourish vigorous and healthy plants that will give rise to orchards, vegetable gardens, and forests of high yield.

In general, growing media are formulated by mixing two or more components, such as peat, pine bark, sphagnum, compost, coconut fiber, or high-chemical activity clay (e.g., perlite and vermiculite). Peat is the most commonly used organic material as a growing medium because it is a standardized, performant, uniform, and cost-effective substrate for horticultural plant growth (Jindo et al, 2020). However, peat is not a renewable material and represents a source of greenhouse gas emissions. Thus, its extraction has been discouraged, causing the need to find renewable substitutes able to provide similar characteristics and be cost-effective. Thus, the priority is to find inexpensive, locally available, longer-lasting, and stable materials that are pathogen-free and with adequate pH, EC, and aeration, bulk density, and balanced in available nutrients that ensure maximum plant growth.

Biochar has been assessed as an additive for growing media or as a peat substitute. However, biochar properties vary greatly according to feedstock or pyrolysis conditions. Thus, wood biochar produced at higher temperatures is potentially the best option to replace peat or other growing media due to its high persistence and low nutrient content and EC (Steiner and Harttung, 2014). Moreover, as opposed to peat, biochar is a renewable source and potentially a C-neutral or C-negative material, which offers an opportunity for C sequestration. Although biochar has been intensively studied in the last two decades, biochar as an additive in growing media has received relatively little attention. Adding up to 25% biochar pellets to peat improved the properties of the substrate for nurseries in small-volume containers (Dumroese et al, 2011). Up to 75% wood biochar could be added to peat as a lime substitute to increase the pH, maintaining similar characteristics for sunflower seedling production (Steiner and Harttung, 2014). Conversely, wood biochar produced at a high pyrolysis temperature (700°C) at a 7.5% mixture with peat was the best ratio for the production of lettuce seedlings, and the application of higher nutrient contents (e.g., N and P) is recommended due to their interaction with biochar (Chrysargyris et al, 2020). Biochar produced from olive stone outperformed perlite or cocopeat materials as solid growth media for tomato seedling production under hydroponic conditions (Karakaş et al, 2017). Across 32 articles dealing with a wide range of feedstocks and pyrolysis temperatures, a range of biochar properties showed similar or positive impacts on plant growth when up to 25% v/v biochar was added as an additive compared with a reference commercial substrate (e.g., peat, perlite, and vermiculite) (Huang and Gu, 2019). Despite the differences among studies with the best mixture ratio of biochar and peat, the advantage of such a mixture goes beyond biochar serving as a substitute for peat in growing media but also creates a synergistic effect. Examples include serving as a pH-controlling agent in acidic peat-based growing media, as a disease-suppressing component at low doses, improving water retention, and adding stable carbon (Kern et al, 2017).

Biochar properties can be modulated during pyrolysis to create conditions for biochar to partially (or even totally) replace peat, perlite, vermiculite, and other traditional growing media. Biochar is also a matrix free of pathogens, persistent over time, of low density, and porous. Biochar also plays a potential role in suppressing plant diseases as well as its biostimulant effect from pyrolysis-derived components (Rasse et al, 2022). Furthermore, carbonized matrices can be produced locally from plant residues, wood residues, postharvest residues, or animal manures. Thus, optimization of pyrolysis conditions and the right choice of feedstock are suitable strategies to increase the proportion of biochar in growing media while improving properties that maximize seedling growth. In this case, it is often better for the substrate to act as a true conditioner that improves aeration, CEC, and pore volume and has adequate pH and EC but low nutrient contents so that the growth rate of plants is controlled according to the supply of nutrients via fertigation and the market demand for seedlings. Currently, biochar is shown to be a suitable additive that might improve growing media quality for better plant growth. As shown in the previous sections, BBF could also be modulated to act as a slow-release nutrient source that might replace chemical fertilizers currently used in growing media to supply nutrients. Research in this direction is definitely needed in the near future, since there is no such research on this subject thus far.

Biochar-based foliar sprays and liquid fertilizers

Biochar foliar sprays (BFS) are produced by micronizing biochar in a liquid solution either at room temperature or at temperatures up to 95°C. The properties of the foliar sprays produced by solubilizing biochar in hot water depend on the type of biochar used (Kumar et al, 2021) and any post-treatment that takes place. During this process, micro- and macronutrients, micron- and submicron-sized minerals, C-bearing particles, and water-soluble organic molecules are released. The liquid extracts are then filtered to allow the particles to go through a specific-sized spray nozzle (usually <100 μm). The organic compounds comprise low molecular weight acids and neutral, polyphenols, polyphenolic acids, and large macromolecules with a high concentration of oxygen functional groups (especially carboxylic acids) and biopolymers. There can be over 200 compounds comprised of alcohols, acids, phenols, ketones, alkanes, alkenes, aromatic and polyaromatic hydrocarbons. The composition of the organic and inorganic compounds and ions includes quartz, diatomite, Al_2O_3, clays, apatite, calcite, TiO_2, Fe oxides, struvite, and dolomite, which are a function of feedstock and pyrolysis and extraction temperature (Lou et al, 2016).

Biomass vinegar (often referred to as wood vinegar, although any type of biomass can be used to produce vinegar) is formed when pyrolysis gases are condensed and then refined through a settling process whereby the heavier tarry fraction settles on the bottom, the water on the top and the wood vinegar is extracted from a middle fraction. These vinegars are sold commercially and are utilized regularly in many Asian countries. Biomass vinegar shares many of the same organic compounds as BFS but has much lower concentrations of salts and minerals. The concentration of the different organic compounds (Lu et al, 2019) and inorganic compounds depends on the feedstock and the temperature at which the gas is collected. For instance, 21 different elements in a wheat straw-derived vinegar were identified with concentrations of Ca and K greater than 5000 mg L^{-1} (Gao et al, 2020).

Few studies have been carried out using biochar extract as a foliar spray. A search in the Web of Science returned only 31 publications using the term "biochar and foliar spray" out of 27,829 publications using the term "biochar" (search on Feb 15, 2023). Positive effects were observed on the yield and quality of wheat and maize straw with BFS used to enhance the growth of Chinese cabbage that was grown in pots with the same addition of chemical fertilizer for all treatments (Lou et al, 2016). The foliar spray was produced by boiling the biochar with water in a ratio of 1:10 (10 g biochar < 2 mm: 200 mL deionized water) to maximize extraction of organic and inorganic compounds and then diluting either to 25, 50, or 100 water to biochar ratio. The control foliar spray was just water. It has been reported that at a dilution of 50:1, extracts of both wheat and maize straw biochar significantly increase yield by approximately 60%. The vitamin C and soluble protein content also increased while decreasing the nitrate content of the cabbage at dilutions of 50 or 100 times (Lou et al, 2016). A similar experiment was carried out with foliar spray obtained from wood:clay:sand (PCS-BC; 70:15:15) and wheat straw:bird manure (WB-BC; 50:50) (Kumar et al, 2021). For such an experiment, a water-to-biochar ratio of 20 was heated to just below the boiling point and simmered for 3 hours. Soilless medium infused with NPK fertilizer was used to grow lettuce. Testing was carried out with 2 dilutions (25:1 and 50:1, water: biochar ratio), and the control was only water. For comparison, chemical fertilizer (CF) foliar spray solutions were prepared to replicate the minor and trace element compositions of the PCS-BC and WB-BC extracts. Major nutrients, N, P, and K were not included in the CF preparations, as these major nutrients were supplied to all the treatments equally in the soil to all treatments. The foliar spray was supplied every 4 days over 32 days. Foliar application of PCS25, WB50, and WB100 led to a significant increase in the plant fresh biomass in comparison to their corresponding chemical fertilizer and to deionized water. Electron microscopy and spectroscopy studies showed the deposition of macro- and nanoscale organomineral particles and agglomerates on leaf surfaces of the examined PCS25-treated plant, and these particles consist of carbon-coated minerals and very fine biochar particles (Kumar et al, 2021).

The micron and mineral nanoparticles (with and without biochar coating) in the foliar spray produced can have a major effect on increasing plant growth, quality, and stress resistance (Shang et al, 2019; Kohatsu et al, 2021). This especially applies to foliar sprays that comprise TiO_2, phosphate, Fe/O, SiO_2 nanoparticles, and aluminosilicate compounds. These minerals could have directly enhanced the biosynthesis of chlorophyll molecules or stabilization of pigment-protein complexes, increasing antioxidant properties by increasing the total phenolic content and biosynthesis of flavonoids and increasing the relative abundance of several metabolites, sugars and sugar alcohols, fatty acids, and small-molecule organic acids (Kumar et al, 2021). These foliar sprays can also affect plant biological functions and soil microbial communities (Tian et al, 2020).

Wood vinegar foliar spray has similar effects (with over 30 peer-reviewed articles published over the last 30 years). Spraying wood vinegar on fertilized rapeseed in comparison to water (i) increased seed yield, leaf area index, and the number of pods per plant by an average of 10%, 23%, and 24% in two years, respectively; (ii) improved the resistance of rapeseed at low ambient temperatures of 2–6°C by increasing the activity of superoxide dismutase, proline, and soluble protein contents; and (iii) reduced the incidence of *Sclerotinia sclerotiorum* and *Peronospora parasitica* (downy mildew) (Zhu et al, 2021). Adding either gibberellin, sodium D-gluconate, or melatonin to wood vinegar further enhanced the rapeseed yield compared with wood vinegar alone. Vinegar produced from walnut shell pyrolysis gas condensed at temperatures between 90 and 480°C was applied to rapeseed at 400 and 800 water dilution ratios and increased soluble protein and SOD after 60 days of foliar application compared with water (Zhu et al, 2021).

Wood vinegar can be used in combination with a solid organic or inorganic fertilizer or biochar amendment. Wheat straw wood vinegar (0 and 90 L ha^{-1}) and urea humate additions increased wheat yields by 12% and maize yields by 7%, but this increased substantially when wood vinegar and urea humate were combined to satisfy N needs at 44% for wheat and 38% for maize (Sun et al, 2021). Nitrogen use efficiency, microbial carbon and nitrogen biomass, and soil inorganic nitrogen (NH_4^+ and NO_3^-) content increased, and urease activity decreased at a depth of 0-0.4 m.

Wheat straw vinegar reduced the activity of *Fusarium graminearum* growth and production of deoxynivalenol (DON) with an EC50 (concentration for 50% of maximal effect) value of 3.1 μL mL^{-1} (Gao et al, 2020). The application of wheat straw vinegar diluted 200-fold significantly decreased the wheat FHB infection rate and DON content by 66% and 69%, respectively. The control efficacy of wheat straw vinegar at a dilution of 200-fold was similar to that of typical chemical fungicide applications.

Although several companies are producing biochar liquid fertilizers, no research data have been published in peer-reviewed journals to date, although some trials are ongoing.

Conclusions and recommendations for future research

There is a growing interest in using biochar to produce enhanced efficiency fertilizers so that locally available waste materials that are challenging in their final use and disposal could be used as feedstock to recycle nutrients and sequester C. New routes for BBF synthesis have already been conceived, and these novel fertilizer formulation processes point to the beginning of a black-green revolution, which indicates the need to optimize the pyrolysis conditions and select feedstocks to produce organomineral fertilizers of high agronomic value.

There has been increasing interest in BBF research in the last five years. However, the development of this research area is still in its infancy. Thus, there is a need to prioritize and focus on research and development, and here, we point out some suggestions (Table 26.3).

Basic research is required to understand the mechanisms by which biochar interacts with nutrients and the microbial community at the rhizosphere level to tailor biochar properties using standardized BBF production. More studies are needed to advance existing technologies for BBF production, such as pelletization, granulation, and encapsulation, as well as the inclusion of selected microorganisms (e.g., phosphate solubilizing microbes) and nanosized biochar. In addition, the delivery and enhanced nutrient uptake from BBF should be investigated. The interaction of bioactive and soluble organic compounds in biochar should be examined which could also be obtained from co-composted biochar as a source for the production of BBF. Easily mineralizable OM compounds extracted from biochar should be studied in their use as foliar spray in fertigation and as hydroponic media, as well as the role these compounds may play in stimulating plant root growth and increasing nutrient uptake, among other important effects on plants. Long-term field experiments should be established to generate the agronomic data for BBF recommendations, including the economics compared with conventional fertilizer conditions. The widespread adoption of biochar as a matrix for BBF production and as an additive in growing media is in line with an emerging interest in the circular economy, and biochar technology and its different uses need to be disseminated to improve the fertilizer sector and many other processes and production chains of services and goods. Finally, risk assessment studies should be carried out to understand the generation and dissipation of PAH compounds as a function of the feedstock used.

Table 26.3 *Ranking of research priorities on biochar-based fertilizers in the coming decade based on current research gaps*

Level of priority	Topic of research
1	Biochar interaction with nutrients and microbial community at the rhizosphere
2	Biochar-based fertilizer production technologies, including the use of nanosized biochar as an additive
3	Bioactivity of soluble organic compounds from biochar, nano-biochar, and co-composted biochar as matrices
4	Extractable compounds from biochar in foliar spray, fertigation, hydroponics, and its biostimulant effect in plants
5	Use of long-term field studies for recommendation and economic feasibility of biochar-based fertilizers
6	Use of biochar as a peat substitute in growing media
7	Risk assessment of biochars from different feedstocks for use in composting or production of biochar-based fertilizers

References

Agegnehu G, Bass AM, Nelson PN, and Bird MI 2016 Benefits of biochar, compost and biochar–compost for soil quality, maize yield and greenhouse gas emissions in a tropical agricultural soil. *Science of the Total Environment* 543, 295–306.

Agyarko-Mintah E, et al 2017 Biochar lowers ammonia emission and improves nitrogen retention in poultry litter composting. *Waste Management* 61, 129–137.

Ahmad Z, Mosa A, Zhan L, and Gao B 2021 Biochar modulates mineral nitrogen dynamics in soil and terrestrial ecosystems: A critical review. *Chemosphere* 278, 130378.

Ahmed M, et al 2021 Valorization of animal bone waste for agricultural use through biomass co-pyrolysis and bio-augmentation. *Biomass Conversion and Biorefinery*, 10.1007/s13399-021-02100-w.

Almanassra IW, Mckay G, Kochkodan V, Ali Atieh M, and Al-Ansari T 2021 A state of the art review on phosphate removal from water by biochars. *Chemical Engineering Journal* 409, 128211.

Antonangelo JA, Sun X, and Zhang H 2021 The roles of co-composted- biochar (COMBI) in improving soil quality, crop productivity, and toxic metal amelioration. *Journal of Environmental Management* 277, 111443.

Bach M, Wilske B, and Breuer L 2016 Current economic obstacles to biochar use in agriculture and climate change mitigation. *Carbon Management* 7, 183–190.

Bai S, et al 2015 Soil and foliar nutrient and nitrogen isotope composition (δ^{15}N) at 5 years after poultry litter and green waste biochar amendment in a macadamia orchard. *Environmental Science and Pollution Research* 22, 3803–3809.

Bai S, et al 2022 Combined effects of biochar and fertilizer applications on yield: A review and meta-analysis. *Science of The Total Environment* 808, 152073.

Barbosa CF, Correa DA, Carneiro JSS, and Melo LCA 2022 Biochar phosphate fertilizer loaded with urea preserves available nitrogen longer than conventional urea. *Sustainability* 14, 686.

Bello A, et al 2021 Composted biochar affects structural dynamics, function and co-occurrence network patterns of fungi community. *Science of the Total Environment*, 775,145672.

Blackwell P, et al 2015 Influences of biochar and biochar-mineral complex on mycorrhizal colonization and nutrition of wheat and sorghum. *Pedosphere* 25, 686–695.

Borges BMMN, Strauss M, Camelo PA, Sohi SP, and Franco HCJ 2020 Re-use of sugarcane residue as a novel biochar fertiliser - Increased phosphorus use efficiency and plant yield. *Journal of Cleaner Production* 262, 121406.

Buss W, Bogush A, and Ignatyev K 2020 Unlocking the fertilizer potential of waste-derived biochar. *ACS Sustainable Chemistry and Engineering* 8, 12295–12303.

Calabi-Floody M, et al 2018 Smart fertilizers as a strategy for sustainable agriculture. *Advances in Agronomy* 147, 119–157.

Carneiro JSS, et al 2021 Long-term effect of biochar-based fertilizers application in tropical soil: Agronomic efficiency and phosphorus availability. *Science of the Total Environment* 760, 143955.

Carneiro JSS, et al 2018 Carbon stability of engineered biochar-based phosphate fertilizers. *ACS Sustainable Chemistry and Engineering* 6, 14203–14212.

Casini D, et al 2021 Production and characterization of co-composted biochar and digestate from biomass anaerobic digestion. *Biomass Conversion and Biorefinery* 11, 2271–2279.

Castejón-Del Pino R, Cayuela ML, Sánchez-García M, and Sánchez-Monedero MA 2023 Nitrogen availability in biochar-based fertilizers depending on activation treatment and nitrogen source. *Waste Management* 158, 76–83.

Chen Z, Pei J, Wei Z, and Ruan X 2021 A novel maize biochar-based compound fertilizer for immobilizing cadmium and improving soil quality and maize growth. *Environmental Pollution* 277, 116455.

Chen P, Shen G, and Liang J 2022 Dissipation and risk assessment of polycyclic aromatic hydrocarbons in industrial-scale biochar composting. *Journal of Soils and Sediments* 22, 1976–1986.

Chew JK, et al 2022 Biochar-based fertiliser enhances nutrient uptake and transport in rice seedlings. *Science of the Total Environment* 826, 154174.

Chew JK, et al 2020 Biochar-based fertilizer: Supercharging root membrane potential and biomass yield of rice. *Science of the Total Environment* 713, 136431.

Chrysargyris A, Prasad M, Kavanagh A, and Tzortzakis N 2020 Biochar type, ratio, and nutrient levels in growing media affects seedling production and plant performance. *Agronomy* 10, 1421.

Clare A, et al 2015 Competing uses for China's straw: The economic and carbon abatement potential of biochar. *Global Change Biology Bioenergy* 7, 1272–1282.

Czekała W, et al 2016 Co-composting of poultry manure mixtures amended with biochar – The effect of biochar on temperature and C-CO_2 emission. *Bioresource Technology* 200, 921–927.

Dai Z, et al 2016 Sensitive responders among bacterial and fungal microbiome to pyrogenic organic matter (biochar) addition differed greatly between rhizosphere and bulk soils. *Scientific Reports* 6, 36101.

Dai Y, Wang W, Lu L, Yan L, and Yu D 2020 Utilization of biochar for the removal of nitrogen and phosphorus. *Journal of Cleaner Production* 257, 120573.

Dickinson D, et al 2015 Cost-benefit analysis of using biochar to improve cereals agriculture. *Global Change Biology Bioenergy* 7, 850–864.

Dil M, Oelbermann M, and Xue W 2014 An evaluation of biochar pre-conditioned- with urea ammonium nitrate on maize (*Zea mays* L.) production and soil biochemical characteristics. *Canadian Journal of Soil Science* 94, 551–562.

Domingues RR, et al 2020 Enhancing cation exchange capacity of weathered soils using biochar: Feedstock, pyrolysis conditions and addition rate. *Agronomy* 10, 824.

Domingues RR, et al 2017 Properties of biochar derived from wood and high-nutrient biomasses with the aim of agronomic and environmental benefits. *PLoS One* 12, 1–19.

Dong D, et al 2020 An effective biochar-based slow-release fertilizer for reducing nitrogen loss in paddy fields. *Journal of Soils and Sediments* 20, 3027–3040.

Du J, et al 2019 Effects of biochar on the microbial activity and community structure during sewage sludge composting. *Bioresource Technology* 272, 171–179.

Duan Y, et al 2019 Positive impact of biochar alone and combined with bacterial consortium amendment on improvement of bacterial community during cow manure composting. *Bioresource Technology* 280, 79–87.

Dumroese RK, Heiskanen J, Englund K, and Tervahauta A 2011 Pelleted biochar: Chemical and physical properties show potential use as a substrate in container nurseries. *Biomass and Bioenergy* 35, 2018–2027.

Fachini J, et al 2021 Novel K-enriched organomineral fertilizer from sewage sludge-biochar: Chemical, physical and mineralogical characterization. *Waste Management* 135, 98–108.

Fachini J, Figueiredo CC, and Vale AT 2022 Assessing potassium release in natural silica sand from novel K-enriched sewage sludge biochar fertilizers. *Journal of Environmental Management* 314, 115080.

Faloye OT, Alatise MO, Ajayi AE, and Ewulo BS 2017 Synergistic effects of biochar and inorganic fertiliser on maize (*zea mays*) yield in an alfisol under drip irrigation. *Soil and Tillage Research* 174, 214–220.

Fischer D and Glaser B 2012 Synergisms between compost and biochar for sustainable soil amelioration. In: Kumar S, Bharti A (Eds) *Management of Organic Waste*. Janeza Trdine: IntechOpen. pp167–198.

Gao T, et al 2020 Wheat straw vinegar: A more cost-effective solution than chemical fungicides for sustainable wheat plant protection. *Science of the Total Environment* 725, 138359.

Gao T, et al 2021 Different responses of soil bacterial and fungal communities to 3 years of biochar amendment in an alkaline soybean soil. *Frontiers in Microbiology* 12, 630418.

Godlewska P, Schmidt HP, Ok YS, and Oleszczuk P 2017 Biochar for composting improvement and contaminants reduction. A review. *Bioresource Technology* 246, 193–202.

Grafmüller J, et al 2022 Wood ash as an additive in biomass pyrolysis: Effects on biochar yield, properties, and agricultural performance. *ACS Sustainable Chemistry and Engineering* 10, 2720–2729.

Hagemann N, et al 2017 Organic coating on biochar explains its nutrient retention and stimulation of soil fertility. *Nature Communications* 8, 1089.

He X, et al 2017 Evaluation of biochar powder on oxygen supply efficiency and global warming potential during mainstream large-scale aerobic composting in China. *Bioresource Technology* 245, 309–317.

He X, Yin H, Sun X, Han L, and Huang G 2018 Effect of different particle-size biochar on methane emissions during pig manure/wheat straw aerobic composting: Insights into pore characterization and microbial mechanisms. *Bioresource Technology* 268, 633–637.

Hestrin R, Enders A, and Lehmann J 2020 Ammonia volatilization from composting with oxidized biochar. *Journal of Environmental Quality* 49, 1690–1702.

Heymann K, et al 2014 Can functional group composition of alkaline isolates from black carbon-rich soils be identified on a sub-100nm scale? *Geoderma* 235–236, 163–169.

Huang L and Gu M 2019 Effects of biochar on container substrate properties and growth of plants—A review. *Horticulturae* 5, 14.

Huang Y, et al 2022 Biochar-based molybdenum slow-release fertilizer enhances nitrogen assimilation in Chinese flowering cabbage (*Brassica parachinensis*). *Chemosphere* 303, 134663.

Jia Y, Hu Z, Ba Y, and Qi W 2021 Application of biochar-coated urea controlled loss of fertilizer nitrogen and increased nitrogen use efficiency. *Chemical and Biological Technologies in Agriculture* 8, 3.

Jiang Z, et al 2019 Exploring the characteristics of dissolved organic matter and succession of bacterial community during composting. *Bioresource Technology* 292, 121942.

Jindo K, et al 2016 Influence of biochar addition on the humic substances of composting manures. *Waste Management* 49, 545–552.

Jindo K, et al 2020 Role of biochar in promoting circular economy in the agriculture sector. Part 2: A review of the biochar roles in growing media, composting and as soil amendment. *Chemical and Biological Technologies in Agriculture*, 7, 16.

Kamau S, Karanja NK, Ayuke FO, and Lehmann J 2019 Short-term influence of biochar and fertilizer-biochar blends on soil nutrients, fauna and maize growth. *Biology and Fertility of Soils* 55, 661–673.

Kammann C, Glaser B, and Schmidt HP 2016 Combining biochar and organic amendments. In: Shackley S, Ruysschaert G, Zwart K, and Glaser B (Eds) *Biochar in European Soils and Agriculture*. London: Routledge. pp 136–164.

Karakaş C, Özçimen D, and İnan B 2017 Potential use of olive stone biochar as a hydroponic growing medium. *Journal of Analytical and Applied Pyrolysis* 125, 17–23

Kern J, et al 2017 Synergistic use of peat and charred material in growing media – An option to reduce the pressure on peatlands? *Journal of Environmental Engineering and Landscape Management* 25, 160–174.

Keske C, Godfrey T, Hoag DLK, and Abedin J 2020 Economic feasibility of biochar and agriculture coproduction from Canadian black spruce forest. *Food and Energy Security* 9, e188.

Khajavi-shojaei S and Moezzi A 2020 Synthesis modified biochar-based slow-release nitrogen fertilizer increases nitrogen use efficiency and corn (*Zea mays* L.) growth. *Biomass Conversion and Biorefinery* 13, 593–601.

Khan N, et al 2014 Maturity indices in co-composting- of chicken manure and sawdust with biochar. *Bioresource Technology* 168, 245–251.

Kohatsu MY, et al 2021 Comparison of foliar spray and soil irrigation of biogenic CuO nanoparticles (NPs) on elemental uptake and accumulation in lettuce. *Environmental Science and Pollution Research* 28, 16350–16367.

Kumar A, et al 2021 Fertilizing behavior of extract of organomineral-activated biochar: Low-dose foliar application for promoting lettuce growth. *Chemical and Biological Technologies in Agriculture* 8, 21.

Lashari MS, et al 2013 Effects of amendment of biochar-manure compost in conjunction with pyroligneous solution on soil quality and wheat yield of a salt-stressed cropland from Central China Great Plain. *Field Crops Research* 144, 113–118.

Lehmann J, et al 2011 Biochar effects on soil biota – A review. *Soil Biology and Biochemistry* 43, 812–1836

Leite A, et al 2020 Selected bacterial strains enhance phosphorus availability from biochar-based rock phosphate fertilizer. *Annals of Microbiology* 70, 6.

Liao J, et al 2020 Effects of biochar-based controlled release nitrogen fertilizer on nitrogen-use efficiency of oilseed rape (*Brassica napus* L.). *Scientific Reports* 10, 11063.

Lou Y, et al 2016 Water extract from straw biochar used for plant growth promotion: An initial test. *Bioresources* 11, 249–266.

Lu X, Jiang J, He J, Sun K, and Sun Y 2019 Effect of pyrolysis temperature on the characteristics of wood vinegar derived from chinese fir waste: A comprehensive study on its growth regulation performance and mechanism. *ACS Omega* 4, 19054–19062.

Lustosa Filho JF, Barbosa CF, Carneiro JSS, and Melo LCA 2019 Diffusion and phosphorus solubility of biochar-based fertilizer: Visualization, chemical assessment and availability to plants. *Soil and Tillage Research* 194, 104298.

Lustosa Filho JF, et al 2020 Aging of biochar-based fertilizers in soil: Effects on phosphorus

pools and availability to *Urochloa brizantha* grass. *Science of The Total Environment* 709, 136028.

Lustosa Filho JF, Penido ES, Castro PP, Silva CA, and Melo LCA 2017 Co-pyrolysis of poultry litter and phosphate and magnesium generates alternative slow-release fertilizer suitable for tropical soils. *ACS Sustainable Chemistry and Engineering* 5, 9043–9052.

Malinowski M, Wolny-Koładka K, and Vaverková MD 2019 Effect of biochar addition on the OFMSW composting process under real conditions. *Waste Management* 84, 364–372.

Marcińczyk M, and Oleszczuk P 2022 Biochar and engineered biochar as slow- and controlled-release fertilizers. *Journal of Cleaner Production* 339, 130685.

Marcińczyk M, Ok YS, and Oleszczuk P 2022 From waste to fertilizer: Nutrient recovery from wastewater by pristine and engineered biochars. *Chemosphere* 306, 135310

Melo LCA, Lehmann J, Carneiro JSS, and Camps-Arbestain M 2022 Biochar-based fertilizer effects on crop productivity: a meta-analysis. *Plant and Soil* 472, 45–58.

Mendes GO, et al 2014 Biochar enhances *Aspergillus niger* rock phosphate solubilization by increasing organic acid production and alleviating fluoride toxicity. *Applied and Environmental Microbiology* 80, 3081–3085.

Mithu MMH, et al 2022 Biochar enriched compost elevates mungbean (*Vigna radiata* L.) yield under different salt stresses. *Crop and Pasture Science* 74, 79–89.

Mosa A, El-Ghamry A, and Tolba M 2018 Functionalized biochar derived from heavy metal rich feedstock: Phosphate recovery and reusing the exhausted biochar as an enriched soil amendment. *Chemosphere* 198, 351–363.

Nardis BO, et al 2022 Production of engineered-biochar under different pyrolysis conditions for phosphorus removal from aqueous solution. *Science of the Total Environment* 816, 151559.

Nardis BO, Carneiro JSS, Souza IMG, Barros RG, and Melo LCA 2020 Phosphorus recovery using magnesium-enriched biochar and its potential use as fertilizer. *Archives of Agronomy and Soil Science* 67, 1017–1033.

Ndoung OCN, Figueiredo CC, and Ramos MLG 2021 A scoping review on biochar-based fertilizers: Enrichment techniques and agro-environmental application. *Heliyon* 7, e08473.

Nguyen TTN, et al 2017 Short-term effects of organo-mineral biochar and organic fertilisers on nitrogen cycling, plant photosynthesis, and nitrogen use efficiency. *Journal of Soils and Sediments* 17, 2763–2774.

Ni J, Pignatello JJ, and Xing B 2011 Adsorption of aromatic carboxylate ions to black carbon (biochar) is accompanied by proton exchange with water. *Environmental Science and Technology* 45, 9240–9248

Nielsen S, et al 2014 Comparative analysis of the microbial communities in agricultural soil amended with enhanced biochars or traditional fertilisers. *Agriculture, Ecosystems and Environment* 191, 73–82.

Patel S, et al 2020 A critical literature review on biosolids to biochar: An alternative biosolids management option. *Reviews in Environmental Science and Bio/Technology* 19, 807–841.

Peng J, et al 2021 Combined application of biochar with fertilizer promotes nitrogen uptake in maize by increasing nitrogen retention in soil. *Biochar* 3, 367–379.

Pogorzelski D, et al 2020 Biochar as composite of phosphate fertilizer: Characterization and agronomic effectiveness. *Science of the Total Environment* 743, 140604.

Prost K, et al 2013 Biochar affected by composting with farmyard manure. *Journal of Environmental Quality* 42, 164–172.

Puga AP, Grutzmacher P, Cerri CEP, Ribeirinho VS, and Andrade CA 2020a Biochar-based nitrogen fertilizers: Greenhouse gas emissions, use efficiency, and maize yield in tropical soils. *Science of the Total Environment* 704, 135375.

Puga AP, et al 2020b Nitrogen availability and ammonia volatilization in biochar-based fertilizers. *Archives of Agronomy and Soil Science* 66, 992–1004.

Qian L, et al 2014 Biochar compound fertilizer as an option to reach high productivity but low carbon intensity in rice agriculture of China. *Carbon Management* 5, 145–154.

Qiu X, Zhou G, Zhang J, and Wang W 2019 Microbial community responses to biochar addition when a green waste and manure mix are composted: A molecular ecological network analysis. *Bioresource Technology* 273, 666–671.

Rashid M, et al 2021 Carbon-based slow-release fertilizers for efficient nutrient management: Synthesis, applications, and future research needs. *Journal of Soil Science and Plant Nutrition* 21, 1144–1169

Rasse DP, et al 2022 Enhancing plant N uptake with biochar-based fertilizers: Limitation of sorption and prospects. *Plant and Soil* 475, 213–236.

Robb S, Joseph S, Abdul Aziz A, Dargusch P, and Tisdell C 2020 Biochar's cost constraints are overcome in small-scale farming on tropical soils in lower-income countries. *Land Degradation and Development* 31, 1713–1726.

Roberts TL 2009 The role of fertilizer in growing the world's food. *Better Crop* 93, 12–15.

Roberts TL, and Johnston AE 2015 Resources, conservation and recycling phosphorus use efficiency and management in agriculture. *Resources, Conservation and Recycling* 105, 275–281.

Rombel A, Krasucka P, and Oleszczuk P 2022 Sustainable biochar-based soil fertilizers and amendments as a new trend in biochar research. *Science of the Total Environment* 816, 151588.

Sánchez-García M, Alburquerque JA, Sánchez-Monedero MA, Roig A, and Cayuela ML 2015 Biochar accelerates organic matter degradation and enhances N mineralisation during composting of poultry manure without a relevant impact on gas emissions. *Bioresource Technology* 192, 272–279.

Sanchez-Monedero MA, et al 2018 Role of biochar as an additive in organic waste composting. *Bioresource Technology* 247, 1155–1164.

Sánchez-Monedero MA, Sánchez-García M, Alburquerque JA, and Cayuela ML 2019 Biochar reduces volatile organic compounds generated during chicken manure composting. *Bioresource Technology* 288, 121584.

Santos SR, Filho JFL, Vergütz L, and Melo LCA 2019 Biochar association with phosphate fertilizer and its influence on phosphorus use efficiency by maize. *Ciência e Agrotecnologia* 43, e025718.

Schmidt HP, et al 2021 Biochar in agriculture – A systematic review of 26 global meta-analyses. *Global Change Biology Bioenergy* 13, 1708–1730.

Shang Y, et al 2019 Applications of nanotechnology in plant growth and crop protection: A review. *Molecules* 24, 2558.

Shi W, et al 2020 Biochar bound urea boosts plant growth and reduces nitrogen leaching. *Science of the Total Environment* 701, 134424.

Sim DHH, Tan IAW, Lim LLP, and Hameed BH 2021 Encapsulated biochar-based sustained release fertilizer for precision agriculture: A review. *Journal of Cleaner Production* 303, 127018.

Sonoki T, et al 2013 Influence of biochar addition on methane metabolism during thermophilic phase of composting. *Journal of Basic Microbiology* 53, 617–621.

Steiner C, and Harttung T 2014 Biochar as a growing media additive and peat substitute. *Solid Earth* 5, 995–999.

Steiner C, Melear N, Harris K, and Das KC 2011 Biochar as bulking agent for poultry litter composting. *Carbon Management* 2, 227–230.

Steiner C, Sánchez-Monedero M, and Kammann C 2015 Biochar as an additive to compost and growing media. In: Lehmann J, Joseph S (Eds). *Biochar for Environmental Management: Science, Technology and Implementation*. London: Routledge. pp 715–735.

Sui F, et al 2021 Effects of iron-modified biochar with S-rich and Si-rich feedstocks on Cd immobilization in the soil-rice system. *Ecotoxicology and Environmental Safety* 225, 112764.

Sun T, et al 2017 Rapid electron transfer by the carbon matrix in natural pyrogenic carbon. *Nature Communications* 8, 14873.

Sun X, et al 2021 Combined urea humate and wood vinegar treatment enhances wheat–maize rotation system yields and nitrogen utilization efficiency through improving the quality of saline–alkali soils. *Journal of Soil Science and Plant Nutrition* 21, 1759–1770.

Sun Y, et al 2019 Assessing key microbial communities determining nitrogen transformation in composting of cow manure using illumina high-throughput sequencing. *Waste Management* 92, 59–67.

Tahery S, et al 2022 A comparison between the characteristics of a biochar-NPK granule and a commercial NPK granule for application in the soil. *Science of the Total Environment* 832, 155021.

Tian L, et al 2020 Foliar Application of SiO_2 nanoparticles alters soil metabolite profiles and microbial community composition in the pakchoi (*Brassica chinensis* L.) rhizosphere grown in contaminated mine soil. *Environmental Science and Technology* 54, 13137–13146.

Tilman D, Cassman KG, Matson PA, Naylor R, and Polasky S 2002 Agricultural sustainability and intensive production practices. *Nature* 418, 671–677.

Vandecasteele B, Sinicco T, D'Hose T, Vanden Nest T, and Mondini C 2016 Biochar amendment before or after composting affects compost quality and N losses, but not P plant uptake. *Journal of Environmental Management* 168, 200–209.

Wang C, et al 2022 Biochar-based slow-release of fertilizers for sustainable agriculture: A mini review. *Environmental Science and Ecotechnology* 10, 100167.

Wang C, et al 2014 Spectroscopic evidence for biochar amendment promoting humic acid synthesis and intensifying humification during composting. *Journal of Hazardous Materials* 280, 409–416.

Wang Y, Villamil MB, Davidson PC, and Akdeniz N 2019 A quantitative understanding of the role of co-composted biochar in plant growth using meta-analysis. *Science of the Total Environment* 685, 741–752.

Wei L, Shutao W, Jin Z, and Tong X 2014 Biochar influences the microbial community structure during tomato stalk composting with chicken manure. *Bioresource Technology* 154, 148–154.

Wei, Y et al 2021 Composting with biochar or woody peat addition reduces phosphorus bioavailability. *Science of the Total Environment* 764, 142841.10.1016/j.scitotenv.2020.142841.

Wiedner K, et al 2015 Acceleration of biochar surface oxidation during composting? *Journal of Agricultural and Food Chemistry* 63, 3830–3837.

Xiang Y, Deng Q, Duan H, and Guo Y 2017 Effects of biochar application on root traits: a meta-analysis. *Global Change Biology Bioenergy* 9, 1563–1572.

Xu N, Tan G, Wang H, and Gai X 2016 Effect of biochar additions to soil on nitrogen leaching, microbial biomass and bacterial community structure. *European Journal of Soil Biology* 74, 1–8.

Yan T, Xue J, Zhou Z, and Wu Y 2021 Biochar-based fertilizer amendments improve the soil microbial community structure in a karst mountainous area. *Science of the Total Environment* 794, 148757.

Yang W, et al 2021 Influence of biochar and biochar-based fertilizer on yield, quality of tea and microbial community in an acid tea orchard soil. *Applied Soil Ecology* 166, 104005.

Yang R, Su Y-Z, Wang T, and Yang Q 2016 Effect of chemical and organic fertilization on soil carbon and nitrogen accumulation in a newly cultivated farmland, *Journal of Integrative Agriculture*, 15, 658–666.

Yao C, et al 2015 Developing more effective enhanced biochar fertilisers for improvement of pepper yield and quality. *Pedosphere* 25, 703–712.

Yao Y, Gao B, Chen J, and Yang L 2013 Engineered biochar reclaiming phosphate from aqueous solutions: Mechanisms and potential application as a slow-release fertilizer. *Environmental Science and Technology* 47, 8700–8708.

Ye J, et al 2016 A Combination of biochar–mineral complexes and compost improves soil bacterial processes, soil quality, and plant properties. *Frontiers in Microbiology* 7, 372.

Ye J, et al 2017 Chemolithotrophic processes in the bacterial communities on the surface of mineral-enriched biochars. *ISME Journal* 11, 1087–1101.

Ye L, et al 2020 Biochar effects on crop yields with and without fertilizer: A meta-analysis of field studies using separate controls. *Soil Use and Management* 36, 2–18.

Zhang H, Voroney RP, Price GW, and White AJ 2017 Sulfur-enriched biochar as a potential soil amendment and fertiliser. *Soil Research* 55, 93–99.

Zhang L, et al 2021a Participation of urea-N absorbed on biochar granules among soil and tobacco plant (*Nicotiana tabacum* L.) and its potential environmental impact. *Agriculture, Ecosystems and Environment* 313, 107371.

Zhao L, et al 2016 Copyrolysis of biomass with phosphate fertilizers to improve biochar carbon retention, slow nutrient release, and stabilize heavy metals in soil. *ACS Sustainable Chemistry and Engineering* 4, 1630–1636.

Zhang Z, et al 2021b Mitigation of carbon and nitrogen losses during pig manure composting: A meta-analysis. *Science of the Total Environment* 783, 147103.

Zhao S, et al 2020 Towards the circular nitrogen economy – A global meta-analysis of composting technologies reveals much potential for mitigating nitrogen losses. *Science of the Total Environment* 704, 135401.

Zheng J, et al 2017 Biochar compound fertilizer increases nitrogen productivity and economic benefits but decreases carbon emission of maize production. *Agriculture, Ecosystems and Environment* 241, 70–78.

Zhou S, Kong F, Lu L, Wang P, and Jiang Z 2022 Biochar - An effective additive for improving quality and reducing ecological risk of compost: A global meta-analysis. *Science of the Total Environment* 806, 151439.

Zhu K, et al 2021 Wood vinegar as a complex growth regulator promotes the growth, yield, and quality of rapeseed. *Agronomy* 11, 510.

Zulfiqar F, et al 2019 Challenges in organic component selection and biochar as an opportunity in potting substrates: A review. *Journal of Plant Nutrition* 42, 1386–1401.

27

Biochar-based materials for environmental remediation

Hailong Wang, Hanbo Chen, Nanthi Bolan, and Shengsen Wang

Introduction

Along with the high level of urbanization and industrialization of modern society, the sustainable development of the ecological environment is undergoing an unprecedented challenge. The increasing load of inorganic and organic contaminants in the environment, soil degradation, and global warming calls for advanced and sustainable approaches to manage these global challenges. In particular, contamination of aqueous and solid phases with (in)organic pollutants is regarded as an intractable issue due to its hazardous impacts on the eco-environment and humans.

Biochar has been proven as an eco-friendly adsorbent/amendment, and as a filler in permeable reactive barriers (PRBs) for the adsorption, immobilization, and degradation of environmental pollutants in soil and water (Lyu et al, 2020a; Chen et al, 2022a). Nevertheless, pristine biochar's application in environmental decontamination has faced some challenges such as separating biochar from soil/water phases (Lyu et al, 2020a), deficient porosity (Kumar et al, 2020), and limited adsorption efficiency (Ahmed et al, 2016). Aiming to address such problems, multiple tailoring techniques including physical, biological, and chemical modification methods (Figure 27.1) have been developed to alter the physicochemical properties of biochar and enhance its application in the environmental decontamination fields. In general, physical activation of biochar possesses advantages such as facilitated operation and is economically viable compared to other methods, yet the energy and time requirements restrict its widespread application. Furthermore, biologically modified biochar can combine the merits of bacteria and biochar to adsorb

DOI: 10.4324/9781003297673-27

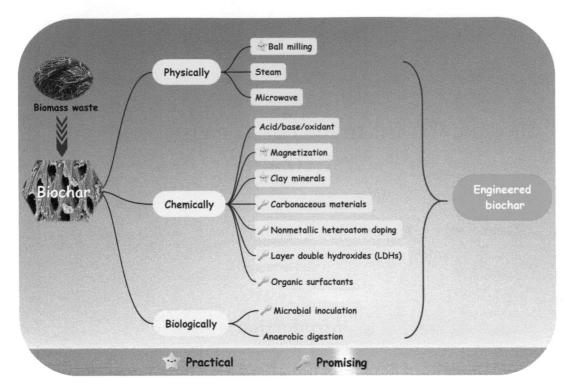

Figure 27.1 *Published methods for biochar modification*

pollutants, which shows great potential for the degradation of organic pollutants in the water and soil systems from an eco-friendly perspective. Related studies using biologically modified biochar as environmental remediation materials are still limited due to its high cost and strict operational requirements. Chemical modification methods simultaneously affect the physical and chemical properties (e.g., surface functional groups, point of zero charges, and electron transfer capacities) of biochars (Medeiros et al, 2022). With abundant resources of modification agents and effective targeted remediation for various pollutants, chemical methods have become the most mature and promising technique for biochar functionalization. Meanwhile, chemical activation still has certain disadvantages, including potential secondary pollution and non-recyclability of resultant biochar.

The purpose of this chapter is to summarize the functionalization of biochar using various modification methods, and the utilization of engineered biochar and biochar-based PRBs for the remediation of toxic (in)organic contaminants in the aquatic and soil systems (Huang et al, 2019; Medeiros et al, 2022; Shaheen et al, 2022). Furthermore, the chapter concludes by identifying knowledge gaps and recommendations for future research.

Pre-and post-treatment methods for biochar

Physical modification

Steam/microwave activation

Steam activation can enhance the porosity and surface area of biochar via the removal of impurities originating from incomplete combustion, and introduce new functional groups (e.g., –OH, C=O) (Ahmed et al, 2016). Recent investigations indicated that steam-activated biochars showed superior performance in the aqueous removal of hazardous compounds (e.g., Pb(II) and antibiotics) (Kwak et al, 2019; Wang et al, 2020), and stabilization of pollutants in the soil (e.g., poly-fluoroalkyl substances (PFAS) and carbamate pesticides) (Sørmo et al, 2021). Microwave activation is an emerging technique based on high-frequency electromagnetic irradiation (Duan et al, 2019). Microwave radiation causes dipole rotation at the atomic level, generating heat energy inside the materials (Yek et al, 2020). This modification approach allows for simultaneous heating of the interior and surface area of biochar without direct contact with a temperature ranging from 200 to 300°C in reaction (Zhou et al, 2021b), resulting in microwave-activated biochar with more functional groups and a larger surface area (Duan et al, 2019). Microwave-activated biochar from waste palm shells proved high removal of chemical oxygen demand (COD) in the hazardous landfill leachate, achieving a maximum removal rate of 65% (Lam et al, 2020). Steam and microwave activation have their limitations including being time-consuming and energy-intensive, this is the direction that needs to be tackled in the future.

Ball milling

Ball milling is an environmentally friendly and cost-effective mechanochemical method to tailor the properties of biochars, such as enhancement of specific surface area and pore volume, and oxygen-containing functional groups (Xiao et al, 2020). The movement of the ball milling process can generate huge kinetic energy to break down the chemical bonds of involved molecules and reduce the particle size of biochar (Kumar et al, 2020). The particle size of biochar after ball milling can be adjusted to the nanoscale (< 1000 nm) to obtain the specific nanobiochar (Lyu et al, 2020b). The operational parameters (e.g., mass ratio of media:biochar, reaction time, milling speed, and reaction atmosphere) during the ball milling process significantly influence the physicochemical properties of resultant biochar, thereby affecting its adsorptive or catalytic performance for contaminants. For example, the composite of biochar and MgAl-layered double hydroxide (LDH) was synthesized through the ball milling method, this composite was identified with enlarged basal spacing, reduced crystallite size of the LDHs, and abundant oxygen-containing functional groups (e.g., C–O and –OH). Meanwhile, the LDHs-biochar composite exhibited superior performance for the removal of Cd(II), achieving a maximum adsorption capacity of 119 mg g^{-1} (Cui et al, 2021b). Ball-milled biochar has exhibited superior adsorptive, electrochemical, and catalytic characteristics, but the time-consuming and energy-intensive production process limited its economical large-scale application. In addition, extensive investigations are required to assess the reusability, cost-effectiveness, and environmental risk of such nano-biochars.

Biological modification

Biological methods were initially developed using anaerobic bacteria to convert biomass into digested residue to produce biochar (Yao et al, 2011). Recently, the biological post-treatment method was developed via the immobilization of microorganisms onto the surface of biochar. The inoculated microorganisms can secrete a diversity of polymers and attach themselves to the biochar skeleton, generating a microbial biofilm that aids in the adsorption and degradation of pollutants (Zheng et al, 2021; An et al, 2022). Immobilized *Bacillus cereus* onto Chinese medicine residue-derived biochar for the removal of chlortetracycline (with an initial concentration of 50 mg L^{-1}) achieved a maximum removal rate of up to 85%, governed by microbial degradation (Zhang and Wang, 2021). By selecting appropriate microorganisms, their colonization onto biochar appears to be an intriguing measure for targeted remediation of pollutants, particularly for organic pollutants. However, such biological modification requires complicated operation procedures with high costs, therefore, a more cost-effective and facile preparation process is needed in the future.

Chemical modification

Acid, base, and oxidant modifications

The primary purpose of acid treatment is to remove the impurities on biochar, and acidic functional groups (e.g., phenolic, carboxylic, and lactonic groups) are also introduced during the process (Wang and Wang, 2019). The specific acid modification (e.g., nitric acid (HNO_3), hydrochloric acid (HCl), and phosphoric acid (H_3PO_4)) renders acid-modified biochars with low pH (Medeiros et al, 2022), and the application

of acidic biochars could aid in the remediation of alkaline calcareous soil (Zhou et al, 2021b). Alkaline modification such as using potassium hydroxide (KOH) and sodium hydroxide (NaOH) renders biochars with higher aromaticity and N/C ratios, yet lower O/C ratios, in contrast to acid-modified biochar (Ahmed et al, 2016). Furthermore, the alkaline treatment can enhance the alkaline hydroxyl groups on biochar, but it can result in a decrease in the acidic functional groups, and the application of alkaline biochar could aid in the remediation of acid soil (Bolan et al, 2023). Oxidizing agents such as H_2O_2 and $KMnO_4$ can also be employed to functionalize the biochar. The resultant biochar was found to be thermostable and effective in adsorbing pollutants (Zhou et al, 2021b). The successful pollutant removal by acid-, base-, or oxidant-modified biochar has been confirmed as mentioned above, yet the potential risks and uncertainty should be considered and investigated over a long-term period when such biochars are introduced to the environment.

Metal oxides and salts

The main purpose of introducing magnetic susceptibility properties into biochar is to enable the separation of the fine biochar from the soil/water phase to recycle and reuse the biochar (Huang et al, 2019). In general, the magnetization of biochar was achieved through the combination of biochar and magnetic or paramagnetic metal oxides and metal salts. Iron (Fe) modification has been the most widely developed and practical technique in the functionalization of biochar due to the resource-abundant, low-cost, and effective decontamination capacity with less environmental risks (Lyu et al, 2020a). To date, iron modification for biochar has been the most mature and practical tailoring technique in the removal of pollutants at lab and pilot scales. Precipitation, co-pyrolysis, thermal reduction, and ball milling are the basic methods to produce Fe-modified biochar (Kumar et al, 2020; Lyu et al, 2020a), which functions as an adsorptive, reductive, and catalytic material in the remediation of environmental contaminants.

Goethite and hematite are the most used Fe minerals to improve the adsorption effectiveness of biochar for metal(loid)s (Lyu et al, 2020a). For instance, α-FeOOH modified biochar derived from wheat straw showed high adsorption capacity for Cd(II) (63 mg g^{-1}) and As(III) (78 mg g^{-1}) (Zhu et al, 2020). In addition, biochars modified by nano zero-valent iron (nZVI), FeS, and FeOOH can provide reductive Fe^0, Fe(II), and S(II) species, which played critical roles in the redox reactions of Cr(VI), U(VI) and organic contaminants, thus mitigating their biotoxicity (Feng et al, 2021; Zhu et al, 2022). For example, the FeS and starch co-modified peanut shell biochar aided the reduction of labile U(VI) to non-labile U(IV) species, and XPS analysis revealed that the reduction process was governed by Fe^0 and S(II) (Liu et al, 2021a). Moreover, Fe-modified biochar has high electron shuttling capacity, abundant persistent free radicals (PFRs), and oxygen-containing functional groups, which can be used as an effective catalyst to generate reactive oxygen species (ROS) (e.g., •OH, $^1O_2$ and $SO_4•^-$) and thus degrade organic contaminants (Lyu et al, 2020a), particularly in the persulfate activation and Fenton-like systems (Feng et al, 2021).

In addition to Fe, other metallic elements such as manganese (Mn), magnesium (Mg), aluminum (Al), copper (Cu), cerium (Ce), lanthanum (La), zirconium (Zr), and bismuth (Bi), etc. were reported as the modifying resources for biochar functionalization (Chen et al, 2022c). For instance, the Mn-modified biochar enhanced the generation of $SO_4•^-$, •OH, and $^1O_2$ in the persulfate activation system and thus facilitated the 4-chloro-3-methyl phenol (CMP) degradation, achieving the removal rate of CMP was nearly 100% (Liu et al, 2021b). Interestingly, recent studies reported the superior degradation efficiency for organic pollutants by the novel biochar-based single-metal-atom catalysts (Cui et al, 2021a, 2022), it might be the new and fascinating option for the functionalization of biochar.

Nonmetallic heteroatom doping

Non-metal heteroatom doping is a new strategy for improving biochar's adsorptive and catalytic capability for pollutant removal by altering its electronic characteristics. Nitrogen (N), sulfur (S), and boron (B) are the most widely employed nonmetallic elements for biochar functionalization. Graphitic N, pyrrolic N, pyridinic N, and amine-N species could be generated on the biochar surface during the N-doping process, and then enhance the electrochemical properties and subsequent environmental performance of biochar (Leng et al, 2020). For example, Xu et al (2022) synthesized N-doped biochar derived from sawdust, and it showed significant adsorption capacity for bisphenol A (54.0 mg g^{-1}). Sulfur (S) can also be used for doping the biochar, and S-doping endows biochar with S-containing functional groups (e.g., C–S, C–S–O, C=S, S–S, and sulfur rings) (Zhang et al, 2021). S-doped wood-waste biochar was an efficient catalyst to activate peroxymonosulfate and finally achieved 91% removal of bisphenol A, which could be ascribed to the introduced phenoxyl radicals (C–O•) and vacancy defects during the S-doping process (Wan et al, 2022). Boron is another excellent prospective heteroatom for modulating electron distribution, tailoring biochar's physicochemical characteristics and providing additional defect sites (Sui et al, 2021). Boron-doping was observed to effectively improve the specific surface area (around 898 m^2 g^{-1}) and surface functional groups of corn-straw biochar, and ultimately enhanced the adsorption capacity for Fe(II) (Sui et al, 2021). A high doping ratio is always adopted to maximize the adsorptive/catalytic abilities of doped biochar which inevitably raises the economic costs. It is necessary to develop a novel and simple doping technique to precisely regulate the N-S-B functional groups with an appropriate doping ratio.

Layered double hydroxides
Layered double hydroxides (LDHs) are effective adsorbents with strong anion exchange capacity (Chen et al, 2022c). In recent years, loading of LDHs onto biochar to synthesize LDHs-engineered biochar has been an emerging functionalization strategy for maximizing the efficiency of both resource materials (Lartey-Young and Ma, 2022). The LDHs-engineered biochar exhibited significant improvement in specific surface area, structure heterogeneity, surface functional groups, and stability (Zubair et al, 2021b). Hydrothermal synthesis, co-precipitation, and co-pyrolysis as the commonly used methods are developed via the reaction of metal cations with hydroxyl in the biochar slurry under certain conditions, rendering the formation of LDHs on the surface of biochar (Zubair et al, 2021b). As a reference, a bamboo-derived ternary Cu-Zn-Fe LDH-engineered biochar was synthesized by Lartey-Young and Ma (2022), and it showed effective adsorption capacity for atrazine (up to 87.04 mg g^{-1}). Another study found that the Mg- or Fe-LDH modified biochar showed better performance in the catalytic degradation of sulfamethoxazole than pristine biochar (Chen et al, 2022c); the Mg- or Fe-LDH modified biochar aided in the activation of urea-hydrogen peroxide to produce •OH, thus ensuring improved removal efficiency of sulfamethoxazole (91%). LDHs-engineered biochar

has shown promise in the treatment of contaminated soils, in addition to its application in aqueous solutions. For instance, the application of Mg or Al LDH-biochar composite significantly decreased the leaching uranium (U) concentration by 54% and the cumulative U loss by 53% in the contaminated soil (total U concentration ≈ 1000 mg kg^{-1}), as compared to the control (Lyu et al, 2021). Biochar-LDHs composites are promising sustainable materials for environmental purification, yet economic methods of mass production that avoid excessive use of chemicals are still lacking, other techniques such as microwave and ball milling could be combined to synthesize the biochar-LDH composites. The commonly used LDHs for biochar engineering are Mg-Al, Mg-Fe, or Ni-Al, future studies using other LDHs such as Zn-Al and Co-Fe may also open up new interesting possibilities to produce novel biochar-LDH composites with high removal efficiency for pollutants.

Clay minerals
Natural clay minerals such as montmorillonite, zeolite, kaolinite, illite, and vermiculite are low-cost adsorbents for various contaminants. The combination of clay minerals and biochar might improve pore structure and the adsorption ability of biochar-clay composite for environmental contaminants. Inexpensive biochar-montmorillonite composites exhibited a substantial adsorption capacity for Pb (II) (140 mg g^{-1}) from wastewater (Fu et al, 2020). Illite-modified biochar (1:4 ratio of illite to walnut shell biochar) had a higher adsorption capacity of metolachlor than pristine biochar (92 mg g^{-1} vs 73 mg g^{-1}) (Liu et al, 2022). Overall, clay-biochar composites are potentially low-cost due to the abundance of both materials, being a promising candidate for the removal of heavy metals, antibiotics, pesticides, and dyes from wastewater and contaminated soils.

Carbonaceous nano-materials

The most widely employed carbonaceous nano-materials are graphene, graphene oxide, and carbon nanotubes. Because these carbonaceous compounds have a great affinity for pollutants in the environment, their loading onto biochar has been extensively researched to attain increased adsorption capabilities for various contaminants. Biochar-graphene oxide composites, exhibited greater adsorption capability of dimethyl phthalate (DMP) (45.65 mg g^{-1}) than that of single biochar due to enhanced hydrophobic effect and π-π interactions (Abdul et al, 2017). Compared to other carbonaceous materials, one of the greatest advantages of biochar is the economic cost. Although loading carbonaceous nanomaterials onto biochar enhanced the removal efficiency of certain contaminants, the extra high cost inevitably rises. This limited the development and practical application of such engineered biochar.

Organic surfactants

Recent research has focused on employing biochar modified with organic surfactants such as polyethyleneimine (PEI) (Li et al, 2020), rhamnolipid (Zhen et al, 2021), chitosan (Chen et al, 2022b), and cetyltrimethylammonium bromide (CTAB) (Murad et al, 2022) for environmental remediation. Chitosan-loaded biochar using glutaraldehyde as the crosslinking agent endowed the modified biochar with more functional groups (e.g., C=O, –NH$_2$, and –OH), showing a significant improvement for the aqueous Sb (III) removal (Chen et al, 2022c). Theoretical calculations verified that the chelation interaction between the chitosan-derived amine group and Sb(III) was the key mechanism during the adsorption process. For surfactant-modified biochar, it is of great importance to further evaluate its potential toxicity before its application in the environment via comprehensive fabrication and detection procedures.

Application of engineered biochar in the aqueous and soil systems

Adsorption of metal(loid)s in aquatic systems

Metal(loid)s adsorption on different engineered biochars has been extensively researched (Table 27.1). In principle, modification techniques, original feedstocks, pyrolysis conditions, and target contaminants are key factors influencing adsorption efficiency. The proposed metal(loid) adsorption mechanisms include physical and chemical processes, such as pore filling, electrostatic effect, ion exchange, complexation, precipitation, and redox (Figure 27.2). For instance, biochar has abundant pore structure, and the metal(loid) ions could be physically adsorbed via pore filling in the mesopore and micropore. Furthermore, biochar is usually negatively charged and can immobilize some cationic metal ions (e.g., Pb (II), Cd(II), and Ni(II)) through electrostatic attraction. Some biochar contains significant rich metal cations (e.g., Ca, Mg, and K), which promote cation exchange reactions with cationic metals (Chen et al, 2022a). More importantly, the functional groups on biochar surfaces such as C=O, C–O–O, and O–H tend to form an inner-sphere complex with metal(loid)s, which might be the most key adsorption mechanism due to the

Table 27.1 Metal(loid) removal and associated mechanisms by engineered biochars in aqueous system

Feedstock	Pyrolysis condition	Modification method	Metal (loid)s	Removal rate or capacity	Mechanisms involved	References
Canola straw	700°C, 2 h	Steam activation	Pb(II)	195 mg g^{-1}	Ion exchange capacity, precipitation, and inner-sphere complexation.	Kwak et al, 2019
Dendro	700°C, –	Ball milling	Cr(VI) Cd(II)	7.4 mg g^{-1} 922 mg g^{-1}	Surface complexation, electrostatic attraction.	Ramanayaka et al, 2020
Corn stalk	800°C, 2 h	S-doping	Fe(II)	50.0 mg g^{-1}	Chemical complexation, ion exchange, and co-precipitation.	Sui et al, 2021
Glucose	800°C, 2 h	KOH and N doping	Cr(VI)	402.9 mg g^{-1}	Physisorption, complexation, reduction.	Liang et al, 2020
Populus L.	600°C, 2 h	FeCl$_3$	As(III) As(V)	87.1% 99.2%	Physical adsorption, Fe-As precipitation, electrostatic interaction, As(III) oxidation	Xu et al, 2020
Ficus microcarpa branch	500°C, 2 h	Chitosan	Sb(III)	86-168 mg g^{-1}	Electrostatic interaction, chelation, surface complexation, π-π interaction, and hydrogen bonding.	Chen et al, 2022b
Phragmites australis	600°C, 25 min	LaCl$_3$	Sb(V)	18.9 mg g^{-1}	Inner-sphere La-O-Sb complex, ligand exchange, electrostatic interaction, H-bonding.	Wang et al, 2018
Peanut shell	250°C, 2 h	FeS and starch	U(VI)	76.3 mg g^{-1}	Electrostatic attraction, surface complexation, precipitation, and reductive reaction.	Liu et al, 2021a
Peanut shell	500°C, 2 h	*Pseudomonas hibiscicola* strain L1 immobilization	Cr(VI) Cu(II) Ni(II)	38.2% 45.8% 81.2%	Complexation, ion exchange, precipitation, reduction.	An et al, 2022

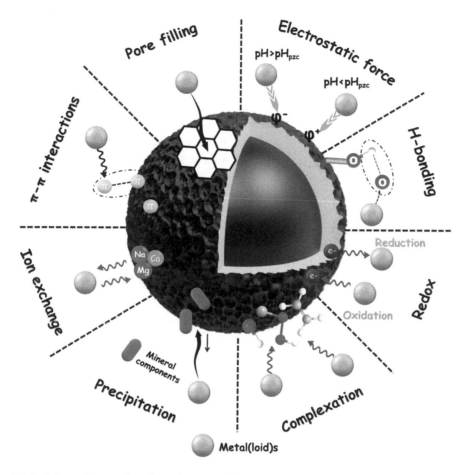

Figure 27.2 *Adsorption mechanisms for metal(loid)s by engineered biochar*

formation of stable chemical bonds. For metal(loid)s with multiple valence states such as Cr, As, and Sb, the adsorption on biochar involves the redox process, which is also a crucial removal mechanism for such elements.

Regarding cationic heavy metals, the adsorption behavior by engineered biochar is greatly governed by biochar's physico-chemical properties and solution conditions. For instance, ball-milled biochar was found to be effective in removing aqueous Cd(II), achieving a maximum adsorption capacity of 922 mg g^{-1} (Ramanayaka et al, 2020). The ball-milled biochar has abundant hydroxyl and carbonyl groups, which facilitated the complexation of Cd ions with the functional groups and enhanced the electrostatic attraction between negatively-charged biochar surface and cationic Cd(II) ions (Ramanayaka et al, 2020). Inoculating *Pseudomonas hibiscicola* strain L1 onto biochar originating from peanut shell led to a removal of Ni(II) and Cu(II) from the aqueous systems of 81.2% and 45.8%, respectively (An et al, 2022). Surface complexation with functional groups, precipitation, ion exchange, and bio-adsorption by strain L1 were proposed as the key mechanisms involved in the removal process (An et al, 2022).

The adsorption of anionic metal(loid)s such as Cr, As, and Sb by pristine biochar in aqueous media was limited due to the electrostatic repulsion effect. Various modification methods exist for biochar to enhance the adsorption of anionic metal(loid)s as shown in Table 27.1. For example, a glucose-derived biochar was activated using KOH and N-doping to remove Cr(VI) in the aqueous solutions, and a significant adsorption (402.9 mg g^{-1}) of Cr(VI) by the modified biochar was noted, mainly attributed to pore filling, electrostatic attraction, surface complexation, and reduction of Cr (VI) to Cr(III) (Liang et al, 2020). Fe-modified biochar derived from *Populus* L. was produced by Xu et al (2020) and achieved 99.9% removal potential for As (III), mainly owing to the formation of Fe-O-As complexes. For Sb, a novel lanthanum (La) doped biochar showed high adsorption performance for Sb(V) 18.9 mg g^{-1} (Wang et al, 2018). Spectroscopic analyses revealed that the formation of La-O-Sb complexes was the crucial mechanism for the enhanced adsorption of Sb(V).

In addition to metal(loid)s mentioned above, many studies have reported the removal of other metal(loid)s such as mercury (Hg) (O'Connor et al, 2018; Lyu et al, 2020b) and U (Liu et al, 2021a; Lyu et al, 2021) by engineered biochar. To maximize the environmental and economic benefits, the choice of modification method for different metal(loid)s should be applied appropriately to improve the removal effectiveness of target pollutants.

Adsorption of nutrients in aquatic systems

Phosphate, nitrate, and ammonium are the most common macronutrients found in the environment, and they all contribute to the eutrophication of waterways. Engineered biochar has also been utilized to improve nutrient adsorption in the aqueous phase (Table 27.2). Loading biochars with different metal salts (i.e., $AlCl_3$, $FeCl_3$, $MnCl_2$, $ZnCl_2$, $MgCl_2$) removed phosphate from water bodies by precipitation between metal ions and phosphate (Sornhiran et al, 2022). Recent research has demonstrated the ability of denitrifying bacteria-inoculated biochar to remove nitrate (Zheng et al, 2021; An et al, 2022). For example, the immobilization of *Pseudomonas hibiscicola* strain L1 on peanut shell biochar resulted in a 65% removal rate for nitrate (An et al, 2022). The enhancement of nitrate removal could be attributed to two reasons: (1) the C-O-containing functional groups (quinone, conjugated ketones, and carboxyl) could act as electron transporters to aid the electron transfer and the subsequent denitrification process; (2) biochar provided a suitable environment for the growth of immobilized strain L1, and thus promoted the immobilization of nitrate substrates by bacterial cells. The functionalization process modified the surface chemistry of biochar, which improved nutrient adsorption through multiple mechanisms including electrostatic interaction, surface complexation, precipitation, ligand exchange, Lewis acid-base interaction, and biological adsorption/immobilization.

Adsorption, degradation, and removal of organic pollutants in the aquatic environment

Dyes, antibiotics, pesticides, plasticizers, phenols, and PAHs are major types of

Table 27.2 *Nutrient removal and associated mechanisms by engineered biochars in aqueous system*

Nutrients	Feedstock	Pyrolysis condition	Modification method	Removal rate or capacity	Mechanisms involved	References
Phosphate	Hickory wood chip	600°C, 1 h	MgO, ball milling	62.9%	Surface precipitation of Mg-P bond and electrostatic interaction.	Zheng et al, 2020
	Bamboo	600°C, 2 h	Polyethyleneimine	9.3 mg g^{-1}	Electrostatic interaction and surface complexation.	Li et al, 2020
	Orange peel	225°C, 2 h	Mg/Al LDH, sodium alginate, *Acinetobacter* sp. FYF8 immobilization	86.1%	Biological adsorption and chemisorption.	Zheng et al, 2021
Nitrate	Soybean straw	500°C, 2 h	AlCl$_3$	40.6 mg g^{-1}	Electrostatic attraction and surface complexation with AlOOH.	Yin et al, 2018
	Orange peel	225°C, 2 h	Mg/Al LDH, sodium alginate, *Acinetobacter* sp. FYF8 immobilization	95.3%	Denitrification and ion exchange.	Zheng et al, 2021
	Peanut shell	500°C, 2 h	*Pseudomonas hibiscicola* strain L1 immobilization	65.3%	Denitrification and surface complexation.	An et al, 2022
Ammonium	Bamboo	500°C, –	Ball milling	22.9 mg g^{-1}	Surface complexation, cation-π interaction, and ion exchange.	Qin et al, 2020
	Paper sludge	200°C, 24 h	H$_2$O$_2$	7.9 mg g^{-1}	π-cation interaction, cation exchange, surface complexation and electrostatic attraction.	Nguyen et al, 2021
	Soybean straw	500°C, 2 h	MgCl$_2$ and AlCl$_3$	0.7 mg g^{-1}	Electrostatic interaction and surface complexation.	Yin et al, 2018

organic contaminants in the aquatic environment (Ahmed et al, 2016). The removal efficiency and associated mechanisms of typical organic contaminants by engineered biochar are summarized in Table 27.3. Multiple mechanisms govern the adsorption of organic pollutants onto biochar, including pore-filling, electrostatic interactions, hydrophobicity, hydrogen bonding, partition, π-π interactions, and biodegradation.

As for dyes, microwaving and steaming to activate the biochar derived from orange peel improved adsorption capacity of Congo red, up to 136 mg g^{-1} (Yek et al, 2020). Additionally, the S-doped biochar originated from tapioca peel waste was effective in adsorbing malachite green (30.2 mg g^{-1}) and rhodamine B (33.1 mg g^{-1}); electrostatic attraction, surface complexation, and hydrogen bonding were proposed as the major removal mechanisms (Vigneshwaran et al, 2021). Many publications have reported using engineered biochar in the removal of aqueous antibiotics. For instance, H$_3$PO$_4$-activated coffee-ground biochar was found to be an effective adsorbent for removing sulfadiazine, the maximum adsorption capacity was up to 139.2 mg g^{-1}, and the adsorption process was governed by hydrophobic effect and π–π EDA interaction (Zeng et al, 2022). Besides, biochar loaded with *Bacillus cereus* LZ01 exhibited a high removal rate (82.3%) for chlortetracycline, which was mainly attributable to biodegradation by LZ01 and biochar adsorption (Zhang and Wang, 2021). Regarding pesticides, iron-modified biochar loaded with *Acinetobacter lwoffii* DNS32 achieved a maximum removal rate achieved 9.71 mg L^{-1} h^{-1} for atrazine (Tao et al, 2019). Two reasons were thought to be responsible for the phenomenon: (1) iron oxide nanoparticles as catalysts to generate •OH and degraded atrazine; and (2) iron-modified biochar as a carrier to promote bacterial growth and biodegradation of atrazine.

Removal of phthalate esters (PAEs) is another research direction of engineered biochar application. For example, a series of hierarchical porous biochars, with a considerably large surface area of up to 2554 m^2 g^{-1}, had a maximum adsorption capacity for DEP of 657 mg g^{-1} governed by pore filling, and potentially involved Lewis acid-base interaction and hydrogen bonding (Cheng et al, 2022). In relation to phenolic substances and PAHs, engineered biochar also possesses a great affinity to such contaminants. For example, Fe-modified biochars (FeBC) can play an important catalytic role in activating ammonium persulfate for the degradation of benzo[a]pyrene (BaP) (Li et al, 2022). The degradation rates of BaP achieved 87.0% with a dose of 0.5 g FeBC L^{-1}. Two pathways were responsible for the enhanced degradation of BaP: (1) radical pathway regulated by SO$_4^{•-}$ and O$_2^{•-}$; (2) non-radical pathway via ^{1}O$_2$ contribution and electron transfer.

Immobilization, degradation, and removal of metal(loid)s and organic pollutants in soils

Intense anthropogenic activities have caused agricultural soils to be polluted with metal(loid)s and organic pollutants in recent decades (Chen et al, 2021). Various engineered biochar has been used as an effective and green amendment to mitigate the environmental risks of such contaminated soils (Table 27.4).

Chitosan-coated biochar treatment significantly decreased the soil-available Cd concentration by 58%, Cd accumulation in plant roots by 54%, and shoots by 73%, compared to the control (Zubair et al, 2021a). 5% S-modified biochar amendment

Table 27.3 *Organic pollutants removal and associated mechanisms by engineered biochars in aqueous system*

Feedstock	Pyrolysis condition	Modification method	Organics	Removal efficiency or capacity	Mechanisms involved	References
Bamboo	500°C, 2 h	Steam activation	Tetracycline	95.8%	Hydrogen bonding, electrostatic interaction, and π-π electron donor-acceptor interactions.	Wang et al, 2020
Coffee ground	700°C, 1 h	H_3PO_4	Sulfadiazine	139.2 mg g^{-1}	Hydrophobic effect and π-π EDA interaction.	Zeng et al, 2022
Glucose	700°C, 3 h	N and Cu doping	Tetracycline	100%	Radical degradation involving •OH and electron transfer.	Zhong et al, 2020
Hickory wood chip	600°C, 1 h	MgO, ball milling	Methylene blue	87.5%	Physical adsorption, electrostatic attractive force, surface complexation.	Zheng et al, 2020
Wood biomass	700°C, 2 h	Graphene oxide	Dimethyl phthalate (DMP)	45.7 mg g^{-1}	π-π interactions and hydrophobicity.	Abdul et al, 2017
Erding	500°C, 1 h	*Bacillus cereus* LZ01 immobilization	Chlortetracycline	82.3%	Biodegradation by LZ01 and biochar adsorption.	Zhang and Wang, 2021
Corn stalk	–	*Vibrio* sp. LQ2 immobilization	Diesel oil	94.7%	Biodegradation and minor physical adsorption.	Zhou et al, 2021a

Table 27.4 *Summary of multiple engineered biochars and their immobilization performance for (in)organic contaminants in soil system*

Contaminants	Feedstock	Pyrolysis condition	Modification method	Contaminants	Addition dose	Main results	References
Metal(loid)s	Coconut shell	800°C, 6 h	Ultrasonication and HCl	Zn (184 mg kg^{-1}), Ni (66 mg kg^{-1}), Cd (0.82 mg kg^{-1})	2.5 and 5%	5% modified biochar application resulted in the soil available Cd, Ni, and Zn decreasing by 12.7%, 57.2%, and 30.1%, respectively.	Liu et al, 2018
	Rice husk	550°C, 2 h	Sulfur powder	Hg ($\approx$1000 mg kg^{-1})	1%, 2% and 5%	Compared to untreated soil, 1%, 2%, and 5% S-modified biochar reduced freely available Hg in TCLP leachates by 95%, 97%, and 99%, respectively.	O'Connor et al, 2018
	Tea branch	500°C, 1.5 h	MnFe$_2$O$_4$	Sb (79 mg kg^{-1}) Cd (696 mg kg^{-1})	0.1%, 1% and 2%	The NH$_4$NO$_3$-extractable concentrations of Sb in soil decreased by 33.8-43.5% with MnFe$_2$O$_4$-modified biochar amendments; the highest decrease of CaCl$_2$-extractable Cd (76.0%) was noted at 2% dose.	Wang et al, 2019
	Corn stalk	350°C, 2 h	Immobilization with *Bacillus subtilis, Bacillus cereus,* and *Citrobacter* sp.	Cd (2.5 mg kg^{-1}) U (29 mg kg^{-1})	3%	The DTPA-extractable U and Cd in the soil decreased by 69% and 56%, respectively; bacteria-loaded biochar reduced metal uptake thus promoting celery growth.	Qi et al, 2021

Organic pollutants	Waste timber	800–900°C, 36 min	Steam/CO_2 activation	PFAS (1200–3800 µg kg^{-1})	0.1–5%	In the soil with low total organic carbon (1.6%), all biochars at a 5% dose strongly reduced leaching concentrations of PFAS, by 98–100%.	Sørmo et al, 2021
	Olive tree pruning residue	400°C, 2 h	$KMnO_4$	Pentachlorophenol (2–30 µg g^{-1})	2.5%	The modified biochar was able to achieve the highest rates of remediation and high removal of extractable PCP under both aerobic and anaerobic conditions.	Chacón et al, 2022
	Biogas residue	700°C, 2 h	Potassium ferrate	Benzo[a]pyrene (8.16 mg kg^{-1})	1%	The Fe-impregnated biochar coupled with ammonium persulfate (FBC-APS) resulted in the overall degradation rate reaching 91.7% after 72 h in contaminated soil.	Li et al, 2022
	Corn stalk	600°C, 2 h	Fe$(NO_3)_3$·$9H_2O$ and loaded with *Acinetobacter lwoffii* DNS32	Atrazine (20 mg kg^{-1})	0.1%	Almost all the atrazine was degraded after the amendment of modified biochar, mainly due to the loaded Fe(II)/(III) enhanced microbial degradation ability as an electron transfer medium.	Tao et al, 2019

reduced freely available Hg in the leachates by 99%, as compared to untreated soil (O'Connor et al, 2018). The impact of engineered biochar on the immobilization of different metal(loid)s in co-contaminated soils has also been observed by many investigations. $MnFe_2O_4$-modified biochar obtained from tea branches at 2% dosage reduced the bioavailable concentration of Sb by 44% and of Cd by 76%, whereas pristine biochar only decreased soil-available Cd concentration by 13-34% (Wang et al, 2019). Coconut shell biochar and biochar modified by ultrasonication and HCl decreased the availability of Cd, Zn, and Ni by 30%, 57%, and 13%, respectively, compared to the control in a co-contaminated soil (at 5% over a 63-d incubation experiment; Liu et al, 2018).

In addition to metal(loid)s, engineered biochars for the remediation of different organic pollutants in soil (e.g., antibiotics, pesticides, PAHs, plasticizers, and phenols) have been examined (Table 27.4). For instance, a sulfidated-nZVI-loaded biochar was synthesized by Gao et al (2022) to degrade nitrobenzene. To be specific, the 1% resultant biochar addition degraded 98% of nitrobenzene within 24 hours. The catalytic reduction of nitrobenzene by FeS_x on biochar was highlighted as the predominant mechanism for enhanced nitrobenzene removal (Gao et al, 2022). Chacón et al (2022) used 0.025 M $KMnO_4$ as the modifying agent to obtain the engineered biochar and tested its effect in the remediation of pentachlorophenol (PCP) contaminated soil. $KMnO_4$-modified biochar exhibited the highest maximum rates of remediation (k_{max}) under aerobic (3.73 μg_{PCP} g_{soil}^{-1} d^{-1}) and anaerobic (2.40 μg_{PCP} g_{soil}^{-1} d^{-1}) conditions, which was much higher than those of raw biochar.

Biochar-based permeable reactive barriers for environmental decontamination

The permeable reactive barrier (PRB) is a passive treatment technology for the decontamination of polluted groundwater. It was defined as an emplacement of reactive media in the subsurface designed to intercept a contaminant plume and provide a flow path through the reactive media, to transform the contaminants to less toxic forms and reduce their concentrations (Thiruvenkatachari et al, 2008). The schematic diagram of biochar-based materials as fillers for the remediation of groundwater in the PRB system is shown in Figure 27.3.

The common reactive materials included ZVI, zeolite, biota, lime, granular activated carbon, phosphate, etc. Recently, engineered biochar-based materials were also used as reactive materials of PRB to decontaminate metal(loid) and organic pollutants in groundwater (Table 27.5). For instance, Mn-coated bone-derived biochar increased the retention ability for As(V), showing an increase in the lifetime of PRB remediation pattern as compared to pristine biochar, due to the formation of manganous arsenate precipitation (Liu et al, 2016). In addition, slowly released nutrient-immobilized biochar as a filling material of PRB is able to remove Cr(VI) from simulated groundwater. In this scenario, *Morganella morganii* subsp thrived as the dominant bacterium and enhanced the bio-reduction of Cr(VI), showing a better

Figure 27.3 *The schematic diagram of the biochar-based permeable reactive barrier*

performance than the traditional PRB filler (such as ZVI) (Hu et al, 2019). Overall, engineered biochar as a filler can enhance the removal rate of potentially toxic elements in the PRB system, showing a high practical application potential. The involved mechanisms for the decontamination of metal (loid)s mainly included surface complexation, adsorption, precipitation, ion exchange, and oxidation-reduction. Moreover, the microbial processes can also participate in the immobilization of metal(loid)s, such as microbially-induced carbonate/phosphate precipitation.

Moreover, biochar was also applied as reactive fillers to remove organic pollutants from polluted groundwater. For example, a treatment column filled with biochar was configured to remediate PAHs-contaminated groundwater, in which coconut shell-derived biochar (CSB) as reactive materials had higher adsorption kinetics and greater removal efficiency for phenanthrene than wheat straw biochar, because PAHs-degraded bacteria (*Pseudomonas* and *Sphingomonas*) became more abundant in CSB filled column (Liu et al, 2019). ZVI supported by microporous biochar (BC/ZVI) as filling materials can also remove trichloroethane from the effluent due to adsorption and catalytic reduction of dechlorination mechanisms (Lawrinenko et al, 2017). Moreover, FeCu bi-functional biochar can significantly remove the chlorobenzene in the effluent from the PRB column compared to the natural sand porous media in the other column, in which chlorobenzene was degraded into some intermediates such as benzene and phenol during the treatment process (Zhang et al, 2018). The removal mechanism of organic pollutants mainly included chemical degradation, biodegradation, pore filling, hydrogen bond, π-π electron donor-acceptor interaction, etc.

Table 27.5 *Biochar as fillers for decontamination of environmental pollutants in the permeable reactive barrier system*

Feedstocks	Pyrolysis condition	Modification method	Contaminant	Removal rate or capacity	Key mechanisms involved	References
Bone	–	MnSO$_4$ Immobilization with	As(V)	9.5 mg g^{-1}	Electrostatic adsorption	Liu et al, 2016
Peanut shell	300°C, 2 h	*Morganella morganii* subsp.	Cr(VI)	80.2 mg g^{-1}	Adsorption and bio-reduction	Hu et al, 2019
Coconut shell	–	Combined with Entisol	Cd(II) Pb(II)	99.9% 92.5%	Surface complexation with hydroxyl functional groups	Paranavithana et al, 2016
Coconut shell	–	–	Phenanthrene	100%	Bio-degradation.	Liu et al, 2019
Herbal residue	600°C, 0.5 h	–	Trichloroethylene	≥ 99.7%	Bio-dechlorination	Siggins et al, 2021
Lignin	900°C, 4 h	nZVI	Trichloroethylene	96% (fast flow) and 99% (slow flow)	Adsorption and reductive dechlorination	Lawrinenko et al, 2017
Bamboo	600°C, –	Fe and Cu powders	Chlorobenzene	61.4% (homogeneous columns) and 68.1% (heterogeneous columns)	Reduction and oxidation effect	Zhang et al, 2018

The way forward

In principle, biochar-based materials have some advantages including inexpensive and widely-sourced feedstocks, and comparatively simple production procedures. They are potentially cost-effective and environmentally friendly adsorbents and amendments that may be utilized to solve a variety of environmental challenges. However, several concerns remain unresolved, and the following points should be considered to achieve a sustainable future for biochar-based materials in environmental remediation.

- Until now, studies involving engineered biochar for environmental remediation were mainly based on laboratory tests, and long-term experiments should be carried out to assess the practical promise at the field scale.
- Cost-benefit analysis of engineered biochar needs to be conducted before its large-scale application. Future studies should endeavor to propose empirical formulae considering the feedstock collection-production-transportation-application-post-treatment of engineered biochar to quantify cost-benefit ratios.
- When exposed to environmental conditions, engineered biochar inevitably undergoes weathering by abiotic and biotic aging processes. However, little information is available on the remediation effect of engineered biochar after aging. The influence of the aging process on the effectiveness of different engineered biochar remediation needs to be further investigated.
- Insufficient information is still available regarding the disposal of spent biochar after the adsorption of toxic pollutants. Related technology should be developed to recover the spent biochar for energy production.
- Some engineered biochars contain hazardous substances (e.g., heavy metals, PAHs, and PFRs) and potentially introduce toxic chemicals during the modification process. The ecotoxicological evaluation of such hazardous biochars should be conducted before their large-scale production and application.
- Advanced spectroscopic techniques (e.g., synchrotron-based X-ray absorption fine structure (XAFS) spectroscopy), theoretical calculations using density functional theory (DFT) and molecular dynamics (MD), artificial intelligence, and machine learning should be considered to fully reveal the decontamination mechanisms for various pollutants.
- To scale up biochar production, support from government agencies may play an important role in collecting biomass feedstock by subsidizing labor and transportation costs for pristine biochar production, thereby decreasing the cost of engineered biochar.

References

Abdul G, Zhu X, and Chen B 2017 Structural characteristics of biochar-graphene nanosheet composites and their adsorption performance for phthalic acid esters. *Chemical Engineering Journal* 319, 9–20.

Ahmed MB, Zhou JL, Ngo HH, Guo W, and Chen M 2016 Progress in the preparation and application of modified biochar for improved contaminant removal from water and wastewater. *Bioresource Technology* 214, 836–851.

An Q, et al 2022 Ni(II), Cr(VI), Cu(II) and nitrate removal by the co-system of *Pseudomonas hibiscicola* strain L1 immobilized on peanut shell biochar. *Science of the Total Environment* 814, 152635.

Bolan NS, et al 2023 Soil acidification and the liming potential of biochar. *Environmental Pollution* 317, 120632.

Chacón FJ, Cayuela ML, Cederlund H, and Sánchez-Monedero MA 2022 Overcoming biochar limitations to remediate pentachlorophenol in soil by modifying its electrochemical properties. *Journal of Hazardous Materials* 426, 127805.

Chen H, et al 2022a Assessing simultaneous immobilization of lead and improvement of phosphorus availability through application of phosphorus-rich biochar in a contaminated soil: A pot experiment. *Chemosphere* 296, 133891.

Chen H, et al 2022b Enhanced sorption of trivalent antimony by chitosan-loaded biochar in aqueous solutions: Characterization, performance and mechanisms. *Journal of Hazardous Materials* 425, 127971.

Chen H, et al 2022c Engineered biochar for environmental decontamination in aquatic and soil systems: A review. *Carbon Research* 1, 4.

Chen H, et al 2021 Sorption of diethyl phthalate and cadmium by pig carcass and green waste-derived biochars under single and binary systems. *Environmental Research* 193, 110594.

Chen Q, et al 2022c Degradation mechanism and QSAR models of antibiotic contaminants in soil by MgFe-LDH engineered biochar activating urea-hydrogen peroxide. *Applied Catalysis B: Environmental* 302, 120866.

Cheng H, et al 2022 Hierarchical porous biochars with controlled pore structures derived from co-pyrolysis of potassium/calcium carbonate with cotton straw for efficient sorption of diethyl phthalate from aqueous solution. *Bioresource Technology* 346, 126604.

Cui P, et al 2022 Atomically dispersed manganese on biochar derived from a hyperaccumulator for photocatalysis in organic pollution remediation. *Environmental Science and Technology* 56, 8034–8042.

Cui P, et al 2021a An N,S-anchored single-atom catalyst derived from domestic waste for environmental remediation. *ACS Environmental Science and Technology Engineering* 1, 1460–1469.

Cui S, et al 2021b New insights into ball milling effects on MgAl-LDHs exfoliation on biochar support: A case study for cadmium adsorption. *Journal of Hazardous Materials* 416, 126258.

Duan W, Oleszczuk P, Pan B, and Xing B 2019 Environmental behavior of engineered biochars and their aging processes in soil. *Biochar* 1, 339–351.

Feng Z, et al 2021 Preparation of magnetic biochar and its application in catalytic degradation of organic pollutants: A review. *Science of the Total Environment* 765, 142673.

Fu C, Zhang H, Xia M, Lei W, and Wang F 2020 The single/co-adsorption characteristics and microscopic adsorption mechanism of biochar-montmorillonite composite adsorbent for pharmaceutical emerging organic contaminant atenolol and lead ions. *Ecotoxicological and Environmental Safety* 187, 109763.

Gao F, et al 2022 Enhanced nitrobenzene removal in soil by biochar supported sulfidated nano zerovalent iron: Solubilization effect and mechanism. *Science of the Total Environment* 826, 153960.

Hu B, et al 2019 Slow released nutrient-immobilized biochar: A novel permeable reactive barrier filler for Cr(VI) removal. *Journal of Molecular Liquids* 286, 110876.

Huang Q, et al 2019 Biochar-based materials and their applications in removal of organic contaminants from wastewater: State-of-the-art review. *Biochar* 1, 45–73.

Kumar M, et al 2020 Ball milling as a mechanochemical technology for fabrication of novel biochar nanomaterials. *Bioresource Technology* 312, 123613.

Kwak JH, et al 2019 Biochar properties and lead (II) adsorption capacity depend on feedstock type, pyrolysis temperature, and steam activation. *Chemosphere* 231, 393–404.

Lam SS, et al 2020 Engineering pyrolysis biochar via single-step microwave steam activation for hazardous landfill leachate treatment. *Journal of Hazardous Materials* 390, 121649.

Lartey-Young G and Ma L 2022 Optimization, equilibrium, adsorption behaviour of Cu/Zn/Fe LDH and LDH/BC composites towards atrazine reclamation in an aqueous environment. *Chemosphere* 293, 133526.

Lawrinenko M, et al 2017 Macroporous carbon supported zerovalent iron for remediation of trichloroethylene. *ACS Sustainable Chemistry and Engineering* 5, 1586–1593.

Leng L, et al 2020 Nitrogen containing functional groups of biochar: An overview. *Bioresource Technology* 298, 122286.

Li T, et al 2020 Polyethyleneimine-modified biochar for enhanced phosphate adsorption. *Environmental Science and Pollution Research* 27, 7420–7429.

Li X, et al 2022 Kill three birds with one stone: Iron-doped graphitic biochar from biogas residues for ammonium persulfate activation to simultaneously degrade benzo[a]pyrene and improve lettuce growth. *Chemical Engineering Journal* 430, 132844.

Liang H, et al 2020 Preparation of nitrogen-doped porous carbon material by a hydrothermal-activation two-step method and its high-efficiency adsorption of Cr(VI). *Journal of Hazardous Materials* 387, 121987.

Liu C, et al 2019 Evaluating a novel permeable reactive bio-barrier to remediate PAH-contaminated groundwater. *Journal of Hazardous Materials* 368, 444–451.

Liu H, et al 2018 Effect of modified coconut shell biochar on availability of heavy metals and biochemical characteristics of soil in multiple heavy metals contaminated soil. *Science of the Total Environment* 645, 702–709.

Liu J, He L, Dong F, and Hudson-Edwards KA 2016 The role of nano-sized manganese coatings on bone char in removing arsenic(V) from solution: Implications for permeable reactive barrier technologies. *Chemosphere* 153, 146–154.

Liu L, et al 2022 Adsorption of metolachlor by a novel magnetic illite-biochar and recovery from soil. *Environmental Research* 204, 111919.

Liu R, Wang H, Hu B, and Qiu M 2021a Reductive and adsorptive elimination of U(VI) ions in aqueous solution by SFeS@Biochar composites. *Environmental Science and Pollution Research* 28, 55176–55185.

Liu T, et al 2021b Removal of chlorophenols in the aquatic environment by activation of peroxymonosulfate with nMnOx@Biochar hybrid composites: Performance and mechanism. *Chemosphere* 283, 131188.

Lyu H, et al 2020a Biochar/iron (BC/Fe) composites for soil and groundwater remediation: Synthesis, applications, and mechanisms. *Chemosphere* 246, 125609.

Lyu H, et al 2020b Thiol-modified biochar synthesized by a facile ball-milling method for enhanced sorption of inorganic Hg^{2+} and organic CH_3Hg^+. *Journal of Hazardous Materials* 384, 121357.

Lyu P, et al 2021 Phosphorus modified biochar cross-linked Mg-Al layered double-hydroxide composite for immobilizing uranium in mining contaminated soil. *Chemosphere* 276, 130116.

Medeiros DCCDS, et al 2022 Pristine and engineered biochar for the removal of contaminants co-existing in several types of industrial wastewaters: A critical review. *Science of the Total Environment* 809, 151120.

Murad HA, et al 2022 A remediation approach to chromium-contaminated water and soil using engineered biochar derived from peanut shell. *Environmental Research* 204, 112125.

Nguyen LH, et al 2021 H_2O_2 modified hydrochar derived from paper waste sludge for enriched surface functional groups and promoted adsorption to ammonium. *Journal of the Taiwan Institute of Chemical Engineers* 126, 119–133.

O'Connor D, et al 2018 Sulfur-modified rice husk biochar: A green method for the remediation of mercury contaminated soil. *Science of the Total Environment* 621, 819–826.

Paranavithana GN, et al 2016 Adsorption of Cd^{2+} and Pb^{2+} onto coconut shell biochar and biochar-mixed soil. *Environmental Earth Sciences* 75, 484.

Qi X, et al 2021 Application of mixed bacteria-loaded biochar to enhance uranium and cadmium immobilization in a co-contaminated soil. *Journal of Hazardous Materials* 401, 123823.

Qin Y, et al 2020 Enhanced removal of ammonium from water by ball-milled biochar. *Environmental Geochemistry and Health* 42, 1579–1587.

Ramanayaka S, Tsang DCW, Hou D, Ok YS, and Vithanage M 2020 Green synthesis of graphitic nanobiochar for the removal of emerging contaminants in aqueous media. *Science of the Total Environment* 706, 135725.

Shaheen SM, et al 2022 Manganese oxide-modified biochar: Production, characterization and applications for the removal of pollutants from aqueous environments – A review. *Bioresource Technology* 346, 126581.

Siggins A, Thorn C, Healy MG, and Abram F 2021 Simultaneous adsorption and biodegradation of trichloroethylene occurs in a biochar packed column treating contaminated landfill leachate. *Journal of Hazardous Materials* 403, 123676.

Sørmo E, et al 2021 Stabilization of PFAS-contaminated soil with activated biochar. *Science of the Total Environment* 763, 144034.

Sornhiran N, Aramrak S, Prakongkep N, and Wisawapipat W 2022 Silicate minerals control the potential uses of phosphorus-laden mineral-engineered biochar as phosphorus fertilizers. *Biochar* 4, 2.

Sui L, et al 2021 Preparation and characterization of boron-doped corn straw biochar: Fe(II) removal equilibrium and kinetics. *Journal of Environmental Science* 106, 116–123.

Tao Q, et al 2019 Copper hexacyanoferrate nanoparticle-decorated biochar produced from pomelo peel for cesium removal from aqueous solution. *Journal of Radioanalytical and Nuclear Chemistry* 322, 791–799.

Thiruvenkatachari R, Vigneswaran S, and Naidu R 2008 Permeable reactive barrier for groundwater remediation. *Journal of Industrial and Engineering Chemistry* 14, 145–156.

Vigneshwaran S, Sirajudheen P, Karthikeyan P, and Meenakshi S 2021 Fabrication of sulfur-doped biochar derived from tapioca peel waste with superior adsorption performance for the

removal of Malachite green and Rhodamine B dyes. *Surface and Interface* 23, 100920.

Wan Z, et al 2022 Stoichiometric carbon catalysis via epoxide-like C-S-O configuration on sulfur-doped biochar for environmental remediation. *Journal of Hazardous Materials* 428, 128223.

Wang J and Wang S 2019 Preparation, modification and environmental application of biochar: A review. *Journal of Cleaner Production* 227, 1002–1022.

Wang L, et al 2018 Enhanced antimonate (Sb (V)) removal from aqueous solution by La-doped magnetic biochars. *Chemical Engineering Journal* 354, 623–632.

Wang R, et al 2020 Synergistic removal of copper and tetracycline from aqueous solution by steam-activated bamboo-derived biochar. *Journal of Hazardous Materials* 384, 121470.

Wang Y, et al 2019 Simultaneous alleviation of Sb and Cd availability in contaminated soil and accumulation in *Lolium multiflorum* Lam. after amendment with Fe-Mn-Modified biochar. *Journal of Cleaner Production* 231, 556–564.

Xiao Y, Lyu H, Tang J, Wang K, and Sun H 2020 Effects of ball milling on the photochemistry of biochar: Enrofloxacin degradation and possible mechanisms. *Chemical Engineering Journal* 384, 123311.

Xu L, et al 2022 Adsorption of micropollutants from wastewater using iron and nitrogen co-doped biochar: Performance, kinetics and mechanism studies. *Journal of Hazardous Materials* 424, 127606.

Xu Y, et al 2020 As(III) and As(V) removal mechanisms by Fe-modified biochar characterized using synchrotron-based X-ray absorption spectroscopy and confocal micro-X-ray fluorescence imaging. *Bioresource Technology* 304, 122978.

Yao, Y, Gao, B, Inyang, M, Zimmerman, AR, Cao, X, Pullammanappallil, P, and Yang, L 2011 Biochar derived from anaerobically digested sugar beet tailings: Characterization and phosphate removal potential. *Bioresource Technology* 102, 6273–6278. 10.1016/j.biortech.2011.03.006.

Yek PNY, et al 2020 Engineered biochar via microwave CO_2 and steam pyrolysis to treat carcinogenic Congo red dye. *Journal of Hazardous Materials* 395, 122636.

Yin Q, Wang R, and Zhao Z 2018 Application of Mg-Al-modified biochar for simultaneous removal of ammonium, nitrate, and phosphate from eutrophic water. *Journal of Cleaner Production* 176, 230–240.

Zeng X, et al 2022 Impacts of temperatures and phosphoric-acid modification to the physicochemical properties of biochar for excellent sulfadiazine adsorption. *Biochar* 4, 14.

Zhang J, et al 2021 Modification of ordered mesoporous carbon for removal of environmental contaminants from aqueous phase: A review. *Journal of Hazardous Materials* 418, 126266.

Zhang S and Wang J 2021 Removal of chlortetracycline from water by *Bacillus cereus* immobilized on Chinese medicine residues biochar. *Environmental Technology and Innovation* 24, 101930.

Zhang X, et al 2018 Application of Fe-Cu/biochar system for chlorobenzene remediation of groundwater in inhomogeneous aquifers. *Water* 10, 13.

Zhen M, Tang J, Li C, and Sun H 2021 Rhamnolipid-modified biochar-enhanced bioremediation of crude oil-contaminated soil and mediated regulation of greenhouse gas emission in soil. *Journal of Soil and Sediments* 21, 123–133.

Zheng Y, Wan Y, Chen J, Chen H, and Gao B 2020 MgO modified biochar produced through ball milling: A dual-functional adsorbent for removal of different contaminants. *Chemosphere* 243, 125344.

Zheng Z, Ali A, Su J, Fan Y, and Zhang S 2021 Layered double hydroxide modified biochar combined with sodium alginate: A powerful biomaterial for enhancing bioreactor performance to remove nitrate. *Bioresource Technology* 323, 124630.

Zhong Q, et al 2020 Oxidative degradation of tetracycline using persulfate activated by N and Cu codoped biochar. *Chemical Engineering Journal* 380, 122608. 10.1016/j.cej.2019.122608.

Zhou H, et al 2021a Enhanced bioremediation of diesel oil-contaminated seawater by a biochar-immobilized biosurfactant-producing bacteria *Vibrio sp.* LQ2 isolated from cold seep sediment. *Science of the Total Environment* 793, 148529.

Zhou Y, et al 2021b Production and beneficial impact of biochar for environmental application: A comprehensive review. *Bioresource Technology* 337, 125451.

Zhu H, et al 2022 Sodium citrate and biochar synergistic improvement of nanoscale zero-valent iron composite for the removal of chromium (VI) in aqueous solutions. *Journal of Environmental Science* 115, 227–239.

Zhu S, Qu T, Irshad NK, and Shang J 2020 Simultaneous removal of Cd(II) and As(III) from co-contaminated aqueous solution by α-FeOOH modified biochar. *Biochar* 2, 81–92.

Zubair M, et al 2021a Efficacy of chitosan-coated textile waste biochar applied to Cd-polluted soil for reducing Cd mobility in soil and its distribution in moringa (*Moringa oleifera* L.). *Journal of Environmental Management* 284, 112047.

Zubair M, et al 2021b Sustainable wastewater treatment by biochar/layered double hydroxide composites: Progress, challenges, and outlook. *Bioresource Technology* 319, 12412

28

Biochar as building and road materials

Harn Wei Kua

Introduction

The mechanical properties of construction materials for buildings and infrastructures are related and substantially affected by their degrees of water sorptivity. Additives for concrete and asphalt, especially additives that are highly porous, can alter the diffusion rate and accumulation of water within the materials, thus modifying these materials' hygro-mechanical properties. Furthermore, the particle size of these additives will determine the density of these materials, as well as provide increased surface area for chemical reactions to take place that control the pace of "maturation" and hardening of these materials.

Biochar is a favorable material to be deployed as an additive to concrete and asphalt for several compelling reasons. Known as a carbon-negative material that is a by-product of the fuel-producing process of gasification or pyrolysis, the properties of biochar can be easily tuned by changing various conditions of the gasification or pyrolysis process, such as residence time, maximum temperature, and heating rate (Ok et al, 2018). These properties include porosity, permeability, and total surface area. Changing the feedstock can also change the morphology of the biochar, and this can be further modified using economical methods, such as grinding and ball-milling. Above all, biochar can be a means of "greening" concrete because, depending on the feedstock used, it can be used to partially replace cement.

This chapter aims to elaborate on how biochar can be used to enhance various properties of concrete, mortar, and asphalt; the limitations of biochar are also shared. An important topic of future research and development in using biochar to increase the sustainability of the built environment is also outlined.

DOI: 10.4324/9781003297673-28

Hygro-mechanical properties of concrete and the critical roles of aggregates

The effect of water on the mechanical properties of concrete

Cement is widely known as the second most utilized material on earth, just behind water. Used in the construction of buildings and infrastructures (such as roads, bridges, and water tanks), cement is mixed with water, fine aggregate (usually sand), coarse aggregate (usually gravel), and special additives to make concrete. When coarse aggregates are left out from the mixture, the mixture is known as mortar. In contemporary architecture and construction, mortar is used for holding masonry units, such as clay bricks and stones, together to form a structural unit. A brick wall is a good example.

Concrete is also known for being "strong in compression but weak in tension". The mechanical strength of concrete 28 days after it is cured under normal conditions (in air but sheltered to prevent excessive water loss through evaporation) is usually taken as its reference strength. The compressive strength of concrete ranges from 2,500 to 9,000 psi (17.2–62.1 MPa), whereas its tensile strength is only about 10–15% of the compressive strength. The reason for this relatively low tensile strength is that under tensile stress, the components of concrete are being pulled apart. The region at the interfaces of aggregates and the surrounding hardened mortar paste matrix – also known as interfacial transition zones (ITZ) – is considerably low in strength. This weakness is partly caused by the formation of plate-like calcium hydroxide (CH) crystals in the ITZ, which will come apart when a concrete sample undergoes tension. Therefore, concrete must be reinforced with additional components, such as steel bars or steel fibers, for its tensile strength to be enhanced.

The mechanical strength of concrete is determined by different factors, including water-cement ratio, level of compaction of the concrete mixture, types and quality of coarse and fine aggregates, types of cement used, and curing conditions. Water is required to initiate the process of cement hydration, with calcium silicate hydrate (commonly written as CSH) and CH as two of its main by-products. CSH can be considered as the "glue" that binds the aggregates and cement paste particles together to form a structural unit. The presence of water in the concrete mixture is measured in terms of water-cement or water-binder (applicable when pozzolanic materials are added to supplement cement as binders) ratio, and the optimal value falls between 0.35 and 0.45. Adding too much water will lower the mechanical strength of concrete; in fact, it is found that when the water-cement ratio changes from 0.45 to 0.60, the concrete porosity increases to 150%, causing strength to reduce by 76% (Kim et al, 2014; Malaiskiene et al, 2017).

Excessive water content (that is, too high a water-cement or water-binder ratio) will result in excess free water in the concrete that can be lost through evaporation and "bleeding" (as shown in Figure 28.1); when that occurs, air voids are left behind. If the volume of such air voids is high enough, they will form an elaborate network of interconnected pores and air voids, leading to increased permeability of the concrete. If acidic water flows through this pore network and comes into contact with the reinforcement bar in the concrete, oxidation of the bar ensues, and this will cause expansion of that particular spot on the bar. This expansion will

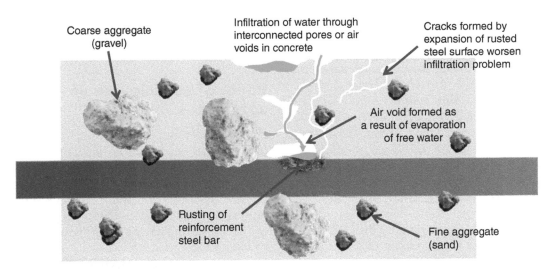

Figure 28.1 *Production of air void network in concrete, as a result of loss of excess free water through evaporation and "bleeding". The infiltration of water and subsequent rusting (oxidation) of the reinforcement bar results in the infamous phenomenon known as concrete spalling*

in turn set off more cracks around the air voids, thus worsening the water infusion problem in a vicious cycle. This progressive deterioration of concrete will ultimately lead to the spalling of concrete and ultimate failure of the structural concrete element.

Mechanical effects of porous aggregates

The mechanical properties of concrete are highly dependent on its moisture content and how it changes with time during the curing process. Assuming that the environment in which a freshly cast sample of concrete is curing is well controlled – for example, covered with polyethylene sheets to reduce evaporation – the distribution of water in the concrete will depend on the water sorption capability of the ingredients in the concrete mix.

Lightweight aggregate (LWA) is a special kind of aggregate used in the production of lightweight concrete products. Popular lightweight aggregates include natural pumice, vermiculite, and perlite. Most LWAs have to be heated to temperatures of about 2,200 °F (1,200°C) so that vaporization of its chemical constituents, including volatile organic compounds (VOCs), expand the volume and create more pores in the aggregate particles, thus lowering the bulk density of the concrete containing these LWAs. However, studies of LWAs also explored their physical and mechanical effects on the ITZ between the aggregates' surfaces and the cement paste matrix. Due to their higher degree of water sorption, it was found that the ITZ in lightweight concrete is different from that of normal concrete. Specifically, the ITZ is found to be more compact than normal concrete (Zhang and Gjorv, 1990; Breton et al, 1993), and mechanical interlocking was observed between the aggregates' surfaces and their surrounding matrix. The detailed morphology of the ITZ depends on the surface morphology of the aggregates (including the surface structure, pore abundance, and structure), the porosity of the cement paste, and the exchange of water between the aggregates' pores and the matrix (Mehta, 1986). The porous and rough surface of the LWA provides sites for the by-products of cement hydration to cross the ITZ and develop "bridges" or "crawls" that mechanically pegged the LWAs to the matrix (Lo and Cui, 2004). This mechanism strengthens the concrete. Additionally, the pores on the LWAs absorb excess free water and hence reduce water loss through evaporation or "bleeding" from the matrix (as shown in Figure 28.1), including from the ITZ. This in turn reduces the formation of an air void network in the vicinity of the LWAs, densifies the ITZ, and results in increased strength.

However, this does not imply that adding more porous aggregates will indefinitely and linearly increase the strength of concrete. A high volume of air void in concrete aggregates will lower the surface energy needed for fracture lines to propagate through the concrete, thus weakening it.

Effect of particle size of aggregates

Even if water infiltration through concrete does not lead to concrete spalling, the ingress and trapping of water within the concrete can cause other problems over time (e.g., mold growth on basement wall). Although the prime cause of this water infiltration problem can be very complex, the absorption of water by concrete components is usually one of the reasons. Although the reduction of water loss (through evaporation and "bleeding") leads to denser and less permeable concrete that absorbs less water when it is used in buildings and infrastructures, the use of fine aggregates as filler to increase the density of the concrete

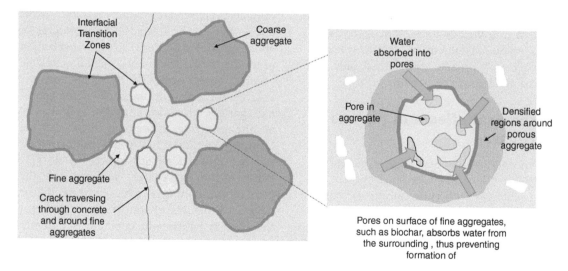

Figure 28.2 *(a) Close-packing with adding more fine aggregates between coarse aggregates strengthens the mortar/concrete because crack lines that are produced have to "travel" around these aggregates, and more energy is required to do so. (b) Pores absorb water from the immediate surroundings and densify these regions because water loss through evaporation and "bleeding" is reduced. When water is used up in hydration, water absorbed in these pores will flow back into the surroundings to sustain the hydration (that is, the "reservoir effect")*

is also a very effective strategy. For example, applying 40% (by mass of cement) of micro-fillers, such as glass powder and clay brick powder, in mortar can increase their compressive strength to about twice that of the control mortar sample within 1 year; correspondingly, the water absorption coefficient due to capillary action decreases by about 32% within this period (Sičáková and Špak, 2019). These results can be attributed to the "close-packing" effect of fine fillers (or micro-fillers), which fill most of the gaps between coarser aggregates and increases the energy needed by any fracture lines to traverse through the material (Figure 28.2a); if there are surface pores on these fine aggregates, free water will also be absorbed into them and so water loss (through evaporation and "bleeding") can also be reduced and this leads to a denser microstructure (Figure 28.2b).

Skid resistance of asphaltic mixture and roles of aggregates

Asphalt is a mixture of coarse aggregates, fine aggregates, and fillers bound with a bitumen binder, and it is used extensively for surfacing roads and airport runways. Known to be a versatile and economical material, the quality and service lifespan of asphalt mixtures can be severely compromised by water infiltration. This will cause a reduction in durability, characterized by a loss in stiffness and structural strength, which can be attributed to a loss of cohesion within the bitumen binder itself and a loss of adhesion between the aggregate and the bitumen (Kennedy et al, 1983; Ruth, 1985; Cui et al, 2014).

Similar to the abovementioned case of concrete, improving the structural integrity, mechanical strength, water resistance, and overall durability of asphalt mainly entails strengthening the bonding between the aggregates and bitumen binder. The chemical and physical properties of the aggregates will also have a strong influence on the final properties of the asphalt. Alkaline aggregates (such as limestone and marble) showed higher water resistance than more acidic aggregates (such as granite) (Cui et al, 2014). Within each of these two groups of aggregates, higher water resistance of aggregates leads to lower porosity. The shape and texture of fine aggregates in an asphaltic mixture also affect the stiffness, fatigue behavior, and durability of the asphaltic pavement (Valdes-Vidal et al, 2019). Specifically, the Particle Index (PI) of the aggregates – derived from a combination of the aggregate's shape, texture, and angularity – has a higher influence on the properties of the asphalt than their chemical composition (Valdes-Vidal et al, 2019). The different types of aggregates had PI values ranging from 14.1 to 23.0, and higher PI values resulted in higher resistance to cracking under fatigue and load-bearing capacity.

Among the different mechanical properties of asphalt, skid resistance is one of the most important performance factors. When the skid resistance of road surface increases by 10%, the traffic accident rate will decrease by about 13% (Ahammed and Tighe, 2009; Wang et al, 2018). The initial skid resistance of new asphalt gradually decreases as the aggregates are continuously worn off (or, polished away) by contact with tires' surfaces. Therefore, the long-term skid resistance of asphalt is determined by the polishing resistance of its aggregates. Their micro-texture typically plays a significant role in affecting the surface texture and skid resistance of pavement. For example, aggregates that can resist polishing will improve skid resistance (Kamel et al, 1982; Dahir, 1993). Although the polishing resistance of aggregates becomes a deciding factor for skid resistance only if the asphalt matrix near the top surface is sufficiently exposed due to prolonged wear and tear (Abdul-Malak et al, 1988; Crouch et al, 2002), it is important even if aggregates are embedded deeper beneath the surface. Specifically, aggregates with high polishing resistance will anchor onto the surrounding binder more strongly, thus causing the matrix layer above the aggregates to apply higher frictional forces on the surface of any object that comes in contact with the matrix layer and exhibit higher skid resistance.

Carbon-based additives in concrete and asphalt

The more commonly used carbon-based additives in construction materials are carbon fibers and carbon black. Carbon fibers are well known to be able to improve the tensile strength of concrete and asphalt. For example, adding fine cured carbon fiber composite materials reduced the porosity of porous hot mix asphalt by 17% and water infiltration rate by up to 20%, and was able to prevent drain-down of asphalt binder at an elevated temperature, while showing improved rutting resistance (Zhang et al, 2019). By adding carbon fibers into asphalt concrete at <2% v/v, the Marshall stability increased from 12.8 kN to 13.5 kN, and residual stability from 91–93% (Liu and Wu, 2011). The rutting dynamic stability also increased from 3,318 times mm^{-1} to 3,403 times mm^{-1}. Similarly, carbon fiber-reinforced polymer composite systems were effective in increasing the flexural strength of reinforced concrete (Brena et al, 2003) but various strengthening configurations were needed to avoid premature de-bonding, including techniques such as placement of transverse straps along the composite laminates or bonding the composites on the side surface of the concrete. In general, incorporating carbon fibers in a medium is challenging because of the electrostatic charges residing on the surfaces of these fibers, which give rise to cohesion forces between carbon fibers, thus causing them to clump and coagulate during mixing. Dispersion of carbon fibers requires special chemicals, and these materials may also affect the mechanical and electrical properties of the composites. Furthermore, the relatively high costs of carbon fibers may also limit their applications in construction.

Besides carbon fibers, other carbon-based materials, such as graphite and carbon black particles had been used to improve the qualities of concrete and asphalt before. For example, for graphite particles, when 0.005–0.025% w/w of cement were added into concrete mixes, compressive strength, frost resistance, heat resistance, and thermo-frost resistance of concrete improved (Yanturina et al, 2017). Similarly, efforts to incorporate carbon black powder in 0.1, 0.2, 0.3, 0.4, and 0.5 carbon black/cement ratio in concrete mixes (containing steel bars) that were later immersed in 3.5% chloride solution for six months showed that with increasing carbon black content, the corrosion rate decreased due to decreased chloride ions permeability (Masadeh, 2015).

Notwithstanding their usefulness, carbon fibers, carbon black, and graphite have their weaknesses. Carbon fibers have very high embodied energy – between 183 and 286 $MJ\,kg^{-1}$; these are close to those of aluminum

$(196–257$ MJ $kg^{-1})$ (Song et al, 2009). Carbon black is derived from incomplete thermal decomposition of petroleum products, such as coal tar and ethylene cracking tar. Production of synthetic graphite is known to be highly energy intensive and, due to about 30–40% mass loss of its reactants as gaseous species, emits large amounts of non-combustion greenhouse gases (Dunn et al, 2015). Comparatively, biochar is a much more sustainable ingredient for "greening"

concrete. This also underlines the importance of employing analytical techniques such as life cycle assessment (LCA) to more holistically evaluate the environmental impacts of construction materials (Hwang et al, 2020). In fact, environmental impacts caused by higher-order consequences of replacing existing materials with a more sustainable alternative should also be evaluated within the LCA (now widely known as consequential LCA) (Kua, 2015).

How biochar affects the mechanical strength and water-tightness of concrete?

Influence of biochar pores and particle size on mechanical strength

In general, the addition of porous biochar increases the air content of concrete. The increase in porosity of the concrete medium will depend on the feedstock and production conditions of the biochar. Due to the capillary absorption of free water, biochar was found to reduce the flowability of cement paste (Choi et al, 2012; Ahmad et al, 2015; Restuccia and Ferro, 2016) and the degree of reduction depends on the porosity of the biochar. For example, food waste (about 80% starch and 20% plant-based) biochar did not significantly reduce flowability up to 2% w/w addition in mortar. In contrast, biochar from wood sawdust results in lower flowability than food waste biochar and rice waste biochar (Gupta et al, 2018; Muthukrishnan et al, 2019). Instead of adding more water to the mixture, higher amounts of superplasticizer are added to increase the flowability in some studies when the flowability becomes too low for the mixture to be useful for casting structures (Gupta and Kua, 2020a).

Adding 1–2% w/w (of the mass of cement in the mixture) biochar from food waste and sawdust significantly improved the compressive strength of mortar by 10–20% (Gupta et al, 2018). However, the strength of mortar decreases when biochar dosage increases beyond 2% w/w because as the volume of biochar increases, the porosity and permeability of the mortar mixture increase as well. With more air voids in the matrix, it becomes easier for fracture lines to traverse through it, thus rendering the mixture weaker. One solution to this dilemma is to ball-mill the biochar particles to bring about two important changes: (1) reducing the average size of the particles; and (2) breaking up of macro-pores and hence increasing the average micro-porosity of the particles. Specific studies on such processed biochar have found that the addition of 1–2% w/w of ball-milled biochar results in higher strength compared to normal biochar (Gupta and Kua, 2019); this can be attributed to a better packing effect from the ball-milled biochar. However, the loss of macro-pores due to ball-milling also means that the biochar cannot absorb as much water as normal biochar, which implies that the densification of the cement paste around the biochar caused by the capillary suction of the pores will also be lower.

Several studies found that biochar addition does not significantly improve flexural strength (Wang et al, 2018). Based on the explanation above, this can be attributed to the incapability of biochar to compensate for the weaknesses of the ITZ. In other words, for biochar to increase the tensile and flexural strength of mortar or concrete, the surface features of biochar particles (including ridges and pores) must be able to anchor them strongly into the surrounding hardened cement paste and/or bridge across the ITZ (thus compensating for the weakness of the CH plates and un-hydrated cement grains in the ITZ) (Figure 28.3). Biochar particles were used in a cementitious coating around polyvinyl alcohol fibers and these coated fibers were mixed in mortar as reinforcement;

the biochar was found to help "anchor" the surface of the fibers to the mortar matrix, thereby increasing the flexural strength of mortar (Kua et al, 2020). This shows that there is potential for rough-surfaced biochar to increase the tensile strength of concrete, as long as it is positioned correctly to enable it to play this role.

Effect of biochar on water permeability and its importance on marine construction

Mortar with 1% w/w of wood-based biochar produced at 300–500°C were shown to have up to 58% and 66% reduction in water absorption and depth of water penetration compared to a control (Gupta et al, 2018). Silica fume (SF) is a widely utilized

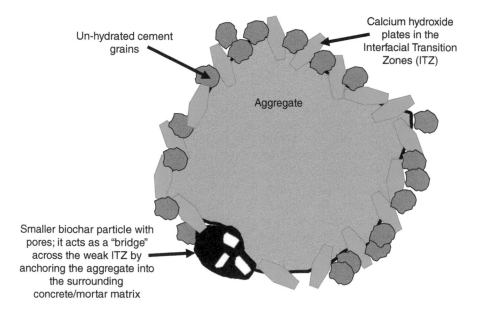

Figure 28.3 *Biochar particle acting as a "bridge" across the weak Interfacial Transition Zone around the aggregate; this will technically increase the tensile strength of the concrete containing this combination of aggregate and biochar additive.*

pozzolanic additive. Biochar–SF mortar was shown to have 65–80% lower water absorption than control mortar, and this is due to the lower water-accessible porosity when biochar and SF are present in the mixture (Gupta and Kua, 2020b). When examining how much SF can be replaced by biochar, up to 60% and 20% w/w of the SF can be reduced by using wood-based biochar and food waste biochar respectively to achieve higher 7-day and 28-day strength than control (by between 10 and 20%). These phenomena can be attributed to the two effects due to biochar that are explained above – the biochar particles act as fillers that keep the aggregates in the hardened mortar paste closely packed together, and absorption of water by biochar pores from the matrix, which densifies the proximity of the biochar.

Low water permeability is a critical quality for concrete used for coastal infrastructure or marine constructions. The presence of chloride and sulfate ions in concrete can result in structural failure triggered by cracks formed by expansive products of chemical reactions involving these ions. Alternatively, these cracks may also be formed from the rust developed on the surface of the reinforcement bars when they come in contact with these ions. Studies on mortar samples submerged in sodium chloride (NaCl) and sodium sulfate (Na_2SO_4) solutions found that by adding 1–2% w/w of wood-based biochar into the samples, they could retain 8–11% higher strength and reduce loss in strength than the control samples after 120-day exposure to NaCl. This is related to the biochar reducing chloride ion absorption after densifying the mortar matrix; furthermore, the pores in the biochar offer additional space for the by-products of these chemical reactions to expand in, without applying damaging pressure and cracking the surrounding matrix. This would result in improved durability and prolonged service lifespan for coastal infrastructures. Furthermore, over a 120-day exposure period in sulfate solution, samples with biochar showed 14–17% higher compressive strength than the control that was also exposed to the solution.

Biochar as a partial cement substitute

Finally, by selecting a specific feedstock for biochar, one can utilize specific chemical composition in the biochar to enhance certain aspects of concrete. For example, rice husk ash (RHA) is commonly used as a pozzolanic substitute for cement in concrete; it is one way of "greening" concrete. However, industrial RHA (iRHA) usually contains traces of unburnt rice husk and its silica content is partially crystalline. For its silica content to be more amorphous, further heat treatment is needed, which will consume more energy. A more sustainable and less energy-intensive alternative is using rice husk biochar (RHB) to partially replace iRHA in concrete. For example, by using a combination of 18% w/w of iRHA and 2% w/w of RHB, the compressive strength of mortar can be increased by about 17% compared to iRHA (Muthukrishnan et al, 2019). In addition to the filler effect and densification of mortar around the biochar caused by capillary absorption, the pozzolanic nature of the amorphous silica found in the biochar enables it to promote the hardening of the mortar. Adding 2% w/w RHB also decreased the primary water absorptivity of the mortar by 27–33%. However, flexural strength was found to be adversely affected by the presence of any form of RHA, regardless of production method. As mentioned above, this inability to increase tensile strength is likely due to a lack of surface undulations on the biochar that enable it to anchor to the surroundings and create bridges across the

ITZ of fine and coarse aggregates to counteract the mechanical weakness of the CH crystals there.

Biochar as "reservoirs" in cementitious mixture

Concrete that undergoes hydration will encounter autogenous shrinkage, where its total volume decreases due to water being dispensed out of the capillary pores in the concrete. Because biochar acts as tiny reservoirs in which the stored water can be released into the matrix to sustain cement hydration, there is less water being taken from the capillary pores, which reduces autogenous shrinkage. This so-called "reservoir effect" was studied by Gupta and Kua (2018) (Figure 28.2b). Mortar samples containing pre-soaked biochar – biochar that was soaked to saturation before being added into the mortar – showed higher internal relative humidity and lower mass loss due to evaporation under air curing conditions than plain mortar and mortar with dry biochar. This is due to water retention in the biochar pores. Furthermore, mixtures with pre-soaked biochar were up to 40% and 30% stronger than plain mortar and mortar with dry biochar, respectively, under air curing. The same trend was found for flexural and split-tensile strength. This shows that the retained moisture was utilized to promote hydration in the surrounding matrix. This process whereby water flows out of the pores into the matrix to sustain hydration is known as internal curing.

How biochar affects the performance of asphalt?

Besides enhancing concrete's properties, biochar is also known to be capable of modifying different mechanical properties of asphalt. The viscosity of biochar-modified asphalt mixtures increased at 135°C and 175°C (Ma et al, 2022), thus enabling it to meet the Superpave standard (<3 Pa·s at 135°C) (Harrigan et al, 1994). As biochar content increased, the viscosity increased, reaching the maximum of 0.85 Pa·s with 15% biochar. It was proposed that the added biochar absorbed a portion of the light components of the binder and hence reduced the binder's fluidity and consequently increased the viscosity. Furthermore, the softening point of the asphalt was also increased. These changes improved the resistance of the asphalt to deformation, thereby improving its mechanical resilience at high temperatures.

Other than the effects on mechanical properties, biochar has been explored as a pollutant remover in asphalt. Heating of asphalt before application is known to cause emissions of VOCs, which, among other effects, leads to the formation of urban haze. These VOCs come in a wide range and each component participates in various physical-chemical activities (Yang et al, 2019). Straw and waste wood-based biochar that was mixed with asphalt adsorbed the 1-pentadecene ($C_{15}H_{30}$), 1-hexadecene ($C_{16}H_{32}$), nonadecane ($C_{19}H_{40}$), and heneicosane ($C_{21}H_{44}$) from the VOCs from asphalt by an average 50% (Zhou et al, 2020), thus making asphalt more sustainable. These studies corroborated the conclusions drawn by Zhang et al (2018) who found that asphalt containing 4% biochar (of average particle size 75 μm) was more resistant to permanent deformation, thermal cracking, and aging when compared to asphalt modified with graphite. Addition of 5% (of weight of binder) was effective in reducing the effect of aging of asphalt binder regardless of the aging method (that is, by oxidation, UV exposure, mixture aging, or weathering) (Rajib et al, 2021). The authors attributed these improvements to the pore structure and functional groups of the biochar, which adsorb reactive species and free radicals generated during the aging process.

Latest development in biochar concrete – accelerated carbonation curing of biochar concrete

One of the ongoing research topics of biochar concrete is the use of biochar to enhance the process of using concrete for carbon dioxide (CO_2) capture; this is known as accelerated carbonation curing (ACC). When left on its own, a concrete sample that is undergoing hydration will react with atmospheric CO_2 to give calcium carbonate; this can be considered as a kind of carbon capture via mineralization. A slow process, by-products of cement hydration such as CH, CSH, and ettringite can be converted into $CaCO_3$ in the following reactions:

$$Ca(OH)_2 + CO_2 \rightarrow CaCO_3 \quad [28.1]$$

$$xCaO \cdot SiO_2 \cdot yH_2O + zCO_2 \rightarrow zCaCO_3 + (x–z)CaO$$
$$\cdot SiO_2 \cdot yH_2O \text{ (decalcified)}$$
$$[28.2]$$

$$xCaO \cdot SiO_2 \cdot yH_2O + xCO_2 \rightarrow xCaCO_3 + SiO_2$$
$$\cdot zH_2O + (y–z)H_2O \quad [28.3]$$

$$3CaO \cdot Al_2O_3 \cdot 3CaSO_4 \cdot 32H_2O \text{ (ettringite)} + 3CO_2$$
$$\rightarrow Al_2O_3 \cdot xH_2O + 3CaCO_3 + 3(CaSO_4 \cdot 2H_2O)$$
$$+ (26–x)H_2O \quad [28.4]$$

ACC is the concept of accelerating this mineralization process by deliberately and carefully exposing the curing concrete to high-concentration CO_2. The degree of carbon capture is quantified by the amount of carbonate formed, and the source of this carbonate can be estimated by comparing concentrations of CH and CSH. Recent studies focused on relating the amount of CO_2 absorbed and any changes in the mechanical properties of the "carbonated concrete". These results show that ACC's effect on concrete microstructure and mechanical properties can vary considerably. Curing mortar samples for 2, 6, 13, and 27 days, in 20,000 ppm of CO_2 resulted in surface precipitation of high concentration $CaCO_3$ that reached a maximum carbonation depth of 5 mm after 28 days, but the compressive strength of CO_2-cured samples was lower than water-cured samples at all ages (Jia et al, 2012). Similarly, carbonation beyond 2 hours reduced the compressive strength of mortar samples compared to the control (Junior et al, 2014) due to carbonation of inner CSH. With only 2 hours of carbonation, a strength improvement of 10% was recorded, corroborated by others (Rostami et al, 2012; Praneeth et al, 2020).

Finally, studies of ACC of biochar mortar or concrete is a comparatively new area of research. Samples containing 1% of biochar showed an 8–10% increase in compressive strength compared to the control even though the CO_2 dosing period was extended to 24 hours (Wang et al, 2020). Increasing the amount of biochar resulted in a decrease in compressive strength for biochar pyrolyzed at 500°C and 700°C; furthermore, when the biochar underwent internal carbonation (in which the biochar was pre-saturated with CO_2 before it was mixed into the mortar), the mortar's compressive strength was lower than those with biochar with external carbonation. In contrast, the internal carbonation strategy can enhance early (7 days) compressive strength by 25 – 30% compared to non-carbonated control, but after 28 days, internally carbonated samples showed similar strength as that of non-carbonated control (Gupta, 2021). On the other hand, biochar that did not undergo internal carbonation became about 5% weaker than control. These results are in contrast with the findings from Praneeth et al (2020), who limited all CO_2 dosing periods to 2 hours and found that all carbonated samples showed significant improvements in compressive strength at 3 days compared to non-carbonated samples.

Future research on the effect of biochar – either with or without internal carbonation – should focus on achieving a balance between prolonging CO_2 capture duration and reducing any drop in compressive strength due to carbonation of CSH or CH. This balance point will also signify a point where ongoing hydration is not compromised by concurrent carbonation reactions (Equations 28.1, 28.2, and 28.3), so that the creation of carbonate is not at the expense of the development of CSH and CH in the matrix. Gupta (2021) postulated that internally carbonated biochar has lower net carbonation; however, it is worthwhile to explore how ACC can be optimized so that internally carbonated biochar can lead to increased carbon capture and strength development at the same time.

Conclusions

The use of biochar to enhance different properties of concrete and asphalt is a relatively young field of research and development. From the first work published by Choi et al (2012) to the active period that saw an acceleration of research activities and innovative discoveries in 2015–2022, valuable lessons have been accrued. This chapter highlighted the key physical properties and effects of biochar that contributed to the results observed so far – namely, macro- and micro-porosity, permeability, biochar particle size, shape, filler effect, and the "reservoir effect". This knowledge will lay the foundation for future work in these fields and establish a stronger case for biochar concrete and asphalt to be deployed in more infrastructure projects.

Given its versatile nature, the fact that it is a by-product of pyrolysis or gasification and the relative ease for its properties to be economically modified, biochar will play an important role as a sustainable and high-performance construction material. The use of biochar to enhance ACC is a good example; although ACC is not a new concept, more effort in understanding how biochar (internally carbonated or otherwise) can affect the effectiveness of carbon capture and strength enhancement is required. As more researchers around the world start working on this topic, the global community may consider collaborating and studying how biochar made from a wide range of feedstock can contribute to ACC as a way of reducing the net greenhouse gas emissions of concrete around the world.

References

Abdul-Malak MAU, Papaleontiou CG, Fowler DW, and Meyer AH 1988 Investigation of the frictional resistance of seal coat pavement surfaces. *Unpublished Report*, https://library.ctr. utexas.edu/digitized/texasarchive/phase2/490-1.pdf (accessed March 15, 2023).

Ahmad S, Khushnood RA, Jagdale P, Tulliani JM, and Ferro GA 2015 High performance self-consolidating cementitious composites by using micro carbonized bamboo particles. *Materials and Design*, 76, 223–229.

Ahammed MA and Tighe SL 2009 Early-life, long-term, and seasonal variations in skid resistance in flexible and rigid pavements. *Transportation Research Record* 2094, 112–120.

Brena SF, Bramblett RM, Wood, SL, and Kreger ME 2003 Increasing flexural capacity of reinforced concrete beams using carbon fiber-reinforced polymer composites. *Structural Journal* 100, 36–46.

Breton D, et al 1993 Contribution to the formation mechanism of the transition zone between rock – cement paste. *Cement and Concrete Research* 23, 335–346.

Choi WC, Yun HD, and Lee JY 2012 Mechanical properties of mortar containing bio-char from pyrolysis. *Journal of the Korea Institute for Structural Maintenance and Inspection* 16, 67–74.

Crouch LK, et al 2002 Determining air void content of compacted hot-mix asphalt mixtures. *Transportation Research Record* 1813, 39–46.

Cui S, Blackman BR, Kinloch AJ, and Taylor AC 2014 Durability of asphalt mixtures: Effect of aggregate type and adhesion promoters. *International Journal of Adhesion and Adhesives* 54, 100–111.

Dahir SH 1993 Criteria for the Selection of Bituminous-Aggregate Combinations to Meet Pavement Surface Requirements. Proceedings of the First Palestinian Convention in Civil Engineering. https://hdl. handle.net/20.500.11888/9177 (accessed March 15 2023).

Dunn JB, et al 2015 *Material and energy flows in the production of cathode and anode materials for lithium ion batteries (No. ANL/ESD-14/10 Rev)*. Argonne National Lab (ANL), Argonne, IL (United States).

Gupta S, Kua HW, and Dai Pang S 2018 Biochar-mortar composite: Manufacturing, evaluation of physical properties and economic viability. *Construction and Building Materials* 167, 874–889.

Gupta S and Kua HW 2019 Carbonaceous micro-filler for cement: Effect of particle size and dosage of biochar on fresh and hardened properties of cement mortar. *Science of The Total Environment* 662, 952–962.

Gupta S and Kua HW 2020a Application of rice husk biochar as filler in cenosphere modified mortar: Preparation, characterization and performance under elevated temperature. *Construction and Building Materials* 253, 119083.

Gupta S and Kua HW 2020b Combination of biochar and silica fume as partial cement replacement in mortar: performance evaluation under normal and elevated temperature. *Waste and Biomass Valorization* 11, 2807–2824.

Gupta S 2021 Carbon sequestration in cementitious matrix containing pyrogenic carbon from waste biomass: A comparison of external and internal carbonation approach. *Journal of Building Engineering* 43, 102910.

Harrigan E, Leahy R, and Youtcheff J 1994 Superpave manual of specifications, test methods and practices. SHRP-A-379. Strategic Highway Research Program.

Huang B, et al 2020 A life cycle thinking framework to mitigate the environmental impact of building materials. *One Earth* 3, 564–573.

Jia Y, Aruhan B, and Yan P 2012 Natural and accelerated carbonation of concrete containing fly ash and GGBS after different initial curing period. *Magazine of Concrete Research* 64, 143–150.

Junior AN, Filho RDT, Fairbairn EDMR, and Dweck J 2014 A study of the carbonation profile of cement pastes by thermogravimetry and its effect on the compressive strength. *Journal of Thermal Analysis and Calorimetry* 116, 69–76.

Kamel N, Musgrove GR, and Rutka A 1982 Design and performance of bituminous friction-course mixes. In: *Asphalts, Asphalt Mixtures, and Additives, Transportation Research Record* (No. 843). pp40–50.

Kennedy TW, Roberts FL, and Lee KW 1983 Evaluation of moisture effects on asphalt concrete mixtures. In: *Asphalt materials, mixtures, construction, moisture effects and sulfur, Transportation Research Record* (No. 911). pp134–143.

Kim YY, Lee KM, Bang JW, and Kwon SJ 2014 Effect of W/C ratio on durability and porosity in cement mortar with constant cement amount. *Advances in Materials Science and Engineering* 2014, 273460.

Kua HW 2015 Integrated policies to promote sustainable use of steel slag for construction—A consequential life cycle embodied energy and greenhouse gas emission perspective. *Energy and Buildings* 101, 133–143.

Kua HW, Gupta S, and Koh ST 2020 Review of biochar as a sustainable mortar admixture and evaluation of its potential as coating for PVA fibers in mortar. In: *Biochar Emerging Applications*, Bristol: OP Publishing Ltd. pp10.1–10.15.

Liu X and Wu S 2011 Study on the graphite and carbon fiber modified asphalt concrete. *Construction and Building Materials* 25, 1807–1811.

Lo TY and Cui HZ 2004 Effect of porous lightweight aggregate on strength of concrete. *Materials Letters* 58, 916–919.

Ma F, et al 2022 Biochar for asphalt modification: A case of high-temperature properties improvement. *Science of the Total Environment* 804, 150194.

Malaiskiene J, et al 2017 The influence of aggregates type on W/C ratio on the strength and other properties of concrete. *IOP Conference Series: Materials Science and Engineering* 251, 012025.

Masadeh S 2015 The effect of added carbon black to concrete mix on corrosion of steel in concrete. *Journal of Minerals and Materials Characterization and Engineering* 3, 271.

Mehta PK 1986 *Concrete: Structure, Properties, and Materials*. Englewood Cliffs: Prentice Hall.

Muthukrishnan S, Gupta S, and Kua HW 2019 Application of rice husk biochar and thermally treated low silica rice husk ash to improve physical properties of cement mortar. *Theoretical and Applied Fracture Mechanics* 104, 102376.

Ok YS, Tsang DC, Bolan N, and Novak JM 2018 *Biochar from Biomass and Waste: Fundamentals and Applications*. Amsterdam: Elsevier.

Praneeth S, Guo R, Wang T, Dubey BK, and Sarmah AK 2020 Accelerated carbonation of biochar reinforced cement-fly ash composites: enhancing and sequestering CO_2 in building materials. *Construction and Building Materials* 244, 118363.

Rajib A, Saadeh S, Katawal P, Mobasher B, and Fini EH 2021 Enhancing biomass value chain by utilizing biochar as a free radical scavenger to delay ultraviolet aging of bituminous composites used in outdoor construction. *Resources, Conservation and Recycling* 168, 105302.

Restuccia L and Ferro GA 2016 Promising low cost carbon-based materials to improve strength and toughness in cement composites. *Construction and Building Materials* 126, 1034–1043.

Rostami V, Shao Y, Boyd AJ, and He Z 2012 Microstructure of cement paste subject to early carbonation curing. *Cement and Concrete Research* 42, 186–193.

Ruth BE 1985 Evaluation and Prevention of Water Damage to Asphalt Pavement Materials: A Symposium. *ASTM International No. 899*, Philadelphia: ASTM.

Sičáková A and Špak M 2019 The effect of a high amount of micro-fillers on the long-term properties of concrete. *Materials* 12, 3421.

Song YS, Youn JR, and Gutowski TG 2009 Life cycle energy analysis of fiber-reinforced composites. *Composites Part A: Applied Science and Manufacturing* 40, 1257–1265.

Valdes-Vidal G, Calabi-Floody A, Sanchez-Alonso E, and Miro R 2019 Effect of aggregate type on the fatigue durability of asphalt mixtures. *Construction and Building Materials* 224, 124–131.

Wang D, Liu P, Xu H, Kollmann J, and Oeser M 2018 Evaluation of the polishing resistance characteristics of fine and coarse aggregate for asphalt pavement using Wehner/Schulze test. *Construction and Building Materials* 163, 742–750.

Wang L, et al 2020 Biochar as green additives in cement-based composites with carbon dioxide curing. *Journal of Cleaner Production* 258, 120678.

Yang C, et al 2019 Abatement of various types of VOCs by adsorption/catalytic oxidation: A review. *Chemical Engineering Journal* 370, 1128–1153.

Yanturina RA, Trofimov BY, and Ahmedjanov RM 2017 The influence of graphite-containing nano-additives on thermo-frost resistance of concrete. *Procedia Engineering* 206, 869–874.

Zhang MH and Gjorv OE 1990 Microstructure of the interfacial zone between lightweight aggregate and cement paste, *Cement and Concrete Research* 20, 610–618.

Zhang R, Dai Q, You Z, Wang H, and Peng C 2018 Rheological performance of bio-char modified asphalt with different particle sizes. *Applied Sciences* 8, 1665.

Zhang K, Lim J, Nassir, S, Englund K, and Li H 2019 Reuse of carbon fiber composite materials in porous hot mix asphalt to enhance strength and durability. *Case Studies in Construction Materials* 11, 00260.

Zhou X, Moghaddam TB, Chen M, Wu S, and Adhikari S 2020 Biochar removes volatile organic compounds generated from asphalt. *Science of the Total Environment* 745, 141096.

29

Biochar as an animal feed ingredient

Roger Hegarty, Stephen Joseph, Melissa Rebbeck,
Sarah Meale, and Nicholas Paul

Introduction

Biochar is increasingly being included in livestock diets in pursuit of productivity improvement and enteric methane mitigation (Kalus et al, 2019; Schmidt et al, 2019; Man, 2021). Results are variable, reflecting the diversity in structure and properties of the C matrix due to variable pyrolysis conditions, as well as by the original composition of polymers and mineral composition of the feedstock. This assessment considers the digestive processes in monogastric and ruminant livestock, and how dietary inclusion of pyrolysis products, principally biochar, can affect the digestive process in ways that may alter productivity and enteric emission from agricultural livestock.

Livestock nutrition and digestive physiology

The growth and productivity of livestock are dependent upon their genotype, their physical and social environment, and their ability to ingest and digest nutrients. Components of the feed may have roles as substrates (e.g., proteins, carbohydrates, fats, and structural minerals), cofactors (such as Mg, Zn), vitamins, or inert carriers (such as indigestible fiber) (Hynd, 2019). As animal nutrition becomes increasingly managed, additives such as pre-, pro-, and anti-biotics and exogenous enzymes have assumed increasing importance in altering gut microbiota to improve nutrient availability and livestock performance (Michalak et al, 2021). In-feed therapeutics or functional feed additives have also become popular, delivered either as pure compounds (essential amino acids, fatty acids, vitamins), crude extracts, or whole natural products (Kasapidou et al, 2015), as is the case for essential oil extracts. Biochar is increasingly being considered as

DOI: 10.4324/9781003297673-29

a functional feed additive to deliver direct health and production advantages, not just to the animal itself, but also indirectly to the farming system by influencing soil attributes and plant growth via manures (Kalus et al, 2019; Schmidt et al, 2019).

Production and health impacts of dietary biochar are moderated by the digestive anatomy and physiology of the animal consuming it. The digestive anatomy of livestock fundamentally divides agricultural livestock into monogastrics (or simple-stomached animals) and ruminants (with a voluminous fermentation vat before the true or acid stomach). All livestock require energy, amino acids, minerals, and vitamins, although some amino acids and vitamins can be produced by gut bacteria, making it not essential that they be included in the diet.

Monogastric digestion

Monogastric livestock such as poultry, pigs, and fish rely on nutrients ingested in the diet, that are dissolved or enzymically liberated after ingestion, before their absorption, primarily from the small intestine (Hynd, 2019). Some monogastrics are omnivorous such as poultry and pigs, and even in industrial agriculture their diet often contains both plant (such as cereal grains) and animal matter (such as fish meal).

In pigs, the mean retention time of digestion through the tract is approximately 40 hours (Wilfart et al, 2007). Digestion starts in the oral cavity where teeth grind the food and saliva is secreted. The oral cavity serves in the prehension of feed and also for particle size reduction through grinding. Once food is chewed and mixed with saliva, it passes through the pharynx and the esophagus to the stomach. The movement through the esophagus involves muscle peristalsis, which is the synchronized contraction and relaxation of muscles to move food. The porcine stomach has four distinct areas which include the esophageal, cardiac, fundic, and pyloric regions. Mucus is secreted and mixed with the digested food in the cardiac region. Food then passes into the fundic region where gastric glands secrete hydrochloric acid, resulting in a pH of 1.5 to 2.5 and the secretion of pepsin, which enzymically hydrolyzes proteins. Finally, the digesta moves to the pyloric region where mucus is secreted to line the digestive tract to prevent damage from low pH digesta as it passes to the small intestine. The small intestine is the major site of nutrient absorption, enabled by pancreatic-sourced amylases, lipases, and proteases, with fats emulsified by bile salts from the liver. Finally, the undigested material and unabsorbed digestion products flow to the large intestines, the principal site from which water is adsorbed, and also volatile fatty acids (VFA) produced by the gut biota.

Chickens are omnivores with a rapid passage of digesta, so non-digested material starts to be voided only 3 hours after ingestion (mean of 177 min), and the average retention of indigestible material is 5–6 hours in commercial poultry production (mean of 340 min; Svihus and Itani, 2019). Food taken in by the beak is blended with saliva in the mouth and passed via the esophagus to the crop, an expandable storage compartment where feed can remain for up to 12 hours. However, this prolonged storage does not occur in industrial poultry raising where feed is always available. The food passes from the crop into the bird's acid stomach (proventriculus) where digestive enzymes are added, then passes to the muscular gizzard where physical grinding of the food occurs, in a manner somewhat analogous to chewing in mammals. The gizzard is a muscular organ that uses retained pebbles or sand particles to grind seeds and fiber into smaller, more digestible particles. From the gizzard, food passes into the small intestine where nutrients are absorbed. The residue then passes into the hindgut and can be diverted into paired cecal blind sacs along the lower intestinal tract, where bacteria ferment undigested food. Cecal reflux back up the gut ensures microbial products can be absorbed from the small intestine. From the caeca, digesta ultimately moves to the large intestine which absorbs water and VFA before the excretion of undigested residue via the cloaca.

The feeding habits of fish vary from predation through omnivory to herbivory, including filter-feeding to sieve plankton, so the digestive anatomy of fish varies accordingly. Consequently, the range of transit times through the gut is large (0.5 to >3 days) and depends on fish size, diet, and temperature. The general structural components of the fish digestive system include the mouth, teeth and gill rakers, esophagus, stomach, pylorus, pyloric caeca, pancreatic tissue (exocrine and endocrine), liver, gall bladder, intestine, and anus. Not all the above components of the digestive system are present in all fish. There is a great diversity in the length and development of the gastrointestinal tract in fish that reflects the nature of the diet. In most carnivorous fish, it is a simple straight or curved tube or pouch with a muscular wall and a glandular lining. Food is largely digested there and leaves the stomach in liquid form into the small intestine where many fish have outpouching into pyloric caeca, which secrete digestive enzymes and absorb nutrients. The distinct large intestine of other livestock is lacking in all fish.

Ruminant digestion

Ruminants have an expanded pre-gastric foregut, with food passing from the mouth to the reticulum via the esophagus and into the rumen or paunch (Van Soest, 1994). The mean retention time of feed particles in the rumen is approximately 24 hours, during which time feed is both chemically degraded by rumen microbes and physically degraded by rumination, which is the return of the digesta to the mouth for further chewing. Microbial fermentation primarily occurs in the rumen (2nd stomach) where the more than 10^{10} bacteria in each milliliter of rumen liquor are largely strict- or facultative-anaerobes in keeping with the low redox potential in the anaerobic rumen (Henderson et al, 2015). Anaerobic protozoa, yeasts, fungi, and a profusion of ruminal viruses are also present. Microbial fermentation converts dietary carbohydrates including cellulose and hemicellulose into VFA or CO_2. Extensive rumen microbial proteolysis causes at least half of dietary protein to be deaminated to ammonium (NH_4^+) in the rumen (Stern et al, 1994). The energy liberated by these catabolic processes is used in the synthesis of new microbial cells.

Microbial fermentation also liberates heat as the feed is biochemically reduced during fermentation in the anaerobic environment. Residual electrons from substrates are transferred via intermediate electron acceptors (NAD^+, FAD^+) into synthetic processes such as lactate or ethanol production, or onto protons to make H_2, both ultimately regenerating NAD^+ to sustain glycolytic ATP production. In the rumen, electrons for microbial methane (CH_4) production are sourced from H_2 or the metabolism of lactate or ethanol to propionate. Approximately 6% of the gross energy of feed eaten is lost as CH_4. Rumen headspace gas is approximately 60% CO_2, 30% CH_4, and less than 2% H_2 by volume (Hegarty and Gerdes, 1999).

However, if H_2 accumulates, feedback inhibition of fermentation can suppress feed intake and animal performance but not always (Martinez-Fernandez et al, 2014) suggesting there may be other routes to regenerate NAD^+ in the rumen, allowing some fermentation pathways to continue. Consequently, at the biochemical level, it is the distribution of electrons released by fermentation into terminal reduction reactions generating CH_4, VFA, and fatty acid saturation (Jenkins et al, 2008) that determines enteric CH_4 production, with CH_4 production being the largest sink for rumen electrons. Methane production has been nature's principal way of disposing of electrons (and H_2) in the rumen so that fermentation can proceed.

Post-rumen, the digesta is passed through the reticulum (1st stomach; honeycomb tripe), exiting via the reticulo-ruminal orifice to enter the omasum (3rd stomach; book tripe) from where water is absorbed, before flowing to the abomasum (4th or acid stomach). Digestive proteases in the abomasum and upper small intestine complete enzymic digestion, with the absorption of amino acids, vitamins, most minerals, and residual sugars in the small intestine. As with most animals, the hindgut is largely for water resorption, although some fermentation and VFA absorption occur.

Historic uses of biochar in animal production

A series of reviews on the use of biochar (often called "charcoal" in this context; here, we use the term "biochar", more explanation in Box 1.1 in Chapter 1) in animal feed have been published since 2019, highlighting the recent interest in animal feed applications (Schmidt et al, 2019; Man et al, 2021; Graves et al, 2022). One of the first references to the use of biochar as an amendment for animal digestive disorders was given in the ancient Roman treatise "On Agriculture" by Cato (Bonner, 2012). He recommended mixing the biochar with wine, salt, and laurel leaves. Over the centuries biochar was administered medicinally for internal disorders in pure form or was mixed with clay, medicinal herbs, butter for cows, eggs for dogs, ash for pigs, and meat for cats (Day, 1906). In the early 1900s, tonics were also prepared by adding biochar with other ingredients, including spices such as cayenne pepper, and digestive bitters like gentian, to reduce digestive disorders, increase appetite, and improve milk production (Pennsylvania State College, 1922). To increase butter-fat, farmers would add charcoal to distillers grain, wheat bran, oats, hominy, and cottonseed meal (Savage, 1922). Biochar and activated carbon (e.g., produced by pyrolyzing biomass at high temperatures and then modifying the carbon either chemically or with steam) were recommended by German veterinarians for reducing and adsorbing pathogenic clostridial toxins from *Clostridium tetani* and *Clostridium botulinum* (Skutetzky and Starkenstein, 1914; Luder, 1950; Mangold, 1936). Similar results were reported for poultry (Totusek and Beeson, 1953; Steinegger and Menzi, 1955). It has also been reported that wild animals, such as deer, eat char left from wildfires (Struhsaker et al, 1997). Similarly, monkeys eat charcoal from traditional charcoal kilns to help digest young Indian Almond (*Terminalia catappa*) or mango (*Mangifera indica*) leaves that contain anti-nutritional phenolic compounds (Cooney and Struhsaker, 1997).

Recent uses of biochar to enhance animal productivity and health

The most recent reviews have shown that biochars (variably called charcoal, activated carbon, activated charcoal, pyrogenic organic matter, pyrogenic carbonaceous matter, etc. see Chapter 1 for nomenclature), can significantly affect animal productivity (weight gain, milk and egg weights and quality, meat quality, feed utilization), as well as manure quality, CH_4 and NH_3 production, and can change microbial communities and resistance to specific pathogens (Schmidt et al, 2019; Man et al, 2021; Graves et al, 2022). The large range of responses observed reflects a broad range of livestock and varied forms of biochar being tested in a relatively limited number of feeding trials (albeit >100). Unlike soil or manure stores, the environment of the gut is dynamic with continuous movement of digesta, and thus reactions may not come to completion before passing further through the gut where conditions differ. The unpredictability of the outcome also highlights that there is much to learn about the functional effects of the different forms of biochar, as well as potential indirect or flow-on effects in different biological systems. The following sections summarize research on biochar use in diets for specific animal production systems and postulate mechanisms for both the positive and sometimes negative effects of biochar.

Monogastrics

Effect of biochar on health, productivity, and product quality in poultry

Several studies have found that biochar made from different plant materials can reduce disease and increase bird and egg weight in poultry. Bamboo biochar with wood vinegar (an acetate-containing, condensation co-product of pyrolysis (Chalermsan and Peerapan, 2009) was fed to 124, four-week-old Betong chickens at 0, 0.5, 1, and 1.5% dietary DM in a corn-soybean ration (57–68% corn, 19–25% soybean meal) *ad libitum* for 16 weeks (Rattanawut et al, 2014). Colony counts of fecal *Escherichia coli* (7.74 v 7.87 $\log_{10}$ CFU g^{-1}) and *Salmonella* spp. (7.29 v 7.39 $\log_{10}$ CFU g^{-1}) were lower (P<0.05) when 1 and 1.5% w/w biochar with wood vinegar were included in the diet for 140 days. The use of wood and green waste biochar was compared with bentonite or zeolite (commercial adsorbents) to suppress pathogen loads in laying chickens while maintaining microbial richness (Prasai et al, 2017). The supplementation rate for each additive was 4% of diet DM, and all three treatments showed potential for the reduction of major poultry zoonotic pathogens without altering the diversity of microbiota. In a further study, commercial green waste wood biochar (76% C, 3.2% H, 0.29% N) included at 4% of total feed increased total egg weight by 5% but bird weight and the number of eggs were unchanged (Prasai et al, 2017). They also reported that biochar reduced the relative abundance of Proteobacteria, Gammaproteobacteria, and Campylobacter bacteria in the chicken cloaca by about half. Some of these bacteria can contaminate eggs and are associated with foodborne illness.

A similar study, where wood biochar was fed at 3% w/w to poultry over 42 days, found that biochar reduced the productive performance of broilers during the starter phase, but enhanced growth during the grower-finisher period, resulting in a better lifetime performance and feed conversion efficiency, with some changes in the gut microbial communities (Goiri et al, 2021). Biochars are also used commercially to remove NH_3 emissions from poultry sheds (Asada et al, 2002) to improve bird health.

Use of biochar for horses

Biochar-based products are now being sold as an amendment to assist horses in digesting high-protein feed. However, there are no independent reports on the efficacy of these commercial products. Only one study has been published on the effects of biochar on total tract apparent digestibility of nutrients, fecal characteristics (pH, concentrations of VFA and NH_3-N), and blood serum parameters (Joch et al, 2022). When horses fed hay and barley (15 g feed kg^{-1} of horse body weight) were supplemented with 10 g biochar kg^{-1} feed DM of high temperature (900°C) spruce wood biochar, the digestibility of all nutrients (except starch) was reduced while fecal pH was higher in 3 of the 8 the biochar-fed horses compared and was higher than from no-biochar controls. There was no effect of biochar on fecal concentration of total VFA, proportions of individual VFA, fecal NH_3-N, or serum concentrations of metabolites, minerals, or vitamins, and no evidence of changed protein digestion.

Use of biochar in aquaculture

Biochars have been used for both nutritional enhancement as well as a method for removing toxins, nutrients, and algae from aquaculture tanks, with several studies reporting that feeding fish with biochar improved growth. Average tilapia fingerling weight increased from 8 to 11 g above control over a 6-week period when water hyacinth biochar was fed at 1% w/w (Najmudeen et al, 2019). Similarly, carp

fingerling weight increased from 6 to 12 g above controls over a 13-week period when manure- and vegetable waste-based biochars were fed at 2 mg kg^{-1} (Khalid et al, 2022), while tiger pufferfish had a weight gain improvement of 122% when 4% w/w bamboo biochar was included in the fish-based diet (Thu et al, 2009). A significantly higher feed intake and feed efficiency were also reported with increased protein efficiency ratios (2.15 versus 1.9 for the control) (Thu et al, 2009). Similar findings using rice husk biochar were reported for trout (Khaki et al, 2017) and for striped catfish (Lan et al, 2021). The mechanism associated with the differences in yield and protein efficiency could be explained by direct changes in microbial communities in the fish's stomach, by changed enzyme activity, especially of lipase, and indirectly by changes to water quality. Amending the soil of tilapia ponds (rather than the diet) with 1% w/w biochar caused a growth benefit (Najmudeen et al, 2019). In contrast, adding up to 4% w/w bamboo biochar did not affect the growth or overall muscle fatty acids of juvenile carp, although blood transaminase activity was reduced and blood glucose increased by 1–4% w/w biochar while changes in intestinal histology were inconsistent (Mabe et al, 2018). The use of biochar to improve the production of other important aquaculture species, such as crustaceans including shrimp and prawns, has not been reported.

There are potentially other indirect benefits from using biochar in aquaculture settings. In fact, most evaluations of biochar in aquaculture are investigations of the use of biochar for water quality improvement, either for production ponds to improve the management of waste (Mahari et al, 2022) or treating soils in preparation for aquaculture (Raul et al, 2021). In one of the few studies partitioning the integrated benefits of biochar as a growth-promoting feed additive with water treatment impacts, Lan et al (2021) highlight total ammoniacal N in tank water for fish was reduced by the inclusion of biochar in feed, more so than if biochar was added directly to the water, which is important as this toxic form of dissolved N can impact larval stages. Removal of heavy metals and other pollutants from the production water to avoid accumulation in the saleable product is another role for biochar in quality improvement in aquaculture (Chen et al, 2022). An additional indirect effect of adding biochar to the water is to induce change in the species composition of microbial aggregates available to, and the gut biota of, filter feeders and omnivores like tilapia, although this biome change did not change levels of digestive amylase or protease activity in the fish (Abakari et al, 2021).

These direct and indirect effects on fish production and the environment mean that more work is needed to identify opportunities to include biochar in diets in appropriate forms, and to explore broader immune responses and health benefits, similar to those in other livestock. Likewise, more research is required in aquaculture to transfer the positive research results for biochar in removing antibiotics or other pollutants reported in laboratory settings (Mrozik et al, 2021) into industry practice. This includes understanding the potential role of biochar in the N cycle of aquaculture by facilitating denitrification in place of existing growth media (Paul et al, 2021). More research also needs to be undertaken on biochar feed additives for other major aquaculture species including shrimp and prawns. It is interesting to note that a small amount of research has been carried out on pyrolyzing crustacean shells to produce biochar for soil applications but not for feed and shrimp feed applications (Mahari et al, 2022).

Ruminants

Use of biochar in the dairy and beef industries

A small number of studies have been carried out feeding a range of biochars to dairy cows and beef cattle. Initial beef growth results on cassava-based diets were positive if not always significant (e.g., Leng et al, 2012; Phongpanith and Preston, 2018), but production may have been due to biochar detoxifying the cyanide in cassava as much as by enhanced nutrient status. The use of 2% dietary biochar over 13 months had no effect on the growth, feed conversion ratio, or carcass attributes of Hanwoo steers on a grain-based diet (Kim and Kim, 2005). Feeding a roughage-based starter diet followed by a corn grain-based finisher diet to well-grown steers found almost no significant effects of biochar on DMI or digestibility on the roughage starter diet or the grain-based finisher diet, except a linear reduction in fiber- and energy-digestibility on the finisher ration from 0–3% pine-trees derived commercial biochar inclusion (85% C, 0.7% N) (Winders et al, 2019).

Supplementing dairy cows with activated carbon (0.5% w/w) did not increase milk production significantly, but significantly increased milk protein and fat content by 2.6% and 6.3% respectively, although the treatment effects were confounded by time (Al-Azzawi et al, 2021). A subjective opinion of 21 dairy farm managers (Gerlach and Schmidt, 2012) identified they believed milk protein and fat were increased by biochar feeding. In contrast, a 9-month study at a 250-head dairy farm where a mixed feedstock biochar (50% eucalyptus wood chip, 25% soybean residue, 350–500°C) was fed at 0.006% of DMI, found no effect on milk fat or milk protein production and a non-significant (2.2%) increase in milk yield compared with the previous 2 years. There was an increase in N, P, and some micro-nutrients in the manure but again, con-founding of biochar provision with time makes interpretation difficult, as all cattle were either on treatment or not on treatment at a given time, so there was no contempora-neous control in comparison with biochar (Taherymoosavi et al, 2022).

Using appropriate controls, the effect of feeding biochar (Commercial product; Norit RBAA-3, rod activated carbon) to Holstein dairy cows at 0, 20, or 40 g day^{-1} was investigated on a fungal-contaminated, poor-quality corn-silage diet (Erickson et al, 2011). Biochar improved feed intake of this potentially toxic (deoxynivalenol containing) diet, and increased milk fat content (quadratic response), suggesting potential protection from ingested mycotoxins, although none of these effects occurred in a subsequent experi-ment on a higher quality diet.

Manure-related impacts have also been associated with biochar feeding to cattle. A farm with 60 head of cattle supplemented with eucalyptus biochar (approximately 600°C) for 3 years, on fields where additional dung beetles had been added, reported an improvement in animal and pasture health (increased soil pH, total P and N), and a buildup of soil organic C. This study also reported that the farmer purchased less phar-maceuticals, fertilizer, or additional feed, and profitability had increased (Joseph et al, 2015). In the 9-month study mentioned pre-viously, the farmer had a 10% increase (AU

$22,000) in total income compared with the previous year, although seasonal effects would be confounded with all reported results (Taherymoosavi et al, 2022). There were increases in manure N, C, Na, Si, Ca, P, and K after feeding biochar to the cows, and increases in the concentrations of soil Ca, K, and P, and plant available N, P, and K were observed following the burial of biochar-infused manure. A dissolution test indicated lower concentrations of leachable Na (3.6%), K (10.3%), Si (0.14%), Ca (0.52%), and Fe (0.21%), but a higher proportion of Mg (5.9%) and P (2.7%) released at the end of the trial. Soil Mg, Na, Fe, extractable NH_4^+ and NO_3^-, and plant-available Ca, Mg, and Na were reduced after biochar was fed over 9 months (Taherymoosavi et al, 2022). The farmer reported that he had an additional two bales of hay as the pasture provided most of the dietary requirements throughout the winter.

Goats

The growth rate of goats increased when bamboo biochar was added to a para grass and acacia diet, at 0.5 g biochar kg^{-1} initial body weight (BW) (Mui and Leden, 2006). The diet was fed at 10 g DM kg^{-1} basal diet. Live weight gain increased by up to 20% day^{-1} compared to the control group and N uptake was greater. As the basal ration included a large proportion of tannin-rich acacia (*Acacia mangium*) leaves, the authors hypothesized that biochar eased the digestion of those leaves by sorption of their tannins, increasing the availability of dietary protein for digestion and absorption. When goats ingesting tannin-rich leaves were supple-mented with 1% w/w biochar, the weight gain of the animals increased by 27% (Silivong and Preston, 2015).

Bioactive properties of biochar

The mechanisms by which different dietary biochars can improve digestion, animal health, and reduce enteric CH_4 emissions within animals have not been examined in detail. Much of the published literature has focused on performance outcomes, including the output of live-weight eggs and milk, reduction in disease, and emissions of NH_3 and other gases in animal sheds. However, there has been little exploration of the attributes of the biochar used to explain the variation in efficacy.

Biochar is produced through the thermal decomposition of biomass in an oxygen-starved environment, or more simply, anaerobic heating (Chapter 3). Its main components are a persistent aromatic C structure, with minerals and some organic molecules that are water or solvent soluble (Joseph et al, 2013). Each of these components can potentially play a role in the efficacy of biochar when it is fed to animals. The effect of biochar on animal health, nutrient uptake, and methane production, are liable to be functions of the relative concentration of these components and the proportion of biochar in feed (Schmidt et al, 2019).

There are 5 classes of soluble organic compounds in biochars:

1   Biopolymers that have a cellulose-type structure which can be used as an energy source for microorganisms;
2   Larger macromolecules (>1000 daltons) with a high concentration of C and O function groups, which can facilitate biotic and abiotic redox reactions;
3   Polyphenols that can act as antioxidants and reduce stress in animals or if at too high a concentration can bind and reduce nutrient digestibility;

4   Low molecular weight acids, comprising acetic, propionic, up to nonadecanoic acid;
5   Other low-molecular-weight compounds.

A list of some of the compounds that have been found in commercial biochar is provided in Table 29.1 (after Fernandez et al, 2022). It should be noted that most biochars have a range of volatile fatty acids from C10 to C20. The concentration of these organic compounds is a function of the biochar feedstock, the final temperature at which pyrolysis occurs, and the temperature-time profile in the pyrolysis kiln. Some gaseous organic molecules found in biochar have been associated with emissions from healthy steers [methyl acetate, heptane, octanal, 2,3-butadione, hexanoic acid, and phenol ($P \leq 0.10$)] and some have been associated with clinically morbid steers [acetaldehyde ($P \leq 0.05$) and decanal ($P \leq 0.10$)] (Spinhirne et al, 2004). Some of these molecules have been found to play a role in reducing enteric CH_4 emission, including lauric (dodecanoic) acid and myristic acids (tetradecanoic acid) (Soliva, 2004). However, the concentration of these compounds in the biochar is very small (typically less than 1 mg kg^{-1}) and when fed at low levels it is very unlikely they would affect animal health, nutrition, or CH_4 production. Low levels of polyaromatic hydrocarbons have been found in biochar from sewage sludge and at high application rates these can be toxic (Zielińska and Oleszczuk, 2016).

The structure and properties of the C matrix are a function of the temperature of pyrolysis and of the original polymeric and ash composition of the feedstock with the degree of graphitization and the graphitic and defect structure likely to affect biochar properties (Chapter 3). High-temperature

Table 29.1 *Some of the organic compounds detected by gas chromatography and mass spectrometry (GC-MS) in a low-temperature wood biochar (after Fernandez et al, 2022)*

2,4-Dimethyl-1-hexene	Nonadecane
2-Propanol, 1,3-dichloro-	9-Octadecenoic acid (Z)-, methyl ester
Benzene, propyl-	9-Octadecenoic acid (Z)-, methyl ester
Furan, 2-pentyl-	9-Octadecenoic acid, 1-
2-Octenal(E)-	Ethyl Oleate
Benzene, butyl-	Octadecanoic acid, ethyl ester
6-Tridecene, (Z)-	9,12-Octadecadienoic acid (Z,Z)-
1-Phenyl-1-butene	Docosanedioic acid, dimethyl ester
Disiloxane, 1,3-diethoxy-1,1,3,3-tetramethyl-	Ethanol, 2-(tetradecyloxy)-
Benzene, pentyl-	9-Octadecenoic acid (Z)-, 2,3-dihydroxypropyl ester
Benzene, hexyl-	Ethanamin', 2,2'-oxybis[N,N-dimethyl-
2,4-Decadienal, (E,E)-	Hexadecanoic acid, 2-(octadecyloxy)ethyl ester
Benzene, heptyl-	Linoleic acid ethyl ester
Phenol, 2,6-dimethoxy-	1-Decanol, 2-hexyl-
Ethanone, 1-[4-(1-methylethyl)phenyl]-	Octanal, 7-methoxy-3,7-dimethyl-
1-Hexanone, 1-phenyl-	9-Octadecenoic acid (Z)-, 2,3-dihydroxypropyl ester
Benzene, octyl-	Heneicosane
Tetradecanoic acid	9-Octadecenoic acid (Z)-, 2,3-dihydroxypropyl ester
Nonadecane	Hexadeca-2,6,10,14-tetraen-1-ol, 3,7,11,16-tetramethyl-, (E,E,E)-
Pentadecanoic acid	Tetratriacontane
Hexadecanoic acid, methyl ester	Tridecanal
Pentadecanoic acid	3Beta-hydroxy-5-cholen-24-oic acid
7-Hexadecenal, (Z)-	Nonadecane
Hexadecanoic acid, ethyl ester	

biochars have a high content of graphitic type structures and a high electrical conductivity (Chapter 6), and therefore can transfer the conjugated π-electron system in the condensed polyaromatic C ring structures in the C matrices and potentially enhance biotic and abiotic redox reactions. Low-temperature biochars are amorphous semi-conductors with a high concentration of surface C and O functional groups and can transfer electrons due to the redox active quinone and hydroquinone groups (Sun et al, 2017). The zeta potential of most small biochar particles is negative and this can result in their ability to adsorb NH_4^+ (Xu et al, 2019) released from protein and urea metabolism. The smaller the particles the higher the zeta potential and the higher the critical coagulation concentration.

All biochars produced under 600°C have persistent free radicals (PFRs) associated with the C structure either as O-centered radicals, C-centered radicals, oxygenated C-centered radicals, or soluble organic compounds (Ruan et al, 2019). These PFRs can stimulate the formation of reactive O species (ROS) that have been associated with the degradation of organic and inorganic toxic compounds, with changes in microbial diversity, and reactions with proteins and hydrogen on the surface of nanoparticles. There may be scope for ROS to oxidize H_2 and CH_4 in the rumen but oxidizable compounds are abundant in the bulk digesta and the majority of PFR could be expected to be reduced in the large mass of anaerobic digesta. Because biochar can adsorb H_2 and CH_4 into the micropores (where liquids are not present) it is possible that these ROS will oxidize these gases.

The surface area and pore volumes of the C matrix of high-temperature woody biochars are much greater than those of the low-temperature, high-mineral ash biochars. These high-temperature biochars have a greater ability to adsorb low-molecular weight compounds into the mesopores of the C matrix. High-temperature biochar has a larger macropore volume than lower-temperature biochar (Chapter 5) but it is not clear whether higher temperature alone is responsible for a greater volume for microbial populations (Chapter 14). Bacteria enveloped in a biofilm can form on the surface and in larger pores of biochars as found after in-vitro fermentation (Figure 29.1).

(a) (b)

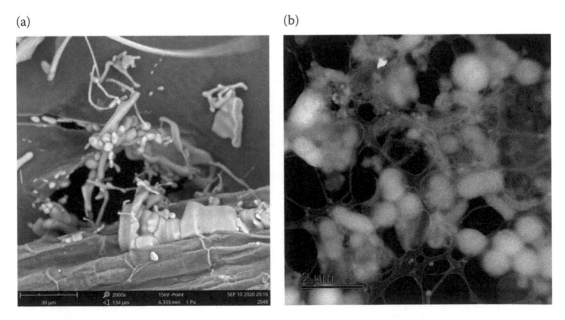

Figure 29.1 *Location of bacteria. (a) Cluster of bacteria enveloped in a biofilm formed on the surface of a wheat straw biochar taken after in-vitro fermentation; (b) bacteria growing in the biochar pore (authors images)*

Biochar can adsorb certain feed and metabolic substances including tannins, phenols, or thionin, which they hypothesized could further increase the electron buffering of biochar particles during their passage through the digestive tract (Schmidt et al, 2019).

The type and concentration of minerals in biochar also depend on feedstock and its pyrolysis temperature. Most minerals in biochars are microns or submicron in size and either exist inside the C lattice or sit on the internal pore surfaces or the external surface of the particle. Some minerals change their structure and oxidation state (e.g., kaolinite, Fe, C, or Mg oxides) during pyrolysis, due to the reaction with the H_2, CO, and CO_2 that is being released. The Eh of minerals is usually higher than that of the C matrix and this means the biochar is composed of domains where there is an electron potential difference that can drive the movement of cations and anions (Joseph et al, 2013). The rates at which these minerals are released are a function of the solution pH as well as of their particle size, and also of where and how the minerals are embedded in the C matrix (Dang et al, 2019). The growth of microorganisms occurs at the interface between the mineral and the C matrix and CO_2-fixing chemolithotrophic bacteria at the interface between a $FeSO_4$ of clays and the C matrix (Ye et al, 2017). Fungi can also grow in pores with a high concentration of Fe-O nanoparticles and Ca phosphate (Joseph et al, 2015). Some of these minerals are mesoporous and can adsorb gases produced during fermentation, and these gases can react or be oxidized or reduced at the surfaces of redox-active minerals (Quin et al, 2015). Some micron- and sub-micron sized mineral particles contain free radicals (especially Fe-O, TiO_2, and ZnO) and can facilitate chain reactions of radicals with organic molecules and oxidation of gases such as H_2 and CH_4, in some environments which may alter the abundance of specific micro-organisms

(Joseph et al, 2010), but there is no evidence of biochar causing oxidation of CH_4 in the rumen. It should be noted that, although the rumen is at a redox potential of around -300 mV, the redox potential inside the pores of the semiconductor minerals that have O in their lattice is positive. Methane and H_2 can be adsorbed into the pores of oxide minerals and potentially can be oxidized (Nan et al, 2022).

While biogas vessels and the rumen differ greatly in their mean retention time, and so in their anaerobic reactions, it is challenging to expect that biochar can stimulate CH_4 production in biogas systems but oxidize CH_4 in the rumen. However, the increased electron-donating capacity of biochars significantly increased CH_4 production in an in-vitro batch (non-rumen) fermentation experiment, but only after 120 days (Viggi et al, 2017). A reduction in CH_4 released in the first 20 days was attributed to the adsorption of CH_4 into the biochar. It should be noted that the ability to significantly adsorb different gases depends on the amount of biochar added in the feed and the speed at which these volatiles could be oxidized. Another mechanism could involve the adsorption of CO_2, H_2, H_2S, NH_3, or CH_4 into the pores of the biochar, along with some of the organic molecules in the ruminal fluids, resulting in a change in the pathway of rumen fermentation. Biochars can have a significant concentration of highly reactive free radicals on their surface and these could disrupt the cell walls of the methanogenic archaea (Zhao et al, 2020).

The properties of fresh biochar change very quickly when mixed in a liquid that is acidic with a high concentration of microorganisms and degrading biomass, such as digesta in the true stomach of animals. Initially, the soluble minerals and organic compounds dissolve, and in low pH environments the surface charge and area or porosity can increase along with the concentration of

C=O functional groups (Joseph et al, 2010). The Eh and pH of the fluid around the particle is likely to change, resulting in the adsorption of gases, the adsorption of anions and cations (Joseph et al, 2013), and the formation of an organic film over part of the biochar's internal and external surfaces (Hagemann et al, 2017). These films are composed of many different organic compounds along with metals and non-metals. Where specific micro-organisms colonize areas and where the local environment is favorable for growth, they may exude acids to dissolve nutrients on the biochar, leaving a residual polysaccharide biofilm. A range of other mechanisms can result in the bonding of cations and anions to biochar, including heavy metals and toxic organic compounds that may be ingested in plants and soil by animals. These mechanisms include ligand exchange, electron donor-acceptor, and electrostatic interactions. Gases that are produced in the fermentation process and specific toxins can also be adsorbed into this organo-mineral layer (Kumar et al, 2020).

In summary, the three biochar components of C backbone, minerals, and embodied organic molecules differ greatly depending on the feedstock material and pyrolysis conditions. When ingested, the biochar can have direct effects on animal production by changing anion and cation availability, redox state, and electron flux, by supporting bioactive organic molecule introduction, as well as having indirect effects through altering the gut biota and potentially the biofilm structure. Further work is required to characterize the detailed biotic and abiotic changes that occur when biochar is added.

Mechanisms underlying dietary biochar amelioration of toxicities and enteric CH$_4$ emissions

As identified above, biochar as a feed ingredient can modify aspects of the chemical and biotic environment in the digestive tract. Below is an assessment of the mechanisms contributing to biochar amelioration of toxicity and potentially of enteric CH$_4$ production in ruminant livestock.

Protection from dietary toxins

Reviews of the use of biochar to passivate heavy metals from wastewater (Wang et al, 2019), in soil remediation (Cheng et al, 2020), and in manure stores (Zeng et al, 2018), identify a key direct chemical detoxifying action. However, its scope to limit heavy metal absorption from the gut digesta in animals is less well studied. The inactivation of organic toxins in the gut has been studied (McKenzie, 1991), suggesting both the direct binding of the toxin to the biochar and indirect inactivation of the toxin by the modified biota associated with biochar-amended soils (Lehmann et al, 2011). The expanded surface area and differing redox environment provided by biochar may support biofilms on feed and digesta particles that have substantially different degradative pathways than digesta or manures without these additives (Leng, 2017).

Reducing enteric CH$_4$ emission from ruminants

There have been more *in vitro* studies of biochar efficacy in enteric CH$_4$ mitigation than studies in animals, and in general biochar efficacy is less in animals than *in vitro*

(Fernandez et al, 2022), as occurs for most additives. Canadian studies with 3 spruce-derived biochars (450°C, zinc chloride, or acid treated) showed no effect on in-vitro fermentation or CH_4 production (Tamayao et al, 2021). In vitro assessment of biochars from 5 substrates across 2 production temperatures (550°C, 700°C) showed neither substrate nor temperature-affected total gas production nor CH_4 content of the total gas, but 550°C supported a lower methane production (ml/incubation) than did 700°C. A recent *in vitro* study of 12 different biochars also showed that CH_4 reduction was determined by thermal treatment procedure, feedstock source, post-treatment (especially with the addition of very small amounts of nitrates (0.1% approx.), and application rate (Fernandez et al, 2022). In their *in vitro* experiment, reduction in CH_4 emissions varied from 18–32% for four of the 12 biochars, with the highest reductions observed at 5% w/w for three of the biochars and 1% w/w for the fourth biochar added to oaten hay. Two of the biochars were then selected for measuring *in vivo* CH_4 emissions with animals held in an open circuit respiration chamber. An acidified wood biochar produced at 450°C for approximately 12 hours decreased CH_4 production (g methane kg^{-1} dry matter intake) by 9–10% compared to the control. Biochar made from wheat straw and wood biomass mixed with minerals, pyrolyzed at 600°C and post-treated with acid and nitrate solutions (appr. 1% of dry matter feed) reduced CH_4 by 10–13% compared to a control in sheep (Fernandez et al, 2022). It should be noted that some of the reduction is due to the nitrate acting as an alternative electron acceptor.

The highest reported mitigation from biochar's fed to cattle have come from studies with very short-term measures (Leng et al, 2012) or where biochar and control diets were not fed concurrently but rather sequentially, confounding biochar and seasonal impacts (Al-Azzawi et al, 2021). These activated carbons are usually expensive (>US$2000 t^{-1}) and thus may only be a commercially viable proposition when policy, programs, or carbon markets reward farmers for reducing CH_4. In the modest amount of animal research that has been conducted using biochar, diversity in the source material, times and temperatures (400–600°C) of pyrolysis, and particle sizes have been used to feed ruminants, which have not shown significant reductions in CH_4. Correspondingly, the mitigation potential of biochar has been judged to be conservative (Hegarty et al, 2021) with the metanalysis of Honan et al. (2021) metanalysis reporting a mean difference from biochar (in 6 studies) of -10 g while Black et al (2021) ascribed 0% mitigation potential to biochar in Australia. However, there is some evidence of efficacy and it is clear that purpose-engineered (and potentially expensive) biochars can result in mitigation.

In pursuing biochar-induced mitigation of enteric CH_4 emissions, three potential mechanisms should be considered:

1   Physical and ecological changes: feeding biochar introduces a new extensive surface area for microbial colonization and thus creates a new habitat in the rumen. This is a new component in the ecosystem and may complement or change the ecology of the biofilms that coat the rumen and feed surfaces in the rumen (Leng, 2017), ultimately changing the net rumen microbial ecology and chemical ecology (Terry et al, 2019; Teoh et al, 2019, Al-Azzawi, 2021). While many feed additives alter the gut microbiome, this is not always reflected in changed methanogenesis (Terry et al, 2019). While little understood, the multi-layer biofilms that build as microbes colonize

substrates and dissipate as the substrate is depleted could be changed by biochars providing not only a new expansive habitat surface, but also a new electron sink.

2   Chemical and metabolic changes: feeding biochars that have a high concentration of microporous mineral oxides or have been treated with ozone or other oxidizing compounds (Huff et al, 2018) may facilitate the adsorption of CH_4 and H_2 into the micro and mesopores of the biochar (Nan et al, 2022). It should be noted that the redox potential on the surface of micropores is positive where either there is a high concentration of O functional groups or where there are mineral oxide particles.

To support CH_4 oxidation or to suppress CH_4 production, the chemistry of abiotic reactions on the surface of biochar in an anaerobic environment that may affect these processes, via electron transfer as hypothesized, must be understood. Fermentation in the O_2-starved rumen involves the breakdown of the hexose and pentose sugars bound up in dietary starch, cellulose, and hemicellulose. In the process, VFA, CO_2, and H_2 are produced, and in the chemically reducing environment, CO_2 and H_2 react to form CH_4 (Equations 29.1, 29.2).

$$CO_2 + 4H_2 \leftrightarrow CH_4 + 2H_2O \quad [29.1]$$

$$CO_2 + 8H^+ + 8e^- \leftrightarrow CH_4 + 2H_2O \quad [29.2]$$

Hydrogen gas is formed through the reduction of H^+ with the involvement of ferredoxins as a catalyst and the enzyme hydrogenase (Equation 29.3). Hydrogen can also be dissociated to H+ using an electron acceptor, typically NAD^+, that is reduced by the electrons derived from H_2 (Equation 29.4).

$$D_{red} + 2H^+ \rightarrow H_2 + D_{ox} \quad [29.3]$$

$$H_2 + A_{ox} \rightarrow A_{red} + 2H^+ \quad [29.4]$$

A = electron acceptor (e.g., NAD^+, O_2, CO_2, nitrate, sulfate, and fumarate)

D = electron donor (coupled with proton reduction and oxidation of electron donors such as ferredoxin; disposes of excess electrons in cells; allowing NAD^+ regeneration and ongoing pyruvate fermentation).

Previous work by Joseph et al (2013, 2015) highlighted the complex set of reactions involving organic C, H_2, O_2, and inorganic N and S compounds that can occur inside the pores and on the surfaces of the biochar (Table 29.2) (Chapter 10). These reactions can also take place within the digestive system of the ruminants (Ungerfeld, 2020) if there is a source of available O and hydronium ions. The biochar can be pretreated with ozone to increase the concentration of reactive O species or the micro and mesopores can be filled with very stable nanobubble O (Wang et al, 2018; Xiong et al, 2023). In flooded paddies, where the redox potential is very low, biochars filled with nanobubble O can increase local Eh and oxidize metals (Chu et al, 2023).

Some of these electrochemical and chemical gas and liquid phase reactions can be catalyzed by porous semiconductor nanoparticles (especially oxides of Cu, Mg, Zn, Fe, and Al). In this case, the C can act as an electron acceptor. As well as adherent nanoparticles catalyzing electrochemical reactions, biochar can also be engineered to contain a high concentration of reactive free radicals (RFR) with O vacancies in the lattice. Some of these (RFR) are associated with the C matrix and many are associated with minerals such as MgO, Fe_3O_4, CuO, and ZnO. Up to 10^{19} unpaired electrons and vacancies both in the C matrix and in specific mineral micron and

Table 29.2 *Reduction potentials (versus standard hydrogen electrodes, SHE) of inorganic redox reactions potentially associated with enteric methanogenesis (Joseph et al, 2013)*

Reaction	E° (Volts vs SHE) Free energy (kJ mol^{-1})
$C + O_2 \leftrightarrow CO_2$	-394
$CH_4 + 3O_2 \leftrightarrow CO_2 + 2H_2O$	-89
$2H_2 + O_2 \leftrightarrow 2H_2O$	-237
$CO_2 + H_2O \leftrightarrow H_2CO_3$	+103
$HCO_3 + 4H_2 + H^+ \leftrightarrow CH4 + 3H_2O$	-136
$Acetate^- + 4H_2O \rightarrow H_2 + 2HCO_3^- + H^+$	+105
$O_2 + 2H_2O + 4e^- \leftrightarrow 4OH^-$	0.40
$O_2 + 4H^+ + 4e^- \leftrightarrow 2H_2O$	1.23
$O_2 + 2H^+ + 2e^- \leftrightarrow H_2O_2$	0.70
$CO_2 + 8H^+ + 8e^- \leftrightarrow CH_4 + 2H_2O$	0.17
$CH_3OH + 2H^+ + 2e^- \leftrightarrow CH_4 + H_2O$	0.50
$CO_2 + 4H^+ + 4e^- \leftrightarrow C + 2H_2O$	0.21
$SO_4^{2-} + 8H^+ + 6e^- \leftrightarrow S^{2-} + 4H_2O$	0.20
$NO_3^- + 10H^+ + 8e^- \leftrightarrow NH_4^+ + 3H_2O$	0.88
$NO_3^- + 2H^+ + 2e^- \leftrightarrow NO_2^- + H_2O$	0.94
$2H^+ + 2e^- \leftrightarrow H_2$	0.00
$H_2O_2 + 2e^- \leftrightarrow 2OH^-$	0.93
$Fe^{3+} + e^- \leftrightarrow Fe^{2+}$	0.77
$Fe(OH)_3 + e^- \leftrightarrow Fe(OH)_2 + OH^-$	-0.55

sub-micron mineral phases can exist in the biochar (Fang et al, 2014). As noted above, the C matrix and the oxide mineral particles act as a redox couple where the biochar accepts excess electrons, which can then form reactive O species with the O functional groups on the surface of the biochar matrix (Joseph et al, 2013, 2015). The lifetime of these free radicals is both a function of the environment that surrounds the biochar and the type of mineral nanoparticles in the biochar (Yuan et al, 2022).

Radicals can be destroyed and formed at the surface of the biochar when it is placed in an environment where there are microbes and organic compounds that can be involved in electron transfer (Huang et al, 2022). Recent studies have highlighted the ability of these free radicals to be involved in the formation of reactive O species, which may oxidize organic molecules, especially where there are both sulfate and nitrate anions, to act as terminal electron acceptors in the pores of the biochar (Yuan et al, 2022). Reactive O species can include H_2O_2, $\cdot OH$, and $ONOO- \cdot O_2$, $NO\cdot$. When biochar is loaded with Fe^{3+} nanoparticles the production of reactive O species can be enhanced through Fenton-type reactions (Yu and Kuzyakov, 2021). These reactions can also be enhanced through ozonation (Zhao et al, 2020) and the addition of H_2O_2. Further research is required to determine if reactive oxygen species have a negative impact on overall digestion.

The mechanisms for oxidation of both H_2 and CH_4 can potentially proceed by both biotic and abiotic mechanisms. In the case of CH_4, the prominent reactions can be (Szécsényi et al, 2018):

$$CH_4 + \cdot OH \leftrightarrow H_2 + \cdot CH_3 \qquad [29.5]$$

$$\cdot CH_3 + O_2 \leftrightarrow CH_3OOH \qquad [29.6]$$

For the case of H_2, the prominent reactions can be (Liu et al, 2018):

$$H_2 + \cdot OH \leftrightarrow H_2O + H. \qquad [29.7]$$

$$H. + \cdot OH \leftrightarrow H_2O \qquad [29.8]$$

Within the rumen, biofilms will form on the biochar and ROS may form in biofilms which can be seeded with metal oxide nanoparticles on biochar (Gold et al, 2018), although there is little experimental evidence in the literature. Recent research (Fernandez et al, 2022; Taherymoosavi et al, 2022) indicates that for biochar to be effective in oxidizing CH_4 and H_2 there must be adsorption of O into the pores of the biochar either in gaseous form (ozone or O_2), as liquids (H_2O_2, peracetic acid), or as salts that can release O. There is no current evidence that oxidation of H_2 or CH_4 will occur in these biofilms because O_2 reactants are naturally limited in the anaerobic rumen unless exogenous H_2O_2 or ozone is added. Both rumen bacteria as well as

Holotrich and Entodiniomorph protozoa rapidly removed dissolved rumen O_2, and the dissolved O_2 concentration in bulk rumen liquor was undetectable (Ellis et al, 1989), even after the introduction of O_2 into headspace gas (Lloyd et al, 1983). Surprisingly, O_2 in the headspace gas has been shown to inhibit ruminal CH_4 oxidation during incubation, not accelerate it (Kajikawa et al, 2003).

Thus, for these oxidation reactions to occur, O must be introduced into the pores of the biochar before feeding to ruminants and then H_2 and the CH_4 will diffuse into the pores and react with the O.

(3) Physical removal of electrons: biochar may store electrons in its matrix and remove them from the rumen when biochar flows into the lower digestive tract, reducing ruminal electron availability for CH_4 production. However, the calculations below suggest this is not likely to be quantitatively important in the rumen. Approximately 60% of dry matter (DM) consumed by the cow is digested by the animal generating approximately 20.7 g CH_4 kg^{-1} DM feed (1.9 mol kg^{-1}) (Charmley, 2015). Formation of 1.9 mol CH_4 requires approximately 16 mol of electrons from the fermentation of the feed (8 mol of electrons mol^{-1} CH_4 [Eqn. 29.1] $\times$ 1.9 mol CH_4 kg^{-1} DM).

Previous measurements have shown that biochars can accept and donate 3–7 mmol electrons g^{-1} (Huang et al, 2022). For this third hypothesis to be true (i.e., biochar to trap all 16 mol electrons kg^{-1} feed, flow from the rumen to starve CH_4-producing organisms of electrons), approximately 3.2 kg biochar would have to be added per kilogram of feed. Since biochar is being added at only 1–5% w/w of the total feed of 5–15 kg DM day^{-1} for growing cattle, this proposed electron export mechanism could not trap sufficient electrons to have a substantial impact on CH_4 reduction.

Conclusions

Biochars can potentially affect productivity, health, and the environmental impact of agriculturally significant livestock, whether ruminant or monogastric, and whether on land or in aquatic or marine environments. These impacts are apparent through both direct effects on the animal, and indirectly through changing the production environment, especially in fish where the role of biochar in improving water quality may find a valuable application. The source of feedstock materials and the selection of pyrolysis conditions greatly affect the impact of these bioactive feed ingredients on animals and their environment. Regarding the direct effects of biochar on livestock, an increasingly clear understanding of the association of the microbiome with biochar surfaces in the gut,

together with the role of biochar in influencing the electrochemical economy of the digestive tract, is needed. Seeding biochars integrating with reactive molecules, such as nitrates and nanoparticles, to increase their capacity to facilitate the diversion of electrons into non-CH_4 electron acceptors warrants attention. Further work is required to determine if loading the micro and mesopores of biochar with oxygen nanobubbles or reactive oxygen species can reduce methane without affecting digestibility and animal health. The indirect impact of dietary biochar on livestock via changing the production environment is likely to be small in land-based systems due to the low quantities of biochar ingested but shows greater promise in sustaining water quality in managed fish production systems.

References

Abakari G, Luo G, Shao L, Abdullateef Y, and Cobbina SJ 2021 Effects of biochar on microbial community in bioflocs and gut of *Oreochromis niloticus* reared in a biofloc system. *Aquaculture International* 29, 1295–1315.

Al-Azzawi M, Bowtell L, Hancock K, and Preston S 2021 Addition of activated carbon into a cattle diet to mitigate GHG emissions and improve production. *Sustainability* 13, 8254.

Asada T, et al 2002 Science of bamboo charcoal: study on carbonizing temperature of bamboo charcoal and removal capability of harmful gases. *Journal of Health Science* 48, 473–479.

Black JL, Davison TM, and Box I 2021 Methane emissions from ruminants in Australia: mitigation potential and applicability of mitigation strategies. *Animals* 11, 951.

Bonner S 2012 *Education in Ancient Rome: From the elder Cato to the younger Pliny*. London: Routledge.

Chalermsan Y, and Peerapan S 2009 Wood vinegar: by-product from rural charcoal kiln

and its role in plant protection. *Asian Journal of Food and Agro-Industry* 2(Special Issue), S189–S195.

Charmley E, et al 2015 A universal equation to predict methane production of forage-fed cattle in Australia. *Animal Production Science* 56, 169–180.

Chen Y, et al 2022 In-situ biochar amendment mitigates dietary risks of heavy metals and PAHs in aquaculture products. *Environmental Pollution* 308, 119615.

Cheng S, et al 2020 Application research of biochar for the remediation of soil heavy metals contamination: a review. *Molecules* 25, 3167.

Cheng Y, et al 2023 In-situ formation of surface reactive oxygen species on defective sites over N-doped biochar in catalytic ozonation. *Chemical Engineering Journal* 454, 140232.

Chu, Q, et al 2023 Oxygen nanobubble-loaded biochars mitigate copper transfer from copper-contaminated soil to rice and improve rice growth. *ACS Sustainable Chemistry and Engineering* 11, 5032–5044.

Cooney DO, and Struhsaker TT 1997 Adsorptive capacity of charcoals eaten by Zanzibar Red Colobus monkeys: implications for reducing dietary toxins. *International Journal of Primatology* 18, 235–246.

Dang VM, et al 2019 Immobilization of heavy metals in contaminated soil after mining activity by using biochar and other industrial by-products: the significant role of minerals on the biochar surfaces. *Environmental Technology* 40, 3200–3215.

Day GE 1906 *Swine: a book for students and farmers.* Des Moines, Iowa: Kenyon Press.

Ellis JE, Williams AG, and Lloyd D 1989 Oxygen consumption by ruminal microorganisms: protozoal and bacterial contributions. *Applied and Environmental Microbiology* 55, 2583–2587.

Erickson PS, Whitehouse NL, and Dunn ML 2011 Activated carbon supplementation of dairy cow diets: effects on apparent total-tract nutrient digestibility and taste preference. *The Professional Animal Scientist* 27, 428–434.

Fang G, et al 2014 Key role of persistent free radicals in hydrogen peroxide activation by biochar: implications to organic contaminant degradation. *Environmental Science and Technology* 48, 1902–1910.

Fernandez GM, Durmic Z, Vercoe P, and Joseph S 2022 *Fit-for-purpose biochar to improve efficiency in ruminants.* MLA peer reviewed report B.GBP.0032. North Sydney: Meat & Livestock Australia Limited. https://www.mla.com.au/contentassets/a28c19322ec049e795b2695613553d26/bgbp_0032-biochar-final-report_mla-website-.pdf Accessed 10/05/2023.

Gerlach A, and Schmidt HP 2012 The use of biochar in cattle farming. *Ithaka Journal* 2012, 281–285.

Goiri I, et al 2021 Assessing the potential use of a feed additive based on biochar on broilers feeding upon productive performance, pH of digestive organs, cecum fermentation and bacterial community. *Animal Feed Science and Technology* 279, 115039.

Gold K, Slay B, Knackstedt M, and Gaharwar AK 2018 Antimicrobial activity of metal and metal-oxide based nanoparticles. *Advanced Therapeutics* 1, 1700033.

Graves C, Kolar P, Shah S, Grimes J, and Sharara M 2022 Can biochar improve the sustainability of animal production? *Applied Sciences* 12, 5042.

Hagemann N, et al 2017 Organic coating on biochar explains its nutrient retention and stimulation of soil fertility. *Nature Communications* 8, 1–11.

Hegarty, RS and Gerdes R 1999 Hydrogen production and transfer in the rumen. *Recent Advances in Animal Nutrition in Australia* 12, 37–44.

Hegarty RS, et al 2021 An evaluation of emerging feed additives to reduce methane emissions from livestock. A report coordinated by Climate Change, Agriculture and Food Security (CCAFS) and the New Zealand Agricultural Greenhouse Gas Research Centre (NZAGRC) initiative of the Global Research Alliance (GRA). https://hdl.handle.net/10568/116489 Accessed 10/05/2023.

Henderson G, et al 2015 Rumen microbial community composition varies with diet and host, but a core microbiome is found across a wide geographical range. *Scientific Reports* 5, 1–15.

Honan M, Feng X, Tricarico JM, and Kebreab E 2021 Feed additives as a strategic approach to reduce enteric methane production in cattle: modes of action, effectiveness and safety. *Animal Production Science* 62, 1303–1317.

Huang C, et al 2023 Effects of heterogeneous metals on the generation of persistent free radicals as critical redox sites in iron-containing biochar for persulfate activation. *ACS EST Water* 3, 298–310.

Huff MD, Marshall S, Saeed HA, and Lee JW 2018 Surface oxygenation of biochar through ozonization for dramatically enhancing cation exchange capacity. *Bioresources and Bioprocessing* 5, 18.

Hynd P 2019 *Animal nutrition: from theory to practice.* Clayton: CSIRO Publishing.

Jenkins TC, Wallace RJ, Moate PJ, and Mosley EE 2008 Board-invited review: Recent advances in biohydrogenation of unsaturated fatty acids within the rumen microbial ecosystem. *Journal of Animal Science* 86, 397–412.

Joch M, et al 2022 Feeding biochar to horses: Effects on nutrient digestibility, fecal characteristics, and blood parameters. *Animal Feed Science and Technology* 285, 115242.

Joseph SD, et al 2010 An investigation into the reactions of biochar in soil. *Soil Research* 48, 501–515.

Joseph S, et al 2013 Shifting paradigms on biochar: micro/nano-structures and soluble components are responsible for its plant-growth promoting ability. *Carbon Management* 4, 323–343.

Joseph S, et al 2015 The electrochemical properties of biochars and how they affect soil redox properties and processes. *Agronomy* 5, 322–340.

Joseph S, et al 2015 Feeding biochar to cows: an innovative solution for improving soil fertility and farm productivity. *Pedosphere* 25, 666–679.

Kajikawa H, et al 2003 Methane oxidation and its coupled electron-sink reactions in ruminal fluid. *Letters in Applied Microbiology* 36, 354–357.

Kalus K, Koziel JA, and Opaliński S, 2019 A review of biochar properties and their utilization in crop agriculture and livestock production. *Applied Sciences* 9,3494.

Kasapidou E, Sossidou E, and Mitlianga P 2015 Fruit and vegetable co-products as functional feed ingredients in farm animal nutrition for improved product quality. *Agriculture* 5, 1020–1034.

Khaki ND, Malcevschi A, Voccia A, and Marzano F 2017 Interaction of dietary biochar (black carbon) on the growth performance and survival rate of early stage larvae of brown trout (*Salmo trutta)*. Aquaculture Europe 2017 Dubrovnik, Croatia.

Khalid MA, Hussain SM, Mahboob S, Al-Ghanim KA, and Riaz MN 2022 Biochar as a feed supplement for nutrient digestibility and growth performance of *Catla catla* fingerlings. *Saudi Journal of Biological Sciences* 29, 103453.

Kim BK, and Kim YJ 2005 Effects of feeding charcoal powder and vitamin A on growth performance, serum profile and carcass characteristics of fattening Hanwoo steers. *Journal of Animal Science and Technology* 47, 233–242.

Kumar A, et al 2020 Mechanistic evaluation of biochar potential for plant growth promotion and alleviation of chromium-induced phytotoxicity in *Ficus elastica. Chemosphere* 243, 125332.

Lan TT, Preston TR, and Leng RA 2021 Feeding biochar or charcoal increased the growth rate of striped catfish (*Pangasius hypophthalmus*) and improved water quality. Livestock Research for Rural *Development* 28, 84.

Lehmann J, et al 2011 Biochar effects on soil biota–a review. *Soil Biology and Biochemistry* 43, 1812–1836.

Leng RA, Preston TR, and Inthapanya S 2012 Biochar reduces enteric methane and improves growth and feed conversion in local "Yellow" cattle fed cassava root chips and fresh cassava foliage. *Livestock Research for Rural Development* 24, 1–7.

Leng RA 2017 Biofilm compartmentalisation of the rumen microbiome: modification of fermentation and degradation of dietary toxins. *Animal Production Science* 57, 2188–2203.

Liu S, Oshita S, Thuyet DQ, Saito M, and Yoshimoto T 2018 Antioxidant activity of hydrogen nanobubbles in water with different reactive oxygen species both in vivo and in vitro. *Langmuir* 34, 11878–11885.

Lloyd D, Scott, RI, and Williams TN 1983 Membrane inlet mass spectrometry—measurement of dissolved gases in fermentation liquids. *Trends in Biotechnology* 1, 60–63.

Luder W 1950 Untersuchungen über die Bakterienadsorption durch Holzkohle. *Schweizer Archiv für Tierheilkunde* 17, 137–153.

Mabe LT, et al 2018 The effect of dietary bamboo charcoal supplementation on growth and serum biochemical parameters of juvenile common carp (*Cyprinus carpio* L.). *Aquaculture Research* 49, 1142–1152.

Mahari WAW, et al 2022 A state-of-the-art review on producing engineered biochar from shellfish waste and its application in aquaculture wastewater treatment. *Chemosphere* 288, 132559.

Man KY, Chow KL, Man YB, Mo W, and Wong MH 2021 Use of biochar as feed supplements for animal farming. *Critical Reviews in Environmental Science and Technology* 51, 187–217.

Mangold E 1936 Die Verdaulichkeit der Futtermittel in ihrer Abhängigkeit von verschiedenen Einflüssen. *Forschungsdienst—Reichsarbeitsgemeinschaften der Landwirtschaftswissenschaft* 1, 862–867.

Martínez-Fernández,G, et al 2014 Effects of ethyl-3-nitrooxy propionate and 3-nitrooxypropanol on ruminal fermentation, microbial abundance, and methane emissions in sheep. *Journal of Dairy Science* 97, 3790–3799.

McKenzie RA, 1991 Bentonite as therapy for Lantana camara poisoning of cattle. *Australian Veterinary Journal* 68, 146–148.

Michalak M, et al 2021 Selected alternative feed additives used to manipulate the rumen microbiome. *Animals* 11, 1542.

Mrozik W, et al 2021 Valorisation of agricultural waste derived biochars in aquaculture to remove organic micropollutants from water–experimental study and molecular dynamics simulations. *Journal of Environmental Management* 300, 113717.

Mui NT, and Ledin I 2006 Effect of method of processing foliage of *Acacia mangium* and inclusion of bamboo charcoal in the diet on performance of growing goats. *Animal Feed Science and Technology* 130, 242–256.

Najmudeen TM, Arakkal Febna MA, Rojith G, and Zacharia PU 2019 Characterisation of biochar from water hyacinth *Eichhornia crassipes* and the effects of biochar on the growth of fish and paddy in integrated culture systems. *Journal of Coastal Research* 86, 225–234.

Nan H, et al 2022 Minerals: A missing role for enhanced biochar carbon sequestration from the thermal conversion of biomass to the application in soil. *Earth-Science Reviews* 234, 104215.

Paul D, and Hall SG 2021 Biochar and zeolite as alternative biofilter media for denitrification of aquaculture effluents. *Water* 13, 2703.

Pennsylvania State University. Agricultural Experiment Station, 1922. Annual Report of the Pennsylvania Agricultural Experiment Station (No. 170). Agricultural Experiment Station, College of Agriculture, Pennsylvania State University.

Peter , A, Chabot , B, & Loranger , E (2021). Pre- and post-pyrolysis effects on iron impregnation of ultrasound pre-treated softwood biochar for potential catalysis applications. *SN Appl Sci*, 3(6), 643–643.10.1007/s42452-021-04636-y.

Phongphanith S, and Preston TR 2018. Effect of rice-wine distillers' byproduct and biochar on growth performance and methane emissions in local "Yellow" cattle fed ensiled cassava root, urea, cassava foliage and rice straw. *Livestock Research for Rural Development* 28, 178.

Prasai TP, Walsh KB, Midmore DJ, and Bhattarai SP 2017 Effect of biochar, zeolite and bentonite feed supplements on egg yield and excreta attributes. *Animal Production Science* 58, 1632–1641.

Quin P, et al 2015 Lowering N_2O emissions from soils using eucalypt biochar: the importance of redox reactions. *Scientific Reports* 5, 16773.

Rattanawut J 2014 Effects of dietary bamboo charcoal powder including bamboo vinegar liquid supplementation on growth performance, fecal microflora population and intestinal morphology in Betong chickens. *The Journal of Poultry Science* 51, 165–171.

Raul C, Bharti VS, Dar Jaffer Y, Lenka S, and Krishna G 2021 Sugarcane bagasse biochar: Suitable amendment for inland aquaculture soils. *Aquaculture Research* 52, 643–654.

Ruan X, et al 2019 Formation, characteristics, and applications of environmentally persistent free radicals in biochars: a review. *Bioresource Technology* 281, 457–468.

Savage ES 1922 Feeding dairy cattle. *Holstein-Friesian World* 19, 13.

Schmidt HP, Hagemann N, Draper K, and Kammann C 2019 The use of biochar in animal feeding. *PeerJ* 7, pe7373.

Silivong P, and Preston TR 2015 Growth performance of goats was improved when a basal diet of foliage of Bauhinia acuminata was supplemented with water spinach and biochar. *Livestock Research for Rural Development* 27, http://www.lrrd.org/lrrd27/3/sili27058.html.

Skutetzky A, and Starkenstein E 1914 Künstliche Nährpräparate. In: Skutetzky A, and Starkenstein E (Eds) *Die neueren Arzneimittel und die pharmakologischen Grundlagen ihrer Anwendung in der ärztlichen Praxis*. Berlin, Heidelberg: Springer. pp189–208.

Soliva CR, Meile L, Hindrichsen IK, Kreuzer M. and Machmüller A 2004 Myristic acid supports the immediate inhibitory effect of lauric acid on ruminal methanogens and methane release. *Anaerobe* 10(5), 269–276.

Spinhirne JP, Kozie J. and Chirase N 2004 Sampling and analysis of volatile organic compounds in bovine breath by solid-phase microextraction and gas chromatography–mass spectrometry. *Journal of Chromatography A* 1025, 63–69.

Steinegger P, and Menzi M 1955 Versuche über die Wirkung von Vitamin-Zusätzen nach Verfütterung von Adsorbentien an mastpoulets. *Gefluegelhof* 18, 165–176.

Stern MD, et al 1994 Evaluation of chemical and physical properties of feeds that affect protein metabolism in the rumen. *Journal of Dairy Science* 77, 2762–2786.

Struhsaker TT, Cooney DO, and Siex KS 1997 Charcoal consumption by Zanzibar Red Colobus Monkeys: its function and its ecological and demographic consequences. *International Journal of Primatology* 18, 61–72.

Sun T, et al 2017 Rapid electron transfer by the carbon matrix in natural pyrogenic carbon. *Nature Communications* 8, 14873.

Svihus B, and Itani K 2019 Intestinal passage and its relation to digestive processes. *Journal of Applied Poultry Research* 28, 546–555.

Szécsényi A, Li G, Gascon J, and Pidko EA 2018 Mechanistic complexity of methane oxidation with H_2O_2 by single-site Fe/ZSM-5 catalyst. *ACS Catalysis* 8, 7961–7972.

Taherymoosavi S, et al 2022 Overall benefits of biochar, fed to dairy cows, for the farming system. *Pedosphere* 33, 225–230.

Tamayao, PJ, Ribeiro, GO, McAllister, TA, Yang, HE, Saleem, AM, Ominski, KH, Okine, EK, and McGeough, EJ 2021 Effects of post-pyrolysis treated biochars on methane production, ruminal fermentation, and rumen microbiota of a silage-based diet in an artificial rumen system (RUSITEC). *Animal Feed Science and Technology* 273, 114802. 10.1016/j.anifeedsci.2020.114802.

Teoh R, et al 2019 Effects of hardwood biochar on methane production, fermentation characteristics, and the rumen microbiota using rumen simulation. *Frontiers in Microbiology* 10, 1534.

Terry SA, et al 2019 A pine enhanced biochar does not decrease enteric CH_4 emissions, but alters the rumen microbiota. *Frontiers in Veterinary Science* 6, 308.

Thu M, Koshio S, Ishikawa M, and Yokoyama S 2009 Effects of dietary bamboo charcoal on growth parameters, apparent digestibility and ammonia nitrogen excretion of tiger puffer fish, *Takifugu rubripes*. *Aquaculture Science* 57, 53–60.

Totusek R, and Beeson WM 1953 The nutritive value of wood charcoal for pigs. *Journal of Animal Science* 12, 271–281.

Ungerfeld EM 2020 Metabolic hydrogen flows in rumen fermentation: principles and possibilities of interventions. *Frontiers in Microbiology* 11, 589.

Van Soest PJ 1994 *Nutritional ecology of the ruminant*. Ithaca: Cornell University Press.

Viggi C, et al 2017 Enhancing methane production from food waste fermentate using biochar: the added value of electrochemical testing in pre-selecting the most effective type of biochar. *Biotechnology for Biofuels* 10, 1–13.

Wang, L, et al 2019 Mechanisms and reutilization of modified biochar used for removal of heavy metals from wastewater: a review. *Science of the Total Environment* 668, 298–1309.

Wang L, Miao X, Ali J, Lyu, T and Pan G 2018 Quantification of oxygen nanobubbles in particulate matters and potential applications in remediation of anaerobic environment. *ACS Omega* 3, 10624–10630.

Wilfart A, Montagne L, Simmins H, Noblet J, and van Milgen 2007 Digesta transit in different segments of the gastrointestinal tract of pigs as affected by insoluble fibre supplied by wheat bran. *British Journal of Nutrition* 98, 54–62.

Winders TM, et al 2019 Evaluation of the effects of biochar on diet digestibility and methane production from growing and finishing steers. *Translational Animal Science* 3, 775–783.

Xiong X, Li Y, Zhou X, and Zhang C 2023 Methane emission mitigation in hypoxic freshwater triggered by oxygen-carrying dual-modified sediment-based biochar. *Journal of Cleaner Production* 394, 136424.

Xu D, Cao J, Li Y, Howard A, and Yu K 2019 Effect of pyrolysis temperature on characteristics of biochars derived from different feedstocks: A case study on ammonium adsorption capacity. *Waste Management* 87, 652–660.

Ye J, et al 2017 Chemolithotrophic processes in the bacterial communities on the surface of mineral-enriched biochars. *The ISME Journal* 11, 1087–1101.

Yu G-H, and Zyakov Y 2021 Fenton chemistry and reactive oxygen species in soil: Abiotic mechanisms of biotic processes, controls and consequences for carbon and nutrient cycling. *Earth-Science Reviews* 214, 103525.

Yuan J, Wen Y, Dionysiou DD, Sharma VK, and Ma X 2022 Biochar as a novel carbon-negative electron source and mediator: electron exchange capacity (EEC) and environmentally persistent free radicals (EPFRs): a review. *Chemical Engineering Journal* 429, 32313.

Zeng, X, et al 2018 Speciation and bioavailability of heavy metals in pyrolytic biochar of swine and goat manures. *Journal of Analytical and Applied Pyrolysis* 132, 82–93.

Zhao L, et al 2020 Ozone decreased enteric methane production by 20% in an in vitro rumen fermentation system. *Frontiers in Microbiology* 11, 571537.

Zielińska A, and Oleszczuk P 2016 Effect of pyrolysis temperatures on freely dissolved polycyclic aromatic hydrocarbon (PAH) concentrations in sewage sludge-derived biochars. *Chemosphere* 153, 68–74.

30

Biochar, greenhouse gas accounting, and climate change mitigation

Annette Cowie, Elias Azzi, Zhe Han Weng, and Dominic Woolf

Introduction: The role of biochar in climate change mitigation

Meeting the goal of the Paris Agreement, to limit global warming to less than 2°C, will require the world to reach the point of net zero CO_2 emissions in the second half of this century, which will necessitate the deployment of carbon dioxide removal (CDR) methods (IPCC, 2022).

Biochar is recognized as a CDR method that combines a biological removal process with a durable storage mechanism (Babiker et al, 2022). Plants take CO_2 from the atmosphere as they grow, retaining the carbon (C) in biomass. When biomass is converted to biochar that is used as a soil amendment or in other applications that prevent combustion of the biochar, C is transferred from the atmosphere to the terrestrial C pool. Biochar is a net negative emissions process if the amount of C transferred is greater than the supply chain greenhouse gas (GHG) emissions associated with its production and use, including indirect impacts. The climate impacts of biochar are determined by comparing the GHG fluxes of a biochar system with a "no biochar" reference scenario.

Besides stabilizing biomass C, the production and use of biochar can provide additional climate change mitigation through the reduction or avoidance of GHG emissions. The following paragraphs summarize the processes contributing to climate change mitigation in biochar systems. Further detail on each of these processes is provided in other chapters. Results of published life cycle assessments are reviewed to quantify the climate change impact of specific biochar systems. Global assessments of biochar mitigation potential and drivers for the adoption of biochar are discussed, including recent initiatives to incentivize biochar through inclusion in carbon markets.

Carbon dioxide removal through biochar

Pyrolysis converts biomass that is readily decomposed by soil microorganisms to a material

DOI: 10.4324/9781003297673-30

that can persist in the soil for hundreds of years, due to its condensed aromatic structure (see Chapter 11). The persistence of biochar C in soil, determined largely by pyrolysis conditions and feedstock, ranges from decades for manures up to several centuries or more for wood pyrolyzed at temperatures above 450-500°C (Joseph et al, 2021; Chapter 11). However, only a fraction of the C in the original biomass is retained in biochar – typically around 40% (assuming 50% C in biomass, biochar yield of 25% with 80% C content), with biochar yield inversely proportional to the pyrolysis temperature (Chapter 3).

The climate benefit of stabilizing biomass C in biochar is dependent on the alternative fate of biomass C. For example, crop residues retained in the field decompose over several months, and burning crop residues returns the C to the atmosphere immediately, so converting to biochar stores this C for much longer than in the "no biochar" scenario. In contrast, coarse branches and stumps remaining after forest harvest decompose slowly on the forest floor, and end-of-life wood products persist for decades in landfill (Ximenes et al, 2018), so the relative benefit of converting these to biochar is smaller.

Adding to the CDR benefit through stabilization of feedstock C, biochar can also enhance soil C sequestration through priming, that is, alteration in the decomposition rate of soil organic matter: while positive priming (accelerated decomposition) is often observed initially, negative priming (slowed decomposition) usually dominates within several months of biochar application (Joseph et al, 2021; Chapter 17). Additionally, biochar can stabilize rhizodeposits (organic compounds exuded by plant roots) and sloughed root material, slowing their decomposition (Weng et al, 2022). Thus, biochar can enhance soil C levels by slowing the rate of turnover of existing soil organic matter, and by stabilizing newly added organic matter. As the application of crop residues or manures can cause positive priming compared with chemical fertilizer (e.g., Abdalla et al, 2022), the negative priming effect of biochar can be greater if the reference scenario involves land-application of these feedstocks, compared with a scenario where they are combusted or landfilled and chemical fertilizers are used. Biochar has also been found to increase inorganic C levels in a calcareous soil (Wang et al, 2023).

Reduction in non-CO_2 greenhouse gas emissions

Biochar commonly reduces N_2O emissions from soil, by decreasing denitrification, or by facilitating reduction of N_2O to N_2 (Chapter 18). Meta-analyses report reductions from 12–50% (Joseph et al, 2021), but the longevity of the effect is uncertain, and it can decline substantially after one or two years (Borchard et al, 2019). Biochar generally decreases CH_4 emissions from flooded soils (Jeffery et al, 2016), although, particularly in the case of high ash biochars, it can decrease the rate of CH_4 uptake in non-flooded soils (Pascual et al, 2020; Chapter 18).

Again, the mitigation benefit is dependent on the "no-biochar" alternative. If biochar is made from manure which is otherwise spread directly on soil, the relative benefit is greater than where chemical fertilizers are displaced, as soil N_2O emissions per unit of applied N tend to be higher for manure than for synthetic fertilizers.

Displacement of fossil fuel

Pyrolysis converts biomass to three products: biochar, bio-oil, and pyrolysis gas. Bio-oil and pyrolysis gas are usually combusted, returning the C to the atmosphere. Excess bio-oil and pyrolysis gas (not required to heat the kiln or dry biomass) can be utilized to displace fossil fuel, for example, in electricity generation or district heating, providing additional climate change mitigation. Co-location of biochar production with activities that can utilize low-grade heat, such as in livestock sheds or glasshouses, can enable beneficial use of process heat. The magnitude of the mitigation benefit depends on the efficiency of conversion to the energy product and the reference energy source and is generally greatest where coal is displaced. As a storable dispatchable energy source, bioenergy from pyrolysis could play a strategic role in supporting energy system transition, enabling more rapid expansion of the intermittent renewables.

It is also relevant to contrast biochar systems with using biomass for energy alone, which can deliver greater mitigation than using biomass for biochar plus energy where the current energy source is coal (Woolf et al, 2016). However, where biochar application substantially reduces N_2O emissions, enhances plant growth, or stimulates negative priming, the climate benefit is generally greater than using biomass solely to displace coal emissions (Cowie et al, 2015). As electricity grids are decarbonized and coal-fired power is phased out, the relative benefit of biochar systems over bioenergy alone is likely to increase (Woolf et al, 2010). Further, the energy supply emissions displaced by a new low-carbon resource depend not on the average GHG intensity of the existing energy system, but rather on the resulting marginal change in GHG intensity.

Avoided emissions

When biomass is used for biochar, GHG emissions from the conventional fate of the biomass are avoided. For example, manures that are stockpiled and spread on soil emit N_2O, CH_4, and NH_3 (leading to indirect emissions of N_2O) during storage and after application to land, particularly in wet climates (van der Weerden et al, 2021). These emissions are substantially reduced if manure is pyrolyzed before application and the pyrolysis plant is co-located with the feedstock source, reducing the storage period (Hou et al, 2017). Putrescible biomass deposited in landfills releases CH_4, which is often not captured and can be avoided if the biomass is diverted for biochar production.

Biochar can reduce the leaching and volatilization of N, increasing the N use efficiency of applied fertilizers (Chapters 18 and 19), thus also reducing fertilizer requirements. As the manufacture of N fertilizer is a GHG-intensive process, biochar use can reduce the emissions from fertilizer manufacture. Likewise, biochar increases soil pH and can partially substitute the use of limestone, another CO_2 source. Furthermore, biochar could reduce fuel use on-farm by increasing moisture retention (Chapter 20) thus reducing irrigation pumping requirements, or by reducing fuel use for cultivation due to reduced soil strength where high rates are applied.

Use of biochar in non-soil applications

Biochar also has many non-agricultural applications, including urban tree planting, landscaping, green roofs, concrete, road construction, and water filters (Chapters 26 and 28). The persistence of biochar C in these applications is less studied but is expected to be as great or greater than in soils. In these applications, biochar can displace the use of other GHG-intensive materials such as peat or plastics, scarce resources such as sand, and improve the performance of the products thereby leading to multiple indirect positive effects (e.g., greater durability of roads reduces maintenance requirements; reduced need for green roof maintenance decreases transportation requirements of operators; improved quality of runoff water reduces the need for stormwater treatment).

Geological storage of pyrolysis co-products

The bio-oil and gas co-products of biochar production are often combusted in the pyrolysis process, used to provide energy for external applications, or flared, returning C to the atmosphere. The CO_2 could be captured at the point of combustion and stored geologically, enhancing the CDR of biochar systems. Carbon capture and storage (CCS) of CO_2 emissions from energy plants is a well-understood process (Bui et al, 2018) in the early stages of commercialization, with around 30 facilities in operation globally, most associated with fossil fuel energy plants. The bio-oil can also be collected, particularly from a fast pyrolysis process that produces bio-oil as the main product, and stored geologically. Pyrogenic C capture and storage (PyCCS),

involving geological storage of bio-oil and pyrolysis gas (Schmidt et al, 2019), could substantially enhance the CDR contribution of biochar systems (Werner et al, 2018; Lehmann et al, 2021).

Quantifying net climate change mitigation

The climate change mitigation benefits of biochar systems are quantified by comparing the biochar system with a "no biochar" scenario in which biomass is not pyrolyzed, conventional soil amendments and fertilizers are applied, and alternative energy sources are used. The reference, which should be explicitly described (Azzi et al, 2021), should represent the most likely scenario(s) in the absence of the biochar system, which might include "business as usual" or alternative

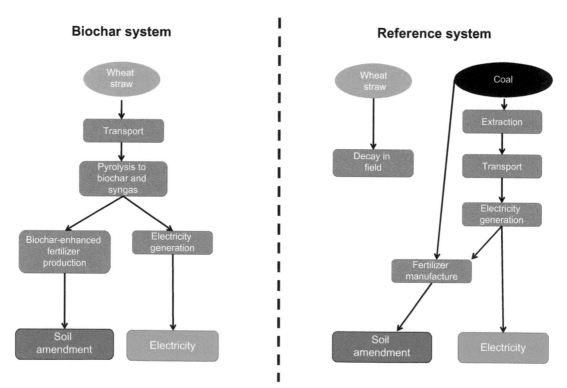

Figure 30.1 *Example of a biochar system, illustrating the components of the biochar life cycle, and the comparison with a reference system*

competing scenarios for use of the biomass (Figure 30.1). Biochar systems provide a climate benefit if the net GHG emissions (emissions minus removals, across the biochar life cycle) are lower than the net GHG emissions of the reference system, when all relevant source and sink processes, including indirect effects, are considered.

Life cycle assessment – A method to quantify the relevance of biochar systems for climate change mitigation

Life cycle assessment (LCA), which quantifies the potential environmental impacts of product systems to enable sound environmental comparisons (Finkbeiner et al, 2006), is used in research, industry, and government. The goal of an LCA study affects the modeling approach and limits the conclusions that can be made. This chapter considers LCAs studying biochar systems to determine whether biochar systems are a relevant climate change mitigation measure, relative to a specific reference or alternative situation, in a specific background context. Chapter 31 addresses LCA for broader environmental impacts.

While results of biochar LCAs can vary widely, this does not imply high uncertainty or lack of consensus, but rather that different studies depict different contexts, rely on different methodologies, or compare biochar to different alternatives. Importantly, the diversity in LCA results relates to the fact that biochar systems are extremely diverse, involving different biomass feedstocks, many types of pyrolysis reactors, and a wide range of applications.

Systematic description of biochar systems from a life cycle perspective

While diverse, biochar production and use are essentially a human enterprise that converts material and energy for some social outcome, which comprises the following stages: biomass production, biomass pyrolysis, use of pyrolysis co-products, and biochar use (Figure 30.2, Azzi et al, 2022). Biochar use, which is usually part of the life cycle of another product or service, can be divided into the production of other materials, and manufacturing, use, and disposal of the biochar-based product. Each stage consumes other products or services, with their respective life cycle (e.g., fuel production, machinery, transportation), and can have direct emissions of environmental stressors, such as CH_4, particles, or N_2O.

The above generic description can be illustrated with some real-life examples:

- A bioenergy facility converts straw harvested from wheat fields (biomass sourcing) to biochar and uses the excess heat for grain drying (biomass pyrolysis and co-product use). The biochar is combined with fertilizers and pelletized (product manufacturing). The biochar-fertilizer pellets are used for the growth of seedlings in a nursery (use-phase), and then transferred to soil (end-of-life).
- A company producing and installing green roofs has invested in a biochar production unit. The biomass feedstock is wood pellets from forest residues (biomass sourcing). Biochar is produced and the pyrolysis gases are combusted for heating greenhouses and office buildings (biomass pyrolysis and co-product use). Biochar is further processed and blended

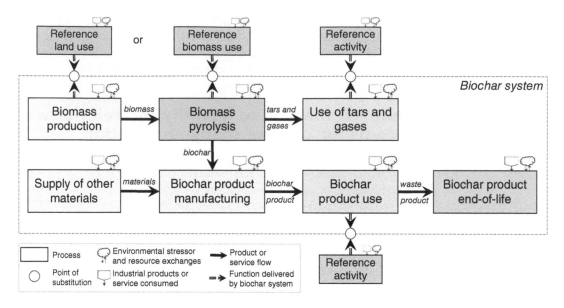

Figure 30.2 *Generic biochar system boundaries for use in life cycle assessment and other systems studies, adapted from Azzi et al (2022)*

with other components to make a mineral soil substrate suited for green roofs (product manufacturing). The soil substrate is cultivated with sedum and then installed on roofs, with an expected lifetime of 50 years (use phase). Then, the substrate can be recycled or repurposed as a landscaping amendment (end-of-life).

• When biochar is directly applied to soil in its use phase, there is no distinction between the use phase and end-of-life.

• In most cases, pyrolysis gases and tars are converted to energy, and therefore, the reference activity is often a reference energy activity. However, in some cases, pyrolysis tars can also be used as a precursor for bio-based materials and chemicals.

Biochar systems are multi-functional, that is, multiple products and services are delivered to society (Figure 30.2). The biochar-based product has a use (e.g., soil amendment, construction material, water filter). During

pyrolysis, energy or chemicals can be valorized. If waste biomass is pyrolyzed, a waste treatment service is provided. If dedicated biomass is utilized as feedstock, land use takes place (this indirect function is particularly important to determine components like land-use change emissions or change in soil organic C stocks). Due to their multi-functionality, biochar systems face the same LCA methodological challenges as biorefineries (Ahlgren et al, 2015).

Thus, to adequately account for multi-functionality and perform sound LCA comparisons, reference activities must be defined for all functions. Together they form a reference system, here depicted outside the boundaries of the biochar system. The reference system is often not unique and multiple reference systems can be used to analyze the relative climate performance of biochar in different contexts. The next paragraphs detail each stage from biomass production to biochar product end-of-life.

Biomass production refers to the cultivation and harvest of biomass dedicated for pyrolysis, or to the collection of biomass residues and organic wastes. Sustainably produced biomass is a limited resource globally and its allocation for biochar raises the same land use and resource competition concerns as for any other bioenergy system (European Commission, 2019; Prade et al, 2017). Suitable biomass types for biochar include woody crops, forestry residues, crop waste, perennial grasses, animal manures, and biosolids (Ippolito et al, 2020). Biochar LCA studies can build on the large body of research into LCA methods for agricultural and forestry systems (Caffrey and Veal, 2013). A specific feature of biochar-biomass systems is the potential feedback loop through which biochar can enhance biomass productivity (Woolf et al, 2010).

Biomass pyrolysis is a thermo-chemical process used to produce biochar (Chapter 3) and can include pre- and post-treatment of the biomass and the pyrolysis products (e.g., drying, steam activation, gas reforming). Reactors vary in configuration and scale, and produce pyrolysis products in different proportions and with different properties (Meyer et al, 2011; Woolf et al, 2014; Cornelissen et al, 2016; Sørmo et al, 2020; Chapter 4). From a life cycle perspective, two additional properties differentiate reactors: the manufacturing and decommissioning requirements and the use phase emissions. These can vary substantially, e.g. between flame-curtain kilns (low capital requirement, high emissions) (Cornelissen et al, 2016), cooking stoves (medium capital requirement, medium emissions) (Gitau et al, 2019), or pyrolysis plants with an advanced combustion chamber (high capital requirement, low emissions) (Sørmo et al, 2020). Data availability is limited, and no widely available LCA database includes datasets for modern pyrolysis reactors.

Use of tar and gases refers to the fate of the pyrolysis gases and tars. They are often co-combusted on-site to provide energy services, including cooking, space heating, and biomass drying (Laird et al, 2009). Production of steam, electricity, and vehicle fuel is less common. Materials such as biochemicals could also be produced from the pyrolysis tars, especially in fast pyrolysis reactors. From a life cycle perspective, quantifying the effects of biochar requires understanding the substitution effects of co-products on energy systems and other supply chains.

Biochar use spans several sectors: agriculture (e.g., soil amendment with biochar or biochar-fertilizer mix), forestry (e.g., seedling growth substrate), urban (e.g., urban greening, construction materials), and industry (filters, bio-materials, site remediation), with possibilities for cascading (sequential) uses (Azzi et al, 2019; Wurzer et al, 2019). Biochar use extends over longer time scales than co-product uses, as it includes e.g. multi-annual crop effects or urban infrastructure that is likely to remain in place for decades. In contrast, the co-products – tar and gases – are generally combusted to provide energy services shortly after production. From a life cycle perspective, the biochar use phase is often included in the life cycle of an existing product or service.

After the use phase, biochar has an ***end-of-life*** that differs from other products, for which common end-of-life processes are disposal in landfills, energy recovery, or recycling. Biochar placed in a given environmental compartment, at a given time, is subject to chemical transformation like aging and decomposition (Sorrenti et al, 2016) and transport to other environmental compartments through various processes. For example, biochar applied to a field could be crushed by cultivation, "age" through interaction with soil minerals and organic matter (Joseph et al, 2021), be washed into a waterway with eroded topsoil, and be deposited in marine sediments, where it is likely to remain for millennia (Major et al, 2010; Haefele et al, 2011; Singh et al, 2015; Kätterer et al, 2019). This biochar end-of-life, which characterizes its CDR function, spans time scales (centuries) that are longer than the time frame of most biochar projects and LCA studies.

Climate change impact of biochar systems in LCA studies

Goal and scope

Most biochar LCA studies have selected a functional unit relating to the biomass or the biochar, e.g. "1 ton of biomass managed through pyrolysis" or "1 ton of biochar produced and used". Both functional units are linked by the biomass to biochar pyrolysis yield and provide equivalent results. Some studies have used units related to the co-products (e.g., 1 MJ energy) or the biochar applications (e.g., 1 m^3 water treated, 1 ton of rice produced). For analyzing biochar as a climate change mitigation option, considering that biomass is a limited resource, the appropriate functional unit is usually based on biomass. This functional unit allows comparison to alternative uses of the biomass in a given context. Another limited resource, land area, can also be a relevant functional unit where biomass is obtained from purpose-grown crops.

The biochar LCA literature has assessed a wide range of feedstock types, including forestry residues, agricultural residues, dedicated crops, manures and sludges, and urban garden waste. This said, few studies have compared multiple feedstocks in a single analysis (Major et al. 2010; Hammond et al, 2011; Azzi et al, 2022). Also, most studies have considered waste-derived biomass feedstocks, rather than feedstock from dedicated

biomass production. Dedicated biomass production is often associated with higher adverse environmental and climate impacts, although a case has been made for the use of carefully selected (often native) plants to simultaneously provide biomass resources and environmental co-benefits (Tilman et al, 2006; Englund et al, 2021). Pyrolysis reactors of all kinds have been studied, from low-tech kon-tiki flame curtain kilns up to prospective large-scale continuous reactors. However, the quality of available data on the manufacturing and operation of different reactors is variable, which partly hampers comparisons.

A range of applications of biochar has been considered in LCA studies, almost all in agriculture, though a few recent studies have analyzed biochar use in urban applications and construction. In agriculture, LCA studies first looked at the direct application of biochar to soil, while newer studies also considered more complex uses like biochar as an animal feed supplement, manure management additive, mixing with fertilizers, or use as a potting medium in plant nurseries (Azzi et al, 2019; Fryda et al, 2019). Non-agricultural applications included use for tree planting in urban areas, production of urban landscaping soils, industrial water filters, and additives to concrete and cement blends (Azzi et al, 2022).

Biochar is often described as a CDR technology with positive side effects. As discussed above biochar side effects are quantified relative to a reference situation, in a given context. The scope of biochar effects included in studies varies significantly (Table 30.1): studies with a focus on agricultural uses of biochar have commonly assumed a reduction in N_2O emissions and increases in fertilizer use efficiency. In other applications, substitution of conventional products was considered (peat in horticulture, sand in water filtering, aggregates in

Table 30.1 *Biochar effects included in 45 biochar LCA studies. If a study modeled both N fertilizer reduction and P fertilizer reduction, the study is counted only once under "Agriculture: fertilizer use reduction". CDR: Carbon dioxide removal; NPP: Net primary productivity; SOC: Soil organic C; NA: Not applicable. Further details are available in Azzi et al. (2021)*

Effect description	No. of studies
Effects included	
CDR: Biochar C sequestration	43
Co-products: avoided heat/power from other fuel	35
Agriculture: fertilizer use reduction	19
Agriculture: soil N_2O emission reduction	19
Pyrolysis: air emissions, relative to reference biomass/land use	12
Agriculture: crop harvest increase	10
Agriculture: biochar-induced SOC change (priming, NPP increase)	7
Agriculture: soil CH_4 emission change	7
Agriculture: avoided nutrient leaching into water	5
Reference biomass/land: land use change emissions	5
Agriculture: avoided limestone production and use	3
Soil toxicity: reduced heavy metal mobility	2
Agriculture: avoided peat use	1
Agriculture: CH_4, N_2O, nutrient flux change in animal husbandry	1
Agriculture: soil albedo changes	1
Other substitutions: clay/gravel/backfill material/landfill space	1
None[a]	1
Effect explicitly not included in the LCA[b]	
Agriculture: crop, NPP, SOC increase	8
Other substitutions: clay or gravel landfill cover substitution	1
Agriculture: soil N_2O emission reduction	1
Other	
Sensitivity on the persistence of biochar effects over time	3

Notes
[a] This study exclusively modeled the material and energy inputs to run a pyrolysis plant.
[b] It is mentioned in the text that this effect exists, but is not included in the analysis for context-specific reasons

construction). Effects such as albedo change or permanence of biochar effects over multiple years are rarely analyzed.

Net climate change effect

To date, several dozen biochar LCA studies have been published, including several reviews (Tisserant and Cherubini, 2019; Matuštík et al, 2020; Azzi et al, 2021; Terlouw et al, 2021). The climate change impact of biochar-to-soil systems in 34 LCA studies (including supply-chain emissions, C sequestration in biochar, substitutions and soil effects) was found to be on average -0.9 ton CO_2e ton^{-1} biomass (range -1.5 to 0 ton CO_2e ton^{-1} biomass) (Tisserant and Cherubini, 2019; Matuštík et al, 2020). This is similar to the range presented in the 2nd edition of this book (-1.2 to 0.4 ton CO_2e ton^{-1} biomass; Cowie et al, 2015) and recent LCAs for biochar use in various urban products (Azzi et al, 2022).

Net climate change impact per mass biomass must be analyzed with care because these results are strongly influenced by the alternative biomass use considered and the background energy context selected by the LCA analyst. In some cases, biochar systems do not contribute to climate change mitigation, for example, when biochar is made from purpose-grown biomass that causes indirect land use change (Roberts et al, 2010), or where the reference use of biomass is bioenergy production replacing fossil fuels in contexts where biochar does not provide agricultural benefits (Woolf et al, 2010).

Contribution analysis

Climate change impacts should not be summarized and presented only as a "net" value. Informative insights are provided by process contribution analyses, illustrated in Figure 30.3, which display the share of emissions or removals associated with different life cycle stages or processes. Contribution analyses can be high-level (e.g. impact divided between biomass supply, pyrolysis, biochar use, and C sequestration) or very detailed (e.g. within each category, showing the contribution of transportation, electricity, thermal energy, direct air emissions, and materials) (Figure 30.3).

Biomass production and supply

In studies of biochar derived from side streams or waste, biomass is often considered burden-free, and only additional transport, handling, and storage are considered. These have relatively small contributions, unless transportation involves long distances (several hundred kilometers) or biomass storage releases CH_4 emissions. In biochar systems utilizing purpose-grown biomass, biomass production impacts are higher due to emissions associated with fertilizers (direct and indirect N_2O emissions), fuel use for cultivation and harvesting, and if any, land use change (direct or indirect).

Biomass pyrolysis

Climate impacts associated with the operation and maintenance of a modern pyrolysis plant are usually small. Plant operation may require electricity, start-up fuel, water for biochar quenching, and bags for biochar bagging. Periodic reactor maintenance may consume cleaning products and energy, and involve disposal of ash, replacement of lubricants, and other mechanical parts. In addition, the stack (chimney) of the pyrolysis plant is a potential source of GHGs and other air pollutants, including traces of CH_4, CO, NO_x, soot, heavy metals, and PAHs. Infrastructure requirements, often neglected in LCA, are another source of impacts that can represent 5% to 15% of supply-chain emissions in some cases.

Significant differences in GHG emissions usually arise between syngas-heated reactors (i.e., pyrolysis reaction is sustained by burning

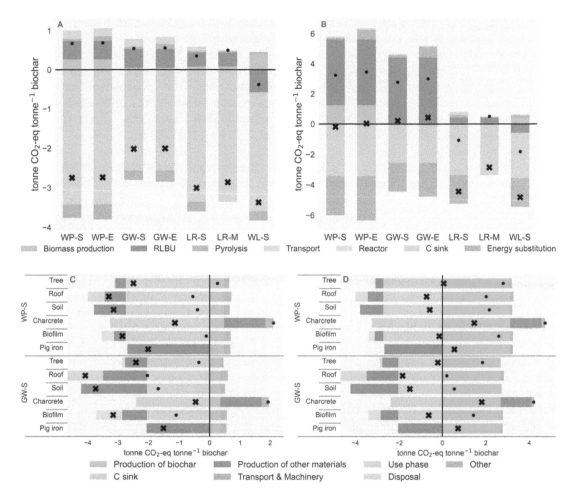

Figure 30.3 *Contribution analysis presenting LCA results for urban biochar applications. Panels A and B: Cradle-to-gate LCA of 7 biochar production supply chains in two energy systems, 2020 Swedish average energy (panel A) and natural-gas-based energy (panel B). Note that cradle-to-gate LCA shows differences between biochar supply chains but does not indicate the net climate change impact of a complete biochar system. Panels C and D: Net climate change mitigation benefits per mass of biochar produced and used, with respect to a reference technology, for each application, and calculated in two energy systems, 2020 Swedish average energy (panel C) and natural-gas-based energy (panel D). WP: wood pellet, GW: garden waste, LR: logging residues, WL: willow woodchips; S: syngas-heated reactor, E: electricity-heated reactor, M: mobile syngas-heated reactor; RLBU: Reference land or biomass use; Tree: street tree establishment; Roof: green roof; Soil: landscaping soil; Charcrete: concrete containing biochar, Biofilm: carrier for biofilm used in water purification; Pig iron: intermediate in steel production. A cross indicates the net impact, while a dot indicates the net impact excluding the biochar C removal (details in Azzi et al, 2022; image reproduced under the terms of the Creative Commons CC BY license, https://link.springer.com/article/10.1007/s42773-022-00144-3#rightslink)*

the pyrolysis gases) and electricity-heated reactors (i.e., pyrolysis reaction is sustained by externally sourced electricity), with the former usually having lower environmental impacts. Construction materials also differ between reactors; utilizing more resource-efficient and durable equipment usually reduces environmental impacts. Importantly, GHG emissions can be substantial for some biomass pyrolysis configurations, e.g. when large quantities of CH_4 are released during pyrolysis without an effective combustion chamber, or when plant operation consumes large amounts of electricity for sustaining the pyrolysis reaction or drying excessively wet biomass using electricity sourced from a fossil-dominated electricity grid. These configurations must be avoided.

Carbon sequestration in biochar

There is consensus in the literature that C stabilization in biochar (delayed oxidation of organic matter) is a major contributor to abatement, usually comprising about 40–60% of the estimated emissions reduction (e.g., Roberts et al, 2010; Hammond et al, 2011; Cowie and Cowie, 2013). Manure biochars, which have a shorter mean residence time in soil than wood biochars (Joseph et al, 2021), make a lower contribution to long-term C storage, (Whitman et al, 2013; Lehmann et al, 2021). The temperature at which the biochar was formed, and the process rate, not only affect the persistence of the biochar but also the yield of biochar vs gas: while at higher temperatures the biochar produced is more persistent in soil (Joseph et al, 2021), the total quantity of biochar produced is lower (Whitman et al, 2013). For more details on biochar permanence and its modeling, refer to Chapter 11.

Many biochar LCA studies assume that the biochar modeled has a high persistence, but few discuss whether the conditions are met for persistence to be high. Sensitivity analyses and contribution analyses displaying results with and without C sequestration in biochar (Figure 31.2), highlight the significance of biochar persistence to the climate outcome, and the consequent need for quality control standards and regulation to ensure biochar meets the criteria for high stability (e.g., IBI, 2015).

Bioenergy from pyrolysis co-products

Pyrolysis generates energy as a co-product, in the form of heat or power. The net climate effect of energy co-production depends on the reference system and the background context. In biochar systems where pyrolysis gas displaces a fossil energy system, the contribution to climate change mitigation of the displacement is of the same order of magnitude as C sequestration in biochar. However, comparing a biochar system to a system where biomass is efficiently used for bioenergy, the biochar system produces less energy, which must be compensated by other means. If this energy is supplied by fossil fuels, the additional emissions commonly negate the benefits from C sequestration in biochar, but if supplied by renewable energy, the penalty is one order of magnitude smaller than C sequestration in biochar.

Significant differences can arise between biomass pyrolysis configurations. The feedstock and the pyrolysis parameters influence the ratio between energy and biochar production (Woolf et al, 2014). For instance, the co-production of electricity is greater from dry feedstocks such as straw that provide a higher net pyrolysis gas output than moist materials such as manure (Cowie and Cowie, 2013).

Biochar use in agriculture

Biochar use in agriculture and animal husbandry can impact GHG fluxes in numerous ways. LCA studies of biochar use in agriculture have commonly found that the effects varied from small to the same order of magnitude as C sequestration in biochar (e.g., Roberts et al, 2010; Hammond et al, 2011; Wang et al, 2014; Azzi et al, 2019). The effects were rarely negative (causing increased emissions), but the longevity of these effects is particularly uncertain.

The contribution of crop yield enhancement through biochar application is generally small, but can be significant if the additional residues are used as a feedstock to increase biochar or bioenergy production (Woolf et al, 2010). Effects of biochar on the use of fertilizer and lime include impacts on pre-farm and on-farm emissions. Production of fertilizer can be avoided when biochar directly supplies nutrients, or where it improves nutrient use efficiency. However, the use of N-rich biomass for biochar that would otherwise have been used as organic fertilizer can lead to increased production of N fertilizer. Avoided production of fertilizer or lime usually has a small contribution. The impacts of avoided N_2O emissions from soil or CO_2 from lime decomposition can be large. However, these fluxes are associated with large variabilities and uncertainties, including over the longevity of the effects, and cannot be generalized.

Some studies have investigated the cascading use of biochar in animal husbandry, composting, or anaerobic digestion before application to soil. Although more complex to model in LCA, cascaded uses of biochar can improve the climate performance of the system by cumulating several smaller benefits. For instance, biochar use on dairy farms both as animal feed and manure management additive was estimated through modeling to yield significant N_2O and CH_4 emission reduction, doubling the benefits from C sequestration in biochar (Azzi et al, 2019).

The contribution to the net GHG outcomes through soil organic matter stabilization or losses (positive or negative priming) that arise as a consequence of biochar incorporation has rarely been included in LCA studies. Under the assumptions applied in their respective studies, Cowie and Cowie (2013) found it to be of minor importance, while Hammond et al (2011) calculated that negative priming contributed up to 27% of the estimated abatement. Significant negative impact due to positive priming is unlikely (Woolf and Lehmann, 2012; Singh and Cowie, 2014). New knowledge of priming (Chapter 17) should be incorporated in future LCA studies, to better reflect this aspect of biochar impacts.

The effects of albedo on the C cycle are rarely considered. Meyer et al (2012) calculated that the reduction in albedo reduced the net climate mitigation benefit of biochar by 13–22%, at a biochar application rate of 30 Mg ha^{-1}, while Genesio et al (2012) found that no statistically significant impact of biochar on soil albedo persisted after two tillage operations.

Biochar use in other products

Recent studies have investigated the use of biochar in urban plantings, construction materials, and water filters, in place of peat, sand, or plastic carriers, respectively. Climate change mitigation through such material substitutions can be substantial though usually of smaller magnitude than C sequestration in biochar. Additional mitigation can arise through indirect benefits. For instance, biochar-based green roofs may reduce fertilizer requirements or reduce plant mortality. However, these use-phase benefits are challenging to include quantitatively in LCA studies due to a lack of data.

Climate metrics

GHGs differ in their efficacy as warming agents and in their atmospheric lifetime. For example, CH_4 is much more powerful than CO_2 but has a short lifetime, of around 12 years, compared with CO_2 that stays in the atmosphere essentially indefinitely. Climate metrics provide a common unit to compare systems that impact fluxes of multiple GHGs. Emissions are commonly expressed as CO_2 equivalent (CO_2e) by multiplying the emissions of a given gas by its scaling coefficient, called its Global Warming Potential (GWP). The GWP expresses the net radiative forcing of a pulse emission of a GHG, integrated (summed) over a specified timeframe, relative to that of an equivalent mass of CO_2. 100-year GWPs are the most commonly applied. Limitations of GWP as a normalization factor have been raised (e.g., Shine, 2009) and alternative metrics have been proposed. While there is ongoing debate over the most appropriate metric to use (Reisinger et al, 2022), there is consensus that no one metric is universally preferable. The Life Cycle Initiative (e.g., Jolliet et al, 2018) has recommended the use of global temperature change potential (GTP) as a measure of longer-term climate-change impacts, to complement the use of GWP100, in LCA.

Timing of emissions

Conventional LCA ignores the timing of GHG fluxes and instead calculates the sum of emissions and removals across the entire life cycle (Brandão et al, 2013). Thus, a process that sequesters and then releases C within the selected time horizon, such as a bioenergy system, is deemed to have no net climate change impact associated with the biogenic C flux. Assuming "carbon neutrality" is valid for systems that cycle C rapidly, such as those utilizing annual crops for bioenergy, where the C is sequestered and released in the same year. This is clearly not the case for biochar: the majority of C in biochar remains in the soil for decades to centuries (Chapter 11). Thus, radiative forcing is reduced for the period that C is retained in the biosphere, and this should be

recognized as a mitigation benefit. However, as biochar production returns a significant portion (usually at least 50%) of the feedstock C to the atmosphere immediately, calculations of the net climate effect should also recognize the hastened release of this fraction of the feedstock C.

Currently, there is no agreed method for quantifying the climate effects of timing of GHG emissions and removals, although several methods have been proposed (Brandão et al, 2013; Parisa et al, 2022), based on the temporal profile of radiative forcing. Some approaches apply a finite assessment period, such as 100 years, and exclude emissions that are delayed beyond the assessment period, while others assign benefits for each year that the C remains stored.

Meyer et al (2012) calculated the time-integrated radiative forcing for two biochar systems and concluded that it is important to consider the temporal shift in the C cycle. More recently, Ericsson et al (2017) applied time-dependent climate metrics to assess the cultivation of willow for biochar or bioenergy, concluding that in a decarbonized energy system, biochar would provide greater climate benefit, even considering biogenic C fluxes.

In summary, LCA studies reveal that, compared with using biomass for conventional bioenergy, biochar systems contribute more efficiently to climate change mitigation if:

1 Biochar persistence in the environment is high, for example favoring applications in which more than 80% of the initial biochar C persists in the environment for more than 100 years.
2 Biochar is used in applications that reduce emissions of non-CO_2 GHGs in agriculture or displace GHG-intensive products.
3 Energy systems are substantially decarbonized through the use of renewable, low-C energy sources such that the heat

and power sources available have a GHG intensity well below that of natural gas energy.

These conditions should be sought and encouraged by a coherent policy that supports (non-biomass-based) renewable energy technologies, energy efficiency, and energy sufficiency alongside the adoption of biochar technologies. Potential emission reduction benefits of biochar should not be overlooked through sole focus on CDR. Rather, optimal outcomes considering both should be sought.

Way forward for future biochar LCA studies

From the current biochar LCA literature, a few recommendations can be made to guide future work:

- Biochar LCA studies should apply broad system boundaries, including all processes across the biochar life cycle and their equivalent in the reference system. Consequential LCA approaches are preferred over attributional, particularly where the results are intended to inform policy development. However, attributional LCA approaches, with a well-defined goal and scope, can also be useful in some applications, for instance in the certification of commercial biochar activities.

- Life cycle data from the manufacturing and operation of commercial-scale pyrolysis reactors should be made available. This would help benchmark different technologies to one another and allow the inclusion of modern pyrolysis data from commercial-scale plants in life cycle databases.

- Most biochar LCAs assume that the pyrolysis oil and gas are valorized as heat or electricity. Advanced uses of

the pyrolysis oil and gas, such as biofuels, hydrogen, or biomaterials, remain under-investigated. Likewise, advanced uses of biochar in high-value applications are not yet well studied. LCAs are needed for these applications.

- Biochar C sequestration is the largest contributor to the climate change benefits of a biochar system. However, some biochar LCAs do not clearly state assumptions on biochar permanence or do not perform sensitivity analysis on these assumptions. This should be improved, considering the latest advances in research on biochar persistence (Chapter 13), in particular the influence of biochar properties and environmental conditions.

- LCA results are context-dependent. This dependence should be better highlighted through scenario analysis, i.e., the inclusion of multiple reference systems or background contexts. Such approaches can make LCA studies easier to compare and lead to more widely applicable results. Ideally, a given biochar system shall be studied: (i) relative to at least two alternative uses of biomass (one non-productive and one productive); and (ii) in at least two background contexts (one based on fossil fuels, and one where renewable energy is available). These are critical factors for comparing the results and conclusions of different studies.

- It is also crucial that LCA results are presented with contribution analysis by life cycle stages, clearly separating the studied system from reference systems; and the contributions of supply-chain, C sequestration, and possible effects like avoided emissions from substitutions or system expansions.

- With the rising interest in CDR technologies and their inclusion in various modeling tools, there is a need for data and models from biochar LCAs to be made available to enable the inclusion of biochar in integrated assessment models, or other modeling on national or continental CDR.

- While many biochar LCA studies have limited the impact assessment to climate change, it should be recognized that other environmental impact categories are also important. Studies that have included multiple impact categories identified both potential trade-offs and synergies from biochar systems, with variations arising from the biochar supply chain, the alternative product considered, or the background context (Azzi et al, 2022). Well-performed multi-criteria LCAs are beneficial for discriminating between biochar systems that would perform equally in terms of climate change mitigation.

The global potential for climate change mitigation through biochar

Published estimates of the global climate change mitigation potential of biochar vary widely. Applying a process-based understanding can help to reconcile the wide range in estimated potential. The processes that contribute to life cycle mitigation can be summarized as: persistent biochar-C (Chapter 11), impact on soil C dynamics (Chapter 17), reduction in emissions of non-CO_2 GHG from soil (Chapter 18), and emissions avoided (displaced fossil fuel, avoided decomposition, displaced materials). Published estimates vary due to including different mitigation processes: all consider C sequestration in biochar; some include reduced non-CO_2 emissions and avoided emissions, while most exclude soil C dynamics. Another source of divergence is the feedstocks considered: some studies consider only biomass residues and wastes, while others include purpose-grown biomass crops. Importantly, some studies assess the technical potential, the theoretical maximum limited by biogeophysical conditions, while others constrain the estimate with economic, sustainability, and feasibility considerations.

Estimates of biochar's global mitigation potential based only on the C sequestered in biochar range from 0.3 Pg CO_2e yr^{-1} (Griscom et al, 2017; Fuss et al, 2018) to 4.9 Pg CO_2e yr^{-1} (Powell and Lenton, 2012). The highest estimate, 11 Pg CO_2e yr^{-1} (Powell and Lenton 2012), combined CDR, fossil fuel displacement, and soil amendment benefits. A comprehensive study that assumed dedicated biomass crops grown on degraded land, applying safeguards to protect biodiversity and ensure food security, estimated sustainable global mitigation potential at 3.7 – 6.6 Pg CO_2e yr^{-1}, including C sequestered in biochar (50% of benefit), avoided emissions of CH_4 and N_2O (20%), and displaced fossil fuel emissions (30%) (Woolf et al, 2010). Many more recent studies based their estimates on Woolf et al (2010). New robust estimates of potential are needed, undertaken through systematic global analyses that apply current knowledge of the various mitigation processes associated with biochar systems. From current studies, it is clear that biochar systems have the potential to contribute substantially to CDR, likely offering gigaton-scale mitigation in 2050.

Drivers for adoption and upscaling of biochar C sequestration

The role of negative emissions in climate change mitigation

The primary focus of climate change mitigation centers on the need for rapid and profound reductions in GHG emissions (Anderson et al, 2019). However, in the face of still-rising emissions (Liu et al, 2022), emissions reduction alone will not suffice to avoid dangerous climate change (IPCC, 2018). Carbon dioxide removal from the atmosphere is therefore expected to be required within a portfolio of mitigation measures and the required volume of CDR to reach global net zero is estimated at up to 10 Pg CO_2 per year (IPCC, 2022). As biochar is one of the few known CDR methods with the potential to contribute substantially to climate stabilization, this need for CDR is expected to be a major driver for its adoption and upscaling.

The extent to which biochar could contribute to climate change mitigation will not only depend on its effectiveness but also on how it compares to other mitigation methods that would compete for limited resources of land and biomass. Other land- and biomass-based climate solutions include reforestation, soil organic C sequestration, and bioenergy (with or without C capture and storage). The efficient allocation of resources among these approaches depends on a multidimensional array of factors including permanence, verifiability, cost, risk, and both trade-offs and synergies with food systems, energy systems, and environmental services, all of which can vary in both space and time (Woolf et al, 2016; Mac Dowell et al, 2022; Smith et al, 2023). Biochar activities could attract a premium in carbon markets, on the basis

that biochar systems can provide net CDR rather than emissions reduction, and that C sequestered in biochar is more secure than CDR through reforestation or soil C management.

Policy mechanisms to drive adoption

Early adoption of biochar may be driven by niche markets such as potting media for high-value crops; waste disposal avoidance; or pollution reduction (such as reducing excess reactive N). Although early adoption in high-value applications may not depend on financial interventions, enabling policies may nonetheless still be required to allow for the use of biochar in soils and crops, or to permit the use of waste streams as feedstocks. For example, the recent European Union regulation 2019/1009 provides minimum quality criteria for the production of biochar to be used for C sequestration in soils (Štrubelj, 2022).

Large-scale adoption of biochar at a rate sufficient to significantly address climate change is not, however, expected to emerge through the free market alone (Xia et al, 2023). To offset costs of feedstock, transport, capital, and running costs, some means to incentivize or compensate producers for the public good of climate-change mitigation would be required to make biochar systems profitable. For example, it has been estimated that biochar may not be competitive with bioenergy until C prices reach an average of around US\$125 Mg^{-1} CO_2 (Woolf et al, 2016).

Interest and investment in climate-change mitigation is currently being driven by a complex suite of measures that vary with scale and sector. The policy options available to governments to reduce GHG emissions range from voluntary measures and incentive schemes to market-based instruments including C taxes and emissions trading schemes. Broadly speaking, policy measures at the national scale are connected to countries' Nationally Determined Contributions (NDCs) under the Paris Agreement. NDCs vary by country both in ambition and type, with countries variously committing to reductions of either absolute emissions or emissions intensity, relative to either a previous point in time or to a business-as-usual baseline, and more recently adopting "net zero" targets (Rogelj et al., 2021).

Internalization of the climate-change mitigation value into a price signal can be achieved through various measures such as voluntary carbon markets, carbon taxes, direct payments, and emissions trading schemes (Chapters 31, 33, 34).

GHG accounting protocols

A common factor of climate policy mechanisms is the need for robust, affordable, and broadly accepted methods (protocols) for quantifying the climate-change impacts of mitigation actions. Protocols guide GHG reporting by nations and companies and serve as an access mechanism to carbon offset markets. Protocols have been developed to provide specific guidance for the quantification of emission reductions from project-based activities to substantiate verifiable claims for offsets and other emission reductions. The use of the protocols is determined by the policy and market frameworks to ensure that the carbon credits generated meet the requirements of the relevant regime.

Many aspects of the avoided emissions that a biochar project would deliver are covered within the scope of existing protocols from other sectors, such as renewable energy

generation, and avoidance of landfill and manure–handling emissions. However, until recently, there were no protocols to quantify the C stabilized in biochar. Gaunt and Driver (2010) highlighted the requirements for biochar protocols to (i) establish the proportion of the C contained in biochar that is more persistent compared to its persistence under its previous management; (ii) provide evidence of its long-term persistence in soil; and (iii) confirm that there are no negative impacts on existing soil C storage when adding biochar to soil. Additionally, protocols should quantify the effects of biochar on other emissions sources, especially if these lead to increased emissions. Protocols for emissions trading share some features with LCA, described above, but differ in some key characteristics. LCA, when applied for research purposes, is commonly comprehensive, including all processes even where data have high uncertainty, low quality, or are laborious or costly to obtain. In contrast, protocols for emissions trading need to be reproducible, auditable, and practical for routine application, and should err on the side of conservatism to avoid over-crediting. Uncertain aspects are often excluded if their omission leads to underestimation of mitigation benefits.

Since the mid-late 2010s, a few GHG accounting methods for biochar have begun to emerge that address these issues. One important step was the publication of a biochar GHG methodology in the 2019 revisions to the IPCC Guidelines for national GHG inventories (Ogle et al, 2019). This method provides a simple Tier 1 or Tier 2 method that can be applied globally with minimal data requirements (broad classification of feedstock material, quantity of biochar produced, classification of pyrolysis temperature into low, medium, or high and, optionally, the $H:C_{org}$ ratio of the biochar). Although material properties of the biochar (such as its $H:C_{org}$ ratio) are considered a more reliable basis on which to estimate biochar longevity than pyrolysis conditions (see Chapter 11), the IPCC method also allows for the use of pyrolysis temperature to allow countries that lack capacity to measure and monitor material properties at scale to include biochar in their national inventories. The method was provided by the IPCC as a good practice guidance annex, meaning that countries can choose to report biochar in their national inventories using the guidance, but are not required to do so. The IPCC methodology has since been expanded (Woolf et al, 2021) to provide improved estimates for a wider range of input feedstocks and to account for differential biochar decomposition rates in different climates. The IPCC method and its extensions (Woolf et al, 2021) have provided a robust basis for the development of accounting methodologies in carbon markets, such as the Puro Standard for Biochar (Puro, 2022) and the Verra VCS Methodology for Biochar Utilization in Soil and Non-Soil Applications (Verra, 2022). Other voluntary market protocols have been developed using different assumptions and methodologies, including the European Biochar Certificate assured by Carbon Standards International (https://www.carbon-standards.com/en/home) and marketed by Carbonfuture (https://www.carbonfuture.earth/). Note that accounting methodologies are continuously reviewed and frequently revised.

It should be noted that the IPCC national inventory guidelines adopt a sectoral approach, such that the emissions and removals associated with an activity are counted in separate parts of the inventory. For example, the IPCC biochar annex does not include emissions associated with feedstock provision, transport, fuel, and electricity used in the pyrolysis plant, avoided waste, or equipment manufacturing, because these are reported in other parts of the inventory. This contrasts with carbon-market methodologies, which are typically practice-based and should, in principle, estimate all lifecycle emissions and removals associated with the activity. Accordingly, many methodologies presently include additional categories of GHG impacts (for example, from the transport of biochar, or CH_4 emissions from pyrolysis), or eligibility rules to minimize upstream and indirect GHG fluxes (e.g., requiring feedstock to be derived from wastes or residues to preclude additional biomass-production emissions). Current methods in the voluntary market count only the C sequestered, adjusted for "permanence" (commonly counting only the C sequestered beyond 100 years) and emissions incurred in the supply chain in comparison with a reference case. Current methodologies differ for the treatment of pyrolysis co-products (syngas and bio-oil, or energy products derived from them): for the sake of simplicity, some methodologies disregard the co-products and allocate all the burdens to the biochar product. In other cases, burdens are allocated between energy products and biochar products, which allows the energy products to be certified under other schemes such as the EU Renewable Energy Directive (which is not a carbon credit scheme). Current methods have not included both fossil-fuel substitutions and C sequestration in biochar into a common carbon credit unit. Future methods may quantify credits for other mitigation contribution processes such as reduction in non-CO_2 GHGs and negative priming, but these are complex to quantify, as they are more variable and require site-specific estimations.

For climate metrics (discussed above), GWP100 is applied in national GHG inventories and protocols for emissions trading. Nonetheless, there is a growing consensus that single metrics are insufficient to fully understand the climate impact of GHG emissions. Separate reporting of short-lived and long-lived GHGs has been suggested (Allen et al, 2022) but, while this could be implemented in national inventories, it is challenging to implement in emissions trading. Further research is needed to inform recommendations for accounting for the different climate impacts of short and long-lived climate pollutants in carbon markets.

Role of carbon markets

Carbon dioxide removal will be required in the short term to reduce net emissions, in the medium term to meet net zero targets, and in the longer term to draw down atmospheric CO_2 to safe levels after overshoot (Babiker et al, 2022). The enormous challenge of delivering the volume of CDR that will be needed, especially in the second half of the century, requires massive investment in innovation and scaling up that is well beyond the resources of governments (Smith et al, 2023). Thus, private-sector investment is critical. Emissions trading, via voluntary and compliance markets, provides an avenue for private investment in the expansion of CDR, including biochar.

Awareness of the need for CDR is stimulating the growing demand for removal credits in the voluntary emission trading market. Biochar projects represent over 20% of CDR through novel C removal methods (Smith et al, 2023). Biochar is favored due to the durability of C storage, the co-benefits of biochar projects, and the relative cost-effectiveness of the technology. The voluntary market is supporting biochar projects across the globe.

The growing attention to CDR should not distract from efforts to reduce GHG emissions. Various initiatives (e.g., Greenhouse Gas Protocol, Science Based Targets Initiative,

and many re-sellers of carbon removal credits) are encouraging companies to adopt a hierarchy in their climate change strategies that prioritizes emissions reduction, followed by removals within their boundary, with removal credits used only to offset residual emissions. Some also push for the creation of financial products that bundle the purchase of removal credits and emission reduction credits.

To facilitate the continued expansion of CDR deployment it is important to ensure the credibility of removal credits. Governments can support the development of effective MRV frameworks that enable GHG accounting for CDR that balances the antithetical needs for accuracy, practicality, and low transaction costs. Effective governance, to manage potential adverse environmental and social effects, and to address the integrity of offset projects and the credits generated, will provide confidence to buyers of credits and certainty to investors in biochar technologies, attracting investment for upscaling biochar.

In the medium term, the voluntary market needs to be better regulated to ensure that credits issued accurately reflect real mitigation impacts. Such regulation can be voluntary (e.g., via certification of standards and registries under the International Carbon Reduction and Offset Alliance (ICROA) or the Integrity Council for the Voluntary Carbon Market (IC-VCM)), or imposed by governments (e.g., the European Union policy on certification of C removals). In the longer term, today's nascent CDR market may be absorbed by compliance markets like the EU Emission Trading Scheme.

The focus of carbon markets on CDR must not obscure the benefits of biochar emissions reduction and various sustainable development objectives (Chapter 31). Focusing on C sequestration only could lead to perverse outcomes, or sub-optimal use of biochar (e.g., landfilling of biochar).

Conclusion and recommendations

Biochar systems can simultaneously provide CDR, emissions reduction, and co-benefits for sustainable development, such as enhancing food security and reducing land degradation. Quantifying the climate change mitigation benefits of biochar requires estimating the emissions from a biochar system in comparison with a "no biochar" reference scenario. The reference includes the conventional use of biomass, the energy system that would be altered, and, in the case of biochar applied to soil, conventional soil management.

Biochar systems can deliver climate change mitigation through several processes. The best recognized is C sequestration in persistent biochar, delaying the decomposition of biomass. Other processes contributing to mitigation include the displacement of fossil fuels through the utilization of gases or bio-oil co-products of pyrolysis for energy; reduced non-CO_2 GHG emissions from soil; negative priming; avoided emissions from biomass decomposition that would otherwise occur; and enhanced plant growth. Indirect benefits can result from reduced need for fertilizer production and lower maintenance costs in non-agricultural applications. Carbon dioxide removal is provided through C sequestration in biochar, negative priming, and enhanced biomass growth while the other processes contribute to emissions reduction.

Life Cycle Assessment has been used to quantify the climate change effects of various biochar systems. LCA studies show that in most cases biochar systems lead to effective climate change mitigation, with C sequestration in biochar being the main contributor. Other effects (displacement of fossil fuels, materials, and emission reductions in agriculture) can double the benefits from C sequestration when these effects are sought and optimized. LCA studies also show that biochar systems perform better than conventional bioenergy if biochar persistence is high, biochar use provides additional climate benefits, and energy systems are already substantially decarbonized.

The use of biomass for biochar is not always the option that gives the greatest mitigation benefit: depending on the reference energy system and the aggregate impact of the various mitigation processes, which in turn depends on the biochar properties and the use of the biochar, using biomass for energy may give greater benefit, particularly where the reference energy system displaced has high GHG intensity. Thus, each situation should be assessed to determine the optimal use of biomass in that context.

The CDR capacity of biochar is a key benefit over use for bioenergy alone. It is now well-recognized that CDR will be required to meet the climate goal of the Paris Agreement. While estimates vary widely, there is strong agreement that biochar has the potential to deliver CDR above one gigaton per year, and thus make a significant contribution to the global CDR requirement. Demand for biochar-based carbon removal credits from companies aiming to achieve carbon neutrality or net zero targets is consequently escalating, thereby supporting biochar deployment globally.

The inclusion of biochar in emissions trading schemes requires scientifically robust yet practical protocols for quantifying the net sequestration of biochar systems as the basis for generating credits. Several methodologies have been recently approved in voluntary schemes. Also, a method has been approved for the voluntary inclusion of biochar in national GHG inventories, paving the way for countries to include biochar in their climate policies.

Current protocols focus on estimating the benefits of retaining C in biochar. Other mitigation processes, which are more variable (dependent on biochar and soil type, for example), are not included in existing protocols, but could be included in the future based on new knowledge that enables sufficiently accurate quantification of these context-specific mitigation benefits.

While LCA is a suitable approach for quantifying the climate effects of individual biochar systems, global-scale modeling is required to provide robust estimates of biochar mitigation potential, from technical and feasible perspectives. Integrated assessment models (IAMs), which combine biophysical and socio-economic drivers, are used to inform international climate policy. IAMs can be used to explore competition for land and biomass between biochar and alternative uses, and ensuing impacts on climate and other societal objectives. These issues must be considered to determine the sustainable potential of biochar at a global scale. However, as most IAMs do not currently consider biochar, there is a critical need to include biochar in IAMs so that the global potential of CDR options can be compared. Several research groups are now working to include biochar in IAMs, using data collected in LCA studies.

Growing recognition of the cost-effective CDR potential of biochar, and its co-benefits for sustainable development, is stimulating interest in biochar from governments and the private sector. Meeting the growing demand for biochar credits, and realizing the mitigation potential, will require significant investment to upscale biochar systems. The voluntary emissions trading market could provide investment funds, particularly if governments implement enabling policies that create incentives while ensuring adequate safeguards to provide confidence to investors and buyers of credits. Informed by the strong scientific evidence base presented in other chapters of this book, governments should devise policy measures to encourage the adoption of biochar systems with the greatest mitigation potential.

References

Abdalla K, et al 2022 Long-term continuous farmyard manure application increases soil carbon when combined with mineral fertilizers due to lower priming effects. *Geoderma* 428, 116216.

Ahlgren S, et al 2015 Review of methodological choices in LCA of biorefinery systems-key issues and recommendations. *Biofuels, Bioproducts and Biorefining* 9, 606–619.

Allen MR, et al 2022 Indicate separate contributions of long-lived and short-lived greenhouse gases in emission targets. *NPJ Climate and Atmospheric Science* 5, 1–4.

Anderson CM, et al 2019 Natural climate solutions are not enough. *Science* 363, 933–934.

Azzi ES, Karltun E, and Sundberg C 2019 Prospective life cycle assessment of large-scale biochar production and use for negative emissions in Stockholm. *Environmental Science and Technology* 53, 8466–8476.

Azzi ES, Karltun E, and Sundberg C 2021 Assessing the diverse environmental effects of biochar systems: An evaluation framework. *Journal of Environmental Management* 286, 112154.

Azzi ES, Karltun E, and Sundberg C 2022 Life cycle assessment of urban uses of biochar and case study in Uppsala, Sweden. *Biochar* 4, 1–17.

Babiker M, et al 2022 Cross-sectoral perspectives. In: Shukla PR et al (Eds) *IPCC, 2022: Climate Change 2022: Mitigation of Climate Change.*

Contribution of Working Group III to the Sixth Assessment Report of the Intergovernmental Panel on Climate Change. Cambridge, UK and New York, NY, USA: Cambridge University Press. 10.1017/9781009157926.005

Borchard N, et al 2019 Biochar, soil and land-use interactions that reduce nitrate leaching and N_2O emissions: a meta-analysis. *Science of the Total Environment* 651, 2354–2364.

Brandão M, et al 2013 Key issues and options in accounting for carbon sequestration and temporary storage in life cycle assessment and carbon footprinting. *The International Journal of Life Cycle Assessment* 18, 230–240.

Bui M, et al 2018. Carbon capture and storage (CCS): the way forward. *Energy and Environmental Science* 11, 1062–1176.

Caffrey KR, and Veal MW 2013 Conducting an agricultural life cycle assessment: challenges and perspectives. *The Scientific World Journal* 2013, 472431. 10.1155/2013/472431. Accessed 26 March 2023.

Cornelissen G, et al 2016 Emissions and char quality of flame-curtain" Kon Tiki" Kilns for Farmer-Scale charcoal/biochar production. *PloS ONE* 11, e0154617.

Cowie AL, and Cowie AJ 2013 Life cycle assessment of greenhouse gas mitigation benefits of biochar. *Case Study Report IEA Bioenergy Task 38* https://www.researchgate. net/publication/369532196_Life_cycle_ assessment_of_greenhouse_gas_mitigation_ benefits_of_biochar Accessed 25 March 2023.

Cowie A, et al 2015 Biochar, carbon accounting and climate change. In: Lehmann J, and Joseph S (Eds) *Biochar for Environmental Management*. London: Routledge. pp795–826.

Englund O, et al 2021 Strategic deployment of riparian buffers and windbreaks in Europe can co-deliver biomass and environmental benefits. *Communications Earth and Environment* 2, 1–18.

Ericsson N, Sundberg C, Nordberg Å, Ahlgren S, and Hansson PA 2017 Time-dependent climate impact and energy efficiency of combined heat and power production from short-rotation coppice willow using pyrolysis or direct combustion. *Global Change Biology Bioenergy* 9, 876–890.

European Commission 2019 Report from the commission to the European parliament, the council, the European economic and social committee and the committee of the regions on the status of production expansion of relevant food and feed crops worldwide. https://eur-lex. europa.eu/legal-content/EN/TXT/HTML/? uri=CELEX:52019DC0142&from=EN Accessed 26 March 2023

Finkbeiner M, Inaba A, Tan R, Christiansen K, and Klüppel H-J 2006 The new international standards for life cycle assessment: ISO 14040 and ISO 14044. *The International Journal of Life Cycle Assessment* 11(2), 80–85.

Fuss S, et al 2018 Negative emissions—Part 2: Costs, potentials and side effects. *Environmental Research Letters* 13, 063002.

Fryda L, Visser R, and Schmidt J 2019 Biochar replaces peat in horticulture: environmental impact assessment of combined biochar and bioenergy production. *Detritus* 05, 132–149.

Gaunt J and Driver K 2010 *Bringing biochar projects into the C marketplace: An introduction to biochar science, feedstocks and technology'*, *Carbon Consulting and Blue Source*. Canada.

Genesio L, Miglietta F, Lugato E, Baronti S, Pieri M, and Vaccari, FP 2012 Surface albedo following biochar application in durum wheat. *Environmental Research Letters* 7, 014025. 10.1088/1748-9326/7/1/014025.

Gitau JK, Sundberg C, Mendum R, Mutune J, and Njenga M 2019 Use of biochar-producing gasifier cookstove improves energy use efficiency and indoor air quality in rural households. *Energies* 12, 4285.

Griscom BW, Adams J, Ellis PW, Houghton RA, and Lomax G 2017 Natural climate solutions. *Proceedings of the National Academy of Sciences* 114, 11645–11650.

Haefele SM, et al 2011. Effects and fate of biochar from rice residues in rice-based systems. *Field Crops Research* 121, 430–440.

Hammond J, Shackley S, Sohi S and Brownsort P 2011 Prospective life cycle carbon abatement for pyrolysis biochar systems in the UK. *Energy policy* 39(5), 2646–2655.

Hou Y, Velthof GL, Lesschen JP, Staritsky IG, and Oenema O 2017 Nutrient recovery and emissions of ammonia, nitrous oxide, and methane from animal manure in Europe: effects of manure treatment technologies. *Environmental Science and Technology* 51, 375–383.

IBI 2015 Standardized product definition and product testing guidelines for biochar that is used in soil. Version 2.1, International Biochar Initiative https://biochar-international.org/wp-content/uploads/2023/01/IBI_Biochar_Standards_V2.1_Final.pdf Accessed 25 March 2023

IPCC 2018 Summary for Policymakers. In: Masson-Delmotte V et al (Eds) *Global Warming of 1.5°C. An IPCC Special Report on the Impacts of Global Warming of 1.5°C above Pre-industrial Levels and Related Global Greenhouse Gas Emission Pathways, in the Context of Strengthening the Global Response to the Threat of Climate Change, Sustainable Development, and Efforts to Eradicate Poverty*. Cambridge, UK and New York, NY, USA: Cambridge University Press, pp3–24.

IPCC 2022 Summary for Policymakers. In: Shukla PR et al (Eds) *Climate Change 2022: Mitigation of Climate Change. Contribution of Working Group III to the Sixth Assessment Report of the Intergovernmental Panel on Climate Change*. Cambridge, UK and New York, NY, USA: Cambridge University Press. 10.1017/9781009157926.001

Ippolito JA, et al 2020 Feedstock choice, pyrolysis temperature and type influence biochar characteristics: a comprehensive meta-data analysis review. *Biochar* 2, 421–438.

Jeffery S, Verheijen FG, Kammann C, and Abalos D 2016 Biochar effects on methane emissions from soils: a meta-analysis. *Soil Biology and Biochemistry* 101, 251–258.

Jolliet O, et al 2018 Global guidance on environmental life cycle impact assessment indicators: impacts of climate change, fine particulate matter formation, water consumption and land use. *The International Journal of Life Cycle Assessment*, 23, 2189–2207. 10.1007/s11367-018-1443-y.

Joseph S, et al 2021 How biochar works, and when it doesn't: A review of mechanisms controlling soil and plant responses to biochar. *Global Change Biology Bioenergy* 13, 1731–1764.

Kätterer T, et al 2019 Biochar addition persistently increased soil fertility and yields in maize-soybean rotations over 10 years in sub-humid regions of Kenya. *Field Crops Research* 235, 18–26.

Laird DA, Brown RC, Amonette JE, and Lehmann J 2009 Review of the pyrolysis platform for coproducing bio-oil and biochar. *Biofuels, Bioproducts and Biorefining* 3, 547–562.

Lehmann J, et al 2021 Biochar in climate change mitigation. *Nature Geoscience* 14, 883–892.

Liu Z, Deng Z, Davis SJ, Giron C, and Ciais P 2022 Monitoring global carbon emissions in 2021. *Nature Reviews Earth and Environment* 3, 217–219.

Mac Dowell, N, Reiner DM, and Haszeldine RS 2022 Comparing approaches for carbon dioxide removal. *Joule* 6, 2233–2239.

Major J, Rondon M, Molina D, Riha SJ, and Lehmann J 2010 Maize yield and nutrition during 4 years after biochar application to a Colombian savanna oxisol. *Plant and Soil* 333, 117–128.

Matuštík J, Hnátková T, and Kočí V 2020 Life cycle assessment of biochar-to-soil systems: A review. *Journal of Cleaner Production* 259, 120998.

Meyer S, Bright RM, Fischer D, Schulz H, and Glaser B 2012 Albedo impact on the suitability of biochar systems to mitigate global warming. *Environmental Science and Technology* 46, 12726–12734.

Meyer S, Glaser B, and Quicker P 2011 Technical, economical, and climate-related aspects of biochar production technologies: a literature review. *Environmental Science and Technology* 45, 9473–9483.

Ogle SM, et al 2019 Generic methodologies applicable to multiple landuse categories. In:

Calvo Buendia E et al (Eds) *2019 Refinement to the 2006 IPCC Guidelines for National Greenhouse Gas Inventories. Vol. IV Intergovernmental Panel on Climate Change Switzerland.* Switzerland: IPCC. pp2.1–2.96.

Parisa Z, Marland E, Sohngen B, Marland G, and Jenkins J 2022 The time value of carbon storage. *Forest Policy and Economics* 144, 102840.

Pascual MB, Sánchez-Monedero MA, Chacón FJ, Sánchez-García M, and Cayuela ML 2020 Linking biochars properties to their capacity to modify aerobic CH_4 oxidation in an upland agricultural soil. *Geoderma* 363, 114179.

Powell TW, and Lenton TM 2012 Future carbon dioxide removal via biomass energy constrained by agricultural efficiency and dietary trends. *Energy and Environmental Science* 5, 8116–8133.

Prade T, Björnsson L, Lantz M, and Ahlgren S 2017 Can domestic production of iLUC-free feedstock from arable land supply Sweden's future demand for biofuels? *Journal of land use Science* 12, 407–441.

Puro 2022 Puro.earth Biochar Methodology https://puro.earth/methodologies/, accessed 25 March 2023

Reisinger A, et al 2022 GHG Emissions Metrics. Cross-Chapter Box 2. In: Shukla PR, et al (Eds) *IPCC, 2022: Climate Change 2022: Mitigation of Climate Change. Contribution of Working Group III to the Sixth Assessment Report of the Intergovernmental Panel on Climate Change.* Cambridge, UK and New York, NY, USA: Cambridge University Press.

Roberts KG, Gloy BA, Joseph S, Scott NR, and Lehmann J 2010 Life cycle assessment of biochar systems: estimating the energetic, economic, and climate change potential. *Environmental Science and Technology* 44, 827–833.

Rogelj J, Geden O, Cowie A, and Reisinger A 2021 Three ways to improve net-zero emissions targets. *Nature* 591, 365–368.

Schmidt HP, et al 2019 Pyrogenic carbon capture and storage. *Global Change Biology Bioenergy* 11, 573–591.

Shine, KP 2009 The global warming potential—the need for an interdisciplinary retrial. *Climatic Change* 96, 467–472. 10.1007/s10584-009-9647-6.

Singh BP, and Cowie AL 2014 Long-term influence of biochar on native organic carbon mineralisation in a low-C clayey soil. *Scientific Reports* 4, 3687.

Singh BP, et al 2015 In situ persistence and migration of biochar carbon and its impact on native carbon emission in contrasting soils under managed temperate pastures. *PloS one* 10, e0141560–e0141560.

Smith SM, et al 2023 The State of Carbon Dioxide Removal. 1st Edition. https://www.stateofcdr.org, accessed 26 March 2023

Sørmo E, et al 2020 Waste timber pyrolysis in a medium-scale unit: Emission budgets and biochar quality. *Science of the Total Environment* 718, 137335.

Sorrenti G, Masiello CA, Dugan B, and Toselli M 2016 Biochar physico-chemical properties as affected by environmental exposure. *Science of the Total Environment* 563, 237–246.

Štrubelj L 2022 Waste, fertilising product, or something else? EU Regulation of Biochar. *Journal of Environmental Law* 34, 529–540.

Terlouw T, Bauer C, Rosa L, and Mazzotti M 2021 Life cycle assessment of carbon dioxide removal technologies: a critical review. *Energy and Environmental Science* 14, 1701–1721.

Tilman D, Hill J, and Lehman C 2006 Carbon-negative biofuels from low-input high-diversity grassland biomass. *Science* 314, 1598–1600.

Tisserant A, and Cherubini F 2019 Potentials, limitations, co-benefits, and trade-offs of biochar applications to soils for climate change mitigation. *Land* 8, 179.

van der Weerden TJ, et al 2021 Ammonia and nitrous oxide emission factors for excreta deposited by livestock and land-applied manure. *Journal of Environmental Quality* 50, 1005–1023.

Verra 2022 VM0044 Methodology for Biochar Utilization in Soil and Non-Soil Applications, v1.0 https://verra.org/methodologies/vm0044-methodology-for-biochar-utilization-in-soil-and-non-soil-applications-v1-0/, accessed 26 March 2023

Wang Y, et al 2023 Inducing inorganic carbon accrual in subsoil through biochar application on calcareous topsoil. *Environmental Science and Technology* 57, 4, 1837–1847.

Wang Z, Dunn, JB, Han, J, and Wang, MQ 2013 Effects of co-produced biochar on life cycle greenhouse gas emissions of pyrolysis-derived renewable fuels. *Biofuels, Bioproducts and Biorefining* 8, 189–204. 10.1002/bbb.1447.

Weng ZH, et al 2022 Microspectroscopic visualization of how biochar lifts the soil organic carbon ceiling. *Nature Communications* 13, 5177.

Werner C, Schmidt H-P, Gerten D, Lucht W, and Kammann C 2018 Biogeochemical potential of biomass pyrolysis systems for limiting global warming to 1.5 C. *Environmental Research Letters* 13, 044036.

Whitman T, Hanley K, Enders A, and Lehmann J 2013 Predicting pyrogenic organic matter mineralization from its initial properties and implications for carbon management. *Organic Geochemistry* 64, 76–83.

Woolf D, Amonette JE, Street-Perrott FA, Lehmann J, and Joseph S 2010 Sustainable biochar to mitigate global climate change. *Nature Communications* 1, 56.

Woolf D, and Lehmann J 2012 Modelling the long-term response to positive and negative priming of soil organic carbon by black carbon. *Biogeochemistry* 111, 83–95.

Woolf D, Lehmann J, Fisher EM, and Angenent LT 2014 Biofuels from pyrolysis in perspective: trade-offs between energy yields and soil-carbon additions. *Environmental Science and Technology* 48, 6492–6499.

Woolf D, Lehmann J, and Lee DR 2016 Optimal bioenergy power generation for climate change mitigation with or without carbon sequestration. *Nature Communications* 7, 13160.

Woolf D, et al 2021 Greenhouse gas inventory model for biochar additions to soil. *Environmental Science and Technology* 55, 14795–14805.

Wurzer C, Sohi S, and Mašek O 2019 Synergies in sequential biochar systems. GreenCarbon ETN Book, *Chapter* 12, 147–160.

Xia L, et al 2023 Integrated biochar solutions can achieve carbon-neutral staple crop production. *Nature Food* 4, 236–246.

Ximenes FA, Cowie AL, and Barlaz MA 2018 The decay of engineered wood products and paper excavated from landfills in Australia. *Waste Management* 74, 312–322.

Biochar sustainability

Cecilia Sundberg and Elias Azzi

Introduction

Sustainability refers to the desirable ability of human societies to be prosperous over long time scales on this planet. Sustainability, as a field of science, has become more complex and transdisciplinary over the years. It builds on almost four decades of work since the Brundtland report and its first definition of sustainable development, "a development that meets the needs of the present without compromising the ability of future generations to meet their own needs" (Brundtland, 1987). Core concepts are the triple-bottom-line concept (social, environmental, and economic dimensions); and the more recent theories of planetary boundaries (Steffen et al, 2015) and social foundation (Leach et al, 2013), where the economy is no longer a pillar but rather a tool for sustainable development. The field's objective is to study the interactions between humans and their environment in order to understand and identify solutions that can tackle the challenges humanity is facing, e.g. climate change, biodiversity losses, pollution,

social injustice, and inequalities. Sustainability also needs to be operational, and as such, it has become a political and societal issue. Policies in many countries and at the international level, aim for sustainability and call for actions to reach common goals. Actions involve both the deployment of new technologies and the modification of human organizational structures, at multiple levels.

This said, while current human activities are clearly unsustainable (Steffen et al, 2015), it is difficult to identify what alternatives could be sustainable. Often, progress toward sustainability is incremental, meaning that changes lead to improvements step-by-step. Also, because sustainability is multi-faceted, involving multiple interconnected social and environmental aspects, it is often split into smaller pieces. The Sustainable Development Goals (SDGs) adopted by the United Nations in 2015 is such a political framework that defines targets to be reached on several interconnected domains.

DOI: 10.4324/9781003297673-31

In that context, it is not obvious how to approach the intersection of biochar and sustainability. One could say that nothing is sustainable in itself, but everything should contribute to sustainability and avoid the pitfalls that undermine it. Biochar is not fundamentally different from any other human enterprise in that respect: pitfalls in biochar systems must be avoided for biochar to deliver its promise of being a carbon dioxide removal method that effectively mitigates climate change with numerous co-benefits. Therefore, when addressing the most important issues for sustainability, aspects that cast doubts and limitations on biochar sustainability are also important to discuss. As biochar practice is becoming more established and markets emerge, it is also important to ask what is needed for biochar to be implemented and grow to scale in ways that contribute to sustainable development. These questions have been guiding the writing.

The main part of this chapter is a description of the potential of biochar to contribute to the fulfillment of the SDGs. This section is organized by following several themes, each covering one or more SDGs. Some SDGs are crosscutting and are described under several themes. Further, some approaches and strategies to sustainable development are described, and the potential connections of biochar to these strategies are outlined. Finally, some sustainability assessment tools are described (life cycle assessment and certification standards among others) and a future outlook is made.

Biochar in the Sustainable Development Goals

The sustainable development goals have been widely recognized and used for sustainability assessment since their introduction in 2015. They outline global objectives for sustainable development with specified goals for 2020 and 2030, with 167 targets in 17 goals covering and integrating social, economic, and environmental sustainability dimensions. The goals were decided by the UN General Assembly and are thus the result of an international consensus process. Consequently, SDGs are inspired by, but are not per se, the result of the state of the art in research on sustainable development.

In the text below, the role of biochar in the SDGs is described in thematic sections (climate, food, health, water, energy, urban development, waste management, partnership, and education), each section focusing on one or two of the SDGs. Indeed, the SDGs are not 17 separate topics, but rather 17 interconnected perspectives on sustainable development. Therefore, some SDG topics do not have separate chapters, but are included in several of the sections. These include poverty, gender, and inequality. Some SDGs, such as Life Below Water (SDG 14) are considered less related to biochar and are not described at all.

Biochar and climate

The effect of biochar on the climate system is extensively described in Chapter 30, with a focus on quantification of net climate change mitigation impact via the use of life cycle assessment. In short, biochar production and use can effectively contribute to climate change mitigation, when properly implemented. Conditions for maximizing climate benefits are: (i) optimal use of the biochar, e.g. selecting biochar types and applications to mitigate soil greenhouse gas emissions, increase crop productivity, or replace other materials with high climate impact; (ii) high persistence of the biochar in soil, achieved through biochar production conditions (Chapter 11); and (iii) clean production principles, i.e. minimizing methane and particle emissions, and utilizing pyrolysis co-products to displace fossil fuels. At a higher systemic level, the availability of biomass and other low-carbon energy sources is important for the trade-off between the use of biomass for energy or biochar. When coal is the main energy source, it is likely that the use of biomass for bioenergy is preferable to biochar (Peters et al, 2015).

Climate impact as estimated in life cycle assessment (LCA) studies (Chapter 30) focuses on climate change mitigation at the global level. However, biochar also has local climate effects of relevance to sustainability which are not captured in LCA studies. For instance, biochar's effects on water fluxes in soil also modify the microclimate. Biochar applied to soils at very high rates, if not incorporated into soil, can lower the soil albedo and thereby lead to local heating. Such local climatic effects can contribute to climate change adaptation and the resilience of food production systems e.g. to drought-related crop failure (Li et al, 2021). These local climatic effects can be measured at the plot level but are difficult to quantify and include in generic sustainability assessments.

Biochar in agriculture and food systems

The ultimate targets of SDG2 are to end hunger (target 2.1) and malnutrition (target 2.2) by 2030. In order to reach these goals, several targets related to agriculture and food systems are included in SDG2. Biochar can contribute to two of these: to increase agricultural productivity and farmer income (target 2.3) and to ensure sustainable food production systems (target 2.4). Target 2.A, to increase investment and international cooperation, is also of relevance for biochar implementation. Other targets that relate to genetic diversity (target 2.5), trade (target 2B), and food commodity markets (target 2 C) will not be discussed further here as they are less relevant to biochar.

After decades of steady improvement in the reduction of hunger globally, there has been a negative trend since 2018. It is estimated that 702–828 million people suffered from hunger in 2021 and the main reasons are conflicts, climate change, and economic downturn (FAO et al, 2022). While technology is not a quick fix to these complex causes, there is nevertheless an opportunity for biochar to contribute to improving the situation, by increasing agricultural productivity, self-sufficiency, and resilience.

How and when biochar contributes to improved agricultural productivity is a result of a range of effects of biochar on plant growth, nutrient availability, water availability, soil physical conditions (Chapter 13), and by affecting plants' resistance to pathogens and insects (Chapter 23).

Biochar implementation in farming requires a systems perspective at several levels (Figure 31.1). At the micro level, there is the integration of biochar in the soil-crop-fertilizer-pest management system. Secondly, there is a need for integration in the farming system of biomass, biochar production, co-product use, and biochar use, which includes the physical and technical resources, but also aspects such as labor, timing, and financing. This in turn requires a macro-level integration with markets, government support as well as access to agronomic and technical knowledge.

One factor that will be of relevance for biochar contribution to SDG2 will be the successful uptake of biochar technologies by

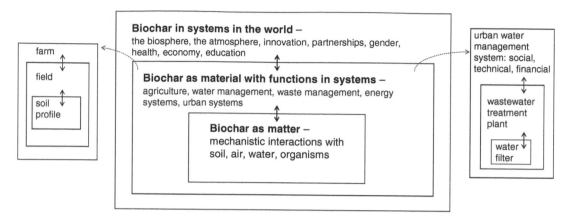

Figure 31.1 *Biochar as nested systems. Innermost are the interactions between biochar and its surroundings, through physical, biological, and chemical processes. Next is biochar providing functions in human-made technical systems which can be described by material, energy, and financial flows. In the outer layer are wider human and planetary system aspects. Each level in the central figure can be described in more detail as described in examples for agriculture and water systems to the left and right, respectively*

smallholder farmers in low- and middle-income countries, who feed a large part of the global population. The SDG target on agricultural productivity also covers the income of small-scale food producers, emphasizing women and family farmers among other groups, and specifying that their access to productive resources and inputs, knowledge, and financial services are important. This shows that an important requirement for biochar to contribute to SDG2 is to make sure that the scientific knowledge on how biochar can increase productivity reaches end users with limited capacity, i.e. those at the highest risk of hunger, who tend to be people with very limited access to agricultural inputs, knowledge, and financial services. There is potential for the uptake of biochar also in communities with limited resources, considering that biochar can be produced from local resources, using low-cost technology and providing direct benefits to farmers (Steiner et al, 2018; Sutradhar et al, 2021). Yet there will be several hurdles,

and the research regarding success factors and constraints for the uptake of agricultural innovations should guide biochar project development (Shilomboleni et al, 2019). This literature suggests that successful dissemination of agricultural innovations requires that the context is well understood, including an understanding of factors behind farmer decision-making (Hermans, 2021), to foster community engagement, build agency and empowerment, and have long-term projects.

In an interdisciplinary project in India, an assessment of soil effects of biochar was combined with an analysis of farmers' local knowledge, practices, and aspirations (Bellè et al, 2022). One finding was that farmers with a stronger financial position showed greater interest in biochar while poorer farmers were more skeptical for various reasons including a lack of biomass resources and labor. This indicates a risk for increased inequalities within rural communities if new technologies are implemented without consideration of the factors behind local inequality. A study of sustained use of biochar in Tanzania observed that the uptake of biochar was very low in the lowest income group, with a lack of access to capital and feedstock for biochar production as well as awareness and knowledge as important factors limiting biochar uptake (Rogers et al, 2021).

The implications of biochar introduction on socio-ecological resilience and vulnerability were assessed in two Indian villages before biochar implementation (Müller et al, 2019). Informed by interviews and focus group discussions, a network was mapped for each village of linked livelihood assets (human, social, financial, physical, and natural assets). Biochar was anticipated to increase the resilience overall by increasing soil quality and yields, and reducing dependence on irrigation systems, manure, and fertilizer. Increased labor demand was identified as a factor reducing the resilience. Small-scale biochar production improved resilience by offering farm management diversity, but had a higher demand on capital, government, scientific knowledge, and land rights. Other important aspects were the market prices and farmers' trust in new technologies and government bodies. In these villages, there was good availability of biomass for biochar production. The barriers to biochar implementation therefore vary depending on the local socio-ecological situation, which not only includes physical and economic aspects but also social and cultural dynamics (Müller et al, 2019). They note that the risks of negative experiences of an inadequate biochar implementation can have wider effects by reducing the community's trust and sense of autonomy.

Although biochar has been presented for nearly a decade as a potential solution to improve food systems in low-income countries, there is still very little literature and documentation on the long-term experiences of biochar projects in such socio-ecological contexts. One rare example is one research group that investigated experiences in Tanzania (Rogers et al, 2021; Fridahl et al, 2022). Participants in 7 projects in Tanzania were interviewed several years after their initiation. Most projects were not viable without external funding and had come to an end. Successfully sustained projects had targeted cash crop farmers who got economic benefits from biochar use. Yield improvements, reduced dependence on external agricultural inputs, and more resilient farming systems were drivers for the smallholder farmers who adopted and sustained using biochar.

There is little research on gender and biochar use. In a study in Tanzania, women farmers used biochar less than men, and it was discussed that women have very little decision-making power over land use in that region (Rogers et al, 2021). Research on gender in agriculture has shown that constraints in resources such as land, labor, and agricultural inputs often cause lower uptake of new technologies by women (Peterman et al, 2014). Moreover, gender aspects are very contextual and vary over time (Doss et al, 2001). There is a need for more investigation of gender aspects in biochar projects.

Biochar and health

Biochar can contribute indirectly to the achievement of SDG3 (Good Health), which covers topics related to diseases, causes of death, and access to healthcare. The specific target where biochar production and use could be beneficial is target 3.9 which aims to reduce the number of deaths and illnesses from hazardous chemicals and air, water and soil pollution and contamination. There are several ways in whereby biochar could contribute to this goal. Two such pathways are the open burning of biomass waste and the traditional burning of biomass for heating and cooking. By shifting practices from open burning to clean pyrolysis processes with biochar production, health effects from air pollution, indoors or outdoors, can be reduced (see section on energy, below). Moreover, by pyrolyzing biomass or by using biochar to treat water or soils, the spread of pollutants in the environment can be reduced, and, as a consequence, human exposure to these pollutants may be reduced (see sections on water and waste management, below).

Target 3.9 is a clear example of the interconnectivity of the SDGs, as it stresses that the human health impacts of environmental pollution should be reduced, and this will primarily be done not by interventions in the health sector but in interventions reducing the emission of pollutants to the environment. Interventions aiming to reduce pollution are also included in some other SDGs related to topics such as clean water (SDG6), energy (SDG7), waste (SDG12), industry, urban areas, etc.

Biochar and water

The adsorption properties of biochar make it useful for many technical applications for clean water and sanitation (SDG6). Biochar can play a role in the treatment of drinking water (target 6.1), wastewater treatment, and other sanitation technologies (targets 6.2 and 6.3). To describe the potential contribution of biochar to SDG6, a hierarchical system perspective can be helpful (Figure 31.1). The hierarchy starts with mechanisms at the molecular level, rises to technical components such as filters, and then to technical systems such as treatment plants. This is followed by socio-techno-economic systems and, finally, wider ecological, economic, and social systems, which affect and are influenced by the technologies.

The mechanisms for biochar interaction with water and pollutants have been described in other chapters in this book. Nutrients such as N and P are described in Chapter 16, heavy metals in Chapter 21, and organic pollutants in Chapter 22. These mechanisms need to be understood for the design of treatment technologies, which need to consider treatment objectives, water quality, biochar properties, and how they change over time (Kamali et al, 2021; Xiang et al, 2020). In addition to drinking water, household wastewater, and industrial water, biochar can be used to reduce nutrient runoff from agricultural land (Rahman et al, 2021).

The potential for biochar to contribute to SDG6 concerns not only the treatment of water, but also the management of nutrients and pollutants that have either been separated from water or kept away from water by source separation to enable nutrient recycling and avoid water contamination. Nutrient-enriched biochar can be used in agriculture after use in water treatment. Biochar can be used in sludge treatment methods such as composting or anaerobic digestion, and sludges can be pyrolyzed and turned into biochar (Gopinath et al, 2021). Source-separated fractions such as urine or fecal sludge can be treated with biochar for reuse in agriculture or other productive uses (Masrura et al, 2021, Pathy et al, 2021 Sutradhar et al, 2021). There are various potential sequential biochar uses involving water treatment or sludge management (Wurzer et al, 2022).

Implementing biochar in water systems also requires the production and availability of biochar with the desired properties. Production of engineered biochar with desired properties, for a treatment method and the final use of the used biochar, places specific requirements on biomass and its pre-treatment, the thermochemical process, and post-treatment of the biochar (Chapter 27).

For biochar technologies to be implemented in practice in water and sanitation systems, biochar must be integrated at the project level, where topics such as stakeholder perspectives, financing, regulations, processes for selection among available technical options, etc. must be resolved. This will require that suitable products based on biochar are developed and made available in the market and in planning tools. It will also require that biochar is integrated at higher system levels (Figure 31.1) described in integrated water resources management (target 6.5) including water ecosystems and catchment areas, policy, and innovation systems.

Meeting the basic needs for clean water (target 6.1) and sanitation (target 6.2) for the millions of people who lack access is a priority in SDG6, and support to developing countries is target 6.7. Moreover, target 6.8 is support to local engagement, emphasizing the need to involve communities in developing technical systems that meet their needs and priorities.

Biochar and energy

SDG7 aims to ensure access to affordable, reliable, sustainable, and modern energy for all, thereby encompassing two major global

challenges; to provide energy to those who lack basic energy services and to transform the global energy system towards sustainability. The targets of SDG7 include providing modern energy services for the large proportion of the human population who lack full access to electricity and clean cooking solutions (target 7.1); increasing the global proportion of renewable energy (target 7.2), and doubling the improvement in energy efficiency (target 7.3). Since biochar is produced through processes that also provide bioenergy, there are close connections between biochar and SDG7, and an opportunity for biochar production to contribute to the achievement of SDG7.

Biochar production is a biomass technology with bioenergy co-products and can conversely be described as a bioenergy technology with a valuable material co-product (Figure 31.2). Modern bioenergy technologies that produce and use biofuels in solid, liquid, and gaseous forms have been implemented and together they are a major renewable energy source globally, supplying more than 5% of total global final energy (IEA et al, 2022). They also have an important role in most scenarios for future renewable energy systems (IEA 2021; Riahi et al, 2022). There is a need for research that compares biochar production and use with other biomass conversion

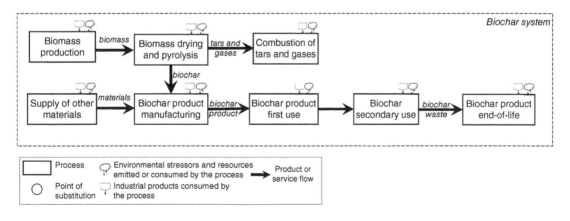

Figure 31.2 *Generic description of a biochar system*

technologies, and that analyses priorities between biochar and its energy co-products. Thereby the relevance of biochar in relation to other priorities of biomass use for energy, material, and land use can be clarified and communicated. Such analyses can be done at different scales, from local case studies to regional and global assessments. In comparison to some established bioenergy technologies, biochar has less competition with food production and more synergies, due to its ability to improve soil quality and productivity.

Bioenergy co-produced during pyrolysis can contribute to SDG7.2 (increase the proportion of renewable energy) if the biomass used for biochar would otherwise not be used for energy. In addition, pyrolysis, with the right technologies to upgrade the pyrolysis gas and oil, can lead to the production of higher-value gaseous or liquid fuels than with conventional bioenergy technologies. There is also a potential for pyrolysis technology to provide a gain in energy efficiency (target 7.3), in case it can replace inefficient bioenergy methods.

About 2.4 billion people lack access to clean cooking technologies, and about 3 million people are estimated to die annually from indoor air pollution (IEA et al, 2022). Reaching SDG target 7.1, ensuring universal access to electricity and clean cooking solutions, is thus a major challenge. There are biochar-producing cookstoves available that could contribute to this goal. These are either pyrolytic or microgasification designs, including top-lit-updraft (TLUD) gasifier cookstoves. In East Africa and South Asia, there have been scientific projects to introduce and test such cookstoves (Whitman et al, 2011; Gitau et al, 2019a) and also several less documented NGO projects. Biochar-producing cookstoves can diversify fuel use by making smaller and lighter pieces of biomass useful for cooking. As they are energy efficient, they can reduce the fuel needed for cooking, in addition to

producing biochar as a valuable by-product. Fuel saving and diversification can reduce the pressure on deforestation. These stoves have been shown to reduce emissions considerably compared to traditional cooking methods (Jetter et al, 2012; Gitau et al, 2019a).

It is well known that there are multiple barriers to the transition from traditional to improved cooking systems (Vigolo et al, 2018; Stanistreet et al, 2014). Barriers occur along the chain of accessing cooking technologies, starting to use them, and sustaining their use. Evidence from studies on the uptake of biochar-producing cookstoves confirms that there are hurdles to the uptake also of this type of stove (Gitau et al, 2019b; Eltigani et al, 2022). Reported challenges include handling difficulties with lighting and refilling fuel, stoves being too small for the cooking pots used, lack of flexibility regarding cooking power, and durability. One difficulty with most designs of biochar-producing cookstoves is that they are batch processes where fuel cannot be added continuously, but needs to be refilled at certain intervals, a process that is time-consuming and interrupts cooking. In theory, biochar cookstoves have a great potential to contribute to the attainment of SDG7 and other SDGs, but there is a need to continue developing the designs of biochar-producing cookstoves interactively with user communities if they are to contribute to the achievement of SDG7. Consideration of gender aspects related to fuels and cookstoves is also crucial for understanding the adoption of biochar-producing cookstoves (Gitau et al, 2018). There is a need for further documentation and sharing of practical experiences of biochar cookstove projects.

While there are many potential synergies between biochar and SDG7, there is also a trade-off situation between biochar and bioenergy. Firstly, biomass can be used for combined biochar and bioenergy production, but another option is to produce bioenergy only from biomass, for example through direct combustion. This is the most energy-efficient use of the biomass. Secondly, the produced char has its uses as biochar in soils, but it also has an energy content and can be used as a fuel. In many countries, there is a large demand for charcoal for household cooking and heating (Box 31.1), and there are also many industrial uses for charcoal.

Biochar and urban development

Biochar contribution to SDG11 (sustainable cities and communities) is possible through various urban uses of biochar and the production of biochar from urban waste. Major applications are planting trees, mixing in landscaping soils, and stormwater management. For the planting of trees, a plant bed that is resistant to compaction and provides space for root growth, air, and water transport, made of a mix of stones and biochar has been developed in the city and Stockholm, Sweden, and is spreading to other cities (Azzi et al, 2021; Saluz et al, 2022). By enhancing tree growth in cities and reducing tree mortality rates, biochar contributes to the benefits of trees in urban areas, which include reducing high temperatures by providing shade and evaporative cooling, improving well-being, and, to some extent, reducing pollution (Pataki et al, 2021).

Box 31.1 Charcoal

Charcoal is a fuel that partially overlaps with biochar in its properties, and the relation between biochar and charcoal will have implications for the sustainability of biochar development. Charcoal is a major fuel in most countries in sub-Saharan Africa and some Asian countries, which is highly problematic from an environmental point of view (Rose et al, 2022). Charcoal is traditionally produced through high-emitting pyrolysis processes with low energy efficiency (Bailis et al, 2015). It contributes to forest degradation or deforestation, often in combination with timber production or conversion of forests to farmland (Sedano et al, 2022). Charcoal production is an important source of income for rural populations in many parts of sub-Saharan Africa (Brobbey et al, 2019; Rose et al, 2022). The main demand for charcoal is as cooking fuel in urban areas, where it is often the most affordable fuel. There is a lack of supporting government policies for improving the sustainability of charcoal production (Doggart and Meschack, 2017) resulting in unsustainable informal value chains (Roos et al, 2021).

In places where charcoal is a major fuel, the introduction of biochar can be expected to interact with the existing charcoal value chains. The question is how? In a negative scenario, biochar demand adds to charcoal demand and becomes a driver of forest and land degradation in the location of production, while it improves soils where biochar is used. In a positive scenario, there is a spillover of clean and efficient biochar production technology, replacing traditional charcoal production methods, thereby reducing the negative environmental impacts of charcoal and increasing rural income from biomass upgrading. Combined value chains are also possible, for example when large pieces are used as fuel and finer pieces as biochar in soil. To develop sustainable biochar markets in areas with large charcoal use, interactions with the charcoal market must be considered.

Mixing biochar in landscape soil not only provides a good substrate for plants but also replaces other substrates, which are often non-renewable virgin resources. In stormwater management, biochar is attractive for its potential for retention of water in soil and removal of pollutants (Mohanty et al, 2018).

There are other uses of biochar in cities where biochar can be an important niche product, but where volumes will be smaller, for example as a substrate material for green roofs (Chen et al, 2018), where the lightweight and good water-holding capacity is an asset (Cao et al, 2014). Biochar also has a potential use as building and road material (Chapter 28). Moreover, biochar can contribute to an urban circular economy, if biochar used in the city is produced from urban waste, for example, park and garden waste, or woody demolition waste. In some countries, the heat from pyrolysis can also be used for heating buildings through district heating systems (Azzi et al, 2019).

Biochar in waste management

Cities and companies all over the world are struggling with the management of waste. Biochar can have a role in waste management, as a minor contribution to the large environmental, technical, financial, and organizational waste management challenges. There are two major routes for biochar in waste management: use of waste biomass as a feedstock for biochar production, and use of biochar as a material in waste management.

Waste management has several goals, which can be summarized as protecting human health, protecting the environment, and using resources efficiently (Wilson et al, 2012). Waste management activities often contribute to all of these goals at the same time. For example, by transforming woody biomass waste to biochar and utilizing the

heat from pyrolysis gases, there are less emissions to the environment, a reduction in exposure of humans to pollutants, and utilization of biomass resources in the form of heat and biochar.

To ensure successful waste management, an integrated approach has been established as the state of the art, whereby many aspects are considered, including technical, organizational, financial, and social (Marshall and Farahbakhsh, 2013). In integrated waste management, the roles and interests of different stakeholders are also considered explicitly. A strategy that is included in any successful waste management is the separation of different waste streams near the point of generation. Thereby the most suitable treatment method can be selected for each waste stream. Another widespread component of integrated waste management is known as the waste hierarchy, or "reduce, reuse, recycle" (3 R) (Redlingshöfer et al, 2020). The waste hierarchy proposes that there is a priority order in waste management with waste prevention at the top, followed by reuse, recycling, energy recovery, and finally, with the lowest priority, safe disposal of waste. Production of biochar from waste can be classified as reuse since a new product is produced from the waste. Biochar is thus at a high level of priority in the waste hierarchy. The waste hierarchy is a central component of the circular economy, which is further described below.

Four targets in the SDGs explicitly concern waste management. Target 11.6, in SDG11 (sustainable cities and communities), focuses on the reduction of the per capita environmental impact of cities, including air pollution and municipal waste management. Targets 12.4 and 12.5 in SDG12 (responsible production and consumption) concern prevention, reduction, reuse, and recycling of waste. Target 12.3

specifically targets the reduction of food waste through prevention, and is thus of less relevance for biochar than the other three targets.

The use of waste for biochar production can be more resource-driven or waste management-driven. In resource-driven biochar production, a waste stream is selected for its characteristics for producing biochar of a certain quality for an intended use. In this case, the demand for biochar of a certain quality will be a major driver for production. In waste-management-driven biochar production, thermochemical conversion with biochar production is identified as a suitable method for managing a certain waste stream. In some cases, pyrolysis of a certain waste stream can present advantages over other waste treatment technologies, e.g. by destruction of microplastics in sludge, or stabilization of certain heavy metals. When waste management is the driver for biochar production, the availability of biomass at low or negative cost is a major driver for production, while the low quality of the biochar may limit its use.

Agro-industrial or forest industry residues are waste streams that can be valuable for conversion to biochar. Potential benefits are the availability of rather homogeneous and clean feedstock for biochar production, often combined with a heat demand on-site. Challenges include the seasonality of production. Production of biochar from different agro-industrial waste streams has been well investigated, while less has been published about the integration in industries, energy considerations, and costs.

Several uses of biochar in waste management have been researched, including its use in composting (Chapter 26; Sanchez-Monedero, 2018), anaerobic digestion (Chen et al, 2023), landfill covers (Qin et al, 2022), and treatment of contaminated soils (Chapters 21 and 27). Biochar has been shown to reduce emissions of methane from composting processes and landfills and to stabilize anaerobic digestion processes.

Biochar partnerships and education

Targets 17.6–17.9 of SDG17 (Partnership for the goals), concern knowledge sharing and cooperation on technology, innovation, and SDG capacity internationally, including the least developed countries. Collaborative projects between academia and practitioners, to learn and co-create biochar systems that live up to sustainability criteria, is a possible way forward for moving biochar from research to large-scale practice with consideration of sustainability. It is also a way to bring practical knowledge into academia.

SDG4 (Quality education) has general targets for access to quality education but also targets that point specifically at topics of relevance for biochar, such as technical and vocational skills (target 4.3), education for sustainable development and global citizenship (target 4.7) and the special needs of developing countries (target 4B). For biochar development, SDG4 is a call for action to integrate biochar into general education as well as training of future and current professionals. This is also a reminder that for biochar to grow and contribute to sustainable development, it needs to be integrated into education worldwide.

The rapidly growing voluntary carbon market for carbon dioxide removal (Arcusa and Sprenkle-Hyppolite, 2022) has the potential to make funding for biochar implementation available globally while setting high sustainability standards. This market could enable global dissemination of high-quality pyrolysis and gasification technology, provided that demand for such credits continues to rise. It also provides global digital infrastructure for verifying projects and disseminating funding, even to smallholder farmers in low-income countries. These are opportunities for partnerships across domains and countries to enable sustainable biochar development. For this emerging market to actually deliver on the SDGs, there are lessons to learn from critical studies of how international carbon offsets have addressed co-benefits such as poverty and gender (Lehmann, 2019; Cavanagh et al, 2021). The carbon offset market has been criticized for claiming poverty reduction co-benefits when marketing its projects, while there is often a lack of consideration of local liveli-hoods in practical implementation (Fischer and Hajdu, 2018). Other criticism of carbon offsets includes that they enable business as usual in terms of climate impacts from pro-duction and consumption while claiming to be acting on climate. For the voluntary carbon

market for biochar to deliver SDG benefits, certification standards should aim at certifying not only carbon removal, but sound carbon removal with all the dimensions that it implies. Sustainability assessment and certification are further described below.

There is limited scientific literature on biochar education or partnerships. Examples include a laboratory exercise where waste-water is filtered in biochar as a way to integrate sustainable development into chemistry education (Arrebola et al, 2020). Extensionists' preliminary screening trials of biochar effects on crops were able to inform future research as well as public outreach education (Hunter et al, 2017). Synergies that emerged as urban carbon sink demon-stration parks with biochar were co-created by scientists, city officials, companies, and citizens (Tammeorg et al, 2021).

Approaches and tools for sustainable biochar development

Circular economy and bioeconomy

Circular economy is a concept that has been put forward as a solution to many sustain-ability problems related to production and consumption. It is a concept that encom-passes many aspects of integrated solid waste management, such as striving to make effi-cient use of resources and promoting cascading uses of materials, but also aspects that go beyond waste management: reducing waste generation, prolonging the lifetime of products, designing products to facilitate recycling, and changing behaviors to mini-mize material needs. The circular economy literature stresses the need to change business models to enable the circular use of materials (Pieroni et al, 2019). Two challenges associ-ated with the circular economy are the

availability of renewable energy to close mate-rial cycles, and the loss of material quality and purity over multiple product cycles.

The bioeconomy is a concept for how biobased products can replace fossil-based products such as bioenergy, biochemicals, and biomaterials. Combined with the cir-cular economy, the circular bioeconomy stresses that biological resources should be used efficiently and in cascading applica-tions, that circular use of nutrients is needed, and that waste should be reduced (Stegmann et al, 2020). Some also stress the need to prioritize the limited available bio-mass resources to essential needs such as food and pharmaceuticals, and that renew-able energy from wind and solar should be prioritized before bioenergy (Muscat et al, 2021). It is acknowledged that implementa-tion of a circular bioeconomy requires a transformation of the economy at large and business models in particular, to enable circularity and thereby reduce the need for virgin resources. Biochar fits well into the circular bioeconomy (Papageorgiou et al, 2022), as it is a long-lived biomaterial that can be produced from waste, and there are applications where biochar can replace other materials with higher environmental impacts (Azzi et al, 2022). There should be ample opportunities for biochar to contribute to circular bioeconomy strategies, policies, and business models.

Sustainability transitions and multi-solving

Sustainability transition theory is an approach that builds on knowledge of how large-scale transitions towards sustainability have happened historically and how markets for innovations emerge. Shared visions among stakeholders have been identified as a success factor, as well as the protection of niche markets to enable sustainable technologies to enter markets, and develop through practical learning to become competitive and grow to scale (Smith and Raven, 2012). As a theory building on sustainable development and innovation systems, transition theory may be useful for guiding biochar development.

Multi-solving is an approach to addressing multiple challenges at the same time through multi-stakeholder collaborations focusing on one solution with several benefits (Sawin, 2018). Some principles and practices characterize successful projects that have managed to address complex interlinked environmental and social challenges. The characteristics are (i) inclusion of everyone, experts, citizens, and politicians, and welcoming new partners; (ii) seeing how certain interventions solve several problems at once; (iii) commitment to experimentation, learning, and documentation of impacts as solutions grow from small to large; and (iv) storytelling to explain what is possible and what has already been achieved. Multi-solving is one of several approaches to resolving complex problems through multi-stakeholder and transdisciplinary collaboration. Such collaboration requires careful stakeholder engagement processes that are sensitive to power imbalances, culture, and gender. As biochar systems typically have multiple functions (e.g., treating problematic biomass waste, while providing heating energy, and replacing non-renewable horticultural substrates) and involve multiple stakeholders (e.g., municipalities, farmers, energy utilities), multi-solving approaches seem very relevant to apply when designing biochar projects.

Environmental life cycle assessment

Life cycle assessment is well established as a tool for quantifying the climate impacts of biochar systems (Chapter 30). LCA in its original form has a broader scope than just climate change and can be used for quantifying the resource use efficiency and a multitude of environmental impacts that a product or service may have along its life cycle, from cradle to grave. When used in an adequate comparative context, such multi-indicator studies can contribute to improved decision-making. However, few studies have considered multiple environmental impact categories when studying biochar (Azzi et al, 2022; Tisserant et al, 2022). The multi-functionality of biochar systems complicates the assessment of biochar's environmental impacts with LCA, but it is nevertheless an appropriate tool for this purpose.

Biochar sustainability assessment

Sustainability assessments can be done qualitatively in ways that aim at covering a broad range of sustainability aspects. A bottom-up approach to biochar sustainability issues can be inspired by life cycle thinking and industrial ecology (Azzi et al, 2021). Biochar effects described in the literature were classified according to their role in the technosphere or the biosphere and then linked to sustainability domains. Twelve categories of biochar effects were identified, covering soil fluxes, soil status, plants, animals, land use, industry, waste, and energy.

The functions of biochar were identified and related to ecosystem services (Nature´s Contribution to People, NCP) and the SDGs (Smith et al, 2019). Nine functions were identified and linked to regulating NCPs (effects on soils and pollutants) material NCPs (biomass and energy) nonmaterial NCPs (learning and inspiration as well as supporting identities by preserving and reviving ancient biochar practices) and overarching NCPs (climate change mitigation). The NCPs were in turn linked to the corresponding SDGs.

Project-level sustainability assessment and certification

Sustainability assessment of biochar systems at the project level is important for one major reason, to mitigate risks, and ultimately to stop projects that will not contribute to sustainability but undermine it. The greatest risks are in unsustainable biomass supply, high emissions from pyrolysis, and soil pollution by contaminated biochar. Twelve biochar sustainability criteria adapted from a review of bioenergy sustainability criteria (Buchholz et al, 2009) were suggested (Verheijen et al, 2015). The list contained environmental criteria: climate change, energy, soil and ecosystem protection, waste, water management, and natural resource efficiency. There were also several social and economic criteria: participation of stakeholders, microeconomic efficiency, compliance with laws, monitoring of performance, and food security. Assessment of greenhouse gases from a lifecycle perspective is also an important part of a sustainability assessment of biochar (Chapter 30).

There are biochar certification systems (IBI Biochar Certification Program; European Biochar Certificate, EBC) that have detailed requirements on the quality of the biochar, including carbon content and content of toxic metals and organic compounds. The EBC covers sustainability aspects such as sourcing of biomass and some requirements on emissions and energy in biochar production. The biochar certifications are intended to provide customers and producers with a reliable quality standard for biochar products. They have been developed to reduce risks of hazards to health and the environment of biochar production and use. Consequently, they cover a broad scope of environmental and toxicity aspects, but they are not full sustainability certifications (Cowie et al, 2012).

As biochar practice is growing, more attention is paid to developing policies to promote sustainable biochar development and reduce risks for negative consequences. So far, biochar certifications have mainly been voluntary instruments developed by market actors and researchers. However, certification can also be mandatory. For example, in Switzerland, the EBC is mandatory for biochar sold for use in agriculture. Biochar legislation is complex and it is connected to laws on topics such as waste, fertilizers, bioenergy, and climate (Štrubelj, 2022).

Experiences from sustainability governance, in particular biofuel governance, should help guide biochar development.

Research stresses the need to build legitimacy and trust through adaptive governance systems (Stupak et al, 2021). Certification systems and other sustainability standards, mandatory or voluntary, have been important for building trust in biomass and bioenergy, though a range of disagreements have emerged (Stupak et al, 2021). Certifications and sustainability criteria are used by market actors and governments to ensure that sustainability criteria are met and that products meet the customers' expectations. There is a wide range of sustainability assessment frameworks and methods for biomass, bioenergy, food, and forestry, with a broad coverage of sustainability aspects, which can be relevant for biochar. The Roundtable for Sustainable Biomaterials (RSB), for example, has a well-established standard with a broad coverage of sustainability aspects (www.rsb.org). There is a growing understanding that product-level sustainability criteria are not necessarily the right tool for reaching macro-level targets such as reducing deforestation and other detrimental land use change (Daioglou et al, 2020). In essence, land and biomass are globally a large but limited resource with multiple competitive uses (Muscat et al, 2020), and it is not possible to quantify the environmental impact of each usage of biomass without considering the overall land use strategy.

While certification systems are important, they are also a cumbersome and costly practice, thus more suitable for large producers. It is equally important to spread guidance on good practices widely so that small producers working in local markets can also provide sustainable products, even though they are not documented and certified. For biochar projects, many sustainability aspects do not need to be assessed quantitatively. A checklist with a traffic-light scale (green-yellow-red) would probably be of use for many purposes.

Passing a sustainability assessment does not mean that it is appropriate to call the project or product "sustainable". Sustainability is a property that emerges at a higher level, at global, regional, or ecosystem levels, and depends on the scale and rate of resource consumption and emissions. To prevent negative outcomes, it is necessary for a product to meet sustainability criteria, but this is never sufficient proof of absolute sustainability.

Outlook: Biochar for a world in crisis

This text has been written within a paradigm of normalcy. For many years, there have been gradual improvements at the global level of most sustainability goals related to social and economic conditions. While global environmental conditions have not improved, there have been improvements in local environmental conditions in many places. As described in this chapter, biochar can play a role in supporting such gradual improvements, while also supporting climate change mitigation and adaptation. However, since the disruption by the global pandemic in early 2020, we are globally no longer in this state of gradual improvements, but rather in a state of consecutive emergencies. While the pandemic is no longer an acute health emergency, its consequences for the global economy and health systems remain strong. In addition, the war by Russia on Ukraine has caused a global energy and food crisis in 2022. Many indicators for the SDGs have turned in the wrong direction during the past few years. Such events and emergencies are taking place in a world that is therefore different from the past century, as the supply of fossil fuels will be structurally constrained and are declining. In fact, the production peak of conventional oil was reached in 2008 (IEA, 2012), and the demand for natural gas is expected to reach a plateau in the 2020–2030 decade (IEA, 2022).

At the same time, there has been a shift in the discourse on climate change. The rise of the global youth climate movement (Sloam et al, 2022) and the rising severity of extreme weather events (www. worldweatherattribution.org), have placed climate change mitigation much higher on political and business agendas. A new focus on net zero emissions has put increased attention on carbon dioxide removal technologies such as biochar. If the high goals set for climate change mitigation including carbon dioxide removals are to be met in the coming decades, there are tremendous technical, political, and organizational challenges at all scales from local to global.

We do not know what lies ahead, but that does not reduce the urgent need for action. It is no coincidence that 'No poverty' and 'Zero hunger' are SDGs number 1 and 2. Even in the best of times, many of the world's poorest have been left behind, and in precarious times, they are the most vulnerable. In theory, biochar provides many opportunities for win-win situations for these SDGs as well as many others. To actually deliver benefits for the SDGs, biochar activities need to consider social aspects and local implications, in addition to scientific and technical knowledge. Practical and theoretical knowledge must be developed and integrated into policies and education systems. In that way, biochar innovations for sustainable development may grow and make a difference.

References

Arrebola JC, Rodríguez-Fernández N, and Caballero Á 2020 Decontamination of wastewater using activated biochar from agricultural waste: a practical experiment for environmental sciences students. *Journal of Chemical Education* 97, 4137–4144.

Arcusa S, and Sprenkle-Hyppolite S 2022 Snapshot of the carbon dioxide removal certification and standards ecosystem (2021–2022). *Climate Policy*, 22 (9–10), 1319–1332.

Azzi ES, Karltun E, and Sundberg C 2021 Assessing the diverse environmental effects of biochar systems: An evaluation framework. *Journal of Environmental Management* 286, 112154.

Azzi ES, Karltun E, and Sundberg C 2019 Prospective life cycle assessment of large scale biochar production and use for negative emissions in Stockholm. *Environmental Science & Technology*, 53, 8466–8476.

Azzi ES, Karltun E, and Sundberg C 2022 Life cycle assessment of urban uses of biochar and case study in Uppsala, Sweden. *Biochar* 4, 18.

Bailis R, Drigo R, Ghilardi A, and Masera O 2015 The carbon footprint of traditional woodfuels. *Nature Climate Change* 5, 266–272.

Bellè SL, Riotte J, Backhaus N, Sekhar M, Jouquet P, and Abiven S 2022 Tailor-made biochar systems: Interdisciplinary evaluations of ecosystem services and farmer livelihoods in tropical agro-ecosystems. *PLOS ONE*, 17, e0263302.

Brobbey LK, Hansen CP, Kyereh B, and Pouliot M 2019 The economic importance of charcoal to rural livelihoods: Evidence from a key charcoal-producing area in Ghana. *Forest Policy and Economics*, 101, 19–31.

Brundtland GH   1987 Our Common Future. *The World Commission on Environment and Development*.

Buchholz TL, Valerie A, and Volk, TA 2009 Sustainability criteria for bioenergy systems: results from an expert survey. *Journal of Cleaner Production*, 17, S86–S98.

Cao CTN, Farrell C, Kristiansen PE, and Rayner JP 2014 Biochar makes green roof substrates lighter and improves water supply to plants. *Ecological Engineering* 71, 368–374.

Cavanagh CJ, Vedeld PO, Petursson JG, and Chemarum AK 2021 Agency, inequality, and additionality: contested assemblages of agricultural carbon finance in Western Kenya. *The Journal of Peasant Studies* 48, 1207–1227.

Chen H, et al 2018 Biochar increases plant growth and alters microbial communities via regulating the moisture and temperature of green roof substrates. *Science of the Total Environment* 635, 333–342.

Chen L, et al 2023 Biochar application in anaerobic digestion: performances, mechanisms, environmental assessment and circular economy. *Resources, Conservation and Recycling* 188, 106720.

Cowie, AL, et al 2012 Is sustainability certification for biochar the answer to environmental risks? *Pesquisa Agropecuária Brasileira* 47, 637–648.

Daioglou V, et al 2020 Progress and barriers in understanding and preventing indirect land-use change. *Biofuels, Bioproducts and Biorefining* 14, 924–934.

Doggart N, and Meshack C, 2017. The marginalization of sustainable charcoal production in the policies of a modernizing African Nation. *Frontiers in Environmental Science* 5, 27.

Doss CR, Eltigani A, Olsson A, Krause A, and Ernest B 2001 Designing agricultural technology for African women farmers: lessons from 25 years of experience. *World Development* 29, 2075–2092.

Eltigani A, et al 2022 Exploring lessons from five years of biochar-producing cookstoves in the Kagera region, Tanzania. *Energy for Sustainable Development* 71, 141–150.

FAO, IFAD, UNICEF, WFP and WHO 2022 The State of Food Security and Nutrition in the World 2022. Repurposing food and agricultural policies to make healthy diets more affordable. Data are available on FAOSTAT (https://www.fao.org/faostat/en/#data/FS)

Fischer K, and Hajdu F 2018 The importance of the will to improve: How 'sustainability'sidelined local livelihoods in a carbon-forestry investment in Uganda. *Journal of Environmental Policy and Planning* 20, 328–341.

Fridahl M, Yanda P, and Hansson A 2022 Exploring lessons from five years of biochar-producing cookstoves in the Kagera Region, Tanzania. *Energy for Sustainable Development* 71, 141–150.

Gitau, JK, Mendum R, and Njenga M 2018 Gender and improvement of cooking systems with biochar-producing gasifier stoves. In: Njenga M, and Mendum R (Eds) *Recovering Bioenergy in Sub-saharan Africa: Gender Dimensions, Lessons and Challenges*. Colombo, Sri Lanka: International Water Management Institute (IWMI). pp49–57.

Gitau JK, Sundberg C, Mendum R, Mutune J, and Njenga M 2019a Use of biochar-producing gasifier cookstove improves energy use efficiency and indoor air quality in rural households. *Energies* 12, 4285.

Gitau JK, Mutune J, Sundberg C, Mendum R, and Njenga, M 2019b Factors influencing the adoption of biochar-producing gasifier cookstoves by households in rural Kenya. *Energy for Sustainable Development* 52, 63–71.

Gopinath A, et al 2021 Conversion of sewage sludge into biochar: a potential resource in water and wastewater treatment. *Environmental Research* 194, 110656.

Hermans TDG, Whitfield S, Dougill AJ, and Thierfelder C 2021 Why we should rethink 'adoption' in agricultural innovation: empirical insights from Malawi. *Land Degradation and Development* 32, 1809–1820.

Hunter B, Cardon GE, Olsen S, Alston DG, and McAvoy D, 2017 Preliminary screening of the effect of biochar properties and soil incorporation rate on lettuce growth to guide research and educate the public through extension. *Journal of Agricultural Extension and Rural Development* 9, 1–4.

IEA, IRENA, UNSD, World Bank, and WHO 2022 *Tracking SDG 7: The Energy Progress Report*. Washington DC: The World Bank.

IEA, 2012. *World Energy Outlook 2012*. Vienna: International Energy Agency.

IEA, 2021. *Net Zero by 2050 A Roadmap for the Global Energy Sector*. Vienna: International Energy Agency.

IEA, 2022. *World Energy Outlook 2022*. Vienna: International Energy Agency.

Jetter J, et al 2012 Pollutant emissions and energy efficiency under controlled conditions for household biomass cookstoves and implications for metrics useful in setting international test standards. *Environmental Science and Technology* 46, 10827–10834.

Kamali M, Appels L, Kwon EE, Aminabhavi, TM, and Dewil R 2021 Biochar in water and wastewater treatment – a sustainability assessment. *Chemical Engineering Journal* 420, 129946.

Leach M, Raworth K, and Rockström J 2013 Between social and planetary boundaries: navigating pathways in the safe and just space for humanity. World Social Science Report: Changing Global Environments, ISSC, UNESCO. pp84–89.

Lehmann I 2019 When cultural political economy meets 'charismatic carbon' marketing: a gender-sensitive view on the limitations of gold standard cookstove offset projects. *Energy Research and Social Science* 55, 146–154.

Li L, Zhang Y-J, Novak A, Yang Y, and Wang J 2021 Role of biochar in improving sandy soil water retention and resilience to drought. *Water* 13, 407.

Marshall RE, and Farahbakhsh F, 2013. Systems approaches to integrated solid waste management in developing countries. *Waste Management* 33, 988–1003.

Masrura SU, et al 2021 Sustainable use of biochar for resource recovery and pharmaceutical removal from human urine: a critical review. *Critical Reviews in Environmental Science and Technology* 51, 3016–3048.

Mohanty SK, et al 2018 Plenty of room for carbon on the ground: Potential applications of biochar for stormwater treatment. *Science of the Total Environment* 625, 1644–1658.

Muscat A, de Olde EM, de Boer IJM, and Ripoll-Bosch R 2020 The battle for biomass: a systematic review of food-feed-fuel competition. *Global Food Security* 25, 100330.

Muscat A, et al 2021 Principles, drivers and opportunities of a circular bioeconomy. *Nature Food* 2, 561–566.

Müller S, Backhaus N, Nagabovanalli P, and Abiven S 2019 A social-ecological system evaluation to implement sustainably a biochar system in South India. *Agronomy for Sustainable Development*, 39, 43.

Papageorgiou A, Sinha R, Azzi ES, Sundberg C and Enell A 2022 The role of biochar systems in the circular economy: biomass waste valorization and soil remediation. In: Zhang T (Ed) *The Circular Economy – Recent Advances in Sustainable Waste Management*. IntechOpen. https://www.intechopen.com/chapters/81303.

Pataki DE, et al 2021 The benefits and limits of urban tree planting for environmental and human health. *Frontiers in Ecology and Evolution* 9, 603757.

Pathy A, Ray J, and Paramasivan B 2021 Challenges and opportunities of nutrient recovery from human urine using biochar for fertilizer applications. *Journal of Cleaner Production* 304, 127019.

Peterman A, Behrman JA, and Quisumbing AR 2014 A review of empirical evidence on gender differences in nonland agricultural inputs, technology, and services in developing countries. In: Quisumbing AR, et al (Eds) *Gender in Agriculture*. Dordrecht: Springer Netherlands. pp145–186.

Peters JF, Iribarren D, and Dufour J 2015 Biomass pyrolysis for biochar or energy applications? A life cycle assessment. *Environmental Science and Technology* 49, 5195–5202.

Pieroni MPP, McAloone TC, and Pigosso DCA 2019 Business modeli for circular economy and sustainability: a review of approaches. *Journal of Cleaner Production* 215, 198–216.

Qin Y, et al 2022 Methane emission reduction and biological characteristics of landfill cover soil amended with hydrophobic biochar. *Frontiers in Bioengineering and Biotechnology* 10, 905466.

Rahman MYA, Cooper R, Truong N, Ergas SJ, and Nachabe MH 2021 Water quality and hydraulic performance of biochar amended biofilters for management of agricultural runoff. *Chemosphere*, 283, 130978.

Redlingshöfer B, Barles S, and Weisz H, 2020 Are waste hierarchies effective in reducing environmental impacts from food waste? A systematic review for OECD countries. *Resources, Conservation and Recycling*, 156, 104723.

Riahi K, et al 2022 Mitigation pathways compatible with long-term goals. In: PR Shukla et al, (Eds), *IPCC, 2022: Climate Change 2022: Mitigation of Climate Change. Contributionof Working Group III to the Sixth Assessment Report of the Intergovernmental Panel on Climate Change.* Cambridge, UK and New York, NY, USA: Cambridge University Press, pp. 295–408.

Rogers PM, et al 2021 Socio-economic determinants for biochar deployment in the southern highlands of Tanzania. *Energies* 15, 144.

Roos A, Mutta D, Larwanou M, Wekesa C, and Kowero G 2021 Operations and improvement needs in the informal charcoal sector: a participatory value stream analysis. *International Forestry Review* 23, 351–364.

Rose J, Bensch G, Munyehirwe A, and Peters J 2022 The forgotten coal: charcoal demand in Sub-Saharan Africa. *World Development Perspectives* 25, 100401.

Saluz AG, Bleuler M, Krähenbühl N and Schönborn A 2022 Quality and suitability of fecal biochar in structurally stable urban tree substrates. *Science of the Total Environment* 838, 156236.

Sanchez-Monedero MA, et al 2018 Role of biochar as an additive in organic waste composting. *Bioresource Technology* 247, 1155–1164.

Sawin E 2018 The Magic of "Multisolving." *Stanford Social Innovation Review* 16, e https:// ssir.org/articles/entry/the_magic_of_ multisolving.

Sedano F, Mizu-Siampale A, Duncanson L, and Liang M 2022 Influence of charcoal production on forest degradation in Zambia: a remote sensing perspective. *Remote Sensing* 14, 3352.

Shilomboleni H, Owaygen M, De Plaen R, Manchur W, and Husak L 2019 Scaling up innovations in smallholder agriculture: Lessons from the Canadian international food security research fund. *Agricultural Systems* 175, 58–65.

Sloam J, Pickard S, and Henn M, 2022 Young people and environmental activism: the transformation of democratic politics. *Journal of Youth Studies* 25, 683–691.

Smith P, et al 2019 Land-management options for greenhouse gas removal and their impacts on ecosystem services and the sustainable development goals. *Annual Review of Environmental Resources* 44, 255–286.

Smith A, and Raven R 2012 What is protective space? Reconsidering niches in transitions to sustainability. *Research Policy* 41, 1025–1036.

Stanistreet D, Puzzolo E, Bruce N, Pope D, and Rehfuess E 2014 Factors influencing household uptake of improved solid fuel stoves in low- and middle-income countries: a qualitative systematic review. *International Journal of Environmental Resarch on Public Health* 11, 8228–8250.

Steffen W, et al 2015 Planetary boundaries: Guiding human development on a changing planet. *Science*, 347, 1259855.

Steiner C, Bellwood-Howard I, Häring V, Tonkudor K, Addai F, Atiah K, Abubakari AH, Kranjac-Berisavljevic G, Marschner B, and Buerkert A 2018 Participatory trials of on-farm biochar production and use in Tamale, Ghana. *Agronomy for Sustainable Development*, 38, 12.

Štrubelj L 2022 Waste, fertilising product, or something else? EU regulation of biochar. *Journal of Environmental Law* 34, 529–540.

Stupak I, Mansoor M, and Smith CT 2021 Conceptual framework for increasing legitimacy and trust of sustainability governance. *Energy, Sustainability and Society* 11, 5.

Stegmann P, Londo M, and Junginger M 2020 The circular bioeconomy: Its elements and role in European bioeconomy clusters. *Resources, Conservation & Recycling: X*, 6, 100029.

Sutradhar I, et al 2021 Introducing urine-enriched biochar-based fertilizer for vegetable production: acceptability and results from rural Bangladesh. *Environment, Development and Sustainability* 23, 12954–12975.

Tammeorg P, et al 2021 Co-designing urban carbon sink parks: case carbon lane in Helsinki. *Frontiers in Environmental Science* 9, 672468.

Tisserant A, et al 2022 Life-cycle assessment to unravel co-benefits and trade-offs of large-scale biochar deployment in Norwegian agriculture. *Resources, Conservation and Recycling* 179, 106030.

Verheijen FG, Bastos AC, Schmidt HP, Brandão M, and Jeffery S 2015 Biochar sustainability and certification. In: Lehmann J and Joseph S (Eds) *Biochar for Environmental Management*. London: Routledge. pp827–844.

Vigolo V, Sallaku R, and Testa F 2018 Drivers and barriers to clean cooking: a systematic literature review from a consumer behavior perspective. *Sustainability* 10, 4322.

Whitman T, Nicholson CF, Torres D, and Lehmann J 2011 Climate change impact of biochar cook stoves in Western Kenyan farm households: system dynamics model analysis. *Environmental Science and Technology* 45, 3687–3694.

Wilson DC, Rodic L, Scheinberg A, Velis CA, and Alabaster G 2012 Comparative analysis of solid waste management in 20 cities. *Waste Management and Research* 30, 237–254.

Wurzer C, Jayakumar A, and Mašek O 2022 Sequential biochar systems in a circular economy. In: Tsang DCW, and Ok YS (Eds) *Biochar in Agriculture for Achieving Sustainable Development Goals*. Dordrecht: Elsevier. pp305–319.

Xiang W, et al 2020 Biochar technology in wastewater treatment: a critical review. *Chemosphere* 252, 126539.

32

Markets for biochar products and services

Josef Maroušek, Anna Maroušková, Otakar Strunecký, and Babak Minofar

The urgency to find markets for biochar

Over the last two decades, biochar has been the subject of intense attention in numerous laboratories all around the world. Such an unprecedented interest resulted in thousands of experiments, academic publications, reports, patents, business opportunities, and even political visions. New possibilities of using biochar in agriculture; food and feed industry; (bio)chemical industry, and environmental engineering are of the highest interest (Kamali et al, 2022). Efforts to date have facilitated a deeper understanding of basic principles and roles of feedstock characteristics; pyrolytic and activation process parameters; pyrolysis unit design; physical and chemical properties of biochar; water, soil, and air interactions; activation, sorption, and filtration dynamics; carbon capture and many others (Wang and Wang, 2019). Although the annual biochar production is rising (as well as its awareness), few biochar businesses are thriving and several countries are already introducing regulations that support low and negative carbon technologies (such as through carbon credits; Chapter 34), all in all, these numbers are far below expectations (Dumortier et al, 2020). Biochar is still perceived as an exotic product and biochar enterprises are, in most countries, still more a matter of enthusiasts with rather locally limited importance.

Most biochar companies nowadays operate small prototypes of biochar production units, face no steady demand, and focus on specific narrow market niches. Thus, from a global perspective, it can be stated that research to date has resulted in only a small number of commercial applications (Thengane et al, 2021). There is a broad consensus that the main reason is the detachment of many environmentally driven biochar visions from economic reality (Vochozka et al, 2016a). However, ignoring economic realities is also spilling over into research that declares its focus on industrial applications (Durana et al, 2021). Among the biggest gaps in science

DOI: 10.4324/9781003297673-32

is the process of "optimization" without taking into account the economic essence of things (Kovacova and Lazaroiu, 2021). Many recent publications describe biochar applications that are rather theoretical with limited market application (Zvarikova et al., 2021). Some authors hesitate to openly communicate that in the current state of the market, investment into the biochar business is risky and so the economic aspects are not adequately addressed (Kovacova et al, 2022). However, being silent about the financial aspects does not help the biochar business (Valaskova et al, 2021). There is a broad consensus that many papers can be suspected of presenting questionable or overestimated financial outcomes. As biochar can also be used for energy generation, fluctuating energy prices play a major role in the orientation of biochar markets (Vochozka et al, 2020). This is a critical challenge because many of the hopes for biochar involve large-scale production which must reflect all economic aspects to be competitive.

Background of economic challenges

Traditional feedstock for charcoal production as a source of energy or other industrial products is oak, beech, or any other hardwood. About the economic aspects, the majority of charcoal and biochar are nowadays made from biowaste to reduce the price as much as possible. Nevertheless, even after two decades of intensive research, finding the balance between biochar´s properties and production costs remains a pressing challenge. One of the main assumptions for cost-competitive biochar production is uncontaminated biowaste which (i) is available all year round in constant quantity and quality; (ii) has a sufficient heating value (to make the process independent of an external power source); and (iii) is declared as waste (so that revenues from its processing can be realized) or at least available free of charge. However, all these requirements can be met at once only under exceptional conditions. An interesting opportunity represents the pyrolysis of withered or otherwise damaged wood in forests. This practice is widely understood (and paid for) as prevention against fires or bark beetles, but it requires a lot of labor and is therefore problematic in countries with high labor costs (Maroušek and Trakal, 2022). Another lucrative opportunity is the processing of postharvest residues, but such a practice is debatable because it reduces the input of organic matter into the soil (Strunecky et al, 2021). Recently, the pyrolysis of various spent grains from distilleries and breweries has been the subject of increased interest (Manolikaki and Diamadopoulos, 2020). Regrettably, these types of feedstock usually have a high water content and low calorific value. It should also be taken into account that other types of valuation, such as bioconversion, usually provide a more cost-effective way of processing. Sludge from sewage treatment plants not only has a high water content but also a high mineral content and thus presents a similar problem of low energy density (Singh et al, 2020). Despite all this, sludge pyrolysis can be economically rational, as many policies support it as the hygienically preferred solution. Interesting synergies can be found regarding the pyrolysis of fermentation residues from biogas plants (also known as digestate) since the production of electricity by burning biogas provides enough waste energy to dry the feedstock (Mardoyan and Braun, 2015) and generates higher-value biogas products than low-value thermal energy through pyrolysis. Coupling with biogas production allows biochar to be produced even at a "negative cost" (Stehel et al, 2020). This is achieved through the accumulation of synergies (Maroušek et al, 2023) where farmers (A) avoid losses associated with digestate management (storage, mixing, loading, transport, application, water and nutrient loss, etc.); (B) waste heat from biogas combustion is used to balance the energy balance of the pyrolysis unit, (C) waste heat from the pyrolysis unit is used to dry the digestate (D) benefits associated with the on-farm application of biochar are generated (nutrient regeneration, process water regeneration, improved soil properties, etc.). The benefits of turning digestate (and similar biowaste) into biochar are supported by the fact that methods to regenerate nutrients in both organic (biochar filters, Maroušek et al, 2019) and mineral form (e.g., through chemical sorption on biochar (Yin et al, 2017; Luo et al, 2023)) are already established on many farms. Even though negligible production costs are a good precondition, they are not a guarantee of the prosperity of a biochar business (Hašková, 2017). Mašek et al (2019) argue that biochar should be applied to arable land in large quantities because it brings a range of direct (improved water and air soil management of soil, increased pH, reduction of density, stimulation of microbial life, etc.) and indirect environmental benefits (C sequestration, improved nutrient dynamics, etc.). However, it cannot be assumed that such action is profitable from the farmer's point of view in each country under every economic and legislative situation, because many of the benefits are long-term or social (Figures 32.1 and 32.2).

However, there is a stormy discussion about the participation of public bodies in the financing of such practices. On the other hand, there is a broad consensus that to achieve higher profitability of biochar, it is advisable

fermentation residues   pyrolysis   activation via $CaCl_2$

soil application   CaP - enriched biochar   sludge water

Figure 32.1 *A common scheme where biochar is produced from the fermentation residues of an agricultural biogas plant by using the waste heat from the conversion of biogas to electricity. The biochar is then activated with calcium chloride to capture P at the wastewater treatment plant, mainly in the form of calcium phosphates, which are well utilized by plants*

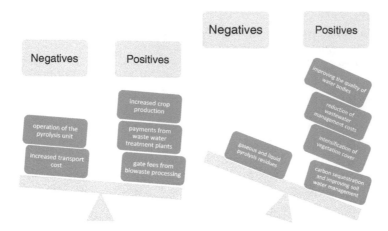

Figure 32.2 *The economic interpretation from the perspective of the farmer (left) and society (right) indicates why there are increasing voices in socially-oriented economies arguing for biochar farming to be supported by public budgets*

to look for more lucrative ways of biochar application than just applying biochar to arable land in conventional agriculture. Unfortunately, not many such markets have yet been found and they are always very specific to a given location and time. The best competitive advantage at the moment seems to be an increased level of knowledge and flexible adaptation to local conditions. Nevertheless, any such proposal should be at least more lucrative than combustion otherwise manufacturers will prefer producing and selling charcoal for energy purposes instead of biochar for agricultural applications (Woolf et al, 2014; Maroušek et al, 2015). So far, a limited number of such applications have been proven commercially.

Searching for profitable uses of biochar products

Predicting future developments regarding carbon materials has in general proven to be challenging. More than a decade ago, Segal et al (2009) claimed that the carbon industry should prepare for a technological leap and a multi-ton-scale mass production of the new generation of carbon materials. To date, however, demand remains negligible as single-layer graphene still costs hundreds of € per square centimeter and graphene oxide costs hundreds of thousands € per kilogram (Bartoli et al, 2020). The market for carbon nanotubes, carbon dots, and carbon black is also stagnant. Given that the filtration capacity of charcoal has been known since antiquity, water purification is one of the most frequently recommended applications for biochar. The literature is rich in claims about high water purification efficiency (Siipola et al, 2020). However, most of these findings were obtained under idealized laboratory parameters (steady level of uniform contaminants, unchanged pH, etc.) and can therefore not be independently achieved at an industrial scale during long-term use. Thus, existing knowledge shows that biochar should be enriched in graphitic carbon forms before it can be used to purify drinking water, and even then, the resulting mixture is hardly competitive with established technologies.

Many independent authors (Manyuchi et al, 2018; Kaetzl et al, 2019) reported that biochar-based filtering material reduces the biological and chemical oxygen demand as well as the concentration of some ions. However, it should be noted that such applications are not straightforward, and with highly variable biochar properties, even contamination may occur instead of purification.

There are efforts to commercialize biochar as a sorbent. Nevertheless, it should be noted that biochar without further modifications is not suitable for saturating sorbates from very dilute solutions (Kolář et al, 2011). Nevertheless, prolonged contact of biochar with atmospheric oxygen, especially in an alkaline environment, results in surface oxidation of the biochar particles, which can form functional groups capable of proton dissociation (Fan et al, 2018). These are mostly carboxylate -COOH and phenolic -OH groups.

Recently, much attention has been paid to the mechanical disintegration of biochar into nanobiochar and its use as a germination accelerator. It has been proven that nanobiochar can not only accelerate plant germination but also increase plant vigor to the same level as silicon nanoparticle preparations (Maroušek et al, 2022). In both cases, however, no commercial interest in this technology currently exists to our knowledge due to disproportionate cost. It has also been confirmed that the black color of biochar warms up the soil and thus allows the growing season to be extended. However, the economic evaluation of these findings is ambiguous because such properties are inappropriate in some areas. Biochar makes it possible to produce different types of special fertilizers (micro granules with complex organo-mineral fertilizers, micronutrients, and biologically active substances) for targeted fertilization to change the characteristics of the crops grown (Maroušek et al, 2018). These findings are still at the early stage of market entry but are attracting a lot of investor interest.

As biochar from some bio-waste contains increased amounts of oxides and carbonates of Ca and Mg, it can be used to raise the pH of the soil, in other words, as a substitute for liming. Although some biochar modifications have been shown to reliably prevent soil acidification, this application is typically not competitive with established lime applications (Gezahegn et al, 2019). The ability of biochar to improve soil aeration and water management is currently considered a key agronomic trait (Obia et al, 2018). However, the use of biochar as a tool for improving soil tilth may also only be profitable in a limited number of cases. Biochar has been repeatedly and independently confirmed to have a wide range of particle size distributions, varying internal pore volumes, and fluctuating porosity, with many researchers promoting the application of biochar as a tool for mitigating erosion, reducing evaporation, reducing soil density, and improving water retention (Smetanova et al, 2013). However, it is still unclear whether these investments represent an attractive investment opportunity (Atukunda et al, 2022). Over time, biochar aging in certain soil types can reduce the physical water-binding properties of the biochar as intrapores, interpores, and capillary structures break down and the overall porosity diminishes (Chapter 20). Taking into account all the above-described soil improvement mechanisms, when biochar investment was considered in terms of its application on arable land, the calculated payback periods were mostly in the range of 13 to 25 years, which is not attractive for most farmers (Vochozka et al, 2016). However, there are few such analyses and these are significantly influenced by both uncertainty and variability with soil type and conditions (Campbell et al, 2018). Biochar properties can be used for the removal of organic pollutants (Chapter 27), such as metabolic residuals of pharmaceuticals, dyes, and polymers residues (Gautam et al, 2021). However, other reactants on the market show similar properties, demand is low and it cannot be argued that the use of biochar for these applications currently represents a significant competitive advantage (Mrozik et al, 2021). Modern carbon materials such as graphene and carbon nanotubes, walls, onions, or dots demonstrate higher efficiency per unit weight in comparison to biochar (Januário et al, 2021). However, when biochar pricing is taken into account, the results are not clear-cut (Jośko et al, 2013; Hale et al, 2015; Godlewska et al, 2017; Bielská et al, 2018). The situation is similar for small organic molecules such as aromatics. However, the high prices of these modern carbon materials represent a commercial advantage for the biochar in sorption and removal of oil spills from water bodies. One advantage of biochar is related to its ability to be tailored by the addition of different chemical functional groups where it can selectively remove some compounds.

Biochar may be used for the production of supercapacitors. Although such a solution is technically possible, the degree of post-treatment of biochar is so demanding that the use of biowaste as a feedstock does not currently appear to represent an economic advantage over conventional materials (Li et al, 2021). The same conclusion may currently be reached when assessing biochar's economic potential as a material for (microbial) fuel cells (Chakraborty et al, 2020). The observations are similar to the use of biochar for battery production. After repeated activations, biochar enhances electronic conductivity, but the cost of the activation procedures and reactants used currently makes the savings on input material negligible.

Environmentally and financially interesting is the use of biochar as a reinforcing agent or filler for (biodegradable) plastics (Kane et al, 2022) and other polymer (bio) composites (Das et al, 2015) as tensile strength, stiffness, and flexural storage modulus are significantly increased. However, this research is at an early stage with market reactions not yet available.

There also seems to be great potential behind the idea of using biochar as a cement substitute and concrete additive (Akhtar and Sarmah, 2018). The technical advantages of this process have been repeatedly demonstrated, and it is considered that biochar, up to a certain limit, can improve many of the characteristics of concrete, in particular, reducing its density without compromising strength. However, if this limit is exceeded, the strength characteristics drop sharply (Gupta et al, 2018). Better results can be achieved by activation, which forms nuclei on the surface of the biochar that accelerate the crystallization of the concrete. Nevertheless, this research is also still in the testing phase. Another disadvantage is that if the proportion of biochar presence exceeds a certain value (usually in the region of 14%) the biochar darkens the concrete surface, which in turn leads to increased heating and the formation of expansion cracks (Maljaee et al, 2021). However, recent research into the formation of crystallization nuclei through the activation of biochar by amorphous structures of Al, Fe, and P may be about to bring a breakthrough.

Biochar could be used as a porous matrix for biodiesel synthesis (Lee et al, 2017). Although the process is feasible under laboratory conditions, it cannot be expected that the synthesis of such biodiesel through pseudo-catalytic transesterification could be commercially viable in the near future. Biochar has been repeatedly and independently verified to improve the population dynamics of anaerobic consortia in anaerobic reactors by accelerating biogas production in multiple laboratories (Khan et al, 2021). It can then be assumed that the subsequent application of fermentation residues enriched with biochar into the soil will bring synergistic effects (Vochozka et al, 2016b). However, despite the technology being declared technically proven, the addition of biochar into biogas plants is rare because commercial-scale verification revealed that this procedure requires further investments to prevent biochar from rising to the surface of the mixing reactor which subsequently manifests itself in difficult pipeline clearance and various blockages.

Another promising application is based on the redox properties of biochar carbonaceous structures that are capable of reducing Cr(VI) toxicity by converting it into Cr(III) while Fe-containing nanoparticles immobilize the Cr(III). It has been shown both experimentally and theoretically that carbonaceous materials can adsorb hexavalent Cr from aqueous solutions where the main interaction in the adsorption process is related to the interaction of $HCrO_4^-$ and CrO_4^{2-} with –OH and –COO- groups. Impregnation of such carbonaceous materials by magnesium ions can improve the adsorption of hexavalent Cr ions by almost 20% which makes the process economically more attractive (Alvarez-Galvan et al, 2022).

Although the use of biochar for the immobilization of other heavy metals such as Cu(II) and As(V) was developed (Inyang et al, 2016), these applications are currently not commercially competitive, as there are other available alternatives.

Crop production relies on regular nutrient additions to compensate for nutrient exports with harvests, as well as for any losses by leaching, erosion, or gaseous evasion and irreversible interactions with soil minerals. Therefore, nutrient recycling is currently considered the biggest market opportunity in the biochar business (Pathy et al, 2021). Much attention is paid to P, which, together with N, are the main cost items in plant nutrition. Technologies are being intensively explored to achieve every potential synergy so that (i) biochar is made of biowaste; (ii) waste heat is utilized; (iii) biochar is activated by waste products or other reactants whose price is irrelevant (because they are used in other operations anyway) and applied sorb nutrients from wastewaters; (iv) savings are realized on wastewater treatment; and (v) biochar is optimized as a fertilizer.

It should be noted that the majority of existing wastewater treatment plants already use P technologies that precipitate phosphates into struvite, hydroxylapatites, or vivianite. However, these technologies are more than century-old concepts (Le Corre et al, 2009) and were designed to reduce the quantity of P released into the water bodies (to reduce eutrophication) not to produce products for plant nutrition. Although great efforts have been made by many companies to increase the nutrient availability of these crystals by deep disintegration and to convince farmers of the efficiency of these byproducts, it was repeatedly and independently proven that the availability of these complex crystals to plant nutrition is low (Stavkova and Maroušek, 2021). Capturing organic forms of P from liquid biowaste via biochar filters is simple (Nardis et al, 2021). Inorganic forms of P are represented by anions (mainly $H_2PO_4^-$, HPO_4^{2-}, and PO_4^{3-}, depending on the pH) and are therefore repelled by negatively charged biochar surfaces. However, adding magnesium ($MgCl_2$) or iron ($FeCl_3$) to feedstock produces biochar-containing mineral forms that can then retain P from a liquid phase (Kopecký et al, 2020; Nardis et al, 2021). Cost breakdown indicates that biochar activation by $CaCl_2$ allows for achieving the best financial indicators because most of the P is transformed into Ca and Mg phosphates, which are the most valuable in terms of plant nutrition.

An assessment of the various applications based on commercial materials, reviewed literature, and practical experience is given in Table 32.1. As is clear from the above, the information is only illustrative with a focus on central Europe in 2022.

Table 32.1 *Financial assessment of markets for biochar, where: IRR = internal rate of return, ECA = experience from a commercial application*

Application	Key risk or competitor	Competitive advantage	IRR	ECA
Recovery of P from wastewater	Fossil P fertilizers	Better availability to plants	high	good
Tailored fertilizers	Agricultural commodity prices	Adaptation to customer needs	high	high
Recovery of P from wastewater via filtration	Phosphorus and N prices	Low acquisition costs	high	good
Biodegradable plastics	Legislation and public opinion	Constant demand for high quantities	medium	good
Cement substitute	Energy prices	Constant demand for high quantities	high	no
Energy source and storage	Charcoal prices	Constant demand for high quantities	medium	low
Water purification	Conventional filtration technology	Residues can be used for soil improvement	low	high
Application of raw biochar to the soil	Agricultural commodity prices	Public opinion on carbon sequestration	low	good
Germination accelerator	Agricultural commodity prices	Public opinion on carbon sequestration	low	low
Removal of toxic pollutants from industrial wastewater	Conventional technologies	Public opinion on carbon sequestration	low	good
Electronics and chemistry	Conventional technologies	Public opinion on carbon sequestration	low	low

Concluding remarks

We have not been able to find a business case in which biochar would be produced from purpose-grown biomass and this would be profitable in the long term. A good prerequisite for cost-competitive biochar marketing is to maximize the synergies between the recovery of waste energy and biowaste valorization. The optimal feedstock for the production of biochar is biowaste that is available in sufficient and stable quality all year round, such as waste from the food industry, distilleries, biogas plants, etc. so that the project's profitability can be significantly improved by the revenues from gate fees. However, pyrolysis of sewage sludge may require an external energy source and the resulting biochar is usually subject to many legislative restrictions, making it difficult to commercialize. On the other hand, biogas plants represent a good synergy because they have enough waste energy from the transformation of biogas into electricity. The competitiveness of the biochar business can be further increased if other (bio)wastes are used to add value to the biochar. The first technologies that make it possible to increase the value of biochar with different types of (bio)waste have already been established and currently include the most profitable ones. An alternative is also to use other wastes to contribute to the energy balance of biochar production and thus achieve even more economic benefits. The application of biochar to arable land has many environmentally positive impacts, but unless the practice is supported by public incentives, the payback periods are long, financial liquidity is practically zero, and the costs present a disproportionately high risk. However, if the rise in food prices continues at its current pace, the situation could change swiftly. In the richest countries,

it can be profitable to use biochar as a feed supplement, customized fertilizer, or soil conditioner in sports fields, gardening, or land remediation.

With regards to non-agricultural and non-soil applications, the largest gains can be made in cosmetics, filtration applications, or construction. Nevertheless, most of these applications represent either low demand or are currently not fully adapted to the market. Notwithstanding the repeatedly and independently proven environmental benefits, it must be stressed that financial sustainability is a prerequisite for environmental sustainability. A viable biochar technology must therefore provide added value that is higher than the price of charcoal as a source of energy. Biochar competitiveness is thus linked to energy prices.

At the current state of knowledge and economy, the most profitable industrially confirmed applications are (i) the recovery of nutrients (depends on current P prices); (ii) the production of tailored-made special fertilizers; and (iii) the production of biodegradable plastics (depending on legislation, purchasing power and public opinion). There are indications that technological breakthroughs can be expected regarding the use of biochar as (iv) a cement substitute; and (v) a material for energy transformations (fuel additive and Power-2-Gas in particular). However, these latter two applications are speculative as they have not yet been sufficiently verified by the market. To increase the profitability of the biochar plant, it is advisable to commercialize all the outputs, i.e., not only biochar but also gaseous and liquid pyrolytic products. The rafination of pyrolytic oil has great economic potential, but such projects require significant investments.

References

Akhtar A, and Sarmah AK 2018 Novel biochar-concrete composites: manufacturing, characterization and evaluation of the mechanical properties. *Science of the Total Environment* 616, 408–416.

Alvarez-Galvan Y, et al 2022 Adsorption of hexavalent chromium using activated carbon produced from *Sargassum* ssp.: comparison between lab experiments and molecular dynamics simulations. *Molecules* 27, 6040.

Atukunda A, Ibrahim MG, Fujii M, Ookawara S, and Nasr M 2022 Dual biogas/biochar production from anaerobic co-digestion of petrochemical and domestic wastewater: a techno-economic and sustainable approach. *Biomass Conversion and Biorefinery*, 10.1007/s13399-022-02944-w.

Bartoli M, Giorcelli M, Jagdale P, Rovere M, and Tagliaferro A 2020 A review of non-soil biochar applications. *Materials* 13, 261.

Bielská L, Škulcová L, Neuwirthová N, Cornelissen G, and Hale SE 2018 Sorption, bioavailability and ecotoxic effects of hydrophobic organic compounds in biochar amended soils. *Science of the Total Environment*, 624, 78–86.

Campbell RM, Anderson NM, Daugaard DE, and Naughton HT 2018 Financial viability of biofuel and biochar production from forest biomass in the face of market price volatility and uncertainty. *Applied Energy* 230, 330–343.

Chakraborty I, Sathe SM, Dubey BK, and Ghangrekar MM 2020 Waste-derived biochar: applications and future perspective in microbial fuel cells. *Bioresource Technology* 312, 123587.

Das O, Sarmah AK, and Bhattacharyya D 2015 A sustainable and resilient approach through biochar addition in wood polymer composites. *Science of the Total Environment* 512, 326–336.

Dumortier J, et al 2020 Global land-use and carbon emission implications from biochar application to cropland in the United States. *Journal of Cleaner Production* 258, 120684.

Durana P, Michalkova L, Privara A, Maroušek J, and Tumpach M 2021 Does the life cycle affect earnings management and bankruptcy? *Oeconomia Copernicana* 12, 425–461.

Fan Q, et al 2018 Effects of chemical oxidation on surface oxygen-containing functional groups and adsorption behavior of biochar. *Chemosphere* 207, 33–40.

Gautam RK, et al 2021 Biochar for remediation of agrochemicals and synthetic organic dyes from environmental samples: A review. *Chemosphere* 272, 129917.

Gezahegn S, Sain M, and Thomas SC 2019 Variation in feedstock wood chemistry strongly influences biochar liming potential. *Soil Systems* 3, 26.

Godlewska P, Schmidt HP, Ok YS, and Oleszczuk P 2017 Biochar for composting improvement and contaminants reduction. A review. *Bioresource Technology* 246, 193–202.

Gupta S, Kua HW, and Low CY 2018 Use of biochar as carbon sequestering additive in cement mortar. *Cement and Concrete Composites* 87, 110–129.

Hale SE, Cornelissen G, and Werner D 2015 Sorption and remediation of organic compounds in s oils and sediments by (activated) biochar. In: Lehmann J, and Joseph S (Eds) *Biochar for Environmental Management*. London: Routledge. pp657–686.

Hašková S 2017 Holistic assessment and ethical disputation on a new trend in solid biofuels. *Science and Engineering Ethics* 23, 509–519.

Inyang MI, et al 2016 A review of biochar as a low-cost adsorbent for aqueous heavy metal removal. *Critical Reviews in Environmental Science and Technology* 46, 406–433.

Januário EFD, et al 2021 Advanced graphene oxide-based membranes as a potential alternative for dyes removal: A review. *Science of the Total Environment* 789, 147957.

Jośko I, et al 2013 Effect of biochars, activated carbon and multiwalled carbon nanotubes on phytotoxicity of sediment contaminated by inorganic and organic pollutants. *Ecological Engineering* 60, 50–59.

Kaetzl K, et al 2019 On-farm wastewater treatment using biochar from local agroresidues reduces pathogens from irrigation water for safer food production in developing countries. *Science of the Total Environment* 682, 601–610.

Kamali M, et al 2022 Biochar for soil applications-sustainability aspects, challenges and future prospects. *Chemical Engineering Journal* 428, 131189.

Kane S, Van Roijen E, Ryan C, and Miller S 2022 Reducing the environmental impacts of plastics while increasing strength: biochar fillers in biodegradable, recycled, and fossil-fuel derived plastics. *Composites Part C: Open Access* 8, 100253.

Khan SA, et al 2021 Mutually trading off biochar and biogas sectors for broadening biomethane applications: A comprehensive review. *Journal of Cleaner Production* 318, 128593.

Kolář L, Kužel S, Peterka J, and Borová-Batt J 2011 Utilisation of waste from digesters for biogas production. In: Bernardes MAS (Ed) *Biofuel's Engineering Process Technology*. Rijeka: InTech. pp191–220.

Kopecký M, et al 2020 Modified biochar—a tool for wastewater treatment. *Energies* 13, 5270.

Kovacova M, and Lazaroiu G 2021 Sustainable organizational performance, cyber-physical production networks, and deep learning-assisted smart process planning in Industry 4.0-based manufacturing systems. *Economics, Management and Financial Markets* 16, 41–54.

Kovacova M, Machova V, and Bennett D 2022 Immersive extended reality technologies, data visualization tools, and customer behavior analytics in the metaverse commerce. *Journal of Self-Governance and Management Economics* 10, 7–21.

Le Corre KS, Valsami-Jones E, Hobbs P, and Parsons SA 2009 Phosphorus recovery from wastewater by struvite crystallization: A review. *Critical Reviews in Environmental Science and Technology* 39, 433–477.

Lee J, et al 2017 Evaluating the effectiveness of various biochars as porous media for biodiesel synthesis via pseudo-catalytic transesterification. *Bioresource Technology* 231, 59–64.

Li X, Zhang J, Liu B, and Su Z 2021 A critical review on the application and recent developments of post-modified biochar in supercapacitors. *Journal of Cleaner Production* 310, 127428.

Luo D, et al 2023 Phosphorus adsorption by functionalized biochar: a review. *Environmental Chemistry Letters* 21, 497–524.

Maljaee H, Madadi R, Paiva H, Tarelho L, and Ferreira VM 2021 Incorporation of biochar in cementitious materials: A roadmap of biochar selection. *Construction and Building Materials* 283, 122757.

Manolikaki I, and Diamadopoulos E 2020 Agronomic potential of biochar prepared from brewery byproducts. *Journal of Environmental Management* 255, 109856.

Manyuchi MM, Mbohwa C, and Muzenda E 2018 Potential to use municipal waste bio char in wastewater treatment for nutrients recovery. *Physics and Chemistry of the Earth, Parts A/B/C* 107, 92–95.

Mardoyan A, and Braun P 2015 Analysis of Czech subsidies for solid biofuels. *International Journal of Green Energy* 12, 405–408.

Maroušek J, Hašková S, Zeman R, Váchal J, and Vaníčková R 2015 Processing of residues from biogas plants for energy purposes. *Clean Technologies and Environmental Policy* 17, 797–801.

Maroušek J, Kolář L, Vochozka M, Stehel V, and Maroušková A 2018 Biochar reduces nitrate level in red beet. *Environmental Science and Pollution Research* 25, 18200–18203.

Maroušek J, Strunecký O, and Stehel V 2019 Biochar farming: defining economically perspective applications. *Clean Technologies and Environmental Policy* 21, 1389–1395.

Maroušek J, et al 2022 Silica nanoparticles from coir pith synthesized by acidic sol-gel method improve germination economics. *Polymers* 14, 266.

Maroušek J, Minofar B, Maroušková A, Strunecký O, and Gavurová B 2023 Environmental and economic advantages of production and application of digestate biochar. *Environmental Technology and Innovation* 30, 103109.

Mašek O, et al 2019 Potassium doping increases biochar carbon sequestration potential by 45%, facilitating decoupling of carbon sequestration from soil improvement. *Scientific Reports* 9, 1–8.

Mrozik W, et al 2021 Valorisation of agricultural waste derived biochars in aquaculture to remove organic micropollutants from water–experimental study and molecular dynamics simulations. *Journal of Environmental Management* 300, 113717.

Nardis BO, et al 2021 Phosphorus recovery using magnesium-enriched biochar and its potential use as fertilizer. *Archives of Agronomy and Soil Science*, 67, 1017–1033.

Obia A, Mulder J, Hale SE, Nurida NL, and Cornelissen G 2018 The potential of biochar in improving drainage, aeration and maize yields in heavy clay soils. *PLoS One* 13, e0196794.

Pathy A, Ray J, and Paramasivan B 2021 Challenges and opportunities of nutrient recovery from human urine using biochar for fertilizer applications. *Journal of Cleaner Production* 304, 127019.

Segal M 2009 Selling graphene by the ton. *Nature Nanotechnology* 4, 612–614.

Siipola V, Pflugmacher S, Romar H, Wendling L, and Koukkari P 2020 Low-cost biochar adsorbents for water purification including microplastics removal. *Applied Sciences* 10, 788.

Singh S, et al 2020 A sustainable paradigm of sewage sludge biochar: valorization, opportunities, challenges and future prospects. *Journal of Cleaner Production* 269, 122259.

Smetanová A, Dotterweich M, Diehl D, Ulrich U, and Dotterweich NF 2013 Influence of biochar and terra preta substrates on wettability and erodibility of soils. *Zeitschrift für Geomorphologie* 57, 111–134.

Stavkova J, and Maroušek J 2021 Novel sorbent shows promising financial results on P recovery from sludge water. *Chemosphere* 276, 130097.

Stehel V, Maroušková A, Kolář L, Strunecký O, and Shreedhar S 2020 Advances in dry fermentation extends biowaste management possibilities. *Energy Sources, Part A: Recovery, Utilization, and Environmental Effects* 42, 212–218.

Strunecký O, Shreedhar S, Kolář L, and Maroušková A 2021 Changes in soil water retention following biochar amendment. *Energy Sources, Part A: Recovery, Utilization, and Environmental Effects*, 10.1080/15567036. 2021.1916652.

Thengane SK, et al 2021 Market prospects for biochar production and application in California. *Biofuels, Bioproducts and Biorefining* 15, 1802–1819.

Valaskova K, Adamko P, Michalikova KF, and Macek J 2021 Quo Vadis, earnings management? Analysis of manipulation determinants in Central European environment. *Oeconomia Copernicana* 12, 631–669.

Vochozka M, Maroušková A, Váchal J, and Straková J 2016a Biochar pricing hampers biochar farming. *Clean Technologies and Environmental Policy* 18, 1225–1231.

Vochozka M, Maroušková A, Váchal J, and Straková J 2016b The economic impact of biochar use in Central Europe. *Energy Sources, Part A: Recovery, Utilization, and Environmental Effects* 38, 2390–2396.

Vochozka M, Horak J, Krulický T, and Pardal P 2020 Predicting future Brent oil price on global markets. *Acta Montanistica Slovaca* 25, 375–392.

Wang J, and Wang S 2019 Preparation, modification and environmental application of biochar: a review. *Journal of Cleaner Production* 227, 1002–1022.

Woolf D, Lehmann J, Fisher E, and Angenent L 2014 Biofuels from pyrolysis in perspective: trade-offs between energy yields and soil-carbon additions. *Environmental Science and Technology* 48, 6492–6499.

Yin Q, Zhang B, Wang R, and Zhao Z 2017 Biochar as an adsorbent for inorganic nitrogen and phosphorus removal from water: a review. *Environmental Science and Pollution Research* 24, 26297–26309.

Zvarikova K, Rowland M, and Krulicky T 2021 Sustainable industry 4.0 wireless networks, smart factory performance, and cognitive automation in cyber-physical system-based manufacturing. *Journal of Self-Governance and Management Economics* 9, 9–20.

33

Economics of biochar production and utilization

*Stephen Joseph, Simon Shackley, Ruy Anaya de la Rosa,
Gerard Cornelissen, Adam O'Toole, and Einar Stuve*

Introduction

Biochar systems are composed of a complex set of operations and processes entailing different costs and benefits, including cultivating and sourcing biomass; feedstock preparation, storage, and transport; capital and operating costs of technologies; yield engineering; post-production processing of biochar and other by-products (bio-liquids and syngas); the packaging, marketing and selling of those products; regulatory costs of both air and liquid emissions and biochar; and the value and persistence of biochar in soil and other applications. This chapter attempts to review and estimate: (i) the net costs and revenues of medium to large-scale biochar projects where more than 1 t (dry) hr^{-1} of biomass is converted to biochar and energy in developed country settings; (ii) the financial, social and environmental costs and benefits of pyrolysis technology, at smaller scales where less than 1 t (dry) hr^{-1} of biomass is converted to biochar and energy, both in developed and developing country settings. Case studies of small-scale biochar technology applications, one from a developed country and one from a developing country, are used to identify where biochar systems might be most appropriate.

Methods and tools for economic analysis

Basic considerations

Biochar technologies and products are currently at different stages of development, and the "true" costs of producing biochar and associated byproducts are often not definitively known (Chapter 33). Estimating the costs and benefits of biochar and its byproducts is thus challenging and subject to significant uncertainty, particularly since

DOI: 10.4324/9781003297673-33

some technologies are not yet mature, and 'dominant designs', accepted by specific markets, have yet to be established (Utterback, 1966).

Many scholars of technological innovation have noted that attempts to predict the costs of yet-to-be-fully-developed or immature technologies are typically highly inaccurate (Collingridge, 1980; Dodgson et al, 2013; IPCC, 2022). First-of-a-kind technologies are more expensive, since all the development, design, and demonstration costs are borne by very few units, and economies of scale in materials and production are not yet realized. To date, there are few examples of fully developed technologies for biochar production, examples including pyrolysis rotary kilns that are mainly built in China (where it is estimated that over 200,000 t of biochar are produced per year, Zheng et al, 2017), screw pyrolyzers (Woolf et al, 2017; Sørmo et al, 2020) and modern container batch kilns. In addition, biomass gasifiers, at a range of scales, are operational in China, the USA, Europe, and Australia, though biochar in some of these plants is a by-product rather than the primary product.

Many of the smaller pyrolyzers (<1 t hr^{-1} biomass input) and the products that are produced from this biochar are currently in the stage of early commercialization, and therefore the 'true' costs of producing biochar and associated by-products may not be known at present. A proviso here is that we assume that the traditional technologies of charcoal production are not the appropriate reference point for considering biochar production costs. To explain why, we need to briefly consider the history, present organization of, and innovation associated with, the charcoal production sector.

Charcoal, hence biochar, can be produced in some countries at low cost due to uncontrolled (frequently non-renewable) removal of wood, low labor costs (e.g.,

$<$US\$ 10 day^{-1}), and low start-up costs for traditional kilns (e.g., simple kilns can be built in sub-Saharan, southern and eastern African countries for ca. US\$150) (Kambewa, 2007). The Freight on Board (FOB) price for charcoal export from such countries can be as low as US\$250 t^{-1} (one-third to one-fifth of the wholesale price of charcoal in industrialized countries) and some biochar retailers in industrialized countries are relying upon such exports. However, such traditional charcoal production does not typically fulfill the requirements of clean efficient production due to: (i) fugitive gaseous and particulate emissions (Pennise et al, 2001; Cornelissen et al, 2016); (ii) the low yield efficiency ranging from 10–14% for earth mounds and pit kilns, and 20–30% for Brazilian beehive, half-orange brick, drum, Casamance, and hot-tail kilns (Kambewa, 2007; Bailis et al, 2013); (iii) under-utilization of 70–80% of the energy produced from the syngas (the produced liquids or heat); and (iv) concerns over the sustainability of the feedstock. Unfortunately, despite the ease and low cost of traditional charcoal production technologies, they are a highly polluting production method (Chapters 3 and 4), potentially hazardous to the health of workers, and therefore should not be considered an appropriate environmental or economic reference point for the goal of clean, efficient production of biochar.

There are lessons to be learned from past experiences of charcoal production in industrialized countries. In the nineteenth and first part of the twentieth century, charcoal was widely produced on large scales, and the pyrolytic liquids were collected and used as chemical feedstocks. Vertical and horizontal continuous production retorts capable of producing 2000 to 6000 t yr^{-1} (7–20 t day^{-1} assuming 80% availability) of high-quality charcoal were developed by companies such as Lambiotte, Lurgi, Mitsubishi E&S, Mitsui E&S, and ITB. The SIFIC Lambiotte retort is still being manufactured in the Baltic region for charcoal production in Eastern Europe, Russia, Australia, and Central Asia (www.lambiotte.com, www.baltcarbon.lv). In Australia, the fines from the kiln are sold as biochar (Joseph et al, 2015). The Lambiotte retort uses pyrolytic vapors for drying the wood and initiating carbonization, while also rinsing the charcoal with hot gases to remove excess volatiles and polycyclic aromatic hydrocarbons (PAHs). The fine charcoal from a Lambiotte kiln has been fed to cattle for over 8 years (Joseph et al, 2015). Balt Carbon is also undertaking research and development (R&D) on the potential for electricity generation from pyrolytic vapors using steam turbines (Roberts et al, 2010). Recovering and using syngas and condensable fractions arising from charcoal making in modern container kilns in Brazil for energy generation (and/or production of tars and creosote replacements) (Bailis et al, 2013). These technologies can be used to inform the production costs of biochar from slow pyrolysis, with some convergence emerging between the separate research, development, and deployment (RD&D) communities involved in charcoal and biochar production.

Another option for considering biochar production costs is to investigate related mature technologies that are already widely deployed. This could include the pyrolysis of coal (for coke production), shale (for oil), and gasification of coal (for syngas and conversion to liquids, etc.). Nowadays, such technologies tend to exist on a large scale because fuels such as coal and oil-bearing shale are energy-dense and spatially concentrated. However, biomass supplies are less energy-dense and much more sparsely distributed. This feedstock configuration results in high transport costs, and a related scaling down of technology size, which has knock-on effects on the efficiency of conversion from biomass to energy. Processing and converting biomass via pyrolysis incurs a higher cost per mass compared to energy-dense fuels such as coal and oil-bearing shale. The production of granular activated carbon (GAC) on an industrial scale started at the beginning of the twentieth century and perhaps provides a closer analog since slow pyrolysis technologies are used and could be adapted to produce biochar. However, obtaining data on activated carbon production costs is challenging due to: (i) varied and multiple sources of supply for the specified carbonized materials which are then activated, operating at a range of scales and using diverse technologies (for instance, village communities may work together to produce coconut shell charcoal that is then sold to the large firms who undertake the activation); (ii) commercial confidentiality; and (iii) the difficulty of separating the costs of slow pyrolysis from the costs of activation and other additional processing stages involved in producing activated carbon.

Thus, overall, it is difficult to find appropriate analogs that may help to predict the future costs and financial benefits of biochar production systems. However, as dominant designs become established, output yield, properties, and costs can be better understood. These processes typically advance iteratively, started by early adopters who are prepared to pay a premium to use new biochar products. This creates 'learning-by-doing' improvement cycles with technology providers and end-consumers. Any current estimates of biochar costs and benefits must therefore be taken as uncertain and conditional on assumptions that are undoubtedly inaccurate, perhaps greatly so. The following section introduces the concept of cost–benefit analysis as a method for evaluating the economics of biochar systems, before presenting the results of a collation and comparison of available data in search of common conclusions regarding profitability, feedstocks, and system design.

Constructing a cost-benefit analysis for biochar systems

There are a range of methodologies and tools for evaluating the economic potential of low-carbon technologies (Stern, 2007). Investors usually identify segments of the value chain and calculate the specific costs and benefits, looking to maximize value in one domain, for instance, technology developers will focus on the potential for specific engineering designs, project developers will consider the profitability of a particular project proposal, whilst marketing and sales experts will be looking at retail and end-consumer markets. Government policy-makers may wish to consider the entire value chain, e.g., to consider comparative indicators such as the marginal abatement cost of carbon. It is important to appreciate the difference between these individual elements of the chain and the system as a whole (Chapter 30). Due to the immaturity of the biochar industry, many of the links in the value chain are immature or under-developed, limiting the confidence with which robust economic analysis of the biochar sector can presently be undertaken. Investors are forced to take on more risk than they would like, unless a public body is willing to step in and underwrite projects through grants or favorable loans.

A simple financial analysis of biochar (Equation 33.1) subtracting costs from benefits (Dutta and Raghavan, 2014; Owsianiak et al., 2021) is:

$$NP = BC + E - F - T - C - O - A \quad [33.1]$$

where NP = net profit; BC = biochar value (e.g., carbon storage, agronomic); E = energy sales (biochar or charcoal, syngas, bio-oil);

F = feedstock costs (production, collection, storage, and processing); T = biochar transport and storage costs; C = capital costs; O = operational costs; A = application costs.

A cost-benefit analysis (CBA) is a bottom-up methodology (Figure 33.1), in that it considers a single (or small number of discrete) investment decisions independent of whatever else is happening in the economy (McCarl et al, 2009). Due to the multi-year character of most biochar projects, a discount factor is applied to account for the time-value of money and to (market perceptions) of relative investment risk (Equation 33.2):

$$\frac{B}{C} = \frac{\Sigma_{T=0}^{N} B/(1+i)^t}{\Sigma_{T=0}^{N} C/(1+i)^t} \quad [33.2]$$

where B = benefit in year t; C = cost in year t; i = discount rate; N = project lifetime.

Net Present Value (NPV) is used to measure cumulative net benefits over the project lifetime. The net cash flow per year is computed using a project discount rate and added up over the project lifetime (Equation 33.3):

$$NPV\,(i, N) = -R_0 + \Sigma_{t=0}^{N} \frac{Rt}{(1+i)^t} \quad [33.3]$$

where NPV = Net Present Value; R_0 = Initial investment; Rt = Cash flow at time t; i = Discount rate; t = time of the cash flow.

Due to the early stage of the industry, most academic studies to date have not included costs associated with project development, engineering commissioning, and procurement, insurance, marketing, and sales. The costs of decommissioning equipment at the end of the project should be accounted for in calculating the NPV. The greater the NPV, the more economically viable the proposed project. While any project with a positive B/C ratio or NPV might be regarded as viable,

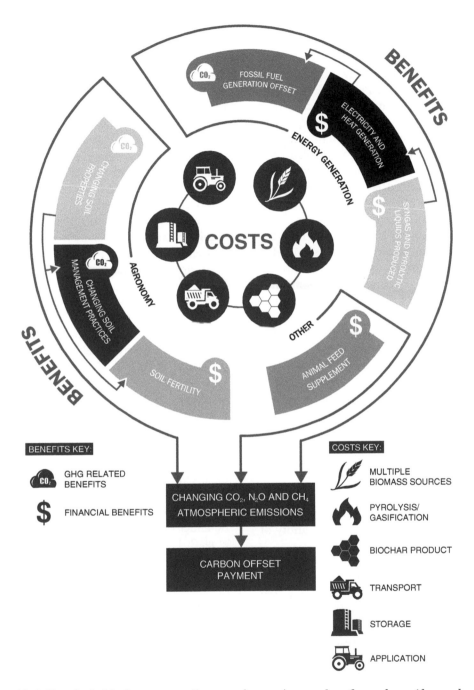

Figure 33.1 *Pyrolysis–biochar system diagram, for use in cost–benefit analyses (drawn by: Jonathan Stevens, Starbit Ltd., with permission)*

whether it will give a 'green light' to a Final Investment Decision (FID) will vary depending on the wider objectives and risk appetite of the investor. Other indicators that investors use include the Payback Period (the time taken for the cumulative net cash flow from the start-up of the plant to equal the depreciable fixed capital investment); Internal Rate of Return (IRR), which calculates the net rate of return on investment (at which the NPV is zero); and Return on Investment (ROI) (the ratio of the net return on investment to the cost of investment).

Levelized Cost of Energy (LCOE) is the ratio of the total discounted costs and discounted energy production over the project lifetime, thereby enabling comparison on a level playing field of different energy generation technologies. Likewise, a Levelized Cost of Biochar (LCOB) for a given application can be defined (Equation 33.4). LCOB could be defined irrespective of application by removing A and T from Equation 33.4. One advantage of LCOB is that it avoids ascribing value to biochar.

$$LCOB = \frac{\sum_n^N \frac{(C_n + O_n + F_n + T_n + A_n + D_n)}{(1 + r)^n}}{\sum_n^N \frac{(B_n)}{(1 + r)^n}} \quad [33.4]$$

where definitions are as in Equation 33.1 and D = decommissioning costs; B = quantity of biochar.

Similar to life cycle assessments (LCA), CBAs and other economic indicators are typically sensitive to where the boundary around the system is drawn (Roberts et al, 2010). Biomass feedstock costs vary due to other demands for the same resource (e.g., as cattle fodder) while the use of biochar for energy or agronomy will depend on charcoal markets, grain prices, carbon price, and user perceptions of biochar's value in soil and their subsequent willingness to pay. Investigating

indirect land-use change (ILUC) (whereby feedstocks for biochar are purpose grown and might displace food production, increase the demand and price of bioresources by increasing competition for land resources) requires a total model of land–food–energy systems (Wise et al, 2009; Henry et al, 2018; Turner et al, 2018; Roe et al, 2021) but their development and use present a conceptual challenge (e.g., regarding whether an equilibrium, quasi-equilibrium or some other response is presupposed).

The agronomic and soil benefits are highly specific to spatial and temporal context as well as uncertain. For example, processes such as the accumulation of soil organic matter (SOM) may have an agronomic value that can be monetized through its positive influence on crop yields. However, uncertainty in our understanding of how SOM relates to crop yields in different soils (Oldfield et al, 2019) and, furthermore, how biochar application relates to (or even influences) SOM levels (Chapter 17), means it is difficult to incorporate such economic relationships into economic analysis at this time, although a considerable volume of research is being undertaken to gather this data. Other approaches are to infer a market value for biochar and then determine whether its agronomic and carbon storage benefits can be anticipated to exceed this price (Galinato et al, 2011).

Some critical questions in evaluating biochar system profitability are:

- Which of these costs are we most certain or uncertain about?
- Which values are likely to change in the future and how might these impact the system?
- Which of the costs and benefits contribute most to overall system profitability, or lack thereof?

Review of studies carried out to date

A small number of economic analyses of the use of biochar have been carried out in agriculture, forestry, animal husbandry, and waste management in developed and developing countries. This section summarizes the results of these studies.

Cost of biochar at factory or farm gate

In this section we look only at the production cost of biochar, intentionally excluding estimates of its potential benefits. With the inclusion of biochar in some voluntary carbon market schemes, a new source of information on biochar costs has become available. Biochar-carbon prices are available (as of mid-2022) at US$100–550 t^{-1} CO_2 removed (based on values provided by the following companies: Husk, Klarna, Carbonfuture, and Mash in 2020; Carbofex, Stripe, Ecoera, Patch, CarboCulture in 2021 (Mocci, 2022; Mocci et al, 2022). These values should encapsulate costs across the supply chain, including marketing and verification, though without more detailed knowledge of the financing arrangements, it is not possible to be certain. The relationship to the cost of biochar *per se* will depend upon the net C removal per mass of biochar. In many cases, this means that the cost per mass of biochar will be higher than the cost per mass of biochar-carbon.

The most comprehensive data on biochar cost at the factory gate, or at the farm gate, exist from North America. A stochastic model was used in one study to estimate production costs and derive a range of US$450–1850 t^{-1}, with a 90% probability that the cost would fall between US$570–1455 t^{-1} (Nematian et al, 2021). An earlier study surveyed 28 biochar producers to generate an average figure of US$ 1500 t^{-1} (Jirka and Tomlinson, 2015). A meta-analysis of biochar prices in North America provided a range from US$800–17,800 t^{-1} (Alhashimi and Aktas, 2017). Other studies in the USA and Canada provide ranges or point values of US$530–1050 t^{-1} and US$1675–1910 t^{-1} (Bergman et al, 2022), US$500–600 t^{-1} (Ahmed et al, 2016; Sessions et al, 2019), and US$500 t^{-1} (Keske et al, 2020). Lower production costs in the region have also been estimated, with values of US$330 t^{-1} (Kim et al, 2015), US$290 t^{-1} (Sorensen and Lamb, 2018), US$240 t^{-1} (Struhs et al, 2020), and as low as US$40–50 t^{-1} (Dutta and Raghavan, 2014). The low cost can partly be explained by the treatment of an existing waste material or other inexpensive feedstocks.

Turning to other parts of the world, a process simulation study in China estimated production costs of US$790–930 t^{-1} when using rice straw or sugarcane bagasse as feedstocks (Lin et al, 2022). However, other studies in China, focused on lowering biochar costs, have provided a figure as low as US$20 t^{-1} (Liang et al, 2019). In Europe, costs have typically been estimated at US$150–600 t^{-1} (Shackley et al, 2011; Dickinson et al, 2015; Aguirre et al, 2021), though with lower costs in some countries (e.g., US$85 t^{-1} for one study in Poland, Latawiec et al, 2021; US$75 t^{-1} in Spain, Ahmed et al, 2016) and where waste feedstocks are utilized. In Australia, costs of between US$265 t^{-1} (Robb et al, 2018) and US$1000 t^{-1} (Wrobel-Tobiszewska et al, 2015) have been estimated. Amongst the few estimates of production costs in developing countries are US$100–165 t^{-1} in sub-Saharan Africa (Dickinson et al, 2015) and US$100 t^{-1} (Ahmed et al, 2016), though much lower costs have also been identified where it is a low-value by-product.

Another approach to estimating the cost of biochar is to identify what the minimum buying price would have to be for the operation to break even. This minimum selling price will be similar to the production cost, though with the inclusion of return on investment. One study in the USA found that the minimum selling price (hence required a willingness to pay by the purchaser) lies at US$75–1280 t^{-1} (Li et al, 2019), while consumers were willing to pay for a peat mix containing biochar to substitute for a proportion of the peat (Thomas et al, 2021).

It can be seen that estimates vary greatly at present, which is a function of the diversity of the supply chains, the large differences between regions, countries, and sub-national areas, and the current lack of maturity of most stages of the biochar supply chain. It is also a function of the diverse methodologies used to estimate production costs in the literature, ranging from limited empirical data to model-based simulations and often a combination of both. In the next section, we delve in more detail into a few of the studies quoted to provide a more nuanced account.

Studies that compare the potential economic returns between developed and developing countries

One-off large applications of biochar may return a positive benefit in developing countries but not in developed countries (Dickinson et al, 2015; Robb et al, 2020), although the use of biochar-based fertilizers could return a net benefit to farmers particularly in field crop production in developed countries (Robb 2020, Robb and Joseph 2019). Assuming a one-off application of 12 t ha^{-1} biochar, biochar application to cereals in Sub Sahara Africa (SSA) returns a positive NPV within 23 years in circumstances of moderate discount rate (5%) and median or better yield increase (Dickinson et al, 2015). For yield increases of 1 t ha^{-1} and a biochar cost of application of US$132 t^{-1}, biochar use obtains a positive NPV in SSA if seven years of yield enhancement can be achieved and reaching an NPV of greater than US$1000 ha^{-1} for 20 years of persistent returns. However, if the one-off effect of biochar does not provide long-term yield benefits beyond six years, then there is a negative return on investment.

Similar results were found when comparing the returns of producing biochar in a simple flame-capped kiln in rural areas and gasifiers in urban areas of five developing countries (Ethiopia, Indonesia, Kenya, Peru, and Vietnam) and using a gasifier in a recently industrialized country (China) (Owsianiak et al, 2021). Quantifying the environmental and economic life cycle impacts of biochar to either home composting or windrow composting as alternative biowaste management systems, both flame cap and gasifier biochar systems performed better than composting and both were expected to bring environmental benefits. The largest environmental benefits were observed for gasifier systems, mainly due to the substitution of electricity production from the grid (resulting in avoided emissions). Damage to ecosystems ranged from -1×10^{-7} to -2×10^{-8} species yr^{-1} and for human health from -1×10^{-5} to -5×10^{-6} (Disability Adjusted Life Years (DALY)) kg^{-1} of biowaste treated, respectively (negative scores indicating environmental benefits). However, net economic benefits were only achieved when low-cost simple kilns were used in countries with low labor costs, such as Ethiopia, Kenya, and Vietnam (net profit from US$0.01 to 0.08 kg^{-1} of biowaste treated). Further, high investment and operating costs and relatively small electricity revenue from substituting the grid electricity resulted in gasifier scenarios being economically unsustainable (net loss US$0.29–1.58 kg^{-1} of biowaste treated). Thus, there are trade-offs between positive environmental impacts for society and net market loss for the individual decision-maker (company or individual). It should be recognized that energy and fertilizer prices have substantially increased since this study was undertaken. Government subsidies for renewable energy have also changed, some increased and some decreased.

A financial analysis and life-cycle carbon emissions assessment of oil palm waste biochar exports from Indonesia for use in Australian field crop agriculture covered application rates of 1–40 t ha^{-1} of empty fruit bunch biochar (Robb and Dargusch, 2018). This biochar could be produced and transported to the farm gate at a minimum sale price of US$266 t^{-1} and the ROI for sugarcane was 67%, irrigated cotton 43%, dry land cotton 22%, and wheat 72%.

Assessment of the costs and benefits in industrialized countries

In China, around 0.9 10^9 t of crop straw is produced every year, of which 77% is collected for use in energy biochar and other on-farm applications. However, due to the rapid development of China's economy and the change in rural residents' lifestyles, crop straw treatment has become a great challenge in agriculture and the rural environment in the past decade. Of the average 0.7 10^9 t of crop straw produced in China in 2016, 54% were directly returned to the field as organic matter, 23% were used as animal feed, and 14% were utilized as fuel, leaving 9% with no clear use. In the last ten years, many studies have been conducted to investigate the effect of applying pyrolyzed straw biochar on soils, crop yield, and greenhouse gas (GHG)

emissions. A meta-analysis of 173 individual studies (Liu et al, 2017) revealed that biochar amendment (average 22.6 t ha^{-1}) increased crop yield by 15% on average for main crops including maize, rice, wheat, soybean, and vegetables.

The high price of biochar (>US$300 t^{-1}) and the large dose (>15 t ha^{-1} either for one-off application or application every 3–5 years) required for yield increases have limited farmer adoption. A financial analysis (Table 33.1) showed that it is unlikely that the farmers in China would buy and use biochar in their fields at application rates greater than 1 t ha^{-1} unless there were subsidies or the farmers could receive a fee for atmospheric carbon removal.

Producing biochar-based fertilizers (Chapter 26) by mixing biochar with inorganic fertilizers at approximately 20% of the total weight of a granule and applying the granule at approximately 500 kg ha^{-1} (equivalent to 100 kg ha^{-1} biochar) can reduce the cost substantially. Field experiments were carried out for 2 years across 146 sites in China (Table 33.2). The addition of biochar-based fertilizers increased median maize yield by 5.3%, rice yield by 9.8%, and soybean yield by 6.8% compared to conventional fertilizers (Table 33.2). Although the price of biochar-based fertilizers is slightly higher than conventional fertilizers (Table 33.3) there was a reasonable return to farmers. Commercial production and sale of this granular biochar-based fertilizer, which can be applied mechanically, is now occurring throughout China.

In Australia, the use of large application rates of low-value wood biochar that is a waste product from a Silicon smelter (delivered at a cost of approximately US$100 t^{-1}) for the growing of avocados resulted in a net benefit of US$ 8,581 ha^{-1} or US$105 t^{-1} of biochar applied (Joseph et al, 2020). Feeding the same biochar to cattle results in a

Table 33.1 *Cost-benefit analysis of a single application of biochar to maize in Xinzhou, Shanxi Province of China (plot size: 4 by 7 m in 3 replicates; maize was air-dried; years: 2011–2012; rainfall: 345–588 mm; soil type: Inceptisol; mean annual temperature:8.0–10.5) (Zhang et al, 2016)*

	Normal biochar price (US$300 t^{-1}) and single application (20 t ha^{-1} that is assumed effective for 5 years)	Biochar price reduced by 30%	Biochar rate reduce to a single application of 1 t ha^{-1} with chemical fertilizer
Maize yield increase of adding biochar (t ha^{-1} yr^{-1})	0.85	0.85	0.85
Financial benefit of biochar addition (assuming 5 years of same effectiveness; US$420 t^{-1} maize)	357	357	357
Additional labor cost for biochar application (US$5 yr^{-1})	5	5	1
Biochar cost (US$ ha^{-1} yr^{-1})	1200	840	60
Farm Revenue (US$ ha^{-1} yr^{-1})	−838	−478	296

Table 33.2 *Changes in crop yield following biochar-based fertilizer application as compared to common inorganic fertilizer (Xu, 2021)*

Crops	Number of experimental sites	Yield increase (%)	Median (%)
Maize	62	1.6–24.3	5.3
Rice	39	0.5–22.5	9.8
Soybean	20	1.6–33.3	6.8
Wheat	5	5.1–14.7	10.8

Table 33.3 *Financial analysis of using biochar-based fertilizers in maize cropping in Anhui Province of China (plot size: 4 by 5 m in 4 replicates; crop yield air-dried; experiments conducted in 2017; rainfall: 840 mm; soil type: Vertisol; mean annual temperature: 14.8°C) (Liu et al, 2017)*

	Biochar-based fertilizer (20% replacement of inorganic fertilizer)
Additional cost compared to the same amount of chemical fertilizer (US$ t^{-1})	84
Maize yield increase (t ha^{-1} yr^{-1})	1.25
Extra labor cost for biochar application (US$ yr^{-1})	0
Additional revenue from increased yield (US$ ha^{-1} yr^{-1})	441

significant financial benefit (net user benefit of approximately US$1000 t^{-1}) (Joseph et al, 2015). A study carried out on a dairy farm (with 250 cows) over 9 months in South Australia indicated that adding approximately 1% of mixed feedstock biochar costing US$800 t^{-1} biochar to dairy cows' feed increased the whole of farm income by more than US$20,000 through reduction of the cost of hay, increased milk yield and improved pasture health (Taherymoosavi et al, 2022). A CBA of the viability of pyrolyzing residues with a mobile kiln from hardwood plantations and using this for growing eucalyptus seedlings indicated that significant savings in site preparation, seedling establishment, and fertilization could be achieved (Wrobel-Tobiszewska et al, 2015). The total benefits depended on the output and capital cost of the pyrolyzer, the amount of biochar that was sold and the selling price, the amount of biochar returned to the forest, and the revenue from selling the carbon credits. If the portable kiln capital cost was US$160,000 which could process 12 t wood day^{-1} then the NPV was calculated at US $100,000.

In North America, the economic value of biochar application on agricultural cropland for carbon sequestration and its soil amendment properties considered the GHG emissions avoided when biochar is applied to agricultural soil, instead of agricultural lime, the amount of C sequestered, and the value of carbon credits (Galinato et al, 2011). Biochar was used to replace lime at an application rate of 6.48 t^{-1} ha^{-1} for winter wheat production in an acidic soil. A sensitivity analysis was carried out at prices for carbon credits of US$1 and US$30 t^{-1} of CO$_2$e and biochar prices from US$0-400 t^{-1}. They reported that, if the carbon offset price is US$ 31 t^{-1} CO$_2$e, the farmer will break even if the price of biochar is approximately US$ 100 t^{-1}. If the carbon offset price

is US$ 1 t^{-1} CO$_2$e, a profit of about US$ 164 t^{-1} is gained when the price of biochar is US$ 10 t^{-1}. At this lower offset price, the farmer will break even if the price of biochar is approximately US$12 t^{-1}.

Later studies concluded that if all the benefits were included higher application rates of biochar could improve farm profitability. When public benefits, such as decreased nitrate leaching and increased soil C level, are included, applying biochar to corn (maize) in the mid-west of the USA at a rate of 22 t ha^{-1} allows for the sustainable annual removal of 50% of corn (maize) residues for 32 years (Aller et al, 2018). This remains profitable for farmers even when biochar has a minimal impact on grain yield. Converting wood waste to biochar, using a portable pyrolyzer for clearing forest trees for the construction of a dam, increased sugar beet yield by 67% (5.59 t ha^{-1} more than farmer practice of 11.4 t ha^{-1}) through the addition of 10 t C ha^{-1} of biochar (cost approximately US$1000 t^{-1}) (Keske et al, 2020). A stochastic analysis with variable prices and yields shows a 0.99 probability of biochar production being profitable when applied to beets at the midline production rate, with an average annualized net return over variable costs of US$4,953 ha^{-1}, and maximum annualized net return of US $11,288 ha^{-1}. Other studies using forest and orchard residues (Eggink et al, 2018) have highlighted the variability of production costs, sales price, and return on investment of applying relatively large rates of biochar to the farmer.

A study carried out on the production of bioenergy and biochar from wood found that, without subsidy, replacing power through the production of biochar was not cost-effective (Garcia-Perez et al, 2019). Reducing power from the plant from 30 MW to 25.5 MW to produce 5.5 Mt hr^{-1} of biochar resulted in a loss of income unless the biochar could be sold for at least US$150 t^{-1}. Including the potential value of biochar based on C sequestration and yield improvement led to the conclusion that without a climate policy compensating farmers for C sequestration, there is only one type of crop (mixed vegetables) which under a fairly optimistic yield improvement assumption (30%), could justify the use of biochar (Garcia-Perez et al, 2019).

Biochar can reduce the cost of above- and below-ground C sequestration by planting eucalyptus in Florida (Rockwood et al, 2022). They reported that planting short rotation woody crops (SRWC) using *E. grandis* and *E. grandis* x *E. urophylla* cultivars with and without soil amendments including biochar, may sequester over 10 t of C ha^{-1} year^{-1}. Under the assumed management costs and market conditions, SRWC management with biochar was found to be more profitable compared to using conventional practice (operational culture) if biochar application costs are ≤US$450 t^{-1}. They also reported that using *E. grandis* cultivars in windbreaks may sequester up to 34 t C ha^{-1} in three years, with additional sequestration by amending soil with biochar.

In Poland, biochar can be produced at relatively low costs of US$85 t^{-1} (Latawiec et al, 2021) and at this price, the payback period for biochar application rates of up to 60 t ha^{-1} was estimated to be three years. Assuming a US$30 t^{-1} CO$_2$e removal value, the use of biochar is profitable in the first year of soybean production.

Interest in the production and use of biochar has been increasing in Norway in recent years. Norway has a large potential for biochar production considering its vast forestry resources and where pyrolysis syngas can be burnt and sold to provide for winter heating needs. Biochar applications under exploration include its use as an animal feed, fertilizer ingredient, peat replacement in soil blends, a substitute for metallurgical coal or imported charcoal, and as a concrete additive. The first commercial biochar plant in Norway was established in 2020 by Oplandske Bioenergi AS (www.oplandske. no/home) situated in the town of Rudshøgda, Innlandet region (Box 33.1). The project achieved significant revenue from carbon markets through both a high carbon price and comparatively high emission reductions.

Studies in developing countries

A study carried out in North Vietnam in a rice-producing village compared the costs and benefits of producing biochar in either a pyrolytic stove or a small drum oven from rice residues and returning these to the fields versus on-field open burning of the residues (Mohammadi et al, 2017). Based on experimental data (Joseph et al, 2015) it was assumed that a household could produce 0.6 t biochar yr^{-1}. Use of this biochar in local cultivation led to the following responses: (i) approximately 10% +/- 6% increase in crop yield; (ii) 30% +/-30% decrease in nitrogen (N) fertilizer use; and (iii) 50%+/-50% decrease in phosphorus (P) and potassium (K) fertilizer use. The modeling assumed that the agronomic impacts of biochar-based fertilizers with increasing biochar applications until reaching maximum benefit at 18 t ha^{-1} which would take 8 years to produce in both these simple devices. This application figure assumed that biochar was added every crop cycle. It was calculated that biochar addition every year for 8 years enhanced the NPV of rice (using a discount rate of 10%) by 12% compared with open field burning. The difference in NPV values between production systems (no addition versus biochar addition) increased to 23% and 71% by crediting GHG

Box 33.1 Case study: slow pyrolysis biochar and district heating in Norway

The biochar plant includes an advanced belt dryer for drying wood chips, a 400 kW C400-I slow pyrolysis screw reactor (Biomacon Gmbh), feedstock storage facilities, and an industrial-sized mixing reactor for pretreatment of feedstocks or biochar (Table 33.4). Approximately 40% of the energy produced is used for drying wood chip (spruce) feedstock, and the excess energy (20%) heats water via a heat exchanger and is sold to a neighboring industrial abattoir and a concrete mixing factory for their heating needs. The company is currently selling three grades of biochar which are intended for 3 applications: OBIO-Agri for general soil applications, OBIO-Förkull for animal feed addition, and OBIO-Urban for landscaping purposes. These products may differ in particle size and requirement for the content of PAHs or heavy metals. The plant is the first in Norway to be certified under the European Biochar Certificate. Its biochar and plant has also undergone an LCA and its biochar is approved for sale for CO_2-equivalent Removal Certificates (CORC) via the puro.earth platform (Puro, 2022) (Table 33.5). Here the CORC is valued at US$137 t^{-1} CO_2e. The biochar made from the wood chips has a C content of 92%, with an emission reduction of 3.01 t CO_2e t^{-1} biochar calculated via LCA.

Table 33.4 *Inputs and outputs of the pyrolysis plant*

Current feedstock capacity (kg hr^{-1})	300 − 400
Duration of operation (hr yr^{-1})	7900
Feedstock capacity (t yr^{-1})	2370 − 3160
Current (potential) biochar production (t yr^{-1})	250 (370)
Exportable energy capacity (kW)	400
Exportable energy generation (MWh yr^{-1})	3160

Table 33.5 *Financial summary of pyrolysis plant at Oplandske Bioenergi*

	US$[d]
Feedstock cost delivered to plant (US$ dry t^{-1})	100
Yearly feedstock costs (US$ yr^{-1})	237,000
Plant CAPEX, including dryer, pyrolysis, buildings etc. (MUS$)	2.2
Public capital subsidy (US$) (32% of investment)	700,000
Total fixed operating expenses (excluding interest and income tax expenses) (US$)	105,000
Heat sales revenue (US$ yr^{-1})	90,000
Biochar factory sale price (US$ dry t^{-1})	1100

Table 33.5 *continued*

	US$[d]
Biochar potential revenue including energy sales (minus variable costs; if 100% sold) (US$ yr^{-1})	498,000
Puro Earth CORC[a] price (US$ CORC^{-1})	137
Potential CORC[a] production (tonnes production yr-1 ×3.01 [t Biochar-C to CO_2e])	1150
Potential biochar CORC[a] net revenue (US$)	138,000
Total revenue potential(heat, biochar, CORC)	636,000
EBITDA[b] (US$)	262,000
IRR[c] over plant lifetime of 20 years with 2.25% inflation rate (%)	5.8

Notes
[a] CORC – CO_2-equivalent Removal certificates
[b] EBITDA = Earnings before taxes, depreciation and amortization
[c] IRR = Internal rate of return
[d] US$ numbers based on US$/NOK exchange rate 20.06.2022

emissions abatement in low (US$15 t^{-1} CO_2e) and high carbon price (US$80 t^{-1} CO_2e) scenarios, respectively.

A study carried out in Brazil on degraded pasture found that the yield of *Brachiaria* spp. increased by 27% through the addition of 15 t ha^{-1} of biochar with fertilization (Latawiec et al, 2019). Biochar addition also increased the contents of macronutrients, soil pH, and CEC. Biochar saved 91 t CO_2e ha^{-1} through a reduction in land clearing. However, the costs of biochar production for smallholder farmers, mostly because of labor costs, outweighed the potential benefits of its use. Biochar in Brazil (at 2018 prices) was 617% more expensive than common fertilizers.

A case study in Indonesia (Box 33.2) showed that the costs of biochar production exceeded the benefits in the form of additional income due to increased maize yields. This was the case even when assuming that the agronomic effects observed in the first season were retained for five years. The yield increase sustained over 5 years offering the same benefit as the cost of biochar making (i.e., the yield increase at break-even) was +26% for Lampung and +22% for Lamongan. Such relative yield increases are not unrealistic for weathered or dry soils such as in much of sub-Saharan Africa, but they are unrealistic for baseline maize yields (9.5 t ha^{-1}) as high as those in Lamongan. The simple cost-benefit analysis thus correlates with the lack of interest in biochar reported by the Lamongan farmers. For the Lampung farmers in Sumatra, a positive gross margin of biochar application could be reached by improved efficiency of biochar production, i.e., reduction of labor costs. This should not be very challenging as the Lampung farmers needed 42 person-days t^{-1} of biochar. Bringing down the labor requirements to 14 person-days t^{-1} would result in benefits equal to costs, even at the modest yield increase rate of 8%.

Box 33.2 Case study: biochar production and soil use in two villages in Indonesia

Baseline agricultural practices, climatic conditions, soil constraints, and socioeconomic characteristics were recorded for 10 farms in Lampung, South Sumatra, and 10 farms in Lamongan, East Java. Farmers were trained to make biochar in flame curtain kilns, enrich it with mineral fertilizer, and apply it to soil. The agronomic and socioeconomic impacts were assessed after one (Lamongan) or two seasons (Lampung) of biochar application, and related to farmers' knowledge and motivation for making and using biochar.

The main crop grown in both sites was maize, followed by rice, watermelon, and peanuts, whereas cassava was only cultivated in Lampung. The planting areas ranged from 0.25 ha to 3 ha. Most respondents were males older than 40 years who held several occupations. The annual income of the participants was between US$1,400 and US$4,700, with Lamongan farmers earning about twice as much as Lampung farmers. Selling prices were about US$190 t^{-1} of maize in Lampung and US$270 t^{-1} of maize in Lamongan. Farmer expenses were dominated by fertilizers in Lampung, whereas expenses were equally distributed over fertilizers, seeds, labor, and pesticides in Lamongan. Cassava stems in Lampung and maize residues in Lamongan were the feedstocks for biochar production. Target crops were cassava, maize, upland rice, and watermelon. Soil conditions were generally poorer (e.g., lower P, K, Ca, Mg, organic C, and pH) in Lampung than in Lamongan. In Lampung, deleterious aluminum was significantly lower in the presence of biochar, whereas K and pH were significantly higher.

The impact of biochar on crop yield was modest for most farmers, with about an 8% increase in Lampung and a 5–10% increase in Lamongan. Baseline maize yields were 6.5–7.5 t ha^{-1} in Lampung and 9.5 t ha^{-1} in Lamongan. 85% of farmers in poorer Lampung (annual income US$2,000) preferred to make biochar by themselves (Table 33.6). In contrast, in Lamongan (annual income US$4,000), 40% of farmers preferred to produce biochar, another 40% preferred to purchase biochar, and 20% neither wanted to make or buy biochar. In Lampung, 90% of farmers were positive about the Kon Tiki kiln and enjoyed operating it, particularly in a group, whereas in the relatively wealthier Lamongan, the majority perceived that this technology was labor-intensive and difficult to operate under the humid conditions in East Java. It should be noted that farmer's experience in biochar production was limited and the efficiency of making biochar would almost certainly increase with more biochar activities. Farmers reported that financial profit rather than climate-change mitigation would be their main motivation to adopt biochar practices. Most carbon certification schemes have valued carbon at a low level compared to the labor investment of US$189 t^{-1} to produce biochar with a Kon Tiki kiln in Lampung. However, the increased value for carbon offsets suggests emerging convergence with the willingness to pay US$58 t^{-1} of biochar (Table 33.6) (Table 33.7).

Table 33.6 *Socio-economic effects of biochar practices (10 farms at each location)*

	Lampung	Lamongan
Biochar production	**% of farmers**	
Prefer to produce biochar	85	40
Prefer to buy biochar	15	40
Neither to produce or buy	0	20

Table 33.6 *continued*

	Lampung	Lamongan
Positive attitude towards Kon Tiki kiln	90	10
Labor costs	**Person-days**	
Person-days needed for 1 t biochar (median)	42	24
Person-days needed for 1 t biochar (IQR)[a]	34–56	16–28
	US$	
Labor costs (US$ t^{-1} of biochar) (IQR)[a]	189[b] (153–252)	470[b] (179–313)
Willingness to pay	**US$**	
Willingness to pay (US$ t^{-1} biochar) (IQR)[a]	58 (38–106)	69 (38–77)
Interest in a more sophisticated biochar technology	**% of farmers**	
Interested	0	10
Not interested	30	60
Interested, but do not have money	70	30
Climate change and carbon credits	**% of farmers**	
Are carbon credits price (US$5 t^{-1} CO_2e) a motivation?	100	30
Is climate-change mitigation a motivation?	0	0
Is money a motivation?	100	30

Notes
[a] IQR, interquartile range.
[b] The average daily income in Lamongan (US$11.2) is higher than that in Lampung (US$4.5).

Table 33.7 *Cost-benefit analysis of biochar application to maize in Indonesia (2018 US$).*

	Lampung	Lamongan
Maize yield increase after adding 10 t biochar ha^{-1} (t ha^{-1} yr^{-1})	0.6	0.0
Financial benefit of biochar addition (assuming 5 years of same effectiveness; US$190 t^{-1} maize)	570	0
Labor cost of making 10 t biochar	1,890	4,700
Willingness to pay for 10 t biochar	580	690
Yield increase needed for break-even at reported labor cost of biochar production (%)	+26	+22
Labor requirement for break-even at the same yield increase as observed in demonstration plots (person-days t^{-1} biochar)	14	n/a[a]

Notes
[a] cannot be calculated for a 0% yield increase.

Farmer livelihoods in Lamongan, East Java, are better than those in Lampung, South Sumatra, with about double the income. Farmers in Lampung were more motivated to continue biochar production and application than in Lamongan because the acidic soils in Lampung were more receptive to biochar amendment, more feedstock was available, and because they were highly motivated to supplement their income. Somewhat surprisingly, the Lamongan farmers indicated a willingness to pay for biochar of US$69 t^{-1} biochar, possibly because they assessed they could sell them for carbon certificates, or because they expected other beneficial long-term effects on their soils.

Comparing biochar amendment with conventional liming, the acid-neutralizing capacity of the biochar in Lampung was promising (Cornelissen et al, 2018). In addition, biochar application adds K, Ca, and Mg to the soil. In comparison, dolomite would add much more of both Ca and Mg, but not K. Small-scale tests with various dolomite additions in Lampung soil showed that around 4 t dolomite ha^{-1} were needed for the same pH increase achieved with 10 t biochar ha^{-1}. At US$250–500 t^{-1} dolomite, this would be a major cost for the small-scale farmers in Lampung. As a comparison, biochar was made for less than US$200 t^{-1} by these farmers in the first demonstrations, a price that might decrease after several rounds of implementation. At this price, the liming effect per US$ invested would be similar for biochar and dolomite lime. Furthermore, compared with liming, biochar offers the additional benefits of K addition (Chapter 16), improvement of soil structure and microbiology (Chapters 10 and 14), nitrous oxide suppression (Chapter 18), and carbon sequestration (Chapter 30).

The voluntary carbon market

The first voluntary carbon credit was traded in 1989 (Trexler et al, 2006) and since then the trend has been to issue carbon credits against the avoidance of GHGs from the combustion of fossil fuels (which may, however, be displaced from one site to another as there are no limits to the growth economy), or to temporary C storage in plants and ecosystems. Carbon trading of this type of credit has expanded but is currently on a very small scale compared to the challenge of net zero CO_2e emissions by 2050. Several voluntary carbon trading platforms and programs now accept biochar applications, including Puro.earth, Carbonfuture, Carbonface, the Verified Carbon Standard (VCS), and the Climate Action Reserve.

As yet, what types of carbon removal such as biochar will be incorporated into compliance emissions trading systems, and how, is unknown. Under the 'corresponding adjustments' requirement of Article 6, UNFCCC Paris Agreement, the use of biochar as a carbon offset might require the addition of an equivalent amount of emission reductions to their Nationally Determined Contribution, thereby increasing the requirement for emissions reduction. Depending upon how corresponding adjustments are implemented and enforced in different countries, the incentives for biochar implementation might vary.

Puro.earth is a CO_2 removal standard, registry, and marketplace for the issuing, trading, and canceling of CORCs. Biochar applications credited by Puro include soil amendments, insulation materials, greenhouse substrates, surface water barriers, landfill and mine absorbers, wastewater treatment components, and animal feed additives. All current CORCs issued are in developed countries, probably due to Puro's requirements for monitoring, reporting, and verification. For example, Puro requires an LCA that shows that the pyrolysis gases have undergone engineered emissions control to reduce methane emissions to negligible levels. The production facility must demonstrate environmental and social safeguards and be capable of metering and quantifying the biochar output reliably. Although the biochar application needs to be proven to be other than energy use, the point of creation of the CORCs is at the production facility and the producer of the biochar is the CORC supplier. This means that Puro favors projects with a small number of engineered biochar reactors that can document and track biochar products accurately. The use of decentralized Kon Tiki kilns by numerous farmers in developing countries is thus not encouraged to obtain CORCs.

Carbonfuture is a platform, registry, and marketplace for permanent C removals with a key focus on biochar. They use a blockchain-based tracking system that follows the biochar materials from production to application. Carbonfuture carbon sink credits are issued based on the European Biochar Certificate (EBC) under the digital registry of Carbon Standards International. One carbon sink credit represents 1 t CO_2e removed from the atmosphere over 100 years.

Carbonface is a new trading platform that provides flexibility and opportunities for small-scale biochar producers in developing countries interested in exploring the carbon markets and getting paid for sequestering relatively small amounts of carbon. Carbonface issues CharTon or CharKilo tokens to "UnMiners" (a term it invented to describe the people who bury biochar in the ground) in developing countries after basic evidence, such as pictures, videos, invoices, or scientific studies, are verified by other "UnMiners" registered in their CarbonFaceSocial network. One CharTon (1 t biochar) is equivalent to 2.5 t CO_2 sequestered, whereas one CharKilo (1 kg biochar) represents 2.5 kg CO_2 sequestered. The price and number of tokens are selected by the UnMiner who created them and are put on sale on the CarbonFace platform.

Verra is in the process of co-developing a "Methodology for biochar utilization in soil and non-soil applications" under their Verified Carbon Standard (VCS) program. The VCS biochar methodology provides a framework for quantifying emission reductions from the diversion of biomass waste to biochar production and carbon removals from the use of biochar in soils and non-soil applications, such as concrete, asphalt, animal bedding, feed supplements, or filtration products. Verra hosted a biochar webinar in August 2021, and the draft methodology went through a 30-day public consultation period ending in September 2021 and was published in 2022.

The Climate Action Reserve is following a multi-stakeholder workgroup process to develop a biochar protocol that aims to guide how to quantify, monitor, report, and verify climate benefits from the production and use of biochar. The biochar protocol will be accompanied by a market analysis including the potential scale and demand for biochar credits.

Conclusions

Although there has been a large increase in the volume of biochar-based products sold in the last five years, the profitability of either making or using biochar depends on several key factors. The most important factors are the cost and quality of the feedstock as delivered to the plant, the capital investment, available subsidies, and favorable loans, labor costs, the value of saleable energy (thermal and electrical), the agronomic or other use benefits and the carbon removal value. Revenues are influenced by the specific application rate, the increase in yields, and the value of carbon credits. Returns are greatest when using biochar with high-value crops such as avocados, and sugar beet and as an animal feed supplement. Large applications of biochar for field crops rarely return a profit to the farmer. Experience from China suggests a need to develop products that combine biochar and nutrients at biochar application rates of less than 1 t ha^{-1} for adoption to accelerate.

References

Aguirre J, Martin M, Gonzalez S, and Peinado M 2021 Effects and economic sustainability of biochar application on corn production in a Mediterranean climate. *Molecules* 26, 3313.

Ahmed M, Zhou J, Huu N, and Guo W 2016 Insight into biochar properties and its cost analysis. *Bioenergy and Biomass* 84, 76–86.

Alhashimi H, and Aktas C 2017 Life cycle environmental and economic performance of biochar compared with activated carbon: a meta-analysis. *Resources, Conservation and Recycling* 118, 13–26.

Aller DM, et al 2018 Long term biochar effects on corn yield, soil quality and profitability in the US Midwest. *Field Crops Research* 227, 30–40.

Bailis R, et al 2013 Innovation in charcoal production: A comparative life-cycle assessment of two kiln technologies in Brazil. *Energy for Sustainable Development* 17, 189–200.

Bergman R, Sahoo K, Englund K, and Mousaua-Awal S 2022 Life-cyle assessment and techno-economic analysis of biochar pellet production from forestry residues and field application. *Energies* 15, 1559.

Collingridge D 1980 *The Social Control of Technology*. London: Pinter.

Cornelissen G, et al 2016. Emissions and char quality of flame-curtain Kon Tiki kilns for farmer-scale charcoal/biochar production. *PloS One* 11, 0154617.

Cornelissen G, et al 2018 Fading positive effect of biochar on crop yield and soil acidity during five growth seasons in an Indonesian Ultisol. *Science of the Total Environment* 634, 561–568.

Dickinson D, et al 2015 Cost-benefit analysis of using biochar to improve cereals agriculture. *Global Change Biology Bioenergy* 7, 850–864.

Dodgson M, Gann D, and Philips N (Eds) 2013 *The Oxford Handbook of Innovation Management*. Oxford: Oxford University Press.

Dutta B, and Raghavan V 2014 A life-cycle assessment of environmental and economic balance of biochar systems in Quebec. *International Journal of Energy and Environmental Engineering* 5, 106.

Eggink A, Palmer K, Severy M, Carter D, and Jacobson A 2018 Utilization of wet forest biomass as both the feedstock and electricity source for an integrated biochar production system. *Applied Engineering in Agriculture* 34, 125.

Galinato SP, Yoder JK, and Granatstein D 2011 The economic value of biochar in crop production and carbon sequestration. *Energy Policy* 39, 6344–6350.

Garcia-Perez M, Brady M, and Tanzil AH 2019 Biochar production in biomass power plants: techno-economic and supply chain analyses. *A report for The Waste to Fuels Technology Partnership 2017–2019 Biennium: Advancing Organics Management in Washington State*, accessed at http://s3-us-west-2.amazonaws.com/wp2.cahnrs.wsu.edu/wp-content/uploads/sites/32/2019/08/Biochar-Production-in-Biomass-Power-Plants-.pdf on July 23 2023/.

Henry RC, et al 2018 Food supply and bioenergy production within the global cropland planetary boundary. *PLoS ONE* 13, e0194695.

IPCC 2022 Mitigation of Climate Change, Working Group III Contribution to the Sixth Assessment Report, IPCC_AR6_WGIII_FullReport.pdf (accessed 07.23.23).

Jirka S, and Tomlinson T 2015 State of the Biochar Industry: A Survey of Commercial Activity in the Biochar Sector, International Biochar Initiative IBI-State-of-the-Industry-2015-final.pdf (biochar-international.org) (accessed 07.23.23).

Joseph S, et al 2020 Biochar increases soil organic carbon, avocado yields and economic return over 4 years of cultivation. *Science of the Total Environment* 724, 138–153.

Joseph S, Lan Anh M, Clare A, and Shackley S 2015 Socioeconomic feasibility, implementation and evaluation of small-scale biochar projects. In: Lehmann J, and Joseph S (Eds) *Biochar for Environmental Management: Science, Technology, and Implementation*. Milton Park: Routledge. pp359–373.

Kambewa P 2007 *Charcoal – The Reality: A Study of Charcoal Consumption, Trade, and Production in Malawi*. London: IIED. Charcoal – the reality: A study of charcoal consumption, trade and production in Malawi | IIED Publications Library (accessed 07.23.23).

Keske C, Godrey T, Hoag D, and Abedin J 2020 Economic feasibility of biochar and agricultural coproduction from Canadian black spruce forest. *Food and Energy Security* 9, e188.

Kim D, Anderson N, and Chung W 2015 Financial performance of a mobile pyrolysis system used to produce biochar from sawmill residues. *Forest Production Journal* 65, 169–197.

Latawiec A, et al 2019 Biochar amendment improves degraded pasturelands in Brazil: environmental and cost-benefit analysis. *Scientific Reports* 9, 11993.

Latawiec A, et al 2021 Economic analysis of biochar use in soybean production in Poland. *Agronomy* 11, 2108.

Li W, et al 2019 Regional techno-economic and life cycle analysis of the pyrolysis-bioenergy-biochar platform for carbon negative energy. *Biofuels, Bioproducts and Biorefining* 13, 1428–1438.

Liang X, Feng L, Yuan G, and Jing W 2019 Low cost field production of biochar and their properties. *Environmental Geochemistry and Health* 42, 1569–1578.

Lin Y, Yong X, Zhang J, and Zhu Z 2022 Process simulation of preparing biochar by biomass pyrolysis via Aspen Plus and its economic evaluation. *Waste and Biomass Valorization* 13, 2609–2622.

Liu X, Pan G, and Li L 2017 Biochar for sustainable soils management in China: from biochar to biochar compound fertilizer. Unpublished report, Biochar for Sustainable Development. Nairobi: UNEP.

McCarl BA, Peacocke C, Chrisman R, Kung CC, and Sands RD 2009 Economics of biochar production, utilization and greenhouse gas offsets. In: Lehmann J, and Joseph S (Eds) *Biochar for Environmental Management: Science and Technology*. London: Routledge. pp341–358.

Mocci B 2022 Staying power. Transform: for Environmental and Sustainability Professionals, IEMA, 24-26, https://s3.eu-west-2.amazonaws.com/iema.net/documents/IEMA-AugSept-FULL-LRC.pdf (accessed 07.23.23).

Mocci B, Brander M, Shackley S, and Tipper R 2022 Is there a link between the price of voluntary offset credits and the permanence of the carbon storage methods? What are the views of the significant groups involved?' (bit.ly/CDRcredits or https://files.edinburgh-innovations.ed.ac.uk/ei-web/production/images/CDR-Credits_Permanence-Report.pdf (accessed 07.23.23).

Mohammadi A, et al 2017 Biochar addition in rice farming systems: economic and energy benefits. *Energy* 140, 415–425.

Nematian M, Keske C, and Ng'ombe J 2021 A techno-economic analysis of biochar production and the bioeconomy for orchard biomass. *Waste Management* 135, 467–477.

Oldfield EE, Bradford MA, and Wood SA 2019 Global meta-analysis of the relationship between soil organic matter and crop yields. *Soil* 5, 15–32.

Owsianiak M, et al 2021 Environmental and economic impacts of biochar production and agricultural use in six developing and middle-income countries. *Science of the Total Environment* 755, 142455.

Pennise DM, et al 2001 Emissions of greenhouse gases and other airborne pollutants from charcoal making in Kenya and Brazil. *Journal of Geophysical Research: Atmospheres* 106, 24143–24155.

Puro 2022 OBIO-biochar from sustainable Norwegian forests. Accessed at https://puro.earth/CORC-co2-removal-certificate/oplandske-bio-biochar-norway-ringsaker-100128 on July 28 2023

Robb S, Joseph S, Aziz A, Dargusch P, and Tisdell C 2020 Biochar's cost constraints are overcome in small-scale farming in tropical soils in lower-income countries. Land Degradation and Development 31, 1713-1726.

Robb S, and Dargusch P 2018 A financial analysis and life-cycle carbon emissions assessment of oil palm waste biochar exports from Indonesia for use in Australian broad-acre agriculture. *Carbon Management* 9, 105–114.

Robb S and Joseph S (2019) A report on the value of biochar and wood vinegar. *ANZBIG* https://anzbig.org/wp-content/uploads/2020/07/ANZBI-2020-A-Report-on-the-Value-of-Biochar-and-Wood-Vinegar-v-1.2.pdf (accessed 07.23.23).

Roberts KG, Gloy BA, Joseph S, Scott NR, and Lehmann J 2010 Life cycle assessment of biochar systems: estimating the energetic, economic, and climate change potential. *Environmental Science and Technology* 44, 827–833.

Rockwood DL, Ellis MF, and Fabbro KW 2022 Economic potential for carbon Sequestration by short rotation eucalypts using biochar in Florida, USA. *Trees, Forests and People* 7, 100187.

Roe S, et al 2021 Land-based measures to mitigate climate change: potential and feasibility by country. *Global Change Biology* 27, 6025–6058.

Sessions J, et al 2019 Can biochar link forest restoration with commercial growth? *Biomass and Bioenergy* 123, 175–185.

Shackley S, Hammond J, Gaunt J, and Ibarrola R 2011 The feasibility and costs of biochar deployment in the UK. *Carbon Management* 2, 335–356.

Sorensen B, and Lamb M 2018 Return on investment from biochar application. *Crop, Forage and Turfgrass Management* 4, 1–6.

Sørmo E, et al 2020 Waste timber pyrolysis in a medium-scale unit: emission budgets and biochar quality. *Science of the Total Environment* 718, 137335.

Stern NH 2007 *The Economics of Climate Change: The Stern Review.* Cambridge: Cambridge University Press.

Struhs E, Mirkouei A, You Y, and Mohajeri A 2020 Techno-economic and environmental assessment of nutrient-rich biochar production from cattle manure: a case-study in Idaho, USA. *Applied Energy* 279, 115782.

Taherymoosavi S, et al 2022 Overall benefits of biochar, fed to dairy cows, for the farming system. *Pedosphere* 33, 225–230.

Thomas, M et al, 2021 Consumer preference and willing-to-pay for potting mixture with biochar. *Energies* 14, 3432.

Trexler MC, Broekhoff DJ, and Kosloff LH 2006 A statistically-driven approach to offset-based GHG additionality determinations: what can we learn? *Sustainable Development Law and Policy* 6, 30–40.

Turner PA, Field CB, Lobell DB, Sanchez DL, and Mach KJ 2018 Unprecedented rates of land-use transformation in modelled climate change mitigation pathways. *Nature Sustainability* 1, 240–245.

Utterback J 1966 *Mastering the Dynamics of Innovation.* Boston: MIT Press.

Wise M, et al 2009 Implications of limiting CO_2 concentrations for land use and energy. *Science* 324, 1183–1186.

Woolf, D, Lehmann, J, Joseph, S., Campbell, C, Christo, FC, and Angenent, LT 2017 An open-source biomass pyrolysis reactor. *Biofuels, Bioproducts and Biorefining* 11, 945–954. 10.1002/bbb.1814.

Wrobel-Tobiszewska A, Boersma M, Sargison J, Adams P, and Jarick S 2015 An economic analysis of biochar production using residues from Eucalypt plantations. *Biomass and Bioenergy* 81, 177–182.

Xu X 2021 Evaluation of Crop Yield Increment and Greenhouse Gas Reduction Potential of Biochar Application in Field Based on Data Integration and Model Simulation. PhD Thesis, Nanjing Agricultural University, China.

Zhang D, et al 2016 Biochar helps enhance maize productivity and reduce greenhouse gas emissions under balanced fertilization in a rainfed low fertility inceptisol. *Chemosphere* 142, 106–113.

Zheng J, et al 2017 Biochar compound fertilizer increases nitrogen productivity and economic benefits but decreases carbon emission of maize production. *Agriculture, Ecosystems and Environment* 241, 70–78.

34

Policy and biochar

Tristan R. Brown and Jenny R. Frank

Introduction

The continued growth of global greenhouse gas (GHG) emissions in the wake of the 2015 Paris Agreement has prompted policymakers to increasingly examine carbon dioxide removal (CDR) strategies as a means of reducing the atmospheric CO_2 concentration and thereby avoiding the worst effects of climate change. Direct air capture (DAC) technologies remain at a very early stage of deployment, however, prompting the policy community to turn to biochar production as a CDR strategy with near-term net sequestration potential. This chapter (1) reviews how biochar has been covered by energy and climate policy historically, (2) examines how biochar is contributing to recent deep decarbonization policies, and (3) considers how biochar may participate in policies implemented in the future.

Motivations for biochar policies

The 2015 Paris Agreement, which entered into force in late 2016 and covered 194 countries (Anon, 2022a) as of 2022, established the twin goals of "limiting global temperature increase to well below 2 degrees Celsius" and "pursuing efforts to limit the increase to 1.5 degrees" (Anon, 2022b). The magnitude of the task of keeping the temperature increase below 2°C, let alone 1.5°C, was starkly presented in the "Summary for Policymakers" section of the Sixth Assessment Report (AR6) of the United Nations Intergovernmental Panel on Climate Change (IPCC) that was released in 2021. Specifically, the Summary showed that, due to a temperature increase of over 1.1°C having already occurred according to the AR6's best estimate, annual net emissions of CO_2 will need to be negative shortly after 2050 if the

DOI: 10.4324/9781003297673-34

temperature increase is to remain below 2°C with a high degree of certainty (IPCC, 2021). Within three decades, in other words, humanity will need to be sequestering more CO_2 than it emits if the Paris Agreement's targets are to be met.

The inadequacies of carbon capture and sequestration

Technologies that capture CO_2 before it enters the atmosphere have long been of interest to policymakers due to their potential to (1) reduce CO_2 emissions from fossil sources when deployed as carbon capture and sequestration (CCS); and (2) reverse legacy CO_2 emissions when paired with atmospheric CO_2 capture mechanisms (e.g., bioenergy with CCS) as CDR (Azar et al, 2010). However, the deployment of CCS systems by large emitters of CO_2 has been limited amid reports of technical challenges and high costs (Baylin-Stern and Berghout, 2021). Only 35 large-scale CCS projects were operational globally as of June 2022 with a total capacity of 44 million t CO_2 yr^{-1} (Anon, 2022c), compared to global annual fossil CO_2 emissions of 36,300 million t (Anon, 2022d). By comparison, the International Energy Agency has estimated that annual removals of 1,286 million t CO_2 will be required by 2030 if the world is to achieve net-zero emissions by 2050 (Anon, 2022c).

CCS has also generated substantial controversy in the climate policy community over its alleged negative effects (Whitmarsh et al, 2019). The pairing of CCS with fossil fuel use is particularly contentious due to concerns that fossil CCS will enable the continued use of fossil fuels, with their attendant negative non-climate impacts on the environment and human health (Batres et al, 2021). In addition to potentially enabling the continued combustion of fossil fuels and consequent emissions

of non-greenhouse gas pollutants such as particulate matter and nitrous oxide, the installation of CCS technologies at existing fossil fuel facilities such as power plants would feasibly extend the lifespan of those facilities. Given that large fossil fuel facilities are disproportionately located in minority and disadvantaged communities, CCS projects are sometimes opposed on environmental justice grounds (Kaswan, 2012). In early 2022, the Biden Administration released guidance on environmental justice concerns and CCS products. Developed by the Council on Environmental Quality, the guidance includes recommendations and actions to be undertaken by federal agencies concerning CCS technologies and their impact on local communities (Anchondo, 2022).

The role and types of CDR technologies

CDR technologies that reduce the level of atmospheric CO_2 rather than capture CO_2 from existing emissions sources have more recently attracted attention from policymakers as a means of combating climate change directly via the reversal of legacy emissions. CDR practices consist of both nature-based solutions (e.g., forest and soil stewardship) and engineered solutions (e.g., biochar production and utilization; direct air capture) (Field and Mach, 2017). Both types of practices face major hurdles to their deployment at the scales that are necessary for them to contribute to the Paris Agreement's targets, however, with the former being constrained by land, other resource availability, and the timing and accuracy of C storage and removal methods and estimates, and the latter being constrained by very high costs (Smith et al, 2016).

CDR technologies differ from conventional CCS technologies in two notable ways. First, CDR technologies are generally unable to be paired with fossil fuel pathways in the manner that CCS technologies are; while CDR technologies reverse past (or future) fossil fuel emissions, they do not reduce those that are actively occurring. Second, this distinction means that CDR technologies do not directly support the incurring of negative externalities (Box 34.1) such as air pollution that results from the combustion of fossil fuels. Instead, CDR technologies provide an important positive externality in the form of reversed CO_2 emissions.

These two differences in CDR technologies compared to CCS technologies are notable at a time when climate policymakers are increasingly imposing restrictions on the production (rather than just the emission) of fossil CO_2 (Erickson et al, 2018). Furthermore, the high cost of existing industrial-scale CDR technologies (US\$600-US\$1,000 t^{-1} CO_2 in 2022) has greatly hindered their deployment (Budinis, 2022) at a time when governments' social costs of carbon, let alone market prices on carbon, are well below CDR costs (Aldy et al, 2021). Taken together, these two factors have prompted climate policymakers in recent years to emphasize the deployment of CDR

technologies that provide positive externalities in addition to the reversal of CO_2 emissions. This new focus reflects a recognition that CDR technologies are more likely to be deployed on the scales and timeframes necessary for the Paris Agreement's targets to be met if they provide benefits to society that increase their overall value under the traditional cost-benefit frameworks that are employed by many policymakers. Recent years have seen a growing number of policies either proposed or adopted by regional and national governments that promote the production, utilization, and research into biochar, both as a CDR technology and as a technology that provides positive externalities.

The last decade has also witnessed a shift in how biochar is perceived as contributing to the larger CDR field. In 2009 biochar was categorized by a The Royal Society report as a "geoengineering" solution alongside practices such as iron fertilization of the ocean and enhanced weathering (The Royal Society, 2009). Biochar differs from most geoengineering solutions by being "low tech", however, as

---

Box 34.1 Definition of externality

An "externality" in economics is an activity by a party that imposes an indirect benefit or (more commonly) cost on an uninvolved third party. Externalities are considered to be "negative" when the activity imposes a cost on the third party and "positive" when it imposes a benefit. Pollution (air, noise, water, etc.) is a classic negative externality in that the party bearing the cost of the polluting activity is frequently not the same party that is creating and benefiting from the pollution (Laffont, 1989). Externalities fall outside of conventional financial transactions, and government action via policy or regulatory action is necessary to align the societal costs or benefits of an externality with its financial impacts (Alesina and Passarelli, 2014). Past successful examples of such government action are the EU's Emissions Trading Scheme, which imposes a cost on GHG emissions, and California's Low Carbon Fuel Standard, which incentivizes low-carbon fuels according to the reduction of their carbon footprint. Several of the beneficial effects of biochar covered in this book, such as reversing legacy pollution, improving soil health, and enhancing biodiversity, are properly classified as positive externalities in certain situations.

evidenced by the large difference in technological complexity and scale between slow pyrolysis and geoengineering technologies (Aquije et al, 2021). The public perception of the consequences and risks of biochar use is also radically different from that of geoengineering technologies (Poumadère et al, 2011; Sweet et al, 2021). Nature-based CDR solutions are broadly more palatable to the public than geoengineering solutions due to these differences, and there is some evidence that public perception of biochar, while limited, views it as existing closer to the nature-based side of the CDR spectrum (Sweet et al, 2021). Furthermore, while biochar production is not a completely natural process in that it requires human intervention, the same is true of other nature-based CDR practices such as afforestation, reforestation, and soil management.

The IPCC's release in early 2022 of the report "Climate Change 2022: Mitigation of Climate Change" placed a strong emphasis on the potential for biochar to be used as a CDR technology (IPCC, 2022). The report determined that up to 6,600 million t CO_2 could be sequestered annually through the global production and utilization of biochar, although it noted a large range of uncertainty around its values for both sequestration potential (300–6,600 million t CO_2) and cost (US$10-345 t^{-1} CO_2) (IPCC, 2022). Notably, the report identified biochar as being "less prone to reversal" than nature-based CDR activities such as forestry and soil management practices. The AR6 is intended to serve as a critical scientific resource for policymakers as they implement the Paris Agreement's targets, illustrating the importance that biochar has attained as a policy solution to catastrophic climate change by 2022.

Biochar as a carbon allowance or offset in emissions trading schemes

The early 21st century saw greenhouse gas (GHG) emissions trading schemes either implemented (e.g., the European Union's Emissions Trading Scheme, or ETS, which was set up in 2005; the Northeastern U.S.'s Regional Greenhouse Gas Initiative, or RGGI, which was started in 2009; and California's Global Warming Solutions Act, or AB 32, which was established in 2013) or almost implemented (e.g., the American Clean Energy and Security Act of 2009, or ACES) by large national emitters of CO_2. A common characteristic of emissions trading schemes is their inclusion of a carbon offset mechanism. A traditional emissions trading scheme imposes a limit on GHG emissions that covered emitters may only exceed if they purchase enough allowances from other emitters. The

carbon offset mechanism enables participating non-covered entities to generate offset credits through qualifying activities such as GHG emission reductions occurring outside of the emissions trading scheme's geographic scope or the establishment of CDR projects. These offset credits may then be sold to covered emitters for use as emissions allowances. The purpose of carbon offsets is to provide covered emitters in hard-to-decarbonize sectors with a means of containing their compliance costs via the indirect reduction of their emissions through the adoption of CDR activities such as afforestation and reforestation. Put another way, carbon offsets increase the supply of allowances that can be acquired by covered emitters by increasing the geographic scope of emission reduction activities.

Biochar has generally struggled to attain government support as a carbon offset activity within the EU. One of the world's largest carbon offset programs in the early 21st century was the UN's Clean Development Mechanism (CDM), which operates under the Kyoto Protocol to provide covered emitters in developed countries with emissions trading schemes (primarily the EU ETS) with the opportunity to obtain allowances in the form of carbon offsets generated by qualifying projects in developing countries. Biochar is not a qualifying offset activity under the CDM, however, for reasons that reflect both its relative novelty and subsequent uncertainty about its impacts (IPCC, 2022) as well as the larger failure of the CDM to increase additionality and prevent fraudulent offset activities (Haya, 2010).

The ACES of 2009 never became law after it failed to be passed by the U.S. Senate, although it was passed by the U.S. House of Representatives (Anon, 2009a). The bill was unique, however, in that it would have tasked the U.S. Department of Agriculture (USDA) with creating a carbon offset program covering the agriculture and forestry sectors (Anon, 2009b). While an initial list of potential offset activities that were listed by the legislation did not include biochar, the USDA would have had the ultimate authority to determine which offset projects qualified for the program. Corresponding legislation that was introduced into but never passed by the U.S. Senate, the Clean Energy Partnerships Act of 2009, did explicitly list biochar production and use as a qualifying offset activity (Johnson, 2009). Biochar has been categorized as being "implicitly" eligible for the offset program that would have been created by ACES of 2009 due to this language (Pourhashem et al, 2019).

In Australia, the Clean Energy Act of 2011 created a carbon pricing mechanism (specifically an emissions trading scheme in which the carbon price was fixed during the first three years) in that country that was implemented the following year (Zeller and Longo, 2012). The new carbon pricing mechanism coincided with the creation of a Carbon Farming Initiative (CFI) in late 2011. The CFI was intended to operate in place of the country's earlier National Carbon Offset Standard by generating carbon offsets for use within the new system. Australia's government addressed the issue of additionality by creating a "positive list" of CDR practices that were neither common practice nor legally required within the country (Macintosh and Waugh, 2012). Biochar used as a soil amendment was one of 14 activities that made the positive list at the time of the pricing mechanism's implementation, and the national government initiated the Biochar Capacity Building Program to facilitate the participation of biochar in the CFI (Anon, 2014). A change of political leadership resulted in the Clean Energy Act of 2011 being repealed just two years after its implementation, however, before biochar and other offset projects had an opportunity to be widely deployed beyond an initial round of projects.

Biochar has achieved greater success as a qualifying carbon offset activity under South Korea's Emissions Trading System (KETS), which began covering the country's large GHG emitters in 2015 (Narassimhan et al, 2018). The KETS allows covered emitters to meet 5–10% of their allowance credit requirements with offsets from qualifying activities. Uniquely, the KETS allows both domestic offset activities and those located in least-developed countries under the CDM (but at least part-owned by a Korean entity) to generate offset credits (Kuneman et al, 2021). Investors in biochar projects that have participated in the voluntary carbon credit markets have also begun to generate offset credits under the KETS, although the

CDM's lack of support for biochar and disruptions to the Paris Agreement's implementation that resulted from the COVID-19 pandemic have limited the participation of biochar projects in the KETS (Vandana, 2021).

Biochar and net-negative CDR

The global community's belated action to stabilize, let alone reduce, its GHG emissions has made it highly likely that various national net-zero emission targets will be insufficient to prevent catastrophic climate change (IPCC, 2021). While global net-zero emissions will still need to be achieved by mid-century, humanity will need to sequester more CO_2 than it emits annually not long after the mid-century mark if warming is to be kept below 2°C. The recognition of this outcome has caused some policymakers to shift their attention away from the use of CDR technologies as a mere decarbonization tool, such as when they are used to generate carbon offsets, in favor of their use as "net-negative CDR" technologies, also known as "negative-emission technologies."

Net-negative emission technologies in policy compared to low- and zero-emission technologies

Net-negative CDR technologies differ from net-zero CDR and low-emission technologies in that they achieve the actual reversal of legacy GHG emissions rather than the mere mitigating or negating of ongoing emissions (Peters and Geden, 2017). Whereas low-emission and net-zero emission technologies are intended to slow the rate of atmospheric CO_2 concentration increase by reducing or offsetting, respectively, fossil fuel emissions, net-negative CDR technologies are employed to reverse legacy emissions. In many cases, the distinction between the two is as much one of policy as it is of technology, with the desired outcome being an explicit reduction in the atmospheric CO_2 concentration. The distinction is more than just technological because many CDR technologies, biochar included, can serve either purpose as is. Consider the use of biochar to offset fossil fuel CO_2 emissions compared to its use to achieve a net reduction to the atmospheric CO_2 level: in both cases, the biochar production pathway is the same, but how that pathway contributes to policy goals is different. Whether biochar production reduces the rate of atmospheric CO_2 increase or the actual atmospheric CO_2 level is therefore the task of policymakers rather than engineers.

National and multinational CDR policies

The Federal Carbon Dioxide Removal Leadership Act of 2022 was introduced into the U.S. House of Representatives in April of that year. The legislation would require the U.S. Department of Energy to remove 10 million t of CO_2 from the atmosphere on a net basis using CDR technologies by 2035 (Anon, 2022e). Similar to New York State's Climate Leadership and Community Protection Act (CLCPA), the bill would also emphasize the mitigation of environmental justice impacts, thereby reinforcing the emphasis on net-negative CDR rather than CDR to generate carbon offsets. While biochar is not explicitly referenced, the bill does classify the pyrolysis of lignocellulosic biomass as a qualifying technology (Anon, 2022e). Similar legislation is being considered in New York State separate from the CLCPA process (Hinchey, 2022).

While a small number of U.S. states have been early promoters of the use of biochar as a net-negative CDR technology due to their status as first movers on economywide deep decarbonization, the growing adoption of similar climate targets by national signatories to the Paris Agreement has also involved biochar. A 2019 report from Switzerland's Federal Office for the Environment "strongly recommended" that the national government declare net-negative CDR technologies to be a critical component of the country's commitment to the Paris Agreement. The report concluded that biochar could achieve annual CO_2 sequestration equal to 18% of Switzerland's annual GHG emissions (Beuttler et al, 2019).

The EU is also in the process of developing a policy framework that would potentially enable biochar to contribute to its member nation's climate targets under the Paris Agreement. Biochar has historically been limited in its ability to participate in EU climate policy because the EU's 2020 emissions target did not account for net GHG emissions from land use, land use change, and forestry (Verde and Chiaramonti, 2021). The years since the Paris Agreement have seen the EU propose the implementation of policies that place a greater emphasis on carbon sequestration and net-negative CDR technologies. These policies are expected to support biochar production and utilization by encouraging the sequestration of C in soil, developing certifications for CDR, and taking net GHG emissions from land use, land use change, and forestry into account when establishing the EU's new 2030 emissions target (Verde and Chiaramonti, 2021). Policies that indirectly support the use of biochar as a CDR technology, such as the inclusion starting in 2022 of biochar as a "fertilizing product" that can be applied to cropland

under EU rules (European Commission, 2022), are also in the process of being implemented.

U.S. State net-negative CDR policies

Biochar has recently attracted the attention of policymakers who are implementing regional and national net-zero emission targets. One of the most explicit such cases occurred during the implementation of New York State's CLCPA. The law, which was passed in 2019, commits the state to achieve, relative to its 1990 levels, at least an 85% reduction, and up to a 100% reduction, to its economywide GHG emissions by 2050 (Anon, 2022f). The CLCPA is unique in that it establishes environmental justice goals such as reductions to criteria pollutant emissions alongside the overarching GHG emission reduction target. Rather than simply treating CDR activities as a source of carbon offsets (although the law does provide flexibility for the generation of offsets), then, it goes further by promoting both net-negative CDR activities that achieve a net reduction to the atmospheric CO_2 concentration and the provision of ecosystem services by CDR activities.

This emphasis on negative-emissions technologies is reflected in the Final Scoping Plan (FSP) that was released in late 2021 as part of the CLCPA's implementation (Anon, 2022f). The FSP contains three policy recommendations that explicitly reference biochar: the first provides for research on the use of biochar to improve soil health; the second provides for research on the use of biochar as a net-negative CDR technology; and the third provides for the funding of demonstration projects that utilize biochar in both capacities (Anon, 2022f). The FSP's policy recommendations will begin to be implemented via regulatory actions in 2023.

State governments in the wildfire-prone Western U.S. have begun to utilize biochar production as a net-negative CDR technology that avoids future GHG emissions and minimizes wildfire risk, in both cases by converting potential wildfire fuel to biochar. In 2021, Colorado passed legislation requiring grants to be awarded to projects, including biochar production, that "demonstrate the utilization of biomass" (Valdez et al, 2021). The California Department of Forestry and Fire Protection has provided funding under its Forest Health Grant program for the development of a protocol for biochar "that will provide guidance on how to quantify, monitor, report, and verify climate benefits from the production and use of biochar" (Climate Action Reserve, 2022). Such efforts by government policymakers illustrate the ability of biochar to provide multiple simultaneous positive externalities.

Biochar and the climate-focused bioeconomy

The commercialization and deployment of biochar production have long been slowed by a lack of government policies that value the many positive externalities, including various ecosystem services, that biochar provides (Thengane et al, 2021). This lack of support is gradually changing with the growing recognition of biochar's ability to stably sequester CO_2 over long time horizons while also contributing to governments' other climate goals. The adoption of economywide deep decarbonization targets by a small but growing number of governments in recent years has accelerated this process due to the large number of economic sectors, many of which have not been the focus of climate policy in the past, that will be impacted. One recent example is New York State's proposal at the end of 2021 to deploy multiple policies in support of the "Climate-Focused Bioeconomy" as part of the implementation of the CLCPA (Anon, 2022f). While the overarching objective of these policies is the deployment of net-negative CDR technologies (see above), biochar is further envisioned as contributing to the sustainability of other sectors while providing long-term carbon sequestration.

For example, one major conclusion of the CLCPA's FSP was that US$30 billion will need to be invested in building upgrades (e.g., building shell and energy efficiency improvements) annually by 2050 for New York State's decarbonization targets to be met (Anon, 2022f). The FSP proposed the implementation of a suite of policies, including research, development, and demonstration (RD&D) support and a low-carbon procurement standard, to increase the development and market access of material products that store atmospheric carbon as one means of directly sequestering CO_2 while also reducing demand for fossil resources in infrastructure and buildings. Other policies in the FSP would utilize technologies such as biochar production to achieve CO_2 sequestration while improving soil health and mitigating methane leakage in the agriculture sector. While biochar is not referenced in all of the recommended policies, their technology-agnostic nature and explicit references to biochar elsewhere in the document make it probable that biochar would be allowed to participate in them if they are adopted in late 2022.

Other policies have focused primarily on the provision of non-climate positive externalities while still enabling biochar to serve as a CDR technology as a co-benefit. Many Chinese cities, especially those in the country's eastern regions, experience very poor air quality from a combination of SO_2, NO_x, particulate matter, and volatile organic compound pollution. By one estimate, the country will need to reduce emissions of these pollutants by 40–60% compared to 2005 levels if it is to reach the World Health Organization's ambient air quality guideline values (Wang and Hao, 2012). Some of this air pollution is the result of the long-standing practice by farmers in China of burning the straw that remains on their fields following a harvest in preparation for the subsequent planting (Liu, 2013). Bans on straw burning have been deployed across much of China's agricultural regions over the last decade, although early compliance rates were hurt due to a lack of alternative means of improving cropland productivity (Huang et al, 2021). The production of biochar from straw feedstock has subsequently emerged as a means of complying with the burn bans while preparing the cropland for the next planting, however, by 2016 the annual production of straw-based biochar in China was 287,647 t (Ren et al, 2019). Utilization of the country's available straw feedstock in that same year was 82% (Shi et al, 2019), and the improvements to China's air quality via biochar production could potentially increase still further as a result.

Policymakers have also recognized biochar for its ability to contribute to water resource availability goals. The Water Efficiency via Carbon Harvesting and Restoration (WECHAR) Act of 2009, which was introduced into the U.S. House of Representatives, would have supported the deployment of pyrolysis facilities to convert water-intensive plant species into biochar (Anon. 2011). A second objective of the legislation was wildfire prevention through the reduction of forest and rangeland fuel loads.

Biochar research

One consequence of the large body of literature discussing biochar's environmental impacts that have appeared over the last decade is a growing recognition that biochar is potentially able to contribute to a variety of policy goals in addition to the climate objectives discussed above. This recognition has in turn led to recommendations in the literature of where additional study is needed to enable biochar to contribute to these policy goals. The utilization of biochar in policy applications necessitates reductions to the uncertain outcomes that have been shown in the literature on biochar over the last decade (or longer) due to the clearly defined goals that policymakers are in the process of establishing. These goals often leave limited room for uncertainty despite the comparative novelty of many of the technologies, such as biochar, that have been identified by policymakers.

For example, while many biochar practitioners take biochar's C sequestration capabilities as a truism, studies have found conflicting results in response to the question of if biochar achieves sequestration on a net basis, with the outcome being dependent on factors such as feedstock, soil fertility, soil chemistry, incubation period, etc. (Wang and Wang, 2019). Findings such as this pose a dilemma to policymakers due to the constraints under which climate policymaking operates. A lack of sufficient financial resources allocated to decarbonization, as evidenced by the continued acceleration of the rise of atmospheric CO_2 concentration, means that those resources that are spent on decarbonization (or, in the immediate case, net-negative sequestration) must achieve the intended outcome. Expending scarce resources on biochar in applications that do not yield such an outcome incurs both an opportunity cost and a heightened risk that global temperatures will exceed the 2°C threshold. Rather than automatically assume that all domestic biochar production represent net sequestration, then, as some proposals such as the Senate version of the ACES of 2009 would have done, these studies **suggest that policymakers should employ narrowly tailored biochar policies while simultaneously supporting research into the conditions that enhance biochar's ability to achieve C sequestration on a net basis.**

A similar dilemma is posed by the literature on biochar's agronomic capabilities, with individual studies providing mixed results that are dependent on many of the same factors that have resulted in the literature's uncertain findings on biochar's C sequestration capabilities (Oni et al, 2019). It could be argued that this agronomic uncertainty is an even greater concern for policymakers than its sequestration uncertainty. After all, whereas the outcome of today's decarbonization and sequestration measures will not be conclusively known for multiple decades, any negative impacts on crop yields resulting from the utilization of biochar will be known as soon as the subsequent growing season. While biochar has been widely identified as one means of mitigating the effects of climate change on agriculture (Stavi and Lal, 2013), the deployment of biochar by policymakers in such a capacity requires assurances that it will

not have the opposite effect. **Additional research into biochar's impacts on nutrient dynamics, soil biota, and diverse soils is therefore an important policy goal rather than merely an interesting research question.**

Finally, while the last decade has seen a wide variety of biochar production systems enter the market across an array of price points, this has not yet coincided with the widespread adoption of these systems by businesses that have abundant access to feedstocks such as farms (Maroušek et al, 2019). This disconnect between technological availability and market adoption represents an additional policy hurdle that must be overcome if biochar is to achieve the goals that policymakers have established for it, as described in this chapter. This hurdle is especially challenging, however, since the lack of market adoption has been attributed in part to the unattractive economics that exist for biochar as a climate solution compared to its agronomic capabilities (Maroušek et al, 2019). Biochar remains broadly unfamiliar to farmers and other agricultural practitioners (if better known than in the past), complicating any effort by policymakers to increase its market adoption on strictly economic grounds. An alternate approach would be for policymakers to **support research that better quantifies biochar's ability to enhance the long-term sustainability of farms and agricultural communities via soil improvement instead of simply treating it as a farming input and output.** Such research is a critical first step in the process of overcoming the socioeconomic and regulatory barriers to novel agronomic practices, such as biochar production and utilization that have resulted from a historical perception of intensive agriculture as a viable recurring practice (Riding et al, 2015).

Conclusion

Policymakers increasingly view biochar as a means of providing a variety of positive externalities including the generation of carbon offsets, the reduction of atmospheric CO_2 levels and criteria pollutant emissions, the prevention of wildfires, and the displacement of fossil products and materials, among others. While care must be taken not to present biochar as a silver bullet capable of solving society's major environmental problems on its own, it is important to note that biochar policy, whether proposed or implemented, often harnesses biochar production and utilization to provide multiple externalities. A common theme among biochar policies is the removal of CO_2 from the atmosphere, but many of these policies also address more localized concerns (e.g., the development of low-carbon building materials). Biochar policies can be categorized according to the geographic scale of the policy objective being pursued (Table 34.1), although these objectives are not necessarily mutually exclusive: many of the policy objectives at the national, regional, and local scales will also contribute to CDR, for example. Furthermore, the level of priority of each objective is properly viewed to be a function of the corresponding policy rather than its geographic scale, since many policymakers prioritize more local concerns even as global concerns such as the atmospheric CO_2 concentration can be expected to affect the greatest number of people.

The existence of a biochar policy can give the misleading impression that there is a reduced need for biochar research. As this chapter shows, however, there is a continued need for research into the mutually supporting benefits (or drawbacks) of biochar production and utilization. Those biochar policies that have been successfully implemented, rather than merely proposed, are characterized by non-global objectives such as wildfire prevention, improved air quality, and fossil resource displacement rather than explicit CDR; the latter has frequently been a supporting objective rather than the primary objective of the policy. To support the global policy objectives that biochar contributes to, then, it is important to further identify biochar's ability to also contribute to non-global policy objectives that are prioritized by policymakers. It is these non-global objectives that, in the absence of a sustained global policy initiative to reduce the atmospheric CO_2 concentration, are the most likely to achieve the objective of CDR.

Table 34.1 *Biochar policy objectives*

Geographic scale	Global	National/regional	Local
Policy objective	Carbon dioxide removal	Fossil resource displacement	Water conservation
		Carbon offset creation	Improved air quality
	Reduced methane leakage	Crop fertility	Soil health
			Wildfire prevention
			Ecosystem protection

References

Aldy JE, Kotchen MJ, Stavins RN, and Stock JH 2021 Keep climate policy focused on the social cost of carbon. *Science* 373, 850–852.

Alesina A and Passarelli F 2014 Regulation versus taxation. *Journal Public Economics* 110, 147–156.

Anchondo C 2022 White House CCS guidance exposes environmental justice rifts. *E&E News*. https://www.eenews.net/articles/white-house-ccs-guidance-exposes-environmental-justice-rifts. Accessed 5 Jan 2023.

Anon 2009a H.R. 2454 – American Clean Energy and Security Act of 2009. In: *Congress.gov*. https://www.congress.gov/bill/111th-congress/house-bill/2454. Accessed 10 May 2022.

Anon 2009b Greenhouse Gas Legislation: Summary and Analysis of H.R. 2454 as Passed by the U.S. House of Representatives. Congressional Research Service, July 27. Washington D.C. https://www.everycrsreport.com/files/2009-07-27_R40643_fcf6e1363093e795404cb5b6d3430b26-cae640de.pdf. Accessed 10 Oct 2022.

Anon 2011 Biochar: Examination of an Emerging Concept to Sequester Carbon. Congressional Research Service, January 11. Washington D.C. https://crsreports.congress.gov/product/pdf/R/R40186. Accessed 10 Mar 2022.

Anon 2014 Biochar Capacity Building Program. In: *Austrailian Government Department of Agriculture, Fisheries and Forestry*. https://www.agriculture.gov.au/agriculture-land/farm-food-drought/climatechange/mitigation/cfi/biochar#the-national-biochar-initiative-ii-a-country-wide-approach-to-biochar-systems–csiro. Accessed 1 Oct 2022.

Anon 2022a Paris Agreement – Status of Ratification. In: *United Nations Framework Convention on Climate Change*. https://unfccc.int/process/the-paris-agreement/status-of-ratification. Accessed 5 Jan 2022.

Anon 2022b Key Aspects of the Paris Agreement. In: *United Nations Framework Convention on Climate Change*. https://unfccc.int/most-requested/key-aspects-of-the-paris-agreement. Accessed 5 Jan 2022.

Anon 2022c Carbon Capture, Utilization, and Storage. In: *International Energy Agency*. https://www.iea.org/reports/carbon-capture-utilisation-and-storage-2. Accessed 1 Aug 2022.

Anon 2022d Global CO2 emissions rebounded to their highest level in history in 2021. In: *International Energy Agency*. https://www.iea.org/news/global-co2-emissions-rebounded-to-their-highest-level-in-history-in-2021. Accessed 1 Aug 2022.

Anon 2022e Federal Carbon Dioxide Removal Leadership Act of 2022. In: *World Resources Institute*. https://www.wri.org/update/federal-carbon-dioxide-removal-leadership-act. Accessed 15 Sep 2022.

Anon 2022f New York State Climate Action Council Final Scoping Plan. Albany. https://climate.ny.gov/resources/scoping-plan/. Accessed 10 Mar 2022.

Aquije C, et al 2021 Low tech biochar production could be a highly effective nature-based solution for climate change mitigation in the developing world. *Plant and Soil* 479, 77–83.

Azar C, et al 2010 The feasibility of low CO_2 concentration targets and the role of bio-energy with carbon capture and storage (BECCS). *Climatic Change* 100, 195–202.

Batres M, et al 2021 Environmental and climate justice and technological carbon removal. *The Electricity Journal* 34, 107002.

Baylin-Stern A, and Berghout N 2021 Is carbon capture too expensive? In: *International Energy Agency*. https://www.iea.org/commentaries/is-carbon-capture-too-expensive. Accessed 5 May 2022.

Beuttler C, et al 2019 The Role of Atmospheric Carbon Dioxide Removal in Swiss Climate Policy. Zurich. https://www.bafu.admin.ch/dam/bafu/en/dokumente/klima/externe-studien-berichte/the-role-of-atmospheric-carbon-dioxide-removal-in-swiss-climate-policy.pdf.download.pdf/The_Role_of_Atmospheric_Carbon_Dioxide_Removal_in_Swiss_Climate_Policy.pdf. Accessed 10 Mar 2022.

Budinis S 2022 Direct Air Capture. In: *International Energy Agency.* https://www.iea.org/reports/direct-air-capture. Accessed 28 Sep 2022.

Climate Action Reserve 2022 Biochar Protocol. In: *Climate Forward.* https://www.climateactionreserve.org/how/protocols/biochar/dev/. Accessed 28 Sep 2022.

Erickson P, Lazarus M, and Piggot G 2018 Limiting fossil fuel production as the next big step in climate policy. *Nature Climate Change* 8, 1037–1043.

European Commission 2022 Fertilising products – pyrolysis and gasification materials. https://ec.europa.eu/info/law/better-regulation/have-your-say/initiatives/12136-Fertilising-products-pyrolysis-and-gasification-materials_en. Accessed 28 Sep 2022.

Field CB, and Mach KJ 2017 Rightsizing carbon dioxide removal. *Science* 356, 706–707.

Haya B 2010 Measuring Emissions Against an Alternative Future: Fundamental Flaws in the Structure of the Kyoto Protocol's Clean Development Mechanism. Berkeley, CA. https://papers.ssrn.com/sol3/papers.cfm?abstract_id=1562065. Accessed 28 Sep 2022.

Hinchey M 2022 *Carbon Dioxide Removal Leadership Act.* New York State Senate, Albany, NY.

Huang L, et al 2021 Assessment of the effects of straw burning bans in China: Emissions, air quality, and health impacts. *Science of the Total Environment* 789, 147935.

IPCC 2021 Summary for Policymakers. In: Masson-Delmotte V et al (Eds) *Climate Change 2021: The Physical Science Basis. Contribution of Working Group I to the Sixth Assessment Report of the Intergovernmental Panel on Climate Change.* Cambridge: Cambridge University Press. pp3–32.

IPCC 2022 *Climate Change 2022: Mitigation of Climate Change. Contribution of Working Group III to the Sixth Assessment Report of the Intergovernmental Panel on Climate Change.* Cambridge, UK and New York, NY, USA: Cambridge University Press.

Johnson R 2009 Agricultural and Forestry Provisions in Climate Legislation in the 111th Congress. *Congressional Research Service.* Washington D.C. https://www.everycrsreport.com/files/20091223_R40994_c35ff967c8a8cd950ccacbdb14f2398889afa580.pdf. Accessed 29 Sep 2022.

Kaswan A 2012 Climate change, the Clean Air Act, and industrial pollution. *UCLA Journal of Environmental Law and Policy* 30, 51.

Kuneman E, Acworth W, Bernstein T, and Boute A 2021 The Korea Emissions Trading System and electricity market. Berlin and Hong Kong. https://www.umweltbundesamt.de/sites/default/files/medien/5750/publikationen/2021-05-19_cc_36-2021_case_study_korea.pdf. Accessed 29 Sep 2022.

Laffont J-J 1989 Externalities. In: Eatwell J, Milagate M, and Newman P (Eds) *Allocation, Information and Markets.* London: Palgrave Macmillan. pp112–116.

Liu C 2013 Can Straw Provide China's Energy Needs? Scientific American, February 20. https://www.scientificamerican.com/article/can-straw-provide-chinas-energy-needs/. Accessed 30 Sep 2022.

Macintosh A, and Waugh L 2012 An Introduction to the Carbon Farming Initiative: Key Principles and Concepts. https://ideas.repec.org/p/een/ccepwp/1203.html. Accessed 30 Sep 2022

Maroušek J, Strunecký O, and Stehel V 2019 Biochar farming: defining economically perspective applications. *Clean Technologies and Environmental Policy* 21, 1389–1395.

Narassimhan E, Gallagher KS, Koester S, and Alejo JR 2018 Carbon pricing in practice: a review of existing emissions trading systems. *Climate Policy* 18, 967–991.

Oni BA, Oziegbe O, and Olawole OO 2019 Significance of biochar application to the environment and economy. *Annals of Agricultural Sciences* 64, 222–236.

Peters GP, and Geden O 2017 Catalysing a political shift from low to negative carbon. *Nature Climate Change* 7, 619–621

Poumadère M, Bertoldo R, and Samadi J 2011 Public perceptions and governance of controversial technologies to tackle climate change: nuclear power, carbon capture and storage, wind, and geoengineering. *Wiley*

Interdisciplinary Reviews Climate Change 2, 712–727.

Pourhashem G, Hung SY, Medlock KB, and Masiello CA 2019 Policy support for biochar: Review and recommendations. *Global Change Biology Bioenergy* 11, 364–380.

Ren J, Yu P, and Xu X 2019 Straw utilization in China – status and recommendations. *Sustainability* 11, 1762.

Riding MJ, et al 2015 Harmonising conflicts between science, regulation, perception and environmental impact: the case of soil conditioners from bioenergy. *Environmental International* 75, 52–67.

Shi Z, et al 2019 Utilization characteristics, technical model and development suggestion on crop straw in China. *Journal of Agricultural Science and Technology* 21, 8–16.

Smith P, et al 2016 Biophysical and economic limits to negative CO_2 emissions. *Nature Climate Change* 6, 42–50.

Stavi I, and Lal R 2013 Agroforestry and biochar to offset climate change: a review. *Agronomy for Sustainable Development* 33, 81–96.

Sweet SK, et al 2021 Perceptions of naturalness predict US public support for Soil Carbon Storage as a climate solution. *Climatic Change* 166, 22.

The Royal Society 2009 Geoengineering the climate: Science, governance, and uncertainty. London. https://royalsociety.org/topics-policy/publications/2009/geoengineering-climate/. Accessed 1 Oct 2022

Thengane SK, et al 2021 Market prospects for biochar production and application in California. *Biofuels, Bioproducts and Biorefining* 15, 1802–1819.

Valdez D, Will P, and Coram D 2021 *Measures to Increase Biomass Utilization.* Colorado General Assembly. https://leg.colorado.gov/bills/hb21-1180. 10 Mar 2023

Vandana S 2021 INTERVIEW: South Korean carbon credit provider looks at voluntary carbon markets. *S&P Global Commodies Insights.* https://www.spglobal.com/commodityinsights/en/market-insights/latest-news/metals/072121-interview-south-korean-carbon-credit-provider-looks-at-voluntary-carbon-markets. Accessed 10 Mar 2022

Verde S, and Chiaramonti D 2021 *The Biochar System in the EU: The Pieces Are Falling into Place, but Key Policy Questions Remain.* Florence: European University Institute. https://cadmus.eui.eu/handle/1814/70349;jsessionid=48196DFD78F4B1E48AEA411127B00A83. Accessed 10 Mar 2022

Wang J, and Wang S 2019 Preparation, modification and environmental application of biochar: A review. *Journal of Cleaner Production* 227, 1002–1022.

Wang S, and Hao J 2012 Air quality management in China: issues, challenges, and options. *Journal of Environmental Sciences* 24, 2–13.

Whitmarsh L, Xenias D, and Jones CR 2019 Framing effects on public support for carbon capture and storage. *Palgrave Communications* 5, 17.

Zeller B, and Longo M 2012 Australia's Clean Energy Act: a new measure in the global carbon market. *Loyola University Chicago International Law Review* 10, 179.

35

Biochar in environmental management

Outlook and conclusion

Yong Sik Ok, Piumi Amasha Withana, Stephen Joseph, and Johannes Lehmann

Introduction

We have acquired a strong understanding of biochar and its associated environmental benefits, as documented throughout the chapters of this book. Biochar is by now more than just a soil amendment, and the research paradigm has shifted as an advanced material in the energy industry and environmental applications and as a beneficial material in the healthcare sector. To improve the intrinsic properties of biochar, its modification for various field applications has recently attracted attention. Studies on purposeful management of biochar have expanded since 2006, and it is essential to commercialize biochar by establishing proper linkages between industrial practitioners, and scientists, with support from the government and policymakers. This will ensure that research data will be converted into real-world applications to obtain the maximum benefits of biochar. Similarly, research directions should be focused on increasing the biochar production capacity and large-scale applications. Life cycle assessment (LCA) of biochar production systems, machine learning, and artificial intelligence are recommended to determine the environmental impacts associated with biochar and to optimize production and applications. Globally available transnational policies and standards are key to broadening the biochar market. Biochar can play an important role in carbon credit programs as it has the potential to withdraw carbon dioxide (CO_2) from the atmosphere in addition to reducing greenhouse gas (GHG) emissions.

DOI: 10.4324/9781003297673-35

What we have learned

What is biochar?

The definition of biochar has evolved over the years (Chapter 1), and the focus on soil amendments has broadened to uses in building materials (Chapter 4, Chapter 28), and animal feeding (Chapter 29), among others. In general, a greater emphasis on environmental management (Chapter 27) reflects an adjustment in what biochar looks like and in what contexts it is investigated and applied. While biochar is primarily known for its use as a soil amendment to improve soil health and plant growth, several new and emerging applications of biochar are being explored. Beyond carbon sequestration, biochar is being used as a filtration medium for water treatment (Chapter 20) to remove contaminants such as heavy metals, organic compounds, and pathogens and also as an adsorbent material to remove pollutants from the air, such as volatile organic compounds and particulate matter. Biochar can be used as a component in supercapacitors and other energy storage devices due to its high surface area and electrical conductivity. Overall, these emerging applications of biochar have the potential to expand its use beyond agriculture and provide new opportunities for sustainable and environmentally friendly solutions in various industries in environmental, social, and governance contexts. At present, biochar has proved its potential to be used as an advanced material in the healthcare sector.

Biochar production systems

Biochar production techniques are continually being modernized to improve their production quality and rate (Chapters 3 and 4). The pyrolysis industry must be well-planned to ensure that long-term sustainability goals are met. Whatever biochar production technology is utilized, it should be able to increase the production capacity. Increasing the production capacity should not deteriorate the quality of biochar, and there should be continuous quality control during biochar production. Technologies with low energy required and low emissions are highly recommended. The success of the growing bioenergy and biorefinery sector depends on interconnection with other industries and energy consumers, as well as state or national household supply programs. Programs aimed at substituting fossil fuels in rural communities can use the heat generated from pyrolysis volatiles to make biochar production economically and environmentally friendly. Additionally, integrating biochar production with carbon capture and sequestration technologies (CCS) and CO_2 transport networks is becoming more viable and should be considered in the planning stages of new installations to accelerate atmospheric CO_2 removal (Chapter 17, Chapter 31).

Biochar modification techniques

There are biochar modification techniques that are currently being researched and developed, each with its own set of benefits and drawbacks (Chapters 6, 22, 27). Overall, the best approach for biochar modification depends on the specific application and desired properties of the modified biochar. A combination of different techniques may be needed to achieve the desired properties and enhance the effectiveness of biochar for a wide range of applications. In this context, the majority of the research has focused on producing modified biochar for the sorption of various contaminants and nutrients (Chapter 27). Yet, for biochar modification focusing on other environmental applications, such as material in energy storage devices, more research needs to be conducted. The choice of biochar modification techniques, whether it be chemical, physical, or biological, depends on the specific application and the desired properties of the modified biochar. In many cases, combining different modification techniques may be the best approach to achieve the desired properties for a specific application. For example, combining physical and chemical modification can enhance the porosity and adsorption capacity of the biochar, while also adding new functionalities to its surface chemistry. Biochar can also be designed to slowly release the available nutrients depending on the nutrient requirements of plants, thereby avoiding nutrient loss via leaching (Chapter 19, Chapter 26). When biochar and compost are combined, the resulting mixture can have several beneficial effects on soil. The use of biochar compost as a soil amendment is a promising approach to improving soil health and fertility while also helping to mitigate climate change. However, further research is needed to better understand the optimum ratios and application rates of biochar and compost for different soil types and crop systems. Many potential biochar modification techniques have not yet been fully explored or developed but hold promise for improving the properties and applications of biochar. Impregnating biochar with nanoparticles such as titanium dioxide or iron oxide could potentially enhance its photocatalytic activity, making it more effective for applications such as water purification or air treatment. Modifying biochar using microorganisms or enzymes could potentially improve its nutrient availability or biodegradability, making it more effective for use as a fertilizer or soil amendment. Overall, these

and other potential biochar modification techniques represent exciting areas of research for improving the properties and applications of biochar. With continued exploration and development, biochar has the potential to become an even more versatile and valuable material for a wide range of industries and applications.

Major environmental applications of biochar and associated benefits

The most frequently studied environmental applications of biochar include its use as an organic amendment to improve soil quality (Chapters 10, 14-16) and crop production (Chapter 13), mitigate GHG emissions (Chapters 18, 31), and retain organic and inorganic contaminants (including potentially toxic heavy metals in soils and water) (Chapters 21, 22, 27). More recently, biochar has been developed as a precursor of the catalyst used in syngas cleaning, as a catalyst for the conversion of syngas into liquid hydrocarbons, and as a solid acid catalyst for biodiesel production (Chen et al, 2019; Wang and Wang, 2019). Its novel applications in environmental systems include usage as a green blend for construction products such as cement (Senadheera et al, 2023), as a gas adsorbent, in fuel cell systems, for the fabrication of biochar-based supercapacitors, and as a raw material for activated carbon that need to be investigated more thoroughly.

Biochar is best known globally as a soil amendment (Chapter 13). Considering its physicochemical and biological properties associated with enhancing soil fertility, biochar is added to soil as an amendment and soil enhancer. The plant-available nutrients (Chapter 8) and carbon content (Chapter 7) that can be achieved in biochar make it a versatile soil amendment. The persistent nature of biochar (Chapter 11), which allows it to be retained in the soil for much longer periods than unpyrolyzed materials, makes it a valuable medium for the removal of atmospheric CO_2. The presence of biochar tends to increase the plant-available nutrient levels in the soils. Very often it remains challenging to distinguish these effects, leading to misinterpretation and misattribution of research results. Similarly, NH_3 and NH_4^+ compounds are retained in the soil by biochar, which can lead to reduced N_2O formation, and, consequently, reduced GHG emissions into the atmosphere even though other effects are likely more important (Chapter 18). Long-term studies are needed to determine the persistence of soil water modifications after biochar application, as most studies are short-term (Chapter 20). The properties of biochar are altered with environmental exposure (Chapter 10), causing long-term shifts in its hydrologic and ecosystem services. Conducting long-term studies will require creative approaches. Field studies need to be well-connected to manipulative lab experiments to reveal mechanisms. There are large discrepancies between field studies and laboratory experiments, constraining the ability to scale up laboratory results. Connecting laboratory, greenhouse, and field trials are therefore necessary. During this scale-up, considering the local environmental conditions is essential to obtain maximum benefits from biochar and to avoid any possible negative impacts. A decision-support system is needed to make robust predictions and recommendations for biochar use in different environments. Now it is high time to bring the biochar research findings into real-world applications as a valuable resource in different industries. The collaboration of global industries in this process is crucial as it helps to bring research findings into large-scale applications. However, there is still a need to create awareness among stakeholders about the benefits of biochar and its potential applications. Furthermore, research is required to optimize the production of biochar and ensure that it meets the required standards.

Adverse effects of biochar on the environment

Biochar has been widely used for soil improvement, pollution control, and climate change mitigation. However, owing to the physicochemical changes occurring in the soil (such as changes in pH) and chemical reactions on the biochar surface (such as oxidation of aromatic carbon rings), biochar may release harmful components into the environment. Adverse effects can also occur as a result of biochar aging in soil, as it can cause changes in soil properties, which have to be considered separately. Soil organisms could suffer from the adverse biological effects of biochar (Chapter 14). Reduced plant-available soil water content, the increased tendency of soil erosion, increased soil salinity, and pH-induced impacts (such as reduced (bio)availability of nutrients and decreased efficiencies of agrochemicals) can reduce the growth and germination of plants that are associated with biochar addition to the soil (Chapter 13). The higher application rates can result in an imbalance in soil nutrients, particularly if the biochar has not been properly processed or is not compatible with the soil type. This can cause a reduction in crop yields, and in some cases, toxic effects on plant growth. In addition, high application rates of biochar can lead to an increase in soil pH, which can negatively impact the availability of nutrients for plant growth. Furthermore, if biochar is not produced sustainably, its production can lead to deforestation, which in turn can have negative impacts on GHG emissions, biodiversity, soil erosion, and social aspects including indigenous rights. Additionally, if the production process is not properly managed, it can result in emissions of GHGs and other air pollutants. Therefore, it is important to carefully consider the appropriate application rate of biochar based on soil type, crop type, and other relevant biochar and environmental factors to avoid negative environmental impacts.

Knowledge gaps

Research studies on biochar utilization optimization using various technological tools

Currently, with the increasing focus on using biochar for environmental management, applications of machine learning (ML) and artificial intelligence (AI) in this field have emerged as novel areas of research. The knowledge gained using ML and AI provides a better understanding of the effectiveness of biochar; the technologies used in biochar production that require minimum energy, less time, and reduced cost; and the feedstock properties that determine the quality and quantity of the biochar. However, despite their immense importance, research on the application of AI and ML in the field of biochar is limited. Most studies have focused on understanding how biochar effects change with feedstock properties and production conditions (such as particle size, heating time, temperature, and heating rate). ML methods can effectively use all of these variables to predict its behavior in the environment with great accuracy, e.g., the heavy metal sorption efficiency of biochar. Hence, other than the discussed aspects, more studies should be conducted with a focus on applying ML and AI to evaluate biochar aging in soil and predict the GHG mitigation potential of biochar systems.

In recent years, increased importance has been given to management strategies for mitigating GHG emissions, and the use of biochar-based carbon management networks (CMNs) is among the top strategies (Ok et al, 2021). Tools, including computer-based system simulations and optimization, remote sensing, and global information systems to acquire and manage relevant biomass sources and biochar are essential for the use of biochar-based CMNs to mitigate GHG emissions. Exploring the use of AI can assist researchers in analyzing large and complex datasets, identifying patterns, making predictions, and identifying research gaps. ML algorithms can be used to predict the physical and chemical properties of biochar, such as its pH, surface area, and porosity, based on its feedstock and pyrolysis conditions. This can help optimize biochar production processes and ensure that biochar produced for a specific application has the desired properties. ML can also be used to analyze the effects of biochar application on soil properties and crop yields and to identify optimum application rates for different crops and soils. This can aid farmers and land managers in making informed decisions about how much biochar to apply to their fields. ML can help analyze long-term data on the effects of biochar application on soil properties, crop yields, and GHG emissions. There is also potential to apply ML to identify new applications for biochar, such as in water treatment, energy storage, or as a feedstock for biofuels. By analyzing large datasets and identifying patterns, ML can assist researchers to explore new possibilities and expand the potential uses of biochar.

LCA studies of biochar applications in soil, water, and air

LCA is a systematic analysis of the environmental impacts of products or services during their life cycles. Very few studies have reported aspects of biochar-related LCA and its environmental applications beyond climate change mitigation. Therefore, it is important to frequently incorporate LCA analysis into the study on biochar, along with its associated environmental implications. LCA studies on biochar have shown that the environmental impacts of biochar production and application can vary depending on a range of factors, including the type of biomass feedstock used to produce biochar, the pyrolysis technology used, and the application rate and timing of biochar in agricultural systems. In many LCA studies, the available information is insufficient to draw a definitive conclusion regarding the best biochar application. Specifically, studies on human toxicity associated with biochar addition to the environment are lacking. Hence, the evaluation of the effect of biochar on impact categories other than climate change should be considered. In addition to using LCA in biochar application in soils and water, other environmental applications, such as using biochar as an additive in the construction industry and as an advanced material, should be linked with LCA to identify the most sustainable and effective ways of producing and using biochar.

Technological advancements associated with biochar

Studies have concentrated more on determining the impact of raw materials and properties on biochar quality and quantity. However, novel biochar production technologies have not been sufficiently explored. Few studies on biochar-metal oxide composites, biochar catalysis development, and engineered biochar have focused on specific environmental applications (Chapter 27). There is a paucity of data regarding methods for improving the reusability of biochar and prolonging its service life. Innovative technologies such as nanotechnology should be adapted to accelerate biochar applications. Nanotechnology can be used to modify biochar surfaces, create composites, develop sensors, and produce carbon-based materials for energy storage (Zhang et al, 2017). Surface modification can improve biochar performance in applications such as water treatment or catalysis. Combining biochar with nanoparticles can create unique properties such as increased strength or catalytic activity. Moreover, new technologies can support the mass production of quality biochar with desired properties.

Next steps in the implementation of biochar

Commercialization of biochar

Biochar has recently gained increasing industry attention owing to its several environmental applications, including soil and water remediation, mitigation of GHGs, and sequestration of soil carbon (Chapters 33, 34). Therefore, it is essential to commercialize biochar to promote its wide application. The commercialization of biochar puts it on the market, which subsequently results in its production, distribution, marketing, and sales.

However, several aspects must be clarified and addressed before commercialization. In addition to the different biochar quality standards in individual countries, global standardization is urgently required. In addition, cheaper, more readily available, and more rapid analytical protocols must be established that are then available for use in all countries worldwide. Countries lacking biochar standards rely on fertilizer and compost standards for biochar application (Chapter 32). Therefore, owing to these limitations and geographical barriers, it is difficult to commercialize biochar production and increase its application. This necessitates the immediate establishment of globally recognized biochar standardization and certification procedures. Similarly, creating a biochar market with a circular economy is essential to attain a sustainable market. Cooperation among biochar market stakeholders, including industry, researchers, consumers, producers, and the government, should be strengthened by collaboratively creating a biochar-based circular economy.

Sustainable solutions for disposal of spent biochar

Although studies on the application of biochar in remediating contaminated water and resource recovery from water have rapidly increased, water–biochar interactions have not yet been fully established. Most studies on biochar-based wastewater remediation have been conducted using simulated wastewater; biochar application in the treatment of wastewater has scarcely been reported. The selective removal of pollutants using biochar should be extensively investigated to develop sustainable remediation processes by scientifically understanding the associated remediation mechanisms. Therefore, it is vital to update the available certification programs by

targeting non-soil applications of biochar (Chapters 28, 29). The use of biochar as an absorbent to remediate contaminated soil and water is increasing rapidly. Managing spent biochar is, therefore, another critical aspect that requires further attention. Laboratory studies commonly focus on applying biochar-based adsorbents, and handling spent biochar disposal is not prioritized, and therefore, no guidelines for end-of-life use are available for implementation. Particular care should be taken when disposing of biochar as an adsorbent for organic and inorganic pollutants, including heavy metals, to avoid secondary contamination. Laboratory studies suggest that biosorbents can be regenerated by using acids, bases, and other chemical agents. However, the practicality of these processes in large-scale applications (such as on an industrial scale) is still questionable. Additionally, the disposal of spent biochar poses social and environmental problems, especially in countries that have minimal or no facilities for waste biochar disposal, such as furnaces and engineered landfills, and no studies have been conducted to overcome this challenge. Incineration, disposal into landfills, reuse, and use as fertilizers are the available options for handling spent biochar. However, sustainable alternatives for the reuse of spent biochar must be further explored. Spent biochar can be recycled to produce raw materials in the construction industry (Chapter 28); however, the mechanical strength of construction may be adversely affected when such recycled materials are used in large quantities. Therefore, improving the mechanical strength of spent biochar for use in the construction industry is another research avenue. Because limited data are available on the performance of biochar-modified cement materials concerning durability in instances of acid attack and fire resistance, further exploration is required before the mass production of biomass-based construction materials. Also, biochar, which is used to retain heavy metals, can be used to produce supercapacitors (Gupta et al, 2020). As data are lacking on how to dispose of or reuse spent biochar, studies on developing sustainable solutions to reuse spent biochar without causing secondary contamination in the environment are required.

Large-scale and long-term field-level studies on biochar application in soils

Comparatively short-term trials can only generate some information on the frequency and quantity of biochar applications. More long-term and large-scale field-level studies are required to investigate the impact of biochar on nutrient availability, soil moisture retention, biochar aging effects on soil properties, microbial community and changes in their activity, and biochar-mediated enhancement of crop production. These aspects will enable farmers to compare biochar with other organic amendments by considering the economic, social, and environmental aspects. Therefore, it is essential to expand the scientific information available regarding the effects of biochar application on soil from short-term laboratory or greenhouse experiments to field-scale experiments. Another important environmental aspect of the sustainable use of biochar is its frequency of application. If biochar remains in the soil for decades, it is important to determine if a one-time application is sufficient; in that scenario, it is also essential to determine the application rate and soil conditions that should be evaluated. Most biochar-related studies have been restricted to the characterization of the material and its application under supervised conditions.

Additional information is required to bridge this knowledge gap by performing long-term analyses of biochar applications under different environmental conditions. These long-term biochar application studies should also focus on the behavior and the fate of biochar in different soil types, including its persistence, mobility in the soil, and disintegration in the soil under different management practices. Therefore, the use of representative pilot areas covering multiple soil regimes with biochar produced from a representative range of feedstocks is an important next step.

Studies on parameters relevant to biochar toxicity

More studies on biochar toxicity, which focus on assessing the bioavailable fractions of the contaminants present in biochar and not on the total amounts, should be conducted (Chapter 22). Among the various quality guidelines proposed by e.g., the IBI, United Kingdom Biochar Center, and European Biochar Certificate, it is important to re-emphasize that a globally applicable biochar standard is needed. To develop this standard, comprehensive studies involving long-term field trials are required to assess the potential bioavailability of all the contaminants in biochar and their combined toxicity effects, the release of these contaminants during field aging, and soil-biochar interactions in removing these contaminants. Analysis of biochar aging in soil and the frequency of the release of associated toxic compounds into the environment is also essential (Chapter 10). There is a need for novel approaches to quantify the extent of contamination in biochar that are rapid, cost-effective, and employ accessible analytics. These methods should be readily available not only to academic research laboratories but also to industrial entities and, potentially, farmers.

Studies focused on low biochar application rates in the soil

The high cost associated with biochar production limits its economic feasibility (Chapter 33). Therefore, low biochar application rates are often preferred over high application rates to manage soil constraints. Optimization of biochar application rates for environmental applications is essential as excess biochar translates to substantial monetary losses and wastage of effort. It is essential to study the optimum rates of biochar application in different soil types (silty, clayey, or loamy). Information on these optimum application rates, in conjunction with that on biochar type and production conditions, would be beneficial, as these parameters strongly influence the quality and quantity of biochar. In addition, co-application of biochar with compost, fertilizer, or manure can reduce labor costs and effort (Chapter 26). The watering frequency can be minimized owing to the high water-holding capacity of biochar (Chapter 20). The method of biochar application, such as incorporation into the topsoil or as a top dressing, can also be influenced by soil conditions and processes, thereby providing information on the benefits of biochar. It is important to identify optimum biochar application practices for specific locations and purposes. Hence, factual information regarding the optimum biochar application rates with the desired soil type, including the type of biochar, production conditions, the quality of biochar (such as particle size, and pyrolysis temperature), the method of application (Chapter 25), and environmental factors (such as avoiding large rain events, storms, extreme winds, and soil moistening) should be available for optimized application of biochar. There is an urgent need to achieve sustainable soil management with

minimum cost associated with biochar production and applications. Because non-soil applications of biochar are increasing rapidly (Chapters 28, 29), it is also essential to focus on optimum biochar application rates in water, for example, to study contaminant removal.

Global, transnational policy development to ensure sustainable growth of the future biochar market

Biochar has received global attention because of its agronomic and environmental benefits. Studies have suggested that biochar applications will gradually increase (Chapter 35). This indicates that biochar production may increase considerably in the future. Therefore, it should be ensured that feedstock for biochar production is sourced from sustainably managed lands and waste materials. Additionally, if biochar is produced from sewage sludge and municipal waste, it should be ensured that the final biochar product is free from contaminants. While biochar should not be considered a standalone solution to environmental challenges, it can contribute to multiple positive externalities (Chapter 35). Biochar is increasingly seen by policymakers as a way to generate positive externalities, such as reducing atmospheric CO_2 levels, preventing wildfires, and displacing fossil products. Biochar policies often aim to provide multiple externalities, with a common theme of removing CO_2 from the atmosphere. These policies can be categorized based on the geographic scale of the objective, but the priority of each objective should be viewed as a function of the corresponding policy rather than its geographic scale. While biochar should not be seen as a silver bullet, it has the potential to provide benefits across different policy objectives.

Inform stakeholders of biochar and its environmental applications

A positive opinion about biochar application among the general public is a key determinant in biochar commercialization, as it can potentially promote the use of biochar. Biochar-related research projects are mostly concentrated in countries such as the USA, China, Australia, and Germany, with little attention in countries with less developed research infrastructure and funding. In some cases, countries with limited resources for research and development may have less emphasis on biochar research compared to countries with more robust research infrastructure and funding. Additionally, some countries may have less agricultural land and therefore less need for soil improvement technologies such as biochar. Cultural, political, and economic factors may also influence the prioritization of biochar research in different countries. By creating adequate awareness about the economic and environmental benefits of biochar application to soil, it might not be difficult to gain the acceptance of farmers, currently the major players in the biochar value chain. As farming is often the primary livelihood in rural regions where a considerable proportion of the residents live below the poverty line, biochar production and its applications may help them obtain a better livelihood and this can be attributed to farming, raw material supply, and biochar production. As biochar can enhance soil productivity in many situations (Chapter 13), it can lead to increased farmers' income. This will ultimately help achieve sustainable development goals, including reducing poverty, while promoting economic growth. An

overall understanding of biochar production and environmental applications among different stakeholder sectors (including the general public, farmers, large-scale and small-scale biochar producers, agronomists, researchers, and soil scientists), along with adequate networking among stakeholders, is essential to maximize the advantages offered by biochar use.

Develop a sustainable role of biochar in mitigating climate change

Biochar could secure its position on the Intergovernmental Panel on Climate Change's list of negative emission technologies owing to its long-term persistence in soil with low biodegradability (Lehmann et al, 2021; Ok et al, 2021). The mechanisms of biochar-induced CH_4 emissions from soil and the influence of biotic and abiotic factors on the underlying processes remain unclear (Chapter 18). One possible reason for this scarcity is the data deficiency in long-term field trials. Information on how biochar aging in soil affects the ability of biochar to reduce GHG emissions is equally scarce. The role of biochar in mitigating GHG emissions varies with the biochar production conditions, its application rates, and environmental conditions (Chapters 17, 18). Therefore, well-designed long-term field experiments are needed to understand the microbial and carbon dynamics associated with biochar application in agricultural soils, as these factors directly govern CO_2, CH_4, and N_2O flux in soils. LCA of biochar-based applications should be conducted to determine the amount of biochar needed to achieve carbon neutrality. Research opportunities include identifying efficacy estimates for specific biochar-soil-climate combinations, developing enhanced efficiency fertilizers using biochar, investigating mechanisms for lowering N_2O emissions, evaluating changes to GHG emissions from manure handling and composting, testing biochar with lower electron-donating capacity in paddy soils to lower methanogenesis, assessing correlations between rice-root growth stimulation by biochar and methanotrophic activity, studying the physical modification of soil, and manipulating changes to soil biological properties for the desired outcome.

Important suggestions for creating a sustainable biochar market

The following suggestions emanate as a priority for research and development (Figure 36.1):

1   Ensuring government support to enhance biochar utilization is essential. Governments can impose policies that promote the utilization of biochar in agriculture. For example, financial aid and incentives can be provided to the families of farmers to encourage biochar application and small-scale biochar production.

2   The creation of a sustainable biochar market cannot be achieved individually by farmers, biochar producers, agronomists, and other stakeholders. Such a market requires effective stakeholder coordination. Studies have indicated that the current trends show a rapid increase in biochar research. However, most of these studies have been conducted by research institutes or universities. Further progress in this field can be achieved by ensuring collaboration between these research institutes and the industries involved in biochar production. They can be commercialized following the necessary evaluations, including long-term field trials.

3   The scarcity of data complicates the establishment of a solid link between biochar application and increased crop production. Therefore, long-term field trials of different environmental applications of biochar are recommended to supplement short-term laboratory studies.

4   Considering the diversity in soil types worldwide and significant variations associated with biochar properties depending on the feedstock and production conditions, best practice guidelines and better accessible material property guidelines (Chapters 9 and 32) for biochar production and intended specific applications are urgently required.

5   Another poorly understood aspect of biochar is its impact on human health during production and handling (Chapter 25). These include the measurement of dioxins, polyaromatic hydrocarbons, heavy metals, and other volatile products generated during pyrolysis. This aspect requires appropriate attention to avoid the unintended effects of biochar on human health, thus creating a safe working environment at the production plant.

6   The creation of a freely available information repository, including global biochar-related projects, is suggested. This will enhance public enthusiasm for initiating such projects on a small scale in different countries. More importantly, in this technology-driven age, the younger generation may be motivated by such information and propose innovative ideas (Figure 35.1).

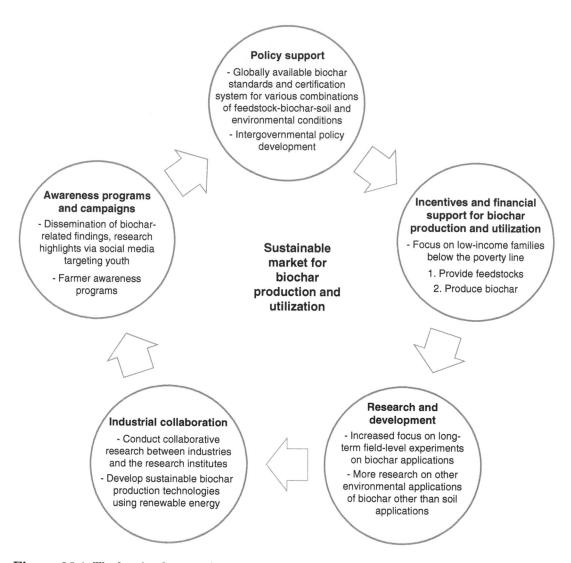

Figure 35.1 *The key implementations to create a sustainable biochar market worldwide.*

References

Chen W, Meng J, Han X, Lan Y, and Zhang W 2019 Past, present, and future of biochar. *Biochar* 1, 75–87.

Gupta S, Sireesha S, Sreedhar I, Patel CM, and Anitha K L 2020 Latest trends in heavy metal removal from wastewater by biochar based sorbents. *Journal of Water Process Engineering* 38, 101561.

Lehmann J, et al 2021 Biochar in climate change mitigation. *Nature Geoscience* 14, 883–892.

Ok Y S, Palansooriya KN, Yuan X, and Rinklebe J 2021 Special issue on biochar technologies, production, and environmental applications in Critical Reviews in Environmental Science & Technology during 2017–2021. *Critical Reviews in Environmental Science and Technology* 52, 3375–3383.

Senadheera SS, et al 2023 Application of biochar in concrete: A review. *Cement and Concrete Composites* 105204.

Wang J, and Wang S 2019 Preparation, modification and environmental application of biochar: A review. *Journal of Cleaner Production* 227, 1002–1022.

Zhang Y, Liu X, Wang S, Li L, and Dou S 2017 Bio-nanotechnology in high-performance supercapacitors. *Advanced Energy Materials* 7, 1700592.

Index

Note: **Bold** page numbers refer to tables and *italic* page numbers refer to figures.